Methods in Enzymology

Volume 128
PLASMA LIPOPROTEINS
Part A
Preparation, Structure, and Molecular Biology

METHODS IN ENZYMOLOGY

EDITORS-IN-CHIEF

Sidney P. Colowick   Nathan O. Kaplan

Methods in Enzymology

Volume 128

Plasma Lipoproteins

Part A

Preparation, Structure, and Molecular Biology

EDITED BY

Jere P. Segrest

DEPARTMENTS OF PATHOLOGY AND BIOCHEMISTRY
UNIVERSITY OF ALABAMA AT BIRMINGHAM
BIRMINGHAM, ALABAMA

John J. Albers

DEPARTMENTS OF MEDICINE AND PATHOLOGY
UNIVERSITY OF WASHINGTON SCHOOL OF MEDICINE
SEATTLE, WASHINGTON

1986

ACADEMIC PRESS, INC.
Harcourt Brace Jovanovich, Publishers
Orlando   San Diego   New York   Austin
London   Montreal   Sydney   Tokyo   Toronto

COPYRIGHT © 1986 BY ACADEMIC PRESS, INC.
ALL RIGHTS RESERVED.
NO PART OF THIS PUBLICATION MAY BE REPRODUCED OR
TRANSMITTED IN ANY FORM OR BY ANY MEANS, ELECTRONIC
OR MECHANICAL, INCLUDING PHOTOCOPY, RECORDING, OR
ANY INFORMATION STORAGE AND RETRIEVAL SYSTEM, WITHOUT
PERMISSION IN WRITING FROM THE PUBLISHER.

ACADEMIC PRESS, INC.
Orlando, Florida 32887

United Kingdom Edition published by
ACADEMIC PRESS INC. (LONDON) LTD.
24–28 Oval Road, London NW1 7DX

LIBRARY OF CONGRESS CATALOG CARD NUMBER: 54-9110

ISBN 0–12–182028–9

PRINTED IN THE UNITED STATES OF AMERICA

86 87 88 89   9 8 7 6 5 4 3 2 1

Table of Contents

Contributors to Volume 128 ix
Preface xv
Volumes in Series xvii

Section I. Overview

1. Introduction to the Plasma Lipoproteins	Antonio M. Gotto, Jr., Henry J. Pownall, and Richard J. Havel	3
2. Molecular and Cell Biology of Lipoprotein Biosynthesis	Donna M. Driscoll and Godfrey S. Getz	41
3. Comparative Analysis of Mammalian Plasma Lipoproteins	M. John Chapman	70
4. Impact of Technology on the Plasma Lipoprotein Field	Angelo M. Scanu	144

Section II. Preparation of Plasma Lipoproteins

5. Precautionary Measures for Collecting Blood Destined for Lipoprotein Isolation	Celina Edelstein and Angelo M. Scanu	151
6. Sequential Flotation Ultracentrifugation	Verne N. Schumaker and Donald L. Puppione	155
7. Density Gradient Ultracentrifugation of Serum Lipoproteins in a Swinging Bucket Rotor	Jim L. Kelley and Arthur W. Kruski	170
8. Single Vertical Spin Density Gradient Ultracentrifugation	Byung H. Chung, Jere P. Segrest, Marjorie J. Ray, John D. Brunzell, John E. Hokanson, Ronald M. Kraus, Ken Beaudrie, and John T. Cone	181

Section III. Isolation and Physical–Chemical Characterization of Plasma Apolipoproteins

9. Delipidation of Plasma Lipoproteins	JAMES C. OSBORNE, JR.	213
10. Isolation and Characterization of Apolipoproteins A-I, A-II, and A-IV	H. BRYAN BREWER, JR., ROSEMARY RONAN, MARTHA MENG, AND CHERI BISHOP	223
11. Isolation and Characterization of Apolipoprotein B-100	WALDO R. FISHER AND VERNE N. SCHUMAKER	247
12. Isolation and Characterization of Apolipoprotein B-48	DAVID A. HARDMAN AND JOHN P. KANE	262
13. Isolation and Characterization of Apolipoprotein E	STANLEY C. RALL, JR., KARL H. WEISGRABER, AND ROBERT W. MAHLEY	273
14. Isolation and Properties of Human Apolipoproteins C-I, C-II, and C-III	RICHARD L. JACKSON AND GEORGE HOLDSWORTH	288
15. Isolation and Characterization of Other Apolipoproteins	W. J. MCCONATHY AND P. ALAUPOVIC	297
16. Serum Amyloid A (ApoSAA) and Lipoproteins	NILS ERIKSEN AND EARL P. BENDITT	311
17. Structure and Lipid Binding Properties of Serum Albumin	ARTHUR A. SPECTOR	320
18. High-Performance Liquid Chromatography of Apolipoproteins	CELINA EDELSTEIN AND ANGELO M. SCANU	339
19. Immunochemical Isolation and Identification of Glucosylated Apolipoproteins	LINDA K. CURTISS, MICHAEL G. PEPE, AND JOSEPH L. WITZTUM	354
20. Isothermal Calorimetry of Apolipoproteins	M. ROSSENEU	365
21. Solution Properties of Apolipoproteins	JAMES C. OSBORNE, JR., NANCY S. LEE, AND GRACE M. POWELL	375
22. Studies of Apolipoproteins at the Air–Water Interface	MICHAEL C. PHILLIPS AND KEITH E. KREBS	387
23. Thermodynamics of Apolipoprotein–Phospholipid Association	JOHN B. MASSEY AND HENRY J. POWNALL	403

Section IV. Structure of Intact and Reconstituted Plasma Lipoproteins

24. Nondenaturing Polyacrylamide Gradient Gel Electrophoresis	ALEX V. NICHOLS, RONALD M. KRAUS, AND THOMAS A. MUSLINER	417

25. Isoelectric Focusing of Plasma Lipoproteins	YVES L. MARCEL AND PHILIP K. WEECH	432
26. Electron Microscopy of Negatively Stained Lipoproteins	TRUDY M. FORTE AND ROBERT W. NORDHAUSEN	442
27. Studies of Lipoproteins by Freeze-Fracture and Etching Electron Microscopy	LAWRENCE P. AGGERBECK AND TADEUSZ GULIK-KRZYWICKI	457
28. Nuclear Magnetic Resonance Studies of Lipoproteins	JAMES A. HAMILTON AND JOEL D. MORRISETT	472
29. Spectroscopic Studies of Lipoproteins	H. J. POWNALL AND J. B. MASSEY	515
30. Circular Dichroism of Lipoprotein Lipids	G. CHI CHEN AND JOHN P. KANE	519
31. Immunochemical Methods for Studying Lipoprotein Structure	ELAINE S. KRUL AND GUSTAV SCHONFELD	527
32. Reconstitution of High-Density Lipoproteins	ANA JONAS	553
33. Reassembly of Low-Density Lipoproteins	MARY T. WALSH AND DAVID ATKINSON	582
34. Reconstitution of the Hydrophobic Core of Low-Density Lipoprotein	MONTY KRIEGER	608
35. Use of Cross-Linking Reagents to Study Lipoprotein Structure	JOHN B. SWANEY	613
36. Synthetic Peptide Analogs of Apolipoproteins	G. M. ANANTHARAMAIAH	627
37. Lipoprotein–Liposome Interactions	ALAN R. TALL, IRA TABAS, AND KEVIN J. WILLIAMS	647
38. Carboxyfluorescein Leakage Assay for Lipoprotein–Liposome Interaction	JOHN N. WEINSTEIN, ROBERT BLUMENTHAL, AND RICHARD D. KLAUSNER	657

Section V. Molecular Biology of Plasma Lipoproteins

39. Measurement of Apolipoprotein mRNA by DNA-Excess Solution Hybridization with Single-Stranded Probes	DAVID L. WILLIAMS, THOMAS C. NEWMAN, GREGORY S. SHELNESS, AND DAVID A. GORDON	671
40. Intra- and Extracellular Modifications of Apolipoproteins	VASSILIS I. ZANNIS, SOTIRIOS K. KARATHANASIS, GAYLE M. FORBES, AND JAN L. BRESLOW	690
41. Characterization of the Apolipoprotein A-I-C-III Gene Complex	SOTIRIOS K. KARATHANASIS, VASSILIS I. ZANNIS, AND JAN L. BRESLOW	712

42. Genetic Polymorphism in the ApoA-I–C-III Complex	CAROL C. SHOULDERS AND F. E. BARALLE	727
43. Molecular Cloning and Sequence Analysis of Human Apolipoprotein A-II cDNA	LAWRENCE CHAN, MARSHA N. MOORE, AND YUAN-KAI TSAO	745
44. Rat Apolipoprotein A-IV: Application of Computational Methods for Studying the Structure, Function, and Evolution of a Protein	MARK S. BOGUSKI, NABIL A. ELSHOURBAGY, JOHN M. TAYLOR, AND JEFFREY I. GORDON	753
45. Detecting Internally Repeated Sequences and Inferring the History of Duplication	WALTER M. FITCH, TEMPLE SMITH, AND JAN L. BRESLOW	773
46. Isolation of cDNA and Genomic Clones for Apolipoprotein C-II	CYNTHIA L. JACKSON, GAIL A. P. BRUNS, AND JAN L. BRESLOW	788
47. Cloning of the cDNA for Rat and Human Apolipoprotein E mRNA	JOHN M. TAYLOR, ROBERT W. MAHLEY, AND CHIKAFUSA FUKAZAWA	801
48. Cloning and Expression of the Human Apolipoprotein E Gene	CATHERINE A. REARDON, YOUNG-KI PAIK, DAVID J. CHANG, GLENN E. DAVIES, ROBERT W. MAHLEY, YUN-FAI LAU, AND JOHN M. TAYLOR	811
49. Genetic Polymorphism in Human Apolipoprotein E	VASSILIS I. ZANNIS	823
50. Genetic Mapping of Apolipoprotein Genes in the Human Genome by Somatic Cell Hybrids	FA-TEN KAO AND LAWRENCE CHAN	851
51. Chromosomal Fine Mapping of Apolipoprotein Genes by *in Situ* Nucleic Acid Hybridization to Metaphase Chromosomes	MARY E. HARPER AND LAWRENCE CHAN	863
52. Genetic Control of Plasma Lipid Transport: Mouse Model	ALDONS J. LUSIS AND RENEE C. LEBOEUF	877
53. Molecular Cloning of Bovine LDL Receptor cDNAs	DAVID W. RUSSELL AND TOKUO YAMAMOTO	895
AUTHOR INDEX		911
SUBJECT INDEX		951

Contributors to Volume 128

Article numbers are in parentheses following the names of contributors.
Affiliations listed are current.

LAWRENCE P. AGGERBECK (27), *Centre de Génétique Moléculaire, Centre National de la Recherche Scientifique, 91190 Gif-sur-Yvette, France*

P. ALAUPOVIC (15), *Lipoprotein and Atherosclerosis Research Program, Oklahoma Medical Research Foundation, Oklahoma City, Oklahoma 73104*

G. M. ANANTHARAMAIAH (36), *Department of Pathology and the Atherosclerosis Research Unit, University of Alabama at Birmingham, Birmingham, Alabama 35294*

DAVID ATKINSON (33), *Biophysics Institute, Housman Medical Research Center, Boston University School of Medicine, Boston, Massachusetts 02118*

F. E. BARALLE (42), *Sir William Dunn School of Pathology, University of Oxford, Oxford OX1 3RE, England*

KEN BEAUDRIE (8), *Department of Pathology, University of Alabama at Birmingham, Birmingham, Alabama 35294*

EARL P. BENDITT (16), *Department of Pathology, University of Washington, Seattle, Washington 98195*

CHERI BISHOP (10), *Molecular Disease Branch, National Heart, Lung, and Blood Institute, National Institutes of Health, Bethesda, Maryland 20205*

ROBERT BLUMENTHAL (38), *Membrane Structure and Function Section, Laboratory of Mathematical Biology, National Cancer Institute, National Institutes of Health, Bethesda, Maryland 20892*

MARK S. BOGUSKI (44), *Department of Biological Chemistry, Washington University School of Medicine, St. Louis, Missouri 63110*

JAN L. BRESLOW (40, 41, 45, 46), *Laboratory of Biochemical Genetics and Metabolism, The Rockefeller University, New York, New York 10021*

H. BRYAN BREWER, JR. (10), *Molecular Disease Branch, National Heart, Lung, and Blood Institute, National Institutes of Health, Bethesda, Maryland 20205*

GAIL A. P. BRUNS (46), *Genetics Division, Children's Hospital, Boston, Massachusetts 02115*

JOHN D. BRUNZELL (8), *Department of Medicine, Division of Metabolism and Endocrinology, University of Washington, Seattle, Washington 98195*

LAWRENCE CHAN (43, 50, 51), *Departments of Cell Biology and Medicine, Baylor College of Medicine, Houston, Texas 77030*

DAVID J. CHANG (48), *Gladstone Foundation Laboratories for Cardiovascular Disease, Cardiovascular Research Institute, University of California, San Francisco, California 94140*

M. JOHN CHAPMAN (3), *Laboratoire de Recherches sur les Lipoprotéines, Institut National de la Santé et de la Recherche Medicale (Inserm), Pavillon Benjamin Delessert, Hôpital de la Pitié, 75671 Paris, France*

G. CHI CHEN (30), *Cardiovascular Research Institute, University of California, San Francisco, California 94143*

BYUNG H. CHUNG (8), *Department of Pathology, University of Alabama at Birmingham, Birmingham, Alabama 35294*

JOHN T. CONE (8), *Department of Pathology, University of Alabama at Birmingham, Birmingham, Alabama 35294*

LINDA K. CURTISS (19), *Department of Immunology, Scripps Clinic and Research Foundation, La Jolla, California 92037*

GLENN E. DAVIES (48), *Gladstone Foundation Laboratories for Cardiovascular Disease, Cardiovascular Research Institute, University of California, San Francisco, California 94140*

DONNA M. DRISCOLL (2), *Imperial Cancer Research Fund, Potters Bar, Hertfordshire EN6 3LD, England*

CELINA EDELSTEIN (5, 18), *Department of Medicine, The Pritzker School of Medicine, The University of Chicago, Chicago, Illinois 60637*

NABIL ELSHOURBAGY (44), *Gladstone Foundation Laboratories, Department of Physiology, Cardiovascular Research Institute, University of California, San Francisco, California 94140*

NILS ERIKSEN (16), *Department of Pathology, University of Washington, Seattle, Washington 98195*

WALDO R. FISHER (11), *Departments of Medicine and Biochemistry, University of Florida College of Medicine, Gainesville, Florida 33610*

WALTER M. FITCH (45), *Department of Physiological Chemistry, Center of Health Sciences, University of Wisconsin-Madison, Madison, Wisconsin 53706*

GAYLE M. FORBES (40), *Section of Molecular Genetics, Cardiovascular Institute, Department of Medicine and Biochemistry, Boston University Medical Center, Boston, Massachusetts 02118, and Metabolism Division, Children's Hospital and Department of Pediatrics, Harvard Medical School, Boston, Massachusetts 02115*

TRUDY M. FORTE (26), *Donner Laboratory, Lawrence Berkeley Laboratory, University of California, Berkeley, California 94720*

CHIKAFUSA FUKAZAWA (47), *National Food Research Institute, Tsukuba-Gun, Ibaraki-Ken, Japan*

GODFREY S. GETZ (2), *Departments of Pathology, Biochemistry and Molecular Biology, and Medicine, The University of Chicago, Chicago, Illinois 60637*

DAVID A. GORDON (39), *Department of Pharmacological Sciences, Health Sciences Center, State University of New York at Stony Brook, Stony Brook, New York 11790*

JEFFREY I. GORDON (44), *Departments of Biological Chemistry and Medicine, Washington University School of Medicine, St. Louis, Missouri 63110*

ANTONIO M. GOTTO, JR. (1), *Baylor College of Medicine and The Methodist Hospital, Houston, Texas 77030*

TADEUSZ GULIK-KRZYWICKI (27), *Centre de Génétique Moléculaire, Centre National de la Recherche Scientifique, 91190 Gif-sur-Yvette, France*

JAMES A. HAMILTON (28), *Biophysics Institute, Housman Medical Research Center, Boston University School of Medicine, Boston, Massachusetts 02118*

DAVID A. HARDMAN (12), *Cardiovascular Research Institute, Department of Physiology, University of California, San Francisco, California 94143*

MARY E. HARPER (51), *Laboratory of Tumor Cell Biology, National Cancer Institute, Bethesda, Maryland 20892*

RICHARD J. HAVEL (1), *Cardiovascular Research Institute, University of California School of Medicine, San Francisco, California 94143*

JOHN E. HOKANSON (8), *Department of Medicine, Division of Metabolism and Endocrinology, University of Washington, Seattle, Washington 98195*

GEORGE HOLDSWORTH (14), *Oak Ridge Research Institute, Oak Ridge, Tennessee 37830*

CYNTHIA L. JACKSON (46), *Center for Cancer Research, Massachusetts Institute of Technology, Cambridge, Massachusetts 02139*

RICHARD L. JACKSON (14), *Merrell Dow Research Institute, Cincinnati, Ohio 45215, and Department of Pharmacology and Cell Biophysics, University of Cincinnati Medical Center, Cincinnati, Ohio 45267*

ANA JONAS (32), *Department of Biochemistry, College of Medicine, University of Illinois at Urbana/Champaign, Urbana, Illinois 61801*

JOHN P. KANE (12, 30), *Cardiovascular Research Institute, Department of Medicine and Department of Biochemistry and Biophysics, University of California, San Francisco, California 94143*

FA-TEN KAO (50), *Eleanor Roosevelt Institute for Cancer Research and Department of Biochemistry, Biophysics, and Genetics, University of Colorado Health Sciences Center, Denver, Colorado 80262*

SOTIRIOS K. KARATHANASIS (40, 41), *Metabolism Division, Children's Hospital, and Department of Pediatrics, Harvard Medical School, Boston, Massachusetts 02115*

JIM L. KELLEY (7), *Department of Pathology, The University of Texas Health Science Center, San Antonio, Texas 78284*

RICHARD D. KLAUSNER (38), *Cell Biology and Metabolism Branch, National Institute for Child Health and Development, National Institutes of Health, Bethesda, Maryland 20892*

RONALD M. KRAUS (8, 24), *Donner Laboratory, Lawrence Berkeley Laboratory, University of California, Berkeley, California 94720*

KEITH E. KREBS (22), *Department of Physiology, The Milton S. Hershey Medical Center, Hershey, Pennsylvania 17033*

MONTY KRIEGER (34), *Health Science, Technology and Management, Walker College, Cambridge, Massachusetts 02139*

ELAINE S. KRUL (31), *Lipid Research Center, Department of Preventive Medicine, Washington University, St. Louis, Missouri 63110*

ARTHUR W. KRUSKI (7), *Department of Pathology, The University of Texas Health Science Center, San Antonio, Texas 78284*

YUN-FAI LAU (48), *The Howard Hughes Medical Institute, Department of Physiology, University of California, San Francisco, California 94143*

RENEE C. LEBOEUF (52), *Research Service, Veterans Administration, Wadsworth Medical Center, Los Angeles, California 90073*

NANCY S. LEE (21), *Molecular Disease Branch, National Heart, Lung, and Blood Institute, National Institutes of Health, Bethesda, Maryland 20205*

ALDONS J. LUSIS (52), *Departments of Medicine and Microbiology, University of California, Los Angeles, California 90024*

ROBERT W. MAHLEY (13, 47, 48), *Gladstone Foundation Laboratories for Cardiovascular Disease, Cardiovascular Research Institute, and Departments of Pathology and Medicine, University of California, San Francisco, California 94140*

YVES L. MARCEL (25), *Laboratory of Lipoprotein Metabolism, Clinical Research Institute of Montreal, Montreal, Quebec H2W 1R7, Canada*

JOHN B. MASSEY (23, 29), *Department of Medicine, Baylor College of Medicine, Houston, Texas 77030*

W. J. MCCONATHY (15), *Lipoprotein and Atherosclerosis Research Program, Oklahoma Medical Research Foundation, Oklahoma City, Oklahoma 73104*

MARTHA MENG (10), *Molecular Disease Branch, National Heart, Lung, and Blood Institute, National Institutes of Health, Bethesda, Maryland 20205*

MARSHA N. MOORE (43), *Departments of Cell Biology and Medicine, Baylor College of Medicine, Texas Medical Center, Houston, Texas 77030*

JOEL D. MORRISETT (28), *Baylor College of Medicine, Methodist Hospital, Houston, Texas 77030*

THOMAS A. MUSLINER (24), *Donner Laboratory, Lawrence Berkeley Laboratory, University of California, Berkeley, California 94720*

THOMAS C. NEWMAN (39), *Department of Biochemical Genetics and Metabolism, Rockefeller University Hospital, The Rockefeller University, New York, New York 10021*

ALEX V. NICHOLS (24), *Donner Laboratory, Lawrence Berkeley Laboratory, University of California, Berkeley, California 94720*

ROBERT W. NORDHAUSEN (26), *Donner Laboratory, Lawrence Berkeley Laboratory, University of California, Berkeley, California 94720*

JAMES C. OSBORNE, JR. (9, 21), *Spinco Division, Beckman Instruments, Palo Alto, California 94304*

YOUNG-KI PAIK (48), *Gladstone Foundation Laboratories for Cardiovascular Disease, Cardiovascular Research Institute, University of California, San Francisco, California 94140*

MICHAEL G. PEPE (19), *Department of Hematology, Stanford University, VA Medical Center, Palo Alto, California 94304*

MICHAEL C. PHILLIPS (22), *Department of Physiology and Biochemistry, The Medical College of Pennsylvania, Philadelphia, Pennsylvania 19129*

GRACE M. POWELL (21), *Molecular Disease Branch, National Heart, Lung, and Blood Institute, National Institutes of Health, Bethesda, Maryland 20205*

HENRY J. POWNALL (1, 23, 29), *Department of Medicine, Baylor College of Medicine, Houston, Texas 77030*

DONALD L. PUPPIONE (6), *Long Marine Laboratory, University of California, Santa Cruz, California 95064*

STANLEY C. RALL, JR. (13), *Gladstone Foundation Laboratories for Cardiovascular Disease, Cardiovascular Research Institute, and Department of Pathology, University of California, San Francisco, California 94140*

MARJORIE J. RAY (8), *Department of Pathology, University of Alabama at Birmingham, Birmingham, Alabama 35294*

CATHERINE A. REARDON (48), *Gladstone Foundation Laboratories for Cardiovascular Disease, Cardiovascular Research Institute, University of California, San Francisco, California 94140*

ROSEMARY RONAN (10), *Molecular Disease Branch, National Heart, Lung, and Blood Institute, National Institutes of Health, Bethesda, Maryland 20205*

M. ROSSENEU (20), *Department of Clinical Chemistry, Algemeen-Ziekenhuis Sin-Jan, 8000 Brugge, Belgium*

DAVID W. RUSSELL (53), *Department of Molecular Genetics, University of Texas Health Science Center at Dallas, Dallas, Texas 75235*

ANGELO M. SCANU (4, 5, 18), *Departments of Medicine, Biochemistry, and Molecular Biology, The Pritzker School of Medicine, The University of Chicago, Chicago, Illinois 60637*

GUSTAV SCHONFELD (31), *Lipid Research Center, Department of Preventive Medicine, Washington University, St. Louis, Missouri 63110*

VERNE N. SCHUMAKER (6, 11), *Departments of Chemistry and Biochemistry and Molecular Biology Institute, University of California, Los Angeles, California 90024*

JERE P. SEGREST (8), *Departments of Pathology and Biochemistry, University of Alabama at Birmingham, Birmingham, Alabama 35294*

GREGORY S. SHELNESS (39), *The Rockefeller University, New York, New York 10021*

CAROL C. SHOULDERS (42), *Sir William Dunn School of Pathology, University of Oxford, Oxford OX1 3RE, England*

TEMPLE SMITH (45), *Department of Biostatistics, Dana Farber Cancer Institute, Boston, Massachusetts 02115*

ARTHUR A. SPECTOR (17), *Department of Biochemistry, University of Iowa, Iowa City, Iowa 52242*

JOHN B. SWANEY (35), *Department of Biochemistry, Hahnemann University, Philadelphia, Pennsylvania 19102*

IRA TABAS (37), *Department of Medicine, College of Physicians and Surgeons, Columbia University, New York, New York 10032*

ALAN R. TALL (37), *Department of Medicine, College of Physicians and Surgeons, Columbia University, New York, New York 10032*

JOHN M. TAYLOR (44, 47, 48), *Gladstone Foundation Laboratories for Cardiovascular Disease, Cardiovascular Research Institute, and Department of Physiology, University of California, San Francisco, California 94140*

YUAN-KAI TSAO (43), *Departments of Cell Biology and Medicine, Baylor College of Medicine, Texas Medical Center, Houston, Texas 77030*

MARY T. WALSH (33), *Biophysics Institute, Housman Medical Research Center, Boston University School of Medicine, Boston, Massachusetts 02118*

PHILIP K. WEECH (25), *Laboratory of Lipoprotein Metabolism, Clinical Research Institute of Montreal, Montreal, Quebec H2W 1R7, Canada*

JOHN N. WEINSTEIN (38), *Theoretical Immunology Section, Laboratory of Mathematical Biology, National Cancer Institute, National Institutes of Health, Bethesda, Maryland 20892*

KARL H. WEISGRABER (13), *Gladstone Foundation Laboratories for Cardiovascular Disease, Cardiovascular Research Institute, and Department of Pathology, University of California, San Francisco, California 94140*

DAVID L. WILLIAMS (39), *Department of Pharmacological Sciences, State University of New York at Stony Brook, Stony Brook, New York 11790*

KEVIN J. WILLIAMS (37), *Department of Medicine, College of Physicians and Surgeons, Columbia University, New York, New York 10032*

JOSEPH L. WITZTUM (19), *Department of Medicine, Division of Metabolic Diseases, University of California at San Diego, La Jolla, California 92093*

TOKUO YAMAMOTO (53), *Department of Molecular Genetics, University of Texas Health Science Center, Dallas, Texas 75235*

VASSILIS I. ZANNIS (40, 41, 49), *Section of Molecular Genetics, Cardiovascular Institute, Boston University Medical Center, Boston, Massachusetts 02118*

Preface

Methodology development has played a central role in understanding the structure, biosynthesis, and physiological functions of the plasma lipoproteins. The ultracentrifuge played a major role in the discovery and characterization of the plasma lipoproteins, and the instrument continues to be a methodologic mainstay. One result has been a progressively better appreciation of the metabolic interrelationships of the different plasma lipoprotein species.

The discovery of the LDL receptor less than fifteen years ago ushered in the molecular era in plasma lipoprotein physiology. Simultaneously, the identification, isolation, and amino acid sequence analyses of the different plasma apolipoproteins laid the foundation for a molecular revolution in our understanding of plasma lipoprotein structure and metabolism. This revolution has been accelerated in the past five years by explosive advances in recombinant DNA technology. The structures of the genes encoding for apolipoproteins, lipoprotein receptors, and enzymes involved in the regulation of lipoprotein metabolism are emerging. With the added availability of new techniques for the isolation and analysis of plasma lipoprotein subspecies, an exciting new era in lipoprotein research beckons. This era promises a detailed understanding of the molecular basis for genetic and metabolic regulation of the plasma lipoproteins.

The exciting state of growth in lipoprotein research, the complete lack of a recent comprehensive treatise on the central methodology presently being used in the field, and the relatively limited treatment of lipoproteins in earlier volumes of *Methods in Enzymology* warranted the assembly of this two-volume work. Volume 128, Part A deals with the preparation, structure, and molecular biology of the plasma lipoproteins; Volume 129, Part B deals with the characterization, cell biology, and metabolism of the plasma lipoproteins. These volumes should serve as convenient handbooks for all investigators involved in lipoprotein research.

We wish to acknowledge our indebtedness to the contributors. We also want to thank Dr. Leon W. Cunningham, Dr. John M. Taylor, and Dr. Antonio M. Gotto, Jr., for their help in the conception and organization of these volumes. We express our appreciation to the staff of Academic Press for their pleasant and efficient assistance.

JERE P. SEGREST
JOHN J. ALBERS

METHODS IN ENZYMOLOGY

EDITED BY

Sidney P. Colowick and Nathan O. Kaplan

VANDERBILT UNIVERSITY
SCHOOL OF MEDICINE
NASHVILLE, TENNESSEE

DEPARTMENT OF CHEMISTRY
UNIVERSITY OF CALIFORNIA
AT SAN DIEGO
LA JOLLA, CALIFORNIA

I. Preparation and Assay of Enzymes
II. Preparation and Assay of Enzymes
III. Preparation and Assay of Substrates
IV. Special Techniques for the Enzymologist
V. Preparation and Assay of Enzymes
VI. Preparation and Assay of Enzymes (*Continued*)
    Preparation and Assay of Substrates
    Special Techniques
VII. Cumulative Subject Index

METHODS IN ENZYMOLOGY

EDITORS-IN-CHIEF

Sidney P. Colowick and Nathan O. Kaplan

VOLUME VIII. Complex Carbohydrates
Edited by ELIZABETH F. NEUFELD AND VICTOR GINSBURG

VOLUME IX. Carbohydrate Metabolism
Edited by WILLIS A. WOOD

VOLUME X. Oxidation and Phosphorylation
Edited by RONALD W. ESTABROOK AND MAYNARD E. PULLMAN

VOLUME XI. Enzyme Structure
Edited by C. H. W. HIRS

VOLUME XII. Nucleic Acids (Parts A and B)
Edited by LAWRENCE GROSSMAN AND KIVIE MOLDAVE

VOLUME XIII. Citric Acid Cycle
Edited by J. M. LOWENSTEIN

VOLUME XIV. Lipids
Edited by J. M. LOWENSTEIN

VOLUME XV. Steroids and Terpenoids
Edited by RAYMOND B. CLAYTON

VOLUME XVI. Fast Reactions
Edited by KENNETH KUSTIN

VOLUME XVII. Metabolism of Amino Acids and Amines (Parts A and B)
Edited by HERBERT TABOR AND CELIA WHITE TABOR

VOLUME XVIII. Vitamins and Coenzymes (Parts A, B, and C)
Edited by DONALD B. MCCORMICK AND LEMUEL D. WRIGHT

VOLUME XIX. Proteolytic Enzymes
Edited by GERTRUDE E. PERLMANN AND LASZLO LORAND

VOLUME XX. Nucleic Acids and Protein Synthesis (Part C)
Edited by KIVIE MOLDAVE AND LAWRENCE GROSSMAN

VOLUME XXI. Nucleic Acids (Part D)
Edited by LAWRENCE GROSSMAN AND KIVIE MOLDAVE

VOLUME XXII. Enzyme Purification and Related Techniques
Edited by WILLIAM B. JAKOBY

VOLUME XXIII. Photosynthesis (Part A)
Edited by ANTHONY SAN PIETRO

VOLUME XXIV. Photosynthesis and Nitrogen Fixation (Part B)
Edited by ANTHONY SAN PIETRO

VOLUME XXV. Enzyme Structure (Part B)
Edited by C. H. W. HIRS AND SERGE N. TIMASHEFF

VOLUME XXVI. Enzyme Structure (Part C)
Edited by C. H. W. HIRS AND SERGE N. TIMASHEFF

VOLUME XXVII. Enzyme Structure (Part D)
Edited by C. H. W. HIRS AND SERGE N. TIMASHEFF

VOLUME XXVIII. Complex Carbohydrates (Part B)
Edited by VICTOR GINSBURG

VOLUME XXIX. Nucleic Acids and Protein Synthesis (Part E)
Edited by LAWRENCE GROSSMAN AND KIVIE MOLDAVE

VOLUME XXX. Nucleic Acids and Protein Synthesis (Part F)
Edited by KIVIE MOLDAVE AND LAWRENCE GROSSMAN

VOLUME XXXI. Biomembranes (Part A)
Edited by SIDNEY FLEISCHER AND LESTER PACKER

VOLUME XXXII. Biomembranes (Part B)
Edited by SIDNEY FLEISCHER AND LESTER PACKER

VOLUME XXXIII. Cumulative Subject Index Volumes I–XXX
Edited by MARTHA G. DENNIS AND EDWARD A. DENNIS

VOLUME XXXIV. Affinity Techniques (Enzyme Purification: Part B)
Edited by WILLIAM B. JAKOBY AND MEIR WILCHEK

VOLUME XXXV. Lipids (Part B)
Edited by JOHN M. LOWENSTEIN

VOLUME XXXVI. Hormone Action (Part A: Steroid Hormones)
Edited by BERT W. O'MALLEY AND JOEL G. HARDMAN

VOLUME XXXVII. Hormone Action (Part B: Peptide Hormones)
Edited by BERT W. O'MALLEY AND JOEL G. HARDMAN

VOLUME XXXVIII. Hormone Action (Part C: Cyclic Nucleotides)
Edited by JOEL G. HARDMAN AND BERT W. O'MALLEY

VOLUME XXXIX. Hormone Action (Part D: Isolated Cells, Tissues, and Organ Systems)
Edited by JOEL G. HARDMAN AND BERT W. O'MALLEY

VOLUME XL. Hormone Action (Part E: Nuclear Structure and Function)
Edited by BERT W. O'MALLEY AND JOEL G. HARDMAN

VOLUME XLI. Carbohydrate Metabolism (Part B)
Edited by W. A. WOOD

VOLUME XLII. Carbohydrate Metabolism (Part C)
Edited by W. A. WOOD

VOLUME XLIII. Antibiotics
Edited by JOHN H. HASH

VOLUME XLIV. Immobilized Enzymes
Edited by KLAUS MOSBACH

VOLUME XLV. Proteolytic Enzymes (Part B)
Edited by LASZLO LORAND

VOLUME XLVI. Affinity Labeling
Edited by WILLIAM B. JAKOBY AND MEIR WILCHEK

VOLUME XLVII. Enzyme Structure (Part E)
Edited by C. H. W. HIRS AND SERGE N. TIMASHEFF

VOLUME XLVIII. Enzyme Structure (Part F)
Edited by C. H. W. HIRS AND SERGE N. TIMASHEFF

VOLUME XLIX. Enzyme Structure (Part G)
Edited by C. H. W. HIRS AND SERGE N. TIMASHEFF

VOLUME L. Complex Carbohydrates (Part C)
Edited by VICTOR GINSBURG

VOLUME LI. Purine and Pyrimidine Nucleotide Metabolism
Edited by PATRICIA A. HOFFEE AND MARY ELLEN JONES

VOLUME LII. Biomembranes (Part C: Biological Oxidations)
Edited by SIDNEY FLEISCHER AND LESTER PACKER

VOLUME LIII. Biomembranes (Part D: Biological Oxidations)
Edited by SIDNEY FLEISCHER AND LESTER PACKER

VOLUME LIV. Biomembranes (Part E: Biological Oxidations)
Edited by SIDNEY FLEISCHER AND LESTER PACKER

VOLUME LV. Biomembranes (Part F: Bioenergetics)
Edited by SIDNEY FLEISCHER AND LESTER PACKER

VOLUME LVI. Biomembranes (Part G: Bioenergetics)
Edited by SIDNEY FLEISCHER AND LESTER PACKER

VOLUME LVII. Bioluminescence and Chemiluminescence
Edited by MARLENE A. DELUCA

VOLUME LVIII. Cell Culture
Edited by WILLIAM B. JAKOBY AND IRA PASTAN

VOLUME LIX. Nucleic Acids and Protein Synthesis (Part G)
Edited by KIVIE MOLDAVE AND LAWRENCE GROSSMAN

VOLUME LX. Nucleic Acids and Protein Synthesis (Part H)
Edited by KIVIE MOLDAVE AND LAWRENCE GROSSMAN

VOLUME 61. Enzyme Structure (Part H)
Edited by C. H. W. HIRS AND SERGE N. TIMASHEFF

VOLUME 62. Vitamins and Coenzymes (Part D)
Edited by DONALD B. MCCORMICK AND LEMUEL D. WRIGHT

VOLUME 63. Enzyme Kinetics and Mechanism (Part A: Initial Rate and Inhibitor Methods)
Edited by DANIEL L. PURICH

VOLUME 64. Enzyme Kinetics and Mechanism (Part B: Isotopic Probes and Complex Enzyme Systems)
Edited by DANIEL L. PURICH

VOLUME 65. Nucleic Acids (Part I)
Edited by LAWRENCE GROSSMAN AND KIVIE MOLDAVE

VOLUME 66. Vitamins and Coenzymes (Part E)
Edited by DONALD B. MCCORMICK AND LEMUEL D. WRIGHT

VOLUME 67. Vitamins and Coenzymes (Part F)
Edited by DONALD B. MCCORMICK AND LEMUEL D. WRIGHT

VOLUME 68. Recombinant DNA
Edited by RAY WU

VOLUME 69. Photosynthesis and Nitrogen Fixation (Part C)
Edited by ANTHONY SAN PIETRO

VOLUME 70. Immunochemical Techniques (Part A)
Edited by HELEN VAN VUNAKIS AND JOHN J. LANGONE

VOLUME 71. Lipids (Part C)
Edited by JOHN M. LOWENSTEIN

VOLUME 72. Lipids (Part D)
Edited by JOHN M. LOWENSTEIN

VOLUME 73. Immunochemical Techniques (Part B)
Edited by JOHN J. LANGONE AND HELEN VAN VUNAKIS

VOLUME 74. Immunochemical Techniques (Part C)
Edited by JOHN J. LANGONE AND HELEN VAN VUNAKIS

VOLUME 75. Cumulative Subject Index Volumes XXXI, XXXII, and XXXIV–LX
Edited by EDWARD A. DENNIS AND MARTHA G. DENNIS

VOLUME 76. Hemoglobins
Edited by ERALDO ANTONINI, LUIGI ROSSI-BERNARDI, AND EMILIA CHIANCONE

VOLUME 77. Detoxication and Drug Metabolism
Edited by WILLIAM B. JAKOBY

VOLUME 78. Interferons (Part A)
Edited by SIDNEY PESTKA

VOLUME 79. Interferons (Part B)
Edited by SIDNEY PESTKA

VOLUME 80. Proteolytic Enzymes (Part C)
Edited by LASZLO LORAND

VOLUME 81. Biomembranes (Part H: Visual Pigments and Purple Membranes, I)
Edited by LESTER PACKER

VOLUME 82. Structural and Contractile Proteins (Part A: Extracellular Matrix)
Edited by LEON W. CUNNINGHAM AND DIXIE W. FREDERIKSEN

VOLUME 83. Complex Carbohydrates (Part D)
Edited by VICTOR GINSBURG

VOLUME 84. Immunochemical Techniques (Part D: Selected Immunoassays)
Edited by JOHN J. LANGONE AND HELEN VAN VUNAKIS

VOLUME 85. Structural and Contractile Proteins (Part B: The Contractile Apparatus and the Cytoskeleton)
Edited by DIXIE W. FREDERIKSEN AND LEON W. CUNNINGHAM

VOLUME 86. Prostaglandins and Arachidonate Metabolites
Edited by WILLIAM E. M. LANDS AND WILLIAM L. SMITH

VOLUME 87. Enzyme Kinetics and Mechanism (Part C: Intermediates, Stereochemistry, and Rate Studies)
Edited by DANIEL L. PURICH

VOLUME 88. Biomembranes (Part I: Visual Pigments and Purple Membranes, II)
Edited by LESTER PACKER

VOLUME 89. Carbohydrate Metabolism (Part D)
Edited by WILLIS A. WOOD

VOLUME 90. Carbohydrate Metabolism (Part E)
Edited by Willis A. Wood

VOLUME 91. Enzyme Structure (Part I)
Edited by C. H. W. HIRS AND SERGE N. TIMASHEFF

VOLUME 92. Immunochemical Techniques (Part E: Monoclonal Antibodies and General Immunoassay Methods)
Edited by JOHN J. LANGONE AND HELEN VAN VUNAKIS

VOLUME 93. Immunochemical Techniques (Part F: Conventional Antibodies, Fc Receptors, and Cytotoxicity)
Edited by JOHN J. LANGONE AND HELEN VAN VUNAKIS

VOLUME 94. Polyamines
Edited by HERBERT TABOR AND CELIA WHITE TABOR

VOLUME 95. Cumulative Subject Index Volumes 61–74 and 76–80
Edited by EDWARD A. DENNIS AND MARTHA G. DENNIS

VOLUME 96. Biomembranes [Part J: Membrane Biogenesis: Assembly and Targeting (General Methods; Eukaryotes)]
Edited by SIDNEY FLEISCHER AND BECCA FLEISCHER

VOLUME 97. Biomembranes [Part K: Membrane Biogenesis: Assembly and Targeting (Prokaryotes, Mitochondria, and Chloroplasts)]
Edited by SIDNEY FLEISCHER AND BECCA FLEISCHER

VOLUME 98. Biomembranes [Part L: Membrane Biogenesis (Processing and Recycling)]
Edited by SIDNEY FLEISCHER AND BECCA FLEISCHER

VOLUME 99. Hormone Action (Part F: Protein Kinases)
Edited by JACKIE D. CORBIN AND JOEL G. HARDMAN

VOLUME 100. Recombinant DNA (Part B)
Edited by RAY WU, LAWRENCE GROSSMAN, AND KIVIE MOLDAVE

VOLUME 101. Recombinant DNA (Part C)
Edited by RAY WU, LAWRENCE GROSSMAN, AND KIVIE MOLDAVE

VOLUME 102. Hormone Action (Part G: Calmodulin and Calcium-Binding Proteins)
Edited by ANTHONY R. MEANS AND BERT W. O'MALLEY

VOLUME 103. Hormone Action (Part H: Neuroendocrine Peptides)
Edited by P. MICHAEL CONN

VOLUME 104. Enzyme Purification and Related Techniques (Part C)
Edited by WILLIAM B. JAKOBY

VOLUME 105. Oxygen Radicals in Biological Systems
Edited by LESTER PACKER

VOLUME 106. Posttranslational Modifications (Part A)
Edited by FINN WOLD AND KIVIE MOLDAVE

VOLUME 107. Posttranslational Modifications (Part B)
Edited by FINN WOLD AND KIVIE MOLDAVE

VOLUME 108. Immunochemical Techniques (Part G: Separation and Characterization of Lymphoid Cells)
Edited by GIOVANNI DI SABATO, JOHN J. LANGONE, AND HELEN VAN VUNAKIS

VOLUME 109. Hormone Action (Part I: Peptide Hormones)
Edited by LUTZ BIRNBAUMER AND BERT W. O'MALLEY

VOLUME 110. Steroids and Isoprenoids (Part A)
Edited by JOHN H. LAW AND HANS C. RILLING

VOLUME 111. Steroids and Isoprenoids (Part B)
Edited by JOHN H. LAW AND HANS C. RILLING

VOLUME 112. Drug and Enzyme Targeting (Part A)
Edited by KENNETH J. WIDDER AND RALPH GREEN

VOLUME 113. Glutamate, Glutamine, Glutathione, and Related Compounds
Edited by ALTON MEISTER

VOLUME 114. Diffraction Methods for Biological Macromolecules (Part A)
Edited by HAROLD W. WYCKOFF, C. H. W. HIRS, AND SERGE N. TIMASHEFF

VOLUME 115. Diffraction Methods for Biological Macromolecules (Part B)
Edited by HAROLD W. WYCKOFF, C. H. W. HIRS, AND SERGE N. TIMASHEFF

VOLUME 116. Immunochemical Techniques (Part H: Effectors and Mediators of Lymphoid Cell Functions)
Edited by GIOVANNI DI SABATO, JOHN J. LANGONE, AND HELEN VAN VUNAKIS

VOLUME 117. Enzyme Structure (Part J)
Edited by C. H. W. HIRS AND SERGE N. TIMASHEFF

VOLUME 118. Plant Molecular Biology
Edited by ARTHUR WEISSBACH AND HERBERT WEISSBACH

VOLUME 119. Interferons (Part C)
Edited by SIDNEY PESTKA

VOLUME 120. Cumulative Subject Index Volumes 81–94, 96–101 (in preparation)

VOLUME 121. Immunochemical Techniques (Part I: Hybridoma Technology and Monoclonal Antibodies)
Edited by JOHN J. LANGONE AND HELEN VAN VUNAKIS

VOLUME 122. Vitamins and Coenzymes (Part G)
Edited by FRANK CHYTIL AND DONALD B. MCCORMICK

VOLUME 123. Vitamins and Coenzymes (Part H)
Edited by FRANK CHYTIL AND DONALD B. MCCORMICK

VOLUME 124. Hormone Action (Part J: Neuroendocrine Peptides)
Edited by P. MICHAEL CONN

VOLUME 125. Biomembranes (Part M: Transport in Bacteria, Mitochondria, and Chloroplasts: General Approaches and Transport Systems)
Edited by SIDNEY FLEISCHER AND BECCA FLEISCHER

VOLUME 126. Biomembranes (Part N: Transport in Bacteria, Mitochondria, and Chloroplasts: Protonmotive Force)
Edited by SIDNEY FLEISCHER AND BECCA FLEISCHER

VOLUME 127. Biomembranes (Part O: Protons and Water: Structure and Translocation)
Edited by LESTER PACKER

VOLUME 128. Plasma Lipoproteins (Part A: Preparation, Structure, and Molecular Biology)
Edited by JERE P. SEGREST AND JOHN J. ALBERS

VOLUME 129. Plasma Lipoproteins (Part B: Characterization, Cell Biology, and Metabolism) (in preparation)
Edited by JOHN J. ALBERS AND JERE P. SEGREST

VOLUME 130. Enzyme Structure (Part K) (in preparation)
Edited by C. H. W. HIRS AND SERGE N. TIMASHEFF

VOLUME 131. Enzyme Structure (Part L) (in preparation)
Edited by C. H. W. HIRS AND SERGE N. TIMASHEFF

VOLUME 132. Immunochemical Techniques (Part J: Phagocytosis and Cell-Mediated Cytotoxicity) (in preparation)
Edited by GIOVANNI DI SABATO AND JOHANNES EVERSE

Section I

Overview

[1] Introduction to the Plasma Lipoproteins

By ANTONIO M. GOTTO, JR., HENRY J. POWNALL, and RICHARD J. HAVEL

Introduction

Although earlier work had suggested a relationship between lipids and plasma proteins, it was Macheboeuf[1] who first isolated and identified a plasma lipoprotein at the Pasteur Institute in the late 1920s. Macheboeuf added ammonium sulfate to 50% saturation and acidified the supernatant to obtain a plasma lipoprotein precipitate. Virtually all of the cholesterol in the precipitate was in the esterified form and we now know that what he actually isolated were the high-density lipoproteins (HDL). He referred to this fraction as C.A. (cenapses precipitated by acid). With the development of electrophoretic methods Macheboeuf later showed that C.A. was an α-globulin. Other contributions were made by Pedersen[2] who separated lipoproteins by magnesium sulfate precipitation and by Cohn[3] and his colleagues who described a large scale procedure for separation of plasma into fractions having α- and β-mobility using ethanol–water mixtures. One of the first major symposia on lipoproteins was held at Birmingham University in England in August of 1949, when there were fewer than 100 published reports on lipoproteins.

More attention was focused on the clinical importance of plasma lipoprotein levels when a strong positive correlation between serum cholesterol levels and the incidence of coronary artery disease (CAD) was found. An intensive study of the role of lipoproteins was begun, as they are the primary carriers of cholesterol in the blood. Gofman and his associates found that plasma low-density lipoproteins (LDL) levels were most strongly correlated with CAD. They also noted a negative correlation between plasma HDL and CAD, an association that initially received relatively little attention. One outgrowth of these early hypotheses was the design and implementation of the now famous Framingham Study that called attention to serum cholesterol levels as a major risk factor. At present, most clinical laboratories perform analyses of serum cholesterol,

---

[1] M. A. Macheboeuf, *Bull. Soc. Chim. Biol.* **11**, 268 (1929).
[2] K. O. Pedersen, *J. Phys. Chem.* **51**, 156 (1947).
[3] E. J. Cohn, F. R. N. Gurd, D. M. Surgenor, B. A. Barnes, R. K. Brown, G. Derouaux, J. M. Gillespie, F. W. Kahnt, W. F. Lever, C. H. Liu, D. Mittleman, R. F. Mouton, K. Schmid, and E. Uroma, *J. Am. Chem. Soc.* **72**, 465 (1950).

triglycerides, and HDL as part of a program designed to identify patients at risk for CAD. The concentration of LDL is estimated from a formula. Some laboratories also offer analyses for apoB and apoA-I, which correlate positively and negatively, respectively, for CAD.

In the late 1940s, John Gofman and his colleagues[4] performed many careful analyses on the analytical ultracentrifuge at the Donner Laboratories. Their early reports and collaborations through the years have helped to make this technique one of the most important standards for the characterization of the physical properties of lipoproteins. In 1955, Havel et al.[5] reported a sequential preparative flotation of lipoproteins in a background density of salt in the ultracentrifuge, which permitted detailed chemical analysis of defined fractions. Alternative methods were reported subsequently, including molecular sieve chromatography by Rudel et al.[6] and others. The ease and economy of scale continues to favor ultracentrifugal flotation of lipoproteins for most studies, but it is now known that some lipoproteins are disrupted or modified during ultracentrifugation.

In 1943, Hahn[7] noted that blood plasma contained a heparin-releasable fraction that rapidly cleared chylomicron triglycerides following a dietary fat load. Subsequent work has shown that this clearing factor is lipoprotein lipase (LPL), an enzyme that is responsible for the clearance of nearly all plasma triglycerides. Since this first observation, LPL has emerged as one of the key enzymes in lipoprotein metabolism. In 1962, Glomset et al.[8] reported that plasma cholesteryl esters are formed by the action of lecithin:cholesterol acyltransferase (LCAT), an enzyme normally present in plasma. This observation led to many further studies of the role of this second key enzyme of plasma lipid transport. Identification of the role of processes catalyzed by these enzymes in atherosclerosis is now an important area of study.

Fredrickson and his colleagues[9] at the National Institutes of Health used preparative ultracentrifugation, precipitation with heparin and manganese, plasma cholesterol and triglyceride assays, and paper electrophoresis to perform a comprehensive study of hyperlipoproteinemia that defined five lipoprotein phenotypes, a sixth being subsequently added. They are summarized in Table I. Although the phenotypes may be secondary to

---

[4] J. W. Gofman, F. T. Lindgren, and H. Elliott, *J. Biol. Chem.* **179**, 973 (1949).
[5] R. J. Havel, H. A. Eder, and J. H. Bragdon, *J. Clin. Invest.* **34**, 1345 (1955).
[6] L. Rudel, J. A. Lee, M. D. Morris, and J. M. Felts, *Biochem. J.* **134**, 89 (1974).
[7] P. F. Hahn, *Science* **98**, 19 (1943).
[8] J. A. Glomset, F. Parker, M. Tjaden, and R. H. Williams, *Biochim. Biophys. Acta* **58**, 398–406 (1962).
[9] D. S. Fredrickson, R. I. Levy, and R. S. Lees, *New Engl. J. Med.* **276**, 34, 94, 148, 215, 273 (1967).

other disorders, certain ones were often found within a given kindred, and this led to speculation that specific phenotypes were associated with genetically determined hyperlipidemia. It was later observed that more than one phenotype may be present in a given genetic form of hyperlipidemia. The recognized genetic hyperlipidemias are also summarized in Table I. In addition to the hyperlipidemias, several lipid-transport disorders characterized by hypolipidemia have been described. These include abetalipoproteinemia in which chylomicrons, VLDL, and LDL are absent and hypobetalipoproteinemia in which LDL is markedly reduced. Tangier disease is characterized by an absence of normal HDL. In familial LCAT deficiency, plasma cholesteryl esters are absent and HDL is severely depressed. Finally, in obstructive liver disease, an abnormal lipoprotein-X is present.

In the mid-1960s the role of apolipoproteins became the focus of interest as methods for their isolation and chemical and physical characterization were reported. These substances proved to be heterogeneous in that each ultracentrifugal class of human plasma lipoproteins contains multiple apolipoprotein components. Gustafson et al.[10] suggested that there are three distinct apolipoprotein patterns: apoA described the proteins of HDL, apoB of LDL, and apoC of VLDL. Independently, Shore and Shore[11] and Brown et al.[12] isolated the individual apoC components. ApoC-III has multiple forms on polyacrylamide gel electrophoresis produced by varying quantities of sialic acid. Shore and Shore[13,14] were the first to separate the major HDL proteins and to establish definitively their chemical identities. These are now designated as apoA-I and apoA-II. The isolation and characterization of these proteins were rapidly confirmed by Scanu et al.,[15] Rudman et al.,[16] and Camejo et al.[17] ApoE was later isolated by the Shores and additional apolipoproteins were identified. Qualitatively, similar distributions of the major apolipoproteins were found in other mammals.

The lipoprotein nomenclature was initially filled with confusion, but a modification of the A, B, C, etc. nomenclature is now generally accepted. The term apoprotein, first coined by Oncley[18] to describe delipidated

---

[10] A. Gustafson, P. Alaupovic, and R. H. Furman, *Biochemistry* **5**, 632 (1966).
[11] B. Shore and V. Shore, *Biochemistry* **7**, 2773 (1968).
[12] W. V. Brown, R. I. Levy, and D. S. Frederickson, *J. Biol. Chem.* **245**, 6588 (1970).
[13] V. Shore and B. Shore, *Biochemistry* **7**, 3396 (1968).
[14] B. Shore and V. Shore, *Biochemistry* **8**, 4510 (1969).
[15] A. Scanu, J. Toth, C. Edelstein, S. Koga, and E. Stiller, *Biochemistry* **8**, 3309 (1969).
[16] R. Rudman, L. Garcia, and C. H. Howard, *J. Clin. Invest.* **49**, 365 (1970).
[17] G. Camejo, Z. M. Suarez, and V. Munoz, *Biochim. Biophys. Acta* **218**, 155 (1970).
[18] J. L. Oncley, in "Brain Lipids and Lipoproteins and Leukodystrophies" (J. Folch Pi and H. Bauer, eds.), p. 5. Elsevier, Amsterdam, 1963.

TABLE I
HYPERLIPOPROTEINEMIA PHENOTYPE DEFINITIONS AND THEIR ASSOCIATION WITH GENETIC AND OTHER DISORDERS[a]

Phenotype	Common name	Laboratory definition	Associated with genetic disorders	Conditions associated with secondary hyperlipoproteinemia
Type I	Exogenous hyperlipemia	Hyperchylomicronemia and absolute deficiency of LPL or PHLA Cholesterol normal Triglycerides greatly increased	Familial LPL deficiency ApoC-II deficiency	Dysglobulinemia, pancreatitis, poorly controlled diabetes mellitus
Type IIa	Hypercholesterolemia	LDL (low-density lipoproteins) increased Cholesterol increased Triglycerides normal	Familial hypercholesterolemia LDL receptor abnormal Familial combined hyperlipidemia Polygenic hypercholesterolemia	Hypothyroidism, acute intermittent porphyria, nephrosis, idiopathic hypercalcemia, dysglobulinemia, anorexia nervosa
Type IIb	Combined hyperlipidemia	LDL increased VLDL (very low-density lipoproteins) increased Cholesterol increased Triglycerides increased	Familial hypercholesterolemia Familial combined hyperlipidemia	

Type III	Dysbetalipoproteinemia	Floating β lipoproteins VLDL cholesterol/VLDL triglyceride >0.35 ApoE-II homozygote on isoelectric focusing Cholesterol increased Triglycerides increased	Familial dysbeta-lipoproteinemia	Diabetes mellitus, hypothyroidism, dysglobulinemia (monoclonal gammopathy)
Type IV	Endogenous hyperlipemia	VLDL increased Cholesterol normal or increased Triglycerides increased	Familial hypertriglyceridemia Familial combined hyperlipidemia	Glycogen storage disease, hypothyroidism, disseminated lupus erythematosus, diabetes mellitus, nephrotic syndrome, renal failure, ethanol abuse
Type V	Mixed hyperlipidemia	Chylomicrons and VLDL increased LDL present but reduced Cholesterol increased Triglycerides greatly increased	Familial hypertriglyceridemia Familial combined hyperlipidemia	Poorly controlled diabetes mellitus, glycogen storage disease, hypothyroidism, nephrotic syndrome, dysglobulinemia, pregnancy, estrogen administration (either contraceptive or therapeutic) in women with familial hypertriglyceridemia

[a] From A. M. Gotto, in "Cardiology Reference Book" (P. D. Kligfield, ed.). Co-Medica, New York, 1984.

lipoproteins, has been used in the past interchangeably with apolipoprotein. In this sense the more definitive term apolipoprotein will be used exclusively. The names and sources of all the major apolipoproteins are given in Table II. Table III lists the physical properties of the major lipoprotein classes defined operationally according to the densities at which they are isolated, as the high (HDL), low (LDL), intermediate (IDL), very low-density lipoproteins (VLDL), and the chylomicrons, which are secreted by the intestine following a dietary fat load.

As progress in the chemistry of lipoproteins progressed, new interest was attracted by two observations. One of these was that of Rothblat *et al.*[19,20] who observed that a factor associated with lipoproteins affected the flux of cholesterol between cells in culture and the medium. The second was the series of studies from the laboratory of Brown and Goldstein[21] which identified a receptor for LDL on the surface of fibroblasts that was responsible for the removal of intact LDL particles and simultaneously for reducing the activity of the rate-limiting enzyme in cholesterol biosynthesis, 3-hydroxy-3-methylglutaryl coenzyme A (HMG-CoA) reductase. In subsequent work, this receptor was isolated[22]; and its detailed structure has recently been reported.[23]

With the addition of new techniques in molecular biology, important contributions have appeared in this research area. Two groups simultaneously cloned HMG-CoA reductase,[24,25] and the LDL receptor has also been cloned.[26] Breslow *et al.*[27,28] and Shoulders *et al.*[29] provided the first reported clones for apoA-I and apoE as well as the partial sequences of

---

[19] G. H. Rothblat, M. K. Buchko, and D. Kritchevsky, *Biochim. Biophys. Acta* **164**, 327 (1968).

[20] C. H. Burns and G. H. Rothblat, *Biochim. Biophys. Acta* **176**, 616 (1969).

[21] M. S. Brown and J. L. Goldstein, *Science* **185**, 61 (1974).

[22] W. J. Schneider, U. Beisiegel, J. L. Goldstein, and M. S. Brown, *J. Biol. Chem.* **257**, 2664 (1982).

[23] T. Yamamoto, C. G. Davis, M. S. Brown, W. J. Schneider, M. L. Casey, J. L. Goldstein, and D. S. Russell, *Cell* **39**, 27 (1984).

[24] L. Liscum, K. L. Luskey, D. J. Chin, Y. K. Ho, J. L. Goldstein, and M. S. Brown, *J. Biol. Chem.* **258**, 8450 (1983).

[25] C. F. Clarke, P. A. Edwards, S. F. Lan, R. D. Tanaka, and A. M. Fogelman, *Proc. Natl. Acad. Sci. U.S.A.* **80**, 3305 (1983).

[26] D. W. Russell, T. Yamamoto, W. J. Schneider, C. J. Slaughter, M. S. Brown, and J. L. Goldstein, *Proc. Natl. Acad. Sci. U.S.A.* **80**, 7501 (1983).

[27] J. L. Breslow, D. Ross, J. McPherson, H. Williams, D. Kurnit, A. L. Nussbaum, S. K. Karathanasis, and V. I. Zannis, *Proc. Natl. Acad. Sci. U.S.A.* **79**, 6861 (1982).

[28] J. L. Breslow, J. McPherson, A. L. Nussbaum, H. W. Williams, F. Lofquist-Kjahl, S. K. Karathanasis, and V. I. Zannis, *J. Biol. Chem.* **257**, 14639 (1982).

[29] C. C. Shoulders, A. R. Kornblitt, B. S. Munro, and F. E. Baralle, *Nucleic Acids Res.* **11**, 2827 (1983).

TABLE II
CHARACTERISTICS OF PLASMA APOLIPOPROTEINS IN NORMAL FASTING HUMANS

	Plasma concentration		Distribution in lipoproteins (mol%)[b]				Major tissue source	Molecular weight
	(mg/dl)	(mol%)[a]	HDL	LDL	IDL	VLDL		
ApoA-I	130	43	100				Liver and intestine	28,016
ApoA-II	40	22	100				Liver and intestine	17,414
ApoA-IV							Liver and intestine	44,465
ApoB-48	80	5		90	8	2	Intestine	264,000
ApoB-100							Liver	550,000
ApoC-I	6	9	97		1	2	Liver	6,630
ApoC-II	3	3	60	10	10	30	Liver	8,900
ApoC-III	12	13	60		10	20	Liver	8,800
ApoD	10	5	100					22,000
ApoE-II								
ApoE-III	5	2	50	10	20	20	Liver	34,145
ApoE-IV								

[a] Based on total plasma concentration.
[b] For each apolipoprotein.

TABLE III
Physical Properties of Human Plasma Lipoprotein Families[a]

	Electrophoretic definition	Particle size (nm)	Molecular weight	Density (g/ml)
Chylomicrons	Remains at origin[b]	75–1200	~400,000,000	0.93
VLDL	Pre-β lipoproteins	30–80	10–80,000,000	0.93 –1.006
IDL	Slow pre-β lipoproteins[c]	25–35	5–10,000,000	1.006–1.019
LDL	β-Lipoproteins	18–25	2,300,000	1.019–1.063
HDL$_2$	α-Lipoproteins	9–12	360,000	1.063–1.125
HDL$_3$	α-Lipoproteins	5–9	175,000	1.125–1.210

[a] From L. C. Smith, J. B. Massey, J. T. Sparrow, A. M. Gotto, Jr., and H. J. Pownall, in "Supramolecular Structure and Function" (G. Pifat and J. N. Herak, eds.), p. 210. Plenum, New York, 1983.
[b] On paper.
[c] On geon pevikon or agarose.

the cDNA. Cheung and Chan[30] provided the first chromosomal localization of an apolipoprotein and the location of several has now been reported.

Lipoprotein Isolation

Lipoproteins may be isolated by ultracentrifugation in a salt solution, gel filtration, or precipitation. They may also be isolated with antibody affinity columns. No method is perfect and few have the economies of scale of ultracentrifugation. Gel filtration appears to cause less protein loss and should be considered when the compositional and structural integrity of the intact lipoprotein is a critical factor. Precipitation by various combinations of reagents is the fastest and most economical method but also appears to alter the structural and biological properties of some lipoproteins. This is a good, fast method, however, when one wants subsequently to isolate the apolipoproteins. Antibody affinity columns are especially useful for small scale analytical procedures in which lipoproteins containing specific apolipoproteins can be removed or isolated.

Purified apolipoproteins are typically obtained from delipidated lipoproteins after one of a number of preparative methods. ApoA-I and

[30] P. Cheung and L. Chan, Proc. Natl. Acad. Sci. U.S.A. **81**, 508 (1984).

apoA-II from apoHDL are separated by gel filtration or DEAE chromatography in guanidine hydrochloride and urea, respectively. ApoB is usually purified by solubilization and chromatography of LDL in sodium dodecyl sulfate or another detergent. ApoC proteins are obtained by ion-exchange chromatography or preparative isoelectric focusing of apoVLDL. The latter method is also used to isolate isoforms of apoE. More recently, high-performance liquid chromatography has emerged as a fast and large scale method for protein purification; we anticipate that it will eventually replace most other methods of apolipoprotein purification.

Composition of Lipoprotein Classes

In Table IV, we list the composition of each lipoprotein class. The larger lipoproteins usually have a higher content of lipids, especially neutral lipids, and as a consequence, their density decreases with increasing size. This may easily be understood in terms of the Shen/Kezdy model of lipoproteins given below, in which the relative amount of the more dense surface components (protein and phospholipid) increases with the square of the lipoprotein radius, and the amount of the less dense core components (triglyceride and cholesteryl esters) increases with the cube of the particle radius.

The fatty acid compositions of the lipids vary in the different human lipoproteins. Usually, a strong similarity exists between the fatty acid

TABLE IV
CHEMICAL COMPOSITION OF NORMAL HUMAN PLASMA LIPOPROTEINS[a]

	Surface components			Core lipids	
	Cholesterol[b]	Phospho-lipids (mol%)	Apolipo-protein	Triglycerides (mol%)	Cholesteryl esters
Chylomicrons	35	63	2	95	5
VLDL	43	55	2	76	24
IDL	38	60	2	78	22
LDL	42	58	0.2	19	81
HDL$_2$	22	75	2	18	82
HDL$_3$	23	72	5	16	84

[a] From L. C. Smith, J. B. Massey, J. T. Sparrow, A. M. Gotto, Jr., and H. J. Pownall, in "Supramolecular Structure and Function" (G. Pifat and J. N. Herak, eds.), p. 213. Plenum, New York, 1983.

[b] May be distributed between the surface and core and in the case of large chylomicrons, more cholesterol may be in the core than on the surface.

compositions of specific lipid classes. This may be due to the effects of several lipid exchange proteins, which have been identified in plasma. Oleic and linoleic are the major fatty acids of cholesteryl esters.[31] These are derived from the action of LCAT on phosphatidylcholine which almost invariably contains saturated and unsaturated fatty acids at the sn-1 and -2 positions, respectively. The unsaturation of the phospholipids in lipoproteins is sufficiently high that they are always in a fluid state at 37°. In contrast many cholesteryl ester-rich lipoproteins undergo thermal transitions around 37°. The exact temperature of these transitions is a function of triglyceride content and the fatty acid composition of the cholesteryl esters.

The complete amino acid sequence of seven of the plasma apolipoproteins is now known. These are apoA-I, apoA-II, apoC-I, apoC-II, apoC-III, apoE, and apoA-IV. With this knowledge, specific regions of these proteins which contain some of the structural and functional determinants of the parent molecules have been synthesized and tested for biological activity. Naturally occurring variants of these structures have been identified so that today there exists a group of apolipoproteinopathies, analogous to the hemoglobinopathies. The synthesis of the first intact apolipoprotein, apoC-I, was simultaneously reported by Sigler *et al.*[32] and by Harding *et al.*[34] Subsequent studies that included peptide synthesis have shown that this method permits a large number of important structural and functional determinants to be systematically tested *in vitro* and *in vivo*.[33] Many of the functions associated with the apolipoproteins are summarized in Table V.

Lipoprotein Structure: Classes and General Concepts

The plasma lipoproteins consist of five major classes and several subclasses. Each class, as separated by ultracentrifugation, is heterogeneous in size and composition. The largest of the particles are chylomicrons, which are synthesized in the gut and carry dietary triglycerides and cholesterol. They contain triglyceride as their major lipid constituent, although their content of cholesteryl ester may be very important in regulat-

---

[31] A. M. Scanu, *in* "The Biochemistry of Atherosclerosis" (A. M. Scanu, ed.), p. 3. Dekker, New York, 1979.
[32] G. Sigler, A. K. Soutar, L. C. Smith, A. M. Gotto, Jr., and J. T. Sparrow, *Proc. Natl. Acad. Sci. U.S.A.* **73**, 1422 (1976).
[33] J. T. Sparrow and A. M. Gotto, Jr., *Ann. N.Y. Acad. Sci.* **348**, 187 (1980).
[34] D. R. K. Harding, J. E. Battersby, D. R. Husbands, and W. S. Hancock, *J. Am. Chem. Soc.* **98**, 2664 (1976).

TABLE V
METABOLIC ROLES OF PLASMA APOLIPOPROTEINS IN LIPID TRANSPORT

Function	Apolipoprotein
Lipoprotein biosynthesis/secretion	B-48 (intestine)
	B-100 (liver)
Enzyme activation	
Lipoprotein lipase	C-II
Lecithin-cholesterol acyltransferase	A-I
	C-I
	A-IV
Interaction of lipoproteins with cellular receptors	
LDL receptor recognition (B/E receptor)	B-100
Chylomicron remnant receptor recognition (E receptor)	E
Inhibition of interaction with hepatic receptors	C-I
	C-II
	C-III

ing the hepatic synthesis of cholesterol. The VLDL, which are synthesized in the liver, carry endogenously synthesized triglyceride as well as cholesterol. The third class, IDL, represents an intermediate in the conversion of VLDL to LDL by lipoprotein lipase. The IDL contain relatively less triglyceride and cholesteryl ester compared to VLDL, but it is important to recognize that VLDL contain particles that may be functionally equivalent to IDL. LDL are the major carriers of cholesterol and cholesteryl ester in the plasma. Approximately 60% of the cholesterol is transported as LDL in man and about three-fourths of this is esterified. The lipids of HDL are primarily phosphatidylcholine and cholesteryl esters. HDL are much richer in protein and contain approximately one-half protein and one-half lipid by weight. HDL are usually subdivided into at least two subclasses, HDL_2 and HDL_3, for two reasons. First, rate zonal ultracentrifugation produces a bimodal distribution of HDL.[35] In contrast, other lipoprotein subclasses are part of an ultracentrifugal continuum. Second, HDL_2 appear to have a stronger inverse statistical relationship with coronary heart disease than HDL_3 (large scale epidemiologic studies are needed to substantiate this point).

Other than albumin-bound fatty acids and a few other specialized carriers such as retinol-binding protein, the plasma lipoproteins are the only recognized transport forms of lipid in the circulation. Lipoprotein

---

[35] W. Patsch, G. Schonfeld, A. M. Gotto, and J. R. Patsch, *J. Biol. Chem.* **255,** 3178 (1980).

particles may be viewed as micellar structures which are in a state of dynamic equilibrium with respect to each other as well as with various tissue and membrane compartments within the body. Not all of the individual components of the plasma lipoproteins are transferred with equal facility; unesterified cholesterol rapidly equilibrates between various lipoprotein particles and cell membranes. However, spontaneous equilibration of cholesteryl esters is very slow and requires a transfer factor; as a general rule, the very insoluble components of lipoproteins require specific transfer factors to facilitate their movement between lipid surfaces.

In 1979, Edelstein et al.,[36] extending earlier models for chylomicrons and VLDL, proposed a general structure for normal plasma lipoprotein particles in which the neutral lipids, cholesteryl esters and/or triglycerides, are separated from the external aqueous environment by a surface monolayer, consisting of the apolipoproteins and the polar lipids, which are mostly phospholipids. However, the distribution of lipids between the surface monolayer and core of these particles, like those of lipid emulsions, is a function of their phase behavior. Thus, appreciable amounts of the cholesterol may be distributed into the core and small amounts of cholesteryl esters and triglycerides are in the surface monolayer.[37]

Each of the normal plasma lipoproteins is spherical when visualized in the electron microscope by shadowing fixed material or by negative staining. One exception to this generalization includes newly synthesized or nascent lipoprotein particles, in particular, the newly synthesized HDL from the liver, which appear as disks that tend to form rouleaux in negatively stained preparations.[38] When a critical amount of the free cholesterol in these nascent particles is converted to cholesteryl ester through the action of the enzyme lecithin : cholesterol acyltransferase (LCAT), the particle assumes a spherical appearance. Thus, a certain proportion of neutral lipid, either cholesteryl ester or triglyceride appears to be necessary in order to maintain the typical spherical structure of the plasma lipoproteins.

A large body of evidence has accumulated over the past decade which indicates that the apolipoproteins contain quasidiscrete regions that have a high affinity for a lipid/water interface. The association of apolipoproteins and phospholipids is driven by the hydrophobic effect, which depends on exclusion from the aqueous phase of apolar amino acid side chains of the apolipoprotein and fatty acyl group of the phospholipid. The

[36] C. Edelstein, F. Kezdy, A. M. Scanu, and B. W. Shen, *J. Lipid Res.* **20**, 143 (1979).
[37] D. M. Small and G. G. Shipley, *Science* **185**, 222 (1974).
[38] R. L. Hamilton, M. Williams, C. Fielding, and R. J. Havel, *J. Clin. Invest.* **58**, 667 (1976).

polar head groups of the phospholipids do not appear to be involved in direct binding to the apolipoproteins. A general structural model of a lipoprotein, for which we use HDL as an example, is given in Fig. 1. The apolipoproteins and the phospholipids are shown as a surface monolayer with the fatty acyl groups of the phospholipids oriented toward the center of the lipoprotein; the cholesteryl esters and triglycerides form a core at the center and the unesterified cholesterol molecules mainly occupy a volume intermediate between the surface and the center.

Structural Determinants of the Mechanism of Lipid Protein Interaction

One of the roles of apolipoproteins is to solubilize lipids for transport from one part of the body to another. Many early studies focused on identification of the structural determinants in an apolipoprotein that are

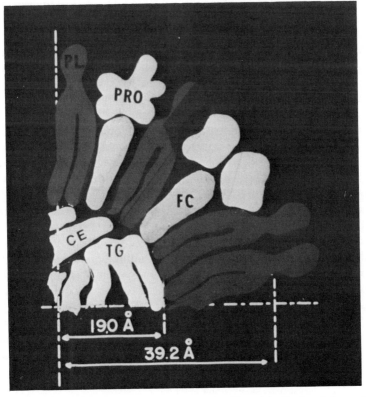

FIG. 1. Model of high density lipoprotein (HDL). (Copyright Baylor College of Medicine.)

involved in association with phospholipid surfaces. The role of helical regions in lipid–apolipoprotein association was first put forth by Segrest et al.[39] Originally it was proposed that specific regions of apolipoproteins had the potential to form what were called amphipathic helices. These regions contain both polar and nonpolar amino acid residues, which are distributed on opposite sides of an α-helix. Based in part upon model building, they found that the acidic and basic amino acid residues, respectively, were located near the center and edge of the polar face; frequently the acidic and basic amino acid residues appeared as pairs that were either adjacent to each other or separated by two nonpolar amino acid residues. Nonpolar amino acids are thought to penetrate part way into the lipid surface: the magnitude of the affinity of an apolipoprotein for phospholipid surfaces is a function of the hydrophobicity of the nonpolar face of the helix. The polar residues on the opposite side of the helix are in contact with the surrounding aqueous phase. It is this structural combination of the polar and nonpolar amino acid residues which leads to their detergent-like properties. We speculate that the role of the polar amino acid residues is to keep the apolipoproteins at the surface of the lipoprotein particle. Such a location would facilitate their transfer between lipoprotein particles and might well be crucial to other specific functions of apolipoproteins, including enzyme activation and the exposure of recognition ligands for binding by receptors on the surface of cells.

Evidence for the amphipathic helix and for its contribution to the stabilization of lipoprotein structure comes from many types of studies. These include protein sequencing, peptide synthesis, preparation of specific synthetic peptide derivatives, and a variety of physicochemical approaches. Other valuable information is derived from the careful investigation of the reassembly of lipoproteins, in particular of HDL. One of the characteristics of the amphipathic helical regions is the tendency to self-associate, a phenomenon which must be taken into account and which can also result in an increase in helical content. From chemical studies involving spectroscopic techniques, the binding of an apolipoprotein to a lipoprotein particle has been shown to involve the transfer of the hydrophobic residues of the amphipathic helix to a hydrocarbon-like phase within the lipoprotein. The polar groups should not contribute significantly to this energy since their environments are similar in the aqueous and lipid-bound forms. Thus, the polar groups and, in particular, the charged groups remain in contact with the aqueous phase surrounding a lipoprotein. This relationship also keeps the hydrophilic residues of the

---

[39] J. P. Segrest, J. D. Morrisett, R. L. Jackson, and A. M. Gotto, *FEBS Lett.* **38**, 274 (1974).

amphipathic helix on the surface of the lipoprotein where their respective physiological functions can be expressed.

A more quantitative model of the surface-seeking properties of the apolipoproteins is provided by the analysis of Eisenberg et al.[40] in which this property is called the helical hydrophobic moment. This algorithm has been applied to the analysis and design of surface-associating peptides. We anticipate that this analysis with additional refinements will greatly aid our understanding of the mechanism of lipid–protein association and the design of active surface-associating peptides.

Lipolytic Enzymes and Transfer Factors

LPL is the enzyme responsible for the hydrolysis of the majority of triglycerides and some of the phospholipid of the triglyceride-rich lipoproteins, VLDL and the chylomicrons. LPL is activated by apoC-II,[41,42] a component of HDL and triglyceride-rich lipoproteins. Presumably, a portion of apoC-II resides in the HDL until secretion of chylomicrons or VLDL, whereupon it is transferred to the triglyceride-rich particles to stimulate lipolysis. A number of cases of apoC-II deficiency have been reported in subjects with severe hypertriglyceridemia.[43,44] Clinically, these resemble lipoprotein lipase deficiency except for the absence of cutaneous xanthomas. A hepatic lipase, which like lipoprotein lipase is released into the blood by heparin, also hydrolyzes phospholipids and triglycerides.[45] It may be distinguished from LPL activity by the lack of activation by C-II and by the absence of inhibition by salt. Its relative importance in the hydrolysis of VLDL and chylomicron lipids is difficult to estimate. A role for hepatic lipase in the degradation of lipids of IDL as well as those of HDL_2 has been speculated.[46]

LCAT is another major enzyme involved in plasma lipid metabolism and is responsible for the formation of nearly all plasma cholesteryl esters in normal persons.[47] The substrates for this enzyme are phosphatidylcho-

---

[40] D. Eisenberg, R. M. White, and T. C. Terwilliger, *Nature (London)* **299**, 371 (1982).
[41] R. J. Havel, V. G. Shore, B. Shore, and D. M. Bier, *Circ. Res.* **27**, 595 (1970).
[42] J. C. LaRosa, R. I. Levy, P. Herbert, S. E. Lux, and D. S. Fredrickson, *Biochem. Biophys. Res. Commun.* **41**, 57 (1970).
[43] D. W. Cox, W. C. Breckenridge, and J. A. Little, *New Engl. J. Med.* **299**, 1421 (1978).
[44] W. C. Breckenridge, J. A. Little, G. Steiner, A. Chow, and M. Poapst, *New Engl. J. Med.* **298**, 1265 (1978).
[45] P. Nilsson-Ehle, A. S. Garfinkel, and M. C. Schotz, *Annu. Rev. Biochem.* **49**, 667 (1980).
[46] R. L. Jackson, *in* "The Enzymes" (P. D. Boyer, ed.), Vol. XVI, p. 49. Academic Press, New York, 1983.
[47] J. A. Glomset, *J. Lipid Res.* **9**, 155 (1968).

line and cholesterol, and the products are cholesteryl esters and lysolecithin. Both hydrolysis and transesterification are activated by apoA-I, which is the major protein component of HDL, the putative physiological substrate.[48] ApoC-I is also an activator and recent evidence suggests that apoA-IV is as well. Fatty acids located at the sn-2 position of phosphatidylcholine are the preferred, although not exclusive, site of cleavage. LCAT preferentially cleaves unsaturated fatty acyl groups and the early reports on its specificity for the sn-2 position must have been influenced by the relative high abundance of sn-1-saturated–sn-2-unsaturated phosphatidylcholines. Further studies on LCAT depend in part on the development of a large-scale purification procedure that yields a homogeneous protein.

Acyl:coenzyme A-O-acyltransferase (ACAT) is an additional source of cholesteryl esters that are formed intracellularly by an enzyme that is located in the endoplasmic reticulum.[49] Following a cholesterol-rich diet, this enzyme forms cholesteryl esters in the intestinal wall; they later appear in the plasma compartment as cholesteryl ester-rich lipoproteins of intestinal origin, and in some mammalian species, of hepatic origin.

A final group of proteins that mobilize lipids from one compartment to another are the lipid-transfer proteins. These are the least well characterized of the proteins involved in lipoprotein metabolism; only one report of a homogeneous protein is available.[50] Lipid transfer among plasma lipoproteins and cell membranes is an important physiological process. Three mechanisms are proposed for these processes. One of these is that lipids are exchanged or transferred during a "sticky" collision between the surfaces of membranes or lipoproteins. Although this mechanism cannot be excluded, little evidence supports it. A second mechanism involves the rate-limiting desorption of single lipid molecules from a membrane or lipoprotein surface into the surrounding aqueous phase from which it rapidly diffuses to an accessible pool of lipid surfaces. This mechanism appears to be applicable to sparingly soluble substances such as cholesterol[51] and short chained phosphatidylcholines,[52] it becomes less important as the chainlength of the lipid, or hydrophobicity of other amphiphiles increases.[53] The relative importance of the desorption process is usually difficult to evaluate *in vivo* because of competing processes, especially lipid hydrolysis. A third mechanism is believed to dominate the transport

[48] L. Aron, S. Jones, and C. J. Fielding, *J. Biol. Chem.* **253**, 7220 (1978).
[49] D. S. Goodman, D. Deykin, and T. Shiratori, *J. Biol. Chem.* **239**, 1335 (1964).
[50] J. J. Albers, J. H. Tollefson, C. H. Chen, A. Steinmetz, *Arteriosclerosis* **4**, 49 (1984).
[51] L. R. McLean and M. C. Phillips, *Biochemistry* **20**, 2893 (1981).
[52] J. B. Massey, A. M. Gotto, and H. J. Pownall, *Biochemistry* **21**, 3630 (1982).
[53] H. J. Pownall, D. L. Hickson, and L. C. Smith, *J. Am. Chem. Soc.* **105**, 2440 (1983).

of very insoluble lipids between lipoprotein or membrane surfaces. This is the transfer of lipid molecules by specific protein factors in plasma. Harmony and co-workers[54] first reported this process for phosphatidylcholine transfer; Zilversmit and co-workers[55] showed that a similar, though not necessarily identical factor is responsible for the transfer of cholesteryl esters. The role of these factors in lipid metabolism is not well understood at this time. Exchange proteins tend to equilibrate the fatty acid compositions of lipids among lipoproteins. This effect may be important in maintaining the "quality" of the interfacial regions of those lipids which are substrates for lipolytic enzymes. In cases where net transfer and not exchange of the same lipid occurs, this mechanism may be an important vehicle for the transfer of lipids from one lipoprotein to another or to cell membranes. The physiological importance of the exchange proteins is presently under intense investigation. Phospholipid exchange proteins seem to be active in all mammals examined to date. By contrast, the activity of cholesteryl ester/triglyceride exchange proteins varies widely among species; in some, little or no activity has been found.

Lipoprotein and Apolipoprotein Synthesis and Secretion

Formation of Triglyceride-Rich Lipoproteins in Intestine and Liver

The formation of triglyceride-rich particles (chylomicrons) in the intestine and their secretion into intestinal lymph during absorption of dietary fat have been known since the pioneering studies of Gage and Fish in 1924.[56] Evidence that the liver secretes triglyceride-rich particles (VLDL) began to accumulate in the late 1950s. Abundant evidence supports the concept that the absorptive cells of the small intestine and hepatocytes are the source of virtually all circulating chylomicrons and VLDL particles.[57] Electron microscopic studies have shown that particles within secretory vesicles of the Golgi apparatus which resemble circulating chylomicrons and VLDL in size and staining characteristics represent nascent lipoprotein particles in these cells.[58,59]

---

[54] M. E. Brewster, J. Ihm, J. R. Brainard, and J. A. K. Harmony, *Biochim. Biophys. Acta* **529**, 147 (1978).
[55] D. B. Zilversmit, L. B. Hughes, and J. Balmer, *Biochim. Biophys. Acta* **409**, 393 (1975).
[56] S. H. Gage and P. A. Fish, *Am. J. Anat.* **34**, 1 (1924).
[57] R. J. Havel, J. L. Goldstein, and M. S. Brown, in "Metabolic Control and Disease" (P. K. Bondy and L. E. Rosenberg, eds.), 8th Ed., p. 393. Saunders, Philadelphia, 1980.
[58] R. L. Hamilton, in "Plasma Protein Secretion by the Liver" (H. Glaumann, T. Peters, Jr., and C. Redman, eds.), p. 357. Academic Press, London, 1984.
[59] S. Sabesin and S. Frase, *J. Lipid Res.* **18**, 496 (1977).

The secretion of triglycerides in chylomicrons by the intestinal epithelium is a direct function of the rate of fat absorption; it is accomplished primarily by an increase in the size of particles secreted, and to a lesser extent by an increase in the number of secreted particles.[60] The former presumably reflects the accommodation of more triglyceride molecules within a single nascent chylomicron particle, whereas the latter may reflect the recruitment of additional cells in the absorptive process. Within a given cell, however, and even within individual secretory vesicles, the size of nascent particles may vary considerably. A similar variation of particle size of VLDL is evident within secretory vesicles of hepatocytes and, as in the intestine, more triglyceride can be secreted when hepatic triglyceride synthesis increases either by secretion of larger particles or of more particles. Variation in hepatic triglyceride synthesis is mainly the result of the rate of uptake of fatty acids or glucose from the blood and the extent to which these precursors are utilized for oxidative metabolism of hepatocytes.[61] It is important to recognize that chylomicrons and VLDL continue to be secreted in the postabsorptive state. Chylomicron triglycerides are then derived from fatty acids of biliary phospholipids or of lipids of cells shed from the intestinal mucosa, whereas VLDL triglycerides are derived mainly from free fatty acids transported to the liver from adipose tissue.[57]

Dietary cholesterol is also transported in chylomicrons, mainly as cholesteryl esters synthesized by ACAT within the absorptive cells.[62] The relative numbers of triglyceride and cholesteryl ester molecules present within chylomicron particles are therefore a function of the rates of absorption of fatty acids and cholesterol. In an analogous manner, VLDL secreted from the liver contain both nonpolar lipids. However, in normal humans, the activity of hepatic ACAT is low[63] and nascent VLDL contain few cholesteryl esters. When ACAT activity and storage of cholesteryl esters in hepatocytes are increased (for example, by cholesterol feeding), nascent VLDL may be greatly enriched in this lipid.[64] Minor nonpolar lipids, such as carotenoids and retinyl and tocopheryl esters, are also incorporated into nascent chylmicrons.

The nonpolar esters of long-chain fatty acids transported in triglyceride-rich lipoproteins are synthesized in the endoplasmic reticulum, as are the phospholipids, which are composed mainly of zwitterionic leci-

[60] R. J. Havel, in "High Density Lipoproteins and Atherosclerosis" (A. M. Gotto, Jr., N. E. Miller, and M. F. Oliver, eds.), p. 21. Elsevier, Amsterdam, 1978.
[61] R. J. Havel, *New Engl. J. Med.* **287,** 1186 (1972).
[62] K. R. Norum, T. Berg, P. Helgerud, and C. A. Drevon, *Physiol. Rev.* **63,** 1343 (1983).
[63] S. Erickson and A. D. Cooper, *Metabolism* **29,** 991 (1980).
[64] L. S. S. Guo, R. L. Hamilton, R. Ostwald, and R. J. Havel, *J. Lipid Res.* **23,** 543 (1982).

thins and sphingomyelins. The requisite enzymes are thought to reside primarily in the outer leaflet of the membrane of this organelle[65] and it is not known how the newly synthesized lipids are transported into the cisternal space where the nascent particles appear to be segregated. One possibility is that the nonpolar lipids are segregated within the bilayer of the membrane and bud off, surrounded by the polar lipids that form the bilayer.[62] Immunocytochemical experiments have failed to demonstrate an association of apolipoproteins (specifically apolipoprotein B) in association with putative nascent chylomicrons or VLDL within cisternae of the smooth endoplasmic reticulum.[66,67] Some observations suggest that the proteins are added to the particle surface at a point where smooth and ribosome-associated (rough) endoplasmic reticulum are joined.[66] As with many other secreted proteins, the apolipoproteins are synthesized on attached ribosomes of the rough endoplasmic reticulum, together with a signal peptide which is rapidly cleaved. They can be visualized on ribosomes and within cisternae of the rough endoplasmic reticulum by immunocytochemical methods.[66] Occasionally, a particle resembling a lipoprotein associated with immunoreactive apolipoprotein can be visualized at the smooth-surfaced termini of the rough endoplasmic reticulum. The nascent particle is thought to be transported to the Golgi apparatus within cisternae of the endoplasmic reticulum, but it is not known whether this occurs by transport within vesicles that bud off from the reticulum or via tubules that connect the endoplasmic reticulum to the Golgi cisternae. Finally, the particles are concentrated within secretory vesicles of the Golgi apparatus, which appear to bud off, migrate to, and fuse with the plasma membrane—in the intestine, with the basolateral membrane of the cell and in the hepatocyte, with the membrane at the sinusoidal front. Thereby, the nascent particles are released into the extracellular space. Secreted nascent chylomicrons have ready access to the lacteals of the intestinal villi via their loose junctional complexes and secreted nascent VLDL have access to the blood via the sieve-plate fenestrae of the sinusoidal endothelium of the liver (Fig. 2).

B apolipoproteins are thought to be essential for the assembly of nascent triglyceride-rich lipoproteins that can be secreted from the cell. This concept is based on the effect of inhibitors of protein synthesis on the secretion of these lipoproteins[68] and the expression of the human genetic

[65] R. Coleman and R. M. Bell, *J. Cell Biol.* **76**, 245 (1978).
[66] C. A. Alexander, R. L. Hamilton, and R. J. Havel, *J. Cell Biol.* **69**, 241 (1976).
[67] N. J. Christiansen, C. E. Rubin, M. C. Cheung, and J. J. Albers, *J. Lipid Res.* **24**, 1229 (1983).
[68] P. Siuta-Mangano, D. R. Janero, and D. M. Lane, *J. Biol. Chem.* **257**, 11463 (1982).

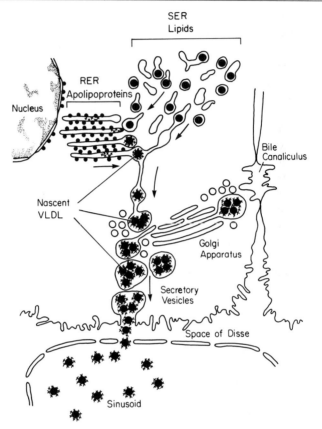

FIG. 2. General scheme for the assembly and secretion of VLDL from hepatocytes. The nascent particle is thought to originate from the membranes of the smooth endoplasmic reticulum (SER) and to be sequestered within the SER lumen. Apolipoprotein B (and other apolipoproteins) are synthesized on ribosomes of the rough endoplasmic reticulum (RER) and after entry into the RER lumen migrate to its smooth-surfaced terminus, where they associate with the nascent particle. The nascent VLDL are then transported to the Golgi apparatus, where the proteins are further glycosylated. The particles are concentrated in secretory vesicles, which bud off and fuse with the plasma membrane, leading to secretion. (From Alexander et al.[66] © 1976 The Rockefeller University Press.)

disorder, abetalipoproteinemia, in which no circulating lipoproteins containing B apolipoproteins are present and triglycerides accumulate in fat droplets in enterocytes and hepatocytes.[69] Circulating HDL containing

[69] P. N. Herbert, A. M. Gotto, and D. S. Fredrickson, in "The Metabolic Basis of Inherited Disease," (J. B. Stanbury, J. B. Wyngaarden, and D. S. Fredrickson, eds.), 4th Ed., p. 544. McGraw-Hill, New York, 1978.

the other recognized apolipoproteins are present in this disorder. Similarly, in orotic acid-fed rats, the liver continues to secrete normal amounts of nascent HDL, but cannot secrete VLDL.[58] In these animals, particles resembling nascent VLDL do not reach the secretory vesicles of the Golgi apparatus, but accumulate within cisternae of the endoplasmic reticulum and a markedly fatty liver develops.[69] The defect in these animals is unknown.

Glycosylation of apolipoprotein B is not essential to secretion of VLDL.[68] The kinetics of assembly of apolipoprotein B and the various lipid components of nascent VLDL are consistent with early formation of a core-glycosylated apoB–phospholipid complex to which a triglyceride-rich particle is then added.[70] Subsequent terminal glycosylation of apolipoproteins occurs in the Golgi apparatus.

The properties of nascent chylomicrons and VLDL separated by rupture of secretory vesicles of Golgi apparatus-rich fractions of intestinal mucosa and liver resemble in many respects those of newly secreted particles in intestinal lymph and liver perfusates. Data obtained from analysis of particles released from hepatic Golgi fractions need to be interpreted with caution, because such fractions may be variably contaminated with multivesicular bodies that contain endocytosed lipoproteins.[71] However, the surface components of nascent rat chylomicrons and VLDL seem to differ in important respects from their counterparts in blood plasma. In general, nascent particles are richer in phospholipids and poorer in cholesterol and apolipoproteins.[58] Presumably, cholesterol and apolipoproteins are added by equilibration with blood cell and lipoprotein surfaces, via small quantities of these components in free solution. The factors responsible for the affinities of the apolipoprotein and phospholipid components to available surfaces are poorly understood, but are thought to involve the composition of the particle and radius of curvature of its surface.

The acquisition of apolipoproteins by the nascent particles is critical to their subsequent metabolism. Rat intestine synthesizes very little of the C apolipoproteins (C-I, C-II, and the several C-IIIs) or of apolipoprotein E.[60] Consequently, these proteins are added after secretion, primarily by transfer from HDL. This occurs to a minor extent in the extracellular space of the intestinal villus, and to a major extent in the blood. In addition to apolipoprotein B-48, which does not exchange with other particles, nascent chylomicrons contain newly synthesized apolipoproteins A-I,

---

[70] D. R. Janero and M. D. Lane, *J. Biol. Chem.* **258**, 14496 (1983).
[71] C. A. Hornick, R. L. Hamilton, E. Spaziani, G. H. Enders, and R. J. Havel, *J. Cell Biol.* **100**, 1558 (1985).

A-II, and A-IV. The A apolipoproteins are transferred in part to HDL in exchange for the added apolipoproteins or phospholipids immediately after secretion or entry into the blood.[60] The liver is the major site of synthesis of the C apolipoproteins; nascent VLDL contain some of these proteins, but considerably more are acquired from HDL after secretion.[58] Nascent VLDL also contain substantial amounts of apolipoprotein E, which is also synthesized in hepatocytes.[58] In the rat, nascent VLDL particles contain either apolipoprotein B-100 or a protein resembling B-48.[72] In other mammals that have been studied (guinea pig, rabbit, dog, and several primates), little or no B-48 is found in VLDL in hepatic perfusates.[73]

Formation of HDL

Unlike triglyceride-rich lipoproteins, nascent HDL have not been successfully identified within subcellular compartments or isolated from Golgi-rich fractions. However, nascent HDL have been consistently found in hepatic perfusates in several mammals (rats, guinea pigs, rabbits, and primates).[58] In the absence of active LCAT (which is also secreted by the liver), nascent HDL appear as discoidal particles about 200 Å in diameter and 45 Å thick. These particles are composed of a bilayer of phospholipids (mainly lecithin). The proteins are thought to be associated mainly with the disk margin.[74] Similar particles can be produced *in vitro* from lecithin liposomes and several apolipoproteins. The origin of the particles in hepatic perfusates is poorly defined, but it is clear that their major proteins (apolipoproteins E and A-I) are synthesized in the liver. In liver perfusates of orotic acid-fed rats, typical discoidal HDL also accumulate in the absence of VLDL, indicating that these particles are formed independently.[58] When LCAT is active in liver perfusates, many of the HDL particles are spherical, contain appreciable amounts of cholesteryl esters, and otherwise resemble HDL in blood plasma.

Particles similar to nascent discoidal HDL have also been found in intestinal lymph of rats, as well as small spherical HDL.[75,76] Mesenteric lymph contains HDL transferred into the extracellular space of the intes-

---

[72] J. P. Kane, *Annu. Rev. Physiol.* **45,** 637 (1983).
[73] R. J. Havel, *in* "Atherosclerosis VI" (F. G. Schettler, ed.), p. 480. Springer-Verlag, Berlin, 1983.
[74] A. Wlodauer, J. P. Segrest, B. H. Chung, R. Chiavetti, Jr., and J. N. Weinstein, *FEBS Lett.* **104,** 231 (1979).
[75] P. H. R. Green and J. Glickman, *J. Lipid Res.* **22,** 1153 (1981).
[76] G. P. Forester, A. P. Tall, C. L. Bisgaier, and R. M. Glickman, *J. Biol. Chem.* **258,** 5938 (1983).

tinal villus, a property which makes it difficult to distinguish nascent from preformed particles. In addition, much of the newly synthesized apolipoprotein A-I in intestinal lymph is usually associated with chylomicrons, and this protein is readily dissociated from the chylomicron surface by centrifugation.[77] Therefore, some "nascent" HDL particles could arise from the chylomicron surface.

The major protein of plasma HDL, apolipoprotein A-I, is secreted from the liver and intestine. The intestine is a major site of synthesis in the rat[78] and apoA-I is a major component of lymph chylomicrons in all mammals studied to date. By contrast, the contribution by the liver seems to vary. In the rabbit, little apolipoprotein A-I appears in hepatic perfusates.[79] Most of the apolipoprotein A-I derived from the intestine reaches HDL indirectly from the chylomicron surface, either as part of the passive exchange of surface components described above or during the formation of chylomicron remnants. As discussed below, a large fraction of the surface components is transferred to HDL during the hydrolysis of chylomicron triglycerides by lipoprotein lipase. This fraction includes essentially all of the A apolipoproteins and a large fraction of the C apolipoproteins and phospholipids.[60,80] Similarly, C apolipoproteins and phospholipids are transferred to HDL during the hydrolysis of VLDL triglycerides by lipoprotein lipase.[80,81]

The major apolipoproteins of HDL, apoA-I and apoA-II, are initially synthesized as preproteins. The prosegment of human apoA-II terminates with typical paired basic amino acids,[82] but that of apoA-I terminates with paired glutamine residues and apoA-I is secreted as a proprotein, the prosegment being subsequently cleaved by a metal-dependent protease.[83] The functional significance of this extracellular cleavage is unknown.

HDL are polymorphic and consist of several more or less discrete subspecies. Many of these subspecies are remodeled by acquisition of surface components of triglyceride-rich lipoproteins.[80,81] In animals in

---

[77] K. Imaizumi, M. Fainaru, and R. J. Havel, *J. Lipid Res.* **19,** 712 (1978).
[78] A. L. Wu and H. G. Windmueller, *J. Biol. Chem.* **254,** 7316 (1979).
[79] L. De Parscau and P. E. Fielding, *J. Lipid Res.* **25,** 721 (1984).
[80] O. D. Mjøs, O. Faergeman, R. L. Hamilton, and R. J. Havel, *J. Clin. Invest.* **56,** 603 (1975).
[81] J. R. Patsch, A. M. Gotto, T. Olivecrona, and S. Eisenberg, *Proc. Natl. Acad. Sci. U.S.A.* **75,** 4519 (1978).
[82] J. I. Gordon, K. A. Dobelier, H. F. Semis, C. Edelstein, A. M. Scanu, and A. W. Strauss, *J. Biol. Chem.* **258,** 14054 (1983).
[83] C. Edelstein, J. Gordon, K. Toscas, H. F. Semis, A. F. Strauss, and A. M. Scanu, *J. Biol. Chem.* **258,** 11430 (1983).

which most of the A apolipoprotein is derived from the intestine, it is likely that HDL arise largely as products of chylomicron metabolism. However, some HDL precursor particles or subspecies evidently can be produced by the liver and, perhaps, by the intestine as independent secretory products.

Synthesis of Apolipoproteins at Sites Other Than Enterocytes and Hepatocytes

Cells, other than those that secrete triglyceride-rich lipoproteins, can synthesize some of the recognized apolipoproteins. Apolipoprotein E is synthesized in human kidney and adrenal glands.[84] Significant levels of apoE mRNA have been found in several tissues of the rat, marmoset, and cynomolgous monkey, among which the highest levels are in the liver, adrenal glands, brain, and spleen.[85,86] This protein is synthesized and secreted from mouse and human macrophages, especially when the cell is activated by certain substances[87] and by cholesterol loading.[88] In those tissues that contain low amounts of the mRNA, resident macrophages could therefore be responsible for synthesis of apoE. However, other cells evidently can synthesize the protein. For example, the apoE in the brain is largely contained within astrocytes or related cells.[89]

In birds, apoA-I and apoB are also synthesized in the kidney[90,91] and in skeletal muscles,[90] especially during late embryonic life.[92]

Relatively little is known about the form in which apoE and apoA-I are secreted from cells other than enterocytes or hepatocytes. A substantial fraction of the apoE accumulating in the medium of cultured macrophages has the density of HDL; some discoidal particles have been found.[93] Given the affinity of apoE and apoA-I for HDL, they probably associate

[84] M. L. Blue, D. L. Williams, S. Zucker, S. A. Khan, and C. B. Blum, *Proc. Natl. Acad. Sci. U.S.A.* **80**, 283 (1983).
[85] N. A. Elshourbagy, W. S. Liao, R. W. Mahley, and J. M. Taylor, *Proc. Natl. Acad. Sci. U.S.A.* **82**, 203 (1985).
[86] T. C. Newman, P. A. Dawson, L. L. Rudel, and D. L. Williams, *Circulation* **II**, 120 (1984).
[87] Z. Werb and J. R. Chen, *J. Biol. Chem.* **258**, 10642 (1983).
[88] S. K. Basu, M. S. Brown, Y. K. Ho, R. J. Havel, and J. L. Goldstein, *Proc. Natl. Acad. Sci. U.S.A.* **78**, 7545 (1981).
[89] J. K. Boyles, R. E. Pitas, and R. W. Mahley, *Circulation* **II**, 120 (1984).
[90] M.-L. Blue, P. Ostapchuk, J. S. Gordon, and D. L. Williams, *J. Biol. Chem.* **257**, 11151 (1982).
[91] M.-L. Blue, A. A. Protter, and D. L. Williams, *J. Biol. Chem.* **255**, 10048 (1980).
[92] J. E. Shackelford and H. G. Lebherz, *J. Biol. Chem.* **258**, 14829 (1983).
[93] S. K. Basu, Y. K. Ho, M. S. Brown, D. W. Bilheimer, R. G. W. Anderson, and J. L. Goldstein, *J. Biol. Chem.* **257**, 9795 (1982).

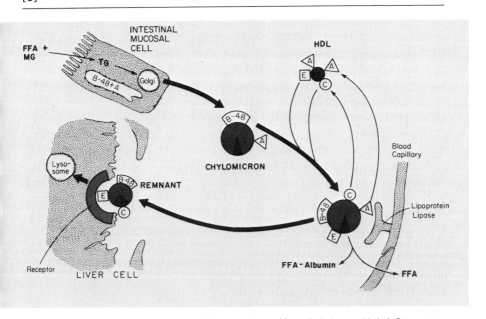

FIG. 3. The pathway for transport of dietary triglyceride and cholesterol is briefly summarized in this diagram, which shows the two steps in chylomicron metabolism following secretion of nascent chylomicrons from intestinal mucosal cells. These steps (lipolysis by lipoprotein lipase in extrahepatic tissues and receptor-mediated uptake of chylomicron remnants into hepatocytes) are discussed in the text. (Modified from R. J. Havel, *in* "Medical Clinics of North America," Vol. 66, No. 2, pp. 319–333, 1982, with permission from W. B. Saunders, Philadelphia.)

in the medium with HDL particles even if they are not secreted as such. In principle, such particles could acquire cholesterol by transfer from cell surfaces and thereby participate in cholesterol transport. Discoidal HDL which have been observed in leg lymph of dogs, especially in animals with diet-induced hypercholesterolemia,[94] could be synthesized in macrophages of the bone marrow or other peripheral cells.

Lipoprotein Processing and Catabolism

Chylomicrons (Fig. 3)

The acquisition of C and E apolipoproteins by nascent chylomicrons in the interstitial fluid of the intestinal villus and in the blood is critical to

---

[94] C. H. Sloop, L. Dory, R. L. Hamilton, B. R. Krause, and P. S. Roheim, *J. Lipid Res.* **24**, 1429 (1983).

their subsequent metabolism, which proceeds in two distinct steps. The chylomicron particles initially bind rapidly to LPL on the endothelial surface of blood capillaries in adipose tissue, cardiac and skeletal muscles, and several other organs, including the brain.[95,96] ApoC-II appears to modify the enzyme–substrate interaction such that the catalytic rate for chylomicron triglycerides is increased many-fold. Most of the triglycerides and some of the component phosphatidylcholine in each chylomicron particle are hydrolyzed by LPL within a few minutes. Comcomitantly, residual A apolipoproteins and most of the C apolipoproteins and remaining phospholipids are transferred to HDL.[80,97] The mode of transfer is uncertain, but may involve micelles containing polar lipids and apolipoproteins as well as transfer of protein monomers through the aqueous phase. The residual chylomicron particle tends to retain its basic structure, which consists of a monolayer, composed mainly of amphiphiles, surrounding a shrunken core that contains the cholesteryl esters, other minor nonpolar lipids and some residual triglycerides. With the loss of apoC-II, together with other changes in the composition of the monolayer, the modified chylomicron can no longer compete effectively with unhydrolyzed chylomicrons for the enzyme, and recirculates in the blood as a "chylomicron remnant." Although most of the component cholesteryl esters of chylomicrons are retained in remnants, some are transferred to the endothelial cells during the period that the chylomicron is closely associated with LPL at the endothelial surface.[98]

In the second step of chylomicron catabolism, the remnant particle is taken up by hepatocytes through receptor-mediated endocytosis. The chylomicron remnant receptor on the sinusoidal surface of hepatocytes appears to recognize apoE.[99] Binding of chylomicron remnants to the receptor is impeded by each of the C apolipoproteins which may prevent premature uptake of small unhydrolyzed chylomicrons by the remnant receptor.[99] Loss of C apolipoproteins during remnant formation not only helps to terminate triglyceride hydrolysis, but permits binding to receptors. For the large chylomicrons (>2000 Å in diameter) that are produced after ingestion of substantial fat loads, the size of fenestrae in the hepatic sinusoidal endothelium presumably limits access of the unhydrolyzed particles to the hepatocyte. The reduced affinity of C apolipoproteins for

---

[95] A. Cryer, *Int. J. Biochem.* **13**, 525 (1981).
[96] R. H. Eckel and R. J. Robbins, *Proc. Natl. Acad. Sci. U.S.A.* **81**, 7604 (1984).
[97] J.-L. Vigne and R. J. Havel, *Can. J. Biochem.* **59**, 613 (1981).
[98] C. J. Fielding, *J. Clin. Invest.* **62**, 141 (1978).
[99] R. J. Havel, in "Treatment of Hyperlipoproteinemia" (L. A. Carlson and A. G. Oleson, eds.), p. 1. Raven, New York, 1984.

chylomicron remnants likely depends on the altered lipid composition of the particle surface. Once the remnant is bound to the receptor, it is taken into the cell through coated pits, which pinch off to form a primary endosome. These endosomes appear to fuse beneath the cell surface to form organelles that migrate to the opposite pole of the cell, near the bile canaliculus, where they are seen as multivesicular bodies.[100] The endosomes and multivesicular bodies contain an ATP-driven proton pump which acidifies the endosomal space.[101] The reduced pH is thought to cause dissociation of the receptor from its ligand. Such dissociation is known in other cases (as with the LDL receptor, described below) to be followed by recycling of the receptor to the cell surface, but this has not been established for the remnant receptor. Primary lysosomes, derived from the nearby Golgi apparatus, fuse with multivescular bodies, releasing acid hydrolases.[102] This process leads to virtually complete lipolysis and proteolysis of remnant components, which presumably pass through the membrane of the secondary lysosomes and are thus made available for metabolic processes within the cell. The released cholesterol can be excreted from the hepatocyte into the nearby biliary canaliculi. Alternatively, the cholesterol may be oxidized to primary bile acids prior to excretion or it can be used in the biosynthesis of lipoproteins.

The first step of chylomicron catabolism, catalyzed by LPL, is highly regulated. The enzyme is thought to be synthesized in parenchymal cells of the tissues that contain it and then secreted for transport to the capillary endothelium, where it appears to be bound by heparan sulfate to the cell surface.[45,95] Although binding to sulfated glycosaminoglycans stabilzies the enzyme *in vitro,* it turns over rapidly *in vivo,* with a half-life of a few hours. The activity of the enzyme is influenced by several hormones, of which insulin seems to be the most important for short-term regulation. In adipose tissue in particular, synthesis of the enzyme is promoted by insulin.[46,95] In the mammary gland, synthesis of the enzyme seems to be coupled to the functional activity of the secretory epithelial cells, as regulated by prolactin.[95] These changes promote the hydrolysis of chylomicron triglycerides in adipose tissue during anabolism or in the mammary gland during lactation. The liberated fatty acids are rapidly taken up into the tissue under these conditions and reesterified to form cellular trigly-

---

[100] A. L. Jones, G. T. Hradek, C. Hornick, G. Renaud, E. E. T. Windler, and R. J. Havel, *J. Lipid Res.* **25,** 1151 (1984).
[101] R. W. Van Dyke, C. A. Hornick, J. Belcher, B. F. Scharschmidt, and R. J. Havel, *J. Cell Biol.* **260,** 11021 (1985).
[102] E. Jost-Vu, R. L. Hamilton, C. A. Hornick, J. D. Belcher, R. J. Havel, *J. Histochem. Cytochem.* **34,** 120 (1986).

cerides, which are stored (in adipose tissue) or subsequently secreted as fatty globules (in the lactating mammary gland). In cardiac and red skeletal muscles, the activity of the enzyme tends to vary in a direction opposite to that found in adipose tissue.[103] In the fasting state, chylomicron triglycerides may therefore be hydrolyzed to a large extent in muscle and the liberated fatty acids oxidized locally.

In contrast to the first step of chylomicron catabolism, which is extrahepatic, the second hepatic step is less closely regulated. Thus, uptake of chylomicron remnants by hepatocytes proceeds efficiently under a variety of conditions.[104] As noted above, most dietary cholesterol is taken up by the liver as a component of remnant particles. Yet even when the liver has become overloaded with chylomicron cholesterol, so that the excess is stored in droplets within the cells, the uptake of remnants continues unabated.

The processing of the surface lipids of chylomicrons that are transferred to HDL is less efficient, and phospholipids in particular accumulate in HDL.[105] The transferred phospholipids associate with existing HDL particles, which become less dense for several hours after ingestion of a fat-rich meal.[106,107] Less cholesterol accumulates in the HDL, presumably because more of it remains with the chylomicron particle. In addition, the HDL may become enriched in triglycerides, in exchange for cholesteryl esters which are transferred from HDL to chylomicrons.[107] Both the phospholipids and triglycerides are gradually hydrolyzed, presumably by hepatic lipase.[46,107]

Both steps of chylomicron catabolism have a large capacity. For example, in humans, up to 300 g of chylomicron triglycerides can be hydrolyzed by lipoprotein lipase daily, but less than 1% of this may be found in the blood at a given time.[108] Likewise, it is difficult to demonstrate accumulation of appreciable quantities of particles with the properties of chylomicron remnants, even in persons who have eaten cholesterol-rich meals. Consistent with such observations, the residence time of triglycerides in chylomicrons is ordinarily less than 10 min, and that of chylomicron remnants appears to be even shorter.

The protein components that are transferred to HDL either remain

---

[103] M. H. Tan, *Can. Med. Assoc. J.* **18**, 675 (1978).
[104] B. Angelin, C. A. Ravida, T. L. Innerarity, and R. Mahley, *J. Clin. Invest.* **71**, 816 (1983).
[105] R. J. Havel, J. P. Kane, and M. D. Kashyap, *J. Clin. Invest.* **52**, 32 (1978).
[106] P. H. E. Groot and L. M. Scheck, *J. Lipid Res.* **25**, 684 (1984).
[107] J. R. Patsch, S. Prasad, A. M. Gotto, Jr., and G. Bengtson-Olivecrona, *J. Clin. Invest.* **74**, 2017 (1984).
[108] J. R. Patsch, J. B. Karlin, L. W. Scott, L. C. Smith, and A. M. Gotto, Jr., *Proc. Natl. Acad. Sci. U.S.A.* **80**, 1449 (1983).

with lipoproteins of this class (as with the major A apolipoproteins) or participate in a process of recycling between HDL and both classes of triglyceride-rich lipoproteins, chylomicrons and VLDL. An appreciable fraction of the apoE transferred from HDL to triglyceride-rich lipoproteins is taken into the liver along with remnant particles.[109] The residence time of C apolipoproteins in humans is about 1.5 days,[110] which substantially exceeds that of apoE, less than 1 day.[111] The residence time of apoB-48 presumably is that of the remnant particle itself, i.e., a few minutes.[112]

VLDL and the Formation of LDL (Fig. 4)

The initial step of VLDL metabolism is basically the same as for chylomicrons—binding to endothelial lipoprotein lipase, followed by hydrolysis of the bulk of the triglycerides. The rate of hydrolysis is a function of the size of VLDL particles. As larger particles contain more molecules of apoC-II, this relationship may be related, at least in part, to the number of productive interactions between lipoprotein lipase molecules with this protein cofactor on the particle surface. As with chylomicrons, phospholipids and C apolipoproteins are transferred to HDL as hydrolysis proceeds; thus the products of lipoprotein lipase action, VLDL remnants, are depleted of triglycerides, phospholipids, and C apolipoproteins, but retain the apoB and apoE.[80,99] In the rat, newly secreted VLDL contain an appreciable complement of cholesteryl esters, synthesized by hepatic ACAT,[113] but acquire little in the plasma, owing to the low cholesteryl ester/triglyceride transfer activity.[114] In other species, such as humans, in which hepatic ACAT activity is low and cholesteryl ester/triglyceride transfer activity is relatively high, VLDL acquire cholesteryl esters from HDL synthesized in the plasma by LCAT, as described below. The amount acquired is in part a function of the residence time of VLDL, which is longer in humans than rats. In many species, most of the VLDL remnants are taken up by the liver by receptor-mediated endocytosis.[115] In humans, a large fraction, perhaps about one-half, is normally pro-

---

[109] F. van't Hooft and R. J. Havel, *J. Biol. Chem.* **256**, 3963 (1981).
[110] M. W. Huff, N. G. Fidge, P. J. Nestel, T. Billington, and B. Watson, *J. Lipid Res.* **22**, 1235 (1981).
[111] R. E. Gregg, L. A. Zech, E. J. Schaefer, and H. B. Brewer, Jr., *J. Lipid Res.* **25**, 1167 (1984).
[112] F. M. van't Hooft, D. A. Hardman, J. P. Kane, and R. J. Havel, *Proc. Natl. Acad. Sci. U.S.A.* **79**, 179 (1982).
[113] O. Faergeman and R. J. Havel, *J. Clin. Invest.* **55**, 1219 (1975).
[114] P. J. Barter and J. S. Lally, *Biochim. Biophys. Acta* **532**, 233 (1978).
[115] R. J. Havel, *J. Lipid Res.* **25**, 1151 (1984).

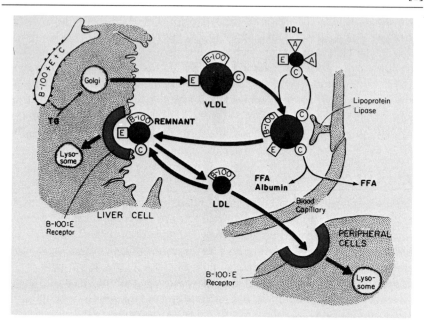

Fig. 4. This diagram summarizes the steps of VLDL metabolism. It shows the similarity of the first step to that for chylomicrons and the partial divergence of the subsequent processing of VLDL remnants to form LDL which, unlike remnants, are catabolized both in the liver and in peripheral cells (see text). (From R. J. Havel, in "Medical Clinics of North America," Vol. 66, No. 2, pp. 319–333, 1982, with permission from W. B. Saunders, Philadelphia.)

cessed further to yield LDL.[115] Small VLDL are thought to yield smaller remnants, which are more likely to form LDL than larger remnants, whereas large VLDL are thought to yield relatively large remnants, which are more likely to be removed rapidly by the liver.[116,117] As the spectrum of VLDL yields a comparable spectrum of remnants, the size and density of the remnant population can overlap that of the precursor VLDL population; however, the remnants are distinguished by a reduced electrophoretic mobility, presumably because of the loss of C apolipoproteins.[118] VLDL remnants with a density greater than 1.006 g/ml are frequently

[116] A. F. H. Stalenhoef, M. J. Malloy, J. P. Kane, and R. J. Havel, *Proc. Natl. Acad. Sci. U.S.A.* **81,** 1839 (1984).

[117] C. J. Packard, A. Munro, A. R. Lorimer, A. M. Gotto, and J. Shepherd, *J. Clin. Invest.* **74,** 2178 (1984).

[118] A. Pagnan, R. J. Havel, J. P. Kane, and L. Kotite, *J. Lipid Res.* **18,** 613 (1977).

referred to as intermediate density lipoproteins (IDL), but there is no evidence that these particles differ qualitatively from VLDL remnants of lower density.

The formation of LDL from VLDL is accompanied by further loss of triglycerides and phospholipids and, perhaps most critically, by the loss of apoE. Only apoB-100 is retained in LDL. The lipid loss is thought to result from the action of hepatic lipase upon the remnant particle[115]; loss of apoE may be a passive consequence of altered composition or curvature of the particle surface. In species with active cholesteryl ester/triglyceride transfer activity, processing of remnants to form LDL may also be accompanied by redistribution of cholesteryl esters to larger particles in exchange for triglycerides[119]; the latter are presumably hydrolyzed by lipoprotein lipase or hepatic lipase.

Terminal Catabolism of VLDL Remnants and LDL

VLDL remnants and LDL, both of which contain apoB-100, are thought to be removed from the blood by interaction with "LDL" receptors.[120] VLDL remnants, which are efficiently taken up by the binding of apoE to LDL receptors on hepatocytes, are removed almost entirely in the liver.[115] By contrast, LDL, which are removed less efficiently by the hepatic receptors through binding to apoB-100, gradually gain access to extravascular compartments of various organs and tissues which contain LDL receptors, so that an appreciable fraction is taken up in extrahepatic tissues as well.[121,122] The distribution of LDL to various tissues presumably depends mainly upon the rate of transcapillary transport and the activity of LDL receptors on cell surfaces. Thus, the adrenal gland, which has a fenestrated endothelium and cortical cells rich in LDL receptors, takes up LDL actively, whereas adipose tissue and muscle, which have nonfenestrated capillaries and few LDL receptors, take up LDL slowly.

ApoE as well as apoB contain a recognition site for the LDL receptor. In apoE, this site is a basic central region of the protein rich in arginyl and lysyl residues.[123] The human LDL receptor is a transmembrane protein of 839 amino acids.[23] Its NH_2-terminal portion, which is rich in cysteine residues, is composed of a 7-fold repeat of about 40 residues. At the

---

[119] R. Deckelbaum, S. Eisenberg, Y. Oschry, E. Butbul, I. Sharon, and T. Olivecrona, *J. Biol. Chem.* **257**, 650 (1982).

[120] J. L. Goldstein, T. Kita, and M. S. Brown, *New Engl. J. Med.* **309**, 288 (1983).

[121] A. D. Attie, R. C. Pittman, and D. Steinberg, *Hepatology* **2**, 269 (1982).

[122] E. F. Stange and J. M. Dietschy, *J. Lipid Res.* **25**, 703 (1984).

[123] T. L. Innerarity, K. H. Weisgraber, K. S. Arnold, S. C. Rall, Jr., and R. W. Mahley, *J. Biol. Chem.* **259**, 7261 (1984).

COOH-terminal end of each repeat is an octapeptide sequence that contains three aspartyl residues and one glutamyl residue. These negatively charged regions presumably constitute the ligand-binding sites of the receptor which interact electrostatically with the positively charged region of the apolipoprotein ligand. Consequently, each receptor has the potential to bind seven ligands. Binding studies have shown that small lipoproteins that contain several apoE molecules can bind to multiple receptor sites,[124] whereas the apoB on an LDL particle presumably interacts with a single binding site. The observation that the affinity of binding of VLDL remnants to the receptor is many-fold greater than that of LDL[125] can plausibly be explained by the binding of the remnant to multiple sites on the receptor through its complement of apoE molecules. This tighter binding may be responsible for the more efficient endocytosis of the remnants than LDL in the liver.

Normal rat VLDL that contain a full complement of C apolipoproteins bind poorly to LDL receptors[125,126] and are taken up slowly by the liver,[127] whereas VLDL remnants bind tightly to the receptor and are rapidly taken up by the liver.[127] Large normal human VLDL, unlike large VLDL from hypertriglyceridemic subjects, are not metabolized by the LDL receptor pathway.[128] Such large VLDL, diameter > 45 nm, bind to the LDL receptor only if they contain apolipoprotein E of an appropriate, accessible conformation which also renders it susceptible to thrombin cleavage.[129] The apoE of normal VLDL is inaccessible both to the LDL receptor and to thrombin.[130] The accessible apoE is acquired by large VLDL in hypertriglyceridemic subjects during its prolonged residence time.

The apoB of large VLDL does not bind to the LDL receptor, suggesting that it is in a different conformation from that of the smaller LDL. This possibility is borne out by studies which show that the immunoreactivity of apoB on large VLDL is different from that of LDL.[131] In contrast to

[124] T. L. Innerarity, R. E. Pitas, and R. W. Mahley, *Biochemistry* **19**, 4359 (1980).
[125] E. T. T. Windler, P. T. Kovanen, Y.-s. Chao, M. S. Brown, R. J. Havel, and J. L. Goldstein, *J. Biol. Chem.* **255**, 10464 (1980).
[126] S. H. Gianturco, F. B. Brown, A. M. Gotto, Jr., and W. A. Bradley, *J. Lipid Res.* **23**, 984 (1982).
[127] E. Windler, Y.-s. Chao, and R. J. Havel, *J. Biol. Chem.* **255**, 5480 (1980).
[128] S. H. Gianturco, A. M. Gotto, Jr., R. L. Jackson, J. R. Patsch, O. D. Tanuton, H. D. Sybers, D. L. Yeshurun, and L. C. Smith, *J. Clin. Invest.* **61**, 320 (1978).
[129] S. H. Gianturco, A. M. Gotto, Jr., S.-L. C. Hwang, J. B. Karlin, A. H. Y. Lin, S. C. Prasad, and W. A. Bradley, *J. Biol. Chem.* **258**, 4526 (1983).
[130] W. A. Bradley, F. B. Brown, A. M. Gotto, Jr., S.-L. C. Hwang, J. B. Karlin, A. H.-Y. Lin, S. C. Prasad, and S. H. Gianturco, *J. Biol. Chem.* **259**, 14728 (1984).
[131] G. Schonfeld, W. Patsch, B. Pfleger, J. L. Witztum, and S. W. Weidman, *J. Clin. Invest.* **64**, 1288 (1979).

large VLDL, small VLDL, 30 to 40 nm in diameter, and IDL, like LDL, bind to the LDL receptor *in vitro* primarily via apoB, not apoE.[130,131] As with chylomicrons, loss of C apolipoproteins or other changes in surface composition of VLDL during remnant formation may render the apolipoprotein E binding site accessible to the LDL receptor. When VLDL remnants are taken up rapidly via the LDL receptor, very little LDL is formed. At least two factors appear to influence the efficiency of hepatic uptake of VLDL remnants: the activity of hepatic LDL receptors and the size of the VLDL precursors.[116,117] The latter relationship may be a function of a greater number of apolipoprotein E molecules on larger remnants,[115] which may lead to tighter binding to the receptor and more rapid endocytosis. Why VLDL remnants are taken up so poorly by hepatic chylomicron remnant receptors is unknown.

Although the receptor responsible for the uptake of particles containing apoB-100 differs from that responsible for uptake of chylomicrons, which contain apoB-48, the pathway of endocytosis and catabolism of each of these particles within the liver is the same as that described above for chylomicron remnants. In cultured cells, LDL receptors recycle within minutes of endocytosis and are reutilized many times before they are eventually catabolized.[132] Presumably, such recycling occurs *in vivo* as well and accounts for the large and continuing capacity of the liver to take up VLDL remnants and LDL when large numbers of hepatic LDL receptors are present.[71]

Some LDL in liver and other tissues are taken up by LDL receptor-independent pathways that also are thought to lead to endocytosis.[121] In the absence of LDL receptors, a large fraction of LDL continues to be taken up by hepatocytes at a rate that exceeds that predicted for fluid phase endocytosis.[121] Hepatocyte surfaces have binding sites other than the LDL receptor that recognize several classes of lipoproteins, including LDL.[133] Binding to such sites could facilitate endocytosis, albeit much less efficiently than that mediated by the LDL receptor. However, receptor-independent uptake of LDL has not invariably been found to exceed that expected for fluid phase endocytosis.[134]

The LDL receptor in the liver and other tissues is subject to close regulation. The activity of hepatic LDL receptors is rapidly reduced when substantial amounts of cholesterol are delivered to the liver in chylomi-

---

[132] M. S. Brown, R. G. W. Anderson, and J. L. Goldstein, *Cell* **32**, 633 (1983).

[133] P. W. Bachorik, F. A. Franklin, D. G. Virgil, and P. O. Kwiterovich, Jr., *Biochemistry* **21**, 5675 (1982).

[134] D. K. Spady, D. W. Bilheimer, and J. M. Dietschy, *Proc. Natl. Acad. Sci. U.S.A.* **80**, 3499 (1983).

cron remnants.[104] Conversely, when the catabolism of cholesterol in hepatocytes is increased by administration of bile acid-binding resins, receptor activity increases.[135] Administration of inhibitors of cholesterol synthesis such as compactin, which competitively inhibits HMG-CoA reductase, also increases the activity of hepatic LDL receptors.[135] In the adrenal cortex, the activity of LDL receptors is stimulated by ACTH, which increases utilization of cholesterol for synthesis of glucocorticosteroids.[136] The behavior of LDL receptors *in vivo* is generally consistent with that observed in cultured cells, as relates to the availability of cholesterol.[104,135,136] In the rat, however, widely varying rates of cholesterol synthesis may not be accompanied by large alterations in the activity of hepatic LDL receptors.[122] Variations in receptor activity are accompanied by corresponding changes in the receptor mRNA,[26] but the molecular events that regulate receptor synthesis are not known.

Two other receptors, the β-VLDL receptor and the "scavenger" receptor, mediate endocytosis of apoB-containing lipoproteins into cells of reticuloendothelial origin, such as macrophages, arterial wall foam cells, and endothelial cells.[137,138] The β-VLDL receptor pathway is specific for chylomicrons,[139,140] large VLDL from hypertriglyceridemic subjects,[139] and the β-migrating, cholesteryl ester rich VLDL from cholesterol fed animals[141] or humans with Type III hyperlipoproteinemia. VLDL, LDL, and HDL from normal animals or humans are not taken up by this pathway.[137,139,141] After binding to the receptor, the lipid-rich particles are internalized, hydrolyzed in the lysosomes, and the lipid is reesterified and accumulates in the cytoplasm, causing the cells to assume a foam cell appearance *in vitro*. The accumulated lipid reflects the predominant lipid carried by the internalized lipoprotein, triglyceride from chylomicrons and hypertriglyceridemic VLDL[139] and cholesteryl ester from cholesteryl ester-rich β-VLDL.[137,141]

The scavenger receptor mediates the uptake of certain chemically

[135] P. T. Kovanen, D. W. Bilheimer, J. L. Goldstein, J. J. Jaramillo, and M. S. Brown, *Proc. Natl. Acad. Sci. U.S.A.* **78**, 1194 (1981).
[136] P. T. Kovanen, J. L. Goldstein, D. A. Chappell, and M. S. Brown, *J. Biol. Chem.* **255**, 5591 (1980).
[137] M. S. Brown and J. L. Goldstein, *Annu. Rev. Biochem.* **52**, 223 (1983).
[138] D. P. Baker, B. J. VanLenten, A. M. Fogelman, P. A. Edwards, C. Kean, and J. A. Berliner, *Arteriosclerosis* **4**, 248 (1984).
[139] S. H. Gianturco, W. A. Bradley, A. M. Gotto, Jr., J. D. Morrisett, and D. L. Peavy, *J. Clin. Invest.* **70**, 168 (1982).
[140] B. J. VanLenten, A. M. Fogelman, M. M. Hokum, L. Benson, M. E. Haberland, and P. A. Edwards, *J. Biol. Chem.* **258**, 5151 (1983).
[141] J. L. Goldstein, M. S. Brown, T. L. Innerarity, and R. W. Mahley, *J. Biol. Chem.* **255**, 1839 (1980).

modified proteins, including LDL.[142] These modifications increase the negative charge of the proteins in various ways, such as by acetylation.[137] It has been noted that the modification of LDL by malondialdehyde is of potential physiological relevance.[143] *In vivo,* such chemically modified proteins are taken up from the blood mainly by hepatic endothelial cells and, to a lesser extent, by Kupffer cells and other macrophage elements.[144–146]

Chemical evidence indicates that the scavenger receptor is a protein as is the LDL receptor.[147,148] Recently, the scavenger receptor has been isolated from tumors produced by injection of a murine macrophage cell line into syngeneic mice. The solubilized tumor receptor has similar chemical properties to those measured in the intact cell; for example, the K_d is $1-2 \times 10^{-8}$ M. It has an apparent molecular weight of 260,000 and an isoelectric point of 6.1. The receptor, purified to homogeneity from human monocytes, bovine endothelial cells, and macrophage tumors, appears to be of identical size.

The cellular expression of the scavenger receptor, in contrast to the LDL receptor, is not a function of cellular cholesterol levels. Furthermore, in human monocytes the acetyl-LDL receptor level is unaffected by either insulin or platelet-derived growth factor, again differing from the LDL receptor.[149] The quantitative role of the β-VLDL receptor and the scavenger receptor in normal lipoprotein catabolism is undetermined, but they may help to clear abnormal lipoproteins from the blood and they may also be involved in the deposition of cholesterol in developing atherosclerotic plaques.

HDL Catabolism and Reverse Cholesterol Transport

The cholesteryl esters that comprise most of the core of HDL particles are produced in blood plasma or lymph through the action of LCAT. When LCAT activity is grossly deficient, HDL particles are mainly lamellar, and unesterified cholesterol accumulates in plasma lipoproteins and

---

[142] R. W. Mahley and T. L. Innerarity, *Biochim. Biophys. Acta* **737**, 197 (1983).
[143] A. M. Fogelman, I. Shechter, J. Jeager, M. Hokum, J. Child, and P. A. Edwards, *Proc. Natl. Acad. Sci. U.S.A.* **77**, 2214 (1980).
[144] J. F. Naglkerke, K. P. Barto, and T. J. C. van Berkel, *J. Biol. Chem.* **258**, 12221 (1983).
[145] R. Blomhoff, C. A. Drevon, W. Eskild, P. Helgerud, K. R. Norum, and T. Berg, *J. Biol. Chem.* **259**, 8898 (1984).
[146] R. E. Pitas, J. Boyles, R. W. Mahley, and D. M. Bissell, *J. Cell Biol.* **100**, 1036 (1985).
[147] J. L. Goldstein, Y. K. Ho, S. K. Basu, and M. S. Brown, *Proc. Natl. Acad. Sci. U.S.A.* **76**, 333 (1979).
[148] D. P. Via, H. A. Dresel, S.-L. Cheng, and A. M. Gotto, Jr., *J. Biol. Chem.* **260**, 7379 (1985).
[149] T. Mazzone and A. Chait, *Arteriosclerosis* **2**, 487 (1982).

some cell membranes.[150] In human plasma, LCAT is largely or entirely associated with a subfraction of HDL that contains apoA-I and apoD, and existing evidence indicates that this fraction is the site of the enzyme's action.[151] The cholesterol substrate for LCAT is derived from the surface of plasma lipoproteins or the plasma membrane of cells. It can reach the site of esterification by diffusion through the aqueous phase.[152] By this means, molecules of cholesterol exchange readily between plasma lipoprotein particles and between lipoproteins and cell membranes. Net movement is a function of the chemical potential of cholesterol at each site. Efflux of cellular cholesterol into human plasma which is coupled to its esterification by LCAT seems to be mediated primarily by a population of HDL particles that contain apoA-I but not apoA-II.[153] LCAT, by consuming cholesterol, promotes net transport of cholesterol from cells into plasma and from other lipoproteins to the site of esterification. In many species, including humans and rabbits, the newly esterified cholesterol is rapidly distributed to other lipoproteins by cholesteryl ester transfer proteins.[62] Blood plasma of rats, dogs, and pigs, however, contains little or no transfer activity, so that cholesterol esterified by LCAT on HDL particles largely remains in this lipoprotein class. The other substrate of the reaction, phosphatidylcholine, can also be derived from lipoproteins, especially those of triglyceride-rich lipoproteins during remnant formation, or cell surfaces; its movement to the site of esterification may be mediated, at least in part, by transfer proteins.[154,155] The product lysophosphatidylcholine is mainly transferred to albumin, from which it is rapidly removed from the blood and reacylated.[156]

In those species with active mechanisms for transfer of cholesteryl esters among lipoproteins, the process may occur as an exchange of nonpolar lipids, mainly cholesteryl esters and triglycerides.[157] Whether the initial movement from the site of esterification occurs as part of an exchange reaction is not known. In any event, those molecules of cholesteryl ester that are transferred to lipoproteins containing apoB-100 or apoE are mainly catabolized during endocytosis of those lipoproteins in the liver, thereby completing the process of reverse cholesterol transport.

[150] J. A. Glomset and K. R. Norum, *Adv. Lipid Res.* **111**, 1 (1973).
[151] P. E. Fielding and C. J. Fielding, *Proc. Natl. Acad. Sci. U.S.A.* **77**, 3327 (1980).
[152] G. H. Rothblat and M. C. Phillips, *J. Biol. Chem.* **257**, 4775 (1982).
[153] C. J. Fielding and P. E. Fielding, *Proc. Natl. Acad. Sci. U.S.A.* **78**, 3911 (1981).
[154] J. Ihm, J. L. Ellsworth, B. Chataing, and J. A. K. Harmony, *J. Biol. Chem.* **257**, 4818 (1982).
[155] A. R. Tall, E. Abreu, and J. Schuman, *J. Biol. Chem.* **258**, 2174 (1983).
[156] Y. Stein and O. Stein, *Biochim. Biophys. Acta* **116**, 95 (1966).
[157] A. V. Nichols and L. Smith, *J. Lipid Res.* **6**, 206 (1965).

In those species in which the transfer mechanism is inactive, the cholesteryl esters must presumably be catabolized by other mechanisms. In the rat, however, a substantial fraction of HDL particles contains apoE and this fraction can be removed by receptors that recognize this protein.[158]

A substantial fraction of the HDL phospholipids consumed by the LCAT reaction may be derived from the surface of triglyceride-rich lipoproteins during remnant formation. HDL become enriched in phospholipids[105] and, in some species, in triglycerides, via cholesteryl ester/triglyceride exchange[107] when lipolysis is stimulated during alimentary lipemia. These lipids are readily hydrolyzed by hepatic lipase *in vitro*.[46,107] Furthermore, in humans, the concentration of phospholipid- and cholesterol-enriched HDL, present in the HDL_2 subfraction, is inversely related to the activity of this enzyme, as measured in postheparin plasma.[159] Therefore, both hepatic lipase and LCAT may participate in the hydrolysis of HDL phospholipids. Hepatic lipase may also participate in the catabolism of HDL triglycerides. These mechanisms[107] could underlie the relationship between HDL_2 levels and the magnitude of postprandial lipemia.[108]

The process whereby cholesteryl esters of HDL are removed appears not to involve the total catabolism of HDL particles. Thus, in the rat, HDL cholesteryl esters are taken up most actively by liver, adrenal gland, and gonads and at a considerably more rapid rate than the major HDL protein, apoA-I.[160–162] This process presumably involves direct association of the particle with the plasma membrane of the cell, but the mechanism of transfer of HDL components into the cell is not known. Both liver and steroid-secreting endocrine glands contain an enzyme that resembles hepatic lipase.[46] Binding of HDL to this enzyme on endothelial cells in the liver could lead to endocytosis, selective loss of lipid, and then retroendocytosis of a lipid-depleted particle.[163] However, cholesteryl ethers, incorporated into rat HDL as slowly metabolized analogs of cholesteryl esters, are taken up almost entirely into hepatocytes.[161]

In surviving hepatocytes and steroidogenic cells, as well as in several other cells in culture, HDL bind to cell surfaces by a mechanism distinct

---

[158] F. M. van't Hooft and R. J. Havel, *J. Biol. Chem.* **257**, 10996 (1982).

[159] T. Kuusi, E. A. Nikkilä, P. Saarinen, P. Varjo, and L. A. Laitinen, *Atherosclerosis* **41**, 209 (1982).

[160] F. M. van't Hooft, T. van Gent, and A. van Tol, *Biochem. J.* **196**, 877 (1981).

[161] C. R. Glass, R. C. Pittman, D. B. Weinstein, and D. Steinberg, *Proc. Natl. Acad. Sci. U.S.A.* **80**, 5435 (1983).

[162] Y. Stein, Y. Dabach, G. Hollander, G. Halperin, and O. Stein, *Biochim. Biophys. Acta* **752**, 98 (1983).

[163] P. K. J. Kinnunen, J. A. Virtanen, and P. Vaino, *in* "Atherosclerosis Reviews" (A. M. Gotto, Jr. and R. Paoletti, eds), Vol. 11, p. 65. Raven, New York, 1983.

from that described for lipoproteins containing apoB-100 or apoE. HDL binding does not appear to depend on calcium ion and is relatively resistant to proteolysis.[62,164,165] In most cases, binding is poorly coupled to lysosomal hydrolytic activities and, at least in some cases, binding is stimulated in cells with increased concentrations of unesterified cholesterol.[166] Evidently, such binding to cells could promote influx or efflux of unesterified cholesterol or uptake of cholesteryl esters. The specificity of the binding site is uncertain. In some, but not all cells or cell membrane preparations, LDL and VLDL have been found to compete with HDL for the same binding site[133]; some evidence supports protein specificity, especially for apoA-I, but other proteins such as the C apolipoproteins may also compete for the site.[165,167,168] Whether these properties of the binding site are tissue specific is unclear.

In vivo, the major apolipoproteins of HDL, apoA-I and apoA-II, are catabolized slowly. In the rat, an appreciable fraction of apoA-I is catabolized in the kidneys.[169] It is postulated that a small fraction of apoA-I present in free solution is filtered by the glomerulus and catabolized in proximal tubular cells. In those tissues that catabolize appreciable amounts of HDL cholesteryl esters, the binding process observed *in vitro* may be involved. In steroidogenic cells, stimulation of hormone synthesis increases both the binding of HDL to the cells *in vitro* and uptake of HDL cholesterol *in vivo.*[165]

It is evident that HDL constitute a varied and plastic class of lipoproteins which actively participate in plasma cholesterol transport. Much remains to be learned about the participation of the discrete subpopulations of HDL in this or other functions of this class of lipoproteins. These functions may differ considerably among species, especially in relation to the varied capacity to transfer core lipids between HDL and other lipoproteins.

Final Perspective

Current knowledge of the structure and function of lipoproteins and apolipoproteins and of the enzymes and receptors that mediate lipoprotein–lipid transport is making it possible to define discrete steps of lipoprotein metabolism at the molecular level. We envision that it may soon

[164] R. J. Havel, this series, Vol. 129 [35].
[165] J. T. Gwynne and J. F. Strauss, III, *Endocrine Rev.* **3,** 299 (1982).
[166] J. F. Oram, E. A. Brinton, and E. L. Bierman, *J. Clin. Invest.* **72,** 1611 (1983).
[167] G. K. Chacko, *Biochim. Biophys. Acta* **795,** (1984).
[168] V. A. Rifici and H. A. Eder, *J. Biol. Chem.* **259,** 13814 (1984).
[169] C. Glass, R. C. Pittman, G. A. Keller, and D. Steinberg, *J. Biol. Chem.* **258,** 7161 (1983).

be possible to understand the association of individual apolipoproteins with specific classes of lipoproteins from the surface properties of lipoprotein particles. In addition, the coupling of cellular regulatory mechanisms to the processes of intracellular lipoprotein biosynthesis and catabolism is beginning to yield to the powerful tools of contemporary cell and molecular biology.

During the past half century, a myriad of methods have been brought to bear upon the biology of the plasma lipoproteins. Most investigators who have entered this field have brought to it specialized techniques and knowledge that they acquired in different disciplinary areas, but no single discipline has dominated. In recent years, the policies of granting agencies have helped to increase communication and collaboration among investigators. Included in this interaction have been clinical scientists and pathologists, whose important contributions have not been described in this survey of lipoprotein structure and metabolism. We believe that the current state of the field is conducive both to further understanding of normal processes and of the role of lipoproteins in atherosclerosis.

[2] Molecular and Cell Biology of Lipoprotein Biosynthesis

By DONNA M. DRISCOLL and GODFREY S. GETZ

Introduction

In the last few years the field of molecular and cellular biology has become one of the most exciting and active areas in experimental biology. Great strides have been made in understanding how eukaryotic genes are organized and regulated, as well as how proteins are synthesized and targeted within the cell. Many of the techniques in molecular biology have been applied recently to the study of plasma lipoproteins. Lipoproteins are lipid–protein complexes which transport lipid in a water-miscible form in the plasma. Structurally all lipoproteins are composed of a hydrophobic core of cholesterol ester or triglyceride and a surface coat of phospholipid, free cholesterol, and specific protein constituents, the apolipoproteins.[1]

[1] L. C. Smith, H. J. Pownall, and A. M. Gotto, *Annu. Rev. Biochem.* **47**, 751 (1978).

Studies in intact animals and in isolated organ systems have established that lipoproteins are synthesized by the liver and intestine. Aside from the biogenesis of subcellular organelles, the production of lipoproteins is one of the most complex biosynthetic processes in eukaryotic cells involving the synthesis of apolipoproteins and lipids and their intracellular assembly into nascent lipoprotein particles.[2] In addition, after secretion the nascent lipoproteins often undergo remodeling in the circulation since many of the apolipoprotein and lipid constituents can exchange between lipoprotein particles in the plasma. The intravascular remodeling also involves modifications of apolipoproteins and of lipids, including the esterification of free cholesterol and the hydrolysis of core triglycerides. The uptake of lipoproteins from the plasma is mediated by several cell-surface receptor systems found in the liver and in extrahepatic tissues, receptors for which apoproteins are the recognized ligands. Thus the spectrum of plasma lipoproteins reflects synthetic as well as catabolic events which are both subject to metabolic regulation by diet and hormones.[1,2]

This review highlights some of the recent findings on the genetics and cell biology of lipoprotein metabolism. We also draw on examples of other well-characterized eukaryotic systems in order to illustrate the potential power of some of the newly developed techniques in molecular biology. We hope to illustrate how the molecular approach is an extremely promising one for investigating the complex processes of lipid transport and lipid metabolism.

Organization and Expression of Eukaryotic Genes

Recent advances in recombinant DNA technology have facilitated the cloning of a large number of eukaryotic genes. The ability to isolate and characterize these genes has greatly advanced our knowledge of gene organization and gene regulation at the molecular level as discussed in the following sections. One of the earliest approaches to this problem involved the cloning of DNA copies of mRNAs as shown in Fig. 1. There are a number of methods currently available for the isolation of mRNA from tissues or cells.[3,4] Double-stranded DNA copies of mRNAs (known

---

[2] G. S. Getz and R. V. Hay, in "Biochemistry of Atherosclerosis" (A. M. Scanu, R. W. Wissler, and G. S. Getz, eds.), p. 151. Dekker, New York, 1979.

[3] J. M. Chirgwin, A. E. Przyalba, R. J. MacDonald, and W. J. Rutter, *Biochemistry* **18**, 5294 (1979).

[4] J. M. Taylor, *Annu. Rev. Biochem.* **48**, 681 (1979).

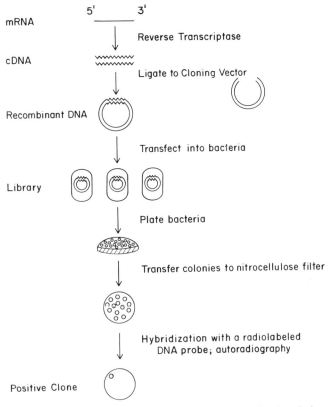

FIG. 1. The construction of a cDNA library. See text for description.

as cDNAs)[4a] can be synthesized *in vitro* using the enzyme reverse transcriptase. These cDNAs can then be ligated to a cloning vector, a DNA molecule which can replicate autonomously in bacteria. The resultant recombinant DNA molecules can be introduced by transfection into bacteria where both the vector and cDNA sequences are replicated and amplified.[5] Libraries of clones containing cDNAs complementary to the

---

[4a] Abbreviations: apo, apolipoprotein; cDNA, double-stranded DNA copy of mRNA; FH, familial hypercholesterolemia; HDL, high-density lipoprotein; HMG CoA reductase, 3-hydroxy-3-methylglutaryl coenzyme A reductase; LDL, low-density lipoprotein; RER, rough endoplasmic reticulum; VLDL, very low-density lipoprotein.

[5] T. Maniatis, E. F. Fritsch, and J. Sambrook, "Molecular Cloning." Cold Spring Harbor Laboratory Press, Cold Spring Harbor, New York, 1982.

entire mRNA population of a particular tissue can be constructed and then screened to identify a particular cDNA clone of interest. The most common screening method involves the transfer of bacterial colonies containing the recombinant DNA from an agar plate to nitrocellulose filter.[6] The filter is then hybridized with a radiolabeled DNA probe to identify clones which contain cDNA sequences complementary to the probe. Several types of probes have been used for this purpose, including radiolabeled cDNA synthesized *in vitro* from a mRNA preparation which has been enriched for a particular mRNA. Several techniques have been developed to enrich for a specific mRNA including polysome immunoprecipitation,[7] sucrose gradient fractionation,[8] or hybridization subtraction chromatography.[9] If the amino acid sequence of a protein is known, the sequence of the cDNA coding for that protein can be predicted although this is limited to some extent by the degeneracy of the genetic code. A radiolabeled oligonucleotide complementary to the cDNA sequence can then be synthesized *in vitro* and used as a probe to screen the library.[10] Once a positive cDNA clone has been identified by nucleic acid hybridization, its identity can be confirmed by hybrid selection[11] or DNA sequencing.[12,13]

Certain cloning vectors have been constructed which allow the expression of eukaryotic genes in bacteria. These "expression vectors" contain a strong promoter sequence which is efficiently recognized by the bacterial RNA polymerase. If a cDNA molecule is inserted behind this promoter in the correct reading frame, the cDNA can be transcribed and translated when transfected into bacteria.[14] Because the gene product is produced, expression libraries can be screened with a radiolabeled antibody directed against a specific protein to identify the cDNA clone encoding it.[15]

[6] M. Grunstein and D. Hogness, *Proc. Natl. Acad. Sci. U.S.A.* **72**, 3961 (1975).
[7] J. P. Kraus and L. E. Rosenberg, *Proc. Natl. Acad. Sci. U.S.A.* **79**, 4015 (1982).
[8] G. N. Buell, M. P. Wickens, F. Payvar, and R. T. Schimke, *J. Biol. Chem.* **253**, 2471 (1978).
[9] M. P. Vitek, S. G. Kreissman, and R. H. Gross, *Nucleic Acids Res.* **9**, 1191 (1981).
[10] B. E. Noyes, M. Meravech, R. Stein, and K. L. Agarwal, *Proc. Natl. Acad. Sci. U.S.A.* **76**, 1770 (1979).
[11] R. P. Ricciardi, J. S. Miller, and B. E. Roberts, *Proc. Natl. Acad. Sci. U.S.A.* **76**, 4927 (1979).
[12] A. W. Maxam and W. Gilbert, *Proc. Natl. Acad. Sci. U.S.A.* **74**, 560 (1977).
[13] F. Sanger, S. Nicklen, and A. R. Coulson, *Proc. Natl. Acad. Sci. U.S.A.* **74**, 5463 (1977).
[14] L. Guarante, T. M. Roberts, and M. Ptashne, *Science* **209**, 1428 (1980).
[15] D. M. Helfman, J. R. Feramisco, J. C. Fiddes, G. P. Thomas, and S. H. Hughes, *Proc. Natl. Acad. Sci. U.S.A.* **80**, 31 (1983).

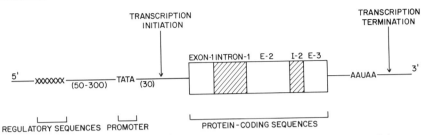

FIG. 2. The organization of a typical eukaryotic gene. The structural element of the gene is composed of protein-coding sequences (exons) which are interrupted by noncoding regions (introns). The TATA box which is the eukaryotic promoter is found 30 nucleotides upstream and the sequences important in the regulation of transcription are found 50 to 300 nucleotides upstream from the site of transcription initiation. The sequence AAUAA which is the signal for polyadenylation is found downstream from the protein-coding sequences and is followed by the site for the termination of transcription.

By definition cDNA clones only contain sequences complementary to the mature processed mRNA. To clone the gene itself, purified chromosomal DNA is fragmented either enzymatically or physically and the DNA framents are ligated to a cloning vector to generate a genomic library. Because of the large size of the eukaryotic genome, cloning vectors derived from bacteriophage λ are used for this purpose since these vectors can accomodate up to 40 kb of DNA.[16] The genomic library can be screened with a radiolabeled cDNA clone to identify the corresponding gene. The studies on cloned eukaryotic genes have culminated in the concept that eukaryotic genes are functionally composed of three elements as illustrated in Fig. 2: a structural element which codes for the gene product, a promoter component which is essential for the transcription of the gene, and regulatory sequences which modulate the expression of the gene.

Structural Elements of Eukaryotic Genes

The development of rapid and simple DNA sequencing techniques has been a particularly important advance in molecular biology.[12,13] In addition to confirming the identity of a particular clone, DNA sequencing can also provide information about the gene product itself. DNA sequencing is much more rapid and easier than peptide sequencing so that deduced amino acid sequence data for "wild type" and mutant proteins can be

---

[16] D. Ish-Horowicz and J. F. Burke, *Nucleic Acids Res.* **9**, 2928 (1981).

very rapidly generated. Structural information about apoE alleles has recently been obtained with this approach[17] but the promise of this methodology in the apolipoprotein field has yet to be fully realized.

The recent analysis of the structure of the LDL receptor provides a signal example of the value of DNA sequencing technology. The LDL receptor is a 160,000 Da glycoprotein found on the plasma membrane of most cells which functions in the receptor-mediated uptake of lipoproteins.[18] Recently a cDNA clone, 5.3 kb in length, for the human LDL receptor was isolated and sequenced.[19,20] The amino acid sequence, deduced from the nucleotide sequence, suggests a division of the protein into five recognizable domains. The LDL receptor precursor includes a signal peptide, 21 amino acids in length. The mature protein contains 839 amino acids. The first domain of the mature protein, consisting of about 300 amino acids made up of multiple 40 amino acid repeats, represents the binding site for apoproteins B and E of LDL and related cholesterol-rich lipoproteins. The repeating elements are rich in cysteine and are strongly homologous to a similar length sequence in human complement component C9. The second domain, containing about 400 amino acids, is homologous to the precursor of mouse epidermal growth factor[19] but not to the mature epidermal growth factor. Domain three of the LDL receptor is represented by a sequence of 48 amino acids, almost 40% of which are serine and threonine residues. These latter residues are the sites of O-glycosylation. A fourth domain of 22 amino acids is the membrane spanning element. The fifth domain contains the COOH-terminal 50 amino acids resident in the cell cytoplasm. Thus a very large amount of information about the LDL receptor was rapidly obtained by sequencing its cDNA. To obtain this information by more conventional protein sequencing methods would have proven time consuming and difficult.

A striking feature of modern molecular biology is how rapidly surprising observations become common knowledge. One such discovery is the finding that the protein-coding regions of most eukaryotic genes (exons) are often interrupted by extensive regions of noncoding sequences (introns). These introns are a part of the primary transcript from which they must be excised before the mRNA is transported to the cytoplasm.[21]

---

[17] J. W. McLean, N. A. Elshourbagy, D. J. Chang, R. W. Mahley, and J. M. Taylor, *J. Biol. Chem.* **259**, 6498 (1984).

[18] M. S. Brown and J. L. Goldstein, *Proc. Natl. Acad. Sci. U.S.A.* **76**, 3330 (1979).

[19] T. Yamamoto, C. G. Davis, M. S. Brown, W. J. Schneider, M. L. Casey, J. L. Goldstein, and D. W. Russell, *Cell* **39**, 27 (1984).

[20] D. W. Russell, W. J. Schneider, T. Yamamoto, K. L. Luskey, M. S. Brown, and J. L. Goldstein, *Cell* **37**, 577 (1984).

[21] P. Breathnach and P. Chambon, *Annu. Rev. Biochem.* **50**, 349 (1981).

Recently genomic clones have been isolated for several of the apolipoproteins including apoA-I, apoC-II, apoC-III, and apoE.[22–29] The organization of exons and introns is very similar in all 4 genes, with exon 1 coding for the 5' untranslated region of the mRNA, exon 2 coding for most of the signal peptide, and exons 3 and 4 coding for the protein sequence. The domain included in the fourth exon of each of these genes codes for the lipid binding element of the apoproteins. It is this fourth exon that includes the series of 66 bp repeats in the apoA-I and apoE genes that encode the 22 amino acid sequence blocks having amphipathic potential.[22] Presumably the same functional organization will be found in the apoA-IV gene, since the apoA-IV cDNA sequence reveals a similar series of 66 bp repeats translating into eight potential amphipathic helices.[30] The fourth exon of the apoC-III gene also codes for the lipid binding domain of this apolipoprotein.[25] This similar pattern of functional gene organization and the striking similarity of these paralogous sequences argue that these genes have evolved from a common ancestral gene.[30] A recent analysis of the gene for the human LDL receptor reveals a 45 kb sequence containing 18 exons, most of which appear to define the functional domains previously recognized in the translated protein sequence deduced from the cDNA sequence. Thus the repeat sequences resembling the C9 component of complement are encoded in distinct exons, while other exons specify sequences similar to the repeat sequences of the precursor to epidermal growth factor. The analyses have led to the notion that the LDL receptor gene belongs to two supergene families, with which exons are shared. These analyses will also facilitate the informed examination of the variety of mutations associated with familial hypercholesterolemia.

[22] S. K. Karathanasis, V. I. Zannis, and J. L. Breslow, *Proc. Natl. Acad. Sci. U.S.A.* **80**, 6147 (1983).
[23] C. C. Shoulders, A. R. Kornblihtt, B. S. Munro, and F. E. Baralle, *Nucleic Acids Res.* **11**, 2827 (1983).
[24] C. R. Sharpe, A. Sidoli, C. S. Shelley, M. A. Lucero, C. C. Shoulders, and F. E. Baralle, *Nucleic Acids Res.* **12**, 3917 (1984).
[25] S. K. Karathanasis, J. McPherson, V. I. Zannis, and J. L. Breslow, *Nature (London)* **304**, 371 (1983).
[26] C. L. Jackson, G. A. P. Bruns, and J. L. Breslow, *Proc. Natl. Acad. Sci. U.S.A.* **81**, 2945 (1984).
[27] S. K. Karathanasis, V. I. Zannis, and J. L. Breslow, this volume [41].
[28] H. K. Das, J. McPherson, G. A. P. Bruns, S. K. Karathanasis, and J. L. Breslow, *J. Biol. Chem.* **260**, 6240 (1985).
[29] Y.-K. Paik, D. J. Chang, C. A. Reardon, G. E. Davies, R. W. Mahley, and J. M. Taylor, *Proc. Natl. Acad Sci. U.S.A.* **82**, 3445 (1985).
[30] M. S. Boguski, N. Elshourbagy, J. M. Taylor, and J. I. Gordon, *Proc. Natl. Acad. Sci. U.S.A.* **81**, 5021 (1984).

One such mutation, representing a 5 kb deletion of several exons near the 3' end of the gene, has already been described.[31]

The development of expression vectors which allow the transcription and translation of eukaryotic genes in bacteria has facilitated the biochemical analysis of rare proteins.[14,32] This methodology has also been of medical importance since interferon, insulin, and growth hormone have all been commercially produced by this method. In addition, nucleotide changes which alter a single amino acid can be introduced into these cloned genes by the technique of oligonucleotide site-directed mutagenesis.[33] The mutant protein which is produced can be analyzed to assess the importance of even single amino acids in the function or conformation of the protein. This novel approach will be extremely powerful in lipoprotein research to help identify the essential structural features in apolipoproteins, their receptors and enzymes of lipid metabolism.

The technique of Southern blotting has proved extremely valuable as a rapid screening method to identify alterations in normal gene structure in genetic diseases. Chromosomal DNA isolated from peripheral blood lymphocytes can be digested with restriction enzymes which cleave DNA at specific nucleotide sequences and the resulting DNA fragments can be separated by electrophoresis in agarose gels. After transfer of these fragments from the gel to nitrocellulose paper, the paper can be probed with ^{32}P-labeled DNA from a cloned gene to detect DNA fragments which contain the gene sequences of interest.[34] Using a battery of restriction enzymes, a detailed restriction map of the normal gene can be constructed and compared with the map of a mutant gene. Genetic polymorphisms such as insertions of DNA, deletions, or even single nucleotide changes have been identified in the human α- and β-globin genes which lead to the production of abnormal globins as in sickle cell anemia[35] or the absence of protein as in β-thalassemia.[36] In the region of apolipoprotein genes such polymorphisms may be associated with disorders of lipoprotein structure and metabolism either by involving the coding sequences of the apolipo-

---

[31] M. A. Lehrman, W. J. Schneider, T. C. Sudhof, M. S. Brown, J. L. Goldstein, and D. W. Russell, *Science* **227**, 140 (1985).

[32] M. J. Casadaban, A. Martinez-Arias, S. K. Shapira, and J. Chou, this series, Vol. 100, p. 293.

[33] G. Dalbadie-McFarland, L. W. Cohen, A. D. Riggs, C. Morin, K. Itakura, and J. H. Richards, *Proc. Natl. Acad. Sci. U.S.A.* **79**, 6409 (1982).

[34] E. Southern, this series, Vol. 69, p. 152.

[35] R. F. Geever, L. B. Wilson, F. S. Nallaseth, P. F. Milner, M. Bittner, and J. T. Wilson, *Proc. Natl. Acad. Sci. U.S.A.* **78**, 5081 (1981).

[36] M. Baird, C. Driscoll, H. Schreiner, G. Sciarratta, G. Sansone, G. Niazi, F. Ramirez, and A. Bank, *Proc. Natl. Acad. Sci. U.S.A.* **78**, 4218 (1981).

proteins or by being linked with disturbances in the regulation of lipoprotein metabolism. The best example of the former relationship is represented by the structural polymorphism of apolipoprotein E which is fundamental to the development of type III dyslipoproteinemia.[37-40] This derives from the poor binding of certain apoE isoforms (E_2) to the LDL receptor and the hepatic apoE receptor. A polymorphism has been found in the apoA-I/C-III gene complex which may be associated with hypertriglyceridemia. This polymorphism involves transversion of CG to GC in the 3' noncoding region of apoC-III mRNA.[41,42] The molecular basis of the association of this polymorphism with hypertriglyceridemia remains to be clarified.

Chromsomal Location of Apolipoprotein Genes

Polymorphisms in the region of apolipoprotein genes, detectable by restriction enzyme analysis or by protein polymorphism, may be used in segregation analysis as genetic markers for the study of linkage with known chromosomal markers. For example, apoE cosegregates with the third component of complement.[43] An alternative powerful method for the assignment of apolipoprotein genes to particular chromosomes rests on the cosegregation of these genes and human chromosomes in interspecific somatic cell hybrids. These two techniques have resulted in the placement of most of the major apoprotein genes on three human chromosomes. Thus apoA-I,[11,44-46] apoC-III,[44] and probably apoA-IV[47] have been placed on chromosome 11. Hybrids that retain only portions of chromosome 11 have permitted the more refined localization of these three genes to the long arm of this chromosome in the region 11q13, and in proximity to the gene for uroporphyrinogen I synthase.[48] In the mouse the latter

---

[37] V. I. Zannis, P. W. Just, and J. L. Breslow, *Am. J. Hum. Genet.* **33**, 11 (1981).
[38] V. I. Zannis and J. L. Breslow, *Biochemistry* **21**, 1033 (1981).
[39] G. Utermann, M. Jaeschke, and J. Mangel, *FEBS Lett.* **56**, 352 (1975).
[40] V. I. Zannis and J. L. Breslow, *J. Biol. Chem.* **255**, 1759 (1980).
[41] S. K. Karathanasis, V. I. Zannis, and J. L. Breslow, *Nature (London)* **305**, 823 (1983).
[42] A. Rees, C. C. Shoulders, J. Stocks, D. J. Dalton, and F. E. Baralle, *Lancet* **1**, 444 (1983).
[43] B. Olaisen, P. Teisberg, and T. Gedde-Dahl, *Hum. Genet.* **62**, 233 (1982).
[44] G. A. P. Bruns, S. K. Karathanasis, and J. L. Breslow, *Ateriosclerosis* **4**, 97 (1984).
[45] S. W. Law, G. Gray, H. B. Brewer, Jr., A. Y. Sakaguchi, and S. L. Naylor, *Biochem. Biophys. Res. Commun.* **118**, 934 (1984).
[46] P. Cheung, F. T. Kao, M. L. Law, C. Jones, T. T. Puck, and L. Chan, *Proc. Natl. Acad. Sci. U.S.A.* **81**, 508 (1984).
[47] S. K. Karathanasis, *Proc. Natl. Acad. Sci. U.S.A.* **82**, 6374 (1985).
[48] M. H. Meisler, L. Wanner, F. T. Kao, and C. Jones, *Cytogenet. Cell Genet.* **31**, 124 (1981).

enzyme and the apoA-I gene are also on the same chromosome, number 9 in this species.[49,50] The gene for human apoA-I, the major apolipoprotein of HDL, is perhaps the best characterized apolipoprotein gene. The apoA-I gene is a single copy gene and is located only 2.5 kb upstream from the 3' end of the gene for apoC-III, a major apolipoprotein of triglyceride-rich lipoproteins.[28,44] These two genes are in reverse orientation, i.e., the coding sequences are on opposite strands of the chromosomal duplex. Karathanasis et al. have examined the structure of the apoA-I gene in chromosomal DNA isolated from patients with a deficiency in plasma HDL, apoA-I, and apoC-III who develop premature atherosclerosis. A restriction enzyme polymorphism in the apoA-I gene of these patients was discovered that was due to the translocation of 6.5 kb of the apoC-III gene into the coding region of the apoA-I gene.[41] The functional significance of this unusual arrangement of these two genes is not clear. The other major HDL apoprotein, apoA-II, has been tentatively assigned to chromosome 1.[51]

Chromosome 19 harbors the remaining major apoproteins other than apolipoprotein B. Thus apoC-I,[52] apoC-II,[26,27] and apoE[28,43] have all been assigned to chromosome 19, as has the gene for the LDL receptor.[53] The genes for apoE and the LDL receptor do not appear to be closely linked. However, this is the second case where the genes for a cell surface receptor and its ligand have been found on the same chromosome. The genes for human transferrin and the transferrin receptor are both on chromosome 3.[54] The genes for apoE and apoC-II are within 2 cM of one another but not closer than 5–7 kb.[55,56]

The functional significance, if any, of the clustering of many of the apolipoprotein genes on two human chromosomes is not clear. Since there is other evidence pointing to the common origin of these genes (vide

---

[49] H. B. Brewer, Jr., T. Fairwell, M. Meng, L. Kay, and R. Ronan, *Biochem. Biophys. Res. Commun.* **113**, 934 (1983).

[50] M. Rosseneu, G. Assmann, M. J. Taveirne, and G. Schmitz, *J. Lipid Res.* **25**, 111 (1984).

[51] M. N. Moore, F. T. Kao, Y. K. Tsao, and L. Chan, *Biochem. Biophys. Res. Commun.* **123**, 1 (1984).

[52] F. Tata, I. Henri, A. Markham, D. Weil, R. Williamson, S. Humphries, and C. Junien, *Hum. Genet.* **69**, 345 (1985).

[53] U. Francke, M. S. Brown, and J. L. Goldstein, *Proc. Natl. Acad. Sci. U.S.A.* **81**, 2826 (1984).

[54] F. Yang, J. Blum, J. R. McGill, C. M. Moore, S. L. Naylor, P. H. Van Brogt, W. D. Baldwin, and B. H. Bowman, *Proc. Natl. Acad. Sci. U.S.A.* **81**, 2752 (1984).

[55] O. Mykelbost, S. Rogne, B. Olaisen, T. Gedde-Dahl, Jr., and H. Prydz, *Hum. Genet.* **67**, 309 (1984).

[56] S. E. Humphries, K. Berg, L. Gill, A. M. Cumming, F. W. Robertson, A. F. H. Stalenhoef, R. Williamson, and A. Borresen, *Clin. Genet.* **26**, 389 (1984).

supra) it is possible that the clustering is an historical vestige. On the other hand it could be argued that apoA-I, apoA-IV, and apoC-III are all involved in triglyceride metabolism and that the clustering of their genes in a small region of chromosome 11 may be of functional significance yet to be discovered. It is not clear that apolipoproteins C-I, C-II, and E bear a simple functional relationship to one another. Nevertheless the presence of the genes for apoE and the LDL receptor on the same chromosome is provocative. However, the major ligand for the LDL receptor is apolipoprotein B, whose gene is located on a different chromosome, number 2.[56a]

Sequences Required for Transcription and Regulation

One of the most important developments in molecular biology has been the establishment of systems which allow cloned eukaryotic genes to be expressed *in vitro* or *in vivo*. Three approaches have been taken to this problem: transformation of cells with cloned genes,[57] microinjection of DNA into cells,[58] and *in vitro* transcription of cloned genes in cell-free systems.[59] Once a gene can be expressed in one of these systems, mutations of the gene can be generated *in vitro* and assayed for expression in order to identify specific nucleotide sequences required for transcription or regulation. As illustrated in Fig. 2, two types of signals important in transcription have been found in the 5' flanking DNA of eukaryotic genes. The TATA box is an 8–10 nucleotide region rich in adenines and thymidines which is found 30–50 base pairs upstream from the mRNA start site[60] and deletion of this sequence will abolish transcription of a gene *in vitro*.[61] The TATA box appears to be the promoter element for many but not all (see section on Regulation of HMG CoA Reductase) eukaryotic genes, i.e., it defines the position in the DNA at which RNA polymerase II will initiate transcription. A consensus nucleotide sequence CCAAT may also be involved in the regulation of transcription.

The second class of signals is found 50–300 base pairs upstream from the gene and is required for the efficient use of the TATA box by RNA

[56a] T. J. Knott, S. C. Rall, T. L. Innerarity, S. F. Jacobson, M. S. Urdea, B. Levy-Wilson, L. M. Powell, R. J. Pease, R. Eddy, H. Nakai, M. Byers, L. M. Priestly, E. Robertson, L. B. Rall, C. Betscholtz, T. B. Shows, R. W. Mahley, and J. Scott, *Science* **230**, 37 (1985).

[57] D. T. Kurtz, *Nature (London)* **291**, 629 (1981).

[58] M. R. Capecchi, *Cell* **22**, 479 (1980).

[59] D. R. Engelke, S. Y. Ng, B. S. Shastry, and R. G. Roeder, *Cell* **19**, 717 (1980).

[60] C. Baker and E. Ziff, *J. Mol. Biol.* **149**, 189 (1981).

[61] N. Proudfoot, M. H. M. Shander, J. L. Manley, N. Gefter, and T. Maniatis, *Science* **209**, 1329 (1980).

polymerase II. These regulatory sequences govern the frequency and efficiency of transcription initiation. The sequences required for the regulation of the gene for metallothionein, a protein synthesized by the liver which functions in the detoxification of heavy metals, have recently been identified. Metallothionein gene expression is normally transcriptionally regulated by cadmium and by glucocorticoid hormones. The technique of gene fusion was used to construct a hybrid gene composed of the 5' flanking region of the metallothionein gene and the structural gene for herpes viral thymidine kinase (TK) which contained only protein-coding sequences.[62] When this fusion gene was injected into heterologous cells or oocytes, the TK gene was expressed and was also subject to regulation by cadmium demonstrating that the promoter and regulatory sequences from the metallothionein gene were present in the fusion gene. Analysis of deletion mutants generated from the fusion gene identified a minimal sequence within 90 nucleotides of the transcription initiation start site which was required for the regulation of the gene by cadmium.[62,63] Another sequence 260 nucleotides upstream from the mRNA start site was required for the regulation of the gene by glucocorticoids.[64]

How a single stimulus can coordinately regulate a number of independent genes has been an intriguing question in cell biology. It has been proposed that groups of genes responsive to the same stimulus may have similar sequences in their 5' flanking DNA which interact with a common regulatory molecule. Recently such shared consensus sequences have been identified in several genes which are coordinately induced, including the heat shock genes of *Drosophila*[65] and the yeast genes for amino acid synthesis.[66] The problem of coordinate regulation is a key one in lipoprotein biosynthesis. Under normal circumstances, each lipoprotein class displays a characteristic apolipoprotein and lipid profile but nothing is known about how this is achieved. How is the synthesis of individual apolipoproteins coordinately regulated? How is apolipoprotein synthesis regulated in relation to lipid synthesis? Despite the close chromosomal linkage of the apoA-I and apoC-III genes, these two genes do not appear to be coordinately regulated in the rat intestine at least in response to a fat

[62] K. E. Mayo, R. Warren, and R. D. Palmiter, *Cell* **29,** 99 (1982).
[63] R. L. Brinster, H. Y. Chen, R. Warren, A. Sarthy, and R. D. Palmiter, *Nature (London)* **296,** 39 (1982).
[64] M. Karin, A. Haslinger, H. Holtgreve, R. I. Richards, P. Kraute, H. M. Westphal, and M. Beato, *Nature (London)* **308,** 513 (1984).
[65] H. R. B. Pelham, *Cell* **30,** 517 (1982).
[66] T. F. Donahue, R. S. Davies, G. Lucchini, and G. R. Fink, *Cell* **32,** 89 (1983).

load.[67] The recent isolation of several of the apolipoprotein genes should enable investigators to pinpoint DNA sequences important in the transcription and regulation of apolipoprotein genes in the near future.

Regulation of Gene Expression

Eukaryotic cells have a remarkable capacity for the selective expression of genetic information, with specific genes being turned on or off in particular cell types or at certain times during development. Although the actual mechanism of gene activation is not understood, it is clear that the physical organization of an active gene differs from that of an inactive one.[68] The globin genes in red blood cell nuclei which are actively transcribed are more sensitive to nuclease digestion and have an altered chromatin configuration, compared to the inactive globin genes in fibroblast nuclei.[69] Presumably both changes reflect an increased accessibility of the gene to the transcriptional complex. Demethylation of certain cytosine residues has been correlated with the expression of developmentally regulated genes.[70] The avian apoVLDL-II gene is an estrogen-responsive gene which encodes the major apoprotein of egg yolk VLDL. The cytosine residues in the 5' flanking region of the gene are initially highly methylated in the embryonic chick liver but become demethylated at day 10 of embryogenesis as the gene becomes estrogen responsive and transcriptionally active.[71]

Transcriptional Regulation of Gene Expression

The expression of active genes can be quantitatively regulated in response to hormonal or environmental stimuli at the transcriptional, posttranscriptional, translational, or even posttranslational levels. However, most eukaryotic gene regulation occurs at the level of transcription as illustrated by the classic studies on the hormonal regulation of the synthesis of vitellogenin, the precursor of egg yolk phosphoprotein. Vitellogenin is normally synthesized in the livers of egg-laying females and can also be artificially induced in the livers of embryos or adult males by the adminis-

---

[67] M. C. Blaufass, J. I. Gordon, G. Schonfeld, A. W. Strauss, and D. H. Alpers, *J. Biol. Chem.* **259**, 2452 (1984).
[68] S. Weisbrod, *Nature (London)* **297**, 289 (1982).
[69] H. Weintraub and M. Groudine, *Science* **193**, 848 (1976).
[70] A. Rhazin and A. D. Riggs, *Science* **210**, 604 (1980).
[71] V. Colgan, A. Elbrecht, P. Goldman, C. B. Lazier, and R. Deeley, *J. Biol. Chem.* **257**, 14453 (1982).

tration of estrogen. The ability to label newly synthesized RNA in isolated nuclei *in vitro* has allowed investigators to measure the transcription rates of specific genes.[72] This technique has been employed to demonstrate that estrogen stimulates the absolute rate of transcription of the vitellogenin gene over a 1000-fold in the *Xenopus* liver.[73]

How can a hormone or other stimulus affect the expression of a particular gene? Some answers to this question may be gleaned from recent experiments on the regulation of the metallothionein gene by glucocorticoids.[64] Similar to other steroid hormones, glucocorticoids act by binding to specific receptors which results in the transport of the hormone/receptor complex to the nucleus. To determine the binding site of this complex, purified glucocorticoid hormone/receptor complex was incubated *in vitro* with DNA from a plasmid containing the metallothionein gene. If the complex bound to a specific region of the gene, that fragment of DNA should theoretically be protected from digestion with DNase. Using this "footprinting" technique, the authors found that the complex protected a region of DNA 260 base pairs upstream from the metallothionein gene. Analysis of deletion mutants demonstrated that this same sequence was functionally required for the regulation of the gene by glucocorticoids.[64] Thus the binding of the hormone/receptor complex to upstream regulatory sequences 260 base pairs away from the mRNA initiation site can apparently enhance the transcription of the gene.

Posttranscriptional Gene Regulation

Gene regulation may also operate at the posttranscriptional level. The generation of a mature cytoplasmic mRNA is a complex process as illustrated in Fig. 3. After transcription by RNA polymerase II is initiated, a methylated guanylate cap (^{7m}Gppp) is added to the 5' end of the primary mRNA transcript. Approximately 200 adenylate (A) residues are added posttranscriptionally to the 3' end of the transcript and this poly(A) tail may play a role in the cytoplasmic stability of the mRNA. In addition the noncoding introns must be removed from the transcript before its transport to the cytoplasm.[74]

Correct splicing of the primary gene transcript is a requirement for normal gene function. Even a single nucleotide change at the splice junction of the α- or β-globin genes can lead to abnormal processing of α- or β-globin mRNA and appears to be the primary genetic defect in some of the

---

[72] E. Hofer and J. E. Darnell, *Cell* **23**, 585 (1981).
[73] M. Brock and D. Shapiro, *J. Biol. Chem.* **258**, 5449 (1983).
[74] J. E. Darnell, *Prog. Nucleic Acid Res. Mol. Biol.* **22**, 327 (1979).

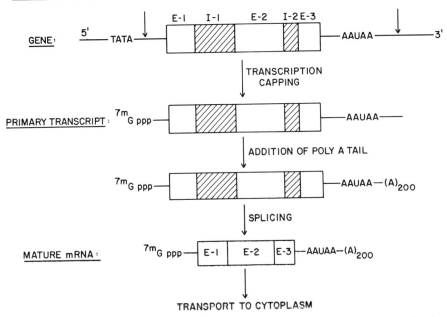

FIG. 3. mRNA processing. A typical eukaryotic gene is illustrated in the upper panel with arrows indicating the sites for the initiation and termination of transcription. As the gene is transcribed by RNA polymerase II, a methylated guanylate cap is added to the 5' end of the primary transcript. This transcript is further processed by the posttranscriptional addition of a poly(A) tail of about 200 adenylate residues. Noncoding introns are removed from the transcript and the remaining exons are spliced together. This mature mRNA is then transported to the cytoplasm for translation.

thalassemias.[75,76] Recently a number of genes have been shown to be regulated at the level of mRNA processing. Multiple mRNAs can be generated from a single gene by differential splicing[77] or by differential polyadenylation of the primary transcript of the gene.[78] Six transcripts are generated from the avian apoVLDL-II gene due to the use of multiple transcription initiation sites and alternative splicing events although the physiologic significance of this is not known.[79] Alternative splicing in different tissues can affect the level of gene expression as described for

[75] R. Treisman, N. J. Proudfoot, M. Shander, and T. Maniatis, *Cell* **29,** 903 (1982).
[76] B. K. Felber, S. H. Orkin, and D. H. Hamer, *Cell* **29,** 895 (1982).
[77] J. E. Schwarzbauer, J. W. Tamnkum, I. R. Lamischka, and R. O. Hynes, *Cell* **35,** 421 (1983).
[78] P. Early, J. Rogers, M. Davis, K. Calame, M. Bond, R. Wall, and L. Hood, *Cell* **20,** 313 (1980).
[79] G. S. Shelness and D. L. Williams, *J. Biol. Chem.* **259,** 9929 (1984).

the murine α-amylase gene[80] or even result in the production of different proteins in different tissues as reported for the calcitonin gene.[81]

The expression of genetic information may also be regulated after the processed mRNA has been transported to the cytoplasm. Changes in the translational activity of a mRNA may be a pivotal control point. The most extreme example of this kind of translational control occurs in *Drosophila* and mammalian cells in response to heat shock. When exposed to elevated temperatures, a series of heat shock proteins are induced and synthesized at the expense of the normal mRNAs which become sequestered in the cytoplasm in a translationally inactive form.[82] Gene expression may also be regulated at the level of mRNA turnover. Estrogen treatment markedly stimulates the production of vitellogenin in rooster and *Xenophus* livers. *Xenopus* liver slices have been pulse-labeled with [^{3}H]uridine and the amount of radiolabeled mRNA which remained during the chase period was determined by DNA excess filter hybridization.[83] Estrogen treatment increased the half-life of vitellogenin mRNA from 16 hr (similar to most poly(A) mRNAs) to 3 weeks.

Regulation of Apolipoprotein Synthesis

Many of the apolipoprotein mRNAs are fairly abundant mRNAs representing 0.1 to 1.0% of the total mRNA from the liver or intestine. Because of their relative abundance, clones of several of the apolipoproteins have been isolated from cDNA librarires constructed from hepatic or intestinal mRNA.[26,84-90] A number of screening approaches have been used to identify these clones, including hybrid selection and hybridization with a synthetic oligonucleotide probe. Several laboratories have investigated changes in apolipoprotein mRNA levels in response to dietary and

---

[80] R. A. Young, O. Hagenbuchle, and U. Schibler, *Cell* **23**, 451 (1981).
[81] M. J. Rosenfeld, J. J. Memod, S. G. Amara, L. W. Swanson, P. G. Sawchenko, J. Rivier, W. W. Vale, and R. M. Evans, *Nature (London)* **304**, 129 (1983).
[82] B. J. DiDomenico, G. E. Bugaisky, and S. Lindquist, *Proc. Natl. Acad. Sci. U.S.A.* **79**, 6181 (1982).
[83] M. Brock and D. Shapiro, *Cell* **34**, 207 (1983).
[84] J. L. Breslow, D. Ross, J. McPherson, H. Williams, D. Kurnit, A. L. Nussbaum, S. K. Karathanasis, and V. I. Zannis, *Proc. Natl. Acad. Sci. U.S.A.* **79**, 6861 (1982).
[85] C. Shoulders and F. E. Baralle, *Nucleic Acids Res.* **10**, 4873 (1983).
[86] P. Cheung and L. Chan, *Nucleic Acids Res.* **11**, 3703 (1983).
[87] S. W. Law and H. B. Brewer, *Proc. Natl. Acad. Sci. U.S.A.* **81**, 66 (1984).
[88] J. C. Ertel-Miller, R. K. Barth, P. H. Shaw, R. W. Elliott, and N. D. Hastie, *Proc. Natl. Acad. Sci. U.S.A.* **80**, 1511 (1983).
[89] J. W. Maclean, C. Fukazawa, and J. M. Taylor, *J. Biol. Chem.* **258**, 8993 (1983).
[90] K. L. Reue, D. H. Quon, K. A. O'Donnell, G. J. Dizikes, G. C. Fareed, and A. J. Lusis, *J. Biol. Chem.* **259**, 2100 (1984).

hormonal manipulation either by *in vitro* translation of RNAs or by hybridization of RNA with a radiolabeled cDNA probe.

ApoE plays a critical role in the recognition and uptake of lipoproteins by the liver. In the normal liver, apoE mRNA represents 0.5–1.0% of the total mRNA.[91] The amount of translatable apoE mRNA in the liver increased 2-fold in rats fed a high cholesterol diet supplemented with propylthiouracil (PTU), an anti-thyroid agent.[92] Whether this increase is in response to the hyperlipidemia, the hypothyroidism, or both is an open question. Plasma levels of apoE increase 20-fold in cholesterol-fed guinea pigs, in part due to a 2-fold increase in apoE secretion by the fatty liver.[92,93] When guinea pigs were fed a cholesterol/corn oil diet, hepatic apoE mRNA levels increased 2- to 3-fold after 4 weeks of diet and then declined to normal levels. Guinea pigs fed a cholesterol/coconut oil diet showed a similar increase in hepatic apoE mRNA only after 12 weeks of diet.[94] In contrast to the guinea pig, apoE mRNA levels did not change in the livers of mice[90] fed a high lipid diet.

Although plasma apoA-I levels decline with cholesterol feeding, there is some evidence that apoA-I mRNA levels are regulated differently in the liver and intestine. In mice fed a diet rich in cholesterol and saturated fat, apoA-I mRNA increased 2-fold in the intestine, perhaps in response to a need to package an increased load of dietary lipid. Hepatic apoA-I mRNA decreased 2-fold in the livers of these animals which led the authors to suggest that apoA-I mRNA synthesis in the liver may be subject to negative feedback control by plasma cholesterol levels.[88] Acute fat feeding increased both apoA-IV and apoC-III mRNA levels but did not affect apoA-I mRNA.[67,95] Despite the linkage of the apoA-I and apoC-III genes in the genome, the regulation of intestinal apoA-I mRNA in the rat appears to be independent of the regulation of the other apolipoproteins synthesized by the intestine. Clearly this issue requires further attention in other physiologic states.

The regulation of apolipoprotein synthesis by dietary cholesterol has just begun to be explored but a number of interesting questions are raised by these studies. How does dietary lipid actually regulate the expression of the individual apolipoprotein genes? One possibility is that cholesterol may modulate apolipoprotein gene transcription either directly or indirectly through a second messenger. Alternatively, intracellular lipid may

[91] C. A. Reardon, R. V. Hay, J. I. Gordon, and G. S. Getz, *J. Lipid Res.* **25**, 348 (1984).
[92] C. Lin-Lee, Y. Tanaka, C. T. Lin, and L. Chan, *Biochemistry* **20**, 6474 (1981).
[93] R. L. Hamilton, J. P. Kane, C. J. Fielding, and L. S. Guo, *J. Lipid Res.* **23**, 531 (1982).
[94] D. M. Driscoll and G. S. Getz, *Circulation* **68**, III-17 (1983).
[95] J. I. Gordon, D. P. Smith, D. H. Alpers, and A. W. Strauss, *J. Biol. Chem.* **257**, 8418 (1982).

alter the lipid composition of cellular membranes, resulting in a change in the processing or stability of a particular apolipoprotein mRNA.

Apolipoprotein E Synthesis in Extrahepatic Tissues

Although lipoproteins are synthesized in the liver and intestine, apoE is synthesized by a broad spectrum of tissues not generally thought to be involved in lipoprotein production in humans, rats, mice, and guinea pigs.[89,94,96,97] ApoE synthesis is fairly prominent in extrahepatic tissues representing between 0.1 and 1.0% of the total protein synthesized. Interestingly very little apoE is synthesized by the intestinal mucosa which is compatible with the finding that the rat apoE gene is more highly methylated in the intestine than in the liver or in extrahepatic tissues.[97] The fact that macrophages can synthesize apoE *in vitro* suggests that tissue macrophages may contribute to the extrahepatic synthesis of apoE.[98] However the magnitude of apoE synthesis in these tissues argues that other cell types are likely to be involved. Indeed Rhesus monkey aortic smooth muscle cells and rat ovarian granulosa cells synthesize and secrete apoE *in vitro*.[97,99]

Although the phenomenon of extrahepatic synthesis of apoE is well documented, little is known about the regulation apoE synthesis in these tissues. When peritoneal macrophages are loaded with cholesteryl ester *in vitro*, the secretion of apoE is markedly stimulated.[98] ApoE synthesis appears to be developmentally regulated in bone marrow-derived macrophages.[100] The functional state of the macrophage may also influence apoE synthesis. ApoE is synthesized by resident peritoneal macrophages and by thioglycollate-elicited macrophages but not by macrophages which are activated immunologically.[101] Because apoE down-regulates lymphocyte proliferation *in vitro*,[102] the decreased production of apoE by immunologically activated macrophages *in vivo* may stimulate the immune response at the site of inflammation. ApoE synthesis can be hormonally regulated in steroidogenic tissues. When rat ovarian granulosa cells are

---

[96] M. L. Blue, D. L. Williams, S. Zucker, S. A. Khan, and C. B. Blum, *Proc. Natl. Acad. Sci. U.S.A.* **80**, 283 (1983).
[97] D. M. Driscoll and G. S. Getz, *J. Lipid Res.* **25**, 1368 (1984).
[98] S. K. Basu, M. S. Brown, Y. K. Ho, R. J. Havel, and J. L. Goldstein, *Proc. Natl. Acad. Sci. U.S.A.* **78**, 7545 (1981).
[99] D. M. Driscoll, J. R. Schreiber, V. M. Schmit, and G. S. Getz, *J. Biol. Chem.* **260**, 9031 (1985).
[100] Z. Werb and J. C. Chin, *J. Cell Biol.* **97**, 1113 (1983).
[101] Z. Werb and J. C. Chin, *J. Exp. Med.* **158**, 1272 (1983).
[102] E. M. Avila, G. Holdsworth, N. Sasaki, R. L. Jackson, and J. A. K. Harmony, *J. Biol. Chem.* **257**, 5900 (1981).

cultured *in vitro,* the secretion of apoE can be stimulated by the addition of follicle-stimulating hormone or cyclic AMP.[99] Clearly further studies are needed to understand the function and regulation of apoE synthesis in extrahepatic tissues.

Regulation of HMG CoA Reductase

In addition to their effects on apolipoprotein synthesis, the uptake of exogenous sterols also controls intracellular cholesterol levels by regulating enzymes involved in cholesterol biosynthesis. HMG CoA reductase is a microsomal enzyme which catalyzes the rate-limiting step in this pathway. Reductase activity is under negative feedback control and transcription of the reductase gene is stimulated *in vivo* by cholesterol deprivation.[103–105] A mutant cell line (UT-1) has been isolated which is resistant to compactin, a competitive inhibitor of reductase. The compactin resistance is due to the overproduction of the enzyme.[105] A cDNA clone for reductase was recently isolated from a cDNA library generated from UT-1 mRNA[106] and hybridization of this clone to UT-1 mRNA demonstrated that reductase mRNA was elevated 100-fold in these cells. The increase in reductase mRNA was in part brought about by gene amplification, a phenomenon observed in other drug-resistant cell lines.[107] Although reductase is normally a single copy gene, hybridization of the reductase clone to UT-1 chromosomal DNA detected 15 copies of the gene in these cells.[108] Presumably there is an increase in the rate of transcription of the reductase gene which also contributes to the 100-fold increase in reductase mRNA.

The hamster gene for HMG CoA reductase has been isolated and consists of 20 exons distributed over 25 kb of DNA.[109] This gene exhibits several unusual features especially in its regulatory region upstream of the coding sequences. It contains no TATA or CCAAT box in this region, but

---

[103] L. Liscum, K. L. Luskey, D. J. Chin, Y. K. Ho, J. L. Goldstein, and M. S. Brown, *J. Biol. Chem.* **258,** 8450 (1983).

[104] C. F. Clarke, P. A. Edwards, S. F. Lan, R. D. Tanaka, and A. M. Fogelman, *Proc. Natl. Acad. Sci. U.S.A.* **80,** 3305 (1983).

[105] D. J. Chin, K. L. Luskey, R. J. Anderson, J. R. Faust, J. L. Goldstein, and M. S. Brown, *Proc. Natl. Acad. Sci. U.S.A.* **79,** 1185 (1982).

[106] K. L. Luskey, D. J. Chin, R. J. MacDonald, L. Liscum, J. L. Goldstein, and M. S. Brown, *Proc. Natl. Acad. Sci. U.S.A.* **79,** 6210 (1983).

[107] R. T. Schimke, R. J. Kaufman, F. W. Alt, and R. F. Kellems, *Science* **202,** 1051 (1978).

[108] K. L. Luskey, J. R. Faust, D. J. Chin, M. S. Brown, and J. L. Goldstein, *J. Biol. Chem.* **258,** 8462 (1983).

[109] G. A. Reynolds, S. K. Basu, T. F. Osborne, D. J. Chin, G. Gil, M. S. Brown, J. L. Goldstein, and K. L. Luskey, *Cell* **38,** 275 (1984).

it does contain five copies of a hexanucleotide sequence CCGCCC or GGGCGG, a sequence thought to be important for the transcription of some viral genes (e.g., ref. 110). Using the powerful techniques of gene fusion and transfection, the importance of these upstream sequences for the control of HMG CoA reductase gene transcription has been established.[111] The 5' noncoding end of the hamster HMG CoA reductase gene was joined to the coding sequence of a marker gene, bacterial chloramphenicol acetyltransferase (CAT). When this chimeric gene was introduced into mouse L cells by transfection,[112] the expression of CAT activity was suppressed by cholesterol as would be the case for HMG CoA reductase itself.[111] Selective removal of portions of the HMG CoA reductase regulatory sequence in these chimeric gene constructs used for transfection revealed that the sequences important for the regulation of HMG CoA reductase expression by sterol was distributed over a 500 bp region placed 300 bp upstream of the reductase transcription initiation sites. This region contains the 5 hexanucleotide repeats that occur in regulatory regions of viral and cellular housekeeping genes.

Cholesterol regulates HMG CoA reductase activity through mechanisms additional to its effects on the transcription of the reductase gene. HMG CoA reductase has a membrane-bound domain and a cytoplasmic domain containing the active site of the enzyme.[113] Cholesterol enhances the degradation of reductase protein and this involves interactions with the membrane-bound domain.[114]

Protein Synthesis and Intracellular Targeting

In the last few years the role of topogenic sequences—sequences which function to direct a precursor protein to its ultimate destination within the cell—has been established in several systems. Many of the mitochondrial proteins are cytoplasmically synthesized as larger precursor molecules containing a sequence which interacts with a receptor on the mitochondrial membrane, resulting in the import of the protein into the mitochondria.[115] Secretory proteins are initially synthesized as larger precursor molecules containing an amino terminal signal peptide not found on the mature plasma protein which directs the nascent chain to the

[110] S. L. McKnight and R. Kingsbury, *Science* **217**, 316 (1982).
[111] T. F. Osborne, J. L. Goldstein, and M. S. Brown, *Cell* **43**, 203 (1985).
[112] C. M. Gorman, L. F. Moffat, and B. H. Howard, *Mol. Cell. Biol.* **2**, 1044 (1982).
[113] L. Liscum, J. Finer-Moore, R. M. Stroud, K. L. Luskey, M. S. Brown, and J. L. Goldstein, *J. Biol. Chem.* **260**, 522 (1985).
[114] G. Gil, J. R. Faust, D. J. Chin, J. L. Goldstein, and M. S. Brown, *Cell* **41**, 249 (1985).
[115] R. V. Hay, P. Boehni, and S. Gasser, *Biochim. Biophys. Acta* **779**, 65 (1984).

membrane of the rough endoplasmic reticulum (RER) as shown in Fig. 4. *In vivo*, the signal peptide is cotranslationally removed by a microsomal enzyme, the signal peptidase, as the nascent protein is sequestered in the lumen of RER.[116] This cleavage of the signal peptide can be reproduced *in vitro* by the addition of canine pancreatic microsomes which have signal peptidase activity to the *in vitro* translation reaction and can be detected by the reduced electrophoretic mobility of the microsome-processed translation product or by sequencing of radiolabeled proteins.[117]

Apolipoprotein Biosynthesis

Although the apolipoproteins are unusual in that they may be secreted from the cell on a particle containing other apolipoproteins and lipid, the initial events in their biosynthesis are similar to those described for other secretory proteins. Many of the apolipoproteins are initially synthesized with an amino terminal signal peptide and their sequences have been determined by sequencing of the primary translation products.[67,91,95,118] The apolipoprotein signal peptides are similar in structure to other known signal peptides, containing a polar amino terminus, a central hydrophobic core, and a charged carboxy terminus often composed of small side chain amino acids. In the face of relatively limited sequence homology between signal peptides, it is notable that there is a high degree of homology between the signal peptides of rat apoA-I and apoA-IV on the one hand and human and rat apoA-I and apoE on the other.[17,91] These relationships suggest that nucleotides encoding the signal peptides of three apoproteins originate from a common ancestral exon.[17] However the precise pattern of amino acid homologies among these three apoproteins raises other possibilities. The signal peptides of apoA-I and apoA-IV are highly homologous in their core regions.[95] In contrast, the apoA-I and apoE signal peptides although not especially homologous in their core regions do exhibit a high degree of homology at both termini.[91] The meaning of these unusual homologies in the signal peptide is not entirely clear but they may be important in the intracellular targeting of the apolipoproteins to the site of their assembly with lipid.

In addition to a signal peptide, the intracellular precursors of apoA-I and apoA-II also contain amino terminal propeptides which are not cotranslationally removed by the signal peptidase. The 5 amino acid propeptide of human apoA-II is structurally similar to other known propeptides

[116] G. Kreil, *Annu. Rev. Biochem.* **50**, 317 (1981).
[117] G. Blobel and B. Dobberstein, *J. Cell Biol.* **67**, 852 (1975).
[118] J. I. Gordon, D. P. Smith, R. Andy, D. H. Alpers, G. Schonfeld, and A. W. Strauss, *J. Biol. Chem.* **257**, 971 (1982).

in that it ends with paired basic amino acids. Using Hep G2 cells, Gordon *et al.* have shown that much but not all of the apoA-II propeptide is cleaved intracellularly prior to secretion, perhaps between the trans-Golgi and plasma membrane as has been shown for other propeptides.[119] The secreted proapoA-II is presumably cleaved extracellularly to generate the mature protein. Both human and rat proapoA-I contain a 6 amino acid propeptide which ends in paired glutamine residues, an unusual carboxy-terminus for a propeptide.[118,120] Due to the presence of an arginine residue in the human propeptide, human proapoA-I can be readily separated from mature apoA-I by isoelectric focusing.[121] Interestingly the rat apoA-I propeptide contains an acidic without a compensating basic residue and therefore in contrast to the human case, rat proapoA-I is more acidic than the mature peptide.[118,122] The apoA-I propeptide is not removed intracellularly and apoA-I is secreted from the liver and intestine mostly as proapoA-I. An enzymatic activity has recently been identified in serum, plasma, and lymph which converts proapoA-I to mature plasma apoA-I.[123] Since the conversion of proapoA-I to mature plasma apoA-I is rapid relative to the half-life of mature plasma apoA-I, the presence of proapoA-I in tissue fluid signifies newly synthesized or secreted apoA-I. Much of the apoA-I in human lymph is proapoA-I suggesting recent secretion by enterocytes.[124] In contrast to the human, about 30% of rat plasma apoA-I is in the form of the propeptide, probably reflecting a rate of conversion of the latter to mature peptide not vastly different than the relatively rapid rate of apoA-I catabolism in the rat.[122]

Unlike the insulin propeptide, the function of the propeptides of apoA-I and apoA-II is not understood. The propeptides may play a role in the intracellular targeting of apoA-I and apoA-II. Because both proteins are extremely lipophilic molecules, the propeptide may function to prevent the nascent proteins from interacting with cellular membranes during their intracellular transport. Tangier's disease is an autosomal recessive disease which is characterized by low plasma levels of HDL and apoA-I in part due to increased catabolism of Tangier apoA-I.[125] Isoelectric fo-

---

[119] J. I. Gordon, K. A. Budelier, H. F. Sims, A. M. Scanu, and A. W. Strauss, *J. Biol. Chem.* **258**, 14054 (1983).
[120] V. I. Zannis, S. K. Karathanasis, H. T. Keutman, G. Goldberger, and J. L. Breslow, *Proc. Natl. Acad. Sci. U.S.A.* **80**, 2574 (1983).
[121] L. L. Kay, R. Ronan, E. J. Schaefer, and H. B. Brewer, *Proc. Natl. Acad. Sci. U.S.A.* **79**, 2485 (1982).
[122] M. B. Sliwkowski and H. G. Windmueller, *J. Biol. Chem.* **259**, 6459 (1984).
[123] C. Edelstein, J. I. Gordon, K. Toscas, H. F. Sims, A. W. Strauss, and A. M. Scanu, *J. Biol. Chem.* **258**, 11430 (1983).
[124] G. Ghiselli, E. J. Schaefer, J. A. Light, and H. B. Brewer, *J. Lipid Res.* **24**, 731 (1984).
[125] E. J. Schaefer, L. L. Kay, L. A. Zech, and H. B. Brewer, *J. Clin. Invest.* **70**, 934 (1982).

cusing and protein sequencing have revealed that 35% of the plasma apoA-I exists as proapoA-I in these patients[126] but this is apparently not due to a deficiency of the proapoA-I converting enzyme.[127]

The biosynthesis of HMG CoA reductase presents an interesting problem since the enzyme is found in the membrane of the RER with the active site facing the cytoplasm. Based on the amino acid sequence derived from the nucleotide sequence of a cDNA clone, reductase contains 7 hydrophobic regions each of which may potentially span the RER membrane.[128] Unlike other transmembrane glycoproteins,[129] reductase does not contain a conventional signal peptide sequence.[128] Further studies are needed to understand how this unusual glycoprotein is inserted into the RER membrane.

Glycosylation

The process of glycosylation is compartmentalized in eukaryotic cells, occurring both in the RER and in the Golgi. High-mannose oligosaccharide chains derived from a dolichol-linked intermediate can be added cotranslationally in the RER by N-linked glycosidic linkages to certain asparagine residues in the nascent protein. These N-linked oligosaccharides can be further processed to complex type oligosaccharides after the protein is transported to the Golgi apparatus. O-linked glycosylation occurs in the Golgi apparatus where carbohydrate moieties donated directly by nucleotide sugars are added to serine or threonine residues in the protein. Sialic acid residues which are sensitive to digestion with neuraminidase are also added in the Golgi to proteins bearing either complex N-linked or O-linked oligosaccharide. The biologic consequences of glycosylation are not understood. The antibiotic tunicamycin which blocks the addition of N-linked oligosaccharides does inhibit the secretion of some glycoproteins but many are secreted normally from hepatocytes in the presence of tunicamycin.[130] Thus N-linked carbohydrate is not required for all glycoprotein secretion.

Many of the apolipoproteins are glycosylated in the RER and/or Golgi during their biosynthesis including apoB, apoE, and apoC-III. The estro-

---

[126] G. Schmitz, G. Assman, S. C. Rall, and R. W. Mahley, *Proc. Natl. Acad. Sci. U.S.A.* **80**, 6081 (1983).
[127] D. Bojanovski, R. E. Gregg, and H. B. Brewer, *J. Biol. Chem.* **259**, 6049 (1984).
[128] D. J. Chin, G. Gil, D. W. Russell, L. Liscum, K. L. Luskey, S. K. Basu, H. Okayama, P. Berg, J. L. Goldstein, and M. S. Brown, *Nature (London)* **308**, 613 (1984).
[129] V. R. Lingappa, F. N. Katz, H. F. Lodish, and G. Blobel, *J. Biol. Chem.* **253**, 8667 (1978).
[130] D. K. Struck, P. B. Siuta, M. D. Lane, and W. J. Lennarz, *J. Biol. Chem.* **253**, 5332 (1978).

gen-stimulated chicken liver has been used to study the biosynthesis and glycosylation of apoB, a glycoprotein containing mannose, galactose, glucose, and sialic acid. The glycosylation of apoB appears to be a two-step process in which high mannose oligosaccharide chains are added cotranslationally in the RER to asparagine residues in the nascent chain; subsequently some of these oligosaccharides are processed to the complex type in the Golgi.[131] Although tunicamycin inhibited apoB glycosylation, the secretion of VLDL from the hepatocyte was not affected, indicating that the N-linked glycosylation of apoB is not necessary for VLDL synthesis, assembly, or secretion.[130,131]

ApoE is also a glycoprotein, containing 2–2.5% carbohydrate. Glycosylation of apoE is not inhibited by tunicamycin, suggesting that the oligosaccharides are O-linked rather than N-linked.[132] In addition the primary amino acid sequence of apoE does not contain the characteristic tripeptide (Asn-X-Ser/Thr) necessary for N-linked glycosylation.[133] The polymorphism of human plasma apoE which is seen on isoelectric focusing is in part due to genetic variation and in part generated by posttranslational modification by the addition of negatively charged sialic acid residues.[134] Human apoE is initially secreted from hepatocytes as a sialo protein but over 80% of plasma apoE is desialylated.[134,135] Desialylation of apoE may be a critical factor in regulating the uptake of apoE-containing lipoproteins by the chylomicron remnant receptor in the liver since there is some evidence that the desialylated forms of apoE are more readily catabolized by the perfused rat liver than the sialylated forms.[136]

Biosynthesis of the LDL Receptor

The LDL receptor is a model for a glycoprotein involved in lipid metabolism which is inserted into the plasma membrane. The physiologic importance of the LDL receptor is best illustrated by patients with the genetic disease familial hypercholesterolemia (FH) who lack functional LDL receptors.[18] FH homozygotes have markedly elevated plasma cholesterol and LDL levels and develop premature atherosclerosis. Tol-

[131] P. Suita-Mangano, S. C. Howard, W. Lennarz, and M. D. Lane, *J. Biol. Chem.* **257**, 4292 (1982).
[132] C. A. Reardon, D. M. Driscoll, R. A. Davis, R. A. Borchardt, and G. S. Getz, *J. Biol. Chem.*, in press (1986).
[133] S. Rall, K. H. Weisgraber, and R. W. Mahley, *J. Biol. Chem.* **257**, 4171 (1982).
[134] V. I. Zannis and J. L. Breslow, *Biochemistry* **20**, 1033 (1981).
[135] V. I. Zannis, D. M. Kurnit, and J. L. Breslow, *J. Biol. Chem.* **257**, 536 (1982).
[136] R. J. Havel, Y. S. Chao, E. Windler, L. Kotike, and L. Guo, *Proc. Natl. Acad. Sci. U.S.A.* **77**, 4349 (1980).

leshaug et al. have used pulse-chase labeling of skin fibroblasts to study the intracellular processing of the receptor in normal and FH patients.[137,138] The LDL receptor contains both N- and O-linked oligosaccharides and is initially synthesized in the RER as a 120,000 Da precursor which is subsequently processed in the Golgi to the mature 160,000 Da receptor. Seven mutant alleles have been identified in FH patients which segregate to the LDL receptor gene locus. Patients homozygous for the null allele are receptor negative and lack any detectable protein. Three alleles were identified which code for abnormal precursor proteins which underwent normal processing whereas the other 3 alleles specified abnormal precursor proteins which were not processed further. Thus FH is a genetically complex disease in that multiple alleles affecting the synthesis, processing, or transport of the LDL receptor can lead to the FH phenotype. The recent isolation of a cDNA clone for the LDL receptor should facilitate understanding of its structure and biosynthesis.

Other Posttranslational Modifications

Recent studies on prepropeptide hormones have demonstrated that gene regulation can occur at the posttranslational level. Preproopiomelanocorticotropin (POMC) is a precursor protein which codes for adrenocorticotropin, a and b melanocyte stimulating hormone, b-lipotropin, and e-endorphin.[139] Although POMC itself is synthesized in a number of tissues, it is processed differently by proteolytic cleavage in each with the result that each tissue secretes a different spectrum of hormones.

Posttranslational processing may play a role in the biosynthesis of apoB, the major apolipoprotein of VLDL and LDL. ApoB is polymorphic in that there are two apoB variants in the plasma: B_h (a 335–590 kDa protein) and B_l (a 240–260 kDa protein).[140-142] Although it was initially believed that B_h was synthesized by the liver and B_l by the intestine, studies with intact animals and liver perfusions have demonstrated that the normal rat liver can synthesize both B_h and B_l.[143,144] The relationship

---

[137] H. Tolleshaug, J. L. Goldstein, W. J. Schneider, and M. S. Brown, *Cell* **30**, 715 (1982).
[138] H. Tolleshaug, K. H. Hobgood, M. S. Brown, and J. L. Goldstein, *Cell* **32**, 941 (1983).
[139] E. Herbert and M. Uhler, *Cell* **30**, 1 (1982).
[140] K. V. Krishnaiah, L. F. Walker, J. Borensztajn, G. Schonfeld, and G. S. Getz, *Proc. Natl. Acad. Sci. U.S.A.* **77**, 3806 (1980).
[141] J. P. Kane, D. A. Hardman, and H. E. Paulus, *Proc. Natl. Acad. Sci. U.S.A.* **77**, 2465 (1980).
[142] J. Elovson, Y. O. Huang, N. Baker, and R. Kannan, *Proc. Natl. Acad. Sci. U.S.A.* **78**, 157 (1981).
[143] L. L. Swift, R. J. Padley, V. S. LeQuire, and G. S. Getz, *Circulation* **64**, IV-101 (1981).
[144] A.-L. Wu and H. G. Windmueller, *J. Biol. Chem.* **256**, 3615 (1982).

between B_h and B_l is not understood. Although the two variants are immunologically cross-reactive and share most peptides in peptide maps,[142] there are some monoclonal antibodies which recognize B_h but not B_l; the converse however is not true.[145] B_h and B_l may be the products of two different genes or they may result from differential processing of the primary mRNA transcript from one gene. Alternatively, there may be a precursor–product relationship between the two variants. Kinetic data are suggestive of the possibility that B_l is generated intracellularly from B_h by proteolytic cleavage.[146]

The posttranslational phosphorylation of proteins is now known to be a generalized mechanism by which cells respond to external stimuli. The activity of HMG CoA reductase, the rate-limiting enzyme in cholesterol biosynthesis, can be acutely regulated by phosphorylation, resulting in inactivation of the enzyme.[147] Phosphorylation may also be important in a number of other cellular processes such as the movement of proteins through or within membranes. Davis et al. have examined the phosphorylation of apoB by radiolabeling newly synthesized phosphoproteins in vitro in cultured hepatocytes or in vivo in rats.[148] Although there is very little if any phosphorylation of apoB_h, apoB_l is secreted from cultured hepatocytes in a phosphorylated form, with most of the phosphate present as phosphoserine. The authors suggest that phosphorylation of the apoB may be important in the intracellular transport of the nascent VLDL during their assembly.

Assembly and Secretion

The classic experiments of Palade used electron microscopy to follow pulse-labeled precursor proteins through organelles in pancreatic acinar cells to study the secretory pathway in eukaryotic cells. It is now well established that secretory proteins follow the morphologic route RER–Golgi–condensing vesicles–secretory granules–exocytosis (Fig. 4).[149] One of the least studied areas in lipoprotein research concerns the intracellular assembly and transport of lipoproteins. The nascent lipoprotein particles appear to follow a route similar to other secretory proteins. Early electron microscopic studies used radiolabeled precursors or perox-

[145] J. P. Kane, Annu. Rev. Physiol. **45**, 673 (1983).
[146] R. J. Padley, L. L. Swift, and G. S. Getz, Circulation **64**, II-101 (1982).
[147] Z. H. Beg, J. A. Stonik, and H. B. Brewer, Proc. Natl. Acad. Sci. U.S.A. **75**, 3678 (1978).
[148] R. A. Davis, G. M. Clinton, R. A. Borchardt, M. Malone-McNeal, T. Tan, and G. Lattier, J. Biol. Chem. **259**, 3383 (1984).
[149] G. Palade, Science **189**, 347 (1975).

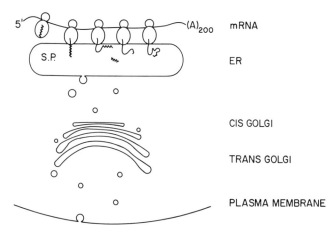

FIG. 4. The secretory pathway. A protein destined for secretion is synthesized on membrane-bound ribosomes and its primary translation product contains an amino terminal signal peptide of 18–24 amino acids which is not found on the mature plasma protein. The signal peptide directs the nascent chain into the lumen of the RER and is then cotranslationally removed by a microsomal enzyme. The completed proteins are transported in vesicles to the Golgi where they are further processed by glycosylation. The trans-Golgi sorts proteins into plasma membrane proteins, lysosomal proteins, and secretory proteins. Secretory proteins are concentrated in condensing vesicles, transported to the plasma membrane by secretory vesicles, and released from the cell by exocytosis.

idase-coupled antibodies to demonstrate that apolipoproteins are synthesized in the RER and triglyceride in the RER or SER.[150,151] Lipidation of the apoproteins has been reported to occur in the RER, in the SER or at the RER/SER junction. Dense osmiophilic particles containing both apolipoproteins and lipid have also been identified in the Golgi and in condensing vesicles.[152]

The biosynthesis of a number of different proteins has been studied by pulse-labeling cells with radiolabeled precursors and following the intracellular movement of proteins by cell fractionation and immunoprecipitation. Studies of this nature have been complicated by the lack of good subcellular fractionation techniques which is a particular problem with lipoproteins since many of the procedures used to disrupt subcellular organelles may also in fact disrupt nascent lipoprotein particles. Recently Howell and Palade have developed a technique which may circumvent

---

[150] H. Glauman, A. Bergstrand, and J. L. Ericsson, *J. Cell Biol.* **66**, 356 (1975).
[151] C. A. Alexander, R. L. Hamilton, and R. J. Havel, *J. Cell Biol.* **69**, 241 (1976).
[152] P. J. Dolphin, *J. Lipid Res.* **22**, 971 (1981).

some of these problems.[153] Golgi fractions were isolated from rat liver homogenates and were treated with Na_2CO_3 at alkaline pH to effectively separate the Golgi membranes from the contents. Approximately 50% of the lipoproteins isolated from the Golgi by this method were similar to plasma VLDL in terms of density (<1.006 g/ml) and biochemical composition.[154] However the remainder of the lipoprotein particles were markedly heterogeneous with respect to size, extent of lipidation, and apolipoprotein profile.

Janero et al. have utilized a kinetic approach to study some of the steps involved in the sequential assembly of VLDL in the chick liver cells.[155] Liver cells were pulse-labeled with radiolabeled precursors and samples of media were analyzed at various times during the chase period for the appearance of radiolabeled constituents in the newly secreted VLDL. The kinetics of the secretion of ^{3}H-labeled apolipoprotein, ^{3}H-labeled triacylglycerol, and ^{3}H-labeled phosphoglyceride indicated that the biosynthesis of VLDL proceeded in a sequential fashion. Based on the kinetic data, the authors suggest that apolipoproteins first associate with phosphoglyceride and then assemble with triacylglycerol to form a triglyceride-rich lipoprotein particle, with additional phosphoglyceride added prior to secretion.

The transport of nascent lipoproteins from the Golgi to the plasma membrane and their secretion from the hepatocyte into the space of Disse requires microtubules since these processes are inhibited by colchicine.[156] The N-linked glycosylation of VLDL apolipoproteins is not necessary for the synthesis, assembly, and secretion of VLDL in the chicken or the rat liver since these processes are not inhibited by tunicamycin.[131,157,158] However, apolipoprotein synthesis is required since inhibition of protein synthesis with cycloheximide will block the secretion of VLDL.[159] ApoB plays a particularly critical role in the intestinal assembly and secretion of the triglyceride-rich chylomicrons and VLDL. Patients with the genetic disease abetalipoproteinemia who lack circulating apoB develop fat malabsorption due to impaired intestinal lipid transport.[160] Although apoA-I is a major apolipoprotein of nascent chylomicrons, apoA-I may not be structurally required for chylomicron assembly and secretion from the

---

[153] K. E. Howell and G. E. Palade, *J. Cell Biol.* **92**, 822 (1982).
[154] K. E. Howell and G. E. Palade, *J. Cell Biol.* **92**, 833 (1982).
[155] D. R. Janero and M. D. Lane, *J. Biol. Chem.* **258**, 14496 (1983).
[156] O. Stein, L. Sanger, and Y. Stein, *J. Cell Biol.* **62**, 90 (1974).
[157] J. Bell-Quint, T. Forte, and P. Graham, *Biochem. J.* **200**, 409 (1981).
[158] P. Siuta-Mangano, D. R. Janero, and M. D. Lane, *J. Biol. Chem.* **257**, 11463 (1982).
[159] R. S. Verbin, P. J. Goldblatt, and E. Farber, *Lab. Invest.* **20**, 529 (1969).
[160] W. O. Dobbins, *Gastroenterology* **50**, 195 (1966).

enterocyte. Patients with genetic defects in apoA-I develop premature atherosclerosis but have normal fat absorption.[41]

Further investigations into the kinetics of lipoprotein assembly and the structure of nascent VLDL and its precursors are needed to answer some very basic questions. How and where are the apolipoproteins lipidated? Is the assembly process affected by posttranslational modifications of the apolipoproteins? How does the assembly process differ in the liver which produces B_h and the intestine which produces B_l? The isolation of intact intracellular precursors of nascent VLDL from RER and Golgi combined with pulse-labeling experiments should help answer some of these questions.

Vistas for the Future

The investigation of lipid transport and lipid metabolism at the molecular level promises to be an exciting and active area of future research. The technology for the cloning of eukaryotic genes is now readily accessible and should be a powerful tool in understanding the structure and function of proteins important in lipid metabolism, particularly proteins such as apoB which have been refractory to chemical analysis. The regulation of apolipoprotein synthesis has just begun to be explored at the molecular level but the search for DNA sequences important in the dietary and hormonal regulation of these genes should be an intense area of investigation in the near future. The isolation of somatic cell mutants in cholesterol biosynthesis (both regulatory and auxotrophic) should facilitate the cloning of the genes for the enzymes and regulatory proteins in the pathway. This will be invaluable in unraveling the mechanisms involved in the regulation of cellular cholesterol metabolism.

Although a great deal has been learned about the structure and function of lipoproteins, very little is known about the biosynthesis of lipoproteins. Some of the early events in apolipoprotein synthesis have been identified but how apolipoproteins are assembled with lipid into lipoprotein particles is an unresolved question. Ultimately one would like to be able to reconstitute the assembly process *in vitro* in order to identify the cellular organelles and factors required. It is likely that the assembly of the two primary lipoprotein particles, VLDL and HDL, follows very different pathways.

Finally much of our knowledge of lipoprotein metabolism and atherogenesis has been derived from individuals with genetic defects. The studies on patients with FH have elucidated the role of the LDL receptor in lipoprotein metabolism and have also shed light on the basic cellular process of receptor-mediated endocytosis. Many of these patients appear to be defective in the glycosylation, intracellular transport, or assembly of the LDL receptor. These studies on FH patients illustrate the value of an intensive combined genetic and biochemical approach. This kind of approach is likely to help decipher the underlying defects in other diseases of lipid metabolism in the near future.

[3] Comparative Analysis of Mammalian Plasma Lipoproteins

By M. JOHN CHAPMAN

Overview

The past decade has seen dramatic developments in our knowledge of lipid transport systems in mammals, to the extent that their genetic regulation is now under intense study.[1-7] Such progress reflects the importance of mammalian species not only as experimental models for lipid and lipoprotein metabolism in diseases afflicting man, such as atherosclerosis, diabetes, and obesity, but also of the direct contribution of certain animals to human nutrition, and therefore of the need to understand processes of lipid absorption and transport in such species during development and adult life. In addition, the availability of genetically defined inbred strains displaying specific defects associated with lipoprotein metabolism, such as the Watanabe-heritable hyperlipidemic (WHHL) rabbit[8-10] which lacks cellular LDL (apoB,E)[10a] receptors, has also stimulated interest in this area.

The aim of this chapter then is to collate available data on the physicochemical characteristics of the plasma lipoproteins and their apolipoproteins in mammals, in such a way as to provide both a rapid access to the

---

[1] M. H. Lin-Su, Y. C. Lin-Lee, W. A. Bradley, and L. Chan, *Biochemistry* **20**, 2470 (1981).
[2] Y. C. Lin-Lee, W. A. Bradley, and L. Chan, *Biochem. Biophys. Res. Commun.* **99**, 654 (1981).
[3] Y. C. Lin-Lee, Y. Tanaka, C. T. Lin, and L. Chan, *Biochemistry* **20**, 6474 (1981).
[4] J. W. McLean, C. Fukazawa, and J. M. Taylor, *J. Biol. Chem.* **258**, 8993 (1983).
[5] R. C. LeBoeuf, D. L. Puppione, V. N. Schumaker, and A. J. Lusis, *J. Biol. Chem.* **258**, 5063 (1983).
[6] A. J. Lusis, B. A. Taylor, R. W. Wangenstein, and R. C. LeBoeuf, *J. Biol. Chem.* **258**, 5071 (1983).
[7] J. C. Ertel Miller, R. K. Barth, P. H. Shaw, R. W. Elliot, and N. D. Hastie, *Proc. Natl. Acad. Sci. U.S.A.* **80**, 1511 (1983).

main features of the lipid transport system in each species, and to facilitate a ready comparison between them. For recent reviews of the chemistry, structure, and comparative aspects of lipoproteins in selected mammals, the reader may also refer to texts on rat,[11] rabbit,[12] canine,[13] porcine,[14] bovine,[15] marine mammal,[16] and certain nonhuman primate lipoproteins,[17-19] and to a general review dealing with a wide range of animal species including mammals.[20]

Nomenclature and Classification of Mammalian Lipoproteins and Apolipoproteins: A Critical Appraisal

Lipoproteins

Of the multiple nomenclatures and classifications proposed for the circulating lipoproteins, that which has seen the widest application re-

---

[8] K. Tanzawa, Y. Shimada, M. Kuroda, Y. Tsujita, M. Arai, and Y. Watanabe, *FEBS Lett.* **118**, 81 (1980).

[9] T. Kita, M. S. Brown, Y. Watanabe, and J. L. Goldstein, *Proc. Natl. Acad Sci. U.S.A.* **78**, 2268 (1981).

[10] A. D. Attie, R. C. Pittman, Y. Watanabe, and D. Steinberg, *J. Biol. Chem.* **256**, 9789 (1981).

[10a] Abbreviations: Apo, apolipoprotein or apoprotein; VLDL, very low-density lipoproteins of $d < 1.006$ g/ml, unless otherwise defined; IDL, intermediate-density lipoproteins, of d 1.006–1.019 g/ml, unless otherwise defined; LDL, low-density lipoproteins, density as defined; HDL, high-density lipoproteins, of d 1.063–1.21 g/ml, unless otherwise defined; VHDL, very high-density lipoproteins, density as defined; LCAT, lecithin:cholesterol acyltransferase; SDS, sodium dodecyl sulfate.

[11] K. H. Weisgraber and R. W. Mahley, *in* "Handbook of Electrophoresis" (L. A. Lewis and H. K. Naito, eds.), Vol. IV, p. 103. CRC Press, Boca Raton, Fla., 1983.

[12] D. C. K. Roberts, *in* "Handbook of Electrophoresis" (L. A. Lewis and H. K. Naito, eds.), Vol. IV, p. 139, CRC Press, Boca Raton, Fla., 1983.

[13] R. W. Mahley and K. H. Weisgraber, *in* "Handbook of Electrophoresis" (L. A. Lewis and H. K. Naito, eds.), Vol. IV, p. 151. CRC Press, Boca Raton, Fla., 1983.

[14] R. L. Jackson, *in* "Handbook of Electrophoresis" (L. A. Lewis and H. K. Naito, eds.), Vol. IV, p. 175. CRC Press, Boca Raton, Fla., 1983.

[15] D. L. Puppione, *in:* "Handbook of Electrophoresis" (L. A. Lewis and H. K. Naito, eds.), Vol. IV, p. 185. CRC Press, Boca Raton, Fla., 1983.

[16] D. L. Puppione, *in* "Handbook of Electrophoresis" (L. A. Lewis and H. K. Naito, eds.), Vol. IV, p. 79. CRC Press, Boca Raton, Fla., 1983.

[17] G. M. Fless, C. A. Rolih, and A. M. Scanu, *in* "Handbook of Electrophoresis" (L. A. Lewis and H. K. Naito, eds.), Vol. IV, p. 17. CRC Press, Boca Raton, Fla., 1983.

[18] R. J. Nicolosi and K. C. Hayes, *in* "Handbook of Electrophoresis" (L. A. Lewis and H. K. Naito, eds.), Vol. IV, p. 33. CRC Press, Boca Raton, Fla., 1983.

[19] A. W. Kruski, *in* "Handbook of Electrophoresis" (L. A. Lewis and H. K. Naito, eds.), Vol. IV, p. 61. CRC Press, Boca Raton, Fla., 1983.

[20] M. J. Chapman, *J. Lipid Res.* **21**, 789 (1980).

sulted directly from the analytical ultracentrifugal studies of Gofman and colleagues in the 1950s.[21] These investigators showed that in human sera, certain maximal and minimal concentrations of lipoproteins occur along a range of density from 0.92 to 1.20 g/ml; using the densities which corresponded to minima as cutoff points to separate one class of particles from another, the human lipoproteins were divided into five major classes: chylomicrons, of $d < 0.94$ g/ml, very low-density lipoproteins (VLDL) of $d < 1.006$ g/ml (i.e., d 0.94–1.006 g/ml), low-density lipoproteins (LDL), d 1.006–1.063 g/ml, high-density lipoproteins (HDL), d 1.063–1.21 g/ml, and very high-density lipoproteins (VHDL), $d > 1.21$ g/ml. This classification recognized the association of variable proportions of lipid and protein components and had the advantage of being an operational one, since Havel and collaborators[22] developed an ultracentrifugal procedure for the preparative isolation of the respective classes. With the advent of interest in mammalian lipoproteins, this nomenclature and classification were tacitly applied, the inherent assumption being that animal lipoproteins displayed similar hydrated densities as the human substances and were distributed in a similar fashion. Indeed this designation has been employed for mammalian lipoproteins to the present day, despite several inadequacies, which include the incomplete separation of low- from high-density particles around the 1.063 g/ml boundary.[23]

The propagation of the density classification was further facilitated by the concomitant formulation of the electrophoretic terminology,[24–26] which is again operational in nature and at its inception, primarily recognized α- and β-migrating lipoproteins (β-migrating lipoproteins are typically of low-density and α-lipoproteins of high-density).[26] The parallel application of the density and electrophoretic classification (and methodologies) is clearly evident in early studies of mammalian lipoproteins,[27,28] although several of these involved use of one or more electrophoretic procedures in semiquantitative studies of a wide variety of species.[29–35]

[21] O. F. DeLalla and J. W. Gofman, *Methods Biochem. Anal.* **1**, 459 (1954).
[22] R. J. Havel, H. A. Eder, and J. H. Bragdon, *J. Clin. Invest.* **34**, 1345 (1955).
[23] G. L. Mills and C. E. Taylaur, *Comp. Biochem. Physiol.* **40B**, 489 (1971).
[24] G. Blix, A. Tiselius, and H. Svensson, *J. Biol. Chem.* **137**, 485 (1941).
[25] E. J. Cohn, L. E. Strong, W. L. Hughes, D. J. Mulford, J. N. Ashworth, M. Melin, and H. L. Taylor, *J. Am. Chem. Soc.* **68**, 459 (1946).
[26] H. G. Kunkel and R. J. Slater, *J. Clin. Invest.* **31**, 677 (1955).
[27] L. A. Lewis, A. A. Green, and I. A. Page, *Am. J. Physiol.* **171**, 391 (1952).
[28] L. A. Lewis and I. H. Page, *Circulation* **14**, 55 (1956).
[29] R. Delcourt, *Acta Zool. Pathol. Antwerp* **48**, 197 (1969).
[30] C. Alexander and C. E. Day, *Comp. Biochem. Physiol.* **46B**, 195 (1973).
[31] I. Hort, *Am. J. Vet. Res.* **29**, 813 (1968).
[32] M. Kalab and W. G. Martin, *Anal. Biochem.* **24**, 218 (1968).

The further use of the electrophoretic classification in the context of mammalian lipoprotein profiles is severely limited, however, since a given density class may display anomalous electrophoretic migration, (e.g., under certain conditions such as cholesterol feeding, VLDL may be of β rather than pre-β in mobility),[36] or may display bands characteristic of more than one density class (bovine LDL may be of both β- and α-mobility upon electrophoresis[15]). These two nomenclatures are therefore not strictly superimposable. Certainly, such designations as "α-migrating LDL," used in the context of bovine lipoproteins,[15] may confuse the unfamiliar reader. The electrophoretic classification has also proved ill adapted to the polydisperse lipoprotein spectrum of mammals and to our ever-increasing knowledge of the structural and metabolic heterogeneity of the particles. Only in rare instances has its use proved indispensable, as in the case of β-VLDL.[36] By contrast, the density classification continues to serve as the cornerstone of mammalian lipoprotein terminology, probably reflecting its inherent flexibility and resolutive potential. For example, HDL was originally divided into two subclasses, HDL_2 (d 1.063–1.125 g/ml) and HDL_3 (d 1.125–1.21 g/ml).[22] A third and less dense HDL subclass, denoted HDL_1, was identified in the analytical ultracentrifuge in analyses of human HDL separated with d 1.063–1.21 g/ml, and defined as those lipoproteins with a flotation rate at d 1.063 g/ml of S_f 0–3 and at d 1.20 g/ml of $F_{1.20}$ 9–20.[21] Such particles, related structurally and electrophoretically (α_2-mobility) to HDL but isolated within the low-density range by ultracentrifugation, were subsequently detected in several species of marine mammal[37,38] and in bovine serum.[37] It has since become apparent that a wide variety of mammals, including such common laboratory species as the rat,[11,39,40] mouse,[5,41] dog,[13,42] as well as certain monkeys and notably the Common marmoset,[43] display HDL_1 particles in the low-density range

---

[33] A. Chalvardjian, *J. Lipid Res.* **12**, 265 (1971).
[34] M. B. N. Johansson and B. W. Karlsson, *Comp. Biochem. Physiol.* **54B**, 495 (1976).
[35] W. G. Dangerfield, R. Finlayson, G. Myatt, and M. G. Mead, *Atherosclerosis* **25**, 95 (1976).
[36] R. W. Mahley, in "Disturbances in Lipid and Lipoprotein Metabolism" (J. M. Dietschy, A. M. Gotto, and J. A. Ontko, eds.), p. 181. Am. Physiol. Soc., Bethesda, Md., 1978.
[37] D. L. Puppione, G. M. Forte, A. V. Nichols, and E. H. Strisower, *Biochim. Biophys. Acta* **202**, 392 (1970).
[38] D. L. Puppione, T. Forte, and A. V. Nichols, *Comp. Biochem. Physiol.* **39B**, 673 (1971).
[39] K. H. Weisgraber, R. W. Mahley, and G. Assmann, *Atherosclerosis* **28**, 121 (1977).
[40] L. T. Lusk, L. F. Walker, L. H. DuBien, and G. S. Getz, *Biochem. J.* **183**, 83 (1979).
[41] M. C. Camus, M. J. Chapman, P. Forgez, and P. M. Laplaud, *J. Lipid Res.* **24**, 1210 (1983).
[42] R. W. Mahley and K. H. Weisgraber, *Circ. Res.* **35**, 713 (1974).
[43] M. J. Chapman, F. McTaggart, and S. Goldstein, *Biochemistry* **18**, 5096 (1979).

under conditions of normal (low-fat) diet. The fact that HDL_1 contain the major protein component of HDL, i.e., apolipoprotein A-I (see Potential Anomalies in Mammalian Lipoprotein Distributions), and that they appear to be metabolically related to HDL,[44] is entirely consistent with their designation: the predominant apoprotein is however apoE.[11,39,40] It remains indeterminate as to whether apoB and apoA-I may reside on the same particle with LDL flotation characteristics, although the badger appears to possess such lipoproteins (P. M. Laplaud, personal communication).

A further and frequent anomaly in mammalian lipoprotein profiles concerns the occurrence of apoB-containing, LDL-like particles within the density range of HDL. Classical examples are the guinea pig,[45] pig,[46-49] and chimpanzee,[50] in which such particles may extend to a higher limiting density of 1.09–1.100 g/ml. Regrettably, no specific nomenclature has been propounded to designate these lipoproteins, for which the terms "LDL" or "dense LDL" are clearly ambiguous.

An additional, distinct molecular form of LDL, first seen in man,[51] and later in other mammals,[52] occurring at densities of ~1.050–1.090 g/ml, and related to the above in possessing apoB as its major protein, is Lp (a); its protein moiety is characterized by the presence of the apo(a) polypeptide in addition to apoB.[53] A mammalian counterpart, termed "LDL-II," and of d 1.050–1.085 g/ml, was characterized in the *Erythrocebus patas* monkey.[54] Immunological studies have revealed a wide distribution of Lp(a) among nonhuman primates (especially Old World species: baboon,[52,55] rhesus,[52,17,56] pigtail monkey,[55] orangutan,[52] and chimpanzee[52]). In this instance then, transposition of the nomenclature from man is entirely satisfactory.

[44] V. Gordon, T. L. Innerarity, and R. W. Mahley, *J. Biol. Chem.* **258**, 6202 (1983).
[45] G. L. Mills, M. J. Chapman, and F. McTaggart, *Biochim. Biophys. Acta* **260**, 401 (1972).
[46] M. Janado, W. G. Martin, and W. H. Cook, *Can. J. Biochem.* **44**, 1201 (1966).
[47] N. Fidge and G. D. Smith, *Artery* **1**, 406 (1975).
[48] R. L. Jackson, O. D. Taunton, R. Segura, J. G. Gallagher, H. F. Hoff, and A. M. Gott, *Comp. Biochem. Physiol.* **53B**, 245 (1976).
[49] R. W. Mahley and K. H. Weisgraber, *Biochemistry* **13**, 1964 (1974).
[50] M. J. Chapman, P. Forgez, D. Lagrange, S. Goldstein, and G. L. Mills, *Atherosclerosis* **52**, 129 (1984).
[51] K. Berg, *Acta Pathol. Microbiol. Scand.* **59**, (Suppl. 166–167), 369 (1963).
[52] K. Berg, *Ser. Haematol.* **1**, 111 (1968).
[53] J. W. Gaubatz, C. Heideman, A. M. Gotto, J. D. Morrisett, and G. H. Dahlen, *J. Biol. Chem.* **258**, 4582 (1983).
[54] R. W. Mahley, K. H. Weisgraber, T. Innerarity, and H. B. Brewer, *Biochemistry* **15**, 1928 (1976).
[55] J. J. Albers, J. L. Adolphson, and W. R. Hazzard, *J. Lipid Res.* **18**, 331 (1977).
[56] L. L. Rudel, D. G. Greene, and R. Shah, *J. Lipid Res.* **18**, 734 (1977).

Development of the classification and nomenclature of mammalian lipoproteins is intimately linked to their protein constituents, the apolipoproteins. Indeed, the most ambitious attempt to update classification of the human lipoproteins, the "family hypothesis" proposed by Alaupovic,[57] directly reflects progress in our knowledge of the molecular organization of lipoproteins and their constitutive apolipoproteins. Advances in mammalian lipoprotein chemistry have occurred more slowly however, thereby limiting the basis upon which a comparable system could be based. Nonetheless, this nomenclature has been extended to the baboon in which a lipoprotein containing a counterpart to human apoD (i.e., lipoprotein D) as its sole protein component, has been identified.[58]

Extensive homology between human and mammalian apolipoproteins is now established[20] (see below), a feature which has prompted fractionation of mammalian lipoproteins by procedures dependent on the physicochemical properties and binding specificities of their protein components. For example, heparin affinity chromatography has been successfully applied to rat lipoproteins, which are separated according to their apoB and apoE contents.[59] Such application paralleled that in man, in which HDL could be separated by a similar procedure[60] into HDL lacking apoE (the major fraction), as well as two minor fractions, HDL with apoE and "HDL" containing apoB. As is often the case, this technique has led to the appearance of a new nomenclature, which represents an amalgam of the density and apolipoprotein terminologies, and of which the terms "apoE HDL" (i.e., HDL_1[13]) and "apoE HDL_c" (a lipoprotein characteristic of cholesterol fed animals[36]) are typical.

Clearly then, no single nomenclature and classification presently lends itself to the comprehensive description of the various anomalies which occur in mammalian lipoprotein profiles as compared to that of man.

Apolipoproteins

The present "alphabetic" nomenclature for the human plasma apolipoproteins was originally proposed by Alaupovic[57] and colleagues, recognizes the A, B, C . . . groups of polypeptides, and has gained wide acceptance. The various constitutive proteins in each group are structurally and metabolically related, and denoted by Roman numerals, e.g.,

[57] P. Alaupovic, *in* "Handbook of Electrophoresis" (L. A. Lewis and J. J. Opplt, eds.), Vol. I, p. 27. CRC Press, Boca Raton, Fla, 1980.
[58] D. Bojanovski, P. Alaupovic, W. J. McConathy, and J. L. Kelley, *FEBS Lett.* **112,** 251 (1980).
[59] E. R. Skinner and J. A. Rooke, *Comp. Biochem. Physiol.* **65B,** 645 (1980).
[60] K. H. Weisgraber and R. W. Mahley, *J. Lipid Res.* **21,** 316 (1980).

A-I, A-II, A-IV, C-I, C-II, C-III, etc. while polymorphic or isomorphic forms of individual apolipoproteins are denoted by either subscript or capital arabic numbers, e.g., A-I$_1$, A-I$_2$, C-III-0, C-III-1, C-III-2, C-III-3, or E-1, E-2, etc. The marked structural homology existing between the apolipoproteins of man and a wide range of mammals[20] has allowed this nomenclature to be transposed successfully. It is also consistent with the existence of common ancestral apolipoprotein genes.[61]

Perhaps the only mammalian apolipoproteins for which the nomenclature is provisional are those of the B group, a situation reflecting our meager knowledge of their primary structure.[62] In order to avoid difficulties associated with the determination of the absolute molecular weight of apoB, Kane and colleagues[63] proposed a centile nomenclature for the human proteins. In this system, the major species are the hepatic form, denoted B-100 ($M_{r\,app}$ 5.4 × 10^5), and the intestinal product of smaller size, B-48 ($M_{r\,app}$ 2.6 × 10^5). The rat possesses counterparts to the human B proteins, and these have also been designated B-100 and B-48.[62] Despite the fact that rat B-100 and B-48 comigrate on SDS gels with their human counterparts,[62] the absolute M_r values of the rat B polypeptides found by some authors tend to differ from those of man,[63] giving rise to such designations as "rat apoB$_{335K}$" and "rat apoB$_{240K}$,"[64] "PI" and "PIII,"[65] and "apoB$_h$" and "apoB$_1$."[66] Analogous proteins have been identified in the mouse,[41,67] rabbit,[68] and pig.[69] Since the structural interrelationships of these proteins in each species remain unresolved, the final choice of a nomenclature must await progress in this direction.

Ultimately, a lipoprotein nomenclature and classification, based on apolipoprotein content, site of formation, and hydrated density would be desirable, a possibility brought closer with the resolution of B-48- and B-100-containing VLDL by immunoadsorbant affinity chromatography using monoclonal antibodies.[70]

[61] W. C. Barker and M. O. Dayhoff, *Comp. Biochem. Physiol.* **57B**, 309 (1977).
[62] J. P. Kane, *Annu. Rev. Physiol.* **45**, 637 (1983).
[63] J. P. Kane, D. A. Hardman, and H. E. Paulus, *Proc. Natl. Acad. Sci. U.S.A.* **77**, 2465 (1980).
[64] A. L. Wu and H. G. Windmueller, *J. Biol. Chem.* **256**, 3615 (1981).
[65] J. Elovson, Y. O. Huang, N. Baker, and R. Kannan, *Proc. Natl. Acad. Sci. U.S.A.* **78**, 157 (1981).
[66] C. E. Sparks and J. B. Marsh, *J. Lipid Res.* **22**, 514 (1981).
[67] P. Forgez, M. J. Chapman, S. C. Rall, and M. C. Camus, *J. Lipid Res.* **25**, 954 (1984).
[68] L. H. Huang, J. S. Jaeger, and D. C. Usher, *J. Lipid Res.* **24**, 1485 (1983).
[69] M. J. Chapman and P. Forgez, *Nutr. Repr. Dev.* **25**, 217 (1985).
[70] R. W. Milne, P. K. Weech, L. Blanchette, J. Davignon, P. Alaupovic, and Y. L. Marcel, *J. Clin. Invest.* **73**, 816 (1984).

Potential Anomalies in Mammalian Lipoprotein Distribution and Their Consequences to the Choice of Methodology for Fractionation

As will be evident from the above section, the principal anomalies met in mammalian lipoprotein profiles concern the presence of HDL particles containing apoA-I and apoE in the density range attributed to LDL (i.e., 1.006–1.063 g/ml) on the one hand, and the occurrence of apoB-containing, LDL-like particles in the density range of HDL on the other. The "dense LDL" have hydrated densities greater than those of human LDL (~1.035 g/ml),[71] overlap the 1.063 g/ml boundary, and are incompletely separated at this solvent density; contamination of HDL results. Two species of Felidae, the lion *(Panthera leo)* and jaguar *(Panthera onca)*, provide an extreme example of this type of anomaly, since in each case the 1.063 g/ml boundary intercepts a continuous lipoprotein profile,[23] and the precise nature of the particles on either side remains to be established.

The degree to which the respective overlaps in density distribution occur may be subject to variation as a function of the strain, age, sex, and nutritional and physiological status of the animal. Indeed, a strain variability in HDL_1 distribution has been indicated in rats of the Sprague–Dawley,[72] Osbourne–Mendel,[73] and Wistar[23] strains.

The complexity of the molecular species which may be present at or close to the 1.063 g/ml boundary is further increased by the occurrence of Lp(a) in nonhuman primates.[17,52,54–56]

Perhaps the most successful methodological approach to the separation of these lipoproteins has been that adopted by Mahley, Weisgraber, and colleagues.[39,42,54] They first isolated lipoproteins of either low-density (d 1.006–1.063[42] or 1.02–1.063[39] g/ml) or of low- and high-density together (d 1.006–1.100 g/ml)[54] by ultracentrifugal flotation, and subsequently resolved the constituent particle species (i.e., LDL, HDL_1,[39,42] and LDL-II/Lp(a)[54] when present) by electrophoresis on a Pevikon block. HDL were isolated by flotation from the ultracentrifugal infranatants at densities of 1.10–1.21,[54] 1.087–1.21,[42] and 1.063–1.21 g/ml.[39]

A similar strategy was successful in isolating LDL over its entire density distribution in the pig.[49] In this instance, a fraction of d 1.006–1.21 g/ml was separated centrifugally and electrophoresed on the starch block, resulting in preparation of purified porcine LDL and HDL of β- and α-mobility, respectively.

[71] D. M. Lee, *in* "Low-Density Lipoproteins" (C. E. Day and R. S. Levy, eds.), p. 3. Plenum, New York, 1976.
[72] S. Koga, D. L. Horwitz, and A. M. Scanu, *J. Lipid Res.* **10**, 577 (1969).
[73] H. G. Windmueller and R. I. Levy, *J. Biol. Chem.* **242**, 2246 (1967).

Other methods have permitted separation of HDL_1, at least in the rat, and these include isopycnic density gradient centrifugation,[40] rate zonal ultracentrifugation,[74] and affinity chromatography on heparin–Sepharose.[59]

A further anomaly which may be encountered involves the isolation of lipoproteins whose triglycerides contain elevated proportions of saturated fatty acids.[75] This phenomenon presently appears restricted to ruminating cattle, in which unusual flattened particles of intermediate density (d 1.006–1.020 g/ml) appear when ultracentrifugal flotation isolation is performed at temperatures below 25°,[15] and is associated with a quasicrystalline state of their core lipids.[15,75] Upon separation at temperatures above that of the core lipid transition (i.e., ~37°), such abnormal lipoproteins are absent from the intermediate density class.[75]

It is noteworthy that when studies of lipoprotein and apolipoprotein profiles are initiated in animals in which little prior information is available, then techniques of isopycnic density gradient ultracentrifugation may profitably be applied. Indeed their potential in resolving complex particle distributions in whole plasma or serum has been amply documented in the mouse,[41] hedgehog,[76] badger,[77] baboon,[19] and Common marmoset.[43]

Identification of Species Groups Based on Lipoprotein Profiles

The "HDL Mammals"

In the vast majority of mammals, high-density lipoproteins are the predominant class, and may account for up to 80% of the total substances of $d < 1.21$ g/ml, as in the random bred, Swiss mouse.[41] Circulating lipoprotein levels in this group are summarized in Table I. Three points are relevant: (1) determination of levels in plasma or serum has not been distinguished as these have been considered equivalent, (2) data are quoted for sexually mature, adult animals, with accompanying data for fetal, newborn, or juvenile animals when available, and (3) since HDL in healthy, normolipidemic man typically represent ~30–32% of the total lipoproteins and LDL preponderate (~50%),[23,78] then any animal in which

---

[74] Y. Oschry and S. Eisenberg, *J. Lipid Res.* **23**, 1099 (1982).
[75] D. L. Puppione, S. T. Kunitake, R. L. Hamilton, M. L. Phillips, V. N. Schumaker, and L. D. Davis, *J. Lipid Res.* **23**, 283 (1982).
[76] P. M. Laplaud, L. Beaubatie, and M. Saboureau, *Biochim. Biophys. Acta* **752**, 396 (1983).
[77] P. M. Laplaud, L. Beaubatie, and D. Maurel, *Biochim. Biophys. Acta* **711**, 213 (1982).
[78] A. V. Nichols, *Adv. Biol. med. Phys.* **11**, 110 (1967).

TABLE I
CIRCULATING LIPOPROTEIN LEVELS IN "HDL MAMMALS"[a]

Mammalian subgroup	Species	Strain/age	Lipoprotein concentration (mg/dl)						
			VLDL	IDL	LDL	HDL$_1$	HDL (total)	HDL$_2$	HDL$_3$
Marsupials	Opossum (*Didelphis virginiana*)		8[b]	11	56–75[a]		660		
Insectivores	Hedgehog (*Erinaceus Europaeus* L.)		~20[c]	9	87	Present	430		
Rodents	Mouse (*Mus musculus*)		13[b]	2	~140		~570		
					27		n.m.		
		Swiss OF 1	41[c]	0	189[f]		206		
		random bred	84[e]	45	69		534	379	155
		Random-bred Swiss	75[f]		65		572	308	264
					42		251		
Rat (*Rattus norvegicus*)		Wistar	66[b]	0	16		246		
		Wistar (young: 1.5 months)	114[x]	24	27		70		
		Wistar (old: 24 months)	93[x]	28	56		126		
		Wistar	36[b]		18		112		
		Sprague–Dawley	107[i]		56		240		
		Sprague–Dawley	46[j]		16		80	220	
		Sprague–Dawley (suckling, 14–15 d)	50[k]		200	125			
		Sprague–Dawley	34[m]	8	51	8–17	~100		
		Holtzmann	70[l]		50		216		
		Zucker (lean)	87[n]	2	62		241		
			27[u]		16		128		
			41[v]		34		125		
	Mongolian gerbil (*Meriones unguiculatus*)		0[b]	0	44		n.m.		
	Porcupine (*Arthenurus macrourus*)								
	Ground squirrel (*Citellus mexicanus*)		0[p]		tr.		60–98		

(*continued*)

TABLE I (continued)

Mammalian subgroup	Species	Strain/age	Lipoprotein concentration (mg/dl)						
			VLDL	IDL	LDL	HDL_1	HDL (total)	HDL_2	HDL_3
Carnivores	Dog (*Canis familiaris*)		15^a		26	13	343		
			73^v		135		479		
	Polar bear		50^v		438		966		
	Dingo (*Canis dingo*)		5^b		99		n.m.		
	Cat (*Felis domesticus*)		n.d.^{a}		70(34–108)		267(94–346)		
	Ferret (*Meles putorius furo* L.)		37^v		112		548		
	Ref fox (*Vulpes Vulpes* L.)		30^v		178		630		
Cetaceans	Killer Whale (*Orcinus orca*)		$60–83^v$	110–104	217–260	68–82	366–590	197	169
								250	340
								270	252
	Bottlenose dolphin (*Tursiops truncatus*)		37^v	33	84	71	522		
	Pacific white-sided dolphin (*Lagenorhynchus obliquidens*)		$3–34^v$	2–39	84–114	2	363	139–202	183–200
Pinnipeds (*Marine carnivores*)	California sea lion (*Zalophus californianus*)		21^v	4	98	84	794	366	428
	Walrus (*Odobenus rosmarus*)	(Pup)	26^v	100	361	172	709	391	318
	Elephant seal (*Mirounga angustirostris*)		$0–47^v$	1	22–186	7	888	450	438
	Harbor seal (*Phoca vitulina*)		$3–17^v$	0–15	133–185	0–68	1203	386	817
Perissodactyls	Domestic horse (*Equus caballus*)		n.m.^{v}		99		423		
	Przewalski horse (*Equus przewalski*)								
	Onager (*Asinus hemionus*)		n.m.^{v}		102		362		
	Mountain zebra (*Equus zebra*)		n.m.^{v}		95		504		
	Common zebra (*Equus burchelli*)		n.m.^{v}		54		606		
			n.m.^{v}		86		819		
Artiodactyls	Malayan tapir (*Tapirus indicus*)		n.m.^{v}	0	269		382		
	Sheep (*ovis aries*)	Columbia–Suffolk	0^b		28		125		
		Clun Forest (lamb 90 d)	77^v		52^v		84		
		Clun Forest (ewe)	15^v		20		466		
					67		262		

Species							
Deer (*Odocoileus virginianus*)		n.d.[A]		66[c]	207		77
Cattle (*Bos sp.*)		10[A]		44	35		
	(Fetus)	21[A]		31	167		
	(Newborn calf)	n.m.[B]		73	365		
	(Steer)	21[b]	0	118	829		
	(Lactating cow)			55	341		
	(Cow)		5	175	640		
Nonhuman primates—New World monkeys							
Squirrel (*Saimiri sciureus*)		5[C]		106	373		
Cebus (*Cebus albifrons*)		3[D]		104	408	296	
Cebus (*Cebus apella*)		31[E]		134	904		
Common marmoset (*Callithrix jacchus*)		2[F]		151	607	200[F]	407
		12[F]		170–280	338–408	81–98	257–310
		50–90[G]					
Old World monkeys							
Rhesus (*Macaca mulatta*)		46[H]		103	395	166	228[F]
Cynomolgus (*Macaca fascicularis*)		42[F]		116	515	199	316
		20[I]	51	257	449	173[F]	276
Stumptail macaque (*Macaca arctoides*)		8[J]	58	116	429	121[F]	308
Pigtail (*Macaca nemestrina*)		7[K]	4	106	345		
African green (*Cercopithecus aethiops*)		22[L]		196	447		
Patas (*Erythrocebus patas*)		22[L]		99	260		
		52[F]		90	271		
		10[M]		150	184		
Mangabey (*Cercocebus albigena*)		34[F]		118	343		
Baboon (*Papio cynocephalus*)		0[N]	3	118	345	211	134
(*Papio anubis*)		11[O]		128	268		
(*Papio papio*)		3[F]		72	475		
Apes							
Chimpanzee (*Pan troglodytes*)		117[F]		234	330		
		40[N]	14	324	424	268	156
		104[P]		215	n.m.	n.m.	341
		119[L]		290	706	158	124
		1[Q]	7	261	282		

(*continued*)

Footnotes to TABLE I

[a] Unless otherwise noted, data are given for adult animals; n.d., not detectable; n.m., not measured.
[a'] I. H. Lewis, A. A. Green, and I. H. Page, *Am. J. Physiol.* **171**, 391 (1952); LDL includes VLDL and IDL.
[b] G. L. Mills and C. E. Taylaur, *Comp. Biochem. Physiol.* **40B**, 489 (1971); IDL determined as lipoproteins of S_f 12–20, LDL of S_f 0–12.
[c] P. M. Laplaud, L. Beaubatie, and M. Saboureau, *Biochim. Biophys. Acta* **752**, 396 (1983); IDL detected in one of five animals; significant amounts of VHDL present.
[d] K. K. Kirkeby, *Scand. J. Clin. Invest.* **18**, 437 (1966); LDL includes VLDL and IDL.
[e] M. C. Camus, M. J. Chapman, P. Forgez, and P. M. Laplaud, *J. Lipid Res.* **24**, 1210 (1983); HDL_2 of $F_{1,20}$ 3.5–9.0, HDL_3 of $F_{1,20}$ 0–3.5.
[e'] VLDL of $d < 1.017$ g/ml, including some IDL; LDL, d 1.023–1.060 g/ml; HDL_2, d 1.060–1.111 g/ml; HDL_3, d 1.111–1.188 g/ml.
[f] R. C. LeBoeuf, D. L. Puppione, V. N. Schumaker, and A. J. Lusis, *J. Biol. Chem.* **258**, 5071 (1983); LDL includes IDL of d 1 006–1.019 g/ml; calculated from chemical composition data.
[g] S. Malhotra and D. Kritchewsky, *Mech. Ageing Dev.* **8**, 445 (1978); determined from lipid concentration data, assuming protein contents of 10, 20, and 40% in VLDL, IDL and LDL, and HDL, respectively.
[h] S. Calandra, I. Pasquali-Ronchetti, E. Gherardi, C. Fornieri, and P. Tarugi, *Atherosclerosis* **28**, 369 (1977).
[i] G. G. de Pury and F. D. Collins, *Lipids* **7**, 225 (1972).
[j] N. L. Lasser, P. S. Roheim, D. Edelstein, and H. A. Eder, *J. Lipid Res.* **14**, 1 (1973).
[k] J. P. Fernando-Warnakulasuriya, M. L. Eckerson, W. A. Clark, and M. A. Wells, *J. Lipid Res.* **24**, 1626 (1983); chylomicrons present in VLDL at a concentration of 85 mg/dl plasma. VLDL of $d < 1.022$ g/ml, including some IDL.
[l] K. A. Narayan, J. J. McMullen, D. P. Butler, T. Wakefield, and W. K. Calhoun, *Atherosclerosis* **23**, 1 (1976); LDL includes HDL_1, animal fed a semisynthetic diet.
[m] K. H. Weisgraber, R. W. Mahley, and G. Assmann, *Atherosclerosis* **28**, 121 (1977); determined from lipid concentration data as in [g].
[n] G. Schonfeld, C. Felski, and M. A. Howald, *J. Lipid Res.* **15**, 457 (1974); LDL of d 1.006–1.063 g/ml, including IDL; concentrations determined separately in four animals.
[n'] VLDL, S_f 20–400; chylomicrons, 15 mg/dl; IDL, d 1.006–1.018 g/ml; LDL, d 1.018–1.050 g/ml; HDL, d 1.050–1.21 g/ml; pooled plasmas.
[o] R. J. Nicolosi, J. A. Marlett, A. M. Morello, S. A. Flanagan, and D. M. Hegsted, *Atherosclerosis* **38**, 359 (1981); animals fed a purified diet containing 15% safflower oil.
[p] H. K. Naito and R. G. Gerrity, *Exp. Mol. Pathol.* **31**, 713 (1974).
[q] R. W. Mahley and K. H. Weisgraber, *Circ. Res.* **35**, 713 (1974).
[r] A. W. Lindall, F. Grande, and A. Schultz, *J. Nutr.* **102**, 515 (1972).
[s] T. L. Kaduce, A. A. Spector, and G. E. Folk, *Comp. Biochem. Physiol.* **69B**, 541 (1981); determined from lipid concentration data as in [g].
[t] A. Cryer and A. M. Sawyerr, *Comp. Biochem. Physiol.* **61B**, 151 (1978).
[u] P. M. Laplaud, L. Beaubatie, and D. Maurel, *Comp. Biochem. Physiol.* **68B**, 125 (1981).

[v] D. L. Puppione, in "Handbook of Electrophoresis" (L. A. Lewis and H. K. Naito, eds.), Vol. IV, p. 79. CRC Press, Boca Raton, Fla., 1983. Levels determined by analytical ultracentrifugation.
[w] W. M. F. Leat, C. A. Northrop, N. Buttress, and D. M. Jones, *Comp. Biochem. Physiol.* **63B**, 275 (1979). Determined by analytical ultracentrifugation.
[x] T. M. Forte, C. E. Cross, R. A. Gunther, and G. C. Kramer, *J. Lipid Res.* **24**, 1358 (1983); LDL includes VLDL and IDL.
[y] W. M. F. Leat, F. O. T. Kubasek, and N. Buttress, *Q. J. Exp. Physiol.* **61**, 193 (1976). Chylomicron: present (14 and 4 mg/dl plasma, respectively, in lambs and ewes); LDL, d 1.006–1.063 g/ml.
[z] M. Fried, H. G. Wilcox, G. R. Faloona, S. P. Eoff, M. S. Hoffman, and D. Zimmerman, *Comp. Biochem. Physiol.* **25**, 651 (1968).
[A] T. M. Forte, J. J. Bell-Quint, and F. Cheng, *Lipids* **16**, 240 (1980). Determined by analytical ultracentrifugation.
[B] D. L. Puppione, in "Handbook of Electrophoresis" (L. A. Lewis and H. K. Naito, eds.), Vol. IV, p. 185. CRC Press, Boca Raton, Fla., 1983.
[C] S. R. Srinivasan, B. Radhakrishnamurthy, L. S. Webber, E. R. Dalferes, M. G. Kokatur, and G. S. Berenson, *Am. J. Clin. Nutr.* **31**, 603 (1978).
[D] D. R. Illingworth, *Biochim. Biophys. Acta* **388**, 38 (1975).
[E] R. J. Nicolosi and K. C. Hayes, in "Handbook of Electrophoresis" (L. A. Lewis and H. K. Naito, eds.), Vol. IV. p. 33. CRC Press, Boca Raton, Fla., 1983.
[F] S. R. Srinivasan, J. R. McBride, B. Radhakrishnamurthy, and G. S. Berenson, *Comp. Biochem. Physiol.* **47B**, 711 (1974). VLDL, LDL, and HDL determined as pre-β-, β- and α-migrating lipoproteins, respectively; HDL level calculated assuming a cholesterol content of 15% by weight.
[F'] Proportions of HDL subclasses calculated from gel scan data. A. M. Morello and R. J. Nicolosi, *Comp. Biochem. Physiol.* **69B**, 291 (1981).
[G] M. J. Chapman, F. McTaggart, and S. Goldstein, *Biochemistry* **18**, 5096 (1979).
[H] S. R. Srinivasan, B. A. Clevidence, P. S. Pargaonkar, B. Radhakrishnamurthy, and G. S. Berenson, *Atherosclerosis* **33**, 301 (1979). VLDL, LDL, and HDL estimated as in F.
[I] A. R. Tall, D. M. Small, D. Atkinson, and L. L. Rudel, *J. Clin. Invest.* **62**, 1354 (1978); concentrations calculated from protein data, assuming contents of 10% in IDL and LDL and 50% in HDL.
[J] T. L. Raymond, A. J. DeLucia, and L. R. Bryant, *Nutr. Rep. Int.* **25**, 75 (1982); concentrations calculated as in I.
[K] R. S. Kuskwaha and W. R. Hazzard, *Biochim. Biophys. Acta* **619**, 142 (1980); calculated from cholesterol concentration data, assuming contents of 20% in VLDL, IDL, and HDL, and 45% in HDL by weight.
[L] S. R. Srinivasan, C. C. Smith, B. Radhakrishnamurthy, R. H. Wolf, and G. S. Berenson, *Adv. Exp. Biol. Med.* **67**, 65 (1972); levels determined as in F.
[M] R. W. Mahley, K. H. Weisgraber, T. Innerarity, and H. B. Brewer, *Biochemistry* **15**, 1928 (1976); levels calculated as in I.
[N] V. Blaton and H. Peeters, *Adv. Exp. Biol. Med.* **67**, 33 (1976).
[O] D. Bojanovski, P. Alaupovic, J. L. Kelley, and C. Stout, *Atherosclerosis* **31**, 481 (1978).
[P] M. Rosseneu, B. Leclercq, D. Vandamme, R. Vercaemst, F. Soeteway, H. Peeters, and V. Blaton, *Atherosclerosis* **32**, 141 (1979).
[Q] M. J. Chapman, P. Forgez, D. Lagrange, S. Goldstein, and G. L. Mills, *Atherosclerosis* **52**, 129 (1984); determined by analytical ultracentrifugation.

HDL accounts for 50% or more of the total $d < 1.21$ g/ml substances has been considered as an "HDL mammal," with a profile distinct from that of urban man.[78]

Upon classification of mammals on the basis of diet and digestive physiology, three major groups appear: herbivores (e.g., rabbit, cow, goat, and sheep), carnivores (e.g., dog, seal, lion), and omnivores (e.g., pig, baboon). Herbivorous species usually possess HDL as their major lipoprotein class (Table I), with some exceptions, such as guinea pig and camel.[23] The same is true of carnivores, whose LDL levels are often low and less than 100 mg/dl serum. Omnivorous mammals present varied lipoprotein profiles, although they tend to display higher LDL concentrations than the two former groups.

Of the three major mammalian subclasses, the Monotremes, Marsupials, and Placental mammals, no data are available on the former, and the opossum appears to be the only marsupial studied to date. The vast majority then of species in Table I are placental mammals. Representatives of almost all of the principal orders of mammals are found in the "HDL" group, with the exception of the Prosimii, Proboscidea (elephants), and Edentata (armadillos and sloths), of whose lipid transport systems we remain largely ignorant.

Table I shows that substantial variation in lipoprotein profile may occur between individuals of the same species, but which differ in strain (e.g., rat and mouse) or stage of development and age (rat, sheep, and cattle). For example, the fetal calf displays LDL as its major lipoprotein but after birth rapidly becomes an "HDL mammal." Variability introduced by the application of distinct methodologies to lipoprotein quantitation in the same species by different investigators should not be underestimated, and may indeed provide at least a partial explanation of certain discrepancies.

Species, such as the chimpanzee, may be considered as borderline "HDL mammals," since in the analyses of some authors, amounts of low-density lipoproteins either predominate over those of high-density, or are approximately the same; in others, HDL clearly dominate. Additional data on anthropoid apes are presently unavailable. A further qualification of data in Table I concerns HDL-related molecules (i.e., HDL_1) which may contribute significantly to "LDL" concentrations, as in the hedgehog[76] and rat,[11] while conversely apoB-containing particles may contribute to HDL (d 1.063–1.21 g/ml), as in hedgehog,[76] pig,[23,49] and chimpanzee.[50] An extreme example of the latter is the guinea pig (Table II).

The availability of quantitative data on lipoprotein levels in only four species of rodent is notable. Observations on lipoprotein patterns obtained by electrophoretic analyses in several other rodents have, however, been described.[30,34,35,79-81]

The "LDL Mammals"

Species characterized by a predominance of lipoproteins of low-density (considered in this instance to be either of $d < 1.063$ or of d 1.006–1.063 g/ml), and which account for more than 50% of the total substances of $d > 1.21$ g/ml, comprise the "LDL mammals." An excellent illustration is the guinea pig, in which HDL are a trace component (~2–6%) of the lipid transport system (Table II). The contribution of "dense LDL" (which may be isolated at densities up to 1.100 g/ml)[82] to guinea pig "HDL" in analyses made in the density interval 1.063–1.21 g/ml[23] is especially noteworthy. A second such example are the Rhinocerotidae, in which HDL are undetectable and LDL appear to be the exclusive lipoprotein class by both electrophoretic and analytical ultracentrifugal criteria.[83]

Like the Rhinoceri, the camel[23] also possesses a single class of lipoproteins, whose density is almost entirely less than 1.063 g/ml. Whether such species possess trace amounts of HDL with a high rate of metabolic turnover, as in the guinea pig,[84] remains indeterminate.

The lipoprotein profile in the rabbit is not only highly dependent on dietary fat and cholesterol,[12,20] but also on strain (Table II). Indeed, whereas the sum of VLDL, IDL, and LDL accounts for 54–65% in Dutch Belt and New Zealand white rabbits, these lipoproteins together represented only 22–45% of the total in Red Burgundy (Fauve de Bourgogne) animals; rabbits of the latter strain could therefore be considered as members of the "HDL group." In the LDL (apoB,E) receptor-defective strain (WHHL, Watanabe heritable hyperlipidemic rabbit)[9] on the other hand, HDL are minor plasma components (~3.2%).[85]

The pig is one of few members of the order Arteriodactyla to present a profile in which apoB-containing, low-density substances are quantita-

[79] H. K. Naito, *in* "Handbook of Electrophoresis" (L. A. Lewis and H. K. Naito, eds.), Vol. IV, p. 255. CRC Press, Boca Raton, Fla., 1983.
[80] H. K. Naito, *in* "Handbook of Electrophoresis" (L. A. Lewis and H. K. Naito, eds.), Vol. IV, p. 263. CRC Press, Boca Raton, Fla., 1983.
[81] W. Galster and P. Morrison, *Comp. Biochem. Physiol.* **58B**, 39 (1977).
[82] M. J. Chapman and G. L. Mills, *Biochem. J.* **167**, 9 (1977).
[83] W. M. F. Leat, C. A. Northrop, N. Buttress, and D. M. Jones, *Comp. Biochem. Physiol.* **63B**, 275 (1979).
[84] P. Barter, O. Faergeman, and R. J. Havel, *Metabolism* **26**, 615 (1977).
[85] R. J. Havel, T. Kita, L. Kotite, J. P. Kane, R. L. Hamilton, J. L. Goldstein, and M. S. Brown, *Arteriosclerosis* **2**, 467 (1982).

TABLE II
CIRCULATING LIPOPROTEIN LEVELS IN "LDL MAMMALS"[a]

Mammalian subgroup	Species	Strain/age	Lipoprotein concentration (mg/dl)							
			VLDL	IDL	LDL	Dense LDL	HDL$_1$	HDL (total)	HDL$_2$	HDL$_3$
Rodents	Guinea pig (*Cavia porcellus*)		0[a']	0	30			64		
			35[b]		127			10		
			28[c]	tr.	83	12.5		2.3		
			4[d]	1	90	90		<5		
		Fetal (60 d)	30[e]		39			3.5		
Lagomorphs	Rabbit (*Oryctolagus cuniculus domesticus*)									
	Dutch belt		25[f]	85	105			140		
	New Zealand White		25[f]	100	115			130		
			19[g]	100	118			160		
			82[g]	38	145			227		
	Fauve de Bourgogne		5[h]		25			53		
			3[h]		15			63		
			52[i]		157			257		
	New Zealand White	Control	206[j]	46	59			248		
		WHHL	366[j]	360	719			48		
Perissodactyls	White rhinoceros (*Ceratotherium simum*)		n.m.[k]		218			n.d.		
	Black rhinoceros (*Diceros bicornis*)		n.m.[k]		127			n.d.		
	Indian rhinoceros (*Rhinoceros unicornis*)		n.m.[k]		210			n.d.		

Artiodactyls	Camel (*Camelus bactrianus*)	10[a]	29		84	11
	Pig (*Sus domesticus*)	0[a']	0	101		103
		30–80[l]		200–300		120–180
		200–350[l]		150–250		80–120
	Miniature swine	54[m]		124		99
New World monkeys	Spider monkey (*Ateles* sp.)	9[n]		219	60	223
	(*Ateles geoffroyi*)	12[o]		242	49	182

[a] n.m., not measured; n.d., not detectable. Tr, traces. VLDL, IDL, LDL, and HDL are fractions of $d < 1.006$ g/ml (S_f 20–400), $1.006–1.019$ (S_f 12–20), $1.019–1.063$ (S_f 0–12), and $1.063–1.21$ g/ml, respectively, unless otherwise stated.

[a'] G. L. Mills and C. E. Taylaur, *Comp. Biochem. Physiol.* **40B**, 489 (1971): LDL of d 1.006–1.063 g/ml.

[b] D. L. Puppione, C. Sardet, W. Yamanaka, R. Ostwald, and A. V. Nichols, *Biochim. Biophys. Acta* **231**, 295 (1971).

[c] C. Sardet, H. Hansma, and R. Ostwald, *J. Lipid Res.* **13**, 624 (1972): LDL, d 1.019–1.063 g/ml; "dense LDL," d 1.063–1.09 g/ml; and HDL, d 1.09–1.21 g/ml.

[d] M. J. Chapman and G. L. Mills, *Biochem. J.* **167**, 9 (1977): LDL and dense LDL, d 1.019–1.100 g/ml; HDL, d 1.00–1.21 g/ml.

[e] T. Bohmer, R. J. Havel, and J. A. Long, *J. Lipid Res.* **13**, 371 (1972); LDL, d 1.006–1.063 g/ml.

[f] R. Shore and V. Shore, *Adv. Exp. Biol. Med.* **67**, 123 (1976): fed. HDL of d 1.081–1.21 g/ml.

[g] H. R. Slater, J. Shepherd, and C. J. Packard, *Biochim. Biophys. Acta* **713**, 435 (1982): HDL calculated from HDL cholesterol level, assuming a 15% cholesterol content.

[h] R. Pescador, *Life Sci.* **23**, 1851 (1978).

[i] J. C. Pinon and A. M. Bridoux, *Artery* **3**, 59 (1977): LDL, d 1.006–1.063 g/ml, fasted.

[j] R. J. Havel, T. Kita, L. Kotite, J. P. Kane, R. L. Hamilton, J. L. Goldstein, and M. S. Brown, *Arteriosclerosis* **2**, 467 (1982); lipoprotein concentrations calculated from triglyceride levels and chemical compositions; 10% triglyceride content assumed for HDL.

[k] W. M. F. Leat, C. A. Northrop, N. Buttress, and D. M. Jones, *Comp. Biochem. Physiol.* **63B**, 275 (1979), small amounts of VLDL.

[l] Fasting, l' postprandial: G. M. Knipping, G. M. Kostner, and A. Holasek, *Biochim. Biophys. Acta* **393**, 88 (1975): VLDL and LDL isolated by phosphotungstate precipitation with subsequent ultracentrifugation at d 1.006 and 1.080 g/ml, respectively.

[m] A. C. Nestruck, S. Lussier-Cacan, M. Bergseth, M. Bidallier, J. Davignon, and Y. L. Marcel, *Biochim. Biophys. Acta* **488**, 43 (1977): LDL, d 1.02–1.07 and HDL, d 1.09–1.21 g/ml, respectively; triglyceride content of HDL not determined.

[n] S. R. Srinivasan, B. Radhakrishnamurthy, L. S. Webber, E. R. Dalferes, M. G. Kokatnur, and G. S. Berenson, *Am. J. Clin. Nutr.* **31**, 603 (1978).

[o] S. R. Srinivasan, J. R. McBride, B. Radhakrishnamurthy and G. S. Berenson, *Comp. Biochem. Physiol.* **47B**, 711 (1974): HDL calculated on the basis of a 15% content of cholesterol. n and o: HDL_2 and HDL_3 levels calculated from densitometric ratios. A. M. Morello and R. J. Nicolosi, *Comp. Biochem. Physiol.* **69B**, 291 (1981).

tively superior to those containing mainly apoA-I and of high-density. Quantitation of porcine LDL as lipoproteins of d 1.006–1.063 g/ml leads to their marked underestimation (Table II; ref. 23).

The porcine pattern is subject to marked change as a function of nutritional state[86] and developmental stage.[87] Thus the lipoprotein profile in the fetal piglet at mid-gestational age (50–70 days) is dominated by LDL.[87] In the newborn animal however, HDL increase rapidly to become the major class some 5 days after birth.[87]

With the possible exception of the chimpanzee (Table I), the spider monkey (*Ateles* sp.) appears alone among the nonhuman primates in displaying levels of low-density lipoproteins ($d < 1.063$ g/ml) which account for 50% or more of the total $d < 1.21$ g/ml substances, tentatively identifying it as an "LDL mammal."[88] Although the Common marmoset possesses LDL levels which in certain animals[43] approached those of man (S_f 0–12; 322–386 mg/dl),[78] high HDL concentrations (~50–60% of total lipoproteins) qualify it as an "HDL mammal" (Table I). Another aspect shared with the human profile concerns the ratio of HDL_2 to HDL_3 (i.e., 1:3.7), which approaches that characteristic of man (~1:3) and differs distinctly from that of many other monkeys in which it typically exceeds 1:2, or 0.5.[89]

Other Groups

The European badger (*Meles meles* L.) is a wild mammal which modulates its activity according to season. As such, marked variation in the absolute concentrations of circulating lipoproteins are seen, to the extent that at certain periods of the year (late autumn and winter), VLDL and LDL predominate, while at others (spring and summer), HDL are the major class.[90] These fluctuations primarily reflect modifications in LDL and VLDL levels (up to a 5-fold change in mean amounts) rather than of HDL (2-fold alteration over the course of the year).[90] Clearly then, such a species cannot be classified according to the criteria applied in the two groups above. Other wild mammals, and notably hibernators such as the hedgehog,[76] may be potential members of this group.

The presence of both a continuous profile and substantial amounts of lipoproteins at and around the d 1.063 g/ml boundary, and a lack of data

---

[86] G. M. Knipping, G. M. Kostner, and A. Holasek, *Biochim. Biophys. Acta* **393**, 88 (1975).
[87] M. B. Johansson and B. W. Karlsson, *Biol. Neonate* **42**, 127 (1982).
[88] S. R. Srinivasan, J. R. McBride, B. Radhakrishnamurthy, and G. S. Berenson, *Comp. Biochem. Physiol.* **47B**, 711 (1974).
[89] A. M. Morello and R. J. Nicolosi, *Comp. Biochem. Physiol.* **69B**, 291 (1981).
[90] P. M. Laplaud, L. Beaubatie, and D. Maurel, *J. Lipid Res.* **21**, 724 (1980).

on high-density components, similarly prevent the classification of two species of Felidae,[23] the lion *(Panthera leo)* and the jaguar *(Panthera onca)* as members of either of the two former species groups.

Chemical and Physical Properties of Lipoprotein Particles

Mammalian species whose lipoproteins and apolipoproteins have been most extensively studied in the normolipidemic, chow-fed state include the mouse, rat, and guinea pig (order Rodentia), rabbit (order Lagomorpha), dog (order Carnivora), pig and cattle (order Artiodactyla), baboon (a nonhuman primate of the family Cercopithecoidea, genus *Papio*), rhesus monkey (family Cercopithecoidea, genus *Macaca*), and chimpanzee (family Pongidae, genus *Troglodytes*). In view therefore of the substantial physicochemical data available, for example on the chemical compositions of the major lipoprotein classes in these animals, one or more representative analyses in each species are presented in the tables; data in less-common species are also included.

Chemical Properties

Lipid and Protein Composition: Chylomicrons and Their Remnants. The chylomicrons, particles of intestinal origin containing triglyceride (up to 95% by weight) as their principal constituent, have a rapid turnover time in the plasma compartment. Under fasting conditions, therefore, they do not constitute a major class in plasma, and must be isolated from either mesenteric or thoracic duct lymph.[91] In the suckling rat, however, appreciable quantities of plasma chylomicrons are found[92] (Table I), permitting their detailed chemical analysis (percentage by weight of lipid and protein components; fatty acid compositions of cholesteryl esters, triglycerides, phosphatidylcholines, phosphatidylethanolamines, sphingomyelins, and lysophosphatidylcholines), and determination of their apolipoprotein content. In chemical composition, plasma chylomicrons were richer in protein, cholesteryl ester, and phospholipid, but poorer in triglyceride, than their lymph counterparts (6.8 vs 1.6, 10.8 vs 2.2, 1.8 vs 0.1, and 79.4 and 95.0% by weight, respectively).[92] The higher content of the 20.4 fatty acid in the cholesteryl esters of plasma chylomicrons as compared to that of both lymph chylomicrons and VLDL was notable. The

---

[91] D. B. Zilversmit, *in* "Disturbances in Lipid and Lipoprotein Metabolism" (J. M. Dietschy, A. M. Gotto, and J. A. Ontko, eds.), p. 169. Am. Physiol. Soc., Bethesda, Md., 1978.

[92] G. J. P. Fernando-Warnakulasuriya, M. L. Eckerson, A. Clark, and M. A. Wells, *J. Lipid Res.* **24,** 1626 (1983).

B_{240}, A-IV, E, A-I, and C apolipoproteins were identified in the protein moiety. Few additional studies of the comparative aspects of mammalian plasma chylomicrons have been described, although chylomicron-like particles of $d < 1.006$ g/ml containing up to 90% by weight of triglyceride were found in fetal guinea pigs,[93] rudimentary data on serum chylomicrons in fat and cholesterol-fed dogs are available,[94] and particles with the physicochemical characteristics of chylomicrons have been detected in the serum of the dairy cow.[95] By contrast, triglyceride-rich lipoproteins were lacking from the sera of newborn and fetal calves.[96]

In spite of a paucity of information on plasma chylomicrons themselves, both the structure and metabolism of their remnants have been investigated exhaustively, and particularly in the rat.[97,98] These particles are enriched in cholesteryl esters and E and B proteins, and are deficient in C peptides and phospholipids.[98,99] They are present only transiently in the circulation due to their rapid and efficient removal by a receptor-mediated pathway in the liver.[99] Indeed, neither chylomicron nor VLDL remnants constitute a quantitatively important class in normal rat plasma. By contrast, it remains indeterminate as to whether the high levels of IDL typical of normolipidemic, control rabbits (Table II) correspond in part to VLDL remnants, as indeed they appear to in the WHHL strain.[85]

VLDL, IDL, LDL, and HDL. As in man,[100] the major lipids of mammalian VLDL, LDL, and HDL are cholesteryl esters, triglycerides, free cholesterol, and phospholipids.[20] Analysis of the lipid esters is incomplete in several mammals. Only in the rat have the major molecular species of each type of phospholipid been evaluated,[101] and that in HDL and a combined VLDL and LDL fraction, while in the sheep *(Ovis aries),* both the contribution of free fatty acids to the lipid content of VLDL, LDL,

---

[93] T. Bohmer, R. J. Havel, and J. A. Long, *J. Lipid Res.* **13,** 371 (1972).

[94] L. A. Hillyard, I. L. Chaikoff, C. Entenman, and W. O. Reinhardt, *J. Biol. Chem.* **233,** 838 (1958).

[95] L. F. Ferreri and R. C. Elbein, *J. Dairy Sci.* **56,** 1025 (1973).

[96] T. M. Forte, J. J. Bell-Quint, and F. Cheng, *Lipids* **16,** 240 (1981).

[97] C. J. Fielding, in "Disturbances in Lipid and Lipoprotein Metabolism" (J. M. Dietschy, A. M. Gotto, and J. A. Ontko, eds.), p. 83. Am. Physiol. Soc., Bethesda, Md., 1978.

[98] O. D. Mjøs, O. Faergeman, R. L. Hamilton, and R. J. Havel, *J. Clin. Invest.* **56,** 603 (1975).

[99] A. D. Cooper, S. K. Erickson, R. Nutik, and M. A. Shrewsbury, *J. Lipid Res.* **23,** 42 (1982).

[100] V. P. Skipski, in "Blood Lipids and Lipoproteins: Quantitation, Composition and Metabolism" (G. J. Nelson, ed.), p. 471. Wiley (Interscience), New York, 1972.

[101] A. Kuksis, W. C. Breckenridge, J. M. Myher, and G. Kakis, *Can. J. Biochem.* **56,** 630 (1978).

"light HDL" (d 1.063–1.075 g/ml), HDL, and VHDL, as well as that of individual phospholipid classes to the total phosphorus-containing lipid in each fraction,[102] have been determined. Data on the contents of free fatty acids and minor lipids (glycolipids, hydrocarbons, and plant sterols) in mammalian lipoproteins are otherwise scant. Indeed most authors have omitted analysis of such components even though they may comprise up to 5–6% by weight of certain human fractions.[100] The transport of composterol and β-sitosterol by rat VLDL, LDL, and HDL is nonetheless established.[103]

The chemical compositions of VLDL, IDL, LDL, and HDL in mammals are compiled in Tables III, IV, V, and VI. Since cholesteryl esters are an essential component of these major lipoprotein classes, analyses which omit this element and which treat the esterified and free sterol together have been omitted. A further omission concerns the lipoproteins of marine mammals, whose lipid composition has been extensively evaluated in the apparent absence of protein determinations.[16]

Moreover, in considering LDL and HDL compositional data in which apoB- and apoA-containing particles contribute significantly to the high- and low-density classes, respectively, and in which one of the limiting densities is 1.063 g/ml, then the highly heterogeneous nature of the lipoproteins analyzed in the whole mixture should not be overlooked. This limitation to chemical data in a number of mammals has been evoked and discussed in previous sections. Clearly then the overall composition of a given lipoprotein fraction will correspond most closely to that of the most abundant particle(s).

The proportion of neutral, core lipids in mammalian VLDL (Table III) resembles that typical of man (~65%),[23,100] with few exceptions. Only in the rat, guinea pig, badger, and dairy cow do the triglycerides and cholesteryl esters together account for more than 75% of the weight of the particles. Certainly, this finding may be at least partially explained by the presence of small chylomicrons and/or chylomicron remnants in the VLDL fraction, in the fetal and in the pregnant guinea pig,[93] in the suckling and in the adult rat,[92,104] and in the dairy cow,[95] particularly when obtained postprandially. The elevated content of cholesteryl ester (>20%) in VLDL from the WHHL rabbit, squirrel monkey, baboon, and in one preparation from the chimpanzee is notable and possibly indicative of the contribution of chylomicron remnants. Equally, the high protein content

---

[102] G. J. Nelson, *Comp. Biochem. Physiol.* **46B**, 81 (1973).
[103] M. Sugano, H. Morioka, Y. Kida, and I. Ikeda, *Lipids* **13**, 427 (1978).
[104] T. G. Redgrave, *J. Clin. Invest.* **49**, 465 (1970).

TABLE III
CHEMICAL COMPOSITION OF MAMMALIAN PLASMA VLDL

Species	Density interval (g/ml)	Cholesteryl ester	Triglyceride	Free cholesterol	Phospholipid	Protein
Man[a]	<1.007	14.9	49.9	6.7	18.6	7.7
Mouse[b]	<1.006	2.3	66.7	5.8	16.9	8.3
Mouse[c]	<1.017	6.3	66.5	6.2	13.2	7.8
Rat[d]	1.014–1.016	3.7	73.3	5.3	11.4	6.3
Rat[e]	<1.006	1.9	73.6	2.7	12.6	9.4
Mongolian gerbil[f]	<1.006	11.8	61.3	2.5	9.8	14.5
Guinea pig[g]	<1.007	8.1	64.8	3.9	12.6	10.7
Guinea pig[h]	<1.006[h']	2.1	90.2	(2.1)[i]	4.2	3.5
Guinea pig[h]	<1.006[h2]	3.8	76	(3.8)[i]	11.2	9.0
Hedgehog[i]	<1.006	11.1	57.3	8.5	14.5	8.5
Dog[j]	<1.006	2	68	6	10	14
Red fox[k]	<1.006	7.4	59.1	3.0	20.1	10.5
Badger[l]	<1.006	9.1	67.8	4.0	9.2	9.9
Rabbit[m]	<1.006[m']	6.6	62.2	3.7	15.6	11.9
Rabbit[m]	<1.006[m2]	21.9	41.9	6.1	17.6	12.2
Camel[o]	<1.007	5.2	55.6	6.4	12.6	20.1
Sheep[n]	<1.006	16.5	41.0	8.0	22.1	11.4
Newborn calf[o]	<1.006	9.8	41.1	6.7	17.9	24.6
Steer[o]	<1.006	2.4	46.8	6.9	21.5	22.5
Dairy cow[p]	<1.006	5	74	7	7	8
Dairy cow[q]	<1.019	3.7	60.2	4.9	25.1	6.0
Pig[r]	<1.006	6.7	63.5	3.8	20.6	5.4
Squirrel monkey[s]	<1.006	34.0	30.5	8.1	14.8	12.6
Cebus monkey[t]	—	3.8	65.9	3.8	15.9	10.7
Common marmoset[u]	<1.017	16.1	57.3	3.7	14.9	8.1
Rhesus monkey[v]	<1.006	13.3	57.6	4.8	15.7	8.6
Baboon[w]	<1.006	27.7	46.2	10.2	6.5	9.2
Chimpanzee[x]	<1.006	4.9	59.3	4.4	20.8	10.6
Chimpanzee[y]	<1.006	22.1	29.9	9.4	17.9	12.9

[a] G. L. Mills and C. E. Taylaur, *Comp. Biochem. Physiol.* **40B**, 489 (1971).
[b] R. C. Le Boeuf, D. L. Puppione, V. N. Schumaker, and A. J. Lusis, *J. Biol. Chem.* **258**, 5063 (1983).
[c] M. C. Camus, M. J. Chapman, P. Forgez, and P. M. Laplaud, *J. Lipid Res.* **24**, 1210 (1983); isolated by density gradient centrifugation.
[d] Suckling rat; G. J. P. Fernando-Warnakulasuriya, M. L. Eckerson, W. A. Clark, and M. A. Wells, *J. Lipid Res.* **24**, 1626 (1983); density limits for the vertical rotor separation taken from B. H. Chung, T. Wilkinson, J. C. Geer, and J. P. Segrest, *J. Lipid Res.* **21**, 284 (1980).
[e] O. D. Mjøs, O. Faergeman, R. L. Hamilton, and R. J. Havel, *J. Clin. Invest.* **56**, 603 (1975).
[f] R. J. Nicolosi, J. A. Marlett, A. M. Morello, S. A. Flanagan, and D. M. Hegsted, *Atherosclerosis* **38**, 359 (1981). Animals fed a semipurified diet containing 15% safflower oil.
[g] M. J. Chapman and G. L. Mills, *Biochem. J.* **167**, 9 (1977).
[h] T. Bohmer, R. J. Havel, and J. A. Long, *J. Lipid Res.* **13**, 371 (1972); [h2] fetal guinea pig at 60 days; [h2] pregnant animals at 60 d.
[i] Percentage cholesteryl esters includes free cholesterol content.
[i2] P. M. Laplaud, L. Beaubatie, and M. Saboureau, *Biochim. Biophys. Acta* **752**, 396 (1983).
[j] J. P. Blomhoff, R. Holme, and J. Ostrem, *Scand. J. Gastroenterol.* **13**, 693 (1978).
[k] P. M. Laplaud, L. Beaubatie, and D. Maurel, *Comp. Biochem. Physiol.* **68B**, 125 (1981).
[l] P. M. Laplaud, L. Beaubatie, and D. Maurel, *J. Lipid Res.* **21**, 724 (1980): analysis of August 1978.
[m] R. J. Havel, T. Kita, L. Kotite, J. P. Kane, R. L. Hamilton, J. L. Goldstein, and M. S. Brown, *Arteriosclerosis* **2**, 467 (1982).
[m'] New Zealand white male control; [m2] WHHL male rabbits.
[n] W. M. F. Leat, F. O. T. Kubasek, and N. Buttress, *Q. J. Exp. Physiol.* **61**, 193 (1976). Neonatal lambs, birth to 90 days postpartum.
[o] T. M. Forte, J. J. Bell-Quint, and F. Cheng, *Lipids* **16**, 240 (1981).
[p] L. F. Ferreri and R. C. Elbein, *J. Dairy Sci.* **65**, 912 (1982).
[q] Data calculated from D. Stead and V. A. Welch, *J. Dairy Sci.* **58**, 122 (1975); minor amounts of IDL present.
[r] N. Fidge, *Biochim. Biophys. Acta* **295**, 258 (1973).
[s] D. R. Illingworth, *Biochim. Biophys. Acta* **388**, 38 (1975).
[t] R. J. Nicolosi and K. C. Hayes, in "Handbook of Electrophoresis" (L. A. Lewis and H. K. Naito, eds.), Vol. IV, p. 33. CRC Press, Boca Raton, Fla., 1983. Density interval not determined. Animals fed a corn oil-supplemented diet.
[u] M. J. Chapman, F. McTaggart, and S. Goldstein, *Biochemistry* **18**, 5096 (1979). Densities are isopycnic banding values from density gradient centrifugation.
[v] L. L. Rudel, D. G. Greene, and R. Shah, *J. Lipid Res.* **18**, 734 (1977); isolated by gel filtration chromatography as fraction IIA.
[w] D. Bojanovski, P. Alaupovic, J. L. Kelley, and C. Stout, *Atherosclerosis* **31**, 481 (1978).
[x] M. Rosseneu, B. Leclercq, D. Vandamme, R. Vercaemst, F. Soetewey, H. Peeters, and V. Blaton, *Atherosclerosis* **32**, 141 (1979); chow diet containing 4% fat.
[y] M. J. Chapman, P. Forgez, D. Lagrange, S. Goldstein, and G. L. Mills, *Atherosclerosis* **52**, 129 (1984).

TABLE IV
CHEMICAL COMPOSITION OF MAMMALIAN IDL

Species	Density interval (g/ml)	Mean % (by weight)				
		Cholesteryl ester	Triglyceride	Free cholesterol	Phospholipid	Protein
Man[a]	1.009–1.019	29.1	20.2	8.4	23.3	19.0
Mouse[b]	1.009–1.019[a']	25.0	17.4	6.8	27.0	23.6
Rat[c]	1.017–1.023	7.2	67.3	5.4	12.0	8.1
Dog[d]	1.006–1.019	12.5	41.1	18.1	12.0	16.3
Cattle: lactating cow[e]	1.006–1.019	3	77	2	6	12
	1.006–1.023[c']	3.1	65.9	7.6	12.6	10.9
	1.006–1.023[e²]	2.3	61.3	7.7	13.2	15.5
Pig[f]	1.006–1.019	12.8	49.2	8.6	17.0	12.4
Squirrel monkey[g]	1.006–1.019	43.8	14.0	11.5	16.1	14.6

[a] D. M. Lee and P. Alaupovic, *Biochemistry* **9**, 2244 (1970); isolated by sequential ultracentrifugation.
[a'] As in *a* with the exception that apoA-containing lipoproteins were removed by immunoprecipitation.
[b] M. C. Camus, M. J. Chapman, P. Forgez, and P. M. Laplaud, *J. Lipid Res.* **24**, 1210 (1983). See *d*, Table VI.
[c] I. Pasquali-Ronchetti, S. Calandra, M. Baccarana-Contri, and M. Montaguti, *J. Ultrastruct. Res.* **53**, 180 (1975).
[d] As *j* in Table III.
[e,e²] D. L. Puppione, S. T. Kunitake, R. L. Hamilton, M. L. Phillips, V. N. Schumaker, and L. D. Davis, *J. Lipid Res.* **23**, 283 (1982); isolated ultracentrifugally at 16°.
[f] N. H. Fidge and G. D. Smith, *Artery* **1**, 406 (1975).
[g] As *s* in Table III.

TABLE V
CHEMICAL COMPOSITION OF MAMMALIAN LDL

Species	Density interval (g/ml)	Cholesteryl ester	Triglyceride	Free cholesterol	Phospholipid	Protein
Man[a]	1.007–1.063	38.0	11.2	9.0	22.1	20.9
Man[b]	1.028–1.050	36.8	4.2	9.4	25.8	23.8
Mouse[c]	1.006–1.063	23.5	19.0	9.5	27.5	20.5
Mouse[d]	1.033–1.060	21.9	28.3	7.4	19.5	24.3
Rat[a]	1.007–1.063	26.8	18.5	8.6	21.2	24.9
Rat[e]		28.8	21.2	13.2	20.4	16.4
Rat[f]	1.02–1.062	27.0	20.5	7.3	25.5	22.1
Rat[g]	"LDL"	31.8	12.8	17.7	19.2	18.5
Mongolian gerbil[h]	1.02–1.05	42.5	11.4	12.2	14.0	21.7
Guinea pig[a]	1.019–1.063	36.3	14.3	4.3	16.3	28.9
Guinea pig[i]	1.007–1.063	35.7	14.3	5.4	12.5	32.1
Guinea pig[j]	1.007–1.100	41.6	15.0	(41.6)[j3]	10.9	32.7
Guinea pig[j]	1.006–1.063[j1]	50	1.7	(50)[j3]	16.3	34.2
Hedgehog[a]	1.006–1.063[j2]	36.3	14.3	4.3	16.3	28.9
Hedgehog[k]	1.007–1.063	22.0	4.0	16.0	27.0	31.0
Dog[l]	1.046–1.055	11	30	11	20	28
Dog[m]	1.019–1.063	24.5	26.6	5.2	22.4	21.3
Badger[n]	1.006–1.063	21.3	15.9	9.5	27.1	26.2
Badger[o]	1.006–1.063	26.3	15.8	9.1	26.2	22.6
Red fox[p]	1.039–1.046	20.9	19.1	8.7	24.1	27.3
Rabbit[q]	1.006–1.063	23.5	30.6	6.3	19.7	19.9
Rabbit	1.019–1.063[q1]	29.6	21.7	8.1	19.1	21.6
Camel[a]	1.019–1.063[q2]	41.2	6.8	11.2	21.5	19.4
Sheep[a]	1.006–1.063	45.1	5.0	9.5	19.6	20.8
Sheep[r]	1.006–1.063	38.6	5.9	8.9	25.8	20.5

(continued)

TABLE V (continued)

Species	Density interval (g/ml)	Cholesteryl ester	Triglyceride	Free cholesterol	Phospholipid	Protein
Cattle						
Fetal calf[a]	1.006–1.063	40.3	0.4	10.3	26.6	22.5
Newborn calf[a]	1.006–1.063	37.1	3.4	7.7	22.3	30.3
Steer[a]	1.006–1.063	39.9	6.4	7.7	21.6	26.0
Dairy cow[b]	1.019–1.039	32.3	0.1	4.9	41.2	21.3
Dairy cow[b]	1.039–1.060	39.3	1.8	5.7	29.2	24.0
Pig[a]	1.019–1.063	43.6	4.2	6.9	24.1	21.2
Pig[a]	1.030–1.040	36.5	6.7	11.4	23.2	22.2
Pig[a]	1.074–1.090	28.1	11.3	6.4	19.6	34.6
Squirrel monkey[b]	1.019–1.063	32.6	8.5	8.5	23.6	27.0
Cebus monkey[b]	n.d.	31.1	10.7	13.1	22.0	23.1
Common marmoset[b]	1.027–1.055	37.4	10.9	7.3	20.9	23.5
Rhesus monkey[b]	1.019–1.050	36.0	2.5	14.0	23.6	24.0
Rhesus monkey[b]	1.036[c] (LDL-II)	37.3	7.6	7.8	25.3	20.8
Baboon[A]	1.006–1.063	37.3	13.1	12.1	12.6	24.7
Baboon[B]	1.024–1.045	30.2	10.8	14.2	21.3	23.5
Cynomolgus monkey[C]	1.019–1.063	42.7	4.0	8.5	22.1	22.7
Chimpanzee[D]	1.006–1.063	27.5	9.6	9.6	25.3	27.8
Chimpanzee[E]	1.024–1.050	38.7	5.3	10.0	22.6	23.1

[a] As Table III.
[b] M. J. Chapman, S. Goldstein, D. Lagrange, and P. M. Laplaud, *J. Lipid Res.* **22**, 339 (1981); isolated by gradient centrifugation; densities are isopycnic banding densities.
[c] R. C. LeBoeuf, D. L. Puppione, V. N. Schumaker, and A. J. Lusis, *J. Biol. Chem.* **258**, 5063 (1983).
[d] M. C. Camus, M. J. Chapman, P. Forgez, and P. M. Laplaud, *J. Lipid Res.* **24**, 1210 (1983); see d, Table VI.
[e] Suckling rat; as d in Table III.

[f] Y. Oschry and S. Eisenberg, *J. Lipid Res.* **23**, 1099 (1982); isolated by rate zonal ultracentrifugation, density undetermined. Fasting animals.
[g] S. Calandra, I. Pasquali-Ronchetti, E. Gherardi, C. Fornieri, and P. Tarugi, *Atherosclerosis* **28**, 369 (1977).
[h] As f in Table III.
[i] As g in Table III.
[j] As h in Table III; j' and j^2 as h' and h^2, respectively.
[j³] Percentage cholesteryl esters includes free cholesterol content.
[k] As i in Table III; densities are isopycnic banding values from density gradient centrifugation.
[l] As j in Table III.
[m] R. W. Mahley and K. H. Weisgraber, *Circ. Res.* **35**, 713 (1974); purified by Pevikon block electrophoresis.
[n] As l in Table III.
[o] P. M. Laplaud, L. Beaubatie, and D. Maurel, *Biochim. Biophys. Acta* **711**, 213 (1982); analysis of December; densities are isopycnic banding values from density gradient centrifugation.
[p] As k in Table III.
[q, q', q²] As m, m', and m^2, respectively, in Table III.
[r] As n in Table III.
[s] As o in Table III.
[t] As q in Table III.
[u] As f in Table IV.
[v] As s in Table III.
[w] As t in Table III.
[x] As u in Table III.
[y] G. M. Fless and A. M. Scanu, *Biochemistry* **14**, 1783 (1975).
[z] G. M. Fless and A. M. Scanu, *J. Biol. Chem.* **254**, 8653 (1979).
[z'] Hydrated density determined by density gradient centrifugation.
[A] As w in Table III.
[B] M. J. Chapman and S. Goldstein, *Atherosclerosis* **25**, 267 (1977).
[C] L. L. Rudel and L. L. Pitts, *J. Lipid Res.* **19**, 992 (1978); control males fed a semipurified diet containing 40% calories as fat; lipoproteins isolated from a $d < 1.22$ g/ml fraction by gel filtration chromatography.
[D] As x in Table III.
[E] As y in Table III.

TABLE VI
CHEMICAL COMPOSITION OF MAMMALIAN HDL AND THE MAJOR SUBCLASSES

Species	Density interval (g/ml)	Mean % (by weight)				
		Cholesteryl ester	Triglyceride	Free cholesterol	Phospholipid	Protein
Man[a]	1.063–1.21	15.0	8.0	2.9	22.7	51.9
Man[b]	1.063–1.125 (HDL$_2$)	16.2	5.7	5.4	29.5	41
	1.125–1.21 (HDL$_3$)	11.7	6.1	2.9	22.5	55
Mouse[c]	1.063–1.21	19.5	1.0	3.0	35.0	41.5
Mouse[d]	1.060–1.163	17.0	4.6	1.9	26.7	50.8
	1.060–1.085 (HDL$_2$)	20.0	11.1	2.6	23.3	42.9
	1.085–1.163 (HDL$_3$)	16.6	4.1	1.8	26.6	50.8
Mouse[e]	1.167 e[1]	20.3	3.7	1.3	28.5	46.2
	1.167 e[2]	24.0	1.8	1.7	26.3	46.2
Rat[f]	1.063–1.21	15.0	2.8	14.8	26.0	41.4
Rat[g]	1.034: "HDL$_1$"	19.6	12.1	16.7	28.1	23.5
	"HDL$_2$"	23.8	0.1	8.2	31.0	36.9
Rat[h]	HDL$_1$	36.8	3.4	11.9	22.2	25.7
Rat[i]	HDL$_1$	28.9	4.1	7.1	33.2	26.1
Rat[i]	"HDL$_2$"	25.8	1.2	3.2	28.1	41.6
Rat[j]	1.063–1.125	27.3	1.8	3.1	29.8	38.0
	1.125–1.21	25.9	3.2	2.7	26.4	46.0
Mongolian gerbil[k]	1.063–1.210	24.6	2.4	5.9	24.2	43.0
Guinea pig[l]	1.063–1.21	31.2	7.5	4.7	13.1	43.4
Guinea pig[m]	1.090–1.21	25	2	9	25	39
Guinea pig[n]	1.063–1.21	22.2	6.5	(22.2)[n']	20.3	51
Hedgehog[o]	1.087–1.100	22.7	2.7	5.4	23.5	45.8
Hedgehog[a]	1.063–1.21	26.4	2.1	6.3	24.4	40.7
Dog[p]	HDL$_1$	36.5	1.9	7.7	34.6	19.2
	1.087–1.21	22.8	0.6	4.9	33.0	38.7
Red fox[q]	1.063–1.21	20.8	0.6	2.8	33.0	42.8
Badger[r]	1.063–1.21	24.3	0.5	4.6	30.5	40.1
Badger[s]	1.087–1.100	27.7	1.4	4.2	30.9	35.8
	1.115–1.130	22.7	1.7	4.7	20.7	50.3

Species	Density					
Rabbit[a]	1.063–1.21	20.8	8.2	2.5	16.5	51.9
Camel[a]	1.063–1.21	37.2	7.3	10.3	13.5	30.4
Sheep[f]	1.063–1.21	26.6	4.9	2.5	19.3	51.2
Sheep[f]	1.063–1.21	25.9	0.6	3.7	23.9	45.5
Cattle						
Fetal calf[u]	1.063–1.21	22.9	0.3	3.0	23.8	50.3
Newborn calf[u]	1.063–1.21	29.0	0.2	2.7	20.6	45.1
Steer[u]	1.063–1.21	37.3	0.1	2.8	22.4	37.5
Dairy cow[a]	1.063–1.21	31.1	4.4	3.7	21.3	43.5
Dairy cow[v]	1.083	36.8	<1	3.7	29.6	29.9
Pig[w]	1.09 –1.21	14.1	1	2.6	27.4	54.8
Pig[x]	1.12 –1.21	19.0	2.9	2.2	33.3	42.6
Squirrel monkey[y]	1.063–1.125 (HDL$_2$)	26.5	0.9	6.1	22.1	44.3
	1.125–1.21 (HDL$_3$)	22.6	0.8	2.0	22.3	52.6
Cebus monkey[z]	—	10.3	10.1	5.6	22.7	51.2
Common marmoset[A]	1.070–1.127 (HDL$_2$)	21.3	8.7	2.2	27.9	39.8
	1.127–1.156 (HDL$_3$)	14.2	6.0	3.8	17.3	58.7
Rhesus monkey[B]	1.063–1.125 (HDL$_2$)	16.2	4.5	5.4	29.5	41.0
	1.125–1.21 (HDL$_3$)	11.7	4.1	2.9	22.5	55.0
Baboon[C]	1.063–1.21	15.4	3.4	3.6	28.8	48.8
African Green monkey[D]	—	19.4	1.9	3.6	30.6	44.5
Cynomolgus monkey[E]	1.063–1.225	20.0	2.0	3.3	29.3	45.3
Chimpanzee[F]	1.063–1.125 (HDL$_2$)	24.9	2.2	6.5	24.6	41.8
	1.125–1.21 (HDL$_3$)	17.0	2.9	3.4	19.0	57.7

(continued)

[a] As Table III.
[b] Adapted from V. P. Skipski, in "Blood Lipids and Lipoproteins: Quantitation, Composition and Metabolism" (G. J. Nelson, ed.), p. 471. Wiley (Interscience), New York, 1972; unidentified lipids omitted.
[c] R. C. LeBoeuf, D. L. Puppione, V. N. Schumaker, and A. J. Lusis, *J. Biol. Chem.* **258**, 5063 (1983); random-bred Swiss mice.
[d] M. C. Camus, M. J. Chapman, P. Forgez, and P. M. Laplaud, *J. Lipid Res.* **24**, 1220 (1983); fractions isolated by density gradient centrifugation; density intervals correspond to isopycnic banding densities. Random-bred Swiss mice (OF 1 strain).
[e] J. D. Morrisett, H. S. Kim, J. R. Patsch, S. K. Datta, and J. J. Trentin, *Arteriosclerosis* **2**, 312 (1982).
[e¹] Animals of the CBA/J strain.
[e²] Mice of the C57Br/cdJ strain. e' and e^2 hydrated density calculated after separation by rate zonal ultracentrifugation.
[f] I. Pasquali-Ronchetti, S. Calandra, M. Baccaran-Contri, and M. Monteguti, *J. Ultrastruct. Res.* **53**, 180 (1975).
[g] Suckling rat; as d in Table III; hydrated density of HDL$_1$ determined by analytical ultracentrifugation. Density of "HDL$_2$" undetermined.

Footnotes to TABLE VI (*continued*)

[h] K. H. Weisgraber, R. W. Mahley, and G. Assmann, *Atherosclerosis* **28**, 121 (1977); isolated by electrophoresis from a d 1.02–1.063 g/ml fraction.
[i] Y. Oshry and S. Eisenberg, *J. Lipid Res.* **23**, 1099 (1982); isolated by rate zonal ultracentrifugation, density undetermined. Fasting animals.
[j] Calculated from the data of A. Kuksis, H. C. Breckenridge, J. M. Myher, and G. Kakis, *Can. J. Biochem.* **56**, 630 (1978).
[k] As f in Table III.
[l] As a in Table III.
[m] C. Sardet, H. Hansma, and R. Ostwald, *J. Lipid Res.* **13**, 624 (1972).
[n] As h' in Table III.
[n'] Percentage cholesteryl esters includes free cholesterol content.
[o] As k in Table V.
[p] As m in Table V.
[q] As k in Table III.
[r] As l in Table III.
[s] As o in Table V.
[t] As n in Table III.
[u] As o in Table III.
[v] D. L. Puppione, S. T. Kunitake, M. L. Toomey, E. Loh, and V. N. Schumaker, *J. Lipid Res.* **23**, 371 (1982); lactating Jersey cow; isopycnic banding density; quantitatively, the peak fraction in 10 density gradient subfractions of HDL (d 1.050–1.21 g/ml).
[w] A. C. Nestruck, S. Lussier-Cacan, M. Bergseth, M. Bidellier, J. Davignon, and Y. Marcel, *Biochim. Biophys. Acta* **188**, 43 (1977).
[x] As r in Table III.
[y] As s in Table III.
[z] As t in Table III.
[A] As u in Table III.
[B] A. M. Scanu, C. Edelstein, L. Vitello, R. Jones, and R. Wissler, *J. Biol. Chem.* **248**, 7648 (1973).
[C] V. Blaton, R. Vercaemst, M. Rosseneu, J. Mortelmans, R. L. Jackson, A. M. Gotto, and H. Peeters, *Biochemistry* **16**, 2157 (1977).
[D] J. S. Parks and L. L. Rudel, *J. Biol. Chem.* **254**, 6716 (1979); isolated by gel filtration chromatography from a d < 1.22 g/ml lipoprotein fraction.
[E] As C in Table V.
[F] V. Blaton, R. Vercaemst, N. Vandecasteele, H. Caster, and H. Peeters, *Biochemistry* **13**, 1127 (1974).

of newborn calf and steer VLDL[96] may represent surface remnant material. Free cholesterol varied over a 4-fold range (2.5–10%), while that of phospholipid was still greater (4.2–25.1%) (Table III). In accordance with the established relationship between particle diameter and the sum of the content of protein, phospholipid and free cholesterol in human VLDL,[105] it may be inferred that differences in mammalian VLDL composition (Table III) reflect marked dissimilarities in their mean particle sizes.

Few analyses of mammalian IDL have appeared (Table IV), and of these, only squirrel monkey IDL is enriched in cholesteryl ester like its human counterpart (44% as compared to 25–29%, respectively). Murine, rat, and porcine IDL are poor in this core lipid (7.2–12.8%), and canine and bovine IDL still poorer (2–3%). By contrast, triglyceride contents are uniformly higher (41–77%) in these mammalian IDL (except for squirrel monkey) than in man (~20%). Like cholesteryl ester, the proportions of both phospholipid and protein in mammalian IDL were rather lower than those in the human fraction. The unusual composition of bovine IDL is related both to the high degree of saturation of its triglyceride fatty acids and to its isolation at a temperature (16°) below their thermal transition.[106]

As evident from Table IV, mammalian LDL have been most often separated in the density interval 1.006–1.063 g/ml, and this despite the inclusion of variable quantities of lipoproteins of intermediate density (d 1.006–1.019 g/ml) (Tables I and II). Amounts of IDL rarely attain concentrations greater than about 30 mg/dl serum, however; nonetheless even such low levels may be of consequence to LDL compositional data when the latter (of S_f 0–12 or d 1.019–1.063 g/ml) are present in similar or slightly greater amounts than IDL, as occurs in the rabbit for instance.[23] For this reason, the composition of one or more of the major LDL subfractions is shown together with that of d 1.006–1.063 g/ml whenever available (e.g., mouse,[41] rat,[107] hedgehog,[76] badger,[77] dairy cow,[108] pig,[47] rhesus monkey,[109] and chimpanzee[50]).

In man, the bulk of the circulating cholesterol is transported in esterified form as LDL. This is not the case in the majority of mammals (i.e., the "HDL mammals," Table I), in which HDL is often the primary cholesterol carrier. Indeed, the relative contributions of HDL and LDL to

[105] T. Sata, R. J. Havel, and A. L. Jones, *J. Lipid Res.* **13**, 757 (1972).
[106] D. M. Small, D. L. Puppione, M. L. Phillips, D. Atkinson, J. A. Hamilton, and V. N. Schumaker, *Circulation* **62**, 111, Abstr. 444 (1980).
[107] S. Calandra, I. Pasquali-Ronchetti, E. Gherardi, C. Fornieri, and P. Tarugi, *Atherosclerosis* **28**, 369 (1977).
[108] D. Stead and V. A. Welch, *J. Dairy Sci.* **58**, 122 (1975).
[109] G. M. Fless and A. M. Scanu, *J. Biol. Chem.* **254**, 8653 (1979).

total cholesterol transport in a given species may readily be calculated from the data in Tables I and II on lipoprotein levels and Tables III, IV, V, and VI on their relative compositions.

The mammals may also be subdivided on the basis of the extent to which their LDL is enriched in cholesteryl ester, as in man (35–40% by weight[100]). Thus, the mouse, rat, rabbit, badger, dog, and red fox all possess LDLs containing 30% or less of this hydrophobic constituent (Table V), distinguishing them from guinea pig, Mongolian gerbil, hedgehog, camel, sheep, bovine, porcine, and New and Old World monkey LDLs containing up to 45%. Conversely, the triglyceride contents in LDL of the former group tended to be elevated (~20–30%), such that the combined proportions of these two apolar lipids represented some 40 to 50% of LDL in species of both groups. The degree to which the relative proportions of cholesteryl esters and triglycerides in mammalian LDLs depend on the rate of cholesteryl ester transfer from HDL is indeterminate. Free cholesterol varied over a 4-fold range (4.3–17.7%); variation in the polar components, i.e., phospholipid and protein, was considerably less (about 2-fold). The elevated phospholipid content of LDL in the dairy cow is noteworthy.[108] Calculation of the average hydrated density of LDLs from the compositional data using published values for partial specific volumes of lipids and protein[105] rapidly reveals the higher average density of LDL enriched in the denser components, primarily phospholipid and protein. In such species, the peaks of the LDL profiles tend to occur at higher densities than those whose LDL are deficient in these components; examples are rabbit (PRO + PL, 39.6%)[85] and rhesus LDL (PRO + PL, 46.1%),[109] whose average hydrated densities are 1.029 (calculated) and 1.064 g/ml (calculated; 1.036 g/ml experimentally determined),[109] respectively. Similar inferences may be made from the data of Mills and Taylaur.[23]

The structural heterogeneity of human HDL has been recognized for some years.[78,100] Many investigators have assumed this to be the case for their mammalian counterparts, and have therefore fractionated HDL into HDL_2 and HDL_3. In several mammals, however, HDL particles display a symmetrical, Gaussian-like distribution unlike that typical of man, and frequently exhibit poorly defined HDL subclasses.[20] Good examples are the rat and mouse, in which the presence of such subclasses remains controversial.[5,11,41,74,107] For comparative purposes, both the compositions of HDL of d 1.063–1.21 g/ml and of its major subclasses have been included in Table V, together with that of HDL_1 when appropriate. Rat and canine HDL_1 are cholesteryl ester-rich and protein-poor, and indeed their contents of these two components are more akin to LDL over whose density range they are primarily found. Upon comparison of data on HDL subclasses, a definite tendency to a reduction in cholesteryl ester and phospholipid and elevation in protein content with increase in density (i.e., HDL_2 to HDL_3) is evident. The combined proportions of protein and phospholipid fall in the range of ~65–70% in HDL_2 and ~70–75% in HDL_3, respectively, a finding consistent with the smaller particle size of the latter. The apparent predominance of one of these subclasses, i.e., HDL_2, in the rat,[74] has been attributed to the absence of a cholesteryl ester transfer protein(s) in this rodent. Cholesteryl esters, formed in the LCAT reaction, cannot therefore be transferred to lighter lipoproteins and accumulate in HDL. The cholesteryl ester fatty acid pattern of rat HDL is thus distinct from that of VLDL and LDL. It will be of interest to determine whether a similar transfer deficiency maintains in the mouse, in which HDL_3-like particles dominate.[41,110] Certainly, the lack of significant cholesteryl ester transfer from HDL to VLDL and LDL in the pig,[111] considered together with an HDL flotation pattern akin to that of human HDL_3,[14] appears inconsistent with such a thesis.[74] Additional factors seem, however, involved in determining the complex structure–composition relationships in both HDL and LDL, and one of these is the plasma triglyceride level.[112]

The major phospholipid subclasses have been separated and quantitated from the lipoproteins of very few mammals; these data are summarized in Table VII. Phosphatidylcholine preponderates in all lipoprotein classes, accounting for ~50 to 88% of the total. Sphingomyelin contents vary considerably in both VLDL and LDL but less in HDL. Levels in rat VLDL and LDL and in rabbit VLDL are markedly lower than those in their human counterparts. Rat LDL and HDL were distinguished by elevated contents (~20%) of lysolecithin. The relevance of these findings to the structure and physiology of the parent lipoprotein particles is presently unknown.

On the contrary, an extensive literature has appeared on the fatty acid distributions of the triglycerides, cholesteryl esters, and phospholipids of mammalian lipoproteins, often in relation to the source of dietary fat. In the case of the phospholipids, it is unfortunate that fatty acid profiles have been frequently assessed on a mixture of the respective subclasses. The

---

[110] J. D. Morrisett, H. S. Kim, J. R. Patsch, S. K. Datta, and J. J. Trentin, *Arteriosclerosis* **2**, 312 (1982).

[111] Y. C. Ha, G. D. Calvert, G. H. McIntosh, and P. J. Barter, *Metabolism* **30**, 380 (1981).

[112] R. J. Deckelbaum, E. Granot, Y. Oschry, L. Rose, and S. Eisenberg, *Arteriosclerosis* **4**, 225 (1984).

TABLE VII
PHOSPHOLIPID DISTRIBUTION IN MAMMALIAN LIPOPROTEINS[a]

Species	VLDL				LDL						HDL				Rhesus monkey[i]		
	Man[a]	Rat[b]	Rabbit[d]	Man[a]	Rat[b]	Dog[c]	Sheep[f]	Rhesus monkey[h]	Man[a]	Rat[b]	Dog[c]	Sheep[f]	Baboon[g]	HDL$_2$	HDL$_3$	Chimpanzee[j]
Phosphatidylcholine	66.2	75.5	72.2	66.3	49.7	80.5	48.6	76.5	74.4	59.1	84.3	66.9	81.2	87.6	88.1	79.8
Sphingomyelin	23.1	4.2	6.7	25.3	5.5	16.5	48.2	14.0	13.2	6.2	6.7	22.9	5.9	3.8	3.8	12.5
Phosphatidylethanolamine		2.5	12.1		3.1		0.4	4.6	3.1	1.5	3.3	0.5	8.3	3.9	3.5	6.4
Phosphatidylserine	5.7	5.5	n.d.	4.4	11.1		n.d.		0.8	8.0	n.d.	n.d.	0.4	1.1	1.1	0.2
Phosphatidylinositol		4.0	3.0		6.3	3.4	0.5	6.5	2.4	1.6	2.3	0.4	3.6	n.d.	n.d.	0.7
Lysolecithin	5.1	7.1	4.2	4.1	20.9		1.9	2.0	2.9	22.4	3.4	7.3	0.7	3.6	3.5	0.4
Polyglycerophosphatides and phosphatidic acid		1.2	n.d.		2.4		n.d.	n.d.	2.2	1.3	n.d.	n.d.	n.d.	n.d.	n.d.	n.d.

[a] n.d., not determined. Values are percentage by weight.
[a'] As b in Table VI.
[b] S. Malhotra and O. Kritchevsky. *Mech. Ageing Dev.* **8**, 445 (1978); data from young (1.5 months) Wistar rats; LDL, d 1.019–1.063 g/ml.
[c] C. Edelstein, L. L. Lewis, J. R. Shainoff, H. Naito, and A. M. Scanu. *Biochemistry* **15**, 1934 (1976).
[d] J. L. Rodriguez, G. C. Ghiselli, D. Torregiani, and C. R. Sirtori. *Atherosclerosis* **23**, 73 (1976).
[e] T. L. Innerarity and R. W. Mahley. *Biochemistry* **17**, 1440 (1978); d 1.02–1.05 g/ml.
[f] G. J. Nelson, *Comp. Biochem. Physiol.* **46B**, 81 (1973).
[g] As C in Table VI; d 1.063–1.21 g/ml.
[h] As z in Table V; LDL-II, hydrated density 1.036 g/ml.
[i] As B in Table VI.
[j] V. Blaton and H. Peeters. *Adv. Exp. Med. Biol.* **67**, 33 (1976); d 1.063–1.21 g/ml.

following reports provide the reader with the basic features of fatty acid profile in the lipoprotein lipids of 12 mammals: suckling rat,[92] mature rat,[74,113,114] guinea pig,[115] Mongolian gerbil,[116] dog,[117,118] rabbit,[85,119] horse,[120] bovine species,[106,108] pig,[18,121] rhesus monkey,[122,123] cebus monkey,[18] baboon,[19,124,125] and chimpanzee.[126,127] Certain of these analyses are, however, incomplete and restricted to cholesteryl ester fatty acids[74,117,118,121] or a single lipoprotein class,[117,118,121] while others pertain to animals fed lipid-supplemented diets.[18,124,125]

The fatty acid profiles in the lipid esters of the nonhuman primate lipoproteins show an overall resemblance to those of the corresponding human fractions,[78] with minor differences. Bovine lipoprotein lipids are distinguished by their marked degree of sturation,[106,108] presumably reflecting hydrogenation in the rumen. The elevated level (50–60%) of linoleate (18:2) in the cholesteryl esters of the IDL, LDL_1, LDL_2, and HDL fractions also appears characteristic[106,108] of this species. Fatty acid analyses of cholesteryl esters in rat lipoproteins have revealed high contents of arachidonate (20:4).[74,113,114] By contrast, the cholesteryl esters of guinea pig lipoproteins lack elevated amounts of this fatty acid,[115] implying that the roles of LCAT, HDL, and the lipid transfer proteins are quite distinct in these two rodents. Thus, in the absence of a cholesteryl ester transfer system,[128] rat HDL function in the transport of essential fatty acids,

[113] G. D. Dunn, H. G. Wilcox, and H. Heimberg, *Lipids* **10**, 773 (1975).
[114] S. Mookerjea, C. E. Park, and A. Kuksis, *Lipids* **10**, 374 (1975).
[115] M. J. Chapman, G. L. Mills, and C. E. Taylaur, *Biochem. J.* **128**, 779 (1972).
[116] R. J. Nicolosi, J. A. Marlett, A. M. Morello, S. A. Flanagan, and D. M. Hegsted, *Atherosclerosis* **38**, 359 (1981).
[117] T. L. Innerarity and R. W. Mahley, *Biochemistry* **17**, 1440 (1978).
[118] C. Edelstein, L. A. Lewis, J. R. Shainoff, H. Naito, and A. M. Scanu, *Biochemistry* **15**, 1934 (1976).
[119] J. L. Rodriguez, G. C. Gishelli, D. Torregiani, and C. R. Sirtori, *Atherosclerosis* **23**, 73 (1976).
[120] M. Yamamoto, Y. Tanaka, and M. Sugano, *Comp. Biochem. Physiol.* **63B**, 441 (1979).
[121] H. J. Pownall, R. L. Jackson, R. I. Roth, A. M. Gotto, J. R. Patsch, and F. A. Kummerow, *J. Lipid Res.* **21**, 1108 (1980).
[122] G. M. Fless, R. W. Wissler, and A. M. Scanu, *Biochemistry* **15**, 5799 (1976).
[123] J. A. Lee and M. D. Morris, *Biochem. Med.* **10**, 245 (1974).
[124] H. Peeters, V. Blaton, B. Leclercq, A. N. Howard, and G. A. Gresham, *Atherosclerosis* **12**, 283 (1970).
[125] A. N. Howard, V. Blaton, N. Vandamme, N. Van Landschoot, and H. Peeters, *Atherosclerosis* **16**, 257 (1972).
[126] V. Blaton, R. Vercaemst, N. Vandecasteele, H. Caster, and H. Peeters, *Biochemistry* **13**, 1127 (1974).
[127] M. Rosseneu, B. Leclercq, D. Vandamme, R. Vercaemst, F. Soeteway, H. Peeters, and V. Blaton, *Atherosclerosis* **32**, 141 (1979).
[128] P. J. Barter and I. J. Lally, *Biochim. Biophys. Acta* **531**, 233 (1978).

whereas the rapid transfer of cholesteryl esters among guinea pig lipoproteins[84] suggests that this function is not the specific property of its HDL.

Carbohydrate Composition. Carbohydrates contribute to both the lipid and protein moieties of mammalian lipoproteins. Carbohydrate side chains are, however, mainly found attached to the principal apoprotein of mammalian LDLs, which are counterparts to human apoB (the B-100 form).[20,63,129] As shown in Table VIII, the apolipoproteins of mammalian LDLs are highly glycosylated, carbohydrate usually representing from 5 to 10% of their weight. Considerable variation in analyses of human LDL is first evident. Galactose, mannose, N-acetylglucosamine, and sialic acid are the major monosaccharides, however, not only in human LDL but also in the animal LDLs examined to date [analyses (a), (c), (d), and (e), Table VIII]. Sialic acid content was most variable (0.3% in marmoset, 0.4% in rat, and 2.1% in rhesus monkey LDL). The data of Pargaonkar *et al.*[130] reveal amounts of carbohydate [total 1.69–3.99%; analyses (f), Table VIII] which appear uniformly low; indeed, these authors detected less than half the amount of monosaccharides found in rhesus LDL by Fless and Scanu.[131] An explanation for this discrepancy may lie in the different methodologies applied.

Small amounts of carbohydrate have also been detected in the apoproteins of porcine and rat HDL (~2% by weight)[132,133] and in those of HDL_2 and HDL_3 (range 1.2–2.2 and 0.9–3.6%, respectively) in spider, rhesus, and patas monkeys, and in the baboon and chimpanzee.[130] The apolipoprotein(s) to which these sugars are attached remains to be established. It is relevant therefore that upon purification, carbohydrate has not been detected in the major protein, apoA-I (see section on Biochemical Characteristics of Mammalian Apolipoproteins), of pig, baboon, rhesus, and patas monkey, and chimpanzee HDL.[20] The presence of minor amounts of apoB could explain these findings, as could that of peptide(s) analogous to human apoC-III_1 and C-III_2.[134]

The carbohydrate component of the lipid moieties of mammalian lipoproteins arises from their content of glycosphingolipids, whose presence in human VLDL, LDL, and HDL was first described by Skipski.[100] The principal lipoprotein glycosphingolipids are glucosylceramide, lactosylceramide, trihexosylceramide, globoside, and hematoside, respec-

[129] S. Goldstein, M. J. Chapman, and G. L. Mills, *Atherosclerosis* **28**, 93 (1977).
[130] P. S. Pargaonkar, B. Radhakrishnamurthy, S. R. Srinivasan, and G. S. Berenson, *Comp. Biochem. Physiol.* **56B**, 293 (1977).
[131] G. M. Fless and A. M. Scanu, *Biochemistry* **14**, 1783 (1975).
[132] M. Janado, Y. Doi, J. Azuma, K. Onodera, and N. Kashimura, *Artery* **1**, 166 (1975).
[133] J. B. Marsh and R. Fritz, *Proc. Soc. Exp. Biol. Med.* **133**, 9 (1970).
[134] P. Vaith, G. Assmann, and G. Uhlenbruck, *Biochim. Biophys. Acta* **541**, 234 (1978).

TABLE VIII
Carbohydrate Composition of Mammalian LDL[a]

Species	Man[a']	Man[b]	Rat[c]	Pig[a']	Rhesus monkey[d]	Common marmoset[e]	Rhesus monkey[f]	Spider monkey[f]	Baboon[f]	Patas monkey[f]	Chimpanzee[f]
Monosaccharide											
Glucose	0.23	—	0.39	0.55	—	n.d.	0.14	0.18	0.13	0.19	0.09
Galactose	1.23	2.13	1.99	1.30	0.82	0.96	0.62	1.11	0.43	1.10	0.77
Mannose	2.38	4.88	1.71	2.18	1.92	1.61	0.60	1.02	0.35	0.87	0.72
N-Acetyl glucosamine	1.77	0.94	1.08	1.39	1.70	1.95	0.85	0.99	0.42	0.85	0.99
Fucose	—	—	4.08[g]	0.14	—	—	0.16	0.13	0.19	0.12	0.14
Sialic acid	0.51	1.73	0.44	0.55	2.13	0.27	0.50	0.56	0.17	0.34	0.70
Total	6.12	9.68	5.60	6.11	6.57	4.79	2.87	3.99	1.69	3.47	3.41

[a] n.d., not determined. Values are percentage monosaccharide per weight of apo-LDL protein.
[a'] A. D. Attie, D. B. Weinstein, H. H. Freeze, R. L. Pittman, and D. Steinberg, *Biochem. J.* **180**, 647 (1979).
[b] N. Swaminathan and F. Aladjem, *Biochemistry* **15**, 1516 (1976).
[c] J. B. Marsh and R. Fritz, *Proc. Soc. Exp. Biol. Med.* **133**, 9 (1970).
[d] G. M. Fless and A. M. Scanu, *Biochemistry* **14**, 1783 (1975).
[e] P. Forgez, M. J. Chapman, and G. L. Mills, *Biochim. Biophys. Acta* **754**, 321 (1983).
[f] P. S. Pargaonkar, B. Radhakrishnamurthy, S. R. Srinivasan, and G. S. Bererson, *Comp. Biochem. Physiol.* **56B**, 293 (1977); LDL of d 1.019–1.063 g/ml.
[g] As neutral hexose.

tively; the neutral forms (the three former) tend to predominate.[135,136] Quantitatively, glycosphingolipids may amount to up to ~9 μmol/g lipoprotein protein, are primarily located in LDL (~60% of total lipoprotein glycosphingolipids),[135,136] but account for only ~1% or less of total lipoprotein lipid.[100] As such, they represent about one-half a mol per mol of HDL and about 5 mol/mol of LDL.[135] To date, the only comparative data available are in the pig,[137] in which plasma glycosphingolipids are essentially equally distributed between LDL (d 1.020–1.060 g/ml) and HDL (d 1.090–1.21 g/ml), and in which galactosylgalactosylceramide and N-acetylgalactosaminylgalactosylgalactosylglucosylceramide predominate. Lower amounts of glycosphingolipids appear to be present in porcine VLDL[137] than in the human fraction.[135,136] Further comparative data on these potentially important constituents are clearly required.

Physical Properties

The physical properties of mammalian lipoproteins, like those of their human counterparts, directly reflect the chemical composition, particle structure, and molecular organization of the particles. Data on several physical properties of lipoproteins from a wide range of mammals are available, and include particle size, molecular weight, isoelectric point, and net (electrical) charge, as well as certain hydrodynamic parameters such as hydrated density, Stokes diameter, partial specific volume, diffusion and frictional coefficients, and flotation and sedimentation rates (S_f and S, respectively) in the analytical ultracentrifuge. Ideally, these parameters relate to the structure of lipoprotein particles in their native, hydrated form. However physicochemical analyses often involve examination in an atypical environment in which the hydration shell is not maintained; such is the case, for example, in negative stain electron microscopy, when particle size is assessed under essentially anhydrous conditions.

Other physicochemical parameters provide information on the physical state or configuration of specific components in the lipoprotein molecule. Such is the case for circular dichroism and optical rotatory dispersion, which may describe the conformation (e.g., α-helix, β-sheet etc.) of the protein moiety, while differential scanning calorimetry, nuclear magnetic resonance (NMR), and electron spin resonance spectroscopy may detail the organization, fluidity, and structural interactions of lipid domains and constituents. In addition, overall particle organization has been

[135] G. Dawson, A. W. Kruski, and A. M. Scanu, *J. Lipid Res.* **17,** 125 (1976).
[136] S. Chatterjee and P. O. Kwiterovich, *Lipids* **11,** 462 (1976).
[137] J. T. R. Clarke and J. M. Stoltz, *Can. J. Biochem.* **57,** 1229 (1979).

usefully studied by small angle X-ray scattering and NMR. Although such procedures have also been applied to structural investigations of recombinants of various lipids with purified mammalian apolipoproteins, and particularly in the case of bovine[138,139] and porcine[140,141] apoA-Is, the present discussion will be restricted to native lipoprotein particles.

Regrettably, comprehensive data on the physical properties of the major lipoprotein classes have been established in only a few mammals, among which the pig, cattle, and rhesus monkey are notable.

The physical properties of mammalian VLDL, LDL, and HDL and their major subfractions are summarized in Tables IX, X, and XI, respectively; each value corresponds to a mean for all of the particles comprising a given preparation, but does not reflect heterogeneity; as in the case of chemical composition then, the values closely reflect the properties of the most abundant molecular species present.

VLDL. Data on the physical properties of VLDL are meager (Table IX), and information on only eight mammals is available; notably absent are dog and rhesus monkey, although canine VLDL are 260–900 Å in diameter.[42] Average particle diameters resemble those in man and are typically in the range 300–500 Å; if one assumes that a similar apolar core/surface monolayer structure maintains in mammalian VLDL as in the human particles,[105] then the hydrated densities of the former suggest their molecular weights fall within the range ~10 to 35×10^6 (Table IX). Exceptions are the rat,[104] whose VLDL are larger (mean diameter 570 Å, MW 56.3×10^6), possibly reflecting the presence of chylomicron and/or VLDL remnants, and the newborn calf and steer whose VLDL are large (>600 Å) and of elevated hydrated density (~1.03 g/ml),[96] a finding apparently explained by the presence of surface remnants enriched in saturated fatty acids.[75,96]

IDL. Physicochemical studies of mammalian IDL have been neglected, and only the rather unusual bovine IDL[75] have been characterized from this viewpoint. Bovine IDL, isolated with d 1.006–1.020 g/ml at 16°, were highly heterogeneous, displaying hydrated densities of 1.013–1.024 g/ml, molecular weights from 19 to 260×10^6, diffusion coefficients of 0.33–0.82, and frictional coefficients of 1.3–1.4. Electron microscopy revealed both typical, spherical IDL particles as well as numerous large, asymmetrical and apparently flattened structures ranging in size from 500

---

[138] A. Jonas, D. J. Krajnovich, and B. W. Patterson, *J. Biol. Chem.* **252**, 2200 (1977).
[139] A. Jonas, S. M. Drengler, and B. W. Patterson, *J. Biol. Chem.* **255**, 2183 (1980).
[140] D. Atkinson, H. M. Smith, J. Dickson, and J. P. Austin, *Eur. J. Biochem.* **64**, 541 (1976).
[141] A. L. Andrews, D. Atkinson, M. D. Barratt, E. G. Finer, H. Hauser, R. Henry, R. B. Leslie, N. L. Owens, M. C. Phillips, and R. N. Robertson, *Eur. J. Biochem.* **64**, 549 (1976).

TABLE IX
PHYSICAL PROPERTIES OF MAMMALIAN VLDL[a]

Species	Density interval (g/ml)	Particle diameter (Å) Mean	Particle diameter (Å) Range	Molecular weight ×10^6	Electrophoretic mobility	Hydrated density (g/ml)	Stokes diameter (Å)	Peak S_f^0 rate	Apparent partial specific volume (cm³/g)
Man[a]	<1.006	364	250–500	—	Pre-β	—	—	—	—
Man[b]		—		10.2	Pre-β	0.998	319	38.6	1.002
		—		21.4	Pre-β	0.978	410	80.9	1.022
		—		37.8	Pre-β	0.962	494	133	1.039
Man[c]	<1.006	—		181.8	—	0.947	850	—	1.056
Mouse[d]	<1.017	494	270–750	—	VLDL/pre-β	0.971	—	—	1.030
Mouse[e]	<1.006	411		20	Pre-β	0.965	404	87	1.036
Rat[f]	<1.006	570	200–900	56.3	Pre-β	0.965	—	—	1.036
Guinea pig[g]	<1.006	463	306–775	—	Pre-β	0.969	—	—	1.032
Rabbit: NZW[h]	<1.006	440		—	Pre-β	0.9759	—	—	1.025
Rabbit: WHHL[h]	<1.006	355		—	Pre-β	0.9904	—	—	1.010
Cattle									
Newborn calf[i]	<1.006	613		—	—	1.031	—	—	0.9699
Steer[i]	<1.006	602		—	—	1.024	—	—	0.9766
Pig[j]	<1.007	—		19.2	—	0.961	—	—	1.041
Pig[k]	<1.007	493	260–780	16–18	VLDL	0.9662	—	68 (20–500)	1.035
Common marmoset[l]	<1.017	318	235–450	—	VLDL	0.9665	—	—	1.035
Chimpanzee[m]	<1.007	417	250–600	—	VLDL	0.9947	—	—	1.005

[a] Molecular weight (MW), hydrated density, Stokes diameter, and peak S_f^* rate determined by analytical ultracentrifugal studies unless otherwise stated. Particle diameter determined by electron microscopy. Partial specific volume calculated as reciprocal of hydrated density ($1/\rho$). Values for the hydrated density of individual components, used in calculation of the hydrated density of lipoprotein fractions from their chemical composition, were taken from T. Sata, R. J. Havel, and A. L. Jones, *J. Lipid Res.* **13**, 757 (1972), with the exception of protein which was taken as 1.373 g/ml [D. L. Puppione, S. T. Kunitake, R. L. Hamilton, M. L. Phillips, V. N. Schumaker, and L. D. Davis, *J. Lipid Res.* **23**, 283 (1982)].

[a] Normolipemic, fasting subjects; J. P. Kane, T. Sata, R. L. Hamilton, and R. J. Havel, *J. Clin. Invest.* **56**, 1622 (1975).

[b] Subfractions isolated from fasting normolipemic males; parameters (including diameter) calculated on the basis of chemical composition; Z. Kuchinskiene and L. A. Carlson, *J. Lipid Res.* **23**, 762 (1982).

[c] Hyperlipemic VLDL; parameters determined by laser scattering from VLDL centrifuged in a density gradient; S. T. Kunitake, E. Loh, V. N. Schumaker, S. K. Ma, C. M. Knobler, J. P. Kane, and R. L. Hamilton, *Biochemistry* **17**, 1936 (1978).

[d] M. C. Camus, M. J. Chapman, P. Forgez, and P. M. Laplaud, *J. Lipid Res.* **24**, 1210 (1983). Hydrated density calculated from chemical composition.

[e] Ascites tumor plasma lipoproteins; S. M. Mathur and A. A. Spector, *Biochim. Biophys. Acta* **424**, 45 (1976).

[f] O. D. Mjøs, O. Faergeman, R. L. Hamilton, and R. J. Havel, *J. Clin. Invest.* **56**, 603 (1975); properties calculated from chemical composition.

[g] M. J. Chapman and G. L. Mills, *Biochem. J.* **167**, 9 (1977); hydrated density calculated from chemical composition.

[h] R. J. Havel, T. Kita, L. Kotite, J. P. Kane, R. L. Hamilton, J. L. Goldstein, and M. S. Brown, *Arteriosclerosis* **2**, 467 (1982). Properties calculated from chemical composition.

[i] T. M. Forte, J. J. Bell-Quint, and F. Cheng, *Lipids* **16**, 240 (1981). Hydrodynamic properties calculated from chemical data.

[j] W. G. Martin and J. Takats, *Biochim. Biophys. Acta* **187**, 328 (1969).

[k] Heterogeneous; M. Janado, W. G. Martin, and W. H. Cook, *Can. J. Biochem.* **44**, 1201 (1966). Particle diameters from J. Azuma and T. Komano, *J. Biochem.* **83**, 1789 (1978).

[l] M. J. Chapman, F. McTaggart, and S. Goldstein, *Biochemistry* **18**, 5096 (1979). Hydrodynamic parameters calculated from chemical data.

[m] M. J. Chapman, P. Forgez, D. Lagrange, S. Goldstein, and G. L. Mills, *Atherosclerosis* **52**, 129 (1984). Hydrodynamic parameters calculated from chemical data.

TABLE X
PHYSICAL PROPERTIES OF MAMMALIAN LDL[a]

Species	Density interval (g/ml)	Particle diameter (Å)		Molecular weight ×10^6	Electrophoretic mobility	Hydrated density (g/ml)	Stokes diameter (Å)	Peak S_f^0 rate	Apparent partial specific volume (cm³/g)	f/f_0	Diffusion coefficient ($D^0_{20,w}$; Ficks)
		Mean	Range								
Man[a']		247	—	2.61	—	1.0492	—	3.71	0.953	1.28	
		203	—	2.89	—	1.0358	—	6.42	0.965	1.24	
		—	—	2.97	—	1.0267	—	9.49	0.974	1.10	
Man[b]	1.030–1.040	244	(d 1.036)[b']	2.45	β	1.0315	156	6.9[b']	0.9695	—	
	1.040–1.053	~306	(d 1.049)[b']	2.38	β	1.0335	154	4.2[b']	0.9676	1.12	
Mouse[c]	1.023–1.060		220–280	—	LDL/β	1.038	—	—	0.9634	—	
Mouse[d]	1.006–1.063	280		6.1	β	1.016	255	17.5	0.9843	—	
Suckling rat[e]	1.020–1.062	219		2.3	LDL	1.028	192	7.0	0.9728	—	
Adult rat[f]		214		2.36	β	1.031	205	—	0.9699	—	
Adult rat[g]	1.030–1.100	~200	178–267	—	—	1.039	—	—	0.9625	—	
Guinea pig[h]	1.02 –1.063	222	125–280	2.19	β	1.056	163	6.3	0.9470	—	
Dog[i]	1.019–1.063	244	160–250	3	β	—	—	—	—	—	
Rabbit: NZW[j]	1.019–1.063	210		—	β	1.056	—	—	0.9470	—	
Rabbit: WHHL[j]	1.02 –1.03	262	200–220	—	β	1.047	—	—	0.9551	—	
Killer whale[k]	<1.063			—	β	—	—	7.5	—	—	
Sheep[l]		260		—	β	1.043	230–300	6.2	0.9588	—	
Cattle											
Fetal calf[m]	1.006–1.063	194		—	—	1.057	—	8.9	0.9461	—	
Newborn calf[n]	1.006–1.063	188		—	—	1.077	—	4.9	0.9285	—	
Steer[m]	1.006–1.063	—		—	—	1.058	—	5.3	0.9452	—	
Dairy cow[n]	1.019–1.039	—		1.14:1.03	β/α	1.032	—	—	0.969	—	
Dairy cow[n]	1.039–1.060			2.37:2.15	β	1.044	—	—	0.958	—	
Pig[o]											
LDL₁	1.020–1.060	221 (274)[p']	195–250	1.60	β	0.9888	173	11.98	1.0113	—	1.57
LDL₂	1.060–1.090	215 (258)[p']	195–250	1.41	β	1.0310	163	4.60	0.9699	—	1.79

Species	Density range	MW	Stokes diameter	S_f^0 rate	Partial sp. vol.	Hydrated density	Mobility	Particle diam.			
Pig[p]		217	210–225	2.6		1.035	β	—	4.9	0.9662	—	—	1.83
		195	193–200	2.0		1.050	β	—	1.8	0.9524	—	—	
Common marmoset[q]	1.027–1.055	210	180–270	—		1.048	β	—	—	0.9542	—	—	2.50
Rhesus monkey[r]	1.019–1.050	—		3.12		1.032	β	222	—	0.969	1.08	1.78
Rhesus monkey[s]	1.019–1.050	196		2.18–2.73 (2.39)		1.042	β	198	—	0.960	1.02	1.07
LDL-I[t]		234		3.32		1.027	β	217	—	0.974	1.07	2.00
LDL-II		212		2.75		1.036	β	205	—	0.965	1.04	1.59
LDL-III		240		3.47		1.050	Pre-β	222	—	0.952	1.20	
Baboon[u]	1.024–1.045	245		—		1.054	β	—	6.8–7.9	0.9488	—	—
Cynomolgus monkey[v]	—	—		3.29		1.054	β	215	—	0.9488	—	—
Chimpanzee[w]	1.024–1.050	220	160–280	—		1.053	β	—	7.7	0.9497	—	—

[a] Molecular weight (MW), hydrated density, Stokes diameter, peak S_f^0 rate, and frictional coefficient (f/f_0) determined by analytical ultracentrifugal studies unless otherwise stated. Particle diameter determined by electron microscopy. Partial specific volume ($\bar{v}$) calculated as reciprocal of hydrated density ($1/\rho$). When unavailable, hydrated densities were calculated from chemical compositions as described in Table IX.

[a'] T. S. Kahlon, G. L. Adamson, M. M. S. Shen, and F. T. Lindgren, *Lipids* **17**, 323 (1982); properties determined by sedimentation equilibrium; subfractions isolated by density gradient centrifugation.

[b] D. M. Lee and P. Alaupovic, *Biochemistry* **9**, 2244 (1970).

[b'] Data from M. M. S. Shen, R. M. Krauss, F. T. Lindgren, and T. M. Forte, *J. Lipid Res.* **22**, 236 (1981) for similar fractions.

[c] M. C. Camus, M. J. Chapman, P. Forgez, and P. M. Laplaud, *J. Lipid Res.* **24**, 1210 (1983); hydrated density calculated from chemical composition.

[d] Ascites tumor plasma lipoproteins; S. N. Mathur and A. A. Spector, *Biochim. Biophys. Acta* **424**, 45 (1976).

[e] G. J. P. Fernando-Warnakulasuriya, M. L. Eckerson, W. A. Clark, and M. A. Wells, *J. Lipid Res.* **24**, 1626 (1983); HDL$_1$ absent.

[f] L. T. Lusk, L. F. Walker, L. H. DuBien, and G. S. Getz, *Biochem. J.* **183**, 83 (1979); HDL$_1$ absent; MW and Stokes diameter determined by agarose gel chromatography, hydrated density from isopycnic gradient centrifugation.

[g] Y. Oschry and S. Eisenberg, *J. Lipid Res.* **23**, 1099 (1982). Hydrodynamic properties calculated from chemical composition; HDL$_1$ absent; isolated by rate zonal ultracentrifugation.

[h] G. L. Mills, M. J. Chapman, and F. McTaggart, *Biochim. Biophys. Acta* **260**, 401 (1972). Particle diameter, for LDL of 1.006–1.100 g/ml, from M. J. Chapman and G. L. Mills, *Biochem. J.* **167**, 9 (1977).

[i] T. L. Innerarity and R. W. Mahley, *Biochemistry* **17**, 1440 (1978). MW determined by gel filtration chromatography; HDL$_1$ removed by Pevikon block electrophoresis.

[j] R. J. Havel, T. Kita, L. Kotite, J. P. Kane, R. L. Hamilton, J. L. Goldstein, and M. S. Brown, *Arteriosclerosis* **2**, 467 (1982). Hydrodynamic properties calculated from chemical composition. NZW, New Zealand White strain; WHHL, Watanabe heritable hyperlipemic rabbit.

(continued)

Footnotes to TABLE X (*continued*)

[k] D. L. Puppione, T. Forte, and A. V. Nichols, *Comp. Biochem. Physiol.* **39B**, 673 (1971).
[l] LDL heterogeneous by gradient gel electrophoresis; hydrated density calculated from chemical composition; data taken from T. M. Forte, C. E. Cross, R. A. Gunther, and G. C. Kramer, *J. Lipid Res.* **24**, 1358 (1983), from W. M. F. Leat, F. O. T. Kubasek, and N. Buttress, *Q. J. Exp. Physiol.* **61**, 193 (1976), and from G. L. Mills and C. E. Taylaur, *Comp. Biochem. Physiol.* **40B**, 489 (1971).
[m] T. M. Forte, J. J. Bell-Quint, and F. Cheng, *Lipids* **16**, 240 (1981): hydrodynamic values calculated from chemical data.
[n] D. Stead and V. A. Welch, *J. Dairy Sci.* **59**, 9 (1976). Values for MW determined by approach to equilibrium and by high-speed equilibrium methodologies in the analytical ultracentrifuge.
[o] R. L. Jackson, O. D. Taunton, R. Segura, J. G. Gallagher, H. F. Hoff, and A. M. Gotto, *Comp. Biochem. Physiol.* **53B**, 245 (1976).
[o'] Determined by laser self-beat spectroscopy.
[p] Subfractions isolated by zonal ultracentrifugation. G. D. Calvert and P. J. Scott, *Atherosclerosis* **22**, 583 (1975). MW estimated on subfractions of d 1.020–1.050 and 1.050–1.090 g/ml, respectively, by X-ray small angle scattering [G. Jurgens, G. M. J. Knipping, P. Zipper, R. Kayushina, G. Degovics, and P. Laggner, *Biochemistry* **20**, 3231 (1981)].
[q] M. J. Chapman, F. McTaggart, and S. Goldstein, *Biochemistry* **18**, 5096 (1979): hydrated density determined from chemical composition.
[r] G. M. Fless, R. W. Wissler, and A. M. Scanu, *Biochemistry* **15**, 5799 (1976): diffusion coefficient determined at infinite dilution, $D^0_{25,b}$.
[s] G. M. Fless and A. M. Scanu, *Biochemistry* **14**, 1783 (1975): mean MW given in parenthesis.
[t] G. M. Fless and A. M. Scanu, *J. Biol. Chem.* **254**, 8653 (1979) subfractions isolated by a combination of isopycnic and rate zonal density gradient ultracentrifugation.
[u] M. J. Chapman and S. Goldstein, *Atherosclerosis* **25**, 267 (1976). Hydrated density calculated from chemical composition.
[v] LDL isolated by agarose gel chromatography: hydrated density calculated from chemical data: A. R. Tall, D. M. Small, D. Atkinson, and L. L. Rudel, *J. Clin. Invest.* **62**, 1354 (1978).
[w] Hydrated density calculated from chemical data: M. J. Chapman, P. Forgez, D. Lagrange, S. Goldstein, and G. L. Mills, *Atherosclerosis* **52**, 129 (1984).

TABLE XI
PHYSICAL PROPERTIES OF MAMMALIAN HDL[a]

Species	Density interval (g/ml)	Particle diameter (Å) Mean	Particle diameter (Å) Range	Molecular weight ×10⁵	Electrophoretic mobility	Hydrated density (g/ml)	Stokes diameter (Å)	Peak $F^0_{1.20}$ rate	Apparent partial specific volume (cm³/g)	f/f_0	Diffusion coefficient ($D^0_{20,w}$; Ficks)
Man[a]											
HDL₂	1.063–1.125	100		3.60	α	1.105	—	5.45a^2	0.905	1.13	3.68
HDL₃	1.125–1.210	75		1.75	α	1.153	—	4.65a^2	0.867	1.31	3.93
Man[b]	1.090–1.098	120		4.10	HDL	—	110/105[b']	5.34	0.914	—	—
	1.108–1.112	101		2.62	HDL	—	102/92[b']	3.14	0.901	—	—
	1.130–1.145	89		1.77	HDL	—	92/79[b']	1.56	0.879	—	—
Mouse[c]	1.060–1.163	—		—	HDL/α	1.154	—	4.2	0.867	—	—
Mouse[d]	1.063–1.21	~131		4.4	α	1.098	109	—	0.911	—	—
Mouse[e]	1.063–1.21	—		4.0	α	~1.100	—	—	0.909	—	—
Mouse[f]	—	—		2.34	α	1.167	—	—	0.857	—	—
Suckling rat[g] HDL₁		—		12.8	HDL	1.034	157	10.5 (3.5–17.5)	0.967	—	—
Adult rat											
HDL₁[h]		200		13.0	α	1.054	168	—	0.9488	—	—
HDL[h']		—		8.8	α	1.156	148	4.4 (0.5–8.5)	0.865	—	—
Adult rat											
HDL₁[i]	1.085–1.21	127	89–178	—	—	1.072	—	—	0.933	—	—
HDL₂[i]	1.063–1.21	92	70–118	—	—	1.124	—	—	0.890	—	—
HDL₂[j]	1.063–1.21	—	—	2.46	α₁	1.102	—	4.0	0.907	—	—
HDL₂[k]	1.063–1.125	136	—	6.03	Prealbumin	1.119	—	7.1	0.894	—	—
HDL₃[l]	1.125–1.21	50–60		—	—	—	—	—	—	—	—
Guinea pig[m]	1.090–1.21	97		—	α	1.115	—	—	0.897	—	—
Dog											
HDL[n]	1.063–1.21	90		2.3(2.13)	α₁	1.117	—	—	0.895	1.108	4.07
HDL₁[o]	1.025–1.10	—	110–350	—	α₂	—	—	—	—	—	—
HDL₂[o]	1.087–1.21	—	55–85	—	α₁	—	—	—	—	—	—

(continued)

TABLE XI (continued)

Species	Density interval (g/ml)	Particle diameter (Å) Mean	Particle diameter (Å) Range	Molecular weight ×10^5	Electrophoretic mobility	Hydrated density (g/ml)	Stokes diameter (Å)	Peak $F^0_{1.20}$ rate	Apparent partial specific volume (cm³/g)	f/f_0	Diffusion coefficient ($D^0_{20,w}$; Ficks)
Rabbit[p]	1.063–1.21	—	90–150	—	α	1.149	—	—	0.870	—	—
Killer whale[q]	1.063–1.21	—	70–130	—	α	—	—	2.8	—	—	—
Harbor seal[r] HDL_1	1.04–1.06	140		12	α	1.138	—	—	—	—	—
Sheep[s]	1.063–1.21	94		2.7	HDL	—	86–89	—	0.879	—	—
Cattle											
Fetal calf[t]	1.063–1.21	82		—	—	1.155	—	2.0	0.866	—	—
Newborn calf[t]	1.063–1.21	90		—	—	1.136	—	3.2	0.880	—	—
Steer[t]	1.063–1.21	126		—	—	1.103	—	6.9	0.907	—	—
Lactating dairy cow[u] HDL_1	1.04–1.06	160		—	α	—	—	—	—	—	—
Dairy cow[v]	1.060–1.21	—		5.76 / 5.67	HDL	1.071	—	—	0.934	—	—
Cow (heifer)[w]	1.063–1.125	—		3.76	HDL	1.099	—	—	0.910	1.08	—
Lactating dairy cow[x]	1.069	—		7.44	α	1.074	65	8.59	0.931	1.15	2.86
	1.083	—		4.89	α	1.082	57	6.72	0.924	1.07	3.53
	1.100	—		3.94	α	1.106	52	5.45	0.904	1.15	3.56
Pig[y] HDL_3	1.11–1.20	100–110	70–130	2.09	α	1.134	—	—	0.882	—	4.9 (D^0_{25})
Pig[z] HDL	1.060–1.17	—		2.54	—	—	—	2.83	—	—	2.0
Common marmoset[A]											
HDL_2	1.070–1.127	100	90–126	—	α	1.111	—	—	0.900	—	—
HDL_3	1.127–1.156	69	57–85	—	α	1.182	—	—	0.846	—	—
Squirrel monkey[B]	1.063–1.21	—		—	—	1.170	—	2.9	0.855	—	—
Cebus monkey[C]	1.068–1.170	40–80 (HDL_3)		—	α	1.154	—	2.3	0.866	—	—
Rhesus monkey[D]											
HDL_2	1.063–1.125	110		3.90	α	1.099	—	—	0.91	1.02	3.60
HDL_3	1.125–1.21	75		1.97	α	1.149	—	—	0.87	1.15	3.96

| Baboon[E] | | | | |
| HDL$_{2b}$/HDL$_{2a}$/HDL$_3$ | 104–120 (HDL$_{2b}$) | — | 1.110/1.125/1.15 | — | 0.90/0.89/0.87 | — |

[a] Molecular weight (MW), hydrated density, Stokes diameter, peak flotation rate, and frictional coefficient (f/f_0) determined by analytical ultracentrifugal studies unless otherwise stated. Particle diameter determined by electron microscopy. Stokes diameter, Stokes diameter, peak flotation rate, and frictional coefficient (f/f_0) calculated as reciprocal of hydrated density ($1/\rho$). When unavailable, hydrated densities calculated from chemical composition as detailed in Table IX.
[a'] A. M. Scanu, L. Vitello, and S. Deganello, *CRC Crit. Rev. Biochem.* **2**, 175 (1974); $\bar{v}$$S^0_{w,20}$.
[b] D. W. Anderson, A. V. Nichols, T. M. Forte, and F. T. Lindgren, *Biochim. Biophys. Acta* **493**, 55 (1977). Density interval values are gradient banding densities. Stokes diameters determined by gradient gel electrophoresis and by gel ultracentrifugation (b'). Partial specific volumes calculated from mean banding densities.
[c] M. C. Camus, M. J. Chapman, P. Forgez, and P. M. Laplaud, *J. Lipid Res.* **24**, 1210 (1983): Random-bred. Swiss mice; hydrated density calculated.
[d] Ascites tumor plasma lipoproteins; S. N. Mathur and A. A. Spector, *Biochim. Biophys. Acta* **424**, 45 (1976).
[e] R. C. LeBoeuf, D. L. Puppione, V. N. Schumaker, and A. J. Lusis, *J. Biol. Chem.* **258**, 5063 (1983). Random-bred. Swiss mice. Isopycnic gradient banding density taken as hydrated density.
[f] Hydrated density calculated from chemical composition; C57BR/cdJ and CBA/J strain mice; J. D. Morrisett, H. S. Kim, J. R. Patsch, S. K. Datta, and J. J. Trentin, *Arteriosclerosis* **2**, 312 (1982). Isolation by rate zonal ultracentrifugation.
[g] J. P. Fernando-Warmakulasuriya, M. L. Eckerson, W. A. Clark, and M. A. Wells, *J. Lipid Res.* **24**, 1626 (1983); isolated by vertical rotor centrifugation; hydrated density from isopycnic banding position.
[h] L. T. Lusk, L. F. Walker, L. H. DuBien, and G. S. Getz, *Biochem. J.* **183**, 83 (1979); isopycnic density gradient separation; hydrated density as in g; Sprague–Dawley strain.
[h'] Hydrated density calculated; peak $F^0_{1.20}$ rate taken from g in Sprague–Dawley strain. MW in h and h' determined by agarose gel chromatography.
[i] Fasted rats; zonal rotor separation: hydrated density calculated; Y. Oschry and S. Eisenberg, *J. Lipid Res.* **23**, 1099 (1982).
[j] G. Camejo, *Biochemistry* **6**, 3228 (1967). Sprague–Dawley strain.
[k] S. Koga, D. L. Horwitz, and A. M. Scanu, *J. Lipid Res.* **10**, 577 (1969).
[l] Wistar strain; I. Pasquali-Ronchetti, S. Calandra, M. Baccarani-Conti, and M. Montaguti, *J. Ultrastruct. Res.* **53**, 180 (1975).
[m] Hydrated density calculated: C. Sardet, H. Hansma, and R. Ostwald, *J. Lipid Res.* **13**, 624 (1972).
[n] C. Edelstein, L. L. Lewis, J. R. Shainoff, H. Naito, and A. M. Scanu, *Biochemistry* **15**, 1934 (1976).
[o] R. W. Mahley and K. H. Weisgraber, *Circ. Res.* **35**, 713 (1974); HDL$_c$ isolated by ultracentrifugation and Pevikon-block electrophoresis.
[p] E. Stange, B. Agostini, and J. Papenberg, *Atherosclerosis* **22**, 125 (1975); hydrated density calculated from the data of G. L. Mills and C. E. Taylaur, *Comp. Biochem. Physiol.* **40B**, 489 (1971).
[q] D. L. Puppione, T. Forte, and A. V. Nichols, *Comp. Biochem. Physiol.* **39B**, 673 (1971); $F^0_{1.20}$ value is mean from males and females.
[r] D. L. Puppione, G. M. Forte, A. V. Nichols, and E. H. Strisower, *Biochim. Biophys. Acta* **202**, 392 (1970).
[s] T. M. Forte, C. E. Cross, R. A. Gunther, and G. C. Kramer, *J. Lipid Res.* **24**, 1358 (1983); MW and hydrated density of lamb HDL from W. M. F. Leat, F. O. T. Kubasek, and N. Buttress, *Q. J. Exp. Physiol.* **61**, 193 (1976).
[t] T. M. Forte, J. J. Bell-Quint, and F. Cheng, *Lipids* **16**, 240 (1981); hydrated density calculated from chemical data.
[u] D. L. Puppione, G. M. Forte, A. V. Nichols, and E. H. Strisower, *Biochim. Biophys. Acta* **202**, 392 (1970); S^0_f rate of peak. 1.0.
[v] D. Stead and V. A. Welch, *J. Dairy Sci.* **59**, 9 (1976); values for MW determined as n in Table X.
[w] A. Jonas, *J. Biol. Chem.* **247**, 7767 (1972).

(*continued*)

Footnotes to TABLE XI (*continued*)

^{x} D. L. Puppione, S. T. Kunitake, M. L. Toomey, E. Loh, and V. N. Schumaker, *J. Lipid Res.* **23**, 371 (1982); fractions 3, 4, and 5 (representing 22.2, 27.9, and 18.3%, respectively, of total bovine HDL of d 1.050–1.21 g/ml) isolated by isopycnic density gradient centrifugation from Holstein cows; banding densities given for "density interval"; $F_{1.20}^0$ values from Guernsey HDL subfractions. Overall particle size range 120–160 Å [A. R. Tall, D. L. Puppione, S. T. Kunitake, D. Atkinson, D. M. Small, and D. Waugh, *J. Biol. Chem.* **256**, 170 (1981)].

^{y} A. C. Cox and C. Tanford, *J. Biol. Chem.* **243**, 3083 (1968); average particle diameter from M. A. F. Davis, R. Henry, and R. B. Leslie, *Comp. Biochem. Physiol.* **47B**, 831 (1974); range of particles of d 1.09–1.21 g/ml HDL from R. W. Mahley and K. W. Weisgraber, *Biochemistry* **13**, 1964 (1974).

^{z} M. Janado, W. G. Martin, and W. H. Cook. *Can. J. Biochem.* **44**, 1201 (1966); $F_{1.20}^0$ rate from T. M. Forte, R. W. Nordhausen, A. V. Nichols, G. Endemann, P. Miljanich, and J. J. Bell-Quint, *Biochim. Biophys. Acta* **573**, 451 (1979).

^{A} M. J. Chapman, F. McTaggart, and S. Goldstein. *Biochemistry* **18**, 5096 (1979); hydrated density (and $\bar{v}$) calculated; density intervals are gradient banding densities.

^{B} $F_{1.20}^0$ rate from P. Hill, W. G. Martin, and J. F. Douglas, *Proc. Soc. Exp. Biol. Med.* **148**, 41 (1975); hydrated density calculated from data of O. W. Portman, M. Alexander, N. Tanaka, and D. R. Illingworth, *Biochim. Biophys. Acta* **486**, 470 (1977).

^{C} $F_{1.20}^0$ rate as in B; density interval values are gradient banding densities from R. J. Nicolosi and K. C. Hayes, *in* "Handbook of Electrophoresis" (L. A. Lewis and H. K. Naito, eds.), Vol. IV, p.33. CRC Press, Boca Raton, Fla., 1983.

^{D} A. M. Scanu, C. Edelstein, L. Vitello, R. Jones, and R. Wissler, *J. Biol. Chem.* **218**, 7648 (1973).

^{E} A. W. Kruski, *in* "Handbook of Electrophoresis," (L. A. Lewis and H. K. Naito, eds.) Vol. IV, p. 61. CRC Press, Boca Raton, Fla., 1983; banding densities for HDI subfractionation in density gradient not reported.

to 2000 Å with average dimension of ~1000 Å. The appearance of such particles appears to be related to their elevated content of saturated triglyceride fatty acids, which may be partially crystallized below body temperature. In this case, crystallization may rupture the "normal" triglyceride-rich particle to give rise to structures akin to surface remnants.

LDL. In normolipidemic man, LDL particles range in diameter from ~270 to 203 Å and in molecular weight from 2.89 to 1.88 × 10^6 over the density range 1.0244 to 1.0597 g/ml.[142] The physical properties of the 14 mammalian LDLs detailed in Table X fall within these limits, with the possible exception of the murine ascites and bovine (d 1.019–1.039 g/ml) fractions. Thus, the molecular weight (6.1 × 10^6) of murine ascites plasma LDL is elevated; the relationship of such particles to normal murine plasma LDL is however indeterminate. By contrast, the molecular weight of bovine (dairy cow) LDL_1 of d 1.019–1.039 g/ml (1.03–1.14 × 10^6) seems about 50% lower than would be anticipated on the basis of its hydrated density and partial specific volume. Some inconsistency is evident in the porcine LDL data, in which case molecular weights are 30–40% less for subfractions of d 1.020–1.060 (LDL_1) and 1.060–1.090 g/ml (LDL_2) (1.60 and 1.41 × 10^6, respectively)[48] than for those of d 1.020–1.050 and 1.050–1.090 g/ml (2.6 and 2.0 × 10^6, respectively)[143] when determined by different procedures (ultracentrifugal flotation and X-ray small angle scattering, respectively). The marked dependence of the molecular diameter of porcine LDL_1 and LDL_2 on analytical methodology (range of means: 173 to 274 and 163 to 258 Å, respectively) is also noteworthy.[48,143] Clearly then, data on the physical parameters of mammalian lipoproteins should be assessed critically in the light of the methods and physical constants employed in their determination.

As originally observed by Mills and Taylaur,[23] the peak S_f^0 rates of mammalian LDLs vary considerably, with the majority between 4 and 9 (Table X); with only rare exception (suckling rat,[92] newborn calf,[96] porcine LDL_1[48]), their hydrated densities range between ~1.030 and 1.055 g/ml.

HDL. In spite of extensive investigation of mammalian HDL, data on their physical properties are often fragmentary. Of the 15 mammals presented in Table XI, only the characterization of bovine, rhesus monkey, and possibly rat HDL is comprehensive. Indeed, even in the latter species, it remains unclear as to whether the physicochemical parameters are

---

[142] M. M. S. Shen, R. M. Krauss, F. T. Lindgren, and T. M. Forte, *J. Lipid Res.* **22,** 236 (1981).

[143] G. Jurgens, G. M. J. Knipping, P. Zipper, R. Kayushina, G. Degovics, and P. Laggner, *Biochemistry* **20,** 3231 (1981).

consistent with a single class of relatively homogeneous HDL particles ("HDL_2"),[74] or whether they are more compatible with two distinct subclasses (HDL_2 and HDL_3)[11,144] as in man and nonhuman primates.[20,78]

In man, high-density lipoproteins are polydisperse macromolecules, ranging in hydrated density from ~1.090 to 1.150 g/ml, in molecular weight from ~1.75 to 4.1×10^5, in molecular diameter from ~75 to 120 Å, and in partial specific volume from ~0.87 to 0.91 g/ml and whose frictional coefficients show them to be slightly asymmetric in shape (f/f_0 1.1–1.3) (Table XI). Among these particles, two major subclasses exist, i.e., HDL_2 and HDL_3.[78] By contrast, the profile of most mammalian HDL tends to be dominated by one class of particle which is relatively homogeneous in properties and structure, and which may be akin to either HDL_2 or HDL_3, or indeed intermediate between them. Murine plasma HDL most resemble human HDL_3 in molecular weight, hydrated density, and $\bar{v}$[41,110] although some findings are at variance with this suggestion.[5] Ovine and porcine HDL appear to be counterparts to HDL_3, as do those of pinnipeds, such as the California sea lion and harbor seal[16]; the limited data in rabbit also tend to this conclusion. Only the lactating dairy cow and steer display HDL with properties distinguishing them as counterparts to HDL_2; indeed, the progressive modification in HDL profile and physical properties with development and growth from HDL_3-type particles in the fetal and newborn calf to HDL_2 in the mature animal is of particular interest. Animals in which the principal HDL particles possess properties intermediate between those of HDL_2 and HDL_3 include the rat, guinea pig, dog, and cetaceans (Killer whale and dolphin).[16] The discordant findings in rat are especially notable, and may be related to strain and hence genetic factors, as recently demonstrated in inbred strains of mice.[6]

Only in the nonhuman primates are discrete particles with the hydrodynamic properties of HDL_2 and HDL_3 consistently seen. In spite, however, of their identification on the basis of flotational and electrophoretic behavior in the Common marmoset, squirrel, cebus, rhesus, cynomolgus, spider, owl, tamarin, pigtail, and African green (grivet) monkeys, baboon, and chimpanzee,[17–19,43,50,89,126,145] their physical properties have been evaluated in detail solely in the rhesus, in which case they are quite comparable to their human counterparts.

Physicochemical values on a further subclass of HDL, i.e., HDL_1, are available in the suckling and adult rat,[40,92] dog,[13] harbor seal,[37] and lactating dairy cow.[37] These particles commonly display lower hydrated density

---

[144] I. Pasquali-Ronchetti, S. Calandra, M. Baccarani-Contri, and M. Montaguti, *J. Ultrastruct. Res.* **53**, 180 (1975).

[145] L. L. Rudel and H. B. Lofland, *Primates Med.* **9**, 244 (1976).

($\sim$1.03–1.07 g/ml) and larger diameters ($\sim$120–300 Å) and molecular weights ($\sim$12–13 × 10^5) (Table XI) than classical HDL and as such are primarily distributed over the "low-density" (1.006–1.063 g/ml) range. Finally, the relatively uniform electrophoretic mobility (i.e., pre-β, β, or α) within each class of mammalian lipoprotein attests to their similarity in net electric charge, but at most represents a semiquantitation estimate of this important parameter. More precise analysis such as isoelectric focusing is required, an approach used by Azuma and Komano[146] to evaluate the surface charge of porcine VLDL, and whose isoelectric points ranged from pH $\sim$4.5 to 5.5.

Physicochemical Studies of Native Lipoprotein Structure

These investigations have been essentially restricted to rat lipoproteins (negative- and positive-stain electron microscopy[144]), rabbit VLDL, LDL, and HDL (fluorescence polarization analysis[147,148]), bovine HDL (small angle X-ray scattering, differential scanning calorimetry, and negative stain electron microscopy,[149] fluorescence polarization analysis,[150,151] intrinsic fluorescence and difference absorption spectra[152]), porcine VLDL, LDL, and HDL (nuclear magnetic resonance studies,[153,154] pyrene fluorescence and electron paramagnetic resonance studies,[154,155] differential scanning calorimetry,[121] X-ray small angle scattering,[143] circular dichorism, and optical rotatory dispersion[156]), rhesus monkey LDL (circular dichroism[109] and negative stain electron microscopy[131]) and HDL (circular dichroism[157]), cynomolgus monkey LDL (calorimetry and X-ray scattering[158]), and chimpanzee lipoproteins (fluorescence polarization

[146] J. Azuma and T. Komano, *J. Biochem.* **83**, 1789 (1978).
[147] F. J. Castellino, J. K. Thomas, and V. A. Ploplis, *Biochem. Biophys. Res. Commun.* **75**, 857 (1977).
[148] E. Berlin and C. Young, *Atherosclerosis* **35**, 229 (1980).
[149] A. R. Tall, D. L. Puppione, S. T. Kunitake, D. Atkinson, D. M. Small, and D. Waugh, *J. Biol. Chem.* **256**, 170 (1981).
[150] A. Jonas and R. W. Jung, *Biochem. Biophys. Res. Commun.* **66**, 651 (1975).
[151] A. Jonas, *J. Biol. Chem.* **247**, 7773 (1972).
[152] A. Jonas, *Biochemsitry* **12**, 4503 (1973).
[153] N. Fidge, *Biochim. Biophys. Acta* **295**, 258 (1973).
[154] E. G. Finer, R. Henry, R. B. Leslie, and R. N. Robertson, *Biochim. Biophys. Acta* **380**, 320 (1075).
[155] R. L. Jackson, J. D. Morrisett, H. J. Pownall, A. M. Gotto, A. Kamio, H. Imai, R. Tracy, and F. A. Kummerow, *J. Lipid Res.* **18**, 182 (1977).
[156] J. Azuma, N. Kashimura, and T. Komano, *J. Biochem.* **83**, 1533 (1978).
[157] A. M. Scanu, C. Edelstein, L. Vitello, R. Jones, and R. Wissler, *J. Biol. Chem.* **248**, 7648 (1973).
[158] A. R. Tall, D. M. Small, D. Atkinson, and L. L. Rudel, *J. Clin. Invest.* **62**, 1354 (1978).

analysis[127]). The microviscosities of the lipid domains of rabbit, bovine, and porcine lipoproteins differed little from those of the corresponding human particles,[159] apart from the elevated microviscosity of bovine HDL, indicative of significant lipid–protein interaction.[150,159] The modulation of lipid fluidity as a function of the fatty acid content of the neutral lipids is relevant.[148,155,160] Both differential scanning calorimetry and X-ray scattering showed that porcine[121,143] and cynomolgus monkey[158] LDLs undergo reversible thermotropic transitions as in man,[161] corresponding to the change in cholesteryl esters from an ordered smectic-like layered structure to a more disordered state; the absolute temperature of the peak of the transition varied according to species and diet, but was in the range ~24–38° for animals on control, low-cholesterol diets. A similar transition was detected in large bovine HDL of diameter 120–160 Å,[149] thereby suggesting that these particles possess a similar core lipid organization to that found in LDL.[161] In sum, these findings argue for a common overall structural organization (surface polar monolayer/apolar lipid core) in corresponding classes of human and mammalian lipoproteins, along the lines proposed for the human substances by Shen et al.,[162] although an alternative and unique "protein-core" model has been suggested for porcine LDL on the basis of NMR studies.[154]

Immunological Properties

Interspecies Relationships. The immunological reactivity of mammalian plasma lipoproteins (as opposed to their constitutive apolipoproteins) was originally examined as a means of evaluating the interspecies relationships of these macromolecules.[29,163] By immunodiffusion and lipid staining, Delcourt[29] demonstrated the extensive cross-reactivity of anti sera to human, to rhesus monkey, and to bovine serum with lipoproteins in sera from a wide range of animals (15 species in the case of anti-human and 12 in the case of anti-rhesus and anti-bovine serum, most of which were ungulates). By contrast, antisera to rabbit, horse, sheep, and goat sera were of restricted cross-reactivity, reacting with lipoproteins in sera from 3 to 6 other mammals in the same series. Clearly then, the lipoproteins of human, rhesus, and cow possess a more extensive and diverse range of antigenic sites than their counterparts in rabbit, horse, sheep, and goat. A rather similar conclusion could be drawn from reaction of 10

---

[159] A. Jonas, *Biochim. Biophys. Acta* **486**, 10 (1977).
[160] F. Schroeder, E. H. Goh, and M. Heimberg, *J. Biol. Chem.* **254**, 2456 (1979).
[161] R. J. Deckelbaum, G. G. Shipley, and D. M. Small, *J. Biol. Chem.* **252**, 744 (1977).
[162] B. W. Shen, A. M. Scanu, and F. J. Kezdy, *Proc. Natl. Acad. Sci. U.S.A.* **74**, 837 (1977).
[163] R. K. W. Walton and S. J. Darke, *Protides Biol. Fluids, Proc. 10th Colloq.* 146 (1969).

mammalian sera with an antiserum to human β-lipoprotein.[163] More recently, the cross-reactivities of LDL from human, rhesus, baboon, pig, and guinea pig were evaluated by immunodiffusion and by a quantitative immunoprecipitation procedure, using antisera to LDL from each species.[164] The cross-reactivity between the human and animal LDLs ranged from 80 to 88% in the Old World monkeys to 36 to 58% in the pig and 26 to 37% in the guinea pig, and were thus consistent with their taxonomic classification. On immunodiffusion, the appearance of a spur beyond the point of coalescence of the precipitin line formed by the animal and by the human fraction attested to their partial immunological identity; this spur became more pronounced as the species became more distant from man. Marmoset LDL and HDL also display a partial immunological identity with their human counterparts.[43] The essentially complete identity (estimated as 85–97% by microimmunoprecipitation) of chimpanzee and human LDL is remarkable, and reflects the marked structural resemblence in their major apoprotein, B-100.[50]

The immunological properties of mammalian lipoproteins have also served as a basis on which to assess the efficacy of procedures used in their separation. Typically, antisera to the whole serum and/or purified lipoproteins from a given mammal have been reacted with lipoproteins freshly isolated from the same species, by double immunodiffusion and/or immunoelectrophoresis, and the purity of the latter evaluated.[42,47,48,54,72,79,153,165] This approach is also particularly useful when exploring the lipoprotein density profile in a particular species for the first time; cases in point are the mouse[41] and Common marmoset.[43]

Allotypy and Immunogenetic Polymorphism. A further aspect of the immunological behavior of mammalian serum lipoproteins concerns their antigenic polymorphism, and indeed a substantial literature has developed on this topic over the past decade (for a review, see Rapacz[166]).

The genetic polymorphism of proteins was first observed by Oudin[167] who found that immunization of a rabbit with serum proteins from another rabbit could elicit production of isoantibodies which reacted with sera from some but not all rabbits. Such genetic variants, identified by their antigenic properties with isoantibodies, are designated allotypes.

The earliest immunogenetically defined polymorphism of a lipoprotein to be recognized was that of human plasma LDL.[168] Subsequently, lipo-

[164] S. Goldstein and M. J. Chapman, *Biochem. Genet.* **14,** 883 (1976).
[165] D. Stead and V. A. Welch, *J. Dairy Sci.* **59,** 1 (1976).
[166] J. Rapacz, *Am. J. Med. Gen.* **1,** 377 (1978).
[167] J. Oudin, *J. Exp. Med.* **112,** 107 (1960).
[168] A. C. Allison and B. S. Blumberg, *Lancet (i),* 634 (1961).

protein allotypes have been detected in several mammals, including mink,[168–171] pig,[172–175] rabbit,[68,176-184] cattle,[185,186] and sheep.[187] Our knowledge of these allotypic systems is most extensive in mink, pig, and rabbit.

In mink, the presence of seven allotypes, Lpm1, Lpm2, Lpm3, Lpm4, Lpm5, Lpm7, and Lpm8, has been established using alloantibodies.[169–171] This "Lpm" system is specifically associated with mink very high-density lipoproteins ($d > 1.21$ g/ml) of α_2-mobility and exhibiting esterase activity. The distribution of Lpm markers among VHDL appears to indicate that each Lpm allotype is under the control of a single structural gene, that the Lpm markers belong to a single immunogenetic system, and are inherited singly or in allogroups (e.g., Lpm4, 7) as allelic codominant characters. Whether any relationship exists between Lpm genotype and relative resistance to atherosclerosis in mink[188] is not as yet elucidated.

The allotype polymorphism in swine is complex, and five distinct systems (Lpp, Lpr, Lps, Lpt, and Lpu) have been identified[175]; these systems involve more than 20 alloantigenic markers. Rapacz and colleagues have demonstrated that the polymorphism in each of these systems is regulated by a set of allelic genes which vary in number from two in the Lpr to seven in the Lpp system.[166,172–175] The Lpp allotypes are specifically associated with LDL of d 1.019–1.063 g/ml[172] and are of particular interest as

---

[169] O. K. Baranov, M. A. Savina, and D. K. Belyaev, *Biochem. Genet.* **14**, 327 (1978).
[170] O. K. Baranov, V. I. Yermolaev, and D. K. Belyaev, *Biochem. Genet.* **16**, 399 (1978).
[171] O. K. Baranov and V. I. Yermolaev, *Immunochemsitry* **15**, 629 (1978).
[172] J. Rapacz, R. H. Grummer, J. Hasler, and R. M. Shackelford, *Nature (London)* **225**, 941 (1970).
[173] J. Rapacz, J. Hasler-Rapacz, W. H. Huo, and D. Li, *Anim. Blood Groups Biochem. Genet.* **7**, 157 (1976).
[174] J. Rapacz, J. Hasler-Rapacz, and W. H. Kuo, *Immunogenetics* **6**, 405 (1978).
[175] J. Rapacz and J. Hasler-Rapacz, *Anim. Blood Groups Biochem. Genet.* **2** (Suppl. 1), 59 (1980).
[176] A. S. Kelus, *Nature (London)* **218**, 595 (1968).
[177] J. J. Albers and S. Dray, *Biochem. Genet.* **2**, 25 (1968).
[178] J. J. Albers and S. Dray, *J. Immunol.* **103**, 155 (1969).
[179] K. Berg, H. Boman, H. Torsvik, and S. M. Walker, *Proc. Natl. Acad. Sci. U.S.A.* **68**, 905 (1971).
[180] A. Gilman-Sachs and K. L. Knight, *Biochem. Genet.* **7**, 177 (1972).
[181] H. Boman, H. Torsvik, and K. Berg, *Clin. Exp. Immunol.* **11**, 297 (1972).
[182] A. L. Børresen, *J. Immunogenet.* **3**, 83 (1976).
[183] A. L. Børresen, *J. Immunogenet.* **3**, 91 (1976).
[184] A. L. Børresen, *J. Immunogenet.* **5**, 13 (1978).
[185] J. Rapacz, N. Korda, and W. H. Stone, *Genetics* **58**, 387 (1968).
[186] J. Wegrzyn, *Anim. Blood Groups Biochem. Genet.* **4**, 15 (1973).
[187] J. Rapacz, J. Hasler-Rapacz, A. L. Page, and R. Antoniewicz, *Anim. Blood Groups Biochem. Genet.* **3**, 31 (Suppl. 1) (1972).
[188] D. B. Zilversmit, T. B. Clarkson, and L. B. Hughes, *Atherosclerosis* **26**, 97 (1977).

certain antigenic markers corresponding to the Lpp3 gene are found in swine resistant to elevation of blood lipid levels, while swine carrying markers corresponding to the Lpp5 gene tend to be susceptible to aortic intimal lipidosis when fed diets enriched in fat.[166,189]

Unfortunately we remain ignorant of the structural basis of the immunogenetic polymorphism of lipoproteins in both the pig and mink, and it will be important to identify and characterize the apolipoprotein(s) which carries the alloantigenic determinant(s). Further data on the mode of inheritance, modulation and sites of formation, and plasma density distribution of the allotypic markers are also needed for the porcine systems.

The allotypy of both low- and high-density lipoproteins is amply documented in the rabbit.[68,176-184] The two allotypic specificities associated with HDL are however controlled by allelic genes situated at an autosomal locus (originally termed Lhj) distinct from that responsible for the polymorphism of LDL.[180] Whether this specific locus is the same as that responsible for the HDL alloantigens described by Berg, Børresen, and colleagues[179,181-184] was not tested. The latter investigators have independently shown that two polymorphic traits of rabbit HDL are inherited in an autosomal dominant mode with no obvious association; these genetic polymorphisms are expressed in the form of the Hl1 and R67 antigens,[179,181] whose concentrations exhibit certain correlations with HDL levels in male and female rabbits of the "Albino" and "Dutch black and white" strains.[183] The two antigens are carried on separate polypeptide chains which are distinct from rabbit apoA-I[182]; tentative estimates of their molecular weights are 40,000 and 17,000 for Hl1 and R67, respectively.[182] Furthermore, as minor apolipoproteins, these antigens reside on separate HDL molecules; on the one hand, the R67 antigen is present on apoA-I-containing HDL, while Hl1 appears to reside on an HDL particle lacking both apoA-I and R67.[184] Additional information on the biosynthesis, metabolism, and primary structure of these proteins is now required. Interestingly, the Hl1 allotypic determinant exists in the pig, at least as judged by the cross-reactivity of pig serum with antiserum to this polypeptide.[190]

Substantial progress has been made in determining the origin of the allotypy of rabbit LDL since the early findings of Kelus[176] and of Albers and Dray.[177,178] Their immunological and genetic studies with four alloantisera[177,178] revealed that two allotypes (denoted Lpq1 and Lpq2) were controlled by one lipoprotein gene, denoted LpqA, and that two other LDL allotypes (Lpq3 and Lpq4) were modulated by alleles of a second

---

[189] J. Rapacz, C. E. Elson, and J. J. Lalich, *Exp. Mol. Pathol.* **27**, 249 (1977).
[190] K. Berg, A. L. Børresen, and H. Bakke, *J. Immunogenet.* **2**, 79 (1975).

gene. More recently, it has been shown that the Lpq3 and Lpq4 genes are not allelic, but that they control the expression of allotypes of distinct gene products.[191] These two latter, genes, designated LpqB and LpqC, are tightly linked to each other and to LpqA.[191] The Lpq gene complex then controls the expression of four different allotypes on rabbit serum LDL and forms part of linkage group VIII.[192]

The nature of the Lpq gene products and thus the structural origin of rabbit LDL allotypy was recently investigated by Huang et al.[68] using immunoblotting techniques. The four alloantisera, directed against the Lpq1, Lpq2, Lpq3, and Lpq4 allotypes, respectively, reacted with two apolipoproteins of high molecular weight (320,000 and 200,000). These proteins cross-reacted with an antiserum to rabbit apoB and were present in VLDL, IDL, and LDL. Nonetheless, the distribution of the allotypes among rabbit lipoproteins was dissimilar, with Lpq1, Lpq2, and Lpq4 being distributed primarily between VLDL, IDL, and LDL, with only traces in HDL. By contrast, the level of the Lpq3 allotype in HDL was 20% of that in LDL. The alloantisera appear then to be specific for rabbit B proteins and as such provide a powerful tool to probe their structure and genetic control, especially as they may allow identification of changes as small as single amino acids in protein sequences.[193]

Physicochemical Characteristics of Mammalian Apolipoproteins

As the now copious literature attests, ever-increasing attention has been focused on mammalian apolipoproteins over the past decade (for recent reviews, see refs. 11, 13, 14, 20, 57, 62, 194, and 195). Thus, these specialized proteins play a major role in determining the structural organization, intravascular metabolism, and tissular fate of lipoprotein particles. As such, they are implicated in specific binding to cellular lipoprotein receptors[196] (apoB and apoE), in the activation or inhibition of certain lipolytic enzymes,[194,195] and in processes of lipid exchange and transfer.[195,197]

---

[191] D. C. Usher, L. S. Huang, B. Soronti, and R. R. Fox, *J. Hered.* **73**, 286 (1982).

[192] R. R. Fox, D. D. Crary, and C. Cohen, in "Inbred and Genetically-Defined Strains of Laboratory Animals" (P. L. Altman and D. D. Katz, eds.), Part 2, p. 570. Fed. Am. Soc. Exp. Biol., Bethesda, Md., 1979.

[193] H. Alexander, D. A. Johnson, J. Rosen, L. Jerabek, N. Green, I. L. Weissman, and R. A. Lerner, *Nature (London)* **306**, 697 (1983).

[194] J. L. Osborne and H. B. Brewer, *Adv. Protein Chem.* **31**, 253 (1977).

[195] G. S. Ott and V. G. Shore, in "Handbook of Electrophoresis" (L. A. Lewis, ed.), Vol. III, p. 105. CRC Press, Boca Raton, Fla., 1983.

[196] R. W. Mahley and T. L. Innerarity, *Biochim. Biophys. Acta* **737**, 197 (1983).

[197] P. E. Fielding and C. J. Fielding, *Proc. Natl. Acad. Sci. U.S.A.* **77**, 3327 (1980).

A large body of structural information has come from application of classical methods of protein chemistry to the purification and characterization of mammalian apolipoproteins. Most recently, however, dramatic progress in our knowledge of apolipoprotein structure has resulted from determination of the nucleotide sequences of their messenger RNAs following cDNA clone analysis. Indeed, such molecular biological studies are now providing insight not only into the amino acid sequences of mammalian apoproteins but also into the proteolytic processing of the nascent polypeptides and into the structure and modulation of their genes. Within the context of the present treatise, only comparative structural aspects of the apoproteins themselves will be considered; their nomenclature and that of the human counterparts was discussed in the section on Apolipoproteins.

The principal physicochemical characteristics of the 10 or more plasma apolipoproteins found to date in mammals are summarized in Table XII, from which it is evident that counterparts to the human A-I, A-II, A-IV, B, C-I, C-II, C-III, E, H,(β_2-glycoprotein I) and threonine-poor/SAA (serum amyloid) apolipoproteins have been purified and characterized in a wide range of species. In addition, apoD (M_r 15,400) was identified in baboon HDL and partially characterized.[58] Moreover, several of the above apoproteins have been detected in mammals by electrophoretic and/or immunological procedures, but have not been purified; these include apoA-I, B, E, and apoCs in sheep,[198] hedgehog,[76] European badger,[90] red fox,[199] and Mongolian gerbil.[116] Still other apolipoproteins have been detected on the basis of their biological activity and notably their capacity to activate lipoprotein lipase; in such cases, either it has proven difficult to isolate them (as in guinea pig[200,201]), or upon isolation their physicochemical properties fail to correspond to those of the supposedly homologous human protein, i.e., apoC-II (as for the bovine C proteins[202-204]).

Most of the information in Table XII pertains to the A-I, A-II, B, and E proteins, each of which is polymorphic, but for which explanations at a structural level are lacking; only in human apoE is this phenomenon fully understood.[205] Amino acid sequences, either partial or complete, have

---

[198] T. M. Forte, C. E. Cross, R. A. Gunther, and G. C. Kramer, *J. Lipid Res.* **24,** 1358 (1983).
[199] P. M. Laplaud, L. Beaubatie, and D. Maurel, *Comp. Biochem. Physiol.* **68B,** 125 (1981).
[200] L. Wallinder, G. Bengtsson, and T. Olivecrona, *Biochim. Biophys. Acta* **575,** 458 (1979).
[201] M. J. Chapman, G. L. Mills, and J. H. Ledford, *Biochem. J.* **149,** 423 (1975).
[202] B. W. Patterson and A. Jonas, *Biochim. Biophys. Acta* **619,** 572 (1980).
[203] C. T. Lim and A. M. Scanu, *Artery* **2,** 483 (1976).
[204] H. N. Astrup and G. Bengtsson, *Comp. Biochem. Physiol.* **72B,** 487 (1982).
[205] K. H. Weisgraber, S. C. Rall, and R. W. Mahley, *J. Biol. Chem.* **256,** 9077 (1981).

TABLE XII
PHYSICAL CHARACTERISTICS OF MAMMALIAN PLASMA APOLIPOPROTEINS[a]

Species	Apolipoprotein	Molecular weight	Apparent pI	Amino acid composition	Sequence	Lipoprotein protein content (%)
Man[a]	A-I	28,300	5.52, 5.65 (major); 5.36, 5.40, 5.85 (minor)	Lacks isoleucine	Determined	HDL (60%); chylos (7%)
Rat[b]	A-I	27,000	5.80, 5.85 (major); 5.69, 5.58 (minor) 6.08, 6.31, 6.44	1.8 mol% isoleucine, 0.4 mol% cysteine	To NH$_2$-terminal residue No. 21	HDL (60%)
Mouse[c]	A-I	25,000–27,000 ~28,000	5.61, 5.54 major; 5.83, 5.49, 5.37 (minor)	0.4 mol% isoleucine, 0.2 mol% cysteine	To NH$_2$-terminal residue No. 30	HDL (~80%), LDL (7%)
Guinea pig[d]	A-I	25,000	5.75, 5.52 (major); 5.67, 5.62, 5.45, 5.40 (minor)	2 mol% isoleucine, 0.8 mol% cysteine	—	HDL
Dog[e]	A-I	~28,000	—	0.8 mol% isoleucine, cysteine absent	Determined	HDL (~90%)
Rabbit[f]	A-I	25,000–27,000	5.50 (major); 5.99, 5.81, 5.66, 5.34 (minor)	1.4 mol% isoleucine, cysteine absent	To NH$_2$-terminal residue No. 29	HDL (>50%)
Cow[g]	A-I	27,000	—	1.7 mol% isoleucine, cysteine absent	—	HDL (88%)

Pig[h]	A-I	26,000;29,000[h]	—	1.2 mol%, isoleucine, cysteine absent	To NH$_2$-terminal residue No. 10	HDL (60%)
Common marmoset[i]	A-I	~27,000	5.02, 4.97, 4.91 (major); 4.85, 4.79 (minor)	1.1 mol% isoleucine, cysteine absent	—	HDL (~60%)
Erythrocebus patas[j]	A-I	28,000	—	Isoleucine and cysteine absent	To NH$_2$-terminal residue No. 20	HDL (>50%)
Baboon[k]	A-I	28,000–29,000	4.67–4.88; 8 isoforms	Isoleucine and cysteine absent	To NH$_2$-terminal residue No. 30	HDL (>60%)
Rhesus monkey[l]	A-I	~27,000	5.21, 5.69	Isoleucine and cysteine absent	—	HDL (70%)
African green monkey (vervet)[m]	A-I	28,000	5.9–6.3; 4 major, ~5 minor isoforms	Isoleucine and cysteine absent	—	HDL (69%)
Chimpanzee[n]	A-I	27,000–29,000	Polymorphic; 5.30, 5.19 major; 5.11 minor	Isoleucine and cysteine absent	To NH$_2$-terminal residue No. 4	HDL (~60%)
Man[o]	A-II	17,400 (dimer)	4.89 (major); 5.16, 4.58, 4.31 (minor)	Lack His, Arg, and tryptophan; 1 mol cysteine per dimer	Determined	HDL (25%); chylos (4%)
Rat[p]	A-II	8,000 (monomer)	4.83	4 mol% Arg; lacks His, cysteine, and tryptophan	—	HDL (≤20%)
Mouse[q]	A-II	8,000–8,400 (monomer)	5.22 (major); 5.33, 5.06 (minor)	1.3 mol% histidine, 1.7 mol% arginine; cysteine absent	—	HDL
Dog[r]	A-II	~8,500 (monomer)	—	—	—	HDL

(continued)

TABLE XII (continued)

Species	Apolipoprotein	Molecular weight	Apparent pI	Amino acid composition	Sequence	Lipoprotein protein content (%)
Cattle[s]	A-II(D$_2$)	8,800 (monomer)	—	Lacks His, Arg, Met, and tryptophan	—	HDL (<10%)
Erythrocebus patas[t]	A-II	8,500 (monomer)	—	Lacks His, Ile, cysteine, and tryptophan; 1.1 mol% Arg	—	HDL
Baboon[u]	A-II	8,500 (monomer)	—	Lacks His, Ile, cysteine, and tryptophan: 1.0 mol% Arg	—	HDL
Rhesus monkey[v]	A-II	8,757 (monomer)	4.65	Lacks His, Ile, cysteine, and tryptophan; 1.3 mol% Arg	Determined	HDL (20%)
African green monkey (vervet)[w]	A-II	9,900 (monomer)	5.17	Lacks His, Ile, cysteine, and tryptophan; 1/3 mol% Arg	—	HDL (8–13%)
Chimpanzee[x]	A-II	17,500 (dimer)	5.02	Lacks His, Arg, and tryptophan; 1.3 mol% isoleucine; 1 mol cystine/dimer	To NH$_2$-terminal residue No. 7	HDL

Man[y]	A-IV	46,000	~5.15	~25 mol% Glu; other amino acids present	To NH$_2$-terminal residue No. 11	$d > 1.21$ g/ml fraction; VLDL, chylos; HDL
Rat[z]	A-IV	46,000	4.95	~23 mol% Glu; tryptophan 0.5 mol%; cysteine 0.5 mol%	To NH$_2$-terminal residue No. 20	HDL, chylos, $d > 1.21$ g/ml fraction
Dog[A]	A-IV	44,000–45,000	—	22 mol% Glu; tryptophan and cysteine not determined	—	HDL, chylos, VLDL
Rat[B]	A-V	59,000	—	14–15 mol% Ser, Glu, and Gly	—	Chylos (lymph.) VLDL ~20%
Man[C]	B-100	549 × 10³	—	~10–12 mol% Asp, Glu, and Leu; ~0.4 mol% cysteine	—	LDL (98%), VLDL (40–60%), chylos (20%)
	B-48	264 × 10³	—	Minor differences in amino acid composition from B-100	—	
Rat[D]	B-100:apoB$_H$:P-1	335 × 10³	—	~10–11 mol% Asp and Leu; 14 mol% Glu; ~0.6 mol% cysteine	—	Chylos, VLDL, LDL

(*continued*)

TABLE XII (continued)

Species	Apolipoprotein	Molecular weight	Apparent pI	Amino acid composition	Sequence	Lipoprotein protein content (%)
	B-48:apoB$_L$: P-III	240×10^3	—	Minor differences from apoB$_H$; contains phosphoserine	—	Chylos (90%), VLDL (70%), LDL (7%)
Mouse[E]	B-100:apoB$_H$: P-I	B$_L$ P-I 534×10^3:320×10^3	—	~10 mol% Asp, Ser, and Leu	—	Chylos, VLDL, LDL
	B-48:apoB$_L$: P-III	B$_L$ P-III 278×10^3:220×10^3		~13 mol% Glu, ~0.5–1 mol% cysteine		Chylos, VLDL, LDL
Guinea pig[F]	B-100	402–422×10^3	—	~11 mol% Asp and Glu; ~14 mol% Leu	—	VLDL, LDL
Dog[G]	B-100 B-48	—	—	—	—	Chylos, VLDL, LDL
Rabbit[H]	B-100:B$_{320K}$ B-48:B$_{220K}$	320×10^3 220×10^3	—	—	—	VLDL, IDL, LDL

Species	Apo	MW		Amino acid composition	CHO side chains	Lipoprotein class
Pig[J]	B-100	$300\text{--}350 \times 10^3$	—	11–12 mol% Asp, Glu, and Leu; ~0.5 mol% cysteine	3 types of CHOate side chains	VLDL, LDL (HDL)
Common marmoset[J]	B-100 B-48 ?	$420\text{--}520 \times 10^3$ B-48:~250×10^3	—	~10–11 mol% Asp and Leu; ~13 mol% Glu; 0.6–0.9 mol% Cys	—	VLDL, IDL, LDL (>80%)
Baboon[K]	B-100	$300\text{--}350 \times 10^3$	—	~10–12 mol% Asp, Glu, and Leu	—	VLDL, LDL
Rhesus monkey[K]	B-100	$300\text{--}350 \times 10^3$	—	~11–12 mol% Asp, Glu, and Leu	—	VLDL, LDL
Chimpanzee[L]	B-100	B-100:$370\text{--}380 \times 10^3$	—	~11–12 mol% Asp, Glu, and Leu	—	VLDL, LDL
Man[M]	C-I	6,600	~7.5	His and Tyr absent, also cysteine; ~15–16 mol% Lys and Glu	Determined	Chylos (14%), VLDL (4%), HDL (1–5%)
Rat[N]	C-I	7,000	—	Lacks Val, Tyr, and cysteine; ~16 mol% lysine and valine	—	VLDL, HDL
Guinea pig[O]	C-I:"Band VI-B"	—	—	11.5 mol% Lys, 14.4 mol% Glu; His 1.8 mol%; Tyr 3.0 mol%	—	HDL

(continued)

TABLE XII (continued)

Species	Apolipoprotein	Molecular weight	Apparent p*I*	Amino acid composition	Sequence	Lipoprotein protein content (%)
Rabbit[P]	C-I: fraction R₁	—	—	Lacks Tyr; ~15–16 mol% Lys and Glu	—	VLDL (<10%)
Man[Q]	C-II	8,800	4.86 (major); 4.69 (minor)	Lacks His and cysteine; ~18 mol% Glu	Determined	Chylos (14%), VLDL (7–10%), HDL (1–3%)
Rat[R]	C-II	8,000	4.74	Lacks cysteine; 14 mol% Leu, 13 mol% Glu	—	VLDL, HDL
Pig[S]	C-II	7,800–10,000	4.82 (major); 4.6, 4.49 (minor)	Lacks His, cysteine, and tryptophan; ~14 mol% Glu	—	VLDL
African green monkey (vervet)[T]	C-II	8,000	5.20	Lacks His and Met; ~16 mol% Glu	—	HDL
Man[U]	C-III	9,200–9,700	5.02; 4.82; 4.62 C-III-0;C-III-1; C-III-2	Lacks Ile and cysteine; ~13 mol% Ser and Glu	Determined: CHOate structure determined	Chylos (34%); VLDL (35–40%), HDL (3–5%)
Rat[V]	C-III	C-III-0:10,000 C-III-3:11,000	C-III-0: 4.67 CIII-3: 4.50 (major isoforms)	Contains 1 mol Ile/mol; lacks His and cysteine; ~15 mol% Glu	To NH₂-terminal residue No. 12	HDL

Mouse[w]	C-III	~9,600	4.74	2.4 mol% Ile; 16.4 mol% Glu; trace, cysteine; ~13 mol% Ser	To NH$_2$-terminal residue No. 13	HDL, VLDL, LDL
Bovine[x]	C-III	D$_2$, D$_3$, D$_4$: 7,100–8,800	—	0.2–1.3 mol% Ile; His, 0–0.2 mol%; 19–21 mol% Glu	—	HDL
African green monkey (vervet)[y]	C-III	9,500	5.05	Lacks Ile and cysteine; 14–15 mol% Glu and Ala	—	HDL
Man[z]	E	35,000–39,000	6.02 (E4), R.89 (E3), 5.78 (E2); 5.64, major; >10 minor	~11 mol% Arg; ~24 mol% Glu; 0–2 mol cysteine/mol E	Determined	VLDL, LDL, HDL
Rat[z1]	E	35,000–37,000	5.31–5.46	10–11 mol% Arg; ~24 mol% Glu	Determined	HDL, VLDL, HDL$_1$
Mouse[z2]	E	35,000–38,000	5.4; 5.7–5.8	—	—	VLDL, HDL
Guinea pig[z3]	E	~34,000	5.24–5.42 (major); 5.17–5.20 (minor)	~10 mol% Arg; ~25 mol% Glu	—	VLDL, LDL, HDL
Guinea pig[z4]	Comigrating polypeptide	31,000–36,000	—	~4 mol% Arg; ~14 mol% Glu	—	VLDL, LDL
Dog[z5]	E	37,000–39,000	~5.0–5.4 (2 major; 2 minor)	10.5 mol% Arg; ~27 mol% Glu; lacks cysteine	To NH$_2$-terminal residue No. 20	VLDL, HDL$_1$
Rabbit[z6]	E	—	Polymorphic (2 major)	~10 mol% Arg; ~23 mol% Glu; lacks cysteine	—	VLDL

(continued)

TABLE XII (*continued*)

Species	Apolipoprotein	Molecular weight	Apparent p*I*	Amino acid composition	Sequence	Lipoprotein protein content (%)
Pig[7]	E	34,000	—	12.4 mol% Agr; ~25 mol% Glu	To NH$_2$-terminal residue No. 13	VLDL, HDL
Erythrocebus patas[8]	E	34,000	—	~11.5 mol% Arg; ~23.5 mol% Glu	—	VLDL, IDL
Man[9]	ApoH β_2-Glycoprotein I	40,000–45,000: 54,000	—	Major discrepancies in the reported amino acid compositions and MW	—	Chylos VLDL (16%), LDL (2%), HDL (17%)
Rat[10]	β_2-Glycoprotein I	58,000	6 major; ~4 minor isoforms	Distinct from human apoH in Arg, Pro, and tryptophan contents	—	VLDL, HDL
Man[11]	Threonine-poor apoproteins ApoSAA	10,000–15,000 40,000	Multiple forms: 5.0–8.0	0.5–1 mol% threonine; 11–13 mol% Asp, Glu, Gly, Ala	Determined for SAA$_1$ to NH$_2$-terminal residue No. 104, and for SAA$_2$ to No. 30	(VLDL, IDL; traces) HDL$_2$, HDl$_3$
Mouse[12]	ApoSAA	10,000	Multiple forms	—	—	HDL
African green monkey (vervet)[13]	Threonine-poor apoproteins	1. 11,500 2. 13,900	6.44 6.04	1–2 mol% threonine; 14–17 mol% Asp, Gly, Ala	—	HDL

[a] Data on apolipoprotein distribution in man taken from E. J. Schaefer, S. Eisenberg, and R. I. Levy, *J. Lipid Res.* **19**, 667 (1978) except where otherwise stated. Values from analyses of purified apoproteins except where stated. While the majority of these mammalian apoproteins were isolated from the lipoproteins of normolipidemic animals, others (notably apoE) were isolated from lipoproteins induced by cholesterol feeding.

[a'] H. B. Brewer, T. Fairwell, A. LaRue, R. Ronan, A. Houser, and T. J. Brorzert, *Biochem. Biophys. Res. Commun.* **80**, 623 (1978). G. S. Oh and V. G. Shore, in "Handbook of Electrophoresis" (L. A. Lewis ed.), Vol. III, p. 105, CRC Press, Boca Raton, Fla. 1983.

[b] J. B. Swaney, F. Braithwaite, and H. A. Eder, *Biochemistry* **16**, 271 (1977); S. Calandra, P. Tarugi, and M. Ghisellini, *Atherosclerosis* **50**, 209 (1984)

[c] M. C. Camus, M. J. Chapman, P. Forgez, and P. M. Laplaud, *J. Lipid Res.* **24**, 1210 (1983); pI values of apoA-I in apo-HDL.

[c'] R. C. LeBoeuf, D. L. Puppione, V. N. Schumaker, and A. J. Lusis, *J. Biol. Chem.* **258**, 5063 (1983).

[d] L. S. S. Guo, R. L. Hamilton, J. P. Kane, C. J. Fielding, and G. C. Chen, *J. Lipid Res.* **23**, 531 (1982). L. S. S. Guo, M. Meng, R. L. Hamilton, and R. Ostwald, *Biochemistry* **16**, 5867 (1977).

[e] C. Edelstein, L. L. Lewis, J. R. Shainoff, H. Naito, and A. M. Scanu, *Biochemistry* **15**, 1934 (1976); H. Chung, A. Randolph, I. Reardon, and R. L. Heinrikson, *J. Biol. Chem.* **257**, 2961 (1982).

[f] S. Calandra, P. Tarugi, and M. Ghisellini, *Atherosclerosis* **50**, 209 (1984); A. L. Børresen and T. J. Kindt, *J. Immunogenetics* **5**, 5 (1978).

[g] A. Jonas, *Biochim. Biophys. Acta* **393**, 460 (1975).

[h] R. L. Jackson, H. W. Baker, O. D. Taunton, L. C. Smith, C. W. Garner, and A. M. Gotto, *J. Biol. Chem.* **248**, 2639 (1973).

[h'] Calculated from amino acid composition. Sequence from R. W. Mahley, K. H. Weisgraber, T. Innerarity, H. B. Brewer, and G. Assmann, *Biochemistry* **14**, 2817 (1975).

[i] P. Forgez, M. J. Chapman, and G. L. Mills, *Biochim. Biophys. Acta* **754**, 321 (1983).

[j] R. W. Mahley, K. H. Weisgraber, T. Innerarity, and H. B. Brewer, *Biochemistry* **15**, 1928 (1976); R. W. Mahley, K. H. Weisgraber, T. Innerarity, H. B. Brewer, and G. Assmann, *Biochemistry* **14**, 2817 (1975).

[k] V. Blaton, R. Vercaemst, M. Rosseneu, J. Mortelmans, R. L. Jackson, A. M. Gotto, and H. Peeters, *Biochemistry* **16**, 2157 (1977); M. Rosseneu, V. Blaton, R. Vercaemst, F. Soeteway, and H. Peeters, *Eur. J. Biochem.* **74**, 83 (1977).

[l] C. Edelstein, C. T. Lim, and A. M. Scanu. *J. Biol. Chem.* **248**, 7653 (1973).

[m] J. S. Parks and L. L. Rudel, *J. Biol. Chem.* **254**, 6716 (1979); found to be immunochemically heterogeneous by J. S. Parks and L. L. Rudel, *Biochim. Biophys. Acta* **618**, 327 (1980).

[n] A. M. Scanu, C. Edelstein, and R. H. Wolf, *Biochim. Biophys. Acta* **351**, 341 (1974); V. Blaton, R. Vercaemst, N. Vandecasteele, H. Caster, and H. Peeters, *Biochemistry* **13**, 1127 (1974); pI values from M. J. Chapman, P. Forgez, D. Lagrange, S. Goldstein, and G. L. Mills, *Atherosclerosis* **52**, 129 (1984).

[o] S. E. Lux, K. John, R. Ronan, and H. B. Brewer, *J. Biol. Chem.* **247**, 7519 (1972); pI values from G. Schmitz, K. Ilsemann, B. Melnik, and G. Assmann, *J. Lipid Res.* **24**, 1021 (1983).

(*continued*)

Footnotes to TABLE XII (continued)

[p] P. N. Herbert, H. G. Windmeller, T. P. Bersot, and R. S. Shulman. *J. Biol. Chem.* **249**, 5718 (1974); J. B. Swaney and L. I. Gidez. *J. Lipid Res.* **18**, 69 (1977).

[q] As c an c'.

[r] R. W. Mahley and K. H. Weisgraber, in "Handbook of Electrophoresis" (L. A. Lewis and H. K. Naito eds.), Vol. IV, p. 151. CRC Press, Boca Raton, Fla. 1983.

[s] C. T. Lim and A. M. Scanu, *Artery* **2**, 483 (1976); B. W. Patterson and A. Jonas, *Biochim. Biophys. Acta* **619**, 572 (1980).

[t] R. W. Mahley, K. H. Weisgraber, T. Innerarity, and H. B. Brewer, *Biochemistry* **15**, 1928 (1976).

[u] V. Blaton, R. Vercaemst, M. Rosseneu, J. Mortelmans, R. L. Jackson, A. M. Gotto, and H. Peeters, *Biochemistry* **16**, 2157 (1977).

[v] C. Edelstein, C. Noyes, P. Keim, R. L. Heinrikson, R. E. Fellows, and A. M. Scanu, *Biochemistry* **15**, 1261 (1976), and as in l.

[w] As m.

[x] As n.

[y] U. Beisiegel and G. Utermann, *Eur. J. Biochem.* **93**, 601 (1979); K. H. Weisgraber, T. P. Bersot, and R. W. Mahley, *Biochem. Biophys. Res. Commun.* **85**, 287 (1978).; tryptophan not determined; J. I. Gordon, C. L. Bisgaier, H. F. Sims, O. P. Sachdev, R. M. Glickman, and A. W. Strauss. *J. Biol. Chem.* **259**, 468 (1984).

[z] J. B. Swaney, F. Braithwaite, and H. A. Eder, *Biochemistry* **16**, 271 (1977); as y; P. S. Roheim, D. Edelstein, and G. G. Pinter, *Proc. Natl. Acad. Sci. U.S.A.* **73**, 1757 (1976); J. I. Gordon, D. P. Smith, D. H. Alpers, and A. W. Strauss, *J. Biol. Chem.* **257**, 8418 (1982).

[A] Isolated from canine HDL_c and amino acid composition determined; K. H. Weisgraber, T. P. Bersot, and R. W. Mahley, *Biochem. Biophys. Res. Commun.* **85**, 287 (1978).

[B] N. H. Fidge and P. J. McCullagh, *J. Lipid Res.* **22**, 138 (1981); trace amounts in plasma HDL: displays similar physicochemical and lipid-binding properties to β_2-glycoprotein I in man; see "Apolipoprotein H"/β_2-glycoprotein I below.

[C] J. P. Kane, D. A. Hardman, and H. E. Paulus, *Proc. Natl. Acad. Sci. U.S.A.* **77**, 2465 (1980). Minor components B-74 (407×10^3) and B-26 (144.5×10^3).

[D] K. V. Krishnaiah, L. F. Walker, J. Borensztajn. G. Schonfeld, and G. S. Getz, *Proc. Natl. Acad. Sci. U.S.A.* **77**, 3806 (1980). J. Elovson, Y. O. Huang, N. Baker, and R. Kannan, *Proc. Natl. Acad. Sci. U.S.A.* **77**, 3806 (1980): The rat B proteins comigrate with human B-100 and B-48. C. E. Sparks and J. B. Marsh, *J. Lipid Res.* **22**, 514 (1981); R. A. Davis, G. M. Clinton, R. A. Borchardt, M. Malone-McNeal, T. Tan, and G. R. Lattier, *J. Biol. Chem.* **259**, 3383 (1984).

[E] As c and c', and P. Forgez, M. J. Chapman, S. C. Rall, and M. C. Camus, *J. Lipid Res.* **25**, 954 (1984); and R. C. LeBoeuf, D. L. Puppione, V. N. Schumaker, and A. J. Lusis. *J. Biol. Chem.* **258**, 5063 (1983).

[F] M. J. Chapman, G. L. Mills, and J. H. Ledford, *Biochem. J.* **149**, 423 (1975).

[G] As r; identification in SDS gels.

[H] Identified by cross-reaction with an antiserum to rabbit LDL apoB; L. S. Huang, J. S. Jaeger, and D. C. Usher, *J. Lipid Res.* **24**, 1485 (1983).

[I] M. J. Chapman and S. Goldstein, *Atherosclerosis* **25**, 267 (1976); G. M. J. Knipping, G. M. Kostner, and A. Holasek. *Biochim. Biophys. Acta* **393**, 88 (1975); R. L. Jackson, O. D. Taunton, R. Segura, J. G. Gallagher, H. F. Hoff, and A. M. Gotto, *Comp. Biochem. Physiol.* **53B**, 245 (1976).

CHOate structure determined by J. Azuma, N. Kashimura, and T. Komarno, *J. Biochem.* **81**, 1613 (1977).

[J] P. Forgez, M. J. Chapman, and G. L. Mills. *Biochim. Biophys. Acta* **754**, 321 (1983); B-48 tentatively identified in SDS gels of apo-VLDL; M. J. Chapman, F. McTaggart, and S. Goldstein, *Biochemistry* **18**, 5096 (1979).

[K] M. J. Chapman and S. Goldstein, *Atherosclerosis* **25**, 267 (1976).

[L] M. J. Chapman, P. Forgez, D. Lagrange, S. Goldstein, and G. L. Mills, *Atherosclerosis* **52**, 129 (1984).

[M] R. S. Shulman, P. Herbert, K. Wehrly, and D. S. Fredrickson. *J. Biol. Chem.* **250**, 182 (1975).

[N] As p.

[O] As F. Identification tentative.

[P] V. G. Shore, B. Shore, and R. G. Hart, *Biochemistry* **13**, 1579 (1974); iso ation under similar conditions as for human C-I.

[Q] R. L. Jackson, H. N. Baker, E. B. Gilliam, and A. M. Gotto, *Proc. Natl. Acad. Sci. U.S.A.* **74**, 1942 (1977). R. J. Havel, L. Kotite, and J. P. Kane, *Biochem. Med.* **21**, 121 (1979); activates lipoprotein lipase.

[R] As p. Activates lipoprotein lipase.

[S] R. L. Jackson, B. H. Chung, L. C. Smith, and O. D. Taunton. *Biochim. Biophys. Acta* **490**, 385 (1977); G. Knipping, E. Steyrer, R. Zechner, and A. Holasek. *J. Lipid Res.* **25**, 86 (1984); activates lipoprotein lipase.

[T] As m; 1 mol sialic acid/mol protein; activates lipoprotein lipase.

[U] H. B. Brewer, R. Shulman, R. Herbert, P. Ronan, and K. Wehrly. *J. Biol. Chem.* **249**, 4975 (1974); pI values from G. S. Ott and V. G. Shore, see a. Each isoform contains 1 mol galactosamine and galactose and 0 to 2 mol sialic acid.

[V] As p: tentative identification of CIII-1, CIII-2 and CIII-4 of pI 4.61, 4.57, and 4.43, respectively; C-III-0 and C-III-3 contain 0 and 3 mol sialic acid/mol protein. Sequence data from M. C. Blaufuss, J. I. Gordon, G. Schonfeld, A. W. Strauss, and D. H. Alpers, *J. Biol. Chem.* **259**, 2452 (1984).

[W] P. Forgez, M. J. Chapman, S. C. Rall, and M. C. Camus, *J. Lipid Res.* **25**, 954 (1984).

[X] C. T. Lim and A. M. Scanu, *Artery* **2**, 483 (1976); B. W. Patterson and A. Jonas, *Biochim. Biophys. Acta* **619**, 572 (1980); tentative identification. Variable sialic acid content.

[Y] As m.

[Z] V. I. Zannis and J. L. Breslow, *Biochemistry* **20**, 1033 (1981); K. H. Weisgraber, S. C. Rall, and R. W. Mahley. *J. Biol. Chem.* **256**, 9077 (1981). S. C. Rall, K. H. Weisgraber, and R. W. Mahley, *J. Biol. Chem.* **257**, 4171 (1982). Major isoforms due to cysteine/arginine interchanges; posttranslational addition of sialic acid.

(continued)

Footnotes to TABLE XII (*continued*)

[z1] J. B. Swaney, F. Braithwaite, and H. A. Eder. *Biochemistry* **16,** 271 (1977); K. H. Weisgraber and G. Assmann, *Atherosclerosis* **28,** 121 (1977); amino acid sequence inferred from nucleotide sequence of apoE mRNA by J. W. McLean, C. Fukazawa, and J. M. Taylor, *J. Biol. Chem.* **258,** 8993 (1983), contains carbohydrate.

[z2] As c and c'; pI 5.4 of apoE secreted from mouse peritoneal macrophages, S. K. Basu, M. S. Brown, Y. K. Ho, R. J. Havel, and J. L. Goldstein, *Proc. Natl. Acad. Sci. U.S.A.* **78,** 7545 (1981).

[z3] L. S. S. Guo, R. L. Hamilton, J. P. Kane, C. J. Fielding, and G. C. Chen, *J. Lipid Res.* **23,** 531 (1982); L. S. S. Guo, M. Meng, R. L. Hamilton, and R. Ostwald, *Biochemistry* **16,** 5807 (1977).

[z4] M. Meng, L. Guo, and R. Ostwald, *Biochim. Biophys. Acta* **576,** 134 (1979); possibly "Band V" in M. J. Chapman, G. L. Mills, and J. H. Ledford, *Biochem. J.* **149,** 423 (1975).

[z5] K. H. Weisgraber, R. F. Troxler, S. C. Rall, and R. W. Mahley, *Biochem. Biophys. Res. Commun.* **95,** 374 (1980), and as r.

[z6] As P: contains glucosamine and galactosamine.

[z7] As Z^5 and R. W. Mahley, K. H. Weisgraber, T. Innerarity, H. B. Brewer, and G. Assmann, *Biochemistry* **14,** 2817 (1975).

[z8] R. W. Mahley, K. H. Weisgraber, and T. Innerarity, *Biochemistry* **15,** 2979 (1976).

[z9] E. Polz and G. M. Kostner, *Biochem. Biophys. Res. Commun.* **90,** 1305 (1979); E. Polz and G. M. Kostner, *FEBS Lett.* **102,** 183 (1979); N. S. Lee, H. B. Brewer, and J. C. Osborne, *J. Biol. Chem.* **258,** 4765 (1983); ~17% by weight carbohydrate.

[z10] E. Polz, H. Wurm, and G. M. Kostner, *Int. J. Biochem.* **11,** 265 (1980).

[z11] V. G. Shore, B. Shore, and S. B. Lewis, *Biochemistry* **17,** 2174 (1978); N. Eriksen and E. P. Benditt, *Proc. Natl. Acad. Sci. U.S.A.* **77,** 6860 (1980); D. C. Parmelee, K. Titani, L. H. Ericsson, N. Eriksen, E. P. Benditt, and K. A. Walsh, *Biochemistry* **21,** 3298 (1982); C. L. Malmendier, P. Paroutaud, and J. P. Ameryckx, *FEBS Lett.* **109,** 43 (1980): isolated from sera with elevated AA-immunoreactivity.

[z12] Endotoxin-stimulated mice: E. P. Benditt, N. Eriksen, and R. H. Hanson, *Proc. Natl. Acad. Sci. U.S.A.* **76,** 4092 (1979).

[z13] As m.

been determined for apoA-I (rat, mouse, dog, rabbit, pig, *Erythrocebus patas* monkey, baboon, and chimpanzee), apoA-II (rhesus monkey and chimpanzee), apoA-IV (rat), apoC-III (rat and mouse), and apoE (rat, dog, and pig). Comparison of the NH_2-terminal sequences of the mammalian apoA-Is reveals a marked degree of homology between species in this section of the protein. Indeed many of the residue differences involve highly conservative mutations and can be accounted for by a single change in a nucleotide base in the codons involved (for apoA-I, see Fig. 5 in ref. 20). Such marked sequence homology is entirely consistent with the extensive (partial) immunological cross-reactivity observed among these proteins, and notably for apoE[206] and apoB.[50,129,164,207] ApoB nonetheless remains the apolipoprotein whose structure is the least understood, not only in man but also in mammals as a whole. Indeed, the molecular weight values shown in Table XII for the human and mammalian B proteins remain controversial since they were determined by polyacrylamide gel electrophoresis in SDS. This methodology is fraught with difficulty and imprecision when applied to studies of labile, carbohydrate-containing proteins of apparent high M_r (>100,000) and a marked tendency to aggregate once devoid of lipid.[62,63,67,207,208] Moreover, it remains to be established as to whether the low M_r form (B-48; apoB$_L$; P-III) seen in mouse, dog, and rabbit[13,67,68] represents a fragment of the larger B protein (B-100, apoB$_H$, P-I) as appears to occur in man and rat.[65,209] Comparative data on counterparts to the human D, F, G, Lp(a), and proline-rich apolipoproteins are also lacking,[195] although, as described above, Lp(a) is widely distributed among nonhuman primates.[17,52,55,56]

Finally, the posttranslational processing of the primary translation products of apolipoproteins, i.e., the preproapolipoprotein, is attracting great interest. Comparative data are however restricted to rat and mouse.[1,2,4,7,210–213] In the former species, it is relevant that the pro-form of apoA-I, containing an additional NH_2-terminal segment of six residues

---

[206] K. H. Weisgraber, R. F. Troxler, S. C. Rall, and R. W. Mahley, *Biochem. Biophys. Res. Commun.* **95**, 374 (1980).

[207] M. J. Chapman and S. Goldstein, *Atherosclerosis* **25**, 267 (1976).

[208] P. Forgez, M. J. Chapman, and G. L. Mills, *Biochim. Biophys. Acta* **754**, 321 (1983).

[209] Y. L. Marcel, M. Hogue, R. Theolis, and R. W. Milne, *J. Biol. Chem.* **257**, 13165 (1982).

[210] J. I. Gordon, D. P. Smith, R. Andy, D. H. Alpers, G. Schonfeld, and A. W. Strauss, *J. Biol. Chem.* **257**, 971 (1982).

[211] J. I. Gordon, D. P. Smith, D. H. Alpers, and A. W. Strauss, *J. Biol. Chem.* **257**, 8418 (1982).

[212] M. C. Blaufuss, J. I. Gordon, G. Schonfeld, A. W. Strauss, and D. H. Alpers, *J. Biol. Chem.* **259**, 2452 (1984).

[213] G. Ghiselli, W. A. Bradley, A. M. Gotto, and B. C. Sherrill, *Biochem. Biophys. Res. Commun.* **116**, 704 (1983).

(X-Glu-Phe-X-Gln-Gln-), circulates at low concentration in rat plasma and presents as an A-I isoprotein with a p*I* of 5.50.[213] This hexapeptide is ultimately cleaved in the plasma to yield the mature protein.

Apart from the rat and rhesus monkey,[214–216] scant information exists on the absolute circulating concentrations of apolipoproteins in mammals. Levels of apoA-I and apoE in the guinea pig[217] (6.2 and 2.2 mg/dl, respectively) are some 15- to 20-fold and some 5-fold lower, respectively, than those of their human counterparts.[216] By contrast, apoC-II concentrations in pig serum (9–13 mg/dl)[218] are some 2- to 3-fold greater than those in man.[219]

Cholesterol-Induced Lipoproteins and Apolipoproteins

The prolonged administration of high-fat, high-cholesterol diets to mammals typically leads to a marked hyperlidemia and to the rapid and premature development of atherosclerosis. Mechanisms of lipid absorption, transport, and metabolism have therefore come under intense study in a variety of animals models, including the rat,[11] dog,[13] guinea pig,[45,82,220] rabbit,[12,119] Erythrocebus patas monkey,[26] rhesus monkey,[121,122] African green monkey,[221,222] cynomolgus monkey,[223] baboon,[124] and chimpanzee.[125] Such investigations have been extensively reviewed in recent years[11,13,36] and have emphasized the appearance of three types of atherogenic lipoprotein particle: β-VLDL, high-molecular-weight LDL and HDL$_c$. All of these particles display elevated cholesteryl ester contents, but while β-VLDL and HDL$_c$ are enriched in apoE, the protein moiety of high M_r LDL appears dominated by B proteins. Marked variation exists, however, in the rate and extent to which such lipoproteins and apoproteins may be induced in the various species, and even between individuals of the same species. Nonetheless, it appears that one of the earliest bio-

---

[214] B. J. Van Lenten, C. H. Jenkins, and P. S. Roheim, *Atherosclerosis* **37**, 569 (1980).
[215] B. J. Van Lenten and P. S. Roheim, *J. Lipid Res.* **23**, 1187 (1982).
[216] J. B. Karlin and A. H. Rubenstein, in "The Biochemistry of Disease" (A. M. Scanu, R. W. Wissler, and G. S. Getz, eds.), Vol. 7, p. 189. Marcel Dekker, New York, 1979.
[217] L. S. S. Guo, R. L. Hamilton, J. P. Kane, C. J. Fielding, and G. C. Chen, *J. Lipid Res.* **23**, 531 (1982).
[218] G. Knipping, E. Steyrer, R. Zechner, and A. Holasek, *J. Lipid Res.* **25**, 86 (1984).
[219] S. I. Barr, B. A. Kottke, J. Y. Chang, and S. J. T. Mao, *Biochim. Biophys. Acta* **663**, 491 (1981).
[220] C. Sardet, H. Hansma, and R. Ostwald, *J. Lipid Res.* **13**, 624 (1972).
[221] L. L. Rudel, L. L. Pitts, and C. A. Nelson, *J. Lipid Res.* **18**, 211 (1977).
[222] L. L. Rudel, J. A. Reynolds, and B. C. Bullock, *J. Lipid Res.* **22**, 278 (1981).
[223] L. L. Rudel and L. L. Pitts, *J. Lipid Res.* **19**, 992 (1978).

chemical events in mammals fed high cholesterol diets may be the induction of hepatic apoE biosynthesis.[3]

Of late, attention has been focused on the cellular degradation of cholesterol-induced lipoproteins, and particularly of the interaction of β-VLDL with monocyte–macrophages.[224] Such interactions may lead to formation of the cholesteryl ester-laden foam cells characteristic of atherosclerotic lesions.[224]

Future Perspectives

It will be evident from this review that much of our knowledge of mammalian lipid transport systems is incomplete and disseminated throughout a vast literature. This remark applies especially to the characterization of mammalian apolipoproteins, and particularly to those present as minor components; their low concentrations in no way diminish their potential physiological importance. It is to be expected therefore that current research directed to the application of molecular biological tools will enhance our fragmentary knowledge of the biosynthesis, chemical structure, and dietary and genetic modulation of apolipoproteins in mammals. Indeed, such investigations may advance our understanding of their molecular evolution, and so provide information on the possible existence of a common ancestral gene(s).[61]

Another notable development concerns determination of the main features of the lipid transport systems in species exhibiting pronounced seasonal variation in activity, such as the hedgehog and European badger.[76,77,90] This approach may well furnish basic data on the role of hormones (e.g., thyroid hormones, androgens, and estrogens) in lipoprotein physiology, a further illustration of the indispensable contribution of comparative studies to our further cognizance of the human lipid transport system.

Acknowledgment

I am indebted to Martine Tassier for preparation of the manuscript.

[224] R. W. Mahley, *Med. Clin. N. Am.* **66,** 375 (1982).

[4] Impact of Technology on the Plasma Lipoprotein Field

By ANGELO M. SCANU

Although it is now generally accepted that the plasma lipids are associated with specific proteins to form lipoprotein complexes of various size and density this notion has slowly evolved through a series of conceptual and technological developments which have taken place independently or in concert with other research areas. The existence of an association between lipids and proteins had been suspected as early as 1914[1] and 1926,[2] but it was not until the studies of Macheboeuf in 1929, at the Pasteur Institute, that well-defined water-soluble lipoproteins were demonstrated.[3] This documentation was derived from a chemical approach in which horse serum was subjected to half-saturated ammonium sulfate at neutral pH and then acidification to pH 3.9 of the globulin fraction which was not precipitated by the salting out procedure. We now know that what had been isolated were the high-density lipoproteins which in horse serum represent the major lipoprotein components. More than 10 years elapsed since this discovery when Blix et al.[4] in Uppsala identified lipoproteins in the α- and β-globulin fractions of blood serum by free boundary electrophoresis, a technique which had been successfully applied to the separation of the other plasma proteins. About 6 years later, Pedersen,[5] in the same laboratory, was the first to make use of the ultracentrifuge to separate a lipoprotein species by flotation in a medium of d 1.04 g/ml. Later De Lalla and Gofman[6] in the early 1950s developed and made wide use of ultracentrifugal flotation techniques and established ultracentrifugation as a reproducible and versatile method for the study of plasma lipoproteins. In its analytical and various preparative modes (flotation, density gradient, zonal, etc.), ultracentrifugation has helped in many important developments in the plasma lipoprotein field both from the structural and metabolic standpoints and still retains a dominant role aided by the developments in ultracentrifuge models and rotors (Chapters [6], [7], and [8], this volume and Chapters [1] and [2], Volume 129).

[1] H. Chick, *Biochem. J.* **8**, 404 (1914).
[2] A. H. T. Theorell, *Biochem. Z.* **175**, 297 (1926).
[3] M. A. Macheboeuf, *Bull. Soc. Chim. Biol.* **11**, 268 (1929).
[4] G. Blix, A. Tiselius, and H. Svensson, *J. Biol. Chem.* **137**, 485 (1941).
[5] K. O. Pedersen, *J. Phys. Chem.* **51**, 156 (1947).
[6] O. De Lalla and J. W. Gofman, in "Methods of Biochemical Analyses" (D. Glick, ed.), Vol. 1. Wiley (Interscience), New York, 1954.

Another physical method which has aided in the studies of plasma lipoproteins has been electrophoresis. From the early use of the free-boundary method, electrophoresis in various supporting media (agarose, cellulose acetate, etc.) has gained a wide acceptance. In fact electrophoresis in albuminated buffers according to the method developed by Lees and Hatch[7] has served as one basis for phenotyping lipoprotein disorders. More specialized applications were made possible by the use of starch, polyacrylamide gel and Geon Pevikon (see for review, ref. 8). More recently, electrophoresis through gradients of polyacrylamide gel has permitted the resolution of lipoprotein subspecies and through changes in gradient geometry has assured versatility both on the analytical and preparative scales generating novel structural and metabolic concepts (Chapter [24], this volume).

Besides the physical methods outlined above, column chromatography in beds of various pore size either alone (Chapter [3], Volume 129) or in combination with ultracentrifugal and electrophoretic techniques has acquired an important place in the study of plasma lipoproteins. Because of the potential artifacts attending the separation of lipoproteins in high salts and in a high gravitational field, relatively "gentle" chromatographic techniques in media of a low ionic strength have provided a useful complement to ultracentrifugation and have also proven to be powerful when used in affinity modes. Ligands such as heparin (Chapter [13], this volume and Chapter [11], Volume 129), antibodies to various apolipoproteins (Chapter [8], Volume 129), etc. immobilized onto a solid support have created efficient systems for either the separation or purification of lipoprotein fractions. Immunoaffinity procedures are seeing increasing application in lipoprotein studies particularly with the use of monoclonal antibodies. Recently high-performance liquid chromatography (Chapter [4], Volume 129) has been introduced to the separation of plasma lipoproteins[9] and more applications in this area are expected through the continuing developments of columns in the molecular exclusion, reverse-phase, and ion-exchange modes. Besides the physical methods outlined above, chemical precipitation techniques (Chapter [5], Volume 129) have also been applied to the study of lipoproteins[10] particularly making use of heparin or other polyelectrolytes in conjunction with metals such as manganese or calcium based on the capacity of these systems to preferentially

[7] R. S. Lees and F. T. Hatch, *J. Lab. Clin. Med.* **61,** 518 (1963).
[8] A. M. Gotto, in "Handbook of Electrophoresis" (L. Lewis, ed.), Vol. III, p. 3. CRC Press, Boca Raton, Fla., 1983.
[9] Y. Ohno, M. Okazaki, and I. Hara, *J. Biochem. (Tokyo)* **89,** 1675 (1981).
[10] M. Burnstein and P. Legmann, *in* "Lipoprotein Precipitation" (T. B. Clarkson, D. Kritchevsky, and O. J. Pollak, eds.). Karger, Basel, 1982.

interact with carbohydrate-containing lipoproteins such as apoB and apoE. In spite of some limitations, techniques of this kind have imposed themselves by their simplicity and have seen particular use in clinical medicine.

The importance that technical advances have made on the lipoprotein field is also well exemplified by the methods which were introduced in the early 1950s for the preparation of apolipoproteins in a lipid-free form. The first successful method dealt with the delipidation of human serum HDL with the generation of a water-soluble protein moiety, apoHDL, amenable to physicochemical characterization.[11] This early procedure and others developed later paved the way to the study of the properties of the various apolipoproteins, an area of research which has seen extraordinary advances in the past few years. It may be pointed out that until the introduction of the delipidation techniques (Chapter [9], this volume), protein chemists were unattracted by lipoprotein research because of interfering lipids. Once this obstacle was overcome, routine methods of protein chemistry could be applied to the lipid-free apolipoproteins. Through the work of several laboratories they were isolated in a pure form and characterized aided by the chromatographic and electrophoretic developments in the general area of protein chemistry (Chapters [10–16], this volume). Urea and SDS–polyacrylamide gel electrophoresis, isoelectric focusing (Chapter [25], this volume), two-dimensional gel electrophoresis (Chapter [10], this volume), conventional column chromatography, and, more recently, high-performance liquid chromatography (Chapter [18], this volume) all have seen extensive use in this area.[12] In addition, the applications of automated sequence techniques have led to the elucidation of the primary structure of these apolipoproteins and the knowledge which was derived has permitted predictions of their secondary structure and identification of amphipathic helical domains involved in lipid binding.[13,14] This information has led to the chemical synthesis of peptides mimicking those present in the naturally occurring apolipoproteins and to a facilitation of our understanding about their interaction with natural and artificial amphipathic surfaces (Chapter [36], this volume).

Of considerable impact has also been the application of immunological techniques made possible by the availability of poly- and monoclonal antibodies against the major apolipoproteins (Chapter [31], this volume).

---

[11] A. M. Scanu, L. A. Lewis, and F. M. Bumpus, *Arch. Biochem. Biophys.* **74**, 390 (1958).
[12] C. Edelstein and A. M. Scanu, this volume [5].
[13] A. M. Scanu, C. Edelstein, and P. Keim, "The Plasma Proteins" (F. Putnam, ed.), 2nd Ed., p. 371. Academic Press, New York, 1975.
[14] J. C. Osborne, Jr. and H. B. Brewer, *Adv. Protein Chem.* **31**, 253 (1977).

The availability of these specific antibodies has been invaluable in the study of the biosynthesis and processing of the apolipoproteins (e.g., apoA-I, apoA-II, apoE, and apoC-III) and have also proven essential in the approaches used and being used in their cloning (Section V, this volume).

In providing an overview on the important role that laboratory techniques have played in the study of lipoproteins and apolipoproteins, mention should also be made about the invaluable aid provided by a number of fascinating human genetic variants such as abeta- and hypobetalipoproteinemia, Tangier disease, and apoC-II deficiency.[15] The studies of these disorders have generated important structural and functional insights and also stimulated technological advances. The field of research on plasma lipoproteins has grown up to be among the exciting ones in modern biology. If we look back on what has happened during the past 60 years we cannot help realize that major conceptual advances have been behind the development of new techniques. The various chapters describing these methods will readily support this conclusion.

---

[15] P. N. Herbert, A. M. Gotto, and D. S. Fredrickson, *in* "The Metabolic Basis of Inherited Disease" (J. B. Stanbury, J. B. Wynguaarden, and D. S. Fredrickson, eds.), p. 545. McGraw-Hill, New York, 1978.

Section II

Preparation of Plasma Lipoproteins

[5] Precautionary Measures for Collecting Blood Destined for Lipoprotein Isolation

By CELINA EDELSTEIN and ANGELO M. SCANU

Introduction

Although the problem of artifacts originating after blood collection has been recognized by some investigators[1] there is no general awareness of this problem among all of the workers in the field. Several studies summarized in a recent review[2] have shown that proteolytic enzymes of different types can affect the cleavage of proapoA-I, apoA-II, apoB, and apoE. For example, the enzyme responsible for the cleavage of proapoA-I to apoA-I is present in circulation and is inhibited by ethylenediaminetetraacetic acid (EDTA).[3] Thus, collection of blood in the presence of EDTA would prevent further processing of proapoA-I. In the case of apoA-II, a protease derived from human polymorphonuclear cells (PMN) has been shown to promote hydrolysis of the HDL-associated apoprotein.[4] Moreover, upon release from the cell the enzyme becomes associated with HDL.[5] This PMN derived enzyme is of the elastase family and is inhibited by diisopropylfluorophosphate (DFP).[6] Addition of this reagent to plasma or serum would thus prevent proteolysis of apoA-II. In the case of apoE addition of DFP would also prevent its degradation by thrombin.[7] ApoB-100 can be degraded by tissue and plasma kallikrein to generate the smaller fragments apoB-74 and apoB-26.[8] A kallikrein inhibitor such as Trasylol (kallikrein inactivator) when added to serum or plasma would remedy this situation.

---

[1] P. S. Bachorik, *Clin. Chem.* **28**, 1375 (1982).
[2] A. M. Scanu, R. E. Byrne, and C. Edelstein, *J. Lipid Res.* **25**, 1593 (1984).
[3] C. Edelstein, J. I. Gordon, K. Toscas, H. F. Sims, A. W. Strauss, and A. M. Scanu, *J. Biol. Chem.* **258**, 11430 (1983).
[4] D. Polacek and A. M. Scanu, *Trans. Assoc. Am. Physicians* **95**, 86 (1982).
[5] D. Polacek, R. E. Byrne, M. Burrous, and A. M. Scanu, *J. Biol. Chem.* **259**, 14531 (1984).
[6] R. E. Byrne, D. Polacek, J. I. Gordon, and A. M. Scanu, *J. Biol. Chem.* **259**, 14537 (1984).
[7] W. A. Bradley, E. B. Gillian, A. M. Gotto, Jr., and S. H. Gianturco, *Biochem. Biophys. Res. Commun.* **109**, 1360 (1982).
[8] A. D. Cardin, R. L. Jackson, J. Chao, V. H. Donaldson, and H. S. Margolius, *J. Biol. Chem.* **259**, 8522 (1984).

TABLE I
PRESERVATIVE "COCKTAIL" FOR BLOOD COLLECTION

Stock solutions	Volume/250 ml	Final concentration
EDTA (sodium salt) 0.2 M, pH 7.4	4 ml	1.2 g/liter
Chloramphenicol 200 mg/ml	0.1 ml	80 mg/liter
Sodium azide 2.5%	1 ml	0.1 g/liter
Gentamicin sulfate 10 mg/ml	2 ml	80 mg/liter
Kallikrein inactivator 20,000 U/ml	0.125 ml	10,000 U/liter
NaCl 0.3 M, pH 7.4	7 ml	0.15 M

Reagent Preparation and Rationale

In view of the multiple enzymatic degradations that can occur during and after withdrawal of blood we have devised a "cocktail" which is added to the bottle before blood collection. The concentrations of the stock solutions and the volumes to be added to a 250 ml collecting bottle are listed in Tables I and II. Stock solutions of each component are kept refrigerated at 4°, except phenylmethylsulfonylfluoride (PMSF) which is kept under a drying agent in the freezer at $-20°$. The "cocktail" is added to the bottle just prior to blood collection. The addition of 0.3 M NaCl ensures that the salt molarity in the "cocktail" volume (14.2 ml) will equal that of the collected blood. The preservatives in Table I will produce plasma which can be an acceptable starting material for lipid and lipoprotein preparations. There are specific circumstances in which serum may

TABLE II
PRESERVATIVE SOLUTIONS FOR LIPOPROTEINS

Stock solution	Volume/ml	Final concentration
Benzamidine 1 M	1 μl	1 mM
PMSF[a] 0.2 M	5 μl	1 mM

[a] PMSF is dissolved in anhydrous methanol and stored under a drying agent at $-20°$.

be preferred to plasma and in such a case the "cocktail" is added after the clot has been removed. Sodium azide has a limited action on strict and facultative anaerobes which, however, are markedly inhibited by gentamicin sulfate and chloramphenicol.[1,3] Although it has been reported that sodium azide promotes the oxidation of lipids and partakes in the cleavage of polypeptide chains,[1,9] the concomitant addition of EDTA prevents these reactions. The choice of EDTA as an anticoagulant is also of importance. Low-molecular-weight anticoagulants such as fluoride, citrate, or oxalate exert osmotic effects that shift large amounts of water from erythrocytes to plasma[1,10] producing artifactually low lipoprotein concentrations. Heparin causes no detectable change in erythrocyte volume and may also be used as an anticoagulant; however EDTA has an added advantage in that it chelates heavy metals, such as Cu^{2+}, that promote autoxidation of unsaturated fatty acids and cholesterol.[1,11] Moreover, EDTA also inhibits bacterial phospholipase C[12] which may produce physicochemical changes of the lipoproteins. Benzamidine is used as a proteolytic inhibitor and is also a specific inhibitor of trypsin.[13] Proper mixing of the "cocktail" with the collected blood is essential since it will ensure that all constituents have come in contact and that preventive measures have begun. In this context, Pflugshaupt and Kurt[14] have measured the appearance of fibrinopeptide A as a result of its cleavage from fibrinogen by thrombin. The appearance of fibrinopeptide A was thus an indicator of thrombin activation. These investigators found that the content of fibrinopeptide A depends strongly on the quality of blood collection and that improved mixing of the anticoagulant with the blood led to a drastic reduction of fibrinopeptide A and therefore activated thrombin. The following procedure is recommended and used routinely in the laboratory for the preparation of lipoproteins.

Procedure

In preparing plasma lipoproteins it is important to utilize known subjects in good condition of health. We generally utilize a young population of male donors between 18 and 24 years of age, blood group A positive.

[9] M. J. Chapman and J. P. Kane, *Biochem. Biophys. Res. Commun.* **66**, 1030 (1975).
[10] C. Alper, in "Clinical Chemistry, Principles and Techniques" (R. J. Henry, D. C. Cannon, and J. Winkelman, eds.), 2nd Ed., p. 373. Harper, New York, 1974.
[11] J. Nishida and F. A. Kummerow, *J. Lipid Res.* **1**, 450 (1960).
[12] H. A. Schwertner and H. S. Friedman, *Am. J. Clin. Pathol.* **59**, 829 (1973).
[13] B. Keil, in "The Enzymes" (P. D. Boyer, ed.), 3rd Ed., Vol. 3, p. 250. Academic Press, New York, 1971.
[14] R. Pflugshaupt and G. Kurt, *Blut* **45**, 73, Abstr. 78 (1982).

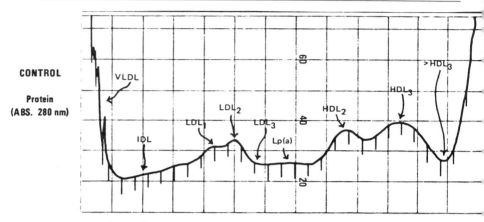

FIG. 1. Single-spin lipoprotein absorption profile of normolipidemic plasma. The gradient consisted of 1 ml of d 1.21 g/ml, 4 ml of 4 M NaCl, 0.2 ml of plasma diluted to 1 ml with 0.15 M NaCl, and 6.8 ml of 0.67 M NaCl. All solutions contained 0.05% EDTA and were adjusted to pH 7.0. The gradient was spun in a swinging bucket rotor at 39,000 rpm for 48 hr at 15°. The abscissa represents the volume collected (0.4 ml/tube) and the ordinate is the absorbance at 280 nm.

Each donor is instructed to fast for at least 12 hr before blood is drawn by trained personnel in the hospital blood bank. In most instances 450 ml of blood is collected into two polycarbonate centrifuge bottles (250 ml each) containing the "cocktail" of stock solutions (Table I) and are placed in a bucket filled with crushed ice. The bottles are manually agitated during collection and within 1 hr the blood is spun in a refrigerated (4°) Sorval RC-2 centrifuge for 20 min at 5000 rpm. Plasma is separated from the sedimented cells and respun if necessary in 35 ml polycarbonate tubes for 15 min at 10,000 rpm. To the clear plasma supernatant benzamidine and PMSF are added to a concentration of 1 mM (Table II) and used immediately for lipoprotein preparation.

On his first visit each donor is phenotyped by isopycnic density gradient ultracentrifugation using a single-spin methodology[15] which permits the identification of the main lipoprotein classes and subclasses and their respective density limits. From this knowledge one is able to classify the donors according to the predominance of a given lipoprotein class or subclasses. For example those donors with relatively high HDL$_2$ are more suited for obtaining apoA-I in high yields whereas HDL$_3$ ensures the preparation of both apoA-I and apoA-II. Moreover, Lp(a)-positive sub-

[15] J. Nilsson, V. Mannickarottu, C. Edelstein, and A. M. Scanu, *Anal. Biochem.* **110,** 342 (1981).

jects can be identified and thus used for the study of these interesting LDL variants. A representative serum profile generated by the single-spin density gradient is shown in Fig. 1. Note the resolution of the major classes of plasma lipoproteins, VLDL, LDL, Lp(a), and HDL and their relative subclasses. Quantitative information can also be obtained from computer-assisted digitization. We find that the lipoprotein distribution varies from individual to individual and is characteristic of each normolipemic subject independent of time.

Acknowledgments

The authors wish to thank the excellent assistance of Mrs. Barbara Kass in typing the manuscript. The work was supported by Program Project USPHS HL-18577.

[6] Sequential Flotation Ultracentrifugation

By VERNE N. SCHUMAKER and DONALD L. PUPPIONE

Sequential Flotation Ultracentrifugation

Because plasma lipoproteins have lower hydrated densities relative to the other plasma proteins, sequential flotation ultracentrifugation has been the principal method used for their isolation and classification. Based on the solvent densities used during ultracentrifugal separation, plasma lipoproteins may be defined as belonging to the very low-density lipoprotein (VLDL) class, the intermediate-density lipoprotein (IDL) class, the low-density lipoprotein (LDL) class, and the high-density lipoprotein (HDL) class. The table[1-4] lists each of these classes together with the density intervals used in their isolations. In addition, the table contains physicochemical data corresponding to each lipoprotein class. Many papers and reviews have been written about the techniques of lipoprotein isolation.[5-13]

[1] D. S. Frederickson, R. I. Levy, and R. S. Lees, *New Engl. J. Med.* **276**, 32 (1967).
[2] D. L. Puppione, G. M. Forte, A. V. Nichols, and E. H. Strisower, *Biochim. Biophys. Acta* **202**, 392 (1970).
[3] V. H. Wieland and D. Siedel, *Z. Klin. Chem. Klin. Biochem.* **10**, 311 (1972).
[4] A. V. Nichols, *Proc. Natl. Acad. Sci. U.S.A.* **64**, 1128 (1969).
[5] O. F. DeLalla and J. W. Gofman, *in* "Methods of Biochemical Analysis" (D. Glick, ed.), Vol. 1, p. 459. Wiley (Interscience), New York, 1954.
[6] R. J. Havel, H. A. Eder, and J. H. Bragdon, *J. Clin. Invest.* **34**, 1345 (1955).

The ability to process simultaneously large volumes or many small samples is one of the principal advantages of sequential flotation ultracentrifugation. For example, lipoproteins from a unit of blood can be readily separated in a rotor, such as a Beckman Ti 70, or as many as 18 separate samples with volumes of 6 ml or less can be processed with a Beckman 40.3 rotor. From 1950 until the present, the majority of ultracentrifugal separations of lipoproteins probably were accomplished with the 40.3 rotor. The fractions obtained have been extensively studied and a considerable amount of concentration and compositional data have been determined for the lipoproteins in each of the density classes.[4,8,12,14]

As a result of these studies, it now is known that some of the apolipoproteins redistribute from one lipoprotein class to another during metabolism, principally between the VLDL and HDL classes.[15] Because centrifugation times are 18 hr or longer, these same types of exchanges or transfers including losses of apoproteins from the lipoprotein surface have been demonstrated to occur during ultracentrifugal separation.[16] This alteration in the protein moiety of the lipoproteins is a major disadvantage of sequential flotation ultracentrifugation. Moreover, during the course of long centrifugation, extensive lipid peroxidation can occur.[17] However, this problem may be minimized by the inclusion of appropriate chelating agents and antioxidants, as recommended below.

Proper use of sequential flotation ultracentrifugation requires attention to some important features, occasionally overlooked. First, in adjusting the density of the plasma to a given volume, calculations must be based on the volume of the background salt solution, and not on the total volume of plasma. Second, if a rotor, other than the two we shall discuss, is to be used, it is important to calculate the effect of differences in

---

[7] F. T. Lindgren, A. V. Nichols, and N. K. Freeman, *J. Phys. Chem.* **59,** 930 (1955).
[8] N. K. Freeman, F. T. Lindgren, and A. V. Nichols, in "Progress in the Chemistry of Fats and other Lipids" (R. T. Holman, W. O. Lundberg, and T. Malkin, eds.), Vol. 6, p. 215. Pergamon, New York, 1963.
[9] F. T. Lindgren, A. V. Nichols, N. K. Freeman, R. D. Wills, L. Wing, and J. E. Gulberg, *J. Lipid Res.* **5,** 68 (1964).
[10] A. M. Ewing, N. K. Freeman, and F. T. Lindgren, *Adv. Lipid Res.* **3,** 25 (1965).
[11] F. T. Hatch and R. S. Lees, *Adv. Lipid Res.* **6,** 1 (1968).
[12] A. V. Nichols, *Adv. Biol. Med. Phys.* **11,** 109 (1967).
[13] F. T. Lindgren, in "Analysis of Lipids and Lipoproteins" (E. G. Perkins, ed.), p. 204. American Oil Chemists Soc., Champaign, Ill., 1975.
[14] A. M. Scanu, C. Edelstein, and B. W. Shen, in "Lipid-Protein Interactions" (P. C. Jost and O. H. Griffith, eds.), Vol. 1, p. 259. Wiley, New York, 1982.
[15] E. J. Schaefer, S. Eisenberg, and R. I. Levy, *J. Lipid Res.* **19,** 667 (1978).
[16] S. T. Kunitake and J. P. Kane, *J. Lipid Res.* **23,** 936 (1982).
[17] B. R. Ray, E. O. Davidson, and H. L. Crespi, *J. Phys. Chem.* **58,** 841 (1954).

APPROXIMATE PHYSIOCHEMICAL PARAMETERS CORRESPONDING TO THE DENSITY CLASSES OF HUMAN LIPOPROTEINS[a–e]

Classes	Density range (g/ml)	Corresponding flotation rates	Corresponding size range (nm)	Corresponding electrophoretic class	Concentration range (mg/dl)[f]	
					Male (35–50 years)	Female
Chylo	<0.940	$S_f^0 >$ 400	80–500	Origin	12 ± 13	2 ± 3
VLDL	0.940–1.006	S_f^0 20–400	30–80	Pre-β	129 ± 122	59 ± 63
IDL	1.006–1.019	S_f^0 12–20	25–30	Pre-β and β	40 ± 23	24 ± 14
LDL	1.019–1.063	S_f^0 0–12	16–25	β	399 ± 81	365 ± 56
HDL	1.063–1.210	$F_{1.20}$ 0–20	9–13	α	300 ± 83	457 ± 115

[a] Chylo, chylomicron; VLDL, very low-density lipoproteins; IDL, intermediate-density lipoproteins; LDL, low-density lipoproteins; HDL, high-density lipoproteins. S_f^0 is used to define the flotation rate as being measured at 26° in a medium of 1.745 molal NaCl (d = 1.063 g/ml). $F_{1.20}$ is used to define the flotation rate as being measured at 26° in a medium of 2.766 molal NaBr, 0.189 molal NaCl (d = 1.200 g/ml).
[b] These relationships are applicable to most, but not all human plasma lipoproteins. For example, the density distribution of α and β lipoproteins can be quite different in individual cases. α Lipoproteins with densities less than 1.063 g/ml are found in persons with hepatocellular disease,[1] abetalipoproteinemia,[1] and extreme hyperalphalipoproteinemia,[2] and β lipoproteins with densities greater than 1.063 g/ml are found in patients with hyperthyroidism.[3] Rather than change the definitions in the table to accommodate such variations, the physicochemical properties of charge and density can be combined to describe these lipoproteins. i.e., α LDL in one case and β HDL in the other. This approach recognizes the polydispersity of these lipoproteins and utilizes these density terms as originally intended, viz. to indicate that lipoproteins within a given density fraction actually have hydrated densities lying within the specified range.
[c] Frederickson et al.[1]
[d] Puppione et al.[2]
[e] Wieland and Siedel.[3]
[f] Nichols.[4]

pathlength. Often separation will not be achieved simply by duplicating the total number of gravity minutes. If conditions necessitate a considerable change in run parameters, particularly a substantial increase in rotor speed, it is strongly recommended that the extent of salt redistribution occurring during a run be determined in a preliminary experiment; some adjustment in density may be necessary to separate the desired class of lipoproteins. Third, it should be kept in mind that the time it takes for lipoproteins to float to the top of the tube is usually less than the time it takes for more dense lipoproteins and plasma proteins to sediment to the bottom, and that both processes must occur to accomplish a separation. Fourth, due to the convection which results during centrifugation with the angle head rotor, it is virtually impossible to reduce the level of contaminating proteins, such as albumin, below detection limits with a single spin. Fifth, although we recommend the use of concentrated salt solutions for the adjustment of density, solid salts, KBr, or NaBr, can be added directly to the plasma, but before weighing these salts, they first must have been dried overnight in an oven. Finally, if the temperature of the run is lowered, the resulting change in the solution viscosity will necessitate increasing the length of the run.

Sample Preparation

Addition of Inhibitors

When chylomicrons are not desired, blood is withdrawn after an overnight fast. While lipoprotein fractions have been isolated satisfactorily from both serum and plasma, collecting blood in tubes containing an anticoagulant is recommended because thrombin can cleave some serum apolipoproteins.[18,19] A 4% w/v solution of Na_2EDTA can be added initially to collecting vials in sufficiently small volumes such that the final concentration of this anticoagulant in blood is 1 mg/ml. Plasma has the added advantage that once the blood is drawn it can be isolated, stored, and refrigerated almost immediately. If citrated blood is used, the Na_2EDTA solution still should be added such that the final concentration is 0.04%; EDTA is required to chelate contaminating heavy metal ions which catalyze lipid peroxidation by molecular oxygen. Recommended additives include the antibiotic gentamicin (which has the advantage of having a low extinction coefficient in the near UV), an organic mercurial

[18] W. A. Bradley, A. J. Gotto, Jr., D. L. Peavy, and S. H. Gianturco, *Arteriosclerosis* **3**, 508a (1983).
[19] D. M. Lee and S. Singh, *Arteriosclerosis* **3**, 511a (1983).

such as thimerosal (merthiolate), and NaN_3. Use of NaN_3 has been criticized,[20] but it is an excellent inhibitor of bacterial and fungal growth. Proteolytic inhibitors may be added as well[21] (see Chapter [5], this volume).

Measure plasma volume and add (final concentration)
 EDTA (4% stock solution) (0.04%)
 Gentamicin (1% stock solution) (0.005%) and/or
 NaN_3 (5% stock solution) (0.05%)
 Optional
  PMSF (0.015%)
  Vitamin E (20 μM)
  Thimerosal (0.01%)
 PMSF (phenylmethylsulfonylfluoride, an inhibitor of serine proteases) 15 mg in 0.5 ml dimethyl sulfoxide per 100 ml plasma
 Aqueous vitamin E (Aquasol E) oil-soluble vitamin E (dl-α-tocopheryl acetate) water solubilized with polysorbate 80. Armour Pharmaceutical Company, Kankakee, Illinois (1 IU = 1 mg)

An extensive discussion of salt solution (NaCl and NaBr) preparation, calculations, and monitoring of density is given, together with tables containing values for density, viscosity, refractive index, and concentration in a useful chapter by Lindgren.[13] For most purposes, however, the tables for NaCl and NaBr solutions at 20° in the "Handbook of Physics and Chemistry" are sufficient,[22] and we have used these in developing the following protocols.

For the isolation of any of the lipoprotein classes with densities less than 1.063 g/ml, NaCl solutions only need to be used to increase solvent density. If the HDL are desired as well, however, it is more convenient to employ NaBr solutions at each step in the isolation procedure, since density calculations become rather complex if salt mixtures are used.

Solutions Required

We recommend an initial preparation of three stock solutions from which all of the other solutions may be prepared, and these are described below. The mock solution is so called because the stated concentration of

---

[20] J. Schuh, G. F. Fairclough, Jr., and R. H. Haschemeyer, *Proc. Natl. Acad. Sci. U.S.A.* **75**, 3173 (1978).
[21] A. D. Cardin, R. L. Jackson, J. Chao, V. H. Donaldson, and H. S. Margolius, *Fed. Proc. Fed. Am. Soc. Exp. Biol.* **43**, 1532 (1984).
[22] Section D, General Chemistry, *in* "Handbook of Chemistry and Physics" (R. Weast and M. J. Astle, eds.). CRC Press, Boca Raton, Fl., 1983.

NaCl has a density equivalent to that of the background salt solution of plasma, although this is not equivalent to the concentration of NaCl in plasma. The two concentrated salt solutions are 25% NaCl and 38% NaBr; these should be prepared with densities close to those suggested below. However, the calculated volumes required for subsequent use should be based on the measured values of the densities of each of these concentrated stock solutions. It is recommended that the NaCl and NaBr be dried overnight in a muffle furnace, and transferred to tightly closed jars prior to weighing. The concentrated salt solutions are more easily prepared if they are gently heated while being stirred. They are conveniently stored in reagent bottles with plastic screw caps. Using volumetric flasks for the final volume adjustment, but not for storage, the following solutions should be prepared:

1. Mock, a solution with the same background salt density as plasma (d = 1.0063 g/ml) 11.42 g NaCl + 100 mg Na_2EDTA per liter.

2. Concentrated NaCl (25 wt%, d = 1.1887 g/ml) 297.2 g NaCl + 100 mg Na_2EDTA per liter.

3. Concentrated NaBr (38 wt%, d = 1.3860 g/ml) 526.7 g NaBr + 100 mg Na_2EDTA per liter.

Sample Calculations

In order to illustrate how these solutions are used, we shall calculate the amount of concentrated NaCl solution and of water to add to 200 ml of plasma, to adjust the final volume to 215 ml and the background salt density to 1.019 g/ml. It will be assumed that plasma has a salt background density of 1.0063 g/ml and is 6% of other solids (mostly protein) by volume. For example, the volume of other solids in 200 ml of plasma would be 200 ml × 6/100 = 12 ml.

The equations which will be employed in all of these calculations are algebraic rearrangements of the mass balance equation for salt:

$$(V_t - V_s) \times d \times (\%A/100) = w$$

where V_t is total volume, V_s is the volume of solids other than the background salt, d is density of the background salt solution, $\%A$ is weight percent of the background salt, and w is the weight of the background salt.

From the "Handbook of Physics and Chemistry" the weight percent of NaCl in a 1.0063 g/ml solution is found by linear interpolation to be 1.1375%, and in a 1.019 g/ml solution it is 2.92%.

Initial weight background NaCl = (200 ml − 12 ml) × 1.0063 g/ml × 1.1375/100 = 2.152 g.

Final weight background NaCl = (215 ml − 12 ml) × 1.019 g/ml × 2.92/100 = 6.040 g.
Amount of salt required = 6.040 g − 2.152 g = 3.888 g.
Amount of 25% concentrated NaCl solution to add = 3.888 g/(1.1887 g/ml × 25/100) = 13.08 ml.
Amount of water to add = 15 ml − 13.08 ml = 1.92 ml.

If the 12 ml volume occupied by protein were not subtracted in making these calculations, the resulting error would cause the addition of too much of the concentrated salt solution, and the final background density would be 1.021 instead of 1.019 g/ml.

Repeating the same problem using NaBr as the density increasing solute requires for simplicity the introduction of a fiction that plasma contains NaBr as its background salt solution of density 1.0063 g/ml. From the salt tables in the "Handbook of Chemistry and Physics" it is found by interpolation that this density corresponds to a weight percent of NaBr of 1.0385%. For the 1.019 g/ml solution, the weight percent of NaBr is 2.65%.

Initial weight of the background "NaBr" = (200 ml − 12 ml) × 1.0063 g/ml × 1.0385/100 = 1.9646 g.
Final weight of the background NaBr = (215 ml − 12 ml) × 1.019 g/ml × 2.65/100 = 5.4817 g.
Additional salt required = 5.4817 g − 1.9646 g = 3.5171 g.
Volume of concentrated NaBr solution to add = 3.5171 g/(1.3860 g/ml × 38/100) = 6.68 ml.
Volume of water to add = 15 ml − 6.68 ml = 8.32 ml.

The dilution of an isolated fraction is another type of problem frequently encountered. For example, to dilute 53 ml of a solution containing a total quantity of 400 mg of low-density lipoproteins (density of LDL = 1.032 g/ml) in a 1.063 g/ml NaBr (8.00% w/v) background solvent to a solution containing a total volume of 175 ml in a 1.019 g/ml NaBr background solvent, the following calculation may be employed:

Volume occupied by the lipoprotein = 0.4 g/1.032 g/ml = 0.39 ml.
Initial weight of the background NaBr = (53 ml − 0.39 ml) × 1.063 g/ml × 8.00/100 = 4.474 g.
Final weight of the background NaBr = (175 ml − 0.39 ml) × 1.019 g/ml × 2.65/100 = 4.7151 g.
Additional salt required = 4.7151 g − 4.474 g = 0.2411 g.
Volume of concentrated NaBr solution to add = 0.2411 g/(1.3860 g/ml × 38/100) = 0.46 ml.
Volume of water to add = 175 ml − (53 ml + 0.46 ml) = 121.54 ml.

1. The plasma protein concentration varies from individual to individual over a range of 6 to 8% by weight, and this affects the volume correction and, thus, the calculated densities. Therefore, it is within experimental accuracy to make the several approximations implicit in these sample calculations. Additivity of volumes is assumed. Linear interpolation of the salt tables given in the Handbook is employed, and the very small effect of the dilute reagents EDTA, NaN_3, antibiotics, or other additives on the final densities is ignored, although the volumes of such added reagent stock solutions must be included in the total volume of the final solution.

2. In order to calculate the volumes occupied by the lipoproteins so that these volumes may be subtracted when adjusting the densities of the background salt for recentrifugation of the isolated lipoproteins, approximate densities and amounts of these lipoproteins must be estimated. Approximate densities which may be used are VLDL, 0.95 g/ml; IDL, 1.012 g/ml; LDL, 1.032 g/ml; HDL, 1.12 g/ml. In normal human plasma, considerable variation in the amounts of these lipoproteins is found, but average concentrations after an overnight fast are listed in the table.

3. When adjusting densities it is crucial to mix all solutions thoroughly but gently to avoid lipoprotein denaturation.

4. Densities are most conveniently measured with the commercial density meters now available. Otherwise pycnometry is recommended, or, for the pure salt solutions, refractive index measurements are a convenient way to estimate densities, using data tabulated in the Handbook.

5. Densities may be measured after each centrifugation on the liquid present in the clear zone which begins just beneath the lipoprotein layer and extends about half way down the tube; essentially all of the protein has been cleared from this zone.

Ultracentrifugation

Choice of Ultracentrifuge Rotors and Tubes

The 40.3 rotor was designed for lipoprotein isolation; this steeply angled rotor with a short path length minimizes run times at 40,000 rpm, and with 18 available holes a number of different samples may be processed simultaneously. Depending upon rotor availability as well as upon the number and size of the samples to be processed, the use of other rotors may be more convenient. Therefore, it is convenient to have a simple way of converting run conditions from one rotor to another. The critical variables are the minimal and maximal radii, r_i and r_f, respectively, the dependence of the viscosity of water, η, upon the temperature, T, and the rotor speed, rpm, and the run time, t. The minimum radius also depends upon the volume placed in the centrifuge tube. The dimensions are provided by the manufacturers in manuals describing their rotors, and the viscosities of NaBr and NaCl solutions, and the dependence of the viscosity of water upon temperature is tabulated in the "Handbook of Physics and Chemistry."[22] The expression relating these variables is

$$\frac{(\text{rpm})_1^2 \, t_1}{(\text{rpm})_2^2 \, t_2} = \frac{\eta_1(T_1) \, \ln(r_f/r_i)_1}{\eta_2(T_2) \, \ln(r_f/r_i)_2}$$

For the 40.3 rotor, the minimal and maximal radii are 48.9 and 79.5 mm, while for the Ti-70, they are 39.5 and 91.9 mm.[23] However, when polycarbonate, thick walled, screw capped bottles are used, the minimal radius becomes 41.5 mm and the maximal radius 89.9 mm for the Ti-70. The number of hours required for a 40,000 rpm centrifugation at 20° in the Ti-70 rotor is calculated from the preceding equation as follows:

$$t_2 = \frac{(40{,}000 \text{ rpm})^2 \times 18 \text{ hr} \times 1.002 \text{ cP} \times \ln(89.9 \text{ mm}/41.5 \text{ mm})}{(40{,}000 \text{ rpm})^2 \times 1.002 \text{ cP} \times \ln(79.5 \text{ mm}/48.9 \text{ mm})}$$

Thus, for the "standard" 40,000 rpm, 18 hr, 20° centrifugation with the 40.3 rotor, the equivalent $\omega^2 t$ conditions would be 23 ml, 40,000 rpm, 28 hr, and 20° using the Ti-70 and the polycarbonate bottles. The run time can be decreased by increasing the rotor speed. But this will result in a more pronounced salt gradient at the top of the centrifuge tube, and speeds above 50,000 rpm are probably excessive when it is desired to duplicate the conventional fractionation procedures. High speeds also may shorten the life of the centrifuge bottles. Temperature is another variable which can have a marked effect on the run conditions by altering the viscosity of the solution. The assumption is made that the temperature dependence of the viscosity of the solution is identical to the temperature dependence of the viscosity of water, which is tabulated in the Handbook. Thus, if the temperature were decreased from 20 to 12°, the viscosity of water would increase from 1.002 to 1.235 cP and the run time using a 40.3 rotor would increase by the ratio of 1.235/1.002 or from 18 to 24 hr.

A wide choice of ultracentrifuge tubes and caps is available. The heat-sealed tubes are convenient for elimination of leakage, sterility, and protection against oxidation if sealed to contain a nitrogen atmosphere.[24] The

---

[23] Rotors Tubes and Accessories, PL-174N, Beckman Instruments, Inc., Palo Alto, Ca. 1984.

[24] D. M. Lee, A. J. Valente, W. H. Kuo, and H. Maeda, *Biochim. Biophys. Acta* **666**, 133 (1981).

"bell-top" design is said to be especially useful for flotation separations of lipoproteins.

The screw-top, thick-walled polycarbonate bottles are especially convenient for sequential flotation ultracentrifugation. After centrifugation, the top fraction containing the lipoproteins is easily seen with these clear bottles, and this fraction is readily and completely drawn off with a capillary pipet after the cap assembly has been removed. For the Ti-70, these polycarbonate bottles are rated to 60,000 rpm, and should last a long time if restricted to use at 40,000 rpm.

Protocol for Preparing Human Plasma Lipoproteins

To one unit of plasma, approximately 200 ml, Na_2EDTA, gentamicin, NaN_3, and PMSF are added as described above. The plasma is then distributed between eight screw-top, thick walled polycarbonate bottles of a Ti-70 rotor, which hold 23 ml each ($8 \times 23 = 184$ ml). If insufficient plasma is available, it may be diluted to the desired volume with Mock. The plasma is centrifuged at 20° and 40,000 rpm for at least 28 hr. The VLDL float to the top of the tube, while the other plasma proteins and lipoproteins sink toward the bottom.

At the conclusion of the centrifugation, the VLDL can be seen as a white, heavily light scattering layer at the meniscus. Below the VLDL will be a clear layer containing very little protein. Normal human LDL and IDL are yellow due to a substantial carotenoid content, and they can be seen toward the bottom half of the tube. A pellet formed at the bottom of the tube will also contain substantial amounts of lipoproteins; it is interesting to note that the sedimentation rate of S_f 6 LDL in a solvent of density of 1.0063 g/ml is about 5 S, thus, the LDL sediment together with the 4.5 S serum albumin in this solvent, and much of the LDL will be found in the protein pellet at the bottom of the tube.

The VLDL are removed in approximately 4 ml. Then, approximately 8 ml of the intermediate, clear layer can be removed and saved for density determination. The remaining subnatent solution is used to resuspend the protein pellet at the bottom of each tube. This can be done using a glass stirring rod to loosen the pellet and then resuspending the pellet in the subnatent solution. After the pellet has been resuspended, the subnatent solutions are pooled together to give about 88 ml. The appropriate amount of concentrated NaBr solution is then added, together with water, to adjust the density to 1.019 g/ml and to bring the final volume to 140 ml. To the water is first added the requisite amounts of the stock solutions of Na_2EDTA, NaN_3, and gentamicin to maintain the concentrations of these reagents in the final volume. This solution is divided between 6 centrifuge tubes and again centrifuged at 40,000 rpm, 20°, for 28 hr to float the IDL to the top.

The IDL are removed in a small volume, the clear zone saved for density determination, and appropriate volumes of concentrated NaBr, water, and additives are mixed with the combined subnatent solutions to adjust the density to 1.063 g/ml and the volume to 140 ml. The solution is divided between 6 tubes and centrifuged at 40,000 rpm, 20°, for 28 hr to float the LDL to the top.

The LDL are removed in a small volume with a capillary pipet, the clear zone saved for density determination, and appropriate volumes of concentrated NaBr, water, and additives are mixed with the combined subnatent solutions to adjust the density to 1.21 g/ml and the volume to 140 ml. The solution is divided between 6 tubes and centrifuged at 40,000 rpm, 20°, for 38 hr to float the HDL to the top. Notice that these smaller particles require longer centrifugation times; moreover, the plasma proteins sediment more slowly in the dense solvent.

The HDL are now removed with a capillary pipet, and the combined HDL fractions are dialyzed against the desired storage buffer, often 1.006 g/ml NaCl containing EDTA, NaN_3, gentamicin, and 10 mM Tris, pH 8.0. LDL and HDL fractions may be sterilized by filtration directly into sterile rubber stoppered bottles and stored at 4°. It is advisable to fill the bottles completely, eliminating most of the air, since the unsaturated lipids are sensitive to oxidation by atmospheric oxygen.

Often a single lipoprotein fraction is desired for analysis; in this case, fewer centrifugations are required for the isolation, although an additional flotation may be required to eliminate traces of contaminating plasma proteins. Thus, to isolate the LDL, the plasma should be initially adjusted to 1.019 g/ml and the VLDL + IDL removed in the first centrifugation, then the density is adjusted to 1.063 g/ml and the LDL are removed after the second centrifugation. Actually, many workers use a density of 1.055 instead of 1.063 g/ml to lessen the contamination of the LDL fraction with Lp(a).[25] Note: The effect of low temperature on the triglyceride-rich lipoproteins has been studied, and it has been found that temperatures below 24° can cause freezing of the triglycerides, if they are enriched in saturated fatty acids.[26] Therefore, in certain metabolic and nutritional studies, it may be desirable to work with the plasma and to perform the initial centrifugation just above room temperature for the isolation of such lipo-

[25] K. Simons, C. Ehnholm, O. Renkonen, and B. Bloth, *Acta Pathol. Microbiol. Scand.* **78B**, 459 (1970).

[26] D. L. Puppione, S. T. Kunitake, R. L. Hamilton, M. L. Phillips, V. N. Schumaker, and L. D. Davis, *J. Lipid Res.* **23**, 283 (1982).

proteins. The artifacts induced by lowering the temperature include both the aggregation of triglyceride-rich lipoprotein and the formation of dense, large, light scattering particles which are separated with the IDL by sequential flotation ultracentrifugation. Compositional analysis indicates that alterations in the apolipoprotein content on the surface of other lipoproteins might also occur to compensate for the structural alteration in the triglyceride-rich lipoproteins.[27]

Fractionation

Removal of the lipoprotein top fraction after centrifugation is readily accomplished with a capillary pipet if the proper lighting and working conditions are provided. A darkened room with a small light source such as a pencil flashlight positioned 6 to 8 in. above and pointed straight down upon the opened centrifuge tube illuminates the light scattering zone of concentrated lipoproteins, and makes them visible during removal. The centrifuge tube should be held in slightly larger, 1.5–2 cm holes drilled in a low plastic block to clearly expose most of the tube from the side; wire racks are not suitable. The plastic block should be mounted about 8 in. above the surfaces of the table, as on a small, sturdy box, with the top of the tube just below eye level. The scientist should be seated comfortably with elbows resting on the table when the fractionation is performed with the pipet. An important advantage of this method is the careful inspection that each tube receives; unusual amounts of lipoproteins, unusual pigmentation, and the presence of extra bands are readily seen and can be separated for analysis.

Although the amount of VLDL can vary a great deal in the normal human, a thin but highly light scattering zone, almost devoid of pigment, is seen after centrifugal flotation of the VLDL into the top fraction. The IDL fraction will contain even less clearly visible material both because the quantity is less, and because these smaller particles scatter less light. Some yellow pigment may be present. The LDL are seen as a distinct yellow band which scatters little light. The HDL are less pigmented and seen as a slightly yellow band. No scattered light is normally seen from the zone containing these small lipoproteins. Below the lipoprotein bands will be a clear zone extending halfway down the tube, from which all the protein has been removed by sedimentation or flotation. At the conclusion of centrifuge runs at 1.0063 and 1.019 g/ml, the yellow pigment will appear about half way down the tube, and it is composed of the pigmented lipoproteins when they are present, as well as some pigmented com-

[27] D. L. Puppione, unpublished observations, 1984.

pounds bound to albumin and to other proteins. The pellet on the tube bottom will contain plasma proteins and more dense lipoproteins, and it must be resuspended between centrifugations or most of the lipoproteins could be lost.

Note: After centrifugation, the salt density will vary substantially near the meniscus and the bottom of the centrifuge tube due to salt redistribution during the course of the centrifuge run. This affects particularly the salt density of the lipoprotein fraction removed from the top. It is good procedure to measure the density on the next milliliter or two of clear zone just beneath the top fraction to estimate the new density prior to calculating the density adjustment if further centrifugation of the isolated lipoprotein fraction is desired. It is also recommended that this density measurement be made to check the previous adjustment. If a density meter is available, the amount of labor is relatively small.

Analysis

The determination of both lipoprotein flotation coefficient distributions and refractometric concentrations by analytical ultracentrifugation probably yields more information than any other analytical technique because it has been employed so extensively in the past, and compositional data usually are given in terms of classes based on flotation rate values.[10] The technique of analytical ultracentrifugation is described in detail by Kahlon et al. in Chapter [2], Volume 129.

Paper electrophoresis and agarose electrophoresis also have been used extensively to characterized human serum lipoproteins.[1,11,28] The isolated fractions of HDL and LDL appear as the faster moving α and the slower migrating β bands, respectively, when studied at pH 8, while between these two the VLDL form a rather broad band, often referred to as a pre-β band.[11] Gradient gel electrophoresis and isoelectric focusing are described in other sections of this volume.

Chemical analyses of the VLDL fraction from normolipidemic individuals show that these particles contain an abundance of triglyceride, which, depending upon the average particle size, will be about 59% of the total mass. About 16% of the mass is phospholipid, 10% is protein, 10% is cholesteryl ester, and 5% is free cholesterol.[29] A typical LDL may contain 42% cholesteryl ester, 9% cholesterol, 8% triglyceride, 21% phospholipid,

[28] F. T. Hatch, F. T. Lindgren, G. L. Adamson, L. C. Jensen, and A. W. Wong, *J. Lab. Chim. Med.* **81,** 946 (1973).
[29] T. Sata, R. J. Havel, and A. L. Jones, *J. Lipid Res.* **13,** 757 (1972).

and 20% protein.[30] The HDL fraction contains approximately 48% protein, 14% cholesteryl ester, 27% phospholipids, 4.2% cholesterol, and 4.3% triglyceride.[14]

Isolation of Nonhuman Lipoproteins

Protocols which we have described are suitable in almost all cases for the isolation of human lipoproteins into relatively pure classes because most triglyceride-rich lipoproteins will float at a density of 1.0063 g/ml and because a density of 1.063 g/ml fortuitously separates the major apolipoprotein A-I-containing lipoproteins from the major apolipoprotein B-containing lipoproteins. Because of their experience in working with human lipoproteins, many investigators refer to all cholesteryl ester-rich lipoproteins containing apolipoprotein B as "LDL" and all cholesteryl ester-rich lipoproteins containing apolipoproteins A-I and A-II as "HDL." In classifying ultracentrifugally isolated nonhuman lipoproteins, however, confusion frequently results because the density intervals as defined in the table may no longer apply to lipoproteins belonging to a single electrophoretic class or having the expected apolipoprotein composition. In characterizing the lipoproteins of certain mammals, a density cut other than 1.063 g/ml may enable a clean separation of the α from the β lipoproteins. However, other animals will have two groups of lipoproteins which although distinct from one another in terms of their size distributions and protein moieties, may be present together within the same density interval. An HDL preparation free of apolipoprotein-B containing lipoproteins often cannot be obtained until a separation is made at 1.090 g/ml. Because human plasma typically contains only Lp(a) between a density of 1.063 and 1.090 g/ml, the majority of the α lipoproteins in human plasma can be isolated between 1.090 and 1.21 g/ml if a preparation free of Lp(a) is desired. However, in other mammalian plasmas the density distribution of α lipoproteins may be so broad that this approach would result in a recovery of only 50% of the lipoproteins containing apolipoproteins A-I and apoA-II. More often than not, the concentration of α lipoproteins greatly exceeds the level of β lipoproteins in the 1.063 to 1.090 g/ml interval. Moreover, obtaining an ultracentrifugally pure preparation of LDL containing apolipoprotein B as the major protein may not be possible, because the density distribution of α lipoproteins extends to a value of 1.020 g/ml in many species. In animals such as the rat, the 1.0063–1.040 g/ml density fraction reportedly yields a relatively pure preparation of β LDL. However, in other mammals such an approach

[30] V. N. Schumaker, *Acc. Chem. Res.* **6**, 398 (1973).

simply leads to a relative enrichment of the β lipoproteins to α lipoproteins. When working with lower vertebrates and invertebrates, specific classes of lipoproteins do not migrate with the electrophoretic mobility of an α- or β-globulin. In characterizing ultracentrifugally the lipoproteins of lower animals, it might simply be clearer to specify the density interval in which they were isolated. Finally, it is also important to determine the background salt density which may differ substantially from mammalian plasma.

Further Purification

Sequential flotation analysis is a separation method based on density alone; therefore, further fractionation can be based on the following:

1. Size, using gradient gels,[31,32] or gel filtration[29] or rate zonal ultracentrifugation techniques.[33]
2. Charge, using electrophoretic techniques such as Geon-Pevicon block electrophoresis.[34]
3. Type of protein or carbohydrate, using affinity columns, such as heparin,[35] or antibodies bound to a solid support.[36]
4. Density, since sequential flotation ultracentrifugation separates lipoproteins into rather broad density classes, it is possible to further fractionate the LDL and HDL into isopycnic classes using continuous density gradients in swinging bucket[37] or zonal rotors.[38]

Advantages and Disadvantages

Among the advantages of sequential flotation ultracentrifugation are that relatively large quantities of plasma or many samples can be processed at one time, that excellent purification of the lipoproteins from other plasma proteins of density >1.21 g/ml can be obtained, and that the

---

[31] D. W. Anderson, A. V. Nichols, T. M. Forte, and F. T. Lindgren, *Biochim. Biophys. Acta* **493**, 55 (1977).
[32] R. M. Krauss and D. J. Burke, *J. Lipid Res.* **22**, 97 (1981).
[33] T. S. Kahlon, L. A. Glines, and F. T. Lindgren, this series, Vol. 129 [2].
[34] R. W. Mahley and K. H. Weisgraber, in "Handbook of Electrophoresis" (L. A. Lewis and H. K. Naito, eds.), Vol. 4, p. 151. CRC Press, Boca Raton, Fl. 1983.
[35] J. Corey-Gibson, H. N. Gingsberg, A. Rubinstein, and W. V. Brown, this series, Vol. 129 [11].
[36] J. P. McVicar, S. T. Kunitake, R. L. Hamilton, and J. P. Kane, *Proc. Natl. Acad. Sci. U.S.A.* **81**, 1356 (1984).
[37] J. L. Kelly and A. W. Kruski, this volume [7].
[38] J. R. Patsch and W. Patsch, this series, Vol. 129 [1].

lipoproteins can be separated into well-studied classes. Moreover, the preparatory centrifuges and rotors employed are available in most modern biology and biochemistry laboratories. Among the disadvantages are the losses of a fraction of some of the apolipoproteins, especially A-I, C, and E. Some aggregation of the TG-rich lipoproteins may result when surface proteins are removed.[39] Full fractionation and purification by ultracentrifugal methods are both lengthy and costly. Also, sequential flotation ultracentrifugation separates lipoproteins according to density only, and the investigator must be aware that particles differing in stage of metabolism, composition, charge, and size may be combined in any of the fractions obtained with this method.

[39] W. J. Lossow, F. T. Lindgren, J. C. Murchio, G. R. Stevens, and L. C. Jensen, *J. Lipid Res.* **10**, 68 (1969).

[7] Density Gradient Ultracentrifugation of Serum Lipoproteins in a Swinging Bucket Rotor

By JIM L. KELLEY and ARTHUR W. KRUSKI

Introduction

Traditionally, human and animal serum lipoproteins have been separated ultracentrifugally by discontinuous stepwise increases in solvent density.[1] Although the methodology for density gradient ultracentrifugation has been available for many years, it did not achieve widespread usage until the mid-1970s when Redgrave *et al.*[2] used a four step density gradient consisting of mixtures of NaBr and KBr to separate human plasma lipoproteins into the major density classes namely, VLDL, LDL, and HDL.[2a] Ultracentrifugal separation was accomplished using a Beckman SW-41 swinging bucket rotor run for 24 hr at 41,000 rpm and 20°. The final salt gradient was continuous and curvelinear. Lipoprotein bands

[1] R. J. Havel, H. A. Eder, and J. H. Bragdon, *J. Clin. Invest.* **34**, 1345 (1955).
[2] T. G. Redgrave, D. C. K. Roberts, and C. E. West, *Anal. Biochem.* **65**, 42 (1975).
[2a] Abbreviations: TC, total cholesterol; VLDL, very low-density lipoproteins, $d < 1.006$ g/ml; IDL, intermediate-density lipoproteins, $d = 1.006-1.019$ g/ml; LDL, low-density lipoproteins, $d = 1.019-1.063$ g/ml; HDL_1, high-density lipoproteins, $d = 1.05-1.063$ g/ml; HDL_2, high-density lipoproteins, $d = 1.063-1.125$ g/ml; HDL_3, high-density lipoproteins, $d = 1.125-1.210$ g/ml.

were aspirated sequentially from the top of the centrifuge tube, saponified, and analyzed for cholesterol content. Foreman et al.[3] used an unusual NaCl step gradient which included solid sucrose crystals in the bottom of the tube and serum at $d = 1.006$ g/ml between two layers of higher density. A centrifugation time of 66 hr was required to ensure equilibrium. Chapman et al.[4] used a five step NaCl/KBr density gradient for separating human serum lipoproteins. Photographs of the separated lipoproteins were scanned for quantitation and fractions were characterized by their chemical, physical, and immunological properties.

An important improvement in the density gradient procedure was introduced by Nilsson et al.[5] A lipoprotein profile was generated by continuously monitoring for protein absorbance at 280 nm as the lipoproteins were pumped from the centrifuge tube. The absorbance was plotted and the area under the curve integrated for quantitation. Groot et al.[6] utilized a four step gradient to separate HDL_2 from HDL_3 by rate zonal centrifugation in a swinging bucket rotor (SW-40).

In this chapter, we describe the density gradient method which has evolved in our laboratory over the past several years[7,8] and will present examples of its application for the separation of human and baboon serum lipoproteins. This method employs a modification of the Redgrave gradient.[2]

Preparation of the Gradient

The stock density solutions used in the preparation of the step gradient are prepared as shown in Table I. The hygroscopic NaBr is dried by heating at 120° for 16 hr and stored in a desiccator. After preparation of the stock solutions, they are filtered and stored in brown bottles at 4°. Density is routinely checked with a Mettler/Paar DMA35 density meter.

A distinct advantage of the density gradient is that volumes from 0.2 to 4.0 ml of serum can be separated in each centrifuge tube. The preparation

---

[3] J. R. Foreman, J. B. Karlin, C. Edelstein, D. J. Juhn, A. H. Rubenstein, and A. M. Scanu, J. Lipid Res. **18**, 759 (1977).
[4] M. J. Chapman, S. Goldstein, D. Lagrange, and P. M. Laplaud, J. Lipid Res. **22**, 339 (1981).
[5] J. Nilsson, V. Mannickarottu, C. Edelstein, and A. M. Scanu, Anal. Biochem. **110**, 342 (1981).
[6] P. H. E. Groot, L. M. Scheek, L. Havekes, W. L. van Noort, and F. M. van't Hooft, J. Lipid Res. **23**, 1342 (1982).
[7] J. L. Kelley and A. W. Kruski, Arteriosclerosis **2**, 443a (1982).
[8] A. W. Kruski, in "Handbook of Electrophoresis" (L. A. Lewis and H. K. Naito, eds.), Vol. IV, p. 61. CRC Press, Boca Raton, Fla., 1983.

TABLE I
Preparation of Stock Sodium Bromide Solutions

	Density (g/ml)				
	1.006[a]	1.019	1.063	1.210	1.386
NaBr (g)[b]	4.5	13.51	42.46	141.7	262.4
10% EDTA (ml)[c]	2.5	2.5	2.5	2.5	2.5
Final volume (ml)	500	500	500	500	500

[a] Correct densities are verified using a Mettler/Paar DMA35 density meter.
[b] Sodium bromide is dried at 120° overnight and stored in a desiccator.
[c] The pH of the ethylenediaminetetraacetic acid (EDTA) is adjusted to 7.0.

of gradients using increasing volumes of serum is shown in Table II. Each stock density solution is layered into the centrifuge tubes manually, using a narrow bore 5 ml glass pipet (Fig. 1).

Centrifugation and Sample Collection

Centrifugation is accomplished using a Beckman SW-40 rotor for 24 hr at 38,000 rpm and 14° in a Beckman Model L5-65 ultracentrifuge. The ultracentrifuge run is allowed to come to a coasting stop (i.e., brake off) in order to minimize disruption of the separated lipoproteins.

Samples are monitored at 280 nm to give a continuous protein absorbance profile of the eluted gradient by means of the apparatus shown in

TABLE II
Preparation of Step Gradients with Different Volumes of Serum

		Stock sodium bromide solutions (ml)				
Serum (ml)	NaBr (g)	1.386	1.210	1.063 (g/ml)	1.019	1.006
0.2	—	0.8	3.2[a]	3.8	3.3	1.2
1.0	0.338[b]	—	3.0	3.8	3.3	1.1
2.0	0.675	—	2.0	3.8	3.3	1.0
4.0	1.350	—	—	3.8	3.3	0.8

[a] The stock density 1.21 g/ml NaBr solution is gently layered over the serum without mixing.
[b] NaBr is added directly to the centrifuge tubes, serum is added, and the salt is dissolved by gently mixing.

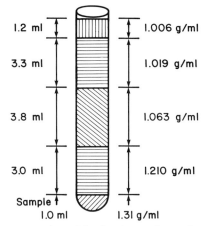

FIG. 1. Schematic representation of the density gradient using 1.0 ml of serum as the sample.

Fig. 2. The centrifuge tube containing the separated lipoproteins is punctured from the bottom and a dense solution is pumped into the bottom of the tube forcing the lipoproteins out of the top and through the UV monitor. A saturated NaBr solution or Fluorinert FC-40 (Sigma Chemical, #F-9755) can be used as the dense solution. Care must be taken not to intro-

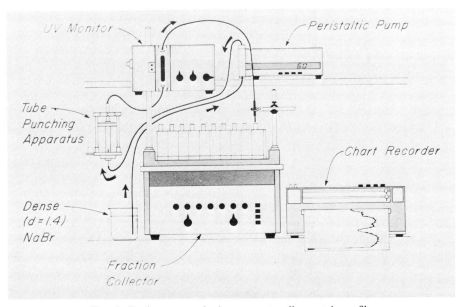

FIG. 2. Equipment required to generate a lipoprotein profile.

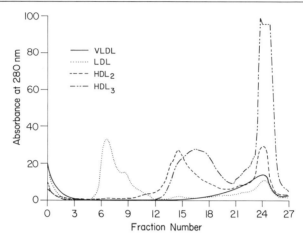

FIG. 3. Lipoprotein profiles of each of the major lipoprotein density classes (VLDL, LDL, HDL$_2$, HDL$_3$) initially prepared by preparative ultracentrifugation then run individually on the density gradient.

duce air bubbles into the gradient when the tube is punctured. This is accomplished by having the pump on during puncture so that the puncture needle is filled with dense solution.

Depending on the volume of serum used, the sensitivity of the UV monitor (LKB, 2138 Uvicord S) will have to be adjusted. If fractions are collected, it will also be necessary to measure and correct for the volume in the tubing from the UV monitor cuvette to the fraction collector in order to correlate the collected fractions with the absorbance profile. A pump speed of 60 ml/hr and a chart speed of 10 mm/min are used.

Fractionation of the 6 tubes from the SW-40 rotor is performed at room temperature. The area under the absorbance profile does not change significantly by diffusion during the 2-hr procedure (variation = 0.43%). The profile is complete when all the lipoproteins have passed through the cuvette and the serum proteins start eluting. The distance between the characteristic initial spike and the off-scale reading of the serum proteins is reproducible. Identification of lipoprotein peaks is accomplished by comparing profiles of lipoprotein fractions initially separated by sequential preparative ultracentrifugation[1,9] (Fig. 3).

In initial studies, we explored different step gradients and determined that the volumes of solutions shown in Table II gave the best separation of lipoproteins in whole serum. The influence of diffusion and centrifugation

[9] J. L. Kelley and P. Alaupovic, *Atherosclerosis* **24**, 155 (1976).

on the density of the fractions was also determined. In Fig. 4, plots of fraction versus density are shown. In these early studies the gradient was separated into 15 fractions. For the diffusion studies, three duplicate sets were placed at 4° (Fig. 4A), without centrifugation, and fractionated at 0, 24, and 48 hr. Identical sets were made up and immediately centrifuged for 6, 24, and 48 hr, and then fractionated (Fig. 4B). It is clear from these studies that diffusion is more important in establishing the gradient than centrifugation time.

In a separate study, the effect of centrifugation time on the separation of human serum lipoproteins by density gradient centrifugation was determined. Aliquots of the same serum were centrifuged for 24, 48, and 66 hr (Fig. 5). The LDL peak moved closer to the top of the tube (less dense)

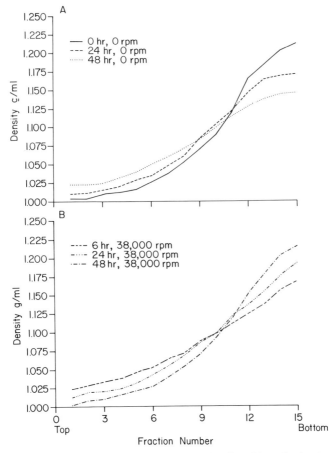

FIG. 4. The influence of diffusion (A) and centrifugation time (B) on the density gradient.

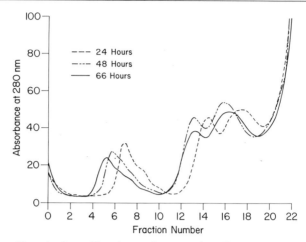

FIG. 5. Effect of length of centrifugation on the separation of human serum lipoproteins.

with increasing centrifugation time. The LDL peak height decreased and became broader, yet the peak area remained the same. The flattening of the gradient at the top of the tube between 24 and 48 hr (Fig. 4B) is in part responsible for the apparent shift in density with increasing time. The HDL peaks also shifted to a lower density between 24 and 48 hr, but they remained in the same relative position between 48 and 66 hr, having appeared to reach isopycnic equilibrium by 48 hr. For routine screening of samples we usually centrifuge for 24 hr, but for quantitation, 48 hr is more appropriate.

Reproducibility was determined by centrifuging aliquots of the same serum in the same run and on different days, with subsequent measurement of the area under the absorbance profile. Areas were measured using a Hewlett Packard Model 9874A digitizer and a Hewlett Packard Model 85 desk top computer. The within run variation was 2.87% and the between run variation was 6.95%.

Application and Analysis

We have used the density gradient procedure for the analysis of lipoproteins in both baboon and human serum. Figure 6 shows selected profiles (24 hr) of several baboons. The abscissa is the fraction number starting from the top to the bottom of the gradient (left to right), and the ordinate is the absorbance at 280 nm. A plot of total cholesterol per fraction versus fraction number would have a similar shape as the absorbance profile at 280 nm; however, the LDL region would be about twice

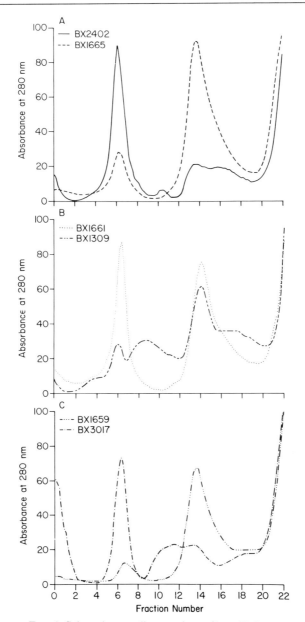

FIG. 6. Selected serum lipoprotein profiles of baboons.

the amplitude while the HDL region would have about half the amplitude. In subsequent descriptions, VLDL, IDL, LDL, HDL_1, HDL_2, and HDL_3 are found in fractions 1–2, 3–4, 5–9, 10–12, 13–17, and 18–20, respectively.

In Fig. 6A, two baboon lipoprotein profiles (BX2402 and BX1665) are compared. As expected, baboon BX2402 with a serum total cholesterol (TC) of 284 mg/dl (typically baboons average about 115 mg/dl)[8] has an elevated LDL peak, while BX1665 with a serum TC of 219 mg/dl has a much smaller LDL peak. The small peak between LDL and HDL_2, which we are designating HDL_1, observed in BX2402, is in the region where Lp(a)[10] has been identified. An interesting result is seen in the profiles when comparing the HDLs. BX2402, with the much higher serum TC, has lower HDL than BX1665. Heterogeneity is clearly observed in the HDL_2 region of BX2402, while it is less obvious in BX1665. Generally, most baboons have a heterogeneous HDL_2 and rarely have a major HDL_3 peak. HDL heterogeneity is sometimes exhibited by a double peak as with BX2402, but most often it occurs as a shoulder on the dense side of the HDL_2 peak as in BX1665.

In Fig. 6B, there is a dramatic difference between the lipoprotein profiles of two baboons with elevated serum TC values (316 mg/dl for BX1661 and 286 for BX1309). Both LDL and HDL peaks are nearly the same height for BX1661. Baboon BX1309 has a very unusual profile, with small IDL and LDL peaks, a moderate HDL_1 peak, and a double HDL_2 peak. The broad peak between LDL and HDL contains a large percentage of cholesterol, as well as apolipoproteins E, A-I, and B.[11] The profiles in Fig. 6C correspond to what might be termed a typical baboon profile (BX1659, TC = 111 mg/dl) and to an animal with an elevated serum TC (BX3017, TC = 307 mg/dl).

We have observed many different profiles with animals having similar serum TC levels as well as profile changes in the same animal due to time, illness, or a change of diet. For this reason we recommend caution when describing any profile as typical, normal, or representative. Having stated this note of caution, the profile of BX1659 with a small LDL peak and larger, heterogenous HDL_2 peak is most typical of chow-fed baboons. Large VLDL and LDL peaks are observed for BX3017 along with a peak between LDL and HDL_2, a small HDL_2 peak, and a rare HDL_3 peak. Much of the information on baboon serum lipoprotein heterogeneity pre-

---

[10] G. M. Fless, K. Fischer-Dzoga, D. J. Juhn, S. Bates, and A. M. Scanu, *Arteriosclerosis* **2**, 475 (1982).

[11] J. L. Kelley, unpublished results, 1983.

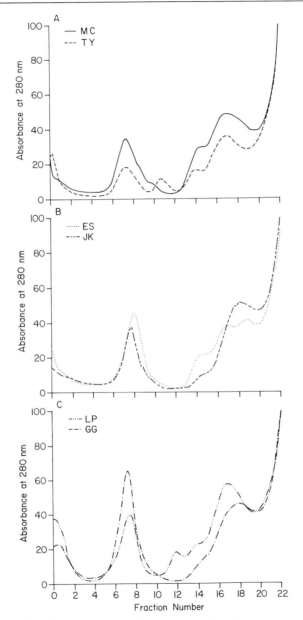

FIG. 7. Selected serum lipoprotein profiles of humans.

sented in Fig. 6 would have been lost, if the discontinuous density cuts for preparative ultracentrifugation were used.

We have now examined more than 100 profiles of human subjects. Considerable variation between subjects was noted, but within subject variation was less. Figure 7 shows several representative, 24 hr, profiles. Subject TY had the lowest serum cholesterol in this series (TC = 151), resulting in a small LDL peak. The peak between LDL and HDL_2 is likely due to the presence of Lp(a) in the HDL_1 fraction.[10] Another subject, MC (TC = 195), showed larger LDL and HDL peaks but the absence of the HDL_1 peak.

The profiles of two normocholesterolemic subjects (JK, TC = 161 and ES, TC = 197) are shown in Fig. 7B. ES has a larger LDL peak than JK and shows considerable heterogeneity in the HDL region. This profile (ES) clearly shows the existence of multiple subfractions in HDL_2.

In Fig. 7C, two profiles of subjects with elevated serum cholesterol (GG, TC = 242 and LP, TC = 277) are shown. LP had moderately elevated VLDL and HDL_1 peaks. GG had a large LDL peak with no HDL_1 and a dense HDL_3 peak. We have noted considerable variation in the density of the HDL_3 peak, with some subjects' HDL_3 barely separating from the serum protein fraction.

Advantages and Disadvantages

Some difficulties may be encountered using density gradient centrifugation for studies of serum from hypertriglyceridimic subjects. Serum samples which have elevated triglycerides can form a cake at the top of the tube which hinders elution of the contents of the centrifuge tube. If chylomicrons or VLDL are present in substantial quantities, they must be removed by aspiration before the fractions are eluted. The method is not appropriate for the screening of large populations because only six samples can be run at a time.

A major advantage of the density gradient procedure is its sensitivity using as little as 0.05 ml of serum. For preparative applications, up to 4.0 ml of serum can be separated in each centrifuge tube. Another important feature is that it provides a visual display of the continuous lipoprotein distribution. There are lipoproteins in the serum of normo- and hypercholesterolemic subjects which overlap the classical density limits, or there may be several peaks within a lipoprotein density class. Separated fractions can be recovered for further analysis. Variations in the step-gradient can be used to separate segments of the density spectrum as Groot *et al.*[6] have done to separate further HDL_2 from HDL_3. We consider that fractionation of serum lipoproteins into a continuous lipoprotein profile as contrasted with discrete density classes gives a more informative picture of the densities and heterogeneity of the lipoprotein species and the appearance of minor or new lipoprotein species.

Acknowledgment

Supported by NHLBI Grants HL-26890 and HL-24752.

[8] Single Vertical Spin Density Gradient Ultracentrifugation

By BYUNG H. CHUNG, JERE P. SEGREST, MARJORIE J. RAY, JOHN D. BRUNZELL, JOHN E. HOKANSON, RONALD M. KRAUSS, KEN BEAUDRIE, and JOHN T. CONE

Lipoproteins are commonly prepared by the differential ultracentrifugation method which makes use of their hydrated density characteristics. In a solvent with a density greater than the lipoprotein density, the lipoprotein floats at a rate dependent on its density, size, and shape. For the preparative isolation of lipoproteins, each lipoprotein class is isolated sequentially by a stepwise increase of the density of serum or plasma with salt, such as KBr or NaBr.[1] The isolation of all major lipoprotein species in plasma by the sequential flotation method requires a minimum of 72 hr of ultracentrifugation (Chapter [6], this volume). Since the sequential flotation method of lipoprotein isolation involves multiple steps, quantitative recovery of all lipoprotein fractions is somewhat difficult.

Another commonly used method of separation is isopycnic (equilibrium) density gradient ultracentrifugation, a method which is somewhat more complicated than differential centrifugation but one with compensating advantages. Advantages of this procedure include preparation of purified lipoprotein fractions in a one step process and the monitoring of lipoprotein density variations which often occur in many dyslipoproteinemic states. This method also permits the analytical measurement of various lipoprotein classes in the plasma.

The most commonly used rotor for isopycnic density gradient separation is the swingout rotor (Chapter [7], this volume). Density gradient ultracentrifugation in the swingout rotor has been successfully applied to

[1] R. J. Havel, H. A. Eder, and H. H. Bragdon, *J. Clin. Invest.* **34,** 1345 (1955).

fractionation of the major lipoprotein species and their subspecies.[2-7] Published density gradient ultracentrifugation methods using swingout rotors, however, require a minimum of 24–36 hr to complete the separation of the major lipoprotein fractions in plasma, and the maximum plasma volume for use in swingout rotors is often too small for preparative purposes.

Recently our laboratory has developed methods to separate plasma lipoproteins using density gradient ultracentrifugation in vertical rotors.[8] The separations of lipoproteins obtained by vertical rotors are comparable to those achieved using swingout rotors and the separation is complete in less than one-tenth the published times for swingout rotors.[8] Since various sizes of vertical rotors and at least two radically different types of ultracentrifuges (standard and tabletop) are currently available, separation of plasma lipoproteins can be achieved over a range of 25–330 min of ultracentrifugation and 4.5–92 ml of total plasma per spin. In this chapter we describe 13 separate vertical spin procedures (as well as 3 adaptations for angled-head rotors) for the preparative separation of a number of different plasma lipoproteins and a method for the quantitative analysis of lipoprotein cholesterol from plasma separated by single vertical spin (SVS) density gradient ultracentrifugation.

Basic Principles

The vertical rotor works on the principle that compression of the gradient geometry by the vertical rotor when at speed shortens spin time over more conventional ultracentrifugation techniques without loss of resolution (Fig. 1). Slow acceleration and deceleration steps are important in horizontal to vertical and vertical to horizontal reorientation of the gradient, respectively. In this regard, the more modern ultracentrifuges with slow start and slow stop modes are necessary to achieve optimal resolution and reproducibility in lipoprotein fraction separation. Combination of vertical geometry compression with inverted rate-zonal ultra-

[2] W. J. Lossow, F. T. Lindgren, J. C. Murchio, G. R. Stevens, and L. C. Jensen, *J. Lipid Res.* **10,** 68 (1969).
[3] R. H. Hinton, J. Kowalski, and A. Mallinson, *Clin. Chim. Acta* **44,** 267 (1973).
[4] F. T. Lindgren, L. C. Jensen, R. D. Will, and G. R. Stevens, *Lipids* **7,** 194 (1972).
[5] J. R. Foreman, J. B. Karlin, C. Edelstein, D. J. Juhn, A. H. Rubenstein, and A. M. Scanu, *J. Lipid Res.* **18,** 759 (1977).
[6] M. J. S. Chapman, D. Goldstein, D. Lagrange, and P. M. Laplaud, *J. Lipid Res.* **22,** 339 (1981).
[7] P. H. E. Groot, L. M. Scheek, L. Havekes, W. L. VanNoart, and F. M. Van't Hooft, *J. Lipid Res.* **23,** 1342 (1982).
[8] B. H. Chung, T. Wilkinson, J. C. Geer, and J. P. Segrest, *J. Lipid Res.* **21,** 284 (1980).

centrifugation allows the achievement of lipoprotein separation with spin times as short as 25 min. This procedure can be considered inverted since the lipoprotein-containing sample is placed at the bottom of the gradient, rather than at the top as is usually the case for rate-zonal centrifugation.

The shortened spin times achieved by the SVS procedure increase the number of samples that can be separated and assayed while minimizing the artifacts due to ultracentrifugal degradation of lipoproteins.

Three separate SVS procedures have been developed. These provide separation and analysis of (1) lipoprotein species [HDL, Lp(a), LDL, IDL, and VLDL] and (2) HDL and LDL subspecies (see Table II). Although we have not yet optimized the ultracentrifugation conditions for separation of IDL and VLDL subspecies, preliminary experiments suggest that these lipoprotein species can also be fractionated into subspecies with SVS ultracentrifugation by suitable modification of gradient and spin times (see, for example, Fig. 7).

For most single spin separations, two-step gradients provide both simplicity in gradient formation with optimal separation of different lipoprotein fractions. In only one case, the complete separation of plasma protein fractions from high-density lipoproteins, has it been necessary to resort to three-step gradients (see Table III).

Gradients and spin times have been determined as follows. The starting plasma is adjusted with KBr (although CsCl or NaBr has also been used) to a density equal to or greater than the densest lipoprotein subspecies to be fractionated and to a volume fraction typically 0.1 to 0.3 of the total centrifuge tube volume. The density-adjusted plasma is then layered under a KBr or NaCl solution adjusted to a density equal to or greater than the least dense lipoproteins subspecies to be fractionated and to a volume equal to the remaining volume of the centrifuge tube. Adjustment of these gradient conditions allows separation by flotation rate-zonal, not isopycnic (equilibrium) centrifugation of the most dense subspecies a short distance away from the bottom, while floating the least dense subspecies very near the top of the tube (Fig. 1). Further fine adjustments in the densities of the bottom and top gradients then provide maximal rapidity and resolution of separation (see Table II). With only two gradients to vary and the short spin times of the SVS procedure, a large number of different experimental conditions can be easily tested.

The conditions developed to fractionate LDL subspecies (Table II) can be used to illustrate this general approach. LDL subspecies vary in density from >1.049 g/ml for the most dense LDL_4 to <1.027 g/ml for the least dense LDL_1.[9] After trying approximately 10 combinations of gradients, 2 ml of 1.08 g/ml density adjusted (KBr) plasma was chosen as the bottom gradient (0.12 volume fraction) and 15 ml of 1.045 g/ml KBr solu-

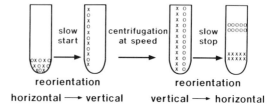

FIG. 1. A schematic diagram illustrating the basic principles of SVS inverted rate-zonal density gradient ultracentrifugation.

tion was chosen as the top gradient (0.88 volume fraction). Optimal separation was achieved by 330 min centrifugation at 65,000 rpm in a TV-865B rotor. A number of other gradient conditions were tried that produced separation but required longer spin times and/or produced less resolution between LDL subspecies.

Materials

Plasma
Ethylenediaminetetraacetate (EDTA)
KBr
Normal saline (0.15 M NaCl–0.01% EDTA, pH 7.4)
Autoflo cholesterol enzymatic reagent consisting of 100 mM Tris buffer, pH 7.4, 50 mM magnesium asparate, 1 mM 4-aminophenazone, 6 mM phenol, 4 mM 3,4-dichlorophenol, 0.3% hydroxypolyethoxyl n-alkane, 250 U/liter cholesterolesterase, and 200 U peroxidase (Boehringer Mannheim Co.)
0.1% Brij solution (Technicon Co.)
Preciset cholesterol standard (Boehringer Mannheim)

Equipment

Ultracentrifuges (Beckman L8-80, L8-80M, Sorvall OTD-2, Beckman TL-100)
Vertical rotors (Sorvall TV-865, TV-865B and TV-850 and Beckman VTi80 and TLV100)
Angled head rotor (Beckman 70Ti)
Gradient Fractionator (Hoeffer Scientific Co.)
Peristaltic pump (Holter Co.)
Technicon Autoanalyzer II consisting of proportioning pump, colorimeter, incubator, and recorder (Technicon Co.)
Apple II e computer equipped with 64K RAM, disc drive, serial interface, silent printer, clock model A102 digital converter, arithmetic processor (model 7811C), parallel triac output board radiofrequency modulator, and color monitor

Preparation of Plasma Samples

Plasma is adjusted to the appropriate density by the addition of solid KBr (Table I). Chylomicrons in the plasma, if any, are removed by centrifugation of the plasma at 30,000 rpm for 30 min at 10°. The chylomicron layer is aspirated with a Pasteur pipet. A discontinuous KBr density gradient is then made by forming a two-layer gradient with the density adjusted plasma forming the bottom layer according to the specifications given in Table II. Formation of the two-layer gradient is easily achieved by overlayering or underlayering. The overlayering method is suitable for the formation of gradients in small centrifuge tubes, where the lower density solution is gently layered over the density adjusted plasma by use of a low-speed peristaltic pump. For the formation of gradients in large centrifuge tubes (TV-865B, TV-850, or 70Ti rotors), the density adjusted plasma is layered under the normal saline by use of a long needle and a peristaltic pump.

Single Vertical Spin Ultracentrifugation

Tubes loaded with sample and gradient are then placed in vertical rotors immediately after formation of the gradient and centrifuged in a Sorvall or Beckman ultracentrifuge at 7° according to the conditions given in Tables II or III. Since the contents of the tubes in a vertical rotor reorient during both the acceleration and deceleration steps of ultracentrifugation, the ultracentrifuge should be set in the slow acceleration and slow deceleration modes in order to avoid disturbance of the gradient during acceleration and deceleration. With a Sorvall OTD-2 ultracentrifuge, we set the ARC-1 automatic controller at rate D with the reograd program activated; with a Beckman L8-80 ultracentrifuge, both slow ac-

TABLE I
KBr REQUIRED FOR DENSITY ADJUSTMENT OF
$d = 1.006$ g/ml SAMPLES

Density (g/ml)	KBr (g added per ml sample)
1.34	0.5739
1.30	0.4946
1.24	0.3816
1.21	0.3265
1.18	0.2654
1.12	0.1722
1.06	0.0834

celeration and slow breaking modes are activated. With the Beckman L8-80M and TL-100 ultracentrifuges acceleration mode 6 and deceleration mode 6 are used.

At the end of ultracentrifugation the tubes are carefully removed from the rotor and the contents are examined. For example, using the conditions of procedure 1 in Table II, each lipoprotein fraction is banded sharply, with VLDL at the top, LDL in the upper middle, and HDL in the lower middle portion of the tube; the plasma protein fraction remains undisturbed in the lower portion of the tube (Fig. 2). The VLDL and LDL bands are readily visible, but the HDL band in many plasmas may be difficult to locate visually. The visibility of the HDL band and other lipoprotein bands can be enhanced by prestaining the plasma lipoproteins with Sudan black[9] prior to ultracentrifugation. The experimenter should be aware, however, that very small changes in flotation density may result from uptake of Sudan black by the lipoprotein particles.

Sudan black (0.1 g) is dissolved in 100 ml ethylene glycol and stored in a brown bottle at room temperature. Depending upon the sample volume, 1–6 drops of the Sudan black solution are added to 2–10 ml of density-adjusted plasma before centrifugation.

The separated lipoprotein fractions can be collected by downward drop fractionation of the gradient sample using a gradient fractionator. Each lipoprotein peak is then pooled following the location of the lipoprotein peaks by assaying the level of cholesterol, triglycerides (Fig. 3), or apolipoproteins in the fractions. Alternatively, each lipoprotein fraction is collected by gentle suction of the lipoprotein band using a long stem pasteur pipet or a syringe with a long needle. The tip of the pasteur pipet or syringe needle is placed just above the lipoprotein band and the lipoproteins are collected sequentially from top to bottom in order to minimize the disturbance of the fractions during collection. Finally, continuous flow cholesterol analysis can be used to quantify the levels of individual lipoprotein species or subspecies (see the following section on analysis).

Ultracentrifugation of a sample under the conditions given in Tables II or III does not produce equilibrium but creates a continuous nonlinear gradient (see for example Fig. 3); the lower half of the gradient is usually two to three times steeper than the upper half. Following procedure 3, Table II, on the basis of a control gradient, VLDL bands in the 1.014–1.016 g/ml density range, LDL in the 1.02–1.062 g/ml density range, and HDL in the 1.062–1.185 g/ml density range. Preisolated authentic

---

[9] A. H. M. Terpsira, C. J. H. Woodward, and F. J. Sanchez-Muntz, *Anal. Biochem.* **III**, 149 (1981).

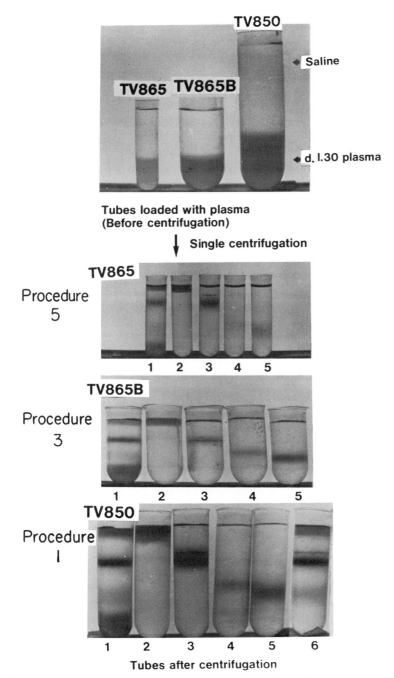

FIG. 2. Separation of plasma lipoprotein species by single vertical spin inverted rate-zonal density gradient ultracentrifugation using three different vertical rotors (TV-865, TV-865B, TV-850 representing procedures 5, 3, 1, Table II, respectively). (1) Plasma, (2) VLDL, (3) LDL, (4) "HDL_2," (5) "HDL_3," and (6) mixture of VLDL, LDL, and HDL.

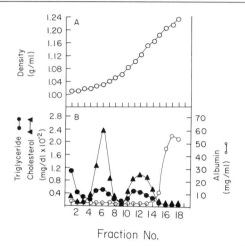

Fig. 3. (A) Distribution of density (g/ml) in the fractions from a gradient of $d = 1.006$ g/ml NaCl and $d = 1.30$ g/ml KBr after SVS centrifugation in a TV-865B rotor at 65,000 rpm for 90 min (procedure 3, Table II). (B) Distribution of cholesterol, triglyceride, and albumin in the fractions following centrifugation of plasma from a normocholesterolemic subject. Albumin measured by the bromcresol green method [B. T. Doumas, W. A. Watson, and H. G. Biggs, *Clin. Chim. Acta* **31**, 87 (1971)].

"HDL$_2$" ($d = 1.063-1.12$ g/ml) and "HDL$_3$" ($d = 1.12-1.2181$ g/ml) have their peak centers at $d = 1.10$ and at $d = 1.15$ g/ml, respectively. If the duration of centrifugation is extended beyond 90 min for procedure 3, Table II, the HDL band spreads wider and the LDL band moves closer to the VLDL band. Collection of lipoproteins by the downward drop fractionation method yields good resolution of VLDL, LDL, and HDL peaks (Fig. 3) and each pooled fraction has the characteristic morphology and size of VLDL, LDL, or HDL.[8]

Analysis of Samples

Analysis of Fractions. Fractions can be collected from a centrifuge tube after SVS ultracentrifugation by either upward or downward flow fractionation. Due to its simplicity, downward flow fractionation is the procedure of choice when the SVS procedure is used analytically, e.g., to measure cholesterol, triglycerides, and apolipoproteins in each fraction. For downward flow fractionation, a Beckman Fraction Recovery System is used with a proportionating pump to control the rate of outflow. The rate of evacuation varies from 0.5 to 8 ml/min depending upon the size of the tube.

TABLE II
GRADIENT CONDITIONS FOR SINGLE VERTICAL[a] SPIN FRACTIONATION OF PLASMA LIPOPROTEINS

Fractions separated	Procedure	Use	Rotor	rpm	Spin time (min)	Total sample volume per rotor (ml)	Bottom gradient (plasma)		Top gradient	
							Density (g/ml)	Volume (ml)	Density (g/ml)	Volume (ml)
Lipoprotein species	1	PREP	TV-850	50,000	150	80	1.30	10	1.006	24
	2	PREP	TV-850	50,000	150	80	1.21	10	1.006	24
	3	PREP	TV-865B	65,000	90	40	1.30	5	1.006	12
	4	PREP	TV-865B	65,000	90	40	1.21	5	1.006	12
	5	ANAL	VTi80	80,000	44	12	1.30	1.5	1.006	3.5
	6	ANAL	VTi80	80,000	44	12	1.21	1.5	1.006	3.5
	7	ANAL	TLV100	100,000	25	4.56	1.21	0.57	1.006	1.33
	8	PREP[b]	70Ti	70,000	210	92	1.21	11.5	1.006	27.5
HDL subspecies	9	PREP	TV-865B	65,000	330	40	1.21	5	1.06	12
	10	PREP[b]	70Ti	70,000	210	64	1.21	8	1.12	31
	11	PREP	TV-865B	65,000	150	28	1.18	3.5	1.12	13.5
	12	ANAL	VTi80	80,000	90	12	1.21	1.5	1.06	3.5
LDL subspecies	13	PREP	TV-865B	65,000	330	16	1.08	2	1.045	15
	14	ANAL	VTi80	80,000	150	4.72	1.08	0.59	1.041	4.41

[a] Angled head rotors can also be used.
[b] Angled head rotor.

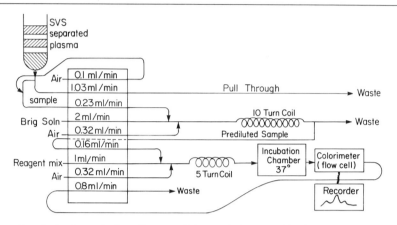

FIG. 4. A schematic flow diagram of the vertical autoprofile (VAP) procedure.

To minimize contamination of lipoprotein fractions with albumin, upward flow fractionation is employed when the SVS procedure is used preparatively. For upward flow fractionation, an ISCO/Instrumentation Specialties Co. Density Gradient Fractionator is used. This instrument contains a syringe volume metering pump. A $d = 1.40$ g/ml cushion of sucrose and NaBr (80 g sucrose and 20 g NaBr dissolved in 100 ml distilled water) is used to displace the KBr gradient upward. This displacement cushion is pumped at a rate of 0.375 ml/min for 5 ml tubes and at 0.750 ml/min for 17, 34, and 40 ml tubes.

Generally a minimum of 20 fractions are collected for the 5 ml tubes and a minimum of 40 fractions collected for the larger tubes. Fractions can be analyzed for total and unesterified cholesterol, triglycerides, phospholipids, apolipoproteins, albumin, and total protein by methods described elsewhere in this series.

Continuous Flow Analysis of Cholesterol. Semiautomated profiles of plasma lipoprotein cholesterol can be developed by the combination of SVS density gradient separation of plasma lipoproteins and continuous flow analysis of cholesterol in the effluent of the gradient sample; this approach has been termed the vertical autoprofile (VAP) procedure.[10] A schematic diagram of the VAP is shown in Fig. 4. In order to achieve maximal peak separations of the lipoprotein species HDL, Lp(a), LDL, IDL, and VLDL, the gradient is slightly modified by adjusting the density of plasma to 1.21 instead of 1.30 g/ml (procedures 2, 4, 6, and 7, Table II).

[10] B. H. Chung, J. P. Segrest, J. T. Cone, P. Pfau, J. C. Geer, and L. A. Duncan, *J. Lipid Res.* **22**, 1003 (1981).

This modified gradient does not produce good separation of HDL from the free plasma proteins but produces a greater peak separation of HDL, Lp(a), and LDL compared to the conditions used for the preparative separation of lipoproteins (procedures 1, 3, and 6, Table II or procedure 16, Table III).

The centrifuged tube containing SVS-separated lipoproteins is placed in a Beckman Fraction Recovery System as described and the bottom of the tube punctured carefully so as not to disturb the gradient formed. The sample effluent is removed downward by a Technicon AA-II pump, and the stream split with use of a T-connector (Fig. 3). Only 18% of the effluent sample is diverted for analysis. This sample is diluted 10-fold with 1% Brij solution, and 7% of the diluted sample is then continuously mixed with enzymatic cholesterol reagent in the ratios shown in the diagram (Fig. 4). In order to prevent the mixing of the samples while they travel through the line, air bubbles are incorporated into the line to separate the samples just before the steps of sample dilution and mixing the diluted sample with the cholesterol reagent (Fig. 4).

Cholesterol in the effluent samples is reacted with cholesteryl esterase and cholesterol oxidase during passage through the incubation chamber, resulting in a stable reddish color. The absorbances of the sample are monitored at 505 nm in a Technicon AAII colorimeter equipped with 1.5 mm flow cells and the values plotted on the recorder. The remaining 82% of the sample effluent (pull through) can be diverted into a fraction collector where the samples are available for other analyses (Fig. 4). A typical lipoprotein cholesterol profile of normolipidemic plasma obtained by the standard VAP method is shown in Fig. 5; the VLDL, LDL, and HDL peaks are separated evenly in the profile.

The total time required to analyze the contents of a small centrifuge tube from a vertical rotor (TV-865 or VTi80) is approximately 9 min and the period of time from the point of puncturing the tube to the point where the absorbance of cholesterol appears on the recorder is approximately 5 min. Continual analysis of samples through the VAP system requires a 1.5 min washing of the system by removal of the emptied tube, mounting a clean, empty tube and filling it half full with distilled water. The empty wash tube is removed and the next gradient-containing tube is punctured. The samples in one rotor (8 tubes) can thus be analyzed by the VAP system in approximately 45 min (5 min per tube).

The positions of a given lipoprotein species or subspecies and thus its relative banding region can be denoted by the time in seconds from the appearance of the leading edge of the HDL cholesterol peak to the peak center of the species in question. The times for the peak centers of the lipoprotein species vary depending upon the subspecies composition of

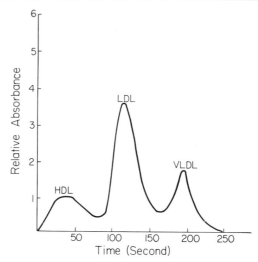

FIG. 5. A typical SVS–VAP cholesterol profile of plasma from a normolipidemic subject (VLDL is slightly elevated). SVS procedure 6, Table II, was used for lipoprotein separation.

each species within the following ranges: HDL, 30–60 sec; Lp(a), 70–100 sec; LDL, 100–140 sec; IDL, 150–180 sec; and VLDL, 200 sec (see Figs. 5–7).

VAP analyses of the contents of larger or smaller tubes can best be performed by varying the pump rate of the pull through line (Fig. 4). For TLV100 rotor analyses (1.9 ml samples) we use a flow rate of 0.23 ml/min for the pull through; for TV-865B rotor analyses (17 ml samples) we use a flow rate of 4.8 ml/min for the pull through. Each produces VAP profiles of approximately 200 sec in length.

The VAP procedure outlined in Fig. 4 can be used directly to analyze SVS separations of HDL and LDL subspecies (procedures 12 and 14, Table II, see Figs. 11–14).

Plasma from several major types of human hyperlipoproteinemia yields characteristically distinct profiles (Fig. 6), reflecting the effectiveness of the VAP method in analysis of hyperlipoproteinemias.

Since the VAP method monitors the cholesterol level along the entire density spectrum of lipoproteins in plasma, the presence of variant forms of lipoproteins is readily detectable as shown in Fig. 7. For example, the presence of Lp(a) in plasma can be monitored by the VAP procedure; this lipoprotein appears as a separate peak between the LDL and HDL peaks in the VAP profile (Fig. 7B).

Quantification of VAP Profiles by Computer-Assisted Curve Decomposition. The area under the lipoprotein cholesterol peaks in the profiles

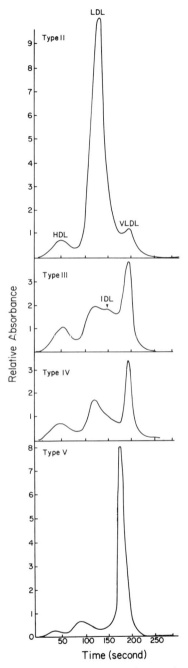

FIG. 6. SVS–VAP cholesterol profiles of plasma from patients with type IIa, type III, type IV, and type V hyperlipoproteinemias [D. S. Fredrickson, R. I. Levy, and R. S. Lees, N. Engl. J. Med. **276,** 34, 94, 148, 218, 273 (1967)]. SVS procedure 6, Table II, was used for lipoprotein separation.

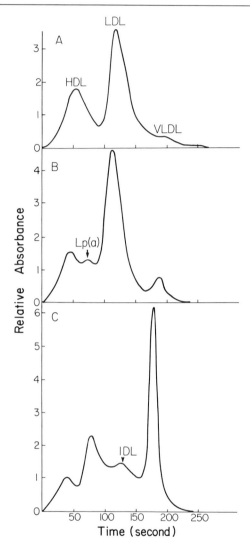

FIG. 7. SVS–VAP cholesterol profiles of plasma from normal (A) and dyslipoproteinemic (B, C) subjects. SVS procedure 6, Table II, was used for lipoprotein separation.

of plasma obtained by the VAP method (procedure 6, Table II) is highly correlated with the level of plasma cholesterol in the tube ($n = 0.998$). The recovery of plasma cholesterol following separation of plasma into individual lipoprotein fractions by SVS is 98.5 ± 3.5% ($n = 10$).[10]

Lipoprotein cholesterol profiles on aliquots of the same plasma sample analyzed by the VAP method are highly reproducible; the coefficients of variation of cholesterol in the VLDL, LDL, and HDL peaks are 4.8, 2.6, and 2.2%, respectively ($n = 8$).[11]

Decomposition of the VAP profile (procedure 6, Table II) into individual lipoprotein species [HDL, Lp(a), LDL, IDL, and VLDL] is done using a strictly linear least-squares technique. With a few assumptions about the shapes of the curves and a fairly accurate knowledge of the times that the curves peak, this can be accomplished.[11]

It is known that the distribution of the various subspecies in a velocity density gradient is Gaussian.[12] The individual peak shapes of isolated species (e.g., LDL) analyzed by VAP is not quite Gaussian due to flow effects in the tubes of the Technicon AA-II analyzer.[11] These effects cause the curves to be slightly skewed, compressed on the leading side and extended on the trailing side. Also a small exponential tail is present due to mixing by (1) wall drag and (2) turbulent flow in the junctions of tubes in the AA-II system.

Once the peak times of all the subspecies [HDL, Lp(a), LDL, and VLDL] are reasonably selected it is now possible to use the linear least-squares algorithm to determine the individual contributions of subspecies to the total cholesterol. This is accomplished as follows. Total cholesterol at each instant of time is the sum of the contributions of the individual subspecies.

$$\text{Total} = a_i e^{-b_i(x-x_0)^2} + c_i e^{-d(x-x_1)}$$

where x_0 is the peak time of the ith curve and x_1 is a set time after the curve peak to start adding the mixing effects.

Assuming that the peak times are known allows the values of the exponentials to be computed at each x point. This means that the values can be considered as a constant and the equations become

$$\text{total} = \Sigma a_i C + D$$

where C is a constant equal to the exponential function, and D is a constant equal to the mixing function, both evaluated at each point.

Another important point is that the b parameters (signifying the width functions) are different on opposite sides of the curves' peak. The algorithm must use the first parameter to calculate the function values until the peak time of the curve is passed and then must use the second b parameter after that. This is done for each curve.

[11] J. T. Cone, J. P. Segrest, B. H. Chung, J. B. Ragland, S. M. Sabesin, and A. Glasscock, *J. Lipid Res.* **23**, 923 (1982).
[12] M. M. S. Shen, R. M. Krauss, F. T. Lindgren, and T. Forte, *J. Lipid Res.* **22**, 236 (1981).
[13] B. H. Chung and J. P. Segrest, unpublished results.

The simplified equation above is linear and can be solved for the various a_i, which are the peak heights of the subspecies. Once the heights are known, the total area under the individual curves can be computed and, using a known standard curve, the absolute total cholesterol can be found.

The linear least-squares method has been applied to HDL–VAP curves (see Fig. 11) but has proved inadequate due to the tendency for large positive/negative oscillations when peak positions are too close. At present we are working on an iterative nonlinear least-squares procedure for the decomposition of HDL–VAP curves. We anticipate that the LDL–VAP curves (see Fig. 14B) can be effectively decomposed with the linear least-squares method since the individual subspecies are reasonably well separated.

Strategies for Preparation of Lipoprotein Species

Single vertical spin ultracentrifugation provides a rapid and effective method of separation of the major lipoprotein species, HDL, LDL, VLDL, IDL, and Lp(a). A wide range of rotors have been adapted to two step gradient single spin separation with spin times from 25 to 210 min (Table II, procedures 1–8). Procedures 1–6, Table II, use vertical rotors and modern floor-mounted ultracentrifuges. Procedure 7 uses a vertical rotor in a Beckman TL-100 tabletop ultracentrifuge and produces a separation with the shortest spin time (25 min). Procedure 8 uses a 70Ti angled-head rotor and modern floor-mounted ultracentrifuges.

Bottom gradient densities of 1.30 g/ml (procedures 1, 3, and 5) provide optimal albumin–HDL separations with vertical rotors and are preferred for preparative use. Bottom gradient densities of 1.21 g/ml (procedures 2, 4, and 6) provide optimal HDL–Lp(a)–LDL separations (see Fig. 7B) with vertical rotors and are preferred for Lp(a) preparation and general analytical use.

The TL-100 table top ultracentrifuge used with the TLV100 vertical rotor (procedure 7) provides the most rapid separation of HDL–LDL–VLDL and gives results comparable to procedure 6.

Angled-head rotors can be adapted to gradient conditions that work with vertical rotors. Procedure 8 was developed to provide optimal LDL, IDL, and VLDL preparations with minimal albumin contamination. Standard slow start and slow stop settings used with the vertical rotors are required. The tubes should be removed from the rotor and the tubes carefully placed in the vertical position before fractionation of lipoproteins is attempted.

LDL and IDL. These lipoprotein species can be prepared by SVS procedures 1–7 (Table II) depending upon the rotors and centrifuges available and the volume of plasma to be centrifuged (for example, see Fig. 3B). However, we presently use an angled-head rotor (procedure 8, Table II) for preparation of LDL and IDL, the reasons being that this procedure allows slightly larger total volumes of plasma per spin (92 ml), provides optimal separation of LDL and IDL, and produces essentially albumin-free LDL and IDL (Fig. 8). A double spin procedure is used in which the LDL and IDL peaks after one spin are isolated by pooling (Fig. 8), are subjected to a second identical ultracentrifugation step, and LDL and IDL pools dialyzed overnight at 4° against two changes of 0.15 M NaCl.

VLDL. This lipoprotein species is also isolated in our laboratory by procedure 8, Table II. It is uncertain at this writing whether the large and small subspecies of VLDL are isolated in equal yields by this procedure (see discussion section on advantages and disadvantages of SVS ultracentrifugation).

Lp(a). The variant lipoprotein, Lp(a), because it has a flotation density between HDL and LDL, requires a procedure for optimizing the separa-

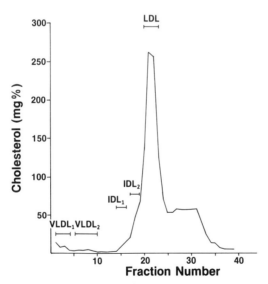

FIG. 8. Distribution in the fraction following centrifugation of normocholesterolemic plasma using a gradient of $d = 1.006$ and $d = 1.21$ g/ml in a 70Ti angled-head rotor at 70,000 rpm for 210 min (procedure 8, Table II). IDL and VLDL fractions were each arbitrarily pooled into the two fractions shown (bars).

tion of HDL and LDL. Procedures 2, 4, and 6 (Table II) are useful in this regard. Subjects with levels of plasma Lp(a) greater than approximately 15 mg/dl cholesterol have well-defined Lp(a) peaks in their VAP profiles (Fig. 7B). The separation between the Lp(a) peak and HDL/LDL is even sharper when procedure 4 (Table II) is used in conjunction with upward flow fractionation (Fig. 9).[13] Isolation of the Lp(a) peak and recentrifugation by procedure 4 produces a single peak that reacts strongly with a monospecific polyclonal antibody to Lp(a).[11]

HDL. This lipoprotein species is presently isolated in our laboratory by procedure 15, Table III, using a 70Ti angled-head rotor. By this procedure, the HDL of whole plasma can be separated almost quantitatively from albumin and LDL by a single 4 hr spin (Fig. 10). The absence of significant recognizable apolipoprotein A-I in the albumin-containing bottom fractions suggests relatively little dissociation of apolipoproteins from the HDL by this procedure. A double spin approach, involving pooling of the apoA-I-containing fractions (HDL), and recentrifugation by procedure 15, produces an HDL free of albumin. An alternative approach is to do a single spin by procedure 15, followed by elution of the pooled HDL fraction through an albumin affinity column.

Yet another approach that allows HDL to be isolated simultaneously with the other major lipoprotein species is to use procedure 16, Table III

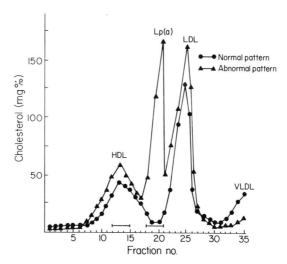

FIG. 9. Distribution of cholesterol in the fractions following centrifugation of normocholesterolemic and an Lp(a)-containing plasma. Centrifugation was with a gradient of $d = 1.006$ g/ml NaCl and $d = 1.21$ g/ml KBr in a TV-865B rotor at 65,000 rpm for 90 min (procedure 4, Table II).

TABLE III
GRADIENT CONDITIONS FOR THREE STEP GRADIENT SINGLE-ANGLED HEAD SPIN FRACTIONATION OF PLASMA LIPOPROTEINS

Procedure	Use	Rotor	rpm	Spin time (min)	Total sample volume per rotor (ml)	Bottom gradient (plasma) Density (g/ml)	Volume (ml)	Middle gradient Density (g/ml)	Volume (ml)	Top gradient Density (g/ml)	Volume (ml)
15	PREP	70Ti	70,000	240	80	1.34	10	1.24	10	1.06	19
16	PREP	70Ti	70,000	260	80	1.35	10	1.20	10	1.006	19

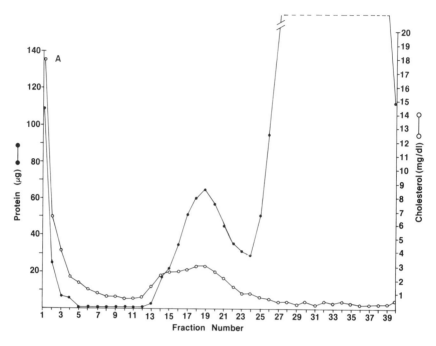
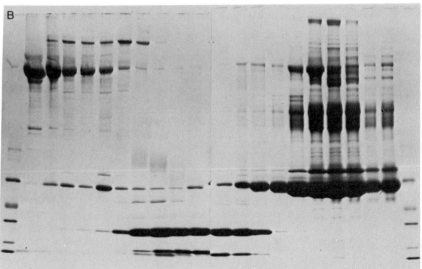

FIG. 10. Separation of HDL from albumin by inverted rate-zonal density gradient ultracentrifugation using a three-step gradient (d = 1.06, 1.24, 1.34 g/ml) in a 70Ti rotor at 70,000 rpm for 240 min (procedure 15, Table III). (A) Distribution of cholesterol and protein (A_{280}) in fractions. (B) Sodium dodecyl sulfate–slab polyacrylamide gel electrophoresis in a gradient of 4–20% polyacrylamide. Odd-numbered fractions have been run and stained for protein with Coomassie brilliant blue. The extreme left and right lanes contain molecular weight standards α-lactalbumin (14,400), soybean trypsin inhibitor (20,000), carbonic anhydrose (30,000), ovalbumin (43,000) bovine serum albumin (67,000), and phosphorylase (94,000).

(70Ti rotor, three step gradient, 4 hr and 20 min spin) to isolate HDL, LDL, IDL, and VLDL. HDL is then subjected to a second spin (procedure 15, Table III) for complete removal of albumin.

Strategies for Preparation of Lipoprotein Subspecies

HDL Subspecies. Reasonable evidence now exists that, based on size and density alone, HDL is composed of at least 11 separate subspecies.[14] In order to study the structural and functional implication of the HDL subspecies, methods for separation are required (see Chapter [24], this volume). One useful approach is single vertical spin ultracentrifugation.

Four separate single spin procedures, three using vertical rotors and one using the 70Ti angled-head rotor, have been developed that provide useful separation of multiple HDL subspecies (procedures 9–12, Table II). Each of these spins spreads HDL throughout most of the gradient while LDL, IDL, and VLDL are concentrated at the top (Fig. 10A and B).

Figure 11A illustrates the results of fractionation by procedure 9 (5.5 hr spin) of HDL from a male subject containing high levels of "HDL_2." Note the results of the radioimmunoassays performed on pooled fractions for apolipoproteins A-II and C-II; apoA-II is localized predominantly in the "HDL_3" region while apoC-II peaks in the left-hand (most dense) region of "HDL_2." Figure 11B illustrates the results of fractionation by procedure 11 (2.5 hr spin) of HDL from the plasma of the same male subject used in Fig. 11A. Again note that apoA-II is localized predominantly in the "HDL_3" region and apoC-II peaks in the left-hand (most dense) region of "HDL_2," showing that procedures 9 and 11 give comparable results. However, procedure 11, compared to procedure 9, provides more rapid separation of HDL subspecies and better resolution, particularly in the "HDL_2" fractions.

The results of a rapid method of HDL analysis applied to the plasma analyzed in Fig. 11A and B are shown in Fig. 11C. This procedure, called an HDL–VAP profile, combines SVS ultracentrifugation by procedure 12, Table I (1.5 hr spin) with VAP continuous flow analysis. In the HDL–VAP profile shown, the LDL/VLDL peak has been mathematically subtracted by computer from the raw profile. Figure 11C shows clearly the two major peaks, "HDL_3" and "HDL_2," seen in Fig. 11A, and B, but, in addition, shows other peaks and shoulders partially resolved at approximately 20, 80, and 90 sec. Each of these five peaks has been identified with one of the 11 known HDL subspecies.[14]

[14] J. P. Segrest, M. Cheung, J. J. Albers, J. T. Cone, B. H. Chung, M. Kashyap, C. G. Brouillette, T. C. Ng, and M. A. Glasscock, submitted for publication (1985).

Six examples of HDL-VAP analyses that illustrate some of the variability seen in the general population are shown in Fig. 12. Even though qualitative, HDL-VAP analyses provide the optimal combination of convenience, rapidity, resolution, and sample volume presently available for analysis of HDL subspecies. Development of a quantitative algorithm for decomposition of HDL-VAP profiles will add further to the scientific and medical value of this procedure.

For those laboratories that do not possess a vertical rotor, Fig. 13 illustrates an adaption of the HDL-SVS procedure to the 70Ti rotor (procedure 10, Table II). The plasma analyzed was from a different sub-

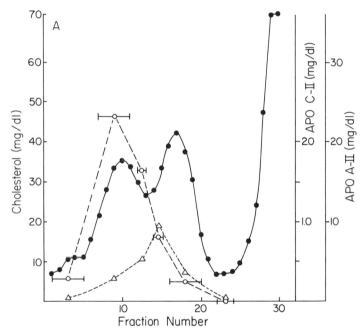

FIG. 11. Fractionation of HDL from the plasma of a male subject into subspecies by SVS inverted rate-zonal density gradient ultracentrifugation. (A) Centrifugation by a two-step gradient (d = 1.06 and 1.21 g/ml) in a TV-865B vertical rotor at 65,000 rpm for 330 min (procedure 9, Table II). The distribution of cholesterol (●), apolipoprotein A-II (○), and apolipoprotein C-II (△), the latter two measured by radioimmunoassay (performed by M. Kasyap, University of Cincinnati School of Medicine) are shown. (B) Centrifugation by a two-step gradient (d = 1.12 and 1.18 g/ml) in a TV-865B vertical rotor at 65,000 rpm for 150 min (procedure 11, Table II). The distributions of cholesterol (●), apolipoprotein A-II (○), apolipoprotein C-II (△), the latter two measured by radioimmunoassay (performed by M. Kasyap, University of Cincinnati School of Medicine), are shown. (C) Analysis by HDL-VAP. Centrifugation was by a two-step gradient (d = 1.06 and 1.21 g/ml) in a VTi80 vertical rotor at 80,000 rpm for 90 min (procedure 12, Table II).

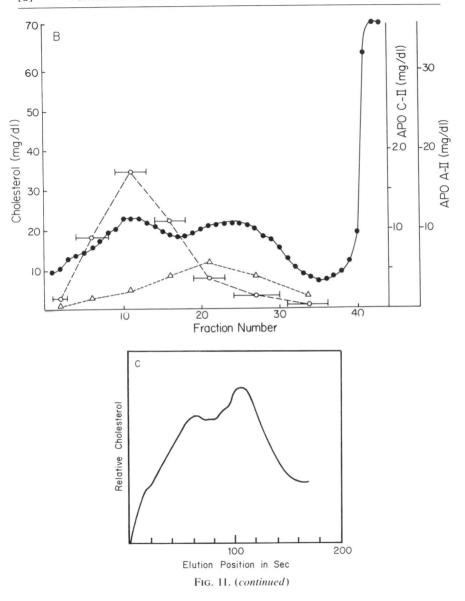

FIG. 11. (*continued*)

ject than used in Fig. 11. Again "HDL$_3$" and "HDL$_2$" peaks can be distinguished. The results of radial immunodiffusion assays for apolipoproteins A-I, A-II, and D and lecithin:cholesterol acyltransferase on pooled fractions are also shown in Fig. 12; these assays indicate differential fractionation of the four proteins by procedure 10.

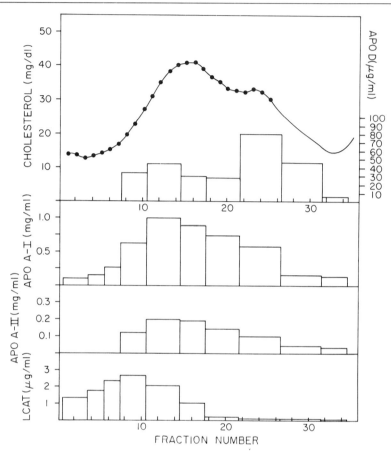

FIG. 12. Fractionation of HDL from the plasma of a female subject into subspecies by inverted rate-zonal density gradient ultracentrifugation in a 70Ti angled-head rotor. Centrifugation was by a two-step gradient (d = 1.06 and 1.21 g/ml) at 70,000 rpm for 210 min (procedure 10, Table II). The distributions of apolipoproteins A-I, A-II, and D and the enzyme lecithin:cholesterol acyltransferase, measured by radial immunodiffusion assay (performed by M. Cheung, University of Washington Health Sciences Center), are shown.

LDL Subspecies. Recent evidence supports the existence, based on size and density, of at least six LDL subspecies[15] (see Chapter [24], this volume). Single vertical spin ultracentrifugation can separate these 6 LDL subspecies with relatively brief spins and thus is a very useful approach to evaluating the structural and functional implications of the multiple LDL subspecies.

[15] R. M. Krauss and D. J. Burke, *J. Lipid Res.* **23,** 97 (1982).

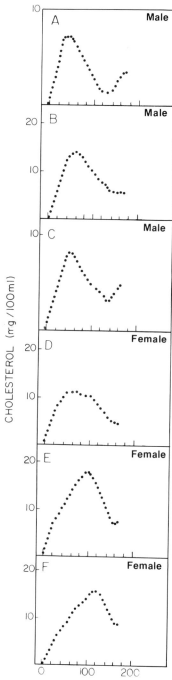

Fig. 13. HDL–VAP analyses on plasma from six subjects.

Two separate single vertical spin procedures, one preparative and one analytical, have been developed that separate the LDL subspecies quite effectively. Figure 14 shows the result of analysis by procedure 13, Table II, of plasma from three subjects known by virtue of nondenaturing polyacrylamide gradient gel analysis (see Chapter [24], this volume) to contain major LDL subspecies of widely differing size and thus widely differing density. Subject 1 shows a major LDL subspecies in the lower LDL density range and several minor subspecies at both lower and higher densities; subject 2 shows a major LDL subspecies in the intermediate LDL density range and several minor subspecies at lower densities; sub-

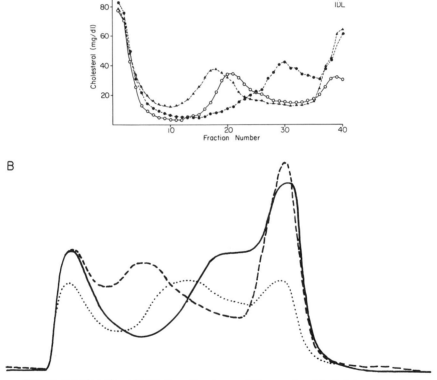

FIG. 14. Fractionation of LDL from the plasma of three male subjects into subspecies by SVS inverted rate-zonal density gradient ultracentrifugation. (A) Centrifugation was by a two-step gradient ($d = 1.045$ and 1.08 g/ml) in a TV-865B vertical rotor at 65,000 rpm for 330 min (procedure 13, Table II). The distribution of cholesterol in the fractions is shown. (B) Analysis by LDL–VAP. Centrifugation was by a two-step gradient ($d = 1.041$ and 1.08 g/ml) in a VTi80 vertical rotor at 80,000 rpm for 150 min (procedure 14, Table II).

ject 3 shows a major LDL subspecies in the higher LDL density range and several minor subspecies at both lower and higher densities. Excellent correspondence in position is seen between the various peaks of the three profiles, indicating a limited number of well-defined LDL subspecies.

The results of an analysis of the plasma from these same three subjects by a rapid VAP adaptation of the LDL–SVS procedure are illustrated in Fig. 14B. The procedure, called an LDL–VAP analysis, combines 1.5 hr spin by procedure 14, Table II, with VAP continuous flow analysis by LDL–VAP. Subjects 1, 2, and 3 show the same major lower, intermediate, and higher density peaks, respectively, seen in Fig. 14A. While some resolution is lost by LDL–VAP analysis compared to analysis of the separate fractions of procedure 13, minor peaks closely corresponding to those seen by procedure 13 are seen for each subject.

LDL–VAP analysis, at the present time, provides the optimal combination of rapidity, resolution, sample volume, and convenience available. Development of a quantitative decomposition algorithm for the LDL–VAP is anticipated soon.

Discussion

Advantages of Single Vertical Spin Ultracentrifugation

The single vertical spin procedure for the isolation and analysis of plasma lipoprotein species and subspecies has a number of advantages over the more conventional methods of sequential flotation, swingout rotor density gradient ultracentrifugation, and zonal ultracentrifugation (see appropriate chapters, this series).

1. *Decreased spin time.* The major advantage of SVS ultracentrifugation is the decreased spin time afforded by the vertical rotor geometry allied with velocity gradient ultracentrifugation. Spin times as short as 25 min produce complete separation of HDL, LDL, and VLDL.

2. *Decreased lipoprotein degradation.* The short spin times not only are more convenient but have an even more important effect; minimization of apolipoprotein dissociation from lipoprotein particles. This dissociation, which is a serious problem for apolipoprotein A-I and E during sequential flotation, is at least partly due to the tendency for lipids and proteins in a density gradient to separate under a centrifugal field and is thus inhibited by shortened spin times.

3. *Continuous spectrum of lipoprotein species and subspecies.* Because of the continuous spectrum of lipoprotein particles separated by SVS, this procedure has a big advantage over sequential flotation; arbitrary density cuts are not used to define lipoprotein species and sub-

species. Rather, positions of continuous peaks and shoulders are available to provide more sophisticated information, such as subspecies composition. As an example, sequential flotation density cuts are ineffective in dealing with Lp(a), whereas Lp(a) appears as a defined peak between HDL and LDL in SVS–VAP profiles (Fig. 7B).

4. *Analysis of lipoprotein subspecies.* Changes in gradients and spin times provide the flexibility to expand a lipoprotein region of interest, e.g., HDL, for more detailed analysis of subspecies.

5. *Simple gradient preparation.* The two step discontinuous nature of most of the gradients used for single spin separation provides gradient simplicity.

6. *Continuous flow monitoring of cholesterol peaks.* The VAP procedure provides a quick and convenient method of both qualitatively and quantitatively assaying the cholesterol profile of a given SVS fractionation of lipoproteins.

7. *Eight samples per spin.* Swingout rotors require a considerable longer time for velocity density gradient analysis of lipoprotein than vertical rotors while allowing only three samples per spin.

8. *Good resolution.* The SVS procedure, compared to the swingout rotor density gradient procedure, provides more rapid lipoprotein separation with equal resolution.

Disadvantages of Single Vertical Spin Ultracentrifugation

1. *Wall adherence of VLDL and albumin.* This is the major drawback of lipoprotein isolation by the vertical rotor. Under extreme conditions of gradient and accumulated g-force, VLDL tends to adhere to the inboard wall of the vertical tube. We find that, for procedure 6, Table II, a constant 40% of total VLDL is lost to the tube wall.[16] Adherence of VLDL does not lead to contamination of the other lipoprotein fractions with VLDL but rather VLDL adherence prevents a quantitative recovery of VLDL and creates the possible differential loss of one VLDL subspecies over another.

The occasional adherence of albumin and other serum proteins to the outboard wall is less of a problem than VLDL adherence but does make it more difficult to remove quantitatively all albumin from the lipoprotein fractions. Use of a denser bottom gradient combined with upward fractionation minimizes this potential problem with vertical rotors.

2. *Resolution.* While resolution between lipoprotein species and subspecies with SVS ultracentrifugation is good, in general resolution is

[16] J. P. Segrest, R. Warnick, and J. J. Albers, unpublished results.

slightly inferior to that achieved by adaptation of the same gradient to swingout rotor ultracentrifugation. Loss of resolution occurs as the result of (1) gradient compression at speed and (2) mixing during reorientation from horizontal to vertical and back again.

3. *Plasma volume limit per rotor.* Due to an upper limit of 80–90 ml of plasma per rotor, sequential flotation might be the procedure of choice for especially large lipoprotein preparations. However, the plasma volume limit is partly compensated for by the short spin times that allow multiple runs.

Solutions to the Problems of Single Vertical Spin Ultracentrifugation

1. *Angled-head rotors.* Use of the 70Ti angled-head rotor as a substitute for a vertical rotor for preparative runs overcomes the problems of VLDL and albumin wall adherence. Angled-head rotors involve some sacrifice of brevity in spin time, however.

2. *Zonal rotors.* Use of the Beckman Ti-15 zonal rotor adapted to the gradient conditions of procedure 1, Table II, overcomes the problem of the plasma volume limit and has been used for routine preparation of LDL (M. Krieger, personal communication).

Section III

Isolation and Physical–Chemical Characterization of Plasma Apolipoproteins

[9] Delipidation of Plasma Lipoproteins

By JAMES C. OSBORNE, JR.

Historical

The fact that lipids, which are only sparingly soluble in aqueous solution, are packaged in plasma as highly soluble lipid protein complexes has intrigued researchers for most of the twentieth century. The initial reports by Hoppe-Seyler[1] that diethyl ether did not extract all of the lipids from plasma have grown into an enormous volume of literature on separation of the protein and lipid components of plasma. Most of the early studies on delipidation were performed with unfractionated plasma or serum and progress in the field was quite slow. This is not surprising in view of our present concept of plasma lipoproteins as a heterogeneous population of micellular-like structures of differing lipid and apolipoprotein content.[2] The quaternary organization of plasma lipoproteins is the result of a fine balance between polar and nonpolar constituents and dissolution of these components can be quite sensitive to temperature and solvent composition. Early studies were concerned primarily with the native structure of plasma lipoproteins. The concept of a protein–lipid interaction which would equate to what we now term a noncovalent bond was first postulated in the early 1900s.[3,4] With the isolation of the first plasma lipoprotein of constant composition in 1929 by Macheboeuf[5] the stage was set for quantitative evaluation of protein–lipid interactions in plasma. Over the years countless solvent systems have been tested and, since the standardization of procedures for the isolation of plasma lipoproteins by ultracentrifugation in the early 1950s,[6] many of these have been applied to separate density classes. In the following overview I shall cover only those solvent systems that result in complete delipidation and that have thus gained wide acceptance in the field. For an excellent review of delipidation procedures used prior to 1969 see the Masters thesis of J. Holland Ledford.[7]

---

[1] F. Hoppe-Seyler, *Handb. Physiol. Chem.* **3**, 196 (1870).
[2] J. C. Osborne, Jr. and H. Bryan Brewer, Jr., *Adv. Protein Chem.* **31**, 253 (1977).
[3] E. G. Young, *Proc. R. Soc. London Ser. B* **93**, 15 (1922).
[4] J. A. Gardner and H. Gainsborough, *Biochem. J.* **21**, 141 (1927).
[5] M. Macheboeuf, *Bull. Soc. Chim. Biol.* **11**, 485 (1929).
[6] J. W. Gofman, F. Lindgren, H. A. Elliott, W. Manty, J. Hewitt, B. Strisower, and V. Herring, *Science* **111**, 166 (1950).
[7] J. Holland Leford, Masters thesis, University of Oklahoma, 1969.

Aqueous Samples

Organic Solvents. Hardy and Gardiner[8] reported on the use of diethyl ether to extract lipids from an alcohol precipitate of serum in 1910. In a more detailed study, Macheboeuf and Sandor[9] reported that diethyl ether alone extracted very little lipid from plasma lipoproteins (horse HDL) but that in combination with ethanol about 50% of the lipids were extracted. The investigations of lipid extractability in the late 1920s and 1930s set the stage for McFarlane's[10] classic letter in *Nature* on the effect of freezing on delipidation of serum. Lowering the temperature of a serum–ether mixture (1:0.3 v/v) to below $-25°$ and rewarming to room temperature allowed the slow (several hours) extraction of a major portion of serum lipids. McFarlane attributed the effect of freezing to the importance of water in the structure of plasma lipoproteins. This was reemphasized in the detailed study by Oncley *et al.*[11] on the properties of LDL.

Over the years many additional factors have been shown to affect the quaternary organization of plasma lipoproteins. The importance of buffer composition, primarily ionic strength, and alcohol to water ratio in the precipitation and extraction of plasma lipoproteins is documented by the studies at Harvard[12] on the fractionation of plasma by ethanol precipitation. The interfacial surface area between aqueous and organic phases has been shown to affect the efficiency of lipid extraction.[13,14] Avigan's[15] report on the extraction of lipids from LDL and HDL emphasizes that the specific properties of different plasma lipoproteins must also be taken into account during delipidation. These studies have been cited in order to emphasize the importance of adhering rigidly to the experimental conditions used in successful delipidation schemes reported in the literature. Changes in buffer composition, especially ionic strength, organic to aqueous volume ratios, temperature, or other experimental conditions may result in incomplete delipidation or formation of aggregates that are difficult to redissolve. This is especially true for plasma lipoproteins, such as LDL, which contain apolipoprotein B.

The delipidation procedures described by Shore and Shore[16] and Scanu

---

[8] W. B. Hardy and S. Gardiner, *J. Physiol.* (*London*) **40,** 68 (1910).
[9] M. Macheboeuf and G. Sandor, *Bull. Soc. Chim. Biol.* **14,** 1168 (1932).
[10] A. S. McFarlane, *Nature* (*London*) **149,** 439 (1942).
[11] J. L. Oncley, F. R. N. Gurd, and M. Melin, *J. Am. Chem. Soc.* **72,** 458 (1950).
[12] E. J. Cohn, L. E. Strong, W. L. Hughes, Jr., D. J. Mulford, J. N. Ashworth, M. Melin, and H. L. Taylor, *J. Am. Chem. Soc.* **72,** 465 (1950).
[13] A. Gustafson, *J. Lipid Res.* **6,** 512 (1965).
[14] D. B. Zilversmit, *J. Lipid Res.* **5,** 300 (1964).
[15] J. Avigan, *J. Biol. Chem.* **226,** 957 (1957).
[16] B. Shore and V. Shore, *Biochemistry* **8,** 4510 (1969).

and Edelstein[17] have gained wide acceptance and form the basis for routinely employed delipidation schemes in many laboratories. With the former procedure, plasma lipoproteins are extracted in the soluble state, whereas the latter is based on a more classical extraction of precipitated plasma lipoproteins. Smaller apolipoproteins, such as the apoC polypeptides, are soluble in the ethanol–ether mixtures used for lipid extraction in both of the above procedures. Loss of total protein is low for HDL but can amount to as much as 20% of the total protein for triglyceride-rich plasma lipoproteins such as VLDL. These peptides are not soluble in ethanol–diethyl ether under anhydrous conditions, or in an alternative solvent system, chloroform–methanol; however, delipidation under these conditions tends to result in an insoluble or poorly soluble residue. Soluble polypeptides may be recovered by an extraction of the combined organic phases.

Detergents. Initial reports on the use of detergents for delipidation were directed primarily at the quaternary organization of plasma lipoproteins. This emphasis on native lipoprotein structure shifted to one of resolubilization with the realization that apoB was irreversibly precipitated with alcohol–ether delipidation.[15] With most delipidation procedures in the literature solubilization of apoB requires high concentrations of denaturants or detergents.

Many schemes that use detergents result in incomplete delipidation. Helenius and Simons[18] have detailed a scheme for the complete delipidation of LDL by sodium deoxycholate, nonidet P40, sodium dodecyl sulfate (SDS), and cetyltrimethylammonium bromide (CTAB). Briefly, the method consisted of incubation of high concentrations of detergent with an aqueous solution of LDL and subsequent separation of the protein and lipid moieties by gel chromatography in the presence of micellular concentrations of detergent. The protein eluted in the void and the lipid in the included volume of the column. All four detergents resulted in complete delipidation, however, SDS and CTAB-treated samples exhibited altered immunochemical properties. This procedure forms the basis for most of the detergent–delipidation schemes reported in the literature.

An interesting modification of the above method is that suggested by Socorro and Camejo[19] in which LDL is delipidated in the presence of an ion-exchange resin. The lipid-free protein is eluted from the resin by high concentrations of inorganic salt. This group recently reported a detailed investigation of the peptides released from LDL upon delipidation with

---

[17] A. M. Scanu and C. Edelstein, *Anal. Biochem.* **44**, 576 (1971).
[18] A. Helenius and K. Simons, *Biochemistry* **10**, 2542 (1971).
[19] L. Socorro and G. Camejo, *J. Lipid Res.* **20**, 631 (1979).

the polyoxyethylene monoether Brij 36T.[20] Gel electrophoresis in SDS of the solubilized protein revealed a heterogeneous population of peptides ranging in apparent molecular weight from 14,000 to 250,000. Other polyoxyethylene monoethers, such as n-dodecyl octaethylene glycol monoether[21] and Triton X-100,[22] are also useful for the total delipidation of LDL. With these latter two detergents, delipidation results in formation of a more homogeneous species of molecular weight 500,000, which is presumably a dimer of the native apolipoprotein.[23] Disparities in the molecular properties of delipidated apoB are a common occurrence in the literature. The resolubilized lipid-free species of apoB is usually highly aggregated, even in the presence of detergents or denaturants, and exhibits nonideal hydrodynamic properties. These complex solution properties of apoB have resulted in major controversies regarding such basic molecular properties as the monomeric molecular weight. For a detailed discussion of the hydrodynamic properties of apoB, consult the chapter by Fisher and Schumaker [11] in this volume.

Tetramethyl Urea. Tetramethyl urea seems to be unique in its effect on plasma lipoproteins and has been used in procedures for purification as well as quantitation of apolipoproteins. The procedure described by Kane and co-workers[24] was quite detailed and has been used by many laboratories. The plasma lipoprotein solution at low ionic strength and pH 6–9 is mixed with an equal volume of tetramethyl urea at 37° and incubated at 37° for 30 min. The dissociated lipids form a separate phase and may be removed after clarification by centrifugation. Apolipoprotein B seems to be selectively precipitated by tetramethyl urea in all plasma lipoprotein density fractions. The tetramethyl urea aqueous phase contains the majority of the remaining apolipoproteins and is a convenient source for isolation of some apolipoproteins, especially apolipoprotein E.[25]

Lyophilized Samples

The importance of water in the structure of plasma lipoproteins has been emphasized repeatedly over the last five decades. It has been estimated that water accounts for greater than 30% of the mass of hydrated LDL.[11] Perturbations which disrupt these waters of hydration have major

---

[20] L. Socorro, F. Lopez, A. Lopez, and G. Camejo, *J. Lipid Res.* **23**, 1982 (1983).
[21] R. M. Watt and J. A. Reynolds, *Biochemistry* **19**, 1593 (1980).
[22] A. I. Kai and M. Hasegawa, *J. Biochem. (Tokyo)* **83**, 755 (1980).
[23] J. C. H. Steele, Jr. and J. A. Reynolds, *J. Biol. Chem.* **254**, 1639 (1979).
[24] J. P. Kane, T. Sato, R. L. Hamilton, and R. J. Havel, *J. Clin. Invest.* **56**, 1622 (1976).
[25] K. H. Weisgraber, R. W. Mahley, and G. Assmann, *Atherosclerosis* **28**, 121 (1977).

effects on the quaternary organization of plasma lipoproteins. The effects of freezing[10] and aqueous–organic interfacial surface area[13,14] on lipid extraction have been attributed classically to disruption of structural water. Nonpolar interactions in aqueous solution are dependent directly upon differences in the structure of waters of hydration and bulk water[26]: given the high percentage of nonpolar moieties in the composition of plasma lipoproteins such as LDL it is not surprising that intermolecular interactions are sensitive to water content.

Lyophilization of plasma lipoproteins prior to delipidation has three potential advantages over delipidation procedures using aqueous samples. (1) The sample size is reduced considerably. This is especially valuable for delipidation of large volumes since most aqueous procedures begin with a dropwise addition of the plasma lipoprotein to a large volume of organic solvent. (2) The small polypeptides are not soluble in the organic phase when delipidation is performed on dehydrated samples. (3) Removal of structural water potentially weakens the nonpolar interactions between components of plasma lipoproteins. This latter effect may be critically dependent upon plasma lipoprotein concentration and solvent composition prior to lyophilization. Formation of intermolecular hydrogen or ionic bonds with dehydration may result in incomplete delipidation and/or residues that are difficult to redissolve. For reproducible results one cannot overstress the importance of adhering strictly to published procedures.

Gustafson[13] was the first to report delipidation of lyophilized plasma lipoproteins. After lyophilization, in the presence of starch, the isolated fractions were stored over P_2O_5 at 4° and subsequently extracted with n-heptane or chloroform–methanol. The latter solvent resulted in complete delipidation whereas the former extracted primarily the neutral lipids. Recovery of soluble protein was essentially quantitative for VLDL and HDL, but was quite poor for LDL. Experience with lyophilized samples thus parallels that obtained with aqueous samples: delipidation of apoB-containing plasma lipoproteins results in a form of apolipoprotein B that is difficult to redissolve.

The procedures described by Brown et al.[27] and Kostner[28] seem to be quite reliable for delipidation of VLDL or HDL. Use of these procedures for delipidation of LDL requires detergents such as decyl sulfate for

[26] H. Edelhoch and J. C. Osborne, Jr., *Adv. Protein Chem.* **30**, 183 (1976).
[27] W. V. Brown, R. I. Levy, and D. S. Fredrickson, *J. Biol. Chem.* **244**, 5687 (1969).
[28] G. M. Kostner, in "The Lipoprotein Molecule" (H. Peeters, ed.), p. 19. Plenum, New York, 1978.

resolubilization of the lipid free residue. Olofsson et al.[29] and Cardin et al.[30] have reported procedures for total delipidation of lyophilized LDL which result in lipid free products that are soluble in aqueous buffers. The former group uses 2 M acetic acid for solubilization whereas the latter group uses the denaturants guanidinium chloride and urea with a subsequent dialysis to a low ionic strength Tris buffer. Unfortunately, these two procedures result in preparations of soluble apoB which differ in molecular weight. For a more detailed discussion of the properties of apoB see the chapter by Fisher and Schumaker (this volume, Chapter [11]).

Sample Preparation

Use of Solvent Additives

As indicated above, the complete delipidation of plasma lipoproteins is quite sensitive to solvent composition, and therefore use of protective agents should be kept to a minimum, especially when following procedures which call for lyophilization. There are, however, several precautions that should be taken to ensure reproducible preparations of lipid-free apolipoproteins. Plasma lipoproteins undergo lipid autoxidation in the presence of oxygen and heavy metals such as copper, and this can result in disruption of the covalent structure of apolipoproteins as has been demonstrated for apolipoprotein B.[31,32] Lipid autoxidation is believed to proceed by a free radical mechanism involving molecular oxygen and is enhanced by elevated temperatures and the presence of heavy metals. The use of EDTA as an anticoagulant serves the second purpose of inhibiting lipid autoxidation; as soon as possible after collection fresh plasma should be made about 1 mM in EDTA. Sodium azide (0.05% w/v) has been reported to enhance lipid autoxidation at room temperature and should be used with caution with lipidated samples.[31] If required, sodium azide should be used at lower concentrations, <0.01% and at lower temperatures, <4°. Finally, serine proteases have been shown to coisolate

---

[29] S.-O. Olofsson, K. Bostrom, U. Svanberg, and G. Bondjers, *Biochemistry* **19**, 1059 (1980).

[30] A. D. Cardin, K. R. Witt, C. L. Barnhart, and R. L. Jackson, *Biochemistry* **21**, 4503 (1982).

[31] J. Schuh, G. F. Fairclough, Jr., and R. H. Haschemeyer, *Proc. Natl. Acad. Sci. U.S.A.* **75**, 3173 (1978).

[32] J. C. H. Steele, Jr. and J. A. Reynolds, *J. Biol. Chem.* **254**, 1633 (1979).

with VLDL[33] and serine protease inhibitors such as phenylmethylsulfonylfluoride (10 μM) should also be added to fresh plasma.

Isolation of Plasma Lipoprotein Fractions

The delipidation schemes described below are general and can be used for delipidation of all plasma lipoprotein density fractions. As indicated above, delipidation of plasma lipoproteins which contain apolipoprotein B tend to result in the formation of residues that are difficult to redissolve. This residue is composed primarily of apolipoprotein B; however, other apolipoproteins which are normally soluble in aqueous buffers may be rendered less soluble in the presence of apolipoprotein B. This is the primary reason for recommending fractionation of plasma lipoproteins prior to delipidation. The schemes described below are designed primarily for delipidation of isolated density fractions of plasma. Fractionation by ultracentrifugation results in losses of some apolipoproteins due to dissociation during centrifugation. Minor modifications may be required prior to application of these schemes to other fractions of plasma, such as those obtained by affinity chromatography.[34]

Dialysis

Buffer composition is an important element in efficient, reproducible delipidation procedures. For delipidation of aqueous samples plasma lipoprotein fractions should be dialyzed exhaustively at 4° against 0.005 M ammonium bicarbonate, 0.15 M sodium chloride, 1 mM ethylenediaminetetraacetic acid, 1 mM sodium azide, 10 μM phenylmethylsulfonylfluoride pH 8.0 buffer. Samples that are to be lyophilized should be dialyzed against 0.005 M ammonium bicarbonate, 1 mM ethylenediaminetetraacetic acid just prior to lyophilization.

Lyophilization

Plasma lipoprotein solutions should be frozen in the form of a thin film (shell freezing) by rotation of the container in an ethanol dry ice mixture prior to lyophilization. Thawing of the sample during lyophilization results in loss of sample and inefficient dehydration and should be avoided.

[33] W. A. Bradley, E. B. Gilliam, A. M. Gotto, Jr., and S. H. Gianturco, *Biochem. Biophys. Res. Commun.* **109**, 1360 (1982).
[34] J. P. McVicar, S. T. Kunitake, R. L. Hamilton, and J. P. Kane, *Proc. Natl. Acad. Sci. U.S.A.* **81**, 1356 (1984).

Delipidation

Aqueous Samples

This procedure can be used for all plasma lipoprotein density fractions and is based primarily on the method described by Herbert and co-workers[35] and was developed in the Molecular Disease Branch under the direction of Thomas J. Bronzert.

Reagents and Apparatus

Absolute methanol (<0.02% water)
Anhydrous diethyl ether (cool to 0° on ice)
Dialyzed lipoprotein solution
Laboratory vortex mixer
Screw cap 16 × 125 mm round bottom test tube

Extraction Procedure

A. Add 250 μl of plasma lipoprotein solution slowly and dropwise to 3 ml of vortexing methanol. Care should be taken that the drops hit the vortexing solvent rather than the sides of the test tube.
B. Add 7 ml of diethyl ether to the vortexing solution and allow the mixture to stand on ice for 10 min.
C. Centrifuge at 1000 rpm for 4 min and remove the supernatant by aspiration.
D. Add 10 ml of diethyl ether while vortexing the slurry from step C and allow to stand on ice for 10 min.
E. Centrifuge at 1000 rpm for 4 min and remove the supernatant by aspiration.
F. Evaporate remaining ether with nitrogen.

Lyophilized Samples

This method is based primarily on the methods proposed by Brown *et al.*[27] and Kostner.[28] It can be used to delipidate all density fractions of plasma.

Reagents

Ethyl alcohol (95%)
Anhydrous diethyl ether

---

[35] P. N. Herbert, L. L. Bausserman, L. O. Henderson, R. J. Heinen, E. C. Church, and R. S. Shulman, in "The Lipoprotein Molecule" (H. Peeters, ed.), p. 35. Plenum, New York, 1978.

Dialyzed lipoprotein solution
Laboratory vortex mixer
Screw cap 16 × 125 mm round bottom test tube

Extraction Procedure

A. Shell freeze approximately 10 mg of lipoprotein in the screw cap vial and lyophilize.
B. Add 10 ml diethyl ether, vortex, and allow to stand, with occasional gentle mixing, for 1 hr on ice.
C. Centrifuge at 1000 rpm for 4 min and remove the supernatant by aspiration.
D. Repeat steps B and C.
E. Add 10 ml of ethanol–diethyl ether (3:1, v/v), vortex, and allow to stand, with occasional gentle mixing, for 1 hr on ice and then repeat centrifugation (step C).
F. Repeat step E four times.
G. Rinse the final pellet twice with diethyl ether and dry with a stream of nitrogen.

Resolubilization

Delipidation of plasma lipoproteins usually results in the formation of residues that are difficult to redissolve. Difficulties with apoB have received most of the attention in the literature, however, other apolipoproteins also tend to form aggregates with delipidation of dehydration. The buffer used to dissolve apolipoproteins should be designed for the specific property under investigation. Simple solubilization to yield a clear solution may not be sufficient for some studies. For instance, soluble aggregates would need to be dissociated prior to evaluating properties that depend directly upon the molecular species in solution. For these types of studies it is a good practice to dissolve the delipidated sample in buffer containing moderate levels of denaturant and then dialyze the solubilized sample against the experimental buffer. We routinely dissolve apolipoproteins in 0.01 M Tris, 0.1 M sodium chloride, 0.001 M sodium azide, 0.001 M ethylenediaminetetraacetic acid, 2 M guanidinium chloride, pH 7.4 buffer. This extra step in preparing samples is time consuming; however, it ensures reproducible solutions of all apolipoproteins except apoB. The solution properties of apoB are not well understood. Detergents such as decyl sulfate, deoxycholate, or Triton X-100 or denaturants such as guanidinium chloride are usually required for solubilization and the resulting solutions usually exhibit nonideal behavior. Consult the chapter by Fisher and Schumaker [11] for details on the properties of apoB. The review by

Kostner[28] is a good starting point for designing resolubilizing buffers for specific apolipoproteins.

Advantages/Disadvantages

This section is included primarily as a brief summary for use in selection of delipidation procedures. The first question for most studies is whether or not one should fractionate plasma prior to delipidation. Fractionation is recommended for most studies; apolipoproteins form mixed aggregates with other components of plasma which may cause inefficient delipidation and/or formation of residues that are difficult to redissolve. The next question is whether or not the sample should be lyophilized prior to delipidation. The major advantage of lyophilization is reduction of sample volume. Loss of polypeptides due to solubility in the organic phase is also minimized with dehydrated samples. Disadvantages include possible incomplete delipidation and formation of apolipoprotein residues that are difficult to redissolve. Incomplete delipidation can usually be avoided by manually disrupting the aggregated apolipoprotein residue at various stages of the delipidation scheme. Resolubilization is not usually a problem when moderate levels of guanidinium chloride are included in the initial solubilization buffer (see above section). Apolipoprotein B, and possibly apolipoprotein E, are best delipidated in the aqueous state, or in the presence of detergents. The molecular properties of these apolipoproteins are sensitive to prior treatment, especially when exposed to procedures such as delipidation which effect their degree of hydration. Detergents used in delipidation procedures may be difficult to remove from the final product, and this can also result in changes in hydrodynamic properties. The delipidation procedure used should be designed for the specific question under study. The published procedures described above are detailed and if followed closely result in efficient reproducible delipidation.

[10] Isolation and Characterization of Apolipoproteins A-I, A-II, and A-IV

By H. BRYAN BREWER, JR., ROSEMARY RONAN, MARTHA MENG, and CHERI BISHOP

Introduction

Human apolipoproteins (apo)A-I and A-II are the major protein (90%) constituents of HDL. HDL as well as apoA-I and apoA-II have been of particular interest since HDL is inversely correlated with the development of premature cardiovascular disease.[1-3] Both apoA-I and apoA-II have been evaluated as potential predictors for the development of premature cardiovascular disease.[4-6] ApoA-IV is also a constituent of HDL; however in human plasma apoA-IV is primarily present unassociated with plasma lipoproteins in the 1.21 g/ml infranate.[7-9] To date apoA-IV has not been identified with any dyslipoproteinemia associated with vascular disease.

ApoA-I, apoA-II, and apoA-IV are synthesized primarily by the liver and intestine.[10-13] ApoA-I has been shown to increase the enzymatic activity of lecithin : cholesterol acyltransferase, the enzyme which catalyzes the esterification of plasma cholesterol.[14] ApoA-II activates hepatic lipase

---

[1] G. J. Miller and N. F. Miller, *Lancet* **1**, 16 (1975).
[2] W. P. Castelli, J. T. Doyle, T. Gordon, C. G. Hames, M. C. Hjortland, S. B. Hulley, A. Kagan, and W. J. Zukel, *Circulation* **55**, 767 (1977).
[3] G. Rhoads, C. L. Gulbrandsen, and A. Kagan, *N. Engl. J. Med.* **294**, 293 (1976).
[4] G. Fager, O. Wiklund, S. O. Olofsson, L. Wilhelmsen, and G. Bondjers, *Arteriosclerosis* **1**, 273 (1981).
[5] P. J. Jenkins, R. W. Harper, and P. J. Nestel, *Br. Med. J.* **II**, 187 (1980).
[6] J. J. Maciejko, D. H. Holmes, B. A. Kottke, A. R. Zinsmeister, D. M. Dinh, and S. J. T. Mao, *New Engl. J. Med.* **309**, 385 (1983).
[7] G. Utermann and U. Beisiegel, *Eur. J. Biochem.* **99**, 333 (1979).
[8] P. H. R. Green, R. M. Glickman, J. W. Riley, and E. Quinet, *J. Clin. Invest.* **65**, 911 (1980).
[9] R. B. Weinberg and A. M. Scanu, *J. Lipid Res.* **24**, 52 (1983).
[10] I. Imaizumi, R. J. Havel, M. Fainaru, and J. L. Vigne, *J. Lipid Res.* **19**, 1038 (1978).
[11] A-L. Wu and H. G. Windmueller, *J. Biol. Chem.* **254**, 7316 (1979).
[12] P. H. R. Green, R. M. Glickman, C. D. Saudek, C. B. Blum, and A. R. Tall, *J. Clin. Invest.* **64**, 233 (1979).
[13] D. W. Anderson, E. J. Schaefer, T. J. Bronzert, F. T. Lindgren, T. Forte, T. E. Starzl, G. D. Niblack, L. A. Zech, and H. B. Brewer, Jr., *J. Clin. Invest.* **67**, 857 (1981).
[14] C. J. Fielding, V. G. Shore, and P. E. Fielding, *Biochem. Biophys. Res. Commun.* **46**, 1493 (1972).

in vitro; however, the physiological significance of this activation remains to be established.[15] There is no known physiological function for apoA-IV; however, recent studies have established that apoA-IV also activates lecithin : cholesterol acyltransferase *in vitro*.[16]

Isolation and Characterization of ApoA-I

Overview

The complete covalent structures of human[17] and dog[18] mature apoA-I have been reported, and are very similar in amino acid sequence. Studies on the biosynthesis of human apoA-I have established that apoA-I is synthesized as a 267 amino acid precursor, preproapoA-I (Fig. 1).[19–23] The human prepeptide contains 18 amino acids, and the propeptide contains a hexapeptide, Arg-His-Phe-Trp-Gln-Gln, attached to the amino terminus of apoA-I. PreproapoA-I undergoes cotranslational cleavage to proapoA-I, and proapoA-I undergoes extracellular conversion to mature apoA-I by an apparent specific calcium-dependent protease (Fig. 2).[24,25] In man the primary form of apoA-I secreted by the cell is proapoA-I, and proapoA-I is quantitatively converted *in vivo* to mature apoA-I with a residence time of approximately 4.5 hr.[26]

ProapoA-I accounts for approximately 4% of fasting plasma apoA-I,[27]

---

[15] C. E. Jahn, J. C. Osborne, Jr., E. J. Schaefer, and H. B. Brewer, Jr., *FEBS Lett.* **131**, 366 (1981).

[16] C.-H. Chen and J. J. Albers, *Arteriosclerosis* **4**, 519a (1984).

[17] H. B. Brewer, Jr., T. Fairwell, A. LaRue, R. Ronan, A. Houser, and T. Bronzert, *Biochem. Biophys. Res. Commun.* **80**, 623 (1978).

[18] H. Chung, A. Randolph, I. Reardon, and R. L. Heinrikson, *J. Biol. Chem.* **257**, 2961 (1982).

[19] S. W. Law, G. Gray, and H. B. Brewer, Jr., *Biochem. Biophys. Res. Commun.* **112**, 257 (1983).

[20] C. C. Shoulders, A. R. Kornblihtt, B. S. Munro, and F. E. Baralle, *Nucleic Acids Res.* **11**, 2827 (1983).

[21] P. Cheung and L. Chan, *Nucleic Acids Res.* **11**, 3703 (1983).

[22] S. K. Karathanasis, V. I. Zannis, and J. L. Breslow, *Proc. Natl. Acad. Sci. U.S.A.* **80**, 6147 (1983).

[23] S. W. Law and H. B. Brewer, Jr., *Proc. Natl. Acad. Sci. U.S.A.* **81**, 66 (1984).

[24] C. Edelstein, J. I. Gordon, K. Toscas, H. F. Sims, A. W. Strauss, and A. M. Scanu, *J. Biol. Chem.* **258**, 11430 (1983).

[25] D. Bojanovski, R. E. Gregg, and H. B. Brewer, Jr., *J. Biol. Chem.* **259**, 6049 (1984).

[26] D. Bojanovski, R. E. Gregg, G. Ghiselli, E. J. Schaefer, J. A. Light, and H. B. Brewer, Jr., *J. Lipid Res.* **Feb.** (1985).

[27] D. L. Sprecher, L. Taam, and H. B. Brewer, Jr., *Clin. Chem.* **30**, 2084 (1984).

Prepeptide

Met Lys Ala Ala Val Leu Thr Leu Ala Val Leu Phe Leu Thr Gly Ser Gln Ala	Arg His Phe Trp Gln Gln					
−24	−7 −6					−1

Mature ApoA-I

Asp Glu Pro Pro Gln Ser Pro Trp Asp Arg Val Lys Asp Leu Ala Thr Val Tyr Val Asp Val Leu Lys Asp Ser Gly Arg Asp Tyr Val
1   10   20   30

Ser Gln Phe Glu Gly Ser Ala Leu Gly Lys Gln Leu Asn Leu Lys Leu Leu Asp Asn Trp Asp Ser Val Thr Ser Thr Phe Ser Lys Leu
31   40   50   60

Arg Glu Gln Leu Gly Pro Val Thr Gln Glu Phe Trp Asp Asn Leu Glu Lys Glu Thr Glu Gly Leu Arg Gln Glu Met Ser Lys Asp Leu
61   70   80   90

Glu Glu Val Lys Ala Lys Val Gln Pro Tyr Leu Asp Asp Phe Gln Lys Lys Trp Gln Glu Glu Met Glu Leu Tyr Arg Gln Lys Val Glu
91   100   110   120

Pro Leu Arg Ala Glu Leu Gln Glu Gly Ala Arg Gln Lys Leu His Glu Leu Gln Glu Lys Leu Ser Pro Leu Gly Glu Glu Met Arg Asp
121   130   140   150

Arg Ala Arg Ala His Val Asp Ala Leu Arg Thr His Leu Ala Pro Tyr Ser Asp Glu Leu Arg Gln Arg Leu Ala Ala Arg Leu Glu Ala
151   160   170   180

Leu Lys Glu Asn Gly Gly Ala Arg Leu Ala Glu Tyr His Ala Lys Ala Thr Glu His Leu Ser Thr Leu Ser Glu Lys Ala Lys Pro Ala
181   190   200   210

Leu Glu Asp Leu Arg Gln Gly Leu Leu Pro Val Leu Glu Ser Phe Lys Val Ser Phe Leu Ser Ala Leu Glu Glu Tyr Thr Lys Lys Leu
211   220   230   240

Asn Thr Gln
241   243

FIG. 1. Amino acid sequence of human preproapoA-I. PreproapoA-I contains an 18 amino acid prepeptide, a 6 amino acid propeptide, and the 243 amino acid mature apoA-I.

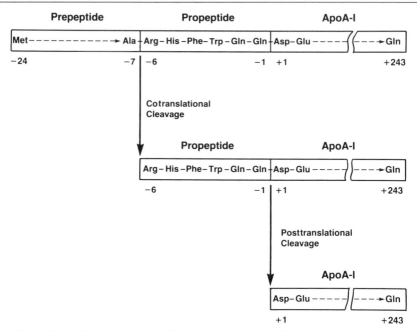

FIG. 2. Proteolytic processing of human preproapoA-I. The cotranslational cleavage of preproapoA-I to proapoA-I occurs intracellularly and the posttranslational cleavage of proapoA-I to mature apoA-I takes place in plasma by a specific calcium-dependent protease.

however, proapoA-I is increased in thoracic duct lymph, and following a meal enriched in lipid.[28] Mature apoA-I is polymorphic in plasma, and has routinely been evaluated by two-dimensional gel electrophoresis (Fig. 3). The nomenclature now utilized for apoA-I isoforms as reviewed previously[27] conforms to the nomenclature standardized for proteins separated by two-dimensional gel electrophoresis. In this system the two-dimensional electrophoretograms are presented with the acidic and basic regions of the gel oriented to the left and right, respectively, and the major unsialylated isoform is designated with a zero subscript. Other isoforms of the protein are codified by unit charge (− or +) from the major unsialylated isoform. Thus, the major mature apoA-I isoform is designated apoA-I_0, and the minor mature isoforms, apoA-I_{-1} and apoA-I_{-2}. ProapoA-I, which contains two additional positive charges when compared to mature apoA-I, is located at the apoA-I_{+2} position on the electrophoretogram

---

[28] G. Ghiselli, E. J. Schaefer, J. A. Light, and H. B. Brewer, Jr., *J. Lipid Res.* **24**, 731 (1983).

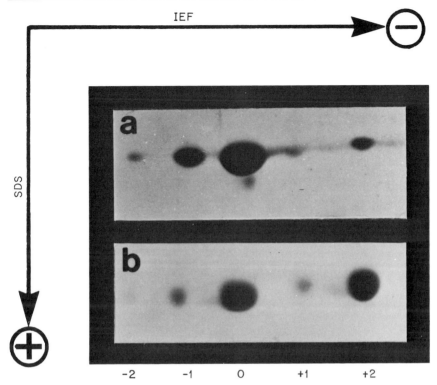

FIG. 3. Two-dimensional gel electrophoretogram of plasma (a) and lymph (b) apoA-I. The major mature apoA-I isoform is designated as apoA-I_0, and the minor isoforms are codified by unit charge shifts by (+) and (−) with respect to the major mature isoform. ApoA-I_{+2} which is selectively increased in lymph apoA-I is proapoA-I.

(Fig. 3). The apoA-I isoform, apoA-I_0, has been previously designated apoA-I_3,[29] apoA-I_4,[30] or apo-A-I_1.[31]

Isolation of ApoA-I-Containing Lipoproteins

Mature apoA-I can be isolated from HDL obtained from plasma of fasting normal subjects. Lipoproteins are separated by sequential ultra-

[29] H. B. Brewer, Jr., T. Fairwell, L. Kay, M. Meng, R. Ronan, S. Law, and J. A. Light, *Biochem. Biophys. Res. Commun.* **113**, 626 (1983).
[30] V. I. Zannis, J. L. Breslow, and A. J. Katz, *J. Biol. Chem.* **255**, 8612 (1980).
[31] G. Utermann, G. Feussner, G. Francecchini, J. Hass, and A. Steinmetz, *J. Biol. Chem.* **257**, 501 (1982).
[32] R. J. Havel, H. A. Eden, and J. H. Bragdon, *J. Clin. Invest.* **34**, 1345 (1955).

centrifugation flotation at various densities. HDL is isolated at density 1.063–1.21 g/ml. Plasma is adjusted to density 1.063 g/ml with solid KBr and centrifuged for 16 hr at 4° (59,000 rpm, Beckman 60 Ti rotor). The infranate, obtained by tube slicing, is adjusted to density 1.21 g/ml and centrifuged for 24 hr at 4° (59,000 rpm, Beckman 60 Ti rotor). The supernate from this second centrifugation is adjusted to density 1.24 g/ml and centrifuged for 48 hr at 4° (39,000 rpm, Beckman 40 rotor). The supernate from this final centrifugation contains the HDL. The isolated HDL is dialyzed extensively against 0.01% EDTA–0.01% NaN_3 (pH 7.4), lyophilized, and delipidated. Delipidation is performed at 5° with chloroform : methanol (2 : 1, v/v; 3–10 mg protein/ml organic solvent) with intermittent vortexing for 1 hr. The sample is centrifuged for 15 min at 2000 rpm, and the delipidation procedure repeated four times.

The isolation of proapoA-I is more readily performed with thoracic duct lymph due to the relative increase in proapoA-I.[28] Lymph (5 ml) is overlayed with 30 ml of 0.9% NaCl (pH 7.0) and centrifuged at 20,000 rpm (Beckman SW 27 rotor) at 5° for 60 min. Chylomicrons are collected in the top 2 ml by tube slicing and recentrifuged under the same conditions. Following the last centrifugation the chylomicrons are dispersed in 0.01 M NH_4HCO_3, lyophilized, and delipidated by three extractions with chloroform : methanol (2 : 1) as outlined above.

Isolation of Polymorphic ApoA-I

The apolipoprotein fractions isolated from plasma and lymph contain a mixture of apolipoproteins in addition to apoA-I. Polymorphic apoA-I can be separated from these other apolipoproteins by size exclusion and affinity chromatography. The isolated apoA-I is homogeneous by analytical sodium dodecyl sulfate (SDS)–gel electrophoresis, urea polyacrylamide gel electrophoresis, Ouchterlony immunodiffusion, Edman amino-terminal analysis, and amino acid analysis.

Size Exclusion. 1. Gel permeation chromatography. ApoA-I can be isolated to virtual homogeneity by gel permeation chromatography. Delipidated HDL apolipoproteins are dissolved (30 mg/ml) in 50 mM glycine, 4 mM NaOH, 0.5 M NaCl, and 6 M urea (pH 8.8) followed by fractionation on a Sephacryl S-200 (Pharmacia) column (4 × 190 cm; 45 ml/hr; 10 ml fractions) equilibrated in the same buffer (Fig. 4). The major peak contains apoA-I (solid bar, Fig. 4), and may be rechromatographed if not completely homogeneous (see below for criteria of purity). The apoA-I pool is dialyzed against 0.01 M NH_4HCO_3, and the concentration adjusted to 1 mg/ml for storage at −20°.

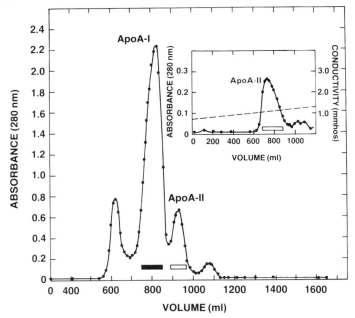

FIG. 4. Gel permeation chromatography of delipidated high-density lipoproteins on Sephacryl S-200. The fractions pooled for apoA-I and apoA-II are indicated by the solid and open bars, respectively. Inset: Partially purified apoA-II separated by Sephacryl S-200 chromatography is further purified by ion-exchange chromatography on DE52. The fractions containing apoA-II (open bar) eluted with a conductivity of 1.1 to 1.4 mmho.

2. *High-pressure liquid chromatography.* Small quantities of highly purified apoA-I may be obtained by gel permeation chromatography with high-pressure liquid chromatography on a TSK SW 3000 (Toyo Soda Manufacturing Co., Ltd.) column (0.75 × 60 cm) equilibrated in 6 M guanidine–HCl and 0.05 M Tris–HCl (pH 7.3) (Fig. 5). Fractions containing apoA-I are dialyzed, and stored at −20° (1 mg/ml).

3. *Preparative sodium dodecyl sulfate–gel electrophoresis.* Delipidated apolipoproteins containing apoA-I are solubilized (2 mg/ml) in 0.1 M Tris–HCl (pH 6.8), 1% SDS, and 10% sucrose followed by incubation for 3 min at 110°. The sample (1.5 mg) is separated by SDS–polyacrylamide gel electrophoresis in a Bio-Rad 220 Dual Slab Cell (Bio-Rad Lab., Richmond, CA) utilizing a 140 × 100 × 3-mm-thick vertical slab gel with a stacking gel containing (per liter) 40 g acrylamide, 1.1 g bisacrylamide, 0.125 mol Tris–HCl (pH 6.8), 2 g SDS, 1.5 ml N,N,N',N'-tetramethylethylenediamine (TEMED), and 0.3 g ammonium persulfate. The running

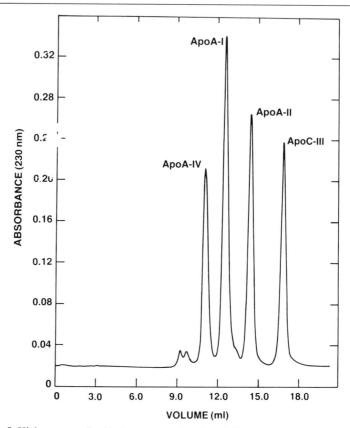

FIG. 5. High-pressure liquid chromatography of apoA-IV (27 μg), apoA-I (27 μg), apoA-II (27 μg), and apoC-III (10 μg) on a TSK SW 3000 column (0.75 × 60 cm, 0.3 ml/min, chart 0.1″/min, AUFS 0.4) equilibrated in 0.05 M Tris–HCl and 6 M guanidine–HCl (pH 7.3).

gel contained (per liter) 150 g acrylamide, 0.8 g bisacrylamide, 0.375 mol Tris–HCl (pH 8.8), 1.0 g SDS, 1.5 ml TEMED, and 0.3 g ammonium persulfate. The tray buffer is 0.05 M Tris, 0.384 M glycine, and 0.1% SDS (pH 8.4). The gels are electrophoresed until the dye front reaches 2 mm from the bottom of the gel (~4 hr; 20 mA for stacking, and 40 mA through the running gel; 4°). After completion of the run a 1-cm section of the edge of the gel is cut and the bands visualized by staining with 0.125% Coomassie Blue R250. The appropriate segments of the gel are cut and the protein eluted by electrophoresis (60 V, 5 hr) into Spectra-Por 3 (Spectrum Medical, Inc., Los Angeles, CA) dialysis bags. The isolated apolipoproteins are desalted on Sephadex G-100 (Pharmacia, 1.2 × 90 cm) in 0.01 M NH_4HCO_3, concentrated to 1 mg/ml, and stored frozen at −20°.

Affinity Chromatography. ApoA-I may be purified from HDL or plasma by affinity chromatography utilizing monospecific polyclonal or monoclonal apoA-I antibodies. ApoA-I antibodies are coupled to CNBr-activated Sepharose 4B (Pharmacia) for 24 hr at 4° with gentle rocking.[33] After coupling the remaining active sites were blocked by gently rocking the resin in 1 M ethanolamine (pH 8.1) for 2 hr at room temperature. Excess blocking agent and adsorbed proteins are removed by washing the resin four times alternately with 0.1 M NaHCO$_3$–0.5 M NaCl (pH 8.0) and 0.1 M CH$_3$COONa–0.5 M NaCl (pH 5.0) buffers. The coupled resin is then equilibrated in 0.01 M sodium phosphate, 0.15 M NaCl, and 0.01% NaN$_3$ (pH 7.5). Delipidated HDL is dissolved or plasma is dialyzed in the equilibration buffer. Delipidated HDL (4 mg/ml coupled resin) or plasma (2 ml/ml coupled resin) is incubated with the anti-apoA-I resin for 18–24 hr at 4° with gentle rocking. Following binding the resin is washed with 12 bed volumes of equilibration buffer. ApoA-I is dissociated and eluted from the resin with 2 bed volumes of the same buffer containing 8 M urea. The isolated apoA-I is dialyzed against 0.01 M NH$_4$HCO$_3$, and stored at 1 mg/ml at −20°.

Isolation of ApoA-I Isoforms

ApoA-I isolated by size exclusion or affinity chromatography is homogeneous by several criteria, however analytical isoelectrofocusing reveals several apoA-I isoforms. The isoforms can be isolated on preparative isoelectrofocusing slab gels. The mature apoA-I isoforms (apoA-I$_0$, apoA-I$_{-1}$, and apoA-I$_{-2}$) and proapoA-I (apoA-I$_{+2}$) are isolated from polymorphic apoA-I obtained from delipidated HDL and lymph chylomicrons, respectively. Lyophilized apolipoprotein samples are dissolved in aqueous 8 M urea (2 mg/ml) and separated by isoelectrofocusing on a Bio-Rad 220 Dual Slab Cell. A 5.0–6.0 pH gradient (Serva Ampholines, Serva AB, Heidelberg, FRG) is employed with a 140 × 100 × 3-mm-thick gel containing (per liter) 75 g acrylamide, 2.0 g bisacrylamide, 480 g urea, 75 ml above mentioned ampholytes, 1.0 ml TEMED, and 0.35 g ammonium persulfate. The upper and lower tray buffers are 0.02 M NaOH and 0.01 M H$_3$PO$_4$, respectively.

The gels are electrophoresed for 16 hr at 250 V. After completion of the electrophoresis, the separated isoforms are localized by cutting a 1-cm section from the side of the slab gel, and staining with a modified Coomassie Blue G250 staining procedure. This Coomassie Blue stain is prepared by adding 5.68 ml of reagent grade H$_2$SO$_4$ to 100 ml of 2% aqueous Coomassie Blue G250. The precipitate formed is allowed to sediment for 1

---

[33] W. B. Jakoby and M. Wilchek, eds., this series, Vol. 34.

hr and the supernatant is filtered through Whatman 1 filter paper. Trichloroacetic acid is added to the filtered supernatant to a final concentration of 12% and titrated with 10 N KOH to pH 5.4–5.6. This staining solution enables the bands to be visualized in minutes without destaining. The appropriate gel segments are cut and the protein is eluted by electrophoresis (60 V, 5 hr) into Spectra-Por 3 dialysis bags. The isolated isoforms are desalted on Sephadex G-100 (Pharmacia, 1.2 × 90 column) in 0.01 M NH_4HCO_3, and concentrated to 1 mg/ml.

Recently, the apoA-I isoforms have been isolated by preparative isoelectrofocusing utilizing an immobline system pH 4.9 to 5.9 (LKB Application Note 321) on a 3.0-mm vertical slab gel. Purified apoA-I or delipidated HDL are solubilized in 0.1 M Tris–HCl, 1% sodium decyl sulfate, 20% sucrose, and 6 M urea (pH 8.0). The upper tray buffer is 0.01 M NaOH and the lower tray buffer 0.01 M glutamic acid. The gels are electrophoresed overnight at 4° at a constant power of 5 W. The focused apolipoproteins are visualized by soaking the gel in water.[34] Individually visualized bands are cut and the protein is eluted by electrophoresis (60 V, 5 hr) into dialysis bags as outlined above. Since the ampholines in immobline are covalently linked to the polyacrylamide this system has the advantage of reducing ampholine contamination during elution, allowing higher salt concentration in the sample, larger sample load, as well as eliminating the problem of pH gradient drift.

Characterization of Isolated ApoA-I and ApoA-I Isoforms

1. Analytical sodium dodecyl sulfate–gel electrophoresis. Lyophilized apoA-I is analyzed by analytical SDS–gel electrophoresis by the methods outlined for preparative SDS–slab gel electrophoresis (Fig. 6). For analytical purposes 2–3 μg of apolipoprotein is electrophoresed in a 140 × 100 × 0.75-mm-thick gel at 4° until the dye reaches 2 mm from the bottom of the gel. Following electrophoresis the gel is fixed and stained for 1 hr with 0.125% Coomassie Blue R 250 in methanol : acetic acid : water (45 : 10 : 45, v/v), and destained in methanol : acetic acid : water (5 : 7.5 : 87.5, v/v) until a clear background is obtained.

2. Urea polyacrylamide gel electrophoresis. Apolipoprotein samples are dissolved (1 mg/ml) in 8 M urea–0.04 M Tris–HCl (pH 8.9). The running gel (0.6 × 10 cm tube) contains (per liter) 100 g acrylamide, 2.0 g bisacrylamide, 0.375 mol Tris–HCl, 0.75 ml TEMED, 480 g urea, and 0.5 g ammonium persulfate (pH 9.1). A 1-cm stacking gel contains (per liter) 25 g acrylamide, 2.0 g bisacrylamide, 0.05 M Tris–H_3PO_4, 0.5 ml TEMED, 480 g urea, 0.2 g ammonium persulfate, and 5.0 mg riboflavin (pH 6.7).

[34] K. von Hunger, R. C. Chin, and C. F. Baxter, *Anal. Biochem.* **128**, 398 (1983).

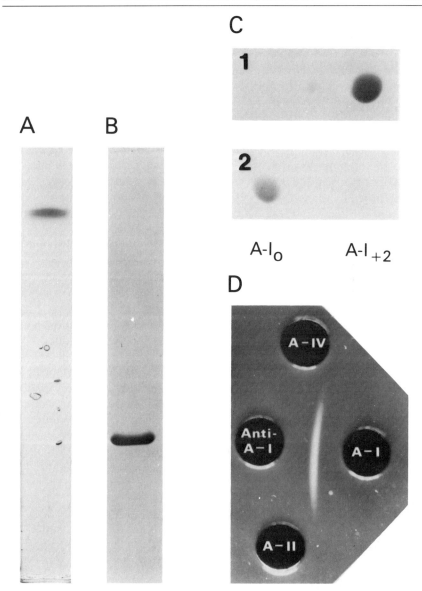

FIG. 6. Characterization of purified apoA-I by urea polyacrylamide gel electrophoresis (A), and analytical SDS–gel electrophoresis (B). C contains proapoA-I (1, apoA-I$_{+2}$) and mature apoA-I (2, apoA-I$_0$) isoform isolated by preparative isoelectrofocusing analyzed by two-dimensional gel electrophoresis. Ouchterlony immunodiffusion of apoA-I utilizing a monospecific antibody against apoA-I is shown in D.

The sample is electrophoresed through the stacking gel at 1.5 mA per gel, and at 2.0 mA per gel in the running gel until the tracking dye reaches 2 mm from the bottom of the gel. The upper and lower tray buffers are 0.04 M Tris–glycine (pH 8.9) and 0.125 M Tris–HCl (pH 8.1), respectively. Following electrophoresis the gels are fixed with 20% TCA for 1 hr, stained in 0.05% Coomassie Blue R 250 in 20% TCA for 1 hr, and destained in 7.5% acetic acid.

3. *Analytical isoelectrofocusing.* Lyophilized samples (3–5 μg protein) are dissolved in 8 M aqueous urea (0.5 mg/ml) and separated according to the method described for preparative isoelectrofocusing. For analytical purposes a 140 × 100 × 0.75-mm-thick slab gel is used. After completion of the electrophoresis, the gel is stained with the modified Coomassie Blue G-250 staining procedure as outlined previously for preparative isoelectrofocusing.

4. *Two-Dimensional gel electrophoresis: First dimension.* Proteins for separation by isoelectrofocusing are dissolved in 8 M aqueous urea. One to 3 μg of protein is focused in 1.5 × 95-mm tube gels utilizing the same gel solutions as previously described for preparative and analytical isoelectrofocusing. The upper and lower tray buffers are 0.02 M NaOH and 0.01 M H_3PO_4, respectively. Following isoelectrofocusing for 18 hr at 200 V at 4° the gel is removed by breaking the tube with a mallet.

Second dimension. The electrofocused gel is placed on the top of a 140 × 100 × 1.5-mm-thick lab gel containing the same gel solutions and buffers as described for preparative and analytical SDS gels, however, no stacking gel is employed. The gel is electrophoresed at 40 mA until the dye front reaches 2 mm from the bottom of the gel. The gel is stained and destained as outlined for analytical SDS–gel electrophoresis (Fig. 6).

5. *Immunoblots.* Apolipoprotein samples separated by SDS–gel electrophoresis or isoelectrofocusing can be further analyzed by immunoblots. The buffers for transfer of isoelectrofocusing and SDS gels to nitrocellulose paper are 0.7% acetic acid and 0.25 M Tris–HCl, 0.192 mM glycine, and 20% methanol (pH 8.3), respectively. Proteins were transferred to nitrocellulose paper in a TE Series Transphor Electrophoresis Unit (Hoeffer Scientific, San Francisco, CA) at 80 V for 2 hr. After the transfer excess protein binding sites on the nitrocellulose paper are blocked by incubation with 0.02 M Tris–HCl and 0.5 M NaCl (pH 7.5) (TBS) containing 3% gelatin for 1 hr. The nitrocellulose blot is then incubated in 1% gelatin in TBS containing a monospecific rabbit antibody to apoA-I for 16–25 hr. After incubation, the nitrocellulose blot is washed three times in TBS (10 min per wash). The final wash is followed by a 4–6 hr incubation of 1% gelatin in TBS containing an affinity purified antirabbit IgG horseradish peroxidase reagent (1/3000 dilution, Bio-Rad Lab, Richmond, CA). The nitrocellulose blot is washed three times in TBS

again, and the peroxidase complex developed with the commercial color reagent (Bio-Rad Labs, Richmond, CA), as outlined by the manufacturer.

6. *Ouchterlony immunodiffusion.* The purity of isolated apolipoproteins may be assessed by Ouchterlony immunodiffusion. Plates (85 × 30 × 1.5-mm-thick gel) are layered with 1.5% agarose in 0.5 M barbital (pH 8.6) or commercially prepared plates may be purchased. Apolipoproteins dissolved in 0.5 M barbital (0.75 μg/5 μl) are added to 5 mm holes and the plates incubated for 24 hr at 25° (Fig. 6).

7. *Edman Amino-terminal analysis.* One to 5 nmol of isolated apoA-I is sequenced by the automated phenylisothiocyanate procedure in a modified Beckman 890B Sequencer utilizing a 0.1 M Quadrol buffer. Polybrene (Aldrich) is used as a carrier in all degradations to reduce protein loss from the cup. The phenylthiohydantoin derivatives are quantitatively identified by high-pressure liquid chromatography on a Zorbax ODS column (0.4 × 35 cm) using a DuPont 8800 LC system equipped with a Perkin-Elmer autosampler (ISS-100) and a Chrom 2 data system.[35] Derivatives are also identified by chemical ionization–mass spectrometry performed on a Finnigan 4510 mass spectrometer using isobutane as the reagent gas.[36]

8. *Amino acid analysis.* Acid hydrolysis of protein samples (0.5 nmol protein) is performed for 24 hr in 6 M HCl (Pierce) containing 2-mercaptoethanol (1 : 2000, v/v) at 108°. Analyses are performed in a Beckman model 6300 amino acid analyzer equipped with a Hewlett Packard Model 3390 A Integrator (see the table).

Isolation and Characterization of ApoA-II

Overview

Human apoA-II is synthesized as a 100 amino acid precursor preproapoA-II, with an 18 amino acid prepeptide and a 5 amino acid propeptide with the sequence Ala-Leu-Val-Arg-Arg (Fig. 7).[37–41] PreproapoA-II is

---

[35] T. Fairwell and H. B. Brewer, Jr., *Anal. Biochem.* **99**, 242 (1979).
[36] C. Zimmerman, J. J. Pisano, and E. Apella, *Anal. Biochem.* **77**, 569 (1977).
[37] J. I. Gordon, K. A. Budelier, H. F. Sims, C. Edelstein, A. M. Scanu, and A. W. Strauss, *J. Biol. Chem.* **258**, 14054 (1983).
[38] C. R. Sharpe, A. Sidoli, C. S. Shelley, M. A. Lucero, C. C. Shoulders, and F. E. Baralle, *Nucleic Acids Res.* **12**, 3917 (1984).
[39] K. J. Lackner, S. W. Law, and H. B. Brewer, Jr., *FEBS Lett.* **175**, 159 (1984).
[40] M. N. Moore, F. T. Kao, Y. K. Tsao, and L. Chan, *Biochem. Biophys. Res. Commun.* **123**, 1 (1984).
[41] T. J. Knott, L. M. Priestley, M. Urdea, and J. Scott, *Biochem. Biophys. Res. Commun.* **120**, 734 (1984).

AMINO ACID COMPOSITION OF HUMAN Apo-A-I,
Apo-A-II, AND ApoA-IV

Amino acids	ApoA-I[a]	ApoA-II	ApoA-IV
Aspartic acid	8.63	4.02	9.04
Theonine	4.08	7.86	4.37
Serine	6.20	7.76	6.04
Glutamic acid	19.23	20.78	20.60
Proline	4.40	5.97	3.69
Glycine	4.83	4.13	7.55
Alanine	8.10	6.88	8.03
Valine	4.90	7.41	5.33
Methionine	1.28	1.11	1.08
Isoleucine	—	1.04	1.56
Leucine	15.68	10.70	12.51
Tyrosine	2.73	5.30	2.03
Phenylalamine	2.58	5.18	3.34
Histidine	2.33	—	2.07
Lysine	8.55	11.82	7.07
Arginine	6.50	—	5.67

[a] Twenty-four hour hydrolysate; data expressed in mol%.

cleaved cotranslationally to proapoA-II. ProapoA-II undergoes posttranslational conversion to mature apoA-II (Fig. 8). A detailed analysis of the *in vivo* conversion of proapoA-II to mature apoA-II in man has not been reported. ProapoA-II is present in human lymph and plasma, however, it represents less than 1% of total apoA-II.[42]

Mature apoA-II in human plasma exists as a dimer of two identical chains of 77 amino acid residues linked by a single disulfide bridge at position 6 in the sequence.[43] In nonhuman primates, as well as in other species, apoA-II is a monomer.[44,45] The primary structure of baboon and rhesus monkey apoA-II has extensive amino acid homology to human apoA-II, however, the cystine at position 6 in human apoA-II is replaced by serine.[46]

[42] K. J. Lackner, S. B. Edge, R. E. Gregg, J. M. Hoeg, and H. B. Brewer, Jr., *J. Biol. Chem.* **260**, 703 (1985).
[43] H. B. Brewer, Jr., S. E. Lux, R. Ronan, and K. M. John, *Proc. Natl. Acad. Sci. U.S.A.* **69**, 1304 (1972).
[44] C. Edelstein, C. Noyes, and A. M. Scanu, *FEBS Lett.* **38**, 166 (1974).
[45] P. N. Herbert, H. G. Windmueller, T. P. Bersot, and R. S. Shulman, *J. Biol. Chem.* **249**, 5718 (1979).
[46] C. Edelstein, C. Noyes, R. Heinrikson, R. Fellows, and A. M. Scanu, *Biochemistry* **15**, 1262 (1976).

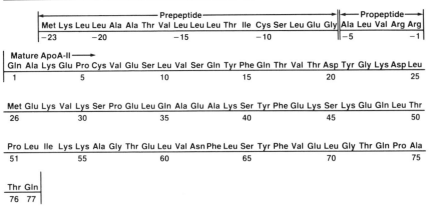

FIG. 7. Amino acid sequence of human preproapoA-II. PreproapoA-II contains an 18 amino acid prepeptide, a 5 amino acid propeptide, and a 77 amino acid mature protein. In plasma apoA-II exists as a dimer of two identical chains linked by a single disulfide bridge at position 6 in the sequence of the mature protein.

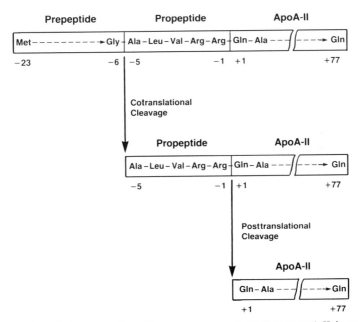

FIG. 8. Proteolytic processing of human preproapoA-II. PreproapoA-II is cotranslationally cleaved to proapoA-II, and proapoA-II undergoes posttranslational cleavage to mature apoA-II.

Mature apoA-II is polymorphic in plasma, and the major apoA-II isoforms are illustrated in Fig. 9. The nomenclature on two-dimensional gel electrophoresis for apoA-II is similar to that outlined above for apoA-I with the major mature isoform designated apoA-II$_0$, and the other isoforms identified by + or − unit charges.[27] Of particular interest is the marked difference in the pI of mature apoA-II (apoA-II$_0$, pI = 4.9), and proapoA-II (apoA-II$_{+3}$, pI 6.79).[41] Sialylated isoforms of apoA-II have also been identified.[42]

Isolation of ApoA-II-Containing Lipoproteins

Human mature dimeric apoA-II is most efficiently purified from plasma HDL as outlined above for apoA-I. ProapoA-II is not enriched in thoracic duct lymph, and therefore may be purified from either plasma HDL, lymph chylomicron–VLDL (density <1.006 g/ml), or lymph HDL. The lipoprotein fractions are isolated and delipidated with chloroform : methanol as outlined above for apoA-I.

Isolation of Polymorphic ApoA-II

Polymorphic apoA-II can be isolated from apolipoprotein fractions by size exclusion, affinity chromatography, and ion-exchange chromatography. The isolated apoA-II is homogeneous by polyacrylamide gel electrophoresis in SDS or urea, Ouchterlony immunodiffusion, and amino acid analysis.

Size Exclusion. 1. Gel permeation chromatography. ApoA-II can be purified from delipidated HDL by gel permeation chromatography on Sephacryl S-200 as outlined above for apoA-I (Fig. 4, see ion-exchange chromatography below).

2. High-pressure liquid chromatography. ApoA-II, as outlined above for apoA-I, may be purified to homogeneity by high-pressure liquid chromatography on TSK SW 3000 gel permeation chromatography (Fig. 5). The isolated apoA-II is pooled, dialyzed in 0.01 M NH$_4$HCO$_3$, and stored frozen at 1 mg/ml for further analysis.

3. Preparative sodium dodecyl sulfate–gel electrophoresis. The isolation of apoA-II by preparative SDS–gel electrophoresis is performed as outlined above for apoA-I except that the running gel contains 18% acrylamide rather than 15% acrylamide due to the smaller molecular weight of apoA-II (17,500) as compared to apoA-I (27,000). Following electrophoresis the gel should be cut rapidly due to the tendency of apoA-II to diffuse in the gel. This latter problem often results in diffuse rather than sharp

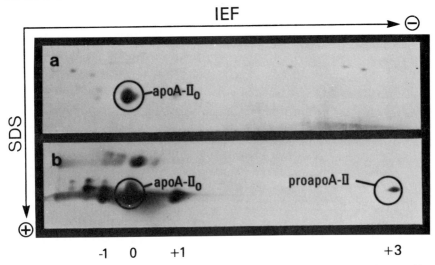

FIG. 9. Two-dimensional electrophoretogram of plasma apoA-II (a) and an immunoblot of the plasma apoA-II electrophoretogram utilizing a monospecific anti-apoA-II antibody (b). The principal apoA-II isoform is designated apoA-II$_0$, and the proapoA-II isoform, apoA-II$_{+3}$. Minor apoA-II isoforms are designated by (+) and (−) unit charges from the major apoA-II$_0$ isoform.

electrophoretic bands for apoA-II. The yield of apoA-II following preparative SDS–gel electrophoresis may be decreased if only the dimer form of apoA-II is isolated. Prior analysis of the sample should be performed to obtain the percentage of apoA-II present as a monomer and dimer.

Affinity Chromatography. The method for the affinity chromatographic isolation of apoA-II is similar to apoA-I, however, the eluted apoA-II may contain a mixture of both monomer and dimer apoA-II. Isolated apoA-II fractions were dialyzed (0.01 M NH$_4$HCO$_3$) and concentrated to 1 mg/ml.

Ion-Exchange Chromatography. Partially purified apoA-II can be isolated to homogeneity by ion-exchange chromatography. ApoA-II initially fractionated by Sephacryl S-200 chromatography is dialyzed, lyophilized, and redissolved in 0.025 M Tris–HCl–6 M urea (pH 8.0). The solubilized protein (325 mg/30 ml) is applied to a Whatman DE 52 column (2.5 × 50 cm) equilibrated in the sample buffer. The column is developed at 40 ml/hr with a 2400 ml linear gradient in 6 M urea from 0.025 M Tris–HCl (pH 8.0) to 0.1 M Tris–HCl (pH 8.0) (Fig. 4 inset). The fractions containing apoA-II are pooled, dialyzed against 0.01 M NH$_4$HCO$_3$, and concentrated to 1 mg/ml prior to storage at −20°.

Isolation of ApoA-II Isoforms

The isoforms of apoA-II can be isolated on preparative isoelectrofocusing gels from partially purified apolipoprotein fractions containing polymorphic apoA-II. The mature apoA-II isoforms (apoA-II$_0$, apoA-II$_{+1}$, and apoA-II$_{-1}$) obtained from plasma HDL are isolated utilizing a 5.0–6.0 pH gradient as outlined above for apoA-I. The proapoA-II isoform (apoA-II$_{+3}$) obtained in plasma or lymph fractions is isolated utilizing a pH gradient of 5.0–7.0 due to the relatively high pI (6.79) of proapoA-II.[42] No fraction of human plasma or lymph contains a relatively increased content of proapoA-II, and the yield of proapoA-II is low necessitating the pooling of a large number of isolated apoA-II$_{+3}$ fractions. The isolated apoA-II isoforms can be desalted on Sephadex G-100 in 0.01 M NH$_4$HCO$_3$, and concentrated to 1 mg/ml for storage at $-20°$.

Characterization of Isolated ApoA-II and ApoA-II Isoforms (Fig. 10)

An assessment of the purity of apoA-II may be obtained by the techniques outlined above for apoA-I (Fig.10). Polyacrylamide gel electrophoresis utilizing SDS or urea should be performed with 18% acrylamide, and the transfer of apoA-II for immunoblot carried out for 1 hr. Edman aminoterminal analysis can not be employed since the amino-terminus of mature apoA-II contains pyrrolidone carboxylic acid and will not react with phenylisothiocyanate.

Isolation and Characterization of ApoA-IV

Overview

ApoA-IV, a major apolipoprotein of lymph chylomicrons, has been extensively studied in both the rat and human.[7,47–49] In the rat apoA-IV is synthesized as a 391 amino acid precursor, preapoA-IV. PreapoA-IV undergoes cotranslational cleavage to mature apoA-IV.[50,51] The structure of human apoA-IV has not been reported. ApoA-IV is found primarily in lymph chylomicrons, however, in human plasma the majority of apoA-IV

---

[47] P. H. R. Green, R. M. Glickman, J. W. Riley, and Quinet, *J. Clin. Invest.* **65,** 911 (1980).
[48] N. Fidge and P. Nestel, *Circulation* **64,** 159A (1981).
[49] R. B. Weinberg and A. M. Scanu, *J. Lipid Res.* **24,** 52 (1983).
[50] J. I. Gordon, D. P. Smith, D. H. Alpers, and A. W. Strauss, *J. Biol. Chem.* **257,** 8418 (1982).
[51] M. S. Boguski, N. Elshourbagy, J. M. Taylor, and J. I. Gordon, *Proc. Natl. Acad. Sci. U.S.A.* **81,** 5021 (1984).

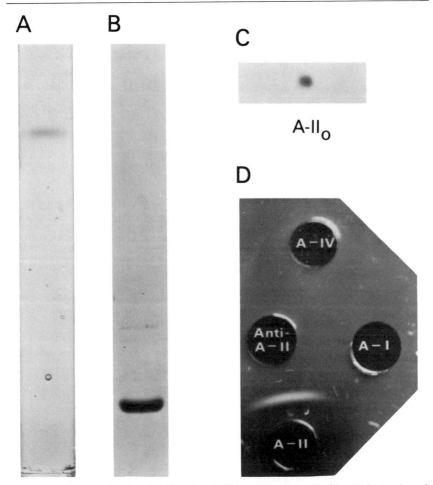

FIG. 10. Characterization of isolated apoA-II by urea polyacrylamide gel electrophoresis (A) and analytical SDS–gel electrophoresis (B). (C) Mature apoA-II$_0$ isolated by preparative isoelectrofocusing, and analyzed by two-dimensional gel electrophoresis; (D) Ouchterlony immunodiffusion of isolated mature apoA-II and a monospecific antibody to apoA-II.

is present unassociated with lipoproteins.[7,47–49] In the rat, a significant fraction of plasma apoA-IV is associated with HDL.[7]

Human apoA-IV is polymorphic in plasma, and the major mature isoform is designated apoA-IV$_0$.[28] A frequent variant of apoA-IV has been designated apoA-IV$_{\text{Marburg}}$.[31] The structural basis for the different isoforms of apoA-IV is as yet unknown.

Isolation of ApoA-IV Containing Lipoproteins

Human apoA-IV can be isolated from the <1.006 g/ml density fraction containing chylomicrons and VLDL of thoracic duct lymph or plasma from patients with type V hyperlipoproteinemia. Chylomicrons and VLDL are isolated by centrifugation at plasma density (1.006 g/ml) for 18 hr at 59,000 rpm (Beckman 60 Ti rotor) at 5°. The lipoproteins are collected by tube slicing, dialyzed in 0.01% EDTA–0.01% NaN_3 (pH 7.4), lyophilized, and delipidated with chloroform : methanol as outlined above.

Isolation of Polymorphic ApoA-IV

Polymorphic apoA-IV can be isolated to homogeneity from apolipoprotein fractions by size exclusion, antibody affinity chromatography, and heparin-Sepharose CL-6B affinity chromatography. Although the latter method produces large yields of apoA-IV the protein must be further purified by gel permeation chromatography.

Size Exclusion. 1. High-pressure liquid chromatography. The isolation of apoA-IV can be achieved by high-pressure liquid chromatography utilizing gel permeation chromatography on a TSK SW 3000 column as outlined above for apoA-I (Fig. 5). Fractions containing apoA-IV are dialyzed, and stored at −20° (1 mg/ml).

2. Preparative sodium dodecyl sulfate–gel electrophoresis. ApoA-IV can be purified from delipidated apolipoproteins by separation on SDS–gel electrophoresis employing a 15% running gel. The isolated apoA-IV is eluted from the gel by electroelution, and desalted as described for apoA-I. The purified apolipoprotein is stored frozen at −20° at 1 mg/ml.

Antibody Affinity Chromatography. ApoA-IV may be isolated by affinity chromatography utilizing monoclonal or polyclonal antibodies to apoA-IV. The methods for coupling and dissociation of the bound apoA-IV by a buffer containing 8 M urea are the same as those detailed above for apoA-I and apoA-II.

Heparin Affinity Chromatography and Gel Permeation Chromatography. ApoA-IV can be purified from delipidated d <1.006 g/ml lipoproteins by heparin-Sepharose CL-6B affinity chromatography followed by gel permeation chromatography. Delipidated apolipoproteins are solubilized (200 mg/80 ml) in 2 mM sodium phosphate–5 M urea (pH 7.4) and fractionated on heparin-Sepharose CL-6B (Pharmacia, 2.5 × 45 cm column, 28 ml/hr) with a 1600 ml linear gradient in sodium phosphate–5 M urea (pH 7.6) from 0.022 M NaCl to 0.25 M NaCl (Fig. 11). The unadsorbed fraction which contains apoA-IV is pooled, dialyzed in 0.01 M NH_4HCO_3, and lyophilized. The apoA-IV containing fraction is dissolved

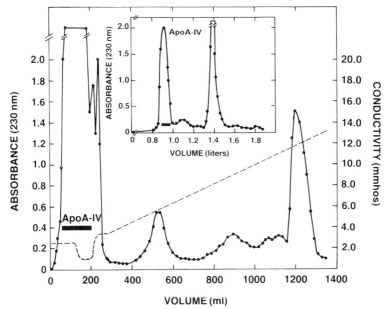

FIG. 11. Heparin-Sepharose CL-6B affinity chromatography of delipidated lipoproteins (d <1.006 g/ml) isolated from thoracic duct lymph (protein load 200 mg). The fractions containing partially purified apoA-IV (solid bar) were pooled, and fractionated on Sephacryl S-200 (inset). The fractions containing apoA-IV are indicated by the bar.

(90 mg/10 ml) in a buffer containing 50 mM glycine, 4 mM NaOH, 0.5 M NaCl, and 6 M urea (pH 8.8) and fractionated on Sephacryl S-200 (4.0 × 190 cm column) equilibrated in the sample buffer (Fig. 11). The fractions containing apoA-IV are pooled, dialyzed in 0.01 M NH$_4$HCO$_3$, and concentrated to 1 mg/ml prior to storage at $-20°$.

Isolation of ApoA-IV Isoforms

The individual isoforms of apoA-IV may be separated by preparative isoelectrofocusing with a pH gradient of 5.0–6.0. The methodology for preparative isoelectrofocusing is included above for apoA-I. The mature apoA-IV$_0$ is the predominate isoform, and no proteolytic processing of apoA-IV has been reported in human plasma.

Characterization of Isolated ApoA-IV and ApoA-IV Isoforms

The general approach to the characterization of purified apoA-IV has been outlined above for apoA-I. The conditions for electrophoresis, and immunoblot are similar to those of apoA-I. The physicochemical proper-

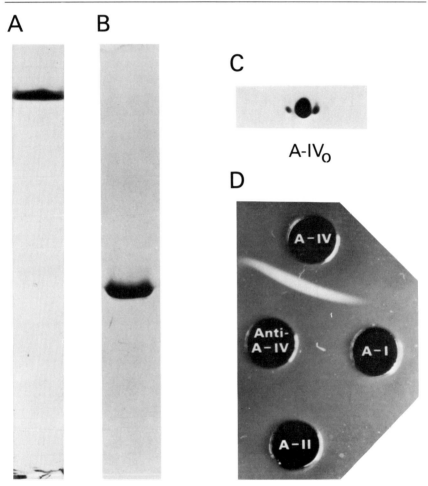

FIG. 12. Characterization of isolated mature apoA-IV by urea polyacrylamide gel electrophoresis (A) and analytical SDS–gel electrophoresis (B). Two-dimensional gel electrophoretogram of Sephacryl S-200 purified apoA-IV (C), and Ouchterlony immunodiffusion of purified mature apoA-IV against a anti-apoA-IV antibody (D).

ties of apoA-IV isolated by heparin-gel permeation chromatography are illustrated in Fig. 12.

Plasma Two-Dimensional Gel Electrophoresis

The evaluation of apoA-I, apoA-II, and apoA-IV as well as the other major apolipoproteins in normal subjects (Fig. 13) and patients with dyslipoproteinemias may be effectively performed by two-dimensional gel

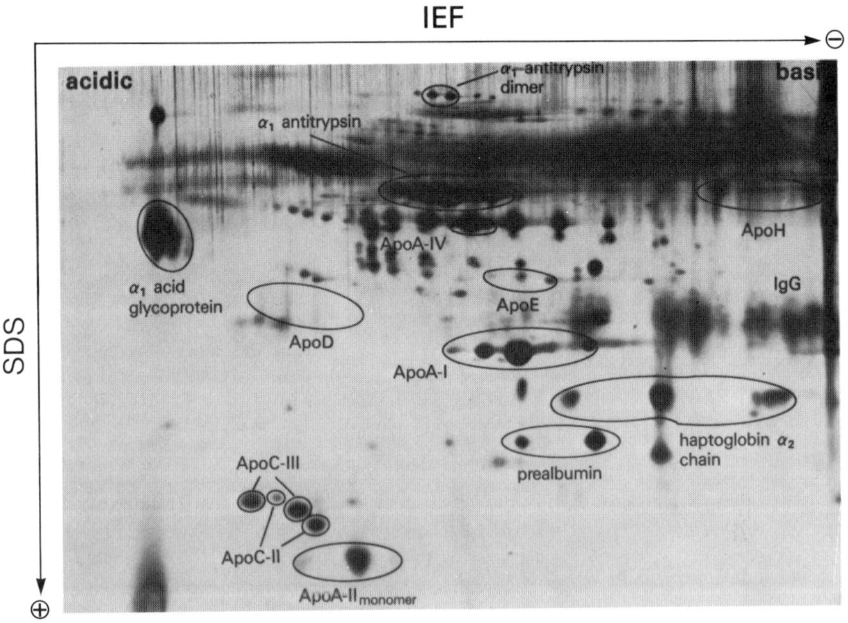

FIG. 13. Two-dimensional gel electrophoretogram of plasma from a normal subject. The major apolipoproteins are circled.[27]

electrophoresis of plasma.[27] This technique is particularly useful in identifying structural mutants of plasma apolipoproteins which may be important in the pathophysiology of diseases associated with a dyslipoproteinemia.

Sample Preparation

Two to 5 µl of plasma is dissolved in 10–12 µl of a detergent solution containing 2% SDS, 5% 2-mercaptoethanol, 20% glycerol, 2% Nonidet P-40, 6% CHAPS, and 5% ampholines (pH 5–7, Serva AB, Heidelberg, FGR) followed by the addition of 8 mg of solid urea. Samples of delipidated lymph chylomicrons (35 µg), VLDL (30 µg), or HDL (50 µg) are redissolved in 6 M urea–5% 2-mercaptoethanol. All samples were intermittently vortexed for 20 min at 25° prior to analysis.

First Dimension

Plasma or delipidated apolipoproteins are separated by isoelectrofocusing in 1.5 × 95-mm glass tubes. The gel composition and tray buffers are identical to those described previously for isoelectrofocusing gels. A pH gradient of 4.0–7.0 was employed. For plasma samples, gels are electrofocused at 4° at 250 V for the first 2 hr, 500 V for the next 15 hr, and 1000 V for the last 2 hr, for a total of 10,000 V-hr. Isolated delipidated lipoprotein fractions are electrofocused at 4° for 16 hr at 250 V. After isoelectrofocusing, gels may be rapidly frozen on solid CO_2 and stored for as long as a month prior to analysis on the second dimension.

Second Dimension

The isoelectrofocused gels are placed for 10 min at 25° in an equilibrium solution containing 1.0% SDS, 2.5% 2-mercaptoethanol, and 0.1 M Tris–HCl (pH 6.8). The gels are then loaded on the top of a 140 × 100 × 1.5-mm-thick polyacrylamide slab gel. A 1-cm stacking gel is used for the analysis of plasma samples. The compositions of the stacking and running gels are identical to those previously described for the isolation of apoA-I by preparative SDS–gel electrophoresis.

The gels are electrophoresed at 4° until the dye front was 2 mm from the bottom of the gel (approximately 3.5 hr; 10 mA per gel for stacking, and 20 mA per gel for the running gel). The gels are stained with either silver stain, similar to that reported by Morrisey[52] or by the Coomassie Blue R250 method described above for SDS–gel electrophoresis.

Immunoblot

Samples separated by two-dimensional gel electrophoresis may be further analyzed by immunoblot utilizing monospecific antibodies to the individual apolipoproteins. Plasma gels are blotted for 1 hr at 80 V to nitrocellulose paper by the techniques outlined above for apoA-I and apoA-II.

Summary

A number of different analytical techniques are now available for the isolation of apoA-I, apoA-II, and apoA-IV. The choice of a particular technique is dependent on the instrumentation available, and the quantity of isolated apolipoprotein required. The isolation and characterization of the separate isoforms and the precursor isoproteins of the individual apolipoproteins are detailed, and methods for the evaluation of the purity of the separate apolipoproteins presented. A method for the evaluation of apolipoproteins in plasma is now available which permits the identification of structural variants of plasma apolipoproteins in patients with dyslipoproteinemias.

[52] J. H. Morrisey, *Anal. Biochem.* **117,** 307 (1981).

[11] Isolation and Characterization of Apolipoprotein B-100

By WALDO R. FISHER and VERNE N. SCHUMAKER

Introduction

Two physically separable macromolecules carry the designation apolipoprotein B. The long recognized form of apoB, large apoB or B-100, is the predominant protein in LDL and the major apoB protein found in human VLDL. Small apoB or apoB-48 was initially described in 1980, and in humans it is produced solely by the intestinal mucosa and is found in chylomicrons and their remnants.[1-3] An arbitrary centile nomenclature, proposed by Kane et al.,[1] has become popular in designating the apoB-related proteins based upon their mobilities in SDS acrylamide gels. According to this nomenclature large apoB, of hepatic origin, is assigned an arbitrary value of 100. The other apoB-related proteins which migrate further into the gel are designated B-76, B-48, and B-26, B-48 being the apolipoprotein produced in the intestine by humans. The properties of apoB-48 are described elsewhere in this volume (Chapter [12], this volume). ApoB-100 is generally an abundant apolipoprotein in plasma being present in normal human plasma in a concentration of about 90 mg/dl.[4] It constitutes over 95% of the protein mass of low-density lipoprotein (LDL) and approximately one-third of the protein mass of very low-density lipoprotein; it is readily isolated from LDL.

A particular concern in working with LDL is the ease with which apoB is proteolytically cleaved,[5] and this has been shown to occur with proteases liberated during blood coagulation or clot lysis.[6-9] The two additional

---

[1] J. P. Kane, D. A. Hardman, and H. E. Paulus, *Proc. Natl. Acad. Sci. U.S.A.* **77**, 2465 (1980).
[2] K. V. Krishnaiah, L. F. Walker, S. Borensztajn, G. Schonfeld, and G. S. Getz, *Proc. Natl. Acad. Sci. U.S.A.* **77**, 3806 (1980).
[3] M. J. Malley, J. P. Kane, D. A. Hardman, R. L. Hamilton, and K. B. Dalal, *J. Clin. Invest.* **67**, 1441 (1981).
[4] J. B. Karlin, D. J. Juhn, A. M. Scanu, and A. H. Rubenstein, *Eur. J. Clin. Invest.* **8**, 19 (1978).
[5] R. B. Triplett and W. R. Fisher, *J. Lipid Res.* **19**, 478 (1978).
[6] J. C. H. Steele, *Thromb. Res.* **15**, 573 (1979).
[7] D. M. Lee and S. Single, *Arteriosclerosis* **3**, 511a (1983).
[8] W. A. Bradley, A. M. Gotto, D. L. Peavy, and S. H. Gianturco, *Arteriosclerosis* **3**, 508a (1983).
[9] A. D. Cardin, R. L. Jackson, J. Chao, V. H. Donaldson, and H. S. Margolins, *Fed. Proc., Fed. Am. Soc. Exp. Biol.* **43**, 1532 (1984).

apoB-associated proteins, B-76 and B-26, are seen with variable frequency in LDL preparations; when proteolysis is carefully blocked they generally, though not always, become very minor components. Whether these presumptive proteolytic fragments are present in circulating LDL or are solely produced following phlebotomy by proteases of the coagulation system is unclear.

Contamination of LDL by B-48 of chylomicron origin is distinctly uncommon in humans; however, Lp(a) may be a contaminating lipoprotein within the 1.006 to 1.063 g/ml density cut commonly used in isolating LDL. In plasma from Lp(a)-positive donors, the amount of Lp(a) present within this cut may approach 10% of total lipoprotein.[10] Accordingly it is suggested that LDL to be used for apoB preparation be screened for contaminating proteins and for proteolysis by SDS-polyacrylamide electrophoresis.

In the rat, the origin, distribution, and varieties of apoprotein B are different than in man. The smaller peptide, which comigrates with human B-48 on SDS gels, appears to be made in both liver and intestine; Sparks and Marsh[11] refer to the smaller peptide as B_l (lower molecular weight), while Elovson et al.[12] use the term apoB-P (peptide) III. The larger peptide, which comigrates with human B-100, is made only in the liver. This large peptide has been called apolipoprotein B-PI by Elovson et al.[12] and B_h (higher) by Sparks and Marsh.[11] In addition, Elovson et al.[12] find a second large rat peptide which migrates very close to the first, which they call apolipoprotein B-PII.

Since the rat has low concentrations of VLDL and LDL, the B apolipoproteins are difficult to isolate in abundance from the normal rat. However, intrajugular injection of the nonionic detergent Triton WR 1339 [0.3 ml/100 g body weight of 20% (w/v) Triton in saline] 5 to 6 hr prior to exsanguination provides a rich yield of VLDL, containing apolipoproteins B-48 and B-100.[12,13] Triton appears to strip off the apolipoprotein Cs, effectively stopping lipolysis, and allowing the VLDL to accumulate in the rat plasma. As much as 10 mg of a mixture of apolipoproteins B-100 and B-48 may be obtained from the plasma of a single rat by this procedure. These two proteins may be separated after delipidation by gel filtration on Sepharose CL-2B in detergent or in 6 M guanidine-hydrochloride, thus, the rat is a convenient source of B-48 for further study.

[10] J. W. Gaubatz, C. Heideman, A. M. Gotto, J. D. Morrisett, and G. H. Dahlen, *J. Biol. Chem.* **258**, 4582 (1983).
[11] C. E. Sparks and J. B. Marsh, *J. Lipid Res.* **22**, 514 (1981).
[12] J. Elovson, Y. O. Huang, N. Baker, and R. Kannan, *Proc. Natl. Acad. Sci. U.S.A.* **78**, 157 (1981).
[13] S. Otway and D. S. Robinson, *J. Physiol. (London)* **190**, 321 (1967).

Inhibition of Proteolysis. LDL is readily proteolyzed both by bacterial enzymes and by the proteases associated with blood coagulation. Since LDL preparation usually requires several days it is important to utilize clean technique, low temperature, and inhibitors of bacterial growth, such as sodium azide and merthiolate.[5] Some investigators also choose to add antibiotics.[14] Most of the enzymes of coagulation are serine proteases and may be inhibited with phenylmethylsulfonyl fluoride (PMSF), aprotinin, or soybean trypsin inhibitor. Because PMSF induces considerable hemolysis when added to whole blood it is more convenient and equally satisfactory to use a protein inhibitor, and the one from soybeans is least expensive. Further inhibition of coagulation is provided by chelating calcium; for this reason and because of concerns of heavy metal catalyzed oxidation of LDL lipids, EDTA is generally added as an inhibitor. A minimal inhibitory cocktail, to which the blood is added when drawn, contains the following ingredients per dl of whole blood: merthiolate 10 mg, sodium azide 20 mg, EDTA 100 mg, soybean trypsin inhibitor 20 mg.

Inhibition of Lipid Peroxidation. There is increasing concern that lipid peroxidation may be associated with chemical modification of apoB, resulting in degradation or chemical cross-linking of the protein.[15,16] Certainly apoB is very sensitive to storage even in the frozen state, particularly if lipid is present. Hence, it has been proposed that apoB should be stored under nitrogen and that, in addition to EDTA, reduced glutathione (0.05%) should be added to plasma and maintained in all solutions containing LDL; glutathione is effective as a heavy metal binder and as an antioxidant. These additional precautions, which are not yet fully evaluated, may be exercised depending upon the judgment of the investigator.

Considerations in Delipidating ApoB. Apolipoprotein B may be freed of natural lipids by delipidation with a number of surfactants, and the physical properties of the resultant apoB–surfactant complexes have been reported.[17-22] All reports indicate that the apoB content of the monomeric complexes ranges between 400 and 500 × 10^3 g.

[14] M. J. Chapman and J. P. Kane, *Biophys. Biochem. Res. Commun.* **66**, 1030 (1975).
[15] J. Schuh, G. F. Fairclough, and R. H. Haschemeyer, *Proc. Natl. Acad. Sci. U.S.A.* **75**, 3173 (1978).
[16] D. M. Lee, A. J. Valente, W. H. Kuo, and H. Maeda, *Biochim. Biophys. Acta* **666**, 133 (1981).
[17] A. Helenius and K. Simon, *J. Biol. Chem.* **247**, 3656 (1972).
[18] J. C. Steele and J. A. Reynolds, *J. Biol. Chem.* **254**, 1633, 1639 (1979).
[19] A. Helenius and K. Simon, *Biochemistry* **10**, 2549 (1971).
[20] A. Ikai, *J. Biochem. (Tokyo)* **88**, 1349 (1980).
[21] R. M. Watt and J. A. Reynolds, *Biochemistry* **19**, 1553 (1980).
[22] B. W. Patterson, L. L. Kilgore, P. W. Chun, and W. R. Fisher, *J. Lipid Res.* **25**, 763 (1984).

When apoB is delipidated with organic solvents in the absence of surfactants, it strongly aggregates even in the presence of denaturing solvents. The properties of the protein in concentrated guanidine–HCl have been examined in a number of laboratories.[18,23–26] Because of the ease with which the protein is cleaved or dissociates, its propensity to aggregate, and its nonideal behavior with physical interaction between particles, reports of its properties have differed. There now seems to be general agreement that the protein is large, probably in the range of 250–400 × 10^3 g, the latter figure corresponding with the value observed in surfactant complexes.

Based on an assumed molecular weight of 250,000, apoB contains 14 half-cysteines, 6 disulfide bonded.[26] Its amino acid composition is unremarkable,[27] and it is a glycoprotein containing about 8% carbohydrate consisting of sialic acid, glucosamine, galactose, and mannose.[28] Recently it has become evident that apoB may undergo nonenzymatic glycolsylation[29] and probably other chemical modification including chemical derivatization with malondialdehyde.[30]

The projected usage of apoB should determine the method of isolation. Surfactant delipidation has provided apoB–surfactant complexes which have been useful in its reconstitution with natural lipids[31] and in immunologic investigations. Once fully denatured in concentrated guanidine the protein may be transferred into urea, but it is poorly solubilized in the absence of denaturants where, though it may be maintained in solution, it is probably heavily aggregated.[26] For tissue culture studies such a preparation may be useful; alternately apoB complexed with bovine serum albumin has been used successfully as a means of introducing apoB into cultures.[32]

The Preparation of LDL for Apolipoprotein B Isolation

The Isolation of LDL. There are a variety of methods for isolating LDL from plasma, the common ones being ultracentrifugation, gel filtra-

---

[23] R. Smith, J. R. Dawson, and C. Tanford, *J. Biol. Chem.* **247**, 3376 (1972).
[24] J. Elovson, J. C. Jacobs, V. N. Schumaker, and D. L. Puppione, *Biochemistry* **24**, 1569 (1985).
[25] B. W. Patterson and W. R. Fisher, Submitted (1986).
[26] A. D. Cardin, K. R. Witt, C. L. Barnhart, and R. L. Jackson, *Biochemistry* **21**, 4503 (1982).
[27] S. Margolis and R. G. Langdon, *J. Biol. Chem.* **241**, 469 (1966).
[28] N. Swaminathan and F. Aladjam, *Biochemistry* **15**, 1516 (1976).
[29] E. Schleicher, T. Deufel, and O. H. Wieland, *FEBS Lett.* **129**, 1 (1981).
[30] A. M. Fogelman, J. Schechter, J. Seager, M. Hokom, J. S. Child, and P. A. Edwards, *Proc. Natl. Acad. Sci. U.S.A.* **77**, 2214 (1980).

tion chromatography, and chemical precipitation.[33–35] For the purpose of apoB isolation the first has some advantage in that the stripping of minor apolipoproteins which occurs during long centrifugation in high salt may reduce protein contamination.[36] Discussions of ultracentrifugal isolation of LDL are in the literature.[33] Briefly, blood drawn from a fasting donor, in the presence of the appropriate inhibitors, is centrifuged at low speed to remove cells; the plasma is then centrifuged at its own density (about 1.006 g/ml) for 18 to 20 hr at 50,000 rpm and 5°. Using a syringe, triglyceride-rich lipoproteins are removed in the supernatant along with about one-half of the clear solution which overlays the yellow lipoproteins in the lower third of the tube. The infranatant solution is recovered and adjusted to a density of 1.063 g/ml by addition of 8 g/dl of KBr, and the solution recentrifuged for 20 hr under the same conditions. LDL in the supernatant is recovered carefully with a syringe so as not to withdraw more than 10% of the tube volume. Such a preparation of LDL contains less than 5% contaminating proteins. If further purification is required the lipoprotein may be redialyzed against a KBr solution of approximately 1.01 g/ml (12.5 g/liter of KBr, 0.01 M phosphate, pH 8 containing the appropriate inhibitors) and taken through the centrifugational sequence again. Alternately the LDL may be directly passed through a 1 m 4 or 6% agarose column at 4° using pH 8 phosphate-buffered saline containing merthiolate, sodium azide, and EDTA. The LDL-containing fractions are visibly yellow and may be quantitated at 456 nm; these are pooled and concentrated by ultrafiltration (e.g., an Amicon XM-100 membrane).

Assessment of Apolipoprotein B Purity of LDL. Since apolipoprotein B is large among the apolipoproteins its purity may be satisfactorily assessed on overloaded SDS acrylamide gels. Because this protein resolves so well from other apolipoproteins, the gel technique is not crucial. By running a nonreduced 4% acrylamide gel one can resolve large apoB or B-100 from small apoB or B-48 and from Lp(a). The degree of irreversible aggregation can be assessed by the amount of protein which does not enter the gel, and the extent of proteolysis can be estimated by the number of bands which fall below the major apoB band.[5] A second gel contain-

---

[31] M. T. Walsh and D. Atkinson, *Biochemistry* **22**, 3170 (1983).
[32] R. Shireman, L. L. Kilgore, and W. R. Fisher, *Proc. Natl. Acad. Sci. U.S.A.* **74**, 5150 (1977).
[33] V. N. Schumaker and D. L. Puppione, this volume [6].
[34] P. S. Bachorik and J. J. Albers, this series, Vol. 129 [5].
[35] L. L. Rudel, C. A. Marzetta, and F. L. Johnson, this series, Vol. 129 [3].
[36] P. N. Herbert, T. M. Forte, R. S. Schulman, M. J. LaPiana, E. L. Gong, R. I. Levy, D. S. Fredrickson, and A. V. Nichols, *Prep. Biochem.* **5**, 93 (1975).

ing 7% acrylamide, into which apoB barely enters, will permit an assessment of the presence of the smaller apolipoproteins which may be visualized though they are not well resolved.

Using overloaded gels it is thus possible to detect levels of a few percent contamination. If a higher level of purity is needed, immunologic methods are required using antisera directed against specific contaminating proteins of concern, and one can use either qualitative Ouchterlony or quantitative immune precipitation methods as described elsewhere.[37]

The Storage of LDL. LDL is a difficult protein to store as it is irreversibly precipitated by lyophilization, and once isolated, freezing denatures the protein. For careful physical studies it is necessary to isolate the lipoprotein and maintain it in solution. The LDL is sterilized by filtration and stored in the presence of bacterial and oxidative inhibitors, under nitrogen and at 4°. It seems to be more stable if left in NaBr or KBr solution.

For most purposes LDL may be isolated from frozen plasma, and for many metabolic, immunologic, and tissue culture experiments the purified lipoprotein can be stored for many months by freezing in the presence of 20% sucrose.[5] Upon thawing, the sample is filtered and the recovered LDL has physical, immunologic, and cellular receptor reactivity (in the fibroblast) which closely resembles the original LDL. VLDL may be similarly preserved, and when recovered it shows only minimal changes when examined by analytical ultracentrifugation or by immune displacement in a radioimmunoassay.

Delipidation Methods

Surfactant Delipidation. The removal of natural lipids from apoB may be accomplished with the use of a variety of surfactants. By contrast to delipidation with organic solvents and with sodium dodecyl sulfate, the primary advantage of nonionic surfactant delipidation is that apoB is never fully denatured or precipitated. The protein retains much of its native immunoreactivity,[38] it may be reconstituted with various lipids,[39] or it may be utilized for physical or chemical studies.[18,22]

Helenius and Simon established the method of surfactant delipidation which has been used, with modification, by subsequent investigators.[19] Three recent reports have described delipidation of apoB using sodium dodecyl sulfate,[18] sodium deoxycholate,[31] and Triton X-100.[22] In each

[37] E. S. Krul and G. Schonfeld, this volume [31].
[38] L. L. Kilgore, B. W. Patterson, and W. R. Fisher, Submitted (1986).
[39] M. T. Walsh and D. Atkinson, this volume [33].

case the procedure is very similar. The LDL is first incubated with a large molar excess of surfactant over lipid to assure that the concentration of free surfactant remains above the critical micellar concentration (CMC). The solution, containing apolipoprotein–surfactant complexes and surfactant–lipid mixed micelles, is then passed over a gel filtration column equilibrated with a buffered solution of surfactant at a concentration again exceeding the CMC. With an appropriately sized column the apolipoprotein elutes well ahead of the lipid–surfactant micelles. The recovered apoB generally contains less than 5% residual natural lipid. After concentration by ultrafiltration, its associated surfactant may be exchanged by passage over an additional agarose column equilibrated with a different surfactant of choice. Some of the considerations and details of surfactant delipidation are illustrated in the following method using Triton X-100, reported by Patterson et al.[22]

Dilute LDL (0.5–1.0 mg/ml), in 0.1 M Na_2CO_3, pH 8.0, is added dropwise with gentle stirring at 24° to a 10% w/v solution of Triton X-100, to provide a final detergent : protein weight ratio of 20 : 1. Upon completion of the addition the solution is chilled and concentrated by ultrafiltration at 4° (Amicon, XM-100 membrane) to a final detergent concentration of about 15%. Triton–apoB complexes are then isolated from surfactant/lipid mixed micelles by gel filtration at 5° using a 1.5 × 100 cm column of Sepharose CL-6B (Pharmacia) resin equilibrated with the Na_2CO_3 buffer containing 0.2% Triton X-100. Since Triton absorbs optically at 280 nm, eluted protein is identified by performing Lowry assays on the fractions,[40] while the yellow, carotenoid-containing lipid micelles are well resolved and can be monitored by absorbance at 456 nm. Also described is a method using ultracentrifugation for isolating Triton–apoB complexes, in which these are separated from lipid micelles by sedimentation in a salt gradient.[22] This procedure permits the isolation of larger quantities of apoB–Triton complexes than can conveniently be separated by gel filtration.

By careful control of the conditions for the delipidation, maintaining low temperature at all subsequent steps, and with appropriate selection of salts and avoidance of disulfide bond reduction, the isolated apoB–Triton complexes are physically homogeneous and immunologically very similar to native apoB in LDL.[38]

Organic Solvent Delipidation of Apolipoprotein B. In planning to isolate apolipoprotein B free of lipids and surfactants, it is important to consider the sulfhydryl chemistry of this protein. There is general agree-

---

[40] O. H. Lowry, N. J. Roseborough, N. J. Farr, and R. J. Randall, *J. Biol. Chem.* **193,** 265 (1951).

ment that there are 14 mol of half-cysteine per 250,000 g of apoB, and that two residues are free, the others being disulfide linked.[26] The protein appears to be constrained by intramolecular disulfide bridging,[25] but following denaturation it is reported to be predisposed to intermolecular disulfide bonding, and concern exists for thiol free-radical induced autoxidation.[26] A number of options are available. The delipidated protein may be maintained in the presence of dithiothreitol (0.01 M). Prior to delipidation, LDL apoB may be carboxymethylated to derivatize the free sulfhydryls, and such preparations are reported to have little intermolecular cross-linking.[18] Alternately, the lipoprotein may be dissolved in a denaturing solvent, such as 6 M guanidine, followed by reduction of disulfides with dithiothreitol and their alkylation. Such a procedure is utilized here.

LDL, either freshly prepared or frozen with sucrose, is dialyzed against 0.1 M Tris–HCl, 0.02% azide, 0.01% merthiolate, 0.05% glutathione, pH 8.9. It is convenient to work with a quantity of between 5 and 100 mg of protein. Upon completion of dialysis solid guanidinium–HCl (GuHCl) is added to 6 M (there is a 75% volume increase upon addition of GuHCl; therefore, add 1 g GuHCl/ml LDL solution); the measured pH is adjusted to 8.9 by addition of NaOH.

The protein is then reduced and alkylated, and the following directions are based on 5 mg of protein. Disulfide bonds are reduced by adding 15 μl of 1 M dithiothreitol (DTT). After 30 min at room temperature the protein is alkylated by addition of 0.1 ml of 0.5 M iodoacetic acid, readjusting the pH at 5 min intervals for 15 min. The completion of alkylation may be demonstrated by use of the Ellman reaction.[41] Excess iodoacetate is destroyed by addition of 50 μl of 1 M DTT, after which the Ellman reaction becomes positive. Solid glutathione is then added to 0.05%.

Delipidation is performed at about $-12°$ using a salt–ice cooling mixture and centrifugations at $-15°$. Diethyl ether is rendered peroxide free by passing it over a column of basic alumina prior to use. The guanidinium solution of LDL is transferred to heavy walled glass stoppered centrifuge tubes or screw capped Corex bottles and, after thorough chilling, 8–10 ml of prechilled 3 : 1 diethyl ether : ethanol is added. Gentle extraction is performed by rotating the tubes in the salt–ice mixture for 5 min followed by centrifugation at 1000 rpm for 10 min and discarding of the upper phase. Four extractions are performed with 3 : 1 ether : ethanol followed by two with ether, after which the protein is gassed with nitrogen to remove excess ether. Throughout this procedure the solutions must be kept very cold to avoid precipitation of the protein. The use of 7 M GuHCl

---

[41] G. L. Ellman, *Arch. Biochem. Biophys.* **82**, 70 (1959).

during the extraction facilitates this objective and permits the extraction to be performed at 0°.

At this stage the protein may be dialyzed against several exchanges of 6 M GuHCl buffered to pH 8 or greater to remove the organic solvents and is then dialyzed into 6 M urea and concentrated using an Amicon XM 100 ultrafiltration cell. Alternately it may be immediately concentrated using Amicon "centricones," 50,000 MW cut-off and applied to a 1 m Sepharose CL-2B column using as a solvent 6 M GuHCl, 0.1 M Tris–HCl, 0.02% EDTA, pH 8 (or greater) and run at 4°. The peak fractions are identified by absorbance at 280 nm.

ApoB delipidated in this manner should be reanalyzed by SDS–acrylamide gel electrophoresis (after replacing the GuHCl with urea) and may be further examined by sedimentation velocity ultracentrifugation or other means. It has a phospholipid and total cholesterol content of about 7% and <2%, respectively. The protein may be rendered virtually lipid free by dialyzing it into 6 M urea, adding 30 mg of SDS per mg apoB, leaving at room temperature of 3 hr, and passing it over a Sepharose CL-6B column in a 2.5 M SDS, 0.01 M Tris–HCl, pH 8.0 buffer. The SDS is then removed by dialyzing the protein against 6 M urea, pH 8, and the protein may be redialyzed into GuHCl. Such a preparation of apoB is very satisfactory for physicochemical studies.

An alternative solvent delipidation procedure using diethyl ether: ethanol has been described by Cardin et al. in which the LDL is initially lyophilized, extracted as a solid, and resolubilized in GuHCl.[26] As measured by viscosity and sedimentation velocity analysis this protein has been found to be aggregated, but it is useful for chemical studies and biological experiments. ApoB prepared by either method may be dialyzed into 6 M urea and then into 3 and 1 M urea followed by 0.01 M Tris–HCl, pH 8.5, in which it remains soluble but aggregated.[26]

Analysis of Apolipoprotein B

Purity of the Product. Assuming that the LDL used in preparing apoB contained no other proteins, then the primary question concerning the isolated apolipoprotein is whether proteolysis has occurred. This is most readily answered using electrophoresis in SDS on 5% acrylamide gels. If the apolipoprotein is in GuHCl it must first be dialyzed into urea since GuHCl precipitates with SDS; however, urea and most surfactants are compatible with SDS buffers used in acrylamide electrophoresis. Usually after organic solvent delipidation in guanidine apoB will demonstrate a few additional faster migrating protein bands. With delipidation in Triton X-100 the delipidated product generally appears unchanged on gels.

Quantitation of Apolipoprotein B. Purified apoB may be quantitated using either a Lowry[40] or a fluorescamine assay[42]; because of the convenience of the latter it is used predominantly in our laboratory. Guanidine interferes with the Lowry assay but does not do so with the fluorescamine assay if the SDS is omitted from the buffer.

The standardization of the fluorescamine assay adapted to run in SDS has been described.[43] The assay is run in new, borosilicate culture tubes or disposable acrylic fluorescence cuvettes, depending upon the available spectrofluorometer. To 1.6 ml of 0.1 M sodium phosphate, pH 8.5, the protein standard or unknown protein is added in a 200 μl volume. The assay is linear over 1 to 100 μg of protein. Fluorescamine solution (0.2 ml) (0.1% in 1,4-dioxane) is added to each sample while vortexing, and after 5 and before 30 min the tubes are read at an emission wavelength of 475 nm while exciting at 395 nm.[43a] The standard curve is determined by calculating a linear regression of fluorescence for the solvent blank and the BSA standards. The protein content of the samples, expressed in terms of BSA, is then obtained by dividing the sample fluorescence by the slope of the BSA standard curve. For apoB the conversion to true from apparent concentration measured using a BSA standard is made by multiplying by 1.50, a factor determined by analysis of amino acid hydrolysates of apoB standards.

The major difficulty encountered with this method is from fluorescent materials present in most commercial detergents. This is resolved by using disposable glassware or, where necessary, washing glassware to be used in the assay with an SDS solution. Thiols also interfere.

Several laboratories have determined conversion factors for Lowry assays when using a BSA standard.[44–47] A factor of 0.78 (SD = 0.03) is an average of measured values for converting apparent protein concentration to true concentration of apoB when using this assay. Certainly acid hydrolysis and amino acid analysis is the method of choice if the concentration must be known accurately.

---

[42] P. Bohlen, S. Stein, W. Dairman, and S. Udenfriend, *Arch. Biochem. Biophys.* **155**, 213 (1973).

[43] L. L. Kilgore, J. L. Rogers, B. W. Patterson, N. H. Miller, and W. R. Fisher, *Anal. Biochem.* **145**, 113 (1985).

[43a] The 1,4-dioxane must be dry. This is accomplished by refluxing overnight in the presence of calcium hydride, followed by distillation. The fluorescamine in dioxane solution should be made fresh daily.

[44] S. Margolis and R. G. Langdon, *J. Biol. Chem.* **241**, 469 (1966).

[45] J. J. Albers, V. G. Cabana, and W. R. Hazzard, *Metabolism* **24**, 1339 (1980).

[46] W. R. Fisher, unpublished data.

[47] V. N. Schumaker, unpublished data.

Measurement of Residual Lipid. Since solubilization of apoB requires the presence of surfactants or denaturants, the newer enzymatic lipid assays cannot be assumed to be valid. Alternately, traditional chemical analyses may be used. Because triglyceride is more easily extracted from lipoproteins than are cholesterol and phospholipid, the major concern is the residual content of the latter lipids. To measure lipid contaminants of the order of a few percent a measured quantity of about 10 mg of apoB, in a urea or Triton X-100 solution, is subjected to vigorous lipid extraction using organic solvents. The method of Bligh and Dyer has proved satisfactory.[48] The single phase system contains proportionately 0.8 : 1 : 2 parts of apoB solution, chloroform, and methanol; it is vortexed for about 5 min at room temperature. A second part of chloroform is added with further vortexing followed by one part of water which is again vortexed; thus the solution is transformed into a two phase system with the extracted lipids separating into the chloroform phase. The lower, chloroform phase is separated, passed through a solvent prewashed Millipore filter, and set aside for lipid analysis. Phospholipids are determined by the Bartlett method which measures total phosphorus, and since they contain approximately 4% phosphorus, multiplying by 25 gives a close estimate of phospholipid content.[49] There are a variety of methods for measuring total cholesterol in a lipid extract; a modification of the Lieberman–Burchardt reaction has proved quite satisfactory.[50]

Characterization of Apolipoprotein B

Generally the presence of contaminating proteins is not a consideration in examining a preparation of apoB since the purity of the LDL selected for the isolation should have previously been assured. Rather the primary concern is with the physical properties of the purified protein. Whether isolated by organic solvent or surfactant delipidation apoB readily aggregates; the protein also has a propensity to dissociate into smaller polypeptides. Certainly the state of the sulfhydryls must be considered if physical studies are planned.

ApoB may be modified chemically with lysines and arginines being readily derivatized in the laboratory, and such derivatization changes the biological properties of the apolipoprotein.[51] *In vivo* lysines are deriva-

---

[48] E. G. Bligh and W. J. Dyer, *Can. J. Biochem. Physiol.* **37,** 911 (1959).
[49] G. R. Bartlett, *J. Biol. Chem.* **234,** 466 (1959).
[50] R. P. Cook, "Cholesterol," p. 484. Academic Press, New York, 1958.
[51] T. L. Innerarity, *et al.,* this series, Vol. 129 [33].

tized by glucose and probably by malondialdehyde again changing the cell receptor reactivity of the protein.[29,30] ApoB is a glycoprotein. While the carbohydrate content and a sequence of several of the carbohydrate chains of the protein have been reported,[28] nothing is known about the differences in carbohydrate which occur when the protein is isolated under different circumstances.

The amino acid analysis of apoB is not particularly useful in characterizing a preparation. It has been shown that up to 30% of the protein may be removed enzymatically with little change in the amino acid composition of the remaining apoB.[5]

If immunologic studies are to be undertaken using a delipidated apoB preparation it must be recognized that the conformation of the delipidated protein may differ strikingly from that of the native protein and that immunologic reactivity need not parallel changes in physical conformation. Thus organic solvent delipidated apoB has lost its native antigens while apoB delipidated with Triton X-100 retains most of them.[38]

A minimal criterion for characterizing an apoB preparation is demonstration of a single band on a 4 or 5% SDS–polyacrylamide gel. A second criterion is the measurement of the sedimentation coefficient.

Two distinguishing features of apoB are its very large size and extreme hydrophobicity. Large molecules are particularly amenable to study by sedimentation velocity techniques, which resolve complex solutions according to the buoyant molecular weights of the constituent macromolecules and do so best when the diffusion coefficients are low, also characteristic of large size. The pronounced hydrophobicity of apoB requires that hydrodynamic experiments be performed in denaturing solvents such as concentrated guanidine–HCl, where CD experiments indicate that some elements of secondary structure may persist until a concentration of 7 M guanidine–HCl is attained.[18] ApoB is also readily solubilized in both ionic and nonionic surfactant solutions, from which it binds substantial quantities of the surfactant. The resulting complexes of protein and detergent may be studied by ultracentrifugation.

Sedimentation Coefficient in Guanidine–HCl. Reduced proteins in 6 M guanidine–HCl generally exist as random coils, and although apoB may not be fully denatured in this solvent, the sedimentation coefficient values determined in 6 M guanidine–HCl agree with each other and are consistent with the values determined in 7.8 M GuHCl when the latter are corrected for viscosity and density to 6 M. Thus, measured values[18] of 1.28 S (7.77 M GuHCl) and 1.36 S (7.8 M GuHCl) become 2.04 S and 2.20 S when converted to 6 M GuHCl at 25°. These values are consistent with values of 2.11 S^{25} and 2.15 S^{24} measured in 6 M GuHCl at 25° and extrapolated to infinite dilution. The dependence of the sedimentation coefficient,

s, upon the concentration is given by the expression $1/s = 1/s_0 (1 + kc)$, where s_0 is the infinite dilution value and k is estimated to be 0.16 ml/mg (unpublished). In making these conversions, a value of 0.703 ml/g was used for the apparent specific volume, as determined by Smith et al.[23] and formulas were employed to determine the solvent viscosity and density as a function of GuHCl concentration.[52]

Although low-molecular-weight material is not easily detected in velocity sedimentation studies of apoB in 6 M GuHCl when the Schlieren optical system is employed, the ultraviolet scanning optical system readily determines nonsedimenting and slowly sedimenting light absorbing material, and allows quantitative measurement. Aggregated material is clearly detected as well; thus Steele and Reynolds report, "Seventy to eighty percent of the initial optical density present sedimented slowly with a sharp boundary, while the remainder of the sample displayed a much broader, faster moving boundary."[18] Samples may be freed of both low-molecular-weight contaminants and aggregated material on a column of Sepharose CL-2B in 6 M GuHCl. The ultracentrifuge scanner patterns of such size-purified material show a single, very sharp peak with a sedimentation cocfficient of 2.15 S.[24]

Sedimentation studies have not yet been reported on the smaller human apolipoprotein B-48; however, they have been performed on the similar rat apolipoprotein B-PIII which migrates very close to B-48 on SDS–polyacrylamide gels. The sedimentation coefficient of rat apolipoprotein B-PIII is 1.47 S at 25° in 6 M GuHCl[53] and 0.915 ± 0.03 S in 7.7 M GuHCl at a concentration of 0.25 mg/ml, which when corrected to 6 M and zero protein concentration, is 1.47 S. In the same solvent and temperature, rabbit muscle myosin has a sedimentation coefficient of 1.42 S ± 0.03. Clearly these two proteins must have about the same molecular weight because they migrate close to each other on SDS gels, which measure the size of the protein–detergent complex, and sediment close to each other in the centrifuge, which measures buoyant molecular weight divided by a size parameter.

Sedimentation measurements may also be performed on microgram quantities of apolipoprotein B using a 5–8 M GuHCl gradient in the presence of dithiothreitol in a swinging bucket rotor. Run conditions are calibrated by a simultaneous measurement using a reference protein. In one study[54] a sedimentation coefficient, corrected to 25° in 6 M GuHCl, of 1.93 S was reported.

[52] K. Kawahara and C. Tanford, *J. Biol. Chem.* **241**, 3228 (1966).
[53] J. Elovson, unpublished data.
[54] B. W. Patterson, V. N. Schumaker, and W. R. Fisher, *Anal. Biochem.* **136**, 347 (1984).

Sedimentation Coefficient in Surfactant Solutions. Sedimentation velocity experiments on apolipoprotein B in the presence of ionic or nonionic detergents reveal a marked dependence upon protein concentration. Thus, in sodium dodecyl sulfate, a value of $s^0_{25,b}$ = 9.07 S has been determined by extrapolation to infinite dilution at 25° in a 2.5 mM sodium phosphate, 1 mM sodium azide buffer, pH 7.4, containing 2.5 mM SDS.[18] In this solvent, the amount of bound SDS was determined to be 1.66 ± 0.05 g/g.[18]

Sucrose density gradient centrifugation in the presence of H_2O and D_2O has been used to measure the binding of Triton X-100 above its critical micellar concentration to human apolipoprotein B-100. A value of $s_{20,w}$ = 8.7 S and an apparent specific volume of 0.84 ml/g has been found for apolipoprotein B in the form of a 0.92 g/g Triton X-100/protein complex.[55]

In more recent investigations, LDL was delipidated with Triton X-100.[22] A major Triton/apoB complex was isolated with a $s^0_{25,b}$ = 12.1 S. This complex had an apparent partial specific volume of 0.823 ml/g, in good agreement with Clarke's value.[55] However, the degree of Triton bound to apoB was considerably higher than Clarke's 1.20 g/g. The complex spontaneously and irreversibly aggregated into 17 S and faster components; however, the extent of such heterogeneity could be controlled.[22] First, gross aggregation resulted when Triton was dissolved directly into LDL; a more homogeneous preparation of 12 S complexes was made when LDL was added slowly to a 10% Triton solution. Second, if samples were not kept refrigerated the 12 S complex aggregated. The rate and extent of aggregation were mediated by temperature. Finally, the presence of disulfide cleavage reagents or low concentrations of chaotropes (perchlorate, thiocyanate) also induced aggregation. The ability to control or eliminate such apoB aggregation is crucial to physical studies on apoB interactions with surfactants or lipids.

The Effects of Solution Nonideality on Molecular Weight Determinations of Apolipoprotein B

The effects of solution nonideality on the determination of the molecular weights of reduced or reduced and alkylated proteins in 6 M guanidine–hydrochloride has been studied.[56–59] The theoretical basis of this

[55] S. Clarke, *J. Biol. Chem* **250**, 5459 (1975).
[56] S. Lapanje and C. Tanford, *J. Am. Chem. Soc.* **89**, 5030 (1967).
[57] F. J. Castellino and R. Barker, *Biochemistry* **7**, 2207 (1968).
[58] J. Gazith, S. Himmelfarb, and W. F. Harrington, *J. Biol. Chem.* **245**, 15 (1970).
[59] P. Munk and D. J. Cox, *Biochemistry* **11**, 687 (1972).

nonideality is explained by Tanford[60]; it is due to the excluded volume occupied by the denatured and unfolded polypeptide chains in this solvent. Nonideality may be described by either an activity coefficient term, or alternatively, by a power-series expansion, which, for dilute solutions, is usually given with sufficient accuracy by a linear expression in macromolecular concentration of the form

$$1/M_{app} = 1/M_{true} + Bc + \cdots$$

or

$$1/M_{app} = 1/M_{true} + 2Bc + \cdots$$

where M_{app} and M_{true} are the apparent and true molecular weights, c is the macromolecular concentration in weight units (such as g/ml), and B is a nonideality term called the second virial coefficient. The first expression shows how the apparent molecular weight varies as a function of concentrations in experiments where the chemical potential of the solvent is being determined, such as osmotic pressure. The second expression describes how the apparent molecular weight varies as a function of concentration in experiments where the chemical potential of the macromolecule is being determined, such as sedimentation equilibrium and light scattering; the nonideality term is exactly twice as great for these latter experiments as for the former.

The magnitude of the nonideality term depends upon the molecular weight of the polypeptide being studied. The largest protein for which data exist is rabbit muscle myosin, and here a value of $B = 0.45 \times 10^{-3}$ mol ml g^{-2} may be calculated from the published data of Gazith et al. in 5 M GuHCl[58] and we have obtained an almost identical value in 6 M GuHCl.[53] For myosin, studied at a protein concentration of 1 mg/ml, an error in molecular weight of about 22% is calculated using the second equation. At 0.1 mg/ml, an error in molecular weight of 3% is calculated. For apoB-48, which appears to have a molecular weight close to that of myosin, the percentage error at these concentrations is probably about the same, while for apoB-100, the error should be substantially larger.

In sedimentation equilibrium experiments, the combined effects of solution nonideality and molecular weight heterogeneity can be seriously misleading, because nonideality causes the log c vs radius squared plot to exhibit negative curvature, that is, to show a decreasing slope at higher concentrations; this effect is in the opposite direction to the positive curvature caused by molecular weight heterogeneity. Therefore, the two effects, solution nonideality and molecular weight heterogeneity, com-

---

[60] C. Tanford, "Physical Chemistry of Macromolecules," p. 192. Wiley, New York, 1961.

bine to yield a log c vs radius squared plot which is deceptively straight. Moreover, the average slope of the resulting curve no longer reflects the molecular weight of the sample, but rather it will be less, sometimes much less. While nonideality effects are serious in 6 M guanidine–hydrochloride, it is likely that similar effects will occur in concentrated detergent solutions, where the apoB molecules may be greatly extended due to the detergent binding. Therefore, published molecular weight values for apoB-100 must be regarded with caution, unless it is shown that the preparation is homogeneous in molecular weight, through sedimentation velocity studies, for example, and a range of concentrations are employed so that the second virial coefficient can be measured and 1/M extrapolated to infinite dilution.

[12] Isolation and Characterization of Apolipoprotein B-48

By DAVID A. HARDMAN and JOHN P. KANE

Introduction

Early studies of the structure of plasma lipoproteins revealed the presence of a protein of very high molecular weight in VLDL and LDL which was termed apolipoprotein B. It was assumed, based on immunochemical cross-reactivity by polyclonal antisera, that this protein was also present in chylomicrons and their remnants. Recently, we and others have shown that the related protein in chylomicrons is distinct from the B protein of VLDL and LDL with respect to both apparent molecular weight[1-3] and amino acid composition.[1] These proteins have been termed apoB-100 and apoB-48, respectively, based on their relative apparent molecular weights in SDS–gel electrophoresis, though other systems of nomenclature have been employed by some authors. It appears that in most animal species the B-100 protein is of hepatic origin and the B-48 protein originates in the intestine. In rats and mice a B-type protein resembling intestinal B-48 appears in hepatogenous VLDL. However, this protein has not been shown to be identical with the intestinal B-48 protein.

[1] J. P. Kane, D. A. Hardman, and A. E. Paulus, *Proc. Natl. Acad. Sci. U.S.A.* **77**, 2465 (1980).
[2] K. V. Krishnaiah, L. F. Walker, J. Borensztajn, G. Schonfeld, and G. S. Getz, *Proc. Natl. Acad. Sci. U.S.A.* **77**, 3806 (1980).
[3] J. Elvoson, Y. O. Huang, N. Baken, and R. Kannan, *Proc. Natl. Acad. Sci. U.S.A.* **78**, 157 (1981).

The B apolipoproteins share certain biochemical properties. Like intrinsic membrane proteins they have very high affinity for the lipid moieties of their respective lipoproteins and do not appear to exchange among lipoprotein particles. They are insoluble in 4.2 M, 1,1',3,3'-tetramethyl urea and they appear to be essential for the secretion of their respective lipoproteins.

The metabolic significance of the speciation of apoB stems from the observations that the different B proteins reside on separate lipoprotein particles,[4-6] that they appear to be under separate genetic control,[7] and that they have different metabolic fates. Thus, further study of apoB-containing lipoproteins must reflect this structural and metabolic heterogeneity. In fact, the individual B apolipoproteins are probably the most reliable markers for subclasses of triglyceride-rich lipoproteins in metabolic studies.

Owing to their very large molecular weights and extensive self-association when separated from their native particles, the resolution of B apolipoproteins has presented unusual technical difficulties. Whereas the B-100 apolipoprotein can be prepared in pure form from LDL, the B-48 protein must be prepared from chylomicrons which are almost invariably contaminated with substantial amounts of B-100-containing VLDL of large diameter. In addition, other proteins, which are not readily soluble in aqueous buffers, are present in chylomicrons, and must be separated from B-48 protein, which constitutes only approximately 5% of chylomicron protein mass. This chapter is devoted to the description of techniques which we have employed in purifying and characterizing the B-48 protein.

Purification of B-48

ApoB-48 is purified from the triglyceride-rich lipoproteins of intestinal origin, the chylomicrons. From humans, the most convenient source of chylomicrons is the serum of individuals with mixed lipemia. An alternate source, where available, is thoracic duct lymph obtained from normolipidemic individuals undergoing lymphophoresis. In either case, samples

---

[4] R. W. Milne, P. K. Welch, L. Blanchetto, J. Davignon, P. Alaupovic, and Y. L. Marcel, *J. Clin. Invest.* **73,** 816 (1984).

[5] J. P. Kane, G. C. Chen, R. L. Hamilton, D. A. Hardman, M. J. Malloy, and R. J. Havel, *Arteriosclerosis* **3,** 47 (1983).

[6] M. Fainaru, R. W. Mahley, R. L. Hamilton, and T. L. Innerarity, *J. Lipid Res.* **23,** 702 (1982).

[7] M. J. Malloy, J. P. Kane, D. A. Hardman, R. L. Hamilton, and K. B. Dalal, *J. Clin. Invest.* **67,** 1441 (1981).

are preserved immediately after they are obtained by the addition of gentamycin sulfate (0.1 mg/ml), NaN_3 (0.05%), and EDTA (0.05%, pH 7.0). These preservatives are present in all solutions that come in contact with the intact chylomicrons. From serum, the largest of the triglyceride-rich lipoproteins are obtained by ultracentrifugation at 20,000 rpm in a 40.3 rotor for 20 min at 12°. Three milliliters of serum is layered under 0.15 M NaCl in each tube. The triglyceride-rich lipoproteins are recovered from the top 1.5 ml and the centrifugation is repeated once. From lymph, the triglyceride-rich lipoproteins are obtained by ultracentrifugation twice at 30,000 rpm for 22 hr.

The isolation of B-48 from these samples is accomplished in three steps. In the first step, the lipoproteins are delipidated using organic solvents and the precipitated proteins are dissolved in aqueous buffer containing SDS. Typically, a lipoprotein sample containing 20 mg of total protein is diluted to a volume of 10 ml using 0.15 M NaCl. This sample is then added, dropwise, with agitation, to 200 ml of ethanol–diethyl ether (3 : 1), and the mixture is allowed to stand at room temperature for 1 hr. The precipitated protein is centrifuged in 50-ml conical tubes (15,000 g min). Next, the bulk solvent is removed from each protein pellet by means of suction and the pellets are flushed briefly with nitrogen to remove excess solvent. Then, the precipitated protein is dispersed in 2% SDS, 0.1 M sodium phosphate to achieve a total protein concentration no greater than 2 mg/ml. The protein is allowed to dissolve at room temperature, and the solution is then heated in a boiling water bath for 15 sec. At this point, the solution may appear turbid due to a trace of remaining lipid. More extensive delipidation with organic solvents will not remove this turbidity, but makes the protein less soluble in SDS. Therefore, we eliminate the turbidity by ultracentrifuging the sample at 38,000 rpm in a 40.3 rotor for 45 min. The floating lipid layer is removed by suction and the clear infranatant is collected.

In the second step, the B proteins are separated from all other apolipoproteins by means of gel permeation chromatography. A volume of 8 ml of the clear infranatant is applied to a 2.2 × 80 cm column packed with Sepharose CL-6B and is eluted at room temperature using buffer containing 0.1% SDS, 0.025 M sodium phosphate, pH 7.0, 0.05% NaN_3, and 0.05% EDTA, at a flow rate of 30 ml/hr. Under these conditions, both B-100 and B-48 elute very close to the void volume. Fractions that contain B-48 are pooled and are concentrated to a volume of 1.5 ml by means of ultrafiltration with an Amicon PM-30 membrane.

In the final step, the B proteins are separated from one another by preparative SDS–PAGE using a modified Shandon Southern instrument.[8]

[8] D. A. Hardman and J. P. Kane, *Anal. Biochem.* **105,** 174 (1980).

The advantage of this system is that it will purify useful quantities of protein in one simple process with virtually complete recovery. In this apparatus, protein separation takes place in a cylindrical slab gel (resolving gel) which is cast within a water jacketed glass column. Below this column is situated a lucite plate incorporating a circular channel (elution chamber) that is aligned with the bottom of the resolving gel. Below this plate is situated another gel (lower gel) which seals the bottom of the elution chamber and also provides electrical contact with the anodic buffer. Each protein travels the entire length of the resolving gel and, as it emerges from the bottom, is swept away by a continuous flow of buffer through the elution chamber.

In order to achieve the resolution and efficiency that we require, we have found it necessary to modify the original apparatus. First, we have discontinued the use of dialysis membrane as a protein barrier beneath the elution chamber. This modification has no effect on the recovery of protein but does eliminate the high resistance and the destructive electroosmotic effects that occur while using the membrane. Second, the glass column, originally designed to support the lower gel, was replaced by a modified structure (lower gel support) machined from lucite. This modification provides a more effective seal around the elution chamber and thus results in a more efficient elution of protein. The new lower gel support is constructed as follows. First, a disk 17 cm in diameter is cut from a sheet of lucite about 1 cm thick. Next, a lucite cylinder, 5 cm i.d. and 2.5 cm long, is centered on one face of the disk and is glued in place. Then, an annular channel, 0.3 cm deep and 2.2 and 2.9 cm inside and outside diameters, respectively, is machined into the opposite face of the disk. Finally, a series of holes 2.5 mm in diameter are drilled through the disk (connecting the channel to the chamber formed by the cylinder). This lower gel support is attached to the lower tank cover as follows. The central opening in the lower tank cover is enlarged to 15 cm diameter. With the annular channel facing upward, the lower gel support is glued to the underside of the lower tank cover.

The assembly and operating procedures for the modified apparatus are as follows. The apparatus is first assembled using a solid lucite disk 75 × 3 mm between the glass column and the lower gel support, in the position that would normally be occupied by the elution chamber during a run. While the assembled unit is inverted, a 10% gel is cast in the lower gel support (10% acrylamide, 0.27% N,N'-methylene bisacrylamide, 0.1 M sodium phosphate, pH 7.2, 0.1% SDS, 0.26% TEMED, and 0.09% potassium persulfate). After this gel has polymerized, the unit is turned upright and a 2.3% resolving gel is cast in the glass column to a height of 10 cm (2.3% acrylamide, 0.06% N,N'-methylene bisacrylamide, etc.). After this gel has polymerized, preferably overnight, the apparatus is again inverted, disassembled, and the lucite disk is carefully removed. The exposed lower surface of the resolving gel is then flooded with water and is covered by a moistened 7-cm-diameter cellulose triacetate filter (0.2 μm pore size). After the excess water is removed by blotting, the filter should adhere firmly to the gel, lying flat with no air bubbles trapped beneath. The elution chamber is then fitted in position above the lower gel, and this assembly is clamped in place on the still inverted column. The fully assembled apparatus is then fitted to the lower tank and both the upper and lower tanks are filled with tank buffer (0.1% SDS, 0.1 M sodium phosphate, pH 7.2, and 0.05% NaN_3).

In order to complete the elution system, a 500 ml reservoir of tank buffer, a flow cell detector, a peristaltic pump, and a fraction collector are connected to the elution chamber as shown (Fig. 1) by means of polyethylene tubing (PE-50). To gain the highest sensitivity and resolution, a flow cell having a path length of 20 mm and a volume of 0.02 ml is used in the detector. Tubing having an i.d. of 1 mm is used in the pump. Note that these components are arranged in such a way that the buffer is always pulled through the elution chamber, never pushed. This is necessary in order to prevent any increase of pressure within the elution chamber that could distort the fragile resolving gel. Also note, however, that while pressure should never be applied to the elution chamber, a considerable amount of suction can be applied since the resolving gel is supported at its base by the cellulose triacetate filter. After all the components of the elution system are in place, buffer is drawn through by means of the pump at 50 ml/hr in order to eliminate bubbles.

The sample of mixed B proteins from the Sepharose column is prepared for electrophoresis by the addition of sucrose and 3-mercapto-1,2-

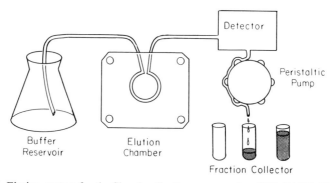

FIG. 1. Elution system for the Shandon Southern preparative SDS–PAGE. The arrangement of components is designed to avoid any increase in pressure within the elution chamber, which could distort the resolving gel.

propanediol to achieve final concentrations of 5 and 1%, respectively. The sample is heated in a boiling water bath for 15 sec and, after it has cooled, is applied to the surface of the resolving gel by layering under the upper tank buffer. Electrophoresis is carried out at a temperature of 25°, a current of 95 mA, and an elution buffer flow of 15 ml/hr. Tank buffer is recirculated at a rate of 300 ml/hr. Fractions are collected at 20 min intervals, B-48 typically elutes after about 10 hr and is collected in a volume of about 15–20 ml (Fig. 2). If necessary, the purified protein may be concentrated by ultrafiltration using an Amicon PM-30 membrane.

Amino Acid Analysis

We have determined the amino acid composition of the purified B-48 as follows. Samples of B-48 containing between 25 and 50 μg of protein are measured into glass tubes (7.5 × 100 mm) that have been washed in chromic acid, rinsed repeatedly in glass distilled water, and rinsed in 1% EDTA. Five separate tubes are needed for a complete analysis. The samples are lyophilized and the residues are extracted with 2 ml of 20% TCA to remove SDS and buffer salts. After each extraction, the precipitated protein is centrifuged (15,000 g min) and the supernatant is discarded. The precipitate is then similarly extracted twice using acetone to remove the

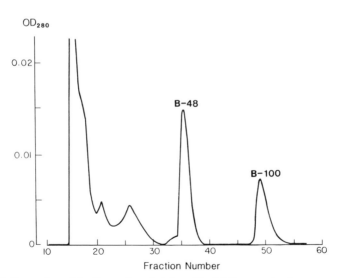

FIG. 2. Separation of the B proteins of triglyceride-rich lipoproteins from serum using the modified Shandon Southern preparative SDS–PAGE apparatus. Current, 95 mA; elution rate, 15 ml/hr; fractions collected at 20 min intervals.

TCA, and is dried under nitrogen. These dry protein samples are then ready for hydrolysis.

The values for all amino acids except Cys, Met, and Trp are determined after hydrolysis in 6 N HCl for three different time periods. One milliliter of 6 N HCl, containing 0.05 mg/ml of phenol, is added to each of three of the extracted protein samples. The tubes are sealed under vacuum and are heated at 105° for periods of 22, 72, and 120 hr. The hydrolysates are lyophilized and the residues are dissolved in 0.2 M sodium citrate (pH 2.2). Analysis is performed on a Beckman 121M amino acid analyzer using a two column program. Values for Thr and Ser are calculated by extrapolation to zero time. The values for Val and Ile are calculated at 120 hr. The remaining values are calculated from the means of all three time points.

Cys and Met are determined as cysteic acid and methionine sulfone after oxidation by performic acid. A volume of 0.1 ml of 88% formic acid is added to one of the extracted protein samples and the dry protein is allowed to dissolve overnight at 23°. Then, 0.3 ml of 30% H_2O_2 is added and the mixture is allowed to react for 6 hr at room temperature. Then, an additional 0.3 ml of formic acid and 0.9 ml of H_2O_2 are added and this mixture is allowed to react for 48 hr at room temperature. The reaction mixture is then lyophilized. In order to reduce frothing during lyophilization, it is helpful to expel dissolved gas by mixing the sample vigorously on a vortex mixer beforehand. After lyophilization, the dried sample is hydrolyzed in 6 N HCl *in vacuo* at 105° for 22 hr.

Analysis of this sample is performed on a Beckman 121M amino acid analyzer using the long column only. However, when samples are analyzed in the routine manner using this column, we have observed a Schlieren peak that coincides with cysteic acid. This artifact is due to the passage of two solvent fronts through the flow cell. The first of these results from the use of application buffer to dissolve the sample and can be eliminated by dissolving the sample in the same buffer that is used to elute the column. The second solvent front results from the injection of a small amount of water onto the column shortly after the sample during the normal injection cycle and can be eliminated by interrupting the injection cycle immediately after the sample is injected (step 4). The injection cycle is then continued after the analysis is complete.

Values for Cys and Met are calculated as ratios to Ala and are then combined with the previous data. As a control to determine the completeness of the oxidation reaction, a sample of lysozyme is treated in parallel.

The value for Trp is determined after hydrolysis in 4 N methanesulfonic acid (MSA) containing 0.2% 3-(2-aminoethyl)indole. From the previous amino acid analysis, the lysine content of the remaining ex-

AMINO ACID COMPOSITIONS OF HUMAN B-48
AND B-100[a] RESIDUES/10^3

	B-100	B-48
Trp	8.76	6.31
Lys	79.15	77.02
His	25.47	24.38
Arg	33.47	38.10
Asp	105.62	106.40
Thr	65.86	67.90
Ser	85.67	89.93
Glu	115.14	109.62
Pro	38.20	38.46
Gly	46.89	55.77
Ala	59.97	65.34
Cys	4.41	6.34
Val	55.11	52.27
Met	15.81	21.11
Ile	59.74	46.42
Leu	117.42	120.55
Tyr	33.28	33.58
Phe	50.02	40.50

[a] B-48 was isolated as described from lymph chylomicrons; B-100 was isolated from LDL.

tracted protein sample is calculated. Then, 4.2 µl of MSA is added for each nanomole of lysine in the sample. Note that it is extremely important to use the correct amount of MSA. On the one hand, if too little MSA is used, Trp may be destroyed during hydrolysis because it is inadequately protected by the indole. On the other hand, if too much MSA is used, the sample could ultimately be too dilute for analysis since, unlike HCl, MSA cannot be removed by lyophilization before analysis. After MSA is added, the sample is frozen and the tube is evacuated for 1 min before it is sealed. The sample is hydrolyzed at 106° for 22 hr. The hydrolysate is diluted with 5.67 volumes of glass distilled water and is applied directly to the short column of a Beckman 121M amino acid analyzer. The value for Trp is calculated as the ratio to Lys and is then combined with the data from the previous analyses. The amino acid compositions for apoB-48 and apoB-100 are compared in the table.

Determination of Molecular Weight

We have also attempted to determine the molecular weight of B-48 using SDS–PAGE. However, as in our earlier work with B-100, we observed that our molecular weight estimates for B-48 were strongly influenced by two factors during electrophoresis: protein load and gel concentration. We believe that this behavior could be partly responsible for the wide variation among estimates of molecular weight that have been reported for both B-48 and B-100 from different laboratories. Based on our analysis of this behavior, we have established a set of conditions for estimating the molecular weights of B proteins, using SDS–PAGE, that forms the basis for our centile system of nomenclature.

The effect of gel concentration on the mobility of any protein during electrophoresis may be shown graphically by plotting log M_r vs gel concentration according to the method of Ferguson.[9] In order to generate the Ferguson plots shown in Fig. 3, each protein sample was subjected to electrophoresis on four separate tube gels (9 × 0.6 cm), each of which contained a different acrylamide concentration (2.5, 3.3, 4, and 5.5%). In all gels the concentration of N,N'-methylene bisacrylamide was 2.6% of the total acrylamide concentration. The tank buffer and gel buffer contained 0.1% SDS and 0.1 M sodium phosphate, pH 7.2. Each sample contained 2 μg of protein in a volume of 100 μl of 0.1% SDS, 0.025 M sodium phosphate, pH 7.2, 5% sucrose, 1% 3-mercapto-1,2-propanediol, and 0.0025% bromophenol blue. Samples were heated in a boiling water bath for 10 sec just before application to the gels. Electrophoresis was at 8 mA/gel throughout. After the tracking dye had migrated 7.5 cm, the gels were removed and stained in 0.025% Serva blue in 30% methanol, 10% acetic acid, and 60% H_2O for 24 hr at 23°. When preparing the staining solution, it is helpful to dissolve the solid dye in the appropriate volume of methanol before adding the other components. The gels were destained in the same solvent. The migration of each protein was measured at the leading edge of the band. The mobility (relative to the tracking dye) for each protein was then calculated in the usual way.

For most proteins on SDS–PAGE, a Ferguson plot yields a straight line (Fig. 3A and B, solid lines); the slope of the line is a function of the size of the protein–SDS complex, while the Y intercept (free electrophoretic mobility) is a function of the charge that it carries. Note that in SDS–PAGE the extrapolated lines corresponding to these typical proteins meet at a single point before they reach the Y axis. This behavior is taken to indicate that all of these proteins adopt a common conformation in the presence of SDS.[10] Under these conditions, all proteins that we have tested, including B-48 and B-100, conform to this pattern.

However, we have observed load-dependent deviation from this pat-

---

[9] K. A Ferguson, *Metabolism* **13**, 985 (1964).
[10] G. A. Banker and C. W. Cotman, *J. Biol. Chem.* **247**, 5856 (1972).

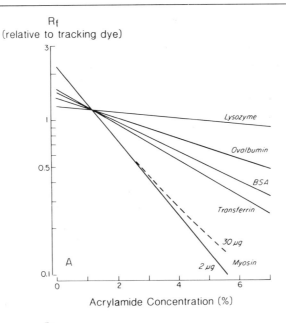

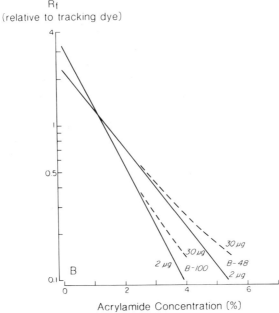

FIG. 3. (A) Ferguson plot of five standard proteins: lysozyme, ovalbumin, bovine serum albumin (BSA), human transferrin, and rabbit back muscle myosin. (B) Ferguson plot of B-48 and B-100. Solid lines represent a protein load of 2 μg/gel; dashed lines represent a protein load of 30 μg/gel.

tern among several high-molecular-weight proteins. Early reports on the usefulness of SDS–PAGE for estimating molecular weights of proteins described the effect of overloading on the migration of protein bands. It was observed at that time that protein bands in SDS–PAGE have a characteristic form consisting of a sharp leading edge and a diffuse trailing edge. It was also observed that increasing the protein load has little, if any, effect on the mobility as measured at the sharp leading edge, but merely broadens the diffuse trailing edge. In contrast, however, we have observed that overloading has a dramatic effect on the migration of very high-molecular-weight proteins.

The effect of increased protein load on the Ferguson plots of three proteins—myosin, B-48, and B-100—is shown in Fig. 3 (dashed lines). In each case, increasing the protein load results in a faster migration of the protein band, with the effect most pronounced at high gel concentration. Thus, for example, the mobility of B-48, as measured on a 5.5% gel, increases by more than 50% when the protein load is increased from 2 to 30 μg. In contrast, in a 2.5% gel, the mobility of B-48 is relatively insensitive to protein load.

Therefore, in order to minimize the effect of protein load and gel concentration on our estimates of the molecular weights of B proteins, SDS–PAGE was performed using very low concentration gels. To this end, mobilities were calculated from the elution profiles of proteins subjected to electrophoresis on 2.3% gels using the preparative electrophoresis apparatus. This apparatus was used because gels of this low concentration are too fragile to handle in the usual way. We have found that the standard curve generated from these profiles by plotting log molecular weight vs M_r is linear over the molecular weight range of 45,000 to 200,000. Linear extrapolation of this standard curve yields molecular weight estimates for B-48 and B-100 of 264,000 and 549,000, respectively.[1] Because of the dangers inherent in extrapolation, we acknowledge that there may be some systematic error in these absolute values. However, the centile designations, which should be relatively insensitive to such errors, would be expected to reflect accurately the relative molecular weights of the B apolipoproteins.

[13] Isolation and Characterization of Apolipoprotein E

By STANLEY C. RALL, JR., KARL H. WEISGRABER, and ROBERT W. MAHLEY

The isolation and purification of human apolipoprotein E (apoE) present a special problem because of its extensive polymorphism, which has been shown to be due to two factors: genetically determined polymorphism resulting from multiple alleles at the single apoE structural gene locus, and polymorphism arising from posttranslational glycosylation (sialylation) of the apoE polypeptide. A concise summary of this polymorphism has been presented by Zannis et al.,[1] who also describe a nomenclature system for human apoE phenotypes and genotypes that will be used in this chapter. Investigators should have a clear understanding of the phenomenon of apoE polymorphism before undertaking the task of isolation of apoE from human sources. Depending on the purpose for which the apoE is intended, the choice of subject(s) and knowledge of the apoE phenotype of individual subjects may be crucial to the investigation. This chapter discusses general and specific procedures for human apoE isolation and outlines procedures for characterization of apoE, including apoE phenotyping.

Isolation

Source

Generally speaking, a single source for human apoE is always preferable to obtaining the protein pooled from several subjects, and the subject should be homozygous to reduce the risk of obtaining heterogeneous apoE, which is often undesirable. For studies on the functional aspects of apoE, this is particularly important because the different apoE polymorphs are heterogeneous in one of their most important functions, which is the ability to bind to lipoprotein receptors (see review by Mahley and Innerarity[2]). Therefore, phenotyping of potential donors should always be carried out before preparative isolation is attempted (see apoE phenotyping section below).

[1] V. I. Zannis, J. L. Breslow, G. Utermann, R. W. Mahley, K. H. Weisgraber, R. J. Havel, J. L. Goldstein, M. S. Brown, G. Schonfeld, W. R. Hazzard, and C. Blum, *J. Lipid Res.* **23**, 911 (1982).
[2] R. W. Mahley and T. L. Innerarity, *Biochim. Biophys. Acta* **737**, 197 (1983).

In a normal healthy individual, plasma levels of apoE are not particularly high, usually about 3–7 mg/dl. To obtain sufficient apoE for most purposes, one and possibly several plasmaphereses will have to be performed. Because of the low levels found in normolipidemic subjects, a better source for human apoE is from subjects who have certain types of hyperlipidemia. Hypertriglyceridemic subjects having either the type IV or V lipoprotein phenotype usually have large amounts of apoE, owing to overproduction or impaired catabolism of very low density lipoproteins (VLDL), and their VLDL tend to be enriched in apoE compared to VLDL from normolipidemic individuals. It is not unusual to obtain 10 mg of apoE from the VLDL contained in 100 ml of plasma from a type V hyperlipoproteinemic subject. Subjects having type III hyperlipoproteinemia are also good sources for apoE, owing largely to accumulation in the plasma of the abnormal lipoproteins, β-VLDL, which are also enriched in apoE. Many of these subjects have plasma apoE levels above 20 mg/dl. However, type III subjects will have one of the abnormal variants of apoE that are known to be variably defective in their interaction with lipoprotein receptors (see review by Mahley and Angelin[3]). This heterogeneity must be taken into account when considering possible donors.

Lipoprotein Isolation and Delipidation

Details of the isolation of apoE-containing lipoproteins from plasma are described by Weisgraber and Mahley.[4] In general, the most convenient method for lipoprotein isolation is ultracentrifugal flotation of d <1.02 g/ml lipoproteins [intermediate-density lipoproteins (IDL) and VLDL] from fresh plasma.[4] The amount of apoE in the d <1.02 fraction is usually increased if the subjects are nonfasted. This is due to the redistribution of apoE from high-density lipoproteins (HDL) to the lower density lipoproteins during fat absorption. It is not necessary to remove apoB by the tetramethylurea technique[5] prior to the apoE isolation. The lipoproteins are dialyzed against 0.01% EDTA (pH 7.4), and then lyophilized in 50-ml conical glass tubes. Delipidation is carried out directly in the same tube containing lyophilized lipoprotein. Chilled (4°) chloroform : methanol (2:1, v/v), ~30 ml, is added to approximately 50 mg lyophilized lipoprotein protein. After standing at 4° for 1 hr, 20 ml methanol is added, and the apolipoproteins are pelleted by low-speed centrifugation after each of three extractions. Following a methanol wash, the *moist*, dispersed apo-

[3] R. W. Mahley and B. Angelin, *Adv. Intern. Med.* **29**, 385 (1984).
[4] K. H. Weisgraber and R. W. Mahley, this series, Vol. 129 [9].
[5] J. P. Kane, *Anal. Biochem.* **53**, 350 (1973).

lipoproteins (they should not be allowed to dry) are then ready for solubilization.

Apolipoprotein E can also be prepared from the apoE-containing subclass of HDL[4] or from the lipoprotein-free infranate ($d > 1.21$ g/ml) after centrifugation. This latter method involves flotation of the apoE from the $d > 1.21$ ultracentrifugal fraction with Intralipid. The procedure has been used primarily for the isolation of apoA-IV but can also be used for apoE.[6] Neither the HDL with apoE subclass nor the $d > 1.21$ fraction can be expected to yield as much apoE as hypertriglyceridemic VLDL does.

Apolipoprotein E General Isolation Techniques

Molecular Sieve Chromatography. Molecular sieve chromatography on Sephacryl S-300 is the method of choice for isolating apoE from the delipidated $d < 1.02$ lipoproteins.[7] The most likely cross-contaminating apolipoproteins, those with molecular weights in the same range as apoE ($M_r = 34,000$), i.e., apoA-I and apoA-IV, tend to be present in relatively low amounts, if at all. The moist, delipidated apolipoproteins (~250 mg total protein) are dissolved in 6 M guanidine, 0.1 M Tris, 0.01% EDTA (pH 7.4), 1% 2-mercaptoethanol, and allowed to stand overnight at room temperature for adequate disulfide reduction and solubilization. Any insoluble material is removed by low-speed centrifugation. The proteins are applied to a 2.5 × 300-cm column equilibrated in 4 M guanidine, 0.1 M Tris, 0.01% EDTA (pH 7.4), 0.1% 2-mercaptoethanol, and eluted with the same buffer at a flow rate of 16–20 ml/hr (peristaltic pump) at room temperature. The column should be packed according to the manufacturer's recommendations and will take 3–4 days for the entire procedure. To maintain proper flow characteristics, it is important not to exceed the stated maximum flow rate. Elution of apoE requires approximately 2–3 days and is monitored by absorbance at 280 nm (the guanidine must be of sufficient quality to allow UV detection). Apolipoprotein E is obtained in good yield, is well separated from apoB and the C apolipoproteins, and is eluted at a volume of about 650 ml, or about 50% of the column volume. Fractions containing the apoE are pooled, exhaustively dialyzed against 5 mM NH$_4$HCO$_3$ at 4°, and lyophilized in acid-washed conical glass centrifuge tubes. Resolubilization of the apoE is usually done with 0.1 M NH$_4$HCO$_3$. The apoE obtained by this procedure is usually of sufficient purity for most purposes, especially if it is obtained from hyperlipidemic $d < 1.02$ lipoproteins. However, an impurity is present that is highly en-

---

[6] R. B Weinberg and A. M. Scanu, *J. Lipid Res.* **24,** 52 (1983).
[7] K. H. Weisgraber, S. C. Rall, Jr., and R. W. Mahley, *J. Biol. Chem.* **256,** 9077 (1981).

riched when normolipidemic subjects are used as the source; this impurity can be readily detected by amino acid analysis of the apoE preparation (see amino acid analysis section below).

Preparative Sodium Dodecyl Sulfate Gels. An alternative general technique for apoE isolation is preparative sodium dodecyl sulfate–polyacrylamide gel electrophoresis (SDS–PAGE).[8] This method is also suitable for mixtures that contain apoA-I and/or apoA-IV (HDL subclasses and $d > 1.21$ fraction). The apoliproteins are incubated at 37° for 2 hr in 2.5 mM Tris–glycine (pH 8.2), 0.3% SDS, and 125 mM 2-mercaptoethanol. A small portion of the sample (1–5%) is dansylated so that the electrophoresis can be monitored directly with UV light.[9] The dansylated and untreated sample are mixed and electrophoresed in a Tris–glycine system (11% acrylamide) according to Stephens.[9] The protein bands are visualized by their UV fluorescence and excised with a razor blade. The gel contents are then electroeluted directly into a dialysis bag with 2.5 mM Tris–glycine (pH 8.2), dialyzed at 4° against 5 mM NH$_4$HCO$_3$, and lyophilized. Recoveries are usually about 50%. Any traces of detergent that might remain do not cause any problems in subsequent procedures or analyses.

Apolipoprotein E Isoform Purification

Apolipoprotein E isoform purification, used in conjunction with the general methods described above, offers several distinct advantages. An investigator may find it desirable to have only a specific isoform of apoE for use in further studies, either because sialylated forms are specifically desired (or undesired), or because the apoE may not be homogeneous (as would occur using a subject with a heterozygous apoE phenotype or multiple subjects as the source). Isoform purification is the surest way to reduce problems caused by apoE polymorphism and, in addition, is almost certain to yield an apoE of higher purity than that which results from the general method.

There are two suitable preparative isoelectric focusing techniques for apoE: flat-bed focusing on a support of Sephadex G-200 and the use of Immobilines on vertical slab gels. The latter method is currently being used in our laboratory and is superior in many regards to the flat-bed method.

Flat-Bed Focusing. The flat-bed technique is performed on an LKB Multiphor apparatus according to Weisgraber *et al.*[7] The gel slurry is prepared according to the manufacturer's recommendation, except that 6

[8] K. H. Weisgraber, R. W. Mahley, and G. Assmann, *Atherosclerosis* **28**, 121 (1977).
[9] R. E. Stephens, *Anal. Biochem.* **65**, 369 (1975).

M urea is included. After pouring the slurry into the tray, dry Sephadex G-200 is sprinkled through fine gauze onto the bed until the gel consistency is such that it does not flow when the tray is tilted 45°. The protein sample in 3–5 ml of 25 mM Tris (pH 7) containing 1% sodium decyl sulfate is applied at the anodic end of the gel. Electrophoresis is carried out for 16 hr at 4° in 8 M urea over the pH range 4–6.5 (2% ampholytes, Pharmacia) at a constant power of 8 W. A paper print technique is used to locate the isoforms.[10] Isoforms are eluted from the gel with 4 M guanidine, 0.1 M Tris (pH 7.4), dialyzed against 5 mM NH$_4$HCO$_3$ at 4°, and lyophilized. Ampholytes are removed from the protein by dissolving the sample in 0.1 M NH$_4$HCO$_3$, adding solid (NH$_4$)$_2$SO$_4$ to 75% saturation, and allowing the solution to stand overnight at 4°. The precipitated protein is pelleted by centrifugation and washed three times with 75% saturated (NH$_4$)$_2$SO$_4$. It is then solubilized in 0.1 M NH$_4$HCO$_3$ and dialyzed against 0.1 M NH$_4$HCO$_3$ at 4° to remove residual (NH$_4$)$_2$SO$_4$. The disadvantages of the above method are as follows: Recoveries are low (25–50%); the isoforms tend to focus as "wavy" bands (which are difficult to excise accurately using the paper print template); and the ampholytes must be removed so that they do not interfere with subsequent analyses (e.g., cell receptor binding studies and amino acid and sequence analyses).

Immobiline Gels. The second method using Immobiline gels obviates all of the above problems. The following technique is not applicable to analytical, multiwell slab gels. The procedure for running horizontal analytical Immobiline gels[11] has been simplified for preparative separation of apoE.[12] Modifications include the use of an isokinetic gradient in place of a linear gradient in pouring the gel (a pH 4.9–5.9 or a pH 5.7–6.7 gel is prepared according to the manufacturer's directions), the inclusion of freshly deionized 5.8 M urea in the gel solutions, and the use of a vertical slab gel apparatus rather than a horizontal gel slab. The 4.9–5.9 gradient is satisfactory for apoE2 and E3 and their sialylated derivatives. For apoE4 it is necessary to use a pH 5.7 to 6.7 gradient. The slabs are 0.15 × 12 × 14 cm. Unlike the case with horizontal analytical gels, both glass plates are left in place after the gel is cast. This provides a trough at the top of the gel for applying the sample. The protein (up to 30 mg) solubilized in 2.0 ml of 0.1 M Tris (pH 8.0) containing 1% sodium decyl sulfate, 20% sucrose, and 6 M urea is applied across the top of the gels. The samples are overlaid with 10% sucrose followed by a layer of 0.01 M NaOH (upper electrolyte buffer). The lower electrolyte buffer is 0.01 M glutamic acid. The gels are

[10] LKB Application Note 198, LKB Instruments, Bromma, Sweden.
[11] LKB Application Note 321, LKB Instruments, Bromma, Sweden.
[12] K. H. Weisgraber, Y. Newhouse, J. Seymour, S. C. Rall, Jr., and R. W. Mahley, *Anal. Biochem.* **151,** 455 (1985).

electrophoresed at 4° for 1 hr at 5 W or until the voltage reaches 800 V, then overnight at 800 V, also at 4°. The voltage is then increased to 1000 V for the last 4–6 hr (total electrophoresis time of 24 hr).

Following electrophoresis, the gel (still attached to one of the glass plates) is immersed in deionized water. The isoforms precipitate and are easily visualized as opaque bands running the entire width of the gel. The bands are excised and eluted from the gel pieces with 4 M guanidine, 0.1 M Tris (pH 7.4), and 1 mM EDTA. The protein is then dialyzed at 4° against 5 mM NH$_4$HCO$_3$, lyophilized, and resolubilized in 0.1 M NH$_4$HCO$_3$. Recoveries (>50%) are higher than with the flat-bed technique, the isoforms can be excised with confidence because they can be totally visualized, the bands are sharp and extremely well separated from one another due to the narrower pH range employed, and ampholytes do not have to be removed after elution because they are covalently attached to the gel support.

Special Purification Techniques

Heparin–Sepharose Chromatography. Heparin–Sepharose chromatography takes advantage of the heparin-binding property of apoE to separate apoE from contaminating proteins. Heparin–Sepharose is prepared in the same manner as described in the isolation of apoE-containing lipoproteins,[4] but the elution procedure is simpler. Apolipoprotein E, isolated by Sephacryl S-300 (1–20 mg in 4 ml of 25 mM NH$_4$HCO$_3$ containing 0.1% 2-mercaptoethanol), is bound to a 1.0 × 15-cm column of heparin–Sepharose equilibrated in the same buffer. The column is washed with 25 ml of starting buffer, and then the apoE is eluted with 0.75 M NH$_4$HCO$_3$. The bound protein eluted at this ionic strength is usually quite pure apoE and recoveries are >80%.

Thiopropyl Chromatography. The second specialized purification method is thiopropyl chromatography on thiopropyl Sepharose 6B as described by Weisgraber et al.[13] This procedure is useful for cleanup of the non-cysteine-containing E-4 isoform, because the contaminating proteins always appear to contain cysteine. The method is particularly suited for easy separation of E-4 from the cysteine-containing isoforms of apoE (E-3 and E-2) without resorting to preparative isoelectric focusing. About 15 mg of column-purified apoE is solubilized in 3 ml of 6 M guanidine, 0.1 M Tris, 0.01% EDTA (pH 7.4), 50 mM dithiothreitol, and allowed to stand overnight at room temperature under a nitrogen atmosphere. Following dialysis against deoxygenated column buffer (4 M guanidine, 0.1 M Tris,

[13] K. H. Weisgraber, S. C. Rall, Jr., T. P. Bersot, R. W. Mahley, G. Franceschini, and C. R. Sirtori, *J. Biol. Chem.* **258**, 2508 (1983).

0.01% EDTA, pH 7.4) to remove the dithiothreitol, the protein is applied to a 1.0 × 10-cm column of thiopropyl Sepharose 6B (4 g) equilibrated with column buffer. The sample is recycled through the column three times to ensure complete reaction. After washing with 40 ml of column buffer to collect the unbound fraction, bound proteins are eluted with column buffer containing 50 mM dithiothreitol. Fractions are dialyzed against 5 mM NH$_4$HCO$_3$ at 4° and lyophilized.

Storage and Handling Precautions

Plasma from which apoE will be isolated should be processed as soon as possible after obtaining fresh plasma, usually within 24 hr. It is best if neither the plasma nor the lipoproteins are frozen at any time, as this can have an adverse effect on both the quality and quantity of apoE ultimately recovered. Purified or partially purified apoE should be stored frozen at −20° in 0.1 M NH$_4$HCO$_3$ solutions. This method is preferable even to storing the apoE lyophilized. Solubilized apoE should never be stored as unfrozen liquid, i.e., never at room temperature or 4° (standard refrigeration).

Some deterioration of apoE caused by the many manipulations during isolation and purification is inevitable and probably unavoidable. This "decomposition" usually manifests itself by the appearance of isoforms that are one and sometimes several charge units more acidic than the starting material. This process is not well understood but is nevertheless a documented phenomenon for both apoE[14] and apoA-I.[13] The presence of the more acidic isoforms becomes more obvious as more manipulations are carried out. Generation of these isoforms will also occur if plasma or VLDL is stored for extended periods of time. The most serious consequence of this phenomenon appears to be in obscuring the true phenotypic apoE pattern in phenotyping studies; this does not affect either structural or functional investigations of the apoE itself. Investigators should be aware of this problem wherever isoelectric focusing patterns are to be interpreted.

Characterization

Apolipoprotein E Phenotyping

When dealing with apoE from human subjects, nothing is likely to be more crucial or informative than accurate phenotyping. In greater than

[14] K. H. Weisgraber, S. C. Rall, Jr., T. L. Innerarity, R. W. Mahley, T. Kuusi, and C. Ehnholm, *J. Clin. Invest.* **73**, 1024 (1984).

99% of the cases, an individual will have one of six phenotypes, displaying either homozygosity (E-2/2, E-3/3, or E-4/4) or heterozygosity (E-4/3, E-3/2, or E-4/2).[1] Each of the apoE polymorphs can (and probably will) demonstrate further complexity upon isoelectric focusing because of sialylation.[15] A few individuals will not fall into one of the six common phenotypic categories. For example, genetically determined (i.e., not sialylated) apoE that focuses in a position corresponding to E-1, E-5, or E-7 has been documented. Furthermore, nonsialylated isoforms focusing in the same position can actually represent several protein structures (proved for E-2 and E-3), so the genotypic picture is also exceedingly complex.

Analytical Isoelectric Focusing. The basic technique for apoE phenotyping is isoelectric focusing. A two-dimensional method, using isoelectric focusing in conjunction with SDS–PAGE, allows the most unequivocal determination of apoE phenotypes,[15] but the one-dimensional focusing method is usually adequate[16] and is considerably more rapid and amenable to screening a large number of subjects. The most convenient procedure is to isolate VLDL ($d < 1.006$) by ultracentrifugal flotation, followed by delipidation of the VLDL and analysis of the apo-VLDL directly by isoelectric focusing in a pH gradient of 4–6.5. Apolipoprotein E isoforms focus with pIs between 5.7 and 6.2; the positions of the more acidic C apolipoproteins provide a convenient reference for each gel (Fig. 1).

Cysteamine Treatment. A companion procedure has been developed that aids in apoE phenotyping.[7] As shown in Fig. 1, the three most common apoE polymorphs can be differentiated by their cysteine content. Apolipoprotein E-2 contains two cysteine residues per mole, while E-3 and E-4 have one and none, respectively. The reagent cysteamine forms a mixed disulfide with cysteine residues in apoE, generating a lysine-like side chain that confers an additional positive charge on the protein for each cysteine residue present. This charge modification can be conveniently monitored by isoelectric focusing. The procedure involves treating the lipoprotein (usually VLDL) with cysteamine (1.0 mg/150 μg lipoprotein protein) for 4 hr at 37°. A companion sample is treated with 2-mercaptoethanol and serves as the control. Both samples are then lyophilized, delipidated with chloroform : methanol, and subjected to isoelectric focusing. An example is shown in Fig. 2. All cysteamine-treated samples appear essentially the same, resembling E-4/4 phenotypes. The important distinction is the charge unit difference between the treated and untreated sample. This allows an assignment of the number of cysteine

---

[15] V. I. Zannis and J. L. Breslow, *Biochemistry* **20**, 1033 (1981).
[16] G. Utermann, A. Steinmetz, and W. Weber, *Hum. Genet.* **60**, 344 (1982).

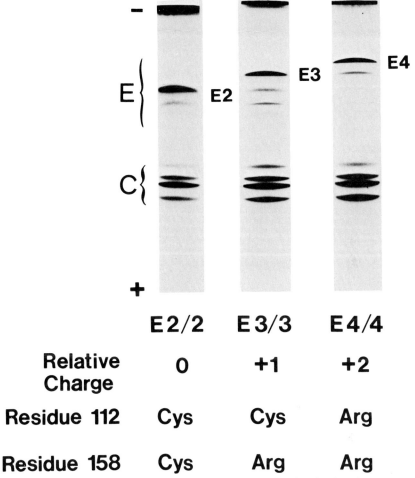

FIG. 1. One-dimensional isoelectric focusing technique showing the three homozygous apoE phenotypes. The amino acid differences among the three major polymorphic forms of apoE are given for comparison. From R. W. Mahley and B. Angelin, *Adv. Intern. Med.* **29,** 385 (1984); reproduced with permission.

residues in the apoE isoforms, which in turn defines the apoE as E-2, E-3, or E-4 (except in a few rare cases). Thus, an unequivocal phenotyping can usually be made without need of a known reference standard.

Neuraminidase Treatment. Another companion procedure is analysis for sialylated forms using neuraminidase. This method is particularly useful in those cases where sialylated isoforms contribute significantly to the total apoE. Neuraminidase (*Clostridium perfringens*) is added to apoE or

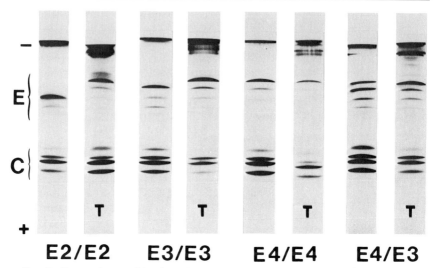

FIG. 2. Cysteamine modification of apoE as an aid in phenotyping. In each pair, the control sample is on the left and the cysteamine-treated sample (T) is on the right. From K. H. Weisgraber, S. C. Rall, Jr., and R. W. Mahley, *J. Biol. Chem.* **256,** 9077 (1981); reproduced with permission.

VLDL at a ratio of about 1 : 20 (w/w) and digestion proceeds at 37° for 4 hr in 0.1 M NH$_4$OAc (pH 4.0 for apoE and pH 6.0 for VLDL). Upon one-dimensional focusing, the treated samples show a marked reduction in the sialylated isoforms, usually sufficient to allow accurate assignment of apoE phenotype. Assessment of sialylated isoforms can also be made without the use of neuraminidase by the two-dimensional method, because the sialylated apoE isoforms migrate with a higher apparent molecular weight in the SDS dimension.[15] The two-dimensional method is actually preferable for this analysis, but is often not as convenient because it is not amenable to analysis of multiple samples.

Sodium Dodecyl Sulfate–Polyacrylamide Gel Electrophoresis. A procedure that has been useful in identifying one particular variant of human apoE is SDS–PAGE.[17] The form of apoE-2 that has cysteine at residue 158 is the apoE-2 with the most severe receptor binding defect and the one most often found associated with the genetic lipid disorder type III hyperlipoproteinemia (familial dysbetalipoproteinemia).[3] On SDS–PAGE (12.5% acrylamide), this apoE migrates with a slightly higher apparent molecular weight than all other human apoE polymorphs, except sialyl-

---

[17] G. Utermann, K. H. Weisgraber, W. Weber, and R. W. Mahley, *J. Lipid Res.* **25,** 378 (1984).

ated species, from which it can be readily distinguished. The key to the success of the method appears to be not only the buffer composition but also the omission of SDS from the upper tray buffer and the gel itself.

Amino Acid Analysis

Of all of the criteria for judging purity of an apoE preparation, the most sensitive is amino acid analysis. This is due in part to the nature of the most frequent contaminant(s), but also to the amino acid composition of human apoE. It is unlikely that impurities will be detected in an apoE preparation (isolated, for example, by Sephacryl S-300 chromatography) by any of the following methods: isoelectric focusing, SDS–PAGE, or amino-terminal analysis. Yet contamination at the level of a few percent is readily detectable by amino acid analysis. The most common contaminant, especially noticeable in apoE preparations from normolipidemic subjects having relatively less apoE in their VLDL, appears to copurify with apoE on column chromatography and goes undetected by isoelectric focusing or SDS–PAGE because it stains poorly in both those procedures (it is probably a glycoprotein). The contaminant cannot be detected by amino-terminal sequence analysis, probably owing to a blocked amino terminus.

As demonstrated in the table, however, amino acid analysis of apoE prepared by various methods gives clearly different results. Apolipoprotein E purified by preparative isoelectric focusing after Sephacryl S-300 chromatography (see the table, columns 4 and 5) yields values closest to the true integral values for each amino acid determined by sequence analysis (see the table, column 6). Owing to the unusual, and therefore characteristic, composition of human apoE, the quantitative values for those amino acids occurring most frequently (particularly arginine and glutamic acid) and least frequently (isoleucine, cysteine, tyrosine, phenylalanine, and histidine) are extremely sensitive to the presence of impurities, which tend to depress the former and increase the latter. It may be significant that a glycoprotein that copurifies with apoE from guinea pig VLDL has a composition relatively enriched in the infrequent apoE amino acids and relatively depleted in the most frequent apoE amino acids.[18]

Sequence Analysis

The amino acid sequence of human apoE is given in Fig. 3. The protein is a single polypeptide of 299 amino acids with a calculated molecular

[18] M. Meng, L. Guo, and R. Ostwald, *Biochim. Biophys. Acta.* **576,** 134 (1979).

AMINO ACID COMPOSITION OF APOLIPOPROTEIN E-3 PREPARED
BY VARIOUS METHODS[a]

Amino acid	1	2	3	4	5	6
Aspartic acid	18.3	14.0	12.8	12.5	12.8	12
Threonine	13.0	11.1	11.2	10.6	10.7	11
Serine	16.3	13.5	13.8	13.6	13.0	14
Glutamic acid	62.5	68.7	69.9	71.0	70.9	71
Proline	9.4	9.5	9.6	8.9	9.2	8
Glycine	17.7	17.5	17.7	17.9	17.9	17
Alanine	30.9	34.1	33.9	36.0	35.2	35
Cysteine	3.0	1.2	1.1	1.3	1.0	1
Valine	19.4	21.7	22.4	22.0	21.8	22
Methionine	5.8	6.3	6.3	6.4	6.0	7
Isoleucine	5.3	2.5	2.2	1.8	2.1	2
Leucine	35.5	37.3	36.6	36.7	37.3	37
Tyrosine	4.7	4.3	4.3	3.4	4.0	4
Phenylalanine	6.0	4.1	3.6	3.1	3.3	3
Lysine	15.1	13.0	11.2	12.1	11.9	12
Histidine	3.6	2.5	2.1	2.1	2.1	2
Arginine	25.8	30.9	33.3	32.7	32.6	34
Tryptophan	ND	ND	ND	ND	ND	7

[a] Analysis of 20-hr hydrolysates (6 N HCl, 110°) with no corrections for hydrolytic destruction or incomplete release. Data expressed as residues/mol, calculated on the basis of 292 residues (tryptophan excluded). Cysteine determined as cysteic acid following performic acid oxidation. Each column represents multiple determinations. Column 1: isolated by Sephacryl S-300 chromatography from $d < 1.02$ g/ml lipoproteins from a normolipidemic subject. Column 2: same as column 1, but from a hypertriglyceridemic (type V) subject. Column 3: same as column 2 plus heparin–Sepharose chromatography. Column 4: same as column 2 plus flat-bed isoelectric focusing. Column 5: same as column 2 plus Immobiline isoelectric focusing. Column 6: composition of apoE-3 from sequence analysis. (ND, not determined.)

weight of 34,200.[19] In apoE, there are several regions that have been proved to be or can be predicted to be functional domains. The receptor binding domain of apoE has been localized to the region encompassing approximately residues 130–160, and a major lipid binding domain resides in the carboxyl-terminal one-third of the polypeptide chain. This latter domain has been predicted based on secondary structure calculations from the amino acid sequence and on the postulate that amphiphilic helices are important lipid binding sites in apolipoproteins. Apolipoprotein E

[19] S. C. Rall, Jr., K. H. Weisgraber, and R. W. Mahley, *J. Biol. Chem.* **257,** 4171 (1982).

```
  1                                                            10                                                           20
Lys-Val-Glu-Gln-Ala-Val-Glu-Thr-Glu-Pro-Glu-Pro-Glu-Leu-Arg-Gln-Gln-Thr-Glu-Trp-
                                                               30                                                           40
Gln-Ser-Gly-Gln-Arg-Trp-Glu-Leu-Ala-Leu-Gly-Arg-Phe-Trp-Asp-Tyr-Leu-Arg-Trp-Val-
                                                               50                                                           60
Gln-Thr-Leu-Ser-Glu-Gln-Val-Gln-Glu-Glu-Leu-Leu-Ser-Ser-Gln-Val-Thr-Gln-Glu-Leu-
                                                               70                                                           80
Arg-Ala-Leu-Met-Asp-Glu-Thr-Met-Lys-Glu-Leu-Lys-Ala-Tyr-Lys-Ser-Glu-Leu-Glu-Glu-
                                                               90                                                          100
Gln-Leu-Thr-Pro-Val-Ala-Glu-Glu-Thr-Arg-Ala-Arg-Leu-Ser-Lys-Glu-Leu-Gln-Ala-Ala-
                                                              110                                                          120
Gln-Ala-Arg-Leu-Gly-Ala-Asp-Met-Glu-Asp-Val-Cys-Gly-Arg-Leu-Val-Gln-Tyr-Arg-Gly-
                                                              130                                                          140
Glu-Val-Gln-Ala-Met-Leu-Gly-Gln-Ser-Thr-Glu-Glu-Leu-Arg-Val-Arg-Leu-Ala-Ser-His-
                                                              150                                                          160
Leu-Arg-Lys-Leu-Arg-Lys-Arg-Leu-Leu-Arg-Asp-Ala-Asp-Asp-Leu-Gln-Lys-Arg-Leu-Ala-
                                                              170                                                          180
Val-Tyr-Gln-Ala-Gly-Ala-Arg-Glu-Gly-Ala-Glu-Arg-Gly-Leu-Ser-Ala-Ile-Arg-Glu-Arg-
                                                              190                                                          200
Leu-Gly-Pro-Leu-Val-Glu-Gln-Gly-Arg-Val-Arg-Ala-Ala-Thr-Val-Gly-Ser-Leu-Ala-Gly-
                                                              210                                                          220
Gln-Pro-Leu-Gln-Glu-Arg-Ala-Gln-Ala-Trp-Gly-Glu-Arg-Leu-Arg-Ala-Arg-Met-Glu-Glu-
                                                              230                                                          240
Met-Gly-Ser-Arg-Thr-Arg-Asp-Arg-Leu-Asp-Glu-Val-Lys-Glu-Gln-Val-Ala-Glu-Val-Arg-
                                                              250                                                          260
Ala-Lys-Leu-Glu-Glu-Gln-Ala-Gln-Gln-Ile-Arg-Leu-Gln-Ala-Glu-Ala-Phe-Gln-Ala-Arg-
                                                              270                                                          280
Leu-Lys-Ser-Trp-Phe-Glu-Pro-Leu-Val-Glu-Asp-Met-Gln-Arg-Gln-Trp-Ala-Gly-Leu-Val-
                                                              290                                                          299
Glu-Lys-Val-Gln-Ala-Ala-Val-Gly-Thr-Ser-Ala-Ala-Pro-Val-Pro-Ser-Asp-Asn-His
```

FIG. 3. The amino acid sequence of human apoE. The most frequently occurring structure (E-3) is shown. Amino acid substitutions have been identified in apoE variants at positions 112, 127, 142, 145, 146, and 158.

has a high degree of secondary structure, particularly α-helix. The calculated α-helical content is 60–70%, depending on the predictive algorithm used; the α-helical content, experimentally determined by circular dichroism, is approximately 45% for human apoE free in solution, and 65% when apoE is bound to phospholipid vesicles.[20]

[20] V. Gordon, K. H. Weisgraber, and R. W. Mahley, unpublished observations (1981).

Intact apoE is rather refractory to Edman degradation by the usual automated method using 0.1 M Quadrol programs. The quality of the degradation deteriorates so rapidly between residues 7 and 15 that the data become uninterpretable by the twentieth cycle. However, when the protein is applied in the presence of 0.5% SDS, the degradation improves enough so that identifications can be made for up to 40–45 cycles. Carboxyl-terminal analysis of apoE is reasonably informative because of the unusual carboxyl-terminal residue (histidine). The terminal histidine can be determined either by hydrazinolysis (although recoveries are low) or by carboxypeptidase digestion. Because of the nature of the carboxyl-terminal sequence of apoE, histidine is the only amino acid released upon digestion with carboxypeptidase A, even at very high enzyme to substrate ratios, which are required for nearly quantitative release of the terminal histidine residue. About 90% of the terminal histidine is released in 4 hr at 37° at enzyme to substrate ratios between 1 : 5 and 1 : 15 in 0.025 M Tris, 0.1 M NaCl (pH 8.5). Quantitation of histidine release is made by direct amino acid analysis of the carboxypeptidase digest.

A useful application of sequencing for human apoE is analysis for the detection and identification of apoE variants. The variants of greatest interest are those having amino acid substitutions that affect the receptor-binding property of the protein.[3] These substitutions almost certainly will occur in the crucial segment of the sequence between residues 130 and 160. This region is contained in a large CNBr peptide of apoE that is relatively easy to isolate.

Purified apoE is cleaved with a 30-fold weight excess of CNBr in 70% HCOOH for 24 hr at 25° (protein concentration of about 5 mg/ml is optimal). After lyophilization, the digestion mixture is redissolved in 2 ml of 20% HCOOH and applied to a 2.5 × 190-cm Sephadex G-50 column (fine beads) and eluted with 0.02 N HCl at room temperature at a flow rate of 15–25 ml/hr. The 93-residue CNBr peptide (residues 126–218) elutes at about 400 ml (45% of the column volume). It may be modestly contaminated (10–20%) with other peptides, but this usually does not interfere with interpretation of subsequent sequence results. The peptide can be conveniently separated from contaminants by repassing the Sephadex G-50 fraction through a 1.2 × 190-cm column of Sephadex G-100, using either 0.02 N HCl or 0.1 M NH_4HCO_3 as eluant. The contaminating peptides elute prior to the main peak containing the 93-residue peptide, which now gives a composition close to the theoretical (sequence-derived) composition: Asp 3.0 (3), Thr 2.0 (2), Ser 3.7 (4), Glu 15.8 (15), Pro 2.2 (2), Gly 9.0 (9), Ala 12.7 (13), Cys, as cysteic acid <0.1 (0), Val 5.4 (5), Met, as homoserine lactone 0.4 (1), Ile 1.0 (1), Leu 14.7 (15), Tyr 1.0 (1), Phe 0 (0), Lys 3.1 (3), His 1.0 (1), Arg 16.5 (17), Trp ND (1). Sequence analysis of this peptide from apoE variants has been used to document amino acid substitutions at residues 127, 142, 145, 146, and 158 (see Fig. 3).

The other apoE CNBr peptides, with one exception, can be isolated in the same fashion (G-50 chromatography). The 51- and 40-residue peptides (residues 222–272 and 69–108, respectively) overlap each other in a broad peak, but reasonably pure peptides can be obtained by judicious pooling of the fractions. The 27- and 17-residue peptides (residues 273–299 and 109–125, respectively) each elute pure and well separated from other peptides. Two small peptides of 4 and 3 residues (residues 65–68 and 219–221, respectively) elute together at the salt (column) volume. The amino-terminal peptide (residues 1–64) cannot be isolated by this method, owing to its tendency to precipitate, to interact with Sephadex beads, and to be susceptible to cleavage by CNBr at its tryptophan residues.

Functional Tests (Receptor Binding)

The most pertinent functional assay of apoE measures its ability to bind to lipoprotein receptors. The most convenient and the most sensitive method is an *in vitro* method for binding to the LDL receptor on cultured human fibroblasts using apoE recombined with phospholipid vesicles. The details of the assay and the preparation of the recombinants are presented elsewhere in this series.[21] Using this assay, dysfunctional variants of human apoE have been identified.[14,22,23] The assay has proved to be sensitive enough and accurate enough that binding defects of less than a factor of 2 are detectable.[23] This powerful tool has been instrumental in making structure–function correlations in human apoE.

[21] T. L. Innerarity, R. E. Pitas, and R. W. Mahley, this series, Vol. 129 [33].
[22] S. C. Rall, Jr., K. H. Weisgraber, T. L. Innerarity, and R. W. Mahley, *Proc. Natl. Acad. Sci. U.S.A.* **79**, 4696 (1982).
[23] S. C. Rall, Jr., K. H. Weisgraber, T. L. Innerarity, T. P. Bersot, R. W. Mahley, and C. B. Blum, *J. Clin. Invest.* **72**, 1288 (1983).

[14] Isolation and Properties of Human Apolipoproteins C-I, C-II, and C-III

By RICHARD L. JACKSON and GEORGE HOLDSWORTH

Introduction

In 1966, Gustafson et al.[1] first reported the isolation of a phospholipid–protein complex from partially delipidated very low-density lipoproteins (VLDL); the protein component was designated "apolipoprotein C" to distinguish it from other proteins present in low- and high-density lipoproteins. Subsequently, the apoCs were isolated from totally delipidated lipoproteins.[2-4] The apoCs, presently designated apoC-I, C-II, and C-III, were isolated by a combination of gel chromatography on Sephadex G-100 and DEAE-cellulose. Since these initial reports, a variety of other methods have been utilized for the preparative isolation of the apoC proteins and their isoforms. These methods include preparative isoelectric focusing (IEF),[5] high-performance liquid chromatography (HPLC),[6-8] and, more recently, chromatofocusing.[9-12]

The purpose of this chapter is to describe in detail the methods used in our laboratory for the isolation of the apoC proteins in a homogeneous state.

Purification

Isolation of Triglyceride-Rich Lipoproteins

Blood from fasting (10 hr) normal subjects or patients with familial hypertriglyceridemia (Types IV and V) is collected with anticoagulant

---

[1] A. Gustafson, P. Alaupovic, and R. H. Furman, *Biochemistry* **5**, 632 (1966).
[2] W. V. Brown, R. I. Levy, and D. S. Fredrickson, *J. Biol. Chem.* **244**, 5687 (1969).
[3] P. N. Herbert, R. S. Shulman, R. I. Levy, and D. S. Fredrickson, *J. Biol. Chem.* **248**, 4941 (1973).
[4] V. G. Shore and B. Shore, *Biochemistry* **12**, 502 (1973).
[5] R. J. Havel, L. Kotite, and J. P. Kane, *Biochem. Med.* **21**, 121 (1979).
[6] G. S. Ott and V. G. Shore, *J. Chromatogr.* **231**, 1 (1982).
[7] W. S. Hancock, C. A. Bishop, A. M. Gotto, D. R. K. Harding, S. M. Lamplugh, and J. T. Sparrow, *Lipids* **16**, 250 (1981).
[8] R. Ronan, L. L. Kay, M. S. Meng, and H. B. Brewer, Jr., *Biochim. Biophys. Acta* **713**, 657 (1982).
[9] G. Knipping, E. Steyrer, R. Zechner, and A. Holasek, *J. Lipid Res.* **25**, 86 (1984).

(EDTA or citrate) by venapuncture or by plasmapheresis. After removal of cells by low-speed centrifugation, EDTA, phenylmethylsulfonylfluoride (PMSF), aprotinin (Tyrsylol), and sodium azide are added to the plasma to final concentrations of 1 mM, 0.5 mM, 50 units/ml, and 0.01%, respectively. The purpose of PMSF and aprotinin is to inhibit proteolysis, particularly the proteolytic activity of thrombin and kallikrein. Sodium azide inhibits bacterial and fungal growth and EDTA prevents oxidative bond cleavage of peptide bonds and lipid peroxidation reactions catalyzed by trace metals.

Triglyceride-rich lipoproteins ($d < 1.02$ g/ml) are isolated by ultracentrifugation in KBr (22.257 g/liter). For the isolation of triglyceride-rich lipoproteins from large volumes, a Beckman 45 Ti rotor which holds 6 tubes (60 ml each) is used; centrifugation is performed at 42,000 rpm for 18 hr at 8°. After centrifugation, the top layer is removed by aspiration; if the subject is hypertriglyceridemic, a gel of lipoprotein may form at the top of the tube. A convenient method for removing this gel is first to remove any solution from the top of the gel and then to remove the gel with a spatula. The triglyceride-rich lipoproteins are resuspended by stirring at 4° in a minimal volume of a standard buffer (10 mM Tris–HCl, pH 7.4, 0.9% NaCl, 0.01% sodium azide, and 1 mM EDTA). The lipoproteins are reisolated by ultracentrifugation: 40 ml of triglyceride-rich lipoproteins is overlayed with 20 ml standard buffer adjusted to d 1.02 g/ml with KBr. After centrifugation as described above, the triglyceride-rich lipoproteins are removed and dialyzed at 4° against standard buffer.

Delipidation of Triglyceride-Rich Lipoproteins

Apolipoproteins are prepared by delipidation with acetone : absolute ethanol. Typically, to triglyceride-rich lipoproteins (150 mg protein in 5 ml of standard buffer) are added 45 ml of acetone : ethanol (1 : 1, v/v); delipidation is performed in 50-ml conical tubes (Kimax) sealed with Teflon caps. After mixing, the delipidation mixture is placed at $-20°$ for 1 hr. The apolipoproteins are then sedimented by low-speed centrifugation for 15 min at 2000 rpm and the solvent is removed by aspiration. The apolipoproteins are redispersed in 45 ml of acetone : ethanol (1 : 1, v/v) by gentle vortexing, and the extraction procedure is repeated until the protein is colorless; typically this requires up to 5 extractions. The sedimented apolipoproteins are next dispersed in anhydrous diethyl ether by gentle vor-

---

[10] J.-F. Tournier, F. Bayard, and J.-P. Tauber, *Biochim. Biophys. Acta* **804**, 216 (1984).
[11] M. Jauhiainen, *Int. J. Biochem.* **14**, 415 (1982).
[12] M. S. Jauhiainen, M. V. Laitinen, I. M. Penttila, and E. V. Puhakainen, *Clin. Chim. Acta* **122**, 85 (1982).

texing at room temperature. After 1 hr at room temperature, the apolipoproteins are pelleted by low-speed centrifugation and the solvent is removed by aspiration. Solubilization buffer is then added to the delipidated protein to give a final concentration of 20 mg/ml (based on the amount of starting lipoproteins). For all purification procedures, fresh solutions of 8 M urea (analytical grade) are prepared weekly, deionized on a mixed-bed resin of Rexyn I-300 (Fisher Scientific Co.), and stored at 4°. After removing small amounts of ether by nitrogen evaporation, the protein mixture (insoluble apoB is present at this stage) is stirred overnight at 4°. The insoluble protein is then removed by centrifugation at 18,000 rpm for 30 min. The urea-soluble proteins (mainly apoCs, apoA-I, and apoE) are removed and stored at 4°. The protein pellet is redissolved in a further volume of the solubilization buffer and the extraction procedure is repeated. The urea-soluble fractions are pooled and the apoCs are further purified as described below.

Sephadex G-75 Gel Filtration of Urea-Soluble Apolipoproteins

The apoCs are separated from higher molecular weight components by chromatography on Sephadex G-75 Superfine (Pharmacia). The column (2.9 × 90 cm) is equilibrated at 4° with 50 mM Tris–HCl, pH 8.6, 6 M urea. The sample (150 mg protein/15 ml) is applied to the column and eluted under gravity with the solubilization buffer. Protein is detected by absorbance at 280 nm. Figure 1 shows a typical Sephadex G-75 elution profile of the urea-soluble apolipoproteins. Fractions containing mainly the apoC proteins are pooled (as indicated in Fig. 1), dialyzed extensively against 10 mM ammonium bicarbonate, using tubing with a 3500 molecular weight cut off, and finally lyophilized. Analytical IEF[13] of the total apoC fraction is shown in Fig. 1, inset.

DEAE-Sephacel Chromatography of ApoCs

The lyophilized apoCs (100–150 mg protein) obtained from chromatography on Sephadex G-75 are dissolved in 6 M urea to give a protein concentration of 10 mg/ml. The pH is adjusted to 8.0 at 25°; the conductivity should be <0.5 mS. If the conductivity is >0.5 mS, the sample is diluted with 6 M urea or dialyzed against the equilibration buffer. The sample is next applied at 4° to a column (2.5 × 35 cm) of DEAE-Sephacel (Pharmacia) previously equilibrated with 10 mM Tris–HCl, pH 8.0, 6 M urea (the pH is 8.0 at 25°). After the sample enters the resin, the column is washed with the equilibration buffer until apoC-I is eluted (Fig. 2); with

---

[13] M. L. Kashyap, B. A. Hynd, K. Robinson, and P. S. Gartside, *Metabolism* **30**, 111 (1981).

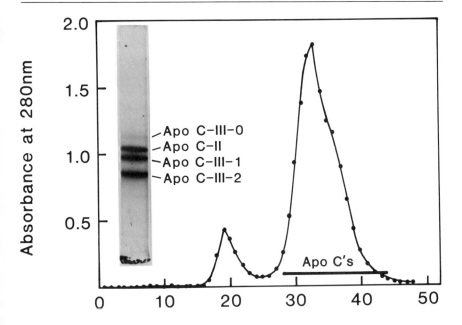

FIG. 1. Sephadex G-75 chromatography of urea-soluble apolipoproteins obtained from triglyceride-rich lipoproteins. The fractions corresponding to the apoCs were pooled and applied to DEAE-Sephacel (Fig. 2). The inset shows an IEF gel (pH 4–8) of the pooled apoCs; apoC-I has a pI > 8 and migrates to the cathode.

these chromatography conditions, apoC-I does not bind to DEAE-Sephacel. A linear 1 liter gradient of 0–0.125 M NaCl in the equilibration buffer is used to elute apoC-II and apoC-III. The flow rate for sample application and elution is 25 ml/hr. Analytical IEF of representative fractions of apoC-II across the elution profile is shown in Fig. 2. The fractions corresponding to apoC-I, C-II, C-III-1, and C-III-2 are pooled, dialyzed against 10 mM NH$_4$HCO$_3$, and lyophilized.

Other Purification Procedures

The criteria of purity for apoC-I is the absence of tyrosine as determined by amino acid analysis. Should the sample contain small amounts of tyrosine, the protein is subjected to rechromatography on Sephadex G-75 in 6 M urea as described above.

As is shown in Fig. 2, the elution profile for apoC-II is complicated by the appearance of multiple bands as determined by analytical IEF. How-

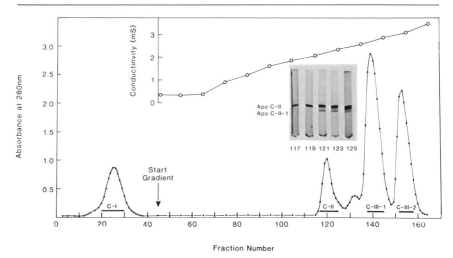

FIG. 2. DEAE-Sephacel chromatography of apoCs. The fractions corresponding to the apoCs (Fig. 1) were pooled and subjected to chromatography on DEAE-Sephacel as described in the text. The insets show the IEF gels (pH 4-8) of the appropriate fractions.

ever, by performing IEF on each fraction across the peak, it is usually possible to pool those fractions which give a single band (pI 4.78), particularly if a single donor is used for isolation of triglyceride-rich lipoproteins. The minor apolipoproteins in the apoC-II peak (Fig. 2) correspond to apoC-III-0, apoC-III-1, and an isoform of apoC-II, termed[5] apoC-II-1. The amount of apoC-II-1 (pI 4.57) is variable and appears to depend on the particular donor and not the purification procedure (unpublished observations). Since apoC-II-1 activates lipoprotein lipase to the same extent as the major forms of apoC-II,[5] absolute purity of apoC-II as determined by IEF is not critical. However, if a single protein is required we routinely use preparative isoelectric focusing or HPLC utilizing a Radial Pak C_{18} column, as described below. In our experience, it is possible to obtain homogeneous apoC-III-1 (pI 4.70) and apoC-III-2 (pI 4.56) by pooling the appropriate fractions from the DEAE-Sephacel column (Fig. 2) that are pure as determined by IEF.

Preparative Isoelectric Focusing. Preparative isoelectric focusing of impure apoC-II is carried out essentially as described in the LKB Application Note 198 (1975). Impure apoC-II obtained from DEAE-Sephacel is equilibrated in 3 ml of 10 mM Tris–HCl, pH 8.2, containing 0.5% sodium decyl sulfate (Eastman Kodak). To load the sample, an 8-cm section of gel is removed 4 cm from the cathode end with the aid of a section cutter. The sample is then mixed with the Ultrodex cut from the gel and the mixture is

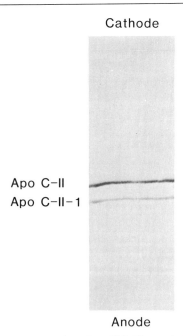

FIG. 3. Preparative isoelectric focusing of apoC-II. The pooled fraction corresponding to apoC-II (Fig. 2) was subjected to preparative isoelectric focusing as described in the text. A paper print of the gel was made and stained with Coomassie brilliant blue R.

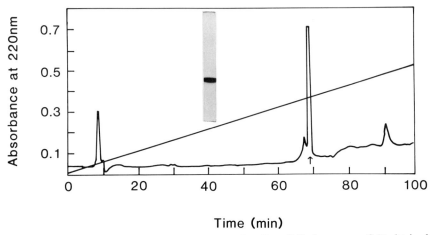

FIG. 4. High-performance liquid chromatography of apoC-II. Impure apoC-II obtained from DEAE-Sephacel (Fig. 2) was applied to a preparative C_{18} Radial Pak column (1.8 × 30 cm) and eluted as described in the text.

```
THR-PRO-ASP-VAL-SER-SER-ALA-LEU-ASP-LYS-LEU-LYS-GLU-PHE-GLY
            5               10              15

ASN-THR-LEU-GLU-ASP-LYS-ALA-ARG-GLU-LEU-ILE-SER-ARG-ILE-LYS
           20               25              30

GLN-SER-GLU-LEU-SER-ALA-LYS-MET-ARG-GLU-TRP-PHE-SER-GLU-THR
           35               40              45

PHE-GLN-LYS-VAL-LYS-GLU-LYS-LEU-LYS-ILE-ASP-SER
           50              55    57
```

FIG. 5. Amino acid sequence of human apoC-I as determined by Shulman et al.[14] and Jackson et al.[15]

poured back into the section cutter; the cutter is then carefully removed from the gel. Electrofocusing is performed for 16 hr at 6 W. After focusing is complete, a surface electrode is used to directly measure the pH of the Ultrodex at 1 cm intervals. If a paper print is required (Fig. 3), the protocol given in the LKB Application Note should be followed. However, since the focused peptides are highly concentrated in the gel, the bands can be visualized directly using a dark background and side lighting. The areas of gel containing the apoC-II bands are carefully removed from the tray and dispersed in a minimum volume of 10 mM NH$_4$HCO$_3$. The Ultrodex slurry is then layered onto the gel surface of a column (2.5 × 35 cm) of Sephadex G-50 (Pharmacia), equilibrated in 10 mM NH$_4$HCO$_3$. The apolipoprotein is then eluted with 10 mM NH$_4$HCO$_3$; fractions containing apoC-II are pooled and lyophilized.

High-Performance Liquid Chromatography. Impure apoC-II from DEAE-Sephacel (pooled peak, Fig. 2) can also be fractionated by HPLC (Fig. 4). The protein is applied to a preparative column (1.8 × 30 cm) of C$_{18}$ Radial Pak (Waters) and eluted with a 100 ml linear gradient of water and n-propanol : methanol (60 : 40, v/v). The flow rate is 1 ml/min. In Fig. 4, the major peak was confirmed to be apoC-II by amino acid analysis and by its ability to activate lipoprotein lipase. It was homogeneous as determined by analytical IEF.

Characterization and Properties

ApoC-I. Human apoC-I contains 57 amino acid residues whose sequence was determined from the protein (Fig. 5).[14,15] Its calculated molec-

---

[14] R. S. Shulman, P. N. Herbert, K. Wehrly, and D. S. Fredrickson, *J. Biol. Chem.* **250**, 182 (1975).
[15] R. L. Jackson, J. T. Sparrow, H. N. Baker, J. D. Morrisett, O. D. Taunton, and A. M. Gotto, *J. Biol. Chem.* **249**, 5308 (1974).

```
THR-GLN-GLN-PRO-GLN-GLN-ASP-GLU-MET-PRO-SER-PRO-THR-PHE-LEU
                5                  10                  15

THR-GLN-VAL-LYS-GLU-SER-LEU-SER-SER-TYR-TRP-GLU-SER-ALA-LYS
               20                  25                  30

THR-ALA-ALA-GLN-ASN-LEU-TYR-GLU-LYS-THR-TYR-LEU-PRO-ALA-VAL
               35                  40                  45

ASP-GLU-LYS-LEU-ARG-ASP-LEU-TYR-SER-LYS-SER-THR-ALA-ALA-MET
               50                  55                  60

SER-THR-TYR-THR-GLY-ILE-PHE-THR-ASP-GLN-VAL-LEU-SER-VAL-LEU
               65                  70                  75

LYS-GLY-GLU-GLU
      79
```

FIG. 6. Amino acid sequence of human apoC-II as determined by Jackson et al.,[16] Hospattankar et al.,[17] and Myklebost et al.[18]

AMINO ACID COMPOSITION OF HUMAN ApoCs

Amino acid	ApoC-I	ApoC-II	ApoC-III
Asp	4	4	7
Asn	1	1	—
Thr	3	9	5
Ser	7	9	11
Glu	7	7	5
Gln	2	7	5
Pro	1	4	2
Gly	1	2	3
Ala	3	6	10
Cys	—	—	—
Val	2	4	6
Met	1	2	2
Ile	3	1	—
Leu	6	8	5
Tyr	—	5	2
Phe	3	2	4
His	—	—	1
Lys	9	6	6
Arg	3	1	2
Trp	1	1	3
Total	57	79	79

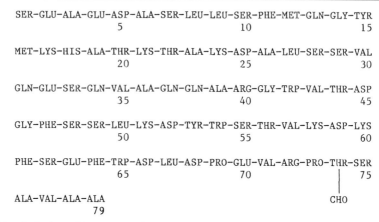

FIG. 7. Amino acid sequence of human apoC-III as reported by Karathanasis *et al.*[19] The carbohydrate attachment at position 74 (Thr) consists of 1 residue each of galactose and galactosamine and either 0, 1, or 2 residues of sialic acid.[20]

ular weight is 6613. The apolipoprotein lacks cysteine, histidine, and tyrosine (see the table).

ApoC-II. Human apoC-II, the activator protein for lipoprotein lipase, contains 79 amino acid residues (Fig. 6).[16–18] The amino acid sequence has been determined from the protein[17] and DNA.[16,18] Its calculated molecular weight is 8826. The protein lacks cysteine and histidine.

ApoC-III. Human apoC-III is a 79 amino acid glycoprotein of known primary structure. The amino acid sequence (Fig. 7) as determined from the DNA[19] is slightly different than that reported earlier from the protein sequence.[20] The protein has an amino acid molecular weight of 8746 and lacks cysteine and isoleucine. The carbohydrate moiety is attached to residue 74 by an O-glycosidic linkage to Thr. The carbohydrate attachment is 1 residue each of galactose and galactosamine and either 0 (apoC-III-0), 1 (apoC-III-1), or 2 (apoC-III-2) residues of sialic acid.

[16] C. L. Jackson, G. A. P. Bruns, and J. L. Breslow, *Proc. Natl. Acad. Sci. U.S.A.* **81,** 2945 (1984).
[17] A. V. Hospattankar, T. Fairwell, R. Ronan, and H. B. Brewer, Jr., *J. Biol. Chem.* **259,** 318 (1984).
[18] O. Myklebost, B. Williamson, A. F. Markham, S. R. Myklebost, J. Rogers, D. E. Woods, and S. E. Humphries, *J. Biol. Chem.* **259,** 4401 (1984).
[19] S. K. Karathanasis, V. I. Zannis, and J. L. Breslow, *J. Lipid Res.* **26,** 451 (1985).
[20] H. B. Brewer, Jr., R. Shulman, P. Herbert, R. Ronan, and K. Wehrly, *J. Biol. Chem.* **249,** 4975 (1974).

Acknowledgments

This research was supported by Public Health Service Grants P01 HL-22619, R01 HL-23019, and General Clinical Research Center and CLINFO Grant NIH RR0-00068. Special thanks go to Ms. Gwen Kraft for preparing the figures and to Ms. Janet Simons for preparing the manuscript for publication.

[15] Isolation and Characterization of Other Apolipoproteins

By W. J. McConathy and P. Alaupovic

Human plasma lipoproteins constitute a structurally complex and metabolically dynamic system of macromolecules. The complexity of lipoproteins is not only reflected in their marked heterogeneity with respect to hydrated density, size, and electrical charge, but also in a number of specific protein moieties or apolipoproteins. To provide an integrated view of this system, we have proposed that apolipoproteins be used as the specific and distinguishing markers for identification and characterization of discrete lipoprotein particles.[1] To express in relatively simple terms the relationship between apolipoproteins, constitutive polypeptides, isomorphic forms, and lipoprotein particles, we have devised a system of nomenclature referred to as the ABC nomenclature.[1] The apolipoproteins are designated by capital letters, the nonidentical polypeptides by Roman numerals, and the polymorphic forms by Arabic numbers. The lipoprotein particles or lipoprotein families are named according to their corresponding apolipoproteins.

The purpose of this chapter is to describe the isolation and physicochemical and immunologic properties of apolipoproteins that are not related to apolipoproteins A-I, A-II, B, C-I, C-II, C-III, and E. This includes minor apolipoproteins D, F, G, and H.

Apolipoprotein D

Apolipoprotein D (apoD) represents a minor protein constituent of the human plasma lipid transport system. Studies in our and other laboratories indicated the presence in HDL of an antigenic determinant that was not related to any of the known apolipoproteins.[2,3] This protein was de-

---

[1] P. Alaupovic, *Ricerca* **12**, 3 (1982).
[2] W. J. McConathy and P. Alaupovic, *FEBS Lett.* **37**, 178 (1973).
[3] W. J. McConathy and P. Alaupovic, *Biochemistry* **15**, 515 (1976).

tected in plasma lipoproteins with both commercial anti-α_1-lipoprotein and anti-HDL_3 sera. Using these antisera, the following scheme was developed for the isolation and characterization of this antigen.

Preparation of ApoHDL[4]

To prepare HDL, 2 liters of pooled fresh or outdated plasma was made 0.01% with respect to sodium azide, EDTA, and thiomerosal. This volume of plasma is necessary in order to obtain sufficient quantities of minor apolipoproteins for further characterization studies. The plasma was adjusted to a solution density of 1.23 g/ml with solid KBr and centrifuged at 45,000 rpm in a Ti60 rotor for 22 hr. The supernate containing the serum lipoproteins was subjected to additional ultracentrifugations with appropriate density adjustments to yield HDL (d 1.073–1.23 g/ml). After several washes at $d = 1.23$ g/ml the HDL were usually free of albumin as demonstrated by double diffusion analysis, though its absence does not appear to be essential for the isolation of either apoD or some other minor apolipoproteins. The composition of HDL isolated by this procedure was indistinguishable from that of normal fasting HDL isolated by standard sequential ultracentrifugation.

HDL were dialyzed against five changes of redistilled water and lyophilized in a number of 50-ml glass-stoppered centrifuge tubes. Lyophilized HDL (1.2–1.6 g of protein, 150 mg/tube) were mixed with 1 volume of chloroform and shaken by inverting the tube several times. Following dispersion of the lyophilized material, 2 volumes of methanol was added. The tubes were shaken several times while stored at $-10°$ for 30 min in chloroform : methanol (1/2, v/v). After low-speed centrifugation at 6°, the solvent was aspirated and the precipitated protein was dispersed in 1 volume of chloroform and 2 volumes of MeOH in the same fashion as the initial step. The residue was extracted four additional times with chloroform : methanol (2/1, v/v) followed by two extractions with peroxide-free diethyl ether. The delipidized HDL (apoHDL) was essentially free of phosphorus and fatty acids.

Isolation of ApoD[2]

To isolate apoD, the apoHDL was dissolved in 10–20 ml of 8 M urea in 1 mM K_2HPO_4, pH 8.0, and the ether was evaporated under a gentle stream of nitrogen. The apoHDL solution was diluted to 2 M with respect to urea by addition of 1 mM K_2HPO_4, pH 8.0, and applied to a hydroxylapatite-cellulose column. The column (30 × 2.2 cm) was packed to a

[4] S.-O. Olofsson, W. J. McConathy, and P. Alaupovic, *Biochemistry* **17**, 1032 (1978).

height of 15 cm with a mixture of 2 volumes of settled hydroxylapatite (Bio-Rad Lab., Richmond, CA) and 1 volume of settled microcrystalline cellulose (Baker, Phillipsburg, NJ). The addition of the microcrystalline cellulose improves the flow rate of hydroxylapatite columns. Prior to sample application, the column was washed with 1 M K_2HPO_4, pH 8.0, followed by equilibration with 1 mM K_2HPO_4, pH 8.0. After application of the sample, the column was eluted with 40–50 ml of 1 mM K_2HPO_4 buffer, pH 8.0. The fraction eluted with 1 mM K_2HPO_4 was rechromatographed under identical conditions on another hydroxylapatite column. After the second chromatography, the eluted fraction was examined by basic polyacrylamide gel electrophoresis (PAGE). The two successive chromatographic steps on hydroxylapatite-cellulose columns were usually sufficient to yield a protein preparation which was characterized by a single band on 7% PAGE. However, in some cases a minor component was present near the junction of the separating and stacking gels. To remove this minor component and urea, gel filtration on Sephadex G-100 equilibrated with 2 M acetic acid was performed and apoD was eluted as a major symmetrical peak. After lyophilization, this major protein fraction only reacted with an antiserum to apoD. On double diffusion analyses, it gave a negative reaction with antibodies to apolipoproteins A-I, A-II, B, C-I, C-II, C-III, E, F, and albumin. On basic 7% PAGE, the isolated polypeptide exhibited a single band with a mobility midway between the bands corresponding to apoA-II and apoC-II. We concluded on the basis of electrophoretic mobility in 7% PAGE, characteristic amino acid and carbohydrate composition, and immunological properties that the isolated protein is a distinct component of the human plasma lipoprotein system. This protein was named apolipoprotein D.

Isolation of Lipoprotein D (LP-D)[3]

In order to define the lipoprotein nature of this isolated protein moiety and to provide evidence that it indeed represents an apolipoprotein according to the criteria proposed by Alaupovic,[1] the following procedure was developed for the isolation of apoD-containing lipoproteins utilizing immunosorber methodology.[3]

Rabbit or goat antiserum specific for apoD was utilized for the isolation of an IgG antibody-containing fraction. The ammonium sulfate-precipitated IgG fraction was dialyzed against 100 mM K_2HPO_4, pH 6.5, and coupled to CNBr-activated Sepharose 4B[5] at pH 6.5. After coupling, the immunosorber was washed extensively with an equilibration buffer (150

---
[5] S. March, I. Parikh, and P. Cuatrecasas, *Anal. Biochem.* **60**, 149 (1974).

mM NaCl, 50 mM Tris–HCl, and 0.01% NaN$_3$, pH 7.5). To avoid prolonged interaction of the dissociating buffer (3 M NaSCN) with lipoproteins eluted from the immunosorber, the chromatography was performed on a column (60 × 2.5 cm) constructed in the following fashion: the column was first packed to a height of 15 cm with medium grade Sephadex G-25 followed by the Sepharose-coupled antibodies (40 cm) and another layer of Sephadex G-25 (4 cm). The G-25 layer at the bottom acts as a molecular sieve and effectively separates the desorbed protein from the dissociation agent, NaSCN. Application of plasma or isolated density classes was followed by extensive washing with the equilibration buffer until the absorbance returned to the baseline. Material bound to the antiapoD-Sepharose column was eluted successively with 50 ml of 3 M NaSCN followed by the equilibration buffer. Approximately 75% of the total bound material was eluted with the equilibration buffer, while the remainder was eluted with NaSCN. Both fractions were first made 1 mM with respect to phosphate by chromatography on a Sephadex G-25 column (60 × 2.5 cm) equilibrated with 1 mM KH$_2$PO$_4$, pH 8.0, and were then applied onto a hydroxylapatite-cellulose column prepared as described for apoD. Fractions eluted with 1 mM KH$_2$PO$_4$, pH 8.0, if not homogeneous, were rechromatographed until a single band was obtained. After purification, the fractions eluted with 1 mM KH$_2$PO$_4$ were desalted by gel filtration on Sephadex G-25 equilibrated with 100 mM (NH$_4$)$_2$CO$_3$ and utilized for characterization studies. Lipoproteins containing apoD, apoA-I, and apoA-II can be eluted from the initial hydroxylapatite column with higher molarities of phosphate buffer.

Characterization of ApoD

ApoD, isolated by a procedure combining hydroxylapatite and Sephadex G-100 column chromatography, migrated on 7% PAGE as a single band with a mobility intermediate between the bands corresponding to apoA-II and apoC-II.[2,3,6–9] On double diffusion and immunoelectrophoresis, apoD only reacted with antiserum to apoD. It was characterized by the presence of all common amino acids including half-cystine (Table I). By analytical isoelectric focusing, apoD was shown to exist in at least three isoforms with pIs of 5.20, 5.08, and 5.00, with the isoform of pI =

[6] S.-O. Olofsson and A. Gustafson, *Scand. J. Clin. Lab. Invest.* **33** (Suppl. 137), 57 (1974).
[7] I. Chajek and C. J. Fielding, *Proc. Natl. Acad. Sci. U.S.A.* **75**, 3445 (1978).
[8] O. Wiklund, G. Fager, S.-O. Olofsson, C. Wilhelmsson, and G. Bondjers, *Atherosclerosis* **37**, 631 (1980).
[9] J. J. Albers, M. C. Cheung, S. L. Ewens, and J. H. Tollefson, *Atherosclerosis* **39**, 395 (1981).

TABLE I
AMINO ACID COMPOSITION OF OTHER APOLIPOPROTEINS

	ApoD[a]	ApoF[b]	ApoG[c]	ApoH (β_2-Glycoprotein-I) Plasma[d]	ApoH (β_2-Glycoprotein-I) Plasma[e]
Lysine	65	33	65	109	106
Histidine	11	2	20	18	26
Arginine	20	17	31	36	40
Aspartic acid	106	70	96	105	116
Threonine	53	44	37	100	105
Serine	36	60	37	82	95
Glutamic acid	100	100	100	100	100
Proline	55	18	49	123	113
Glycine	34	87	50	82	91
Alanine	49	90	66	64	79
Half-cystine	20	10	N.D.	82	78
Valine	58	54	55	68	67
Methionine	10	12	Tr.	14	14
Isoleucine	51	28	30	45	54
Leucine	68	69	80	68	70
Tyrosine	28	31	19	54	50
Phenylalanine	28	4	36	64	75
Tryptophan	19	N.D.	N.D.	18	6

[a] W. J. McConathy and P. Alaupovic, *Biochemistry* **15**, 515 (1976).
[b] S. O. Olofsson, W. J. McConathy, and P. Alaupovic, *Biochemistry* **17**, 1032 (1978).
[c] M. Ayrault-Jarrier, J. F. Alix, and J. Polanovski, *Biochimie* **60**, 65 (1978).
[d] N. Heimberger, K. Heide, H. Haupt, and H. E. Schultze, *Clin. Chim. Acta* **10**, 293 (1964).
[e] E. Polz, H. Wurm, and G. M. Kostner, *Artery* **9**, 305 (1981).

5.08 representing the major form.[9] The chemical basis for this heterogeneity remains unknown. The amino terminal acid was blocked.[3] Carbohydrate analysis demonstrated that apoD is a glycoprotein with hexose, glucosamine, and sialic acid accounting for 18% of its dry weight (Table II). By applying similar procedures, several groups have isolated a protein constituent from HDL with similar characteristics.[6–9] The major reported discrepancy between these and the initial report[3] is the apparent molecular weight as estimated by SDS–PAGE. Our reevaluation of the molecular weight of apoD by SDS–PAGE now agrees with the recent reports of an estimated molecular weight in the range of 32,000–34,000.[7,9] The reason for the original discrepancy remains unclear but may be related to the high carbohydrate content of apoD. However, the estimated molecular weight of apoD from the amino acid and carbohydrate composition is 22,100, a

TABLE II
CARBOHYDRATE CONTENT OF APOLIPOPROTEIN
D AND APOLIPOPROTEIN H (β_2-GLYCOPROTEIN-I)

	ApoD[a] (% weight)	ApoH[b] (% weight)
Hexose	9.3	6.7
Hexosamine	4.5	5.8
Neuraminic acid	4.8	4.4

[a] W. J. McConathy and P. Alaupovic, *Biochemistry* **15**, 15 (1976).
[b] N. Heimburger, K. Heide, H. Haupt, and H. E. Schultze, *Clin. Chim. Acta* **10**, 293 (1964).

value consistent with the elution volume of apoD when sized by molecular sieve chromatography in the presence of 6 M guanidine–HCl on Sephadex G-100 column chromatography.

Lipoprotein D (LP-D) was isolated by a procedure combining chromatography of HDL or whole serum on an immunosorber containing antibodies to apoD, and hydroxylapatite column chromatography. LP-D displayed a single, symmetrical boundary in the analytical ultracentrifuge and a single band on 7% PAGE. When injected into rabbits it produced antisera that only reacted with apoD. LP-D consists of 65–75% protein and 25–35% lipid. The lipid moiety contains cholesterol, cholesteryl ester, triglyceride, and phospholipid (Table III). The phospholipid composition is characterized by a relatively high content of lysolecithin and sphingomyelin and a relatively low content of lecithin. By the same proce-

TABLE III
LIPID COMPOSITION OF LP-D AND LP-F

	LP-D[a] (%)	LP-F[b] (%)
Triglyceride	8.2 (6.2)	5.2 (1.0)
Cholesterol	18.1 (1.0)	21.1 (0.6)
Cholesteryl ester	27.7 (7.6)	63.3 (3.2)
Phospholipid	46.5 (8.2)	11.5 (2.5)

[a] W. J. McConathy and P. Alaupovic, *Biochemistry* **15**, 515 (1976).
[b] E. Koren, W. J. McConathy, and P. Alaupovic, *Biochemistry* **21**, 5347 (1982).

dures, a similar lipoprotein species was isolated from baboon plasma.[10] These results demonstrated that other species contain a plasma apolipoprotein analogous to human apoD.

Distribution and Levels of ApoD

Plasma concentrations and distribution of apoD in ultracentrifugally defined lipoprotein density classes (VLDL, LDL, and HDL) were determined by various immunoassays. Studies on plasma levels have shown that apoD represents a minor apolipoprotein with reported levels for normolipidemic subjects ranging from 6 to 10 mg/dl.[8,9,11] Distribution studies indicated that apoD is primarily localized in HDL (60–65%) with only trace amounts present in VLDL and LDL, and the remainder in VHDL.[11] Such analyses are consistent with HDL as the starting material for isolating apoD or LP-D, though apoD can also be isolated from VLDL and LDL.[3]

We concluded from these studies that apoD is a unique apolipoprotein that exists in the form of a distinct lipoprotein family with a macromolecular distribution extending from very low-density lipoproteins into very high-density lipoproteins, but with a maximum concentration in high-density lipoproteins.[3]

Functional Aspects of ApoD

The functional role of apoD in the metabolism of plasma lipoproteins remains unclear. Several different lines of evidence have linked apoD with lecithin: cholesterol acyltransferase (LCAT), as a component of the same macromolecular complex,[11,12] an activator of LCAT,[13] and a substrate/product lipoprotein.[6] In addition, apoD has been characterized as a transfer protein mediating the movement of cholesterol ester from HDL to LDL or VLDL accompanied by a reciprocal reverse transfer of triglyceride from VLDL and LDL to HDL.[7] This observation has been disputed by other investigators,[9,14] though it seems plausible that there may be several different proteins participating in the movement of lipids between different lipoprotein species. Confirmatory evidence from other

---

[10] D. Bojanovski, P. Alaupovic, W. J. McConathy, and J. L. Kelly, *FEBS Lett.* **112,** 251 (1980).
[11] M. D. Curry, W. J. McConathy, and P. Alaupovic, *Biochim. Biophys. Acta* **491,** 232 (1977).
[12] P. E. Fielding and C. J. Fielding, *Proc. Natl. Acad. Sci. U.S.A.* **77,** 3327 (1980).
[13] G. Kostner, *Scand. J. Clin. Lab. Invest.* **33** (Suppl. 137), 19 (1974).
[14] R. E. Morton and D. B. Zilversmit, *Biochim. Biophys. Acta* **663,** 350 (1981).

investigative groups will be necessary to substantiate these different views on the role of apoD in lipoprotein metabolism.

Apolipoprotein F

In addition to the well-characterized apolipoproteins present in HDL, electrophoretic and immunochemical analyses of apoHDL revealed the occurrence of protein constituents with properties differing from those of the known apolipoproteins. In order to clarify the nature of one of these proteins, the following procedures were utilized to isolate and partially characterize one of these constituents which we designated apolipoprotein F (apoF).

Isolation of ApoF [4]

The apoHDL, prepared as described under the apoD section, was solubilized in 2 M acetic acid and the diethyl ether was evaporated under a gentle stream of nitrogen. Essentially all of the apoHDL was soluble after the addition of ultrapure urea to approximately 2 M with respect to urea. ApoHDL (1.2–1.6 g) in a volume of 50–60 ml was applied to a Sephadex G-100 column (5.0 × 150 cm) equilibrated with 2 M acetic acid (flow rate, 30 ml/hr). Thirteen-milliliter fractions were collected and monitored at 280 nm. Fractions were combined on the basis of elution pattern and lyophilized. The lyophilized void volume fraction was utilized for the next fractionation step on carboxymethylcellulose (CM52, Whatman). The microcrystalline CM-cellulose was washed with 5 mM phosphoric acid, pH 3.5, until the slurry had a pH of 3.5. The CM-cellulose was poured into a column (1.5 × 25 cm). Prior to use, the ion-exchange bed was equilibrated with the eluting buffer 1 mM KH_2PO_4, pH 3.5, in 6 M urea. The urea solution was deionized on a mixed bed ion-exchange resin prior to addition of the KH_2PO_4 and adjustment of pH.

The lyophilized void volume fraction was dissolved in 1 mM KH_2PO_4, pH 3.5, containing 6 M urea, and chromatographed at 6° on carboxymethylcellulose (flow rate, 25 ml/hr) using the solubilizing buffer to elute the unretained fraction. The unretained fraction was eluted in a volume of 60–80 ml. Chromatography of this fraction was repeated under identical conditions until it was free of apoA-I and apoA-II as demonstrated by double diffusion analyses. Two chromatographies over carboxymethylcellulose were sufficient to yield a homogeneous preparation of apoF. This protein preparation gave no reaction with antisera to apolipoproteins A-I, A-II, B, C-I, C-II, C-III, D, or E. The apoF was desalted on a Sephadex G-25 column equilibrated with 2 M acetic acid. This material was lyophilized and used for all subsequent analyses. The yield of apoF from 2 liters of plasma was approximately 2–4 mg.

Characterization of ApoF[4]

The described isolation procedure yielded a protein which migrated as a single band on basic PAGE in a position similar to apoD. Amino acid analysis demonstrated the presence of all common amino acids except tryptophan (Table I). The molecular weight was estimated to be 28,000. Isoelectric focusing gave a relatively low isoelectric point ($pI = 3.7$) with no indication of microheterogeneity by this technique. This low isoelectric point suggests that apoF may represent the most acidic apolipoprotein of the human plasma lipoprotein system. A monospecific antiserum to apoF only reacted with apoF. The apolipoprotein nature of this polypeptide was indicated by the uptake of Oil Red O by the precipitin arcs formed when anti-apoF serum was reacted against LDL and HDL. Nonidentity reactions between the known lipoprotein families and the lipoprotein form of apoF were indications that apoF is the protein moiety of a distinct lipoprotein family designated according to the ABC nomenclature lipoprotein F (LP-F).

Isolation and Characterization of Lipoprotein F (LP-F)[15]

In order to provide more direct evidence for the lipoprotein nature of apoF, the lipoprotein forms of apoF were investigated by the use of immunosorbers. The IgG fractions of antisera to apolipoproteins A-I, A-II, F, and apoF-free plasma were isolated, coupled to Sepharose, and the corresponding immunosorbers were constructed as described in the apoD section. Using an anti-apoF immunosorber, the apoF-containing lipoproteins were present in the retained fraction eluted from the immunosorber. Crossed immunoelectrophoretic patterns with an antiserum to apoF showed no difference between the retained fraction and whole plasma. In addition to apoF, only small but constant amounts of apoA-I and apoA-II were always found in the retained fraction.

By passing this retained fraction over the immunosorbers constructed with antibodies to apoA-I, apoA-II, and apoF-free plasma, it was possible to demonstrate in the unretained fraction by double diffusion analyses, electroimmunoassay, and basic PAGE a lipoprotein species that only contained apoF as the protein moiety. The determination of the lipid composition showed the prevalence of cholesteryl esters (Table III). On

---

[15] E. Koren, W. J. McConathy, and P. Alaupovic, *Biochemistry* **21**, 5347 (1982).

crossed immunoelectrophoresis against an anti-apoF serum, lipoprotein F showed β-lipoprotein mobility.

Distribution and Functional Aspects of ApoF

There are no published reports on the plasma concentrations of apoF though preliminary results from this laboratory by electroimmunoassay indicate apoF levels of 2.7 mg/100 ml with the major amount occurring in HDL (75–80%) followed by LDL (15–20%) and traces in VLDL and VHDL ($d > 1.21$ g/ml). Currently, no information is available on the functional role of apoF.

Apolipoprotein H (β_2-Glycoprotein-I)

In the early 1960s, a previously unrecognized β_2-globulin was isolated from human serum.[16] Further studies led to the designation of this component as β_2-glycoprotein-I which distinguished it from a similar protein of lower carbohydrate content, β_2-glycoprotein-II.[17] Recent investigations by several groups have described the interaction of β_2-glycoprotein-I with lipoproteins, particularly the triglyceride-rich lipoproteins.[18–20] Based on these observations β_2-glycoprotein-I was designated as apolipoprotein H (apoH).[20] In the following presentation, β_2-glycoprotein-I will be referred to as apoH.

Isolation of ApoH

Several different variations of similar procedures have been used to isolate apoH from both plasma and triglyceride-rich lipoproteins. Such procedures have included precipitation of serum proteins with Rivanol (2-ethoxy-6,9-diaminoacridine lactate) or various acids,[21] ultracentrifugation, and a variety of chromatographic steps taking advantage of the size and charge characteristics of apoH.[19,20]

As a first step in the isolation of apoH from human plasma, most procedures have utilized the precipitation of the bulk of plasma proteins with 0.2 N perchloric acid which leaves a substantial quantity of plasma glycoproteins, including apoH, in the supernate. As previously outlined,[21] 200 ml of acid–citrate–dextrose (ACD)-treated plasma is acidified with 25

---

[16] H. E. Schultze, K. Heide, and H. Haupt, *Naturwissenschaften* **48**, 719 (1961).
[17] H. Haupt and K. Heide, *Clin. Chim. Acta* **12**, 419 (1965).
[18] M. Burstein and P. Legmann, *Protides Biol. Fluids* **25**, 407 (1977).
[19] E. Polz, H. Wurm, and G. M. Kostner, *Artery* **9**, 305 (1981).
[20] N. S. Lee, H. B. Brewer, Jr., and J. C. Osborne, Jr., *J. Biol. Chem.* **258**, 4765 (1983).
[21] J. S. Finlayson and J. F. Muskinski, *Biochim. Biophys. Acta* **147**, 413 (1967).

ml of 1.8 N perchloric acid at $-2°$ followed by low-speed centrifugation at 4° to remove the precipitated proteins. The turbid supernate is adjusted to pH 7 by dropwise addition of 12 N NaOH which clarifies the solution. To the supernate is added solid $(NH_4)_2SO_4$ (380 g/liter) with stirring. Following the solubilization of ammonium sulfate, the solution is allowed to stand for 30 min at 4° followed by collection of the precipitate by low-speed centrifugation. The precipitate is solubilized and dialyzed against 50 mM Tris–HCl, pH 8.0. The dialyzed fraction is applied to a DEAE-cellulose column (Whatman DE52, 1.5 × 30 cm) equilibrated with the same buffer. After application of the sample and elution of the unretained fraction, the remaining proteins are eluted with a linear salt gradient from 0 (200 ml) to 300 mM NaCl (200 ml) in 50 mM Tris–HCl, pH 8.0. The apoH is eluted in the unretained fraction and at the beginning of the gradient using commercially available antisera to β_2-glycoprotein-I as the monitoring tool. The fractions containing apoH are pooled, dialyzed exhaustively against distilled water, and lyophilized. As a final step in the purification procedure, the lyophilized fraction is solubilized in 10 mM Tris–HCl buffer, pH 8.0, and applied to a heparin-Sepharose column (1.5 × 25 cm) as previously described.[20] The column is eluted with 100 ml of 10 mM Tris–HCl, pH 8, followed by a linear gradient consisting of 300 ml solutions of the 10 mM buffer and the same buffer containing 400 mM NaCl. The major portion of retained apoH is eluted as a symmetrical peak and its presence is detected by the use of commercially available anti-β_2-glycoprotein-I. Fractions reacting with anti-β_2-glycoprotein-I serum are monitored by basic PAGE and those appearing homogeneous are dialyzed and lyophilized. The purity of apoH is tested by both basic PAGE and reactivity with antisera to apolipoproteins and other serum proteins, including the commercial antiserum specific for β_2-glycoprotein-I.

As an alternative procedure,[19] serum (100 ml) is mixed with an equal volume of a 1.68% (w/v) Rivanol solution (Serva). The pH is adjusted to 8.0 by adding 10% Na_2CO_3, followed by stirring for 10 min at room temperature. This mixture is centrifuged and to the supernate is added solid NaCl to a final concentration of 5 g/100 ml. Addition of 1 M HCl to give a pH of 7.0 is followed by stirring for 10 min and then low-speed centrifugation to remove the precipitate. The supernate is cooled to $-2°$ and 3 ml of 70% (w/v) perchloric acid is added per 100 ml. After stirring for 10 min at $-2°$, the material is centrifuged in the cold and the precipitate is discarded. The supernate is immediately neutralized with 10% Na_2CO_3, dialyzed exhaustively against distilled water, and lyophilized. Solubilization of the lyophilized crude apoH preparation in 10 mM Tris–HCl, pH 8.0, is followed by chromatography on a heparin-Sepharose (Pharmacia) column (25 × 1 cm). The column is eluted in steps with the initial buffer (10 mM

Tris) containing 200 mM NaCl, 300 mM NaCl, and 1 M NaCl. The majority of apoH is eluted in the 300 mM NaCl fraction and is judged homogeneous on the basis of basic PAGE and double diffusion analyses. Differences in the molarity of NaCl required to elute apoH from heparin-Sepharose have been observed. For this reason, the molarity of NaCl solutions for elution of apoH may require adjustments, depending on the source of the heparin-Sepharose column.

Procedures for isolating apoH from VLDL have been previously outlined and have included heparin-Sepharose and gel permeation column chromatography.[19,20] Analyses of apoH isolated from plasma and VLDL have demonstrated a charge heterogeneity by both DEAE-cellulose column chromatography and isoelectric focusing.[19,21]

Characterization of ApoH

On basic PAGE, the isolated apoH has a mobility similar to that of apoE.[19] Amino acid analyses of apoH isolated from plasma and VLDL demonstrated the occurrence of all common amino acids, with a relative enrichment in proline and half-cystine, and similar amino acid compositions for serum (Table I) and the isomorphic forms isolated from VLDL.[19] ApoH is a glycoprotein with hexoses, hexosamine, and sialic acid accounting for approximately 16% of the dry weight (Table II). The estimated molecular weight of apoH is 54,000 by SDS–PAGE,[19] while the weight average molecular weight by sedimentation equilibrium has been shown to be 43,000 in the presence and absence of denaturing solvents.[20] The difference between the two methods is probably due to the anomalous behavior of glycoproteins on SDS–PAGE. Studies on the isomorphic forms of apoH (pI range, 5.6–6.4)[22] have suggested that the polymorphism is due to the oligosaccharide side chains rather than to a variation in the sialic acid content in each of these various forms.[17,21,22]

Distribution of ApoH[23]

The identification of apoH (β_2-glycoprotein-I) as a constituent of plasma lipoproteins has led to studies on its concentration and distribution in lipoprotein density classes. Determinations were performed by radial immunodiffusion using commercially available antisera and 1,1,3,3-tetramethyl urea-treated lipoprotein fractions. Plasma levels were in the range of 16–30 mg/100 ml with 70–75% in the 1.21 g/ml infranate, 15–18%

---

[22] I. Schousboe, *Int. J. Biochem.* **15**, 35 (1983).
[23] E. Polz and G. M. Kostner, *FEBS Lett.* **102**, 183 (1979).

in HDL, 8–10% in VLDL, and 1% in LDL. Five hours after a heavy fat load, apoH can be detected in chylomicrons.

Function of ApoH

Several different lines of evidence have indicated a role of apoH in the metabolism of triglyceride-rich lipoproteins. One includes the detection of β_2-glycoprotein-I as a constituent of both chylomicrons and VLDL.[18–20] Other studies have shown that the addition of apoH increases the enzymatic activity of lipoprotein lipase in the presence of apoC-II, suggesting that the enzymatic activity of LPL in triglyceride metabolism may be modulated by apoH.[24] Additional experiments using the rat as an *in vivo* model have shown that the infusion of apoH increases the clearance rate of intralipid triglycerides.[25] In addition to its possible role in the metabolism of triglyceride-rich lipoproteins, apoH has been shown to have an affinity for both mitochondria and platelets.[22]

Other Constituents

One additional protein of human plasma has been designated as an apolipoprotein. This component was isolated from VHDL (d 1.21–1.25 g/ml) by a column chromatographic procedure similar to that outlined for apoD using hydroxylapatite.[26] This protein, however, was eluted with 100 mM K_2HPO_4, pH 8.0. It had an apparent molecular weight of 72,000, exhibited no reactivity with antisera directed against known apolipoproteins, and had a distinct amino acid composition (Table I). Based on these observations and its reactivity with anti-apoHDL serum, this protein component was designated apolipoprotein G (apoG). However, it has not yet been established whether this apolipoprotein occurs in lipoprotein particles as a sole protein or in combination with other apolipoproteins. No information is available either on the lipoprotein forms or the function of apoG.

In addition to the protein components discussed in this and other chapters of this volume, a number of other protein components have been reported to be associated with human plasma lipoproteins. The amino acid analyses of some of these proteins isolated from HDL are shown in Table IV. Due to insufficient information related to the chemical, physical, and immunologic properties of these various protein components,

[24] Y. Nakaya, E. J. Schaefer, and H. B. Brewer, Jr., *Biochem. Biophys. Res. Commun.* **95**, 1168 (1980).
[25] H. Wurm, E. Beubler, E. Polz, A. Holasek, and G. Kostner, *Metabolism* **31**, 484 (1982).
[26] M. Ayrault-Jarrier, J.-F. Alix, and J. Polonovski, *Biochimie* **60**, 65 (1978).

TABLE IV
MINOR POLYPEPTIDE CHAINS PRESENT IN ApoHDL OR ApoVHDL

	Glycine-serine-rich polypeptide[a]	D-2[b]	Proline-rich polypeptide[c]	Threonine-poor serum amyloid A (SAA) polypeptides[d]	
				pI = 6.5	pI = 6.0
Lysine	29	59	46	41	40
Histidine	15	1	21	23	20
Arginine	17	3	40	86	70
Aspartic acid	54	24	71	144	116
Threonine	29	32	57	5	9
Serine	125	38	72	85	77
Glutamic acid	100	100	100	100	100
Proline	28	27	70	42	35
Glycine	117	25	68	110	90
Alanine	54	34	32	154	123
Half-cystine	9	8	27	N.D.	N.D.
Valine	25	35	36	10	17
Methionine	0	5	5	11	17
Isoleucine	16	7	32	22	18
Leucine	38	52	50	42	40
Tyrosine	13	23	30	43	38
Phenylalanine	15	25	30	75	60
Tryptophan	0	6	N.D.	33	25

[a] S. O. Olofsson, G. Fager, and A. Gustafson, *Scand. J. Clin. Lab. Invest.* **37**, 749 (1977).

[b] C. T. Lim, J. Chung, H. J. Kayden, and A. M. Scanu, *Biochim. Biophys. Acta* **420**, 332 (1976).

[c] T. Sata, R. J. Havel, L. Kotite, and J. P. Kane, *Proc. Natl. Acad. Sci. U.S.A.* **73**, 1063 (1976).

[d] V. G. Shore, B. Shore, and S. B. Lewis, *Biochemistry* **17**, 2174 (1978).

including capacity to form distinct lipoproteins and circumstances of their appearance within the lipoprotein density spectrum, it is still not possible to recognize these proteins as integral components of the plasma lipoprotein system. The common characteristic of all of these proteins is an affinity for some portion of the lipoprotein molecule. It is not known what function they may play in the transport or metabolism of lipids. Although there is no available evidence to indicate a direct role for these proteins in the transport of lipids, they may have some auxiliary structural or metabolic functions. We have suggested, therefore, that some of these proteins, if not recognized eventually as apolipoproteins, may be considered and classified as auxiliary proteins of the lipid transport system.[1]

[16] Serum Amyloid A (ApoSAA) and Lipoproteins

By NILS ERIKSEN and EARL P. BENDITT

The discovery of a unique protein (AA) characteristic of amyloid deposits in tissues of patients with chronic inflammatory diseases[1] led to the immunological recognition of a structurally related protein (SAA) in human serum. Normally at a trace level, the concentration of SAA increases markedly in a variety of pathological conditions,[2,3] the bulk of it circulating as an apolipoprotein of the high-density lipoproteins (HDL), not only in human subjects but also in experimental animals, such as endotoxin-stimulated mice.[4,5] Several apoSAA isotypes have been found in human HDL; one of them, named by us apoSAA$_1$, consists of 104 amino acid residues, the first 76 being homologous with the sequence of tissue amyloid protein AA.[6] Because of this amino acid sequence homology, apoSAA has come to be regarded as a likely precursor of the related but somewhat smaller protein extractable from amyloidotic tissue. If a precursor–product relationship does exist between apoSAA and AA, it has yet to be ascertained whether all apoSAA isotypes are amyloidogenic. According to the evidence of a study of amyloidogenesis in several strains of mice, only one of the two major circulating murine apoSAA isotypes is amyloidogenic: on the basis of NH_2-terminal sequence identity, only apoSAA$_2$ was represented in the tissue amyloid protein AA.[7] Since spleen, a major site of amyloid deposition, does not produce apoSAA message during amyloidogenesis,[7a] it seems reasonable to conjecture that the deposition of protein AA results from an aberration in local metabolism of apoSAA. These facts indicate at least a site to be investigated further for clues to the function(s) of the apoSAA molecules.

---

[1] E. P. Benditt and N. Eriksen, *Am. J. Pathol.* **65,** 231 (1971).
[2] G. Husby and J. B. Natvig, *J. Clin. Invest.* **53,** 1054 (1974).
[3] C. J. Rosenthal and E. C. Franklin, *J. Clin. Invest.* **55,** 746 (1975).
[4] E. P. Benditt and N. Eriksen, *Proc. Natl. Acad. Sci. U.S.A.* **74,** 4025 (1977).
[5] E. P. Benditt, N. Eriksen, and R. H. Hanson, *Proc. Natl. Acad. Sci. U.S.A.* **76,** 4092 (1979).
[6] D. C. Parmelee, K. Titani, L. H. Ericsson, N. Eriksen, E. P. Benditt, and K. A. Walsh, *Biochemistry* **21,** 3298 (1982).
[7] J. S. Hoffman, L. H. Ericsson, N. Eriksen, K. A. Walsh, and E. P. Benditt, *J. Exp. Med.* **159,** 641 (1984).
[7a] R. L. Meek, J. S. Hoffman, and E. P. Benditt, *J. Exp. Med.*, in press (1986).

Preparation of ApoSAA

Initially, apoSAA was prepared by sequential molecular-sieve chromatography of high-SAA-level serum under dissociating conditions, as summarized immediately hereafter. Serum or defibrinated plasma, after dialysis against distilled water and acidification to a final concentration of 10% (v/v) HCOOH, is passed through a BioGel P-60 column equilibrated with 10% HCOOH. Collected fractions reactive with antiserum to AA (10,000–15,000 D) are passed sequentially through Sephadex G-75 and G-50 superfine columns likewise equilibrated. Pooled fractions with the just-mentioned immunoreactivity are lyophilized. The product has been shown to contain significant quantities of albumin fragments and prealbumin[8]; also, contamination with β_2-microglobulin has been mentioned.[9]

Preferably, apoSAA is isolated from HDL that have been separated from serum or plasma by sequential-flotation centrifugation according to established methods.[9-11] After being washed by recentrifugation at density 1.21 g/ml and concentrated by lyophilization, ultrafiltration, or dehydration through a dialysis membrane in contact with a water-absorbing material, the HDL are delipidated with organic solvents (e.g., methanol and diethyl ether, 1:3). The HDL apolipoproteins are fractionated under dissociating conditions (5 M guanidine/0.1 M CH_3COOH, 6 M urea/HCOOH, pH 3, or 5 M guanidine–HCl) on a column of Sephadex G-100, Sephacryl S-200, or BioGel A 0.5 M; chromatographic components of interest (10,000–15,000 D) are rechromatographed on the same column, collected, dialyzed, and lyophilized. The product consists of SAA apolipoproteins contaminated principally with the C apolipoproteins.

Further purification of SAA may be achieved by ion-exchange chromatography on DEAE-cellulose or DEAE-Sephacel in a salt gradient at pH 8.2 in the presence of 6 M urea,[11,12] whereby several apoSAA isotypes are resolved. As a final step in the purification, the individual isotypes, dissolved in 6 M guanidine at pH 7, may be subjected to high-performance liquid chromatography (HPLC) in an acetonitrile gradient[6]; alternatively, preparative isoelectric focusing[9,12] or chromatofocusing[13] may be used.

Physical and Chemical Characteristics of ApoSAA

Six apoSAA isotypes have been resolved by ion-exchange chromatography on DEAE-cellulose and partially characterized[11]; two more have

[8] G. Marhaug and G. Husby, *Clin. Exp. Immunol.* **45**, 97 (1981).
[9] G. Marhaug, K. Sletten, and G. Husby, *Clin. Exp. Immunol.* **50**, 382 (1982).
[10] N. Eriksen and E. P. Benditt, *Proc. Natl. Acad. Sci. U.S.A.* **77**, 6860 (1980).
[11] L. L. Bausserman, P. N. Herbert, and K. P. W. J. McAdam, *J. Exp. Med.* **152**, 641 (1980).
[12] B. Skogen, K. Sletten, T. Lea, and J. B. Natvig, *Scand. J. Immunol.* **17**, 83 (1983).
[13] C. L. Malmendier and J. P. Ameryckx, *Atherosclerosis* **42**, 161 (1982).

been recognized by similar chromatography on DEAE-Sephacel.[12] Two consistently major isotypes, apoSAA$_1$ and apoSAA$_2$, have identical amino acid sequences as far as compared (30 residues), except that the apoSAA$_2$ sequence begins with serine, lacking the N-terminal arginine of apoSAA$_1$.[10] ApoSAA$_1$ and apoSAA$_2$ have isoelectric points at approximately pH 6.1 and 5.7, respectively.[14] A third, sometimes prominent, isotype is distinguished by its practical insolubility above pH 3 and a more basic nature than any of the others, and a fourth lacks the N-terminal tripeptide (Arg-Ser-Phe) of apoSAA$_1$.[15] The remaining isotypes have been seen only in minor or trace amounts.

Generally somewhat sparingly soluble in neutral salt solutions, the apoSAA isotypes require acid or agents such as SDS, urea, or guanidine for ready solubility. Observed differences among the several isotypes in regard to solubility, and also to electrophoretic mobility and isoelectric point, are probably attributable in most instances to slight structural variations resulting from posttranslational modifications.

The molecular mass of apoSAA$_1$ has been calculated to be 11,685 D from its amino acid sequence. This figure may be assumed to be very close to the maximal mass for at least six apoSAA isotypes, inasmuch as the six have been reported to have identical C-terminal sequences, no attached carbohydrate, and chain-length variations restricted to shortenings by one or more residues at the N-terminus.[11,14] Molecular mass determinations by sedimentation equilibrium have yielded mean values of 11,640 and 11,840 D for two of the isotypes.[16] Estimates based on SDS–polyacrylamide gel electrophoresis in several laboratories place the mean molecular mass of the unresolved isotypes in the range 11,000–14,000 D.

The apoSAA isotypes are characterized by a high content of alanine, glycine, arginine, and aspartic acid plus asparagine, a relatively high content of phenylalanine and tyrosine, generally an absence of threonine, and without exception an absence of cystine. A potentiality for the formation of amphipathic α-helices by residues 1–24 and 50–74, and therefore a basis for association with HDL, is revealed by an examination of the amino acid sequence of apoSAA$_1$; indications of structural regularity are lacking in the intervening segment (residues 25–49) and in the C-terminal segment (residues 75–104), which respectively contain one and three of the four proline residues in the molecule.[6] The sequence analysis of apoSAA$_1$ revealed two forms of the protein, one containing valine and alanine at positions 52 and 57, respectively, and the other containing alanine and

---

[14] N. Eriksen and E. P. Benditt, *Clin. Chim. Acta* **140**, 139 (1984).
[15] L. L. Bausserman, A. L. Saritelli, P. N. Herbert, K. P. W. J. McAdam, and R. S. Shulman, *Biochim. Biophys. Acta* **704**, 556 (1982).
[16] L. L. Bausserman, P. N. Herbert, T. Forte, R. D. Klausner, K. P. W. J. McAdam, J. C. Osborne, Jr., and M. Rosseneu, *J. Biol. Chem.* **258**, 10681 (1983).

valine at these positions. These reciprocal substitutions resulted in identical compositions. Whether the two forms represent allotypes or products of separate genes is not clear. The only other indication of a genetic basis for the existence of apoSAA isotypes is the presence of a single, internal threonine residue in the sequences of two isotypes and the absence of threonine in the other isotypes[11,15]; a deletion or a substitution must have occurred at an as yet unknown position.

The unusual amino acid composition of the apoSAA isotypes renders unreliable commonly used colorimetric methods when applied without correction factors to the estimation of total protein in purified apoSAA preparations. It is preferable for this purpose to rely on an extinction coefficient; based on the recovery of amino acids liberated by hydrolysis of a sample of the protein, a value of 26 has been calculated for the absorbance of a 1% solution (1 cm) of apoSAA$_1$ in 0.1 M NaHCO$_3$ at 280 nm.[17]

Assay of SAA

General Considerations. Soon after the discovery of SAA several methods for its quantification came into use; all were based on immunoreactivity of SAA with antibodies to AA, and results were expressed as AA-equivalent concentrations in the assayed samples. As purified apoSAA became available, immunoassays based on hybrid systems or a complete apoSAA/anti-apoSAA system were devised. All assay systems apparently are capable of distinguishing between high and low levels of SAA, but wide interlaboratory discrepancies exist with respect to SAA levels in normal serum; reported mean normal levels range from less than 0.1 μg AA-equivalent/ml to as much as 20 μg apoSAA/ml.[3,18,19] It makes little difference whether concentrations are expressed in mass units of apoSAA or AA per unit volume, the molecular mass ratio being only about 1.35. Other factors must therefore account for the bulk of the discrepancy. The most obvious possibilities are mentioned in the remainder of this section.

If a ^{125}I-labeled AA/anti-AA radioimmunoassay system is used, substantially higher SAA levels will be obtained with reference to apoSAA standards as opposed to AA standards because AA is much more effective than apoSAA in displacing ^{125}I-labeled AA from binding by anti-AA. Also to be considered are the purity of the standard, the specificity of the

---

[17] G. Marhaug, *Scand. J. Immunol.* **18**, 329 (1983).
[18] M. H. van Rijswijk, Ph.D. thesis, State University of Groningen, The Netherlands, 1981.
[19] W. Hijmans and J. D. Sipe, *Clin. Exp. Immunol.* **35**, 96 (1979).

antibodies, and the availability of antigenic sites on the apoSAA molecules associated with HDL or other serum proteins. The latter consideration raises the question of sample denaturation before assay. In some laboratories it has been found that sample denaturation does not consistently increase SAA immunoreactivity and in fact may decrease it.[17,20] In other laboratories it has been found that acidic (10% HCOOH, 37°, 24 hr)[21] or alkaline (0.1 M NaOH, room temperature, 6 hr)[18] denaturation of samples substantially increases SAA immunoreactivity. We routinely heat our samples for 1 hr at 60° before assay; such thermal denaturation has consistently yielded 2-fold to 3-fold increases in the measured apoSAA concentrations. Heat treatment has the additional potentiality of protecting SAA against digestion by serum proteases.[22] The interlaboratory differences in regard to the need for denaturation may well be due to the fact that some antibodies are directed toward exposed epitopes on the native SAA-containing complexes, the need for sample denaturation when such antibodies are used in the immunoassay of SAA thus being obviated.[20] A comprehensive investigation of the denaturation question is obviously needed. A related matter is the stability of SAA during storage of specimens. Considerable increases were observed in SAA concentrations of serum specimens stored several months at 4° with thymol as a preservative[21]; it may be that during storage the same epitopes were unmasked as those that deliberate denaturation would have revealed. Elsewhere, the mean SAA level in 50 normal serum specimens that had been stored frozen for 1 year was found not to differ significantly from that of 50 other normal serum specimens that were analyzed 1 day after sampling.[17] Our practice is to subject serum samples to thermal denaturation on receipt and store them frozen until a convenient time for assay.

An additional and important consideration is the background against which SAA measurements are often made: a large amount of extraneous material of which apoSAA represents only a very small percentage. It has been our experience that serum samples with low concentrations of SAA may yield radioactivity-binding values (B) that equal or even exceed the 100% level of binding (B_0). This excess-binding phenomenon can be eliminated by the addition of an SAA-depleted serum specimen to all assay tubes except those containing the samples to be assayed. Such a specimen

---

[20] R. E. Chambers and J. T. Whicher, *J. Immunol. Methods* **59**, 95 (1983).
[21] J. D. Sipe, K. P. W. J. McAdam, B. F. Torain, and G. G. Glenner, *Br. J. Exp. Pathol.* **57**, 582 (1976).
[21a] We have carried out sets of assays at total reaction volumes as large as 150 μl to accommodate samples with low SAA levels but have not systematically studied the effect of total reaction volume on apparent SAA levels.
[22] L. L. Baussermann and P. N. Herbert, *Biochemistry* **23**, 2241 (1984).

can be prepared by ultracentrifugal delipidation (at density 1.21 g/ml) of a normal human serum, passage of the delipidated serum through a Sephadex G-200 column in phosphate-buffered saline (PBS), and pooling of the effluent fractions representing the major chromatographic peaks (the two following the void-volume peak). By this procedure, any lipoprotein associated, macrocomplexed, or free apoSAA is removed from the bulk serum proteins, which can then be concentrated by any of several suitable methods and used in the assay of low-SAA-level serum samples to provide an appropriate protein background at a volume equivalent to that of the serum being assayed (see Fig. 1).

A report comparing three different methods for assaying SAA[17] contains much useful information concerning methodology as well as problems encountered in quantification of SAA. The author recommended a double-antibody radioimmunoassay with apoSAA$_1$ as standard and labeled antigen, and polyclonal rabbit anti-SAA as first antibody.

Suggested Method for Assay of SAA. The method used in our laboratory for quantification of SAA is basically the same double-antibody procedure previously described,[4] with modifications introduced at different times since the initial description. HPLC-purified apoSAA$_1$ (concentration in stock solution estimated from extinction coefficient) is the standard, ^{125}I-labeled apoSAA$_1$ is the labeled antigen, and anti-apoSAA$_1$ is the primary antibody, which is isolated from antiserum produced in a rabbit by injection of apoSAA$_1$ purified by ion-exchange chromatography on DEAE-cellulose. The isolation is achieved by affinity chromatography on Sepharose 4B (Pharmacia) to which HPLC-purified apoSAA has been covalently linked according to instructions provided by the manufacturer. The antibody is eluted with 0.1 M CH$_3$COOH/0.5 M NaCl, collected in tubes containing sufficient 1 M NaHCO$_3$/NaOH (pH 8.5) to neutralize the acid, stored frozen, and subsequently used in the assay without further manipulation. The radiolabeling with ^{125}I is done enzymatically with immobilized lactoperoxidase, as before, but the reactants are contained in a 12 × 75-mm glass tube on a vortex mixer; after 5 min the reaction mixture is transferred to the G-50 Sephadex column, the termination step with NaN$_3$ being eliminated. The two or three effluent fractions (in 0.01 M SDS/0.01 M phosphate, pH 8.6; 0.7 ml per fraction) representing the coincident peaks of ultraviolet absorption (278 nm) and radioactivity are pooled, stored frozen, and subsequently used in the assay. The iodinated antigen gives no indication of radiation damage for a period exceeding 3 months.

A description of a typical assay for SAA in our laboratory follows. The reaction is carried out in a 400-μl polyethylene tube with attached cap, at a total volume of 40 μl.[21a] Additions of reagents are made with Carlsberg constriction micropipets. Reagents containing particulate matter are clarified by centrifugation in a microfuge. Each tube receives 10 μl each of (1) heat-denatured (60°, 1 hr) and diluted (e.g., 7-fold in PBS) sample serum, or PBS only; (2) apoSAA$_1$ standard (5 to 1000 ng) dissolved in 0.1% Tween 20/0.01% SDS/0.01 M phosphate, pH 7.4, or the solvent only; (3) stock ^{125}I-labeled apoSAA$_1$ diluted, e.g., 4-fold with 0.25% Tween 20/ 0.01% SDS/0.01 M phosphate, pH 7.4; and (4) anti-apoSAA$_1$ preparations diluted, e.g., 2.5-fold with bicarbonate-neutralized acetic acid elution medium (see above) and containing normal rabbit serum at 16-fold dilution as carrier, or, for the determination of nonspecific binding (N), a 14-fold dilution of normal rabbit serum in bicarbonate-neutralized acetic acid elution medium. The amount of antibody used is sufficient to bind 40–50% of the total radioactivity added to the tubes containing antibody. The total Tween 20 concentration, 0.07% (v/v), is effective in reducing nonspecific binding of radioactivity without noticeably affecting the antigen/antibody interaction. The reactants are mixed, the tubes are incubated for 1 hr at 37° and a 0.5 hr at 22° (room temperature), and then sufficient goat antiserum to rabbit IgG for maximal precipitation (e.g., 5 μl containing 25 mg antibody/ml) is added to each tube. After an additional hour at 22°, during which initial radioactivity counts may be made to assess uniformity of addition of labeled antigen (extreme variation not exceeding 5% of mean count per tube considered satisfactory), the secondary antigen/antibody precipitate is washed by centrifugation as described previously.[4] If the final rinse and centrifugation are omitted and the washed pellet is counted in the intact tube, a considerable saving in time will be achieved at the expense of only a slight increase in the nonspecific binding value.

Representative displacement curves are shown in Fig. 1. The standard curve is essentially linear over the range 10–200 ng apoSAA$_1$ and approximately parallel to the displacement curve for the high-SAA-level serum ($\sim$160 μg apoSAA/ml). The two displacement curves for the low-SAA-level serum ($\sim$5 μg apoSAA/ml) illustrate the effect of substituting, for PBS in the control, B_0, and standard tubes, SAA-depleted serum in amounts equivalent to the different volumes of low-SAA-level serum being tested; the downward shift in the displacement curve is equivalent to an increase of a few micrograms of apoSAA per milliliter in the apparent SAA concentrations, a negligible difference with respect to high levels of SAA but an approximate doubling in this instance. The indication of nonparallelism between the standard and low-SAA-level displacement curves suggests that what is being measured in such specimens is not completely identical to apoSAA. Supporting evidence for this view is

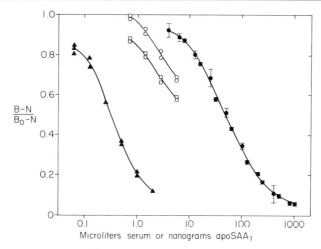

FIG. 1. Displacement of ^{125}I-labeled apoSAA$_1$ from binding to anti-apoSAA$_1$: (●) apoSAA$_1$ standard, serially diluted with 0.1% Tween 20/0.01% SDS/0.01 M phosphate, pH 7.4, mean ± SD of 2 to 6 replicates per data point; (▲) high-SAA-level serum, serially diluted with PBS; (○) low-SAA-level serum, serially diluted with PBS; (□) same low-SAA-level serum, serially diluted with PBS and referred to B_0 tubes containing SAA-depleted serum specimen (see text). B, B_0, and N defined in text.

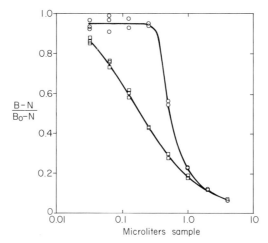

FIG. 2. Displacement of ^{125}I-labeled apoSAA$_1$ from binding to anti-apoSAA$_1$: (○) high-SAA-level HDL$_3$ preparation, serially diluted with PBS; (□) same preparation, serially diluted with 0.1% Tween 20/0.01% SDS/0.01 M phosphate, pH 7.4.

provided by our data on the distribution of apoSAA among the lipoprotein density classes of normal human serum[14]; only a small part of the apparent total apoSAA was found in the HDL fraction, the bulk of it having been found in the density 1.21 bottom fraction. This nonlipoprotein-associated material with SAA-like immunoreactivity has not been further characterized.

A property of apoSAA that may have an effect on its quantification is illustrated in Fig. 2, which shows displacement curves for a high-SAA-level HDL_3 preparation (~250 μg apoSAA/ml) serially diluted in glass tubes with PBS or 0.1% Tween 20/0.01% SDS/0.01 M phosphate, pH 7.4. Dilution with PBS resulted in a curve exhibiting almost no displacement at the lowest concentrations, whereas dilution with the detergent-containing solvent resulted in a displacement curve resembling that of the serum from which the HDL_3 was isolated (Fig. 1). At the highest concentrations, the two displacement curves merged completely. Adsorption of apoSAA to the glass surface that presumably occurred in PBS was apparently prevented by the detergents. Serial dilution with PBS of the high-SAA-level serum without apparent adsorption losses (Fig. 1) is attributed to the presence of sufficient extraneous protein to cover adsorption sites.

Concluding Remarks

The claim cannot be justifiably made that SAA is measured with completely satisfactory reliability by any of the methods of assay in use at this time. As in the quantification of other apolipoproteins, a number of difficulties remain to be overcome, particularly in regard to specimens with low SAA levels. ApoSAA might participate in a variety of molecular associations, any one of which could interfere to some extent with its detection. It is also possible that what is being measured includes entities that are structurally related but not identical. Limited proteolysis of apoSAA *in vivo* or during storage of specimens, for example, could result in the appearance of AA, which itself could participate in various associations and thus further complicate the quantification of apoSAA.

What role, if any, apoSAA plays among the variety of normal metabolic functions served by the plasma apolipoproteins and whether it is an essential determinant of structure in a particular subset of HDL particles are questions that remain to be answered. It has been suggested that apoSAA must serve an important function in vertebrate physiology, inasmuch as human and murine apoSAA show perfect amino acid sequence homology with residues 33 through 45 (numbering system for human AA) of AA proteins from man, monkey, mink, and duck, this partial sequence having been conserved for an evolutionary period estimated to be about

270 million years.[23] Moreover, methionine residues have been found to occupy positions 17 and 24 in all AA and SAA proteins that have been sequenced for the requisite distance. Whatever its function may be, apo-SAA is already included in that group of proteins (acute-phase reactants) whose synthesis is stimulated in the response of mammals to injury. As a sensitive indicator of tissue damage, it could serve a clinically useful purpose.

Acknowledgments

We express our appreciation to Ronald H. Hanson, Marlene Wambach, and Philip Lu for technical assistance during various phases of our investigations, and to Virginia Wejak for preparation of the manuscript. The work was supported by U.S. Public Health Service Grant HL-03174.

[23] R. S. Stearman, C. A. Lowell, W. R. Pearson, and J. F. Morrow, *Ann. N.Y. Acad. Sci.* **389**, 106 (1982).

[17] Structure and Lipid Binding Properties of Serum Albumin

By ARTHUR A. SPECTOR

One of the main physiologic functions of albumin, the most abundant protein in the blood plasma, is to transport long-chain fatty acids. The fatty acid that is carried by albumin is for the most part released from adipocyte triacylglycerol stores in response to hormonal, nervous, or nutritional stimuli. These fatty acids contain 16- or 18-carbon atoms and even though they exist predominantly in anionic form, they are very poorly soluble in water. Through binding to albumin, the amount of fatty acid that can be present in plasma exceeds the aqueous solubility by as much as 1000-fold. The binding is a physical process that occurs through a combination of nonpolar and electrostatic interactions.[1] Albumin transports the fatty acid to tissues where it is either oxidized as a source of energy or esterified into complex lipids. Because the fatty acid transported by albumin is in unesterified form, it is called free fatty acid. This is to signify that it is not covalently bonded and to contrast it with the bulk of the plasma fatty acid which exists in ester linkage as components of

[1] A. A. Spector, *J. Lipid Res.* **16**, 165 (1975).

phospholipids, triacylglycerols, and cholesteryl esters. Although present in relatively low concentrations, the plasma free fatty acid turns over very rapidly and, therefore, is a very important source of lipid for many tissues.[2]

Information regarding fatty acid binding to albumin is of interest for several reasons. The most common is to obtain specific details about the fatty acid transport process. Another involves the need to add a long-chain fatty acid to an aqueous solution. Since long-chain fatty acids have a very low water solubility, it usually is necessary to use some type of carrier in order to get enough into the incubation mixture for a particular experiment. Furthermore, in those cases where high concentrations can be achieved without a carrier, the fatty acid often exerts detergent effects and is toxic.[3] Because it is the physiologic fatty acid transport protein in plasma, albumin usually is used as the carrier, even for studies involving intracellular organelles and enzymes. When used in this way, albumin acts as a fatty acid buffer. Although retaining most of the fatty acid in bound form, it rapidly and continuously allows enough to dissociate to replace the unbound material as it is utilized, thereby maintaining a relatively constant supply of unbound fatty acid for the reaction. Albumin is not unique in this function. Other proteins that bind fatty acid can be substituted; for example, β-lactoglobulin or low-density lipoproteins.[4,5] The other reason for interest in fatty acid binding to albumin is that this simplified system has provided fundamental insight into lipid–protein interactions, helping to understand the more complex processes that occur in plasma lipoproteins and membranes.

Preparation of Fatty Acid–Albumin Solutions

Methods for adding long-chain fatty acids to aqueous solution will be described in detail even though this may seem to be a trivial issue. There are numerous instances, however, where fatty acids have been added incorrectly to incubation media. A strong possibility exists that some conclusions regarding fatty acid effects on cells and enzymes, or lack thereof, either are incorrect or of little biologic importance because the medium contained toxic amounts of fatty acid or, conversely, the fatty acid was not adequately solubilized.[3] It is critical to remember that under the usual physiologic conditions, the molar ratio of fatty acid to albumin rarely exceeds 2, more than 99% of the fatty acid is bound, and the

[2] D. S. Fredrickson and R. S. Gordon, Jr., *J. Clin. Invest.* **37,** 1504 (1958).
[3] A. A. Spector and J. C. Hoak, *Science* **190,** 490 (1975).
[4] A. A. Spector and J. E. Fletcher, *Lipids* **5,** 403 (1970).
[5] A. A. Spector and J. M. Soboroff, *J. Lipid Res.* **12,** 545 (1971).

unbound concentration to which tissues are exposed usually is between 5 nM and 1 µM and probably never exceeds 10 µM.[1,2]

Delipidation of Albumin

Because it is available commercially at relatively low cost, bovine albumin is used for most experimental studies. Two types of bovine preparations are offered, crystalline albumin and Fraction V albumin. The latter contains some plasma protein impurities, but albumin usually accounts for about 95% of the protein in most preparations. Bovine Fraction V albumin is 5 to 10 times less expensive than crystalline albumin and, therefore, is used for most experimental purposes. Unless specifically treated, commercial albumins contain free fatty acid in amounts between 0.1 and 2.4 mol/mol protein,[6] the usual value for bovine Fraction V albumin being about 0.3 to 0.5 mol/mol protein. Fatty acid-poor or fatty acid-free bovine Fraction V albumin, which contain only 0.05 to 0.1 mol/mol protein, can be purchased at a somewhat higher price. Alternatively, delipidation can be performed in about 2 hr by the method of Chen in which the fatty acid is removed by activated charcoal during a 1 hr incubation at pH 3.0 in an ice bath.[6] Lowering the pH unfolds the protein and removes the negative charge from the fatty acid carboxyl group. Both of these changes facilitate removal of the fatty acid from albumin. Albumin is not denatured by this treatment, and its capacity to bind fatty acid remains intact when the pH is returned to the neutral range.[6]

Our laboratory routinely adds two steps to the published procedure of Chen.[6,7] After the 20,000 g centrifugation to sediment the bulk of the charcoal, the solution is adjusted to pH 7.4, transferred to ultracentrifuge tubes, placed in a fixed angle rotor, and centrifuged at 5° for 45 min at 100,000 g. Any small charcoal particles remaining after the low-speed centrifugation are sedimented in this step. Subsequently, the protein solution is dialyzed against 40 volumes of a buffered salt solution such as Krebs Ringer phosphate to remove any low-molecular-weight impurities. The dialysis solution is changed three times, and the total time of dialysis is 18 to 24 hr. The procedure of Chen removes essentially all of the inherent free fatty acid from the albumin.[6]

Addition of Fatty Acid Salts

This procedure is the quickest, but it requires skill and experience in order to be successful. An albumin solution is prepared to contain twice

---

[6] R. F. Chen, *J. Biol. Chem.* **242**, 173 (1967).
[7] A. A. Spector, K. John, and J. E. Fletcher, *J. Lipid Res.* **10**, 56 (1969).

the concentration of protein, salt, and buffer that will be required in the final preparation. The concentration of albumin should not exceed 0.5 mM. If necessary, fatty acids can be purified by either gas–liquid or high-performance liquid chromatography and collection of the column effluent.[8,9]

A measured amount of fatty acid is dissolved in a very volatile solvent such as hexane and placed in a small beaker. Since there usually is incomplete transfer of the fatty acid to albumin by this method of addition, it is best to begin with about 10 to 20% more fatty acid than is actually needed. A small excess of 1 N NaOH is added to the fatty acid dissolved in hexane; KOH or NH_4OH can be substituted for studies involving intracellular organelles or enzymes. The hexane is evaporated under N_2; the beaker should be warmed gently to facilitate removal of the solvent. It is critical to dry the fatty acid salt thoroughly at this point, until it gives the appearance of a chalky, white residue. This must be done carefully, especially with unsaturated fatty acids, because excessive heat or exposure to air can cause oxidative damage. After drying, the fatty acid salt is dissolved in 1 to 3 ml of highly purified distilled water. The mixture is warmed gently until a clear solution, often containing a soap film at the surface, is produced. This is the most delicate step. Excessive heating will damage the fatty acid, whereas failure to produce a clear solution interferes with the preparation of a tightly bound complex with albumin. Unless this solution is somewhat alkaline, it will not clarify sufficiently on warming. A simple test is that unless it is possible to read printed material held behind the solution, it will not form a good complex with albumin.

The warm, clear fatty acid salt solution immediately is added dropwise over 30 to 40 sec to the albumin solution, which is stirred magnetically. It is best to warm the pipet used for the dropwise addition to prevent fatty acid from precipitating on the glass surface. Subsequently, some of the albumin solution is drawn into the pipet, and the beaker is rinsed with it several times to remove any residual fatty acid. The solution is then adjusted with 0.2 N HCl to the required pH, usually 7.4, and then diluted with distilled water to twice the volume of the initial albumin solution. The pH adjustment should be done immediately because even though the solution contains albumin and buffer, it usually becomes slightly alkaline when the fatty acid solution is added. Since albumin, salt, and buffer were added initially at twice the desired final amounts, dilution with distilled water brings these concentrations to the proper level.

It is important to measure the free fatty acid concentration at this

[8] C. P. Burns, S.-P. L. Wei, and A. A. Spector, *Lipids* **13**, 666 (1978).
[9] M. I. Aveldano, M. VanRollins, and L. A. Horrocks, *J. Lipid Res.* **24**, 83 (1983).

stage,[10] for the exact amount that is transferred into the albumin solution is unpredictable. For this reason, it is best to prepare a solution containing more fatty acid than actually is required. A second albumin solution is prepared containing the same final concentration of protein, salt, and buffer, but no fatty acid. By adding a measured amount of this solution to the albumin solution containing fatty acid, the fatty acid concentration is adjusted downward to the required level without changing the protein, salt, or buffer concentrations. Since the binding and dissociation of fatty acid occur very quickly,[11,12] the fatty acid rapidly redistributes over the entire population of albumin molecules to a uniform, equilibrium mixture. The final fatty acid–albumin solution should be optically clear, and its absorbance should not be more than 0.01 optical density units above that of a corresponding albumin solution containing no added fatty acid. If the solution is cloudy, it likely will be toxic to cells, act as a detergent on membranes and denature enzymes. This is due to the fact that the solution contains an excessive concentration of unbound fatty acid in micellar form. A cloudy solution cannot be salvaged by filtration or centrifugation, and it should be discarded.

The salt loading method works well with saturated fatty acids containing up to 16 carbon atoms and for all unsaturated fatty acids. With practice, it even is possible to combine stearic acid (octadecanoic) with albumin. If the final albumin concentration is in the range of 0.1 to 0.2 mM, solutions containing at least 5 mol of fatty acid per mol of protein can be prepared, except in the case of stearic acid where the maximum is about 3 mol/mol protein. Many variations of this general method are in current usage.[13] Because of its complexity and the attention that is required to minor details, however, the investigator should gain some experience with this technique before preparing solutions for definitive experiments.

Celite Method[14]

Another procedure used to add fatty acid to albumin solutions is to first coat the fatty acid on a particulate material. Transfer will occur even if the fatty acid is coated on the surface of a beaker, but the process is more efficient if the particles are small and can be dispersed in the al-

---

[10] S. Laurell and G. Tibbling, *Clin. Chim. Acta* **16**, 57 (1967).
[11] W. Scheider, *Biophys. J.* **24**, 260 (1978).
[12] W. Scheider, *Proc. Natl. Acad. Sci. U.S.A.* **76**, 2283 (1979).
[13] B. R. Lokesh and M. Wrann, *Biochim. Biophys. Acta* **792**, 141 (1984).
[14] A. A. Spector and J. C. Hoak, *Anal. Biochem.* **32**, 297 (1969).

bumin solution. Celite, amberlite, or microthene are satisfactory,[14] with Celite being used most commonly.

The fatty acid is dissolved in a volatile organic solvent such as hexane, Celite particles are added, and the solvent is evaporated under N_2 with gentle warming. It is important to dry the Celite well at this point, so that is has a white, talc-like appearance, and then to thoroughly mix the dry powder. The fatty acid-coated Celite can be stored and used as needed, although precautions to prevent oxidation should be followed if the fatty acid is unsaturated. Routinely, 1 mmol of fatty acid is added per 10 g of Celite.

A weighed amount of fatty acid-coated Celite is incubated with an albumin solution in a beaker having a large surface area, usually for 30 min to 1 hr at room temperature with gentle magnetic stirring. If the fatty acid is unsaturated, the incubation is done under N_2. The contents then are transferred to a plastic centrifuge tube, and the Celite is sedimented by centrifugation at 2° for 10 min at 15,000 g. Centrifugation is repeated once, the clear supernatant solution is passed through a 1.2-μm filter, and the pH is adjusted. The amount of fatty acid transferred into the albumin solution can vary to some extent from one preparation to the next. Because of this, the actual free fatty acid concentration of the solution must be measured in each preparation. It is best to prepare a solution having somewhat more fatty acid than is ultimately desired and, as described in the method for loading fatty acid salts, dilute with a corresponding fatty acid–free albumin solution to achieve the desired final concentration.

Figure 1 shows the effect of fatty acid structure on the concentration that can be produced in a bovine albumin solution by the Celite method under a fixed set of conditions. Less saturated fatty acid is loaded into the solution as the length of the acyl chain increases. With the 18-carbon atom fatty acids, unsaturation increases the concentration that can be achieved. However, there is little difference whether the fatty acid has either one or two cis double bonds. The differences probably reflect the different binding affinities of albumin for these long-chain fatty acids,[1] as well as differences in the unbound concentrations of anion monomer than can be achieved with each fatty acid.

The Celite method has several important advantages. It is straightforward, and none of the steps requires any special skill or training. Furthermore, overloading does not occur. If the capacity of albumin to tightly bind the fatty acid is exceeded, transfer reaches a steady state, and large amounts of fatty acid do not accumulate in the aqueous phase as micelles or aggregates. Therefore, solutions that are overtly toxic to cells or injurious to enzymes are not produced, a problem often encountered when the

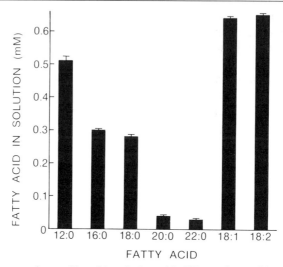

FIG. 1. Concentrations achieved in solution with different fatty acids. Each incubation contained 50 mg of fatty acid-coated Celite, and the Celite contained 1 mmol of fatty acid per 10 g in each case. The time of incubation was 30 min. Abbreviations used for the fatty acids are 12:0, lauric; 16:0, palmitic; 18:0, stearic; 20:0, arachidic; 22:0, behenic; 18:1, oleic; 18:2, linoleic.

fatty acids are added by the salt method. Two potentially serious problems, however, should be recognized. One is the cost factor with certain fatty acids, especially radioactive polyunsaturates such as arachidonic acid. Since only half or less of the material added to Celite usually is transferred into solution, there is considerable waste of these expensive compounds. For reasons that are not completely clear, the Celite containing residual fatty acid does not serve as a good donor if it is recovered and added to a fresh albumin solution. Another problem concerns the possibility that the fatty acid-coated Celite might bind and thereby remove substances from the aqueous solution. This is less serious with albumin solutions, although even here some divalent cation such as calcium may possibly complex with the particles. It is a concern when a medium containing serum must be loaded, as is often the case in tissue culture studies.[15]

Other Approaches. Fatty acids have been added to aqueous solutions by first dissolving them in a small amount of an organic solvent such as

[15] A. A. Spector, R. E. Kiser, G. M. Denning, S.-W. M. Koh, and L. E. DeBault, *J. Lipid Res.* **20,** 536 (1979).

ethanol.[16] Another approach is to incubate the albumin solution with an *n*-heptane solution containing fatty acid and then remove the hydrocarbon phase.[17] Fatty acids also have been sonified for long periods in a solution containing albumin.[18] Alternatively, the albumin solution can be incubated with a hormone-stimulated fat pad or a triglyceride emulsion containing activated lipase.[17] Each of these procedures has certain advantages but at the same time, serious disadvantages, and none of them has come into general usage.

Measurement of Fatty Acid Binding to Albumin

A number of physical techniques have provided important insights into the binding of fatty acids to albumin, particularly the molecular interactions and the localization of the binding sites. For example, long-chain fatty acids quench the tryptophan fluorescence of bovine albumin and cause a decrease in the wavelength of maximum emission.[17] Fluorescent fatty acid analogs such as *cis*-parinaric and *cis*-eleostearic acids are available, and their binding can be followed by fluorescence spectroscopy.[19] Alternatively, the addition of fatty acids will perturb the emission of other bound fluorescent ligands such as 1-anilino-8-naphthalenesulfonate.[20,21] They also perturb the absorbance of bound organic ligands such as trinitrobenzenesulfonate.[22] Another approach is to follow the binding of stearic acid nitroxides by electron spin resonance,[23] or to measure the perturbation in the spectrum of other anionic spin-labeled compounds when fatty acids are added.[24] Nuclear magnetic resonance has been used to study the binding of [1-^{13}C]oleic acid to albumin.[25] Finally, gel filtration columns have been employed to study fatty acid binding to fatty acid binding protein,[26] and such a procedure can be adopted for albumin binding studies. While each of these methods has contributed to an under-

---

[16] L. W. Daniel, B. A. Beaudry, L. King, and M. Waite, *Biochim. Biophys. Acta* **792**, 33 (1984).
[17] A. A. Spector and K. M. John, *Arch. Biochem. Biophys.* **127**, 65 (1968).
[18] H. W. Cook, J. T. R. Clarke, and M. W. Spence, *J. Lipid Res.* **23**, 1292 (1982).
[19] C. B. Berde, B. S. Hudson, R. D. Simoni, and L. A. Sklar, *J. Biol. Chem.* **254**, 391 (1979).
[20] E. C. Santos and A. A. Spector, *Biochemistry* **11**, 2299 (1972).
[21] E. C. Santos and A. A. Spector, *Mol. Pharmacol.* **10**, 519, (1974).
[22] L.-O. Anderson, J. Brandt, and S. Johansson, *Arch. Biochem. Biophys.* **146**, 428 (1971).
[23] S. J. Rehfeld, D. J. Eatough, and W. Z. Plasky, *J. Lipid Res.* **19**, 841 (1978).
[24] B. J. Soltys and J. C. Hsia, *J. Biol. Chem.* **253**, 3023 (1978).
[25] J. S. Parks, D. P. Cistola, D. M. Small, and J. A. Hamilton, *J. Biol. Chem.* **258**, 9262 (1983).
[26] R. K. Ockner and J. A. Manning, *J. Clin. Invest.* **54**, 326 (1974).

standing of fatty acid binding, values for the number of albumin binding sites and binding constants are based primarily on either equilibrium dialysis or equilibrium partition measurements.

Equilibrium Dialysis[27]

Fatty acids containing 14 or more carbon atoms will not freely pass through ordinary dialysis membranes in the physiologic pH range. Therefore, equilibrium dialysis only can be employed to study the binding of fatty acids containing up to 12-carbon atoms. Commercially available plastic dialysis cells and membranes are ideal for this purpose (Bel-Art Products). These cells have two compartments, each 1 ml, separated by the dialysis membrane. Prior to insertion, sheets of dialysis membrane are placed in boiling distilled water for 15 min and then soaked in fresh distilled water for 24 to 48 hr at room temperature. After the cell is assembled, albumin and fatty acid are added to one compartment, and the same buffered salt solution is added to the other. The compartments are sealed, and the cells are incubated in a temperature-controlled water bath covered with a hood and shaken at 40 oscillations per min. Equilibration of the fatty acid is reached with 16 hr at 37°; no albumin crosses the membrane. The fatty acid concentration in both compartments is determined after equilibrium is reached, usually by measuring radioactivity. The unbound fatty acid concentration, F, is obtained from the compartment having no albumin. By subtracting this value from the total fatty acid concentration of the compartment containing albumin, the amount of fatty acid bound to albumin, B, is obtained. For binding experiments, a set of 10 to 25 cells is set up, each having the same albumin concentration, usually between 10 and 100 μM, and increasing amounts of fatty acid, usually from 5 μM to 2 mM. A binding isotherm is constructed from the values of B, in moles of fatty acid per mole protein, and the corresponding value of F, in moles/liter. In the most widely used graphical display, a Scatchard plot, the term B/F is plotted on the y axis and B on the x axis.

Equilibrium Partition[7]

This method is employed to measure the binding of long-chain fatty acids containing at least 12-carbon atoms. Two sets of flasks are required, one containing a heptane solution of a fatty acid and an albumin solution, and the other containing the heptane–fatty acid solution and the buffered salt solution without albumin. As originally designed by Goodman,[28] the

[27] J. D. Ashbrook, A. A. Spector, and J. E. Fletcher, *J. Biol. Chem.* **247,** 7038 (1972).
[28] D. S. Goodman, *J. Am. Chem. Soc.* **80,** 3892 (1958).

method utilized separatory funnels and relatively large volumes of solution. This has been modified, and the procedure now is done in specially constructed small screw-capped vials that contain a tubular sampling device.[7] A binding experiment requires about 15 to 20 vials containing albumin and an equal number containing no albumin. In both cases, equilibrium is reached within 24 hr if the vials are shaken at 40 oscillations per min in a water bath at 37°. Before the vials are loaded, a tightly fitting glass rod is inserted into the sampling tube, touching the bottom of the vial, and left in place throughout the incubation. Either 1 ml of buffered salt solution, or 1 ml of this solution containing albumin is added to each vial, and this is covered with 1 ml of n-heptane containing radioactive fatty acid. No solution will enter the sampling tube as long as the glass rod is left in place. A single albumin concentration is used in each vial, usually 100 μM. The fatty acid contained in heptane is varied over a range of 20 μM to 5 mM. After the system comes to equilibrium, the screw cap is removed and the amount of radioactive fatty acid in the heptane and aqueous phases of each flask is measured. First, the heptane is sampled with a micropipet. Next, the glass plug is removed from the sampling tube, allowing some of the aqueous solution to enter. The aqueous phase is then sampled by inserting a micropipet through the glass tube, being careful to avoid touching the heptane solution. It is best to draw a sample into the micropipet, discard it, and then take a second sample for isotope counting.

From the set of vials containing albumin, one obtains the equilibrium concentrations of fatty acid in the heptane and aqueous phases. As indicated by Eq. (1), the concentration in this aqueous phase represents the sum of the unbound concentration, F, and the amount bound to albumin, B.

$$FA_{heptane} \rightleftharpoons FA_{unbound} + FA_{albumin} \quad (1)$$

It is necessary to measure B and F separately, for these are the values required to assess binding. The information to calculate B and F from the total value is obtained from the second set of vials, those that contain no albumin in the aqueous solution. Each of these flasks gives a partition ratio, the distribution of the fatty acid at equilibrium between heptane and the aqueous phase.[28] At equivalent fatty acid concentrations in heptane, the aqueous concentration in these vials must be equal to the unbound fatty acid concentration in the aqueous phase in Eq. (1). This is the value of F. By subtracting this value from the total fatty acid present in the aqueous phase containing albumin, the amount bound to albumin, B, is obtained.

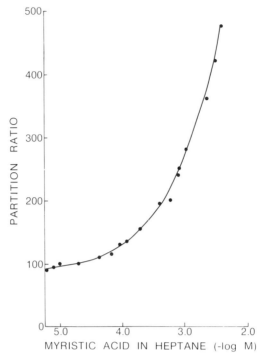

FIG. 2. Partition of myristic acid (14:0) between n-heptane and a buffer solution containing 16 mM Na$_2$HPO$_4$, pH 7.4. The incubation temperature was 37°. In order to spread out the plot evenly, the content of myristic acid in the heptane solution is given as the negative logarithm of the molar concentration.

Partition Ratio

The partition of a fatty acid between a hydrocarbon and aqueous solution is not a constant value; it depends on the fatty acid concentration. This is illustrated by the data shown in Fig. 2, a plot of the partition ratio relative to the fatty acid concentration in n-heptane for myristic (tetradecanoic) acid at 37°. The results should be plotted showing the concentration in heptane because this is the value that is needed to apply these results to the corresponding flasks containing albumin. As shown in Fig. 2, the fatty acid concentration in heptane usually is plotted as a logarithmic function in order to spread out the values evenly over the entire range that is examined. From these results for myristic acid, a value of 86 is calculated for the partition coefficient, the distribution at infinite dilution.[29] Using this value and the partition curve, the concentration of

[29] R. B. Simpson, J. D. Ashbrook, E. C. Santos, and A. A. Spector, *J. Lipid Res.* **15,** 415 (1974).

myristic acid in the aqueous solution can be calculated at all concentrations that are likely to be present in the heptane solution of the incubations containing albumin using Eq. (2).

$$[FA]_{aqueous} = \frac{(FA)_{heptane}}{\text{Partition ratio}} \quad (2)$$

Binding Isotherm. The next step is to utilize the partition results and the values obtained from the set of vials containing albumin to construct a binding isotherm. For each fatty acid concentration in the heptane phase of the set of vials containing albumin, the unbound aqueous fatty acid concentration can be calculated from the partition graph. This value is F, the unbound aqueous concentration of fatty acid in moles/liter. To obtain the quantity of fatty acid bound to albumin, the value of F is subtracted from the total fatty acid concentration in the corresponding aqueous solution containing albumin. Since the albumin concentration is known, the value of B, the fatty acid bound to albumin in moles/mole protein, can be calculated. This is done for all of the flasks containing albumin, and an additional value, B/F, is calculated in each case. The series of values are plotted in the Scatchard format, B/F as a function of B, as shown in Fig. 3 for myristic acid binding to bovine albumin.

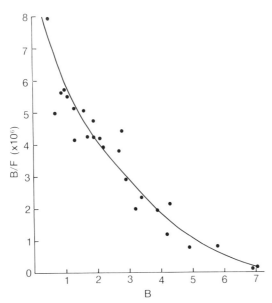

FIG. 3. Scatchard plot of myristic acid binding to bovine plasma albumin. The conditions are similar to those given in Fig. 2.

Unbound Fatty Acid Association. The fatty acid anion binds to albumin. However, in constructing a binding isotherm such as Fig. 3, the total fatty acid concentration, not the fatty anion concentration actually is used. This approximation is introduced because it is difficult to measure only the anion concentration. Initially, the artifact caused by this approximation was thought to be insignificant because the fatty acid pK_a is about 4.8. Since binding measurements usually are made at pH 7.0 or above, almost all of the fatty acid in the aqueous solution should be in anionic form. This assumption is not strictly valid for fatty acids containing 12 or more carbon atoms, at least when the heptane–water partition system is employed.[30] The probable explanation is the formation of fatty anion dimers in the aqueous solution,[31] although other associated forms of fatty acid also may be present.[32,33]

The partition function takes all of these possibilities into account, as is shown in Eq. (3)

$$PR = \frac{C_h}{C_w} = \frac{[HA]_h + 2[(HA)_2]_h}{[HA]_w + [A^-]_w + \sum_{m=0}\sum_{n=2} n[H_mA_n^{-(n-m)}]_w} \quad (3)$$

where C_h is the total fatty acid concentration in heptane, C_w is the total fatty acid concentration in water, $[HA]_h$ is the acid monomer concentration in heptane, $[(HA)_2]_h$ is the acid dimer concentration in heptane, $[HA]_w$ is the acid monomer concentration in water, $[A^-]_w$ is the anion monomer concentration in water, and $n[H_mA_n^{-(n-m)}]_w$ is the concentration of all of the various associated forms in water.[29] It is possible to calculate $[A^-]_w$ for each value of C_h, which would provide the actual unbound fatty anion monomer concentration for use in constructing a binding isotherm. To simplify the analysis, however, one usually calculates C_w as a function of C_h and, for the reasons indicated above, assumes that C_w is a reasonable approximation of $[A^-]_w$. This was done in plotting Fig. 2 and in subsequently using the results to calculate the binding isotherm shown for myristic acid in Fig. 3.

There is uncertainty as to whether premicellar association is a real phenomena or simply an artifact of the aqueous–hydrocarbon partition system. One reason for this uncertainty is that no premicellar association is detected by electron spin resonance spectroscopy when spin-labeled stearic acid probes are dissolved in an aqueous buffer.[23]

[30] D. S. Goodman, *J. Am. Chem. Soc.* **80**, 3887 (1958).
[31] P. Mukerjee, *J. Phys. Chem.* **69**, 2821 (1965).
[32] R. Smith and C. Tanford, *Proc. Natl. Acad. Sci. U.S.A.* **70**, 289 (1973).
[33] J. D. Ashbrook, A. A. Spector, E. C. Santos, and J. E. Fletcher, *J. Biol. Chem.* **250**, 2333 (1975).

Correction of Binding Data for Fatty Acid Association. The extent of premicellar association in the biphasic heptane–water system is dependent on fatty acid concentration and becomes appreciable when the concentration in the aqueous solution exceeds 1 μM.[29-32] At a given concentration, the degree of association of saturated fatty acids increases as the chain length increases.[29,31] To compensate for premicellar association, the binding isotherm for long-chain fatty acids obtained by the equilibrium partition procedure often is corrected by calculating the fraction of the total unbound fatty acid concentration that is in anionic form.[33] The correction can be considerable. For example, the first association constant (K_1) for stearic acid binding to human albumin increases from 1.5×10^8 to $9.1 \times 10^8\ M^{-1}$ when a correction for unbound stearic acid association is applied to the binding results.[33] By contrast, application of this correction only increases K_1 for myristic acid binding to human albumin from 2.1×10^7 to $2.5 \times 10^7\ M^{-1}$.[33]

Analysis of Fatty Acid Binding Data

Data for myristic acid binding to albumin, such as in Fig. 3, indicate that a single molecule of albumin must contain multiple binding sites for fatty acid. The reason that the plot is not linear is that the binding sites are heterogeneous.[25] They have been grouped into three different classes: strong sites that are specific for long-chain fatty acids, intermediate sites that are generally available to all organic anions, and weak sites that initially interact with the anion and facilitate its entry to the stronger sites.[28] To complete the binding analysis, it is necessary to estimate the number of binding sites and calculate the binding constants using an appropriate analytical method.

Scatchard Model

This analysis can be done graphically[28] or with computer assistance.[7] The binding data in the form of B and F are fitted to Eq. (4)

$$B = \frac{n_1 k_1 F}{1 + k_1 F} + \frac{n_2 k_2 F}{1 + k_2 F} + \cdots + \frac{n_n k_n F}{1 + k_n F} \qquad (4)$$

where each term represents a separate class of binding sites, n_i is the number of sites in the class, and k_i is the average binding constant of these sites. The simplest Scatchard model that provides a good fit for fatty acid binding data, such as for the results shown in Fig. 3, contains three classes of sites.[7,28] The primary class contains either 2 or 3 sites, and the k_1 is in the range of $10^7\ M^{-1}$. These sites are specific for long-chain fatty acids. There are 3 to 5 sites in the secondary class, the k_2 is in the range of 10^{-6} to 10^{-5} M^{-1}, and these sites are generally available to organic compounds, including many drugs. The third class is large, containing between 20 and 60 sites. They bind fatty acids very weakly, the k_3 being in the range of 10^{-3} to 10^{-2} M^{-1}, and these sites do not contribute appreciably to the binding capacity of albumin under physiologic condition. Rather, they are readily accessible for fatty acid attachment on the surface of the globular structure. The tertiary sites interact rapidly with fatty acids and then facilitate their entry to the primary and secondary binding sites which are located deeper within the globular structure.[12] The kinetics of this rearrangement are first order, and it is limited by a negative entropy of activation.[12]

A major advantage of the Scatchard model is that good estimates of n_i or k_i can be obtained by graphic extrapolation. Since the only terms that are of real physiologic importance are n_1, n_2, k_1, and k_2, the information is easy to utilize. This model, however, has several disadvantages. Often, the separation of sites into the n_1 and n_2 categories is somewhat arbitrary, and the value selected for n_i has a big influence on the value of the corresponding k_i. In addition, the model does not take cooperative binding effects into account. Notwithstanding, the Scatchard model is the most widely used formulation for fatty acid binding data.

Multiple Binding Model

This analysis is more complicated and can only be carried out efficiently with computer assistance.[34] The values of B and F are fitted by a nonlinear least-squares procedure to Eq. (5)

$$B = \frac{K_1 F + 2K_1 K_2 F^2 + \cdots + nK_1 K_2 \cdots K_n F^n}{1 + K_1 F + K_1 K_2 F^2 + \cdots + K_1 K_2 \cdots K_n F^n} \tag{5}$$

where K_i are the binding constants for each sequential fatty acid that associates with the protein. The big advantage of the multiple binding analysis is that the values of K_i do not depend on a somewhat arbitrary subdivision of the binding sites into classes. Analysis of fatty acid binding by this model indicates that the K_i occur in decreasing order, indicating that major positive cooperative effects do not accompany multiple binding.[7] Moreover, as predicted from the graphic analysis, the K_i do not separate into distinct classes. For long-chain fatty acids, K_1 is in the range of 10^{-8} M^{-1}, K_2 between 10^{-8} and 10^{-7} M^{-1}, and K_3 between 5×10^{-7} and 5×10^{-6} M^{-1}.

Distribution Analysis. The values for the individual binding constants calculated by the multiple binding analysis are not widely separated.

---
[34] J. E. Fletcher, A. A. Spector, and J. D. Ashbrook, *Biochemistry* **9**, 4580 (1970).

Therefore, binding is not localized to only one site at a time. Instead, each mole of fatty acid is spread over several binding sites having K_i that differ by only 5- to 10-fold. The distribution can be calculated from Eq. (6)

$$P_i = \frac{[PF_i]}{[P_T]} = \frac{K_1 K_2 \cdots K_i F^i}{1 + K_1 F + K_1 K_2 F^2 + \cdots + K_1 \cdots K_n F^n} \quad (6)$$

where P_i is the fraction of the albumin that has i molecules of fatty acid, PF_i is the albumin that contains fatty acid, P_T is the total albumin, and K_i and F are the same as in Eq. (5). Several results that are not intuitively obvious become apparent from a distribution analysis. For example, not all albumin molecules contain one bound fatty acid when the molar ratio of fatty acid to albumin is 1, only 43% do. There are other species of albumin in such a solution; 31% do not contain fatty acid, 22% contain two fatty acid molecules, 3.6% contain 3, and 0.4% contain 4.[35] This is important regarding physiologic considerations. Even though the molar ratio of free fatty acid to albumin in blood plasma usually is between 0.3 and 1.0 and rarely exceeds 2, more than the two primary binding sites are involved in fatty acid transport. Small quantities of fatty acid are present at the third, fourth, and probably fifth binding site so that these sites play a biologic role. Furthermore, because some fatty acid is present at these secondary sites, there is a possibility for competitive or even complementary binding interactions with other types of organic ligands at these secondary sites. Heretofore, such interactions were explained entirely on the basis of conformational changes.[7,36]

Effect of Fatty Acid Structure on Binding. Fatty acid containing less than 12-carbon atoms bind only to the secondary and tertiary sites; they apparently do not associate with the two primary sites.[27,28] The longer fatty acids bind to all three classes of sites.[28] A considerable decrease in binding occurs when the carboxyl group is converted to a methyl ester, a primary alcohol, or a methyl group.[7] Likewise, binding is reduced when the positively charged amino acid side chains of albumin are modified.[7] This has been interpreted to indicate that ionic interactions between the carboxylate anion of the fatty acid and positively charged lysine or arginine residues of the protein are important in the binding process.[7] Likewise, the strength of association increases as the chain length of a saturated fatty acid increases. The log K_i increases by about 5.0 units as the chain-length increases from 6- to 18-carbon atoms.[33] Oleic acid, which

---

[35] A. A. Spector and J. E. Fletcher, *in* "Disturbances in Lipid and Lipoprotein Metabolism" (J. M. Dietschy, A. M. Gotto, Jr., and J. A. Ontko, eds.), p. 236. Amer. Physiol. Soc., Washington, D.C., 1978.

[36] A. A. Spector, E. C. Santos, J. D. Ashbrook, and J. E. Fletcher, *Ann. N.Y. Acad. Sci.* **226**, 247 (1973).

contains a single cis double bond in the 9,10 position, is bound more tightly than the corresponding 18-carbon atom saturated fatty acid, stearic acid.[33] The presence of a second double bond, however, appears to weaken the association. Linoleic acid, an 18-carbon atom dienoic fatty acid containing cis double bonds in the 9,10 and 12,13 positions, is bound less tightly than stearic acid. The K_1 for linoleic acid is about $8 \times 10^7 \, M^{-1}$, about one-half as large as the K_1 for oleic acid.[33] For arachidonic acid, a 20-carbon atom tetraenoic acid, K_1 is about $3 \times 10^7 \, M^{-1}$.[37] This suggests that increases in the number of cis double bonds above two produces a smaller reduction in the strength of binding than the introduction of the second double bond.

Albumin Structure and Lipid Transport

Bovine albumin is composed of a single polypeptide chain containing 581 amino acid residues and having a molecular weight of about 66,200. There are many regions of homology in the amino acid sequence, and the polypeptide chain is folded into three similar globular domains, designated I, II, and III, composed of 185, 192, and 204 amino acid residues, respectively.[38] A schematic representation of the structure of bovine albumin, adapted from the work of Brown,[38] is shown in Fig. 4. Each domain contains three loops of the polypeptide chain, and these are held together by disulfide bonds. There are a total of nine such loops, and they are held in place by 17 disulfide bonds. One cysteine residue, contained in domain I in loop 1, in present in the reduced form. Bovine albumin contains two tryptophan residues: one in domain I in loop 3, and another in domain II in loop 4 that is located in a crevice between the first two domains. This is a major difference from human plasma albumin, which contains only the tryptophan residue in loop 4. When fatty acids bind to bovine albumin, the tryptophan fluorescence is quenched by as much as 44%, and there is a shift in the maximum fluorescence to lower wavelengths.[17] A similar effect does not occur with human albumin,[17] suggesting that quenching involves the tryptophan residue of bovine albumin that is located in the first domain. Fatty acid binding is thought to cause a conformational change in the region of this tryptophan. This leads to a change in the ionization state of an adjacent lysine residue, producing the quenching.[39] The decrease in the wavelength of maximum fluorescence probably is due to movement of one of the tryptophan residues into a less polar environment.[1]

[37] L. Savu, C. Benassayag, G. Vallette, N. Christeff, and E. Nunez, *J. Biol. Chem.* **256**, 9414 (1981).
[38] J. R. Brown, *Fed. Proc., Fed. Am. Soc. Exp. Biol.* **35**, 2141 (1976).
[39] C. J. Halfman and T. Nishida, *Biochim. Biophys. Acta* **243**, 294 (1971).

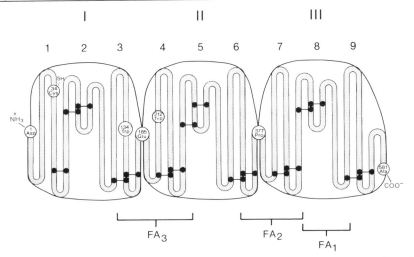

FIG. 4. Schematic representation of the structure of bovine plasma albumin. This figure is adapted from the studies of Brown.[38] The domains are designated as I, II, and III, beginning at the amino terminal residue of the polypeptide chain. The loops of the polypeptide chain are designated 1–9, again beginning at the amino terminus. Each of the disulfide bonds that hold the looped structure together is indicated by the heavy cross-link. The approximate localization of the three strongest fatty acid binding sites, designated as FA_1, FA_2, and FA_3, respectively, also is shown.

The primary and secondary fatty acid binding sites are located between or within loops lined with hydrophobic amino acid side chains. The hydrocarbon tail of the fatty acid interacts with these nonpolar regions, providing most of the binding energy.[7] In addition, the ionized fatty acyl carboxyl group interacts with a positively charged lysine or arginine residue situated near the surface of the protein at the entrance to the binding site.[7,25] Two methods have been employed to localize the binding sites within the albumin structure. One is fluorescence spectroscopy; the other is proteolytic cleavage followed by equilibrium binding to the resulting protein fragments.[40] Both methods of analysis indicate that the strongest fatty acid binding site is located within the third domain, in the region of polypeptide loop 8, about 100 amino acid residues removed from the carboxyl terminus of the protein.[18,40] The second primary binding site is located either between loops 7 and 8, adjacent and antiparallel to the strongest site,[19] or between loops 6 and 7 near the juncture of the second and third domains.[40] The third strongest binding site, which is shared by fatty acids and other organic ligands, is situated in the region of loops 3

[40] R. G. Reed, R. C. Feldhoff, O. L. Clute, and T. Peters, Jr., *Biochemistry* **15**, 5394 (1976).

and 4 at the junction of the first and second domains.[40] No other fatty acid binding sites have been localized, although kinetic measurements suggest that the weak class of tertiary sites is present at or near the surface of the globular structure.[12] Fatty acid binding is associated with small conformational changes that allow the fatty acid to penetrate into the interior of the globular protein structure and enable the albumin binding sites to adapt to the configuration of the fatty acyl chain.[41]

Tissue Uptake. A small amount of unbound fatty acid, less than 0.1% of the total free fatty acid under physiologic conditions, is in equilibrium with the fatty acid bound to albumin as indicated in Eq. (1).[7] Almost all of this is in ionized form. Even though the unbound concentration under physiologic conditions is in the range of 10^{-8} to 10^{-6} M, it probably is the species that is taken up by the tissues.[42] As uptake occurs, bound fatty acid rapidly dissociates to maintain the equilibrium, replacing the pool of unbound fatty acid and providing a continuous supply for uptake. An alternative mechanism has been suggested, based on the finding that the liver contains an albumin receptor.[43] According to this view, the dissociation of fatty acid from albumin is facilitated when albumin binds to this plasma membrane receptor. Even in this case, however, albumin does not appear to be involved in the actual transfer of fatty acid across the cell membrane. There is general agreement that almost all of the fatty acid dissociates from albumin when it leaves the capillary or is taken up by a cell and that albumin itself does not act as a transmembrane carrier for fatty acid.

Comparison with Plasma Lipoproteins. Like albumin, the lipoproteins play an important role in plasma lipid transport. Yet, there are fundamental differences in the structure of these two types of lipid transport complexes and in the mechanisms by which cells utilize their lipid payloads. Plasma lipoproteins contain from 50 to 98% lipid, and they have a micellar structure into which one or more proteins components are embedded. By contrast, the fatty acid–albumin complex consists at most of about 1% lipid. In lipoproteins, lipid solubilization is conferred by the micellar structure, with the surface phospholipids playing a major role. The various protein components function as enzyme cofactors, receptor recognition factors, or lipid transfer factors. Some of these apolipoproteins probably also are involved in the structural organization of the mixed micelle. By contrast, in a fatty acid–albumin complex, the protein moiety serves primarily to solubilize the lipid. It is the reservoir that enables a relatively

---

[41] W. Scheider, *Ann. N.Y. Acad. Sci.* **303**, 47 (1977).
[42] A. A. Spector, *Ann. N.Y. Acad. Sci.* **149**, 768 (1968).
[43] R. Weisiger, J. Gollan, and R. Ockner, *Science* **211**, 1048 (1981).

high concentration of long-chain fatty acid to exist in the plasma in unesterified form. Major differences also exist with respect to the role of the protein in metabolism of the lipid payload. Albumin plays a passive role in free fatty acid utilization, whereas the protein constituents of the plasma lipoproteins direct and regulate the metabolism of the lipoprotein lipids. This probably reflects that fact that albumin-bound fatty acid is utilized by nearly all tissues, whereas lipoproteins are targeted to specific tissues that contain the appropriate receptors or lipolytic enzymes. Except through differential release of individual fatty acids because of different binding affinities,[33] albumin probably does not play an analogous role in directing fatty acids to either specific tissues or metabolic pathways.

Acknowledgment

These studies were supported by Arteriosclerosis Specialized Center of Research Grant HL14,230 from the National Heart, Lung and Blood Institute, National Institutes of Health.

[18] High-Performance Liquid Chromatography of Apolipoproteins

By CELINA EDELSTEIN and ANGELO M. SCANU

From the preparative viewpoint, procedures to separate apolipoproteins in their lipid-free form have mainly been based on conventional molecular sieving and/or ion-exchange chromatography. Successful examples have been the isolation of the apolipoproteins obtained from HDL, i.e., apoA-I and apoA-II as well as the apoCs obtained either from apoHDL or apoVLDL. More recently techniques based on discrete differences in charge such as preparative gel isoelectric focusing and chromatofocusing have also been applied. The table[1–20] summarizes the gen-

[1] A. C. Nestruck, G. Suzue, and Y. L. Marcel, *Biochim. Biophys. Acta* **617**, 110 (1980).
[2] A. M. Scanu, J. Toth, C. Edelstein, S. Koga, and E. Stiller, *Biochemistry* **8**, 3309 (1969).
[3] D. Polacek, C. Edelstein, and A. M. Scanu, *Lipids* **16**, 927 (1981).
[4] C. Edelstein, C. T. Lim, and A. M. Scanu, *J. Biol. Chem.* **247**, 5842 (1972).
[5] G. S. Ott and V. G. Shore, *J. Chromatogr.* **231**, 1 (1982).
[6] W. S. Hancock, H. J. Pownall, A. M. Gotto, and J. T. Sparrow, *J. Chromatogr.* **216**, 285 (1981).
[7] A. M. Scanu, C. T. Lim, and C. Edelstein, *J. Biol. Chem.* **247**, 5850 (1972).
[8] C. T. Lim, J. Chung, H. J. Kayden, and A. M. Scanu, *Biochim. Biophys. Acta* **420**, 332 (1976).

eral procedures for apolipoprotein separation and also lists the various techniques for further subfractionation of the isoforms of some of these proteins.

In an effort to improve both the speed and resolution of the fractionation of peptides and proteins, a new and rapidly developing chromatographic technique, high-performance liquid chromatography (HPLC), has emerged. This type of chromatography has been made possible by the development of support materials characterized by a high mechanical stability under pressure. HPLC has recently been extended to the fractionation of lipoproteins and apolipoproteins and currently used in lieu of and as a complement to previous analytical and preparative separation techniques. In this chapter we will focus on the utilization of HPLC for the separation of the apolipoproteins contained in HDL and VLDL with particular regard to the gel permeation, anion exchange, and reverse phase modes. HPLC of lipoproteins will be discussed in Vol. 129 [4].

General Comments on the Separation of Apolipoproteins

In general, the apolipoproteins derived from HDL, apoA-I, apoA-II, and the C peptides are readily solubilized in aqueous buffers of low molarity (5 to 50 mM) at neutral to basic pH (pH 7–8.5).[21] Although higher ranges of buffer molarity can be used, we have observed that above 50 mM at pH 7.5 the solubility of apoHDL increases as a function of time and that above a protein concentration of 20 mg/ml complete dissolution is not possible. Solubility of apoHDL decreases dramatically below pH 7;

---

[9] Y. L. Marcel, M. Bergseth, and A. C. Nestruck, *Biochim. Biophys. Acta* **573**, 175 (1979).
[10] W. S. Hancock, C. A. Bishop, A. M. Gotto, D. R. K. Harding, S. M. Lamplugh, and J. T. Sparrow, *Lipids* **16**, 250 (1981).
[11] C. E. Sparks and J. B. Marsh, *J. Lipid Res.* **22**, 514 (1981).
[12] B. Shore and V. G. Shore, *Biochemistry* **12**, 502 (1973).
[13] K. H. Weisgraber and R. W. Mahley, *J. Biol. Chem.* **253**, 6281 (1978).
[14] P. Weisweiler and P. Schmidt, *Clin. Chim. Acta* **124**, 45 (1982).
[15] D. Pfaffinger, C. Edelstein, and A. M. Scanu, *J. Lipid Res.* **24**, 796 (1983).
[16] V. I. Zannis, J. L. Breslow, and A. J. Katz, *J. Biol. Chem.* **255**, 8612 (1980).
[17] G. Schmitz, K. Ilsemann, B. Melrik, and G. Assman, *J. Lipid Res.* **24**, 1021 (1983).
[18] V. I. Zannis and J. L. Breslow, *Biochemistry* **20**, 1033 (1981).
[19] G. S. Ott and V. G. Shore, *in* "Handbook of Electrophoresis" (L. A. Lewis, ed.), Vol. 3, p. 106. CRC Press, Boca Raton, Fla., 1983.
[20] A. M. Scanu, C. Edelstein, and B. W. Shen, *in* "Lipid Protein Interactions," (P. Jost and O. Hayes Griffith, eds.), Vol. 1, p. 260. Wiley, New York, 1982.
[21] D. M. Lee, A. J. Valente, W. H. Kuo, and H. Maeda, *Biochim. Biophys. Acta* **666**, 133 (1981).

Protein	MW[a]	Distribution	pI	Separation method – Conventional	Separation method – HPLC
ApoA-I	28,331	HDL Lymph Chyl.	>6.5, 5.85, 5.74, 5.64 (major), 5.52, 5.40[16]	Preparative isoelectric focusing in 7 M urea[1] Gel filtration on Sephadex G-200 in 6 M urea[2] Anion-exchange chromatography in 6 M urea[4]	Molecular sieving on TSK 3000 columns in 6 M urea[3] Anion-exchange on Synchropak AX 300 in 6 M urea[5] Ion-paired reversed-phase chromatography[6]
ApoA-II	17,400	HDL Lymph Chyl.	5.16, 4.89 (major), 4.58, 4.31[17]	Gel filtration on Sephadex G-200 in 8 M urea[2] Anion-exchange chromatography in 6 M urea[7]	Molecular sieving on TSK 3000 columns in 6 M urea[3] Anion-exchange on Synchropak AX 300 in 6 M urea[5] Ion-paired reversed-phase chromatography[6]
ApoC-I	6,630	Chyl. VLDL HDL	6.5[9]	Anion-exchange chromatography in 6 M urea[8] Preparative isoelectric focusing[9]	Anion-exchange on Synchropak AX 300 in 6 M urea[5] Ion-paired reversed-phase chromatography[10]
ApoC-II	8,500	Chyl. VLDL HDL	4.86 (major)[9], 4.69 (minor)	Same as for apoC-I	Same as for apoC-I
ApoC-III	8,764	Chyl. VLDL HDL	5.02, 4.82, 4.62[9]	Same as for apoC-I	Same as for apoC-I
ApoE	34,145	Chyl. VLDL IDL	6.02, 5.89, 5.78, 5.70, 5.61[18]	Gel filtration in SDS[11] Ion-exchange chromatography in 8 M urea[12] Preparative polyacrylamide gel electrophoresis[13] Chromatofocusing[14]	Molecular sieving on BioGel TSK 50, Bio Sil TSK 400, and TSK 3000 in 4 M GdmCl[15]
ApoB	?	VLDL IDL LDL	3.9, 4.75, 5.27, 5.60[19]	Gel filtration in SDS[11]	Same as for apoE

[a] The molecular weight was calculated from the covalent structure of the apolipoprotein.

above pH 8.5 the dissolved protein may undergo deamidation. Contrary to apoA-I, apoA-II, and apoC, apoB and apoE can only be solubilized in the presence of denaturing agents, i.e., urea, guanidine–hydrochloride (GdmCl), or detergents, i.e., sodium dodecyl sulfate (SDS).[20,22] ApoE may be solubilized in basic buffers (100 mM, pH 8) but its solubility is limited to a protein concentration in the range of 1 mg/ml. Thus in a mixture of proteins derived from HDL and VLDL, a gross separation among them may be achieved based on their solubilities in nondenaturing buffers.

Although apoHDL is readily soluble in nondenaturing buffers, the individual proteins self-associate in a concentration-dependent manner.[22,23] In physiological buffer this property hinders the separation of the apolipoproteins into single molecular weight species unless urea or GdmCl is used during the separation procedure. In the presence of denaturants, apoB is soluble but it is still unclear whether one is dealing with a monomer or a highly aggregated form. ApoE is dissociated into monomers in 4 M GdmCl or 6 M urea solution.

The table lists the molecular weights of the apolipoproteins derived from HDL and VLDL. Except for apoB, whose molecular weight remains controversial, all of the apolipoproteins listed in the table have been sequenced.[20,23] Under denaturing conditions the apolipoproteins derived from HDL (apoA-I, apoA-II, apoC) acquire their monomer molecular weight and can be separated from each other using molecular sieve chromatography.[2] If separation cannot be accomplished in this manner, exploitation of their charge differences can accomplish the task.

The isoelectric points of the various apolipoproteins listed in the table are good indicators as to how separation based on ionic charge can be achieved. ApoHDL can be fractionated using anion-exchange chromatography in the presence of urea. In this procedure the C peptides are separated as well as the isoforms of apoC-III and those of apoA-I. ApoVLDL under similar conditions also yields the apoC-III isoforms which are in turn separated from apoE.[12] Due to its high resolution (0.01 pH unit), preparative isoelectric focusing on granular beds has been used frequently to isolate isomorphic forms of apoC-III, apoA-I, apoA-II, and apoE.[1,9,13] For analytical purposes, two-dimensional gel electrophoresis appears to be the most powerful tool.[24] This method combines the features of charge and size separation with the resulting fingerprint of the individual pro-

[22] J. C. Osborne, Jr. and H. B. Brewer, Jr., *Adv. Protein Chem.* **31**, 253 (1977).
[23] S. C. Rall, K. H. Weisgraber, and R. W. Mahley, *J. Biol. Chem.* **25**, 4171 (1982).
[24] P. H. O'Farrel, *J. Biol. Chem.* **250**, 4007 (1975).

teins. As with other electrophoretic techniques, the protein bands can be recovered by electroelution.

Methodology

Gel Permeation High-Performance Liquid Chromatography

Gel permeation chromatography permits the separation of components according to their differential ability to permeate a porous packing, the larger molecular components eluting ahead of the smaller ones. The technique of HPLC as applied in the molecular sieving mode[25] has been invaluable in the isolation, purification, and quantitative analysis of proteins and peptides. As with any chromatographic method attention to elution conditions and sample preparation will result in successful and reproducible separations. In the case of HPLC, reproducibility is also strongly dependent on column maintenance and regeneration. In contrast to the compressible gels used in conventional steric exclusion chromatography, supports for HPLC are rigid[25,26] permitting operation at higher flow rates thus enabling shorter analysis times. The experimental procedures described below are an outgrowth of the experience gained in our laboratory in separating the apolipoproteins of apoHDL and apoVLDL using molecular sieve HPLC.[3,15]

ApoHDL. Delipidation. HDL of d 1.063–1.21 g/ml is dialyzed against 0.15 M NaCl, pH 7.0 solution containing 0.02% sodium azide, and 1 mM EDTA. Delipidation with a mixture of ethanol/diethyl ether, 3:2 v:v is carried out at −20° as described by Scanu and Edelstein.[27] The protein precipitate is washed with anhydrous diethyl ether overnight and the final ether powder is dried and stored under nitrogen or preferably argon gas at −80°.

Preparation of Sample. Ten milligrams of apoHDL is dissolved at room temperature in 1 ml of 6 M urea solution previously adjusted to pH 3.15 with formic acid. Solubilization is rapid, and the sample is filtered thru a 0.2-μm polycarbonate filter (Nucleopore, Pleasanton, CA). Each injection volume is 200 μl and contains 2 mg of apoHDL.

Preparation of Mobile Phase (6 M Urea, pH 3.15). Urea is crystallized from ethanol and dissolved in deionized and double distilled water to a concentration of 6 M. This solution is then adjusted to pH 3.15 with

---

[25] F. E. Regnier and K. M. Gooding, *Anal. Biochem.* **103**, 1 (1980).
[26] F. E. Regnier, *Science* **222**, 245 (1983).
[27] A. M. Scanu and C. Edelstein, *Anal. Biochem.* **44**, 576 (1971).

formic acid. The final solution is filtered right before use thru a 0.4-μm polyester filter (Nucleopore Co., Pleasanton, CA).

Chromatographic System. Chromatography is performed in a Varian Model 5000 liquid chromatograph equipped with a Valco Model C6U loop injector and Varian Model UV-100 variable wavelength detector connected to a Vista CDS 401 data collector (Varian Instruments Group, Walnut Creek, CA). Two analytical columns (300 × 7.5 mm) TSK gel 3000 SW are connected in series and are preceded by a guard column (75 × 7.5 mm) TSK guard column SW. The columns are manufactured by Toyo Soda Manufacturing Co., LTD., Tokyo, Japan and are purchased prepacked. These columns are size exclusion supports which consist of a hydrophilic phase chemically bonded to microparticulate silica. The columns permit the permeation of proteins and dextran in the molecular weight range of 1000–300,000 and 1000–100,000, respectively.

Chromatographic Conditions. The columns are equilibrated with the mobile phase prior to the first protein sample injection, at a flow rate of 0.5 ml/min. ApoHDL is injected in a volume of 200 μl at a concentration of 10 mg/ml. The flow rate is maintained at 0.5 ml/min at a pressure of 18 atm and 26°. The eluted fractions are continuously monitored at 280 nm, recorded, and collected by means of a fraction collector. Selected fractions are pooled and either desalted through Sephadex G-25 (Pharmacia) or dialyzed at 4° against 5 mm NH_4HCO_3. A typical fractionation of apoHDL is shown in Fig. 1. ApoA-I and apoA-II are well separated from each other and from the C peptides. A shoulder on the descending portion of apoA-II can be resolved upon rechromatography on the same column system (Fig. 1, inset). This protein has not been identified but is under investigation.

Comments. Under the chromatographic conditions described, maximum resolution can be attained with a 90% mass recovery up to a protein load of 3 mg. Good resolution is obtained at a flow rate of 1 ml/min, however, for a quantitative study where high yields of pure protein fractions are needed, a flow rate of 0.5 ml/min is recommended. In this context, particular attention should also be paid to the delipidation step. Completeness of delipidation is essential. Poor delipidation will result in lower yields of apoA-I and incomplete resolution between apoA-I and apoA-II. The use of urea in acidic solution prevents cyanate formation which otherwise may lead to the carbamylation of lysine residues in the protein resulting in the production of artifactual isoforms.[28] As an alternative, GdmCl (4 *M*) may be used as the denaturant in a buffer at pH 7.0.

[28] J. J. T. Gerding, A. Kappers, P. Hagel, and H. Bloemendal, *Biochim. Biophys. Acta* **243**, 374 (1971).

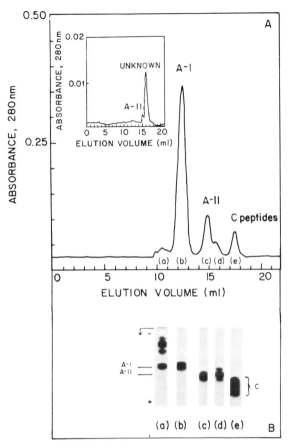

FIG. 1. (A) Elution pattern from the HPLC column of apoHDL. Conditions: injected volume, 100 μl at 10 mg/ml of apoHDL; flow rate, 1 ml/min; chart speed, 1 cm/min. (B) SDS gel electrophoretic patterns of eluted fractions from a thorough e. Inset: elution profile of rechromatographed fraction d (shoulder of apoA-II peak). Data from D. Polacek, C. Edelstein, and A. M. Scanu, *Lipids* **16**, 927 (1981).

ApoVLDL. Delipidation. Two milliliters (2–4 mg/ml) VLDL of d 1.006g/ml is delipidated with 50 ml of a 2 : 1 chloroform–methanol mixture at room temperature followed by five peroxide-free ethyl ether washes at 4°. The final apoVLDL preparation is dried under nitrogen and the powder stored at −70°.

Preparation of Sample. Ten milligrams of apoVLDL is dissolved at room temperature in 1 ml of 0.01M Tris–HCl, 0.01 M dithiothreitol (DTT), 6 M GdmCl solution, pH 7.0 containing 1 mM phenylmethylsulfonylfluoride. After a clear solution is obtained (2–3 hr) the sample is filtered through a 0.2-μm polycarbonate filter (Nucleopore) and 200-μl aliquots containing 2 mg of protein are used for each chromatographic injection.

Mobile Phase. The columns are preequilibrated and eluted with a 4 M GdmCl solution in 0.01 M Tris, 0.01 M DTT, pH 7.0, which has been filtered through a 0.4-μm Millipore filter. The inclusion of DTT is to promote solubilization of apoB and the dissociation of apoE, a portion of which is disulfide-linked to apoA-II.

Chromatographic System. The column system is composed of a BioGel TSK Guard column, 75 × 7.5 mm (Bio-Rad, Richmond, CA), followed by a BioGel TSK 50, 300 × 7.5 mm column (Bio-Rad), a BioSil TSK 400, 300 × 7.5 mm column (Bio-Rad), and a Spherogel TSK 3000, 300 × 7.5 mm column (Altex, Berkely, CA), all connected in series. BioSil TSK 400 and Spherogel TSK 3000 are composed of spherical porous silica gel. The BioGel TSK 50 is a hydroxylated polyether-based material. The liquid chromatographic instrument and peripherals are the same as previously described for the apoHDL system.

Chromatographic Conditions. The columns are preequilibrated for 3 hr with the mobile phase at a flow rate of 0.5 ml/min. The sample is injected in a volume of 200 μl at a concentration of 10 mg/ml. The back pressure with the three column system is 40 atm. The effluents are continuously monitored at 280 nm and fractions of 0.25 ml/tube are collected. Samples from each peak are pooled and can be desalted immediately for use or stored at −70°.

A typical HPLC profile of apoVLDL is shown in Fig. 2. The 34, 48.7, and 58.3 min components correspond to apoB, apoE, and apoC peptides, respectively. The 44 min component is dependent upon the length of incubation of the apoVLDL in the solubilizing buffer. A 3 hr incubation time is sufficient to eliminate this shoulder and suggests that it corresponds to the apoE–A-II disulfide-linked complex.

Comments. As noted for apoHDL, the success of the fractionation is strongly dependent on the completeness of the delipidation step. Repetitive rinsing with peroxide-free diethyl ether ensures complete removal of residual lipids and results in a white powder which is easily solubilized in the 6 M GdmCl solution. Filtration through the 0.22-μm polycarbonate filter results in a 10% mass loss mainly due to apoB. A loading of 2 mg of apoVLDL is considered to be ideal because the resolution is at an optimum and the total column recovery for apoVLDL is 93%. Under identical conditions, rechromatography of apoE results in a recovery of 80%. Moreover, pure samples of apoB and apoC can be obtained.

Maintenance of Columns. The lifetime of columns of 6 to 10 months can be prolonged if certain precautions are followed. In the presence of

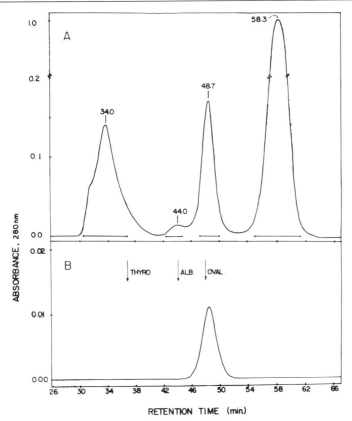

FIG. 2. (A) HPLC profile of apoVLDL. Conditions: injected volume, 200 µl at 10 mg/ml of apoVLDL; flow rate, 0.5 ml/min; chart speed, 1 cm/min. The number of each peak maximum indicates the retention time. (B) Rechromatography of the 48.7 min fraction in A. The arrows represent the elution position of standards. Data from D. Pfaffinger, C. Edelstein, and A. M. Scanu, *J. Lipid Res.* **24,** 796 (1983).

denaturants such as urea or GdmCl, the pH should not exceed 7 and the columns should be subjected to thorough rinsing with distilled water at the end of the working day. This can be accomplished by constant rinsing at 0.5 ml/min overnight. All parts that come in contact with urea or GdmCl, i.e., solvent reservoir filters, pump, loop, and pump valves are also separately flushed with water. The columns must never be left standing in buffers without flow. The operating pressure should also be guarded carefully. When an increase in pressure is observed (20–30% increase), this may be an indication that the frit of the guard column is obstructed and needs attention. Cleaning of the frit is accomplished by disconnecting the top of the guard column which holds the frit, pumping distilled water through the frit, and observing the pressure. If the pressure is still increased then a quick soak (5–10 min) of the frit in 6 N nitric acid, followed by flushing with distilled water will remedy the problem. Column problems can be minimized by using buffers prepared with ultra pure double-distilled water and passed through a 0.4-μm filter. Also ultrafiltration of the sample using either a 0.4- or 0.2-μm filter is recommended.

Ion-exchange High-Pressure Liquid Chromatography

In contrast to gel permeation chromatography, ion-exchange chromatography is dependent on the ionic interactions between species on the support surface and charged groups in the protein. Ott and Shore[5] have taken advantage of this technique and have separated two forms of apoA-I, apoA-II, and the individual C peptides.

ApoHDL. Delipidation and Preparation of Sample. HDL of d 1.065–1.210g/ml is delipidated with ethanol/diethyl ether[29] and the apoHDL is dissolved in the starting buffer (6 M urea, 0.01 M Tris–HCl, pH 7.9).

Chromatographic System. Anion-exchange HPLC is performed with SynChropak AX300, an oxirane-cross-linked polyethyleneimine-coated 10-μm macroporous spherical silica support (SynChrom, Linden, IN). Both semipreparative (250 × 9 mm) and analytical scale (300 × 4.5 mm) columns can be used. The columns are fitted with SynChropak GPC 100 guard columns (60 × 2 mm).

Chromatographic Conditions. After 10 min preequilibration with starting buffer, the columns are eluted at 1 ml/min at room temperature with a linear 180 min gradient of 0.02–0.20 M Tris–HCl in 6 M urea at pH 7.9 for the semipreparative column, and a 45 min linear gradient of 0.02 0.15 M Tris–HCl in 6 M urea at pH 7.9 for the analytical scale column system.

Figure 3a illustrates the separation of apoHDL on a semipreparative scale. Anion-exchange HPLC resolves apoA-I, apoA-II, apoC-I, C-III$_1$, and C-III$_2$. ApoC-II is not well resolved from apoA-II in the 45 min gradient and requires a 120 min run with a shallower gradient (0.02–0.225 M Tris–HCl at pH 7.9 in 6 M urea). Two isoforms of apoA-I can also be separated if chromatographed on an analytical column in the latter gradient.

Comments. The maximum efficient loading of the semipreparative column is approximately 20 mg with yields in the range of 65 to 85%. Although commercially packed columns are available, supports such as

---

[29] V. Shore and B. Shore, *Biochemistry* **7,** 3396 (1968).

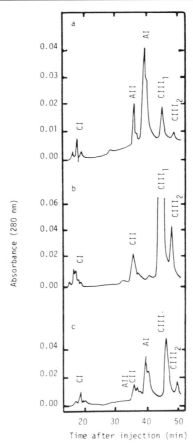

FIG. 3. Semipreparative scale AX 300 chromatography of urea-soluble VLDL and HDL apolipoproteins. Apolipoproteins were eluted with a linear 45 min gradient of 0.02–0.15 M Tris–HCl in 6 M urea at pH 7.9 and a flow rate of 1 ml/min. (a) HDL apolipoproteins; (b) urea-soluble VLDL apolipoproteins; (c) urea-soluble VLDL apolipoproteins plus HDL apolipoproteins. Data from G. S. Ott and V. G. Shore, *J. Chromatogr.* **231**, 1 (1982).

the one described can be packed with the use of a slurry packer (Model CP-111, Jones Chromatography, Inc., Columbus, OH).

ApoVLDL. Delipidation. VLDL ($d < 1.007$ g/ml) is delipidated by the tetramethylurea extraction method[30] which precipitates apoB and releases apoE and the apoC peptides in soluble form.

The chromatography of the urea-soluble VLDL apolipoproteins is identical to the conditions employed for separation of the HDL apolipoproteins described above.

[30] J. P. Kane, T. Sata, R. L. Hamilton, and R. J. Havel, *J. Clin. Invest.* **56**, 1622 (1975).

Figure 3b shows the semipreparative scale AX 300 chromatography of the urea-soluble apoVLDL. As expected, the C apolipoproteins separate well, however, apoE is not resolved.

Reversed-Phase High-Performance Liquid Chromatography

The variation in hydrophobicity of both peptides and proteins provides a means by which polypeptides may be fractionated. Reversed-phase chromatographic systems are based on the partitioning of solutes between hydrocarbon stationary phases and polar mobile phases. The hydrocarbon stationary phase is chemically bonded to a porous microparticulate silica support. The hydrocarbon phase is usually bonded as a monolayer with ~5 to 12% carbon loading. This process leaves residual silanols which can interact with basic groups on peptides. Thus endcapped supports are available in which a secondary silanization reaction has been carried out to eliminate this problem.

ApoHDL. The separation of apoA-I and apoA-II by ion-paired reversed-phase HPLC was achieved by Hancock *et al.*[6] The technique called ion pair chromatography involves the production of neutral species by pairing the ionized species of interest with a counterion. For large peptides and particularly those with strong hydrophobic interactions such as the apolipoproteins, good results are obtained with phosphate buffers between pH 2 and pH 4. Under these conditions, protonated amine groups are able to form hydrophilic ion pairs with phosphate ions, reducing affinity of the peptides for the stationary phase. In addition, by using an amine phosphate, irreversible binding of the apolipoproteins to free silanol groups on the support is eliminated.

Delipidation and Preparation of Sample. The apolipoproteins are isolated from HDL by delipidation with ether–ethanol (3:1). A partially purified apoHDL fraction is initially obtained by conventional gel filtration and dissolved in 1% triethylammonium phosphate, 6 M GdmCl, pH 7.5 at a concentration of 0.125 mg/ml. Immediately before use, the sample (50–200 μl) is treated with 10 μl of orthophosphoric acid to decrease the pH and to ensure a large excess of the ion pairing reagent.

Preparation of Mobile Phase (1% Triethylammonium Phosphate, pH 3.2). The mobile phase is prepared by addition of 10 ml of orthophosphoric acid to a liter of deionized, distilled water. The pH is adjusted to 3.2 with triethylamine to make the ion pair reagent about 0.17 M, filtered with a Fluoropore (FH) 0.5-μm filter (Millipore, Bedford, MA) and degassed with helium.

Chromatographic System. The column is a μBondapak-alkylphenyl column, 250 × 4 mm (Waters Associates, Millford, MA) containing 10-μm microparticulate silica particles with a surface coating of ~3 μmol/m^2, and a pore diameter of ~60 Å. Unreacted silanol groups are minimized by secondary capping with trimethylchlorosilane.

Chromatographic Conditions. The proteins are fractionated at room temperature at a flow rate of 1.5 ml/min using the mobile phase with a 2 hr gradient of 0 to 50% acetonitrile followed by a 30 min gradient of 50 to 80% acetonitrile. The eluted fractions are monitored at 220 nm, collected, and adjusted to pH 6 with triethylamine to prevent decomposition and lyophilized.

Figure 4 shows the analysis of a partially purified sample of apoHDL. Peaks 1 to 3 are the C peptides and peaks 4 and 5 are apoA-I and apoA-II, respectively.

Comments. The method described accomplishes the separation of the C peptides from the A apolipoproteins, but the resolution between apoA-I and apoA-II is not satisfactory. However, these proteins may be resolved to a greater extent by using a 2 hr gradient of 0 to 80% acetonitrile or a Zorbax-C8 column 250 × 4.6 mm (Dupont, Wilmington, DE) with the same gradient. At least two isoforms of apoA-I can be obtained when this protein is eluted from a Radial-Pak (C$_{18}$) cartridge, 100 × 8 mm (Waters Associates) using the same mobile phase but a 30 min linear gradient of 0 to 30% acetonitrile.

ApoVLDL. Delipidation and Preparation of Sample. VLDL of $d <$ 1.006 g/ml is delipidated with ether/ethanol (3 : 1) and initially processed by conventional gel filtration to isolate the C apolipoprotein mixture. The

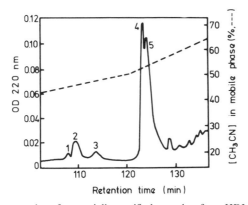

FIG. 4. Chromatography of a partially purified sample of apoHDL on a μBondapak-alkylphenyl column. A 2 hr gradient of 0 to 50% was used and was followed by a 30 min gradient of 50 to 80% acetonitrile. Peaks 1 to 3 are the C apolipoproteins; peaks 4 and 5 are apoA-I and apoA-II, respectively. Data from W. S. Hancock, H. J. Pownall, A. M. Gotto, and J. T. Sparrow, *J. Chromatogr.* **216**, 285 (1981).

C apolipoproteins are dissolved in 1% triethylammonium phosphate, 6 M GdmCl, pH 7.5 and treated with orthophosphoric acid precisely as for apoHDL. The mobile phase and the chromatographic system employed is identical to that for apoHDL.

Chromatographic Conditions. The column is equilibrated with the mobile phase and 100 μg of the C apolipoproteins is injected. The elution gradients at a flow rate of 1.5 ml/min consist of a 10 min concave gradient of 0 to 37% acetonitrile and a 20 min convex gradient of 37 to 42% acetonitrile. After chromatography, the collected proteins are brought to pH 6 as outlined for apoHDL.

The elution profile for a mixture of the C apolipoproteins is shown in Fig. 5. All the C apoproteins are well resolved including some minor components which may be due to apoB and apoE.

Comments. The recovery of the individual apolipoproteins is 80 to 95%. To obtain good reproducibility between injections, it is necessary to follow the analysis by equilibration of the reversed phase with a high

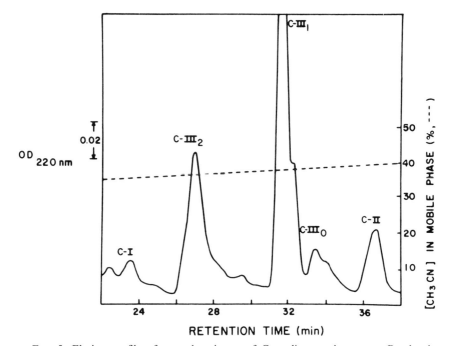

FIG. 5. Elution profile of a crude mixture of C apolipoproteins on a μBondapakalkylphenyl column with a mobile phase of 1% triethylammonium phosphate and a flow rate of 1.5 ml/min. The gradient consisted of a 10 min concave gradient of 0 to 37% acetonitrile and 20 min convex gradient of 37 to 42% acetonitrile. Data from W. S. Hancock, C. A. Bishop, A. M. Gotto, D. R. K. Harding, S. M. Lamplugh, and J. T. Sparrow, *Lipids* **16,** 250 (1981).

concentration of the organic modifier (2 min gradient to 80% acetonitrile, and 2 min more at the same concentration) and then equilibration with the aqueous phase (5 min gradient to 0% acetonitrile, followed by a 10 min equilibration).

Advantages and Disadvantages of HPLC

The major advantages of HPLC in the separation of the apolipoproteins are rapidity, high resolution, and recoveries. It can also be used as a sensitive analytical probe for assessing apolipoprotein purity. The sensitivity can be controlled by the detection system, e.g., by measuring the absorbance at 220 nm, fluorescence intensity, and radioactivity. Due to its rapidity, proteins which may be unstable can be quickly isolated with minimum loss of structural properties or enzymatic activity if applicable.

A drawback to the use of HPLC is the high cost of the columns. In particular, the molecular sieving columns have relatively short life times (up to 1 year). But if careful attention is paid to their care as outlined, they can be functional over a period of 2 years. Packing material is, however, available for the ion-exchange and reverse-phase columns. This makes their use more economical.

Concluding Remarks

The methods which were described in the previous sections provide guidelines which should enable researchers to tailor their chromatographic system to their needs. Moreover, with the development of appropriate HPLC molecular sieving columns, it should be possible to approach the problem of the isolation of the large- and small-molecular-weight forms of apoB. In the area of ion-exchange or reversed-phase modes, a ready separation of the various isoforms of apoE and apoA-I may also be achieved. It is apparent that HPLC has simplified the isolation and quantitation of apolipoproteins making them a readily available reagent in the laboratory.

Acknowledgments

The authors wish to thank the excellent assistance of R. Scott in typing the manuscript. The work was supported by Program Project USPHS HL 18577.

[19] Immunochemical Isolation and Identification of Glucosylated Apolipoproteins

By LINDA K. CURTISS, MICHAEL G. PEPE, and JOSEPH L. WITZTUM

A significant accompaniment of hyperglycemica in diabetes is the resultant posttranslational, nonenzymatic glucosylation of plasma and cellular proteins. As illustrated in Fig. 1 nonenzymatic glucosylation of proteins leads to the attachment of glucose to free amino groups of proteins to form a labile Schiff base. Free amines available for reaction with glucose are the ε-amino group of lysine and all unblocked N-terminal α-amines. The labile Schiff base, once formed, undergoes an Amadori rearrangement to form the more stable ketoamine. The ketoamine, in turn, is in equilibrium with a hemiketal or ring form. As a result of this reaction a bulky glyco group is introduced onto a free amine, leading to changes in charge as well as spacial conformation. At physiologic pH and temperature, the rate-limiting step is the formation of the Schiff base. This occurs very slowly in normoglycemics (glucose <110 mg/dl); however in hyperglycemic conditions this rate can increase significantly such that 2- to 12-fold increases in the extent of specific protein glucosylation have been reported.[1,2]

We have demonstrated that all of the major isolated lipoprotein classes undergo this posttranslational change even in euglycemic individuals, and that the extent of glucosylation of lipoproteins in diabetics is 2- to 10-fold higher.[3] The functional consequences of the attachment of glucose to apoB have been investigated. Witztum *et al.*[4] have demonstrated that the metabolism and regulatory activity of *in vitro* heavily glucosylated low-density lipoproteins (LDL) (in which 50% or more of the lysine residues are blocked) differ from that of control LDL in that (1) it is not taken up and degraded by the high-affinity LDL pathway in normal skin fibroblasts; (2) its uptake and degradation are not accompanied by changes in cellular HMG-CoA reductase or ACAT activity; and (3) its fractional catabolic rate *in vivo* is markedly reduced. These studies suggest that the normal cellular regulatory activities of LDL are blocked by the lack of recogni-

[1] E. G. Means and M. K. Chang, *Diabetes* **31** (Suppl. 3), 1 (1982).
[2] L. Kennedy, T. D. Mehl, E. Elder, M. Varghese, and T. J. Merimee, *Diabetes* **31** (Suppl. 3), 52 (1982).
[3] L. K. Curtiss and J. L. Witztum, *J. Clin. Invest.* **72**, 1427 (1983).
[4] J. L. Witztum, E. M. Mahoney, J. M. Banks, M. Fisher, R. Elam, and D. Steinberg, *Diabetes* **31**, 382 (1982).

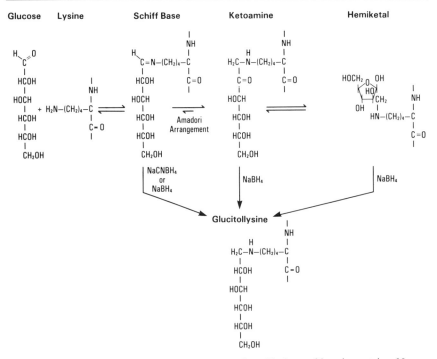

FIG. 1. Nonenzymatic glucosylation of the ε-amine of lysine residues in proteins. Nonreduced glucosylated adducts include the labile Schiff base and the more stable ketoamine and hemiketal Amadori rearrangement products. The reduced glucosylated adduct is glucitollysine. Reduction of the hemiketal by NaBH$_4$ also can result in the generation of mannositollysine, an epimer of glucitollysine.

tion of LDL by the cellular LDL receptors. This loss of recognition is a direct result of the modification of the lysine residues of apoB in a manner similar to that observed by other lysine modifications of LDL such as acetylation, reductive methylation, or carbamylation.[5] In some diabetics up to 6% of the lysine residues of LDL are glucosylated, and Steinbrecker and Witztum[6] have recently shown that modification of as few as 2–6% of the lysine residues can inhibit *in vivo* LDL clearance in the guinea pig by 5–25%. The role that nonenzymatic glucosylation of other apolipoproteins may play in the ability of these lipoproteins to regulate cell function is unknown; however, there is evidence to suggest that it could be important. Structure–function studies have identified a number of lysine resi-

[5] R. W. Mahley, T. L. Innerarity, K. H. Weisgraber, and S. Y. Oh, *J. Clin. Invest.* **64**, 743 (1979).
[6] U. P. Steinbrecher and J. L. Witztum, *Diabetes* **33**, 130 (1984).

dues on different apolipoproteins that are required for various functional activities including receptor recognition, enzyme activation, enzyme interaction, lipid binding, and the regulation of cellular proliferation. Thus, glucosylation of specific lysine residues of the various apolipoproteins could have a profound influence on the function of that apolipoprotein.

In the past, confirming that specific proteins of interest are glucosylated *in vivo* has been a technically difficult and arduous task, and one that has usually involved specific isolation of the protein of interest followed by various chemical and physical demonstrations that some of the amino acids are glucosylated. Recently an immunochemical approach has been successfully applied to the problem of the identification of glucosylated apolipoproteins in diabetes. In succeeding portions of this chapter, these immunochemical approaches will be described.

Preparation of in Vitro Glucosylated Protein Immunogens and Antigens. Proteins are glucosylated *in vitro* either in the presence or absence of a reducing agent. In the absence of a reducing agent three products are found, the labile Shiff base and the more stable Amadori rearrangement products (Fig. 1). In the presence of a reducing agent, all of the labile Schiff base that is formed is immediately converted to glucitollysine (Fig. 1). Therefore, higher degrees of glucosylation are obtained *in vitro* if glucosylation is carried out in the presence of a reducing agent. In addition the reduced adducts are highly immunogenic as described below. Glucosylation is accomplished by incubating the proteins in phosphate-buffered saline (PBS) with 1 mM EDTA at 37° in the presence of 80 mM glucose and in the presence or absence of 12.5 mg/ml NaCNBH$_3$.[3,4] Because incubations are often carried out for 5 to 7 days, the reaction mixtures are sterilized by passage through a 0.45-μm filter. After incubation, the glucosylated proteins are exhaustively dialyzed against a physiologic buffer (usually PBS containing 1 mM EDTA). The degree of nonenzymatic glucosylation that can be attained is proportional to the time of incubation. In the presence of a reducing agent up to 50% of the lysine residues of LDL can be modified in 7 days by the covalent attachment of glucose. In the absence of a reducing agent the degree of modification is more variable but usually less than 6 or 7% of the lysine residues of LDL are modified in 7 days. Similar degrees of glucosylation of proteins other than LDL are attained, however individual differences exist. For example, the incubation of albumin, hemoglobin, transferrin, and high-density lipoproteins for 7 days with 80 mM glucose in the presence of NaCNBH$_3$ resulted in the modification of 57.1, 85, 51.4, and 29.2%, respectively, of the available lysine residues of each of these proteins. Because of this variability, proteins used as immunogens or antigens should be individually checked to assess the degree of derivitization.

Two independent nonimmunochemical procedures are used to assess the extent of lysine modification of *in vitro* glucosylated proteins. They are amino acid composition analysis[4] and the trinitrobenzenesulfonic acid (TNBS) assay.[3,4,6] For low degrees of protein modification like those obtained without a reducing agent, amino acid composition analysis is preferred because of its sensitivity.[6]

Antibody Preparation. Either polyclonal or monoclonal antibodies can be generated for use.[3,7] The majority of the approaches described below have been performed with monoclonal antibodies; however immunopurified polyclonal antibodies are equally useful. Antibodies specific for the reduced conjugate of glucose and lysine are prepared by immunizing the animals with autologous reduced and glucosylated LDL (i.e., rabbits are immunized with glucosylated and reduced rabbit LDL and mice are immunized with glucosylated and reduced mouse LDL). The choice of LDL as the homologous protein for modification appears to be an important one. For example, we have immunized guinea pigs with autologous glucosylated and reduced albumin and compared the antibodies obtained with those obtained from guinea pigs immunized with autologous glucosylated and reduced LDL.[8] Antibodies generated against glucosylated and reduced LDL are directed against glucitollysine and therefore react with other glucosylated proteins that contain the same lysine derivative. However, antibodies generated against glucosylated and reduced guinea pig albumin react only with glucosylated and reduced guinea pig albumin, and not with glucitollysine or with other glucosylated and reduced proteins. Therefore, although antibodies can be obtained that will bind glucosylated albumin and not native albumin by immunizing with autologous albumin, the antibodies generated appear to be less restricted in their specificity and as a consequence are not useful for detecting glucosylation in a series of structurally unrelated proteins such as the apolipoproteins. On the other hand, such an antibody ought to be useful for specifically detecting glucosylated albumin.

An immunization schedule which was used to generate a panel of mouse monoclonal antibodies that bind glucosylated and reduced human LDL[3] is shown in the table. Key points in this scheme are (1) the use of an autologous modified protein as immunogen, (2) multiple immunization steps, (3) a minimum interval between immunization of 2 weeks, and (4) a prefusion immunization with human glucosylated LDL. This perfusion

---

[7] J. L. Witztum, U. P. Steinbrecher, M. Fisher, and A. Kesaniemi, *Proc. Natl. Acad. Sci. U.S.A.* **80**, 2757 (1983).

[8] U. P. Steinbrecher, M. Fisher, J. L. Witztum, and L. K. Curtiss, *J. Lipid Res.* **25**, 1109 (1984).

IMMUNIZATION SCHEDULE FOR THE GENERATION OF GLUCITOLLYSINE-SPECIFIC
MOUSE ANTIBODIES

Day 0	Prebleed, then inject 20 μg mouse glucosylated and reduced LDL intraperitoneally in complete Freund's adjuvant
Day 14	Inject 20 μg of mouse glucosylated and reduced LDL intraperitoneally in incomplete Freund's adjuvant
Day 28	Same as day 14
Day 38	Bleed and check serum antibody titers
Day 51	Inject 20 μg of human glucosylated and reduced LDL intravenously in saline
Day 55	Bleed and sacrifice for fusion

immunization is used to specifically propagate *in vivo* the clones of B-lymphocytes which are making antibodies against glucosylated and reduced mouse LDL that will cross-react with glucosylated and reduced human LDL. B-lymphocytes that have the highest probability of contributing to the generation of successful hybridomas are those that are actively proliferating at the time of fusion. The short interval between the prefusion injection and sacrifice for fusion is insufficient for primary stimulation of human LDL-specific B cell clones, but sufficient to permit clonal expansion of the desired memory B cell clones. Recently 500 units of gamma interferon have been included with both the primary and secondary immunizations and vigorous antibody responses in these immunized animals have been observed.[9]

Preparation of in Vivo Glucosylated Protein Antigens for Assay. Protein or lipoprotein glucosylation that occurs *in vivo* in either euglycemic or hyperglycemic conditions is limited to the formation of the labile Schiff base and the more stable Amadori rearrangement products of the Schiff base, the ketoamine or hemiketal adducts (Fig. 1). Antibodies prepared by immunization with glucosylated and reduced LDL do not bind these nonreduced forms.[3,7] Therefore, all plasma samples and lipoproteins obtained for assay must first be reduced to convert the Schiff base and the Amadori rearrangement products to glucitollysine (Fig. 1). $NaCNBH_3$ will reduce the Schiff base but not the ketoamine or hemiketal forms, whereas $NaBH_4$ a more general reducing agent will reduce each of these adducts. We have demonstrated[3] that quantitation of glucitollysine residues of plasma proteins after reduction with $NaCNBH_3$ (after removal of free glucose) gives a minimum estimate of glucose adducts in the Schiff base form, and quantitation of glucitollysine residues in plasma proteins reduced with $NaBH_4$ (after removal of glucose) gives a minimum estimate of total glu-

[9] M. Nakamura, T. Mauser, G. D. N. Pearson, M. J. Daley, and M. L. Gifter. *Nature (London)* **307**, 381 (1984).

cose adducts. Therefore, for routine measurements all plasma samples and isolated proteins are reduced with NaBH$_4$ after removal of free glucose. Plasma is collected in EDTA and kept at 4°. Free glucose is removed by chromatography on Sephadex G-25 in PBS containing 1 mM EDTA, or by extensive dialysis against PBS containing 1 mM EDTA. Plasma or isolated proteins are reduced by incubation with 10 mM NaBH$_4$ for 4 hr at 37° and then dialyzed against PBS containing 1 mM EDTA.

Solid-Phase Radioimmunoassay for Glucosylated Proteins. Solid phase radioimmunoassays are used to monitor antibody titers, screen for positive hybridoma clones, and demonstrate antibody specificity. Direct (noncompetitive) binding assays are used when a large number of culture supernatants or serum samples are screened for their ability to react with a single or a limited number of antigens. Competitive binding assays are used when the binding of a limited number of antibodies to a panel of antigens is assessed. Both assays are performed in flexible round bottom polyvinyl chloride 96-well microtiter plates. For assay the wells are coated with antigen by adding 0.05 ml of antigen in PBS and incubating the plates at room temperature for 3 hr. Longer coating times can be used but are not necessary, and shorter coating times are feasible if a higher antigen concentration or a higher temperature (37°) is used. The binding of a protein antigen to the microtiter plate is a unique property of the antigen, and differences exist in binding efficiency. For example to bind 50 ng of protein per well (a maximum amount of antigen bound for most proteins), VLDL is used at 50 µg protein/ml, IDL at 6.2 µg/ml, LDL at 4.4 µg/ml, and HDL at 24 µg/ml.[10] However, more dilute antigen solutions can be used. *In vitro* glucosylated and reduced proteins are routinely used at 1 µg/ml for the immunoassays. After antigen coating, the antigen solution is removed and the remaining adsorptive sites on the plastic wells are blocked by postcoating the wells with a nonreactive protein. In most instances this is 0.2 ml of PBS containing 3% bovine serum albumin (BSA) and 3% normal goat serum. Goat serum is used because the second antibody is a goat anti-mouse Ig antibody.

For noncompetitive binding assays, 0.05 ml of solution containing first antibody diluted in PBS containing 3% BSA and 0.02% Tween-20 is added to the wells (first antibody refers to the antibody that binds the antigen immobilized on the plate). Incubations can range from 2–3 hr at 37° for nonequilibrium conditions to 18–24 hr at 4° to approach equilibrium conditions. After this incubation the first antibody solution is removed, all unbound antibody removed by repeated washing of the wells in PBS containing 3% BSA, and bound antibody detected by incubation with

[10] L. K. Curtiss and T. S. Edgington, *J. Biol. Chem.* **257**, 15213 (1982).

10 ng/well of a second antibody dilution. In most cases, this second antibody is radioiodinated goat anti-mouse Ig. The antibody is radioiodinated enzymatically using immobilized lactoperoxidase and glucose oxidase (Enzymobeads, Bio-Rad) to specific activities of 3–4 μCi/μg.[10] Because the final step involves a second antibody detection system, this incubation can be adjusted to give good binding of the second antibody in a minimum amount of time (usually 2–4 hr).

Competitive assays are performed in an identical manner except that they contain 0.025 ml of competitor diluted in PBS (with 3% bovine albumin, 3% goat serum, and 0.02% Tween-20) and 0.025 ml of first antibody. The amount of first antibody added in 0.025 ml is determined in preliminary direct binding studies. Because antibody must be limiting in a competitive assay, a dilution is used that will give between 30 and 50% of maximum binding. Quantitation of glucitollysine residues isolated on lipoproteins and apolipoproteins using the competitive RIA is achieved as follows.[3] The wells are coated with antigen (*in vitro* glucosylated and reduced LDL) at 1 μg/ml for 1 hr at room temperature and postcoated for 30 min. First antibody (0.025 ml of hybridoma culture supernatant diluted to give half maximal binding) is incubated for 18 hr at 4° with 0.025 ml of standard or unknown. Glucitollysine is used as standard and is used between 10 and 0.05 nmol/ml. Background is determined by replacing the first antibody with an irrelevant antibody of the same immunoglobulin heavy chain type. The extent of first antibody binding is determined by the degree of binding of ^{125}I-labeled goat anti-mouse Ig in 4 hr at 4°. Data are expressed as the ratio of B/B_0, where B equals the counts bound in the presence of competitor and B_0 is the amount of radioiodinated second antibody bound in the absence of competitor.

The competitive assay can be used to measure the molar concentration of glucitollysine residues in a given mass of protein. If isolated NaBH$_4$ reduced lipoproteins or apolipoproteins are used, these assays are used to assess the number of glucitollysine residues/mg of isolated protein. However, it should be pointed out that these assays do not identify which apolipoproteins on a given lipoprotein are glucosylated. Therefore to identify specific apolipoprotein glucosylation in the lipoproteins of diabetic subjects, additional immunochemical approaches are used.

Apolipoprotein Glucosylation by Western Blot Analyses. For analysis the apolipoproteins of the isolated and reduced lipoproteins are separated by polyacrylamide gel electrophoresis in the presence of SDS. A vertical slab gel apparatus (14 × 12 × 0.15 cm) is used (Hoeffer Scientific Instruments). The gels are prepared according to Laemmli[11] using a 24 mM

[11] U. K. Laemmli, *Nature (London)* **227**, 680 (1970).

Tris–glycine buffer at pH 8.6. The upper stacking gel is 3% acrylamide and 1% SDS, and the lower running gel is a gradient gel that also contains 1% SDS. The acrylamide gradients chosen depend upon which apolipoproteins are to be separated. Gradients of 3–20% acrylamide separate a wide range of apolipoproteins and can be used to visualize both apoB and apoC. But these 3–20% gradient gels do not resolve individual apoC proteins or individual apoB species. A 3–6% acrylamide gradient will resolve the apoB species[10] and a 7.5–10% acrylamide gradient will resolve the lower molecular weight apolipoproteins.[12] Lipoproteins (1–50 μg protein per lane) are delipidated by boiling for 3 min in electrophoresis sample buffer that contains 1% SDS, 10 mM Tris, 0.24 mM EDTA. Electrophoresis is carried out at constant current overnight for approximately 200 mA-hr until the dye front begins to exit the gel.

Immediately after electrophoresis, the separated apolipoproteins are transferred electrophoretically to nitrocellulose paper.[10,13] Before transfer the gels are washed in distilled H_2O for 10 min, then for an additional 10 min in 25 mM Tris, 192 mM glycine, pH 8.3, that contains 20% v/v ethanol. Transfer is accomplished with a Hoeffer TE-40 series transfer unit at 400 mA for 2 hr. The efficiency of transfer of individual apolipoproteins varies and in general parallels the apparent solubility of the apolipoproteins in aqueous buffers.[10] Apolipoproteins B and E are electrophoretically transferred to nitrocellulose with low efficiency, whereas apolipoproteins A-I, A-II, and C are transferred with good efficiency. It has been determined that the limiting step is solubilization and transfer of apolipoproteins E and especially B out of the polyacrylamide gels, rather than binding to the nitrocellulose. After transfer, all remaining active sites on the nitrocellulose paper are blocked by soaking the paper overnight in PBS containing 3% BSA and 3% normal goat serum.

The postcoated nitrocellulose paper transfers are incubated for 6 hr with first antibody which is diluted in PBS containing 3% BSA and 3% normal goat serum. After repeated washing antibody binding to the immobilized apolipoproteins is detected with a second 4 hr incubation with 0.5 μCi/ml of radioiodinated goat anti mouse Ig antibody. This is the same second antibody that is used in the radioimmunoassays. A common problem with Western blotting techniques is the high background caused by nonspecific binding to the nitrocellulose paper. This can be reduced in two ways: (1) use minimum amounts of first antibody, and (2) wash the nitrocellulose with PBS that contains 3% BSA and 0.05% Tween-20 alternated with a solution of 0.5 M LiCl that contains 0.1% SDS.

[12] L. K. Curtiss and E. F. Plow, *Blood* **64**, 365 (1984).
[13] H. Towbin, T. Staehelin, and J. Gordon, *Proc. Natl. Acad. Sci. U.S.A.* **76**, 4350 (1979).

After washing and air drying the paper transfers, binding of second antibody is detected by autoradiography. X-Omat film (Eastern Kodak) is exposed for 1 to 24 hr at $-20°$. If desired the autoradiograph can be quantitated by soft laser densitometric scanning. Specific apolipoproteins can be tentatively identified by their relative mobilities, and positively identified by reaction with apolipoprotein-specific antibody if necessary.

Apolipoprotein Glucosylation by Column Chromatography. In this approach glucosylation is identified directly in the apolipoproteins after separation by molecular sieve column chromatography. As for the other procedures lipoproteins are first reduced by incubation for 4 hr at 37° in the presence of 10 mM NaBH$_4$. This is followed by dialysis into 1.0 mM EDTA, pH 7.4 and lyophilization. Lipoproteins are delipidated in absolute ethanol followed by three washes of absolute ethyl ether at $-20°$. Proteins are dried under N$_2$ and solubilized by incubation overnight at 4° in 0.2 M Tris, pH 8.0, containing 6 M guanidine–HCl, and 0.1% mercaptoethanol. The insoluble apoB is removed by low-speed centrifugation, and the soluble proteins diluted 1:2 with 4 M guanidine and applied to a 100 × 2.5 cm column of Sephacryl S-300 equilibrated in 0.2 M Tris, pH 8.0, containing 4 M guanidine and 0.1% mercaptoethanol. Approximately 60, 2 ml fractions are collected. Apolipoproteins and glucitollysine residues on the apolipoproteins are identified directly in each column fraction by a modified solid phase immunoassay. Samples (0.05 ml) of the column fractions are incubated for 2 hr at room temperature directly in the polyvinyl chloride microtiter plate. The 4 M guanidine does not interfere with the binding of apolipoproteins to the assay plates. Following washing, the apolipoproteins or plasma proteins are identified with apolipoprotein-specific first antibodies followed by radioiodinated second antibody. Minimum detectable levels for apoA-I, A-II, B, C-I, E, and albumin in this assay are 500, 20, 10, 20, 500, and 10 ng, respectively. Glucosylation of the apolipoproteins is identified in similar assays using glucitollysine-specific antibodies in place of the apolipoprotein-specific antibodies, and the chromatographic behavior of the glucitollysine residues compared with the chromatographic behavior of the individual apolipoproteins.

Apolipoprotein Glucosylation by Immunoadsorption. Because most apolipoproteins are soluble in aqueous buffers after delipidization, apolipoprotein glucosylation can be studied also using immunoadsorption techniques. This approach can not be used to measure apoB glucosylation. Unlike the previous approaches, this procedure can be made quantitative if the proper bookkeeping is performed. Apolipoproteins containing glucitollysine residues are separated from nonglucosylated apolipoproteins by preparative affinity chromatography on a glucitollysine-specific immunoadsorbent. The glucitollysine-specific antibodies are first immunopurified by chromatography over immobilized bovine albumin that has been maximally glucosylated *in vitro* in the presence of NaCNBH$_3$. Glucosylated albumin is preferred because albumin is a more homogeneous molecule and it will simplify the problems of leaching and dissociation associated with the use of lipoprotein immunoabsorbent. The purified glucitollysine-specific antibody is coupled to CNBr-activated Sepharose 4B using standard procedures (25 mg protein/15 ml of swollen gel) and the immunoadsorbent equilibrated in PBS containing 1 mM EDTA and 5 mM benzamidine, pH 7.4.

Lipoproteins are delipidated for immunoaffinity chromatography with 3:1 (v/v) anhydrous ether/ethanol and washed with anhydrous ether. After drying under N$_2$ the proteins (10 mg) are solubilized in 100 ml of 6 M urea, and dialyzed exhaustively at 4° overnight into PBS containing 1 mM EDTA and 5 mM benzamidine. Soluble proteins are separated by centrifugation and applied to the immunoadsorbent. Chromatography is performed at 4° and the nonretained fraction (e.g., soluble apoproteins that do not contain glucitollysine residues) collected. The retained fraction is eluted with a bed volume of 3 M KI and the fractions dialyzed immediately into PBS to remove the KI. Both the retained and nonretained fractions are dialyzed into 1 mM EDTA, lyophilized, and resolubilized in a minimum volume of PBS containing 5 mM benzamidine for further characterization.

The apolipoprotein composition of the retained and nonretained fractions from the immunoadsorbent columns are identified by SDS–polyacrylamide electrophoresis and/or Western blot analysis, and quantitated if necessary with apolipoprotein-specific immunoassays. The retained portion of all lipoprotein fractions from most diabetic subjects studied to date represents between 5 and 10% of the nonretained fractions, whereas the retained portions of identical lipoprotein fractions from normal subjects represent less than 1% of the nonretained fraction. Immunoadsorption as described results in a remarkable enrichment for glucosylated proteins, and should permit further biochemical characterization of these glucosylated apolipoproteins.

Tandem Assay for Apolipoprotein Glucosylation. To determine the prevalence of apolipoprotein glucosylation in diabetes, population studies must be done. Isolation of specific lipoprotein classes and further separation of the apolipoproteins are not feasible for large numbers of subjects. Therefore, a tandem immunoassay was developed to make use of both glucitollysine-specific and apolipoprotein-specific antibodies. The principle of the assay is that a monoclonal antibody directed against a lipoprotein (or protein) in question is first used to separate the desired protein from plasma. Then a glucitollysine-specific antibody is used to quantitate the glucitollysine residues on that lipoprotein. This assay can be performed directly on a small volume of plasma. The assay has the following components. First, a monoclonal antibody that will bind a specific lipoprotein or protein and is capable of binding equally well to either the glucosylated or the native protein. For example, to measure glucosylation of LDL, an apoB-specific antibody is immunopurified and coupled to a solid support such as a Sepharose bead, a plastic plate, or a plastic ball. The immobilized antibody is incubated with plasma and a fixed quantity of LDL is removed. It is important to document that this antibody removes the same amount of LDL from all plasmas. To verify this, a second monoclonal antibody is used that recognizes a site on apoB that is distinct from the site recognized by the first apoB antibody. This antibody also binds glucosylated or native lipoprotein with equal efficiency. This second antibody is immunopurified, radioiodinated, and used to quantitate LDL removal. The third component of this tandem assay is an antibody that can identify glucitollysine. This third antibody also is immunopurified as described earlier and radioiodinated.

To assay plasma, 0.5 ml is placed in a dialysis bag and to remove free glucose, dialyzed overnight against PBS containing 10 mM EDTA. After dialysis the bags are placed in a solution of 10 mM NaBH$_4$ and incubated for 6 hr at 37° to reduce all glucose adducts to glucitollysine. The bags are then dialyzed against PBS containing 10 mM EDTA and 5 mM benzamidine to remove the reducing agent. Plasma diluted 1 : 5 with dialysis buffer (0.1 ml) is incubated with the solid phase or immobilized apolipoprotein-specific antibody for 4 hr at 20°. Nonbound plasma proteins are removed by extensive washing and the immobilized antibodies (containing the bound lipoproteins) are incubated a second time with the two types of radioiodinated second antibodies. Incubation with a radioiodinated antibody of the same specificity as the immobilized antibody verifies that the same amount of lipoprotein antigen is removed from all plasmas. Incubation of a parallel tube with a radioiodinated glucitollysine-specific antibody quantitates the number of glucitollysine residues in the immobilized lipoprotein. Preliminary evidence suggests that this or similar tandem assays can be used to identify and quantitate the extent of glucosylation of apoB lipoproteins containing apoB or A-I as well as extent of glucosylation of albumin.

Summary. Confirming that specific proteins or apolipoproteins are glycosylated *in vivo* has been a technically difficult task, and one that has usually involved isolation of the protein of interest followed by various chemical and physical demonstrations that some of the lysine residues of the proteins are glucosylated. The availability of a panel of antibodies that are specific for the reduced conjugate of glucose and lysine has provided a new immunochemical approach to identifying the extent to which apolipoproteins are glucosylated in the hyperglycemic state.

[20] Isothermal Calorimetry of Apolipoproteins

By M. ROSSENEU

Introduction

The quantitation of the binding enthalpy between protein and lipids can usefully contribute to the evaluation of the driving forces in the association process and of the stability of the complex.[1]

The binding enthalpy can be either measured directly by calorimetry or calculated from free energy measurements at different temperatures using the Van t'Hoff equation.[2] The first approach offers the advantages of a direct measurement in terms of accuracy and sensitivity. It usually requires a smaller amount of material and can be carried out under a variety of experimental conditions.

The enthalpy measurements are usually carried out in a microcalorimeter, given the small quantities of apolipoproteins available, and will be referred to as "microcalorimetric" measurements. In this chapter, we will try to summarize several applications of this technique and analyze the information in terms of apolipoprotein structure and especially in terms of their association properties with lipids.

Experimental Set-up

Isothermal microcalorimetry consists in the measurement of the enthalpy released or absorbed on mixing two reagents. These enthalpy changes can be due to the dilution of a solute, to a titration with an acid or base or to solute–solute interactions.

Two major types of calorimeters, the adiabatic and the heat conduction calorimeter, can be used for these isothermal measurements. In the adiabatic calorimeter there is no heat exchange between the calorimeter and its surroundings while in the heat conduction calorimeter, the heat of reaction is quantitatively transferred to the surroundings, the "heat-sink" through the thermopiles.

The LKB batch calorimeter type 10.007 is a thermopile conduction calorimeter in a twin arrangement (Fig. 1). The reaction vessels consist of

---

[1] M. Rosseneu, F. Soetewey, V. Blaton, J. Lievens, and H. Peeters, *Chem. Phys. Lipids* **13**, 203 (1974).
[2] C. Bjurulf, J. Layuez, and I. Wadsö, *Eur. J. Biochem.* **14**, 47 (1970).

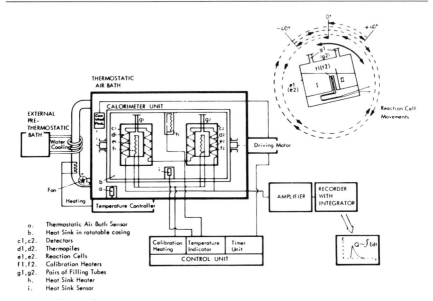

FIG. 1. Schematic representation of the LKB batch microcalorimeter.

two 18-carat gold cells divided by a partition wall into two compartments of 4 and 2 ml, respectively.

The reaction cells are on both sites, in intimate thermal contact with thermocouple plates mounted in opposition to form a sandwich-like construction. The metal block surrounding the two calorimeter units is insulated by styrofoam and enclosed in an aluminum cover. The calorimeter block is suspended in a thermostated air bath. Mixing of the reagents is achieved by rotation of the calorimeter assembly. This type of calorimeter is suitable for measurement of both fast and slow (hours or days) processes.

Thermostatization at 0.001° is obtained by means of an air-thermostat regulated by a water thermostat. Temperature differences of around 0.001° are measured by the thermocouples and amplified by means of a Keithley microvoltmeter. The signal generated during the reaction is recorded on a Philips recorder (Fig. 2). The area under the reaction curve (Fig. 2) is quantitated by means of an integrator and compared to the amount of Joule heat generated by the electrical calibration. This calibration was achieved by circulating a current intensity of 1–5 mA through the 50 Ω resistance located inside the measurement cell for about 10 sec.

The heat of friction generated by the rotation of the calorimetric unit around its axis is subtracted from the heat of reaction. This effect

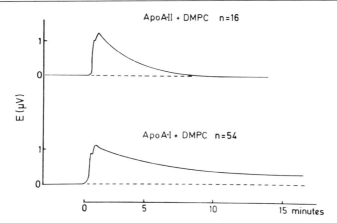

FIG. 2. Heat signal obtained on mixing apoA-I and apoA-II with DMPC. n = molar apolipoprotein/DMPC ratio.

amounts to 2.5% of the reaction heat on the 10 μV scale and is negligible at the 30 μV sensitivity.

The protein amount used in most experiments was 2 ml at a final concentration of 1×10^{-5} M for apolipoproteins. The final phospholipid concentration was varied between 2×10^{-4} and 2×10^{-3} M. The heat released amounted to 1 to 5 mcal depending on the extent of reaction. Special care was taken to ensure thorough and reproducible cleaning of the cells. They were rinsed with 1 N NaOH subsequently flushed 3 times with distilled water and finally with the buffer used in the particular experiments. This procedure efficiently removed any reagent adsorbed on the cell surface.

Enthalpy of Association between Apolipoproteins and Phospholipids

The enthalpy of association between apolipoproteins and phospholipids has been measured for several soluble plasma apolipoproteins of human and nonhuman primates.[3-10] The association enthalpies of human

---

[3] M. Rosseneu, F. Soetewey, G. Middelhoff, H. Peeters, and M. V. Brown, *Biochim. Biophys. Acta* **441,** 68 (1976).

[4] M. Rosseneu, V. Blaton, R. Vercaemst, F. Soetewey, and H. Peeters, *Eur. J. Biochem.* **74,** 83 (1977).

[5] M. Rosseneu, in "The Lipoprotein Molecule" (H. Peeters, ed.), p. 129. Plenum, New York, 1978.

[6] V. Blaton, M. Rosseneu, R. Vercaemst, R. Jackson, and A. M. Gotto, *Biochemistry* **16,** 2157 (1977).

apoA-I, apoA-II, apoC-I, and apoC-III with synthetic lecithins have been reported.[3-8] The association enthalpy for fragments of apoC-III obtained by thrombin cleavage was also measured.[9] The binding enthalpy of DMPC (dimyristoyl phosphatidylcholine) to baboon apoA-I and apoA-II[5,6] and to rabbit apoE was also determined.[10]

At 25° all apolipoproteins bind to lecithins with a highly exothermal enthalpy; these association processes are very rapid and the magnitude of the enthalpy changes is dependent on the apolipoprotein. A comparison of the binding enthalpies expressed as kcal/amino acid residue (see the table) shows that they become more endothermal in the sequence apoC-III ≃ apoA-II > apoC-I > apoA-I.[7]

The highly exothermal enthalpy of binding can be attributed to several factors: the conformational changes which accompany lipid binding to all apolipoproteins,[11] the changes in the crystalline structure of the phospholipid bilayer, involving a decrease in the size of the cooperative units within the liposomes,[3] and finally the state of self-association of the apolipoproteins.[4]

We will successively analyze the contribution of those factors to the total enthalpy of association and to the mechanism of lipid–apolipoprotein binding.

Enthalpy of Protein Conformational Changes

An important contribution to the enthalpy of lipid–apolipoprotein association arises from the conformational changes occurring to the apolipoproteins upon lipid binding. A helicity increase of 20 to 30% has been observed as a common feature associated with apolipoprotein–lipid complex formation.[11] This conformational change is likely to be associated with some exothermal effect as observed with synthetic polypeptides.[12] This could be demonstrated for the apolipoproteins apoA-I, A-II, and C-III using different experimental conditions.

For apoA-I we compared the structure and phospholipid-binding prop-

---

[7] M. Rosseneu, F. Soetewey, H. Peeters, L. Bausserman, and P. N. Herbert, *Eur. J. Biochem.* **70,** 285 (1976).

[8] H. Pownall, J. Massey, J. Hsu, and A. M. Gotto, *Can. J. Biochem.* **59,** 700 (1981).

[9] J. Sparrow, H. Pownall, F. Hsu, L. Blumenthal, A. Culwell, and A. M. Gotto, *Biochemistry* **16,** 5427 (1977).

[10] R. Roth, R. Jackson, H. Pownall, and A. M. Gotto, *Biochemistry* **16,** 5030 (1977).

[11] J. Morrisett, R. Jackson, and A. Gotto, *Annu. Rev. Biochem.* **44,** 183 (1975).

[12] H. Pownall, F. J. Hsu, M. Rosseneu, H. Peeters, A. M. Gotto, and R. Jackson, *Biochim. Biophys. Acta* **488,** 190 (1977).

ENTHALPY CHANGES ON THE ASSOCIATION OF PHOSPHOLIPIDS TO PLASMA APOLIPOPROTEINS

Reference	Apolipoprotein	Species	Phospholipid	T (°C)	Molar complex composition	ΔH/mol protein (kcal/mol)	ΔH/mol PL (kcal/mol)
3	ApoHDL	Human	Lyso-PC	25	72	−33	−0.46
3	ApoA-I	Human	Lyso-PC	25	70	−26	−0.37
3	ApoA-II	Human	Lyso-PC	25	61	−34	−0.56
14	ApoA-II	Human	Lyso-PC	30	—	49	—
3	ApoHDL	Human	DMPC	28	80	−220	−2.75
3	ApoA-I	Human	DMPC	28	140	−215	−1.53
3	ApoA-II	Human	DMPC	28	65	−230	−3.54
14	ApoA-II	Human	DMPC	23.5	75	+90	+1.20
14	ApoA-II	Human	DMPC	24.5	75	−260	−3.48
14	ApoA-II	Human	DMPC	30	45	−62	−1.4
7	ApoC-III	Human	DMPC	24.9	90	−250	−2.8
8	ApoC-III	Human	DMPC	28	260	−130	−2.0
8	ApoC-III	Human	DMPC	42.0	92	−490	−4.5
7	ApoC-I	Human	DMPC	28	130	−65	−0.50
4	ApoA-I	Baboon	DMPC	28	120	−180	−1.5
4	ApoA-II	Baboon	DMPC	28	35	120	−3.42
10	ApoE	Rabbit	DMPC	28	625	−614	−0.98

erties of this apoprotein at pH 7.4 and 3.1 by microcalorimetry, circular dichroism, and density gradient ultracentrifugation.[12] At pH values of 7.4 and 3.1, apoA-I binds to DMPC to form complexes of similar composition and helical content. The difference in the enthalpies of association at the two pHs was attributed to differences in the increase of helical content at the lower pH. This corroborated the direct measurement of pH-induced helix coil transition between the two pHs, both in aqueous and in guanidine–chloride solutions. Comparable values were obtained for the helix–coil formation in apoA-I and in poly-L-lysine amounting to 1.2–2.0 kcal/residue.[12] Massey et al.[13] have shown that for apoA-II and C-III there is a linear correlation of the increase in helical structure and the exothermic enthalpy of association. The measured value of -1.3 kcal/mol of amino acid residue converted from random coil to α-helical structure is close to that of α-helix formation by charged amino acids (-1.2 kcal/mol) and suggests that amino acid side chains contribute little heat to the random coil–helix transition in the plasma apolipoproteins.

This enthalpy of α-helix formation is the major source of enthalpy of association between apoA-II and DMPC at 30° and with lysomyristoyl phosphatidylcholine (LMPC) at all temperatures.[14] This hypothesis is also valid for apoA-I.[3]

An endothermal contribution of about -20 kcal/mol apoA-I was also observed upon phospholipid binding to human and baboon apoA-I.[3,4] This effect was correlated with the extent of self-association of the apolipoprotein and could be reversed by preincubation of the protein with 10 mol lysolecithin/mol apoA-I.[4]

Enthalpy Changes from the Phospholipid Crystalline Structure

The enthalpy of phospholipid–apolipoprotein association has been measured mainly using micellar or bilayer phospholipid dispersions.[1] Due to their partly hydrophilic character, lysolecithin and short-chain lecithins (up to C_{12}) exist in water as micellar structures. Longer chain lecithins tend to spontaneously form bilayer structures in aqueous solutions; these consist of either concentric phospholipid bilayers or closed vesicles after input of sonication energy. Within the bilayer the phospholipids can exist in either a crystalline or a liquid-crystalline structure dependent upon the temperature. Below the transition temperature, the phospholipid acyl chains are highly ordered in crystalline state whereas their mobility increases to a liquid-crystalline state above the transition. The presence of

[13] J. Massey, A. M. Gotto, and H. Pownall, *J. Biol. Chem.* **254**, 9359 (1979).
[14] J. Massey, A. M. Gotto, and H. Pownall, *Biochemistry* **20**, 1575 (1981).

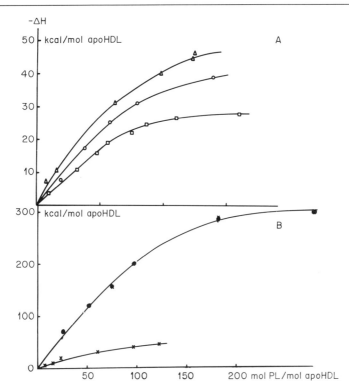

FIG. 3. Enthalpy changes on binding lysolecithin and lecithins to apoHDL. (A) Lysolecithin (□), dicaproyl lecithin (○), and dilauroyl lecithin (△). (B) Dilauroyl lecithin (×) and dimyristoyl lecithin (●).

other lipids such as phospholipids or cholesterol inside a mixed bilayer also influences the crystalline state of the phospholipids.[15,16]

We have compared the enthalpy of association of apoHDL and apoA-I with micellar phospholipid and with bilayer phospholipids (Fig. 3). The comparison of the binding enthalpies indicate that the enthalpy is significantly lower for micellar than for bilayer phospholipids around the transition temperature.

The formation of apolipoprotein–phospholipid complex shifts the transition temperature of the lipid toward higher temperatures,[5] and decreases the mobility of the phospholipid acyl chains. These changes in the lipid crystalline state account for part of the difference in binding enthalpy

[15] J. Massey, A. M. Gotto, and H. Pownall, *Biochim. Biophys. Acta* **794,** 137 (1984).
[16] F. Soetewey, J. Lievens, R. Vercaemst, M. Rosseneu, H. Peeters, and W. V. Brown, *Eur. J. Biochem.* **79,** 259 (1977).

between apolipoprotein micellar and bilayer phospholipids. The table lists such enthalpies for different phospholipids and apolipoproteins.

Massey et al.[15] have investigated the effect of temperature on the enthalpy of association of apoA-II with DMPC and LMPC. They observed that at temperatures above 30° for DMPC, and at all temperatures for LMPC, the association is exothermic. The major source of enthalpy arises from the increase in helical structure of the apolipoprotein. Between T_c of DMPC and that of its complexes with apoA-II, the enthalpy of association is highly exothermic (-260 kcal/mol of apoA-II); the enthalpy in this temperature range is assignated to the sum of protein-induced, isothermal acyl chain crystallization and coil–α-helix formation. Below T_c the association is endothermic but occurs spontaneously ($+90$ kcal/mol of apoA-II), suggesting that the entropy contribution to the free energy of association is greater than -90 kcal/mol of apoA-II. The crystalline state of the lipid phase can also be modulated by the addition of increasing amounts of cholesterol. This gradually decreases the amplitude of the phase transition which eventually disappears above 20 mol% of cholesterol.[5]

With apoA-II,[15] the enthalpy of lipid–protein association is a function of cholesterol content and at 25° increases linearly with the mol% cholesterol in the reaction mixture until it becomes endothermic between 15 and 20 mol% cholesterol. These data fit a model in which cholesterol is excluded from phospholipids in the "boundary" layer, which is perturbed by the protein. At high cholesterol concentrations the formation of a lipid–protein complex is thermodynamically unfavorable. Upon binding apoA-I to either DMPC bilayers or to mixed bilayers containing 10 mol% cholesterol we did not observe any significant difference in the association enthalpies, suggesting a different behavior for apoA-I and apoA-II in the association with phospholipid–cholesterol mixtures.[16]

Enthalpy of Ionization of the Apolipoproteins

According to the model of Segrest et al.,[17] the lipid-binding sites on the plasma apolipoprotein are located in the helical amphipathic segments of the amino acid sequence. Hydrophobic forces are mainly responsible for the cohesion of the complexes, though ionic contribution cannot be excluded. The contribution of the ionic forces towards apolipoprotein–phospholipid association can be derived from the comparison of the ionization patterns of the native apolipoproteins to those of their complexes with phospholipids. Microcalorimetric measurements provide a direct

[17] J. Segrest, R. Jackson, J. Morrisett, and A. M. Gotto, FEBS Lett. 38, 247 (1974).

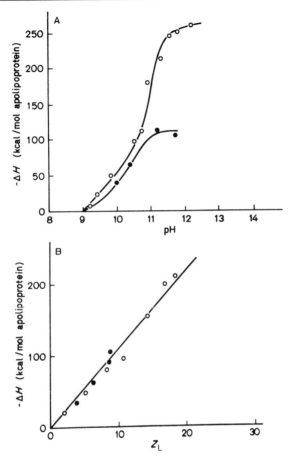

FIG. 4. Enthalpy titration curves for apoA-I (○) and apoA-I–DMPC complex (●). The enthalpy changes are plotted as a function of pH (A) and of the number of charged residues (B).

measure of the ionization enthalpy while the number of charged residues is obtained by potentiometric titration.[18,19]

The experimental results corrected for the heat of dilution of the components and adjusted such that $\Delta H_T = 0$ at pH 9.0 are summarized in Fig. 4. ΔH_T at any pH is the molar enthalpy change associated with the titration of either apoA-I protein or the apoA-I protein/lecithin complex from

[18] M. Rosseneu, P. Van Tornout, J. Lievens, and G. Assmann, *Eur. J. Biochem.* **117**, 347 (1981).
[19] M. Rosseneu, F. Soetewey, J. Lievens, R. Vercaemst, and H. Peeters, *Eur. J. Biochem.* **79**, 251 (1977).

pH 9.0 to that pH. In order to provide a thermodynamic interpretation of the ionization processes in the native apoA-I protein and its complex with dimyristoyl lecithin, we combined the calorimetric and the potentiometric titration data.

From the comparison of the calorimetric titration curves for the native apoA-I protein and for the apoA-I protein–lecithin complex (Fig. 4) it appears that the enthalpy of titration in the complex amounts to about half of the maximal enthalpy change evolved for the native apoA-I protein. Both curves level around pH 11.5 indicating that most of the groups are titrated at that particular pH. The titration curve for the native apoA-I protein consists of two parts: one from pH 9 to pH 10.8 with an enthalpy change of -110 kcal/mol, followed by a second process (pH 10.8–11.8) characterized by a larger enthalpy change per pH unit leveling at -260 kcal. From the amino acid composition of the native apoA-I protein the titration of 21 lysines ($\Delta H_T = -11$ to -14 kcal/mol) and 7 tyrosines ($\Delta H_T = -5$ to -6 kcal/mol) is expected. The magnitude of the enthalpy of ionization observed experimentally agrees with the predicted value.

From the potentiometric titration curves, the number of charged residues could be calculated at each pH. These data indicated that less charged groups are available for titration in the complex, accounting for the lower ionization enthalpy. The enthalpies of titration of the lysines of native apoA-I was identical to that of the complex and amounted to -11 kcal/mol lysine.[18]

Assuming that both hydrophobic and ionic binding are involved in the phospholipid–apolipoprotein association, the binding sites in the apoA-I protein should consist of sequential segments, providing apolar regions in close proximity to charged residues. A computer program was used to locate these sequences in the apoA-I, apoA-II, apoC-I, and apoC-III proteins.[19] For apoA-I these lipid-binding segments should include 10 lysines, 7 arginines, and 22 acidic residues to agree with the titration data. The observed pK shift of about 1 pH unit for 39 charged residues in the apoA-I–DMPC complex corresponds to a free energy charge of -0.2 kcal/phospholipid, compared to -2.4 kcal/phospholipid for a pure hydrophobic interaction. These titration data therefore suggest that only weak ionic interactions exist between charged residues and phospholipids at discrete binding sites on the apolipoproteins, while the major interaction forces are hydrophobic.

Conclusion

The direct measurement of the enthalpy of association between lipoprotein and lipids, by microcalorimetry, provides information about the thermodynamics of the complex formation. These measurements can be performed on low apolipoprotein concentrations and with several phospholipids. The effect of temperature, pH, and denaturing agents can be investigated to probe the mechanism of apolipoprotein–lipid association. The exothermal reactions observed between apolipoproteins and phospholipids suggest that the enthalpy is a driving force in the association process. These enthalpy changes have been attributed mainly to protein conformational changes involving coil to helix transition and to changes in the crystalline structure of the phospholipid.

[21] Solution Properties of Apolipoproteins

By JAMES C. OSBORNE, JR., NANCY S. LEE, and GRACE M. POWELL

Overview

In this chapter we shall detail methods used in our laboratory for reproducible measurements of the solution properties of plasma apolipoproteins. This methodology is based primarily upon the observed hydrodynamic properties of apolipoproteins and thus, although reference is made to specific instrumentation and/or techniques, the procedures are applicable and recommended for all studies on the behavior of apolipoproteins in aqueous solution. In line with the policy of this volume, the emphasis is on sample preparation and experimental conditions rather than specific molecular properties of each apolipoprotein, which have been detailed in several recent reviews.[1-3] In order to avoid unnecessary confusion we shall begin by defining several useful terms.

Definition of Terms

Apolipoprotein. An apolipoprotein is a homogeneous protein composed of a single polypeptide chain, or several polypeptide chains held together by covalent bonds, which contains no detectable noncovalently bound lipid and which associates with or forms an integral component of plasma lipoproteins. This definition is general and does not depend upon

---

[1] A. M. Scanu, C. Edelstein, and B. W. Shen, *in* "Lipid Protein Interactions" (P. C. Jost and C. H. Griffith, eds.), Vol. 1, p. 259. Wiley, New York, 1982.
[2] J. A. Reynolds, *in* "Lipid Protein Interactions" (P. C. Jost and C. H. Griffith, eds.), Vol. 2, p. 293. Wiley, New York, 1982.
[3] J. C. Osborne, Jr., and H. B. Brewer, Jr., *Adv. Protein Chem.* **31**, 253 (1977).

the method of isolation. Many apolipoproteins associate reversibly with lipoprotein particles and isolation by classical techniques, such as density centrifugation, often results in dissociation of lipid–apolipoprotein complexes.

Protomer/Oligomer. The majority of purified proteins form intermolecular complexes with themselves and other macromolecules in aqueous solution. This interaction is governed by the laws of mass action and in buffers of appropriate pH and sufficient ionic strength is, in most cases, quite specific. A protomer is the lowest molecular weight species undergoing complex formation and need not necessarily be a monomer. Oligomers, i.e., dimers, trimers, etc., are the higher molecular weight species in solution. Apolipoprotein–apolipoprotein interactions are subdivided further on the basis of rate of complex dissociation by the following two definitions.

Aggregation. Aggregation is defined as the "irreversible" formation of oligomers. In other words, upon dilution the higher molecular weight aggregates in solution do not dissociate to protomeric forms during the time scale of the experimental observation. Aggregates of apolipoproteins form with delipidation, lyophilization, and/or storage for long periods at high protein concentrations. By definition, aggregates are noncovalent complexes of apolipoproteins; however, in some cases disulfide exchange within aggregates can result in covalent bond formation between protomers.[4]

Association. Associating systems are those in which the rates of intermolecular interaction, i.e., protomer association and oligomer dissociation, are rapid enough for the system to be analyzed in terms of equilibrium thermodynamics. Thus, upon dilution the oligomers of a self-associating system dissociate to their corresponding protomeric forms and a new equilibrium distribution of species is established within the time scale of the experimental observations. At equilibrium the rates of association and dissociation are equal and the system can be described in terms of equilibrium constants. Self-association is defined as the reversible formation of oligomers from identical protomers and mixed-association is the reversible formation of oligomers from nonidentical protomers.

Molecular Species in Solution

The presence of both aggregated and associated species of apolipoproteins in solution accounts, most probably, for the majority of differing

[4] A. D. Cardin, K. R. Witt, C. L. Barnhart, and R. L. Jackson, *Biochemistry* **21**, 4503 (1982).

experimental observations in the literature. Aggregates of apolipoproteins form with delipidation and lyophilization and also form slowly with time in most aqueous systems. Thus, the concentration of aggregates in solution is variable depending upon the history of the sample. In order to obtain reproducible results any aggregates must be dissociated or removed from the samples prior to analysis. With the exception of apoB, aggregates have been removed/dissociated prior to analyzing the systems described below.

Nonassociating. Apolipoprotein H (β_2-glycoprotein I) satisfies all of the criteria to be classified as an apolipoprotein, however, its secondary, tertiary, and quaternary organization is quite different from that of other well-characterized apolipoproteins.[5] The secondary and tertiary structures of apoH are more resistant to the effects of denaturants than that of other apolipoproteins and this is the only apolipoprotein described thus far that does not self-associate in aqueous solution. Thus, the molecular weight of apoH is 43,000 in the presence and absence of denaturants.

Aggregating. The tendency of apoB to aggregate, even in the presence of denaturants and detergents, and to exhibit nonideal behavior in aqueous solution has resulted in a vast, and more often than not, conflicting literature on the molecular species in solution. Estimates of the monomeric molecular weight of apoB have ranged from about 8,000 to over 250,000; in almost all reports the presence of higher molecular weight aggregates has been noted by the authors.[4,6] Needless to say the solution properties of apoB are not well understood at the present time. For a detailed, up to date discussion of the properties of apoB consult Fisher and Schumaker, this volume [11].

There have been two detailed reports on the hydrodynamic properties of apoC-II and both groups classify this apolipoprotein as an aggregating system.[7,8] An upper limit of the size of the aggregated species was not estimated, however, both groups noted time-dependent precipitation of apoC-II from solution.

Associating. Apolipoproteins A-I, A-II, C-I, C-III, and E self-associate and also form mixed-associated complexes in aqueous solution at neutral pH (9). The table lists the molecular species for self- and mixed-associations as predicted from studies in our laboratory. For a detailed discussion of association schemes consult recent reviews.[1-3]

---

[5] N. S. Lee, H. B. Brewer, Jr., and J. C. Osborne, Jr., *J. Biol. Chem.* **258**, 4765 (1983).
[6] J. C. H. Steele, Jr. and J. A. Reynolds, *J. Biol. Chem.* **254**, 1639 (1979).
[7] W. W. Mantulin, M. F. Rohde, A. M. Gotto, Jr., and H. J. Pownall, *J. Biol. Chem.* **255**, 8185 (1980).
[8] S. Tajima, S. Yokoyama, Y. Kawai, and A. Yamamoto, *J. Biochem.* **91**, 1273 (1982).

MOLECULAR SPECIES FOR SELF- AND MIXED-ASSOCIATIONS

	Self-association	
Apolipoprotein	Monomer molecular weight	Association scheme
A-I	28,016	Monomer–dimer–tetramer–octamer
A-II	17,380	Monomer–dimer
C-I	6,630	Monomer–dimer–tetramer
C-III	8,764	Monomer–dimer–trimer
E	34,000	Not known; oligomer size $>10^6$

Mixed-association	
Apolipoproteins	Oligomer stoichiometry
ApoA-I : apoA-II	1 : 1, 2 : 2
ApoA-II : apoC-I	2 : 4
ApoA-I : apoC-I	2 : 4

Molecular Conformation

Apolipoproteins A-I, A-II, C-I, and C-II undergo structural transitions with association/aggregation. Changes in secondary structure can be quite large and in the case of apoC-I approach that obtained by interaction with lipid.[9] These transitions seem to be rapid and reversible and have been characterized in terms of equilibrium constants. Perturbations, such as dilution or the addition of low levels of denaturants, that effect the equilibrium distribution of species in solution also effect secondary structure of these apolipoproteins. The secondary structure of apolipoproteins B, C-III, E, and H is more resistant to the effects of mild perturbations such as the addition of low levels of denaturants and in this respect resemble classical gobular proteins.

Preparation of Samples

Use of Solvent Additives

As soon as possible after collection of plasma, ethylenediaminetetraacetic acid, to about 1 mM, should be added in order to inhibit lipid autoxidation, which is believed to cause covalent modification of apolipo-

[9] J. C. Osborne, Jr. and H. Bryan Brewer, Jr., *Ann. N.Y. Acad. Sci.* **348,** 104 (1980).

proteins.[10] The use of a serine protease inhibitor such as phenylmethylsulfonylfluoride (10 μM) is recommended[11] as well as low levels, <0.01%, of sodium azide. The temperature should be kept below 4° and if possible the plasma should be purged with dry nitrogen.

Delipidation

Numerous delipidation procedures are available in the literature, however, the majority result either in incomplete removal of lipids, especially phospholipids, or formation of protein residues that are difficult to redissolve. The procedures described by Shore and Shore,[12] Scanu and Edelstein,[13] Brown et al.,[14] and Herbert et al.[15] have gained wide acceptance and are recommended for routine delipidations. The efficiency of delipidation and solubility of the lipid free product is dependent critically upon many factors such as temperature, buffer composition, and ratio of organic to aqueous phase. For reproducible results the published procedures should be followed closely. For additional details see Chapter [9], this volume.

Solubilization

The title of this section should actually be "solubilization and dissociation of any aggregated apolipoprotein." Aggregation of apolipoproteins may result in changes in molecular and/or functional properties and thus cause experimental variability. Aggregates need not be of high molecular weight or be less soluble than nonaggregated material. A dimer that did not dissociate upon dilution would be classified as an aggregate. The presence of aggregates in a self-associating system complicates considerably the evaluation of association scheme and corresponding equilibrium constants. Aggregates of apolipoproteins should be removed or dissociated prior to experimentation. We routinely dissolve precipitated or lyophilized apolipoproteins in 0.01 M Tris, 0.1 M sodium chloride, 0.001 M sodium azide, 0.001 M ethylenediaminetetraacetic acid, 2 M guanidinium

[10] J. Schuh, G. F. Fairclough, Jr., and R. H. Haschemeyer, *Proc. Natl. Acad. Sci. U.S.A.* **75,** 3173 (1978).
[11] W. A. Bradley, E. B. Gilliam, A. M. Gotto, Jr., and S. H. Gianturco, *Biochem. Biophys. Res. Commun.* **109,** 1360 (1982).
[12] B. Shore and V. Shore, *Biochemistry* **8,** 4510 (1969).
[13] A. M. Scanu and C. Edelstein, *Anal. Biochem.* **44,** 576 (1971).
[14] W. V. Brown, R. I. Levy, and D. S. Fredrickson, *J. Biol. Chem.* **244,** 5687 (1969).
[15] P. N. Herbert, L. L., Bausserman, L. O. Henderson, R. J. Heinen, E. C. Church, and R. S. Shulman, *in* "The Lipoprotein Molecule" (H. Peeters, ed.), p. 35. Plenum, New York, 1978.

chloride, pH 7.4 buffer and then dialyze the resulting solution against the desired final buffer. This procedure usually results in aggregate-free solutions of apolipoproteins. Exceptions are apoB, which resists solubilization in the absence of detergents, and preparations of other apolipoproteins that have been stored frozen or lyophilized for long periods (1–2 years). In the latter case, higher levels of guanidinium chloride or repurification may be required to free the preparation from aggregated material.

Covalent Modification

Covalent modification or labeling of apolipoproteins may result in changes in physical properties and almost always results in generation of aggregated material. Perturbations in hydrodynamic properties may be specific for a particular amino acid in the polypeptide backbone and occur even with low levels of labeling. For instance, iodination of apoA-I can result in generation of an incompetent species with properties that are quite different from those of the unlabeled material.[16] This also occurs with nitration and is due presumably to modification of a specific tyrosine residue on apoA-I. The incompetent material in this case may be removed by chromatography. We recommend the following two steps prior to using covalently modified apolipoproteins for experimental studies:

Repurification. At a minimum, the labeled material should be subjected to column chromatography in order to detect and remove aggregated material. We routinely chromatograph solutions of apolipoproteins using a high-performance liquid chromatograph (HPLC) and TSK 2000-3000 SW columns (Toyo Soda, Tokyo, Japan). In buffers of neutral pH and sufficient ionic strength ($>0.1\ M$) the profiles are characteristic of a given apolipoprotein and yields are essentially quantitative. The chapter by Scanu and Edelstein in this volume [18] should be consulted for additional details on HPLC of apolipoproteins.

Evaluation of Molecular Properties. The extent of this step depends upon the question under study and may vary considerably from laboratory to laboratory. For instance, if labeled apolipoproteins are used to quantitate mixed-association between apolipoproteins then the association scheme and corresponding equilibrium constants should be evaluated independently for the labeled material and compared to those obtained with the unlabeled species. The majority of experimental studies require much less rigorous characterization. We have found HPLC chromatogra-

---

[16] J. C. Osborne, Jr., E. J. Schaefer, G. M. Powell, N. S. Lee, and L. A. Zech, *J. Biol. Chem.* **259**, 347 (1984).

phy of labeled material in the presence of an excess of unlabeled species to be useful in comparing hydrodynamic properties. The analysis is rapid (~1 hr) and for self-associating systems, such as apoA-I, can be quite sensitive to differences in molecular properties.[16]

Storage

We routinely store apolipoproteins at low concentrations (~0.1 mg/ml) in ammonium bicarbonate buffer or as a lyophilized powder at −20°. Aggregates may form under these conditions and samples are routinely processed for analysis by first dissolving lyophilized apolipoprotein in buffer containing 2 M guanidinium chloride and then dialyzing exhaustively against appropriate experimental buffer.

Methods of Study

Although the solution properties of apolipoproteins are quite complex, they are predictable in most cases and correspond to that expected of interacting protein systems. The major difference in the behavior of apolipoproteins as compared to classical globular proteins is the ease with which apolipoproteins, especially apoA-I, apoA-II, and apoC-I, refold to accommodate a given environment. In the monomeric form apolipoproteins are folded loosely with a high degree of exposure of nonpolar groups to solvent.[17] Presumably the amino acid sequence of apolipoproteins is such that, in the monomeric state, little free energy is gained by intramolecular folding to sequester nonpolar residues from solvent. Upon interaction with other compounds the additional degrees of freedom (i.e., the additional nonpolar surface area of the added ligand) is used to shield solvent from nonpolar residues. This is accomplished, in many cases, by a major refolding of the polypeptide backbone. Reactions involving changes in the degree of exposure of nonpolar compounds to solvent show a characteristic dependence upon experimental variables such as temperature, pressure, and solvent composition. In this section we have chosen several experimental techniques in order to illustrate some precautions that should be taken into account in the design of experiments on apolipoproteins in aqueous solution. Earlier chapters in this series should be consulted for details on each of the techniques discussed in the following subsections.[18]

[17] J. C. Osborne, Jr., H. B. Brewer, Jr., T. J. Bronzert, E. J. Schaefer, and R. L. Tate, NIH Publication No. 83-1266, 178 (1983).
[18] C. H. W. Hirs and S. N. Timasheff, eds., this series, Vol. 27.

Sedimentation Equilibrium/Velocity Measurements

The method of choice for quantitating associating systems is sedimentation equilibrium. Molecular weights are obtained directly from first principles and are independent of shape. In addition, equations for describing the primary data in terms of various association schemes are on firm theoretical ground. One possible drawback of this technique is that the pressure in the sample cell increases with increasing radius (i.e., distance from the center of rotation), and this can complicate analysis of interacting systems due to the pressure dependence of equilibrium constants.

$$(\partial \ln k/\partial P)_T = -\Delta V/RT \qquad (1)$$

where k is the equilibrium constant, P the pressure, T the temperature, R the gas constant, and ΔV the change in volume upon interaction. Changes in partial specific volume with association can be quite large for apolipoproteins, due primarily to large conformational changes, and this results in molecular weights (or more specifically equilibrium constants) that depend upon radius and rotor speed. Collection of equilibrium data at several different rotor speeds and initial concentrations of protein combined with appropriate data analysis yields equilibrium constants that correspond to atmospheric pressure.[19] Pressure effects are more difficult to detect with sedimentation velocity measurements especially when the number of protomers in the highest oligomer is small. Increases in pressure obey the following equation

$$P(r) = \frac{\rho\omega^2}{2}(r^2 - r_m^2) + P_m \qquad (2)$$

where P is the pressure, ρ the solvent density, ω the angular velocity, r the radial position in the cell, and m the meniscus. Therefore, near the meniscus, where sedimentation coefficients are usually measured, the pressure is essentially atmospheric and pressure effects are small. Near the bottom of the cell, pressure effects are larger, however, they are more difficult to quantitate due to the effects of diffusion and radial dilution on the associated species.

In addition to pressure effects, the presence of aggregates and/or nonideal behavior also greatly affect the evaluation of interacting systems. If possible, aggregates should be removed or dissociated and studies should be performed in buffers of sufficient ionic strength (usually 0.1 M inorganic salt) to minimize nonideal behavior due to charge effects.[18]

---

[19] S. Formisano, H. B. Brewer, Jr., and J. C. Osborne, Jr., *J. Biol. Chem.* **253**, 354 (1978).

Circular Dichroic/Fluorescence Measurements

This section was included in order to emphasize three properties of apolipoproteins in solution. The first is that minor changes in the environment of an apolipoprotein can result in major changes in structure. Self- and mixed-association, changes in temperature, buffer composition (especially ionic strength), as well as the pressence of other amphiphiles have been shown to cause refolding of the polypeptide backbone of apolipoproteins.[9] Therefore spectroscopic properties should be equated with specific experimental conditions, especially the concentration of apolipoproteins and other molecules in the system.

The second point in this section deals with circular dichroic measurements in the far ultraviolet and is directed primarily at solutions of apolipoprotein B. Artifacts due to light scattering are magnified in the far ultraviolet region of the spectrum. Solutions that are not visibly turbid may exhibit distorted spectra of decreased intensity due to light scattering. Decreases in signal become much stronger with decreasing wavelength and in some cases the spectrum of a protein with a high α-helical content can resemble that expected of a β-sheet configuration. In this situation the minimum in negative ellipticity is usually shifted to longer wavelengths and the signal-to-noise ratio decreases as the intensity of positive or negative ellipticity decreases.

The final point in this section deals with measurement of spectroscopic properties as a function of apolipoprotein concentration in order to monitor molecular interaction. For instance, the secondary structure of apolipoproteins A-I, A-II, C-I, and apoC-II decreases with dissociation of oligomers and measurements of circular dichroism in the far ultraviolet have been used to characterize this transition.[9] For these studies artifacts that perturb signal intensity, such as nonspecific absorption, must be monitored carefully. Nonspecific absorption to cuvette walls and air–buffer interfaces occurs at all apolipoprotein concentrations: losses due to transfer of solution from one cuvette to another are immeasurable at high concentrations (~1 mg/ml) but may become substantial in the region of dissociation of most apolipoprotein oligomers (<0.1 mg/ml). We have found that it is a good practice to dilute apolipoprotein solutions in a single cuvette by replacing a given volume of sample with buffer.

Chromatography

High-performance liquid chromatography (HPLC) is a powerful tool for processing and evaluating apolipoproteins.[20] We have found conven-

[20] A. M. Scanu, M. Halari, and C. Edelstein, NIH Publication No. 83-1266, 230 (1983).

tional gel permeation chromatography using TSK 2000-5000 SW columns to be an efficient and time-saving method for changing buffer composition and for monitoring sample heterogeneity, the presence of aggregates, and molecular associations.[20] Chromatography time is about 1 hr and we have found that the recovery of apolipoprotein in usual buffers of sufficient ionic strength (~0.1 M inorganic salt) to be excellent. Sample heterogeneity can be evaluated in buffer containing 6 M guanidinium chloride: we normally use 0.01 M Tris, 0.1 M sodium chloride, 0.001 M sodium azide, 6 M guanidinium chloride, pH 7.4. An added benefit is that elution position under these conditions is porportional to molecular weight and less sensitive, compared to SDS–gel electrophoresis, to the presence of carbohydrate. Chromatography in the absence of denaturants can be used to monitor self- and mixed-associations and to evaluate the integrity of covalently modified species. These topics are treated in detail in Chapter [18] by Edelstein and Scanu.

Light Scattering Measurements

Light scattering measurements compliment sedimentation equilibrium measurements in that both yield weight average molecular weights from first principles and the former analysis is not complicated by pressure effects. Therefore, molecular weight versus concentration profiles predicted at atmospheric pressure by sedimentation equilibrium measurements as a function of rotor speed can be verified directly by light scattering measurements. In addition, whereas sedimentation equilibrium is of limited use for proteins with molecular weights above 250,000, light scattering, especially using laser light sources at angles below 5°, can be used for macromolecules with molecular weights greater than 10^7. The primary technical drawbacks of light scattering measurements is sample clarification. The intensity of scattered light is proportional to molecular size and low levels of large contaminates, such as dust, result in high estimates of molecular weight. We routinely clarify solutions of apolipoproteins by ultrafiltration directly into sample cells. There are two primary artifacts that should be taken into account when using this technique of clarification. First, some filters contain detergents as "wetting agents" and the apolipoprotein may leach these out of the filter and into the sample cell. Second, loss of apolipoprotein due to nonspecific absorption during filtration may result in an overestimation of the actual concentration of sample in the cell. Both of these artifacts can cause substantial over/underestimates of molecular weight. We usually process more solution than required and discard the first volumes of sample passed through the filter system.

Potential Problems

We have included this section primarily to summarize the unique properties of apolipoproteins. Assumptions used in experimental design and data analysis that would be acceptable for classical globular proteins could cause substantial problems in the evaluation of solution properties of apolipoproteins.

Sample Purity

Contaminating apolipoproteins contribute to the signal being observed and also compete with the apolipoprotein under study for intermolecular interactions. Methods for the routine isolation and purification of most of the known apolipoproteins are available (see appropriate chapters in this series) and the presence of contaminating apolipoproteins is not usually a major obstacle in evaluating solution properties of apolipoproteins. On less firm ground is the presence of lipid, especially phospholipid, resulting from incomplete delipidation procedures. Lipids can cause conformational transitions in apolipoproteins and stabilize some molecular species, and thereby affect evaluation of intermolecular interactions, especially when using spectroscopic techniques such as fluorescence or circular dichroic measurements. Bound lipid also effects the partial specific volume and thus may perturb density gradient centrifugation and sedimentation equilibrium measurements. Contaminating lipid can be minimized by following closely the published delipidation procedures (see Chapter [9] in this volume).

Experimental Design

Apolipoproteins, like most classical globular proteins, self-associate in aqueous solution and thus have properties that depend upon protein concentration as well as solvent environment. The major difference between apolipoproteins and classical globular proteins is that the former are more flexible molecules and as such their secondary and tertiary structure responds more dramatically to a given perturbation in environment. This flexibility leads to characteristic responses to changes in temperature, pressure, and solvent composition, all of which must be monitored carefully for reproducible results. As polyelectrolytes, apolipoproteins exhibit nonideal behavior in buffers of low ionic strength. Moderate concentrations of inorganic salt are recommended for most studies; we routinely use 100 mM sodium chloride in all buffers. Finally, aggregates of apolipoproteins form with lyophilization and with storage in most buffers. The experimental protocol should include processing steps to ensure dissociation of aggregated material. This step can be combined with the recommended repurification steps when using covalently modified apolipoproteins.

Data Analysis

This section is included primarily for analysis of sedimentation equilibrium measurements; however, the general principles are also applicable to other types of studies in which theoretical models are fit to experimental data. First, the primary data should be analyzed directly, rather than in the form of linear or other types of transformations. This is especially important for ligand binding[2,21] and sedimentation equilibrium measurements.[22,23] Transformation of primary data can lead to distortions of experimental error and inadequate statistical correlations between the parameters of the various models being tested. Equations for self- and mixed-associating systems in terms of the primary data from sedimentation equilibrium measurements are available in the literature.[24,25] Finally, all of the assumptions that apply to a particular model should be kept in mind when interpreting "best fit" parameter values. If a system is irreversible or nonideal, then estimates of parameter values using models that assume reversibility or ideality are necessarily inaccurate and may be meaningless.

Relevance to Native Lipoprotein Structure and Metabolism

The unusual properties of apolipoproteins, especially the ability to adapt to a given environment by rapid polypeptide backbone refolding, allow a unique view of the forces involved in the structural stability of proteins and macromolecules in general. Changes in conformation, especially secondary and tertiary structure, can be induced by minor changes in environment and the thermodynamics of unfolding/refolding processes can be measured without resorting to extremes in pH or temperature or high levels of denaturant. This fact alone probably explains the time, manpower, and literature that has been devoted to understanding the solution properties of apolipoproteins. The usefulness of these studies actually goes far beyond the theoretical interests of classical protein

[21] P. J. F. Henderson, *in* "Techniques of Protein and Enzyme Biochemistry." Vol. BI/II, p. 113, Elsevier, Amsterdam, 1978.
[22] H. Fujita, *in* "Foundations of Ultracentrifugal Analysis" (P. J. Elving and J. D. Winefordner, eds.) (Chemical Analysis, Vol. 42). Wiley, New York, 1975.
[23] J. C. Osborne, Jr., T. J. Bronzert, and H. B. Brewer, Jr., *J. Biol. Chem.* **252,** 5756 (1977).
[24] J. C. Osborne, Jr., *J. Biol. Chem.* **253,** 359 (1978).
[25] L. Servillo and J. C. Osborne, Jr., *Biophys. Chem.* **13,** 35 (1981).

chemists. It is easy to understand why a knowledge of hydrodynamic properties is useful and in fact fundamental to *in vitro* studies with apolipoproteins. Changes in a given parameter may depend critically on all interacting species in solution and computed theoretical parameters, such as binding affinity or number of binding sites, may be a complex product of several terms. Since apolipoproteins interact with most nonpolar compounds, their behavior in solution is basic to the design and interpretation of *in vitro* studies ranging from immunoquantitation to molecular interactions with cells in culture. The solution properties of apolipoproteins are also a useful framework for evaluating the integrity of covalently modified apolipoproteins which are often assumed to act as tracers of unlabeled species in experimental studies.

If all apolipoproteins behaved as apoB and remained with the same particle through all stages of metabolism in plasma then the usefulness of studies on solution properties would not be extended to the *in vivo* state. However, the apolipoprotein composition of plasma lipoproteins changes dramatically with metabolism. The rate of apolipoprotein transfer between particles is governed by the laws of mass action and in some cases apolipoprotein composition can be described in terms of equilibrium thermodynamics. In other cases transfer between particles may be on the same time scale as particle metabolism and may actually be the rate-limiting step in metabolism. In either case, transfer between particles occurs in all probability through water-soluble species. These soluble species may contain a few molecules of lipid or may actually be lipid free. Measurements of solution properties of apolipoproteins *in vitro* bear directly on the properties of these water-soluble species. Thus, knowledge of the hydrodynamic properties and modes of interaction with small amphiphiles and other macromolecules can be used as a framework for evaluating one possible mechanism of *in vivo* communication between plasma lipoproteins.

[22] Studies of Apolipoproteins at the Air–Water Interface

By MICHAEL C. PHILLIPS and KEITH E. KREBS

Introduction

There is a long history to the use of the surface balance for studying biological molecules. The investigation of the properties of proteins at the air–water interface also extends back to the earliest days of surface balances (for reviews, see Neurath and Bull[1]; MacRitchie[2]). Insights into the properties and packing of protein molecules at biological interfaces have been derived, and the high surface activity of serum apolipoproteins makes them ideal candidates for such surface balance studies. The literature on the surface properties of apolipoproteins has been reviewed recently.[3]

The parameter which can be measured most easily is surface pressure, but if the surface concentration is also known, a much more detailed analysis of the surface activity of apolipoproteins is possible. Here we describe the procedures for measuring these parameters for spread and adsorbed apolipoprotein films at the air–water and lipid monolayer–water interfaces. In order to study the surface concentration of adsorbed films, [^{14}C]apolipoprotein is required and the methods for such labeling are outlined first. The use of the surface balance to evaluate the surface behavior of apolipoproteins is then treated.

Materials

Apolipoproteins

Chromatographically pure apolipoproteins may be isolated by the procedures outlined in other chapters in this volume. The purity of the sample may be verified by sodium dodecyl sulfate–polyacrylamide gel electrophoresis, isoelectric focusing, or high-performance liquid chromatography. The protein samples may be stored in 3 M guanidine–hydrochloride at $-20°$, and desalted on a BioGel P-2, or equivalent, column immediately prior to use so that all apolipoprotein molecules are present as monomers. It is not recommended that the protein be lyophilized and stored, since it has been demonstrated that it can be difficult to obtain the apolipoprotein in a monomeric state by dissolving the lyophilized sample.[4]

A precise determination of protein concentration is essential for interpretation of results obtained in interfacial studies. This is most easily accomplished by means of a Lowry protein analysis incorporating sodium dodecyl sulfate to dissolve the protein,[5] taking care that the protein sam-

---

[1] H. Neurath and H. B. Bull, *Chem. Rev.* **23**, 391 (1938).
[2] F. MacRitchie, *Adv. Protein Chem.* **32**, 283 (1978).
[3] G. Camejo and V. Muñoz, in "High Density Lipoproteins" (C. E. Day, ed.), p. 131. Dekker, New York, 1981.
[4] S. Yokoyama, S. Tajima, and A. Yamamoto, *J. Biochem.* **91**, 1267 (1982).
[5] M. K. Markwell, S. M. Haas, L. L. Bieber, and N. E. Tolbert, *Anal. Biochem.* **87**, 206 (1978).

ple buffer does not contain agents that may interfere with the reaction.[6] By performing a Lowry determination, and measuring an absorbance value for the protein solution at 280 nm, an extinction coefficient may be determined for rapid measurement of the protein concentration during later use. Alternatively, an accurate determination of protein concentration may be obtained from a Kjeldahl nitrogen analysis.

For quantitating adsorption and desorption at the air–water interface, surface radioactivity techniques can be employed to determine the surface concentration of protein. To perform these experiments it is necessary to radioactively label the purified apolipoprotein. Reductive methylation of the apolipoprotein using the procedure described by Jentoft and Dearborn[7,8] results in ^{14}C-labeled methyl groups being substituted into the amino group of lysine residues. In a typical reaction, a 1 ml total volume of 50 mM HEPES buffer at pH 7.5 containing 1 mg of human apoA-I, 0.4 mM [^{14}C]formaldehyde (specific activity = 32 Ci/mol), and 4 mM sodium cyanoborohydride is incubated at room temperature for 18 hr. This results in an apoA-I preparation with a specific activity of about 5 μCi/mg, in which less than 5 of the 22 lysine residues present in each molecule are methylated. The degree of chemical modification can be determined by amino acid analysis using lysine, and mono- and dimethyllysine as standards. The fraction of lysine residues modified may also be determined by reaction of the protein with ninhydrin, which yields the phenylthiohydantoin derivative upon reaction with a primary amine. The methylation of lysine residues prevents reaction with ninhydrin and hence the methylated protein has decreased absorbance at 550 nm compared to the unmodified protein.

As an alternative to reductive methylation, reaction of protein with [^{14}C]acetic anhydride[9] leads to acetylation of lysine residues. The advantage of reductive methylation over acetylation is that the former does not eliminate the positive charge on the lysine residues. For the detection of apolipoprotein at the interface by gas flow counter (see below), a ^{14}C specific activity of at least 1 μCi/mg protein is required.

Reagents for Surface Chemistry

Because of the sensitivity of the surface chemistry techniques to impurities, reagent grade salts must be employed (it may be necessary to roast the salts to remove surface active contaminants). In addition, good qual-

---

[6] A. Bensadoun and D. Weinstein, *Anal. Biochem.* **70**, 241 (1976).
[7] N. Jentoft and D. G. Dearborn, *J. Biol. Chem.* **254**, 4359 (1979).
[8] N. Jentoft and D. G. Dearborn, this series, Vol. 91, p. 570.
[9] D. J. Adams, M. T. A. Evans, J. R. Mitchell, M. C. Phillips, and P. M. Rees, *J. Polymer Sci. (Part C)* **34**, 167 (1971).

ity distilled or deionized water is necessary so that contaminants do not influence surface tension measurements.[10]

Surface Pressure–Molecular Area Isotherms for Spread Monolayers

Measurement of Surface Pressure

Surface pressure (π) is the decrease in surface tension of an air–water interface due to the presence of a monolayer, and is defined as $\pi = (\gamma_0 - \gamma_{monolayer})$ where γ_0 and $\gamma_{monolayer}$ are the surface tensions of the clean air–water interface and the monolayer-covered interface, respectively. Surface pressure measurements may be carried out using a variety of techniques that have been described in a number of monographs.[10,11] In this chapter, the more common techniques which have been applied to the study of apolipoproteins at the air–water interface are presented.

Surface pressure–molecular area isotherms of spread films are measured on surface balances which are comprised of a hydrophobic Teflon rectangular trough filled with a suitable aqueous substrate (Fig. 1). The insoluble film is spread quantitatively onto the surface at the desired temperature between a moveable Teflon barrier which is used for film compression and a Teflon float which is used to monitor changes in surface pressure with changes in surface area (Langmuir balance). There is a clean air–water interface behind the float, and the lateral pressure exerted by the protein film in the surface displaces the float thereby rotating a torsion wire which is connected to the float. The torsion wire in turn is interfaced to a force detection device.

The surface pressure of spread films may also be measured by the Wilhelmy plate technique.[10,12] This procedure utilizes a thin plate which dips into the aqueous substrate when suspended from the arm of a recording electrobalance, as illustrated in Fig. 1. The change in surface pressure may be followed over time by interfacing the balance to a recorder. The downwards force exerted by the meniscus on the plate is $F = (\gamma \cos \theta)$ where γ = surface tension and θ = the contact angle between the plate and the monolayer-covered aqueous phase. By the appropriate selection of plate material, a contact angle of zero is achieved (i.e., $\cos \theta = 1$) so that $F = \gamma$. A decrease in γ (increase in π) leads to a decrease in F which is detected by a plate of perimeter P as a weight loss, $(-)\Delta W = PF$. It

---

[10] G. L. Gaines Jr., "Insoluble Monolayers at Liquid–Gas Interfaces." Wiley (Interscience), New York, 1966.
[11] A. W. Adamson, "Physical Chemistry of Surfaces," 4th Ed. Wiley, New York, 1982.
[12] L. Wilhelmy, *Ann. Phys.* **119,** 177 (1863).

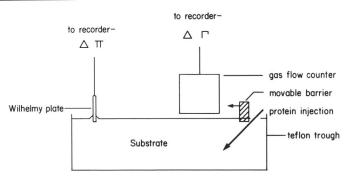

FIG. 1. Configuration of surface balance for measuring surface pressure and surface concentration. Protein adsorption is monitored by following the changes in surface pressure (π) with the Wilhelmy plate, and surface radioactivity (counts per minute) with the gas flow counter, after injection of a given concentration of ^{14}C-labeled protein beneath the surface into a suitable substrate buffer. The surface concentration (Γ) is derived from a calibration curve relating surface radioactivity (cpm) to a known protein surface concentration as described in the text. Determination of the surface pressure exerted by a protein monolayer as a function of molecular area is obtained by spreading the protein as an insoluble film and decreasing the surface area with the moveable barrier. Alternatively, increasing quantities of protein can be spread at a constant surface area.

follows that for a wetted plate π (mN m^{-1}) = $(-)\Delta W/P$ where the units of ΔW and P are mg and cm, respectively.[10]

Wilhelmy plates can be cut from glass, platinum, or mica because when these materials are clean they are wetted by aqueous solutions (i.e., $\theta = 0°$). Roughening the surface of the plate with an abrasive promotes wetting. The ease with which mica plates can be created in various dimensions makes this material a good choice; it can be readily obtained from suppliers of electron microscopy accessories. An additional advantage of the use of mica as the plate material is the fact that phospholipid and protein do not readily adsorb to wetted mica, decreasing the probability of contact angle changes during the experiment. Between experiments the plate can be cleaned in chromic acid solution and thoroughly rinsed in distilled water. Typical plate dimensions are 0.006 cm in thickness, 2 cm in height, and 1 cm in width to give a plate with a perimeter (P) of approximately 2 cm. The sensitivity in π for a plate of the above dimensions is 0.1 mN m^{-1}, with an error of ± 1 mN m^{-1}.

Another technique, related to the Wilhelmy plate, which has been used to study apolipoproteins at the air–water interface is that employing the du Noüy ring.[13]

---

[13] B. W. Shen and A. M. Scanu, *Biochemistry* **19**, 3643 (1980).

Spreading of Insoluble Apolipoprotein Monolayer

Typical experiments with spread monomolecular protein films involve quantitative spreading at the air–water interface, compression of the monolayer to known areas, and the measurement of corresponding π values by the above procedures to obtain π-molecular area (A) isotherms. The quantitative spreading of protein at the clean interface with a minimum loss into the substrate to produce the monomolecular film is a critical step in such monolayer experiments. The air–water interface is initially cleaned by compressing the surface and sucking away the surface layer which contains any surface active contaminants. This procedure is repeated until there is no increase in π on sweeping the full length of the trough. It is difficult to spread aqueous protein solutions directly onto the cleaned surface without some of the protein dissolving in the subphase. In order to avoid this problem of protein loss, three alternative procedures are available. (1) The apolipoprotein at a concentration of less than 0.06 mg/ml in the appropriate buffer solution is spread as a film down a water-wetted, 5-mm-diameter glass rod onto the interface at a flow rate of less than 0.2 ml/min.[14] The acid-cleaned glass rod is positioned in the interface such that the distance between the point of delivery of the protein solution to the rod and the interface does not exceed 5 cm. Spreading the protein from dilute solution decreases the probability of aggregation. (2) Improved protein spreading is achieved by adding 30% w/v isopropanol to the spreading solution which is applied dropwise directly to the surface.[15] (3) If the substrate contains a very high salt concentration (e.g., 3.5 M KCl), the apolipoproteins are insoluble and will remain in the interface when added directly from buffer solution. A disadvantage of this approach is the difficulty of eliminating all surface-active impurities from the large amount of salt added to the substrate.

Following the quantitative spreading of apolipoprotein at the air–water interface, the monolayer is compressed and the π–A isotherm recorded. Examples of typical π–A isotherms are given in Figs. 2 and 3, respectively, for rat apoHDL[16] and a series of purified apolipoproteins.[17]

Surface Pressures of Adsorbed Monolayers

The adsorption of protein to an interface may be monitored by following the increase in π after injection of protein into the subphase. Typi-

---

[14] H. J. Trurnit, *J. Colloid Sci.* **15**, 1 (1960).
[15] C. W. N. Cumper and A. E. Alexander, *Rev. Pure Appl. Chem.* **1**, 121 (1951).
[16] G. Camejo, G. Colacicco, and M. M. Rapport, *J. Lipid Res.* **9**, 562 (1968).
[17] K. E. Krebs, M. C. Phillips, and C. E. Sparks, *Biochim. Biophys. Acta* **751**, 470 (1983).

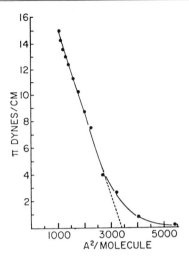

FIG. 2. Surface pressure–molecular area isotherm for a spread monolayer of rat high-density lipoprotein apolipoprotein. Rat apoHDL was spread as a film on 0.01 M phosphate buffer, pH 7.6, ionic strength 0.1, at 24 ± 1°. The apolipoprotein solution (conc. = 200 μg/ml) was spread dropwise at the air–water interface in 0.5-μl aliquots, and the subsequent increase in surface pressure as a function of molecular area was measured by a Wilhelmy plate (roughened platinum). The limiting molecular area, given by extrapolation of the linear region of the isotherm to $\pi = 0$, is 3400 Å^2/molecule, which corresponds to approximately 15 Å^2/amino acid residue. [Reprinted with permission from G. Camejo, G. Colacicco, and M. M. Rapport, *J. Lipid Res.* **9**, 562 (1968).]

cally, these experiments are carried out using the Wilhelmy plate technique or a du Noüy ring. For dilute protein solutions, the rate-limiting step in adsorption is the diffusion of protein to the interface,[18] and this may be speeded up by constant stirring of the substrate following protein injection. This stirring enhances protein mixing in the substrate and reduces the thickness of the unstirred layer beneath the air–water interface.

Adsorption kinetics for proteins are characteristically measured as a change in π as a function of the initial or final substrate protein concentration (C_p) in weight percent (g/100 ml). Protein concentration is expressed in weight units rather than in moles because adsorption of polymers is mediated by the independent kinetic units within the polymer chain rather than the whole molecule. Use of mass units facilitates comparison of the surface activities of proteins of different molecular weights because equal masses of polypeptide are present.

[18] D. E. Graham and M. C. Phillips, *J. Colloid Interface Sci.* **70**, 403 (1979).

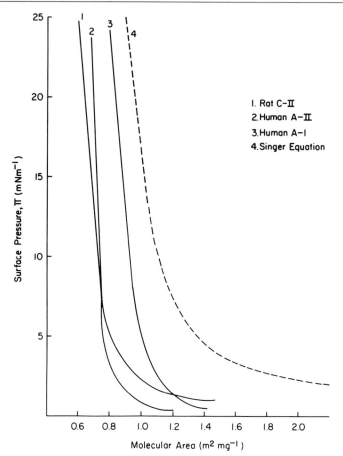

FIG. 3. Surface pressure–molecular area isotherms for purified apolipoproteins. Purified rat [K. Krebs, M. C. Phillips, and C. E. Sparks, *Biochim. Biophys. Acta* **751**, 470 (1983)] and human apolipoproteins were spread on a phosphate buffer (pH 7.0, 22°) and compressed with a movable barrier to generate the π–A isotherms. Extrapolation of the linear region of the isotherms to $\pi = 0$ gives limiting molecular areas of between 15 and 18 Å²/amino acid residue, which are consistent with close packed α-helical rods lying in the plane of the interface. The isotherm given by the Singer equation [S. J. Singer, *J. Chem. Phys.* **16**, 872 (1948)] represents the more expanded interfacial conformation assumed by a random coil protein.

The time courses of the increase in π due to adsorption at the air–water interface for rat apoHDL and two globular proteins, serum albumin and ribonuclease, are compared in Fig. 4. After 30 min, π for apoHDL attains a steady state value of $\sim$15 mN m^{-1} whereas at the same initial C_p, the globular proteins give $\pi < 1$ mN m^{-1}.

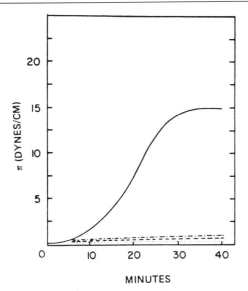

FIG. 4. Comparison of rates of adsorption at the air–water interface of various proteins. π was measured as a function of time using a Wilhelmy plate (roughened platinum). Rat apoHDL (——), rat serum albumin (···), ribonuclease (– – –), and trypsinized rat apoHDL (·—·—·) were injected into a phosphate buffer (pH 7.6, ionic strength 0.1, 24 ± 1°) to give an initial substrate protein concentration of 2×10^{-4} wt%. Rat apoHDL adsorbed from the substrate to give a steady state pressure of 15 mN m^{-1} at 40 min whereas the trypsinized apoHDL, and the globular proteins gave little increase in π. [Reproduced with permission from G. Camejo and V. Muñoz, in "High Density Lipoproteins" (C. E. Day, ed.), p. 131. Dekker, New York, 1981.]

Surface Concentration of Adsorbed Films

Surface Radioactivity

The surface concentration (Γ) of a protein monolayer at the air–water interface may be measured with radiolabeled protein samples by the surface radioactivity technique.[18,19] The type of instrumentation used to monitor ^{14}C-labeled protein adsorption has been described[9,18]; it is important that the rate meter has a long time constant (e.g., 50 sec) so that surface counts can be integrated to give a stable signal. The thin window gas flow detector is positioned a few millimeters above the surface (Fig. 1), and the counts in the surface are monitored with the rate meter as described in the rate meter instruction manual. The range of the soft β

---

[19] M. Muramatusu, in "Surface and Colloid Science" (E. Matijevic, ed.), Vol. 6, p. 101. Wiley, New York, 1973.

radiation emitted by the ^{14}C nucleus is approximately 300 μm in water, so that only counts from ^{14}C-labeled protein molecules near the surface are detected. The efficiency of counting is enhanced by using a thin mylar window in the gas flow counter that facilitates the passage of the soft β radiation to the detector (e.g., gold-coated mylar of density 0.9 mg/cm^2).

In a typical experiment in which the ^{14}C-labeled protein has been radiolabeled to a specific activity of 5 μCi/mg, a gas flow counter of window radius 1.7 cm, positioned 3 mm above the surface, will detect counts in the range 500–5000 cpm depending on the quantity of protein in the surface. When $\Gamma \sim 1$ mg m^{-2}, the surface radioactivity would be about 1000 cpm. The efficiency of the counter is approximately 15%, and the error is ±100 cpm for a counting range of 5000 cpm. A calibration curve relating counts to protein mass per unit area can be generated by spreading known quantities of radioactive protein at different areas; to ensure that all of the protein molecules remain at the interface as an insoluble monolayer, the protein may be spread onto a substrate of 3.5 M KCl, as described above. Molecules near the surface but not actually adsorbed contribute background radioactivity; this can be determined by counting the surface radioactivity when a substrate contains the same concentration of ^{14}C atoms in a soluble, non-surface-active, molecule such as [^{14}C]sucrose. For the conditions specified here, this substrate contribution amounts to ~1% of the protein surface radioactivity. If necessary this can be subtracted from the calibration line relating surface counts to Γ.

The radiolabeled apolipoprotein is injected into the substrate beneath the interface (Fig. 1) and the change in surface concentration can be monitored simultaneously with the change in surface pressure. Representative adsorption isotherms [π–substrate concentration (C_p) and $\Gamma - C_p$] are depicted in Fig. 5. The reversibility of adsorption may be examined by injecting cold apolipoprotein beneath an adsorbed monolayer of radiolabeled apolipoprotein (e.g., Fig. 6). The surface radioactivity procedure outlined above has the advantage of giving a direct measure of the surface concentration of labeled protein. However, the technique has the disadvantage that the labeling procedure may alter the surface properties of the protein. This can be monitored by comparing π at given C_p values for the native and labeled protein.

An indirect method for the determination of surface concentration involves the removal of an aliquot of substrate after labeled protein of known specific activity has adsorbed. The decrease in radioactivity gives a measure of the decrease in substrate protein concentration due to protein adsorption to the air–water interface, and Γ can be calculated.[20] Apolipoproteins in which tyrosine residues are iodinated with ^{125}ICl by

[20] M. C. Phillips and C. E. Sparks, *Ann. N.Y. Acad. Sci.* **348**, 122 (1980).

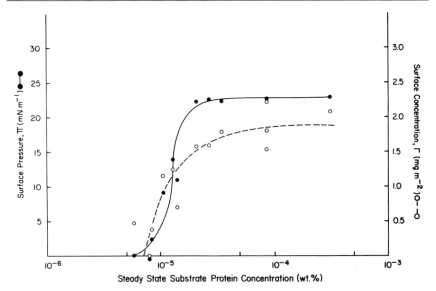

FIG. 5. Surface pressure and surface concentration isotherms for human apolipoprotein A-I adsorbed at the air–water interface. ^{14}C-labeled human apoA-I (4–7 μCi/mg) was injected into 80 ml of a phosphate buffer (pH 7.0, 5.65 mM Na$_2$HPO$_4$/3.05 mM NaH$_2$PO$_4$/0.08 M NaCl, 22 ± 2°) beneath an air–water interface, and the changes in π and surface radioactivity (cpm) were measured simultaneously until a steady-state π was attained. Measurement of π (●) was made using a roughened mica Wilhelmy plate. The change in surface concentration (○) was derived from a calibration curve relating cpm to a known Γ. The steady-state substrate protein concentration (C_p) was determined by subtracting the quantity of protein present in the surface when π reached a steady-state value from the initial quantity of protein injected into the substrate. The times required to achieve steady-state π and Γ values ranged from approximately 4 hr for high C_p to 10 hr for low C_p.

the modified McFarlane procedure[21] can be utilized in such experiments; the γ-radiation from ^{125}I is too long range for use with the surface radioactivity method, but facilitates detection of changes in protein concentration of the substrate. A serious drawback of this procedure is that the calculation of Γ is indirect, and it is assumed that no protein adsorbs to the walls of the container. As a result, this procedure is relatively inaccurate, as is the approach of attempting to quantitatively remove an adsorbed protein film for mass determination.[13]

Surface Potential

In principle, the surface potential at an air–water interface can be used to monitor the adsorption of an apolipoprotein but, to date, there are no

---

[21] D. W. Bilheimer, S. Eisenberg, and R. I. Levy, *Biochim. Biophys. Acta* **260,** 212 (1972).

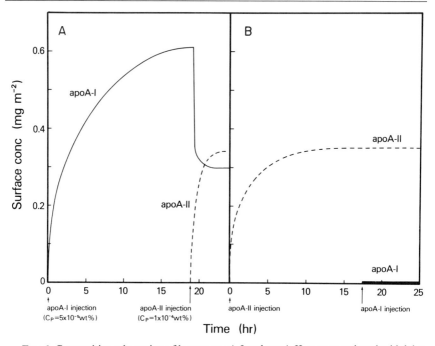

FIG. 6. Competitive adsorption of human apoA-I and apoA-II to an egg phosphatidylcholine monolayer. Egg phosphatidylcholine (PC) was spread from 9:1 v/v hexane/ethanol onto 80 ml of a phosphate buffer (pH 7.0, 5.65 mM Na$_2$HPO$_4$/3.05 mM NaH$_2$PO$_4$/0.08 M NaCl, 22 ± 2°) to give an initial pressure π_i of 10 mN m^{-1} as measured by a roughened mica Wilhelmy plate. (A) ^{14}C-labeled human apoA-I (4 μCi/mg), was injected beneath the lipid–water interface at 5 × 10^{-5} wt% and allowed to adsorb to give a steady-state π = 26 ± 0.5 mN m^{-1} and the surface concentration depicted (as determined by the gas-flow counter technique). Unlabeled human apoA-II was then injected beneath the ^{14}C-labeled apoA-I/egg PC mixed monolayer, and the displacement of ^{14}C-labeled apoA-I monitored as a decrease in surface radioactivity using a gas-flow counter. The concomitant adsorption of apoA-II to the apoA-I/egg PC monolayer gives steady-state π = 29 ± 0.5 mN m^{-1}. The increase in Γ of apoA-II was monitored in a parallel experiment using unlabeled apoA-I and ^{14}C-labeled apoA-II (6 μCi/mg). Two molecules of apoA-II adsorb for every apoA-I molecule desorbed. (B) Reciprocal experiments demonstrated that apoA-II is not displaced from the lipid–water interface by apoA-I.

reports of this approach in the literature. The surface potential of a film-covered surface is different from that of a clean air–water interface and is a function of Γ. The apparatus includes a surface electrode which is coated with an air-ionizing radionuclide, such as [241]Am, and positioned a few millimeters above the interface. This surface electrode is connected to a reference electrode in the substrate in a potentiometer circuit. The radiation from the surface electrode ionizes the air gap between the surface electrode and interface, creating a conducting medium across which the potential difference may be measured. For a more detailed discussion of the technique and materials, the reader is directed to one of the monographs on the subject.[10,11] In order to use surface potential values measured for an adsorbed film to estimate Γ, calibration curves would have to be derived as described above for the ^{14}C surface radioactivity technique. An advantage is that no labeling of the protein is required to measure surface potentials.

Protein Adsorption to Lipid Monolayers

The adsorption of protein to an insoluble lipid monolayer is studied by first spreading lipid at the air–water interface and then injecting the apolipoprotein into the substrate beneath the lipid monolayer.[22,23] Changes in surface pressure, surface concentration, and surface potential are monitored by the techniques described above.

In a typical experiment, chromatographically pure lipid is spread from a volatile organic solvent (e.g., egg yolk phosphatidylcholine is spread from 9:1 v/v hexane/ethanol) by applying the solution dropwise from a pipet onto the clean surface to attain the appropriate initial surface pressure. Prior to spreading the lipid monolayer, the surface pressure recording device is nulled at the surface tension of the clean air–water interface; the Wilhelmy plate should be removed during lipid spreading to avoid contact angle changes due to lipid adsorption on the plate. Protein injections are made through a polypropylene tube projecting through the monolayer into the subphase, or through an injection port built into the trough, to avoid perturbation of the lipid monolayer during the injection. The changes in π and Γ of the protein are then followed during adsorption of protein to the lipid–water interface. Typical data are presented in Fig. 7.

Changes in lipid monolayer surface concentration may be followed by means of the gas-flow counter technique with ^{14}C-labeled lipid and unlabeled proteins.

Interpretation of Data

Surface Conformation of Apolipoproteins

Spread monolayers of α-helical homopolypeptides at the air–water interface give limiting areas of 13–19 Å^2/amino acid residue, with the

[22] G. Colacicco, *J. Colloid Interface Sci.* **29**, 345 (1969).
[23] M. C. Phillips, M. T. A. Evans, and H. Hauser, *Adv. Chem.* **144**, 217 (1975).

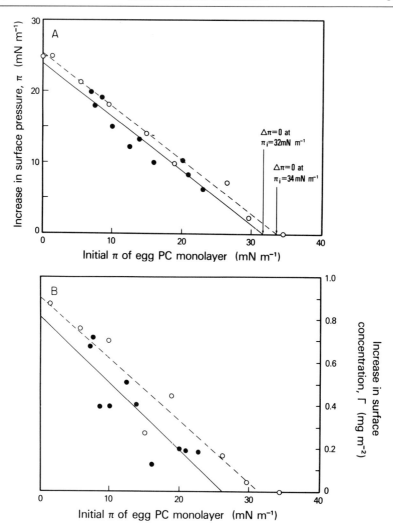

FIG. 7. Apolipoprotein adsorption to an egg phosphatidylcholine monolayer–water interface: comparison of human apoA-I and apoA-II. Egg PC was spread from 9:1 v/v hexane/ethanol onto 80 ml of a phosphate buffer (pH 7.0, 5.65 mM Na$_2$HPO$_4$/3.05 mM NaH$_2$PO$_4$/0.08 M NaCl, 22 ± 2°) to give initial surface pressure (π_i) values of between 0 and 40 mN m^{-1}, as determined by a roughened mica Wilhelmy plate. The ^{14}C-labeled apolipoproteins (4–7 μCi/mg) were injected beneath the lipid–water interface through a polypropylene tube to give an initial substrate protein concentration of 5 × 10^{-5} wt%. The change in surface pressure (A), and the change in surface concentration (B), detected as the change in cpm by gas-flow counter, were measured simultaneously during protein adsorption: (●) apoA-I and (○) apoA-II. The lipid packing density at which an apolipoprotein is excluded from the interface is determined by extrapolation to the π_i axis of the linear regression lines for π and Γ as a function of π_i.

variation in area being due to differences in the amino acid side chain.[24] Random coil proteins yield a more expanded isotherm[25] which is close to the isotherm expected from a statistical thermodynamic treatment of a random two-dimensional polymer.[26] The π–A isotherms in Figs. 2 and 3 indicate that the apolipoproteins studied assume a limiting interfacial area of 15–17.5 Å^2/amino acid residue (obtained by extrapolation of the steep part of the isotherm to $\pi = 0$) which is consistent with close-packed α-helices lying in the plane of the air–water interface. These monolayer studies are in agreement with circular dichroism studies in which the apolipoproteins are recombined with phospholipid and found to contain a considerable amount of amphipathic α-helix.

The molecular conformation of apolipoprotein monolayers formed by adsorption from the subphase can also be deduced from consideration of molecular areas. Adsorption isotherms such as that shown in Fig. 5 can be combined to give π as a function of A ($=1/\Gamma$) for the adsorbed film.

Adsorption Kinetics

The molecular processes involved in the adsorption and conformational rearrangement of apolipoproteins at the surface may be studied via the measurement of π and Γ as a function of time. The adsorption of a protein to an air–water interface takes place in two phases: initially, when the surface concentration approximates zero, adsorption is diffusion controlled, and later when the surface concentration approaches a steady-state value, the rate of adsorption is related to molecular rearrangements within the plane of the surface.[25] The rates of the processes involved in adsorption may be analyzed by the first-order equation

$$\ln\left(\frac{\pi_{ss} - \pi_t}{\pi_{ss} - \pi_0}\right) = -\frac{t}{\tau}$$

where π_{ss}, π_0, and π_t are the surface pressure values at steady-state conditions, $t = 0$, and at any time, t, respectively, and τ is the relaxation time (k^{-1}, where k is the rate constant).

Typically, during protein adsorption π continues to increase after Γ has attained a steady-state value. The application of the above equation to a protein adsorption experiment results in two linear regions, with the change in slope occurring at the time that Γ attains steady state. The two linear regions may be described by τ_1, in which both π and Γ are increasing due to protein molecules penetrating the interface, and τ_2, in which Γ

[24] B. R. Malcolm, *Prog. Surf. Membr. Sci.* **7**, 183 (1973).
[25] D. E. Graham and M. C. Phillips, *J. Colloid Interface Sci.* **70**, 415 (1979).
[26] S. J. Singer, *J. Chem. Phys.* **16**, 872 (1948).

is constant and changes in π take place as a result of molecular rearrangements within the protein film after adsorption has ceased. τ_1 and τ_2 have values of about 1 and 4 hr for apoA-I adsorption and adsorption experiments must be continued for periods of up to 16 hr, depending upon the initial substrate protein concentration (C_p).

Surface Activity

The interaction of apolipoproteins with hydrophobic surfaces is thought to be mediated by amphipathic α-helices within the protein, such that the interaction energy of a helix with an interface is a function of its amphipathicity. Eisenberg and co-workers[27] have described the concept of a mean helical hydrophobic moment (μ_H) which quantitates the degree of amphipathicity of a given helical segment (see Chapter [23], this volume).

A good prediction of surface activity is obtained by averaging μ_H over all helices in a protein molecule using the relationship

$$\bar{\mu}_H = \sum_{i=1}^{n} \mu_H/n$$

where n = number of helical segments in the protein. To permit comparison of different types of proteins, the value of $\bar{\mu}_H$ is multiplied by the fraction of helix in the protein (F). The surface pressures exerted by apolipoproteins at a given C_p are proportional to the product ($\bar{\mu}_H F$); a linear correlation coefficient of 0.98 is obtained for apolipoproteins as well as for surface-seeking peptides and globular, highly structured proteins.[28]

Displacement Experiments

The gas-flow counter technique is well suited for studies of the interfacial exchange of radiolabeled apolipoprotein molecules while monitoring changes in surface pressure. An example of an application of these interfacial studies is the displacement of apoA-I by apoA-II from the air–water and lipid monolayer–water interface. Previous studies have demonstrated that apoA-II can displace apoA-I from the surface of canine high-density lipoprotein (HDL),[29] and from phosphatidylcholine–cholesterol complexes[30] in the ratio of 2 molecules of apoA-II adsorbed for 1 molecule of apoA-I displaced. It is therefore of interest to examine the changes in

[27] D. Eisenberg, R. M. Weiss, and T. C. Terwilliger, *Nature (London)* **299**, 371 (1982).
[28] K. E. Krebs and M. C. Phillips, *Biochim. Biophys. Acta* **754**, 227 (1983).
[29] P. A. Lagocki and A. M. Scanu, *J. Biol. Chem.* **255**, 3701 (1979).
[30] M. Rosseneu, R. Van Tornout, M. J. Lievens, and G. Assmann, *Eur. J. Biochem.* **117**, 347 (1981).

surface pressure and the kinetics of adsorption/desorption that accompany this phenomenon in monolayers, and indeed, to see if the same stoichiometry applies at the lipid monolayer–water interface.

Experiments have been performed at the egg phosphatidylcholine monolayer–water interface, in which apoA-I is first adsorbed to the lipid monolayer, and apoA-II is injected beneath the mixed monolayer to induce displacement. The data in Fig. 6 indicate that two molecules of apoA-II displace one molecule of A-I in this model system, which is similar to the results obtained with canine HDL and apolipoprotein/lipid complexes. ApoA-I does not displace apoA-II from the lipid monolayer–water interface. Similar results are observed in analogous experiments with apoA-I and apoA-II carried out at the clean air–water interface. By the use of radiolabeled lipid and apolipoprotein, independently, the surface balance provides a suitable model system for studying the molecular exchange events that take place on the surface of a lipoprotein particle during metabolism.

[23] Thermodynamics of Apolipoprotein–Phospholipid Association

By JOHN B. MASSEY and HENRY J. POWNALL

Apolipoproteins perform several tasks that are a function of their affinity for a lipoprotein surface. Certain apolipoproteins contain the determinants for receptor-mediated endocytosis; others are specific activators for lipolytic enzymes. Thus, apolipoproteins are multifunctional in that they contain some protein sequences that associate with a phospholipid surface and others that mediate a biological activity.

The amphipathic helical model of the lipid-associating regions of plasma apolipoproteins[1] has been useful in understanding the structure and molecular dynamics of plasma lipoproteins. More recently, there have been several refinements[2-5] of the model and some experimental

[1] J. P. Segrest, R. L. Jackson, J. D. Morrisett, and A. M. Gotto, Jr., *FEBS Lett.* **38**, 247 (1974).
[2] H. J. Pownall, R. D. Knapp, A. M. Gotto, Jr., and J. B. Massey, *FEBS Lett.* **159**, 17 (1983).
[3] D. Eisenberg, R. Weiss, and T. C. Terwilliger, *Nature (London)* **299**, 371 (1982).
[4] D. Eisenberg, R. M. Weiss, and T. C. Terwilliger, *Proc. Natl. Acad. Sci. U.S.A.* **81**, 140 (1984).
[5] K. E. Krebs and M. C. Phillips, *Biochim. Biophys. Acta* **729**, 227 (1983).

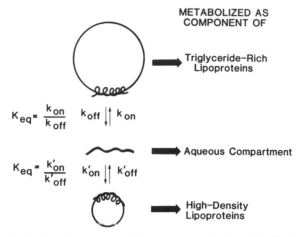

FIG. 1. A physicochemical concept of lipoprotein metabolism. The distribution of apolipoproteins is determined by the relative affinity for a lipoprotein. The equilibrium constants (K'_{eq} and K_{eq}) for apolipoprotein distribution between various lipoprotein classes that have different catabolic routes will determine the fraction of the apolipoprotein that is removed from the plasma compartment by that route.

tests that have provided a quantitative basis that permits one to estimate the relative affinity of amphipathic helical regions of proteins for lipid surfaces from their primary structure. The affinity of an apolipoprotein for a lipoprotein surface is a function of the qualities of both the surface and the protein. The equilibrium constant for the distribution of an apolipoprotein between various lipoproteins and the aqueous phase is an important factor in determining whether it will be catabolized as a part of a lipoprotein or as a component of the aqueous phase; moreover, the equilibrium constant for apolipoprotein distribution among various lipoprotein classes that have different catabolic routes will determine the fraction of that apolipoprotein that is removed from the plasma compartment by that route (Fig. 1). Thus, there are two important areas in which there are unanswered questions: (1) what are the physicochemical factors that regulate apolipoprotein–lipid association *in vitro;* (2) are the *in vitro* equilibrium measurements consistent with a predictable pattern of *in vivo* turnover of apolipoproteins?

To systematically investigate what structural determinants regulate protein–lipid association, several laboratories have synthesized peptide analogs of lipid-associating regions of plasma apolipoproteins.[6–9] Figure 2

[6] H. J. Pownall, A. M. Gotto, Jr., and J. T. Sparrow, *Biochim. Biophys. Acta* **793,** 149 (1984).

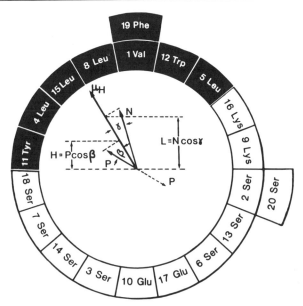

FIG. 2. The peptide (LAP-20) is represented to show an amphipathic α-helical segment projected as an Edmundson helical wheel. The hydrophobic residues are shown as white on a black background. The vector analysis to determine the helical amphipathic moment is demonstrated. Schematic representation of the resolution of the polar and nonpolar components of the individual δG_i into a resultant vector corresponding to the helical amphipathic moment, $\langle \mu_H \rangle$. The polar vector (--→ P) is always a positive free energy. This is converted to a negative free energy contribution to $\langle \mu_H \rangle$ which is always negative, by rotation through 180° as shown. H and L are the respective projections of the polar and nonpolar vectors onto the helical amphipathic moment. The value H/L is a measure of hydrophile–lipophile balance. N and P are the respective resultants of the nonpolar and polar amino acid residues.

gives an illustration of one such model peptide designed by Sparrow and his colleagues.[6]

To correlate structure and function for the interaction of peptides with phospholipids, we require a method that directly quantifies changes in protein structure with measurable thermodynamic quantity. The association of apolipoproteins with surfaces occurs with the development of an

---

[7] D. Fukushima, S. Yokoyama, D. J. Kroon, F. T. Kezdy, and E. T. Kaiser, *J. Biol. Chem.* **255**, 10651 (1980).

[8] S. Yokoyama, D. Fukushima, J. P. Kupfenberg, F. T. Kezdy, E. T. Kaiser, *J. Biol. Chem.* **255**, 7333 (1980).

[9] J. P. Segrest, B. H. Chang, G. G. Brouillette, P. Kanellis, and R. McGahan, *J. Biol. Chem.* **258**, 2790 (1983).

α-helix in which polar and nonpolar faces are formed. The energetics of this arrangement can be quantified by the vector addition of the free energies of transfer of amino acids from water to hydrocarbon to give a resultant helical amphipathic moment. The helical amphipathic moment can be used to directly compare peptide secondary structure to measured equilibrium constants.

Helical Amphipathic Moment

To identify correlations between primary and secondary structure for phospholipid interface-associating proteins, we have used the following method.[2] The method demonstrated in Fig. 2, consists of 3 steps: (1) the free energy of transfer to amino acids from water to hydrocarbon δG_i, is obtained from known hydrophobicity scales[10]; (2) assuming a helical arrangement of the peptide the δG_i for each of N amino acids is represented as a vector directed from the helical axis through its position on the circumference of the helix; (3) the calculated vector sum of the δG_i, the helical amphipathic moment $\langle \mu_H \rangle$, is obtained on a mean residue basis as

$$\langle \mu_H \rangle = \sum_{i=1}^{N} \delta \mathbf{G}_i / N \qquad (1)$$

and the mean residue hydrophobicity is given by

$$\langle H \rangle = \sum_{i=1}^{N} \delta G_i / N \qquad (2)$$

The point along the helical axis at which $\langle \mu_H \rangle$ is exerted is obtained from calculations in the same way that a center of gravity is calculated; $\langle \mu_H \rangle$ was also calculated separately for all polar and nonpolar residues. The hydrophile–lipophile balance (H/L) is obtained as the ratio of the projections of the polar and nonpolar vectors onto the resultant as shown in Fig. 2.

The mean residue helical amphipathic moments of apolipoproteins are consistently higher than those of membrane spanning proteins. Surface-associating peptides have a low mean residue hydrophobicity and a high helical amphipathic moment. Integral membrane proteins have a high hydrophobicity and a low helical amphipathic moment.[2,11]

[10] M. Levitt, *J. Mol. Biol.* **104**, 59 (1976).
[11] D. Eisenberg, *Annu. Rev. Biochem.* **53**, 595 (1984).

Measurement of Thermodynamic Parameters

It is now well recognized that many sparingly soluble components of lipoproteins can be transferred between surfaces as monomers. Specifically, the mechanism involves rate-limiting desorption of a molecule from the surface followed by rapid diffusion of the monomer through water to another membrane surface. Molecules that fit into this category are amphipathic in that they have a measurable solubility in both water and lipid surfaces. The affinity of the amphipath for phospholipids can be expressed as the free energy of association, ΔG_{eq}. ΔG_{eq} can be calculated from the distributions of the peptide between phospholipid and aqueous phase according to

$$\Delta G_{eq} = -RT \ln K_{eq} \tag{3}$$

and

$$K_{eq} = X_w/X_l = [n_p^w/(n_p^w + n_w)]/[n_p^l/(n_p^l + n_l)] \tag{4}$$

where X_w and X_l represent the respective mole fractions of peptides in the aqueous and lipid phase; n_p^w and n_p^l are the numbers of moles of peptide in water and lipid, respectively, and n_l and n_w are the respective numbers of moles of lipid and water.

Assuming that $n_p^w \ll n_w$ and that $n_p^l \ll n_l$ gives

$$K_{eq} = (n_p^w/n_w)/(n_p^l/n_l) \tag{5}$$

Experimentally K_{eq} can be calculated from

$$K_{eq} = (B/F)([W]/[P]) \tag{6}$$

where B and F represent the concentration of phospholipid-associated and free peptide. W and P represent the molar concentration of water and of the phospholipid, respectively. Under conditions where the moles of peptide $\ll$ moles of lipid, a linear relationship between [P]/[W] and (B/F) has been found demonstrating that the K_{eq}s were independent of peptide concentration.

When spectroscopic measurements are used, at a lipid molar concentration corresponding to the midpoint of the spectroscopic change produced by dilution of the peptide–phospholipid complexes, equal amounts of the peptide are in the aqueous and lipid phases so that $n_p^l = n_p^l$. Then the equilibrium constant is simply the ratio of the molarities of water and the phospholipid at which the midpoint in the shift is observed, thus

$$K_{eq} = [PC]/55 \tag{7}$$

Methods

The methods to determine K_{eq} may be divided into two types. In one the concentrations of free and bound peptide are determined directly and in the second, spectroscopic changes are monitored. Methods that determine free and bound peptide concentrations are equilibrium dialysis, ultrafiltration, ultracentrifugation, and equilibrium gel filtration.

Equilibrium dialysis is conducted by placing 1 ml of the apolipoprotein and phospholipid inside regular dialysis tubing and dialyzing against a 50 ml solution of buffer within sealed Nalgene test tubes. To verify that equilibrium is obtained, experiments are performed in duplicate where the peptide is inside or outside the dialysis bag. At equilibrium the same K_{eq} is observed irrespective of the locations of the peptide at the beginning of the dialysis. The amount of bound peptide is obtained from the difference between the amount on the inside and outside of the dialysis tubing. The amount on the outside is the free peptide. Figure 3 gives an example of a typical experiment as analyzed by Eq. (6). The disadvantages of this technique are that long times are needed for equilibration and only low-molecular-weight peptides readily dialyze so that most native apolipoproteins are too large.

Pownall et al.[12,13] have developed a technique based on ultracentrifugation. Protein and lipid are placed in polyallomar centrifuge tubes, placed in the cups of a Beckman SW 50 rotor, and centrifuged for 18 hr at 45,000 rpm in a thermostated Beckman Model L8-70 ultracentrifuge. In each experiment, the mixing incubation and centrifugation are conducted at the same temperature. At the end of each spin, each tube is fractionated with a Buchler Instruments Densiflow connected to a Pharmacia peristaltic pump. Typically, 6–8 fractions and a lipid–protein pellet are collected. Each sample is assayed for lipid and protein. With this binding assay, unsaturated lipids are not dense enough to pellet by ultracentrifugation. This technique works for large-molecular-weight apolipoproteins; however, it is not necessarily an equilibrium technique.

Yokoyama et al.[8] have used ultrafiltration and gel filtration for the rapid separation of bound and free protein. The ultrafiltration method and gel permeation method yielded binding constants which were within experimental error. Because of self aggregation of apolipoproteins at high concentration, the gel permeation method was preferred for the determination of its binding characteristics. This technique assumes that the equilibrium is not perturbed during separation.

[12] H. J. Pownall, D. Hickson, and A. M. Gotto, Jr., *J. Biol. Chem.* **256**, 9849 (1981).
[13] H. J. Pownall, Q. Pao, D. Hickson, J. T. Sparrow, and A. M. Gotto, Jr., *Biophys. J.* **37**, 175 (1982).

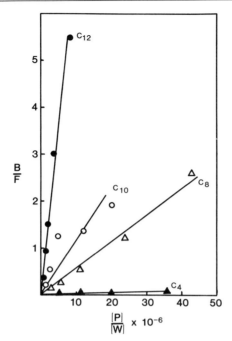

FIG. 3. Determination of the constant of equilibrium ($K_{eq}'s$) for the binding of C_4, C_8, C_{10}, and C_{12} LAP[14] to apolipoprotein A-I/1-palmitoyl-2-oleyl-phosphatidylcholine recombinants (R-HDL) at 37°. The model apolipoprotein consists of a 15 residue peptide which is acylated on the amino terminus with a 4 (C_4), 8 (C_8), 10 (C_{10}), or 12 (C_{12}) carbon saturated fatty acid. Measurements of the ratio of R-HDL-bound to free LAPs (B/F) were performed by equilibrium dialysis. Several concentrations of R-HDL were used with each LAP. The $K_{eq}'s$ are given by the slopes of the regression lines. The values for C_0 LAP almost coincided with the x axis and are not shown for clarity of the figure. The high degree of self-association of C_{16} and C_{18} LAP precluded their use in equilibrium dialysis experiments. The peptide has the following sequence:

C_1-Ser-Ser-Leu-Lys-Glu-Tyr-Trp-Ser-Ser-Leu-Lys-Glu-Ser-Phe-Ser
C_2
C_n

The association of apolipoproteins and phospholipids are associated with several spectroscopic changes. Far UV circular dichroism has been used to measure changes in protein secondary structure, mainly formation of an α-helix. Changes in the environment of aromatic residues, predominantly tryptophan, on going from an aqueous to nonpolar envi-

---

[14] G. Ponsin, K. Strong, A. M. Gotto, Jr., J. T. Sparrow, and H. J. Pownall, *Biochemistry* **23**, 5337 (1984).

TABLE I
MEASURED FREE ENERGY VALUES FOR THE ASSOCIATION OF
APOLIPOPROTEINS WITH PHOSPHOLIPIDS

Apolipoprotein	$-\Delta G_{eq}$ (kcal/mol)	Reference
RCM-apoA-II	7–9	12
ApoC-III	9.3	13
ApoA-I	10.6	8
LAP-16	<6.5	6
LAP-20	8.9	6
LAP-24	9.6	6
LAP-16-C_8	6.6	14
LAP-16-C_{12}	8.0	14
Kanellis–Segrest Peptide 1	10.6	9
Kaiser–Kezdy Peptide 1	10.1	8
ApoA-I 121–164	9.3	7
ApoA-I 144–165	7.8	7

ronment can be followed by UV difference spectra, near UV circular dichroism, and fluorescence spectroscopy; changes in the emission maxima of the tryptophan residue have been used to determine equilibrium constants for apolipoprotein–lipid association.[6,9]

The enthalpic contribution to the free energy of association has been measured by several laboratories on a LKB 10070 batch calorimeter[15,16] (see Chapter [20], this volume). Enthalpy values are connected for the heat of dilution of lipid and protein and for the friction of mixing. The heat of mixing is minimized by dialyzing the lipid and protein against a common buffer at room temperature prior to the calorimetric experiment. In a typical experiment, the total heat released is 1 to 6 mcal. In practice, this instrument has poor stability at temperatures far removed from room temperature.

Comparison of Results

Table I gives a summary of the measured free energy values for the association of apolipoproteins and synthetic peptide analogs for phospholipid surfaces. The most surprising fact is that there is no correlation between free energy with size of the apolipoprotein. Small peptides (MW ~2000) have free energy values similar to those of native apolipoproteins

[15] J. B. Massey, A. M. Gotto, Jr., and H. J. Pownall, *Biochemistry* **20**, 1575 (1981).
[16] J. B. Massey, A. M. Gotto, Jr., and H. J. Pownall, *J. Biol. Chem.* **254**, 9359 (1979).

(MW 7,000–28,000). This is perplexing since the native apolipoproteins are made of several amphipathic helical regions on which the synthetic peptides are based. Evidently, the sum of the free energies of the individual helical regions is much less than the free energy for the native protein.

The free energy can be broken down into enthalpic and entropic components. The enthalpy of apolipoprotein association with phospholipids is a function of accompanying structural changes in both the lipid and the protein.[15] The association of apolipoproteins with fluid lipids results in an enthalpy change of −1.3 kcal/mol amino acids converted from a random coil to an α-helical structure.[16] In the absence of a change in α-helical structure, the enthalpy of association is practically nil. The apolipoproteins have a low free energy (ΔG = 2–3 kcal/mol) difference between native and unfolded states[17,18] which is similar to the low free energy values for the stability of α-helices.[19] These results suggest that the enthalpy of helix formation is compensated by a loss in conformational entropy such that the contribution of protein structural changes to the free energy of lipid–protein association is small.

The major contribution to the free energy of association has been generally accepted to be the transfer of hydrophobic amino acids from an aqueous to lipophilic environment. However, the sum of the free energies of transfer of the individual amino acid side chains on the nonpolar face of an α-helix is much larger than the measured free energy of association. This discrepancy has been hypothesized to be due to the hydrophobic amino acid side chains associating with a region of the lipid of a lipoprotein that is much more polar than the interior of a lipid bilayer, such that the magnitude of the hydrophobic effect is less. Also, the addition of an apolipoprotein increases the acyl-chain order of the phospholipid matrix.[20] Thermodynamically this would be expressed as a loss in conformational entropy which would be entropically unfavorable.

Comparisons of the free energies of transfer of several peptides (Table II) suggest that changes in the nature of the polar face are not important in the formation of a stable helix at the lipid–water interface. The hydrophobic face and the hydrophobicity thereof is considered more important since the energetics of transfer from water to an interface involves the transfer of these residues from water to an apolar environment. This is not

---

[17] A. R. Tall, D. M. Small, G. G. Shipley, and R. S. Lees, *Proc. Natl. Acad. Sci. U.S.A.* **72**, 4940 (1975).
[18] W. W. Mantulin, M. F. Rohde, A. M. Gotto, Jr., and H. J. Pownall, *J. Biol. Chem.* **255**, 8185 (1980).
[19] J. Hermans, Jr., *J. Phys. Chem.* **70**, 510 (1966).
[20] W. W. Mantulin, J. B. Massey, A. M. Gotto, Jr., and H. J. Pownall, *J. Biol. Chem.* **256**, 10815 (1981).

TABLE II
HELICAL MOMENT ANALYSIS FOR SYNTHETIC APOLIPOPROTEINS

	$-\Delta G_{eq}$ (kcal/mol)	$\langle H \rangle$ (kcal/mol)	$\langle \mu H \rangle$ (kcal/mol/AA)	Vector components (kcal/mol)		
				Polar	Nonpolar	Resultant
LAP-16	<6.5	−0.04	1.01	7.4	10.5	16.3
LAP-20	8.9	−0.18	0.98	7.7	13.4	19.6
LAP-24	9.6	−0.28	0.93	8.1	15.3	22.4
Kanellis–Segrest Peptide 1	10.5	0.26	1.13	11.7	8.8	20.4
Kaiser–Kezdy Peptide 1	10.1	0.97	0.89	11.8	8.9	20.6
Kaiser–Kezdy apoA-I 144–165	7.8	0.45	0.48	6.2	4.1	10.1

true for the polar residues; after transfer from water to the surface they remain in contact with the aqueous phase. Therefore, it is probable that the polar face does not contribute much to the free energy of transfer of apolipoproteins from water to a phospholipid surface. (However, see Chapter [36], this volume.)

Figure 4 shows a comparison of the measured free energies of association of several peptides with the magnitude of the hydrophobic vector.

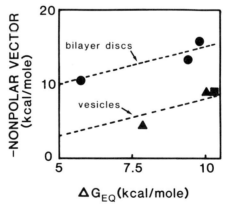

FIG. 4. The measured free energies for the association of model apolipoproteins with single bilayer phospholipid vesicles or micellar bilayers disks are compared with the magnitude of the nonpolar vector determined by the helical moment analysis (Table II). The dashed lines are to indicate what would be expected for a linear correlation between the two measurements.

The free energies have been determined for peptide association with single bilayer vesicles and dissociation from small micellar lipid–protein complexes. The only comment to be made is that the trend of increasing hydrophobicity appears to correspond with increasing free energy of association. Additionally, the intrinsic surface activities of proteins and peptides, as determined by surface pressure at the air–water interface (Chapter [22], this volume), correlates with the product ($\bar{\mu}_H F$) where $\bar{\mu}_H$ is the average value of μ_H for all helices in the molecule, and F is the fraction of α-helical structure in the protein.[5] Currently, the use of the amphipathic helical moment analysis has proved promising in correlating structure with function and in the design of model lipid-binding peptides.

Section IV

Structure of Intact and Reconstituted Plasma Lipoproteins

[24] Nondenaturing Polyacrylamide Gradient Gel Electrophoresis

By ALEX V. NICHOLS, RONALD M. KRAUSS, and THOMAS A. MUSLINER

Human plasma lipoproteins comprise a spectrum of macromolecules which exhibit both a wide range of particle size (diameter, 5–120 nm), hydrated density (0.93–1.16 g/ml), molecular weight (0.15 × 10^6–500 × 10^6), and a broad diversity of lipid and protein composition. Within this spectrum, major classes of lipoproteins have been identified according to size, density, and ultracentrifugal flotation rate. These classes include chylomicrons, very low-density lipoproteins (VLDL), intermediate-density lipoproteins (IDL), low-density lipoproteins (LDL), and high-density lipoproteins (HDL). Additional polydispersity in size and density within each of the major lipoprotein classes has been described and the presence of unique subpopulations has been established. Highly informative in this regard have been studies utilizing gradient gel electrophoresis to describe the particle size distributions and subpopulations of HDL[1] and LDL[2] classes. Characterization of subpopulations is important to the interpretation of lipoprotein spectra in health and disease and to basic studies of metabolic origins and interconversions of subspecies of lipoproteins.

Since the functional specificity of plasma lipoproteins resides in major part in the apolipoprotein moieties associated with the lipoproteins, techniques that combine size analysis, such as gradient gel electrophoresis, with apolipoprotein identification, either directly by immunoblotting or indirectly by immunoaffinity chromatography,[3] appear highly promising for investigation of lipoprotein structure and function. This chapter will consider the application of gradient gel electrophoresis primarily to the analysis of HDL and LDL subpopulations and will indicate approaches for more detailed identification of these subpopulations.

Principles

Electrophoresis According to Particle Size

Gradient gel electrophoresis entails the migration of charged particles through a matrix comprised of increasing concentrations of polyacryl-

---

[1] A. V. Nichols, P. J. Blanche, and E. L. Gong, in "Handbook of Electrophoresis" (L. Lewis and J. Opplt, eds.), Vol. III, p. 29. CRC Press, Boca Raton, Fla., 1983.
[2] R. M. Krauss and D. J. Burke, *J. Lipid Res.* **23**, 97 (1982).
[3] M. C. Cheung and J. J. Albers, *J. Biol. Chem.* **259**, 12201 (1984).

amide gel.[4] The effective pore size of the matrix is progressively reduced as the gel concentration increases, resulting in differential retardation of the migrating charged particles. With increasing retardation, the charged particles approach an effective "exclusion limit" which is a function of the size and shape of the particles. The gel concentration gradient and the time required for approach to zero migration rate is established by consideration of the size range of the specific particle mixture under investigation. Calibration is usually performed by use of standards comprised of protein mixtures of known physical properties. Validation of particle size calibration of lipoproteins can be made by direct analysis of particle size by alternative techniques, such as electron microscopy.

Gel Gradients and Lipoprotein Particle Size Ranges

Effective resolution of subpopulations of high-density lipoproteins (HDL) (particle size 7.0–12.0 nm) can be obtained utilizing a linear gel gradient in the range of 4–30% polyacrylamide concentration. Time of electrophoresis for HDL is generally 24 hr. The 4–30% gel gradient is also effective for analysis of lipoprotein particles (designated HDL_c) in the size range (12.5–22.0 nm) intermediate to LDL and HDL.

LDL subpopulations in the particle size range of 21.5–27.5 nm can be resolved by use of gel gradients in the range of 2–16% polyacrylamide concentration with an electrophoresis time of 24 hr. Particle sizes of intermediate-density lipoproteins (IDL) can also be analyzed by this gel gradient. While penetration of larger very low-density lipoproteins (VLDL) is limited in this gel gradient, estimates of particle size of major VLDL components of diameter <40.0 nm can still be made.

Localization and Identification of Lipoprotein Subpopulations: Protein vs Lipid Staining

By virtue of their content of both protein and lipid constituents, lipoproteins within the gel matrix can be stained by either protein- or lipid-specific stains. Protein-specific stains (such as Coomassie G-250) are most effective when electrophoretic analysis is performed on lipoproteins previously isolated from interfering proteins present in whole human plasma. Such isolation can be made by techniques including ultracentrifugation and immunoaffinity chromatography.

Lipid-specific stains (such as Oil Red O) can be used both in the presence and absence of proteins, which would ordinarily interfere when using protein-specific staining. Thus, lipoprotein subpopulations in whole

[4] D. Rodbard, G. Kapadia, and A. Chrambach, *Anal. Biochem.* **40**, 135 (1971).

plasma, as well as in isolated fractions, can be analyzed by use of lipid-specific stains.[5]

Stains specific for other chemical groups within lipoprotein structures can also be used (including periodic acid–Schiff reagent for interaction with carbohydrate groups in apolipoproteins).

Localization and Identification of Lipoprotein Subpopulations: Autoradiography and Immunoblotting

Lipoprotein subpopulations containing radiolabeled constituents, such as radioiodinated apolipoproteins, can be localized and identified by autoradiography. With standard 3-mm-thick gels, resolution can be significantly improved by electrophoretic transfer of radiolabeled apolipoproteins to nitrocellulose sheets prior to autoradiography. Different variants of immunoblotting methodology can be applied to localization of specific apolipoproteins within lipoprotein subpopulations resolved by gradient gel electrophoresis.

Calibration

When calibration is performed using a mixture of proteins of known molecular weight, it is generally assumed that the proteins are globular and hence can serve as calibration standards for apparently spherical lipoprotein particles. The size calibration can be based either on hydrated particle sizes of the standard proteins calculated from the Stokes equation and diffusion constant of the particular protein, or on nonhydrated particle sizes of the standard proteins calculated from the molecular weight and partial specific volume of the protein, assuming spherical shape.[6] Direct calibration of particle size can also be performed by gradient gel electrophoresis of purified lipoprotein fractions whose particle sizes have been determined by other techniques, such as electron microscopy.

Procedure

Supplies and Reagents

Basic equipment for gradient gel electrophoresis includes slab gel electrophoresis chamber, constant voltage power supply, and refrigerated (temperature-controlled) circulating water bath. Polyacrylamide gradient gels appropriate for HDL analysis consist either of a 4–30% or 2.5–27%

---

[5] G. J. Bautovich, M. J. Dash, W. J. Hensley, and J. R. Turtle, *Clin. Chem.* **19**, 415 (1973).
[6] D. Rodbard and A. Chrambach, *Anal. Biochem.* **40**, 95 (1971).

gradient; a gel effective for LDL–IDL–VLDL analysis provides a 2–16% gradient. The electrophoresis buffer is generally Tris base (90 mM), boric acid (80 mM), sodium azide (3 mM), and EDTA (3 mM), pH 8.35.

Sample Preparation

Gradient gel electrophoresis of the major lipoprotein classes can be performed directly on plasma when a lipid-specific stain is used to visualize the lipoprotein bands. Plasma preparation should be expeditious and include measures to minimize oxidative degradation, lipoprotein transformation, or proteolysis by plasma enzymes, and bacterial activity. A variety of additives are used to control the above and include EDTA with storage under nitrogen (oxidative alteration); 5,5'-dithiobis-(2-nitrobenzoic acid) (inhibition of lecithin:cholesterol acyltransferase); phenylmethylsulfonylfluoride or aprotinin (inhibition of protease activity); streptomycin, penicillin, ethylmercurithiosalicylic acid, sodium azide, gentamicin, and/or chloramphenicol (bacteria-associated activity). Freezing of plasma can be deleterious to lipoprotein integrity and effects of specific procedures should be evaluated.

Lipoprotein fractions isolated from interfering plasma proteins can be analyzed by gradient gel electrophoresis using either lipid- or protein-specific stains. Isolation of lipoprotein fractions from plasma can be performed by standard ultracentrifugal procedures which take into account the precautionary measures cited above as well as precautions to limit prolonged exposure of lipoproteins to centrifugal forces. The latter precaution is especially applicable to lipoproteins containing quite readily dissociable apolipoproteins such as apolipoproteins A-I and E.

All of the major lipoprotein classes of human plasma can be analyzed by gradient gel electrophoresis of the plasma $d < 1.20$ g/ml fraction obtained after single-spin ultracentrifugation (114,000 g, 24 hr, 15°, Beckman 40.3 rotor). LDL, IDL, and VLDL classes of human plasma can also be analyzed by gradient gel electrophoresis of the $d \leq 1.063$ g/ml fraction obtained after single-spin ultracentrifugation (114,000 g, 24 hr, 15°, Beckman 40.3 rotor). Lipoprotein fractions obtained by other ultracentrifugal procedures or by other separation methods can be analyzed by gradient gel electrophoresis provided precautions have been taken with respect to maintaining lipoprotein integrity.

Ultracentrifugal fractions containing lipoproteins in an elevated salt background solution can be electrophoresed directly without dialysis to electrophoresis buffer, if "spacer solution" is applied to gel lanes immediately adjacent to the salt-containing lipoprotein fraction. The salt content of the "spacer solution" should match that of the lipoprotein sample to avoid streaking and distortion of the lipoprotein bands.

Electrophoresis

Prior to sample and "spacer solution" application to the gel, the gels should be preconditioned in electrophoresis buffer for 20 min at 125 V in the electrophoresis chamber maintained at 10°. The plasma or lipoprotein fraction is mixed (4:1 by volume) with solution consisting of sucrose (40%) and bromophenol blue (0.05%). When application volume is monitored, thyroglobulin (1.75 mg/ml) is incorporated into the diluting solution above. Application volume of the lipoprotein-containing mixture is usually in the range of 10–20 μl (sufficient volume to provide an amount of total lipoprotein protein of approximately 10–15 μg). For lipoproteins rich in lipid content (LDL, IDL, and VLDL), the amount applied should reflect the expected intensity of lipid staining, previously established by calibration.

For particle size calibration in the size range of HDL diameters, a standard protein mixture of the following proteins is prepared in appropriate spacer solution: thyroglobulin (hydrated: 17.0; nonhydrated: 11.6 nm), ferritin (12.2; 10.2 nm), catalase (10.4; 8.2 nm), lactate dehydrogenase (8.1; 6.9 nm), and bovine serum albumin (7.1; 5.4 nm). When applied to the gel, 6 μl of the mixture provides 2.5 μg of each of the above proteins.

For particle size calibration in the size range of the LDL and IDL, two separate calibration preparations are applied to two separate lanes in the gel. One calibration preparation is the same standard protein mixture as used for HDL particle size determination. The second preparation consists of a solution of latex particles (Dow Chemical, 38.0 nm, used at 1:10 dilution). By combining the results after electrophoresis of these two preparations, a calibration curve can be drawn that includes particle sizes on either side of the LDL and IDL particle size range. Separate electrophoresis is performed on the above calibration preparations to avoid abnormalities in migration arising from interaction of latex particles with the proteins of the standard mixture.

"Spacer solution," prior to application to alternate lanes on the gel, is diluted (4:1 by volume) with the same sucrose-containing solution used to dilute the lipoprotein samples prior to application. The application volume of the diluted "spacer solution" is the same as that of the diluted lipoprotein sample.

To initiate electrophoresis, the chamber voltage is applied in the following sequence: 15 V for 15 min, 70 V for 20 min, and 125 V for 24 hr. The final voltage is maintained, giving 3000 V-hr.

Fixing and Staining

Protein Stain. For fixation prior to protein staining, gels are exposed to sulfosalicylic acid (10%) for 1 hr immediately after electrophoresis. For protein staining using Coomassie G-250, the fixed gels are exposed for 1.5 hr to a solution comprised of the stain (0.04%) in perchloric acid (3.5%). Destaining is accomplished by exposure of the stained gel to acetic acid (5%) with gentle agitation until gel background is clear. Alternatively, gels may be stained without prior fixation in 0.05% Coomassie R-250 in 50% methanol, 10% acetic acid, and destained in 20% methanol, 9% acetic acid.

Lipid Stain. For lipid staining using Oil Red O, gels are exposed (without prior fixation) for at least 24 hr at 55–60° to a solution comprised of the stain (0.04%) in ethanol (60%). Gels are destained and rehydrated in acetic acid (5%).

Densitometry

Densitometry of stained gels can be performed utilizing standard equipment commercially available. For Coomassie G-250-stained gels, lipoprotein and calibration protein densitometry is usually performed at 596 nm; for Coomassie R-250-stained gels, 555 nm is used; for Oil Red O-stained gels, lipoprotein densitometry is performed at 530 nm. Calibration protein and latex particle densitometry on lipid-stained gels is performed at 280 nm. Alternatively, lanes containing standard proteins can be selectively stained with Coomassie R-250 in 50% methanol, 10% acetic acid, by soaking a narrow strip of filter paper placed over the appropriate lane.

Computer-assisted densitometry can facilitate determination of particle sizes of major components as well as areas under peaks associated with the major components. Computer-assisted resolution of multicomponent electrophoresis patterns has been reported[7] and should provide more detailed information on constituent species within the major classes of lipoproteins.

Electrophoretic Transfer to Nitrocellulose Sheets

Electrophoretic transfer of lipoproteins from gradient gels occurs with considerably lower efficiency than the transfer of proteins from linear

[7] C. Chen, K. Applegate, W. C. King, J. A. Glomset, K. R. Norum, and E. Gjone, *J. Lipid Res.* **25**, 269 (1984).

polyacrylamide gels,[8] presumably because of migration to limiting pore size. Transfer efficiency can be considerably improved by performing the procedure in the presence of detergents capable of releasing apolipoproteins from associated lipid components. Efficient transfer can be obtained with the following procedure.

After electrophoresis under standard conditions outlined above, the gel (either 2–16 or 4–30%) is transferred to 0.05% acetic acid (v/v) and gently agitated at room temperature for 90–120 min with three changes of solution. The gel is then placed in a "sandwich" between nitrocellulose paper and Whatman No. 1 filter paper, presoaked in "blotting buffer" consisting of 2% Triton X-100 and 0.05% acetic acid. The sandwich is placed in an appropriate electrophoretic chamber (such as a Pharmacia GD-4 II destainer) containing the same buffer, with the nitrocellulose paper facing the cathode. Electrophoresis is carried out for 24 hr at 24 V (current of 170–190 mA), with the chamber temperature maintained at 19–20°. The nitrocellulose blot is then rinsed in 0.05% acetic acid and either dried or stored in this solution.

When autoradiography is to be performed, the boundaries of the gel and the ferritin standard (which can be visualized as a sharp yellow band on the nitrocellulose blot) are marked with traces of radioactive protein. Alternatively, radiolabeled standard proteins may be used. Autoradiography is performed at −70° using Kodak XAR-5 film and appropriate intensifier screens. Additional calibration standards and the origin of the gel may be marked on the developed autoradiograms by aligning them (using the ferritin standards) with residual bands on the original gel stained with Coomassie blue, and reequilibrated with the blotting buffer at blotting temperature. Computer-assisted densitometry with calculation of particle diameters of major components can be performed on autoradiograms in a fashion similar to that described above for stained gels.

The blot may also be directly stained for protein using amido black (0.1% in 45% methanol/10% acetic acid) and destained with 20% methanol, 9% acetic acid. Immunostaining using antisera directed against specific apolipoproteins and peroxidase-conjugated second antibodies can be performed following the procedure of Towbin et al.[8]

Discussion

Normal Human Plasma HDL Distribution

Utilizing the procedures described above for electrophoretic analysis of particle size distributions of human plasma HDL, additional subpopu-

[8] H. Towbin, T. Staehelin, and J. Gordon, *Proc. Natl. Acad. Sci. U.S.A.* **76**, 4350 (1979).

lations have been identified within major lipoprotein classes characterized earlier by ultracentrifugal procedures. At present, five subpopulations of plasma HDL with reproducible properties have been identified by gradient gel electrophoresis: $(HDL_{2b})_{gge}$, $(HDL_{2a})_{gge}$, $(HDL_{3a})_{gge}$, $(HDL_{3b})_{gge}$, and $(HDL_{3c})_{gge}$. The size and density ranges of these subpopulations are presented in Table I. A representative pattern of an HDL distribution enriched in $(HDL_3)_{gge}$ subpopulations is shown in Fig. 1A; a representative pattern of an HDL distribution enriched in $(HDL_2)_{gge}$ subpopulations is shown in Fig. 1B.

Distribution of Normal Human Lipoprotein Subpopulations of d < 1.063 g/ml

Electrophoresis on 2–16% gradient gels of the LDL fractions (1.019 < d < 1.063 g/ml) from normal human subjects reveals multiple discrete bands in the size range from 22.0 to 27.5 nm (Fig. 2). Similar banding is seen on lipid-stained gels electrophoresed with nonultracentrifuged plasma. Analyses of the distribution of gradient gel electrophoretic peaks in LDL from a large group of healthy men and women revealed seven modes, from the largest to smallest diameter.[2] LDL from individual subjects may contain from two to five peaks corresponding to these different

TABLE I
HDL SUBPOPULATIONS AND THEIR PROPERTIES

HDL Subpopulation	$(HDL_{2b})_{gge}$	$(HDL_{2a})_{gge}$	$(HDL_{3a})_{gge}$	$(HDL_{3b})_{gge}$	$(HDL_{3c})_{gge}$
Particle size interval (nm)	12.9–9.7	9.7–8.8	8.8–8.2	8.2–7.8	7.8–7.2
R_f interval[a]	0.445–0.627	0.627–0.711	0.711–0.781	0.781–0.841	0.841–0.962
Particle size at mode of size frequency distribution within subpopulation interval (nm)	10.6	9.2	8.4	8.0	7.6
Density interval (g/ml)	1.063–1.100	1.100–1.125	1.125–1.147	1.147–1.167	1.167–1.200
Density at mode of density-frequency distribution within subpopulation interval (g/ml)	1.085	1.115	1.136	1.154	1.171

[a] R_f interval: R_f designates migration distance of an HDL component's peak relative to migration distance of the peak of bovine serum albumin in the standard protein mixture.

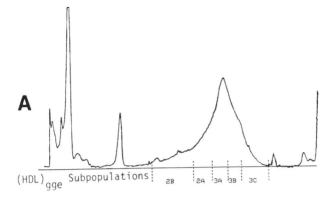

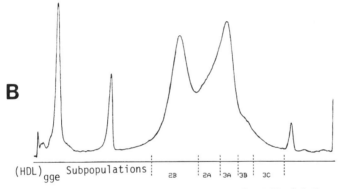

FIG. 1. Densitometric scans of electrophoresed plasma $d \leq 1.20$ g/ml ultracentrifugal fractions from two normal subjects: (A) (HDL)$_{gge}$ pattern enriched in (HDL$_3$)$_{gge}$ and (B) in (HDL$_2$)$_{gge}$. Electrophoresis was performed on 4–30% gradient gels at 125 V, 10° for 24 hr, followed by protein staining with Coomassie Blue G-250. Particle sizes and R_f values corresponding to (HDL)$_{gge}$ subpopulation boundaries are listed in Table I. The scan pattern, reading from left to right, shows the presence of peaks corresponding to low-density lipoprotein (LDL), thyroglobulin (added as an internal standard), and the HDL subpopulations. The minor peak to the right is albumin, present as a contaminant in the preparation of the plasma $d \leq 1.20$ g/ml ultracentrifugal fraction.

size groups, and there is further size heterogeneity among individuals. The sizes of these subpopulations are summarized in Table II, along with the density intervals in which the bulk of these species may be isolated. There is, however, considerable overlap of size and density among these subpopulations.[2]

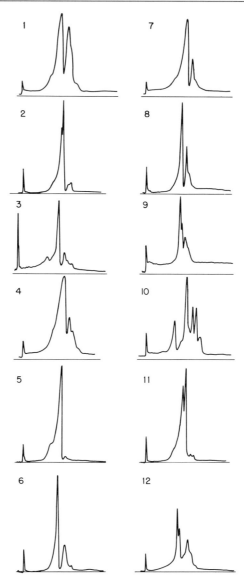

FIG. 2. Coomassie Blue R-250-stained 2–16% gradient gel electrophoretograms of total LDL from 12 normal subjects. The sharp peak at the left end of each scan represents the top of the gel. Particle diameters of the major peaks (not shown) vary from 22 to 27.5 nm.

TABLE II
VLDL, IDL, AND LDL SUBPOPULATIONS

Subpopulation	Peak size range (nm)	Peak density interval (g/ml)
VLDL		
Large	>35.0	<1.006
Small	30.0–35.0	<1.006–1.010
IDL 1	28.0–30.0	1.008–1.022
IDL 2	27.0–28.5	1.013–1.028
LDL I	26.0–27.5	1.025–1.032
LDL IIA	25.5–26.4	1.030–1.038
LDL IIB	25.0–25.5	1.035–1.040
LDL IIIA	24.7–25.2	1.038–1.048
LDL IIIB	24.2–24.6	1.038–1.048
LDL IVA	23.3–24.1	1.048–1.065
LDL IVB	22.0–23.2	1.048–1.065

At least two distinct subpopulations are also evident by gradient gel electrophoresis in both VLDL ($d < 1.006$ g/ml) and IDL ($1.006 < d < 1.019$ g/ml) fractions from normal subjects (Table II).[9] The smaller of the VLDL subspecies (30.0–33.0 nm) resembles β-VLDL in several respects, and is relatively cholesterol-enriched and triglyceride-depleted compared with the larger VLDL (peak diameter > 35.0 nm). The two IDL subspecies overlap in size and density, yet may be observed by gradient gel electrophoresis of unspun plasma as well as isolated IDL fractions, implying they are not simply ultracentrifugation artefact. IDL subspecies, however, must be distinguished from bands corresponding to Lp(a) which are also found in the size range of 30.0–32.0 nm.

Particle Size Distributions of Lipoproteins in States of Abnormal Lipoprotein Metabolism

Several examples of HDL particle size distributions in states of abnormal lipoprotein metabolism are presented in Fig. 3. These patterns indicate multicomponent particle size distributions within particle size ranges of normally occurring HDL as well as in size ranges outside those of normal HDL species. Some components within these patterns [e.g., particles with sizes > $(HDL_{2b})_{gge}$ in Fig. 3a] are not spherical in shape but exhibit discoidal or vesicular morphology. When such abnormal lipoproteins are present in HDL samples, it is necessary to establish their size

[9] T. A. Musliner, C. Giotas, and R. M. Krauss, *Arteriosclerosis* **6**, 79 (1986).

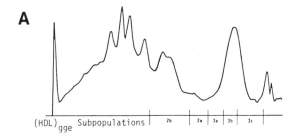
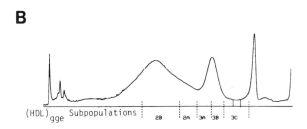
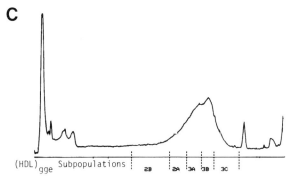

FIG. 3. Densitometric scans of electrophoresed plasma $d \leq 1.20$ g/ml ultracentrifugal fractions from three subjects with abnormal lipoprotein metabolism: (A) lecithin : cholesterol acyltransferase deficiency, (B) apolipoprotein A-I–C-III deficiency, and (C) hypertriglyceridemia.

characteristics by separate calibration since their migration properties may differ considerably from those of spherical particles.

Examples of abnormal distributions of $d < 1.063$ g/ml lipoproteins are presented in Fig. 4. Subjects with severe hypertriglyceridemia typically have increased levels of a discrete and quite small (22.0–23.0 nm) LDL subspecies in addition to variable amounts of larger species and pronounced increases in large VLDL. Dysbetalipoproteinemia is accompa-

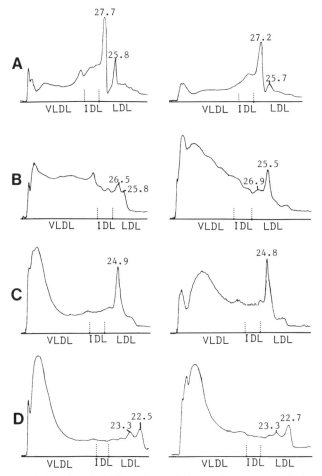

FIG. 4. Oil Red O-stained 2–16% gradient gel electrophoretograms of plasma from patients with hyperlipoproteinemia. (A) Familial hypercholesterolemia: left, female homozygote; right, female heterozygote. (B) Type III hyperlipoproteinemia: left, female; right, male. (C) Moderate hypertriglyceridemia (plasma triglyceride, 500–1000 mg/dl): left, female; right, male. (D) Severe hypertriglyceridemia (plasma triglyceride > 1000 mg/dl): left, female; right, male. Numbers above peaks refer to particle diameters (nm) calculated from calibration standards.

nied by markedly increased amounts of lipoproteins of the size of small VLDL and IDL in normals. Patients with heterozygous familial hypercholesterolemia demonstrate sharp LDL peaks of unusually large particle diameter, commonly approaching the size range of IDL components in normal subjects.

Molecular Weight Determination

While protein molecular weight determination is possible by gradient gel electrophoresis by use of appropriate calibration standards, such determination of lipoprotein molecular weights is still only approximate and requires validation by use of lipoprotein species with molecular weights determined by other methods.

Electrophoretic Blotting following Gradient Gel Electrophoresis

Examples of autoradiograms following electrophoretic transfer of radioiodinated LDL ($1.019 < d < 1.063$ g/ml) from a 2–16% gradient gel and radioiodinated HDL ($1.063 < d < 1.21$ g/ml) from a 4–30% gel are shown in Fig. 5. The LDL fraction was isolated from a subject with hypertrigly-

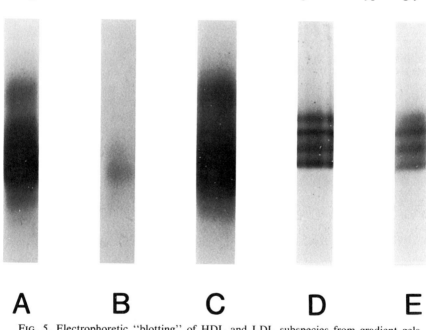

FIG. 5. Electrophoretic "blotting" of HDL and LDL subspecies from gradient gels. Conditions were as outlined in the text. (A) HDL from a normolipidemic subject electrophoresed on a 4–30% gradient gel and stained with Coomassie Blue R-250; (B) a similar gel to that shown in (A), but stained with Coomassie Blue R-250 after electrophoretic blotting; (C) autoradiogram of the nitrocellulose sheet following electrophoretic blotting of ^{125}I-labeled HDL from the same subject; (D) LDL from a hypertriglyceridemic subject electrophoresed on a 2–16% gradient gel and stained with Oil Red O; and (E) autoradiogram of the nitrocellulose sheet following electrophoretic blotting of ^{125}I-labeled LDL from the same hypertriglyceridemic subject as in (D).

ceridemia who demonstrated multiple, closely sized LDL subspecies; the HDL was derived from a normolipidemic subject.

The efficiency of transfer of HDL is greater than 90% (Fig. 5), and densitometric scans of the autoradiograms show that the peaks are similar to those in the protein-stained gradient gels. LDL, IDL, and VLDL leave the gel somewhat less efficiently (60–70%, depending on conditions), presumably because of the larger size or lower solubility of apoB. Nevertheless, because the efficiency of transfer of different subfractions of IDL and LDL has been found to be similar, densitometric scans of autoradiograms of preparations including multiple subspecies closely resemble those of protein-stained lipoproteins in the original gradient gel.

A variety of factors can influence the size of peaks on autoradiograms or immunoblots of gradient gels, and must be controlled for when the procedure is used quantitatively. These include overloading of the binding capacity of nitrocellulose, decreased efficiency of transfer at very low lipoprotein loads (perhaps due to nonspecific binding of apolipoproteins within the gels) and potential competition for binding to nitrocellulose among apolipoproteins and between apolipoproteins and other proteins within the preparation. The latter may particularly pose a problem with whole plasma, which contains significant amounts of proteins which overlap in size with HDL.

Within these limitations, the blotting procedure has varied applications. It may be used to follow size changes in radiolabeled lipoproteins as they are metabolized *in vivo*. The distribution of apolipoproteins among lipoprotein subfractions can be assessed by immunostaining of blots from gradient gels on which appropriate subfractions have been resolved.

Gradient Gel Electrophoresis following Immunoaffinity Chromatography

Identification of lipoprotein subpopulations according to both their particle size and the presence of specific apolipoproteins can also be performed by a sequence of immunoaffinity chromatography followed by gradient gel electrophoresis. HDL subpopulations observed by gradient gel electrophoresis fall within two major apolipoprotein-specific populations [population Lp(A-I − A-II), containing apoA-I but no apoA-II, and population Lp(A-I + A-II), containing both apoA-I and apoA-II] in human plasma.[3]

[25] Isoelectric Focusing of Plasma Lipoproteins

By YVES L. MARCEL and PHILIP K. WEECH

Charge Difference as a Basis for Lipoprotein Heterogeneity

The apolipoproteins and lipids which constitute the plasma lipoproteins are heterogeneous and carry different charges. The apparent pI of most apolipoproteins and of their isoforms are known and cover a wide range of pI from 3.7 for apoF to about 6.5 for apoC-I. Polar lipids are located on the surface of lipoproteins and are thought to contribute to the net charge of the lipoproteins. However, little is known about their actual charge in soluble lipoproteins. In addition, the pI of an apolipoprotein may be modified by the protein's conformation and its binding to lipids when it is present in a lipoprotein. Indeed the equilibration of HDL with dimyristoyl phosphatidylcholine liposomes resulted in a modification of the isoelectric focusing profile of HDL with a decrease in a fraction banding at pH 5.1 and a concomitant increase in a fraction centered at about pH 5.6.[1]

Lipoprotein particles carry different complements of apolipoproteins, which differ not only in nature but also in the number of each individual protein carried on each particle. Lipoprotein particles are also characterized by their lipid molecules which vary in nature as well as in number. Accordingly lipoproteins are expected to be heterogeneous in their net charge and in fact this heterogeneity can be demonstrated by a variety of separation techniques based on charge, of which isoelectric focusing has the highest charge resolution.

Charge-Dependent Separations of Lipoproteins

Ion-exchange chromatography has demonstrated the charge heterogeneity of the main lipoprotein density classes. Whether on DEAE-cellulose or DEAE-agarose, VLDL,[2] LDL,[3] and HDL[4] are eluted as broad asymmetrical peaks and in each case analyses of fractions taken across the peaks have shown different complements of apolipoproteins as well as

[1] J. Damen, M. Waite, and G. Seherphof, *FEBS Lett.* **105**, 115 (1979).
[2] W. K. K. Ho, *FEBS Lett.* **50**, 175 (1975).
[3] B. Rubenstein and G. Steiner, *Can. J. Biochem.* **22**, 1023 (1976).
[4] B. Rubenstein, S. Evans, and G. Steiner, *Can. J. Biochem.* **55**, 766 (1977).

lipids. However the resolution of this chromatographic technique is insufficient for the isolation of discrete lipoprotein subclasses.

Electrophoresis on polyacrylamide gel systems permits the separation of human HDL into 8 to 9 components, the proportions of which differ between individuals and especially between males and females.[5] However the potentially high resolution of this method is hampered by the molecular sieving effect of the polyacrylamide gel. Of great interest is the cellulose acetate gel electrophoresis which has been used to characterize VLDL and which can be interfaced with an immunoelectrophoresis on agarose gel to yield a two-dimensional analysis of these lipoproteins.[6]

Isoelectric focusing and other related techniques, such as isotachophoresis[7,8] and chromatofocusing,[9] which are the methods of highest resolution for charge separations, have also been applied to the analysis of lipoproteins and have clearly demonstrated their heterogeneity. We will review the different support phases which have been used in isoelectric focusing of lipoproteins and discuss their respective advantages and inconveniences.

Isoelectric Focusing in Liquid Phase

Liquid phase isoelectric focusing is achieved on vertical columns using classically a sucrose density gradient to stabilize the separated fractions and to allow their elution.

Kostner et al.[10] were the first to report the isoelectric heterogeneity of human serum lipoproteins prestained with Sudan black and focused in pH 3–10 or pH 4–6 gradients in the presence of 33% ethylene glycol. However, the subclasses thus separated were not characterized.

Subsequently, Eggena et al.[11] have also documented the isoelectric heterogeneity of human serum HDL in the presence of 33% ethylene glycol. In this system, HDL_2 is resolved into 7 different fractions with the major bands focused at pH 5.1 to 5.4, each of which differs with respect to apolipoprotein composition.

[5] G. Utermann, *Clin. Chim. Acta* **36**, 521 (1972).
[6] L. Holmquist, *FEBS Lett.* **111**, 162 (1980).
[7] G. Bittolo Bon, G. Cazzolato, and P. Avogaro, *J. Lipid Res.* **22**, 998 (1981).
[8] M. Bojanovsky and C. J. Holloway, *in* "Electrophoresis '81" (R. C. Allenand and P. Arnaud, eds.), p. 809. De Gruyter, Berlin, 1981.
[9] A. C. Nestruck, P. D. Niedman, H. Wieland, and D. Seidel, *Biochim. Biophys. Acta* **753**, 65 (1983).
[10] G. Kostner, W. Albert, and A. Holasek, *Hoppe-Seyler's Z. Physiol. Chem.* **350**, 1347 (1969).
[11] P. Eggena, W. Tivol, and F. Aladjem, *Biochem. Med.* **6**, 184 (1972).

Similar separations have been obtained by another group[12,13] who focused on a pH 3–6 gradient HDL_2 and HDL_3, which were separated into 5 and 6 subfractions, respectively. These fractions were also shown to differ in lipid and apolipoprotein composition. Recombination of these subfractions yielded a product indistinguishable from HDL by analytical ultracentrifugation and agarose gel electrophoresis. These observations are important as they demonstrate that ampholytes do not associate irreversibly with the lipoproteins and do not modify the lipoprotein density and electrophoretic mobility. Therefore it appears that lipoproteins do not undergo significant irreversible denaturation during isoelectric focusing.

These studies have shown that HDL_2[11] and HDL_3[13] consist of several discrete lipoprotein species which differ in their apolipoprotein composition and most notably in the apoA-I to apoA-II ratio. Eggena et al.[11] have also noted the absence of apoC-III in their HDL_2 subfractions with the most cathodic migration.

Heterogeneity of VLDL and LDL has also been noted upon isoelectric focusing either in the presence of 33% ethylene glycol[14] or 10% glycerol.[15] However precipitation of these lipoproteins during focusing could not be totally prevented with either ethylene glycol or glycerol. Scanu et al.[16] also noted this problem with HDL as well as LDL and concluded that, as a preparative method, isoelectric focusing in liquid phase possessed limited value because of the tendency of the separated products to precipitate at their isoelectric point. Indeed very sharply delineated rings can be observed on the isoelectric focusing column as focusing of HDL or LDL progresses but these rings of floculated or aggregated lipoproteins collapse as they reach their apparent pI and mix with one another, thus preventing collection of the fractions (unpublished observations).

Isoelectric Focusing on Polyacrylamide Gel

Isoelectric focusing on polyacrylamide gel is limited in its application to lipoproteins by the pore size of the gel which restricts the mobility of the larger lipoproteins. Initially 5% polyacrylamide gels have been used for the focusing of serum and serum lipoprotein fractions prestained with Sudan black using 33% ethylene glycol to minimize electroendoosmosis.[10] Although the system afforded a good resolution of HDL species, the

[12] S. L. MacKenzie, G. S. Sundaram, and H. S. Sodhi, Clin. Chim. Acta 43, 223 (1973).
[13] G. S. Sundaram, S. L. MacKenzie, and H. S. Sodhi, Biochim. Biophys. Acta 337, 196 (1974).
[14] E. Pearlstein and F. Aladjem, Biochemistry 11, 2553 (1972).
[15] S. G. Sundaram, K. M. M. Shakir, and S. Margolis, Anal. Biochem. 88, 425 (1978).
[16] A. M. Scanu, C. Edelstein, and L. Aggerbeck, Ann. N.Y. Acad. Sci. 209, 311 (1973).

migration of VLDL was not demonstrated. Electrofocusing in 3.5% polyacrylamide gel allowed the migration of VLDL in the gel, although interfacing of the focusing gel with a nondenaturing polyacrylamide-gradient gel electrophoresis, to give a size separation in a second dimension, indicated that VLDL appeared as a streak rather than defined spots.[17] This appearance could have been due either to an incomplete focusing caused by molecular sieving effects and aggregation of the particles or to an intrinsic charge heterogeneity of the VLDL, but this issue has not been resolved.

In conclusion, neither in liquid phase nor in polyacrylamide gels has isoelectric focusing provided a good method for the analysis and preparation of lipoproteins.

Isoelectric Focusing on Agarose Gel

While, early on, agarose appeared to be the support of choice for isoelectric focusing of large molecules such as lipoproteins, the presence on agarose of charged groups, heterogeneous in nature as in distribution, conferred to this material high electroendoosmotic properties which prevented its use in isoelectric focusing. However, chemically modified agarose with low endoosmosis has become available in which either positive and negative charges have been balanced or in which stable charges independent of pH have been introduced. Several manufacturers produce these agarose specially for isoelectric focusing and the gels made with these products are best cast on hydrophilic plastic backing sheets. These agarose isoelectric focusing systems have proven to be of high resolution and reproducibility for the analysis of human serum lipoproteins and the application of this technique to the separation of HDL[18] is described below. For this lipoprotein class, narrow range ampholytes (pH 4.60–5.85), which yield an optimum separation, have been prepared by electrofocusing of commercial ampholytes (pH 4–6) in an Ultrodex gel system. The fractions of appropriate pH are pooled and the ampholytes are extracted with water and concentrated to their original volume by distillation under vacuum.

Preparation of Isoelectric Focusing Gel Films. The isoelectric focusing gel was prepared by dissolving 280 mg agarose (IEF grade) and 720 mg sorbitol in 28 ml water in a boiling water bath for about 15 min. The

[17] A. V. Emes, A. L. Latner, M. Rahbani-Nobar, and B. H. A. Tan, *Clin. Chim. Acta* **71**, 293 (1976).
[18] Y. L. Marcel, P. K. Weech, T.-D. N'guyen, R. W. Milne, and W. J. McConathy, *Eur. J. Biochem.* **143**, 467 (1984).

quality of the commercial IEF grade agarose was found to vary between manufacturers and sometimes between batches from the same brand, therefore requiring pretesting of this product. The solution was allowed to cool to 78° and then maintained at that temperature. Ampholytes were added to the hot solution and mixed by swirling: 28 µl ampholytes, pH 3.5–10, and 1.4 ml ampholytes, pH 4.60–5.85.

A glass plate was warmed in a glassware drying oven, placed on a leveling table, and kept warm with an infrared lamp. A sheet of gel-bond film (12.5 × 11 cm) was fixed onto the warm glass plate with a drop of ethanol or glycerol, with its hydrophilic surface up. Using a fast-flow pipet warmed in hot water, the agarose–ampholyte solution (10 ml) was quickly pipetted over the surface of the gel-bond to form a thin film with a uniform, smooth surface. Rapid pipetting together with the use of the infrared lamp was essential to avoid gelation before the entire 10 ml had been applied and the liquid surface had found its level, since surface irregularities cause distortion of the lipoprotein bands during focusing. After a few minutes, when the agarose had gelled, the glass plates carrying the films were placed in a humid, plastic box with tight-closing lid, and stored at 4° for 2 hr to 5 days maximum before use.

Isoelectric Focusing Conditions. The gel-bond carrying the agarose-IEF film was placed on the cooling plate (9°) of the electrophoresis chamber. Paper electrofocusing wicks were wetted with acetic acid (0.2 M) for the anode and with ethanolamine (0.2 M) for the cathode. The wicks were briefly blotted on a sheet of absorbant paper and then laid along the long edges of the agarose film, on the gel surface. To serve as a sample applicator a glass capillary of appropriate length was laid on the surface of the gel, parallel to the electrode wicks, 2.5 cm from the cathodic edge of the film. This capillary was cut from a 1–5 µl disposable, capillary pipet and had its free openings sealed in a flame to avoid entry of solutions. Lipoprotein samples were dialyzed against 5 mM Tris–HCl, 1 mM disodium EDTA, 0.02% NaN_3, pH 7.4 before isoelectric focusing. The sample of HDL was pipetted along the cathodic side of the capillary: 20–30 µl per cm at a concentration ranging from 1 to 17 mg protein/ml. This method of application was found to be the best of those tested, which included paper applicator, plastic template, and various troughs in the agarose gel. The sample was electrophoresed immediately at 1200 V for 50 min. The current was initially about 6 mA, rising quickly to about 8 mA, then falling to 2–3 mA toward the end of the 50-min period. Cooling was essential to avoid drying of the gel; although the gel surface may be blotted with Whatman 1 paper before applying the sample, this is unnecessary and is better avoided to obtain the best resolution of HDL bands.

Staining of IEF Films. HDL bands were visualized by soaking the electrophoresed agarose-IEF film in a bath of 10% trichloroacetic acid and 5% sulfosalicylic acid, then viewing the strip by indirect light against a dark background. The bands were stained as follows. For protein staining the film was washed twice (30 min each) in ethanol, acetic acid, water (35:10:55, v/v), covered with a sheet of Whatman paper and allowed to dry slowly at room temperature then stained for 15 min in Coomassie blue G-250 0.2 g/100 ml dissolved in the same solution. The film was destained in the same solvent for a few minutes, rinsed with water, then dried in air. For lipid staining, the trichloroacetic/sulfosalicylic acid-containing film was washed twice (30 min each) in ethanol : isopropanol : acetic acid : water (57:3:5:100, v/v), covered with a sheet of Whatman paper and allowed to dry slowly at room temperature then stained in a freshly prepared solution of Sudan black B (0.2%, w/v) in ethanol (58%) for 15 min. The film was destained briefly in ethanol : isopropanol : water (19:1:20, v/v), rinsed in water again, and allowed to dry. Sudan black B precipitates on the surface of the film, especially the hydrophobic under surface of gelbond film. This precipitate can be removed during destaining by wiping with paper and destaining solution or chloroform, but on no account allow the chloroform to come into contact with the agarose surface.

Considerations for Optimization of the Method. The quality of the agarose gel is of utmost importance to ensure regular migration of the lipoprotein bands. The agarose gel must be homogeneous and especially complete dissolution of the gel beads must be ensured. The surface of the agarose film needs to be perfectly plane and free of dust. Prolonged storage of the agarose film results in superficial drying and formation of a denser layer which prevents the normal migration of the lipoproteins. Some electroendoosmosis always occurs, leading to drying of the gel on the anodic side and accumulation of water next to the cathode. This problem can never be totally eliminated although blotting the excess liquid off the wicks helps. In addition, the time and power used in the focusing were chosen to ensure maximum migration and separation of the bands while minimizing the endoosmotic effects which increase with time.

The choice of the pH range of the main ampholytes depends upon the classes of lipoprotein which are to be analyzed. In the case of HDL, a pH range of 4.60 to 5.85 was considered optimum and the use of a narrower pH range improved somewhat the separation but at the expense of the resolution since the bands became wider. This phenomenon is in contrast with the sharp resolution which is obtained with simple proteins, and it is related either to a lipoprotein pI heterogeneity greater than the number of ampholyte species or to an unknown artifact as the lipoproteins approach their pI. Increasing the ampholyte concentration did not improve the separation but small volumes of wide-range ampholytes were added to stabilize the pH gradient close to the electrodes. Improved separation of lipoproteins can be achieved either with narrow range ampholytes over a short physical distance[18] or with conventional pH 4 to 6 range over a longer physical distance.[19]

Agarose Gel Isoelectric Focusing of Plasma Lipoproteins

Isoelectric focusing on agarose gels has proven to be a simple and flexible method for the analysis of lipoproteins provided that an appropriate pH gradient for the optimum resolution of lipoprotein classes can be obtained. Normal plasma lipoproteins including VLDL, LDL, and HDL separated by single spin discontinuous gradient ultracentrifugation[21] (without prestaining) have been focused in 3 different pH gradients. A wide gradient (pH 4–8) allowed the migration of the major plasma lipoproteins as illustrated in Fig. 1A. In this system VLDL was resolved into about 9 subfractions of different apparent pI which overlapped but did not coincide with most of the HDL subfractions. Accordingly, direct focusing of the native plasma lipoproteins in the same gradient resulted in a profile which is a composite picture of the subfractions present in isolated VLDL and HDL. In the same gradient, LDL also exhibited considerable charge heterogeneity and was resolved into about 5 distinct subfractions, the pI of which were less acidic than those of VLDL and HDL subfractions. However, the presence of albumin, which stained with Sudan black due to the presence of free fatty acids and overlapped with LDL subfractions, prevented observation of LDL heterogeneity directly in plasma. This LDL heterogeneity has also been observed earlier in pH 3.5–10 using the same system.[22]

As narrower and more acidic pH gradients were selected (Fig. 1B and C), the resolution of HDL subfractions increased using pH 4–6 to pH 4.6–5.9. Using the narrowest pH range (4.6–5.9) the VLDL and LDL migrated very little, neither as isolated fractions nor when present in the plasma sample. This diminished their effect on the pH gradient in the center of the gel where HDL was focused, and consequently there was a good concordance between the lipoprotein bands observed when isolated HDL and plasma were focused side by side (Fig. 1C). It is intriguing to note that VLDL and HDL subfractions focused in the same range of pI when applied in a pH 4–8 gradient (Fig. 1A) but not when applied in the narrow and more acidic pH gradient (Fig. 1B and C). This appears to be

[19] C. Luley, H. Watanabe, and K. U. Kloer, *J. Chromatogr.* **278**, 412 (1983).
[20] H. Towbin, T. Staehelin, and J. Gordon, *Proc. Natl. Acad. Sci. U.S.A.* **76**, 4350 (1979).
[21] A. M. M. Terpstra, C. S. H. Woodward, and F. J. Sanchez-Muniz, *Anal. Biochem.* **111**, 149 (1981).
[22] R. W. Milne and Y. L. Marcel, *FEBS Lett.* **146**, 97 (1982).

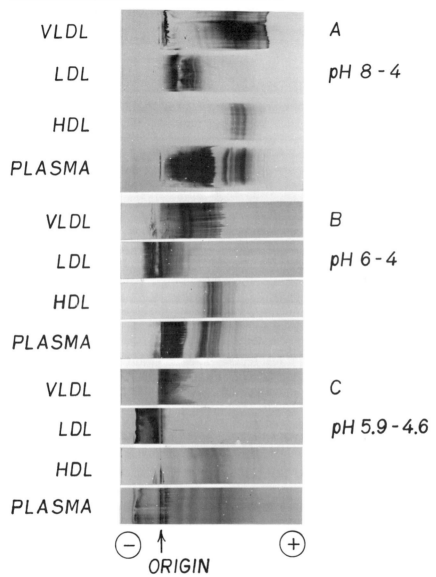

Fig. 1. Isoelectric focusing of human plasma lipoproteins. VLDL, LDL, and HDL were isolated by a single, density-gradient ultracentrifugation[21] from postprandial plasma of a normolipemic person and dialyzed against 5 mM Tris–HCl, 1 mM EDTA, 0.02% NaN$_3$, pH 7.4. Isoelectric focusing gels were prepared as described in the text using as main ampholytes: (A) pH 8–4 (equal volumes of pH 8–5 and 6–4); (B) pH 6–4; (C) pH 5.9–4.6. The protein concentrations of the samples were VLDL 1.24 mg/ml, LDL 1.25 mg/ml, HDL 1.36 mg/ml. They were applied to the gels at 20 μl/cm (VLDL, LDL, plasma), or 30 μl/cm (HDL). After isoelectric focusing the gels were stained with Sudan Black B.

related to the precipitation of VLDL which occurs increasingly when this lipoprotein class is focused in pH gradients that are more and more acidic.

Two-dimensional Analysis with Agarose Isoelectric Focusing and Other Electrophoretic Techniques

Strips of agarose gel can easily be cut out after isoelectric focusing and interfaced with other electrophoretic media to yield two-dimensional analysis of lipoproteins. Owing to the compatibility of the supporting media, agarose isoelectric focusing strips can easily be imbedded into agarose gels containing various antisera to give a two-dimensional immunoelectrophoretic profile of the separated lipoproteins. This technique has been applied to the analysis of plasma lipoproteins as well as HDL_2 and HDL_3 using antisera against apoA-I, apoA-II, and apoC-III which were included in the gel either alone or in combinations of two.[18] Luley et al.[19] have also used two-dimensional isoelectric focusing–immunoelectrophoresis with antiserum to apoA-I to characterize the lipoproteins present in the supernatant of dextran sulfate precipitation.

Agarose isoelectric focusing gels have also been interfaced in a second dimension with SDS–polyacrylamide gel electrophoresis.[18] The latter gel cast on a gel-bond film was placed on a horizontal electrophoretic system. The strip of agarose isoelectric focusing gel on which HDL had been focused was placed gel down on the exposed surface of the stacking gel and electrophoresed in the second dimension. After migration, the separated apolipoproteins could be either stained or transferred onto nitrocellulose paper[20] and identified by reaction with various specific antibodies.[18]

Heterogeneity of HDL upon Isoelectric Focusing

Human HDL_2 and HDL_3 that were homogeneous and polydisperse by the usual criteria of agarose gel electrophoresis and electron microscopy have been found to be heterogeneous by isoelectric focusing. The heterogeneity was evident as about 10 narrow bands after staining for either lipid or protein.[18] This indicates that, although HDL are homogeneous and polydisperse by some physical characteristics and techniques, they are actually a complex mixture of lipoproteins which can be separated by their electrical properties on isoelectric focusing into a series of fractions which contain both lipid and protein. A similar number of bands were seen in both HDL_2 and HDL_3, and these bands migrated at the same rates indicating identical charges. The same heterogeneity was also evident in whole plasma which demonstrates that the bands observed are a characteristic of plasma lipoproteins and do not result from ultracentrifugation.

ApoA-I and apoA-II are present in all fractions isolated after focusing of HDL_2 and HDL_3 and in both HDL classes acidic lipoprotein bands are enriched in apoA-II, apoC-III, and apoD. In contrast, alkaline lipoproteins are enriched in apoC-I and apoE. ApoC-II and apoC-III are not similarly distributed in HDL_2 and HDL_3 whereas apoC-I distribution is similar in both classes. The difference in the distributions of apoC-I, apoC-II, and apoC-III in HDL_2 and HDL_3 indicates that the existence of a lipoprotein containing simultaneously C-I, C-II, and C-III can account for only a small fraction of these apolipoproteins.

Therefore these experiments substantiate the theory of the protein basis of HDL heterogeneity and suggest that the majority of apolipoproteins are present in complexes which, upon isoelectric focusing, result in lipoprotein fractions of identical pI for both HDL_2 and HDL_3. In addition, subtle differences in the distribution of apolipoproteins in HDL_2 and HDL_3 indicate the occurrence of some different lipoprotein species in these 2 subclasses of HDL.

There is now ample evidence for the charge heterogeneity of human HDL using the techniques of isoelectric focusing, isotachophoresis, and chromatofocusing. The former technique has been employed and developed in many laboratories, has been shown to be a reliable technique for the fractionation of HDL without their denaturation, and shows a concordance between the HDL and homologous apolipoprotein A-I-containing lipoproteins in plasma. Very little study has been devoted to the VLDL and LDL using charge separation techniques and, although we present evidence in Fig. 1 for charge heterogeneity of these lipoproteins, it is obvious that the separation of VLDL and LDL fractions depends to a great extent on the ampholytes used. The technique of isoelectric focusing of VLDL and LDL is now in its infancy and needs to be optimized in the same way as the isoelectric focusing of HDL. While there is general agreement in the published studies that the heterogeneous protein composition of HDL particles is seminal to their lipoprotein heterogeneity, we have no information available for the source of VLDL and LDL charge heterogeneity. Development of isoelectric focusing for these lipoprotein classes may give us new insight to the structure and composition of these lipoprotein classes and permit their analysis directly in plasma, as is possible for the HDL[18,19] (Fig. 1C) using narrow-range ampholytes.

Acknowledgment

The studies from the authors' laboratory presented here were supported by grants from the Medical Research Council of Canada (Program Grant PG-27) and the Quebec Heart Foundation. Mrs. L. Lalonde gave excellent secretarial assistance.

[26] Electron Microscopy of Negatively Stained Lipoproteins

By TRUDY M. FORTE and ROBERT W. NORDHAUSEN

The technique of negative staining was initially used by electron microscopists to study viruses, subcellular fractions, proteins, and lipid dispersions,[1,2] and subsequently the technique was applied to the study of lipoproteins.[3] Negative staining was first utilized in light microscopy to study bacterial structure. Essentially, the translucent bacteria on a glass slide were surrounded by India ink before viewing. Light passed through the bacteria and not through the dense stain, thus revealing the specimen by negative contrast. The principle of embedding biological samples in a dense matrix was later applied to electron microscopy. Biological materials examined by electron microscopy generally have poor inherent electron scattering characteristics and hence must be visualized either by means of positive staining (e.g., by OsO_4), which interacts directly with specific chemical groups in the specimen, or by negative contrast. The latter technique, called negative staining, is extremely useful in examining small macromolecular structures and, specifically, does not require interaction between the stain and the specimen. Criteria for an ideal negative stain for electron microscopy are (1) it must not react with the specimen to be examined; (2) it must have high electron density; (3) it must resist sublimation in the electron beam; and (4) it must have minimal granularity when examined in the electron microscope.

Negative stain techniques applied to lipoproteins are especially useful for defining pathologic conditions and examining lipoprotein fractions derived from all manner of *in vivo* and *in vitro* studies. It is an extremely useful technique for obtaining insight into particle size distribution and morphology under conditions where sample size is extremely small and/or dilute. Negative staining lends itself extremely well to the examination of smaller, more dense lipoproteins such as low-density lipoproteins (LDL) and high-density lipoproteins (HDL). Larger, triglyceride-rich particles do not lend themselves as readily to negative stain techniques because of their tendency to flatten and/or coalesce during sample preparation.

---

[1] A. M. Glauert, *Lab. Invest.* **14** (Part 2), 331 (1965).
[2] R. W. Horne, *Lab. Invest.* **14** (Part 2), 316 (1965).
[3] T. M. Forte and A. V. Nichols, *Adv. Lipid Res.* **10**, 1 (1972).

Stains

Negative stains commonly used in electron microscopy are uranyl acetate or uranyl formate, ammonium molybdate, and phosphotungstate. Uranyl acetate and uranyl formate have generally been favored by electron microscopists examining protein structure. The pH of this stain is around 3.5 to 4.0 and is too low for application to lipoproteins. At this pH, lipoproteins aggregate and precipitate. An alternative uranyl stain is uranyl oxalate, which can be adjusted to pH 6.5; however, in our experience, this stain appears to be more granular than phosphotungstate and therefore has little advantage over phosphotungstate. Ammonium molybdate has been used for negative staining lipoproteins, however, this stain is not so dense as tungstate. In our experience, phosphotungstate has the most desirable characteristics for routine use. It can be adjusted over a wide pH range, is very electron dense, and has fine granularity. For use with lipoproteins, phosphotungstate is adjusted to pH 7.0–7.5 with NaOH or KOH. We use the former basic solution since the sodium ion is the major extracellular cation. The stock staining solution consists of 2% sodium phosphotungstate, pH 7.4. After the final pH is established, the solution is filtered through a 0.45-μm Nalgene filter (Nalge Co., Rochester, NY) and stored in a clean glass bottle. If care is taken so that only clean pipets are introduced into the stain bottle, the stain is usable for months.

Preparation of Sample

For best results in negative staining of lipoproteins, the specimens should be fresh. Isolated very low-density lipoproteins (VLDL) rapidly deteriorate and form aggregates which are impossible to dissociate if the sample is "old." VLDL should ideally be examined within 1 to 2 days of preparation. At the other end of the spectrum, HDL appear to be more resistant and can be stored from 1 to 2 weeks without serious deterioration of ultrastructure. Deterioration of HDL ultrastructure in electron microscopy can be recognized by two obvious difficulties: (1) the particles form clumps, and (2) the particles tend to lose contrast and become fuzzy. We feel that the latter problem is related to either a loosening of the protein structure and hence penetration of stain or that aging leads to certain degree of positive staining effect. The net result is that the crispness of the negative image is lost.

Fractions isolated by ultracentrifugation are best preserved and stored in their high salt background; however, EDTA (1 mg/ml) and gentamicin sulfate (0.1 mg/ml) or an equivalent antibacterial agent should be present.

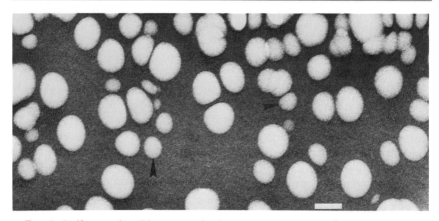

FIG. 1. Artifact produced by sucrose in the sample. The electron-lucent structures are not lipoprotein particles. Bar represents 100 nm.

If the storage effects of a particular buffer or elution solution are unknown, it is best to store the particles in 0.9% sodium chloride plus EDTA and gentamicin. Prior to examination in the electron microscope, the sample is dialyzed to a volatile buffer such as ammonium acetate or ammonium bicarbonate. A volatile buffer is used so that the thin film of stain on the grid can rapidly set into a "glass," thus locking in the particle structure. Volatile buffers assure rapid drying and leave no salt residue which can be troublesome. Even small amounts of sodium chloride, other salts, or organic material left in the sample can seriously affect the quality of the pictures taken. An example of an artifact introduced by small amounts of sucrose in the lipoprotein sample is shown in Fig. 1. The large, round, lucent structures almost look like VLDL but are, in fact, due to sucrose. The electron beam volatilizes the sucrose and leaves a clear or lucent image. This process can be seen on the viewing screen of the microscope, and it appears as if the material is "boiling" in the beam. One should be suspicious of final images if such a phenomenon occurs in the microscope.

Dialysis of Samples

In our laboratory we routinely use a dialysis buffer consisting of 0.125 M ammonium acetate, 2.6 mM ammonium carbonate, and 0.26 mM tetrasodium EDTA at pH 7.4. If large volumes or small numbers of samples are to be dialyzed, the most efficient method is to use 1/4 to 1/2 in. dialysis tubing (Spectropor dialysis membrane) and dialyze overnight in the cold against a large excess of dialysis fluid. If 5 or more samples require dialysis, a multiwell dialyzer unit is extremely useful. The unit we recommend is the MRA Corporation M145-1 lucite chamber (MRA Corp., 1058 Cephas Rd., Clearwater, FL) which dialyzes up to 15 samples with volumes of 0.1 to 1 ml.

Grids and Specimen Support Films

Copper mesh grids (200, 300, or 400 mesh) which have a matte finish on one side and a shiny surface on the other are used in conjunction with either a Formvar-carbon substrate or carbon-only substrate. Before grids are used, they should be thoroughly cleaned in acetone and dried on filter paper. We typically maintain a supply of Formvar-carbon-coated grids as well as girds with thin carbon films.

Standard Formvar-Carbon-Coated Grids

This substrate is useful where the negative staining workload is large. Large numbers of these grids can be made up ahead of time and they are very stable. The larger mesh (200 and 300) grids are typically used for Formvar-carbon-coated grids. Description of support films are adequately covered by Kay[4] and Hayat.[5] The major steps in making Formvar-carbon-coated support films are shown in Fig. 2. Formvar (polyvinyl formal) at a concentration of 0.3% is made up in ethylene dichloride. A clean microscope slide is dipped into the Formvar solution, withdrawn, and allowed to air dry. The dried film is scored with a razor blade on the flat side of the slide (Fig. 2A) and then the film is slowly floated off onto distilled water in a crystallizing dish (Fig. 2B). The film can be seen on the H_2O surface by reflected light from a desk top lamp. Grids are then placed matte side down in rows on the film (Fig. 2C). The support film with grids is picked up by overlaying it with clean Parafilm (Fig. 2D). Air dry the grids on the Parafilm. To carbon coat the grids, first place two narrow strips of double sticky tape onto a glass slide, then carefully attach grids, *Formvar-side down,* by their edges to the sticky tape as shown in Fig. 2E. Slides with attached grids are placed in a clean vacuum evaporator, and carbon evaporation is carried out by using 3-mm-diameter carbon rods whose tips are turned down to 1 × 2.4 mm. Approximately half of this tip is evaporated at a distance of 80 mm from the rotating stage. When a sample is to be applied to the coated grid, the grid should be turned so that the Formvar

---

[4] D. H. Kay, "Techniques for Electron Microscopy." F. A. Davis, Co., Philadelphia, 1965.
[5] M. A. Hayat, "Principles and Techniques of Electron Microscopy. Biological Applications," Vol. 1. Van Nostrand-Reinhold, New York, 1970.

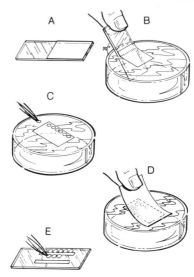

FIG. 2. Steps in preparing Formvar-carbon-coated grids. See text for full description.

surface is uppermost. This latter surface has far better wetting properties than the carbon surface. On rare occasions one may find that samples spread very unevenly. If this occurs with several different preparations, one usually suspects that some oil has gotten into the vacuum evaporator and that the grids are contaminated with oil. It is best to throw out all such grids and start anew.

Thin Carbon Support Films

Thin carbon films are used where the work requires greater resolution and contrast, such as with small HDL particles or small lipid–protein complexes. Two approaches can be used in making up such grids. In the first case, 400-mesh grids are covered with a film of thin carbon. The carbon films are made by cleaving mica and depositing a thin layer of carbon onto the cleaved surface; the thin film is then floated onto a clean water surface. To cover the grid, firmly grasp the grid edge with fine forceps and lower it into the water at an angle and then come up under the floating carbon film and lift the grid to the surface. Air dry grids and store; the easiest and cheapest way to store them is to use a clean slide with two strips of double sticky tape similar to that shown in Fig. 2E. An alternative method for preparing thin carbon support films is to mount them onto

200-mesh grids first covered with holey triafol (cellulose acetate butyrate) films. The procedure we use is that of Fukami and Adachi[6] as modified by Reichelt et al.,[7] which utilizes 1% glycerol and 0.25% triafol in ethyl acetate. The triafol films are cast in the same fashion as those for standard Formvar-coated grids. The holey, triafol covered 200-mesh grids are then backed with a heavy layer of carbon which gives maximum stability to the holey substrate. Thin carbon films are then picked up onto the holey supports as outlined above. The advantage of this system is that it produces a support with various sized holes which can be covered with extremely thin carbon films.

Application of Sample

The simplest and most straightforward method of applying sample to the grid is to use small droplets of fluid. With standard Formvar-carbon-coated grids, be sure the Formvar side is uppermost. One approach with the droplet method (favored in our laboratory) is to mix equal proportions of NaPTA and sample in a conical test tube. The final concentration of HDL and LDL samples should be approximately 100 to 250 μg lipoprotein/ml in 1% NaPTA. When working with VLDL, the best dilution criterion is that the diluted mixture should be just slightly opalescent. The sample is gently mixed by tapping the side of the tube. A small aliquot of lipoprotein plus stain is taken up in a 10-μl disposable microcapillary and a small droplet placed on the grid held in locking fine forceps. Let stand for 20–30 sec and then remove excess fluid with filter paper. A word of caution in sample application: the forceps should be rinsed in ethanol after each sample is applied in order to avoid dragover of the previous sample, which tends to adhere to the tip of the forceps. Immediately after preparing the sample, place the grid into the microscope specimen chamber and examine. After examining the grid, do not save it for future use; the sample rehydrates when reintroduced to ambient conditions and results in rehydration–dehydration artifacts.[8] The single droplet method is ideal for determining the presence and morphology of lipoproteins in very dilute samples. A very small (1–2 mm diameter) droplet is placed on the grid and is either allowed to dry down completely or is blotted, and the edge of the droplet is examined in the electron microscope. This procedure effectively concentrates particles at the edge while the droplet is

[6] A. Fukami and H. Adachi, *J. Electron Microsc.* **14**, 112 (1965).
[7] R. Reichelt, T. König, and G. Wangermann, *Micron* **8**, 29 (1977).
[8] E. C. Gong, G. M. Forte, and A. V. Nichols, *Physiol. Chem. Phys.* **2**, 180 (1970).

drying. An alternative method for staining the specimen used in some laboratories is what we call the 2-drop method. First place a droplet of lipoprotein onto the grid (wait 15–20 sec); then remove excess lipoprotein fluid and immediately add a droplet of stain and remove excess stain. One must be sure with this method that the lipoprotein sample on the grid does not dry down prior to adding stain.

Another approach to sample application is to spray the sample onto the grid. This method of application has great value especially, as noted below, when the distribution of particles is to be determined. Two simple methods for spraying samples are (1) De Vilbiss nebulizer and (2) Effa Spray mounter (Ernest F. Fullam, Inc., Schenectady, NY). The De Vilbiss nebulizer requires approximately 0.7–0.8 ml of sample (lipoproteins in stain). Hold the grid perpendicular to the orifice of the nebulizer at a distance of 6–8 in. and atomize the sample twice. The Effa spray assembly has the advantage of using a smaller sample size and disposable capillary tubes. The sprayed samples should deposit small round droplets onto the grids; in the electron microscope look for droplets that have a uniform gray appearance.

Generally lipoprotein samples are maintained at 4° and mixed with stain which is at room temperature. There are certain instances, however, when samples should be maintained at approximately 37° throughout the staining procedure. This is particularly true for some chylomicron, VLDL, and IDL preparations which contain highly saturated fatty acids. Puppione et al.[9] and Clark et al.[10] showed that the lipids in such particles undergo transitions at low temperatures and become less fluid. This alteration in the physical state of the lipids also produces unusual flattened, irregular morphologies in the particles. To avoid such a phenomenon, the lipoprotein sample, solutions, grids, and forceps are maintained at 37° (or other desired temperature) with a slide warmer during the entire negative staining procedure.

Examination of Sample

Instrument magnifications generally used for LDL are 30,000–40,000×, while those used for HDL are 40,000–60,000×. Chylomicron and VLDL examination requires low magnifications (10,000–25,000×).

[9] D. L. Puppione, S. T. Kunitake, R. L. Hamilton, M. L. Phillips, V. N. Schumaker, and L. D. Davis, *J. Lipid Res.* **23**, 283 (1982).
[10] S. B. Clark, D. Atkinson, J. A. Hamilton, T. Forte, B. Russell, E. B. Feldman, and D. M. Small, *J. Lipid Res.* **23**, 28 (1982).

Since we are interested primarily in high resolution images at high magnification, we use an accelerating voltage of 100 kV. The latter sacrifices the contrast obtained by using lower accelerating voltages; however, contrast is effectively regained by using a small objective aperature (50 μm). Care should be taken to minimize radiation damage and contamination of the sample. This is best achieved by keeping the illumination reduced when scanning the grid, and when an area of interest is chosen for photography, keep beam exposure down to a few seconds. Focusing of the sample can be a problem, particularly to the novice. The best rule-of-thumb is to adjust the focus of the background to minimal granularity. The tendency is to defocus too far so that the sample is very contrasty to the eye. An example of good focus is shown in Fig. 3a and poor focus in Fig. 3b.

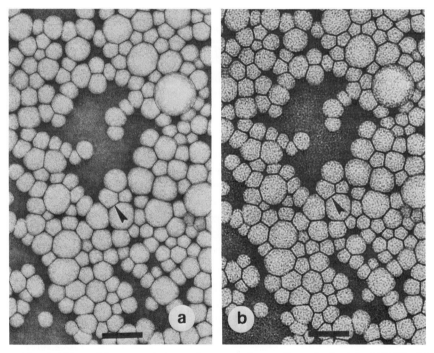

FIG. 3. Effect of focus on electron microscopic image of negatively stained VLDL. (a) This micrograph is near focus; note smooth stain background. Arrow indicates interparticle distance, which is 1.5 nm. (b) Micrograph of same area, 1.2 μm under focus. Note the gross granularity of the stain background and of the particles. Arrow indicates identical region as in micrograph (a) and shows that the interparticle distance is two times greater (3.0 nm) than in (a). Bars indicate 100 nm.

Figure 3a has a smooth looking stain background, while Fig. 3b has a very grainy background and the VLDL particles begin to appear to have substructure. The latter leads to inaccurate assumptions about particle substructure and inaccurate measurement of particle size (note the increase in interparticle space in Fig. 3b versus Fig. 3a, arrows).

Some thought must be given to adequacy of sampling, i.e., are the photographs taken representative of the sample? It is best to take random photographs within a grid square and to use random grid squares. When more than one type of particle appears in the sample or when the sample is extremely heterogeneous (e.g., VLDL) a more accurate determination of morphology and size distribution can be obtained by using the spraying technique. A photo(s) of the entire droplet should be taken for the most accurate results.

Sizing of particles can be carried out either by using (1) an ocular micrometer or (2) a digitizing tablet interfaced with a computer. The ocular micrometer (Bausch and Lomb Micrometer with 0.1 mm divisions is excellent) requires no expensive equipment but is labor intensive. With the micrometer one can work directly with the negative, but great care must be taken so that the emulsion is not scratched since the negatives are the primary data, or one can work from prints enlarged 3-fold. The digitizing tablet is useful because a single individual can size large numbers of particles in a short period of time and a computer program can store the information and/or generate histograms as needed on particle size distribution. Since many laboratories have existing computers, the most inexpensive way of collecting data is to purchase a digitizer such as the sonic digitizer (Graf-pen, Science Accessories Corp., Southport, Conn.) used in our laboratory and to write the appropriate programs. It is now possible to purchase computerized image analyzers which come with software packages for analyzing many of the functions needed in electron microscopy of lipoproteins. The resolution and reproducibility of data obtained by these digitizers should be checked against the values obtained with a micrometer. The error can be considerable when extremely small particles are measured by the digitizer.

Artifacts

Accurate sizing and interpretation of data depends a great deal on knowledge of possible artifacts associated with negative staining. Common artifacts encountered in negative staining of lipoproteins and their products are summarized in the table. Clumping of particles (see Fig. 5 for an example) clearly makes it difficult to assess particle size and size distribution. Clumping of particles is often indicative of sample aging and,

COMMON ARTIFACTS IN NEGATIVE STAINING

Phenomenon	Cause	Remedy
1. Clumping and/or aggregation	Aged sample Low pH Too concentrated	Fresh sample Check stain Dilution
2. Grossly distorted particles	Aged sample Dehydration and rehydration	Fresh sample Check sample application
3. Spotty or uneven background	Contaminated support film	Use new Formvar-carbon grids or carbon supports
	Suspension fluid or stain contaminated	Check suspension fluid and stain in EM; made up new solutions if necessary
4. Flattening which introduces error into size measurements	Particularly severe with triglyceride-rich particles	Prefix particles with OsO_4[a]

[a] J. A. Glomset, K. Applegate, T. Forte, W. C. King, C. D. Mitchell, K. R. Norum, and E. Gjone, *J. Lipid Res.* **21**, 1116 (1980).

as indicated earlier, is more evident with triglyceride-rich particles. The best way to avoid this problem is to examine the samples promptly. Occasionally the stain will produce clumping; this suggests that the pH may not be adjusted properly. Aggregation of particles, as opposed to nondescript clumping, may not necessarily be totally bad. Under certain conditions aggregation of particles can, in fact, provide additional information on the physical properties of the particles. The classical example of the aggregation artifact providing additional insights into lipoprotein structure was our early observation on HDL from lecithin:cholesterol acyltransferase deficient patients.[11] Under our negative staining conditions, the major HDL constituent formed rouleaux during drying, thus providing the clue that the individual particles were discoidal in shape and had the thickness of a phospholipid bilayer. If such specimens are diluted a great deal, one perceives round profiles and no rouleaux. In such a case, one is likely to interpret the en face particles as "round" or "spherical" structures. It is important to remember that the structure of the negative image is seen only in two dimensions. Figure 4a and b provide further evidence that aggregation of particles in more concentrated samples can

[11] T. Forte, K. R. Norum, J. A. Glomset, and A. V. Nichols, *J. Clin. Invest.* **50**, 1141 (1971).

be of use to the investigator. The HDL in Fig. 4a are ideally spread so that particle diameters of individual freestanding structures are easy to determine. This photograph, however, tells little about the geometry of the particle; from what we know about the chemistry of the particles, we would probably decide that they are quasi spheres. Figure 4b is of the same sample in higher concentration; here the particles are aggregated in a highly organized way which shows hexagonal packing. Hexagonal packing of particles is additional evidence for spherical geometry of the HDL particles. Yet another view of an unusual HDL morphology seen upon negative staining is shown in Fig. 4c. This micrograph is of HDL isolated from the $d < 1.063$ g/ml fraction from an abetalipoproteinemic patient; however, similar structures are also found in human interstitial fluid[12] and in sheep lung lymph.[13] The particles in this case pack in square arrays.[3,14] This is not to say that the freestanding particles are cubes, but rather it suggests that these unusual HDL may have a unique distribution of protein and/or surface lipid components so that they align in square arrays when packed. It is apparent from such examples that aggregation artifact can often be "meaningful artifact" by supplying greater insights into the physical and chemical structure of the particle. Hence, it is often useful to take photographs of more concentrated samples as well as dilute ones.

Particles which have undergone dehydration in the negative staining process and are allowed to rehydrate will show great distortion of particle shape. This is the prime reason that one discards the grid after examining the sample. One will also find distortion when samples are extremely old. Distorted particles often appear to have protuberances or have an angular appearance. As mentioned earlier, triglyceride-rich particles enriched in saturated fatty acids are distorted in shape when maintained at low temperatures, due to crystallization of the core lipids. To avoid this phenomenon, as mentioned earlier, the samples must be kept above their transition temperature during all handling steps.[10]

An insidious artifact which is sometimes difficult to assess and overcome in seen in Fig. 5a and b. It consists of a spotty background which appears as electron lucent small round structures. These are, in fact, not particles but hydrophobic regions on the grid which do not stain evenly. Figure 5a shows areas of normal smooth stain with clumped VLDL; the same grid area shows small lucent structures. The fact that there are

[12] T. M. Forte, unpublished observation (1984).
[13] T. M. Forte, C. E. Cross, R. A. Gunther, and G. C. Kramer, *J. Lipid Res.* **24**, 1358 (1983).
[14] A. M. Scanu, L. P. Aggerbeck, A. W. Kruski, C. T. Lim, and H. J. Kayden, *J. Clin. Invest.* **53**, 440 (1974).

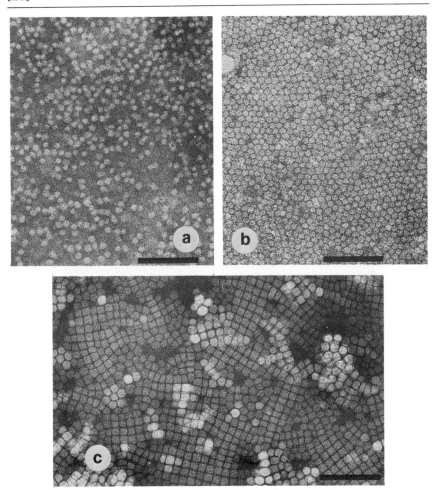

FIG. 4. Examples of aggregation artifact during negative staining which can provide additional information on the physical characteristics of the particle. (a) and (b) are the same HDL preparations seen at different concentrations. (a) shows freestanding round profiles, while (b) shows that the particles tend to hexagonally pack during aggregation and are therefore homogeneous spherical structures. (c) shows an unusual HDL fraction (d 1.006–1.063) isolated from a patient with abetalipoproteinemia; upon aggregation numerous particles form square arrays, suggesting that the protein and/or lipid arrangement in these particles is different from that of normal HDL. Bar represents 100 nm.

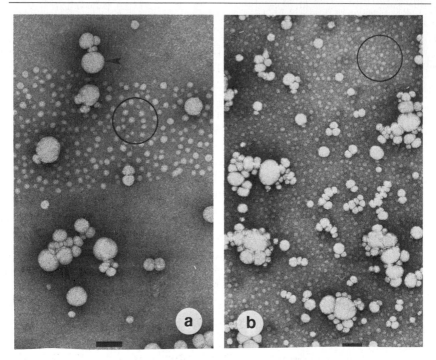

FIG. 5. Examples of background artifact. (a) VLDL sample. VLDL particles are large and show some clumping (arrows). Small lucent round structures (outlined in circle) form a band across the EM field, while the rest of field has a uniform, smooth stain background. The small lucent regions are generated by hydrophobic regions on the grid substrate. (b) VLDL sample with a more generalized stippled background; artifact indicated in circled region. Grids should be suspected for possibility of contamination. Another source would be a contaminant in either the stain or sample buffer. Bars represent 100 nm.

adjacent normal and abnormal staining areas suggests that there are contaminated regions on the grid which repel the stain and thus create the appearance of negatively stained particles. In Fig. 5b the lucent areas are more subtle and blend into the background. However, in this case the spotty background is more uniform over the entire grid square. When one suspects such artifact, the stain and dialysis solution, without lipoproteins present, should be tested in order to determine whether the grids or one of the solutions are responsible for the problem. It will be necessary to make up new coated grids if the problem is attributable to contaminated grid surfaces. In some instances it may be necessary to make up new stain or dialysis buffer.

Flattening of lipoprotein particles is an obvious problem with the larger, less dense particles and can lead to serious problems in calculating particle size and volume. We encountered this problem in earlier studies with negatively stained VLDL[15] and adjusted for flattening of the VLDL by carrying out parallel studies with osmicated VLDL. The fixed VLDL or chylomicrons are rigid and their size can be determined more accurately. To fix the triglyceride-rich particles, place the lipoprotein solution into dialysis tubing and suspend in a cylinder containing a 2% solution of OsO_4 in 0.15 M sodium cacodylate buffer, pH 7.4, for several hours at room temperature. The fixed particles turn brown-to-black in color, depending on concentration and size of the particles. To remove excess OsO_4, take the dialysis bag and dialyze extensively against distilled water. The fixed particles are sprayed onto grids and are readily seen as positive images and can be accurately measured. A flattening factor can thus be calculated for correcting VLDL and/or chylomicron size in negatively stained images.

Triglyceride-rich particles on occasion will appear to have a "halo" around the periphery; this can be incorrectly interpreted as substructure. Examination of micrographs where this appears reveals that this phenomenon is artifact due to the rupture of a thin layer of stain which covered these large particles. The "halo" is usually irregular in width and shows increased density on the edges which have rolled back (Fig. 6).

Agreement between Electron Microscopy and Other Physical–Chemical Techniques

Each of the lipoprotein classes constitutes a polydisperse spectrum of particle sizes and in certain pathological conditions, such as LCAT deficiency,[11] Tangier disease,[16] and fish eye disease,[17] they are, as well, heterogeneous in geometry. Heterogeneity of particle size, and possibly morphology, must be kept in mind when comparing electron microscopic parameters with other techniques. Human VLDL size (35–85 nm) and morphology (spherical) obtained by negative staining techniques corrected for flattening are in good agreement with the early observations of Lossow et al.[18] using preparative ultracentrifugal isolation and analytic ultracentrifugal analysis of isolated VLDL fractions. The mean diameters

---

[15] J. A. Glomset, K. Applegate, T. Forte, W. C. King, C. D. Mitchell, K. R. Norum, and E. Gjone, *J. Lipid Res.* **21**, 1116 (1980).

[16] G. Assmann, P. N. Herbert, D. S. Fredrickson, and T. Forte, *J. Clin. Invest.* **60**, 242 (1977).

[17] T. M. Forte and L. A. Carlson, *Arteriosclerosis* **4**, 130 (1984).

[18] W. J. Lossow, F. T. Lindgren, J. C. Murchio, G. R. Stevens, and L. C. Jensen, *J. Lipid Res.* **10**, 68 (1969).

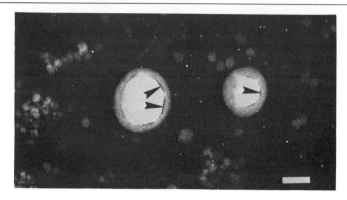

Fig. 6. Artifact that suggests substructure on VLDL surface. The thin layer of stain which covered the particles has ruptured and gives the appearance of a "halo." Frequently one can see a slightly denser region at the inner edge of the halo which is evidence that the phenomenon is an artifact resulting from splitting and rolling back of the stain film. This artifact is associated only with large, triglyceride-rich particles. Bar represents 100 nm.

of LDL by negative staining techniques range from 20 to 27 nm in diameter. The mean particle diameter of LDL varies considerably from individual to individual; and furthermore, these lipoproteins show polydispersity of particle size. The size of the LDL determined by the negative stain technique[19] is in good agreement with gradient gel electrophoresis data.[20] Others have found good agreement between electron microscopic data and that obtained by laser light scattering.[21] A discrepancy between LDL diameter determined by electron microscopy and that determined by sedimentation equilibrium ultracentrifugation has been noted.[22]

Part of the discrepancy between various techniques may depend upon basic assumptions such as particle hydration and sphericity. It is generally assumed that LDL are nearly spherical in morphology; however, it has been suggested by several laboratories that the particle may not be a true sphere.[23,24] A small degree of flattening of LDL, which can lead to overestimation of the diameter, may also occur during negative staining. Spherical HDL size ranges from 6 to 12 nm. The mean particle diameter

[19] M. M. S. Shen, R. M. Krauss, F. T. Lindgren, and T. M. Forte, *J. Lipid Res.* **22,** 236 (1981).
[20] R. M. Krauss and D. J. Burke, *J. Lipid Res.* **23,** 97 (1982).
[21] C. J. Packard, J. Shepherd, S. Joerns, A. M. Gotto, and O. D. Taunton, *Biochim. Biophys. Acta* **572,** 269 (1979).
[22] T. S. Kahlon, G. L. Adamson, M. M. S. Shen, and F. T. Lindgren, *Lipids* **17,** 323 (1982).
[23] H. Pollard, A. M. Scanu, and E. W. Taylor, *Proc. Natl. Acad. Sci. U.S.A.* **64,** 304 (1969).
[24] T. Gulik-Krzywicki, M. Yates, and L. P. Aggerbeck, *J. Mol. Biol.* **131,** 475 (1979).

of normal HDL or HDL fractions agrees well with that obtained by analytic ultracentrifugation and gradient gel electrophoresis.[25,26]

Acknowledgments

This work was performed under NIH NHLBI Program Project Grant HL 18574.

[25] D. W. Anderson, A. V. Nichols, T. M. Forte, and F. T. Lindgren, *Biochim. Biophys. Acta* **493**, 55 (1977).
[26] P. J. Blanche, E. L. Gong, T. M. Forte, and A. V. Nichols, *Biochim. Biophys. Acta* **665**, 408 (1981).

[27] Studies of Lipoproteins by Freeze-Fracture and Etching Electron Microscopy

By LAWRENCE P. AGGERBECK and TADEUSZ GULIK-KRZYWICKI

Introduction

Electron microscopic studies of biological ultrastructures generally employ fixation steps to preserve the sizes, shapes, interactions, and orientations of the macromolecules or the supramolecular complexes to be observed. Chemical fixation and dehydration procedures may, however, introduce extensive structural artifacts; this is particularly so in the case of lipid-containing material. Cryofixation is one alternative for avoiding some of these problems. Freeze-drying, freeze-fracturing and etching, and low temperature electron microscopy of frozen hydrated specimens are methods for studying cryofixed material. The freeze-fracture (and etching) electron microscopic technique consists of observing and photographing the metallique replica of the fractured (and etched) surface of the cryofixed sample (solution). The technique is based upon the early work of Steere,[1] Moor *et al.*,[2] Hall,[3] and Meryman[4] and was applied to the study of whole cells early on. Following subsequent improvements, freeze-fracture electron microscopy has since become the preeminent

[1] R. L. Steere, *J. Biophys. Biochem. Cytol.* **3**, 45 (1957).
[2] H. Moor, K. Muhlethaler, H. Waldner, and A. Frey-Wyssling, *J. Biophys. Biochem. Cytol.* **10**, 1 (1961).
[3] C. E. Hall, *J. Appl. Phys.* **21**, 61 (1950).
[4] H. T. Meryman, *J. Appl. Phys.* **21**, 68 (1950).

technique for the structural study of cell membranes and lipid phases and has been applied to the study of biological macromolecules in solution. In the case of serum lipoproteins, human chylomicrons,[5] very low-density lipoprotein (VLDL),[5,6] low-density lipoprotein (LDL),[6] apoLDL,[7] high-density lipoprotein (HDL),[6,8] and normolipidemic[9] and hyperlipidemic[10] rhesus monkey LDL have been the subject of freeze-fracture and freeze-etching electron microscopic studies. The technique consists of four essential steps: cryofixation of the sample solution, either in the presence or absence of antifreeze agents, fracture of the cryofixed sample solution followed by etching (sublimation of ice) of the fractured surface and finally, heavy metal replication of the fractured and etched surface. Following cleaning, the replica is examined and photographed in the electron microscope. To obtain meaningful results, careful attention must be paid to optimize each of these steps. We describe here a simple and inexpensive procedure for the study of plasma lipoproteins using a Balzer's freeze-etching unit. Several technical variations exist and references will be made to some of these alternative procedures which are well described in the literature.

Sample Preparation

Prior to cryofixation, lipoproteins are dialyzed in the dark against several 200-fold volumes of buffer solution. Volatile salt solutions (ammonium acetate or ammonium bicarbonate, for example) or dilute (10 mM) sodium phosphate solutions at alkaline pH (7–8) are appropriate. For convenient observation and analysis, the lipoprotein concentrations are adjusted, by dilution with dialysis buffer, to approximately 10–50 μM VLDL or LDL and 100–200 μM HDL. Following adjustment of the concentration, the samples are filtered (Millipore filter, type HA, 0.2 μm pore size, or equivalent).

---

[5] T. Forte and A. V. Nichols, *Adv. Lipid Res.* **10**, 1 (1972).
[6] L. P. Aggerbeck, M. Yatès, and T. Gulik-Krzywicki, *Ann. N.Y. Acad. Sci.* **348**, 352 (1980).
[7] A. Gulik, L. P. Aggerbeck, J. C. Dedieu, and T. Gulik-Krzywicki, *J. Microsc.* **125**, 207 (1981).
[8] M. Ohtsuki, C. Edelstein, M. Sogard, and A. M. Scanu, *Proc. Natl. Acad. Sci. U.S.A.* **74**, 5001 (1977).
[9] L. P. Aggerbeck, T. Gulik-Krzywicki, V. Luzzati, J. L. Ranck, A. Tardieu, G. Fless, and A. Scanu, *in* "Latent Dyslipoproteinemias and Atherosclerosis" (J. L. DeGennes, J. Polonovski, and R. Paoletti, eds.), p. 109. Raven, New York, 1984.
[10] T. Gulik-Krzywicki, M. Yatès, and L. P. Aggerbeck, *J. Mol. Biol.* **131**, 475 (1979).

Cryofixation of the Sample

The goal of this step is to quench the sample rapidly enough so that after cryofixation the structure is the same as it was at the initial temperature. Many factors, however, such as ice crystal formation, temperature-induced structural transitions, changes in partial specific volume, etc., have the potential to alter the sample structure during quenching. One approach to minimize such perturbations is to pretreat the sample by chemical fixation and/or by adding cryoprotectants.[11] Such pretreatment, however, may itself alter the sample structure. Alternatively, freezing-induced perturbations may be minimized by using rapid cryofixation procedures. Copper block,[12] spray,[13] propane jet,[14] and "sandwich"[15] freezing are, presently, the most appropriate for ultrarapid cryofixation. Ultrarapid cryofixation of plasma lipoproteins may easily be achieved using the procedure shown in Fig. 1. A thin copper specimen holder is freshly washed with 10% nitric acid solution followed by several washes with distilled water and a final wash with ethanol. After drying, the upper surface of the holder is scratched with a fine emory paper to increase the sample adherence and the holder is rinsed one final time with ethanol and dried. A small drop (about 0.1 μl) of the solution is deposited on the specimen holder (Fig. 1, part 1). The drop is then squeezed between this sample holder and a thin copper plate (Fig. 1, parts 1 and 2) previously cleaned in the same way as the sample holder but without the emory-paper treatment. It is also possible to prepare the sample by squeezing the solution between two identically treated sample holders if the fracture is to be performed with a double replica device (see below). Cryofixation of the sample is obtained by rapid plunging into liquid propane, as illustrated in Fig. 1, part 3, or by projecting, at high speed, a stream of liquid propane onto the metal surfaces.[14] The quenched specimens are then stored under liquid nitrogen.

Fracture and Etching of the Cryofixed Sample

The quenched sample is mounted on a cold table (123 K) in a Balzer's (or equivalent) freeze-etching unit and the chamber is evacuated to a

---

[11] F. Franks, *J. Microsc.* **111**, 3 (1977).
[12] J. E. Heuser, T. S. Reese, M. J. Denis, Y. Jan, L. Jan, and L. Evans, *J. Cell Biol.* **81**, 275 (1979).
[13] L. Bachmann and W. W. Schmitt-Fumian, in "Freeze-Etching Techniques and Applications" (E. L. Benedetti and P. Favard, eds.), p. 73. Société Française de Microscopie Electronique, Paris, 1973.
[14] M. Mueller, N. Meister, and H. Moor, *Mikroscopie (Vienna)* **36**, 129 (1980).
[15] T. Gulik-Krzywicki and M. J. Costello, *J. Microsc.* **112**, 103 (1978).

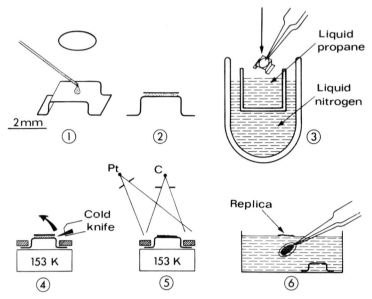

FIG. 1. Schematic representation of the freeze-fracture/etching technique. (1) A small drop of sample solution (about 0.1 µl) is placed upon a clean thin copper specimen holder. (2) The drop is squeezed between the specimen holder and a thin copper disk (hat) and then (3) plunged into a container of liquid propane cooled by liquid nitrogen after which the ensemble is stored in liquid nitrogen until (4) the specimen holder (to which the cryofixed sample solution and copper hat are still attached) is mounted upon the cold stage of the freeze-etching apparatus at 123 K (for the Balzer's freeze-etching unit). The fixed sample solution is fractured at 153 K by removing the hat with the cold knife of the apparatus. Immediately or following a period of etching, the fractured (and etched) surface is (5) shadowed with platinum (or another heavy metal) and then covered with a thin layer of carbon to increase the strength of the replica. Finally, (6) the replica is floated off the sample holder, cleaned, and deposited on an electron microscope grid for examination and photography in the electron microscope.

vacuum better than 10^{-6} Torr. The sample can be fractured at 123 K if the vacuum is better than 10^{-8} Torr, otherwise the temperature of the table should be increased (153 K for 10^{-6} Torr), in order to avoid the contamination of the cold, fracture-exposed surface. The fracturing is performed by removing the upper plate with the liquid nitrogen cooled knife, as illustrated in Fig. 1, part 4, and in Fig. 2, part 2. When the sample is frozen between two identical holders, the fracturing can be obtained also using a Balzer's (Balzer's Union BB187260T, BB172138T) or any home-made double replica device.

Parts of the sample situated below the fracture-exposed surface may be further exposed by sublimation of the ice, as illustrated in Fig. 2, part

3. This sublimation (etching) is obtained by increasing the sample temperature to values between 158 and 173 K, depending on the local vacuum conditions around the sample. In order to avoid any contamination of the fractured surface and to improve the local vacuum, the sample should be protected during the time of etching by a cold shroud, such as the liquid nitrogen-cooled knife in the Balzer's freeze-etching unit.

Replication of the Exposed Surface

The replication of the surface exposed by fracture (freeze-fracture) or by fracture and etching (freeze-etching) is performed by evaporation of heavy metals (Fig. 1, part 5 and Fig. 2, parts 4 and 5). Platinum–carbon is commonly used but a finer grain size is obtained with tungsten or tungsten–tantalum alloys.[16]

The metal deposit, which has a mean thickness of 1–2 nm, is too fragile to manipulate or observe in the electron microscope. To strengthen the replica, a 10- to 20-nm-thick layer of carbon is evaporated onto the surface. The carbon layer, at this thickness, does not contribute appreciably to the final electron microscope image. Since evaporation of the metals during the replication step requires high temperatures, heating and possible distortion of the exposed surface must be carefully avoided. This is particularly true when metals other than platinum are used. Surface heating can be minimized by working at the optimal conditions for a given source of evaporation and by using a multistep evaporation procedure allowing cooling of the partially shadowed surface between steps.[7] The best revelation of the various aspects of the sample ultrastructure frequently needs to be determined empirically by using unidirectional (Fig. 2, part 4) or circular (obtained by rotating the sample, Fig. 2, part 5) shadowing or combinations thereof at different shadowing angles.

Cleaning and Observation of the Replica

After replication, the sample is thawed and then solubilized or digested with an appropriate solvent system such as distilled water followed by chromic acid diluted one to one. After elimination of the sample, the replica is washed with distilled water, deposited on an electron microscope grid (Fig. 1, part 6), and dried prior to observation in the electron microscope. Incomplete elimination of the sample material giving a

---

[16] R. Abermann, M. M. Salpeter, and L. Bachmann, in "Principles and Techniques of Electron Microscopy" (M. A. Hayat, ed.), Vol. 2, p. 197. Van Nostrand-Reinhold, New York, 1972.

"dirty" replica can, on occasion, raise a problem. This can often be resolved by letting such replicas soak overnight in diluted chromic acid. The copper electron microscopic grid dissolves and the clean platinum–carbon replica can be washed again with distilled water and deposited on a new electron microscope grid. Finally, it must be stressed that since the replica is an object having three dimensions, stereo electron microscopy is invaluable for properly determining the sample ultrastructure. Further, for ordered samples, optical or electron diffraction may provide information about the lattice parameters and symmetry of the sample ultrastructure revealed by different fracture planes.

Interpretation of the Images

Ideally, when properly quenched, lipoproteins remain randomly distributed and randomly oriented in the cryofixed solvent as shown schematically in Fig. 2, part 1. When the cryofixed solvent is fractured, the fracture plane may pass across the surfaces of the lipoproteins leaving some lipoproteins exposed to varying degrees above the fixed solvent while others are removed leaving impressions of varying depths. In some cases, the fracture plane may pass through the particle (Fig. 2, part 2). The particles observed on the freeze-fracture replica (Fig. 2, part 6) reflect these varying degrees of exposure and orientation of the lipoproteins. The freeze-fracture particles and their shadows indicate whether the form of the object under study is globular or elongated, whether subunits are present, and, if so, their symmetry, and, in the case of fractured particles, the internal structure of the object may be apparent. A more quantitative description of the morphology may be given by analyzing histograms of the sizes of the freeze-fracture particles measured perpendicular to the direction of shadowing. For several standard proteins, LeMaire et al.[17] and Gulik et al.[7] have shown that, when unidirectional shadowing is used, such histograms can be related to the sizes and the shapes of the proteins. Although not yet applied extensively to the investigation of serum lipoproteins and apolipoproteins, a study of apoLDL in sodium deoxycholate solutions has appeared.[7]

Although lipoproteins (and other macromolecules in solution) may be studied by freeze-fracture alone, etching prior to shadowing may be very useful. The etching step (Fig. 2, part 3) further exposes the lipoproteins revealed by fracture as well as some of those which were situated immediately below the fracture surface. This may result in some particles coming to rest upon or right next to others (Fig. 2, part 3). In the case of either the

[17] M. LeMaire, J. V. Moller, and T. Gulik-Krzywicki, *Biochim. Biophys. Acta* **643**, 115 (1981).

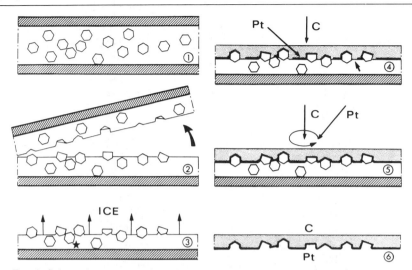

FIG. 2. Schematic representation of the freeze-fracture/etching technique at the molecular level. (1) The lipoproteins have random positions and orientations in the sample solution squeezed between the specimen holder and its hat and this disposition is maintained following cryofixation. (2) Upon removal of the hat a fracture plane is established through the fixed sample solution with part of the material being removed with the hat and the rest adhering to the specimen holder. The lipoproteins or the parts of the lipoproteins that remain behind are exposed to varying degrees above the cryofixed solvent as the fracture plane passes around or through them. (3) Etching or sublimation of the cryofixed solvent further exposes the lipoproteins above the surface of the cryofixed solvent. Some particles may come to rest upon others during this process (∗). The fractured and etched surface is shadowed with a heavy metal (platinum for example) either unidirectionally (4) giving rise to particle shadows (arrow) or circularly (5) entirely covering the lipoprotein to varying degrees. The replica (6) maintains the three-dimensional aspects of the surface of the fractured and etched sample solution. The sample has been completely removed during the cleaning procedure leaving behind only its outlines.

freeze-fracture or the freeze-etching technique, unidirectional (Fig. 2, part 4) or circular (Fig. 2, part 5) shadowing permits the preparation of very thin (about 1–2 nm) heavy metal replicas of the exposed surfaces which display high contrast (particularly in the case of unidirectional shadowing) due to the differences in the thickness of the deposited metal over the uneven surface.

Figure 3 shows freeze-etching electron microscopic images of unidirectionally shadowed HDL_3, LDL_2, and VLDL which illustrate several aspects of the freeze-etching technique previously discussed. Replicas of HDL_3 show particles that are roughly globular in form having an indented surface further evidenced by slightly irregular shadows but without any obvious subunit structure. A range of particle sizes is apparent, reflecting the varying degrees of exposure of the lipoproteins. In marked contrast,

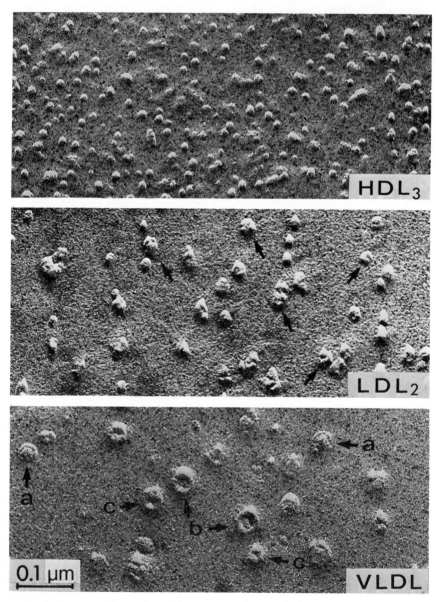

FIG. 3. Freeze-etching replicas of human serum lipoproteins in 10 mM sodium phosphate buffer pH 7.4, etched at 168 K, 10^{-6} to 10^{-7} Torr, for approximately 1 min, unidirectionally shadowed with platinum and coated with carbon. The shadowing direction is from the bottom of the page toward the top. The bar represents 0.1 μm. HDL$_3$, human serum HDL$_3$, concentration 100 μM; note the somewhat irregular particle shapes and shadows. LDL$_2$, human serum LDL$_2$, concentration 50 μM. Note the irregular outline of the shadows and the presence of globules of about 0.8 nm in size consistent with a knobby surface (arrows). VLDL, human serum very low-density lipoprotein, concentration 30 μM. Some particles appear to have been fractured with removal of a layer (a) or a large portion (b) of the particle while others appear to possess globular subunits (c).

replicas of LDL_2 look markedly knobby and appear to contain a small number of globules (approximately 8 nm in diameter) either isolated or clustered in small bunches approximately 30 nm wide. The particle shadows are markedly irregular. As for HDL, varying particle sizes, reflecting varying degrees of exposure of the lipoproteins, are present. Replicas of VLDL are different from those of HDL or LDL and are more complex. Many particles are roughly spherical in shape and some seem to have been fractured with removal of a layer or a large part of the interior of the particle. Other particles appear to possess globular subunits approximately 9 nm in diameter either isolated or clustered in small bunches. As for the other lipoproteins, there are varying particle sizes.

Freeze-fracture or freeze-etching studies employing circular shadowing (Fig. 2, part 5) may, in some cases, provide a clearer impression of the lipoprotein morphology. In the case of LDL_2, rotary shadowing shows most clearly the details of the lipoprotein morphology. Many of the LDL_2 particles show coarse 2- or 3-fold symmetry and the maximum number of globules apparent at any one time appears to be less than four (Fig. 4). This morphology is different from those of other macromolecules of similar sizes, such as the turnip yellow mosaic virus shown at the bottom of Fig. 4. The latter, in contrast to LDL_2, display a "soccer ball" appearance, reflecting the spherical shape and symmetrical subunit structure of the virus.

Finally, since the three-dimensional nature of the replicated surface is retained by the replica itself, three-dimensional reconstruction of the lipoprotein from the electron microscope images can be envisioned.

Although the freeze-fracture/etching technique may provide detailed information on lipoprotein morphology, the chemical nature of the various components observed on the replica cannot be determined by the freeze-fracture/etching technique alone. A detailed structural interpretation can be made only in the context of complementary information provided by other techniques. As an example we may cite the study of LDL_2 structure using solution X-ray scattering[18] and freeze-etching electron microscopic[10] techniques. Based upon these data, a model of the low-density lipoprotein has been proposed[18] which details the positions of the various chemical constituents within the particle.

Origin of Artifacts

The structural information that may be obtained by electron microscopy may be limited, in addition to the usual theoretical factors, by a

---

[18] V. Luzzati, A. Tardieu, and L. P. Aggerbeck, *J. Mol. Biol.* **131**, 435 (1979).

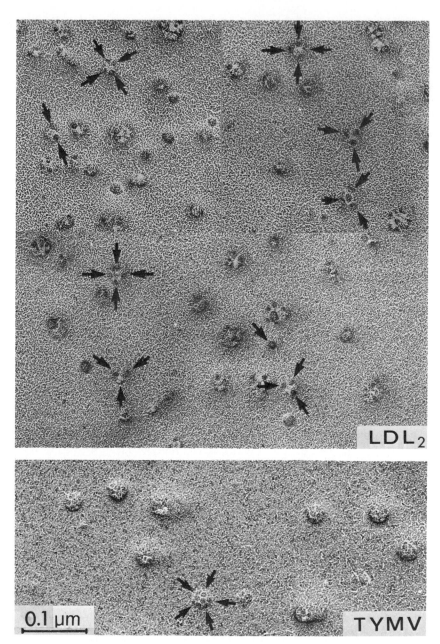

FIG. 4. Montage of freeze-etching replicas of human LDL$_2$ and turnip yellow mosaic virus (TYMV), prepared as in Fig. 3 except rotary shadowed and coated with carbon. The bar represents 0.1 μm. The LDL$_2$ particles are knobby and look like clusters of a small number of globules. Some of the particles display coarse 2- and 3-fold symmetry. The TYMV particles, in contrast, have a soccer ball appearance reflecting the spherical shape of the particles. There is a regular distribution of subunits over the surface with the presence of 2-, 3-, and 5-fold symmetry [see also T. Gulik-Krzywicki, M. Yatès, and L. P. Aggerbeck, *J. Mol. Biol.* **131**, 475 (1979)].

variety of practical aspects related to the particular technique employed. In the case of freeze-fracture and etching electron microscopy these practical aspects concern the sample preparation, cryofixation, etching, and replication steps.

Sample homogeneity is of crucial importance for obtaining structural information. The particles on the freeze-fracture replica represent portions of the macromolecule exposed by the fracture above the surface of the cryofixed solvent. Quantitative analysis of the distributions of these particle sizes, which reflect the varying degrees of exposure and orientation of the macromolecule, provides information on its morphology. Heterogeneity or the presence of aggregates distorts the distribution of the particle sizes and makes quantitative analysis and deduction of morphological parameters very complicated if not impossible.

The cryofixation step may alter the sample structure in a variety of ways. Less than optimal quenching rates result in crystallization of the solvent which leads to aggregation of the solute and possibly to morphological damage. Even when obvious aggregation of the solute does not occur, small ice crystals (on the order of a few tens of nanometers) may alter the sample structure. Independent of these changes due to solvent crystallization, there may be changes in the sample itself resulting from the change in temperature and associated changes in the partial specific volumes of the sample components or due to sample thermal structural transitions. These effects are of particular importance in the case of LDL where a thermal structural transition involves the surface as well as the core components.[18] In such cases, an independent assessment of the structural state of the lipoprotein after cryofixation may be necessary (low temperature X-ray or neutron scattering may be the best techniques for this).

The effect of cryofixation upon the sample solution may be determined quantitatively by examining the sample before and after fixation by X-ray diffraction techniques. The experimental procedure has been described in detail by Gulik-Krzywicki and Costello[15] and has been used by Aggerbeck and Gulik-Krzywicki[19] to evaluate the effect of cryofixation upon human serum low-density lipoprotein. "Slow cryofixation" of a solution of LDL (50–100 μm thick samples on gold planchettes are plunged into liquid freon) results in large domains of ice separated by domains of aggregated particles (Fig. 5C). The domains of aggregated particles may appear heterogeneous with globular structures ranging in size from about 8 to 30 nm but individual LDL particles are not readily identified. Slowly cryofixed LDL solutions also exhibit marked changes in their low temperature X-

[19] L. P. Aggerbeck and T. Gulik-Krzywicki, *J. Microsc.* **126**, 243 (1982).

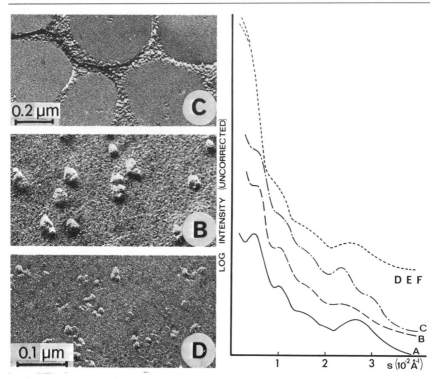

FIG. 5. Freeze-etching electron micrographs of human serum LDL_2 solutions (in 150 mM NaCl, pH 7.4) slowly cryofixed, rapidly cryofixed, and cryofixed in the presence of glycerol. (C) LDL_2 solution slowly cryofixed by plunging into liquid freon 22. (B) LDL_2 solution cryofixed by spray freezing, 2 min etching at 168 K, 10^{-6} to 10^{-7} Torr. (D) LDL_2 solution plus 75% glycerol cryofixed in liquid freon 22, fracture only and corresponding low temperature (153 K) X-ray scattering curves: (C,B,D) as above; (A) control LDL_2 solution at 294 K; (E) control LDL_2 solution plus 75% glycerol at 294 K; (F) control LDL_2 solution (3.3 M NaBr) at 294 K. $s = 2 \sin \theta/\lambda$ (2θ is the scattering angle, λ is the wavelength in Å [see V. Luzzati, A. Tardieu, and L. P. Aggerbeck, *J. Mol. Biol.* **131,** 435 (1979) for details].

ray scattering curves (153 K) as compared to the control LDL curve (Fig. 5, compare curve C with curve A).

Freeze-fracture electron micrographs of rapidly cryofixed (spray frozen or sandwich technique) and fractured LDL show particles that look markedly knobby and appear to contain small globules of 8 to 10 nm in diameter either isolated or clustered in small bunches having a total width of about 30 nm (Fig. 5B). The low temperature X-ray scattering curve of the rapidly fixed LDL resembles much more closely the control LDL curve although there is a marked increase in scattered intensity throughout the angular range and less marked maxima and minima (Fig. 5, curve

B). In contrast to these results, the scattering curve of LDL cryofixed in 75% glycerol (Fig. 5, curve D) is identical to the control LDL curve in 75% glycerol obtained at 294 K (Fig. 5, curve E) and to a control LDL curve at 294 K in the presence of 3.3 M NaBr, a solvent of identical electron density as 75% glycerol (Fig. 5, curve F). The freeze-fracture electron micrograph resembled that of the rapidly fixed LDL appearing markedly knobby (Fig. 5D). These results illustrate several points regarding the cryofixation procedure. First, in the presence of 75% glycerol, cryofixation results in minimal, if any, structural alteration of, at least, the LDL lipid moiety as assessed by X-ray scattering. Further, it shows that, in this particular case, glycerol, as a cryoprotectant, is innocuous and, indeed, that it may be important for adequate preservation of LDL under certain conditions. In the case of rapidly cryofixed LDL samples, the alterations in the low temperature X-ray scattering curves may well be correlated, in part, with the large scattered intensity of the cryofixed solvent indicating the presence of electron density fluctuations in the range of dimensions typical of LDL probably due to the quenching process (small ice crystals). Finally, for the slowly cryofixed preparations, in addition to the effects of large ice crystal formation, there is a marked change in the particle morphology in the freeze-etching electron micrographs and marked alteration in the low temperature X-ray scattering curve.

Correlation of the X-ray scattering data with the electron microscopic images must be made with the awareness that the X-ray data are obtained using cryofixed material that has not undergone fracturing, etching, or shadowing. Each of these subsequent steps can introduce artifacts, as well. Plastic deformation may distort the sample structure during the fracturing step.[20] This type of distortion may be reduced by fracturing at lower temperatures. Deposition of residual water vapor may contaminate the newly exposed fractured surface giving rise to small spherical particles or crystalline appearing material. This type of contamination may be reduced by operating under a higher, cleaner vacuum.

Sublimation of the ice during the etching step partially dehydrates the sample and, particularly for low-density lipoproteins, may markedly affect the particle morphology. Immediately following fracture and up to 4 min of etching at 10^{-6} to 10^{-7} Torr at 168 K the LDL particles appear markedly knobby as previously described (Figs. 3 and 6). However, with increasing periods of etching up to 30 min, the particles become increasingly spherical. Further, there is a rather homogeneous size distribution for the majority of the particles. This is due to the fact that most of the

[20] V. B. Sleytr and A. W. Robards, *J. Microsc.* **110**, 1 (1972).

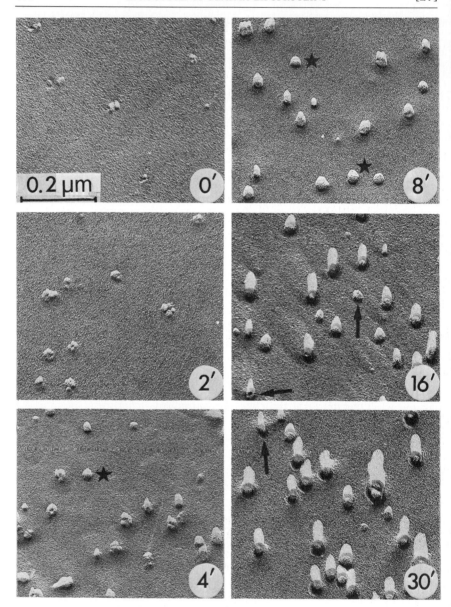

FIG. 6. Freeze-etching electron micrographs of human serum LDL_2 solutions in 10 mM sodium phosphate, pH 7.4, etched for different periods of time at 168 K, 10^{-6} to 10^{-7} Torr, unidirectionally shadowed with platinum and coated with carbon. The shadowing is from the bottom of the page toward the top. The bar represents 0.2 μm. (0′) fracture only, no etching; (2′) 2 min etching; (4′) 4 min etching; (8′) 8 min etching; (16′) 16 min etching; (30′) 30 min

lipoproteins have been completely exposed during the etching process. The few particles of varying smaller sizes presumably represent lipoproteins which have become only partially exposed, to varying degrees, during the etching process. The morphological fragility of the LDL during the etching process is of note and its origin is under investigation. Although this effect has been noted only with low-density lipoproteins while several other proteins and viruses do not display similar behavior, the possibility of such etching effects needs to be evaluated for each lipoprotein studied. Finally, the replication step may introduce artifacts if care is not taken to avoid excessive heating of the surface during heavy metal deposition or if too little or too much metal is applied.

Other Methods Employing Cryofixation

As mentioned in the introduction, two other methods permit the examination of cryofixed material, freeze-drying and low temperature electron microscopy of frozen hydrated samples. Freeze-drying has been extensively used for the study of biological macromolecules in solution. The technique has been well described[21] but applied rarely to the study of serum lipoproteins.[22] Low temperature electron microscopy of frozen hydrated biological macromolecules has developed more recently[23] and presents several advantages. In particular, the sample solution is not fractured and the sample is observed in its hydrated state. Preliminary studies suggest that this technique may be very useful for the study of serum lipoproteins.

[21] M. V. Nermut, in "Freeze-Etching Techniques and Applications" (E. L. Benedetti and P. Favard, eds.), p. 135. Société Française de Microscopie Electronique, Paris, 1973.
[22] Freeze-drying images which resemble closely deeply etched freeze-fracture images (Fig. 6, 15 to 30 min etching) have been obtained in the case of human serum LDL_2. (Jacques Escaig, personal communication, 1982.)
[23] M. Adrian, J. Dubochet, J. Lepault, and A. W. McDowall, *Nature (London)* **308**, 32 (1984).

---

etching. Note (1) The density of the particles on the replica increases with time as the etching process exposes more and more of the lipoproteins. (2) The particles have varying sizes, exhibit the presence of globules of about 8 nm in size and have very irregular shadow outlines consistent with a knobby surface at 0, 2, and 4 min of etching. (3) With increasing periods of etching (4', 8', 16') the proportion of particles exhibiting the knobby surface progressively decreases with the appearance of more spherical appearing particles (*). (4) At long etching times (16', 30') the particles are primarily spherical in morphology with a rather homogeneous size distribution with only a few knobby particles apparent (arrows).

Acknowledgments

The authors would like to thank Dr. V. Luzzati for useful discussions and J. C. Dedieu for excellent technical assistance and artwork. This work was supported, in part, by grants from the Délégation Générale à la Recherche Scientifique et Technique (Contract Number 80.70386), Ministère de la Recherche et Industrie (Contract Number 82.L.0752), the Institut National de la Santé et Recherche Mèdicale (Contract Number 817001), and the National Institutes of Health (Contract Number HL-18577).

[28] Nuclear Magnetic Resonance Studies of Lipoproteins

By JAMES A. HAMILTON and JOEL D. MORRISETT

Over the past 30 years nuclear magnetic resonance (NMR) spectroscopy has evolved into a highly sophisticated, extremely versatile methodology. In its formative stage it was of primary interest to physicists, but as the technique matured it became an important tool of the chemist. Now it is used extensively in biology and medicine. NMR has had reasonably extensive application to lipoproteins, and we have used examples from published studies in conjunction with basic principles to illustrate appropriate methodology. However, even with its limitation to high-resolution NMR techniques as applied to plasma lipoproteins and related systems, this chapter is intended to be representative rather than comprehensive.

NMR spectroscopy is a collection of techniques based on magnetic properties of nuclei. Certain nuclei have a magnetic moment (μ) and a nuclear spin (I) which are related by the proportionality $\mu = \gamma \hbar I$, where γ is the gyromagnetic ratio. Nuclear resonance is observed when a nucleus with $I \neq 0$ is placed in a strong, uniform static magnetic field, producing different nuclear energy states, and rf energy of the appropriate frequency is applied to induce transitions between energy states. In the earliest NMR spectrometers, the sample was placed in a permanent magnet or electromagnet and the radiofrequency energy was varied continuously through a small frequency range to produce an absorption spectrum. This type of spectroscopy (continuous wave or CW) has largely been supplanted by pulsed Fourier transform (FT) NMR spectroscopy, a newer methodology which is dependent on more highly sophisticated electronics and computers. In pulsed FT NMR, a high-power rf pulse is applied for a very short time (μsec) to excite all the nuclear spins simultaneously. Following the rf pulse, the decaying amplitudes of the magnetization vector are measured as a function of time, as the excited nuclei return to equilibrium via various relaxation processes. The time domain data (the free induction decay, or FID) are converted by a computer-calculated Fourier transformation into the more conventional form of an amplitude vs frequency plot, or spectrum. Although FT NMR is a more complex and costly approach than CW NMR, it offers numerous advantages, including improved sensitivity and the capability of making a variety of NMR measurements that are difficult, if not impossible, with CW NMR. Most FT NMR instruments are now equipped with superconducting magnets which produce magnetic fields significantly higher than those obtainable with iron-core magnets and generally afford increased sensitivity and resolution. Except as noted, all methodological discussions herein pertain to pulsed FT NMR spectroscopy.

In spite of the large number of nuclei which have magnetic moments and the subtle as well as gross differences in their NMR properties, there are several basic NMR measurements which are common to these nuclei. The most basic is *chemical shift,* which is the absorption frequency of a particular nucleus normalized to the magnetic field strength and referenced to some standard. Different nuclei of the same magnetic isotope may exhibit peaks or "resonances" at different chemical shifts, depending on their local magnetic environment. The chemical shift of a particular nucleus is the combined effect of *intrinsic factors* such as chemical bonding and conformation and *extrinsic factors* such as hydrogen bonding and local ionic environment. A second general NMR measurement is that of *peak intensity.* Although in some cases peak height may be a reasonable or necessary approximation, the integrated area under the peak is the preferred measure of peak intensity. Under appropriate experimental conditions (see below), the peak area is directly proportional to the number of nuclei absorbing the applied rf energy.

Two other quantities commonly measured in NMR spectra, *linewidth* and *spin-lattice relaxation time,* can be related to the motion(s) of a particular nucleus. The linewidth ($\nu_{1/2}$) is defined as the width of a resonance (line) at one-half its maximum height. For a Lorentzian line (the type generally found in lipoprotein spectra) which is well resolved, $\nu_{1/2}$ is inversely proportional to T_2, the spin-spin relaxation time: $\nu_{1/2} \leq 1/\pi T_2$. T_2 can be quantitatively related to molecular motions, and is particularly sensitive to anisotropic motions and the slower components of complex motions. The measured $\nu_{1/2}$ may also include significant contributions from such sources as instrumental field inhomogeneity and chemical exchange; it therefore provides only a lower-limit estimation of T_2. The spin-lattice relaxation time (T_1) cannot be measured directly from a single time domain or frequency domain spectrum, but rather is measured from a series of spectra. It is the time constant for the equilibration of an excited nuclear spin with its surroundings ("lattice") and thus is of fundamental importance in the FT NMR experiment, in which time-domain spectra are accumulated in regular short intervals and summed in the computer in order to provide adequate signal-to-noise ratios (S/N). The T_1 of each resonance must be known to determine the total time interval between pulses (pulse interval or recycle time) which is appropriate for obtaining the greatest S/N per unit time (1.3 times T_1) or for measuring equilibrium peak intensities,[1] which reach 95% of the equilibrium value at 3 times T_1 and 99.3% at 5 times T_1. T_1 can be quantitatively related to molecular motions, particularly isotropic motions and the faster components of complex motions.

For a detailed development of basic NMR theory, the reader is referred to definitive basic texts by Carrington and McLachlan,[2] Farrar and Becker,[3] Jardetzky and Roberts,[4] and Shaw.[5] Practical aspects of FT NMR in general are discussed in some of the above[3-5] as well as in Becker et al.[1] We have focused on basic NMR measurements, experimental designs, practical aspects, and interpretational caveats of NMR spectroscopy of lipoproteins and closely related systems. In some cases, results of NMR studies have been discussed within the framework of current knowledge of lipoprotein structure but without extensive cross-reference to other studies. A previous review of NMR spectroscopy of lipoproteins provides a detailed correlation of NMR results with results of other physical measurements.[6]

Several nuclei in lipoproteins are magnetically active and are thus potential NMR probes of lipoprotein structure. Table I lists the magnetic isotopes present in the covalent structures of the molecular constituents of lipoproteins: lipids, proteins, and carbohydrates. Every type of nucleus that is part of the endogenous structure of these molecules has at least one magnetic isotope. Each magnetic nucleus represents an intrinsic and completely nonperturbing probe (when at the natural abundance level) of local molecular motion and magnetic environment. The NMR experiment itself is also nonperturbing and nondestructive. Table I also lists for each nucleus its nuclear spin, its natural isotopic abundance, its sensitivity, and its resonance frequency at two commonly employed magnetic fields, one

[1] E. D. Becker, J. A. Ferretti, and P. N. Gambhir, *Anal. Chem.* **51**, 1413 (1979).
[2] A. Carrington and A. D. McLachlan, "Introduction to Magnetic Resonance." Harper, New York, 1967.
[3] T. C. Farrar and E. D. Becker, "Pulse and Fourier Transform NMR." Academic Press, New York, 1971.
[4] O. Jardetzky and G. C. K. Roberts, "NMR in Molecular Biology," Academic Press, New York, 1981.
[5] D. Shaw, "Fourier Transform NMR Spectroscopy." Elsevier, New York, 1976.
[6] P. Keim, in "The Biochemistry of Atherosclerosis" (A. M. Scanu, R. W. Wissler, and G. S. Getz, eds.), p. 9. Dekker, New York, 1979.

TABLE I
PROPERTIES OF MAGNETICALLY ACTIVE NUCLEI IN LIPOPROTEINS[a]

Nucleus	Nuclear spin[b]	Natural isotopic abundance	Sensitivity[c]	Resonance frequency (MHz)	
				21.14 kG[d]	47.0 kG[d]
^{1}H	1/2	99.98	1.00	90.0	200.0
^{2}H	1	0.0156	9.65×10^{-3}	13.82	30.7
^{13}C	1/2	1.108	1.59×10^{-2}	22.63	50.3
^{14}N	1	99.64	1.01×10^{-3}	6.50	14.4
^{15}N	1/2	0.365	1.04×10^{-3}	9.12	20.3
^{17}O	5/2	0.037	2.9×10^{-2}	12.20	27.1
^{31}P	1/2	100	6.63×10^{-2}	36.44	81.0
^{33}S	3/2	0.74	2.26×10^{-3}	6.90	15.3

[a] Taken from Bruker Instruments NMR Periodic Table; Magnetic nuclei which may be present in ionic form in aqueous lipoprotein samples include ^{23}Na, ^{39}K, ^{41}K, ^{43}Ca, ^{35}Cl, and ^{37}Cl.
[b] Units of $h/2\pi$.
[c] Relative to sensitivity of ^{1}H as 1.0, with equal numbers of nuclei and at constant field.
[d] Magnetic field in kilogauss (kG).

in the low field range (21.14 kG or 2.11 Tesla) and the other in the high field range (47.0 kG or 4.70 Tesla). Of the nuclei listed in Table I, ^{1}H, ^{13}C, and ^{31}P have been the primary ones studied in lipoproteins. The general advantages and disadvantages afforded by these and other nuclei as probes of lipoprotein structure will be discussed. ^{13}C NMR spectroscopy, the method which has had the most extensive application (and probably has the greatest future potential) to lipoproteins, will be treated in greatest detail, but many of the principles described apply to other nuclei as well.

^{31}P NMR

^{31}P is a spin $\frac{1}{2}$ nucleus with intermediate intrinsic sensitivity and with the highest possible natural abundance (Table I) which together make it a nucleus of good overall sensitivity (detectability). However, the chemical abundance of ^{31}P in lipoproteins is low since it is present only in the phosphate groups of phospholipids. In all classes of human plasma lipoproteins, phosphatidylcholine (PC) is by far the most abundant phospholipid and sphingomyelin (SM) the next most abundant.[7] The limited number of different chemical environments for phosphorus in lipoprotein constit-

[7] A. M. Scanu, *Ciba Founda. Symp.* **12**, 223 (1973).

uents is advantageous in that it affords spectral simplicity, but is disadvantageous in that it limits the scope of ^{31}P NMR spectroscopy of lipoproteins.

Published applications of ^{31}P NMR to lipoproteins illustrate several basic NMR experimental approaches, including spectral assignments, intensity measurements, and the use of shift reagents. The ^{31}P spectra of lipoproteins are quite simple, as exemplified[8] by those of HDL$_3$ in Fig. 1B–E. Assignment of the major resonances was made by comparison with a simple model system (PC and SM in H$_2$O, Fig. 1A). The large chemical shift range of the ^{31}P nucleus together with the small $\nu_{1/2}$'s of the resonances in the HDL$_3$ spectrum allowed clear resolution of a resonance for PC for SM (Fig. 1A) and other phospholipids, as well as resolution of a lyso PC resonance in aged HDL$_3$ (Fig. 1C). Figure 1 is also informative because it illustrates spectra obtained by the CW method (A,B,D,E) and the pulsed FT method (C). Spectrum C has a major resonance with the same $\nu_{1/2}$ as that in spectrum B and S/N which appears to be comparable to or better than that for spectrum B. However, the CW spectrum required "2–4 days" of signal averaging, whereas the FT spectrum required only $\frac{1}{2}$ day. Significantly, sweep time per CW spectrum was 32–64 sec, while the pulse interval for the FT spectrum was 4 sec. Subsequent ^{31}P FT NMR studies of lipoproteins (e.g., Brainard et al.[9]) showed that spectra with good S/N can be obtained in 30 min or less using a total volume of 2.0–2.5 ml at 10–20 mg lipoprotein/ml in 10-mm sample tubes. In the early 1970s the transition from CW to pulsed FT NMR proceeded rapidly, in part because of the great time savings afforded by the latter.

A more rigorous approach for assignment of ^{31}P resonances in lipoprotein spectra determined ^{31}P chemical shifts for aqueous dispersions of each phospholipid present in lipoproteins and for recombinant particles of apoHDL and PC or SM.[10] Only two ^{31}P resonances were detected in spectra of HDL and LDL; these corresponded to PC and SM, as in Fig. 1. A single, relatively broad resonance was observed for VLDL at the chemical shift for PC. The identity of PC and SM chemical shifts in aqueous (sonicated) dispersions and in lipid–protein recombinants aided the assignment of lipoprotein spectra and also showed that the PC and SM phosphorus magnetic environments were not detectably perturbed by the binding of apolipoprotein. This observation does not necessarily indicate

---

[8] T. Glonek, T. O. Henderson, A. W. Kruski, and A. M. Scanu, *Biochim. Biophys. Acta* **348**, 155 (1974).

[9] J. R. Brainard, E. H. Cordes, A. M. Gotto, Jr., J. R. Patsch, and J. D. Morrisett, *Biochemistry* **19**, 4273 (1980).

[10] G. Assmann, E. A. Sokoloski, and H. B. Brewer, Jr., *Proc. Natl. Acad. Sci. U.S.A.* **71**, 549 (1974).

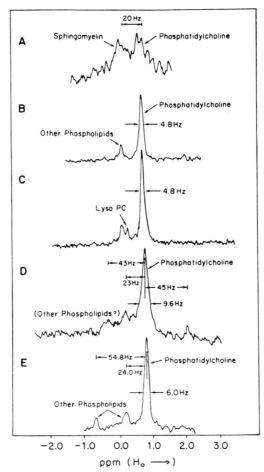

FIG. 1. ^{31}P NMR spectra (40.5 MHz) from a phospholipid dispersion (A) and from samples of human HDL$_3$ (~15 mg/ml) in the presence of various metal ions (B–E): (A) spectrum from a phospholipid dispersion of phosphatidylcholine and sphingomyelin (1 : 1, w/w) with Na$^+$; (B) HDL$_3$ with Na$^+$; (C) aged HDL$_3$ with Na$^+$; (D) HDL$_3$ with the tetramethylammonium countercation; (E) HDL$_3$ with Na$^+$ to which MgCl$_2$ has been added (ratio of Mg^{2+} to phosphorus is 1 : 5). Spectra A, B, D, and E are signal-averaged continuous-wave recordings taken during broad-band ^{1}H decoupling; 2–4 days of signal averaging were required using a sweep width of 120 or 240 Hz and sweep rate of 3.7 Hz/sec. Spectrum C is a Fourier-transform spectrum taken during broad-band ^{1}H decoupling after 12 hr signal averaging using a pulse width of 4.5 μsec, a delay time of 238 μsec, a sweep width 2100 Hz (238 μsec per data point, 16,384 data points), and a recycle time of 4 sec. All spectra are displayed with the same horizontal scale. From T. Glonek, T. O. Henderson, A. W. Kruski, and A. M. Scanu, *Biochim. Biophys. Acta* **348**, 155 (1974).

a lack of strong interactions between the phosphate region and protein groups, but possibly an insensitivity of the ^{31}P chemical shift to these interactions.[10]

The above study and others[8,11,12] have attempted to correlate *ratios* of ^{31}P peak intensities with their chemical abundance in lipoproteins. In pulsed FT NMR experiments, obtaining peak intensities that accurately represent the number of nuclei contributing to the resonance can be fraught with difficulties. First, the T_1 of each resonance must be known. If "flip angles" (pulses) of <90° are used[1] or if pulse intervals which do not give equilibrium intensities are employed, then the appropriate corrections to the measured intensities must be calculated. In practice, it is best to use a 90° flip angle and a pulse interval ≥ 5 times T_1 to obtain the equilibrium peak intensity.[1] Low S/N of peaks in lipoprotein spectra may place a severe limitation on the accuracy of peak area measurements, as may the presence of partially overlapping resonances. Another complicating factor in peak intensity measurements in ^{31}P NMR spectra is the nuclear Overhauser effect. ^{31}P NMR spectra are routinely acquired with ^{1}H irradiation to remove ^{1}H–^{31}P couplings and collapse each ^{31}P multiplet into a singlet. However, the populations of the ground and excited nuclear states are perturbed by this procedure, such that the true peak intensities are increased, an effect called nuclear Overhauser enhancement (NOE).[4] Thus, to relate ^{31}P intensities to the number of contributing nuclei, the NOE must either be known, or spectra must be acquired under gated decoupling conditions which suppress the NOE.[13,14] Measurements of NOE values may involve rather large errors unless S/N is very good, because the NOE is measured as a ratio of intensities of the peak in two different spectra acquired under different conditions (with broad-band decoupling and gated decoupling).[15]

Some of the difficulties described above are illustrated by published studies. One ^{31}P NMR study[10] reported that the peak height ratio of the SM and PC resonances in spectra of several HDL and LDL samples corresponded closely (within ~20%) to the chemical ratio of the two species. However, the reported pulse interval used to acquire spectra for quantitating peak intensities (1.3 sec) was short compared to the T_1 values of the two peaks (1.7 and 2.3 sec). Thus, equilibrium intensities were not obtained, and the peak with the longer T_1 would be attenuated more than

---

[11] T. O. Henderson, A. W. Kruski, L. G. Davis, T. Glonek, and A. M. Scanu, *Biochemistry* **14**, 1915 (1975).

[12] E. B. Brasure, T. O. Henderson, T. Glonek, N. M. Pattniak, and A. M. Scanu, *Biochemistry* **19**, 3934 (1978).

[13] S. J. Opella, D. J. Nelson, and O. Jardetzky, *J. Chem. Phys.* **64**, 2533 (1976).

[14] P. L. Yeagle, R. G. Langdon, and R. B. Martin, *Biochemistry* **16**, 3487 (1977).

the peak with the shorter T_1. In addition, the possible effect of differential NOEs on peak intensities was not considered in this study[10] or in others.[8,11] The NOEs for the SM and PC resonances of LDL have been measured and are different, although barely outside of the estimated experimental error.[14] Interestingly, the T_1 values of the two resonances measured in LDL[14] were about the same (1.6 and 1.7 sec). Thus, it is not clear how accurately the ^{31}P NMR peak area ratios reflect the chemical ratios in lipoproteins.

In addition to quantitation of *relative* intensities of different lipoprotein ^{31}P resonances, intensities have also been quantitated in an *absolute* sense; i.e., the percentage of the ^{31}P nuclei in the NMR sample that contributes to its NMR spectrum.[11,14] In this case the above guidelines for intensity measurements must be extended to a suitable reference peak (e.g., inorganic phosphate). To obtain the fraction of lipoprotein ^{31}P contributing to the spectrum, the lipoprotein/reference ^{31}P peak area ratio is compared to the lipoprotein phosphorus/reference phosphorus chemical ratio. One study[11] determined that all ($\pm 2.5\%$) of the phosphorus in HDL and LDL contributed to the ^{31}P spectrum; however, the spectra were not obtained under conditions of NOE suppression and the effects of NOE were not considered. Using gated decoupling to eliminate NOEs, another study found that all the phosphorus of HDL but only ~80% of the phosphorus of LDL contributed to the ^{31}P spectra.[14]

The usefulness of NMR spectroscopy for monitoring lipoprotein metabolism was first demonstrated in a ^{31}P NMR study of the hydrolysis *in vitro* of HDL phospholipids by phospholipase A_2.[12] Because of the long time course of the reaction (~12 hr) and the relatively good sensitivity of the ^{31}P nucleus, spectra could be accumulated for 30 min intervals to provide information at "points" along the course of the reaction. From intensities of PC and lyso PC peaks, the average amount of the major reactant (PC) and one of major products (lyso PC) in each time interval was determined; this quantitation agreed closely with the chemical analysis.[12] More important, this type of study can provide information about the structural organization of products and reactants during the reaction by monitoring other spectral properties such as chemical shift and $\nu_{1/2}$.[12]

^{31}P NMR studies of lipoproteins also illustrate the use of NMR shift reagents to assess features of structural organization. Figure 2 shows the effect of the lanthanide ion Pr^{3+} on the ^{31}P NMR spectrum of lipoprotein X (LP-X).[9] The single resonance in the spectrum of native LP-X (Fig. 2A) is split into two resonances after addition of small amounts of Pr^{3+}. For example, 1.6 mM Pr^{3+} results in a shifted and an unshifted resonance (Fig. 2B) and 2.2 mM Pr^{3+} (Fig. 2C) gives the complete resolution of the two peaks needed for accurate quantitation of peak intensities. Additional

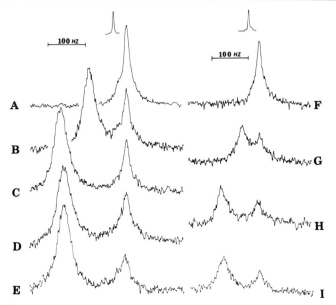

FIG. 2. ^{31}P NMR spectra (40.5 MHz) of native and model lipoprotein-X (LP-X) in the presence of increasing concentrations of Pr^{3+}. Left (A–E): native LP-X_1 at 27.3 mg in 2.5 ml of buffer was titrated with 17.8 mM Pr(NO$_3$)$_3$ at 34°. Conditions: (A) 0 mM Pr^{3+}; (B) 1.6 mM Pr^{3+}; (C) 2.2 mM Pr^{3+}; (D and E) 2.2 mM Pr^{3+}, recorded after 4 days at 40° (D) and after an additional 1 hr at 55° (E). An external reference signal of 85% H$_3$PO$_4$ is shown at the top of the figure. Spectra are from 1024 scans obtained at a 1.21 sec recycle time. Right (F–I): model LP-X consisting of phosphatidylcholine, cholesterol, human serum albumin, and apolipoprotein C. Pr^{3+} concentrations were (F) 0, (G) 0.63, (H) 0.95, and (I) 0.95 mM after an additional hour at 55°. From J. R. Brainard, E. H. Cordes, A. M. Gotto, Jr., J. R. Patsch, and J. D. Morrisett, *Biochemistry* **19**, 4273 (1980). © 1980 American Chemical Society.

spectra (Fig. 2D and E) show the stability of the LP-X in the presence of Pr^{3+}. The NMR effects result from binding of the paramagnetic ion Pr^{3+} to the phosphate group of LP-X phospholipids and the consequent alteration of the local magnetic field of the ^{31}P nuclei.[15] Because of rapid exchange of the ion among different phosphates, the molar ratio of shift reagent to phospholipid can be ≪1. However, a significant portion of the ^{31}P nuclei of LP-X were unaffected by the shift reagent.[9] By analogy with model phospholipid vesicle systems (Fig. 2F–I), the unaffected ^{31}P nuclei may be located on the inner leaflet of an impermeable spherical phospholipid bilayer and hence be inaccessible to the externally added ions.[9]

Depending on the particular experimental requirement or goal, spe-

---

[15] W. C. Hutton, P. L. Yeagle, and R. B. Martin, *Chem. Phys. Lipids* **19**, 255 (1977).

cific paramagnetic ions may be chosen to shift resonances upfield or downfield, to broaden resonances, or to alter the spin lattice relaxation.[16] These effects are transmitted beyond the ^{31}P nucleus of the surface-located phospholipids to other nuclei (notably, ^{1}H and ^{13}C) on groups proximal to the phosphate, such as the phospholipid choline methyl and methylenes, glyceryl backbone, and carbonyl groups. Hence, these reagents affect ^{1}H and ^{13}C resonances in lipoprotein spectra (see below).

The effects of several paramagnetic reagents on the ^{31}P resonances in spectra of HDL and LDL have been studied.[11] Simple hydrated transition metal ions caused complete broadening beyond the point of detection ("quenching") of all ^{31}P resonances. Low levels of hydrated lanthanide ions led to sample precipitation, and previous results[10] suggesting that all ^{31}P nuclei in HDL spectra were accessible to Eu^{3+} could not be duplicated.[11] Of the reagents tested, only Mn^{2+} in the presence of a molar excess of EDTA showed selective effects, quenching 50% of the intensity of LDL ^{31}P resonances and 80% of the intensity of HDL ^{31}P resonances.[11] A later ^{31}P NMR study of LDL using Pr^{3+} (a combination not reported by Henderson et al.[11]) found that the ^{31}P resonances could be completely shifted at a molar ratio of Pr^{3+} to phospholipid of 0.2.[17] Differences between the results on LDL[10,17] may be rationalized by postulating that the MnEDTA complex is too large to perturb the environments of all phosphates[17] or that the Pr^{3+} perturbs the native environment to a greater extent than the MnEDTA complex.

Thus, the use of shift reagents with lipids and lipoproteins and the interpretation of resultant NMR changes are not necessarily straightforward. The paramagnetic reagent is an additional component in an already complex system, and one which may be structurally perturbing. In addition to the desired spectral perturbation, these reagents may, as noted above, cause gross alteration (disruption or precipitation) of the sample. Localized structural changes, such as conformation changes, may also occur.[18] Reagents which induce a change in chemical shift may also cause broadening of a resonance, a change in the T_1 value, and/or a change in the NOE.[16,19] The latter two effects can complicate intensity measurements. It is always advisable, when possible, to use independent methods to aid in the interpretation of results obtained with NMR paramagnetic reagents.

[16] G. Royden, A. Hunt, and L. R. H. Tipping, *J. Inorg. Biochem.* **12**, 17 (1980).
[17] P. L. Yeagle, R. B. Martin, L. Pottenger, and R. G. Langdon, *Biochemistry* **17**, 2707 (1978).
[18] H. Hauser, W. Guyer, B. Levine, P. Skrabal, and R. J. P. Williams, *Biochim. Biophys. Acta* **503**, 450 (1978).
[19] B. C. Mayo, *Chem. Soc. Rev.* **2**, 49 (1973).

¹H NMR

In contrast to the ^{31}P nucleus, the ^{1}H nucleus is present in all lipids and proteins of lipoproteins and in nearly all moieties of a particular molecule. Thus, ^{1}H NMR spectroscopy can, in principle, be used to probe structural features and motions of many different molecular groups. In practice, however, the large amount of information often contained in the ^{1}H NMR spectrum is of limited accessibility. The relatively small chemical shift range of protons (~10 ppm) and the presence of spin–spin couplings between nearby protons, which result in peak multiplicity[20] and broadening, diminish resolution in complex systems such as lipoproteins. However, because ^{1}H–^{1}H couplings are nearly ubiquitous and are difficult to remove (except selectively), ^{1}H NMR spectra are routinely acquired without decoupling, with the result that NOEs are not present and peak area quantitation is simplified. ^{1}H has the highest intrinsic sensitivity of any nucleus (e.g., Table I) and a natural abundance of nearly 100%. These factors together with its high chemical abundance make it the easiest nucleus to detect in lipoproteins and related systems.

The first published NMR spectra of lipoproteins were low field ^{1}H NMR spectra of LDL and HDL shown in Fig. 3B and C.[21] These spectra appear rather simple, since they consist of only six resolved peaks. Several additional peaks are seen at higher fields, as exemplified by spectra of human HDL at 200 and 400 MHz (Fig. 4), but the difference in resolution between these two spectra is modest. Both contain very few resolved resonances considering the great number of chemically different protons. It is essential to determine whether only certain groups with sufficiently rapid molecular motions contribute to the spectrum or whether there is still inadequate resolution of many different ^{1}H resonances. This question can be addressed by measuring absolute peak intensities. For both LDL and HDL, the total intensity of the aliphatic spectral region (measured against an external reference) was the same in the native spectrum and the spectrum of the sonicated total lipid extract; the relative peak intensities were also the same (Fig. 3A and B).[21] Thus, most molecular groups (at least with respect to the lipid components) contribute to the spectrum. Absolute intensity measurements in spectra of porcine LDL and HDL showed that all the choline methyl protons of HDL phospholipids and ~70% of those of LDL phospholipids contribute to the ^{1}H NMR spectrum.[22] With the exception of the choline methyl resonance, the accuracy

---

[20] For small molecules in solution, analysis of multiplet patterns and coupling constants can of course provide valuable structural information.

[21] J. M. Steim, O. J. Edner, and F. G. Bargoot, *Science* **162**, 909 (1968).

[22] E. G. Finer, R. Henry, R. B. Leslie, and R. N. Robertson, *Biochim. Biophys. Acta* **380**, 320 (1975).

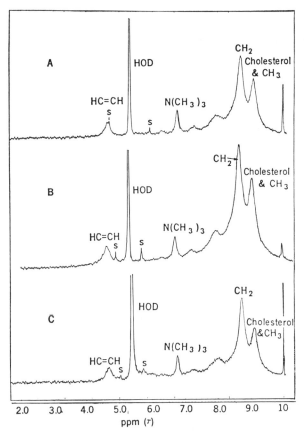

FIG. 3. ^{1}H NMR spectra (60 MHz) of (A) LDL lipids dispersed in D$_2$O by sonication at 512 μg phosphorus/ml; (B) LDL in 0.1 M NaCl in D$_2$O at 625 μg lipid phosphorus/ml; (C) HDL in 0.1 M NaCl in D$_2$O at 832 μg lipid phosphorus/ml. The spectrum for HDL lipids dispersed in D$_2$O was essentially identical to (A). Spinning side bands are labeled S. Note that in this study and in some of the other early ^{1}H NMR studies, the τ scale for chemical shift was used. We will use the δ scale, which is 10-τ. From J. M. Steim, O. J. Edner, and F. G. Bargoot, *Science* **162**, 909 (1968). © 1968 by The American Association for the Advancement of Science.

of integration will not be sufficient to determine the contribution of minor constituents or to determine whether all of a given constituent contributes to a peak. In porcine LDL spectra[22] the integral of the 0 to 3 ppm region but not that of the −2 to 3 ppm region varied with temperature because of a marked broadening of aliphatic resonances below ~50°. Thus, when reporting integration results, the sample temperature and the region of integration of broad resonances must be specified. Even for a well-resolved broad resonance, substantial integration errors may result from

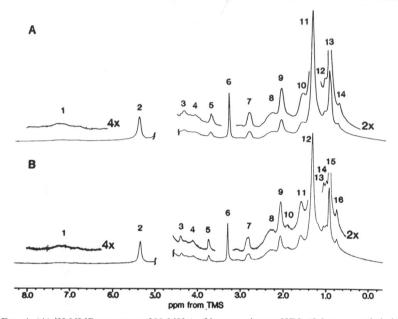

FIG. 4. (A) ¹H NMR spectrum (200 MHz) of human plasma HDL (8.3 mg protein/ml in D₂O, pD = 6.3) at 38° after 250 spectral accumulations using a pulse interval of 1.86 sec, 8192 time domain points, and a spectral width of 2200 Hz with 0.1 Hz linebroadening. (B) ¹H NMR spectrum (400 MHz) of the same sample (obtained 2 hr after the above spectrum) at 38° after 1000 spectral accumulations using a pulse interval of 1.86 sec, 16,384 time domain points, and a spectral width of 4400 Hz with 0.2 Hz linebroadening. The HDO peak has been deleted. Insets are printed with the indicated (2× or 4×) vertical expansion. In a comparison of resolution between these two spectra many variables that may affect resolution, such as sample composition, concentration, viscosity and temperature, digitizer resolution, and time-dependent sample changes are strictly controlled. Chemical shifts are in ppm downfield from tetramethylsilane (TMS). Assignments of numbered peaks are presented in Table II. Note that the aliphatic region (~0–3 ppm) is dominated by fatty acyl resonances which obscure weaker resonances. J. A. Hamilton, unpublished results, 1984. The authors thank Dr. Michael Blumenstein for help in obtaining the 400 MHz spectrum and Mr. Howard Lilly for the 200 MHz spectrum.

difficulty in determining where the Lorentzian tails merge with the baseline (and hence the proper limits of integration), and computer simulation of the peak may be of value.[23]

The poor resolution of some ¹H NMR lipoprotein spectra precludes complete and accurate assignments by conventional methods. The most fundamental approach is to compare the chemical shifts in lipoprotein

[23] W. Dietrich and R. Gerhards, *J. Magn. Reson.* **44,** 229 (1981).

spectra with those for the lipid and protein constituents.[24] However, complex regions like the aliphatic region cannot be assigned solely by chemical shift comparisons. Since many peaks may be masked by more intense peaks, an important strategy is to locate suitable marker peaks for different constituents. In LDL and HDL spectra, the only markers for different lipids are the choline methylene and methyl peaks for choline-containing phospholipids, and the C-18 and C-19 methyl peaks for cholesterol and/or cholesteryl esters. Because these groups can have rapid molecular motions and hence narrow resonances while other parts of the same molecule have restricted motions and consequently broad resonances (e.g., the choline peaks remain narrow in the gel phase of PC[25]), their presence in the ^{1}H spectrum does not necessarily indicate that other groups in the same molecule contribute to the spectrum. Similarly, the presence (or absence) of protein resonances in spectral regions free of lipid resonances (such as the aromatic region) does not necessarily imply the presence (or absence) of protein resonances in regions of poor spectral resolution like the aliphatic region,[26] since different amino acid groups may have markedly different molecular motions. Thus, absolute and relative intensity data must be evaluated carefully to deduce the contribution of different groups in regions with many overlapping peaks. The complexity of ^{1}H spectra of lipoproteins has been most clearly demonstrated in a recent two-dimensional (2D) NMR study of bovine HDL.[27] 2D heteronuclear ^{13}C–^{1}H chemical-shift correlation spectra resolved, for example, several cholesterol ring ^{13}C resonances corresponding to 1.3 ppm in the conventional ^{1}H spectrum[27] (Table II). This technique is probably the most promising approach for assigning complex ^{1}H NMR spectra.

Assignments for the HDL spectra in Fig. 4 are presented in Table II. They are based on a critical survey by the authors of all published assignments and integration results.[21,22,24,26,27] Note that there is very little contribution of protein to the spectrum of HDL. In cases of overlapping resonances, assignments are based on those of the above 2D study.[27]

Inadequate resolution in conventional ^{1}H NMR spectra of LDL and HDL can complicate the interpretation of chemical shift, T_1, and $\nu_{1/2}$ measurements. For example, a direct comparison of methylene chemical shifts, particularly those for the fatty acyl C-2′ and C-3′ methylenes, in

---

[24] D. Chapman, R. B. Leslie, R. Hirz, and A. M. Scanu, *Biochim. Biophys. Acta* **176**, 524 (1969).
[25] A. G. Lee, N. J. M. Birdsall, Y. K. Levine, and J. C. Metcalfe, *Biochim. Biophys. Acta* **255**, 43 (1973).
[26] R. B. Leslie and D. Chapman, *Chem. Phys. Lipids* **3**, 152 (1969).
[27] S. Coffin, M. Limm, and D. Cowburn, *J. Magn. Reson.* **59**, 268 (1984).

TABLE II
CHEMICAL SHIFTS AND ASSIGNMENTS OF ^{1}H RESONANCES OF
HUMAN SERUM HDL AT 47 kG (200 MHz) AND 94 kG
(400 MHz)[a]

	Chemical shift (ppm)	
Assignment	94 kG (400 MHz)	47 kG (200 MHz)
Protein aromatic	7.2 (1)	~7.2 (1)
CH=CH	5.31 (2)	5.32 (2)
Choline CH_2O	4.34 (3)	4.33 (3)
Glycerol CH_2O[b]	4.06 (4)	~4.1 (4)
Choline CH_2N	3.70 (5)	3.70 (5)
Choline $(CH_3)_3N$	3.25 (6)	3.25 (6)
=CCH_2C=	2.8 (7)	2.77 (7)
CH_2CO	2.24 (8)	2.25 (8)
CH_2—C=C	2.02 (9)	2.02 (9)
Unassigned[c]	1.85 (10)	
CH_2CCO[c]	1.54 (11)	1.53 (10)
$(CH_2)_n$[d]	1.27 (12)	1.27 (11)
Cholesterol C-19[e]	1.01 (13)	1.01 (12)
Cholesterol C-21[e]	0.95 (14)	
CH_3CH_2[c,f]	0.88 (15)	0.89 (13)
Cholesterol C-18	0.70 (16)	0.70 (14)

[a] Chemical shifts are in ppm downfield from tetramethylsilane. The phospholipid choline methyl group at 3.25 ppm was used as an internal reference. The numbers in parentheses after the chemical shifts are the peak designations in the spectra of HDL shown in Fig. 4A (200 MHz) and Fig. 4B (400 MHz). H indicates specially assigned ^{1}H (J. A. Hamilton, unpublished results, 1984).

[b] Both glyceryl CH_2O groups are probably contained in this broad resonance.

[c] May contain some contribution from protein [D. Chapman, R. B. Leslie, R. Hirz, and A. M. Scanu, *Biochim. Biophys. Acta* **176**, 524 (1969)].

[d] In bovine HDL, contributions to this resonance from CH_3CH_2, $CH_3CH_2CH_2$, and cholesterol C-9 have been detected in 2D spectra [S. Coffin, M. Limm, and D. Cowburn, *J. Magn. Reson.* **59**, 268 (1984)].

[e] These assignments may be reversed.

[f] In bovine HDL, contributions to this resonance from the cholesterol C-26 and C-27 methyls have been detected [S. Coffin, M. Limm, and D. Cowburn, *J. Magn. Reson.* **59**, 268 (1984)].

spectra of HDL and PC[28] cannot be made because of the inherent ambiguity created by overlapping resonances from different lipid types. Most ^{1}H peaks contain significant contributions from chemical shift inhomogeneity, and extrapolation of peak width to zero magnetic field[22] will not necessarily give an accurate $\nu_{1/2}$ of a single well-resolved peak (which in the case of methylene and methyl peaks will include *at least* the contribution of different fatty acyl esters). The splitting of peaks by scalar couplings also makes quantitative interpretation of ^{1}H $\nu_{1/2}$ results difficult. T_1 values of the aliphatic peaks[22] are difficult to interpret unambiguously because they are a composite of (potentially) several different T_1's. Another complication is that intermolecular as well as intramolecular interactions may contribute to ^{1}H relaxation processes,[29] making it more difficult to relate proton T_1's to intramolecular motions.

In ^{1}H NMR spectra of LDL and HDL, the peak which is most amenable to straightforward study is the choline methyl resonance, since it is well resolved at all magnetic fields and is relatively narrow and intense. It represents primarily only two chemical species, PC and SM. Reliable intensity, chemical shift, T_1, and $\nu_{1/2}$ data can be obtained for this peak.[22] The ^{1}H choline methyl resonance is also responsive to the presence of NMR shift reagents, as demonstrated by studies of porcine LDL and HDL[22] with $Fe(CN)_6^{3-}$ and of porcine HDL[30] with $Fe(CN)_6^{3-}$ and the lanthanide ion, Gd^{3+}.

Lipoproteins which have a more homogeneous composition or a vast predominance of one type of lipid may yield simpler ^{1}H spectra. Whether either of these factors translates into spectra from which reliable quantitative information can be obtained depends on factors other than the composition. LP-X, for example, is comprised mainly of phospholipid and cholesterol, with very small amounts of other lipids and proteins. The ^{1}H NMR spectrum at 220 MHz contains no resonances from cholesterol (they are too broad to observe) and an aliphatic region with broad, overlapping peaks.[9] Thus, the structural organization of LP-X imposes severe restrictions on molecular motions of the lipids, and the spectrum is poorly resolved even though there are few contributing species. In contrast, ^{1}H NMR spectra of chylomicrons[31] have several well-resolved peaks, some of which show splittings (Fig. 5). In this case resonances from triglycerides dominate the spectra, and the low relative abundance of lipids

[28] H. Hauser, *FEBS Lett.* **60,** 71 (1975).
[29] P. A. Kroon, M. Kainosho, and S. I. Chan, *Biochim. Biophys. Acta* **433,** 282 (1976).
[30] H. Hauser, in "The Lipoprotein Molecule" (H. Peeters, ed.). NATO Advanced Study Institutes Series, Plenum, New York, 1978.
[31] J. A. Hamilton, D. M. Small, and J. S. Parks, *J. Biol. Chem.* **258,** 1172 (1983).

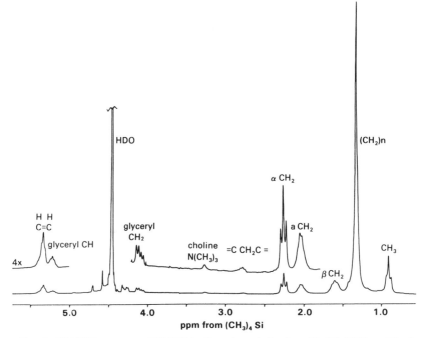

FIG. 5. ^{1}H NMR spectrum (200 MHz) of monkey chylomicra (S_f 400–2000) at 55° after four spectral accumulations with a pulse interval of 12.0 sec, 8192 time domain points, and a 2000 Hz spectral width. Peaks corresponding to fatty acyl proton groups are labeled as follows: HC=CH, olefinic protons; =CCH$_2$C=, methylene protons between olefinic groups; αCH$_2$, methylene protons of the C-2' carbon; aCH$_2$, allylic protons; βCH$_2$, methylene protons of the C-3' carbon; (CH$_2$)$_n$, methylene groups except for those designated individually; and CH$_3$, terminal methyl protons. The glyceryl protons of the triglyceride glyceryl backbone are indicated, and protons of the choline methyl groups of phospholipids are designated as choline N(CH$_3$)$_3$. The inset is printed with a 4-fold (4×) vertical expansion. Side bands of the strong HDO resonance overlap the low field half of the glyceryl CH$_2$ resonance not shown in the expansion. To minimize the signal from water protons, the sample was dialyzed exhaustively against 99.8% D$_2$O. From J. A. Hamilton, D. M. Small, and J. S. Parks, *J. Biol. Chem.* **258**, 1172 (1983).

other than triglycerides and the rapid molecular motions of the liquid triglycerides are independent factors favoring good resolution. However, below the liquid–solid phase transition of the triglycerides, fatty acyl resonances broadened with a loss of resolution in the aliphatic region.[31] Nevertheless, reliable intensity data for the CH$_3$, allylic CH$_2$, and αCH$_2$ peaks were obtained at temperatures down to 5°, providing unique data on the differential melting of saturated and unsaturated fatty acyl chains of triglycerides.[31]

^{1}H is the most sensitive NMR nucleus, giving ^{1}H NMR spectroscopy an important versatility. For example, numerous spectra can be acquired as a function of a given variable such as temperature[22,31] in a short period of time. It is possible to study purified subfractions of lipoproteins using small amounts of material that may not meet the minimum material requirement for NMR studies of other nuclei. High-quality ^{1}H NMR spectra of lipoproteins can be obtained in minutes with <1 mg lipid/ml using ~0.2 ml sample volume in 5-mm-diameter tubes.[32] Thus, ^{1}H NMR is the method best suited (where appropriate for the research question) for small amounts of sample, whether the limitation is encountered by low-yield subfractionation procedures or other experimental conditions, by special dietary studies, or by disease states.

^{13}C NMR

Relative to ^{1}H NMR, ^{13}C NMR spectroscopy offers distinct advantages for the study of complex systems like plasma lipoproteins. These include (1) increased resolution, (2) more readily assignable resonances, and (3) relative ease of obtaining information about molecular motions.[33] Increased spectral resolution results in large part from the much wider dispersion of ^{13}C chemical shifts; the usual range is ~200 ppm compared with ~10 ppm for ^{1}H. In addition, spin–spin couplings between ^{13}C nuclei and ^{1}H are generally removed by the experimental procedure of complete ("broad-band") ^{1}H decoupling. Spin–spin couplings between ^{13}C nuclei are not observed because of the low natural abundance (1.1%) of the ^{13}C nucleus and therefore the very low abundance (0.01%) of molecules with two ^{13}C isotopes on adjacent carbons. Thus, natural abundance ^{13}C NMR spectra acquired under routine conditions are comprised of singlets, a feature which contributes to all three advantages cited above.

Like ^{1}H, the ^{13}C nucleus is ubiquitous in lipids and proteins. However, the inherently poor sensitivity and low natural abundance of the ^{13}C nucleus (Table I) combine to make it ~10,000 times less sensitive than the ^{1}H nucleus. This disadvantage has been partially overcome by the use of the pulsed Fourier transform technique, the digital accumulation of many individual scans (signal averaging), and data manipulation techniques (e.g., use of an exponential multiplier) to improve S/N.[33] In addition, broad-band ^{1}H decoupling increases S/N by collapsing the multiplet into a singlet of narrower width and greater amplitude and by producing an

---

[32] J. A. Hamilton, unpublished results, 1982.
[33] J. A. Hamilton, C. Talkowski, R. F. Childers, E. Williams, E. M. Avlia, A. Allerhand, and E. H. Cordes, *Science* **180**, 193 (1973).

NOE of 1.15 to 2.98.[34] The latter effect complicates intensity measurements in ^{13}C spectra, since NOE values for different resonances may be different.[13,34]

The major advantages of ^{13}C NMR in comparison to ^{1}H NMR spectroscopy were evident in the first published lipoprotein ^{13}C spectra.[33] Natural abundance FT NMR ^{13}C spectra of VLDL, LDL, and HDL at 15.2 MHz revealed up to 27 resonances (HDL spectrum). Thus the resolution was better in these ^{13}C spectra, which were obtained at the lowest field used for lipoprotein studies, than in ^{1}H spectra at any magnetic field. All of the ^{13}C resonances were assigned; there are several suitable marker peaks for the different constituents (see below). However, the low S/N of the ^{13}C nucleus required use of all the methods noted above for improving S/N and the use of a large (20-mm) sample tube containing ~12 ml of sample (>30 times that required with the 5-mm tube) which gave a 3-fold increase in sensitivity over that afforded by 12- and 13-mm tubes. Still, high concentrations (80–125 mg lipoprotein/ml) and long accumulation times (10–20 hr per spectrum) were required.

In cases where the experimental goal is to observe the maximum number of ^{13}C peaks in a sample of low sensitivity it is important to optimize spectrometer parameters, particularly the pulse interval, to obtain the best S/N per unit time. Measurements of T_1 values of selected lipoprotein peaks revealed a large range of values (36 msec–1.2 sec[35]), so that a single pulse interval would not optimize S/N for all peaks. Nevertheless, using a shorter pulse interval than in the first study, improved S/N was obtained for most peaks in about the same total time as above,[33] and several additional resonances were detected in the 15 MHz spectra of VLDL, LDL, and HDL (Fig. 6). These spectra contain numerous well-resolved resonances, including several single-carbon resonances, from which valuable quantitative information can be obtained. The spectra of LDL and HDL are similar in overall appearance, but differ significantly in detail. The spectrum of VLDL differs in appearance from LDL and HDL spectra in most regions.

Our discussion of ^{13}C methodology will be divided into six sections: (1) chemical shifts and assignments, (2) quantitation, (3) spin-lattice relaxation time, (4) linewidth, (5) sample temperature, and (6) optimizing sample volume and signal-to-noise ratio. Most sections will be prefaced by a general discussion followed by a discussion of strategies and problems in lipoprotein studies.

---

[34] D. Doddrell, V. Glushko, and A. Allerhand, *J. Chem. Phys.* **56**, 3683 (1972).

[35] J. A. Hamilton, C. Talkowski, R. F. Childers, E. Williams, A. Allerhand, and E. H. Cordes, *J. Biol. Chem.* **249**, 4872 (1974).

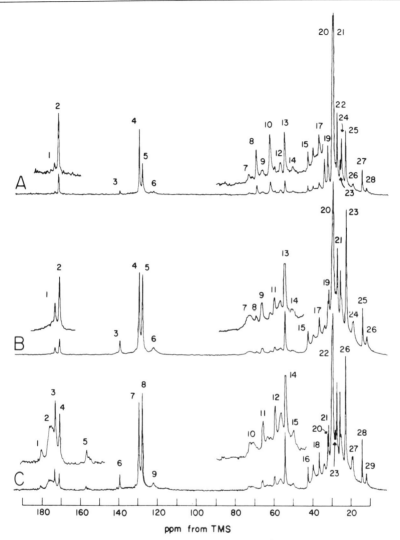

FIG. 6. Natural abundance ^{13}C Fourier transform NMR spectra (15.2 MHz) of human serum lipoproteins at 36° recorded with ^{1}H decoupling in 20-mm sample tubes with the use of 250-ppm spectral widths, 4096 time domain points, and a recycle time of 0.555 sec. The insets were printed with a 4× vertical expansion. Digital broadening of 0.3 Hz (main spectra) and 2.4 Hz (insets) was introduced to improve S/N. Peak numbers are the same as those shown in parentheses in Table II. (A) VLDL, 37 mg/ml, pH 7.5, after 165,254 accumulations (25 hr); (B) LDL, 110 mg/ml, pH 7.5, after 65,536 accumulations (10 hr); (C) HDL, 70 mg/ml, pH 8.6, after 65,536 accumulations (10 hr). From J. A. Hamilton, C. Talkowski, R. F. Childers, E. Williams, A. Allerhand, and E. H. Cordes, *J. Biol. Chem.* **249,** 4872 (1974).

Chemical Shifts and Assignments

The chemical shift of a resonance provides the most fundamental (and sometimes the only) data required for making correct spectral assignments. The general prerequisites for measurement of precise chemical shifts are the same ones that apply to measurement of T_1, $\nu_{1/2}$, NOE, and intensity: adequate spectral resolution, S/N, and digital resolution (or data point density). The meaning of "adequate" will depend somewhat on the particular experimental goal. For example, consider typical accumulation conditions for obtaining lipoprotein ^{13}C spectra at 50 MHz. Using 32K data points and a 10,000 Hz spectral width, the digital resolution is 16,384 points/10,000 Hz (since only half the data points are transformed into a real spectrum), or 1.6 pts/Hz (0.6 Hz/pt). With this digital resolution, changes in chemical shift would be significant only if they are $\geqslant 0.6$ Hz (or 0.01 ppm). On the other hand, this digital resolution provides adequate precision for peak assignments.

In most cases chemical shift is not measured as an absolute frequency but relative to a reference peak. Thus, the total error will be greater than the error for measurement of one peak, and hence the precision of measurement of the reference peak must be considered. The reference may be external or internal; the internal reference may be a peak from a chemical additive or from the sample. There is no generally ideal internal chemical reference for aqueous systems,[4] and any chemical additive should be considered in terms of its potential interaction with lipoprotein components. Use of an external shift reference carries an inherent *inaccuracy* because of bulk susceptibility differences,[4] but can still allow *precise* chemical shift measurements and is particularly appropriate when the effect of an additional component, such as a paramagnetic ion, on the spectrum is being studied. In addition, the concentration, temperature (and in some cases, the pH) dependence of the chemical shift of any reference peak requires evaluation. For most ^{13}C NMR studies of lipoproteins, the fatty acyl methyl resonance has been used as an internal reference. It is a well-resolved, narrow (<10 Hz), symmetric peak whose chemical shift (14.1 ppm, the value for the fatty acyl methyl of lipoprotein lipids dissolved in CDCl$_3$ and referenced to internal tetramethylsilane) does not usually change with environmental perturbations.[36]

A general strategy in assigning complex spectra is to obtain the best possible resolution, and this can be achieved in two straightforward ways in lipoprotein and lipid ^{13}C NMR studies:[37] by increasing the magnetic

---

[36] J. A. Hamilton, unpublished results, 1974.

[37] Data manipulation techniques may be also used to enhance resoltuion. A method commonly used is convolution–difference spectroscopy [I. D. Campbell, C. M. Dobson,

field and by increasing the temperature. In the latter case, resolution increases because most ^{13}C lipid resonances exhibit a decrease in $\nu_{1/2}$, an effect observed particularly with LDL (see below). Increasing the field strength improves resolution because $\nu_{1/2}$'s (in Hz) of most lipid resonances do not increase. A comparison of HDL spectra at 14 (Fig. 6C), 47,[38] and 94 kG[39] reveals a more modest improvement between the latter two fields than the first two fields and shows that magnetic fields of intermediate strength will provide adequate resolution for most ^{13}C lipoprotein studies. The same conclusion may apply to ^{1}H NMR, based on comparison of HDL spectra at three fields (Figs. 3C and 4).

The peak assignments of the Fig. 6 spectra are given in Table III. The vast majority of peaks were assigned to lipid carbons[33,35] by comparison of their chemical shifts with those of individual lipids in solvents (triolein, trilinolein, and cholesteryl acetate in organic solvents, and dipalmitoyl lecithin in D_2O). Assignment of protein, cholesterol, and lipid carbonyl resonances will be presented as examples of strategies for ^{13}C assignments. Proteins contain many more classes of carbon atoms than do lipids; hence, in lipoproteins their *molar* concentration is lower, a factor which could account for the lack of prominent protein signals. ^{13}C NMR spectra of lipoproteins contain large regions devoid of lipid resonances, but only the HDL spectrum (Fig. 6C) contains resonances which were clearly attributable to protein carbons (peaks 1, 2, and 5). To clarify the contribution of protein to the HDL spectrum, the lipoprotein was extensively hydrolyzed by trypsin, and the protein-depleted lipid complex was separated from soluble, lipid-free protein fragments.[40] The ^{13}C NMR spectrum of unfractionated, trypsinized HDL showed numerous additional narrow resonances which corresponded to peaks in the spectrum of the lipid-free protein fragments.[40] The spectrum of the protein-depleted lipid complex (from which the cleaved peptides had been removed) was similar to that of intact HDL (Fig. 6C) except that peaks 1, 2, and 5 were absent, and the relative intensity of peak 17 was lower.[40] Thus, the *primary* reason for observing few protein resonances in the HDL spectrum is that the rotational and segmental motions of most protein carbons are restricted compared to those of lipid carbons. The carbons which have observable resonances have a predictably high mobility. Even with improved S/N

R. J. P. Williams, and A. V. Xavier, *J. Magn. Reson.* **11**, 172 (1973)]. However, since this technique significantly degrades S/N, it has not been applied to lipoprotein studies because spectral regions of low S/N are often ones of considerable interest.

[38] J. A. Hamilton and E. H. Cordes, *J. Biol. Chem.* **253**, 5193 (1978).
[39] H. Hauser and G. M. Kostner, *Biochim. Biophys. Acta* **573**, 375 (1979).
[40] J. A. Hamilton, Ph.D. thesis, Indiana University (1974).

TABLE III
Chemical Shifts and Assignments of ^{13}C Resonances of Human Serum Lipoproteins[a]

Assignment	Chemical shift (ppm)		
	VLDL	LDL	HDL
Glu carboxyl			181.0 (1)
Protein carbonyl			175 (2)
Carbonyl	173.7 (1)	173.6 (1)	173.6 (3)
Carbonyl	171.7 (2)	171.2 (2)	171.3 (4)
Arg C-6			157.2 (5)
Tyr C-6			b
Cholesterol C-5	139.8 (3)	139.8 (3)	139.8 (6)
—CH=CH—CH$_2$—CH$_2$	129.7 (4)	129.7 (4)	129.7 (7)
—CH=CH—CH$_2$—CH=CH—	128.1 (5)	128.0 (5)	128.0 (8)
Cholesterol C-6	122.2 (6)	122.3 (6)	122.3 (9)
Cholesterol C-3	73.2 (7)	72.4 (7)	72.4 (10)
Glycerol CH	69.1 (8)	69.0 (8)	c
Choline CH$_2$N	66.2 (9)	66.1 (9)	66.1 (11)
Glycerol CH$_2$	62.0 (10)	61.9 (10)	c
Choline CH$_2$O	59.6 (11)	59.6 (11)	59.6 (12)
Cholesterol C-14, C-17	56.6 (12)	56.6 (12)	56.8 (13)
Choline (CH$_3$)$_3$N	54.3 (13)	54.2 (13)	54.2 (14)
Cholesterol C-9	50.2 (14)	50.2 (14)	50.2 (15)
Cholesterol C-13	42.4 (15)	42.4 (15)	42.4 (15)
Cholesterol C-24[d]	39.7 (16)	39.6 (16)	39.7 (17)
Cholesterol C-10	36.6 (17)	36.6 (17)	36.6 (18)
—CH$_2$—CH$_2$—CO—	33.8 (18)	33.7 (18)	34.0 (19)
CH$_3$—CH$_2$—CH$_2$—CH$_2$—	32.1 (19)	32.1[e]	32.1 (20)
	31.7[e]	31.6 (19)	31.6 (21)
Fatty acyl (CH$_2$)$_n$	29.9 (20)		29.9[e]
	29.5 (21)	29.5 (20)	29.5 (22)
Cholesterol C-25	f	f	28.1 (23)
—CH$_2$—CH$_2$—CH=CH—	27.3 (22)	27.3 (21)	27.3 (24)
—CH=CH—CH$_2$—CH=CH—	25.7 (23)	25.7 (22)	25.7 (25)
—CH$_2$—CH$_2$—CO—	24.9 (24)	25.0[e]	25.1[e]
CH$_3$—CH$_2$—CH$_2$—	22.8 (25)	22.7 (23)	22.8 (26)
Cholesterol C-26, C-27		19.2[e]	19.3[e]
Cholesterol C-19, C-21	19.1 (26)	18.9 (24)	18.9 (27)
CH$_3$—CH$_2$—	14.1 (27)	14.1 (25)	14.1 (28)
Cholesterol C-18	12.0 (28)	11.9 (26)	12.0 (29)

[a] Chemical shifts are in ppm downfield from tetramethylsilane. The terminal methyl group of fatty acyl chains at 14.1 ppm was used as an internal reference. Numbers in parentheses after chemical shifts are the peak designations in the spectra of VLDL (Fig. 6A), LDL (Fig. 6B), and HDL (Fig. 6C). Unless otherwise noted, the assignments refer to the fatty acyl chain. C indicates the specifically assigned type of carbon. Chemical

and use of higher fields, subsequent ^{13}C NMR studies of HDL have detected only two additional (weak) protein resonances, at 115[38,41] and 17 ppm.[38] Since so few protein resonances can be observed, it is clear that detailed studies of protein structure and dynamics in intact lipoproteins by conventional ^{13}C NMR methods are not possible.

Cholesterol and cholesteryl ester resonances in lipoprotein spectra were assigned by comparison with spectra of unsolvated (neat) lipids (e.g., cholesteryl esters and triglycerides) and neat lipid mixtures found in lipoproteins.[38,41] In spectra of cholesterol dissolved in cholesteryl oleate or in spectra of cholesterol oleate with triolein, the only steroid resonances which could be distinguished for the esterified and unesterified forms of cholesterol were the C-3, C-5, and C-6.[38,41] Their chemical shifts (in ppm) are tabulated below.

Assignment	Cholesterol	Cholesteryl ester
C-3	71.0	73.1
C-5	141.2	139.7
C-6	120.7	122.3

In spectra of HDL, cholesterol was first detected at low field by the C-5 peak at 141.3 ppm,[41] and subsequently also by the C-6 peak at 120.7 ppm at higher fields, which resolved the small C-6 unesterified cholesterol peak from the more intense cholesteryl ester C-6 peak.[38] Even at 100.6 MHz, the C-3 region in the HDL spectrum contains incompletely resolved resonances.[39]

In some instances the chemical shift, because of its sensitivity to extrinsic effects, provides structural information. For example, ^{13}C NMR spectra of lipoproteins contain two narrow lipid carbonyl peaks (~171.2 and ~173.6 ppm) which have a chemical shift *difference* that is much larger than that for lipids in an organic solvent.[35] Studies of the carbonyl

[41] E. M. Avila, J. A. Hamilton, J. A. K. Harmony, A. Allerhand, and E. H. Cordes, *J. Biol. Chem.* **253**, 3983 (1978).

---

shifts accurate to ±0.2 ppm [J. A. Hamilton, C. Talkowski, R. F. Childers, E. Williams, A. Allerhand, and E. H. Cordes, *J. Biol. Chem.* **249**, 4872 (1974)].
[b] Shoulders upfield of peak 5 in Fig. 1C.
[c] Could not be identified unambiguously.
[d] May also contain ε carbon of lysine and β carbon of leucine in HDL spectrum.
[e] Shoulder in Fig. 1.
[f] Barely discernible at about 28 ppm.

region of spectra of lipids in anhydrous and hydrated environments suggested that these magnetically inequivalent environments originate from hydrated (173.6 ppm) and unhydrated (171.2 ppm) carbonyls.[42,43] The hydrated site is presumably occupied by phospholipids and the anhydrous site(s) by cholesteryl esters and triglycerides.[43] Another study approached the assignment of the lipid carbonyl resonances in a novel way. HDL enriched with 1,2-[dioleoyl-1-^{13}C,1-^{14}C]-sn-phosphatidylcholine was subjected to the action of lecithin : cholesterol acyltransferase (LCAT) to transfer the sn-2 acyl chain of lecithin to cholesterol, thereby forming labeled cholesteryl oleate.[44] By comparison of the ratio of the radioactivity in the two lipids with the ratio of ^{13}C peak heights, the downfield peak was assigned to phospholipid carbonyls. This analysis assumes that the NMR peak intensity accurately reflects the chemical composition, which is not always a valid assumption [e.g., the case of cholesterol (see below)].

Another example in ^{13}C NMR studies of lipoproteins of how the chemical shift has provided structural information is the C-4 resonance of cholesterol. This resonance is environmentally sensitive, having a significantly different chemical shift in PC vesicles from that in a hydrocarbon environment.[45] In natural abundance ^{13}C NMR spectra of lipoproteins, the cholesterol C-4 resonance is in a crowded spectral region (~42 ppm) and cannot be assigned. However, complete replacement of HDL cholesterol with 90% ^{13}C-enriched C-4 cholesterol by total reconstitution of the HDL particle or by specific exchange of cholesterol yielded cholesterol signal(s) of very good S/N.[45] The major cholesterol C-4 peak is a broad (~55 Hz) resonance with a chemical shift characteristic of cholesterol in vesicles. Somewhat more than 50% of the total unesterified cholesterol contributes to this peak[45]; cholesterol in this pool probably yields resonances too broad to observe in the natural abundance spectrum of native HDL (see below). The results for a second resonance identified as core-located cholesterol[45] are not as clear. The chemical shift of this narrower peak does not coincide precisely with that of cholesterol in triolein and is very close to the chemical shift of the nonprotonated C-13 carbon of cholesteryl ester.[42,45] Thus, identification of, and any quantitative measurements of, this peak may be complicated by the natural abundance ^{13}C spectrum.

In general, the chemical shift of a lipid resonance in lipoprotein ^{13}C spectra agrees closely (within 0.1–0.2 ppm) with the chemical shift of the

---

[42] J. A. Hamilton, N. Oppenheimer, and E. H. Cordes, *J. Biol. Chem.* **252**, 8071 (1977).
[43] J. A. Hamilton, E. H. Cordes, and C. J. Glueck, *J. Biol. Chem.* **254**, 5435 (1979).
[44] G. Assmann, R. J. Highet, E. A. Sokoloski, and H. B. Brewer, Jr., *Proc. Natl. Acad. Sci. U.S.A.* **71**, 3701 (1974).
[45] S. Lund-Katz and M. C. Phillips, *Biochemistry* **23**, 1130 (1984).

corresponding resonance in simple model systems (without organic solvents). Thus, except for the examples cited above, novel structural information has not been derived from ^{13}C chemical shift measurements in lipoprotein spectra. Important findings from the assignments are that nearly all resonances are attributable to lipid carbons and that fatty acyl resonances, which are generally the most dominant ones, are not distinguishable for the different ester types (i.e., cholesteryl ester, triglyceride, and phospholipid) except for carbonyl and, in some cases,[43] the adjacent C-2' methylene carbon. In high-field ($\geq$47 kG) spectra, at least one well-resolved *single carbon* resonance for each major lipid type is observed.[38,39]

Quantitation

General guidelines for accurate intensity measurements have been presented in the ^{31}P and ^{1}H sections and apply to ^{13}C spectra as well. The problems and assumptions attendant to the quantitation of ^{13}C peaks with very low S/N are illustrated by early efforts to quantitate the contribution of unesterified cholesterol to the spectrum of HDL.[41] In HDL samples for which the chemical composition was determined, the cholesterol C-5 to cholesteryl ester C-5 ratio (peak heights and/or integrated areas) in the NMR spectrum was ~50% lower than the chemical ratio of cholesterol to cholesteryl ester. Since the cholesterol resonances were too weak to quantitate accurately against an external absolute standard, an indirect approach was adopted. First, the ratio of intensities of the cholesteryl ester C-5 and the PC choline methyl resonances were measured under conditions of NOE suppression and found to correspond to the ratio calculated from the chemical analysis of the NMR sample.[41] If the ^{13}C choline resonance represents all of the choline groups in HDL (a reasonable extrapolation of the ^{1}H data[22]), it can be inferred that all the cholesteryl esters in HDL contribute to the ^{13}C NMR spectrum, at least with respect to resonances from the fused ring. Thus, the cholesterol (and not the cholesteryl ester) C-5 peak intensity is attenuated, but the cause *could* be differences in the T_1 or NOE values of these two C-5 peaks. Although the cholesterol C-5 peak was too weak to allow measurement of accurate T_1 values, spectra obtained at different pulse intervals showed that this peak was not being selectively saturated (compared with the cholesteryl ester C-5 peak) by the pulse interval employed for peak area measurements.[41] It was also not possible to measure an accurate NOE of the cholesterol peaks or to obtain intensities with NOE suppressed, since a spectrum with NOE suppressed requires a total pulse interval of 8–10 times the T_1 value.[13] However, because the cholesterol C-5 and the cholesteryl ester C-5 are both nonprotonated, structurally similar carbons, and because the T_1 and linewidth values of the two C-5 peaks are similar, it is reasonable to assume that the NOEs are also similar.[34,38,41] Therefore, it was concluded that the narrow (observed) peaks from unesterified cholesterol accounted for <50% of this lipid in HDL, and that the resonances corresponding to the remaining cholesterol were much broader and not observed. Subsequently, similar conclusions were reached in a study of HDL at 67.9 MHz which quantitated the cholesterol C-6 resonance using a similar strategy[38] and in a study of HDL at 100.6 MHz which compared relative intensities of the cholesterol and cholesterol ester C-5 and C-6 peaks obtained under conditions of NOE suppression.[39] These studies[38,39,41] also show that in the majority of instances where steroid resonances of the unesterified and esterified cholesterol are indistinguishable by the criterion of chemical shift, the peak intensity will reflect primarily the ester (75–80% of the total intensity).

Spin-Lattice Relaxation Time (T_1)

T_1 is not measured directly from a single spectrum but from a series of spectra obtained under controlled conditions using a specific pulse sequence in which nonequilibrium intensities are obtained. The method most commonly used in lipoprotein studies is the "inversion-recovery" method[46] which utilizes the pulse sequence: $[180°-\tau-90° \text{ (FID)}-T]_n$. After a 180° pulse to invert the nuclear populations, a delay time τ is varied to allow the system to relax partially before applying a 90° pulse to bring the magnetization into the x–y plane for measurement. T is a fixed delay set to >5 times the estimated T_1 value, and a group of n ($n = 8$–10) spectra is obtained. Thus, T_1 measurement of resonances from dilute samples can be very time consuming and may be highly impractical, if not impossible, for weak resonances. It is possible to shorten appreciably the total time required to determine T_1 by using a "fast inversion-recovery" method, which uses a shorter fixed delay (T) and relaxes the requirement that the longest τ value permit full recovery of the magnetization from inversion.[47] Based on analyses of ideal cases[48,49] and of neat lipid systems,[50] we recommend use of the fast inversion recovery method with $T \sim 2$ times T_1, and the longest $\tau \sim 2$ times T_1. Data analysis must then be done with a three-parameter exponential, and not a linear least-squares fitting rou-

---

[46] R. L. Vold, J. S. Waugh, M. P. Klein, and D. E. Phelps, *J. Chem. Phys.* **48**, 3831 (1968).
[47] D. Canet, G. C. Levy, and I. R. Peat, *J. Magn. Reson.* **18**, 199 (1975).
[48] T. P. Pitner and J. F. Whidby, *Anal. Chem.* **51**, 2203 (1979).
[49] H. Hanssum, *J. Magn. Reson.* **45**, 461 (1981).
[50] J. A. Hamilton, unpublished results, 1982.

tine[51] to obtain accurate and reproducible T_1 values.[48] This method is also advantageous because the estimate of T_1 needed to choose appropriate τ values and the T value does not have to be as accurate as with other methods of measuring T_1.

Considerations for measuring reliable T_1's have been discussed in detail elsewhere.[49–52] A few pertinent recommendations for measuring and interpreting T_1's will be presented here. When accumulating spectra for T_1 measurements, the digital resolution must be adequate to define the lineshape clearly, since peak heights are generally used instead of peak areas for T_1 calculations. When analyzing data, note that the standard deviation of the T_1 calculation is often much smaller than the true error involved in reproducing the data. This error, the standard deviation of separate measurements of T_1 on the same or a similar sample,[53] is often not evaluated because of the time-consuming nature of T_1 measurements. T_1 data may show a good fit to an exponential function but be highly inaccurate, as in the case where the delay times are too short. Errors in T_1 should not be evaluated solely from the errors of the fitting routine, but also from the highest S/N in the set of partially relaxed spectra. In general, obtaining T_1 values in lipoprotein systems to a precision of better than ±5% is neither required nor feasible. T_1 differences this small are almost always impossible to interpret meaningfully in terms of molecular motions (see below) and seldom provide an empirical means for distinguishing different systems. Such factors as sample temperature, concentration, volume, compositional variability, and viscosity may affect T_1 values. Therefore, to make detailed comparisons of T_1 values in different experiments, these and other[52] variables must be strictly controlled.

Knowledge of T_1 values is essential for determining pulse intervals for maximizing S/N and for measuring equilibrium peak intensities and NOE values. In addition, T_1 values of protonated carbons are sensitive probes for studying rotational and segmental motions of unassociated molecules in solution and molecules in biological aggregates such as lipoproteins. The ^{13}C relaxation of *protonated* carbons of large molecules is dominated by $^{13}C-^1H$ dipolar interactions with directly bonded hydrogens,[34] and T_1 is related to the rate of reorientation of the C—H vectors. In the case of isotropic rotation of an entire molecule having no internal motions, a single correlation time (τ_R) describes its reorientation. The equation relating T_1 to τ_R, and plots of the dependence of T_1 on τ_R for a methine carbon

---

[51] M. Sass and D. Ziessow, *J. Magn. Reson.* **25**, 263 (1977).
[52] G. C. Levy and I. R. Peat, *J. Magn. Reson.* **18**, 500 (1975).
[53] J. R. Brainard, R. D. Knapp, J. R. Patsch, A. M. Gotto, Jr., and J. D. Morrisett, *Ann. N.Y. Acad. Sci.* **348**, 299 (1980).

at three magnetic fields have been described by Doddrell et al.[34] T_1 is a double-valued function which takes on minimal values at $\tau_R \sim 10^{-8}$–10^{-9} sec and becomes progressively larger at shorter and longer correlation times. T_1 is field-dependent except at $\tau_R \leq 10^{-10}$ sec.[34] An analysis of the T_1 value of the cholesteryl ester C-6 resonance of LDL and HDL was made by Hamilton et al.[35] based on the assumption of isotropic rotation, since the fused ring does not undergo internal motions; two solutions for τ_R (2.0 × 10^{-9} and 2.5 × 10^{-8} sec) were obtained. In principle, experimental values of NOE and $\nu_{1/2}$ can be used to select the correct value of τ_R.[34] In this case, however, a clear choice between the two solutions could not be made on the basis of $\nu_{1/2}$, probably because the steroid ring reorientations are (detectably) anisotropic. However, these correlation times are within about an order of magnitude of the correlation times of cholesteryl esters calculated using an anisotropic model (see below).

When the correlation time is sufficiently small ($\leq$ about 10^{-9} sec) to fall in the "extreme narrowing region,"[34] the relationship between T_1 and τ_R becomes

$$1/NT_1 = \hbar^2 \gamma_h^2 \gamma_c^2 r^{-6} \tau_R \tag{1}$$

where N is the number of directly bonded hydrogens, γ_h and γ_c are the gyromagnetic ratios of ^{1}H and ^{13}C, respectively, and r is the C—H bond length. Equation (1), or the more general form of this equation,[34] will not apply to carbons of lipids in lipoproteins (except in the extreme case if they were anchored rigidly in the particle) because they have anisotropic motions and/or several degrees of rotational motion. However, in some cases it can be useful to combine the different correlation times into an "effective" rotational correlation time τ_{eff} and rewrite Eq. (1) as

$$1/NT_1 = \hbar^2 \gamma_h^2 \gamma_c^2 r^{-6} \tau_{eff} \tag{2}$$

or

$$\tau_{eff} \propto 1/NT_1 \tag{3}$$

for a given C—H vector. Thus, as the effective correlation time decreases (the rate of rotation increases), T_1 is predicted to increase. Note these equations apply only to (1) carbons with directly bonded hydrogens, (2) carbons relaxing by a dipolar mechanism, and (3) T_1 values obtained with complete decoupling of the ^{1}H.[34]

A comparison of NT_1 values for several corresponding fatty acyl carbons in VLDL, LDL, and HDL showed that HDL and LDL values were indistinguishable, and that VLDL values were slightly longer, suggesting a somewhat greater segmental mobility of these groups in VLDL.[35] However, the small differences would be difficult to explain in a rigorous, quantitative way because of the complexities of fatty acyl chain motions.[6,54,55] The similar NT_1's for HDL and LDL do not necessarily imply structural similarities,[35] particularly since the resonances do not distinguish different types of acyl chain esters. The relative contribution of different esters will differ in spectra of the two lipoproteins. For example, based on the "average" composition of LDL and HDL_3 given in Scanu,[7] in LDL phospholipids will contribute 35% and the cholesteryl esters, 40% of the total intensity of each fatty acyl resonance (except for the carbonyls) and in HDL_3 phospholipids will contribute 63% and the cholesteryl esters, 25%. An important consideration for ^{13}C fatty acyl resonances of lipoproteins is whether T_1 values of the core constituents (cholesteryl ester and/or triglyceride) are different from those of the surface phospholipids. If they are not, specific structural models of fatty acyl chain interactions will not be testable using T_1 values.

The use of T_1 data to elucidate features of structural organization in lipoproteins requires careful evaluation. For example, the result that the phospholipid choline methyl ^{13}C T_1 is the same in egg PC vesicles, in egg PC recombined with apolipoproteins, and in native HDL led to the conclusion that ionic interactions between the choline group and protein groups in HDL are of minor significance.[44] However, the sensitivity of the NMR quantity measured to the structural feature under scrutiny was not established; i.e., it was not shown that the choline T_1 is sensitive to different types of ionic interactions. The rapid rotation of the methyl groups, as reflected in the T_1 value, may be insensitive to changes in ionic interactions that occur (presumably) through the hydration layer. Hence, the conclusion may be invalid.

In lipoproteins and related systems, T_1 data alone will not provide sufficient information to describe the complex molecular motions of most carbons. A more comprehensive approach, entailing $\nu_{1/2}$ and NOE measurements (sometimes at different magnetic fields) is necessary to test different models of molecular motions.[34,54,55] However, T_1 will often provide a good estimate of the shortest correlation time (i.e., most rapid motion) in a complex system.[54,55]

Table IV shows the NT_1 values at four magnetic fields for selected well-resolved resonances in HDL spectra. Most resonances have T_1 values which are significantly field dependent. Those resonances which have little or no field dependence are from nonprotonated carbons (e.g., the carbonyl carbons) or from groups with extra degrees of internal rotation (the choline side chain and fatty acyl chain methyls). Thus, T_1's of lipopro-

---

[54] R. E. London and J. Avitable, *J. Am. Chem. Soc.* **99**, 7765 (1977).
[55] M. F. Brown, *J. Chem. Phys.* **80**, 2808 (1984).

TABLE IV
SELECTED ^{13}C NT$_1$ VALUES OF HDL RESONANCES AT DIFFERENT MAGNETIC FIELDS (IN kG)[a]

Peak	14.2 kG[b]	23.2 kG[c]	47 kG[d]	63.4 kG[c]
CH=CH (7)	0.22	0.29	0.37	0.43
CH=CH—CH$_2$—CH=CH— (8)	0.27	0.33	0.41	0.53
Cholesteryl ester C-6 (9)	0.036	0.050	0.12	0.16
Cholesteryl ester C-3 (10)	—	—	0.15	0.20
Choline CH$_2$N (11)	—	—	0.56	0.72
Choline CH$_2$O (12)	—	—	0.48	0.68
Choline (CH$_3$)$_3$N (14)	1.2	1.3	1.4	1.5
CH$_3$CH$_2$ (28)	3.3	4.8	4.5	—
Cholesteryl ester C-18 (29)	—	0.81	1.7	—

[a] Values of T_1 for HDL or HDL$_3$ were measured at 37 ± 2°. Each T_1 value (in seconds) has been multiplied by the number of directly bonded hydrogens (N). The number in parentheses after the peak assignments are the peak designations in Fig. 6C. C indicates the specifically assigned carbon.
[b] J. A. Hamilton, C. Talkowski, R. F. Childers, E. Williams, A. Allerhand, and E. H. Cordes, *J. Biol. Chem.* **249**, 4872 (1974).
[c] J. R. Brainard, R. D. Knapp, J. R. Patsch, A. M. Gotto, Jr., and J. D. Morrisett, *Ann. N.Y. Acad. Sci.* **348**, 299 (1980).
[d] J. A. Hamilton, unpublished results, 1980.

teins must be measured at the specific field of interest, and model system data must be acquired at the same field.

Linewidth ($\nu_{1/2}$)

The experimentally measured $\nu_{1/2}$ of a ^{13}C peak always has a contribution from molecular motions and may include significant contributions from instrumental broadening, chemical shift effects, scalar couplings, and chemical exchange. Instrumental broadening (e.g., field inhomogeneities) will be uniform for all resonances and thus can be evaluated from the $\nu_{1/2}$ of the narrowest peak in the spectrum, which in the case of lipoproteins will generally be that from a nonprotonated carbon or the fatty acyl methyl carbon (e.g., Fig. 6). Experimental problems such as poor shimming can contribute several Hz to $\nu_{1/2}$ and should be assessed carefully. Chemical shift effects may cause an apparent broadening of a peak and complicate the relationship of $\nu_{1/2}$ to molecular motions. For example, the C-14,17 resonance at 56.6 ppm (Table III) appears as a single resonance in the low field spectrum of HDL (Fig. 6C) but exhibits a splitting in high field HDL spectra.[38] The fatty acyl C-2' peak also appears as a single peak at low fields (e.g., Fig. 6) but shows a splitting in high field spectra of arterial plaques and LDL.[42] Other fatty acyl resonances which appear as single peaks may include contributions from chemical shift effects because of the different ester linkages or possibly because of structural features. These contributions must be small on an absolute basis, since most fatty acyl resonances are narrow, but could be significant on a relative basis. Chemical exchange has not been shown to affect the $\nu_{1/2}$ of any resonance in lipoprotein spectra. In two cases where exchange between more than one environment may take place (lipid carbonyl resonances[43] and the cholesterol C-4 resonances[45]), the exchange is slow on the NMR time scale. Although scalar couplings between ^{13}C and ^{1}H are removed by broadband ^{1}H decoupling, scalar couplings of ^{13}C with ^{31}P and ^{14}N are not. The choline methylene carbons and the glyceryl carbons of phospholipids are coupled to the ^{31}P nucleus, and the choline CH$_2$N and the quaternary methyl carbon to the ^{14}N nucleus; thus, the $\nu_{1/2}$ of these peaks includes contributions from scalar coupling.[56]

In the case of isotropic reorientation, the spin–spin relaxation time T_2 is inversely proportional to τ_R. If factors other than molecular motions are insignificant, $\nu_{1/2} = 1/\pi T_2$, and $\nu_{1/2}$ is directly proportional to τ_R; the $\nu_{1/2}$ of a methine ^{13}C resonance increases monotonically from <0.1 Hz at $\tau_R = 10^{-11}$ sec to >100 Hz at $\tau_R = 10^{-7}$ sec.[34] In the extreme narrowing limit, $T_2 = T_1$; except in the range of $\tau_R = 8 \times 10^{-9}$ to 5×10^{-7} sec, $\nu_{1/2}$ or T_2 does not have a field dependence.[34] The equation for isotropic rotation can be used to predict the $\nu_{1/2}$ for a carbon rigidly attached to a spherical lipoprotein particle. The rotational correlation time of HDL$_3$ is 1.3×10^{-7} sec at 36° in a viscosity of 0.012 P.[35] The $\nu_{1/2}$ for a carbon with no internal rotations would be ~100 Hz (methine), ~200 Hz (methylene), or ~300 Hz (methyl). ^{13}C peaks of >100 Hz are difficult to detect on high-resolution NMR instruments unless the resonance is well isolated and S/N is excellent. Since most lipid resonances in HDL spectra are much narrower than 100 Hz[38] and can be narrower in spectra of large lipoprotein particles (e.g., VLDL[57]) than for HDL, $\nu_{1/2}$ of these resonances must be dominated by internal motions that are much faster ($\ll 10^{-7}$ sec) than the correlation time for particle rotation ($\geq 10^{-7}$ sec). That isotropic particle reorientation is not sufficient to produce narrow resonances is shown by the fact that lipoproteins which contain solid[31] or liquid-crystalline lipids (see below) do not exhibit narrow ^{13}C (or ^{1}H) peaks for these moieties.

The usefulness of the ^{13}C linewidth for detailed studies of the dynamics of different carbon atoms in the same molecule is exemplified by

[56] R. E. London, T. E. Walker, D. M. Wilson, and N. A. Matwiyoff, *Chem. Phys. Lipids* **25**, 7 (1979).
[57] J. A. Hamilton, N. J. Oppenheimer, R. Addelman, A. O. Clouse, E. H. Cordes, P. M. Steiner, and C. J. Glueck, *Science* **194**, 1424 (1976).

the temperature-dependent spectral changes of cholesteryl esters. Figure 7 shows ^{13}C NMR spectra of neat cholesteryl oleate at various temperatures. In the isotropic liquid phase, narrow resonances are observed for steroid ring and acyl chain carbons. With decreasing temperature in the isotropic phase, there is selective broadening of resonances: most fatty acyl resonances remain narrow while most steroid ring resonances broaden significantly. In addition, the steroid ring resonances broaden differentially. For example, the C-6 $\nu_{1/2}$ increases from 10 to 18 Hz and the C-3 $\nu_{1/2}$ from 20 to 55 Hz between 46 and 37°. Thus, the C-3/C-6 $\nu_{1/2}$ increases from 2.0 to 3.0 and is largest near the liquid → liquid crystalline phase transition. Since the C-6 and C-3 are both methine carbons in a rigid ring system undergoing no internal motions, they would have equivalent $\nu_{1/2}$'s in the case of isotropic reorientation.[34] The $\nu_{1/2}$ results show that steroid ring motions are anisotropic and become increasingly anisotropic near (within 10°) the phase transition.[42] Below the phase transition temperature, most steroid ring resonances are broadened beyond detection, as are fatty acyl resonances proximal to the steroid ring (e.g., the carbonyl, C-2′, and C-3′); several other fatty acyl resonances remain relatively narrow. Thus, qualitatively, the $\nu_{1/2}$ results demonstrate a marked restriction of steroid ring motions in the cholesteric and smectic liquid crystalline phases, but a much lower restriction of the hydrophobic terminal portion of fatty acyl chains. Quantitatively, steroid ring motions of cholesteryl esters have been described using a model of anisotropic motion with two correlation times, one for the long axis of the ring and another for the two short axes.[58] This analysis showed that $\nu_{\frac{1}{2}}$ was much more sensitive to changes in anisotropic motions of the steroid ring than T_1 or NOE. The same quantitative approach has been applied to neat cholesteryl esters and to cholesteryl esters in lipoproteins and in rabbit atherosclerotic plaques.[59]

Even without a rigorous analysis relating $\nu_{1/2}$ to correlation times, $\nu_{1/2}$ data can provide an empirical means of distinguishing cholesteryl esters in the different microenvironments in various plasma lipoproteins and in atherosclerotic plaques. Linewidths can also provide a quantitative *relative* measure of cholesteryl ester anisotropic motions in these systems.[38,43,58,60] These types of analyses are based on predictions from model systems, such as those in Table V and those described by Hamilton *et al.*[42] Table V shows $\nu_{1/2}$ data at ~37° for the C-3 and C-6 cholesteryl ester resonances in spectra of cholesteryl esters alone and of cholesteryl

[58] D. M. Quinn, *Biochemistry* **21**, 3548 (1982).
[59] P. A. Kroon, D. M. Quinn, and E. H. Cordes, *Biochemistry* **21**, 2745 (1982).
[60] J. D. Morrisett, J. W. Gaubatz, A. P. Tarver, J. K. Allen, H. J. Pownall, P. Laggner, and J. A. Hamilton, *Biochemistry* **23**, 5343 (1984).

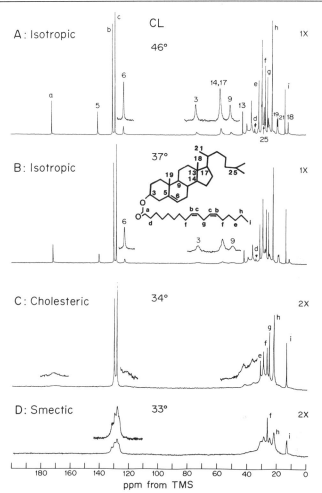

FIG. 7. Natural abundance ^{13}C Fourier transform NMR spectra (25.2 MHz) with ^{1}H decoupling at various temperatures of neat cholesteryl linoleate (CL) recorded after 2048 accumulations with 250-ppm spectral widths, 8192 time domain addresses, and a recycle time of 1.50 sec. The main spectra in C and D were printed with a 2-fold vertical expansion relative to the main spectra in A and B, as indicated in the upper right-hand corner of each spectrum. The insets in A and B were printed with a 4-fold vertical expansion relative to the main spectra; those in C and D with a 2-fold vertical expansion. The temperature at which each spectrum was obtained and the type of phase present is indicated for each spectrum. Assignments are indicated as follows: the numbered peaks correspond to carbon atoms of cholesterol as numbered by the standard system and shown in the inset in B; lettered peaks correspond to fatty acyl carbons as indicated [J. A. Hamilton, N. Oppenheimer, and E. H. Cordes, *J. Biol. Chem.* **252**, 8071 (1977)].

TABLE V
LINEWIDTHS AND SPIN-LATTICE RELAXATION TIMES[a]

System[b]	Temperature	C-6 $\nu_{1/2}$[c]	C-6 T_1[d] (msec)	C-3 $\nu_{1/2}$[e]	$\dfrac{\text{C-3 }\nu_{1/2}}{\text{C-6 }\nu_{1/2}}$[f]
76% TO, 18% CL, 6% CO	36°	6.5		7.5	1.15
58% TO, 31.5% CL, 10.5% CO	36°	6.0	62	8.0	1.3
22% TO, 58.5% CL, 19.5% CO	36°	13		21	1.6
13% TO, 65% CL, 22% CO	37°	13		28	2.2
4% TO, 72% CL, 24% CO	36.5°	18	62	65	3.6
75% CL, 25% CO	37°	21		70	3.3
Neat CL	38°	18	60	50	2.8

[a] Linewidths ($\nu_{1/2}$) and spin-lattice relaxation times (T_1) at 23.4 kG (25.2 MHz) at ~37° for the cholesterol ring C-6 and C-3 resonances in cholesteryl linoleate (neat) and in a 3/1 cholesteryl linoleate/cholesteryl oleate mixture containing various amounts of triolein (J. A. Hamilton, unpublished results, 1977).
[b] Composition by weight percent of triolein (TO), cholesteryl linoleate (CL), and cholesteryl oleate (CO); each system (except neat CL) contains a 3/1 CL/CO ratio and was an isotropic liquid at the indicated temperature. The mixture containing 13% TO exhibited an isotropic → smectic-like transition at ~26° on rapid cooling. The three systems with >22% TO exhibited only an isotropic → crystal transition, and the three systems with <4% exhibited isotropic → liquid crystalline phase transitions, as reported in J. A. Hamilton, N. J. Oppenheimer, and E. H. Cordes, *J. Biol. Chem.* **252**, 8071 (1977).
[c] Linewidth of the cholesterol ring C-6 resonance in Hz.
[d] Spin-lattice relaxation time (T_1) in msec for the C-6 carbon for selected systems. For each system, the T_1 value of the C-6 resonance was equal or nearly equal to the T_1 value of the C-3 resonance.
[e] Linewidth of the cholesterol ring C-3 resonance in Hz.
[f] Ratio of the C-3 $\nu_{1/2}$ to the C-6 $\nu_{1/2}$.

esters with added triolein. The C-3 and C-6 $\nu_{1/2}$'s decrease with increasing triolein content; $\nu_{1/2}$'s and the C-3/C-6 $\nu_{1/2}$ ratio are smallest for the systems with the greatest triolein content (58 and 76%). The individual $\nu_{1/2}$'s and C-3/C-6 $\nu_{1/2}$ are largest for systems which at ~37° are close to an isotropic → liquid crystalline transition (neat cholesteryl linoleate, cholesteryl linoleate/cholesteryl oleate, and the latter mixture with 4% triolein). The T_1 of the C-6 resonance is insensitive to the changes in composition and consequent changes in anisotropic motions. Thus, in this case, $\nu_{1/2}$ but not T_1 provides a means of distinguishing cholesteryl esters in a wide range of physiologically relevant environments.

The application of ^{13}C linewidth measurements to complex lipid systems is exemplified by the use of $\nu_{1/2}$ to probe temperature-dependent changes in the structural organization of human LDL. This lipoprotein exhibits a thermotropic transition just below body temperature as a result of a phase transition in the core of the particle.[61] Natural-abundance ^{13}C FT NMR spectra of LDL at 36 and 45° (Fig. 8A and B) exhibit narrow resonances from all major lipid types. Notably, the intense C-6 and C-3 resonances do not include a contribution from unesterified cholesterol, as shown by the resolution of the unesterified cholesterol C-6 peak (peak 6') from the more intense cholesteryl ester C-6 resonance (peak 6), and thus provide a probe specific for cholesteryl esters. The C-3 resonance is broader than the C-6 resonance at all temperatures, indicating that the steroid ring motions are anisotropic. With decreasing temperature the C-6 and C-3 peaks broaden differentially; the C-3/C-6 $\nu_{1/2}$ ratio increases from 1.4 at 45° to 1.8 at 32°, showing that the steroid ring motions become increasingly anisotropic with decreasing temperature.[43] At 12° (Fig 8D) most cholesteryl ester steroid ring resonances are broadened beyond detection. Although the fatty acyl resonances include an appreciable contribution from phospholipids and triglycerides as well as cholesteryl esters (see above), comparison of the LDL spectrum at 12° with spectra of cholesteryl linoleate (Fig. 7) shows that the LDL spectra do not contain broad components in the olefinic and aliphatic regions that are characteristic of the smectic mesophase of neat cholesteryl linoleate. In contrast, several fatty acyl resonances remain narrow at 12°, as do those resonances for the choline side chain. In fact, the $\nu_{1/2}$ of the choline methyl resonance does not change significantly between 45 and 12°, and thus serves as an internal $\nu_{1/2}$ standard. Several weaker resonances seen at 45° (peak 6' and "PL" and "TG" peaks) broaden with decreasing temperature and disappear by 25°; however, the S/N is very low and the degree of broadening does not have to be large before these peaks become "unobservable." Thus, the temperature-dependent $\nu_{1/2}$ changes in the LDL spectrum are mainly attributable to cholesteryl ester resonances and specifically indicate changes in steroid ring anisotropy characteristic of a liquid → liquid crystalline transition. No effects of this transition are detected by the $\nu_{1/2}$'s of the choline side chain resonances or resonances from fatty acyl carbons in the distal portion of the chain.

Sample Temperature

Temperature is a critical variable in NMR studies of lipids and lipoproteins, particularly since many of these systems exhibit thermotropic phase transitions. Linewidth, T_1, and NOE values and intensities of lipid resonances may exhibit a marked temperature dependence. Two aspects

---

[61] R. J. Deckelbaum, G. G. Shipley, and D. M. Small, *J. Biol. Chem.* **252**, 744 (1977).

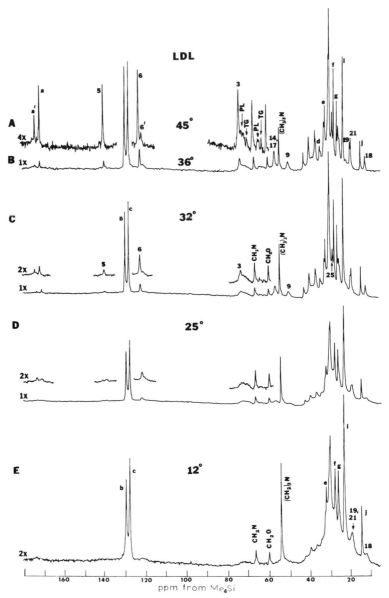

FIG. 8. Natural abundance ^{13}C Fourier transform NMR spectra (67.9 MHz) with ^{1}H decoupling of LDL (ρ 1.025–1.050) at 30 mg protein/ml at various temperatures: (A) incomplete spectrum at 45°, 29,873 accumulations with a pulse interval of 1.20 sec and 16,384 time domain points. Spectrum is printed with a 4-fold vertical expansion relative to the main spectra in B–D; (B) complete spectrum at 36°, 32,768 accumulations with a pulse interval of

of temperature will be emphasized: thermal inhomogeneities and measurement of sample temperature. In most commercial high-resolution NMR spectrometer systems sample temperature is controlled by passing a stream of air by the sample tube (e.g., Fig. 9). The temperature is measured by a thermocouple located near the bottom of the NMR tube (Fig. 9) and displayed on the temperature control unit. In ^{1}H NMR studies using a 5-mm NMR tube, the sample temperature will be similar but not necessarily identical to the external air temperature. In ^{13}C NMR studies, use of broadband ^{1}H decoupling irradiation increases the sample temperature and causes temperature gradients by inductive dielectric heating.[62,63] The extent of the thermal perturbations depends on the magnetic field strength, the decoupler power, and the ionic strength of the sample; internal sample temperatures can increase by several degrees for aqueous samples of low ionic strength (≤ 0.05 M NaCl) with continuously applied broad band decoupling power.[62] Although the sample temperature reaches a steady-state value, it may be significantly higher than that of the external air temperature. It is thus imperative to measure the temperature in the sample, particularly in ^{13}C NMR experiments using large diameter (≥ 10 mm) NMR tubes. Since there is no accurate and convenient internal chemical "thermometer" for ^{13}C NMR samples, we measure the temperature in the sample by a thermocouple. After 15–20 min in the NMR probe under the *exact* accumulation conditions (except with the vortex plug removed) to reach the steady-state temperature, the sample is ejected. The tip of a thin thermocouple supported in a long glass capillary is placed in the center of the sample (without touching the sides of the NMR tube), and temperature is recorded as a function of time. Extrapolation to time zero gives a temperature which is generally accurate to $\pm 0.5°$ in a range of about $\pm 25°$ from room temperature. A nonmagnetic optic fiber thermometer (Luxtron) is now commercially available and permits measurement of

[62] J. J. Led and S. B. Petersen, *J. Magn. Reson.* **32**, 1 (1978).
[63] D. S. McNair, *J. Magn. Reson.* **45**, 490 (1981).

0.501 sec; (C) 32° and (D) 25°, conditions as in B; insets are printed with a 2-fold vertical expansion; (E) 12°, 32,768 accumulations with a pulse interval of 0.401 sec; spectrum is printed with a 2-fold vertical expansion relative to spectrum B and main spectra in C and D. Peaks are labeled as in Fig. 7, except for the following additions: peak a', phospholipid carbonyl carbons; peak b, as in Fig. 7 inset, also contains the methine (unsaturated) carbons of oleate esters; peak c, as in Fig. 7 inset; PL and TG, glyceryl backbone carbons of phospholipids (PL) and triglycerides (TG). A resonance for unesterified cholesterol C-6 is designated 6'; the intensity of this peak relative to the cholesteryl ester C-6 peak is low, as in the case of HDL, and this narrow component probably represents only the fraction of the cholesterol which is dissolved in the core of LDL. The chemical shift sale is in ppm downfield from tetramethylsilane (Me$_4$Si) (J. A. Hamilton, unpublished data, 1978).

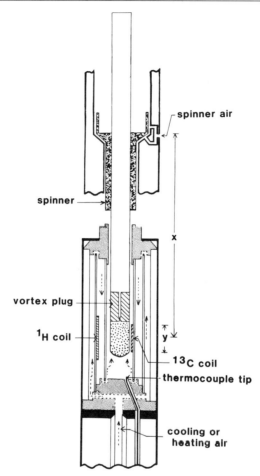

FIG. 9. Schematic partial cross-section of the Bruker 10 mm ^{13}C NMR probe showing a 10-mm sample tube in place. The ^{13}C transmitter/receiver coil and the ^{1}H decoupling coil are located on concentric glass cylinders, as shown. To localize the sample (shaded area in NMR tube) precisely in the ^{13}C coil, the distance between the spinner and the ^{13}C coil (for example, distance x) and the length (distance y) of the coil must be known. For Bruker narrow-bore probes for any nucleus, the distance between the top of the flange on the spinner and the center of the ^{13}C coil (distance x) is *generally* 100 mm; the length of the transmitter/receiver coil (distance y) is variable. The cooling/heating air enters the chamber below the sample (and does not contact the sample), passes by the ^{1}H decoupling coil and then passes the thermocouple, the ^{13}C coil, and the sample, as indicated by the dashed arrows.

temperature in the sample when in the probe. The temperature control unit can be calibrated by measuring the temperature in a range of settings using the lipid-free aqueous buffer to avoid sample contamination and loss. However, the calibration must be repeated if the ionic strength of the sample or if the air flow is changed. Temperature gradients within the sample can be reduced by using the minimum required sample volume (see below), reducing the decoupling power or gating the decoupler,[64] reducing the ionic strength of aqueous samples, or by using highly conductive material for the NMR tube.[63] Improved temperature uniformity can also be achieved by using smaller sample tubes or by placing the sample in a coaxial insert immersed in the lock solvent contained in the NMR tube.[65]

Optimizing Sample Volume and Signal/Noise Ratio

Sample volume considerations are especially important in ^{13}C NMR spectroscopy because of the low sensitivity of the ^{13}C nucleus. Two fundamental methods for improving the S/N of a peak (all other factors constant) are to increase the number of nuclei in the volume sampled by the detection coil and to signal average. S/N is directly proportional to the number of nuclei but only to the square root of the number of spectral accumulations. Thus, in ideal cases, the S/N can be doubled by concentrating the sample by a factor of two or by increasing the number of scans by a factor of four. (A notable exception to this ideal case is that of a small signal in a spectrum containing a very intense signal.[66]) Thus, the best strategy for ^{13}C NMR studies where sensitivity is important is to use the minimum required volume of the most concentrated sample. (In some cases, however, concentrating a sample may change its bulk or microscopic properties in a manner which decreases S/N or alters undesirably a property being investigated.) As shown in Fig. 9, the location of the detection coil relative to the spinner must be determined for precise placement of the sample within the coil. The sample volume should extend slightly (~1 mm) above the coil, and the rounded part of the NMR tube should extend slightly below the bottom of the coil (Fig. 9). In this scheme, a vortex plug is used and all trapped bubbles must be removed; otherwise both S/N and field homogeneity will decrease. A small savings in sample volume can be realized by using a flat-bottom NMR tube. If sample sensitivity is not particularly critical, a slight sample excess can be

[64] T-M. Chan and J. L. Markley, *Biochemistry* **22**, 5982 (1983).
[65] G. S. Ginsburg, D. M. Small, and J. A. Hamilton, *Biochemistry* **21**, 6857 (1982).
[66] B. J. Kimber, K. Roth, and J. Feeney, *Anal. Chem.* **53**, 1026 (1981).

used for convenience, but a large excess will increase thermal gradients. When only small amounts of sample are available, use of a cylindrical or spherical sample tube insert will optimize S/N by placing all the sample within the coils (provided again that the probe geometry is accurately known). This configuration also obviates the use of a vortex plug and permits sample volume changes in temperature studies. However, the field homogeneity must be checked carefully when such sample configurations are used. When larger amounts of sample are available in a suitably concentrated form, larger sample tubes can be used, depending on the probe size available. The increase in S/N may be somewhat less than that expected on the basis of the volume increase. If conditions are optimized on a Bruker WP200 spectrometer equipped with a 10 mm probe, lipoprotein spectra of quality (i.e., S/N per unit time) comparable to those shown in Fig. 6 are obtained with only a 1.2 ml sample using about the same concentration of lipoprotein.[67]

Model Systems

Judicious choice of model systems is essential for lipoprotein NMR studies, as exemplified in the above discussions. Model systems range from very simple ones like neat cholesteryl esters to very complex ones, such as total lipoprotein recombinants. An important advantage provided by models is the capability of using lipid or protein components in which specific nuclei are enriched. For example, ^{13}C-enriched lipids may provide information which cannot be gleaned from natural abundance ^{13}C spectra of native lipoproteins because of inadequate resolution, low detectability, or inability to distinguish between similar structural moieties (e.g., fatty acyl chain carbons of phospholipids, cholesteryl esters, and triglycerides). The low natural abundance of the ^{13}C nucleus, which is generally viewed as a liability, becomes an asset when selective enrichment is used, since this allows observation of a particular carbon with greatly enhanced sensitivity. Some advantages and problems of using ^{13}C enrichment in lipoprotein studies have been noted above. In studies of modified or reassembled lipoproteins containing ^{13}C-enriched moieties, it is essential to determine the degree to which specific structural features of the model lipoprotein resemble those of the native counterpart. Also, it is useful to obtain a background natural abundance spectrum with as good S/N as feasible to help validate the comparison of native and model systems.

There are a variety of model system studies which have been directly

---

[67] J. A. Hamilton, unpublished results, 1981.

or peripherally related to lipoprotein structure. Selected examples not cited above are given below in brief outline form to illustrate different strategies:

1. Use of specifically ^{13}C-enriched lipids to measure T_1 values of fatty acyl resonances of different lipids in lipid–protein recombinants.[68]
2. Use of specifically ^{13}C-enriched apolipoprotein C-I ([^{13}C]methyl enrichment of the methionine-38 residue) to compare molecular motions of the apolipoprotein in aqueous solution and in complexes with dimyristoyl PC.[69]
3. Use of [^{13}C]carbonyl-enriched triglycerides and cholesteryl esters to determine the solubility and conformation of these lipids in phospholipid bilayers.[70]
4. Use of the ^{1}H chemical shift of the choline methyl resonance to monitor structural changes in complexes of apolipoprotein A-I with phospholipids as a function of the lipid/protein ratio.[71]
5. Use of ^{1}H spin–spin couplings to determine the effect of apolipoprotein A-I on the PC polar group conformation.[72]
6. Use of specifically ^{2}H-enriched cholesteryl palmitate to study motions and conformations of this ester in recombinant HDL.[73]

Other Nuclei

Several magnetically active nuclei present in lipoproteins have received very little, if any, attention to date. These include ^{2}H, ^{14}N, ^{15}N, ^{17}O, and ^{33}S (Table I). A primary limitation of all these nuclei is their inherently low sensitivity and, except for ^{14}N, their low isotopic abundance (Table I). In addition, except for ^{2}H, these nuclei have a limited chemical distribution in lipoprotein lipids. However, each nucleus may provide unique information and may have special applications to lipoproteins.

[68] W. Stoffel, O. Zierenberg, B. Tunggal, and E. Schreiber, *Proc. Natl. Acad. Sci. U.S.A.* **71**, 3696 (1974).
[69] T-C. Chen, R. D. Knapp, M. F. Rohde, J. R. Brainard, A. M. Gotto, Jr., J. T. Sparrow, and J. D. Morrisett, *Biochemistry* **29**, 5140 (1980).
[70] J. A. Hamilton, K. W. Miller, and D. M. Small, *J. Biol. Chem.* **258**, 12821 (1983).
[71] C. G. Brouilette, J. L. Jones, T. C. Ng, H. Kercret, B. H. Chung, and J. P. Segrest, *Biochemistry* **23**, 359 (1984).
[72] D.-J. Reijnoud, S. Lund-Katz, H. Hauser, and M. C. Phillips, *Biochemistry* **21**, 2977 (1982).
[73] Y. I. Parmar, H. Gorrissen, S. R. Wassall, and R. J. Cushley, *J. Biol. Chem.* **258**, 2001 (1983).

Deuterium is a nucleus with spin $I = 1$ and with an electric quadrupole moment; thus, ^{2}H NMR spectra may exhibit quadrupolar splittings from which order parameters can be calculated.[4] The very low overall NMR sensitivity of ^{2}H can be improved markedly by isotopic enrichment, as exemplified by a study of HDL reconstituted with specifically deuterated cholesteryl palmitate.[73] In this study, molecular motions at the enrichment sites were probed by T_1 and $\nu_{1/2}$ measurements, but quadrupole splittings were not observed because of rapid particle tumbling which averaged the quadrupolar interactions and produced narrow Lorentzian resonances.[73] It might be possible to observe quadrupole couplings in ^{2}H NMR spectra of lipoproteins that have been concentrated to a viscous gel or immobilized by attachment to a solid matrix. Cholesteryl esters in liquid crystalline phases can be aligned in strong magnetic fields to yield ^{2}H NMR spectra with quadrupole splittings.[74] NMR spectroscopy of ^{2}H may have good potential for studying the apolipoproteins in a lipid matrix since the molecular motions of many protein groups are greatly restricted compared to those of the lipids.[6,33,35,40] In addition, the incorporation of selectively deuterated amino acids, especially aromatic ones, into synthetic peptides that mimic the lipid-binding behavior of apolipoproteins may provide information about local motions in hydrophobic regions[75] which are thought to have strong interactions with phospholipid acyl chains.

Oxygen is found in the polar regions of all lipoprotein lipids. The ^{17}O nucleus has a large chemical shift range,[4] and ^{17}O NMR[76] could be useful for studying hydrogen-bonding interactions that might occur between proton donor and acceptor moieties on molecules such as PC, SM, and cholesterol. However, the low sensitivity and the very broad resonances seen in general for ^{17}O will limit its application to lipoproteins.

Sulfur does not occur naturally in lipoprotein lipids, although thio analogs of cholesterol and phospholipids have been synthesized. Sulfur is present in methionine, cysteine, and cystine residues in several human apolipoproteins (e.g., apolipoprotein A-II, apolipoprotein B, apolipoprotein E, and apolipoprotein [a]). Some thiol or disulfide moieties play important structural and/or functional roles, and ^{33}S NMR spectroscopy could be useful in detailing these features. However, the low molar concentration of these nuclei combined with their very low isotopic abundance and low sensitivity greatly limit the feasibility of ^{33}S NMR experiments.

[74] Z. Luz, R. Poupko, and E. T. Samulski, *J. Chem. Phys.* **74**, 5825 (1981).
[75] M. A. Keniry, H. S. Gutowsky, and E. Oldfield, *Nature (London)* **307**, 383 (1984).
[76] J.-P. Kintzinger, *in* "NMR Basic Principles and Progress" (P. Diehl, E. Fluck, and R. Kosfeld, eds.), Vol. 17, p. 50. Springer-Verlag, Berlin, 1981.

Nitrogen has a limited distribution in lipoprotein lipids, since it is present mainly in choline or ethanolamine phospholipids and only in the polar regions of these lipids. Nitrogen is much more abundant and widely distributed in proteins. ^{14}N is the more abundant of the two nitrogen isotopes and has a spin $I = 1$ (Table I) with a generally quite large quadrupole coupling constant. Even for molecules in solution with rapid components of motion, ^{14}N resonances may be very broad.[77] In contrast, ^{15}N is a rare nitrogen isotope and has a spin $I = \frac{1}{2}$ (Table I). The combination of low abundance, low intrinsic sensitivity, and negative NOEs makes NMR of even highly ^{15}N-enriched samples difficult. However, the very large chemical shift range of the ^{15}N nucleus (~1000 ppm) makes it a good probe of molecular interactions involving ionic and hydrogen bonds.[78] In addition, the capability of interpreting T_1 and $\nu_{1/2}$ data in terms of molecular motions (in a manner similar to ^{13}C) could make it attractive for studying dynamics of various nitrogen-containing moieties in lipoproteins.

[77] T. T. Cross, J. A. DiVerdi, and S. J. Opella, *J. Am. Chem. Soc.* **104**, 1759 (1982).
[78] F. Blomberg and H. Ruterjans, in "Biological Magnetic Resonance" (L. J. Berliner and T. Reuben, eds.), Chap. 2. Plenum, New York, 1983.

[29] Spectroscopic Studies of Lipoproteins

By H. J. POWNALL and J. B. MASSEY

Introduction

All molecules absorb light with an absorbtivity and frequency that is a function of their electronic structure. Following light absorption the molecule remains in the excited state for a time interval that is a function of the rates of several processes that deactivate the excited state. These processes include both radiative and nonradiative transitions. A major radiative mechanism is that of fluorescence, a process that involves a transition of the molecule from its lowest excited singlet state to the singlet groundstate. Both absorption and emission spectra are sensitive to the environment of the molecule that is irradiated. For this reason electronic spectroscopy provides an important technique for assessing protein concentration and structure. These two areas can be divided into intrinsic and extrinsic methods. In the former one can, in favorable situations, utilize the known spectroscopic properties of fluorescent moieties within

TABLE I
EXTINCTION COEFFICIENTS OF SEQUENCED APOLIPOPROTEINS

Protein	Molecular weight	Molar extinction coefficient[a] (M^{-1} cm^{-1})	Weight extinction coefficient (ml/mg-cm)
ApoA-I	28,016	31,720 (280 nm)	1.13
ApoA-II	17,414	12,000 (276 nm)	0.69
ApoC-I	6,630	5,690 (280 nm)	0.86
ApoC-II	8,916	12,090 (280 nm)	1.36
ApoC-III[b]	8,764	19,630 (280 nm)	2.24
ApoE	34,145	44,950 (280 nm)	1.32

[a] Values are calculated for their near ultraviolet maxima given in parentheses.
[b] Protein only; weight coefficient varies with sialic acid content.

a large molecule such as a protein to address structural questions concerning the protein and its environment. In proteins the most important intrinsic structural probes are the aromatic amino acids phenylalanine, tyrosine, and tryptophan, with the latter being most important. With either extrinsic or intrinsic probes one can study protein structure by several methods including measuring the fluorescence spectrum, polarization, quenching of intensity and lifetime, and energy transfer to suitable donors.

Absorption Spectroscopy

The absorption spectra of purified proteins may be used to quantify their concentration according to the summation method of Edelhoch.[1] This procedure states that the absorption spectrum of a protein is the arithmetic sum of the spectra of its aromatic amino acid residues. From this one can estimate the molar extinction coefficient Σ of a protein from $\varepsilon = \Sigma_i \varepsilon_i$, where ε_i is the molar extinction coefficient of the ith aromatic amino acid residue. If the absorption (A) is measured in a denaturant such as guanidinium chloride (GdmCl), the concentration (c) calculated from Beers law ($c = A/El$, where l = cell pathlength) is within 5% of that determined by an independent method. This permits us to calculate the A_{280} of any apolipoprotein from its amino acid composition. Table I contains a list of the molar and weight extinction coefficients of a number of apolipoproteins whose compositions are known with certainty. These values are useful for the rapid and accurate determination of protein concentration and are most reliable when the absorbance is measured in 3 M

[1] H. Edelhoch, *Biochemistry* **6**, 1948 (1967).

GdmCl. The greatest risk in making absorbance measurements of lipoproteins and apolipoproteins is the contribution of scattered light to the attenuation of the incident light intensity. This may be obviated, in part, by the use of detergents with lipoproteins or the addition of GdmCl to apolipoproteins. In spite of these precautions, light scattering from macromolecules such as proteins and especially the larger lipoproteins can present a problem. If the light scattering is significant but not overwhelming, one can calculate the true absorbance from the known λ^{-4} variation of scattered light intensity. Proteins absorb light below 300 nm and above 300 nm most of the light attenuation is due to Rayleigh scattering. From this relationship we can determine that the light scattering at 280 nm is twice that observed at 332 nm. The calculated value is then subtracted from the observed absorbance. In every case, a spectrum of the buffer should be collected to ensure that the actual light signal at 280 nm is optimized.

Fluorescence Properties of Apolipoproteins

All of the A and C apolipoproteins contain aromatic amino acids that contribute to their fluorescence and phosphorescence. Only the former is easily measured at ambient temperatures and will be discussed here. The fluorescence of a protein is usually characteristic of the amino acid that has the lowest lying excited singlet state. Illumination of proteins that contain phenylalanine, tyrosine, and tryptophan almost always produces fluorescence from tryptophan, which has the lowest lying excited singlet state. This group includes apoA-I, apoB, apoC-II, and apoC-III; apoC-I

TABLE II
WAVELENGTH MAXIMA FOR THE FLUORESCENCE OF
SELECTED APOLIPOPROTEINS

Protein	Concentration (μM)	λ_{max} (nm)	
		Tris-buffered saline	3 M GdmCl
ApoA-I	3	333	350
ApoA-II[a]	14	304	304
ApoC-I	2	347	350
ApoC-II[b]	7	343	350
ApoC-III	6	343	350

[a] Contains tyrosine but no tryptophan.
[b] Dialyzed from GdmCl into 150 mM NaCl, 10 mM Tris, pH 7.4 and measured within 6 hr of start of dialysis. Due to self-association some of these values are concentration dependent unless a denaturing salt such as guanidine hydrochloride (GdmCl) is employed.

contains phenylalanine and tryptophan but no tyrosine. All four of these proteins exhibit tryptophan fluorescence (Table II). In denaturing solvents such as guanidine hydrochloride the emission, which is maximal at 350 nm, is characteristic of tryptophan in an aqueous environment. If the tryptophan is buried in a hydrophobic "pocket" such as one might expect in lipoproteins and in solutions of self-associated or folded proteins, the emission maximum of the tryptophan is shifted to shorter wavelengths. The magnitude of the blue shift is a qualitative measure of the hydrophobicity of the environment surrounding this amino acid. With due caution, this observation is sometimes taken to represent the environment of all hydrophobic amino acids. ApoA-II contains phenylalanine and tyrosine but no tryptophan. Since the lowest excited singlet state of tyrosine is lower than that of phenylalanine, apoA-II exhibits only tyrosine fluorescence. The fluorescence of tyrosine is not nearly as sensitive to solvent perturbations as tryptophan and its fluorescence is therefore rarely useful for structural studies except for quenching and pH studies.

Fluorescence is a very sensitive method and the level of detection is largely a function of the type of instrument employed and the purity of the solvent. The apolipoprotein fluorescence intensity may, with a suitable calibration curve based upon the fluorescence of standards of known intensity, be used to quantify concentration. Absorbance is defined by $A = -\ln T$ where T is the percentage transmission. Alternatively,

$$A = \log I_0/(I_0 - I_a) \quad \text{and} \quad I_a = 1 - 10^{-A}$$

where I_0 and I_a are the intensities of the incident and the absorbed light, respectively. The intensity of the emitted light (I_f) is directly proportional to I_a. On the basis of this equation, and the absorbance, one can calculate the percent of light that is absorbed (I_a). At infinite dilution, the change in I_a is a linear function of the absorbance. In practice if one uses fluorescence to quantify concentration, the intrinsic error is less than 5% if $A = 0.05$.

Conclusions

The characteristic absorption and fluorescence of apolipoproteins are useful in their detection, quantification, characterization, and structural analysis. The rapidity with which photometric measurements can be made favors their use in the routine analysis of apolipoproteins.

[30] Circular Dichroism of Lipoprotein Lipids

By G. CHI CHEN and JOHN P. KANE

Introduction

The application of circular dichroism (CD) in the ultraviolet wavelength region to the study of protein conformation was extensively reviewed in this series.[1] The lipid contents of different classes of human serum lipoproteins range from over 90% of the total mass for very low-density lipoproteins (VLDL) to about 80% for low-density lipoproteins (LDL) and 50% for high-density lipoproteins (HDL).[2] Such high contents of lipids make it important to ascertain the lipid contribution when analyzing the protein conformation from the CD spectrum of lipoproteins. In this chapter, we will discuss first the CD contributions in the ultraviolet wavelength region of constituent lipids of lipoproteins and their effects on the conformational analysis of the protein moieties. Second, we will discuss the application of a probe technique utilizing the induced CD of β-carotene, a normal constituent of lipoproteins, in the study of order in the lipid cores of those macromolecular complexes.

CD Contribution of Lipids in the Ultraviolet Wavelength Region

Intrinsic CD of Constituent Lipids of Lipoproteins

Several of the constituent lipids of lipoproteins are optically active molecules. We have shown that pure component lipids, namely cholesterol, cholesteryl esters, and phospholipids, when dissolved in organic solvents of spectral quality show appreciable CD bands in the ultraviolet wavelength region.[3] Among them, sphingomyelin shows the most prominent CD spectrum at 190–200 nm. Moreover, the observed CD are thermally dependent. Similarly, lipid extracts of LDL in trifluoroethanol–ethyl ether (3:1, v/v) or the fractions of extracted lipids soluble in trifluoroethanol or hexane show appreciable temperature-dependent CD bands in the same wavelength region.[3] Correction for the CD contribution of lipids in LDL, by subtracting the CD intensity corresponding to

---

[1] J. T. Yang, C.-S. C. Wu, and H. M. Martinez, this series, in press.
[2] V. P. Skipski, in "Blood Lipids and Lipoproteins: Quantitation, Composition, and Metabolism" (G. J. Nelson, ed.), p. 471. Wiley (Interscience), New York, 1972.
[3] G. C. Chen and J. P. Kane, *Biochemistry* **14**, 3357 (1975).

amounts of lipids in organic solvents equivalent to those found in LDL, shows that a substantial part of the thermal change in CD spectrum at 208–210 nm of LDL may be attributable to lipids.[3] However, correcting for the lipid contribution based on the studies of CD in organic solvents may be complicated by potential effects of those solvents on CD behavior. Still more important may be the organized state of lipids in the lipoprotein particles. As an approach to the evaluation of this question, we have studied liposomes and microemulsion particles as models for lipid structure in lipoproteins and found CD contribution due to organized lipids, especially in the far ultraviolet wavelength region.[4]

CD of Organized Lipids in Model Systems

Liposomes Containing Phospholipids Alone and in Combination with Cholesterol. Liposomes of egg phosphatidylcholine (egg PC) alone and liposomes of egg PC containing different amounts of cholesterol were prepared by passing aqueous lipid suspensions through a French pressure cell (at 20,000 psi) and were centrifuged at 35,000 rpm for 1 hr at 4° to remove small amounts of multilamellar liposomes.[5] Liposomes of egg PC alone showed a positive CD band between 210 and 230 nm (Fig. 1, curve 1). Below 210 nm, no CD signal could be detected on the most sensitive scale of the instrument. On the other hand, liposomes of egg PC containing cholesterol showed intense negative CD bands below 210 nm (Fig. 1, curves 2 and 3). Incorporation of cholesterol into the liposomes appears to be responsible for the intense negative CD bands but contributes little to the positive CD band. The specific ellipticity of the cholesterol-containing liposomes was then calculated, assuming that only egg PC contributed to the positive CD around 220 nm and only cholesterol contributed to the negative CD around 205 nm. The resulting magnitude of specific ellipticity at 205 nm was about 40 times that at 220 nm (Fig. 2A).

Microemulsion Particles Containing Phospholipids and Cholesteryl Esters. Microemulsion particles were prepared by injecting a mixture of egg PC and cholesteryl oleate dissolved in 2-propanol into aqueous buffer.[4] In a typical experiment, 20 mg of egg PC and 10 mg of cholesteryl oleate in 2.3 ml of 2-propanol at 50° were rapidly injected through a syringe fitted with a 27-gauge needle into 40 ml of rapidly stirred aqueous solution. The microemulsion solution was concentrated about five times in an ultrafiltration cell with the XM-50 membrane (Amicon Corp., Lex-

[4] G. C. Chen, L. S. S. Guo, R. L. Hamilton, V. Gordon, E. G. Richards, and J. P. Kane, *Biochemistry* **23**, 6530 (1984).
[5] R. L. Hamilton, J. Goerke, L. S. S. Guo, M. C. Williams, and R. J. Havel, *J. Lipid Res.* **21**, 981 (1980).

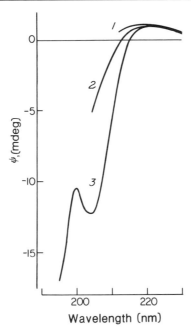

FIG. 1. CD spectra of egg PC liposomes containing various amounts of cholesterol at 25°. Suspensions of 10 mg egg PC containing 0, 25, and 50 mol% of cholesterol, respectively, in 0.02 M phosphate buffer and 0.001 M EDTA (pH 7.5) were passed twice through the French pressure cell (20,000 psi) at room temperature. The liposomes were centrifuged at 35,000 rpm for 1 hr at 4°. The concentration of egg PC and cholesterol (mg/ml), respectively, in the three preparations of liposomes measured were as follows: curve 1 (3.6, 0); curve 2 (3.5, 0.3); and curve 3 (2.2, 0.7), corresponding to 0, 13, and 37.5 mol% of cholesterol. CD signal, ψ, in millidegree was normalized as in a cell of 1 mm path length. [From G. C. Chen, L. S. S. Guo, R. L. Hamilton, V. Gordon, E. G. Richards, and J. P. Kane, *Biochemistry* **23**, 6530 (1984); adapted with the permission of the American Chemical Society (copyright 1984).]

ington, MA). The concentrated microemulsion preparation was centrifuged twice at 35,000 rpm for 16 hr to remove liposomes and then was separated on a 2% agarose column (1.2 × 90 cm) at room temperature. The microemulsion particles in the included volume fractions formed a peak containing nearly equal amounts of egg PC and cholesteryl oleate with particle diameters of 200–600 Å (mean 320 Å).[4]

These microemulsion particles showed CD spectra between 190 and 230 nm similar to those of the cholesterol-containing liposomes. Cholesteryl oleate appeared to have an intensity about 10 times higher in the positive band than egg PC, as indicated by the magnitude of specific ellipticity at 222 nm of about 40° cm²/dg for egg PC liposomes and about 300–400° cm²/dg for microemulsion particles (Fig. 2B).

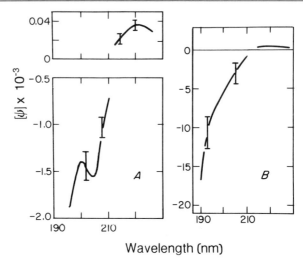

Wavelength (nm)

FIG. 2. Specific ellipticity, [ψ] in degrees centimeter squared per decagram at room temperature. (A) Liposomes of egg PC and cholesterol. The magnitude above 210 nm was calculated based on the concentration of egg PC and below 210 nm based on the concentration of cholesterol. Bars indicate the range of 3 preparations. (B) Microemulsion particles of egg PC and cholesteryl oleate. The magnitude was calculated based on the concentration of cholesteryl oleate. Bars indicate the range of seven samples from separate microemulsion preparations. [From G. C. Chen, L. S. S. Guo, R. L. Hamilton, V. Gordon, E. G. Richards, and J. P. Kane, *Biochemistry* **23**, 6530 (1984); adapted with the permission of the American Chemical Society (copyright 1984).]

Correction of CD Spectra of Lipoproteins for the Contribution of Lipids. The calculation of CD contribution of lipids in intact lipoproteins was based on the specific ellipticity of lipids in liposomes and microemulsion particles (Fig. 2). The CD due to lipids was then subtracted at 1-nm wavelength interval from the CD of the intact lipoproteins. Substantial differences were observed between the corrected CD spectrum and the original CD spectrum of the intact lipoproteins, as shown in Fig. 3. Specifically, the deep trough at 208 nm diminished to almost the same intensity as that at 222 nm. This was accompanied by a markedly increased peak with a red-shift of the crossover point, reflecting the strong negative CD contribution of lipids, as shown in liposomes and microemulsion particles. A CD spectrum with a deep 208 nm trough and a peak of relatively small magnitude, such as the uncorrected curve in Fig. 3, has been observed in VLDL of the hypercholesterolemic rabbit,[6] in the β-migrating

---

[6] G. C. Chen and J. P. Kane, *J. Lipid Res.* **20**, 481 (1979).

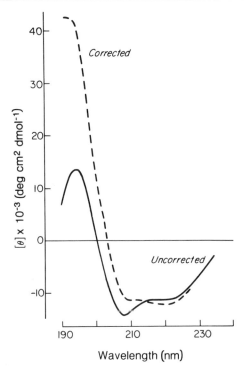

FIG. 3. Correction for the contribution of lipid to the CD spectrum of β-VLDL from patients with familial dysbetalipoproteinemia. CD data were expressed in terms of mean residue ellipticity, [θ] in degrees centimeter squared per decimole using a mean residue mass of 112 for the protein moiety. [Unpublished results by G. C. Chen (1984).]

VLDL of patients with familial dysbetalipoproteinemia,[7] and in canine apoE HDL_c.[4] This spectral feature thus appears characteristic of cholesterol- (or cholesteryl ester-) rich lipoproteins having a low ratio of protein to cholesterol.

Because of the small CD contribution of lipids around 222 nm[4] and the high CD intensity of helical conformation of the peptide bond[1] in this wavelength region, estimation of helix content for the protein moiety of lipoproteins based on the reference value of ellipticity at 222 nm[8] may be accepted with some confidence. On the other hand, estimation of various conformations of lipoproteins from the whole CD spectrum between 190

[7] J. P. Kane, G. C. Chen, R. L. Hamilton, D. A. Hardman, M. J. Malloy, and R. J. Havel, *Arteriosclerosis* **3**, 47 (1983).
[8] Y.-H. Chen, J. T. Yang, and H. M. Martinez, *Biochemistry* **11**, 4120 (1972).

and 240 nm[1] could be very uncertain. The uncertainty is due not only to the lipid contribution at the shorter wavelength region but also to the smaller signal-to-noise ratio in the shorter wavelength region. Although liposomes do not resemble closely the structure of lipids in lipoproteins, they are reasonable models at least for studying the CD behavior of lipids of the surface monolayers of lipoproteins. Microemulsion particles allow study of the CD contribution of both monolayer and core lipids at least approaching the probable arrangement of lipids in soluble lipoproteins. Though much is still uncertain with respect to quantitative details, this CD study of organized lipids appears to explain the unusual spectral feature observed in the cholesterol-rich lipoproteins.

β-Carotene as a Spectroscopic Probe of Order in Lipoprotein Lipids

Carotenoids, which absorb light in the visible wavelength region, are found in all classes of human serum lipoproteins, but the major portion is carried in LDL. β-Carotene constitutes about 90% of the native carotenoids of LDL, with lycopene and lutein comprising most of the remainder. We have shown that when the temperature of LDL is lowered below 37°, β-carotene, a symmetrical chromophore, displays a CD spectrum between 350 and 550 nm.[9] This temperature-dependent CD appears to be induced by environmental constraints in the lipoprotein complex because β-carotene lacks intrinsic optical activity. The cholesteryl esters in the core of LDL may provide the asymmetric environment when they are cooled below their unordered isotropic liquid to ordered liquid crystalline phase transition temperature. This induced optical activity of carotenoids in LDL is the basis of an abnormality in the visible wavelength region found in earlier optical studies of LDL, and originally attributed to protein.[9]

Validation in Reconstituted LDL

Reconstitution of LDL has provided the means for validation of β-carotene as a spectroscopic probe for determining the changes in the physical state of core lipids of lipoproteins, since the induced CD of β-carotene accurately reflects thermal transitions of cholesteryl esters placed in reconstituted LDL.[10] Mixtures of β-carotene, cholesteryl esters, triolein, and unesterified cholesterol were incorporated into LDL. The reconstituted lipoproteins showed induced CD similar to that of native

[9] G. C. Chen and J. P. Kane, *Biochemistry* **13**, 3330 (1974).
[10] G. C. Chen, M. Krieger, J. P. Kane, C.-S. C. Wu, M. S. Brown, and J. L. Goldstein, *Biochemistry* **19**, 4706 (1980).

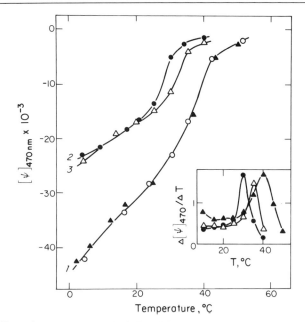

FIG. 4. Effect of temperature on the induced CD of β-carotene at 470 nm of reconstituted LDL. Curves 1–3 are reconstituted LDL with the cores replaced with cholesteryl oleate, cholesteryl linoleate, and cholesteryl linolenate, respectively. (Inset) The data from the main figure were plotted as first derivatives to show transition temperatures. (▲,○) preparations containing cholesteryl oleate, (△) containing cholesteryl linoleate, and (●) containing cholesteryl linolenate. [From G. C. Chen, M. Krieger, J. P. Kane, C.-S. C. Wu, M. S. Brown, and J. L. Goldstein, *Biochemistry* **19,** 4706 (1980); reproduced with the permission of the American Chemical Society (copyright 1980).]

LDL. Based on the thermal dependence of the induced CD, the transition temperatures of approximately 40, 35, and 30° (Fig. 4) for the cholesteryl oleate-, cholesteryl linoleate-, and cholesteryl linolenate-containing preparations, respectively, are very similar to those for pure cholesteryl ester model systems undergoing liquid crystalline to isotropic liquid phase transition determined by differential scanning calorimetry. Thus, the thermal transitions detected by using the induced CD of β-carotene reflect the ordered–unordered phase transition of the cholesteryl esters in the core of reconstituted LDL.

Native Lipoproteins

We have used the induced CD of β-carotene as a spectroscopic probe to determine whether the cholesteryl esters in the cores of a variety of native lipoproteins exist in an ordered state. In addition to native and

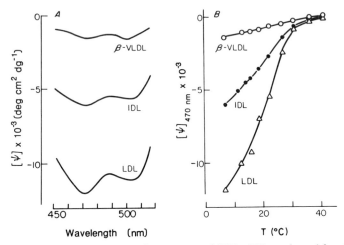

FIG. 5. (A) Induced CD spectra of β-carotene of LDL, IDL, and a subfraction of β-migrating VLDL at 7° from a patient with familial dysbetalipoproteinemia on a β-carotene-enriched diet. (B) Temperature dependence of induced CD of β-carotene at 470 nm of LDL, IDL, and a subfraction of β-migrating VLDL. [From G. C. Chen, J. P. Kane, and R. L. Hamilton, *Biochemistry* **23**, 1119 (1984); adapted with the permission of the American Chemical Society (copyright 1984).]

reconstituted LDL, the intermediate-density lipoproteins (IDL) and subfractions of the β-migrating $d <1.006$ g/ml lipoproteins (β-VLDL) showed induced CD of β-carotene between 7 and 30°, with contours similar to that of LDL but with smaller magnitudes (Fig. 5A).[11] In contrast, HDL[12] and subfractions of the pre-β-migrating $d <1.006$ g/ml lipoproteins (pre-β-VLDL)[11] showed no detectable induced CD of β carotene under the same conditions. These results indicates that, as in the core of LDL, the β-carotene molecule in the cores of IDL and the major portion of the β-VLDL becomes optically active by virtue of its local environmental constraint, probably arising from the formation of some ordered state of cholesteryl esters in the core. The temperature dependence of the induced CD spectra of β-carotene (Fig. 5B) also reflects the thermal-dependent changes in the cores, though the broader thermal dependence in β-VLDL and IDL suggests less cooperativity than in LDL.

The occurrence of induced CD of β-carotene in lipoproteins is found to depend upon the composition of core lipids and the diameter of the lipoprotein particles.[11] These two determinants also influence the thermal

---

[11] G. C. Chen, J. P. Kane, and R. L. Hamilton, *Biochemistry* **23**, 1119 (1984).
[12] G. C. Chen and J. P. Kane, unpublished observation (1983).

behavior of the cores in lipoproteins.[13] Thus, the absence of a thermal transition in human VLDL has been attributed to their high content of triglyceride and the absence of a transition in human HDL to their small diameter. Our finding of no detectable induced CD of β-carotene in human HDL and pre-β-VLDL is consistent with the absence of a thermal transition in these lipoproteins. Based on the carotene probe studies, we have found that the cores of triglyceride-rich lipoproteins can exist in some ordered state below 30° if they have a relatively low ratio of triglycerides to cholesteryl esters (mass ratio <1.6) and relatively small particle diameter (<60 nm).[11] On the other hand, the cores of cholesteryl ester-rich lipoproteins can exist in some ordered state over a wide range of particle diameter.

The high specific ellipticities of the induced CD of β-carotene (strong CD signals are observed even though only 4–6 molecules of β-carotene are dispersed among 1300–1500 molecules of cholesteryl esters in a particle of LDL[11]) and the absence of CD signal in an unconstrained environment make β-carotene an exquisitely sensitive probe of order in lipoprotein core regions. Thus, this technique can detect lipid order at a level which would likely escape detection by differential scanning calorimetry. The fact that β-carotene is normally present in lipoproteins allows such detections without the perturbations of structure which might result from the introduction of an extrinsic probe molecule.

Acknowledgment

This work was supported by U.S. Public Health Service Grant HL 14237 (Arteriosclerosis SCOR).

[13] A. R. Tall, *Ann. N.Y. Acad. Sci.* **348**, 335 (1980).

[31] Immunochemical Methods for Studying Lipoprotein Structure

By ELAINE S. KRUL and GUSTAV SCHONFELD

Introduction

The lipoproteins in plasma represent several sets of discrete particles whose structures are constantly changing because of their interactions with each other and with several enzymes in plasma. In addition to these interactions, the entrance into plasma of nascent particles and the removal from the plasma of "old" particles gives rise to great heterogeneities of structure. Subtle alterations in the dispositions of functionally important domains of the apolipoproteins on the surfaces of lipoproteins occur which can be detected only by sensitive immunologic techniques or by tests of function.[1-3] Appropriate changes of apolipoprotein conformations probably are needed for normal function. Abnormal conformations of apolipoproteins arising from defective structures acquired during biosynthesis or posttranslational modification can result in abnormal function.[4] Alterations in diet, hormonal, or health status also can give rise to changes in the intracellular assembly or the intravascular catabolism of the lipoproteins, creating subtle differences in lipoprotein organization which in turn affect metabolism.

The high degree of specificity of antibodies has made them useful tools for probing lipoprotein structures, for documenting apolipoproteins dispositions on lipoproteins, and for detecting physiologic or pathologic changes.

Production of Polyclonal Antisera

The methods for raising polyclonal antisera in a variety of animals have been described earlier in this series.[5] Antisera directed toward several lipoproteins and apolipoproteins have been successfully raised in rabbits in our laboratory using the following immunization protocol.[6]

Two to three kilogram female New Zealand White rabbits are injected with 0.5–1.0 mg of lipoprotein protein or purified apolipoprotein. The antigen is emulsified in complete Freund's adjuvant and is injected subcutaneously in multiple sites along the back of the rabbits. Usually 2–3 subsequent injections are given at 2–3 week intervals using 0.2–0.5 mg protein per rabbit in incomplete Freund's adjuvant. Seven to ten days after the final "boost," blood is drawn from the ear vein or artery and antibody activity is evaluated.

---

[1] G. Schonfeld, W. Patsch, B. Pfleger, J. L. Witztum, and S. W. Weidman, *J. Clin. Invest.* **64,** 1288 (1979).

[2] T. L. Innerarity, K. H. Weisgraber, K. S. Arnold, S. C. Rall, Jr., and R. W. Mahley, Jr., *J. Biol. Chem.* **259,** 7261 (1984).

[3] S. H. Gianturco, A. M. Gotto, Jr., S-L. C. Hwang, J. B. Karlin, A. H. Y. Lin, S. C. Prasad, and W. A. Bradley, *J. Biol. Chem.* **258,** 4256 (1983).

[4] T. L. Innerarity, E. J. Friedlander, S. C. Rall, Jr., K. H. Weisgraber, and R. W. Mahley, *J. Biol. Chem.* **258,** 12341 (1983).

[5] B. A. L. Hurn and S. M. Chantler, this series, Vol. 70, p. 104.

[6] G. Schonfeld and B. Pfleger, *J. Clin. Invest.* **54,** 236 (1974).

Production of Monoclonal Antibodies

Immunogens and Immunizations. The use of purified apolipoproteins as immunogens has resulted in the production of monoclonal antibodies that recognize a number of "native" epitopes on hololipoproteins.[7-9] Antibodies also may be produced against "native" epitopes present but not detectable (e.g., masked) in hololipoproteins, or "artifactual" epitopes present only in the lipid-free immunogen. Collection of lipoproteins (or apolipoproteins) in the presence of antimicrobial agents[10] and anti-proteolytic agents[3] reduces the potential of generating antibodies to degraded products or "artifactual" epitopes. Indeed, since the physical state of the immunogen may be important in determining the range of specificities of the antibodies produced by the hybridomas, antibodies against sufficient numbers of native expressed epitopes may not be produced by using only isolated apolipoproteins. Therefore, native hololipoproteins may need to be used as immunogens. It would appear that lipoprotein conformation is not destroyed by emulsifying in the adjuvant.[11] Antibodies which recognize "native" determinants on LDL (or HDL) particles have been produced using intact LDL (or HDL) as immunogens.[12-16] There are two potential disadvantages in using intact lipoproteins as immunogens. First, antibodies may be produced against irrelevant but more immunodominant apolipoprotein determinants on the lipoprotein particles. Second, antibodies against "masked" yet native epitopes may not be produced.

Another alternative is to immunize with human apolipoproteins reconstituted with lipid vesicles or microemulsions, or with mouse lipoproteins. Apolipoproteins in apolipoprotein–lipid recombinants are present in configurations that are closer to native than are lipid-free apolipoproteins,[17] therefore antibodies against some native configurations could be

---

[7] E. S. Krul, M. J. Tikkanen, T. G. Cole, J. M. Davie, and G. Schonfeld, *J. Clin. Invest.* **75**, 361 (1985).

[8] R. W. Milne, Ph. Douste-Blazy, Y. L. Marcel, and L. Retegui, *J. Clin. Invest.* **68**, 111 (1981).

[9] G. Schonfeld, R. Dargar, and T. Kitchens, *Circulation* **70** (Supp. II), 315 (1984).

[10] J. P. Kane, D. A. Hardman, and H. E. Paulus, *Proc. Natl. Acad. Sci. U.S.A.* **77**, 2465 (1980).

[11] Y. L. Marcel, M. Hogue, P. K. Weech, and R. W. Milne, *J. Biol. Chem.* **259**, 6952 (1984).

[12] M. J. Tikkanen, R. Dargar, B. Pfleger, B. Gonen, J. M. Davie, and G. Schonfeld, *J. Lipid Res.* **23**, 1032 (1982).

[13] R. W. Milne, R. Theolis, Jr., R. D. Verdery, and Y. L. Marcel, *Arteriosclerosis* **3**, 23 (1983).

[14] L. K. Curtiss and T. S. Edgington, *J. Biol. Chem.* **257**, 15213 (1982).

[15] S. J. T. Mao, R. E. Kazmar, J. C. Silverfield, M. C. Alley, K. Kluge, and C. G. Fathman, *Biochim. Biophys. Acta* **713**, 365 (1982).

[16] T. S. Watt and R. M. Watt, *Proc. Natl. Acad. Sci. U.S.A.* **80**, 124 (1983).

[17] A. Jonas, this series, Vol. 128, p. 553.

elicited. Human apolipoproteins complexed with appropriate mouse lipoproteins may be even more desirable immunogens since the configurations of the human apolipoproteins on these "hybrid" particles may be even closer to the native. If mouse apolipoproteins are not immunogenic, antibodies would be produced only against the expressed human apolipoprotein determinants. To obtain antibodies that recognize only selected regions of apolipoproteins it may be necessary to immunize with either lipid-free peptide fragments[18] or fragments reconstituted with lipids.

Immunizations of mice require 50–100 μg protein per injection. The antigen is injected subcutaneously initially with complete Freund's adjuvant followed 2–3 weeks later by antigen in incomplete Freund's adjuvant. Antihuman lipoprotein monoclonal antibodies have been produced also after intraperitoneal[14,15] and intravenous[16] immunizations. In most cases, a final booster dose of antigen (~10 μg) in phosphate-buffered saline (PBS) is injected intraperitoneally or intravenously 3 days prior to spleen cell fusion.

As lipoprotein surface probes, antibodies of the IgG class are preferable in view of their smaller size and greater stability. To ensure a population of hybridomas secreting this immunoglobulin class it may be necessary to carry out immunizations over periods of up to 6 months using multiple injections and multiple sites. Before fusion, antibody activity can be sought in the 19 S and 7 S fractions of the immunized mouse's serum.

Cell Fusions and Expansion of Clones. Methods for producing murine hybridomas have been described extensively elsewhere.[19]

Assays for Detection of Antibody Activity. Several procedures for detection of antibody activity in spent hybridoma culture media or mouse ascitic fluids have been described,[19] but two methods which have been proven particularly useful in our laboratory will be described. The presence of antibody activity should be sought in assays using both purified apolipoproteins and native lipoproteins to ensure detection of all possible antibodies directed against linear or configurational specificities.

Sandwich Plate Assay

Materials

0.01 *M* phosphate-buffered saline, pH 7.2–7.5
BSA–PBS: 3% BSA in PBS (w/v)

---

[18] J.-C. Fruchart, J.-C. Ghesquiere, C. Delpierre, C. Cachera, and A. Tartar, *Clin. Chem.* **30**, 992 (1984).

[19] G. Galfre and C. Milstein, this series, Vol. 73, p. 3.

96-well microtiter plates (Dynatech Laboratories Incorporated, Alexandria, VA)
Anti-mouse IgG (Cappell, Malvern, PA) radiolabeled with ^{125}I[20]

Procedure

1. The microtiter plate can be marked with a porous tip pen to ensure proper alignment of plate, antibodies used, etc. and rinsed with water to remove any electrostatic charge. The plate is flicked dry and placed in a humidified chamber (a tray lined with damp paper towels). A volume of 150 μl of apolipoprotein or lipoprotein (1–100 μg/ml in PBS) is dispensed into each well except for those that are to be used as controls.
2. Cover the tray and incubate the plate overnight at room temperature.
3. Empty the wells. Wash the wells three times with PBS.
4. Fill each well (300 μl) with BSA–PBS and let incubate at room temperature for at least 1 hr.
5. Remove the BSA–PBS.
6. Hybridoma culture supernates to be tested can be added (150 μl) to each well. Tissue culture medium alone is added as a negative control. Dilutions of the serum of immune mice or known positive monoclonal antibodies are added as positive controls. When ascites fluid is to be tested, appropriate dilutions of the ammonium sulfate precipitate (in 1% BSA–PBS) are added to each well.
7. The tray is covered and the plates allowed to incubate for 4 hr at room temperature.
8. The culture supernates or ascites fluids are removed and the plate rinsed three times with PBS.
9. Appropriate dilutions of the second antibody (1000 cpm/μl) in 150 μl of 1% BSA–PBS are added to each well.
10. The tray is covered and the plate incubated for 2–4 hr at room temperature.
11. The plate is washed carefully at least three times with PBS and dried.
12. Individual wells are obtained by cutting the plate and they are carefully transferred to clean tubes for counting.

Dot Immunoblotting.[21] Immunoblotting is a rapid, visual technique for testing antibody activity toward small amounts (<1 μg) of antigen.

---

[20] J. J. Marchalonis, *Biochem. J.* **113**, 299 (1969).
[21] R. Hawkes, E. Niday, and J. Gordon, *Anal. Biochem.* **119**, 142 (1982).

Materials

PBS

BSA–PBS

BSA–PBS–NIRS: BSA–PBS containing 3% nonimmune rabbit serum (w/v)

Anti-mouse IgG (Cappell, Malvern, PA) radiolabeled with ^{125}I

Nitrocellulose paper (Bio-Rad, Richmond, VA)

Kodak X-Omat AR X-ray film

Procedure

1. Gloves free of dust or talc should be worn when handling nitrocellulose paper. With a pencil or pen, draw a rectangular grid onto a nitrocellulose paper (5 × 5 mm per square is a convenient size).

2. Wash the nitrocellulose paper for 5–10 min by gentle agitation in distilled water. Allow the paper to dry at room temperature.

3. When the nitrocellulose paper is dry apply a small drop (<10 µl) of antigen solution (0.1–1 mg/ml protein in PBS) to the center of each square. A capillary pipet is useful for this.

4. The spot is allowed to dry thoroughly at this point and can be stored dry for periods of time (several weeks).

5. If a series of apolipoproteins or lipoproteins is repeated in rows the nitrocellulose paper can be cut with clean, dry scissors into strips each having an identical pattern of antigen to be tested against different antibodies. One end of the strips should be left blank so as to facilitate handling of the strips with forceps.

6. The nitrocellulose paper is blocked with BSA–PBS–NIRS for at least 1 hr at room temperature with shaking. If a sheet of nitrocellulose is used this is accomplished by placing the paper into a Seal-a-Meal bag (Dazey Products, Industrial Airport, KA) with 10 ml of blocking solution being sure to remove air bubbles. Alternatively, strips of nitrocellulose paper are placed into scintillation vials (twisting helically so that paper does not overlap) with 2 ml blocking solution.

7. The BSA–PBS–NIRS is removed and replaced with an equal volume of appropriately diluted ammonium sulfate precipitated ascites in BSA–PBS. The solutions are allowed to incubate overnight with gentle agitation at room temperature.

8. The antibody solutions are decanted and the nitrocellulose papers are washed 4 times × 15 min with equal or larger volumes of PBS.

9. Radiolabeled second antibody is added to the nitrocellulse papers in BSA–PBS (10 ml for sheets, 2 ml for strips) at a concentration of 7.5 ×

10^5 cpm/ml. The second antibody is allowed to incubate with the nitrocellulose for 2–4 hr with gentle agitation.

10. The radiolabeled antibody is removed and the nitrocellulose papers are washed 2 × 15 min with PBS, 1 × 15 min with PBS containing 0.02% Tween-20, and finally 2 × 15 min with PBS.

11. The nitrocellulose papers are allowed to dry thoroughly. Sheets or strips of nitrocellulose are then mounted on plain white paper by means of clear tape.

12. The nitrocellulose is then exposed to X-ray film at −70°.

An example of a dot immunoblot assay is shown in Fig. 1.

Interpretation of Results. This method like many other visual methods bears a degree of subjectivity. Therefore, the following controls should be run simultaneously on any dot immunoassay. Hybridoma antibodies of known specificity toward several of the "test" antigens should be used as

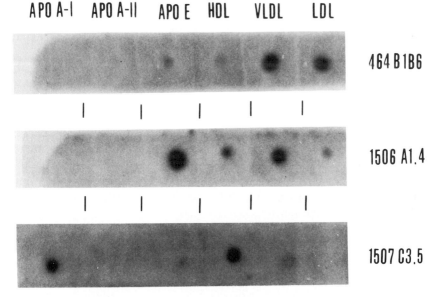

FIG. 1. Autoradiograms of dot immunoblots incubated with three different monoclonal antibodies. Approximately 1 μg of each apolipoprotein or lipoprotein was applied to the nitrocellulose strips. Strips then were incubated with antibodies 464 B1B6 (immunogen was LDL), 1506 A1.4, or 1507 C3.5 (immunogen for both was apoHDL). Radiolabeled goat antimouse IgG was used to detect monoclonal antibody binding to the antigens. Antibody B1B6 reacts with the apoB-containing lipoproteins. Although the two other antibodies were generated with a single immunogen (apoHDL), 1506 A1.4 is clearly directed toward apoE and apoE-containing lipoproteins whereas 1507 C3.5 recognizes apoA-I and apoA-I-containing lipoproteins.

positive controls. Irrelevant proteins should be spotted on the nitrocellulose as negative controls in Step 3.

Purification of Antibodies. Ascites fluids are generally subjected to centrifugation to remove particulate matter and stored in aliquots at $-70°$ until needed. Due to the presence of irrelevant proteins in these fluids it is often desirable to purify the monoclonal antibodies. Several methods employed in our laboratory are listed below. Readers are referred to the indicated references for descriptions of general methodology.

1. Ammonium sulfate precipitation.[22,23]
2. DEAE Affi-Gel Blue chromatography.[24]
3. Protein-A Sepharose chromatography.[25,26]
4. Apolipoprotein (lipoprotein) affinity chromatography. Isolation of a specific monoclonal antibody can be achieved by chromatographing the ascites fluid or the salt fractionated immunoglobulin on Sepharose coupled to the immunogen apolipoprotein or lipoprotein. This allows for purification of IgG or IgM antibodies. Methods for preparing protein coupled matrices for affinity chromatography have been described extensively elsewhere.[27]

Antihuman apoB monoclonal antibodies have been purified on LDL-coupled matrices[12,15,16,28] and in one case over insoluble apoB.[28] ApoA-I-coupled Sepharose-4B has been used in this laboratory to purify antihuman apoA-I antibodies successfully.

5. Purification of IgM monoclonal antibodies. Aside from affinity chromatography mentioned above, other methods applicable to monoclonal IgM purification have been described and they include gel filtration,[29] high-performance hydroxylapatite (HPHT) chromatography,[30] and sucrose density ultracentrifugation.[31] The latter method has been used in this laboratory for the purification of several IgM monoclonal antibodies. Precipitation methods for preparing IgM monoclonals often generated irreversibly denatured antibody.

[22] K. Heide and H. G. Schwick, *in* "Handbook of Experimental Immunology" (D. M. Weir, ed.), p. 7.1. Blackwell, London, 1978.
[23] P. Parham, this series, Vol. 92, p. 110.
[24] C. Bruck, D. Portetelle, C. Glineur, and A. Bollen, *J. Immunol. Methods* **53,** 313 (1982).
[25] P. L. Ey, S. J. Prowse, and C. R. Jenkin, *Immunochemistry* **15,** 429 (1978).
[26] I. Seppala, E. Sarvas, F. Peterfy, and O. Makela, *Scand. J. Immunol.* **14,** 335 (1981).
[27] P. Cuatrecasa and C. B. Anfinsen, this series, Vol. XXII, p. 345.
[28] B. P. Tsao, L. K. Curtiss, and T. S. Edgington, *J. Biol. Chem.* **257,** 15222 (1982).
[29] J. L. Fahey and E. W. Terry, *in* "Handbook in Experimental Immunology" (D. M. Weir, ed.), p. 8.1. Blackwell, London, 1978.
[30] H. Juarez-Salinas, S. C. Engelhorn, W. L. Bigbee, M. A. Lowry, and L. H. Stanker, *Biotechniques* **2,** 164 (1984).
[31] T. Pearson, G. Galfre, A. Ziegler, and C. Milstein, *Eur. J. Immunol.* **7,** 684 (1977).

Purification of Fab Fragments. The smaller size of these specific antigen binding molecules makes them potentially more useful as probes of epitope expression than the parent IgG molecules. Fab fragments would be expected to have fewer steric restrictions on lipoprotein binding.[12,32] The preparation of Fab fragments is described elsewhere.[33,34]

Assignment of Epitopes to Individual Apolipoproteins

When hololipoproteins containing several apolipoproteins are used as immunogens and screening of hybridomas also is performed with hololipoproteins, one can identify antibody activity in culture media or in ascites fluids but the individual monoclonal antibodies may be directed against any of several individual apolipoproteins. Therefore, it becomes important to identify the apolipoprotein chain specificities of the antibodies. Dot immunoblotting using a "panel" of all the probable apolipoproteins against which the antibodies are likely to be directed provides a facile means of assigning chain specificities. Alternatively, immunoelectrotransfer or nonelectrophoretic protein transfer methods may be used. Here the apolipoproteins of the lipoprotein are separated from each other by gel electrophoresis, transferred to nitrocellulose paper, and incubated with the antibodies. The binding of antibodies to apolipoproteins then is determined.

Dot Immunoblotting. Protocols for this procedure are discussed above.

Protein Transfer Blots. Methods for electrophoretically transferring proteins from polyacrylamide gels to nitrocellulose sheets using SDS gels[35,36] or nonelectrophoretic transfer for IEF gels[37] have gained considerable popularity.[38] The combined advantages of the ability to resolve complex mixtures of proteins (or peptide fragments) and the accessibility of the proteins on nitrocellulose for binding antibodies has made this a useful tool for the assignment of epitopes to apolipoproteins chains.[14,39,40]

Separation of apolipoproteins on polyacrylamide gels can be achieved

---

[32] R. W. Milne and Y. L. Marcel, *FEBS Lett.* **97** (1982).
[33] G. Gorini, G. A. Medgysi, and G. Doria, *J. Immunol.* **103**, 1132 (1969).
[34] J. L. Fahey and B. A. Askonas, *J. Exp. Med.* **115**, 623 (1962).
[35] H. Towbin, T. Staehelin, and J. Gordon, *Proc. Natl. Acad. U.S.A.* **76**, 4350 (1979).
[36] V. C. W. Tsang, J. M. Paralta, and A. R. Simons, this series, Vol. 92, p. 377.
[37] M. P. Reinhart and D. Malamud, *Anal. Biochem.* **123**, 229 (1982).
[38] J. M. Gershoni and G. E. Palade, *Anal. Biochem.* **131**, 1 (1983).
[39] K.-S. Hahm, M. J. Tikkanen, R. Dargar, T. G. Cole, J. M. Davie, and G. Schonfeld, *J. Lipid Res.* **24**, 877 (1983).
[40] Y. L. Marcel, M. Hogue, R. Theolis, Jr., and R. W. Milne, *J. Biol. Chem.* **257**, 13165 (1982).

in several ways prior to protein electrotransfer. Gradient polyacrylamide gels containing sodium dodecyl sulfate[41,42] are useful for resolving apolipoproteins of widely variant molecular weights simultaneously. For the separation of the high-molecular-weight apoB chains, 3% SDS–polyacrylamide gels have been used[10] and the addition of 0.5% agarose[40] helps to improve gel resiliency in handling during subsequent protein electrotransfer. When running gradient or regular SDS–PAGE it is advisable to include 5.6 mM $CaCl_2$ in the sample buffer (or at least twice the molar concentration of any EDTA present) to promote efficient electrotransfer of the higher molecular weight apolipoprotein chains to nitrocellulose.[14] Analysis of the lower molecular weight apolipoproteins is conveniently performed by polyacrylamide gel electrophoresis in the presence of 6–8 M urea[43] after delipidation of the lipoprotein with TMU[44] or organic solvents.[45] Isoelectric focusing (IEF[46]) permits the separation of apolipoproteins having genetic or sialated isoforms. As methods of polyacrylamide gel electrophoresis are described elsewhere in this volume only the protein transfer to nitrocellulose and subsequent incubation with antibodies will be discussed here.

Electrotransfer after SDS–PAGE or Urea–PAGE[35,36]

Materials

25 mM sodium phosphate, pH 6.5 (SDS–PAGE) or (transfer buffer) 0.7% acetic acid (Urea–PAGE)
Nitrocellulose paper (Bio-Rad, Richmond, VA) cut to size to match gel pieces and marked with pencil to indicate gel edges
Trans-Blot electrophoretic Transfer Cell (Bio-Rad, Richmond, VA)

Procedure

1. Run SDS–PAGE or urea–PAGE slab gels (14 × 12 × 0.15 cm).
2. Remove gel from electrophoresis apparatus and wash the SDS–PAGE gel in transfer buffer 3 × 20 min with gentle shaking. Urea–PAGE gels do not require this wash step.
3. The gel may be cut longitudinally (e.g., if a series of lanes is repli-

---

[41] J. B. Swaney and K. S. Kuehl, *Biochim. Biophys. Acta* **446**, 561 (1976).
[42] H. Aburatani, T. Kodama, A. Ikai, H. Itakura, Y. Akanuma, and F. Takaku, *J. Biochem.* **94**, 1241 (1983).
[43] B. J. Davis, *Ann. N.Y. Acad. Sci.* **121**, 404 (1965).
[44] J. P. Kane, *Anal. Biochem.* **53**, 350 (1973).
[45] V. Shore and B. Shore, *Biochemistry* **7**, 3396 (1968).
[46] S. W. Weidman, B. Suarez, J. M. Falko, J. L. Witztum, J. Kolar, M. Raben, and G. Schonfeld, *J. Lab. Clin. Med.* **93**, 549 (1979).

cated on the gel). One piece may be stained at this point and used as a control for the other pieces to be used for protein transfer.

4. A sheet of filter paper (3 mm) is moistened in transfer buffer and placed on top of the porous pad on one side of the sandwich assembly of the Trans-Blot Cell. One or several gel pieces are placed on top, followed by approximately cut moistened, nitrocellulose sheets. As transfer of some protein occurs immediately it is best not to lift and replace nitrocellulose from gel surface too often. The nitrocellulose is covered with a second sheet of moistened filter paper, and the second porous pad. Care should be taken so that there are no air bubbles between surfaces.

5. The entire assembly is placed into the Trans-Blot Cell filled with transfer buffer. For SDS–PAGE gels, the nitrocellulose paper should be closest to the anode with the gel facing the cathode. For urea–PAGE gels, the gel should be facing the anode with the nitrocellulose sheet closest to the cathode.

6. The voltage is set at 27 V for 2–3 hr on the Trans-Blot Cell to effect protein transfer.

7. After transferring, the nitrocellulose sheets are dried. (Staining of the nitrocellulose sheets can be carried out at this point with 0.1% Amido Black in 45% methanol/10% acetic acid.)

8. Nitrocellulose sheets to be used for immunoblotting can be placed dry between sheets of filter paper and wrapped in plastic wrap and stored at 4° for extended periods (several weeks).

9. Polyacrylamide gels that were used for the protein transfer can be stained to assess the efficiency of transfer by comparison to the stained control gel piece.

Protein Transfer after IEF–PAGE[37]

Materials

5 mM Tris, 0.15 M NaCl, pH 8.15 (transfer buffer)
Nitrocellulose paper (Bio-Rad, Richmond, VA)

Procedure

1. Run IEF–PAGE gel (14 × 12 × 01.15 cm).
2. Remove gel from electrophoresis apparatus. The gel can be cut longitudinally into strips as described for Step 3 above.
3. A piece of filter paper is placed on a solid support so that its ends are immersed in troughs containing transfer buffer. A piece of moistened nitrocellulose is placed on to the filter paper, followed by the gel (being careful to remove any air bubbles) and a second piece of nitrocellulose.

This is covered by a second piece of moistened filter paper. A stack of paper towels is placed on top of this assembly to ensure buffer flow through the layers of nitrocellulose and gel. A glass plate holding a weight of 200–400 g is placed over the towels to provide an even distribution of weight to allow all the surfaces to be in close contact. (Note that two nitrocellulose transfers are obtained with this method.)

4. Transfer is allowed to proceed for 3–4 hr.

5. Subsequent steps are the same as for Steps 7–9 for SDS–PAGE and urea–PAGE electrotransfer.

Immunological Detection of Proteins on Nitrocellulose

Materials

(Same as for dot immunoblotting.)

Procedure

1. These steps are the same as Steps 6–12 described for dot immunoblotting.

Interpretation of Results. At least one group has reported the loss of some monoclonal antibody detectable epitopes following electrophoresis and electrotransfer of apoB.[14] This is a potential problem in the analysis of lipophilic apolipoproteins. We have tried to minimize protein denaturing conditions by using nonmethanolic buffers for electrotransfer, namely 25 mM phosphate[47] or the "native" nonelectrophoretic blot method.[37] The presence of methanol also has been shown to reduce the efficiency of transfer of high-molecular-weight proteins.[48] Other considerations in protein transfer are discussed in a review by Gershoni and Palade.[38]

Immunoelectrotransfer blots of VLDL, LDL, and HDL incubated with monoclonal antibodies directed toward apoB, apoE, or apoA-I are shown in Fig. 2.

Identification and Enumeration of Distinct Epitopes on Apolipoproteins. When a series of monoclonal antibodies known to react with a given apolipoprotein are available, it becomes important to know how many different epitopes can be identified on that apolipoprotein. It also becomes important to be able to distinguish between native expressed, native "masked" or artifactual epitopes. Native expressed epitopes may be defined operationally as those detectable on native hololipoproteins. Native "masked" epitopes are detectable after treatment of native lipoproteins with lipolytic enzymes, neutral detergents, or other agents which do

---

[47] M. Bittner, P. Kupferer, and C. F. Morris, *Anal. Biochem.* **102**, 459 (1980).
[48] W. N. Burnette, *Anal. Biochem.* **112**, 195 (1981).

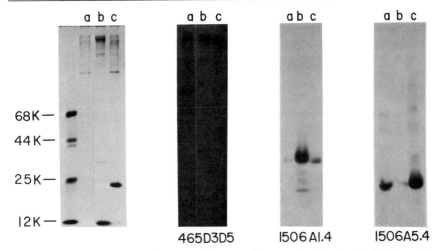

FIG. 2. Immunoelectrotransfer autoradiograms of lipoproteins incubated with three different monoclonal antibodies. Molecular weight standard proteins, delipidated LDL (a), VLDL (b), and HDL (c) were electrophoresed on 3–20% SDS–polyacrylamide gels and gels were stained with Coomassie Blue (shown on the far left). The proteins on the gels were transferred electrophoretically to nitrocellulose and incubated with the indicated monoclonal antibodies. The binding of the antibodies was monitored by incubating the transfers with labeled goat anti-mouse IgG. The 465 D3D5 (immunogen was LDL) reacted only with the higher molecular weight bands. Monoclonal anti-apoHDL antibody 1506 A1.4 bound to protein bands corresponding to apoE (the 22K band in VLDL is a fragment of apoE). Anti-apoHDL antibody 1506 A5.4 demonstrated high affinity for bands corresponding to apoA-I.

not grossly alter apolipoprotein conformation. Artifactual epitopes may be defined as those which are detected only on isolated apolipoproteins, denatured hololipoproteins or even on reconstituted apolipoprotein–lipid complexes, but never in native hololipoproteins.

Antibody Competition. In these experiments purified radiolabeled and unlabeled antibodies are made to compete against each other for occupancy of the antigenic binding sites. The assumption is that if two antibodies bind to the same or closely spaced epitopes competition will occur but if binding is to distant epitopes the unlabeled antibody will fail to reduce the binding of the labeled antibody.

Materials

0.01 M phosphate-buffered saline, pH 7.2–7.5
BSA–PBS: 3% BSA in PBS (w/v)
96-well microtiter plates (Dynatech Laboratories Incorporated, Alexandria, VA)
Purified monoclonal antibody radiolabeled with ^{125}I[20]

Procedure

1. Follow the procedure for the Sandwich plate assay above up to and including Step 5.

6. Serial dilutions of each monoclonal antibody in BSA–PBS to be used as competitor are added to each well.

7. Radiolabeled monoclonal antibody in BSA–PBS is added to the dilutions of competitor antibody in a volume such that the final volume in the wells is 150 μl with a total ~150,000 cpm.

8. The tray is covered and the plate is incubated for 2–4 hr at room temperature.

9. The plate is washed at least three times with PBS and dried. The individual wells are counted as for the sandwich plate assay.

An example of antibody competition is depicted in Fig. 3. In this case the two antibodies 464 B1B3 and 464 B1B6 bind to closely spaced epitopes on LDL. When labeled 464 B1B3 was used as tracer, 464 B1B6 was shown to compete as effectively for binding as unlabeled 464 B1B3 (not shown).

Interpretation of Results. When antibodies are observed to compete for binding to an apolipoprotein or lipoprotein, several explanations are possible.

1. The antibodies may be recognizing identical or structurally overlapping epitopes.

2. Antibodies may be directed against nonoverlapping epitopes but mutual binding of the two antibodies may be prevented merely by their physical bulk (steric hindrance). To minimize this "artifact" monovalent Fab fragments can be used.

3. Conformational changes induced by the binding of one antibody may alter the epitope recognized by the second antibody.

Failure to reduce binding of one antibody by another usually is due to their binding to distinct epitopes, but it may be observed also in cases where the labeled and competitor antibodies bind to overlapping, nonidentical epitopes but the labeled antibody binds the antigen with much higher avidity than does the competitor. One way to distinguish between these possibilities is to conduct reciprocal assays in which the radiolabeled and unlabeled antibodies are reversed.

Antibody Cotitration.[49] This assay is based on the principle that any antibody will be bound to a certain degree by a limiting concentration of antigen and if a second antibody is present the amounts of both antibodies

---

[49] H. G. Fisher and G. Brown, *J. Immunol. Methods* **39,** 377 (1980).

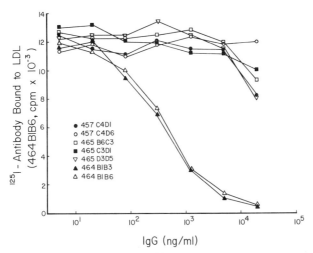

FIG. 3. Inhibition of binding to LDL of ^{125}I-labeled monoclonal anti-LDL antibody 464 B1B6 by unlabeled antibodies. A series of monoclonal antibodies were generated using LDL as immunogen. In the experiment depicted here, LDL was used to coat microtiter plate wells. After blocking with BSA–PBS, the wells received increasing amounts of the indicated unlabeled antibodies while the amount of ^{125}I-labeled 464 B1B6 was held constant. After the plates had incubated, the wells were rinsed with PBS, dried, and counted. Determinations carried out in triplicate had coefficients of variation of ~5%.

bound will be additive if they recognize completely independent epitopes. If they recognize closely linked epitopes then the maximum amount of antibody bound cannot be any higher than that bound when either of the antibodies were titrated separately. Partially overlapping epitopes would be expected to yield amounts of antibody bound that are intermediate.

Materials

(Same as for sandwich assay.)

Procedure

1. Follow steps for sandwich plate assay up to and including Step 5.
6. Serial dilutions in BSA–PBS of individual antibodies or combinations of two antibodies in 1:1 ratios are added to the wells. (When two antibodies are titrated together, the antibodies are initially combined in a 1:1 ratio and subsequently serially diluted.) Total volume added to the wells is 150 μl.
7–12. Same as for sandwich plate assay.

Interpretation of Results. The results can be expressed as follows:[49]

$$[(A + B)/(A) + (B)] \times 100/50$$

where (A), (B) are the counts bound by individual antibodies at *saturation* and $(A + B)$ are the counts bound at saturating levels of a 1 : 1 mixture of both antibodies.

Ideally, a value of 1 would be obtained if the antibodies giving rise to (A) and (B), respectively, bound epitopes that were very close spatially. Where the two antibodies bind two clearly distinct epitopes the total amount of antibody bound in a mixture would be equivalent to the addition of the two individual saturation binding levels and would result in a value of 2 for the above equation.

Interpretation of the results obtained with cotitration is subject to the same pitfalls as the antibody competition assays, however, the former method offers several advantages in that only a single radiolabeled second antibody is required in order to compare a series of monoclonal antibodies, crude antibody preparations can be used and generally less antibody is required than for antibody competition assays.

The use of cotitration in detecting unique epitopes on human LDL has been used by Milne *et al.*[13]

Assignment of Epitopes to Apolipoprotein Domains

The general approach in assigning epitopes to specific regions of an apolipoprotein is to fragment the protein, isolate the fragments, and assess their immunologic reactivities with the available panel of antibodies in a variety of assays. As epitopes frequently depend on protein conformation[50] as well as amino acid sequence, it may not be possible by this method unequivocally to assign every epitope. However, assignment should be easier for those epitopes which form parts of larger fragments where more of the native structure is likely to be retained.

Cyanogen Bromide Fragmentation of Proteins. Cyanogen bromide fragmentation of proteins produces fairly large size fragments and generally serves as a starting point in the analysis of protein fragments. Fragmentation of apoB,[51] apoE,[52] apoA-I,[53] apoA-II,[54] apoC-I,[55] and apoC-II[56]

---

[50] M. Z. Atassi, *Immunochemistry* **12**, 423 (1975).
[51] D. G. Deutsch, R. L. Heinrikson, J. Foreman, and A. M. Scanu, *Biochim. Biophys. Acta* **529**, 342 (1978).
[52] K. H. Weisgraber, S. C. Rall, Jr., and R. W. Mahley, *J. Biol. Chem.* **256**, 9077 (1981).
[53] H. N. Baker, R. L. Jackson, and A. M. Gotto, Jr., *Biochemistry* **12**, 3866 (1973).
[54] H. B. Brewer, Jr., S. E. Lux, R. Ronan, and K. M. John, *Proc. Natl. Acad. Sci. U.S.A.* **69**, 1304 (1972).

by this method has been reported. Polyclonal antisera have been used to identify epitopes on COOH-terminal and NH_2-terminal CNBr fragments of apoA-I and in competition assays those fragments were used successfully in assessing apoA-I immunoreactivity on the surface of native HDL particles.[57]

Enzymatic Cleavage of Apolipoproteins. Plasmin has been used to digest apoA-I,[58] and trypsin to digest apoA-I,[59,60] apoA-II,[61] apoB,[14,39,62] apoC-I,[63] apoC-II,[56] apoC-III,[60,64] and apoE.[60,65] Thrombin has also been used to cleave apoC-III[66] and apoE[3,67] into fragments with distinct physiologic and immunologic characteristics.

The choice of a particular protease depends on the state of knowledge of the protein's amino acid sequence, and the region of the protein under study. The conditions of incubation are determined for each enzyme used and each protein to be fragmented. This topic is dealt with earlier in this series.[68]

Chemical Modification of Amino Acid Residues. Since modifications may alter immunoreactivity either by altering the epitopes themselves chemically or by disrupting secondary structure, changes in immunoreactivity could be subject to more than one interpretation. However, chemical modification-induced changes in immunoreactivity would confirm the presence of an epitope on a protein or its fragment. If none of the modifications produce changes in immunoreactivity, the involvement of the altered residues in the formation of an epitope can be ruled out.

---

[55] R. L. Jackson, J. T. Sparrow, H. N. Baker, J. D. Morrisett, O. D. Taunton, and A. M. Gotto, *J. Biol. Chem.* **249,** 5308 (1974).
[56] R. L. Jackson, H. N. Baker, E. B. Gilliam, and A. M. Gotto, Jr., *Proc. Natl. Acad. Sci. U.S.A.* **74,** 1942 (1977).
[57] G. Schonfeld, R. A. Bradshaw, and J-S. Chen, *J. Biol. Chem.* **251,** 3921 (1979).
[58] H. R. Lijnen and D. Collen, *Thromb. Res.* **24,** 151 (1981).
[59] H. B. Brewer, Jr., T. Fairwell, A. LaRue, R. Ronan, A Houser, and T. J. Bronzert, *Biochem. Biophys. Res. Commun.* **80,** 623 (1978).
[60] L. V. Pereira and P. J. Dolphin, *Can. J. Biochem.* **60,** 790 (1982).
[61] S. J. T. Mao, A. M. Gotto, Jr., and R. L. Jackson, *Biochemistry* **14,** 4127 (1975).
[62] S. Goldstein and M. J. Chapman, *Biochem. Biophys. Res. Commun.* **87,** 121 (1979).
[63] R. S. Shulman, P. N. Herbert, K. Wehrly, and D. S. Fredrickson, *J. Biol. Chem.* **250,** 182 (1975).
[64] R. S. Shulman, P. N. Herbert, D. S. Fredrickson, K. Wehrly, and H. B. Brewer, Jr., *J. Biol. Chem.* **249,** 4969 (1974).
[65] S. C. Rall, Jr., K. H. Weisgraber, and R. W. Mahley, *J. Biol. Chem.* **257,** 4171 (1982).
[66] S. J. T. Mao, P. K. Bhatnager, A. M. Gotto, Jr., and J. T. Sparrow, *Biochem.* **19,** 315 (1980).
[67] K. H. Weisgraber, T. L. Innerarity, K. J. Harder, R. W. Mahley, R. W. Milne, Y. L. Marcel, and J. T. Sparrow, *J. Biol. Chem.* **258,** 12348 (1983).
[68] D. G. Smith, this series, Vol. XI, p. 214.

Particular attention has been paid by lipoprotein researchers to lysine and arginine residues on apolipoproteins B and E as chemical modification of these amino acids leads to reduction of lipoprotein receptor interactions.[69] Abolition of the positive charge of as few as 20% of the ε-amino lysine residues by acetylation,[70] acetoacetylation,[71] malondialdehyde treatment,[72] or carbamylation[73] suffices to reduce receptor recognition.[74] Recently, it has been shown that chemical modification of the single arginyl residue of apoC-II with cyclohexanedione or butanedione resulted in a loss of the protein's ability to activate lipoprotein lipase.[75] Several investigators have demonstrated altered immunoreactivities of lipoproteins following specific chemical modification of amino acid residues.[8,12,13,76] These experiments provided information on the specificities of epitope expression as well as enabling a distinction between monoclonal antibodies on the basis of their characteristic reactions with the modified lipoprotein. The reversibility of certain chemical modifications such as cyclohexadione treatment of arginyl residues,[77,78] citraconylation,[59,79] and acetoacetylation[80] of lysyl residues and cysteamine charge modification of cysteine residues[52] allows one to assess whether changes in antibody reactivity are due to irreversible alterations (e.g., protein unfolding induced by manipulation) or due to an alteration of the epitope associated with the specific amino acid residue.

Reversible modifications also are useful in protein fragmentation because they can narrow the specificities of some proteases by blocking their access to certain cleavage sites thereby reducing the numbers and increasing the sizes of fragments yielded by a given enzyme. This is useful for epitope mapping.

Synthetic Peptides. The laboratory synthesis of apolipoprotein fragments has proven to be useful in identifying the lipid binding regions of

[69] R. W. Mahley, *Klin. Wochenschr.* **61**, 225 (1983).
[70] H. Fraenkel-Conrat, this series, Vol. IV, p. 247.
[71] A. Marzotto, P. Pajetta, L. Galzigna, and E. Scoffone, *Biochim. Biophys. Acta* **154**, 450 (1968).
[72] A. M. Fogelman, I. Shechter, J. Seager, M. Hokom, J. S. Child, and P. A. Edwards, *Proc. Natl. Acad. Sci. U.S.A.* **77**, 2214 (1980).
[73] K. H. Weisgraber, T. L. Innerarity, and R. W. Mahley, *J. Biol. Chem.* **253**, 9053 (1978).
[74] R. W. Mahley and T. L. Innerarity, *Biochim. Biophys. Acta* **737**, 197 (1983).
[75] G. Holdsworth, J. G. Noel, K. Stedje, M. Shinomiya, and R. L. Jackson, *Biochim. Biophys. Acta* **794**, 472 (1984).
[76] A. Scanu, H. Pollard, and W. Reader, *J. Lipid Res.* **9**, 342 (1968).
[77] R. W. Mahley, T. L. Innerarity, R. E. Pitas, K. H. Weisgraber, J. H. Brown, and E. Gross, *J. Biol. Chem.* **252**, 7279 (1977).
[78] L. Patthy and E. L. Smith, *J. Biol. Chem.* **250**, 557 (1975).
[79] H. B. F. Dixon and R. N. Perham, *Biochem. J.* **109**, 312 (1968).
[80] G. Schonfeld, W. Patsch, B. Pfleger, and M. Abrams, *Adv. Physiol. Sci.* **12**, 515 (1980).

several apolipoproteins and in assessing the effects of the addition or substitution of amino acid residues.[81] Solid phase radioimmunoassays with synthetic peptidyl resins have been able to identify a major antigenic determinant of apoA-II.[82] The synthesis of apolipoprotein peptides is discussed in Chapter [36], this volume.

Expression of Epitopes on Hololipoproteins

Having assigned epitopes to specific regions of apolipoproteins one can proceed to assess how uniformly a given epitope is expressed in a given population of lipoproteins, how many epitopes are expressed per particle, compare epitope expression on different populations of particles (e.g., VLDL, LDL, IDL, and native or reconstituted HDL) with respect to avidity and number, and determine the effects of physiologic and pathologic perturbations of lipoprotein structure on epitope expression.

Assessment of the binding of antibody to lipoproteins in sandwich plate assays (described above) or in competitive binding assays (described below) provides the information on the degree of homogeneity of apolipoprotein expression[8,16,28,83,84] in a given lipoprotein population and whether any heterogeneity is due to altered affinity or altered numbers of epitopes expressed or both.

Quantitation of binding of Fab fragments of monoclonal antibodies to a lipoprotein by conventional Scatchard analysis can provide information as to how many times per particle an epitope is expressed.[32] By affinity chromatography of lipoproteins on antibody columns it is possible to identify and isolate a subpopulation of particles on which given epitopes are expressed. Especially when dealing with hololipoproteins, the possibility of different subpopulations of particles expressing different epitopes can drastically alter any conclusions regarding epitope sequence. This problem can be evaluated by demonstrating that immunoprecipitation of labeled antigen (or immunoaffinity chromatography) by one monoclonal antibody removes all antigen molecules capable of reacting with another antibody and vice versa.

Competitive Binding Assay. A variety of methods to isolate lipoproteins and their subfractions exists (see other chapters in this volume). In order to compare these particles in competitive binding assays, the meth-

[81] J. T. Sparrow and A. M. Gotto, Jr., *CRC Crit. Rev. Biochem.* **13**, 87 (1982).
[82] P. K. Bhatnagar, S. J. T. Mao, A. M. Gotto, Jr., and J. T. Sparrow, *Peptides* **4**, 343 (1983).
[83] M. J. Tikkanen, T. G. Cole, and G. Schonfeld, *J. Lipid Res.* **24**, 1494 (1983).
[84] M. J. Tikkanen, T. G. Cole, K.-S. Hahm, E. S. Krul, and G. Schonfeld, *Arteriosclerosis* **4**, 138 (1984).

ods of isolation must be considered. Distinct VLDL subpopulations can be isolated either by their ultracentrifugal properties[85] or by affinity chromatography on heparin[86] or antibodies. Another consideration is the choice of tracer or reference ligand, since studies with polyclonal antisera have yielded distinct displacement patterns with different peptide fragments of apoA-I used as tracer.[57] When using radioiodinated proteins as tracers, the choice of iodination method is important so as not to destroy the epitope of interest.[87]

Materials

Monoclonal (or polyclonal) antibody
96-well microtiter plates
0.01 M phosphate-buffered saline, pH 7.2–7.5
BSA–PBS: 3% BSA in PBS (w/v)
Reference ligand to be used as tracer radiolabeled with ^{125}I

Procedure

1. The microtiter plate is prepared as for the sandwich plate assay. A volume of 150 μl of purified antibody in PBS (1–25 μg/ml) is dispensed into each well except for those that are to be used as controls.

2–5. Same as Steps 2–5 of sandwich plate assay.

6. Serial dilutions of the competitor apolipoprotein or lipoprotein in 1% BSA–PBS are added to the wells in duplicate or triplicate in a total volume of 100–130 μl.

7. Radiolabeled tracer diluted in 1% BSA–PBS is added in an appropriate volume so that the final volume in the wells is 150 μl and the total radioactivity per well is ~1000 cpm/μl.

8. The tray is covered and incubated for 2–4 hr at room temperature.

9. The plate is washed, dried, and counted as described for the sandwich plate assay.

Interpretation of Results. Displacement curves generated by the competitor species can be plotted as B/B_0 (bound over total counts) versus log competitor dose. Where displacement curves are widely separated for individual competitor apolipoproteins or lipoproteins, this purely graphical analysis can provide much information as to relative potencies of epitope binding to any given antibody. Linear log–logit transformation[88]

---

[85] S. H. Gianturco, A. M. Gotto, Jr., R. L. Jackson, J. R. Patsch, H. D. Sybers, O. D. Taunton, D. L. Yeshurun, and L. C. Smith, *J. Clin. Invest.* **61**, 320 (1978).
[86] P. Nestel, T. Billington, T. Nada, P. Nugent, and N. Fidge, *Metabolism* **32**, 810 (1983).
[87] R. E. Gregg, D. Wilson, E. Rubalcaba, R. Ronan, and H. B. Brewer, Jr., Proc. Workshop on Apolipoprotein Quantification. NIH Publication 83-1266, 383 (1983).
[88] D. Rodbard, *Clin. Chem.* **20**, 1255 (1974).

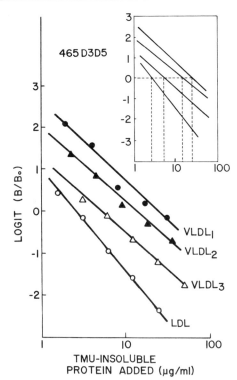

FIG. 4. Log–logit transformation of competitive displacement curves produced by VLDL subfractions and LDL using monoclonal anti-LDL antibody 465 D3D5. A competitive binding assay was carried out using immobilized antibody, VLDL subfractions, isolated as previously described,[84] LDL, and tracer ^{125}I-labeled LDL. The results indicate that LDL > VLDL$_3$ > VLDL$_2$ > VLDL$_1$ with respect to competitive ability versus ^{125}I-labeled LDL. Relative displacing potencies can be calculated from the doses required to achieve 50% displacement of label (logit = 0) shown in the inset. Regression analyses can provide values for the slopes of the displacement curves. Slopes were −2.71, −1.72, −1.82, and −1.88 for LDL, VLDL$_3$, VLDL$_2$, and VLDL$_1$, respectively.

simplifies interpretation of the data and allows estimations of the relative binding affinities of various competitors by slope comparisons.[84] This is exemplified in Fig. 4. Curve fitting by computer-aided four parameter logistic modeling[89] generates theoretical binding curves by best fit analysis on families of competition curves which can be compared statistically on the basis of slope (affinity) or potency.

Antibody Affinity Chromatography. The coupling of monoclonal antibodies of known specificity to chromatographic matrices permits the iso-

[89] A. DeLean, P. J. Munson, and D. Rodbard, *Am. J. Physiol.* **235**, E97 (1978).

lation of antigen with specific epitope expression. As a result, complex mixtures of lipoproteins can be separated into subpopulations or purified on the basis of subtle epitope determinants. For any particular monoclonal antibody that is to be used for affinity chromatography, optimal conditions must be found for binding and elution of the antigen. Foremost among the considerations is the capacity of the immunoadsorbent. Especially when dealing with hololipoproteins, which may potentially share similar epitopes but vary widely in size, binding to immobilized antibody may be sterically restricted for larger particles when, in fact, equimolar concentrations of antigen on smaller particles would have achieved "saturation."[90] To circumvent this possibility one can perform an analytical chromatographic run using a trace amount of radiolabeled antigen. Should the elution profile obtained using labeling antigen correspond to that obtained using preparative quantities of the same antigen then one can be assured that column overloading did not occur. Some investigators routinely add purified labeled antigen to crude preparations to monitor purification over monoclonal antibody affinity columns.[91]

The use of monoclonal antibodies to isolate subpopulations of apolipoproteins or complex lipoproteins relies on the specificity of binding to a single epitope on the antigen. Therefore, any nonspecific binding due to hydrophobic or electrostatic forces can interfere with this process. To circumvent this problem one can pass the antigen-containing mixture over a "virgin" sepharose column[7,23,91] and/or column containing nonspecific mouse IgG[23,92] prior to passage over the monoclonal affinity column to remove nonspecifically bound substances.

Methods for coupling monoclonal antibodies to Sepharose and for affinity chromatography are described extensively elsewhere.[23,91] Affinity chromatography has been used to detect subpopulations of VLDL having LDL–receptor binding activity (see the table).[7] Milne et al.[92] have used this methodology preparatively to separate apoB-48 VLDL from apoB-100 VLDL.

In Vitro Perturbation of Lipoprotein Structure and Analysis with Antibodies

Enzymatic Treatment of Lipoproteins. The *in vivo* process of lipolysis gives rise to much of the size, change, and compositional heterogeneity in the VLDL ($d < 1.006$) range. Lipolysis induced modulation of apoB and

[90] J. Turkova, *J. Chromatogr. Libr.* **12**, 1 (1978).
[91] M. F. Mescher, K. L. Stallcup, C. P. Sullivan, A. P. Turkewitz, and S. H. Herrmann, this series, Vol. 92, p. 86.
[92] R. Milne, P. K. Weech, L. Blanchette, J. Davignon, P. Alaupovic, and Y. L. Marcel, *J. Clin. Invest.* **73**, 816 (1984).

BINDING OF ^{125}I-LABELED LIPOPROTEINS TO A MONOCLONAL ANTI-LDL IMMUNOAFFINITY COLUMN[a]

^{125}I-labeled ligands	Counts bound to column (%)	Counts precipitable by TMU (%)	Counts bound by (%) control column
VLDL ($d < 1.006$)	34	50	1.0
VLDL$_1$	23	37	1.0
VLDL$_2$	42	49	0.2
VLDL$_3$	53	69	0.1
LDL	96	97	N.D.[b]
HDL	<0.5	<0.1	N.D.

[a] Monoclonal antibody 464 B1B6 was coupled to Sepharose 4B as described by others.[23] This antibody has been shown to bind near the LDL receptor recognition site on human apoB.[12] Various ^{125}I-labeled lipoproteins were applied to the column ($\sim 2.5 \times 10^6$ cpm corresponding to 10–25 μg of total protein) in 3% BSA–PBS pH 7.4 at 4°. VLDL subfractions were isolated as previously described.[84] The control column consisted of uncoupled Sepharose 4B. HDL was used as a negative binding control. VLDL and VLDL subfractions demonstrate heterogeneity in the ability to bind to 464 B1B6, indicating that the receptor recognition site is expressed differentially on VLDL particles.[7]

[b] N.D., Not detectable.

apoC-II epitope expression has been demonstrated with monoclonal[84,93] and polyclonal[1,94] antibodies. An example of this is shown in Fig. 5 where lipolysis of VLDL *in vitro* increased its immunoreactivity toward two anti-apoB monoclonal antibodies so that it more closely resembled LDL in terms of apoB epitope expression.

Digestion by proteolytic enzymes of intact hololipoproteins has been carried out.[39,62,95] Digestion of human LDL by enzymes of differing specificity has demonstrated with the use of monoclonal antibodies unique patterns of residual immunoreactivity on the core particle.[39] Glycosidases and neuraminidase have been used to assess the role of carbohydrate moieties on lipoprotein epitope expression.[96]

Chemical Delipidation of Lipoproteins. The specific role of lipid–protein interactions on the immunoreactivity of apolipoproteins can be evalu-

[93] J. G. Patton, M. C. Alley, and S. J. T. Mao, *J. Immunol. Methods* **55,** 193 (1982).
[94] S. I. Barr, B. A. Kottke, J. Y. Chang, and S. J. T. Mao, *Biochim. Biophys. Acta* **663,** 491 (1981).
[95] H. Maeda, N. Nakamura, and H. Uzawa, *J. Biochem.* **92,** 1213 (1982).
[96] V. N. Schumaker, M. T. Robinson, L. K. Curtiss, R. Butler, and R. S. Sparks, *J. Biol. Chem.* **259,** 6423 (1984).

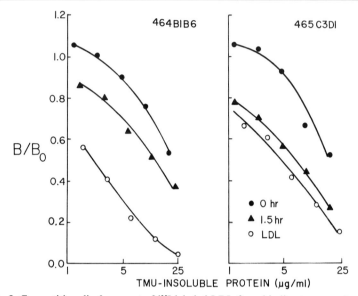

FIG. 5. Competitive displacement of ^{125}I-labeled LDL from binding to monoclonal anti-LDL antibodies 464 B1B6 or 465 C3D1 by LDL or VLDL$_1$ incubated *in vitro* with milk lipoprotein lipase. Incubation with lipase for 1.5 hr increased the immunoreactivity of the VLDL$_1$ subfraction so that it more effectively competed for LDL indicating an increased expression of antibody-specific epitope.

ated by using chemical means to delipidate lipoproteins. A wide variety of effects on immunoreactivity have been reported.[6,16,84,94]

Reconstituted Lipoprotein Models. The ability to synthesize apolipoprotein–lipid vesicles or microemulsions with defined composition is a useful tool to analyze lipoprotein structure. The demonstrated biological activity of these particles (see other chapters in this volume) points to their suitability in use in experiments with antibodies as probes. Polyclonal[66] and monoclonal[11] antibodies have been used with reconstituted apolipoprotein–lipid complexes in competitive binding assays to delineate epitope expression.

Naturally Occurring Sources of Lipoprotein Variation and Analysis with Antibodies

Allelic Variations. Slaughter *et al.*[97] devised a scheme for the detection of enzyme polymorphism with monoclonal antibodies. By construction of binding ratio profiles, that is, the amounts of a series of specific

[97] C. A. Slaughter, M. C. Coseo, M. P. Cancro, and H. Harris, *Proc. Natl. Acad. Sci. U.S.A.* **78,** 1124 (1981).

monoclonal antibodies bound to assay apolipoprotein versus reference apolipoprotein determined in sandwich plate assays, Schumaker et al.[96] were able to detect polymorphisms in human LDL epitope expression.

Role of Other Lipoprotein Components on Epitope Expression. The role of carbohydrates in epitope expression of human LDL has been assessed in competitive binding assays with polyclonal and monoclonal antibodies using a variety of monosaccharides or glycopeptides as competitors.[96,98]

Linking Epitopes to Apolipoprotein Function

As lipoproteins are studied for their expression of different epitopes, attempts can be made to relate given epitopes to specific metabolic functions.

Receptor Binding and Uptake Studies. In view of the roles of apoB and apoE in lipoprotein receptor recognition, much interest has been generated in trying to elucidate the mechanisms and modulation of their recognition by a variety of cells.[74] Studies using monoclonal antibodies have provided information in the relative contribution of apoB or apoE in the binding and uptake of various lipoproteins (Fig. 6)[7,99]

For these experiments it is important that the affinities of the antibodies to be used as probes to block cellular uptake are of the same order of magnitude as the receptor, otherwise negative results may be interpreted as the target apolipoprotein epitope not having any role in receptor recognition. A wide range of antibody concentrations should at first be tested to ensure "saturation" of all epitope binding sites on the lipoprotein (or lipoprotein analog). A preincubation of the antibody with lipoprotein at room temperature or 37° prior to the addition to cells (or cell membranes) is advised for similar reasons.

Epitopes Conserved through Evolution. That natural selection favors functionally important proteins is a widely accepted idea in evolutionary theory. Similarly, epitopes deemed to have important metabolic functions are conserved in a variety of species. Conversely, Goldstein and Chapman[100] discovered a relatively high conservation of LDL structure related to taxonomic classification by immunochemical means. Similar studies with highly sensitive monoclonal antibodies have confirmed these findings[96,101] and indicate a potential of monoclonal antibodies in the detection

---

[98] S. Goldstein and M. J. Chapman, *Biochemistry* **20**, 1025 (1981).
[99] D. Y. Hui, T. L. Innerarity, and R. W. Mahley, *J. Biol. Chem.* **259**, 860 (1984).
[100] S. Goldstein and M. J. Chapman, *Biochem. Gen.* **14**, 883 (1976).
[101] C. A. Nelson, M. A. Tasch, M. J. Tikkanen, R. Dargar, and G. Schonfeld, *J. Lipid Res.* **25**, 821 (1984).

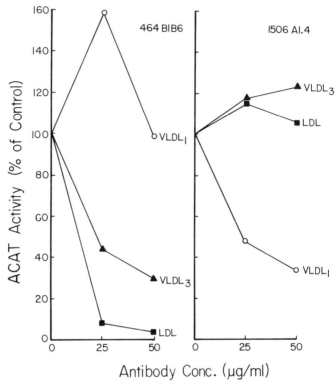

FIG. 6. Inhibition by monoclonal anti-LDL (464 B1B6) or anti-apoE (1506 A1.4) antibodies of the lipoprotein-stimulated esterification of cholesterol in cultured human fibroblasts. Indicated doses of antibodies were incubated with 25 μg/ml of hypertriglyceridemic lipoproteins for 30 min at 37° before being added to the fibroblasts. Inhibition of specific binding and uptake of the lipoproteins are reflected by the antibody concentration-dependent inhibition of lipoprotein-stimulated cholesterol esterification. This experiment supports the concept that the role of apoB in mediating lipoprotein binding and uptake decreases in the order LDL > $VLDL_3$ > $VLDL_1$. Conversely, the role of apoE in mediating cellular uptake of lipoproteins decreases in the order $VLDL_1$ > $VLDL_3$.

of conserved sequences or structures. Nelson et al.[101] noted that epitopes spatially located near the cellular recognition site of apoB have a greater tendency to be conserved in a variety of animal species.

Enzyme Catalysis. Polyclonal antibodies have been used to effectively block activation of lecithin:cholesterol acyltransferase enzyme by apoA-I,[102] and the activation of lipoprotein lipase by apoC-II.[103]

[102] S. Yokoyama, T. Murase, and Y. Akanuma, *Biochim. Biophys. Acta* **530**, 258 (1978).
[103] B. W. Liu, B. A. Hynd, and M. L. Kashyap, *Biochem. Biophys. Res. Commun.* **121**, 946 (1984).

Region Specific Immunoassays

The identification, localization, and characterization of a specific epitope recognizable by a monoclonal antibody can finally lead to region specific immunoassays which would be invaluable as clinical tools to quantitate genetic apolipoprotein isoforms, enzyme activator levels, receptor defective proteins, and abnormal or modified lipoprotein structures.

Acknowledgments

The authors would like to thank Dr. Koji Oida for helpful suggestions concerning dot immunoblotting and immunoelectrotransfer. E. K. acknowledges support by a Medical Research Council of Canada Fellowship. The typing of this manuscript by Lois Weismantle was appreciated.

[32] Reconstitution of High-Density Lipoproteins

By ANA JONAS

Introduction

The detailed investigation at the molecular level of protein–lipid interactions, functions, and metabolism of high-density lipoproteins (HDL),[1] led during the past decade to the development of several reconstitution methods for micellar particles of defined chemical composition, having the overall size and density of HDL. The main components for most reconstitution procedures have been purified apolipoproteins (A-I, A-II, E, or Cs), synthetic or natural phosphatidylcholines (PC), and free (i.e., unesterified) cholesterol. In a few cases partly isolated HDL components have been used, neutral lipids have been added, or other phospholipids

[1] Abbreviations: HDL, high-density lipoprotein; apoA-I, apoA-II, apolipoproteins isolated from HDL; apoE, apoC's(C-I, C-II, C-III), apolipoproteins isolated from very low-density lipoprotein; PC, phosphatidylcholine, DLPL, dilauroyl-PC; DMPL, dimyristoyl-PC; DPPC, dipalmitoyl-PC; VLDL, very low-density lipoprotein; SDS, sodium dodecyl sulfate; DEAE-cellulose, diethyl aminoethyl cellulose; TLC, thin-layer chromatography; T_c, gel to liquid-crystalline phase transition temperature; MLV, multilamellar vesicles (also multibilayer liposomes); SUV, small unilamellar vesicles; LUV, large unilamellar vesicles; DSPC, distearoyl-PC; POPC, palmitoyloleoyl-PC; LCAT, lecithin : cholesterol acyltransferase; DPH, diphenylhexatrine; Dns-Cl, dansyl-Cl.

(e.g., phosphatidylethanolamine, sphingomyelin, or ether PC analogs) have been included in the reconstitution of HDL-like particles.

Three major methods for the codispersion and interaction of apolipoproteins with various lipids have been employed: (1) cosonication of components, (2) spontaneous interaction of apolipoproteins with lipid vesicles, and (3) detergent-mediated reconstitution. The cosonication method has not gained wide acceptance because of reproducibility and yield problems. In this chapter only the latter two approaches: spontaneous reaction of vesicles with apolipoproteins, and the detergent-mediated reconstitution methods will be described in detail. Both of these methods give discoidal particles with apolipoproteins, PC, and free cholesterol, of dimensions and composition analogous to nascent, discoidal HDL particles; however, spontaneous reaction of PC vesicles with apolipoproteins is kinetically controlled by the physical state and composition of the lipid vesicles, and is restricted to only a few PC types [e.g., dilauroyl PC (DLPC), dimyristoyl PC (DMPC), dipalmitoyl PC (DPPC), and their mixtures]. The detergent reconstitution method, on the other hand, is based on the dispersion of lipids in sodium cholate, which allows the incorporation of a much larger variety of phospholipids into discoidal, HDL-like particles. Unless stated otherwise all the references to apolipoproteins in this chapter are to human apolipoproteins.

Preparation of Materials

Apolipoproteins

The isolation and purification of the HDL-associated apolipoproteins is described in Chapters [10]–[16] of this volume. The reader is referred to the appropriate chapters for details of the procedures.

Routinely, the purity of apolipoproteins (A-I, A-II, E) is assessed by polyacrylamide slab–gel electrophoresis using 10% gels, about 50 μg of apolipoproteins, and 0.1% SDS. Electrophoresis of the C apolipoproteins is performed in 8 M urea rather than SDS to allow separation by charge of these small-molecular-weight proteins. The purified apolipoproteins are stored in lyophilized or frozen-solution form, at −15°, for periods not exceeding 1 year. In order to prepare stock solutions with reproducible properties, the stored apolipoproteins (lyophilized) are dissolved in a small volume of 4 M guanidine hydrochloride in the buffer of choice, to give concentrations in the range of 4–10 mg/ml of protein. After thorough dialysis against the selected buffer, at 4°, these apolipoprotein solutions can be stored for at least 2 months at 4° until use in reconstitution experiments. During the dialysis steps it is important to note the pore size of the dialysis tubing: the low-molecular-weight of C apolipoproteins requires small pore size dialysis tubing, such as Spectropor 6.

Apolipoproteins in solution can be quantitated by several methods: (1) from absorbance at 280 nm and the known percentage extinction coefficients (see Chapter [29] of this volume); (2) Lowry et al.[1a] assay using either bovine serum albumin or apoA-I as the standards; (3) the fluorescamine assay[2] using a standard solution of the corresponding apolipoprotein. For the precise determination of absolute concentrations, or comparison of different apolipoproteins it should be noted that the different amino acid composition of the apolipoproteins will affect the results of the Lowry et al.[1a] and the fluorescamine assays.[2]

Lipids

Most phosphatidylcholines (PC) used in reconstitution experiments are obtained from commercial sources (all saturated, unsaturated synthetic PCs, and egg-PC); PC from human HDL is extracted from HDL by the Folch et al.[3] method and is recovered in pure form by chromatography on a silicic acid column eluted with a chloroform/methanol/H_2O gradient.[4] The purity of the commercial PCs is normally checked by one-dimensional thin-layer chromatography (TLC) on silica gel-G plates developed in chloroform/methanol/H_2O (65/25/5, v/v); natural PC purity is assessed by two-dimensional TLC in chloroform/methanol/28% ammonia (65/35/5, v/v) and in chloroform/acetone/methanol/acetic acid/H_2O (5/2/1/1/1.5, v/v). Cholesterol is a commercial, crystalline product which is tested for purity by one-dimensional TLC using cyclohexane/ethyl acetate (60/40, v/v) or petroleum ether/diethyl ether/acetic acid (90/12/1.5, v/v) solvents. Lipid spots are visualized by exposure to iodine vapors. Radiolabeled (^3H or ^{14}C) cholesterol or PC (^3H or ^{14}C, DMPC or DPPC) purity is examined in the same TLC solvent systems as the unlabeled lipids, adding small amounts of cold carrier lipids. Radiolabeled spots are detected by autoradiography or by scraping the TLC plate in thin sections and by scintillation counting.

Lipids are stored either in the solid state or in organic solvent at $-15°$, desiccated over Drierite or silica gel. Periodic checks of purity are con-

---

[1a] N. H. Lowry, N. J. Rosebrough, A. L. Farr, and R. J. Randall, *J. Biol. Chem.* **193**, 265 (1951).
[2] P. Böhlen, S. Stein, W. Dairman, and S. Undenfriend, *Arch. Biochem. Biophys.* **155**, 213 (1973).
[3] J. Folch, M. Lees, and G. H. Sloane-Stanley, *J. Biol. Chem.* **226**, 497 (1957).
[4] G. Rouser, G. Kritchevsky, D. Heller, and E. Lieber, *J. Am. Oil Chem. Soc.* **40**, 425 (1963).

ducted, particularly of the radiolabeled lipids, which are purified by TLC chromatography and organic solvent extraction, if impurities exceed 5%. Dry lipids are weighed accurately and are dissolved in cold redistilled chloroform/methanol (1/1, v/v) to prepare concentrated stock solutions, just before use in reconstitution experiments. Lipids stored in organic solvent require an accurate determination of concentration prior to use. The Chen et al.[5] assay is used in the quantitation of phosphate in phospholipids, and the enzymatic assay kit of Boehringer-Mannheim for the determination of cholesterol. We have determined that stock solutions prepared by weighing solid lipids agree within 10% with the concentrations determined by the chemical assay methods. However, when very hygroscopic lipids are used or a precise starting concentration is required, a chemical assay of stock solutions is advisable. Lipid mixtures (e.g., cholesterol and PC, various PCs, or mixtures containing tracer ^{14}C or ^{3}H lipids) are prepared by mixing thoroughly the appropriate volumes of stock solutions (in redistilled organic solvents) at room temperature; they are usually prepared immediately before use in teflon-lined screw cap test tubes. Such containers are the most convenient for the long- and short-term storage and handling of lipids in organic solvents.

Pure lipid solutions or lipid mixtures are dried, and the solvent is completely evaporated by heating at less than 60° in a heating block under a stream of N_2. Periodic vortexing spreads the lipid on the walls of the tubes, increases the efficiency of solvent removal, and facilitates subsequent dispersion in aqueous solvents. After organic solvent removal, the appropriate aqueous buffer is added to give final dispersions containing from 2 to 20 mg lipid per ml. Although most of our work has been performed using 10 mM to 0.1 M Tris–HCl, pH 7.6 to 8.2, 0.001 to 0.01% EDTA, 1 mM NaN_3 buffers, with or without 0.150 M NaCl, lipid dispersions and complex preparations in phosphate and other buffers, in the neutral pH range (pH 6 to 9), appear to work equally well. Lipids are then dispersed by vortexing the samples, adjusted to temperatures higher than the gel to liquid crystalline phase transition temperature (T_c) of the highest melting phospholipid in the mixture (e.g., mixtures of DMPC and DPPC are dispersed above 41°). The dispersion is complete when lipid no longer clings to the walls of the test tube and the milky suspension has a uniform consistency. At this point the preparation contains very large multilamellar vesicles (MLV) of phospholipid with or without cholesterol, which intercalates in the phospholipid bilayers. These MLV can be used directly in spontaneous interactions with apolipoproteins, under the re-

[5] P. S. Chen, T. Y. Toribara, and H. Warner, *Anal. Chem.* **28**, 1756 (1965).

strictions indicated in the next section. Most commonly the MLV are processed further, by sonication, to give small unilamellar vesicles (SUV) or by dispersion with sodium cholate, to form mixed micelles of lipids and detergent.

Small unilamellar vesicles of PCs are prepared by the sonication of MLV dispersions, as described by Huang.[6] In our experience the Heat Systems, Inc. sonifier, Model W185, operating at a power of 80 to 90 W for 30 min to 1 hr, in 3 min bursts, at 1 to 2 min intervals, with probe immersion, gives mostly SUV. Shorter sonication times (~5 min) give more heterogeneous vesicle preparations with larger average diameters. Sonication of unsaturated PCs is usually performed under a stream of N_2, cooling the sonication vessel with ice-water and protecting from light, in order to minimize lipid autoxidation. With saturated PCs the precautions are less stringent and consist of cooling with ice-water to keep the temperature below 50°. Thin-layer chromatography of lipids extracted after sonication, and gas–liquid chromatography of fatty acid methyl esters, do not indicate any significant lipid degradation during the period of sonication. After sonication, metal probe fragments and very large lipid particles (MLV) are removed by centrifugation at 15,000 rpm, 15°, for 1 hr. If homogeneous SUV preparations are desired the supernatant of the centrifugation step is passed through a Sepharose CL-4B column, and the fractions on the trailing edge of the included lipid peak[6] (detected by absorbance due to light scattering, radioactivity of tracer lipids, or phosphate determination) are collected. These stock solutions of SUV are assayed for lipid concentrations and are used within short periods in reconstitution experiments. Storage of SUV will result in their aggregation and fusion into large unilamellar or multilamellar vesicles (LUV and MLV); the rates of these processes are dependent on the type of PC and the temperature of storage. For example egg-PC SUV can be stored for about 1 week at 4° without major changes in their properties, but DPPC SUV aggregate and fuse after several hours at room temperature or at 4°.

Sodium cholate (commercial, high purity) dispersions of lipids are prepared from MLVs by adding a concentrated sodium cholate solution in buffer (30 mg/ml) to give molar ratios from 1/1 to 2/1, sodium cholate/PC, with a final sodium cholate concentration of at least 3 mg/ml. Under these conditions the preparations become translucent. Incubations for 1 to 2 hr near or slightly above the T_c of the PC with the highest T_c in the mixture, usually improve the micellization process and subsequent yields of reconstituted complexes with apolipoproteins.

[6] C. Huang, *Biochemistry* **8**, 344 (1969).

Preparation of Apolipoprotein–Lipid Complexes

Spontaneous Interaction of Apolipoproteins with Lipid Vesicles

Under appropriate conditions the incubation of MLV, LUV, SUV, of PCs with apolipoprotein A-I, A-II, E, or C solutions will result in the disruption of the lipid vesicles and the formation of micellar, discoidal products containing apolipoprotein and PC. The term "micellar" is used in this review to designate small lipid particles spherical or discoidal, which have all the polar lipid head groups on the surface of the particle, and which are devoid of an internal aqueous compartment. The latter property distinguishes micelles from lipid vesicles or "vesicular" complexes containing apolipoproteins.

The rates of discoidal, micellar complex formation from PC vesicles and apolipoproteins are determined by a variety of factors: (1) the type of apolipoprotein, (2) the type of PC, (3) the phase state of the PC vesicles as determined by the temperature of the reaction, and (4) the PC vesicle size and composition. The morphology and size of the product particles are also influenced by the above factors, but the preponderant factor affecting the size and shape of the product is the PC/apolipoprotein molar ratio in the reaction mixture. It has been shown that apoA-II reacts more rapidly with DMPC vesicles than apoA-I,[7] but no comparative kinetic studies have been conducted with the other apolipoproteins. The following discussion of the lipid vesicle properties governing the kinetics of discoidal, micellar complex formation of PC with apolipoproteins will be based mostly on results obtained with human apoA-I.

It is evident that PC vesicle bilayer disruption and/or penetration by the apolipoprotein is involved in the formation of discoidal, micellar complexes of PC and apolipoprotein. Thus any conditions which stabilize the vesicle bilayer will, in turn, decrease the reaction rates. Planar bilayers of MLV or LUV react with apoA-I much more slowly than the highly curved, inherently less stable bilayers of SUV.[8] Phosphatidylcholine vesicles at their T_c, where lipid lattice defects occur most frequently, react much more readily than vesicles in their gel or liquid-crystalline states.[9] This temperature dependence also correlates with the type of vesicle: SUV have the highest reaction rates at the T_c of the PC, but also appreciable rates 5 to 10° on either side of the T_c; MLVs only react in a very

[7] P. Van Tornout, R. Vercaemst, M. J. Lievens, H. Caster, M. Rosseneu, and G. Assmann, *Biochem. Biophys. Acta* **601,** 509 (1980).
[8] J. R. Wetterau and A. Jonas, *J. Biol. Chem.* **257,** 10961 (1982).
[9] H. J. Pownall, J. B. Massey, S. K. Kusserow, and A. M. Gotto, Jr., *Biochemistry* **17,** 1183 (1978).

narrow temperature window around the T_c.[8] Long chain saturated PCs, DLPC, DMPC, DPPC, and distearoyl-PC (DSPC), react with apoA-I with rates that are inversely proportional to the acyl chain length, due to increasing hydrophobic stabilization of the bilayers.[10] Cholesterol inclusion into DMPC bilayers increases reaction rates up to 12 mol% cholesterol contents, but essentially abolishes reaction above 20 mol% in MLV or 30 mol% cholesterol contents in SUV.[11,12]

Spontaneous reaction of SUV of saturated PCs (DMPC or DPPC) with apolipoproteins can give rise to different products, depending on the ratios of PC to apolipoprotein in the reaction mixture.[13–17] As indicated in Fig. 1, apoA-I reacts with DMPC SUV to form vesicular complexes at PC/apoA-I molar ratios exceeding 1000/1, large discoidal complexes around 300/1, and small discoidal complexes around 100/1 molar ratios of PC/apoA-I. At even lower PC/apoA-I molar ratios free apoA-I and small discoidal complexes are recovered. The large and small discs contain 3 and 2 apoA-I molecules per particle, respectively.[14,16] It is not completely clear whether or not even larger discs with more apoA-I molecules are formed, but vesicular complexes with only a few apoA-I molecules (1 to 3) per vesicle have been reported using DMPC or DPPC vesicles at very high PC/apoA-I molar ratios.[13] Within each complex class there is some compositional and/or size heterogeneity.[17]

Table I lists selected apolipoprotein–lipid systems, which have been used in the spontaneous recombination of HDL-like discoidal complexes, and gives the conditions of the reactions. A specific example is the synthesis of small discoidal complexes of human apoA-I with DMPC performed routinely in our laboratory. A solution of SUV of DMPC containing 10 mg DMPC in 1 ml of buffer is equilibrated to 25°; to this solution 3.6 mg of apoA-I is added from a concentrated stock solution (~10 mg apoA-I/ml). The mixture is equilibrated, with periodic stirring, for 3 to 5 hr at 25°, to ensure complete reaction. The course of the reaction to equilibrium can be followed nondestructively by observing the change in the

---

[10] A. Jonas and W. R. Mason, *Biochemistry* **20**, 3801 (1981).
[11] A. Jonas and D. J. Krajnovich, *J. Biol. Chem.* **253**, 5758 (1978).
[12] H. J. Pownall, J. B. Massey, S. K. Kusserow, and A. M. Gotto, Jr., *Biochemistry* **18**, 574 (1979).
[13] A. Jonas, S. M. Drengler, and B. W. Patterson, *J. Biol. Chem.* **255**, 2183 (1980).
[14] J. B. Swaney, *J. Biol. Chem.* **255**, 8791 (1980).
[15] J. B. Massey, M. F. Rohde, W. B. Van Winkle, A. M. Gotto, Jr., and H. J. Pownall, *Biochemistry* **20**, 1569 (1981).
[16] J. R. Wetterau and A. Jonas, *J. Biol. Chem.* **258**, 2637 (1983).
[17] C. G. Brouillette, J. L. Jones, T. C. Ng, H. Kercret, B. H. Chung, and J. P. Segrest, *Biochemistry* **23**, (1984).

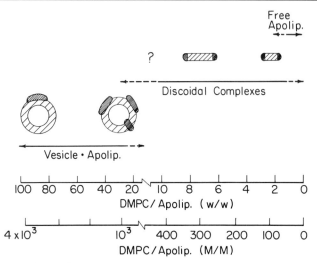

FIG. 1. Types of DMPC–apoA-I complexes formed as a function of the starting DMPC/apoA-I ratios in molar (M/M) or in weight (w/w) terms. The vesicular complexes include up to 3 apoA-I molecules per particle and have molecular weights around 2×10^6. Discoidal, micellar complexes contain 3 or 2 apoA-I molecules per particle, and have molecular weights in the vicinity of 8×10^5 or 2×10^5, respectively. Within the large and small discoidal complex classes there are subsets of particles with slightly different lipid contents.

absorbance of the solution at 350 nm; as vesicles are disrupted the absorbance decreases to a constant low value. If DMPC MLV instead of SUV are used in the reaction it is important to equilibrate at the T_c of the lipid (24°) and to stir the reaction mixtures to prevent settling of the MLV. Alternatively a buffer containing 8.5% NaBr may be used to keep the MLVs in suspension.[9] Evidently this is a very simple method to prepare discoidal, HDL-like complexes of apolipoproteins with some saturated PCs (particularly DMPC) or sphingomyelin; the complexes may include up to 20 mol% cholesterol. However, the rates of reaction and the yields of apoA-I discoidal complexes with DPPC are quite low, and reactions with more physiological lipids such as egg-PC, palmitoyloleoyl PC (POPC), or naturally occurring HDL PCs, are practically undetectable under normal conditions. Apolipoprotein A-II appears more reactive than apoA-I with some of the long chain unsaturated PCs and with DPPC (see Table 1).

Detergent-Mediated Synthesis of Complexes

The kinetic block in the spontaneous reaction of PC vesicles with apolipoproteins, alluded to in the previous section, can be bypassed by

TABLE I
SPONTANEOUS REACTION OF HUMAN APOLIPOPROTEINS WITH PHOSPHOLIPID VESICLES TO FORM MICELLAR, DISCOIDAL COMPLEXES[a]

Apolipoprotein	Phospholipids	Reaction conditions[b] (vesicle, temp, time, PL/apolip)	Complex properties[c] (size and shape, PL/apolip)
ApoA-I[d]	DLPC	MLV, 0°, 18 hr, 2.5/1 (w/w)	Small particles
ApoA-I[e]	DMPC	SUV, 23–37°, 3 hr, 130/1 (M/M)	Small discs, 90/1 (M/M)
ApoA-I[f]	DMPC	MLV, 37°, >12 hr, 20/1 to 780/1 (M/M)	Small discs below 80/1 (M/M) Excess MLV at >100/1 (M/M)
ApoA-I[g]	DMPC	MLV, 24°, ~15 min, 206/1 (M/M)	Small particles
ApoA-I[h]	DMPC	SUV, 25°, 20 hr, 50/1 to 4000/1 (M/M)	Small discs, ~100/1 (M/M) Vesicle–apoA-I complexes at >900/1 (M/M)
ApoA-I[i]	DPPC	SUV, 42°, 24 hr, 50/1 to 500/1 (M/M)	Small discs, ~100/1 (M/M) Large discs, ~280/1 (M/M) (yields >80%)
ApoA-I[j]	DPPC	LUV, MLV, 41°, 24 hr, 150/1 (M/M)	Small discs, ~100/1 (M/M) (yields <40%)
ApoA-I[k]	Sphingomyelin (beef brain)	MLV, 28°, 24 hr, 2.5/1 to 15/1 (w/w)	Large discs, 300/1 to 500/1 (M/M)
ApoA-I[l]	DMPC-DPPC; DMPC-DSPC	MLV, 26 to 39°, 18–40 hr, 2.5/1 to 15/1 (w/w)	Small discs, ~140/1 (M/M) Large discs, ~300/1 (M/M)
ApoA-I[m]	DMPC-POPC	MLV, 10 to 20°, ~1 hr, 2.4/1 (w/w)	Small particles (yields <50%)
ApoA-I[n]	DMPC-cholesterol	SUV, 37°, 2 hr, 200/1 (M/M), up to 33 mol% cholesterol	Small discs, ~100/1 (M/M)
ApoA-I[o]	DMPC-cholesterol	MLV, 24°, 12 hr, 210/1 (M/M), up to 24 mol% cholesterol	Small particles, ~170/1 (M/M)

(continued)

TABLE I (continued)

Apolipoprotein	Phospholipids	Reaction conditions[b] (vesicle, temp, time, PL/apolip)	Complex properties[c] (size and shape, PL/apolip)
ApoA-II[p]	DMPC	MLV, 20, 24, 30°, 24 hr, 4/1 to 12/1 (w/w)	Small and large discs, 45/1, 75/1, and 240/1 (M/M)
ApoA-II[q]	DPPC	SUV, MLV, 25°, 12 hr, >30/1 (M/M)	Small particles, 20/1 (M/M)
ApoA-II[r]	Egg-PC; sphingo-myelin (egg)	SUV, ambient temp., 1 hr, 1/1 to 7/1 (w/w)	Small particles, 20/1 to 40/1 (M/M)
ApoA-II[s]	DMPC-cholesterol	SUV, 25°, 3 hr, 2/1 (w/w)	Small particles, ~100/1 (M/M)
ApoE[t]	DMPC	SUV, 24°, 3 hr, 3.8/1 (w/w)	Small discs, 4.4/1 (w/w)
ApoC-I[u]	DMPC	SUV, 25°, 5 hr, 3.6/1 (w/w)	Small discs, 25/1 (M/M)
ApoC-III[v]	DMPC	MLV, 30°, ? min, 20/1 (M/M)	Small particles, ~20/1 (M/M)
ApoC-III[w,x]	DMPC	SUV, 28°, 30 min, 4/1 (w/w)	Small discs, ~4/1 (w/w)
ApoA-II[y] (50-77 fragment)	DMPC	MLV, 24.5°, 12 hr, 5/1 (w/w)	Small particles, 40/1 (M/M)
LAP-20[z] (synthetic peptide)	DMPC	MLV, 24°, 12 hr, 2/1 (w/w)	Small particles, 20/1 (M/M)

[a] Only *selected* references, using pure human apolipoproteins, and illustrating best the reaction conditions and the characteristics of the discoidal complex products are given in this table.

[b] The phospholipid vesicle types are multilamellar (MLV), large unilamellar (LUV), and small unilamellar (SUV) vesicles. The ratios of phospholipid/apolipoprotein in the reaction mixtures are given in terms of weight ratios (w/w) or molar ratios (M/M).

[c] Small discs have molecular weights in the range from 10^5 to 4×10^5, axial ratios of 2/1 to 3/1, diameters from 100 to 140 Å, and the thickness of a lipid bilayer. Large discs have molecular weights ~10^6, axial ratios ~4/1 or larger, and diameters around 200 Å. When "small particles" is entered in the table, it means that the product characterization in the reference was not complete, but that these are probably small discoidal particles.

[d] J. B. Swaney and B. C. Chang, *Biochemistry* **19**, 5637 (1980).
[e] A. Jonas, D. J. Krajnovich, and B. W. Patterson, *J. Biol. Chem.* **252**, 2200 (1977).
[f] A. R. Tall, D. M. Small, R. J. Deckelbaum, and G. G. Shipley, *J. Biol. Chem.* **252**, 4701 (1977).
[g] H. J. Pownall, J. B. Massey, S. K. Kusserow, and A. M. Gotto, Jr., *Biochemistry* **17**, 1183 (1978).
[h] A. Jonas, S. M. Drengler, and B. W. Patterson, *J. Biol. Chem.* **255**, 2183 (1980).
[i] J. R. Wetterau and A. Jonas, *J. Biol. Chem.* **258**, 2637 (1983).
[j] J. R. Wetterau and A. Jonas, *J. Biol. Chem.* **257**, 10961 (1982).
[k] J. B. Swaney, *J. Biol. Chem.* **258**, 1254 (1983).
[l] J. B. Swaney, *J. Biol. Chem.* **255**, 8798 (1980).
[m] J. B. Swaney, *J. Biol. Chem.* **255**, 8791 (1980).
[n] A. Jonas and D. J. Krajnovich, *J. Biol. Chem.* **253**, 5758 (1978).
[o] H. J. Pownall, J. B. Massey, S. K. Kusserow, and A. M. Gotto, Jr., *Biochemistry* **18**, 574 (1979).
[p] J. B. Massey, M. F. Rohde, W. B. Van Winkle, A. M. Gotto, Jr., and H. J. Pownall, *Biochemistry* **20**, 1569 (1981).
[q] G. Middelhoff, M. Rosseneu, H. Peeters, and W. V. Brown, *Biochim. Biophys. Acta* **441**, 57 (1976).
[r] G. Assmann and H. B. Brewer, Jr., *Proc. Natl. Acad. Sci. U.S.A.* **71**, 989 (1974).
[s] P. Van Tornout, R. Vercaemst, M. J. Lievens, H. Caster, M. Rosseneu, and G. Assmann, *Biochim. Biophys. Acta* **601**, 509 (1980).
[t] T. L. Innerarity, R. E. Pitas, and R. W. Mahley, *J. Biol. Chem.* **254**, 4186 (1979).
[u] A. Jonas, J.-P. Privat, P. Wahl, and J. C. Osborne, Jr., *Biochemistry* **24**, 5205 (1982).
[v] H. Träuble, G. Middelhoff, and V. W. Brown, *FEBS Lett.* **49**, 269 (1974).
[w] K. C Aune, J. G. Gallagher, A. M. Gotto, Jr., and J. D. Morrisett, *Biochemistry* **16**, 2151 (1977).
[x] P. Laggner, A. M. Gotto, Jr., and J. D. Morrisett, *Biochemistry* **18**, 164 (1979).
[y] S. J. T. Mao, R. L. Jackson, A. M. Gotto, Jr., and J. T. Sparrow, *Biochemistry* **20**, 1676 (1981).
[z] H. J. Pownall, Q. Pao, D. Hickson, J. T. Sparrow, S. K. Kusserow, and J. B. Massey, *Biochemistry* **20**, 6630 (1981).

using micellar dispersions of lipids with sodium cholate or deoxycholate.[17-22]

Sodium cholate micelles with PC will form at sodium cholate/PC molar ratios from 1/2 to 2/1. A typical reaction mixture contains 5.6 mg of egg-PC and 3.1 mg of sodium cholate (1/1, mol/mol) in 0.5 ml of buffer (10 mM Tris–HCl, pH 8.0, 0.15 M NaCl, 0.005% EDTA, 1 mM NaN$_3$). The lipid dispersion should become translucent to clear when thoroughly mixed and incubated for 1 to 2 hr at 4°. If the dispersion remains milky, more sodium cholate may be added but not to exceed the 2/1, sodium cholate/PC molar ratio. Apolipoprotein in concentrated solution (~10 mg/ml) is added next: 2.0 mg of apoA-I to give a PC/apoA-I, 100/1 molar ratio, in 0.7 ml of solution. The reaction mixture is stirred thoroughly and is incubated for 12 to 16 hr at 4°. At the end of the equilibration period the sample is transferred to an appropriate dialysis bag and is extensively dialyzed at 4°. At least 5 buffer changes in a volume excess of 500-fold each time, over 3 days will essentially remove all of the sodium cholate. Less than 0.2 mol% of sodium cholate could be detected at the end of the dialysis period.[19]

To generalize the procedure to other apolipoproteins and PCs it should be noted that incubation of the PC with sodium cholate, the equilibration of the lipid with apolipoprotein, and the early stages of the dialysis should be performed at temperatures as close to the T_c of the PC as practical (e.g., for PCs with T_c below 0°, 4° is an appropriate temperature; and dialysis at warm room 37–40°, temperature works well with DPPC). Since different apolipoproteins and PCs may give discoidal complexes with stoichiometries distinct from 100/1, PC/apolipoprotein (mol/mol), several molar ratios should be tried in pilot experiments.[23,24] A good starting point, however, is a weight ratio equivalent to 100/1, egg-PC/apoA-I molar ratio. As in the case of spontaneous reaction of lipid vesicles with apolipoproteins, widely different PC/apolipoprotein molar ratios of the same components can yield discoidal complex classes with different sizes and stoichiometries.[22,25] In the dialysis step the choice of the dialysis tubing is dictated by the size of the apolipoprotein in the reaction mixture.

At the end of the dialysis for the removal of sodium cholate, any

[18] C. E. Matz and A. Jonas, *Fed. Proc., Fed. Am. Soc. Exp. Biol.* **40**, 1634 (1981).
[19] C. E. Matz and A. Jonas, *J. Biol. Chem.* **257**, 4535 (1982).
[20] C. Chen and J. J. Albers, *J. Lipid Res.* **23**, 680 (1982).
[21] H. J. Pownall, W. B. Van Winkle, O. Pao, M. Rohde, and A. M. Gotto, Jr., *Biochim. Biophys. Acta* **713**, 494 (1982).
[22] A. V. Nichols, E. L. Gong, P. J. Blanche, T. M. Forte, *Biochim. Biophys. Acta* **750**, 353 (1983).
[23] A. Jonas and C. E. Matz, *Biochemistry* **21**, 6867 (1982).
[24] A. Jonas, S. A. Sweeny, and P. N. Herbert, *J. Biol. Chem.* **259**, 6369 (1984).
[25] A. Jonas and H. T. McHugh, *Biochim. Biophys. Acta* **794**, 361 (1984).

unreacted or precipitated lipid (normally not exceeding 20% of the total) can be removed by centrifugation at 15,000 rpm, 15°, for 1 hr. The supernatant contains all of the apolipoprotein in complexes with lipid. Isolation of relatively homogeneous complexes requires gel filtration on a BioGel A-5m column (50 × 1.8 cm) equilibrated with buffer, and selection of appropriate column fractions.[19,23] Higher resolution of complex classes can be achieved by using a longer column.[22] After dialysis or column fractionation all complexes are very stable and can be stored at 4° for at least 2 months.

Table II lists selected lipid–apolipoprotein systems which have been used to date to produce discoidal HDL-like complexes by the sodium cholate method. Similarly to the spontaneous reaction of PC vesicles with apolipoproteins, different PC to apolipoprotein molar ratios can yield discoidal particles of different sizes.[22,25] The smallest discs of apoA-I with various PCs contain two apolipoprotein molecules,[19] but larger particles with three or more apolipoprotein molecules can be prepared by using higher PC/apolipoprotein molar ratios.[22,25,26] As indicated in Table II this method of reconstitution of HDL-like particles can be used with a variety of apolipoproteins, PCs (synthetic or natural), and possibly with other phospholipids or glycolipids. Up to 30 mol% of cholesterol can be incorporated into the large discs, but only 20 mol% cholesterol can be recovered in the small discs of egg-PC and apoA-I. Higher levels of cholesterol in reaction mixtures result in very poor yields and highly heterogeneous complexes.[26] It is not yet clear why higher cholesterol amounts cannot be introduced, especially into the larger complexes. The boundary layer of PC molecules adjoining the apolipoprotein in the discoidal complexes apparently excludes cholesterol, which would partially explain why cholesterol incorporation is less than the 50 mol% maximum observed in egg-PC vesicles. However, in the larger discs the boundary PC represents only about 10% of the total particle PC, so that overall cholesterol content could be as high as 45 mol%, if the bulk PC could take up as much cholesterol as vesicles of the same lipid (e.g., egg-PC). At present it is not clear whether physical factors or chemical composition limit the level of cholesterol incorporation. In spite of this limitation, the detergent-mediated reconstitution method is very useful for the synthesis of discoidal HDL-like particles of defined size and composition, using a wide variety of lipid and apolipoprotein components.

Synthesis of Spheroidal HDL-Like Complexes

Spherical HDL-like particles can be prepared by cosonication of HDL components, including neutral lipids; however, the yields and reproduc-

[26] A. Jonas and H. T. McHugh, *J. Biol. Chem.* **258**, 10335 (1983).

TABLE II
SODIUM CHOLATE-MEDIATED FORMATION OF MICELLAR, DISCOIDAL COMPLEXES OF HUMAN APOLIPOPROTEINS WITH PHOSPHATIDYLCHOLINES

Apolipoprotein	PC	Reaction conditions[a] [temp, time, PC/apolip, PC/cholate (± cholesterol)]	Complex properties[b] (size and shape, PC/apolip)
ApoA-I[d]	Egg-PC, DPPC	5°, 37°, 12 hr, 75/1 to 200/1 (M/M), 4/1 to 1/4 (M/M) (± cholesterol)	Small discs, ~100/1 (M/M)
ApoA-I[e]	Egg-PC	24°, 20 min, 312/1 (M/M), 1/3.3 (M/M) (+ 5 mol% cholesterol)	Vesicles ?, 312/1 (M/M)
ApoA-I[f]	Egg-PC	4°, 18 hr, 25/1 to 150/1 (M/M), 1/20 to 2/1 (M/M) (± cholesterol)	Small and large discs, 85/1 to 190/1 (M/M)
ApoA-I[g]	PPOPC[c]	Ambient temp, ? min, 100/1 (M/M), 1/4.5 (M/M) (+ 2 mol% cholesterol)	Small particles, 100/1 (M/M)
ApoA-I[h]	POPC[c]	4°, 12 hr, 150/1 (M/M), 1/1 (M/M) (+ 15 mol% cholesterol)	Small discs, 94/1 (M/M)
ApoA-I[i]	Egg-PC-variable cholesterol	4°, 12 hr, 80/1 (M/M), 2/3 (M/M), (3/1 to 100/1, PC/chol, M/M)	Small discs and large discs, 70/1 to 170/1 (M/M)
ApoA-I[j]	DPPC	41 to 45°, 12 hr, 70/1 to 500/1 (M/M), 2/3 (M/M) (+ 10 mol% cholesterol)	Small, large, and very large discs, 70/1, 280/1, 470/1 (M/M)
ApoA-II, apoC-I, apoC-II, apoC-III[k]	Egg-PC	4°, 12 hr, 2.8/1 (w/w), 1/1 (M/M) (+ 10 mol% cholesterol)	Small and large discs, 1.8/1 (w/w)

[a] The PC/apolipoprotein and PC/cholate ratios in the reaction mixtures are expressed as molar (M/M) or as weight (w/w) ratios. Cholesterol was included in most preparations and is indicated in parenthesis as mol% of the total lipid in the reaction mixture.
[b] Small and large discs have the characteristics described in footnote [c] of Table I. Very large discs have axial ratios around 8/1. The particles described by Pownall et al., and Chen and Albers appeared round by negative stain electron microscopy.
[c] PPOPC, palmitoylpalmitoleoyl-PC; POPC palmitoyloleoyl-PC.
[d] C. E. Matz and A. Jonas, J. Biol. Chem. **257**, 4535 (1982).
[e] C.-H. Chen and J. J. Albers, J. Lipid Res. **23**, 680 (1982).
[f] A. V. Nichols, E. L. Gong, P. J. Blanche, and T. M. Forte, Biochim. Biophys. Acta **750**, 353 (1983).
[g] H. J. Pownall, W. B. Van Winkle, Q. Pao, M. Rohde, and A. M. Gotto, Jr., Biochim. Biophys. Acta **713**, 494 (1982).
[h] A. Jonas and C. E. Matz, Biochemistry **21**, 6867 (1982).
[i] A. Jonas and H. T. McHugh, J. Biol. Chem. **258**, 10335 (1983).
[j] A. Jonas and H. T. McHugh, Biochim. Biophys. Acta **794**, 361 (1984).
[k] A. Jonas, S. A. Sweeny, and P. N. Herbert, J. Biol. Chem. **259**, 6369 (1984).

ibility of such preparations have not been optimized.[27–30] Recently, in a different approach, HDL has been partially delipidated by extraction of neutral lipids with heptane and reconstituted with cholesteryl linoleate plus radiolabeled cholesteryl linoleyl ether. The exogenous lipids were added in heptane, and the mixture was solubilized in buffer after organic solvent removal. These particles were not characterized in detail, but they behaved as native HDL in functional tests.[31]

A promising approach is the enzymatic transformation of discoidal complexes containing apoA-I, PC, and cholesterol by lecithin:cholesterol acyltransferase (LCAT). Discoidal complexes prepared by the sodium cholate reconstitution method react very efficiently with purified LCAT (see Chapter [45], Volume 129, for the purification and assay methods of this enzyme). The cholesteryl ester products of the reaction accumulate inside the particles whereas much of the lyso-PC is removed by binding to serum albumin, which is included in the reaction mixture. The resultant particles lose their discoidal morphology, become less asymmetric, and probably approach a spherical shape.[32] A typical reaction mixture contains 1.0 mg of egg-PC, 0.45 mg of apoA-I, and 0.124 mg cholesterol (i.e., 80/1/20, mol/mol) in small discoidal complexes, plus 14 mg of defatted bovine serum albumin, and approximately 50 units of enzyme (1 unit = 1 nmol of cholesteryl ester formed/hr). The final volume is 4.0 ml in 10 mM Tris–HCl, pH 7.6, 5 mM EDTA, and 4 mM 2-mercaptoethanol. The reaction is carried out at 37° for 24 hr, and then the mixture is fractionated on a NaBr linear density gradient by centrifugation in an SW 41 rotor at 41,000 rpm, 20°, for 84 hr. Evidently shorter centrifugation times could be used at higher centrifugal fields. The isolated complexes have the following composition in terms of molar ratios: 65/1/4/16, PC/apoA-I/cholesterol/cholesteryl ester; they appear more symmetrical than the starting discoidal complexes.[32] Potentially, this method could be adapted for the enzymatic synthesis of spherical, chemically defined HDL-like particles, provided that enough free cholesterol could be incorporated into the starting discs or could be made otherwise available for the LCAT reaction. At the present time, as indicated above, there are

[27] A. Scanu, E. Cump, J. Toth, S. Koga, E. Stiller, and L. Albers, *Biochemistry* **9**, 1327 (1970).
[28] T. M. Forte, A. V. Nichols, E. L. Gong, S. Lux, and R. I. Levy, *Biochim. Biophys. Acta* **248**, 381 (1971).
[29] G. Schonfeld, B. Pfleger, and R. Roy, *J. Biol. Chem.* **250**, 7943 (1975).
[30] M. C. Ritter and A. M. Scanu, *J. Biol. Chem.* **252**, 1208 (1977).
[31] C. Glass, R. C. Pittman, D. B. Weinstein, and D. Steinberg, *Proc. Natl. Acad. Sci. U.S.A.* **80**, 5435 (1983).
[32] C. E. Matz and A. Jonas, *J. Biol. Chem.* **257**, 4541 (1982).

limits to the incorporation of cholesterol into the discoidal complexes which, in turn, restricts the amount of cholesterol ester products available to make up a spherical core for the product particles.

Characterization of Apolipoprotein–Lipid Complexes

Isolation

The discoidal complexes prepared by spontaneous reaction of PC vesicles with apolipoproteins or by the sodium cholate dialysis method can be isolated from unreacted lipid or apolipoprotein by two methods: (1) gel filtration on large pore gels (e.g., BioGel A-5m, Sepharose CL-4B), and (2) isopycnic or gradient density ultracentrifugation.

1. In most preparations of complexes where free apolipoprotein is not present, the *gel filtration* method on columns of BioGel A-5m about 1.8 cm in diameter and around 50–70 cm in length, equilibrated with common aqueous buffers, affords good separations of unreacted lipid and complexes.[16,23] Separations of free apoA-I and complexes, even on long columns, are usually poor due to apolipoprotein self-association and elution near the small complexes. Prior to sample application to the column, any solid lipid should be removed by centrifugation at 15,000 rpm, 15°, for 1 hr. The elution of complexes is monitored by absorbance at 280 nm or by protein fluorescence to detect protein elution, and by tracer lipid radioactivity or chemical analysis to detect lipid elution. The complex peaks are occasionally broad and often the protein and lipid elutions do not coincide exactly, indicating the presence of chemical and size heterogeneity in the complexes. If high homogeneity of complexes is essential, then the use of longer columns with better resolution is indicated.[22] Gel filtration isolation of the complexes can be performed at ambient temperature or at 4° but storage for prolonged periods should be at 4°. In work involving the spontaneous reaction of DPPC vesicles with apoA-I we used jacketed columns which were maintained at 45° in order to prevent vesicle aggregation which occurs at lower temperatures. Under such conditions we observed a much better resolution of unreacted apoA-I from small discoidal complexes than at the lower temperatures.[16]

2. *Isopycnic or gradient density centrifugation* methods are most useful when free apolipoproteins must be separated from complexes. Since small discoidal complexes have densities around 1.10 g/ml, a background density of 1.15 g/ml or higher (up to 1.25 g/ml) will result in the flotation of complexes (and free lipid vesicles, if present) and sedimentation of apolipoproteins upon centrifugation at 50,000 rpm, 15°, for 24 hr.[24] Ultracentrifugation in a linear NaBr gradient between densities of 1.04 and 1.17 g/ml

at 50,000 rpm, 25°, for 96 hr can separate free DMPC vesicles, small discoidal complexes, and free apoA-I.[13,19] In ultracentrifugal separations, particularly by the density gradient approach, one must keep in mind the sizes and partial specific volumes of all the mixture components, as well as possible partial specific volume changes with temperature. The differences can be significant for lipids undergoing a phase transition or for different apolipoproteins; for example, the partial specific volume of DMPC is 0.933 ml/g at 10° and 0.970 ml/g at 25°.[33] Apolipoproteins A-I and A-II have quite different partial specific volumes at 20°: 0.737[34] and 0.892 ml/g,[35] respectively. Note that these values were obtained by different methods. Isolated discoidal complexes of apolipoproteins with lipids can be stored at 4°, in 10 mM Tris–HCl, pH 8.0, 0.15 M NaCl, 0.001% EDTA, 1 mM NaN$_3$ buffer for at least 2 months without undergoing changes in their properties.

Chemical Composition

The chemical composition of isolated complexes can be determined using a total sample of 0.3 mg apolipoprotein in small discoidal complex form. Protein concentrations are determined by the Markwell *et al.*[36] method, a modification of the Lowry *et al.*[1a] assay which includes 1% SDS for the solubilization of lipid. Prior to this assay, interfering Tris buffer must be removed and replaced with deionized water. The usual standards are bovine serum albumin or human apoA-I; several sample aliquots must be used to increase precision. For apolipoproteins with known extinction coefficients (see the "Preparation of Materials" section) absorbance at 280 nm can serve as a good relative measure of protein content down to about 30 µg/ml (for apoA-I). As complexes are stored there is a slight, gradual increase in absorbance at 280 nm which has not been explained satisfactorily. In any event, even fresh complexes give about 10 to 15% higher protein contents from absorbance measurements when compared to the results from the Markwell *et al.*[36] assay. The fluorescamine[2] assay is quite sensitive and accurate for relative measurements using the same apolipoprotein. For absolute measurements a standard of the protein being determined should be employed because this method depends on the number of free amino groups in proteins. Obviously, buffers devoid of free amino groups are required for this assay. Phosphate and cholesterol

---

[33] A. Watts, D. Marsh, and P. F. Knowles, *Biochemistry* **17**, 1792 (1978).
[34] S. Formisano, H. B. Brewer, Jr., and J. C. Osborne, Jr., *J. Biol. Chem.* **253**, 354 (1978).
[35] C. Edelstein, M. Halari, and A. M. Scanu, *J. Biol. Chem.* **257**, 7189 (1982).
[36] M. K. Markwell, S. M. Haas, L. L. Bieber, and N. E. Tolbert, *Anal. Biochem.* **87**, 206 (1978).

assays are the same as for the free lipids (see "Preparation of Materials" section), since there is no interference from the protein in the complexes. Frequently, when using radiolabeled cholesterol, we determine a working specific activity from the total dry weight of cholesterol and the total radioactivity in the stock cholesterol solution; subsequently we use this specific activity value to determine cholesterol contents in complex samples. Quenching effects in various preparations are not significant when using [^{14}C]cholesterol and scintillation fluids with high capacity for water.

Size and Shape Determination

The size of discoidal complexes can be determined by a number of methods: (1) gel filtration, (2) electron microscopy, (3) gradient gel electrophoresis, (4) analytical ultracentrifugation, (5) fluorescence polarization, and (6) dynamic light scattering. Several of these methods can also provide information on the asymmetry (i.e., shape) of the particles.

1. Size estimates by *gel filtration* can be made by elution of complexes through a BioGel A-5m column, of the size used in the isolation of complexes, calibrated with standard proteins or vesicles of known size.[16,25,26] Since the shape and partial specific volume of discoidal complexes are quite different from the corresponding properties of standard proteins, Stokes radii rather than molecular weights should be used in the calibration. Convenient markers are thyroglobulin (Stokes radius, 85 Å), ferritin (Stokes radius, 61.0 Å), catalase (Stokes radius, 41 Å), serum albumin (Stokes radius 35.5 Å), ovalbumin (Stokes radius, 30.5 Å), and chymotrypsinogen A (Stokes radius, 20.9 Å).

2. Transmission *electron microscopy*, using negative staining with phosphotungstic acid, provides not only an average measure of particle diameters and thicknesses, but also an estimate of particle heterogeneity in terms of the calculated standard deviation from the mean dimension[16,17,22] (see Chapter [26], this volume). Electron micrographs are obtained with instruments such as the JOEL-100C electron microscope. Samples of complexes at protein concentrations of 0.1 to 1.0 mg/ml, in the usual buffers, are mixed with an equal volume of 2% phosphotungstic acid adjusted to pH 7 with NaOH, and are placed on copper grids (200 or 300 mesh) coated with carbon and Formvar. Excess solution is removed after 1 min with filter paper and the dry grids are examined at instrument magnifications of 80,000- to 100,000-fold, at accelerating voltages of 80 kV. Several prints of different particle fields are made using 2- to 3-fold magnifications of the negatives. Usually discoidal complexes appear as stacks of particles of similar diameters and thicknesses, but frequently particles lying flat on the grid can also be seen. For good statistical analy-

sis of the dimensions from electron micrographs, at least 50 particles should be measured, manually or with a microcomparator. Some selectivity during measurement is warranted since quite often there is apparent fusion of particle stacks lying side-by-side or there are particles which are out of focus or have fuzzy edges. Evidently, stack formation is an artifact of the staining and drying procedure because hydrodynamic methods indicate that free, individual complex particles exist in solution.

3. *Gradient gel electrophoresis* (see Chapter [9] of this volume) is performed in 4 to 30% slab gradient acrylamide gels, pH 8.3. A stacking 3% gel can be used to increase resolution. Electrophoresis is performed at 150 V, for 24 to 36 hr, at 10 to 15°, in 50 mM Tris and 0.38 M glycine, pH 8.3. Gel fixing and staining are performed in 0.2% Coomassie Brilliant blue, 5% methanol, and 7.5% acetic acid. Proteins used in the calibration of gel filtration columns are also useful as standards in gradient gel electrophoresis.[17,22]

4. The weight average molecular weight of homogeneous complex preparations can be determined by *analytical ultracentrifugation,* including equilibrium and sedimentation methods (see Chapter [21] of this volume). The latter requires an independent measurement of the diffusion coefficient of the particles, and has not been applied to the determination of apolipoprotein–lipid complex molecular weights. The low-speed equilibrium method[37] at centrifugation speeds of 10,000 to 12,000 rpm, at 18°, using a Beckman Model E969 ultracentrifuge with scanner or interference optics, gives after 48 hr of equilibration the concentrations of small discoidal complexes as a function of the distance from the center of rotation. These data are analyzed using the weight average molecular weight expression:

$$M_w = [2RT/(1 - \bar{v}\rho)\omega^2]d \ln c/dr^2$$

where R is the gas constant, T is absolute temperature, $\bar{v}$ is partial specific volume, ρ is the density of the solution, ω is angular velocity, c is the concentration of the solute, and r is the distance from the center of rotation. Partial specific volumes can be measured with a density meter, such as the Anton Paar, DMA-02 C Precision Density Meter, but high solute concentrations and precise temperature control are required for this method. Alternatively, $\bar{v}$ can be obtained either from equilibrium or from sedimentation measurements in solvents of varying densities.[33,35] The $\bar{v}$ parameter can also be estimated from the chemical composition of the complexes and the $\bar{v}$ values of the isolated components, assuming that volume changes do not occur during complex formation.

[37] K. E. Van Holde, *in* "The Proteins" (H. Neurath and R. L. Hill, eds.), p. 256. Academic Press, New York, 1975.

5. Rotational relaxation times have been used in our laboratory to determine the size and shape of discoidal complexes from *static fluorescence polarization* measurements of fluorescently labeled complexes.[38] Fluorescent probes with lifetimes longer than 10 nsec (e.g., dansyl covalent groups) must be employed because the rotational relaxation times of small discoidal complexes are in the range of 300 nsec at 25°. Ideally the probe must be tightly bound to the complexes, but if local rotational motions contribute significantly to the fluorescence depolarization, the overall rotational relaxation time of the particles can still be determined by varying the viscosity of the solution and measuring fluorescence polarization at constant temperature.[39] The relationship between static fluorescence polarization and the rotational relaxation time or molar particle volume is given by the equations:[40]

$$(1/p - 1/3) = (1/p_0 - 1/3)(1 + 3\tau/\rho_h); \quad \rho_0 = 3V\eta/RT$$

where p is fluorescence polarization, p_0 is intrinsic polarization in the absence of Brownian rotations, τ is the fluorescence lifetime, ρ_h is the mean rotational relaxation time, ρ_0 is the rotational relaxation time of a sphere, V is the molar volume of the particle (including hydration), η is the viscosity of the solution, and R, T have been defined previously. If the molecular weight, partial specific volume, and hydration of the complex are known then a ρ_0 for an equivalent sphere can be calculated. Since ρ_h/ρ_0 ratios are correlated with particle asymmetry, in terms of ellipsoid axial ratios, an estimate of particle shape can be made.[40]

The preparation of fluorescent conjugates with Dns-Cl (dansyl chloride) is usually limited to a few probe molecules per complex. The chemical reaction can best be carried out on the free apolipoprotein before reconstitution with lipid. We have shown that the kinetics of spontaneous reconstitution, but not the nature of the products are altered by Dns labeling of apoA-I.[16,38] A solution of apoA-I in 1% NaHCO$_3$, at a concentration from 1 to 5 mg/ml, is kept near 0°; Dns-Cl (4 mg/ml) in cold acetone is added in small aliquots with vigorous stirring, until about 0.2 mg of Dns-Cl is added over 1 to 2 hr. Following an incubation of 2 hr at 4°, the reaction mixture is dialyzed against Dowex 2-X (initially in OH$^-$ form) and buffer in order to remove unreacted and hydrolyzed Dns-Cl. The fluorescence characteristics of Dns-labeled apolipoproteins and complexes are very convenient for the detection of small amounts of protein, as well as for the measurement of fluorescence spectra (excitation wave-

---

[38] A. Jonas and D. Krajnovich, *J. Biol. Chem.* **252**, 2194 (1977).
[39] P. Wahl and G. Weber, *J. Mol. Biol.* **30**, 371 (1967).
[40] G. Weber, *Adv. Protein Chem.* **8**, 415 (1953).

length 340 nm, maximum emission wavelength ~500 nm, but varies with the probe environment) and fluorescence polarization of solutions. Although polarizers are available with many commercial spectrofluorometers, precise determination of fluorescence polarization values requires the simultaneous measurement of parallel and perpendicularly polarized fluorescent light, which requires two detectors and signal averaging capability, such as that available in the SLM Model 400 polarization instrument. In addition, fluorescence polarization measurements require precise control of solution temperature ($\pm 0.1°$).

6. *Quasielastic light scattering* can be used to determine translational diffusion coefficients and Stokes radii for homogeneous complex solutions. Light scattering measurements are performed with an argon laser (Spectra Physics Model 165) as the source of incident light (wavelength of 4579 Å) which is focused at the center of a scattering cell (1.0 × 0.4 × 4.5 cm). Polydisperse solutions, where the autocorrelation function does not decay exponentially, cannot be analyzed by this method.[41]

7. *Other methods.* Although NMR methods (see Chapter [28], this volume) cannot be used routinely for the characterization of complexes of apolipoproteins with lipids because of the large amounts of material required and the complexity of the instrumentation, it should be noted that ^{1}H or ^{31}P NMR provides an unambiguous method to distinguish between vesicle–apolipoprotein complexes and micellar complexes. Since vesicles have inner aqueous compartments, inner phospholipids are protected from external chemical shift reagents, and, in SUV, have an altered head group ^{1}H environment due to crowding. With micellar, discoidal complexes the characteristic spectra of vesicles are not observed because all the phospholipid head groups are exposed and are essentially equivalent.[13,17]

Small angle X-ray scattering and neutron diffraction have been used to determine the size, shape, and apolipoprotein localization of discoidal complexes of apolipoproteins with DMPC; such studies obviously require highly specialized expertise and equipment.

Lipid Phase Behavior in Discoidal Complexes

In the discoidal complexes the lipids are organized in a bilayer surrounded by an apolipoprotein ring.[42] The effect of the apolipoprotein is to slightly increase the mobility and decrease the order of the PC in the gel

[41] K. C. Aune, J. G. Gallagher, A. M. Gotto, Jr., and J. D. Morrisett, *Biochemistry* **16**, 2151 (1977).
[42] A. R. Tall, D. M. Small, R. J. Deckelbaum, and G. G. Shipley, *J. Biol. Chem.* **252**, 4701 (1977).

state, and to decrease mobility and increase order in the liquid crystalline state of the PC. The apolipoprotein effect on the liquid crystalline state of the lipids is probably responsible for the increase of 2 to 4° in their T_c. The broadening of the transition reflects the relatively small size of the cooperative lipid unit, particularly in the small discs. The large discs have lipid transitions resembling more those of pure lipid vesicles (MLV).[8,10,43]

A very convenient method to observe the phase behavior of lipids in discoidal complexes is the measurement of the fluorescence polarization of a lipophilic fluorescent probe dissolved in the lipid domain of the complexes.[8,10,43] Diphenylhexatriene (DPH) has proved to be an excellent probe which is essentially quenched in water but becomes highly fluorescent in lipid. Usually PC/DPH molar ratios in excess of 200/1 are used in order to minimize phase perturbations caused by the probe itself. The DPH probe is added in a very small volume of tetrahydrofuran to the discoidal complex solutions in buffer; the mixtures are allowed to incubate for about 1 hr prior to measurements of fluorescence polarization over a temperature span from 60 to 5°. Removal of tetrahydrofuran by bubbling N_2 through the solution does not change the results. With DPH the exciting light is 366 nm and the fluorescent light is filtered through Corning Glass C-74 filters which pass light above 400 nm.

Several other methods have been used routinely to measure the phase transition behavior of lipids in discoidal complexes as well as in native lipoproteins; these include electron spin resonance employing spin probes, light scattering, and calorimetric methods.

Apolipoprotein Structure in Discoidal Complexes

Some aspects of apolipoprotein structure in complexes can be examined by spectroscopic methods. Circular dichroism spectra and intrinsic fluorescence intensity and polarization measurements in the static or dynamic mode are the most common approaches.

Circular dichroism spectra recorded with an automatic spectropolarimeter, such as the Jasco J-40A instrument, using 1 mm cells, and complex solutions containing about 0.1 mg/ml apolipoprotein at ambient temperature, show marked ellipticity minima at 222 and 208 nm. After baseline corrections with buffer blanks the molar ellipticities $[\theta]$ can be determined from

$$[\theta] = \mathrm{MRW}\theta°/10lc$$

where MRW is the mean residue weight, obtained from the amino acid composition of the apolipoprotein, $\theta°$ is the measured ellipticity in de-

---

[43] A. Jonas, J.-P. Privat, and P. Wahl, *Eur. J. Biochem.* **133**, 173 (1983).

grees, l is the light path in cm, and c is the apolipoprotein concentration in g/ml. The molar ellipticity values can then be used to estimate the secondary structure of the apolipoprotein, particularly the percentage α-helix content by the empirical approaches of Greenfield and Fasman:[44]

$$\% \ \alpha\text{-helix} = ([\theta]_{208} - 4000)/29000$$

With the small discoidal complexes spectral distortions due to light scattering do not appear to be significant.

Various fluorescence methods can provide average structural information about apolipoproteins in complexes, particularly by comparison with the same fluorescence parameters in free apolipoproteins or in other complexes. Intrinsic fluorescence spectra excited at 280 nm and recorded between 290 and 400 nm with a recording spectrofluorometer, such as the Hitachi-Perkin Elmer MPF III instrument, can give information on the polarity and changes in environment of the aromatic residues of the apolipoprotein. In apolipoproteins containing tryptophan (Trp) residues, fluorescence spectra are usually dominated by the Trp fluorescence. In fact, choosing an excitation wavelength of 295 nm, where only Trp absorbs, will eliminate the contribution of tyrosine fluorescence which normally occurs with a maximum around 305 nm. Phenylalanine fluorescence is very weak and can be disregarded for all exciting wavelengths. The wavelength maxima of Trp in proteins occur between 320 and 355 nm and represent very nonpolar (e.g., hydrocarbon-like) to very polar (aqueous) environments, respectively. Tyrosine fluorescence does not have a similar sensitivity to solvent polarity. Fluorescence intensity changes represent changes in the quenching and the environment of the aromatic residues, but cannot be easily interpreted in physical terms because of the complicated and largely unexplained fluorescence quenching effects that occur in proteins.

The use of quenching molecules or ions, in conjunction with the Stern–Volmer[45] equation, can provide information on quenching constants and on the degree of aromatic residue exposure to solvent. Common quenching agents are iodide (KI), pyridinium chloride,[46] and acrylamide. For example, a stock KI solution (5 M) is stabilized with 1.0 mM $Na_2S_2O_3$ in the usual buffers. The complex solution having an absorbance of less than 0.1 at 280 nm is placed in a fluorescence cuvette, and the initial fluorescence is recorded. Subsequent titration with the KI solution

[44] N. Greenfield and G. D. Fasman, *Biochemistry* **8,** 4108 (1969).
[45] O. Stern and M. Volmer, *Phys. Z.* **20,** 183 (1919).
[46] H. J. Pownall and L. C. Smith, *Biochemistry* **13,** 2590 (1974).

results in progressive quenching of fluorescence. Analysis of the results is performed by the Stern–Volmer[45] expression:

$$F_0/F = 1 + K_Q(X)$$

and/or the modified equation:[47]

$$F_0/\Delta F = 1/(X)f_a K_Q + 1/f_a$$

where F_0 is the initial fluorescence intensity, F is the measured fluorescence intensity, K_Q is the quenching Stern–Volmer constant, X is the molar concentration of quencher, ΔF is $F_0 - F$, and f_a is the fractional maximum accessible protein fluorescence.

Static fluorescence polarization measurements of Trp residues in apolipoprotein–lipid complexes reflect the local rotational motions of the Trp groups, and not whole particle rotations, since the overall rotational relaxation times of small discoidal complexes are much longer (~300 nsec) than the average fluorescence lifetime of Trp (~1 to 4 nsec). At 25° the intrinsic polarization of Trp is 0.21 at an exciting wavelength of 280 nm; the measured fluorescence polarization values in complexes are of the order of 0.10 to 0.14, indicating that there is considerable depolarization due to local Trp rotational motions, which evidently depend on the environment of these amino acid residues.[25,48] Because of the relatively small values and changes in Trp fluorescence polarization, accurate measurements require the use of a fluorescence polarization instrument such as the SLM Model 400, using 280 or 295 nm exciting light and Corning Glass 0-54 filters in the path of the fluorescent light.

Time-dependent fluorescence methods allow the determination of average fluorescence lifetimes, as well as the resolution of up to three different lifetimes; similarly two to three rotational relaxation times can be resolved directly from fluorescence anisotropy (or polarization) decay measurements. Two distinct methods, the pulse or the phase-modulation method, are employed for this purpose and require specialized instrumentation. With the pulse method about 2 ml of complex sample containing 0.5 mg/ml apolipoprotein is required. Temperature is closely controlled over the prolonged periods of data accumulation with the photon counting detection system. The fluorescence lifetime and rotational relaxation time results are obtained by deconvolution of fluorescence decay curves and fitting to sums of exponential functions. The fluorescence lifetimes obtained in this manner give an indication of fluorophore heterogeneity and

[47] S. L. Lehrer, *Biochemistry* **10**, 3254 (1971).
[48] A. Jonas, J.-P. Privat, P. Wahl, and J. C. Osborne, Jr., *Biochemistry* **24**, 6205 (1982).

help distinguish static from dynamic quenching processes. Resolution of rotational relaxation times provides information on local fluorophore rotations as well as on backbone motions involving the fluorophore.[43,48] Time resolved fluorescence anisotropy measurements of probes such as diphenylhexatriene (DPH) in bilayer lipid regions give not only rotational relaxation times but also order parameters.[43]

Comparison of Reconstituted Discoidal Complexes with Native HDL

Structure and Composition

Human HDL of discoidal structure, as observed by electron microscopy, have been detected in the plasma of patients with familial lecithin: cholesterol acyltransferase (LCAT) deficiency[49,50] and in patients with alcoholic hepatitis.[51,52] Liver perfusion studies of animal species[53] in the presence of LCAT inhibitors have indicated the presence of discoidal, nascent HDL particles; and similar structures were demonstrated in rat intestinal lymph,[54] the peripheral lymph of cholesterol-fed dogs,[55] and in the tissue culture medium of rat macrophages.[56] The human and animal nascent HDL particles isolated between densities of 1.063 and 1.210 g/ml have the characteristic discoidal morphology with diameters in the range from 130 to 240 Å and widths from 40 to 50 Å. Isolated apolipoprotein E-containing discoidal particles have been reported to have more heterogeneous sizes, with disc diameters from 140 to 400 Å.[50] Disc stacks are usually observed by negative stain electron microscopy. The human discoidal HDL contain 28% by weight protein and 72% lipid.[49,51] The protein composition includes about 37% apoE, 35% apoA-I, and 15% apoCs and

---

[49] K. R. Norum, J. A. Glomset, A. Y. Nichols, T. Forte, J. J. Albers, W. C. King, C. D. Mitchell, K. R. Applegate, E. L. Gong, V. Cabana, and E. Gjone, *Scand. J. Clin. Lab. Invest.* **35** (Suppl. 142), 31 (1975).

[50] C. D. Mitchell, W. C. King, K. R. Applegate, T. Forte, J. A. Glomset, K. R. Norum, and E. Gjone, *J. Lipid Res.* **21,** 625 (1980).

[51] S. M. Sabesin, H. L. Hawkins, L. Kuiken, and J. B. Ragland, *Gastroenterology* **72,** 510 (1977).

[52] J. B. Ragland, P. D. Bertram, and S. M. Sabesin, *Biochem. Biophys. Res. Commun.* **80,** 81 (1978).

[53] R. L. Hamilton, M. C. Williams, C. J. Fielding, and R. J. Havel, *J. Clin. Invest.* **58,** 667 (1976).

[54] R. H. R. Green, A. R. Tall, and R. M. Glickman, *J. Clin. Invest.* **61,** 528 (1978).

[55] C. H. Sloop, L. Dory, R. Hamilton, B. R. Krause, and P. S. Roheim, *J. Lipid Res.* **24,** 1429 (1983).

[56] S. K. Basu, Y. K. Ho, M. S. Brown, D. W. Bilheimer, R. G. W. Anderson, and J. L. Goldstein, *J. Biol. Chem.* **257,** 9788 (1982).

apoA-II, in terms of stain intensity on polyacrylamide gel electrophoresis.[51] On the average, the weight percent composition of the lipids is 62% phospholipids, 24% unesterified cholesterol, 11% triglycerides, and 3% esterified cholesterol, with considerable individual donor variability. The average molar phospholipid to unesterified cholesterol ratio is 1.3/1 in these particles. The well-characterized rat nascent HDL[53] have disc diameters of 190 ± 25 Å and widths of 46 ± 5 Å; 38 wt% of protein and 62 wt% of lipid. Phospholipids account for 65%, unesterified cholesterol 19%, triglyceride 9%, and cholesteryl esters for 7% of all the lipid weight; and the molar phospholipid to free cholesterol ratio is 1.7/1 in these particles. In the case of human nascent HDL it is not yet clear whether hybrid particles containing both apoE and apoA-I exist or if mixtures of particles each with a single apolipoprotein component are present.

The synthetic HDL-like complexes of apolipoproteins and lipids described in this chapter have the same discoidal morphology observed for the native discoidal HDL, and sizes in the same range: 110 to 350 Å diameters and 40 to 50 Å widths. It is worth noting that synthetic discoidal particles with diameters around 200 Å and apolipoprotein only in the peripheral ring in α-helical configuration would have a chemical composition of around 10 wt% protein and 90 wt% lipid. In fact apoA-I complexes with DPPC in this size range have such a composition.[16,25] On the other hand, recombinant complexes of apoA-I with egg-PC and nascent human HDL, with diameters around 200 Å, have protein contents of 22 and 28 wt%, respectively, indicating that not all the apolipoprotein can be present in the peripheral ring structure. Some apolipoprotein is probably associated with other parts of the discoidal surface, such as the face of the disc. In contrast, the small recombinant discs of apoA-I with a variety of PC appear to have all the apolipoprotein in the peripheral ring structure.

In terms of apolipoprotein composition the synthetic complexes can be prepared with a variety of apolipoproteins including apoA-I, apoE, apoA-II, and the C apolipoproteins. The phospholipid content and composition can be easily manipulated in the complexes prepared by the sodium cholate method, and unesterified cholesterol can be incorporated. However, the present recombination methods, as indicated earlier, allow the preparation of discoidal complexes containing only 4/1, PC/cholesterol molar ratios in small discs (i.e., 20 mol% cholesterol), or 2/1, PC/cholesterol molar ratios in large discs (i.e., 33 mol% cholesterol). These are considerably larger molar ratios than the 1.3/1 observed with the native, discoidal HDL. It remains to be shown whether the particle compositions (apolipoprotein and/or phospholipid) or the cholesterol incorporation mechanism are the reason for this discrepancy in cholesterol content between the synthetic and native discoidal HDL.

Reactions with LCAT

Native HDL are the best lipoprotein substrates for LCAT:[57] discoidal HDL react most efficiently with LCAT followed by small spherical HDL, and then by the larger spherical HDL subclasses. In all cases the reactivity is related directly to the apoA-I content of the particles.[58] It is well known that in terms of lipid substrates LCAT preferentially transfers long chain unsaturated acyl chains from the sn-2 position of PC to cholesterol.[57]

Recent work with discoidal complexes of PC, cholesterol, and apoA-I showed that these synthetic HDL-like particles are excellent substrates for purified LCAT.[18,23,32] Particles containing PCs with long chain unsaturated fatty acyl groups such as egg-PC or POPC are about 5- to 10-fold more reactive than discoidal complexes prepared with DPPC, and about 50- to 100-fold more reactive than particles containing DMPC.[23,32] Although the reactivity of native discoidal HDL and synthetic discoidal complexes has not been compared under the same solution conditions and using the same enzyme preparations, it is clear that qualitatively the synthetic discs are comparable in their enzyme reaction rates and lipid substrate specificity to the native, discoidal HDL. Furthermore, studies of the enzymatic mechanism which are precluded by the compositional heterogeneity and scarcity of the native discoidal HDL can be easily performed using the synthetic discoidal complexes. For example, it has been shown that PC/cholesterol ratios and cholesterol contents of the complexes do not affect enzyme reaction rates;[26] that apoA-I and apoC-I are the best activators for LCAT but that other apolipoproteins (apoC-III, apoC-II, and apoA-II) can activate to lesser extents;[24] and that enzyme activation by apoA-I depends on the structure of the apolipoprotein in the complexes.[24,25]

Extensive reaction of LCAT with discoidal complexes of egg-PC and apoA-I, containing 20 mol% cholesterol, can lead to essentially complete transformation (90%) of the cholesterol into cholesteryl esters, with loss of PC and conversion to a particle less asymmetrical than the starting disc.[32] A similar enzymatic transformation of native discoidal HDL has been postulated to occur in circulation, during the maturation of nascent HDL into spherical HDL structures. Thus synthetic discoidal complexes can serve as models of nascent HDL structural and functional changes in plasma.

[57] J. A. Glomset, *in* "Blood Lipids and Lipoproteins: Quantitation, Composition, and Metabolism" (G. J. Nelson, ed.), p. 745. Wiley (Interscience), New York, 1972.

[58] Y. L. Marcel, C. Vezina, D. Emond, and G. Suzue, *Proc. Natl. Acad. Sci. U.S.A.* **77**, 2969 (1980).

Interactions with Cells and Lipoproteins

In the important areas of HDL function, involving cell surface (receptor or membrane) interactions and lipid transfers and exchanges with other lipoproteins or with membranes, synthetic discoidal complexes have not yet been sufficiently exploited. The probable reason is that until recently only DMPC-containing complexes could be easily prepared; however, such particles are quite unphysiological and are unstable in the presence of lipoproteins.[59,60] Increased use of synthetic complexes can be anticipated with the advent of the cholate reconstitution method, which allows the incorporation of a variety of PCs into discoidal particles.

Human and canine apoE reconstituted with DMPC bind to the apoB, E receptors of human fibroblasts with the same or greater affinity as the native canine HDL_c lipoproteins. The synthetic particles are internalized and degraded in a manner similar to the native lipoproteins.[61] Analogous experiments performed with human apoE3 fragments show that the NH_2-terminal region of the protein (residues 1–191) binds DMPC and interacts with the fibroblast receptors.[62]

Presumably spherical particles, reconstituted from rat HDL by partial delipidation and incorporation of labeled cholesteryl ethers, deliver cholesteryl esters to liver, adrenal gland, and gonads of the rat in a manner similar to the specific mechanism of native HDL.[31] Thus reconstituted HDL, discoidal and spherical, apparently retain the apolipoprotein characteristics necessary for the recognition of specific receptors.

High-density lipoproteins are known to act as acceptors of cholesterol from various cell types in tissue culture. When the cells are exposed to isolated apolipoproteins, combined with phospholipids under conditions that could give rise to the formation of discoidal complexes, efficient cholesterol removal also occurs.[63–65] Although the complexes have not been characterized in detail before and after exposure to cells, and quantitative results are not available, it appears likely that cholesterol transfer between cells and native HDLs or discoidal complexes of similar size, could be comparable.

---

[59] A. V. Nichols, E. L. Gong, P. J. Blanche, and T. M. Forte, *Biochim. Biophys. Acta* **617**, 480 (1980).
[60] R. L. Jackson, A. D. Cardin, R. L. Barnhart, M. Ashraf, and J. D. Johnson, *Lipids* **17**, 338 (1982).
[61] T. L. Innerarity, R. E. Pitas, and R. W. Mahley, *J. Biol. Chem.* **254**, 4186 (1979).
[62] T. L. Innerarity, E. J. Friedlander, S. C. Rall, Jr., K. H. Weisgraber, and R. W. Mahley, *J. Biol. Chem.* **258**, 12341 (1983).
[63] R. L. Jackson, A. M. Gotto, Jr., O. Stein, and Y. Stein, *J. Biol. Chem.* **250**, 7204 (1975).
[64] O. Stein, J. Vanderhoek, and Y. Stein, *Biochim. Biophys. Acta* **431**, 347 (1976).
[65] Y. K. Ho, M. S. Brown, and J. L. Goldstein, *J. Lipid Res.* **21**, 391 (1980).

Discoidal complexes of egg-PC with apoA-I can act as acceptors of radiolabeled PC from egg-PC SUV membranes, similarly to native HDL particles; however, the transfer rate constants are 22-fold higher with the native spherical HDL than with the discoidal complexes.[66] When partially purified phospholipid transfer protein is added to the above system, significant enhancement is seen in the transfer rates with the native HDL but no change in transfer rate can be detected with the discoidal complexes.[67] Thus it appears that PC desorption rates from native, spherical HDL or from discoidal complexes are significantly different due to particle curvature or composition.

In summary, the synthetic discoidal complexes can be used as chemically and physically defined structural and functional models of nascent HDL with confidence; as models for mature HDL, they seem to mimic the apolipoprotein functional properties quite well, but should be used with caution in lipid transfer studies, or in detailed structural investigations.

Acknowledgments

The work from A. Jonas' laboratory cited in this review was supported by NIH Grants HL-16059, HL-29939, and American Heart Association Grant-in-Aid 80-753.

[66] G. E. Petrie and A. Jonas, *Biochemistry* **23**, 720 (1984).
[67] S. A. Sweeny and A. Jonas, unpublished results (1983).

[33] Reassembly of Low-Density Lipoproteins

By MARY T. WALSH and DAVID ATKINSON

Introduction

Over the last 10 years our understanding of the assembly, disassembly, interconversions, and molecular exchange processes occurring with plasma lipoproteins has developed rapidly, together with the general picture of lipoprotein metabolism. Biochemical defects in lipid and lipoprotein metabolism are expressed at different levels of the physiological processes and lead to alterations in the composition, structure, and properties of the plasma lipoproteins. Ultimately these processes must be determined by the physicochemical properties, molecular interactions, and structural organization of the lipids and specific apolipoproteins of the plasma lipoproteins.

We now have a low resolution understanding of lipoprotein structure and molecular interactions (primarily HDL and LDL)[1] derived from electron microscopy,[1a] X-ray and neutron[2] structural studies, and nuclear magnetic resonance.[3] However, the structural information at the current resolution is insufficient to explain the molecular basis of these metabolic processes. This low resolution picture of lipoprotein structure describes lipoproteins as emulsion or microemulsion particles consisting of a neutral lipid core (triglycerides, cholesteryl esters) stablilized by a surface monolayer of polar phospholipids, cholesterol, and specific apolipoproteins. Consequently lipoprotein molecular organization can be considered in terms of (1) lipid structure and lipid–lipid interactions relevant to the lipid core of the particles, (2) lipid–lipid and lipid–protein interactions relevant to the surface structure of the lipoprotein particle, and (3) the interrelationship between the lipid core and surface interactions which stabilize the intact lipoprotein particle.

The molecular complexity of native lipoprotein both in terms of the species complexity of the lipids and their fatty acyl distribution, together with apolipoprotein complexity, in many respects hinders further progress using direct approaches to these questions. Thus the disassembly of lipoproteins into their constituent apolipoprotein classes and individual lipid species followed by sequential and selective reassembly of model lipoprotein systems provides an important approach to the investigation of the molecular architecture and interactions in lipoproteins. Additionally these model systems provide a mechanism by which to establish the consequences of changes in lipoprotein composition and structure on the metabolic processes in which lipoproteins are involved. In the case of HDL, the major advances in describing the primary structure of lipoproteins A-I, A-II, E, and C peptides have led to a conceptual description of the secondary structure of these proteins and their interaction with the lipid components. This sequence information and structure prediction

---

[1] Abbreviations: HDL, high-density lipoprotein; LDL, low-density lipoprotein; apoB, apolipoprotein B-100; SDS, sodium dodecyl sulfate; CMC, critical micelle concentration; CD, circular dichroism; NaDC, sodium deoxycholate; TES, N-tris(hydroxymethyl) methyl-2-aminoethanesulfonic acid; SDS–PAGE, SDS–polyacrylamide gel electrophoresis; $C_{12}E_8$ n-dodecyloctaethyleneglycol monoether; RCAM-apoB, reduced carboxyamidomethylated apoprotein B; EPC, EYPC, egg yolk phosphatidylcholine; DMPC, dimyristoyl phosphatidylcholine; CO, cholesteryl oleate; m-LDL, model LDL as described by Lundberg and Suominen[37]; r-LDL, reconstituted LDL as described by Krieger et al.[38]; DSC, differential scanning calorimetry; V_0, void volume; V_t, total volume; ΔH, enthalpy of transition, cal/g.

[1a] T. M. Forte and A. V. Nichols, Adv. Lipid Res. **10**, 1 (1972).

[2] P. Laggner and K. W. Muller, Q. Rev. Biophys. **II:3**, 371 (1978).

[3] W. G. Bradley and A. M. Gotto, in "Disturbance in Lipid Metabolism," p. 111. Am. Physiol. Soc., 1978.

methods[4,5] have led to the concept of amphipathic helices.[6-8] A plethora of studies have investigated the structure and interactions of these apolipoproteins with phospholipids in discoidal analogs of "nascent" HDL and in HDL recombinant systems. From these studies, a fairly detailed picture has emerged of the molecular and structural basis of the interactions between this class of apolipoprotein and lipids.

The major apolipoprotein of LDL, apoB, has proved much more difficult to isolate and characterize. A thorough description of the composition and properties of apoB is not yet possible since many of the conventional methods for protein handling have been hampered by the extreme insolubility of this protein in aqueous buffers and its tendency to form aggregates.[9] Nonetheless, a variety of solubilization methods employing chemical modification,[10,11] denaturants,[10,12-15] or detergents[15-20] have been developed in order to maintain apoB in solution. The use of denaturants and detergents to dissociate apoB from the lipids of LDL suggests that strong hydrophobic lipid–protein interactions maintain the protein in its native state in LDL.

At best, information concerning apoB "in solution" is really information concerning the interaction of apoB with amphipathic detergents. A variety of detergent–apoB complexes have been described[15,17,20-22] in all of which the protein maintains a high molecular weight (MW) form. This

[4] P. Y. Chou and G. D. Fasman, *Biochemistry* **13**, 211 (1974).
[5] P. Y. Chou and G. D. Fasman, *Biochemistry* **13**, 222 (1974).
[6] A. L. Andrews, D. Atkinson, M. D. Barratt, E. G. Finer, H. Hauser, R. Henry, R. B. Leslie, N. L. Owens, M. C. Phillips, and R. N. Robertson, *Eur. J. Biochem.* **64**, 549 (1976).
[7] J. P. Segrest, R. L. Jackson, J. D. Morrisett, and A. M. Gotto, *FEBS Lett.* **39**, 247 (1974).
[8] J. D. Morrisett, R. L. Jackson, and A. M. Gotto, *Annu. Rev. Biochem.* **44**, 183 (1975).
[9] A. M. Scanu and W. L. Hughes, *J. Biol. Chem.* **235**, 2876 (1960).
[10] J. P. Kane, E. G. Richards, and R. J. Havel, *Proc. Natl. Acad. Sci. U.S.A.* **66**, 1075 (1970).
[11] A. Scanu, H. Pollard, and W. Reader, *J. Lipid Res.* **9**, 342 (1968).
[12] R. Smith, R. Dawson, and C. Tanford, *J. Biol. Chem.* **247**, 3376 (1972).
[13] B. Shore and V. Shore, *Biochemistry* **8**, 4510 (1969).
[14] R. Shireman, L. L. Kilgore, and W. Fisher, *Proc. Natl. Acad. Sci. U.S.A.* **74**, 5150 (1977).
[15] J. C. H. Steele, Jr. and J. A. Reynolds, *J. Biol. Chem.* **254**, 1633 (1979).
[16] J. L. Granda and A. M. Scanu, *Biochemistry* **5**, 3301 (1966).
[17] A. Helenius and K. Simons, *Biochemistry* **10**, 2542 (1971).
[18] L. Socorro and G. Camejo, *J. Lipid Res.* **20**, 631 (1979).
[19] A. D. Cardin, K. R. Witt, C. L. Barnhart, and R. L. Jackson, *Biochemistry* **12**, 4503 (1982).
[20] M. T. Walsh and D. Atkinson, *Biochemistry* **22**, 3170 (1983).
[21] A. Ikai and M. Hasegawa, *J. Biochem. (Tokyo)* **83**, 755 (1978).
[22] R. M. Watt and J. A. Reynolds, *Biochemistry* **19**, 1593 (1980).

description of current methodology for the reassembly of model LDL particles will first focus on methods for the solubilization of LDL and apoB which are potentially useful for reassembly studies.

Experimental Methods

Solubilization of ApoB.

Detergents. Sodium dodecyl sulfate. Sodium dodecyl sulfate (SDS), an ionic detergent which has been widely used in the solubilization of intrinsic membrane proteins, has been used by several groups to solubilize apoB of LDL.[15,17,23,24] In the methodology described by Steele and Reynolds,[15] LDL (d 1.019–1.050 g/ml) was dialyzed against 0.1 M sodium bicarbonate, pH 8.6. The two free sulfhydryl groups per 250,000 g of protein were blocked with a 1.5 molar excess of iodoacetamide; LDL was further dialyzed against 117 mM sodium phosphate, 1 mM sodium azide, pH 7.4 at an ionic strength of 0.3. Twenty grams of SDS was added per g of apoB. Following incubation for 3 hr at room temperature, 10–25 mg of protein was applied to a 1.5 × 90 cm column of Sepharose 4B, equilibrated and eluted with 117 mM sodium phosphate, 1 mM sodium azide, 1–2.5 mM SDS, pH 7.4. Small fractions were collected and monitored for protein and lipid. ApoB in the presence of SDS above its critical micellar concentration (CMC = 0.91/mM)[25] eluted as a symmetrical peak, which was well included in the column, and was completely delipidated.[15,17]

Circular dichroic spectra of apoB in 0.1–2.5 mM SDS showed that the secondary structure of apoB was 25% α-helix, 45% β-sheet, and 35% random coil, which is different from that of native LDL. At maximal SDS binding (1.66 g SDS/g of apoB) there was a dramatic change in the CD spectrum: 30% α-helix, 20% β-sheet, and 50% random coil.[15]

In addition Helenius and Simons[17] have shown that apoB in 3.5 mM SDS yields two precipitin lines against anti-LDL serum, neither of which fused with native LDL, suggesting that SDS at this concentration alters the immunological reactivity of apoB.

Triton X-100. Triton X-100, a nonionic detergent which has proven useful for the characterization of membrane proteins in the native state, has also been utilized in the solubilization of apoB by Socorro and Camejo.[18] Ten to twenty milligrams of LDL (d = 1.019–1.063 g/ml) in 1–4 ml was dialyzed against 50 mM Tris, pH 8.8 and applied to a 1.5 × 15 cm

---

[23] A. Ikai, *J. Biochem. (Tokyo)* **79,** 679 (1976).
[24] K. Simons and A. Helenius, *FEBS Lett.* **7,** 59 (1970).
[25] A. Helenius, D. R. McCaslin, E. Fries, and C. Tanford, this series, Vol. 56, p. 734.

column of DEAE-Sepharose CL-6B. A linear gradient of 60 ml of 0–2% (w/v) Triton X-100 was then pumped through the column at a rate of 20 ml/hr (yielding lipid-Triton micelles). The column was washed with 10 column volumes of detergent-free buffer (yielding Triton in the eluate). ApoB which was bound to the column was eluted with 2 column volumes of 1 M NaCl buffered to pH 7.4 with 50 mM Tris–HCl. ApoB solubilized in Triton X-100 retains the immunological properties of intact LDL. However apoB isolated under these conditions appears to be made of oligomers of two monomeric subunits, one with MW ~22,700 and a smaller one of MW ~8000.

Ikai and Hasegawa[21] have also utilized Triton X-100 in the delipidation and solubilization of apoB. Utilizing the method of Helenius and Simons[17] LDL was dissolved in 2% Triton X-100. ApoB–Triton X-100 mixed micelles eluted as a discrete peak at the void volume of a 1.2 × 75 cm column of Sepharose CL-6B which had been equilibrated with buffer + Triton X-100. The apoB–Triton X-100 mixed micelles isolated by this method have been shown by sedimentation equilibrium[21] to have f/f_{min} of 2.2 and thus are asymmetric, not compact, and the apoB is an elongated or a loosely expanded dimer of MW 570,000.

Sodium deoxycholate. Sodium deoxycholate (NaDC), a bile salt and weakly ionic detergent, has been extremely useful in the solubilization of apoB,[17] since it dissociates interactions between protein molecules effectively. Based on the method of solubilization described by Helenius and Simons,[17] recent modifications[20] have been employed which improve the recovery and prolong the stability of apoB.

LDL donors were "prescreened" for the presence of B-74 and B-26[26] prior to utilization of their LDLs for solubilization and subsequent reassembly experiments. Only those donors whose LDL showed one high MW Coomassie blue staining band (B-100) on 3% SDS–PAGE were used for subsequent studies.

LDL (d 1.025–1.050 g/ml) was dialyzed against 0.1 M sodium bicarbonate, pH 8.6, and following the procedure described by Steele and Reynolds was alkylated with a 1.5 molar excess of iodoacetamide to block the two free sulfhydryl groups.[15,19] The LDL was then dialyzed against 20 mM TES, 0.15 M sodium chloride, pH 7.4 and finally against 0.05 M sodium carbonate, 0.05 M sodium chloride, pH 10. Solubilization of apoB was accomplished by the slow addition of 1.10 g of NaDC to LDL (20 mg of LDL protein in a volume of 2 ml) and slow stirring for 3 hr at room

[26] J. P. Kane, D. A. Hardman, and H. F. Paulus, *Proc. Natl. Acad. Sci. U.S.A.* **77**, 2465 (1980).
[27] G. Zampighi, J. A. Reynolds, and R. M. Watt, *J. Cell Biol.* **87**, 555 (1980).

temperature in the dark. The solution was applied to a 2.5 × 40 cm column of Sepharose CL-4B which had been equilibrated and eluted with buffer plus 10 mM NaDC. The NaDC-solubilized apoB eluted as a single peak completely delipidated (see Fig. 1) which was included in the column, with protein recovery of ~70%. The protein-containing fractions from the column separation were pooled and concentrated to 2 mg protein/ml by ultrafiltration using Amicon YM-10 filters (Lexington, MA).

Far UV CD spectra of apoB–NaDC at ambient temperature showed apoB to have a secondary structure similar to apoB in native LDL[20] (42.1 ± 0.7 α-helix, 17 ± 5% β-sheet, 40 ± 5% random coil).

Electrophoresis on 3.0% SDS–polyacrylamide gels performed on preparations of LDL and NaDC-solubilized apoB showed one band which corresponded to an apparent MW of 366,000 ± 10,000 (Fig. 2). No additional bands were observed at lower MWs on similar gels even after prolonged sample storage (up to 6 weeks at 4°). Over the time course of detergent removal and chromatography, no degradation of apoB had occurred.

$C_{12}E_8$ *(n-dodecyloctaethylene glycol monoether)*. $C_{12}E_8$, another nonionic detergent, has also been used as a solubilizing and delipidating agent

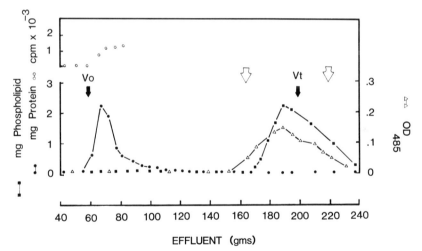

FIG. 1. Gel filtration chromatography of 1.025 < d < 1.050 g/ml LDL solubilized with NaDC. LDL at 20 mg of LDL protein in 2 ml of 50 mM sodium carbonate, 50 mM sodium chloride, 0.02% sodium azide, pH 10 buffer was incubated with 55 mg of NaDC/mg of protein and applied to a Sepharose CL-4B column of dimensions 2.5 × 40 cm and eluted downward with the above buffer plus 10 mM NaDC at room temperature. Solid arrows mark V_0 (Blue Dextran 2000) and V_t (tryptophan) of the column. Open arrows mark elution position of [^{14}C]NaDC micelles. (○) cpm ([^{14}C]NaDC); (●) protein (mg/fraction); (■) phospholipid (mg/fraction); (△) OD$_{485}$.[20]

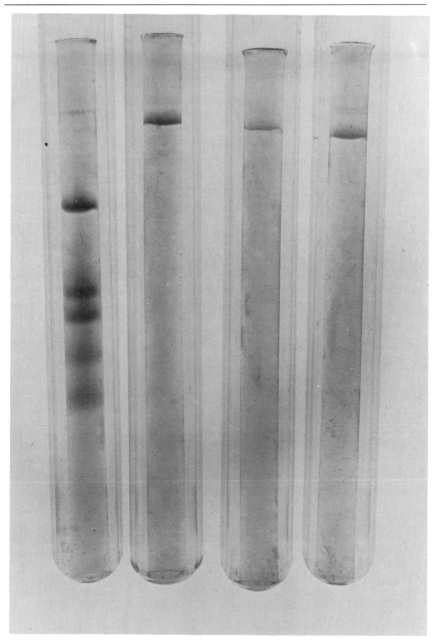

FIG. 2. Three percent SDS–polyacrylamide gels. (Left–right) High MW standard (Bio-Rad high MW protein standards (top–bottom); myosin (200,000), β-galactosidase (116,250), phosphorylase B (92,500), bovine serum albumin (66,200), and ovalbumin (45,000); LDL; NaDC-solubilized apoB; DMPC–apoB 4:1 w/w.[20]

for apoB.[22] $C_{12}E_8$ (30 mg/mg of LDL protein) was added to LDL[22] (2–3 mg of LDL protein in 1.5 ml). The solution was gently stirred for 3 hr at room temperature in the dark and applied to a 1.5 × 90 cm column of Sepharose CL-4B equilibrated in buffer containing 1 mM $C_{12}E_8$. The major portion of the protein which had been applied to the column eluted as a defined fraction which was included in the column. The shape of the elution peak has been shown to be pH dependent. As pH is increased the peak becomes more symmetrical. ApoB of the apoB–$C_{12}E_8$ complex at pHs above 7.4 retains a similar secondary structure to native LDL, as shown by CD.[22] Sedimentation velocity and sedimentation equilibrium measurements show the apoB–$C_{12}E_8$ complex to be highly asymmetric with f/f_{min} equal to 2.0–2.3. ApoB–$C_{12}E_8$ complex has been shown by negative stain electron microscopy to be a rod-shaped particle 75–80 nm in diameter and 4.5–5.5 nm in thickness.[27]

Organic Solvents: Ethanol-Diethyl Ether. Organic solvents have been utilized by several investigators[12,19,28] in the delipidation of LDL. Briefly, LDL was washed several times with 3 : 1 (v/v) ethanol/diethyl ether over 2 days and further extracted with several changes of ether over another 2 days. Recently Lee *et al.*[28] and Cardin *et al.*[19] have modified and refined this method to ensure stability and purity of apoB. The resultant protein is free of lipids, contaminating minor apolipoproteins, and may be subsequently solubilized with SDS, guanidine–HCl, or urea.

Denaturants. Lee *et al.*[28] have developed a method to isolate apoB from human LDL which minimizes the possibility of oxidation throughout isolation, fractionation, and delipidation. This procedure ultimately results in apoB being soluble in 6 M urea. Briefly, LDL apolipoprotein obtained by ethanol–ether extraction was solubilized in 6 M guanidine–HCl containing the reducing agent dithiothreitol. ApoB solubilized in 6 M guanidine–HCl was then carboxymethylated with recrystallized iodoacetic acid. The reaction was terminated after 10 min by the addition of an excess of dithiothreitol. Reduced carboxylmethylated apoB (RCAM-apoB) was obtained by gel filtration chromatography on a 2 × 160 cm column of Sepharose CL-6B equilibrated with buffer containing 6 M guanidine–HCl. RCAM-apoB in 6 M guanidine–HCl eluted near the void volume on this column and was dialyzed into 6 M urea. At low protein concentrations apoB may be dialyzed into other aqueous buffers or to a lower concentration of urea.

Smith *et al.*[12] and Steele and Reynolds[29] have utilized guanidine HCl in the solubilization and analysis of apoB. LDL delipidated with ethanol/

[28] D. M. Lee, A. J. Valente, W. H. Kuo, and H. Maeda, *Biochim. Biophys. Acta* **666**, 133 (1981).
[29] J. C. H. Steele, Jr. and J. A. Reynolds, *J. Biol. Chem.* **254**, 1639 (1979).

diethyl ether[12] or SDS-solubilized apoB[29] were reduced with 2-mercaptoethanol and carboxymethylated with iodoacetamide. ApoB was either resolubilized in guanidine–HCl solution or dialyzed against 7 M guanidine–HCl.

SDS-solubilized apoB which has been dialyzed into guanidine–HCl exists predominantly as a monomeric species of MW 250,000 ± 10%, shows no residual secondary structure by CD criteria, and has a Stokes radius of 147 Å which is close to the value of 156 Å expected for a random coil of this MW, as well as the value reported by Smith *et al.*[12] for LDL delipidated with organic solvents. ApoLDL in guanidine–HCl exists as a random coil of 200,000–290,000, as shown by gel filtration. Sedimentation equilibrium studies[12,30] have shown apoB to have a MW of 250,000 as calculated from S^0.

Role of Sulfhydryl Blocking. Following delipidation of LDL, apoB has been shown to form covalent aggregates. This process has been shown to be reversed by the addition of 2-mercaptoethanol, suggesting that free sulfhydryl groups may play a role in this process. Cardin *et al.*[19] have shown that each polypeptide chain of LDL contains 2 mol of free sulfhydryl and 6 intramolecular disulfide bonds. As shown by Steele and Reynolds,[15] if these 2 free sulfhydryl groups of apoB are not first blocked, covalent aggregates form. Cardin *et al.*[19] have recently shown by gel electrophoresis that the free sulfhydryl groups of apoB mediate this covalent aggregation. Carboxyamidomethylation of LDL in the absence of glutathione inhibits aggregation, while in the presence of glutathione covalent aggregation occurs. This suggests that a free sulfhydryl group is necessary to initiate the cross-linking reaction. If the sulfhydryl groups of apoB are not first blocked, covalent aggregation occurs. The data suggest that this aggregation occurs by a sulfhydryl–disulfide exchange mechanism, which occurs through the 2 free sulfhydryl groups.

Reassembly Methods

Phospholipid–ApoB Complexes

Two studies have been published concerning the reassembly of LDL apoB with phosphatidylcholine, the major surface lipid in native LDL. The first describes the reassembly of egg yolk PC with SDS-solubilized apoB,[31] and the second, developed in our laboratory, describes the reas-

---

[30] B. W. Patterson, V. N. Schumaker, and W. R. Fisher, *Anal. Biochem.* **136**, 347 (1984).
[31] R. M. Watt and J. A. Reynolds, *Biochemistry* **20**, 3897 (1981).

sembly of dimyristoyl phosphatidylcholine with NaDC-solubilized apoB.[20]

ApoB–Egg Yolk PC Complexes.[31] In this method apoB was solubilized with SDS by the method of Steele and Reynolds[15] described above, and interacted with EYPC which had been deposited as a thin dried film on the walls of a conical test tube.

[35]S-labeled SDS was added to the SDS-solubilized apoB (2 mg protein/ml) and the concentration of SDS, monitored by liquid scintillation counting, was lowered from 2.5 to 0.25–0.5 mM by dialysis. SDS–apoB was added to a dried film of egg PC and allowed to stand 6–8 hr at room temperature and then stirred for 3–4 hr. The PC : apoB ratio was a 2000 : 1 molar excess of lipid : apoB. Excess SDS was removed from the PC–apoB–SDS mixture by dialysis against 2 liters of 20 mM carbonate buffer, pH 10 for 24 hr.

Solid sucrose was added to a final density of 1.1 g/cm^3 and the solution was spun at top speed in a Beckman airfuge for 1 hr in cellulose nitrate tubes that had been presaturated with phospholipid. Free phospholipid vesicles containing no protein floated to the top, and the infranate contained >95% of the apoB complexed to lipid. The infranate was collected from the bottom of the centrifuge tube by using a Hamilton syringe.

Gel filtration chromatography of the EYPC/apoB complex was performed on a 0.7 × 50 cm column of Sepharose CL-4B equilibrated and eluted with 20 mM carbonate buffer at an ionic strength of 0.3 at pH 10. As shown in Fig. 3, phospholipid and protein coeluted from the column in a region that was well included in the column. A small amount of residual SDS elutes close to the total volume. On occasion a small amount of phospholipid was detected in the void volume. However, protein was not found in association with these liposomes. Throughout the leading two-thirds of the coeluting protein and phospholipid peaks the ratio of phospholipid to apoB was constant at 0.5 g/g (320 mol EYPC/500,000 g apoB). On the descending side of the peak the binding levels were slightly lower.

Sedimentation equilibrium studies at 5200 and 5600 rpm in a double sector cell were performed on the EYPC/apoB particle (0.36 g of lipid/g of protein) following gel filtration chromatography. A nonlinear ln OD vs r^2 plot indicated that two species are present. Of the particles 65% have M_ρ $(1 - \phi'\rho)$ of 1.43 × 10^5 corresponding to ~535,000 for the MW of the protein. This fraction does not exhibit concentration-dependent aggregation during centrifugation and is stable for a long period of time. The authors conclude that the second component, which corresponds to 35% of the sample, is formed during the detergent removal process.[31]

Electron micrographs of negatively stained samples of the EYPC/

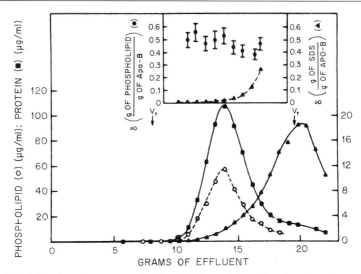

FIG. 3. Gel filtration chromatography of apoB-PL particles. After flotation of contaminating phospholipid vesicles at $d = 1.10$ g/cm^3, gel filtration chromatography was employed to free the preparation of both sucrose and residual detergent; 420 µg of apoB-egg yolk phosphatidylcholine particles in 530 µl of solution was applied to a 50 × 0.7 cm Sepharose CL-4B column. The eluant was 20 mM sodium carbonate–bicarbonate, pH 10.0, adjusted to an ionic strength of 0.3 with NaCl. Arrows mark the void (V_0) and total (V_t) volumes of the column (adapted from Ref. 31). © 1983 American Chemical Society.

apoB complex show it to be 140 Å in diameter, smaller than native LDL, and suggest that the complex may be a flattened ellipsoid rather than a sphere. Five percent SDS–PAGE shows apoB of the EYPC/apoB complex to migrate as one band.[31]

The EYPC–apoB complex prepared by this method cross-reacts with rabbit antibodies to native LDL and with apoB in either SDS or $C_{12}E_8$ (below their critical micellar concentration).

ApoB–DMPC Complex. Walsh and Atkinson[20] have described the recombination of apoB with DMPC at a series of ratios (10:1–1:1 w/w DMPC/apoB). Gel chromatography and density gradient ultracentrifugation have shown that at all ratios examined a reasonably homogeneous complex of DMPC : apoB 4 : 1 w/w is obtained. The following summarizes the method which is currently in use for the production of a high yield of the DMPC–apoB 4 : 1 w/w complex from a starting incubation mixture of 2.5 : 1 w/w DMPC to apoB.

Mixed micelles of DMPC and NaDC are prepared by the following procedure. A ^{14}C-labeled DMPC stock solution in CHCl$_3$: MeOH (2 : 1 v/v) is made by adding L-[α-dipalmitoyl-^{14}C]phosphatidylcholine (100 mCi/

mmol) (New England Nuclear, Boston, MA) to DMPC in quantities which total less than 0.01 mol% of the total lipid mixture. The specific activity in disintegration per min per mg is determined by dividing the disintegrations per min per unit volume by the dry weight (milligrams) per unit volume. The specific activity is adjusted to approximately 100,000 dpm/mg. The solution can then be stored at $-20°$ under N_2, in a sealed vial.

DMPC of the desired mass (25 mg) is obtained from a radiolabeled "stock solution," dried under N_2, and desiccated *in vacuo* at 4° overnight; 12.5 mg of solid NaDC and 2 ml of 50 mM NaCl, 50 mM sodium carbonate, pH 10 are then added to the dried film. The vial is capped and vigorously shaken at room temperature for 2 hr. An optically clear mixed micellar solution results under these conditions.

DMPC–NaDC mixed micellar solution (2.0 ml) is added to 10 mg of NaDC-solubilized apoB and incubated without agitation at room temperature for 1 hr. Detergent is removed by dialysis at 4° against 3–4 liter changes of 0.05 M sodium chloride, 0.05 M sodium carbonate, pH 10. The optically clear solution is concentrated to 2 ml by ultrafiltration and fractionated by density gradient ultracentrifugation.

When [^{14}C]NaDC was included in the incubation mixture 99.8% of the detergent was found in the dialysate after 36 hr and no radioactivity was detectable in association with the protein after density gradient ultracentrifugation.

Density gradient fractionation of a DMPC–apoB mixture prepared at 2.5:1 w/w[20] (Fig. 4) shows that a stable complex with a lipid to protein ratio of 4:1 w/w is obtained.[20] The DMPC–apoB 4:1 w/w complex is separated from either uncomplexed protein or uncomplexed lipid and may be isolated over the density range of d 1.074–1.115 g/ml (median density = 1.095 g/ml).

Electron micrographs of negatively stained particles of DMPC/apoB 4:1 w/w (Fig. 5) show the structural organization of this complex to be similar to a single bilayer phospholipid vesicle of particle diameter 210 ± 20 Å.[20] Three percent SDS–PAGE[32] of apoB of the DMPC–apoB 4:1 w/w complex performed after isolation showed apoB to migrate as one high MW band; 7.5% SDS–PAGE of apoB of the same complex after several weeks of storage at 4° showed no smaller MW bands suggesting that no degradation of apoB has occurred over the time course of storage. DMPC/apoB 4:1 w/w exhibits β-migration on agarose gels, similar to native LDL.[33]

---

[32] K. Weber and M. J. Osborne, *J. Biol. Chem.* **244**, 4406 (1969).
[33] R. P. Noble, *J. Lipid Res.* **9**, 693 (1968).

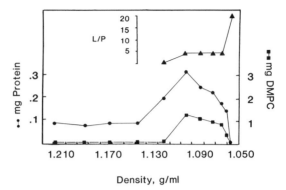

FIG. 4. Density gradient centrifugation of DMPC–apoB 2.5 : 1 w/w incubation mixture on a gradient of potassium bromide.[20] The bottom of the tube is at the left, and the top of the tube is at the right. (●) mg of protein/fraction; (■) mg of DMPC/fraction; (▲) L : P (mg of DMPC to mg of protein) per fraction.

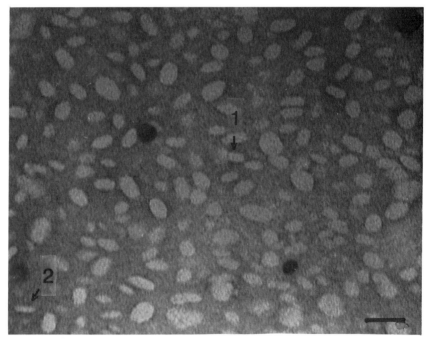

FIG. 5. DMPC–apoB 4 : 1 w/w complex isolated by density gradient ultracentrifugation, negatively stained with uranyl oxalate. Isolated from 1 : 1 w/w incubation ratio at $d = 1.098$ g/ml. (Arrows) (1) flattened elongated particle; (2) completely flattened particle. Bar = 500 Å.

Phospholipid–Cholesteryl Ester–ApoB Complexes (Reassembled LDL)

Microemulsion–ApoB Complexes: Models for Totally Reassembled LDL. Reassembled model LDL complexes have been prepared by the interaction of lipid-free NaDC-solubilized apoB[20] of native human LDL with preformed 200 Å diameter microemulsions of cholesteryl oleate (CO) surface stabilized by either EYPC or DMPC.[34] These microemulsion systems which represent protein free lipid models for LDL have been extensively studied and characterized in this laboratory. Here we describe briefly the methodology for producing these systems which serve as "precursor" lipid particles with which to interact apoB.

[^{3}H]Cholesteryl oleate, synthesized by the method of Lentz *et al.*,[35] was added to solutions in $CHCl_3$/MeOH of the unlabeled cholesteryl oleate. ^{14}C-labeled phospholipid solutions in $CHCl_3$/MeOH were made by adding [1-^{14}C]DPPC (100.0 mCi/mmol) to unlabeled phosphatidylcholines in quantities which totaled <0.01 mol% of the total lipid mixture. Specific activity in disintegrations per min per mg are determined as described above for DMPC.

Aliquots of [^{3}H]cholesteryl oleate and [^{14}C]phospholipid in designated amounts appropriate to give the desired molar ratio of the two components were taken from stock solutions and the initial total amount and starting molar ratio of the two lipids was verified by liquid scintillation counting of an aliquot from the mixture. The solvent was removed by evaporation under a stream of N_2 followed by vacuum desiccation at 4° for 12–16 hr. The dried lipids were resuspended in 10 ml of 0.1 M KCl, 0.01 M Tris–HCl, 0.025% NaN_3 at pH 8.0. Total lipid concentrations were approximately 10% w/v in all experiments. The cloudy suspension was sonicated for 300 min under an N_2 atmosphere at a temperature at which both components were in a liquid or liquid crystalline state. The temperature was maintained above 51° for CO systems. Temperature was monitored by a thermocouple inserted directly into the sonication vessel. Sonication was performed using a Heat Systems Sonifier (W-350) equipped with a standard 0.5 in. horn at a power setting of 4 (50 W output) in the "continuous" operating mode. After sonication the solution was centrifuged at 195,000 g for 30 min in a Beckman SW 41 rotor at 4°. All centrifugation was done without braking so as to prevent intermixing of fractions. The upper 10% of the solution, containing particles which float at a background density of 1.006 g/ml was removed. The remaining infranatant was adjusted to a background density of 1.25 g/ml with KBr and recentrifuged at 195,000 g for 2 hr at 4°. The top 20% of the tube volume was removed

---

[34] G. S. Ginsburg, D. M. Small, and D. Atkinson, *J. Biol. Chem.* **257**, 8216 (1982).
[35] B. R. Lentz, Y. Barenholz, and T. E. Thompson, *Chem. Phys. Lipids* **15**, 216 (1975).

including the gelatinous layer of lipids which forms in the top 1–2 mm at 25°. The lipid was readily resuspended. Gel filtration chromatography of the PC/CO microemulsions is illustrated in Fig. 6.

As shown schematically in Fig. 7, NaDC-solubilized apoB was interacted with the microemulsion particles at room temperature by placing 16 mg of either EYPC/CO or DMPC/CO microemulsions into an open-ended dialysis bag (16 mg of total lipid–8 mg of EYPC or DMPC plus 8 mg CO in 2 ml of 50 mM sodium chloride, 50 mM sodium carbonate, 0.02% sodium azide, pH 10) which was suspended in 6 liters of the above buffer. Two milligrams of NaDC-solubilized apoB[20] in a 2 ml volume was introduced to the microemulsions in the dialysis bag at a very slow rate (0.5 ml/hr) by means of a peristaltic pump. Two 6-liter changes of buffer were made over

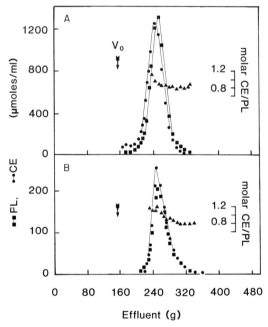

FIG. 6. Column chromatography of PL/CO microemulsions. PC/CO microemulsions were prepared at an initial molar ratio of 1:1 by sonication for 300 min. Fraction S_1 ($\rho <$ 1.006) was isolated and removed and fraction S_2 (1.006 $< \rho <$ 1.220) was applied to a Sepharose 4B column (2.6 × 100 cm) at room temperature. (A) Elution profile of ultracentrifugation fraction S_2 resulting from the sonication of CO/DMPC (initial molar ratio 1:1) for 300 min. (B) Elution profile of concentrated fractions comprising the effluent between 220 and 300 g from (A). Fractions were pooled, concentrated by ultracentrifugation, and applied to the same column. V_0 indicates the void volume; (●) cholesteryl ester (CE) in nmol per ml of effluent; (■) phospholipid (PL) in nmol per ml of effluent; CE/PL is the *molar* cholesterol ester/phospholipid ratio indicated on the descending portion of the elution curves as ▲.[34]

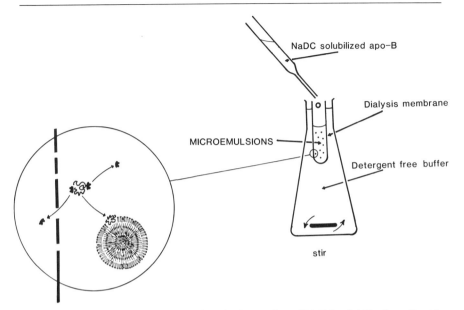

FIG. 7. Schematic diagram illustrating the interaction of NaDC solubilized apoB with PC/CO microemulsions with detergent removal by dialysis.

a 24-hr period to complete the removal of NaDC. After extensive dialysis DMPC/CO/apoB or EYPC/CO/apoB was concentrated to 2 ml and applied to a 2.5 × 40 cm column of Sepharose CL-4B equilibrated with 50 mM sodium chloride, 50 mM sodium carbonate, 0.02% sodium azide, pH 10.

Gel filtration chromatography of EYPC/CO/apoB or DMPC/CO/apoB[36] (Fig. 8) showed that phospholipid, cholesteryl ester, and protein coeluted at a position that was included in the column, indicative of a stable complex. In the central region where all components coeluted, the molar ratio of CO/PC is 1.05, comparable to that observed in the initial microemulsions and the apoB/PC ratio is 0.32 w/w. Both ratios were relatively constant.

EYPC/CO/apoB and DMPC/CO/apoB complexes have a circular morphology with a diameter of 219 ± 24 Å (Fig. 9B and C) similar to LDL, as determined by negative stain electron microscopy.[36] Similar to native LDL, both complexes exhibit β-migration on agarose gels.[33] ApoB of EYPC/CO/apoB and DMPC/CO/apoB complexes migrated as one high MW band on 3% SDS–PAGE.

[36] G. S. Ginsburg, M. T. Walsh, D. M. Small, and D. Atkinson, *J. Biol. Chem.* **259**, 6667 (1984).

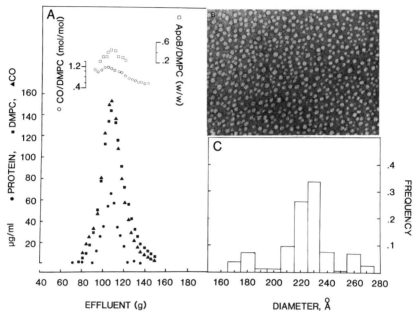

FIG. 8. Gel chromatography and electron microscopy of DMPC/CO/apoB complexes. (A) Gel filtration chromatography of DMPC/CO/apoB. The contents of the dialysis bag were concentrated to 2 ml by ultrafiltration and applied to a column (2.5 × 40 cm) of Sepharose CL-4B which was eluted downward with 50 mM sodium carbonate, 50 mM sodium chloride, 0.02% sodium azide, pH 10 buffer. (●) Protein, μg/ml; (■) DMPC, μg/ml; (▲) CO, μg/ml; (○) CO/DMPC, mol/mol; (□) apoB/DMPC, w/w. (B) Electron microscopy of DMPC/CO/apoB from a fraction in the central region of coeluting DMPC, CO, and apoB (see A) negative staining was carried out using sodium phosphotungstate, pH 7.4. Bar = 500 Å. (C) Size distribution histogram of DMPC/CO/apoB. Diameters of 200 randomly chosen circular particles were measured from representative fields of the electron micrographs. Average values were 219 ± 24 Å.[36]

A variation in the above method has been recently published by Lundberg and Suominen.[37] For the preparation of microemulsions, 4 mg of [^{14}C]cholesteryl oleate in 4 ml of 0.01 M Tris–HCl, pH 9 was prepared by injection of CO in 400 μl of diethyl ether–chloroform 2:1 through the bottom of a thermostatted sonication vessel. Simultaneous with injection, sonication was conducted at 30 sec intervals for a period of 30 min under a nitrogen atmosphere at 52° (MSE sonifier, titanium probe, setting 6, 100 W output). EYPC was dried as a thin film on the walls of a glass ampoule. An aliquot of the sonicated [^{14}C]CO suspension was transferred immediately to the EYPC-containing ampoule (CO:EYPC = 2:1 w/w). The

[37] B. Lundberg and L. Suominen, *J. Lipid Res.* **25**, 550 (1984).

ampoule was immediately flushed with N_2, sealed, and sonicated at 52° for 20 min (Branson bath-type).

ApoB was solubilized from native human LDL (d 1.025–1.050 g/ml) with sodium deoxycholate essentially by the method of Helenius and Simons.[17] ApoB–NaDC mixed micelles were separated from lipid–NaDC mixed micelles by gel filtration chromatography on a 1.6 × 60 cm column of Sepharose CL-6B equilibrated and eluted with 50 mM sodium chloride, 50 mM sodium carbonate, 10 mM sodium deoxycholate, pH 10. To remove detergent, the apoB–NaDC mixed micelles (concentrated by ultrafiltration on Amicon PM-30 filters) were passed through a Sephadex G-75 column (1.6 × 30 cm) and eluted with 0.01 M Tris–HCl buffer, pH 9.0. ApoB eluted in the void volume and the bile salt close to the total volume of the column.

The preparation of model LDL (m-LDL) was accomplished by addition of the apoB to the CO–EYPC microemulsion at a weight ratio of 1:1.2 between EYPC and apoB. The mixture was incubated for 10 min at 52° and sonicated in the bath sonifier for 15 min. An optically clear preparation of m-LDL was obtained.

The m-LDL solution was then subjected to density gradient centrifugation using a 0–40% (w/v) linear sucrose gradient. The m-LDL samples (0.5 ml) were layered on top of the gradient and centrifuged at 50,000 rpm for 24 hr at 10° in a Beckman SW-60 Ti rotor. Small samples were collected by tube puncture and analyzed for protein and radioactivity. Sample densities were determined by refractometry.

Density gradient ultracentrifugation of m-LDL solutions shows the particles were concentrated to a maximum at d 1.07 g/ml. Over the major portion of the peak the weight ratios for the components were quite constant (CO/EYPC = 1.5, protein/EYPC = 0.8). For all further studies the d 1.04–1.09 fractions from the density gradient were pooled and dialyzed against 0.01 M Tris–HCl, pH 9.0 buffer.

Gel filtration chromatography of m-LDL on Sepharose CL-4B (Fig. 9) showed the particles to elute as a symmetrical peak (K_{av} = 0.50) with fairly constant component ratios over the central region of the peak corresponding to a MW of 2.2 × 10^6. Electron microscopy of negatively stained particles showed spherical structures (d 160–290 Å) with a mean particle diameter of 210 Å. Agarose electrophoresis of m-LDL showed the particles to exhibit β-migration.

Reconstitution of LDL by Cholesteryl Ester Replacement. Krieger *et al.*[38] have described a procedure by which >99% of the core-located

---

[38] M. Krieger, M. S. Brown, J. R. Faust, and J. L. Goldstein, *J. Biol. Chem.* **253,** 4093 (1978).

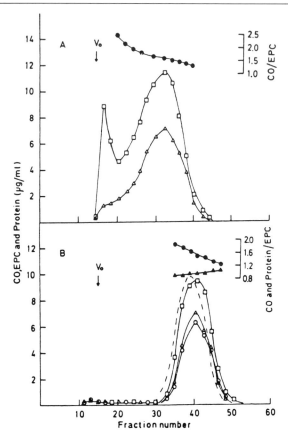

FIG. 9. Sepharose CL-4B gel filtration of (A) CO/EPC microemulsion (initial weight ratio 2:1) and (B) reconstituted m-LDL after separation by sucrose gradient centrifugation. (○) Protein; (△) EPC; (□) CO. Insets: (●) CO/EPC weight ratio; (▲) protein/EPC weight ratio. The dotted line in (B) represents native LDL.[37]

cholesteryl esters are removed from LDL and the particles are subsequently reconstituted by the addition of exogenous neutral lipid. This procedure is described in detail in Chapter [34], this volume.

Physical Characterization of Model Particles

Physical–chemical studies (calorimetry, NMR, X-ray scattering, etc.) of LDL have established our current understanding of the molecular interactions and structure of the native particle. Several chapters in this series describe the details of these studies. Similar studies of model reassembled LDL particles serve two functions. The first of these is to demonstrate similarities between the model particles and the native system. The second and perhaps more important function is to use the systematic compositional variation possible in these model systems coupled with physical–chemical studies to extend the detail of our understanding of molecular interactions and structure of lipoproteins.

Since the reassembly of LDL is still in its infancy the number of detailed physical–chemical studies on model particles is limited and these studies are currently primarily of a comparative nature. The main features of these studies are summarized here.

Differential Scanning Calorimetry (DSC)

Thermodynamic characterization by differential scanning calorimetry has shown that LDL undergoes two distinct thermal transitions.[39] As shown in Fig. 10 the first transition occurs reversibly over the temperature range of 20–40° in normal LDL and has been shown to be associated with an order–disorder transition of the core-located cholesteryl esters.[39] A number of studies have demonstrated that the temperature at which this transition takes place is determined by the fatty acid composition of the cholesterol esters and by the triglyceride content of the LDL particle. The second transition occurs at ~85° and is associated with disruption of the LDL particle.

The reassembled model LDL particle developed in this laboratory from sodium deoxycholate-solubilized apoB and the model lipid microemulsion particles show many structural and physical properties similar to native LDL.[36] In addition, the model LDL particles retain the thermodynamic characteristics unique to their precursor protein-free phospholipid–cholesteryl ester microemulsions.

In the model LDL particle with egg PC forming the surface monolayer the order–disorder transition of the cholesterol oleate in the core occurs at 38° with an enthalpy of 0.3 cal/g CO. A high temperature transition at 85° corresponding to particle disruption and protein unfolding is also observed. Thus this model particle shows similar thermal behavior to the native LDL particle. These two transitions are also evident in the DMPC/CO/apoB complex. In addition, a high enthalpy transition occurs centered at 22° with an enthalpy of 2.6 cal/g DMPC, and has been shown to be associated with the order–disorder transition of the fatty acyl chains of the surface monolayer of DMPC.[34,36]

Interestingly the transition temperatures and enthalpies of the core-located cholesterol esters and, in the case of the particle formed with

[39] R. J. Deckelbaum, G. G. Shipley, and D. M. Small, *J. Biol. Chem.* **252**, 744 (1977).

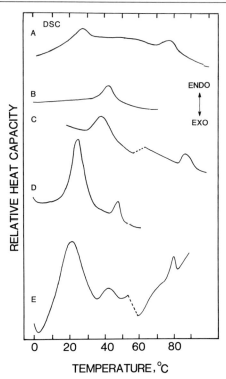

FIG. 10. Differential scanning calorimetry. Measurements were made in a Perkin-Elmer DSC-2 (Perkin-Elmer, Norwalk, CT) at heating and cooling rates of 5°/min and a range of 0.2 mcal/sec. Samples were concentrated by vacuum dialysis immediately and hermetically sealed in 75 μl sample pans. The reference pan contained an equal mass of buffer. Sample masses were determined either by Lowry protein [O. H. Lowry, N. J. Rosebrough, A. L. Farr, and R. J. Randall, *J. Biol. Chem.* **193**, 265 (1951)] or liquid scintillation counting. Enthalpies of transition (ΔH) were calculated from the areas under the peaks as measured by planimetry and related to the area of the crystal-liquid melting transition of an indium standard (A) LDL; (B) EYPC/CO; (C) EYPC/CO/apoB; (D) DMPC/CO; (E) DMPC/CO/apoB.

DMPC, the transition temperature and enthalpy of the surface phospholipid monolayer are lower than those observed in the precursor protein-free microemulsion particles. These observations suggest that apoB either directly or indirectly influences the properties of both the core and surface components.

DSC of reconstituted LDL prepared by the method of Krieger et al.[38] have been reported by Tall and Robinson.[40] LDL was reconstituted utilizing three different ratios of cholesterol ester (either cholesteryl linoleate

[40] A. R. Tall and L. A. Robinson, *FEBS Lett.* **107**, 222 (1979).

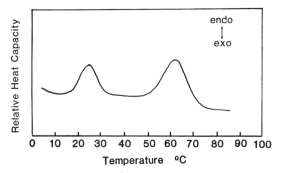

FIG. 11. Differential scanning calorimetry of DMPC/apoB 4:1 w/w.

or cholesteryl oleate). In these reconstituted LDLs at all ratios and compositions examined there was no evidence of the reversible thermal transition associated with the order–disorder transition of the core-located cholesteryl esters over the temperature range 0–40°. These results suggested the presence of an impurity, perhaps heptane, in the core of the reconstituted particle. Thus the core-located cholesteryl ester of LDL reconstituted by this method does not exhibit similar thermal behavior as the core lipids of native LDL.

The vesicular DMPC–apoB complex developed in this laboratory exhibits two thermal transitions[41] (Fig. 11). The first is reversible and occurs with a peak temperature of 24° and is attributed to the gel to liquid crystalline transition of the hydrocarbon chains of the DMPC bilayer, with an enthalpy of 3.34 kcal/mol of DMPC. At 62° the complex undergoes a thermal transition which is associated with particle disruption and protein unfolding. This transition temperature is ~20° lower than that of the native LDL particles or the reassembled model LDL particles suggesting that apoB is less thermally stable in association with phospholipid alone.

Circular Dichroism

ApoB of EYPC/apoB complexes described by Watt and Reynolds[31] and isolated by gel filtration chromatography containing either 0.5 g of EYPC/g of apoB (leading edge) or 0.35 g of EYPC/g (trailing edge) of apoB exhibits a pronounced minimum at 217 nm in CD spectra, indicative of a significant amount of β-pleated sheet structure (Fig. 12). Overall, there are no significant differences in these spectra with respect to those of LDL. There is no change in the shape of the spectra and the minima do not shift. Only the magnitude of the minimum is altered.[31]

[41] M. T. Walsh and D. Atkinson, *J. Lipid Res.*, in press (1986).

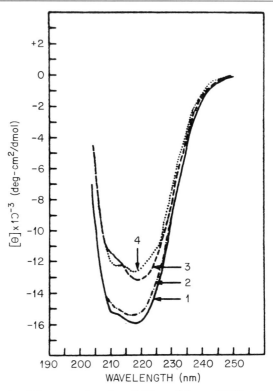

FIG. 12. Circular dichroic spectra of apoB-PL particles and LDL_2. Spectra of fractions taken from the gel filtration column shown in Fig. 3 were recorded. Samples eluting at 12.5 (curve 1), 14.0 (curve 2), and 16.4 (curve 4) g of effluent were compared to holo-LDL (curve 3). Spectra were taken at room temperature in 1 mm rectangular cells at protein concentrations of 0.05–0.1 mg/ml.[31]

The far UV CD spectra of apoB of LDL, of the vesicular complex of DMPC–apoB and of apoB in complexes with microemulsions of DMPC/CO and EYPC/CO developed in this laboratory have been presented previously.[20,36,41] Spectra of apoB of LDL and the complexes are shown in Fig. 13. These spectra were recorded at 50°, a temperature at which all the lipids of LDL and the three complexes are above their order–disorder transitions. The secondary structure of apoB as estimated from molar ellipticities at 217 and 222 nm by the method of Greenfield and Fasman[42] and Morrisett et al.[43] is summarized in the table.

[42] N. Greenfield and G. D. Fasman, Biochemistry **8,** 4108 (1969).
[43] J. D. Morrisett, J. A. K. David, H. Pownall, and A. M. Gotto, Biochemistry **12,** 1290 (1973).

SECONDARY STRUCTURE OF ApoB IN LDL AND REASSEMBLED
COMPLEXES AT 50°

	% α-Helix (%)	β-Sheet (%)	Random coil (%)
LDL	40	20	40
DMPC–apoB	48	12	40
EYPC/CO/apoB	37	3	60
DMPC/CO/apoB	9	11	80

The CD spectrum of apoB of LDL at 50° is characterized by negative minima at 217 and 222 nm, indicative of a large amount of α-helix and β-structure. In the vesicular complex with DMPC, apoB is characterized by a negative minimum at 227.5 nm and a high overall ellipticity. ApoB on

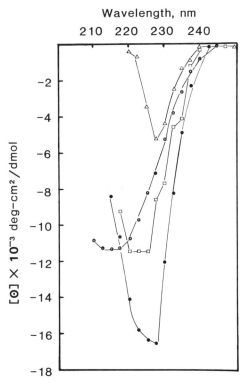

FIG. 13. Circular dichroic spectra. (○) LDL; (□) EYPC/CO/apoB; (△) DMPC/CO/apoB; (●) DMPC/apoB. Spectra were recorded at 50° in a Cary Model 61 spectropolarimeter in a 1-cm quartz cell at protein concentrations of 0.03–0.09 mg/ml. All samples were dialyzed against 0.005 M sodium tetraborate, pH 10 prior to use.[20,36,41]

the EYPC/CO/apoB complex exhibits its minimum from 220 to 225 nm, which suggests that on this complex, apoB exhibits a similar α-helical content as apoB on LDL. However, the lower value of molar ellipticity at 217 nm would suggest that there is less β-structure. ApoB on the complex of DMPC/CO exhibits a very low overall ellipticity, with a minimum at 228 nm, suggesting a great deal of random coil and a small amount of α-helix.

Biological Characterization

Preliminary investigations have been carried out testing the ability of microemulsion/apoB complexes (EYPC/CO/apoB and DMPC/CO/apoB) to compete with human ^{125}I-labeled LDL for the binding to the apoB and E receptors of normal human fibroblasts.[44,45] As shown in Fig. 14 displacement of ^{125}I-labeled LDL from the receptors requires a greater amount of protein for both of the complexes than for normal LDL. In both cases, however, competitive binding is achieved, thus demonstrating the biological activity of both types of microemulsion/apoB complexes. Experiments performed with the vesicular DMPC/apoB complex show relatively little binding to the receptor.

Similarly, Lundberg and Suominen[37] have studied in a more extensive manner the ability of m-LDL to compete with ^{125}I-labeled LDL for binding, incorporation, and degradation to the apoB and E receptors on cultured human fibroblasts. m-LDL competes with LDL for the receptor, indicating that m-LDL is recognized and taken up by the receptor pathway.

The metabolism of ^{125}I-labeled LDL and m-LDL were compared in binding and degradation studies with fibroblasts at 37°. For ^{125}I-labeled LDL, a steady state for binding and cellular content of radioactivity was achieved in 2 hr. Assay of TCA-soluble radioactivity in the medium showed that almost all of the ^{125}I-labeled LDL that was hydrolyzed was released into the medium and the rate of appearance is time dependent. Metabolism of ^{125}I-labeled m-LDL is similar to that of ^{125}I-labeled LDL. Binding and cellular content of unhydrolyzed ^{125}I-labeled m-LDL reaches a plateau at 2.5 hr. Degradation of ^{125}I-labeled m-LDL was slightly slower than that for ^{125}I-labeled LDL. These data were interpreted to indicate that this reconstituted LDL is bound to the LDL receptor with the same affinity as native LDL. In addition, the particles are taken up and hydrolyzed like native LDL.

[44] J. L. Goldstein and M. S. Brown, *Annu. Rev. Biochem.* **46**, 897 (1977).
[45] T. L. Innerarity, R. E. Pitas, and R. W. Mahley, *J. Biol. Chem.* **254**, 4186 (1979).

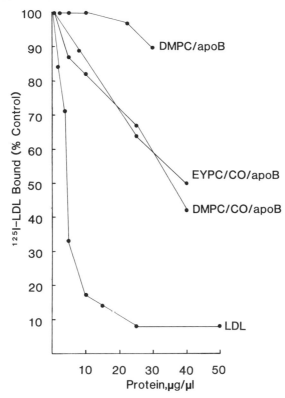

FIG. 14. Ability of microemulsion/apoB complexes to compete with human ^{125}I—labeled LDL for binding to receptors on normal human fibroblasts.[44,45] Cells were incubated for 2 hr on ice in the presence of ^{125}I-labeled LDL (2 mg/ml) and the microemulsion/apoB complex or DMPC/apoB complex. The 100% control value was 15 ng of ^{125}I-labeled LDL protein bound per mg of cellular protein.

The biological activity of LDL reconstituted by the method of Krieger et al.[38] has been reported in detail. r-LDL is reported to bind to the LDL receptor in cultured fibroblasts with the same affinity as native LDL and is taken up and hydrolyzed like native LDL. The cholesterol released from lysosomal hydrolysis of r-LDL modulates intracellular cholesterol metabolism.

Summary

The methodologies described here for the selective and sequential reassembly of model LDL particles, although in many instances still in the developmental stages, will undoubtedly provide a basis on which further advances in LDL reassembly will be made.

Reassembled LDL complexes of defined lipids and apoB provide well-defined model systems in which to study the molecular interactions and structural organization of LDL, including the lipid–lipid interactions in the particle core, the lipid–lipid and lipid–protein interactions which determine the surface organization and protein conformation, and the interactions between the core and surface components. These reassembled LDL complexes should serve as important models to study the delivery of isotopically labeled lipids with differing physical properties to cells in order to investigate the metabolic complexity of intracellular LDL catabolism and its relationship to positive cholesterol balance and atherogenesis.

[34] Reconstitution of the Hydrophobic Core of Low-Density Lipoprotein

By MONTY KRIEGER

Low-density lipoprotein (LDL), the major cholesterol transport vehicle in human plasma,[1] is a large particle (approximately 2.5×10^6 Da) which contains a core of neutral lipid surrounded by an amphiphatic shell.[2] LDL's core is composed mainly of fatty acyl esters of cholesterol (primarily cholesteryl linoleate) and lesser amounts of triglyceride. The amphipathic shell contains phospholipid, unesterified cholesterol, and a large protein called apolipoprotein B. The cholesterol in LDL is delivered to cells by receptor-mediated endocytosis.[1] LDL binds to cell surface LDL receptors which are located in specialized regions of the plasma membrane called coated pits. The invagination and pinching off of coated pits leads to the formation of coated endocytic vesicles containing LDL. Subsequently, LDL dissociates from its receptor and is transported to lysosomes wherein the LDL particle is disassembled by lysosomal enzymes. These enzymes include lipases which hydrolyze the core cholesteryl esters to unesterified cholesterol. The unesterified cholesterol is then available to satisfy the cholesterol needs of the cell, including membrane biosynthesis and the regulation of cellular sterol metabolism.

[1] J. L. Goldstein, S. K. Basu, and M. S. Brown, this series, Vol. 98, p. 241.
[2] R. J. Deckelbaum, G. G. Shipley, and D. M. Small, *J. Biol. Chem.* **252**, 744 (1977).

The finding that the apolipoprotein B component of LDL was responsible for the lipoprotein's binding to surface LDL receptors[3] raised the possibility that the core cholesteryl esters could be extracted and replaced by other molecules without interfering with the ability of the lipoprotein to bind to LDL receptors and be transported to lysosomes. We have developed a method to replace or reconstitute the core of LDL with a variety of compounds and thus convert LDL into a general purpose delivery vehicle for the receptor-mediated transport of molecules into cells.[4-6] It has been possible to use the reconstitution method to introduce a variety of spectroscopic probe molecules into the core and thus study the lipoprotein's structure and its relation to its function.[7-9] Reconstitution of LDL has also permitted the incorporation of radiolabeled cholesteryl esters, fluorescent lipids, and toxic lipids into the core. These reconstituted LDLs have been used to study cellular cholesterol metabolism[10] and the genetics of receptor-mediated endocytosis.[11,12]

Reconstitution of LDL

Principle. To remove the apolar core of LDL, the lipoprotein is first lyophilized in the presence of starch and then the core is selectively extracted into heptane. This method was originally described by Gustafson.[13] The starch presumably serves to stabilize the lipoprotein to dehydration and prevent aggregation of the particles. Dehydration is apparently required to permit access of the very apolar extraction solvent to the core lipids. If solvents of significantly greater polarity are used, the phospholipids are also extracted and the apolipoprotein apparently denatures.

[3] R. W. Mahley, T. L. Innerarity, R. E. Pitas, K. H. Weisgraber, J. H. Brown, and E. Gross, *J. Biol. Chem.* **252,** 7279 (1977).
[4] M. Krieger, M. S. Brown, J. R. Faust, and J. L. Goldstein, *J. Biol. Chem.* **253,** 4093 (1978).
[5] M. Krieger, M. J. McPhaul, J. L. Goldstein, and M. S. Brown, *J. Biol. Chem.* **254,** 3845 (1979).
[6] M. Krieger, L. C. Smith, R. G. W. Anderson, J. L. Goldstein, Y. J. Kao, H. J. Pownall, A. M. Gotto, Jr., and M. S. Brown, *J. Supramol. Struct.* **10,** 467 (1979).
[7] M. Krieger, J. Peterson, J. L. Goldstein, and M. S. Brown, *J. Biol. Chem.* **255,** 3330 (1980).
[8] G. C. Chen, M. Krieger, M. S. Brown, J. L. Goldstein, and J. P. Kane, *Biochemistry* **19,** 4706 (1980).
[9] P. Kroon and M. Krieger, *J. Biol. Chem.* **256,** 5340 (1981).
[10] M. S. Brown, J. L. Goldstein, M. Krieger, Y. K. Ho, and R. G. W. Anderson, *J. Cell. Biol.* **82,** 597 (1979).
[11] M. Krieger, M. S. Brown, and J. L. Goldstein, *J. Mol. Biol.* **150,** 167 (1981).
[12] M. Krieger, this series, Vol. 129 [14].
[13] A. Gustafson, *J. Lipid Res.* **6,** 512 (1965).

It appears likely that the core of the heptane-extracted lipoprotein is filled with heptane.

To introduce into the core an exogenous compound, the compound, dissolved in an apolar solvent is added to the LDL starch residue and the solvent is evaporated with an inert gas (e.g., nitrogen or argon). During the course of the evaporation, the compound is concentrated in the core. If the evaporation is not complete, solvent may remain trapped in the core and reduce the amount of incorporated compound. This could present particular difficulties if the reconstituted LDL is to be used for structural studies. The reconstituted LDL, designated r-(compound)LDL, can then be solubilized by the addition of an aqueous buffer (pH 7–9) and isolated by low-speed centrifugation.

Reconstituted LDL resembles native LDL in many of its chemical and physical properties (e.g., lipid: protein mass ratio, electrophoretic mobility, buoyant density, core thermotropic transitions, electron microscopic appearance, etc.). The biological activity of r-(cholesteryl ester)LDL as judged by its ability to be processed by the LDL pathway of endocytosis in cultured cells is virtually identical to that of native LDL [e.g., binding affinity, rate of internalization, ability of r-(cholesteryl ester)LDL to delivery metabolically active cholesterol[4]].

Using this procedure, LDL can only be reconstituted by apolar molecules and the reconstitution process appears to be facilitated if the compound contains a long hydrocarbon chain ($>C_{10}$) which is either bent by the presence of at least one cis double bond (e.g., oleic acid) or branched by side chains (e.g., polyisoprenes such as phytol).[5] The sterol moiety is not required; triglycerides and fatty acyl methyl esters can be reconstituted into the core.[5] A variety of toxins (25-hydroxycholesteryl oleate, N-[[[4-[3β-(oleoyloxy)androst-5-en-17β-yl]pentyl]oxy]-carbonyl]-N,N-bis-(2-chloroethyl)amine[14]), spectroscopic probes (e.g., β-carotene,[8] 12-doxylstearate methyl ester[7]), and fluorescent probes (e.g., pyrene[6] and fluorescein derivatives of cholesteryl esters[15]) have been reconstituted into the core of LDL and used to study the structure of LDL and the mechanism of endocytosis[11] and to develop a diagnostic test for familial hypercholesterolemia.[15] However, caution must be exercised in using reconstituted LDL. Not all reconstituted LDLs are suitable for examining receptor-mediated endocytosis in cultured cells. Some molecules can reconstitute LDL yet leak out of the lipoprotein and enter cells by receptor-independent pathways.[14] Structural features of lipids which permit this

---

[14] R. Firestone, J. Pisano, J. Falck, M. McPhaul, and M. Krieger, *J. Med. Chem.* **27**, 1037 (1984).

[15] M. Krieger, Y. K. Ho, and J. R. Falck, *J. Receptor Res.* **3**, 361 (1983).

nonspecific leaking have not been defined; however, this leaking has not been observed when the reconstituted lipid contains a sterol moiety. Thus, the sterol structure may serve to anchor lipids in the core.[14] Methods for testing the receptor-specific delivery of reconstituted lipids to cells have been described previously.[1,14]

Reagents

    LDL at a protein concentration of 4–6 mg/ml
    Dialysis buffer: 0.3 mM EDTA, pH 7.0
    Potato starch, purified powder (Fischer Scientific)
    Prosil-28 (American Scientific Products) or equivalent glassware antiwetting agent
    Liquid nitrogen
    Heptane, reagent grade or better
    Compound to be incorporated into LDL, 30 mg/ml in heptane or benzene
    Nitrogen or argon gas
    Tricine buffer, 10 mM, pH 8.4, sterile

Preparation of LDL. Human LDL may be isolated from freshly drawn blood by KBr density centrifugation using either the sequential flotation method[1] with fixed angle rotors or the density gradient method using a zonal rotor with a reorienting core.[16] Prior to reconstitution, the LDL (4–6 mg protein/ml) should be extensively dialyzed against 0.3 mM EDTA, pH 7.0 (4 changes of 6 liters at least 6 hr between changes). If the pH falls more than a few tenths of a unit below 7, the LDL may precipitate. The pH of the dialysis buffer should be checked immediately before use. Protein content is measured by the Lowry method.[17] The dialyzed LDL may be stored in plastic or Prosil-28-treated glass tubes under inert gas at 4° in the dark for at least 2 weeks.

Preparation of Starch Tubes. Reconstitution is performed in 13 × 100 mm disposable glass tubes. To prevent LDL from nonspecifically sticking to the glass, the tubes are first treated with Prosil-28 or an equivalent antiwetting agent according to the manufacturer's instructions. Potato starch (25 mg) is added to each tube. To simplify the addition of starch to multiple tubes, a 50-μl aliquot from a freshly mixed and continuously stirring slurry of starch (0.03 g/ml of distilled water) may be added to each tube. The starch must be dried (e.g., by evaporation at room temperature overnight) and the tubes may be stored at room temperature.

[16] B. H. Chung, J. P. Segrest, M. J. Roy, K. Beaudru, and J. T. Cone, this volume [8].
[17] O. H. Lowry, N. J. Rosebrough, A. L. Farr, and R. J. Randall, *J. Biol. Chem.* **193**, 265, (1951).

Procedure

1. Lyophilization of LDL. Add 1.9 mg protein of dialyzed LDL (approximately 0.3–0.5 ml) to each starch tube, cover with two layers of parafilm, and punch holes in the parafilm with a dissecting needle or pipet tip. Form an even suspension by gently vortexing the mixture and rapidly freeze in a thin shell with liquid nitrogen (or a dry ice slurry). Lyophilize for 4–6 hr until the sample is completely dry.

2. Extraction of endogenous core lipids. The extraction may be more conveniently conducted at room temperature than at $-10°$ as previously reported.[4,5] Add 5 ml of heptane to each lyophilized LDL starch tube, vortex vigorously for 30–60 sec, and pellet the LDL starch residue by centrifugation at 2000 rpm for 10 min in a clinical centrifuge (e.g., Beckman TJ-6R). Discard the yellow supernatant after aspiration and repeat the extraction two more times. The last supernatant should be colorless. Immediately proceed to the next step after removing the last supernatant. Do not allow the heptane covering the LDL starch residue to completely evaporate at this stage.

3. Reconstitution with exogenous lipid. Add 6 mg of lipid in 200 μl of apolar solvent to the LDL starch residue. Heptane, petroleum ether, carbon tetrachloride, and benzene can be used as the solvent. After gentle shaking the mixture of LDL starch residue : lipid, allow to incubate for 5–10 min in a standard freezer ($<-10°$) and then slowly evaporate the solvent under a stream of inert gas (nitrogen or argon) at room temperature until the sample is completely dry (usually 30–60 min). Multiple samples may be evaporated using a N-EVAP evaporator (Organomation) or its equivalent.

4. Solubilization and isolation. All subsequent steps are performed at $4°$ and all liquid transfers are made using either plastic pipet tips or Prosil-28-treated glass pasteur pipets. Add 1 ml of sterile buffer (10 mM tricine, pH 8.4) to each tube, vortex gently to resuspend the dry pellet, cover (parafilm) and incubate for at least 10 hr but no more than 24 hr. During this time, the reconstituted LDL is released from the starch while any denatured LDL and excess lipid remain bound to the starch or the glass. Lipids that are liquid at $4°$ may form droplets that float on the buffer.

The solubilized reconstituted LDL is separated from the bulk of the starch and excess lipid by low-speed centrifugation (2000 rpm, in a TJ-6R centrifuge or equivalent). The specimen should be further clarified by at least one or two additional centrifugations (approximately 10,000 rpm for 10 min) using a Beckman microfuge or its equivalent. The supernatant may contain a swirling very fine powdery substance, especially if polyunsaturated lipids are used. Most of this substance, which will float on the buffer, may be removed by carefully withdrawing the supernatant after each centrifugation. If necessary, the samples may be clarified by filtration through a 0.8-μm Unipore filter (Bio-Rad). Little or no insoluble material is observed when the lipid contains a polycyclic aromatic hydrocarbon moiety, such as a pyrene group. Protein recovery, an excellent assay for the success of the reconstitution, can be determined by the Lowry method. The protein content of the specimens could be as much as 1.8 mg/ml (95% yield). At these low concentrations a substantial fraction of the lipoprotein will stick to standard Millipore HA filters and their use is not recommended. Methods used to determine the amount of lipid incorporated will vary depending on the lipid used for the reconstitution. For example, gas chromatography[4] or enzymatic assay (kits available from several companies, e.g., Boehringer) may be used to detect incorporation of cholesteryl esters or triglycerides. The reconstituted LDL should be stored under an inert gas at 4°.

Acknowledgment

Preparation of this manuscript was supported by a NIH Research Career Development Award and by NIH research grants.

[35] Use of Cross-Linking Reagents to Study Lipoprotein Structure

By JOHN B. SWANEY

Scope

In 1970 it was shown by Davies and Stark[1] that the introduction of covalent bonds between protein chains through the use of bifunctional reagents was useful for analysis of the quaternary structure of complex enzymes. Since that time this technique has been applied to the structural analysis of a wide variety of macromolecular assemblies, both for the purpose of determining the number of protomers in an assembly and for ascertaining nearest-neighbor relationships among different proteins in a multimeric macromolecule. Application of this approach to the study of lipoprotein structure has been relatively recent and, as yet, inadequately exploited. The cross-linking methodology has been used most extensively

[1] G. E. Davies and G. R. Stark, *Proc. Natl. Acad. Sci. U.S.A.* **66**, 651 (1970).

for the study of apolipoprotein self-association and recombinant lipoprotein structure, but recent studies have also utilized this technique for structural studies of the high- and low-density lipoproteins.

Criteria for the Selection of a Cross-Linking Reagent

A wide variety of cross-linking reagents have been utilized for studies of protein structure and many of these have been reviewed earlier in this series.[2] What follows is a brief discussion of reagents which have been applied specifically for the study of lipoproteins, or which may prove useful for such studies.

Diimidoester Reagents. The majority of cross-linking studies which have been performed on intact lipoproteins and on apolipoproteins have utilized bifunctional reagents with imidoesters as the active functional groups (see the table). Of this reagent category, the most widely used is dimethyl suberimidate (DMS).[3] The diimidoester reagents, as a class, possess several advantages over other cross-linking reagents, the most important of which is their high level of solubility in aqueous media, thus avoiding organic solvents which might disrupt lipoprotein structure. In addition, these reagents are specific for primary amino groups and react under mild conditions of solvent, pH, and ionic strength. Lysine side chains react in their unprotonated form, so that reaction can proceed readily at pH values of 8.5 or higher. The result of reaction with the imidoester function is an amidine linkage; the amidine group possesses similar charge characteristics to the original amine, which appears to be the explanation for the observation that amidination leads to only minimal perturbations in protein structure.

This class of reagents includes a number of compounds which vary mainly in the length of the spacer alkyl chain which separates the imidoester functions (see the table). This results in a choice of cross-linking distances ranging from about 5 to as great as 12 Å, or more. Since the reactive protein functionality is the amino group of lysine residues, flexibility of this amino acid side chain creates an effective span which is even greater than that implied by the length of the cross-linking reagent per se.

One of the longest of these reagents, dimethyl suberimidate, has been used most extensively because of its ability to cross-link widely separated proteins and thereby to maximize the probability of covalently linking all protein components on a lipoprotein particle. Analysis of nearest-neigh-

---

[2] F. Wold, this series, Vol. 25, p. 623.
[3] Abbreviations: DMS, dimethyl suberimidate; SDS, sodium dodecyl sulfate; DFDNB, difluorodinitrobenzene; HDL, high-density lipoprotein; LDL, low-density lipoprotein; DMPC, dimyristoyl phosphatidylcholine.

CROSS-LINKING REAGENTS

REAGENT	SPAN

I. Diimidoesters

$$Cl^-H_2{}^+N=C(OCH_3)-(CH_2)_n-C(OCH_3)=N^+H_2Cl^-$$

n = 6 Dimethyl suberimidate (DMS)	11 Å
n = 4 Dimethyl adipimidate (DMA)	8.6 Å
n = 2 Dimethyl malonimidate (DMM)	6.2 Å

II. Bis(sulfosuccinimidyl)suberate

[structure: $^+Na^-O_3S$ and $SO_3{}^-Na^+$ substituted succinimidyl groups linked via $N-O-C(=O)-(CH_2)_6-C(=O)-O-N$]

~11 Å

III. 1,5-Difluoro-2,4-dinitrobenzene (DFDNB)

[structure: benzene ring with F, NO_2, F, NO_2 substituents]

~3 Å

IV. N-Hydroxysuccinimidyl-4-azidobenzoate

[structure: N_3–phenyl–$C(=O)-O-N$(succinimidyl)]

~6-7 Å

bor proteins, on the other hand, requires a reagent with a short span. Although short span diimidoester reagents can be prepared to suit this purpose (e.g., dimethyl malonimidate) it has been this author's experience that these reagents have an increased rate of hydrolysis, rendering the cross-linking reaction highly inefficient. It should be noted that some members of this class of reagents are commercially available with internal disulfide bonds, allowing convenient chemical cleavage of the cross-links.

Difluorodinitrobenzene (DFDNB). The reagent of choice for analysis of nearest-neighbor proteins in lipoproteins has been DFDNB (see the table). This reagent reacts with primary amino groups at elevated pH, but may also react with tyrosine side chains. The two fluorine groups are not equal in reactivity, with monosubstitution being favored over disubstitution. Also, the reagent is sparingly soluble in water, adding to the incompleteness of the cross-linking reaction. This is not a disadvantage, however, since limited reaction assures that only proteins which are in close proximity will be covalently linked. An additional advantage of this reagent is that the products possess characteristic visible and ultraviolet spectra, facilitating quantitative assessment of the reaction.[4] This reagent has a maximum bridge span of 3–5 Å.

Other Reagents. In recent years a number of new cross-linking reagents have been reported and many are commercially available. One particularly attractive alternative to the diimidoester reagents is bis(sulfosuccinimidyl)suberate (see the table), which was recently described by Staros.[5] This reagent is reported to be less susceptible to hydrolysis than the diimidoesters, resulting in more efficient cross-linking. An additional advantage is that the reaction can be performed in aqueous buffers at neutral pH. Both this reagent and a disulfide-containing analog are commercially available.

In addition to these, there have appeared in recent years a number of heterobifunctional reagents, many containing photoactivatable groups, which have been designed for special purposes. One example is the use of N-hydroxysuccinimidyl-4-azidobenzoate (see the table) for the photoaffinity labeling of receptors.[6] Such reagents are first reacted (in the dark) with a ligand molecule, mixed with a binding component, and photoactivated to induce covalent linkage via a nitrene radical. While this approach has not yet been applied to lipoproteins, the utility of this approach for the study of lipoprotein receptors is evident.

Performing the Cross-Linking Reaction

Reaction with Dimethyl Suberimidate. Dimethyl suberimidate (DMS) is available from a number of commercial suppliers. Alternatively, it can be readily prepared by methylation of suberonitrile by the procedure of McElvain and Schroeder.[7]

Reaction of this reagent with lipoproteins or apolipoproteins is performed as follows. A solution of DMS in 1 M triethanolamine–HCl, pH 9.7, is freshly prepared and 1 volume is added to 10 volumes of the apoprotein or lipoprotein solution (protein concentration between 0.1 and

---

[4] H. Zahn and J. Meierhofer, *Makromol. Chem.* **26**, (1958).
[5] J. V. Staros, *Biochemistry* **21**, 3950 (1982).
[6] G. L. Johnson, M. I. MacAndrew, Jr., and P. F. Pilch, *Proc. Natl. Acad. Sci. U.S.A.* **78**, 875 (1981).
[7] S. M. McElvain and J. P. Schroeder, *J. Am. Chem. Soc.* **71**, 40 (1949).

1.0 mg/ml). In studies of apolipoprotein self-association it was found that the protein should be renatured by dissolution in 3 M guanidine hydrochloride and the denaturant slowly removed by dialysis.

The reaction rate was found to be relatively independent of temperature over the range of 4 to 30°. Buffers containing compounds with reactive groups, such a thiols or primary amino groups, must be avoided. The reaction can, however, be performed over a wide range of ionic strengths and in the presence of protein denaturants like guanidine hydrochloride. The reaction with lysine side chains is optimized at elevated pH values, due to deprotonation of this group, but hydrolysis of the reagent is also increased. Practical pH limits for the reaction are between 8.5 and 9.5, since it has been shown that cross-linking is highly inefficient at pH values near or below 8.

The reaction can be monitored as a function of time by adding either ammonium ions (to react with the reagent) or acetic acid (to lower the pH) to terminate the cross-linking reaction. The samples are then subjected to SDS–gel electrophoresis and compared with unmodified material. Time-course studies of the cross-linking reaction with apolipoprotein A-I showed that the reaction was complete within 90 min.[8]

It should be emphasized that a control for the effectiveness of the cross-linking reaction should be performed during each experiment. This is most easily accomplished by performing cross-linking of a protein with a known quaternary structure, such as the tetrameric protein aldolase. As with the unknown samples, cross-linking is monitored by performing SDS gel electrophoresis and quantitating the extent of oligomer formation by scanning stained gels.

Reaction with DFDNB. For cross-linking with DFDNB the lipoprotein is prepared at a concentration of 1–2 mg/ml in 0.025 M NaHCO$_3$, pH 8.8. To this solution is added DFDNB dissolved in ethanol to yield a final reagent concentration of 20 μg/ml. The sample is incubated at room temperature for 6 hr and the reaction stopped by adding SDS to a concentration of 1%.

Reaction with Bis(sulfosuccinimidyl)suberate. The reagent is freshly prepared at a concentration of 10 mM in a solution of 50 mM sodium phosphate, pH 7.4. This solution is added to the protein, which is dissolved in the same buffer, to achieve a final reagent concentration of 1–2 mM. After incubating at room temperature for 6 hr, the reaction is stopped by the addition of 1/6 volume 50 mM ethanolamine in 50 mM sodium phosphate, pH 7.4.

[8] J. B. Swaney and K. J. O'Brien, *Biochem. Biophys. Res. Commun.* **71**, 636 (1976).

Analysis of Results

Analysis of the results of chemical cross-linking experiments has generally been accomplished by performing SDS–gel electrophoresis of the cross-linked samples in order to resolve the oligomeric species produced. Although a wide variety of procedures have been used for this purpose, consideration must be given to the range of molecular weights which can result from the cross-linking reaction. For this reason, we have utilized an acrylamide gradient gel with acrylamide concentrations ranging from 3 to 27%. Cross-linked oligomers tend to produce more diffuse bands than unmodified proteins, perhaps due to heterogeneity of cross-linking, and the gradient gel technique helps to minimize this effect, as well as to extend the size range of the species which can be resolved.

An important aspect of the analysis is the identification of the oligomeric species produced on the basis of their molecular weights. Since the presence of either intrachain or interchain cross-links (e.g., disulfides) is known to affect mobility on SDS gels, an appropriate selection of molecular weight standards is imperative. Steele and Nielsen[9] have shown that proteins cross-linked with diethyl pyrocarbonate yield linear relationships between molecular weight and electrophoretic mobility, although the observed mobilities were somewhat greater than for uncross-linked proteins. Consequently, it is evident that cross-linked proteins should be employed as molecular weight standards.

A convenient means for generating a series of molecular weight standards is to cross-link a pure protein at a high concentration where interchain cross-links become prominent. For example, serum albumin cross-linked at a concentration of 50 mg/ml yields a series of oligomers ranging from monomer (65,000) through pentamer (325,000). This approach can be used with any highly soluble monomeric protein to obtain a series of oligomers spanning any desired range of molecular sizes.

It should be noted that there has been one report of an artifactual band on SDS gels when a control sample, containing no protein, was applied to an SDS gel.[10] This Coomassie blue staining material migrated in the position of a 40,000 MW protein. However, such an artifactual band has never been observed in the author's laboratory, and may be a contaminant of the dimethyl suberimidate reagent which was used.

An alternative, but more laborious, means of analyzing the molecular sizes of the cross-linked oligomers is by performing gel filtration in the presence of denaturing agents. When cross-linked oligomers of apoA-I were chromatographed on Sepharose Cl-4B in the presence of 6 M guani-

---

[9] J. C. Steele, Jr. and T. B. Nielsen, *Anal. Biochem.* **84,** 218 (1978).
[10] W. F. Burke and H. C. Reeves, *Anal. Biochem.* **63,** 267 (1975).

dine hydrochloride the molecular weights obtained were very similar to those measured by SDS gel electrophoresis.[11] In the case where a relatively monodisperse product is obtained in the cross-linking reaction, as with the cross-linking of HDL, analytical ultracentrifugation could be employed to establish the size of the product.

Quantitation of Extent of Modification

The extent of reaction with cross-linking reagents can be established by one of several different procedures, depending upon the nature of the reagent. The number of amino acid side chains modified can often be quantitated by amino acid analysis. In the case of DFDNB, the lysine and tyrosine derivatives are stable to normal acid hydrolysis and can be quantitated by a normal amino acid analyzer short column run.[12] In the case of the diimidoester reagents, modification can be quantitated by the loss of free lysine residues, although the small conversion of modified lysine to free lysine requires correction. Reaction of apoA-I and apoA-II with dimethyl suberimidate resulted in nearly quantitative modification of the lysine residues.[11]

Extent of modification cannot be equated with cross-link formation, however, since reaction at only one end of the reagent may occur. Determination of the number of cross-links requires knowledge of the number of reagent molecules incorporated into the protein. This can be determined through the use of a radioactive cross-linking agent, or spectroscopically, if the reagent molecule possesses a characteristic spectrum.

Interparticle Cross-Linking

One concern in cross-linking studies is that proteins located on separate particles might cross-link due to random collisions. In the case of stable quaternary structures, this concern can be addressed by cross-linking at various protein concentrations. Interparticle cross-linking is highly concentration dependent and should be evident only at higher protein concentrations, whereas intraparticle cross-linking should be relatively concentration independent. This is not the situation, however, with studies of apolipoprotein self-association, since the proportion of oligomeric species is, itself, concentration dependent. As a general guideline, we have found that interparticle cross-linking is not significant at protein concentrations below 1 mg/ml.

An additional control for this possibility is to add a marker protein at

---

[11] J. B. Swaney and K. C. O'Brien, *J. Biol. Chem.* **253,** 7069 (1978).
[12] L. C. Jen, Ph.D. thesis, University of Illinois, Urbana, Illinois, 1967.

an equivalent concentration to the protein of interest. If this protein becomes cross-linked to the protein of interest, then interparticle cross-linking must be suspected. Of course, a marker protein which is known not to bind to the sample protein being investigated must be selected. This approach has been described in studies of apolipoprotein self-association.[13]

Applications

High-Density Lipoprotein

The first application of the cross-linking methodology to analysis of lipoprotein structure was by Grow and Fried,[14] who used DFDNB to determine the spatial relationships of A-I and A-II on human HDL. Reaction of this reagent with HDL yielded a 46,000 MW oligomer (Fig. 1C). This species was tentatively identified as an A-I/A-II dimer, since this molecular weight corresponds to the sum of A-I (28,000) and A-II (17,000). Because A-II is a disulfide-bonded dimer of identical monomers, a disulfide reductant was included in a separate sample (Fig. 1D). The molecular weight of the cross-linked species was reduced by approximately 8500, indicating that the product was an A-I/A-II dimer in which the A-I was cross-linked to only one of the two A-II chains.

These cross-linking data allow several conclusions. First, some HDL particles must contain both A-I and A-II. Second, the A-I and A-II proteins lie in close proximity on the surface of the HDL particle. Third, either the topographical relationships or the amino group orientations are such that homologous cross-linking (A-I/A-I or A-II/A-II) occurs with a low frequency. Although this suggests that on the surface of the intact HDL there may be particular affinity of A-I for A-II, recent data with discoidal recombinant particles suggest that under conditions of limited surface area there is little evidence for such specific interactions.[13]

More recent studies have utilized longer cross-linking agents to elucidate protein stoichiometry of the HDL particle. Initial studies were performed on the HDL from dog, chicken, and cow, which possess only the A-I protein, since the absence of multiple protein types simplifies the interpretation of cross-linking results.[15] When cross-linked dog or chicken HDL were compared against cross-linked apoA-I, as a molecular weight reference, each showed a major band corresponding to the A-I trimer;

[13] J. B. Swaney and E. Palmieri, *Biochim. Biophys. Acta* **792**, 164 (1984).
[14] T. E. Grow and M. Fried, *Biochem. Biophys. Res. Commun.* **66**, 352 (1975).
[15] J. B. Swaney, *Biochim. Biophys. Acta* **617**, 489 (1980).

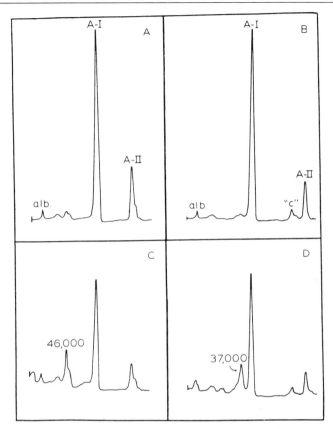

FIG. 1. SDS–polyacrylamide gel electrophoresis of SDS solubilized apoproteins of HDL. (A) Control, without reducing agent. (B) Control, with 1% 2-mercaptoethanol. (C) DFDNB-cross-linked HDL, no reducing agent. (D) DFDNB-cross-linked HDL with 1% 2-mercaptoethanol. Reproduced with permission of T. Grow and M. Fried, *Biochem. Biophys. Res. Commun.* **66**, 354 (1975).

from this it was concluded that the HDL of these species contain three copies of the A-I protein per HDL particle. These results were in good agreement with compositional data. Bovine HDL, on the other hand, yielded a cross-linked oligomer which was twice as large as that produced by the HDL of either dog or chicken.

Studies with human HDL_2 and HDL_3 have been complicated by the presence of substantial amounts of A-II and C proteins on this particle, in addition to A-I, and by the known heterogeneity of these fractions. However, HDL_3 and HDL_2 each yields a single, diffuse, cross-linked oligomeric band with estimated molecular weights of 90,000 and 120,000, re-

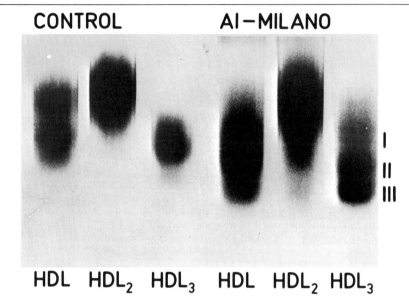

FIG. 2. SDS–gel electrophoretic separations of cross-linked lipoproteins from control and A-I$_{Milano}$ subjects. Purified lipoproteins (30 μg of lipoprotein–protein) were cross-linked with DMS and applied to 5% acrylamide SDS gels. A similar mobility is shown for HDL$_2$ particles, whereas three different subpopulations of HDL$_3$ (I, II, III) are noted in the A-I$_{Milano}$ subjects. Reproduced with permission of G. Franceschini, T. Frosi, C. Manzoni, G. Gianfranceschi, and C. Sirtori, *J. Biol. Chem.* **257,** 9926 (1982).

spectively.[15] These results are roughly in agreement with compositional data.

An exciting application of the cross-linking methodology to the analysis of abnormal HDL structures was reported by Franceschini *et al.*[16] These workers have identified patients who possess an HDL of abnormal morphology and compositional properties, and with an A-I protein (A-I$_{Milano}$) which contains a cysteine residue; this thiol group allows this protein to enter into disulfide linkages with itself and with the A-II protein.

When the HDL$_3$ fraction from such individuals was cross-linked with DMS, three oligomeric bands were produced (Fig. 2). The first of these (I) has a molecular weight comparable to that found with HDL$_3$ from control subjects (90,000). In addition, however, two major bands of molecular weight 70,000 and 58,000 are observed. This increased degree of heteroge-

[16] G. Franceschini, T. G. Frosi, C. Manzoni, G. Gianfranceschi, and C. Sirtori, *J. Biol. Chem.* **257,** 9926 (1982).

neity was supported by the results of gel filtration chromatography. The correlation of these unusual subcomponents of HDL_3 with a reduced incidence of coronary heart disease raises interesting questions with respect to their functional role in tissue cholesterol homeostasis.

Other recent studies have taken the heterogeneity of HDL_2 and HDL_3 into account by performing a separation of subpopulations by hydroxyapatite chromatography prior to cross-linking.[17] In both cases subpopulations were isolated which contained various combinations of the A-I, A-II, and C proteins; in some cases albumin contamination caused unnecessary ambiguity in interpreting the gel patterns. These subpopulations were cross-linked with either dithiobispropionimidate dihydrochloride or DFDNB.

From their data, Yachida and Minari[17] concluded that 75% of the HDL_3 material consisted of particles with a stoichiometry of $(A-I)_2$-$(A-II)_2$. Of the remaining particles, 10% possessed a stoichiometry of $(A-I)_3(C)_2$ and the rest consisted of $(A-I)_4$ or $(A-I)_3$.

On the other hand, 70% of the HDL_2 was found to possess particles containing $(A-I)_4$, with the remaining material possessing an equal proportion of $(A-I)_4(C)_2$ and $(A-I)_2(A-II)_2$ stoichiometries. These results are in accord with previous results, both in terms of the heterogeneity of molecular sizes represented and in the molecular sizes of the major components of HDL_2 and HDL_3.

The fact that $(A-I)_2(A-II)_2$ particles were found in both density fractions, but that $(A-I)_2(A-II)$ was found in neither is consistent with the results of limited DFDNB cross-linking observed by Grow and Fried.[14] The observation that two copies each of A-I and A-II reside on a major component fraction of HDL_3 also supports the early prediction of Scanu.[18] However, it does not support the conclusion of Friedberg and Reynolds[19] that HDL_3 contains a stoichiometry of $(A-I)_2(A-II)$.

Low-Density Lipoproteins

A limited amount of cross-linking data is also available for LDL.[20] Several studies have reported that apoB in LDL has a molecular weight of 250,000 and that LDL contains two B chains per lipoprotein particle. Ikai and Yanagita cross-linked human LDL with dimethyl suberimidate in an attempt to obtain supportive evidence for this conclusion. However, these workers found that this bifunctional reagent did not induce signifi-

---

[17] Y. Yachida and O. Minari, *J. Biochem. (Tokyo)* **94**, 459 (1983).
[18] A. M. Scanu, *Biochim. Biophys. Acta* **265**, 471 (1972).
[19] S. J. Friedberg and J. A. Reynolds, *J. Biol. Chem.* **251**, 4005 (1976).
[20] A. Ikai and Y. Yanagita, *J. Biochem. (Tokyo)* **88**, 1359 (1980).

cant amounts of dimer formation. They conclude that the hydrophilic domains of apoB are segregated from each other, resulting in a separation between the two B chains which is too great to allow cross-linking. Recent studies which have reported a molecular weight of 400,000–500,000 for the B-100 form of this protein suggest an alternative explanation; namely, that LDL contains only a single apoB per particle, in accordance with its chemical composition, and that this is the reason that no dimer forms of apoB are seen by chemical cross-linking.

Recombinant Lipoproteins

Another area of usefulness for the cross-linking methodology is in the characterization of recombinant lipoproteins prepared with single apolipoproteins and pure phospholipids. The apolipoproteins in solution have been shown to yield characteristic cross-linking profiles which result from various degrees of self-association.[11] However, when multilamellar dispersions of particular phospholipids are added to the proteins and incubated at the phase transition temperature of the phospholipid, cross-linking shows dramatic changes in the pattern of oligomers produced. This was first shown for the formation of recombinants between apoA-I and dimyristoyl phosphatidylcholine.[21] Figure 3 shows that cross-linked apoA-I yields a pattern with bands ranging in molecular weight from 28,000 to 140,000 when cross-linked with DMS. When this protein is combined with dimyristoylphosphatidylcholine at a lipid : protein ratio of 5 : 1, two major bands are observed and these correspond to dimer and trimer forms of A-I. Figure 3 shows that similar results are obtained with DMS, bis(sulfosuccinimidyl)suberate, and DFDNB, but that significantly less trimer form is observed with the latter reagent. It was concluded that synthetic recombinants containing phosphatidylcholines can be readily formed which contain two or three molecules of A-I, but that there is little tendency to form particles containing less than two or more than three A-I chains. Similar results have been obtained with DMPC and pig A-I.[22] It has been found, however, that when apoA-I is recombined with bovine brain sphingomyelin, larger structures are formed which contain either three of four A-I chains per particle.[23]

Little work has been performed on cross-linking of recombinants formed using apoA-II. One report indicates that association of A-II with egg yolk lecithin yields mainly 51,000 MW (trimer) bands upon cross-linking,[21] with a small amount of tetramer formation. However, studies by

[21] J. B. Swaney, *J. Biol. Chem.* **255,** 877 (1980).
[22] A. Mougin-Schutz and A. Girard-Globa, *Biochimie* **65,** 485 (1983).
[23] J. B. Swaney, *J. Biol. Chem.* **258,** 1254 (1983).

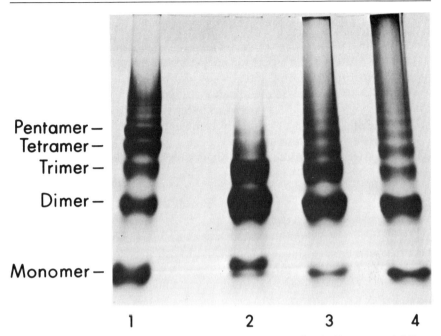

FIG. 3. SDS slab gel electrophoresis (3–27% acrylamide gradient) of human apoA-I (lane 1) and a DMPC : A-I recombinant complex (5 : 1 w/w) (lanes 2–4). Samples were cross-linked with DMS (lanes 1 and 2), BS (lane 3), or DFDNB (lane 4).

Massey et al. concluded that A-II forms several complexes with DMPC which range in stoichiometry from 5 to 8 mol of A-II per complex, based upon molecular size and composition.[24] These differences may be attributable to the different lipid used in each study, but further investigation into the quaternary structure of these complexes is warranted.

Additional studies of A-II/phospholipid recombinants were performed by Stoffel and Preissner.[25] Although cross-linking was not used to interpret quaternary structure of the particles, it was employed to obtain information on the tertiary structure of this protein. Based on cross-link formation between specific lysine pairs it was found that the two monomers were not oriented as to allow interchain cross-link formation. However, based on interchain cross-link formation a model for A-II folding has been proposed.[25]

More recently, studies have been performed on recombinants pre-

[24] J. B. Massey, M. F. Rohde, W. B. Van Winkle, A. M. Gotto, Jr., and H. J. Pownall, Biochemistry **20**, 1569 (1981).
[25] W. Stoffel and K. Preissner, Hoppe Seyler's Physiol. Chem. **360**, 691 (1979).

pared by incubating apoA-I with subcellular membranes from rat liver.[26] Under conditions of elevated pH the microsomal membrane fraction can serve as a lipid donor to form an A-I-containing recombinant which possesses two protein chains per particle, as shown by cross-linking with dimethyl suberimidate. Also in this study a novel use for the cross-linking reaction in analyzing the time-course for association of the A-I with membrane lipids was reported. This approach took advantage of the change in protein quaternary structure which accompanies lipid association, so that by cross-linking after various periods of incubation between the A-I and membranes, the formation of the dimeric A-I species was taken as a measure of the extent of reaction.

Future Directions

There are a substantial number of areas in which chemical cross-linking can be expected to add useful insights. One of these is in elucidating the structure of the larger lipoprotein classes. For example, cross-linking of very low-density lipoprotein might indicate the topographical proximity of the numerous protein components to each other; the same would be true of the chylomicrons.

Another question which can be addressed is whether the family concept of apolipoproteins can be substantiated in lipoproteins from other animal species. For example, since rat HDL contains a more complex protein complement than HDL from other animal species, it is of interest to know whether each of the component proteins (A-I, A-IV, E, C-I, C-II, and C-III) resides on a single class of particles or whether the rat HDL density class contains two or more classes of lipoproteins which differ in their protein composition.

A third area where cross-linking should prove useful is in aiding in the identification and isolation of specific lipoprotein receptors. Heterobifunctional reagents have proved extremely useful in identifying various hormonal receptors and should prove equally useful for lipoprotein receptors.

[26] J. Nunez and J. B. Swaney, *J. Biol. Chem.* **259,** 9141–9148 (1984).

[36] Synthetic Peptide Analogs of Apolipoproteins

By G. M. Anantharamaiah

Introduction

The amphipathic helix hypothesis for apolipoproteins, in its original form, was first proposed by Segrest *et al.*[1] This model has been generally presumed to be the structural form of the lipid-associating domains of the exchangeable apolipoproteins A, C, and E from the plasma lipoproteins. The amphipathic helix model is defined by an arrangement of amino acid residues which result in helical domains containing opposing polar and nonpolar (hydrophobic) faces. A specific distribution of charged residues was proposed, with positively charged amino acids occurring at the polar and nonpolar interface and the negatively charged amino acids occurring at the center of the polar face. The alkyl side chains from the positively charged Lys or Arg residues at the polar and nonpolar interface are believed to contribute to the hydrophobicity of the hydrophobic face.[2] This arrangement of the charged residues also seems to form topographically close complimentary ion pairs, the number of which may prove to be significant. Further, this model allows for ionic interactions between the positively charged side chains and the phosphate group of the phospholipid as well as negatively charged residues and positive substituents of the phospholipid. Such interactions may also play a role in initiating or contributing to the stability of the protein–lipid complex. Since this theory has been proposed, many smaller peptide analogs of apolipoproteins have been synthesized and studied (Table I). Recently a model for the lipid-associating domain of apolipoprotein B in the form of a β-strand structure has also been synthesized and studied.[3]

An ideal synthetic peptide analog of apolipoproteins should mimic many properties of apolipoproteins and compete for binding with the apolipoproteins present in lipoproteins. Among the properties of apolipoprotein A-I, the major protein component of HDL, lecithin: cholesterol acyltransferase (LCAT) activation, which is involved in the reverse-cholesterol transport, is important.

Although the general features of this theory have been supported by some experimental evidence, the proof of such a theory comes from

---

[1] J. P. Segrest, R. L. Jackson, J. D. Morrisett, and A. M. Gotto, *FEBS Lett.* **38**, 247 (1973).
[2] J. P. Segrest and R. J. Feldmann, *Biopolymers* **16**, 2053 (1977).
[3] D. Osterman, R. Mova, F. J. Kezdy, E. T. Kaiser, and S. C. Meredith, *J. Am. Chem. Soc.* **106**, 6845 (1984).

TABLE I
PEPTIDE ANALOGS OF APOLIPOPROTEINS

Peptide sequence (abbreviation)	LCAT activation (%)	Fluorescence max with DMPC	α-Helicity (%) Buffer	α-Helicity (%) DMPC complex	α-Helicity (%) TFE	Reference
DWLKAFYDKVAEKLKEAF (18A)	18	352	15	30	39	6
SSSDWLKAFYDKAEKLKEAFSS (18As)	Not studied		9	46		17
AADWLKAFYDKVAEKLKEAFAAA (18Aa)	Not studied		12	20	33	17
DWLKAFYDKAEKLKEAF (desVal[10]18A)	0	362	11	12	17	6
KWLDAFYKDVAKELEKAF (reverse 18A)	0	361	49	53	41	6
DWLKAFYDKVAEKLKEAFPDWLKAFYDKVAEKLKEAF (18A-Pro-18A)	140	358				6
PKLEELKEKLKELLEKLKEKLA (Ap)			50			3
VSSLKEYWSSLKESFS (LAP-16)	18	350				4
VSSLLSSLLSSLKEYWSSLKESES (LAP-20)	0	334				4
VSSLLSSLLSSLKEYWSSLKESES (LAP-24)	Activator	338				4
SSLEKEYWSSLKESFS (Co-LAP)	Activator	356	Nil		Nil	5
CH$_3$(CH$_2$)$_7$-CO-SSLKEYWSSLKESFS (C$_8$-LAP)	0	343	Nil		35	5
CH$_3$(CH$_2$)$_{17}$-CO-SSLKEYWSSLKESFS (C$_{18}$-LAP)	20	345				5
CH$_3$(CH$_2$)$_3$-CO-SSLKEYWSSLKESFS (C$_4$-LAP)	50		Nil		10	5
CH$_3$(CH$_2$)$_{11}$-CO-SSLKEYWSSLKESFS (C$_{12}$-LAP)			9		35	5
CH$_3$(CH$_2$)$_{15}$-CO-SSLKEYWSSLKESES (C$_{16}$-LAP)			23		36	5

peptide synthesis. Due to a vast improvement in the techniques of solid-phase peptide synthesis, and purification techniques such as high-performance liquid chromatography (HPLC), it has been possible to provide experimental proof for the theory of the amphipathic helix as a mechanism to explain the association phenomena between lipids and apolipoproteins that occurs in plasma lipoproteins.

Design of Peptides

The peptide synthesized by Yokoyama et al.,[4] peptide Ap in Table I, was designed for optimizing the amphipathic helix. This 22-peptide possesses Lys and Glu as the charged amino acids and Leu as the hydrophobic amino acid. The peptide was designed to be capable of forming an α-helix with a hydrophobic face covering about half of the surface and the hydrophilic face covering the remaining half. This peptide was also designed to have minimal homology to naturally occurring amphipathic α-helical domains of apolipoproteins (Fig. 1). Sparrow et al.[5] designed three peptides: LAP-16, LAP-20, and LAP-24 (Table I). Here, one can note that although there is a hydrophobic region and a polar region (Fig. 1), importance was not given to the position of charges. Recently a variation of LAP-16 has been reported from the same laboratory in which saturated fatty acyl chains of various lengths have been coupled to the N-terminus of LAP-16.[6] These were designed to study the quantitative effect of hydrophobicity on lipid-protein interaction.

The peptides designed from our laboratory[7] were directed toward proving not only the importance of opposing polar and hydrophobic regions in plasma apolipoproteins, but also a particular arrangement of polar amino acids in their lipid-associating ability. We believe this is important for the stability of the protein-lipid complex.

Peptides have been designed with the following features: (1) an idealized amphipathic peptide based on the general features of the amphipathic helix; (2) various positions for the charged groups (for instance reversing the positively and negatively charged amino acid position) to assess their importance in lipid binding; (3) deletion of an amino acid at the center of

---

[4] S. Yokoyama, D. Fukushima, J. P. Kupferberg, F. J. Kezdy, and E. T. Kaiser, *J. Biol. Chem.* **255**, 7333 (1980).
[5] J. T. Sparrow, C. R. Ferenz, A. M. Gotto, Jr., and H. J. Pownall, in "Peptides" (D. Rich and E. Gros, eds.), p. 253. Pierce Chem. Co., Rockford, Ill., 1981.
[6] G. Ponsin, K. Strong, A. M. Gotto, J. T. Sparrow, and H. J. Pawnall, *Biochemistry* **23**, 5337 (1984).
[7] G. M. Anantharamaiah, J. L. Jones, C. G. Brouillette, C. F. Schmidt, B. H. Chung, T. A. Hughes, A. S. Bhown, and J. P. Segrest, *J. Biol. Chem.*, **260**, 10248 (1985).

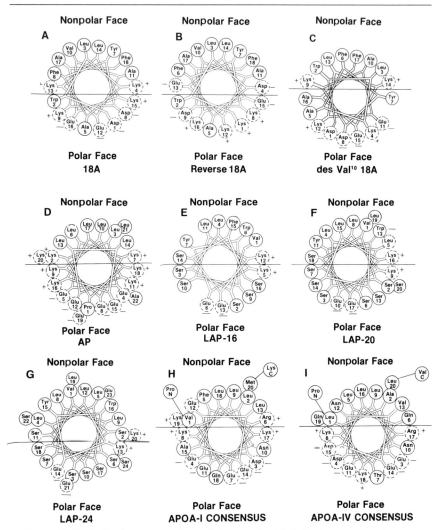

FIG. 1. Helical wheel representations of the amphipathic helical structures of synthetic peptides.

the helix. Such a deletion causes the rotation of polar and nonpolar interface by 100°; and (4) covalent linkage of multiple helical domains. The latter modification would be predicted to better mimic apolipoprotein lipid association due to the cooperativity of amphipathic helical domains. Peptides were also designed to represent the consensus domains of apoA-I or

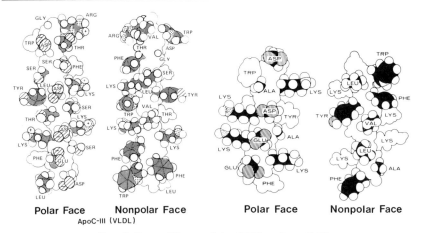

FIG. 2. Space filling models of 18A and apoC-III.

apoA-IV which are shown in Fig. 1.[8] Key features of the apoA-I consensus domains are the strong localization of positively charged residues at the polar-nonpolar interface, and the central polar location of four of the five negatively charged residues. The key difference in the apoA-IV consensus domain, compared to A-I, is the positioning of one of the positive charges in the center of the polar face.

Figure 2 shows a comparison of the space filling models of 18A and apolipoprotein C-III. Peptide 18A is designed to have a polar and a nonpolar face of approximately equal surface areas. Positively charged Lys are present at the polar-nonpolar interface. The side chains of Lys may thus contribute to the hydrophobicity of the hydrophobic face and also possess charges. Negatively charged Asp and Glu are present at the center of the nonpolar face. This is thus an idealized amphipathic helical peptide and is expected to mimic the properties of apoA-I. Peptide 18A is also a reasonable analog of the consensus domain of apoA-I. Peptide reverse 18A is 18A with its charged residue positions reversed. As shown in Fig. 1B, this results in negative charges at the polar and nonpolar interface. This was designed to investigate the importance of the position of charges for lipid association.

A helical wheel representation of the peptide desVal^{10}18A, which is the result of the deletion of Val from 18A, is shown in Fig. 1C. It is clear from the figure that this is less hydrophobic than 18A. Such a deletion causes not only the rotation of the polar and nonpolar interface of the

[8] J. L. Gorden, personal communication.

helix by 100°, but also reduces the hydrophobicity of the hydrophobic face. This peptide is comparable to the consensus domain of apoA-IV. Peptide 18A-Pro-18A has two 18A peptides bridged by a proline. This was designed to test the effect of multiple amphipathic helical domains.

Peptide Synthesis

Most of the peptides reported in the literature have been synthesized by the solid phase method of peptide synthesis.[9] There are many reviews of solid phase peptide synthesis, including a recent one by Barany and Merrifield.[10] In this method the C-terminal N-protected amino acid of the sequence to be synthesized is attached to an insoluble polymer (polystyrene cross-linked with divinyl benzene is the most commonly used polymer). Then the N-protection is removed and the next amino acid residue is introduced using the corresponding N-protected amino acid in the presence of a condensing agent. Deprotection and condensations are repeated until the desired amino acid residues are incorporated. Since reactions are carried out on an insoluble polymer, excess of reagents and by-products can be removed just by washing with solvents. The schedule that is commonly used in most of the laboratories is shown in Table II. Due to the repetitive nature of the steps involved and since all the reactions are carried out in one reaction vessel, the method is amenable to automation. The release of peptide from the polymer however is done by acidolytic cleavage (HBr-TFA or HF) or base treatment (NaOH) or ammonolysis if the C-terminal amide is required. Under these conditions many side reactions have been reported.[10] A modification of the HF procedure which reduces the chances of side reactions[11] is as follows. The reactions are carried out in a special all fluorocarbon apparatus. Peptide resin (500 mg) is mixed with dimethyl sulfide/p-cresol/p-thiocresol (0.25 ml/0.65 ml/0.55 ml). Anhydrous HF is condensed over the resin at $-78°$. The temperature of the reaction vessel is raised to $-10°$ and stirred at this temperature for 30 min. It is then stirred for 30 min at $0°$ and HF is evaporated at $0°$ under reduced pressure. Ether is then added and liberated peptide along with the resin is filtered and washed with ether. The peptide is then separated by stirring the mixture of resin and peptide in aqueous acetic acid (10 to 50%) and lyophilized.

[9] R. B. Merrifield, *J. Am. Chem. Soc.* **86**, 304 (1964).
[10] G. Barany and R. B. Merrifield, *in* "The Peptides" (E. Gross and J. Meienhofer, eds.), Vol. 2, p. 1. Academic Press, New York, 1980.
[11] J. P. Tam, T. N. Wong, M. W. Riemen, F. S. Tjoeng, and R. B. Merrifield, *Tetrahedron Lett.* 4033 (1979).

TABLE II
Washing Schedule

Step	Reagents	Number of cycles	Volume (ml)	Time (min)
Wash	CH$_2$Cl$_2$	2	30	2
Deblock	40% TFA in CH$_2$Cl$_2$ (10% anisole: 1% mercaptoethanol)	1	30	5
	40% TFA in CH$_2$Cl$_2$ (10% anisole: 1% mercaptoethanol)	1	30	25
Wash	CH$_2$Cl$_2$	3	30	2
Neutralize	5% DIEA in CH$_2$Cl$_2$	2	40	5
Wash	CH$_2$Cl$_2$	3	30	2
Couple	Boc amino acid + HOBt (3 equivalents each in 1:1 CH$_2$Cl$_2$:DMF) + DCC (3 equivalents in 1:1 CH$_2$Cl$_2$ DMF)		20	2
				120–360
Wash	CH$_2$Cl$_2$	2	30	2
Wash	t-Butanol (10% isoamyl alcohol)	3	30	2
Wash	CH$_2$Cl$_2$	2	30	2

If benzyl-type protecting groups are used, the hydrogenolysis method can theoretically be used to cleave the peptide from the solid support while simultaneously removing the protecting groups. Since this procedure utilizes an insoluble catalyst and is a surface phenomenon, it cannot be used for the solid phase peptide synthesis in which the substrate is insoluble. A modification of this procedure, catalytic transfer hydrogenation,[12] is becoming a useful alternative in peptide synthesis and has been used in solid phase peptide synthesis.[13]

Many laboratories have applied this procedure to cleave the protecting groups and also release the peptide from the solid support.[13–15] The solid phase peptide synthesis procedure in which benzyl-based protecting groups are used for the side chain protection and the procedure of transfer hydrogenation, as used by us, is described here.

[12] G. M. Anantharamaiah and K. M. Sivanandaiah, *J. Chem. Soc. Perkin I* **5**, 490 (1977).
[13] S. A. Khan and K. M. Sivarandaiah, *Synthesis* 750 (1978).
[14] K. M. Sivanandaiah and S. Gurusiddaappa, *J. Chem. Res. (S)* 108 (1979).
[15] M. K. Anwar and A. F. Spatola, *Synthesis* **929** (1980).

Benzhydrylamine resin (Beckman Bioproducts, 0.59 mM NH$_2$/g of resin) was used as the solid support. The COOH-terminal amino acid (Boc-Phe) was attached to the solid support through a 4-(oxymethyl)-phenacetyl group.[16] This is a more stable linkage than the conventional benzyl ester linkage, yet the finished peptide can be cleaved by hydrogenation. The following protecting groups were used for the side chains: 2,6-dichlorobenzyl for tyrosine, benzyl ester for aspartic and glutamic acids, 2-chlorocarbobenzoxy for lysine, and formyl for the indole of tryptophan. The schedule used for the synthesis is shown in Table II.

Removal of the α-Boc was achieved at each stage by treatment with 40% trifluoroacetic acid in the presence of 10% anisole and 1% mercaptoethanol for 30 min. Neutralization was achieved by treatment with 5% DIEA in methylene chloride. All the reagents and solvents were distilled prior to their use except for dichloromethane (Burdick and Jackson, "glass-distilled").

Stepwise coupling of each Boc-amino acid was carried out using three equivalents of each Boc-amino acid, dicyclohexylcarbodiimide and 1-hydroxybenzotriazole. The couplings were monitored by the Kaiser test.[17] Whenever the test was positive, recoupling was done by the symmetrical anhydride method as described below: Boc-amino acid (4 equivalents) was dissolved in dichloromethane and cooled to 0°. To this was added two equivalents of DCC and after 10 min the liberated dicyclohexylurea was filtered off and the solvent evaporated. The residue was taken in DMF (15 ml/g of resin), mixed with the resin, and the reaction was done for 30 min.

Release of peptides from the solid support was achieved by using transfer hydrogeneration. The peptide resin was mixed with equal amount of palladium acetate and stirred for 1 hr in the presence of DMF (20 ml/g). Formic acid (5% total concentration with respect to DMF) was used as the hydrogen donor. Reactions were carried out at room temperature for 15 to 24 hr. Ether was added to the reaction mixture and filtered. The peptide was separated from the catalyst and resin by stirring the mixture with 50% acetonitrile (0.1% TFA) and filtering. This solvent has been found to be suitable in our case although acetic acid (10–50%) is usually used to extract the released peptide. The filtrate was lyophilized to obtain the crude peptide.

For the removal of the formyl group of Trp, peptide was dissolved in 0.1 M hydroxylamine hydrochloride solution (pH 9.5, approximately 100 ml/100 mg peptide) and the pH was adjusted to 9.5 and stirred overnight.

---

[16] A. R. Michel, S. B. H. Kent, and R. B. Merrifield, *J. Org. Chem.* **43**, 2845 (1978).
[17] E. Kaiser, R. L. Colescott, C. D. Bosinger, and P. I. Cook, *Anal. Biochem.* **23**, 595 (1970).

After readjusting the pH to 7.0 the solution was lyophilized. In many cases peptide separated out as a solid which was isolated.

Peptides 18A, desVal^{10}18A, and reverse 18A were purified by high-performance liquid chromatography (HPLC) on C_{18} silica (particle size 13–25 μm) using Michel Miller glass columns. Acetonitrile (35 to 40% with 0.1% TFA) was used as the eluent. The individual fractions from the main peak were subjected to thin-layer chromatography and pure fractions were pooled and lyophilized. Peptide 18A-Pro-18A was purified on a preparative HPLC column (particle size 5 μm) using the Perkin Elmer HPLC system. Fractions from the main peak were subjected to analytical HPLC and thin-layer chromatographic analyses and pure fractions were pooled.

In a typical experiment, 1.0 g of 18A-Pro-18A resin was mixed with the same amount of palladium acetate and stirred with DMF (15 ml) for 1 hr. Formic acid (total concentration of 5% with respect to DMF) was added and stirred for 15 hr. Ether was added to the reaction mixture and filtered. The peptide was separated from the resin and the catalyst by dissolving in 50% acetonitrile (0.1% TFA) and filtering. After dilution and lyophilization of the filtrate, removal of the formyl group was achieved as described earlier during which the peptide (250 mg) precipitated out. This was purified on a preparative Perkin Elmer HPLC system (particle size 5 μm, column size 2.5 × 25 cm, monitored at 213 nm), using acetonitrile–water with 0.1% TFA (40 to 60% gradient), as eluent. Fractions of 1.2 ml (flow rate = 1.2 ml/min) were collected. Individual fractions were inspected by analytical HPLC and thin-layer chromatography for the presence of pure material and fractions 63 to 72 were pooled and lyophilized. From 40 mg of crude material, 17 mg of pure 18A-Pro-18A was obtained.

Peptides were subjected to amino acid sequencing analysis on a modified Beckman automated sequencer and the phenylthiohydantoin derivatives of the released amino acids were identified by HPLC. Each peptide was found to have the desired sequence. Amino acid analyses of acid hydrolysates of the peptides gave the expected results.

For determining the complete removal of protecting groups, peptides were dissolved in DMSO and ^{1}H NMR (300 MHz) spectra were run and carefully examined for the absence of benzyl protons characteristic of benzyl-based protecting groups. These signals appear in a region (5.0–5.2 ppm) where there are no signals due to peptide protons. The spectra were compared with the spectra of individual amino acid derivatives with such protecting groups. In all cases there was no detectable peak in this region indicating essentially complete deblocking of the side chain protecting groups, although the possibility that less than 2% of the peptides still have a blocking group cannot be ruled out by this technique.

Peptide–Lipid Interactions

Various physical methods have been used to establish the interaction of synthetic amphipathic peptide analogs with lipids. Some of them are density gradient ultracentrifugation, negative stain electron microscopy (EM), nondenaturing gradient gel electrophoresis (GGE), proton NMR, differential scanning calorimetry (DSC), intrinsic tryptophan fluorescence, and circular dichroism (CD). In many cases the complexes formed by the association of apoA-I with the lipid have been compared with the peptide–lipid complexes by EM, proton NMR, ultracentrifugation, and GGE. Because of the absence of Trp and Tyr in the peptide analog (Ap) synthesized by Yokoyama et al., the intrinsic tryptophan fluorescence technique could not be applied.[4] In this case the binding of peptide Ap or apoA-I to vesicles was quantitated by rapid ultrafiltration and by gel permeation chromatography using Sepharose CL-4B. Typically 4–25 μg peptide in 0.025 M 3-(N-morpholino)propanesulfonic acid in 1.6 M KCl (pH 7.4) was incubated with 2.4×10^{-4} to 2.4×10^{-3} M lecithin in the same buffer at 22–37.5° for 3 hr. The binding of synthetic peptide to vesicles was quantitated by rapid ultrafiltration sampling of the peptide. For binding of apoA-I to vesicles, incubation was done under the same conditions but for 1 hr and 500-μl aliquots were subjected to sepharose CL-4B column (1.5 × 5 cm) chromatography and fractions were treated with fluorescamin. The fluorescence intensity was measured which is proportional to the protein concentration. Results indicated that peptide Ap and apoA-I bind to egg PC in a similar manner. The binding constant $K_d = 1.92 \times 10^{-6}$ M and the binding capacity $N = 1.51 \times 10^{-2}$ peptides/lecithin, and for apoA-I, $K_d = 9.0 \times 10^{-7}$ M and $N = 1.74 \times 10^{-3}$ proteins/lecithin. In the presence of cholesterol, the lipid binding capacity, N, per apoA-I is 1.95×10^{-3} protein/molecule and $N = 1.21 \times 10^{-2}$ peptide/molecule of lecithin (an increase in the number of molecules bound by 113 and 80%, respectively), and K_d, the dissociation constant, increases for peptide binding, whereas for apoA-I binding, the K_d decreases.

Essentially all the physical techniques described earlier have been used in this laboratory to investigate peptide–lipid interactions.

Electron Microscopy

Peptide–DMPC complexes are stained with 2% (w/v) potassium phosphotungstate, pH 5.9, and examined with a Philips EM 400 microscope on carbon-coated, Formvar grids (see Chapter [26], this volume). Grids are prepared for electron microscopy by the following procedure. Approximately 20 μl of the sample is placed on the coated grid for 3–5 sec, followed by blotting excess liquid with bibulous paper. This is followed by

20 μl drop of the stain for 2 min after which the excess stain is floated. The grid can be used immediately or stored for several days at room temperature. Figure 3 shows that at 1:1 weight ratios 18A/DMPC and apoA-I/DMPC complexes are similar-sized protein annulus-bilayer disc structures with approximately 100 Å diameter and 50 Å thick. This has been further confirmed by density gradient ultracentrifugation (see Chapter [7], this volume for details) which showed that the band formed by the peptide–DMPC complex was at the same density range as the band that was formed by the apoA-I/DMPC complex.

Detailed EM studies performed on the complexes formed between DMPC and the four analog peptides, 18A, 18A-Pro-18A, desVal^{10}18A, and reverse 18A, at different ratios have been compared with the results of GGE and CD studies.

Nondenaturing Gradient Gel Electrophoresis (GGE)

This method (described in detail in Chapter [24], this volume) is useful for determining the size and stability of the complexes formed by amphipathic peptides and DMPC. In this method peptides are incubated with DMPC in the presence of a trace amount of 4-nitro-1,2,3-benzotriazole (NBD) as a fluorescence marker and subjected to GGE; 12 × 14 cm gradient slab gel was formed from a 4–25% linear gradient of acrylamide, pH 8.3. A 3% stacking gel, pH 8.3, was poured after resolving gel had

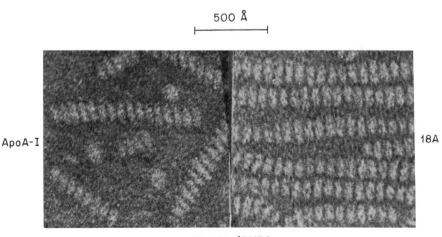

FIG. 3. Negative stain electron microscopy of apoA-I and 18A complexes with DMPC at a 1:1 weight ratio.

polymerized. Electrophoresis was performed at 15° in 0.05 M Tris–0.38 M glycine, pH 8.3, at 150 V constant voltage for 36 hr. It was determined that the pore limiting migration position of DMPC/apoA-I complexes was established by 24 hr. The gel was fixed and stained with 0.2% Coomassie Brilliant blue, 5% methanol, and 7.5% acetic acid in presence of 5% formaldehyde. The protein standards thyroglobulin, ferritin, catalase, lactate dehydrogenase, and bovine serum albumin were used to generate a calibration curve relating Stokes diameter to the electrophoretic R_f, the latter defined as the equilibrium migration distance relative to bovine serum albumin.

The assessment of the size of complexes formed at different ratios of peptide and lipid by EM and GGE is given in Table III. From these results one can conclude that peptides 18A and 18A-Pro-18A form stable particles with DMPC at all ratios studied, whereas reverse 18A and desVal[10]18A form unstable complexes with DMPC.

Intrinsic Tryptophan Fluorescence

Based on the shift in the tryptophan fluorescence maxima with and without lipid, one can determine the increase in α-helicity of the peptide. In the case of LAP peptides, LAP-16 and LAP-20 in solution showed fluorescence maxima near 350 nm.

The addition of DMPC to LAP-20 and LAP-24 has been found to produce a large blue shift in the peptide fluorescence. This shift has been assigned to the association of DMPC with these peptides. When N-acyl-

TABLE III
COMPARISON OF RESULTS OF EM AND GGE STUDIES

Lipid/peptide ratio used[a]		Stokes diameter (Å)			
	Method	18A	18A-Pro-18A	desVal[10]18A	Reverse 18A
1	EM	85 + 9	91 + 20	152 ± 20	298 ± 25
	GGE	97 (75–140)[b]	103 (85–149)	219	225[c]
2.5	EM	122 ± 20	117 ± 13	180 ± 34	233 ± 24
	GGE	127 (89–180)	112 (84–154)	222	225
5	EM	206 ± 30	148 ± 19	224 ± 42	237 ± 26
	GGE	179 (106–222)	140 (88–165)	224	225
7.5	EM	239 ± 36	202 ± 29	244 ± 37	264 ± 45
	GGE	192 (131–222)	152 (114–177)	225	225

[a] EM, electron microscopy; GGE, nondenaturing gradient gel electrophoresis.
[b] Stokes diameter range.
[c] Did not enter the gel.

ated LAP-16 peptides were subjected to tryptophan fluorescence studies in the presence and absence of DMPC, the fluorescence maxima of C_0-, C_4-, and C_8-LAP peptides corresponded to that of tryptophan in aqueous solution.[6] The blue shift of C_{12}- and C_{16}-LAP peptides has been assigned to self-association of peptides. The blue shift and the α-helicity increase with increasing acyl chain length.

Peptide 18A shows the greatest blue shift with a max of 361 nm in solution and 352 nm in the presence of DMPC, while 18A-Pro-18A has the smallest blue shift. The degree of blue shift and relative fluorescence increase with increasing DMPC concentration relative to the peptide 18A in solution. This indicates that with the addition of more lipid there is increase in interaction between the tryptophan and the acyl chain of the fatty acid. These studies further indicate that reverse 18A and desVal¹⁰18A have less avidity as there is no considerable shift in fluorescence maxima with DMPC, which is in accordance with EM and GGE results. A smaller blue shift with peptide 18A-Pro-18A has been attributed to the presence of one tryptophan residue close to the proline which may be involved in formation of a helical hairpin structure.

Circular Dichroism

The CD studies for peptides and proteins have been very useful for determining the secondary structural features. The peptides and peptide–lipid complexes were placed in 10 mM phosphate buffer, 150 mM NaCl, pH 8.0 in 0.1 cm cells and CD spectra were obtained on a JASCO J-500A spectropolarimeter containing a DPN 500 data processor. CD studies of our peptides indicated that 18A is helical in solution and 18A-Pro-18A is highly helical, whereas desVal¹⁰18A is slightly helical and reverse 18A is not helical. An increase in α-helicity with the addition of lipid has been observed by the CD studies of 18A and desVal¹⁰18A, whereas already highly helical 18A-Pro-18A had no further increase in helicity with the addition of lipid. The CD studies clearly show that the rank order percentage α-helicity of the peptide in its DMPC complex is 18A-Pro-18A > 18A > desVal¹⁰18A > reverse 18A, an order inversely proportional to the complex diameter (18A-Pro-18A < 18A < desVal¹⁰18A < reverse 18A). These results suggest a lipid-associating affinity of the four peptides as follows: 18A-Pro-18A > 18A > desVal¹⁰18A > reverse 18A. The tryptophan blue shift data are roughly consistent with this postulated order for lipid-associating affinity.

In case of N-acylated LAP-16 peptides, the α-helicity increased with the increase in acyl chain length indicating the importance of hydrophobicity of the amphipathic peptide for lipid association.[6]

Proton NMR

Proton NMR has been employed to explore the phospholipid head group structure in discrete DMPC/ApoA-I complexes.[17] The chemical shift of the phospholipid N-methyl resonance is a very sensitive indicator of the head group surface area, which, in turn, is related to a combination of the head group conformation and/or hydration. NMR has been used to prove that the phospholipids around the perimeter of a discoidal micelle are in a different chemical environment than the core phospholipid due to the presence of protein.[17]

It is known that high field (400 MHz) proton NMR of apolipoprotein A-I/phosphotidylcholine discoidal complexes produces a single $N(CH_3)_3$ resonance (as compared to a bimodal resonance for sonicated PC vesicles). The chemical shift produced is a function of the apoA-I/PC ratio, and thus the diameter of the complex.[18]

This method can therefore be used to compare the sizes of the complexes produced with peptides/DMPC and apoA-I/DMPC. Figure 4 is a plot of chemical shift versus discoidal diameter (measured by EM) for three separate 18A/DMPC weight ratios (1:1, 1:2.5, 1:5), for two apoA-I/egg PC weight ratios (1:1 and 1:2) and for apoA-I/DMPC weight ratios (1:2.5 and 1:17.5). Even though the scale of the theoretical curve is forced to fit the extremes of the chemical shifts, the general shape of the predicted fit corresponds nicely to the intermediate points. This is in accordance with the earlier postulate of a single two state fast exchange model for all three types of complexes, where one state is lipid associated with the peptide.[18]

Differential Scanning Calorimetry (DSC)

Differential scanning calorimetry measures the difference in heat flow into a sample compared to a standard or reference buffer as the two solutions are heated. Protein conformational changes (usually denaturation) and lipid phase transitions (e.g., gel → liquid-crystalline) are endothermic transitions easily detected by DSC. For protein–lipid mixtures, the magnitude of the interaction between the proteins and lipid is most commonly assessed by the changes observed in the midpoint temperature, enthalpy and/or cooperativity of the lipid phase transition upon the addition of protein.

These studies can be conducted to understand the lipid phase transition in the presence of peptides. The following results were obtained with

---

[18] C. G. Brouillette, J. L. Jones, T. C. Ng, H. Kercret, B. H. Chung, and J. P. Segrest, *Biochemistry* **23**, 359 (1984).

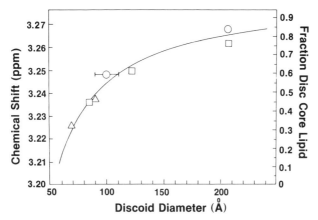

FIG. 4. Chemical shift of —N(CH$_3$)$_3$ proton NMR spectra of apoA-I or 18A complexes with phospholipid versus discoidal diameter measured by electron microscopy.

the DSC studies of the peptide–lipid complexes formed by 18A, reverse 18A, 18A-Pro-18A, and desVal^{10}18A. From transition parameters one can observe several trends. (1) Two lipid transitions for each complex studied can be one occurring at or below (T_m − 1) and one occurring well above (T_m − 2) the transitions T_m of pure DMPC multilamellar vesicles (23.6°). (2) When heated to just above lipid phase transition, the bimodal transition appears reversible. (3) There is a clear trend in the half height width of the transition at the 1:5/peptide:lipid ratios as follows: 18A-Pro-18A > 18A > desVal^{10}18A > reverse 18A. There is considerably less variation in the half height width of transition 2. (4) The smaller 18A/DMPC complex (1:2.5 ratio) has a broader transition 1 than the larger 18A/DMPC complex (1:5 ratio). (5) The two smallest complexes have the lowest total enthalpies, 3.1 and 2.6 kcal/mol, respectively, compared to 3.9 ± 0.1 kcal/mol for the three remaining complexes. These data have been explained with a model as shown in Fig. 5, in which T_m − 1 and T_m − 2 represent disc core DMPC and disc edge DMPC, respectively. The core DMPC (associated with transition 1) should have a phase transition temperature close to that of pure DMPC. Also, transition 1 is expected to be more cooperative in larger discoidal complexes. This has been found to be true as seen by a sharp increase of T_m − 1 with increasing disc size. At peptide/DMPC ratios of 1:5 the rank order of size of complexes is 18A-Pro-18A > 18A > desVal^{10}18A > reverse 18A and the rank order of half height. Width of transition 1 is 18A-Pro-18A < 18A < desVal^{10}18A < reverse 18A.

A modification of the edge-core model of Fig. 5A, in which peptide

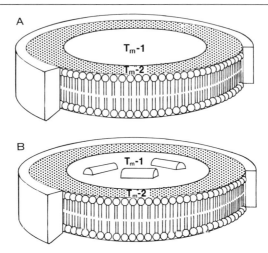

FIG. 5. Schematic diagrams of two possible edge-core lipid domain models to explain the differential calorimetry data. (A) Protein annulus-bilayer disc with independent edge ($T_m - 2$) and core ($T_m - 1$) lipid transitions. (B) Same model containing amphipathic peptides associated with core lipid region.

associates with the core DMPC with varying concentrations that depend upon the lipid affinity of the peptide and affect the size of DMPC core (Fig. 5B), has been invoked as an alternative to explain the DSC data. These results agree with the results from CD and tryptophan blue shift in suggesting the lipid associating affinity of the four peptides as follows: 18A-Pro-18A > 18A > desVal^{10}18A > reverse 18A.

Competition Experiments

Binding Studies

These studies have been done by Ponsin et al.,[6] who studied the effect of hydrophobicity of their alkylated LAP peptides in their capacity to bind to HDL. In these experiments they measured the equilibrium constants K_{eq} for the association of ^{125}I-labeled acylated LAP to reassembled HDL by equilibrium dialysis at several temperatures. At 37°, K_{eq} increased by 3 orders of magnitude as the carbon chain is increased from 0 to 16, also showing a linear relationship between K_{eq} and the acyl chain lengths. The free energy of association (ΔG) decreased by a constant value for each methylene unit added.

LAP peptides labeled with ^{125}I were also incubated overnight at 4° with purified human HDL or rat serum, and the incubation mixtures were separated on a Sepharose CL-4B column. It was found that the unalkylated peptide was recovered as a free peptide, where as the C_6-LAP associated with HDL. The peptide was incubated overnight with a mixture of purified HDL, LDL, and VLDL, having the same concentrations of phospholipid and the sample was chromatographed on a Sepharose CL-4B column. The results showed that approximately 74% of the peptide was associated with HDL and approximately 8 and 5% with VLDL and LDL, respectively. These results indicate that the binding of amphipathic peptide to lipoprotein is governed by its hydrophobicity.

Displacement Studies

Systematic studies were conducted in our laboratory to correlate structure of amphipathic peptide analogs with function.[16] In these studies, radial immunodiffusion technique was used to quantitate the amount of apoA-I displaced by amphipathic peptide analogs from native HDL. Earlier it was shown that an analog of 18A, containing triserine at both NH_2- and COOH-terminal ends, displaced apolipoproteins from HDL and VLDL.[19] With a view to quantitate the native apolipoproteins, displaced HDL was incubated with four peptide analogs, 18A, 18A-Pro-18A, reverse 18A, and desVal^{10}18A, at a peptide to HDL weight ratio of 1:2. Two milligrams (protein) of isolated HDL was placed in a glass test tube (12 × 100 mm) and 0.5–4 mg of synthetic peptide was added. The volume of the mixture was adjusted to 1.4 ml by the addition of Tris buffer. The lipoprotein–peptide mixtures were incubated overnight (18–20 hr) at 7° and subsequently were subjected to single vertical spin ultracentrifugation at 80,000 rpm for 30 min employing a VTi-80 rotor and a Beckman L8-80 ultracentrifuge. HDL without peptide was used as a control. A total of 25 fractions was collected from each tube by puncturing the bottom with a gradient fractionator (Hoefer Scientific Instruments, San Francisco, California). The levels of cholesterol in each gradient fraction were measured by using an enzymatic cholesterol assay kit (Biodynamic/Boehringer Co.) and protein was measured by OD 280. The cholesterol-free bottom fractions (free protein) and the fraction containing cholesterol (HDL) were pooled and dialyzed against Tris buffer by using spectrapor membrane tubing (Fisher Scientific) with a 3500 MW cutoff. The identity of displaced apolipoproteins recovered in the free protein fraction and the change of apolipoprotein composition were determined by urea gel electrophoresis. The major changes are a decrease in the apoA-I/apoA-II ratio for HDL modified by incubation with 18A and especially 18A-Pro-

[19] B. H. Chung, G. M. Anantharamaiah, C. G. Brouillette, T. Nishida, and J. P. Segrest, *J. Biol. Chem.* **260,** 10256 (1985).

18A. The levels of A-II did not appear to change after incubation with any of the four peptides.

For quantitating the apparent decrease in apoA-I following incubation with 18A, 18A-Pro-18A, reverse 18A, and desVal^{10}18A radial immunodiffusion was used. Agarose, 1% containing a suitable amount of purified monospecific rabbit antibody to apoA-I, was prepared and poured into plastic circular plates. After puncturing the sample wells, 5 μg of suitably diluted apoA-I as standard, and 5 μg of buffer as control and peptides, incubated HDL samples were placed in the wells of the radial immunodiffusion plates. The levels of HDL were standardized based on cholesterol content, and HDL was treated with tetramethyl urea before diluting with Tris buffer. The immunodiffusion plates were then incubated in humid chambers until the precipitate rings attained final sizes (48–72 hr). The diameters of the precipitate rings were measured in 0.1 mm units using a calibrated immunodiffusion viewer. In this assay system, the concentration of apoA-I is proportional to the square of the radius of the precipitation ring produced. The assays were reproducible to within 3% error. The results showed that reverse 18A displaces only minimal amount of apoA-I compared to control, desVal^{10}18A displaces 0.5 mg of apoA-I/mol of HDL or approximately 20% of the total apoA-I. 18A displaces twice as much as desVal^{10}18A or 40% of total apoA-I from HDL. Remarkably, 18A-Pro-18A displaces twice as much as 18A or 80% of apoA-I from HDL.

When HDL was incubated with 18A at three different peptide/HDL ratios, increasing 18A concentrations displaced increasing amounts of apoA-I. However, even at 1:1 weight ratio, 18A failed to displace as much of apoA-I as 18A-Pro-18A at 1:2 weight ratio. This shows that 18A-Pro-18A, the dimer of 18A with a proline at the center, shows cooperativity in its ability to displace apoA-I.

LCAT Activation

One of the major functions of apoA-I, the main constituent of HDL, appears to be to act as a cofactor in the LCAT mediated catalysis of cholesteryl ester formation. The conversion of the amphipathic surface lipid, cholesterol, to the nonpolar core lipid, cholesteryl ester, has been suggested to trap cholesterol in HDL, thus lowering the concentrations of free cholesterol in peripheral cell membranes. This process is known as reverse cholesterol transport. Most of the synthetic peptides have been tested for their ability to serve as LCAT cofactors.

The first synthetic peptide to be tested for this activity as a cofactor for LCAT activation was by Yokoyama et al.[4] in which they found that the peptide Ap was 18% as active as apoA-I. In this assay system, as conducted in our laboratory, an egg PC:[^{3}H]cholesterol mixture (90 nM:20 nM) was sonicated and to the resulting small unilamellar vesicles was added 0.13 μg of a highly purified LCAT preparation (0.207 units/ng). Varying amounts of activator peptide (18A-Pro-18A, desVal^{10}18A, and reverse 18A) or protein (apoA-I or A-II) were added, and the mixture incubated for 60 min at 37°. The reaction was stopped by extraction of free and esterified cholesterol. Formation of cholesteryl ester was determined by the counting of spots cut from plastic thin-layer chromatography sheets using CHCl$_3$:MeOH (2:1) solvent system.

When C_8- and C_{18}-LAP peptides were tested for LCAT activating activity they activated the enzyme in a concentration-dependent manner. When compared on a mass concentration bases, apoA-I was 50 and 20% times more potent than C_{18}- and C_8-LAP, respectively.[6]

The results of LCAT activity experiments with various peptide analogs synthesized by us are shown in Fig. 6.

As shown in Fig. 6, the peptide analog 18A-Pro-18A has been found to be a powerful activator of LCAT when incubated with unilamillar egg PC

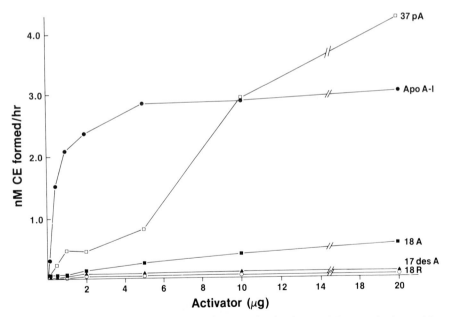

FIG. 6. Lecithin:cholesterol acyltransferase activation by apoA-I or synthetic peptide analogs as measured by the small unilamellar egg PC vesicles as a substrate for lecithin:cholesterol acyltransferase: Activation by 18A, reverse 18A, desVal^{10}18A, 18A-Pro-18A, or apoA-I.

vesicles, reaching 140% of the activity of apoA-I at a 1:1.75 ratio of peptide egg PC ratio.

Measurement of cholesteryl ester formation as a function of time for 10 mg of 18A-Pro-18A and apoA-I essentially showed parallel curves, confirming the results obtained by 18A-Pro-18A. This ability of 18A-Pro-18A to activate LCAT has been attributed to the spontaneous conversion of preformed egg PC vesicles to protein annulus: bilayer discs which are better LCAT substrates than vesicles.

Neither apoA-I, apoA-II, nor any of the other peptides, 18A, reverse 18A, and desVAL^{10}18A, has been found to convert preformed egg PC vesicles to discoidal structures and this perhaps explains the ability of 18A-Pro-18A to exceed even apoA-I in its ability to activate LCAT in the egg PC vesicular assay system.

Advantages

Synthetic peptide analogs of apolipoproteins have shed light on our understanding of the importance of the secondary structural features of apolipoprotein. Many questions such as the importance of position of charges, hydrophobicity, and cooperativity, which cannot otherwise be answered, have been answered by the synthetic peptide analogs. At least in one instance a synthetic peptide (18A-Pro-18A) has been found more active in activating LCAT than apoA-I. A systematic study to investigate the mechanism of LCAT activation is possible by synthetic peptide analogs. It is thus possible to provide answers with further design and synthesis of peptide analogs. Synthetic peptide analogs have given good support to the protein annulus bilayer disc model for HDL. We have some preliminary results indicating that the highly conserved region of apoA-I, (8–33) apoA-I, when synthesized, does not bind to lipid at all. Now the question is, what part is played by this highly conserved region in apoA-I? We are investigating the possibility of this as a binding site for LCAT. It may even be possible in the future to design a peptide to effectively elevate the levels of HDL by administering a synthetic HDL. Another important application is in understanding the different types of HDL present in plasma by using displacement studies. Displacement studies have already indicated that there is a certain amount of displaceable apoA-I in HDL which is not needed for LCAT activating activity. Synthetic peptides have been useful in equilibrium studies of apolipoproteins between HDL and the aqueous phase.[20,21]

[20] J. P. Segrest, B. H. Chung, C. G. Brouillette, P. Kanellis, and R. McGahan, *J. Biol. Chem.* **258**, 2290 (1983).

[21] C. G. Brouillette, J. P. Segrest, T. C. Ng, and J. L. Jones, *Biochemistry* **21**, 4569 (1982).

Disadvantages

Native apolipoproteins possess multiple helical domains along with tertiary structure. The importance of a tertiary structure for various functions of apolipoproteins cannot be understood by the synthetic peptides approach, as techniques for protein synthesis are still far from being developed. Semisynthesis may be the answer to this problem but it is still in a rudimentary stage of development.

[37] Lipoprotein–Liposome Interactions

By ALAN R. TALL, IRA TABAS, and KEVIN J. WILLIAMS

Introduction

The interaction of lipoproteins with phospholipid liposomes has been studied as a model of the transfer of phospholipids from triglyceride-rich lipoproteins into high-density lipoproteins (HDL) during lipolysis.[1] These studies may also have relevance to lipoprotein–cell interactions, the potential use of liposomes as carriers of drugs,[1a] and to the enhanced regression of atherosclerosis induced by infusion of phospholipid liposomes into experimental animals.[2,3] Although all lipoprotein species may be involved, the major interaction of liposomes with lipoproteins in plasma involves high-density lipoproteins (HDL), and results in dissolution of the liposome structure and transfer of liposomal phospholipid into the HDL density range.[4] Under certain circumstances liposomes remain in plasma as intact particles, resembling lipoprotein-X.[5] The outcome of specific experiments where phospholipid liposomes are incubated with isolated lipoproteins or plasma or are injected intravenously is markedly influenced by a number of variables, such as the amount and type of phospholipid, the cholesterol/phospholipid ratio, the temperature, and the species of lipoproteins. The transfer of vesicle phospholipid into HDL is greatly enhanced by plasma phospholipid transfer protein(s).[6,7] This chapter will

---

[1] A. R. Tall, C. B. Blum, G. P. Forester, and C. Nelson, *J. Biol. Chem.* **257**, 198 (1982).
[1a] F. Szoku and D. Papanadjopoulos, *Annu. Rev. Biophys. Bioeng.* **9**, 467 (1980).
[2] S. O. Byers and M. Friedman, *J. Lipid Res.* **1**, 343 (1960).
[3] K. J. Williams, V. P. Werth, and J. A. Wolff, *Perspect. Biol. Med.* **27**, 417 (1984).
[4] A. R. Tall and P. H. R. Green, *J. Biol. Chem.* **256**, 2035 (1981).
[5] K. J. Williams and A. M. Scanu, *Biochim. Biophys. Acta,* in press.
[6] A. R. Tall, L. Forester, and G. Bongiovanni, *J. Lipid Res.* **24**, 277 (1983).
[7] A. R. Tall, E. Abreu, and J. Shuman, *J. Biol. Chem.* **258**, 2174 (1983).

cover the effects of different variables or liposome–lipoprotein interactions, methods used to separate the different products of this interaction, the preparation of phospholipid transfer protein(s), and the preparation of HDL containing radiolabeled phospholipids of high specific activity. We will also survey the biological relevance of lipoprotein–liposome interactions.

Preparation of Liposomes

Multilamellar liposomes may be prepared by drying down phosphatidylcholine (PC) from organic solvent in a round bottomed flask, followed by agitation with buffer above the gel to liquid crystalline transition temperature of the phospholipid. Unilamellar vesicles of phosphatidylcholine have been prepared by sonication (15–30 × 30 sec bursts) to clarify above the phospholipid transition temperature followed by purification on a Sepharose CL4B column.[1a,4] Other methods for preparing unilamellar vesicles include mixing of phospholipids with a detergent followed by removal of the detergent, injection of lipids dissolved in ethanol into an aqueous buffer solution, extrusion of multilamellar vesicles through a French press, and formation of water in oil emulsions of phospholipids and buffer in an excess organic phase followed by evaporation of the organic solvent.[1a] Each of these methods yields liposomes of different physical properties, yet little is known about how these differences could potentially affect lipoprotein–liposome interactions. The inclusion of cholesterol in phospholipid vesicles has several effects on the organization and physical properties of the phospholipids; for instance, cholesterol decreases the permeability of the vesicles to ions and small molecules and above 30 mol%, increases the size of small unilamellar vesicles.[1a] The preparation of cholesterol/phospholipid liposomes requires considerably longer sonication times, especially at greater than 33 mol% cholesterol.

Effects of Liposome Composition on Lipoprotein–Liposome Interactions

Unilamellar vs Multilamellar Liposomes. Multilamellar liposomes of saturated phospholipids (C_8–C_{14}) display clearing of turbidity upon incubation with HDL, indicating incorporation of the phospholipid into smaller particles which do not scatter light. However, multilamellar liposomes of egg phosphatidylcholine do not undergo dissolution when incubated in plasma.[8]

[8] G. Scherphof, H. Morsett, J. Regts, and J. C. Wilschut, *Biochim. Biophys. Acta* **556**, 196 (1979).

Phospholipid Composition. The interaction of dimyristoyl PC with isolated HDL proceeds more rapidly than that of egg PC, possibly reflecting the greater aqueous solubility of the dimyristoyl PC molecule. Also, the products of interaction of dimyristoyl PC with HDL are somewhat different from those of egg PC (see below).

Cholesterol/Phospholipid Ratio. The presence of cholesterol has a profound influence on the interaction of vesicles with isolated HDL, with HDL in plasma, and with HDL in the intact animal.[8-10] With increasing cholesterol/phospholipid ratio there is a decrease in the net transfer of phospholipid molecules from vesicles into HDL. This is particularly pronounced as the ratio of cholesterol/egg PC is increased above 33 mol%. The presence of cholesterol also results in a decrease in the lipoprotein or plasma-induced leakiness of phospholipid vesicles, as assessed by release of carboxyfluorescein.[10] However, cholesterol/phospholipid vesicles still display active exchange of phospholipid molecules with HDL.[11] In incubations of egg PC/cholesterol vesicles in human plasma we have found that with increasing cholesterol content, there is first a decreased incorporation of phospholipids into preexisting HDL particles, but cholesterol/phospholipid/apolipoprotein discs are still formed. At higher ratios of cholesterol/phospholipid, vesicles remain intact, but adsorb HDL apolipoproteins such as apoA-I and apoE. When injected into the rat, vesicles of low cholesterol/PC ratio are broken down and the vesicle phospholipid is incorporated into HDL; as the cholesterol/PC ratio is increased, progressively less PC is incorporated into HDL, the vesicles retain entrapped solutes, and are preferentially taken up by the liver.[9] The latter event may reflect interaction of adsorbed apolipoproteins with the hepatic apoB,E or E receptors.[3]

Preparation of Lipoproteins

All plasma lipoproteins may incorporate phospholipid when incubated with phospholipid vesicles.[12-14] However, the uptake by HDL is most pronounced. Within HDL, the more dense subpopulations show greater capacity to incorporate phospholipid (i.e., following incubation with vesicles they show a greater increase in phospholipid/protein ratio), though

---

[9] A. R. Tall, *J. Lipid Res.* **21**, 354 (1980).
[10] L. S. Guo, R. J. Hamilton, J. Goerke, J. N. Weinstein, and R. J. Havel, *J. Lipid Res.* **21**, 993 (1980).
[11] A. Jonas and G. T. Maine, *Biochemistry* **18**, 1722 (1979).
[12] A. R. Tall and D. M. Small, *Nature (London)* **265**, 163 (1977).
[13] J. V. Chobanian, A. R. Tall, and P. I. Brecher, *Biochemistry* **18**, 180 (1979).
[14] Z. Shahrokh and A. V. Nichols, *Biochem. Biophys. Res. Commun.* **108**, 888 (1982).

on a molar basis a large HDL_2 particle can probably take up more phospholipid molecules than a smaller HDL_3 particle.[4] These results may reflect a greater intrinsic capacity of smaller HDL particles to take up phospholipid molecules and/or the presence in smaller HDL particles of increased amounts of phospholipid transfer proteins.[6] In general, lipoproteins used in studies of lipoprotein–liposome interactions have been obtained by preparative ultracentrifugation. It is notable that the ability of the $d < 1.25$ fraction to promote incorporation of vesicle phospholipid into HDL is reduced in proportion to the period of ultracentrifugal flotation. Flotation for 20 hr resulted in a fraction with 2.5-fold greater ability to stimulate transfer of phospholipid radioactivity, compared to a 60 hr flotation.[4] This reduced ability to incorporate phospholipid reflects the centrifugal separation from HDL of phospholipid transfer protein(s) which are subsequently recovered in the $d > 1.21$ fraction.[4,6] The activity of plasma phospholipid transfer protein(s) accounts for the more rapid transfer of phospholipid into HDL when vesicles are incubated in plasma, compared to when they are incubated with centrifugally isolated HDL.[6]

Incubation Conditions

Temperature. The dissolution of phospholipid liposomes in plasma is markedly dependent on the incubation temperature. Multilamellar liposomes of saturated phospholipids are broken down in plasma at temperatures close to the gel to liquid crystalline phase transition temperature of the phospholipid. For example, multilamellar liposomes of dimyristoyl PC were readily solubilized in plasma at the phase transition temperature of 24°, but were quite resistant to dissolution at 15 or 37°.[8] It is likely that unilamellar vesicles of saturated PC show a similar temperature dependence. However, unilamellar vesicles of egg PC do undergo dissolution in plasma at 37°, even though the phase transition temperature of egg PC is about $-10°$.

Vesicle/HDL or Vesicle/Plasma Ratio. The plasma HDL has a finite capacity to accommodate additional PC molecules. When egg PC vesicles were incubated with HDL_3 there was increasing incorporation of vesicle PC into preexisting HDL particles up to a limit of about 0.25 mg PC/mg HDL protein.[4] At this point the HDL_3 particles had a mean density of about 1.11 g/ml and were increased in diameter from 8.0 to 9.2 nm, as determined by negative stain electron microscopy. With increasing vesicle/HDL ratios there was also increased incorporation of PC into discoidal complexes containing PC, apoA-I, and apoA-II. As the vesicle/HDL ratio is increased further (>2/1 PC/HDL protein), intact vesicles with adsorbed HDL proteins are probably formed.

When small amounts of vesicles were incubated in plasma the majority of the vesicle phospholipid was rapidly incorporated into preexisting spherical particles or discs in the HDL density range.[4] There was an associated net transfer of cholesterol from other lipoproteins into HDL.[4] When these incubations were conducted in the absence of an inhibitor of plasma lecithin : cholesterol acyltransferase the discs were converted into cholesteryl ester-containing spherical lipoproteins.[4] With larger amounts of vesicles, phospholipids are also transferred into LDL and probably also VLDL particles, resulting in the formation of larger less dense particles.[13,14] With further increase in the amount of vesicles, the capacity of the lipoproteins to incorporate phospholipids is exceeded and intact vesicles of modified composition remain.[5] There is a net transfer of cholesterol from HDL and other lipoproteins into the vesicles.[5] The vesicles probably adsorb apoE, apoA-I, and small amounts of C apolipoproteins.[5] These vesicles are very similar to lipoprotein-X, the abnormal particle formed during cholestasis or intralipid infusion.

The results obtained with vesicles of saturated PC were somewhat different.[15,16] At lower ratios of PC/HDL, there was incorporation of PC into preexisting HDL particles, associated with the appearance of free apoA-I in the lipoprotein-free fraction. At higher ratios of dimyristoyl PC/HDL, there was also formation of discoidal complexes containing PC and apoA-I. These results suggested that dimyristoyl PC inserted into the surface of HDL displacing apoA-I, which was then available for formation of discoidal particles.[15,16] A possible role of preparative ultracentrifugation below the gel to liquid crystalline transition temperature of dimyristoyl PC in liberation of apoA-I from dimyristoyl PC/apoA-I discs or dimyristoyl PC-enriched HDL has not been assessed.

Separation and Analysis of Phospholipid-Enriched HDL Species

The products of vesicle–HDL interaction are heterogeneous and their separation may be problematic. Differences in density have been used, but sometimes may only provide partial resolution of the different phospholipid-enriched lipoproteins. Thus, when liposomes of dimyristoyl PC were incubated with HDL_3, then analyzed by isopycnic density gradient ultracentrifugation, the peak of the dimyristoyl PC/apoA-I discs was only partially separated from that of the phospholipid-enriched HDL_3 particles.[17] When egg PC vesicles were incubated in plasma then analyzed by

---

[15] A. V. Nichols, E. L. Gong, T. M. Forte, and P. J. Blanche, *Lipids* **13**, 943 (1978).
[16] E. L. Gong and A. V. Nichols, *Lipids* **15**, 86 (1980).
[17] A. R. Tall, V. Hogan, L. Askinazi, and D. M. Small, *Biochemistry* **17**, 322 (1978).

density gradient ultracentrifugation, the discs were at slightly lower density than the major peak of phospholipid-enriched HDL_3, but there was considerable overlap.[4] When egg PC vesicles were incubated with isolated HDL_3, the discoidal particles were of relatively low density (1.04–1.06 g/ml). Thus, they could be readily separated from the phospholipid-enriched HDL by appropriate density gradients, but were close in density to intact vesicles containing adsorbed apolipoproteins.

Size has also been used as a criterion to separate the products of incubation of liposomes with HDL. Phospholipid-enriched spherical HDL can be separated from intact vesicles by agarose chromatography.[18] However, discs, though morphologically distinct, may be similar in diameter to unreacted phospholipid vesicles, and therefore may elute in the same position as unreacted vesicles.[4] These results indicate that a combined approach employing density gradient and column chromatographic techniques, with careful monitoring of fractions by negative stain electron microscopy, is essential for the careful definition of products.

Role of Plasma Phospholipid Transfer Protein(s) in Promoting Exchange and Transfer of Phospholipids between Vesicles and HDL

The transfer of phospholipids from vesicles into HDL may be promoted by the plasma $d > 1.21$ fraction or by protein(s) purified from this fraction.[6,7] The $d > 1.21$ fraction and partially purified plasma phospholipid transfer proteins enhance the transfer of phospholipids from vesicles into preexisting HDL, resulting in the formation of larger and less dense HDL particles.[6,7] The plasma phospholipid transfer protein(s) are normally isolated in the density 1.19–1.25 g/ml fraction and elute in HDL-sized particles when plasma is analyzed by agarose chromatography. With prolonged centrifugation, the phospholipid transfer proteins appear to dissociate from HDL particles.[6] Procedures for isolation of phospholipid transfer proteins are described in Chapter [48], Volume 129.

Methods for Phospholipid Enrichment or Radiolabeling of HDL

The interaction of phospholipid vesicles with HDL can be exploited as a means of enriching HDL with phospholipids and of introducing radiolabeled phospholipids into HDL. The transfer/exchange activity of the plasma phospholipid transfer protein(s) is used in this procedure. In order to obtain a phospholipid-enriched HDL_3 particle, plasma is raised to a

[18] A. Jonas, *J. Lipid Res.* **20**, 817 (1979).

background salt density of 1.125 g/ml then centrifuged for 48 hr at 45,000 rpm in a Beckman Ti 50.3 rotor. The 1.125 infranatant is then dialyzed against 0.15 M NaCl, and made 2 mM in dithionitrobenzoic acid to inhibit endogenous lecithin:cholesterol acyltransferase activity. The $d > 1.125$ fraction is incubated with unilamellar vesicles of egg phosphatidylcholine (the vesicle/HDL ratio will determine the degree of phospholipid enrichment of HDL) for 3 hr at 37° in a shaking water bath. Subsequently, the $d > 1.125$ fraction is readjusted to density 1.105 g/ml, overlaid with density 1.090 g/ml NaBr solution, and centrifuged for 24 hr at 45,000 rpm. This procedure results in flotation of discoidal phospholipid/apoA-I complexes and also unreacted phospholipid vesicles. The 1.090 infranatant containing the phospholipid-enriched HDL is adjusted to density 1.25 g/ml, overlaid with density 1.21 g/ml NaBr solution, then centrifuged for 48 hr at 45,000 rpm. The top 2 ml of the tubes contains the phospholipid-enriched HDL_3 particles in which the additional phospholipid molecules have been introduced into preexisting spherical HDL particles.

In order to obtain HDL containing high specific activity radiolabeled phospholipids, a similar procedure to that outlined above is followed. However, the radiolabeled phospholipids are dissolved in a small volume of ethanol, then introduced into the $d > 1.125$ fraction by slow injection through a fine-gauge needle positioned just beneath the surface of the stirred solution. The solution is placed under a stream of N_2 then incubated for 1 hr at 37°. The HDL containing radiolabeled phospholipids is isolated by preparative ultracentrifugation as indicated above.

Physical Studies of Phospholipid-Enriched HDL

The phospholipid-enriched HDL particles are suitable for study by a variety of spectroscopic methods. We have found very little evidence of a conformational change of the HDL apolipoproteins associated with uptake of phospholipids. Thus, the intrinsic protein fluorescence of HDL apolipoproteins was found at the same wavelength maximum in native HDL_3 and phospholipid enriched HDL_3. Also, spectrophotometric titration showed identical curves for HDL_3 and phospholipid-enriched HDL_3 between pH 8 to 12.[4] Finally, we have found no change in the circular dichroic spectra of HDL_3 as a result of phospholipid enrichment, suggesting a similar content of protein secondary structure in the native and phospholipid-enriched HDL particles. However, evidence of a more fluid lipid environment in phospholipid-enriched HDL was obtained from the fluorescence polarization of diphenylhexatriene (shortening of the rotational relaxation time).[18] Also, differential scanning calorimetry showed an altered thermal denaturation pattern of the phospholipid enriched HDL particles, suggesting a lower thermal stability of the phospholipid-enriched particle.[4]

Molecular Basis of the Transfer of Phospholipids into HDL

The incubation of phospholipid liposomes in plasma or with isolated HDL_3 results in incorporation of phospholipids into HDL. All studies are in agreement that there is insertion of phospholipid molecules into preexisting HDL, resulting in an increase in size and decrease in density of the HDL particles. In addition, we and others[4,15,16] have shown that phospholipid/apoA-I/apoA-II discs are formed.

The discs that are formed during incubation of vesicles with HDL_3 result from the interaction of apoA-I or apoA-II released from HDL with vesicle phospholipid. After incubation of vesicles containing radioactive phospholipid with isolated HDL, the discs that are formed have much higher phospholipid specific activity than the phospholipid-enriched spherical HDL, indicating that vesicles do not first combine with HDL and then give rise to discs.[4] Also, when dimyristoyl PC vesicles are incubated with HDL, apoA-I is released at lower vesicle/HDL ratios, and discs are formed only at higher ratios, suggesting that apoA-I is first released from HDL, then combines with the vesicles to form discs.[15,16]

The uptake of phospholipid by spherical HDL may involve at least two different mechanisms. The first mechanism involves phospholipid-apolipoprotein substitution in the surface of HDL. Thus, for both egg PC and dimyristoyl PC, the uptake of phospholipid by HDL at all ratios is accompanied by release of HDL apolipoproteins, giving rise to free apolipoproteins or forming phospholipid/apolipoprotein discs. In the case of egg PC vesicle/HDL_3 interactions, the amount of released apolipoprotein occurs in a fixed stoichiometry with the amount of added vesicle phospholipid, as shown by the formation of discs of constant density and lipid/protein ratio as the vesicle/HDL was increased; also the amount of HDL phospholipid increased linearly with increasing vesicle/HDL ratio.[4] These observations suggest that phospholipid–apolipoprotein substitution is a major mechanism of phospholipid uptake at all vesicle/HDL ratios. An additional mechanism of uptake may involve "flexibility" of the HDL surface, such that it has the ability to incorporate additional phospholipid molecules without the loss of apolipoproteins.[18] This molecular property of HDL could result from a decrease in the average surface area per phospholipid molecule in the lipoprotein surface, or from convolution of the lipoprotein surface. Also, it is conceivable that the HDL apolipoproteins could undergo a conformational change allowing the lipids to occupy a greater amount of the surface. However, as noted above, there is little evidence to support a conformational rearrangement of HDL apolipoproteins in phospholipid-enriched particles (see, however, Chapter [32], this volume). When phospholipid molecules are incorporated into HDL in concert with active lecithin : cholesterol acyltransferase, both the surface and the core of HDL particles are increased; thus, additional phospholipid molecules can be accommodated in the HDL surface, without phospholipid–apolipoprotein substitution, or without a need for "flexibility" of the HDL surface. It is likely that such a simultaneous increase in core and surface lipids occurs as phospholipids are transferred into HDL under physiological circumstances, such as during alimentary lipemia.[1]

There are two more potential molecular rearrangements that may occur when liposomes are incubated with lipoproteins. Fusion of individual lipoprotein particles leading to quantum increases in lipoprotein size may result from the depletion of surface apolipoproteins of spherical lipoproteins when incubated with liposomes.[17] The conditions promoting this potential mode of lipoprotein transformation are poorly understood. A second potential mechanism has recently been elucidated by Nichols et al.[19] When phospholipid/apoA-I discs were incubated with LDL or VLDL in the presence of the $d > 1.21$ fraction and an inhibitor of lecithin : cholesterol acyltransferase, there was formation of small, spherical lipoprotein particles, consisting of phospholipid, cholesterol, and apoA-I. It is conceivable that a similar particle might be formed when vesicles containing small amounts of cholesterol are incubated with HDL in the presence of the $d > 1.21$ fraction.

Biological Relevance of Liposome–Lipoprotein Interactions

There is evidence that HDL becomes enriched with phospholipids under physiological circumstances. The transfer of the surface phospholipids of triglyceride-rich lipoproteins into HDL probably accounts for the increase in HDL phospholipids during alimentary lipemia.[1] Also, it is likely that lipolysis of VLDL is associated with transfer of phospholipids into HDL. The physiological transfer of phospholipids from triglyceride-rich lipoproteins into HDL involves insertion into preexisting particles simultaneous with an LCAT-mediated increase in core constituents.

[19] A. V. Nichols, E. L. Gong, P. J. Blanche, T. M. Forte, and V. S. Shore, *Biochim. Biophys. Acta* **793,** 325 (1984).

Whether discoidal particles are also generated transiently during this process, as postulated,[20] is unknown.

Although there is not much direct experimental evidence, it is also likely that HDL can incorporate phospholipids derived from cell membranes. Thus, in lymph collected from rat intestine, sheep lung, or dog foot, the HDL contains both phospholipid-enriched spherical HDL particles and also phospholipid/apolipoprotein discs. It is probable that the formation of these particles may result in part from an interaction of filtered spherical HDL with plasma membranes. The discs may be apparent in these situations because of the low activity of lecithin : cholesterol acyltransferase in lymph. The discs may also be derived as a direct result of the cellular secretion of apolipoproteins.

Liposomes have been used to encapsulate drugs, thereby altering their delivery through the blood to target tissues. The interactions of liposomes with lipoproteins can result in disruption of the liposomes and release of the drugs. However, the dissolution of liposomes in plasma may be minimized by the incorporation of sufficient unesterified cholesterol into the liposomes, or by the use of high melting phospholipids (above 37°), or by giving larger doses of liposomes. Interactions of liposomes with lipoproteins can also result in adsorption of apolipoproteins by otherwise intact liposomes. This may be of great importance since the adsorption of apoE could lead to the uptake of liposomes via the cellular apoB,E receptor.[3] Since the expression of the latter can be regulated in the liver, this could have important consequences for the ultimate disposal of liposomes containing drugs.

Finally, the ability of repeated injections of phospholipids to produce regression of experimental atherosclerosis may reflect interactions between lipoproteins and the injected phospholipids.[3] Injection of phospholipids may produce phospholipid enrichment of HDL and influx of cholesterol from the tissues or other lipoproteins into HDL, with subsequent esterification of LCAT. Circulating liposomes adsorb apolipoproteins from lipoproteins. Complexes of apolipoproteins and phospholipids may be more effective in extracting cholesterol from cultured cells than protein-free phospholipid liposomes. Also, these apolipoproteins, especially apoE, may enhance uptake of cholesterol-bearing liposomes by the liver. Presumably, the liver would excrete cholesterol delivered to it by these particles. With repeated infusions of relatively large amounts of phospholipids, the major cholesterol-transporting particle in plasma is probably a cholesterol/phospholipid vesicle with adsorbed apolipoproteins, includ-

---

[20] A. R. Tall and D. M. Small, *Adv. Lipid Res.* **17**, 1 (1980).

ing apoE. The metabolism of such a particle, though poorly understood at present, may hold the key to understanding the antiatherogenic effect of liposome infusions.

Acknowledgments

Supported by NIH research Grants Hl22682 and T-07343. Alan Tall is an Established Investigator of the American Heart Association.

[38] Carboxyfluorescein Leakage Assay for Lipoprotein–Liposome Interaction

By JOHN N. WEINSTEIN, ROBERT BLUMENTHAL, and RICHARD D. KLAUSNER

Lipoproteins and apolipoproteins can exchange components with liposomes in a number of different ways, as described in Chapter [37], this volume. Some of those interactions result in penetration or disruption of liposomal bilayers sufficient to permit escape of entrapped hydrophilic solutes. In this chapter we describe a technique (termed the "fluorescence self-quenching," or FSQ, method) by which such solute escape can be continuously monitored and the mechanism(s) of liposome–lipoprotein interaction studied. The FSQ method[1] was initially developed for studies of liposome–cell interaction[2-10] but later applied to the interaction of lipo-

[1] J. N. Weinstein, E. Ralston, L. D. Leserman, R. D. Klausner, P. Dragsten, P. Henkart, and R. Blumenthal, in "Liposome Technology" (G. Gregoriadis, ed.), Vol. 3, p. 183. CRC Press, Boca Raton, Florida, 1984.
[2] J. N. Weinstein, S. Yoshikami, P. Henkart, R. Blumenthal, and W. A. Hagins, *Science* **195,** 489 (1977).
[3] W. A. Hagins and S. Yoshikami, in "Vertebrate Receptors" (P. Fatt and H. B. Barlow, eds.), p. 97. Academic Press, New York, 1978.
[4] J. N. Weinstein, R. Blumenthal, S. O. Sharrow, and P. Henkart, *Biochim. Biophys. Acta* **509,** 272 (1978).
[5] J. T. Lewis and H. M. McConnell, *Ann. N.Y. Acad. Sci.* **308,** 124 (1978).
[6] R. E. Pagano, A. Sandra, and M. Takeichi, *Ann. N.Y. Acad. Sci.* **308,** 185 (1978).
[7] F. C. Szoka, Jr., K. Jacobson, and D. Papahadopoulos, *Biochim. Biophys. Acta* **551,** 295 (1979).
[8] M. W. Fountain, R. Chiovetti, Jr., H. Kercret, D. O. Parrish, and J. P. Segrest, *Biochim. Biophys. Acta* **597,** 543 (1980).
[9] J. Van Renswoude and D. Hoekstra, *Biochemistry* **20,** 540 (1981).
[10] R. Blumenthal, E. Ralston, P. Dragsten, L. D. Leserman, and J. N. Weinstein, *Membr. Biochem.* **4,** 283 (1982).

somes with lipoproteins,[11-14] serum,[15-19] and other macromolecular species.[20-21a] (Table 1 of ref. 1 contains a fuller listing of past applications.)

Principle

If the water-soluble fluorophore carboxyfluorescein (CF) is encapsulated in liposomes (lipid vesicles) at sufficiently high concentration (e.g., 100 mM), its fluorescence is almost completely quenched by interaction between neighboring fluorophore molecules as shown in Fig. 1. Release of CF from the liposomes can therefore be assessed directly from the observed increase in fluorescence.

Characteristics of Carboxyfluorescein

CF has the structure of fluorescein but with an extra carboxyl group located at the 5- or 6-position (the product from Eastman is a mixture of the two isomers[22]). Like fluorescein, CF has an excitation maximum at approximately 492 nm and an emission maximum at approximately 520 nm. The additional carboxyl group decreases the butanol–water partition coefficient by about three orders of magnitude over a broad range of

[11] L. S. S. Guo, R. Hamilton, J. Goerke, J. N. Weinstein, and R. J. Havel, *J. Lipid Res.* **21**, 993 (1980).
[12] J. N. Weinstein, R. D. Klausner, T. Innerarity, E. Ralston, and R. Blumenthal, *Biochim. Biophys. Acta* **647**, 270 (1981).
[13] C. Kirby, J. Clarke, and G. Gregoriadis, *FEBS Lett.* **111**, 324 (1980).
[14] R. D Klausner, R. Blumenthal, T. Innerarity, and J. N. Weinstein, *J. Biol. Chem.*, in press (1985).
[15] T. M. Allen and L. G. Cleland, *Biochim. Biophys. Acta* **597**, 418 (1980).
[16] M. Yatvin, J. N. Weinstein, W. H. Dennis, and R. Blumenthal, *Science* **202**, 1290 (1978).
[17] J. N. Weinstein, R. L. Magin, R. L. Cysyk, and D. S. Zaharko, *Cancer Res.* **40**, 1388 (1980).
[18] C. Kirby and G. Gregoriadis, *Biochem. J.* **109**, 251 (1981).
[19] R. L. Magin and J. N. Weinstein, in "Liposome Technology" (G. Gregoriadis, ed.), Vol. 3, p. 137. CRC Press, Boca Raton, Florida, 1984.
[20] R. Chen, *Anal. Lett.* **10**, 787 (1977).
[21] R. D. Klausner, N. Kumar, J. N. Weinstein, R. Blumenthal, and M. Flavin, *J. Biol. Chem.* **256**, 5879 (1981).
[21a] P. I. Lelkes and H. B. Tandeter, *Biochim. Biophys. Acta* **716**, 410 (1982).
[22] There has been confusion about the chemical name of CF. Originally provided as "6-carboxyfluorescein" (Eastman catalog No. 49), it was renamed "4(5)-carboxyfluorescein" (Eastman catalog No. 50) after it was found to consist of two isomers. The correct name appears to be "5(6)-carboxyfluorescein." The 6-isomer of the closed, lactone form is listed (Chemical Abstracts) as 3',6'-dihydroxy-3-oxo-spiro[iso-benzofuran-1(3H),9'-xanthen]-6-carboxylic acid (registry number 3301-79-9).

 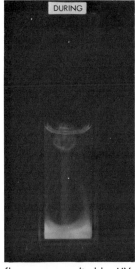 

fluorescence excited by UV

FIG. 1. Fluorescence self-quenching of CF in small unilamellar vesicles. (A) A suspension of vesicles containing 200 mM CF. (B) Triton X-100 is added to disrupt vesicles. (C) Released CF fluoresces with approximately 20 times the original yield. The suspension contained 70 μM CF and 300 μM dioleoyl phosphatidylcholine (adapted from Ref. 1).

concentrations and pHs.[23] Hence, CF is retained in liposomes much longer than is fluorescein itself. The most water-soluble and fluorescent form of CF is the trivalent anion predominant at neutral and alkaline pH (see Fig. 2). In practice, it is important to calibrate fluorescence data for the ambient pH and to be aware that the less soluble forms present in acid medium have a greater tendency to bind to proteins, detergent molecules, and other organic materials.[21a] Solutions should be well buffered at pH 7 or above.

Purification of Carboxyfluorescein

To prepare a stock of purified CF,[24] 35 g of practical grade CF (Eastman, Rochester, NY) is dissolved in 200 ml absolute ethanol in an Erlenmeyer flask. Two grams of activated charcoal (Norit A) is added and the

[23] P. A. Grimes, R. A. Stone, A. M. Laties, and W. Li, Arch. Ophthalmol. **100,** 635 (1982).
[24] E. Ralston, L. M. Hjelmeland, R. D. Klausner, J. N. Weinstein, and R. Blumenthal, Biochim. Biophys. Acta **649,** 133 (1981). The method of purification is based on a protocol developed by W. A. Hagins and S. Yoshikami.

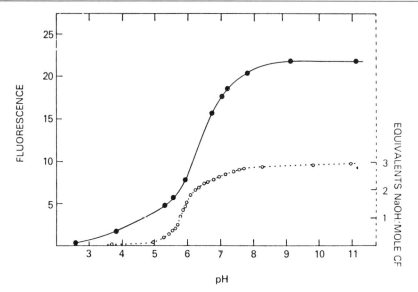

FIG. 2. Titration of CF at 23°. Closed circles: fluorescence of 100 nM dye (excitation 470 nm; emission 520 nm). At high pH the dominant species is the highly fluorescent trivalent anion. Open circles: NaOH required for titration of 2 mM dye from the acid form to the indicated pH (adapted from Ref. 1).

solution boiled for 5 min. Brown material will be left on the bottom of the flask. The solution is decanted and vacuum filtered through a Buchner funnel lined with Whatman No. 50 filter paper. Cold distilled water (about 400 ml) is slowly added with magnetic stirring until cloudiness persists. The solution is cooled slowly and then placed in a $-20°$ freezer overnight. CF precipitates as a light orange powder. The powder is placed on a Buchner funnel and washed extensively with iced distilled water. No ethanol odor should remain. A solution of approximately 250 mM CF is made up in distilled water by titrating to pH 7.4 with 1 N NaOH. The acid form will not dissolve. The amount of CF to be added can be determined by weight if the powder has been thoroughly dried, but it is simpler to make up a concentrated solution from wet precipitate and determine its concentration by optical density at 492 nm. The molar extinction coefficients obtained at the 492 nm peak for 5-CF and 6-CF isomers, respectively, are 74,000 and 76,900 M^{-1}/cm.[24]

Fifty milliliters of 250 mM CF is passed over a 40 × 5 cm column of LH20 Sephadex (Pharmacia, Piscataway, NJ), eluting with distilled water. This step removes principally the hydrophobic impurities. Small samples from the resulting broad red band of CF are examined for impurities by thin-layer chromatography on silica gel plates, eluting with butanol/ acetic acid/water (80:20:20, v/v). Appropriate fractions are pooled and the concentration determined by optical density. CF solution can be stored in the refrigerator (protected from light) for months or years with little deterioration. Lelkes[25] has developed a somewhat different method for purification of CF, also using an LH20 column. Chromatographically purified (99+%) 5(6)-CF can be obtained from Molecular Probes (Junction City, OR). Separated 5- and 6-isomers (purity greater than 95%) can be obtained from Calbiochem-Behring (San Diego, CA).

Preparation of Liposomes

Lipids can be purified and liposomes prepared in a large number of ways, as reviewed elsewhere[26,27] and described in Chapter [37] of this volume. The lipid or lipid mixture chosen will depend on the nature of the intended experiments. If interaction with lipoproteins at the liquid–crystalline phase transition temperature is to be studied, then dimyristoyl phosphatidylcholine (DMPC; T_c 23°), dipalmitoyl phosphatidylcholine (DPPC; T_c 42°), and distearoyl phosphatidylcholine (DSPC; T_c 54°) are natural choices. Mixtures of DMPC and DPPC or DPPC and DSPC give intermediate temperatures of transition. It is often useful to add lipid species labeled with ^{3}H or ^{14}C in order to monitor the amount of lipid in each sample.

The following is one useful method for making small unilamellar vesicles (SUV). The lipids, typically 25–50 mg, are dried from organic solvent onto a glass scintillation vial (60 mm high, 24 mm internal diameter) under a stream of argon or nitrogen gas, and lyophilized overnight in a freeze dryer. Five milliliters of 100 mM CF (diluted from stock solution with distilled water) is added to the vial, which is then vortex-mixed to remove lipid from the glass walls. The suspension is sonicated under argon or nitrogen gas with a titanium probe (Heat-Systems sonicator, model W350, Plainview, NY) at power level 2–4. Typically, the turbid solution will clarify in about 5 min. Sonication should be continued for about 4 times the clearing period to ensure obtaining predominantly small vesicles. Temperature is controlled by a water bath. Both vortex mixing and sonication must be performed above the transition temperature. The suspension is then maintained above transition for 15–30 min (this step may not

[25] P. I. Lelkes, *in* "Liposome Technology" (G. Gregoriadis, ed.), Vol. 3, p. 225. CRC Press, Boca Raton, Florida, 1984.
[26] F. Szoka and D. Papahadjopoulos, *Annu. Rev. Biophys. Bioeng.* **9**, 465 (1980).
[27] G. Gregoriadis, ed., "Liposome Technology," Vol. 1. CRC Press, Boca Raton, Florida, 1984.

be necessary) and quickly cooled to ice temperatures. Titanium particles are removed by centrifugation (3000 g for 1 min).

At this point, the suspension should consist principally of small unilamellar vesicles, but there will also be a fraction of larger forms with different CF release characteristics. A more homogeneous fraction of SUV can be obtained by chromatography on Sepharose 4B (Pharmacia, Piscataway, NJ). Five milliliters of solution is eluted on a 2.6 × 30 cm column with phosphate-buffered saline solution, pH 7.4.

It should be noted that small unilamellar vesicles are not thermodynamically stable structures, and there is still "magic" to their preparation. Because of stress arising from the low radius of curvature, the transition temperature is several degrees lower than that of multilamellar vesicles of the same composition. SUV whose lipids are in the solid phase often change phase transition characteristics within hours.[28] Therefore, the vesicles should be used as soon as possible after preparation.

Lipoproteins

Preparation of lipoproteins and apolipoproteins is described elsewhere in this volume.

Continuous Monitoring of CF Release

Slow release of dye can be monitored by mixing the liposome suspension and lipoprotein preparation in a cuvette, then placing the cuvette in the excitation beam of a standard fluorometer. The fluorescence signal is read continuously on a y-t recorder. An electrically driven sample changer can be used to monitor multiple incubations. Experiments are typically performed as follows. Two milliliters of buffer containing lipoprotein or apolipoprotein is added to each cuvette. The cuvettes are then brought to the desired temperature in a sample turret within the fluorometer and a small volume of vesicle suspension is added. The cuvettes are mixed by covering them with pieces of Parafilm and gently inverting. The cuvettes are returned to the turret for fluorescence measurement over a period of minutes or hours. Emission is monitored at 515 nm; the excitation is generally at 470 nm. The excitation wavelength chosen is below the excitation maximum in order to increase the excitation/emission separation and thus reduce the contribution of light scattering to the emission reading. Further reduction of the signal from scattering can be achieved

---

[28] E. Schullery, C. F. Schmidt, P. Felgner, T. W. Tillack, and T. E. Thompson, *Biochemistry* **19**, 3919 (1980).

by use of a polarizing filter in the excitation beam. At the end of each run, 100 μl of 10% Triton X-100 is added to each cuvette for "post-Triton" measurement of total CF fluorescence. Triton has no significant intrinsic fluorescence, and at low concentration its effect on the fluorescence efficiency of CF is negligible at neutral pH. However, at low pH or in the presence of serum, there is a reduction in fluorescence.[1,21a] A free-CF standard should always be checked for fluorescence quenching by the combination of detergent and lipoprotein preparation to be used in the experiment. [We have recently noted that the detergent $C_{12}E_9$ (Calbiochem-Behring), used at a detergent/lipid ratio of 10:1 (M/M), has less effect on CF fluorescent than does Triton X-100.]

Phase Transition Release

The apparatus diagrammed in Fig. 3 is useful for measurements of rapid CF release at the lipid phase transition. The fluorometer sample chamber is heated to a temperature above T_c, e.g., 47° in the case of dipalmitoyl phosphatidylcholine vesicles. A 135-μl portion of vesicle suspension (generally about 30–60 μM lipid) is mixed on ice with 15 μl of the appropriate dilution of lipoprotein, and 135 μl of the resulting mixture is pipetted into a cuvette for measurement. To accelerate temperature equilibration, small cuvettes (3 × 3 mm internal diameter, 5 × 5 mm outer diameter, 24 mm high; Precision Cells, Hicksville, NY) are used in place of the standard size cuvettes. The cuvette is placed in an adaptor (No. J-6114, American Instrument Co.) at room temperature. The cuvette and adaptor are then transferred to the heated sample chamber, and fluores-

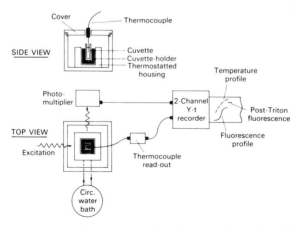

FIG. 3. Schematic diagram of the apparatus for phase transition release (from Ref. 12).

cence is monitored continuously on a y-t recorder with excitation at 470 nm and emission at 515 nm. A copper-constantan thermocouple probe (Type PT-6, Bailey Instruments, Inc., Saddle Brook, NJ) is inserted to lie within the sample, approximately 1 mm above the lightpath. Typically, heating rates of 10–15°/min are obtained at T_c. At these rates, the temperature readings are accurate to within 1°. For routine quantitation of release, the thermocouple probe can be omitted. More nearly constant heating rates can be obtained using a Peltier cell-heating unit. After the temperature scan, 5 μl of 10% Triton X-100 is added and the suspension carefully mixed for measurement of "post-Triton" fluorescence. It is feasible to scan one sample every 90 sec.

Calculations

For calculations of stoichiometry,[12] we assume the following values: DSPC, 789 Da; DPPC, 733 Da; DMPC, 677 D; apolipoprotein A-I, 28,300 Da; apolipoprotein A-II, 17,400 Da; HDL apolipoprotein, 23,000 Da; 5000 lipid molecules per small unilamellar vesicle (this last is an approximation). The molecular weight of HDL apolipoprotein was calculated assuming a 4.4:2:1 molar ratio of apolipoproteins A-I, A-II, and C. The extent of release with time or at the phase transition can be estimated well enough for many purposes by simple examination of the y-t recording. The following, however, are explicit calculations from the data in a form appropriate for phase transition release. Each fluorescence reading (F) is first corrected to 45° by an empirical expression obtained from fluorescence measurements on free CF: $F' = F(1.437 - 0.00977T)$, where T is the temperature. The temperature-corrected post-Triton fluorescence is multiplied by an additional factor of 1.04 (to account for the dilution with Triton) to obtain F_t'. The CF fluorescence F' observed at any moment during a scan is

$$F' = [\beta + \alpha(1 - \beta)]F_t' \quad (1)$$

where β is the fraction of CF outside of liposomes and α is the quenching factor (see Fig. 4) for CF in the liposomes. Rearranging Eq. (1),

$$\beta = (F'/F_t - \alpha)/(1 - \alpha) \quad (2)$$

In order to use Eq. (2), it is necessary to decide on a model for the CF release process. If the mechanism were all-or-none, α would remain constant (at about 0.05 for vesicles containing 100 mM CF), since dye remaining in the unperturbed vesicles would still be at 100 mM. If, on the other hand, the process were a permeation involving all vesicles equally, α

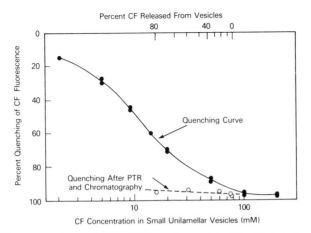

FIG. 4. All-or-none release of CF by HDL apolipoprotein. The filled circles represent the quenching of CF in vesicles. Open circles represent quenching ratios after phase transition release with apolipoprotein, followed by chromatographic separation of released CF (from Ref. 12).

would change as a function of the fractional dye release; an iterative calculation would be required. Experiments suggest that spontaneous release away from the transition in the absence of lipoprotein is a permeation, but that release induced by apolipoprotein at T_c is all-or-none (see Fig. 4).

Phase Transition Release Profiles

Figure 5A shows a series of fluorescence profiles indicating CF release at different HDL concentrations. There is little or no release in the absence of lipoprotein but essentially 100% release from 60 μM liposomes at T_c within a few seconds for HDL concentrations above 20 μg/ml. On this fast time scale, there was little release at temperatures well below T_c. Above T_c, the release rate takes on intermediate values. To explore the kinetics of interaction, it would often be necessary to use capillary tubes for even faster temperature equilibration or to employ fast-flow techniques. In performing phase transition release experiments with dilute apolipoproteins and liposomes, it is important to remember that binding steps prior to passage through T_c may be rate limiting.[14] The method of calculation just described is used to translate the experimental curves in Fig. 5A into a titration profile such as that in Fig. 5B. Similar profiles have been obtained for mixed HDL apolipoproteins,[12] apoA-I,[14] and other lipoprotein and apolipoprotein fractions.

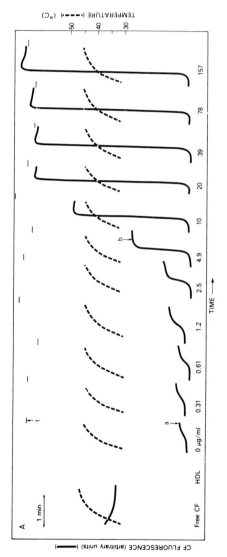
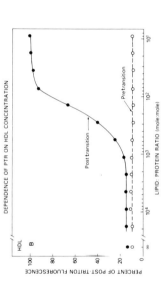

FIG. 5. Phase transition release for small unilamellar DPPC/DSPC vesicles as a function of HDL concentration. (A) Release profiles. Temperature measurements with a thermocouple probe are shown as dashed lines. The first profile is that for free CF. The downward slant results from the temperature dependence of CF fluorescence. Scans 2–11 are for vesicles. The lipid concentration was 60 μM for each scan, and the HDL protein concentration were varied as indicated. The letters a, b, and t are as described in ref. 12. (B) Phase transition release titration curve calculated from the profiles in (A) (from Ref. 12).

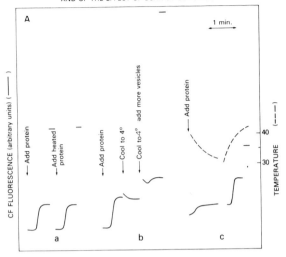

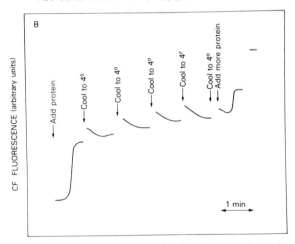

FIG. 6. Studies of the mechanism of interaction between DPPC vesicles and HDL apolipoprotein. (A) Repeated temperature scans with small unilamellar DPPC vesicles and 0.029 μM HDL apolipoprotein indicate irreversibility of interaction. After the initial scan, there was little additional release of CF until more apolipoprotein was added to bring the final concentration to 0.058 μM. (B) Two cycles of the protein through T_c before addition to the vesicles made no difference (a). After the combining power of the apoprotein had been "used up" in an initial scan, addition of more CF liposomes led to little or no further CF release (b). Downward temperature scan produced little release; upward scan then produced the expected release (c). The lipid and protein concentrations were 40 and 0.027 μM, respectively (from Ref. 12).

Mechanisms of Liposome–Apolipoprotein Interaction

Phase transition release has been used in conjunction with other biochemical and spectroscopic techniques to obtain information on the mechanisms of interaction.[12,14] Those references should be consulted for detailed descriptions and results. The approach is illustrated for HDL apolipoprotein by the sequence of temperature scans in Fig. 6A and B. These scans indicate an irreversible interaction in which lipid and protein are used up in a recombination process and become unavailable for further interaction. The open circles in Fig. 4 indicate that the release process is all-or-none. For that experiment, small unilamellar vesicles were formed in 92 mM CF. Portions of the suspension were passed through T_c, with or without sufficient HDL apolipoprotein to produce partial CF release. Each sample was then cooled to room temperature and eluted on a Sephadex G-25 column. As the vesicle peak eluted from the column, a portion was dripped directly into a cuvette for reading of fluorescence. Triton was then added and a second reading taken. The percentages of quenching thus obtained after partial release and chromatographic separation of released CF could be compared with those found immediately after formation of vesicles in different concentrations of CF. The quenching changed little with partial release, consistent with an all-or-none process in which vesicles not perturbed still contained their full complement of dye after passage through T_c.

Section V

Molecular Biology of Plasma Lipoproteins

[39] Measurement of Apolipoprotein mRNA by DNA-Excess Solution Hybridization with Single-Stranded Probes

By DAVID L. WILLIAMS, THOMAS C. NEWMAN, GREGORY S. SHELNESS, and DAVID A. GORDON

Introduction

Nucleic acid hybridization provides an invaluable method for the detection and measurement of specific mRNA molecules. There are basically three hybridization methods one can use: RNA-excess solution hybridization or R_0t analysis, hybridization to immobilized RNA or blot hybridizations, and DNA-excess solution hybridization. Detailed discussions of these techniques and their applications have been presented.[1-3] R_0t analysis is based on the kinetics of hybridization of a tracer cDNA probe to excess RNA. This method will measure relative amounts of the specific mRNA in different samples and will give absolute amounts if a pure mRNA standard is available. The disadvantages of R_0t analysis are that it requires very large amounts of RNA to measure low abundance mRNAs and often shows nonideal kinetics at high RNA concentrations. R_0t analysis is generally regarded as unreliable for measurements of less than about 10-20 mRNA molecules/cell.[4,5] Blot hybridization is usually carried out with RNAs which have been resolved by gel electrophoresis and transferred to nitrocellulose paper (Northern analysis) or after direct application to the paper (dot blot analysis). Northern analysis is primarily a qualitative method although quantitative data can be obtained if appropriate standards are included with each blot. The dot blot procedure is simple, requires a minimum of probe preparation, and is a quantitative method. The disadvantages of dot blot analysis are that only a fraction of the immobilized RNA is available for hybridization and the total amount of RNA that can be analyzed is usually limited to a few micrograms.

DNA-excess solution hybridization provides the greatest sensitivity and reliability for mRNA measurement. This method is analogous to a receptor binding assay in which the receptor (the mRNA) is reacted to

---

[1] J. O. Bishop, *Acta Endocrinol. Suppl.* **161**, 247 (1972).
[2] S. J. Flint, *in* "Genetic Engineering" (J. K. Setlow and A. Hollaender, eds.), Vol. 2, p. 47. Plenum, New York, 1981.
[3] S. S. Longacre and B. Mach, *Nucleic Acids Res.* **6**, 1241 (1979).
[4] R. G. Deeley, J. I. Gordon, A. T. H. Burns, K. P. Mullinix, M. Bina-Stein, and R. F. Goldberger, *J. Biol. Chem.* **252**, 8310 (1977).
[5] J. P. Jost, T. Ohno, S. Panyim, and A. R. Schuerch, *Eur. J. Biochem.* **84**, 355 (1978).

completion with an excess of ligand (single-strand cDNA). In practice, total cellular RNA is incubated with an excess of radiolabeled single-stranded probe complimentary to a specific mRNA. Probe which remains unhybridized is subsequently digested with S1 nuclease, a single-strand-specific nuclease. Hybridized probe is collected by acid precipitation and measured by scintillation spectrometry. This procedure has several advantages. First, the assay is carried out in solution such that all the mRNA and probe are available to react. Second, the reaction proceeds to completion and within broad limits is independent of factors that can influence hybridization kinetics. Third, this method provides absolute values for mRNA measurements. Fourth, the assay is exquisitely sensitive. Routine sensitivity is at the level of 1 mRNA molecule/cell, and the sensitivity can be greatly increased if necessary. Fifth, the assay requires small amounts of total cellular RNA. Sixth, the assay procedure itself is very simple as is the preparation of single-stranded probe. Until recently, the preparation of such probes has been difficult. However, recent developments in labeling techniques and the use of single-stranded cloning vectors have made the preparation of probes routine. We describe here the hybridization assay and two methods for the preparation of high specific activity single-stranded probes. These methods are used routinely in this laboratory for the measurement of mRNAs for human apolipoprotein (apo)E and apoA-I and for avian apo II and vitellogenin. The cDNA clones employed here for apoE and apoA-I are those isolated by Breslow et al.[6,7] The avian cDNAs were isolated and characterized in this laboratory.[8,9] Nucleotide numbers and restriction sites refer to the original publications.

Preparation of Single-Stranded Probes

Selection of an Appropriate Probe

Two methods are described for the preparation of single-stranded probe. In the first, a cDNA fragment is subcloned into bacteriophage M13 to facilitate probe synthesis. In the second, double-stranded probe is isolated and the strands separated. In either case it is important to select an

---

[6] J. L. Breslow, J. McPherson, A. L. Nussbaum, H. W. Williams, F. Lofquist-Kahl, S. K. Karathanasis, and V. I. Zannis, *J. Biol. Chem.* **257**, 14639 (1982).

[7] J. L. Breslow, D. Ross, J. McPherson, H. Williams, D. Kurnit, A. L. Nussbaum, S. K. Karathanasis, and V. I. Zannis, *Proc. Natl. Acad. Sci. U.S.A.* **79**, 6861 (1982).

[8] A. A. Protter, S.-Y. Wang, G. S. Shelness, P. Ostapchuk, and D. L. Williams, *Nucleic Acids Res.* **10**, 4935 (1982).

[9] G. S. Shelness and D. L. Williams, *J. Biol. Chem.* **259**, 9929 (1984).

appropriate region of the cDNA clone as the probe. The requirements to be met by a suitable probe include ease of isolation from the vector and a minimal amount of internal complimentarity in order that background resistance to S1 nuclease is low. In addition, one should avoid DNA fragments containing homopolymer tracks due to the initial cloning method or the poly(A) region of the mRNA. Although the effects of homopolymer tracks have not been tested, such regions might influence probe availability or hybridization kinetics. In general we have found that cDNA fragments of 150–300 nucleotides make the best probes. The most efficient way to identify suitable fragments is with one of the many sequence analysis programs available for microcomputers. The January 1982 and January 1984 issues of *Nucleic Acids Research* describe such programs for many computers. We have employed the IBM PC with the program of Conrad and Mount[10] which allows the user to define the window size for comparison and the extent of base pairing which must occur for the sequence to be scored as complimentary. From measurements with apoE cDNA fragments, we find that a window of 12 bases and a matching of 10/12 bases will provide a good indication of resistance or sensitivity to S1 nuclease. For fragments of 150–300 bases, 5 or fewer regions of internal homology resulted in negligible nuclease resistance (<1%) while one fragment with more than 40 regions of homology was 30% resistant to S1 nuclease.

Probe Preparation with Bacteriophage M13

Single-stranded probe is most easily prepared with the use of a single-stranded template as provided by the male specific bacteriophage M13. The appropriate cDNA fragment is first cloned into the double-stranded M13 replicative form at a restriction site distal to the priming site for *in vitro* DNA synthesis. One may then isolate the single-stranded M13 template containing the correct strand of the cDNA insert. The synthesis of single stranded probe from this M13 template is illustrated in Fig. 1. A commercially available 15-mer (P in Fig. 1) is hybridized to the template and the complementary strand of DNA is synthesized *in vitro* under conditions which control both the specific activity and the total amount of radiolabeled nucleotide. After synthesis past the inserted cDNA, excess unlabeled deoxynucleotide triphosphates are added to permit further elongation of the newly synthesized DNA. This facilitates subsequent isolation of the small probe DNA (region A) from the larger M13 DNA fragments (region B). The partially double-stranded structure resulting

[10] B. Conrad and D. W. Mount, *Nucleic Acids Res.* **10**, 31 (1982).

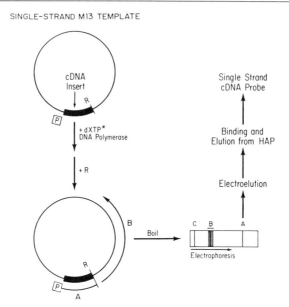

FIG. 1. Preparation of single-stranded cDNA probe by the M13 method.

from this reaction is digested at a unique restriction site at or near the 3' end of the newly synthesized cDNA (R in Fig. 1). The strands are separated by boiling, and the fragments are electrophoresed on a denaturing polyacrylamide gel to separate the probe fragment (A) from the labeled M13 fragments (B) and the unlabeled M13 template (C). After recovery from the gel by electroelution, the probe fragment is adsorbed to and eluted from hydroxylapatite (Fig. 1, HAP) to remove a small amount (1–5%) of double-stranded material. The final probe shows only 0.1–0.5% resistance to S1 nuclease and is suitable for DNA excess hybridizations. Note that the probe will contain the unlabeled M13 15-mer primer and a small region of labeled M13 sequence at the 5' end and may contain some M13 sequence at the 3' end depending on the choice of the R restriction site (Fig. 1).

Specific Probes for Apolipoprotein mRNAs

Suitable probes for human apoE mRNA have been prepared from *Pst*I fragments 29–222 and 514–671[6] cloned in M13 mp 9. In these cases R (Fig. 1) is the *Hin*dIII site in the adjacent M13 sequence. Probe prepared from apoE *Pst*I fragment 223–513 is not suitable due to high resistance to S1 nuclease. A suitable probe for human apoA-I mRNA has been prepared from the *Pst*I–*Msp*I fragment 1–480 (1 is the first coding nucleotide of the reported clone)[7] cloned into *Acc*I–*Pst*I cut M13 mp9. In this case R is the *Bgl*I site within the apoA-I fragment; cleavage at this site yields a 191 nucleotide *Msp*I–*Bgl*I apoA-I probe and a smaller 133 nucleotide apoA-I *Bgl*I fragment from a second *Bgl*I site in the insert. In the case of avian apoII, a suitable probe is provided by *Pst*I fragment 130–369[11]; R is the *Hin*dIII site in the adjacent M13 sequence. Strategies and detailed protocols for cloning cDNA fragments into M13 mp 8 and mp 9 or similar derivatives are presented elsewhere.[12,13]

Probe Synthesis

The following is an example of a reaction to synthesize an apoE probe from the M13 clone containing *Pst*I fragment 514–671. The reaction is sufficient to make approximately 10 ng of probe at a specific activity of 640 cpm/pg. The size of the reaction and the specific activity can be easily modified if desired. M13 DNA (3.75 µg) is mixed with the 15-mer sequencing primer (45 ng) (New England Biolabs) in 7.5 µl of 0.05 M NaCl, 0.01 M Tris–HCl, pH 7.5, in a 1.5 ml polypropylene centrifuge tube. The tube is sealed with teflon tape to prevent evaporation and placed in a preheated 15 ml water filled flask in a 65° oven for 15 min. Slow cool the sample by placing the flask at room temperature for 30 min and microfuge for 5 sec. This mixture and a 1.5 µl wash of the tube with 0.009 M DTT are added to another 1.5 ml centrifuge tube in which radiolabeled and unlabeled deoxynucleotide triphosphates and buffer components have been dried. A Speed Vac Concentrator (Savant Instruments, Inc., Hicksville, NY 11801) is preferred over lyophilization because the reagents are concentrated in a small area of the tube and can be dissolved in a volume of only a few microliters. After the addition of the Klenow fragment of *E. coli* DNA polymerase (5 units) (New England Biolabs), the final concentrations are 250 µM dATP, 250 µM dGTP, 250 µM dTTP, 81 µM [α-^{32}P]dCTP (222 Ci/mmol, Amersham), 0.009 M Tris–HCl, pH 7.5, 0.05 M NaCl, 0.008 M MgCl$_2$, 0.0015 M DTT, plus components from the Klenow storage buffer. The sample is incubated for 1 hr at 37° after which 1 µl of a chase mixture containing each of the four dXTPs at 5 mM is added. Incubation is continued for 15 min at 37°. Synthesis is terminated by heating for 10 min at 65°.

For the restriction digest at R1 (Fig. 1), add to the sample 2 µl 0.5 M Tris–HCl, pH 8, 0.5 M NaCl, 0.1 M MgCl$_2$, 40 units *Hin*dIII (Bethesda

---

[11] B. Wieringa, G. Ab, and M. Gruber, *Nucleic Acids. Res.* **9**, 489 (1981).
[12] J. Messing, B. Gronenborn, B. Muller-Hill, and P. W. Hofschneider, *Proc. Natl. Acad. Sci. U.S.A.* **74**, 3642 (1977).
[13] B. Gronenborn and J. Messing, *Nature (London)* **272**, 375 (1978).

Research Laboratories), and water to a final volume of 30 µl. Incubate for 2 hr at 37°. Ethanol precipitate by the addition of 3 µl 3 M Na-acetate and 2.5 vol ethanol and incubation at $-70°$ for 30 min. After centrifugation for 5 min in a microfuge, the pellet is rinsed with 100% ethanol, recentrifuged, and lyophilized for 5 min. The ethanol supernatants should be discarded as radioactive waste.

The DNA pellet is dissolved in 50 µl 90% deionized formamide, 1 × TBE (0.089 M Tris, 0.089 M boric acid, 0.002 M EDTA, pH 8.3), 0.02% bromophenol blue, 0.004 M NaOH, 0.02% xylene cyanol, denatured by boiling for 5 min, and electrophoresed on a prerun (30 min at 45 mA to warm gel) 7 M urea–5% polyacrylamide gel (2–2.5 mm spacers) to resolve the probe fragment (A, Fig. 1) from the labeled M13 fragments (B, Fig. 1) and the M13 template (C, Fig. 1). The labeled probe fragment is located by a 15 sec X-ray film exposure and recovered from the gel by electroelution in 0.5 × TBE. The electrophoresis and electroelution procedures are described in detail in the Cold Spring Harbor cloning manual.[14] The electroeluted DNA may be loaded directly onto hydroxylapatite. Efficient electroelution also is achieved with the Model UEA Unidirectional Electroelutor (cat. No. 46000, International Biotechnologies, Inc., PO Box 1565, New Haven, CT 06506). With this apparatus the probe is recovered in 7 M ammonium acetate. After a 3-fold dilution, the sample can be loaded directly onto hydroxylapatite.

Hydroxylapatite Fractionation

Hydroxylapatite (HAP) (Bio-Rad HTP) is suspended in 0.02 M Na-phosphate, pH 7.1 (NaP$_i$) and allowed to settle for a few minutes before removing the fines. This is repeated three times, and the HAP is suspended such that 1 ml of slurry yields approximately 0.3 ml of packed HAP after a 5 sec centrifugation in a microfuge. Probe is added to 0.3 ml of packed HAP in a 1.5 ml centrifuge tube and mixed at room temperature for 5 min. The tube is centrifuged for 5 sec in a microfuge and the supernatant removed. This process is repeated until all the sample from the electroelution has been loaded onto the HAP. The supernatants from the loadings should be saved at each step and checked with a geiger counter. If a significant amount of probe remains in the supernatant, additional HAP (0.1 ml) should be added to the supernatant and mixed as above. After centrifugation, this HAP should be added to the first batch. The HAP is washed sequentially with increasing NaP$_i$ concentrations to elute

---

[14] T. Maniatis, E. F. Fritsch, and J. Sambrook, "Molecular Cloning." Cold Spring Harbor Press, Cold Spring Harbor, New York, 1982.

the single and double-stranded material. It is useful to do an entire series of NaP$_i$ washes the first time a new probe is made in order to determine the complete elution profile. Once the profile is established, only the washes to elute single-strand material need be done. The NaP$_i$ washes are done by suspending the HAP in 0.75 ml, incubating for 8 min at 55° in a water bath, and centrifuging 5 sec in a microfuge. The number of washes at each NaP$_i$ concentration are given as (1×, 2×, etc.). The NaP$_i$ washes are 0.02 M (2×), 0.07 M (2×), 0.12 M (3×), 0.17 M (3×), 0.22 M (3×), 0.27 M (2×), 0.32 M (2×), 0.37 M (2×), 0.42 M (2×), and 0.47 M (2×). The single-stranded probe elutes between 0.12 and 0.22 M NaP$_i$ while the double-stranded material elutes between 0.37 and 0.42 M NaP$_i$. It is important to do the 0.02 and 0.07 M washes because very small double-stranded DNA fragments which do not bind strongly to HAP are removed. The single-strand fractions are pooled and dialyzed against two changes of 0.01 M Tris–HCl, pH 7.5, 0.001 M EDTA to remove phosphate which inhibits S1 nuclease. An alternative to dialysis is a vacuum dialysis procedure using collodion membranes (Model UH 020/2a, Schleicher and Schuell, Inc., Keene, New Hampshire, 03431). With this inexpensive apparatus the sample is dialyzed and concentrated simultaneously without the need to handle a dialysis bag containing a highly radioactive sample. This procedure is also faster than normal dialysis generally requiring only 2–3 hr. The colloidion membrane is presoaked in 200 μg/ml herring sperm DNA to reduce nonspecific loss of probe. The membrane may be rinsed after use, stored in 20% ethanol, and reused many times. After dialysis, a sample is taken for the measurement of radioactivity, and probe is stored at 0–4°. Probe preparation by this method can be carried out in 1 day.

Probe Preparation by Replacement Synthesis

An alternate method for preparing single-stranded DNA probe involves radiolabeling of the desired double-stranded fragment followed by strand separation. Two labeling methods have been employed. Nick translation[15] does not appear suitable for this purpose since the resultant single-stranded probes contain high levels (5–10%) of S1 nuclease-resistant material which is not removed by HAP.[16] In contrast, replacement synthesis with T$_4$ DNA polymerase[17] yields probes with very low S1 nuclease resistance. Figure 2 illustrates this method for labeling the avian

---

[15] P. W. J. Rigby, M. Dieckmann, C. Rhoades, and P. Berg, *J. Mol. Biol.* **113**, 237 (1977).
[16] A. A. Protter, Ph.D. thesis, State University of New York at Stony Brook, New York, 1982.
[17] P. H. O'Farrell, E. Kutter, and M. Nakanishi, *Mol. Gen. Genet.* **179**, 421 (1980).

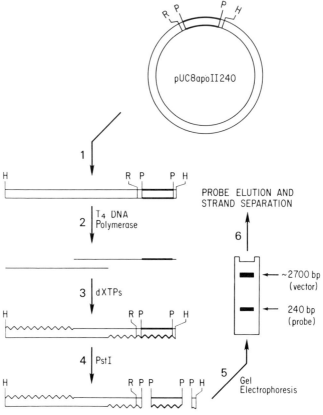

FIG. 2. Preparation of single stranded apo II cDNA probe by replacement synthesis and strand separation. R, EcoRI; P, PstI; H, HindIII.

apoII PstI fragment 130–369 which was cloned into the PstI site of the pUC8 vector. The plasmid is first linearized by digestion at the unique HindIII site (Fig. 2, step 1, H) which lies just outside the cDNA insert. In the absence of deoxynucleotide triphosphates the strong 3'-5' exonuclease activity of T_4 DNA polymerase resects the entire noncoding strand of the apoII cDNA in a timed reaction (Fig. 2, step 2). After the addition of unlabeled and labeled deoxynucleotide triphosphates, replacement synthesis results in uniform labeling of one strand of the cDNA insert (Fig. 2, step 3). The DNA is then digested with PstI (Fig. 2, step 4), and the 240 base pair probe is purified by gel electrophoresis under nondenaturing conditions (Fig. 2, step 5). After elution from the gel, the probe fragment is denatured and strand separated (Fig. 2, step 6) by hybridization to excess rooster liver RNA under conditions which permit the formation of RNA:DNA hybrids but not DNA:DNA duplexes.[18] Following digestion with S1 nuclease to eliminate the unwanted DNA strand, RNA is eliminated by alkaline hydrolysis. Probe prepared in this fashion shows about 1% resistance to S1 nuclease; this can be further reduced to 0.05–0.1% by a second round of hybridization to liver RNA if desired.

Probe Synthesis

Apo II cDNA plasmid (3.5 µg) is digested with HindIII (25 U) in 30 µl 0.033 M Tris-acetate, pH 7.9, 0.066 M K-acetate, 0.01 M MgCl$_2$ for 2 hr at 37°. Subsequent incubations are also at 37°. T$_4$ DNA polymerase (Bethesda Research Laboratories) (36 U) is added and the incubation continued for 45 min. The reaction is then supplemented with 2.8 nmol [α-^{32}P]dCTP (50–200 Ci/mmol) (Amersham) and 20 nmol each of dTTP, dGTP, and dATP in the above buffer to give a final volume of 60 µl, and incubation is continued for 30 min. PstI (50 U) is added, and the sample is incubated for 2 hr. The sample is then loaded on a neutral 4% polyacrylamide-TBE 14 cm slab gel[14] and electrophoresed until the bromophenol blue tracking dye has migrated 2/3 the length of the gel. The 240 basepair probe fragment is located by autoradiography and recovered by electroelution.[14] After the addition of 100 µg carrier DNA, the probe is ethanol precipitated as described above.

Strand Separation

Double-stranded probe (100 ng) is mixed with 50 µg poly(A)$^+$ RNA prepared from estrogen-stimulated rooster liver,[8] and ethanol precipitated as described above. The pellet is dissolved in 100 µl 0.1 M N-2-hydroxyethylpiperazine-N'-2-ethanesulfonic acid, pH 6.8, 0.5 M NaCl, 0.0025 M EDTA, 70% deionized formamide, and incubated at 75° for 3 min to denature the probe. After a further incubation at 51° for 3 hr, the sample is added to 4 ml of 0.03 M Na-acetate, pH 4.6, 0.3 M NaCl, 0.006 M MgCl$_2$ containing 3200 U S1 nuclease (Miles Laboratories), and incubated for 1 hr at 44°. Carrier DNA (100 µg) is added, and the probe is ethanol precipitated as above. A second round of hybridization to poly(A)$^+$ RNA is carried out as above. The final DNA pellet is dissolved in 0.3 M NaOH, incubated for 12–18 hr at 37°, and chromatographed on a 10 ml column of Sephadex G-100 equilibrated with 0.02 M NaOH. The void volume is neutralized with HCl, and the probe is ethanol precipitated as above, and washed once with 100% ethanol. Probe is dissolved in 0.01 M Tris–HCl, pH 7.5, 0.001 M EDTA and stored at 4°.

[18] J. Casey and N. Davidson, *Nucleic Acids Res.* **4**, 1539 (1977).

Comments

1. The use of a unique restriction site adjacent to the cDNA insert as illustrated in Fig. 2 simplifies the protocol since timing of the exonuclease step is not critical. Nevertheless, one should determine empirically that the digestion proceeds beyond the cDNA fragment in order that the probe fragment is uniformly labeled. This is most easily accomplished by a restriction digest, after replacement synthesis, with an enzyme that cuts at numerous sites in the vector. Inspection of the radioactivity in each band after electrophoresis and autoradiography will show how far the exonuclease digestion proceeded. The T_4 DNA polymerase labeling procedure can also be used to label small DNA fragments (200–400 base pairs) directly, although in this case the timing of digestion is critical since digestion past the midpoint of the molecule leads to strand separation and more rapid digestion of the individual strands.

2. In the above protocol, strand separation is achieved by hybridization to excess poly(A)$^+$ RNA containing the mRNA complimentary to the cDNA probe. This approach is suitable only if the mRNA of interest is moderately to highly abundant as is the case with apo II mRNA. We have also used this approach to prepare single-stranded probes to avian vitellogenin and serum albumin mRNAs. With less abundant mRNAs, the long hybridization times and high RNA concentrations make this approach impractical. Several alternatives are available for strand separation including strand separation gels[14] or denaturing gel electrophoresis of complimentary DNA strands which are of slightly different lengths due to the ends left by particular restriction enzymes. With fragments up to 175 nucleotides in length, a difference of a few nucleotides will permit strand separation in DNA sequencing gels.

3. The protocol for labeling the apo II cDNA probe has been designed to minimize buffer changes and manipulation of the DNA. Since most of the steps are carried out in the same buffer, the enzyme reactions include more activity than would be needed under optimal buffer conditions for each enzyme.

Hybridization Assay

RNA Preparation

The recovery of intact RNA free of contaminating ribonuclease is essential. While a variety of methods are available for RNA purification, two methods, in particular, yield high quality RNA even from tissues such as liver that are rich in ribonuclease. The first is the guanidine–HCl

TABLE I
CONCENTRATION RANGES FOR mRNA MEASUREMENT

mRNA (molecules/cell)	Assay range (μg)	Example
1–10	10–200	ApoE, monkey thymus
10–100	1–10	ApoE, monkey kidney
100–1000	0.1–1	ApoE, monkey liver
1000–10,000+	0.01–0.1	Apo II, chick liver

method of Cox[19] followed by phenol-chloroform extractions as described.[8] The second is the guanidine thiocyanate method[20] combined with the CsCl$_2$ centrifugation procedure of Glisin et al.[21] A detailed protocol for the second method is given in the Cold Spring Harbor cloning manual.[14] Purified RNA may be stored in 0.02 M Tris–HCl, pH 7.5, 0.001 M EDTA at $-70°$ or as an ethanol precipitate at $-20°$. One problem encountered with liver RNA purified by the guanidine–HCl method of Cox[19] is contamination with large amounts of polysaccharide which interfere with the measurement of RNA by UV absorbance. Accurate measurement of RNA in such samples can be made by acid precipitation of the RNA in expendable aliquots of the sample. For this purpose a sample containing 5–100 μg RNA is adjusted to 1.25% trichloroacetic acid and kept on ice for 10 min. After centrifugation in a microfuge for 5 min at 4°, the supernatant containing the polysaccharide is removed. The RNA pellet is washed once with 66% ethanol, 0.15 M NaCl and once with 100% ethanol to remove trichloroacetic acid, and the pellet is dried by lyophilization for 5 min. The pellet is dissolved in 1 ml 0.02 M Tris–HCl, pH 7.5, 0.001 M EDTA for the measurement of absorbance at 260 nm. An absorbance of 1 in a 1 cm pathlength cuvette is equivalent to 40 μg/ml RNA.

The RNA to be assayed is prepared in 0.02 M Tris–HCl, pH 7.5, 0.001 M EDTA and several dilutions are made in duplicate directly in the 1.5 ml microfuge tubes in which the hybridizations will be performed. The RNA concentration range to cover depends on the abundance of the mRNA species in question. Table I shows the approximate RNA input ranges for mRNAs of different abundances with several examples. For each sample several RNA dilutions are assayed in duplicate within the linear range of the assay. If no information is available as to the abundance of the

---

[19] R. A. Cox, this series, Vol. 12, p. 120.
[20] J. M. Chirgwin, A. E. Przybyla, R. J. MacDonald, and W. J. Rutter, *Biochemistry* **18**, 5294 (1979).
[21] V. Glisin, R. Crkvenjakov, and C. Byus, *Biochemistry* **13**, 2633 (1974).

mRNA, a preliminary assay with RNA dilutions in 10-fold steps is advisable. If the approximate range is known, dilutions are in 3-fold steps.

Assay Procedure

The RNA sample (20 µl) is mixed with the probe (80 µl) containing the buffer components in a 1.5 ml polypropylene centrifuge tube. The final concentrations are 100 pg probe, 0.03 M Tris–HCl, pH 7.0, 0.3 M NaCl, 0.02 M EDTA, 5 µg/ml yeast RNA, and 100 µg/ml denatured herring sperm DNA. After mixing, the tube is centrifuged in a microfuge for 5 sec and overlaid with 50 µl paraffin oil. The sample is incubated for 60 hr in a 68° oven. The sample is then diluted with 0.9 ml 0.03 M Na-acetate, pH 4.5, 0.3 M NaCl, 0.006 M $ZnCl_2$, 20 µg/ml denatured calf thymus DNA, containing 3000 U S1 nuclease (Miles Laboratories) and incubated for 2 hr at 45°. After the addition of 100 µg carrier DNA and 200 µl 50% trichloroacetic acid, the sample is kept on ice for 10 min, filtered through a G F/C glass fiber filter (Whatman), washed twice with 20 ml 7.5% trichloroacetic acid, and dried under an infrared lamp. The filter is counted in a scintillation cocktail containing 3.8 g Omnifluor (New England Nuclear), 25 ml Protosol (New England Nuclear), and 3.5 ml water per liter of toluene. The DNA is solubilized from the filter in this cocktail.

Comments

1. In principle, the hybridization probe should be in a sufficiently high concentration that the reaction is driven to completion within the assay period. In practice, the hybridization reaction will approach completion asymptotically as a function of time or probe concentration. The influence of probe concentration on the extent of completion of the hybridization reaction has been estimated by examining the reassociation of double-stranded probes under the standard assay conditions given above. Since the rate constants for RNA : DNA hybridization and DNA : DNA reassociation are very similar (1,2), the extent of DNA : DNA reassociation for a particular probe concentration under standard assay conditions (0.3 M NaCl, 68°) should approximate the extent of the hybridization reaction at the same probe concentration when the probe is in excess to the hybridizing mRNA species. Table II illustrates the influence of probe concentration on the reassociation of a 200 base pair double-strand fragment. The rate constant, K, for these calculations was determined experimentally as described in Table II. These values show approximately 90% completion of the reaction in 60 hr at a probe concentration of 100 pg/100 µl. This probe concentration was selected for the standard assay as a compromise that permits near completion of the reaction without using prohibitive

TABLE II
CALCULATED REASSOCIATION VALUES AS A
FUNCTION OF FRAGMENT CONCENTRATION

Probe[a] (pg/100 μl)	Completion[b] (%)
50	81.2
75	86.6
100	89.6
150	92.8
200	94.5
250	95.6
300	96.3

[a] Calculations are for a fragment size of 200 base pairs.
[b] This is the extent of reassociation of DNA strands under standard assay conditions (0.3 M NaCl, 68°, 60 hr) calculated from $C_0/C = 1 + KC_0t$ [R. J. Britten and D. E. Kohne, Science **161**, 529 (1968)]. The value of K (2.5×10^4 LM^{-1} sec^{-1} per 100 bp fragment length) was determined experimentally by monitoring the reassociation of a 240 bp apo II cDNA fragment and a 580 bp vitellogenin cDNA fragment under standard assay conditions.

amounts of probe. Table III shows similar values which illustrate the influence of fragment size on the reassociation of DNA strands under standard assay conditions. Note that in extrapolating these values to the hybridization reaction, the concentration of the mRNA species would also contribute to the extent of the reaction. However, this effect would be small since the probe is in excess to the hybridizing mRNA. The basic point is that even in RNA samples with very low amounts of the mRNA of interest, the probe concentration alone is sufficient to drive the hybridization reaction to completion.

2. The typical assay includes several dilutions of each RNA sample as well as samples that contain probe but not RNA. One set of samples lacking RNA are used to determine the S1 nuclease-resistant background of the probe; this is usually less than 1% of the input. This background value is subtracted from the experimental samples. The other set of samples lacking RNA are carried through the assay but are not treated with S1 nuclease. These samples confirm the amount of probe used in the assay. Samples containing up to 200 μg of RNA may be assayed as described

TABLE III
CALCULATED REASSOCIATION VALUES AS A FUNCTION OF FRAGMENT SIZE

Probe (base pairs)	$C_0 t_{1/2}{}^a$ (ML^{-1} sec × 10^{-4})	$C_0 t / C_0 t_{1/2}{}^b$	Completion[c] (%)
100	0.4	17	94.6
150	0.6	11.5	92.0
200	0.8	8.6	89.6
250	1.0	6.9	87.3
300	1.2	5.8	85.1
400	1.6	4.3	81.2
500	2.0	3.4	77.5

[a] $C_0 t_{1/2}$ values were calculated for the indicated fragment size using the equation and K value as described in Table II.
[b] This is the ratio of the $C_0 t$ achieved in 60 hr with 100 pg/100 μl probe to the $C_0 t_{1/2}$ for the indicated probe size.
[c] This is the percentage completion of the reassociation reaction achieved in 60 hr at 100 pg/100 μl probe calculated as described in Table II. The K value for each probe is the reciprocal of the $C_0 t_{1/2}$.

above. If larger RNA samples are to be assayed, it may be necessary to increase the amount of S1 nuclease. Note that all suppliers do not have the same unit definition for S1 nuclease activity. The unit referred to here is that amount of enzyme which produces 1 μg of acid soluble deoxypentose in 30 min.

3. Calculation of mRNA values from the hybridization data may be done in two ways. First, one may calculate how much mRNA is present on the basis of the hybridization value, the probe specific activity, the size of the probe compared to the mRNA, and the extent of completion of the reaction. This method relies heavily on the absolute specific activity of the radiolabeled precursor as supplied by the manufacturer, the accuracy of any changes one may make in the specific activity, and the absolute recovery of the sample in the assay. For example, we normally make probe with a single precursor at about 200 Ci/mmol which is a 4-fold change from the specific activity as supplied by the manufacturer. Errors in the mixing of labeled and unlabeled nucleotide, in the measurement of the unlabeled nucleotide, or variations in the quality of labeled or unlabeled nucleotide may lead to substantial errors in the calculation of mRNA values.

Second, one may determine the mRNA value by comparison to an hybridization standard. This method is reliable and gives consistent values from one assay to the next. In the case of probes made by the M13 method, the template M13 DNA serves as an appropriate standard which is single stranded, pure, and can be accurately measured by UV absorbance. In addition, since the assay proceeds to completion, any minor differences in the kinetics of DNA:DNA versus DNA:RNA annealing are eliminated. Figure 3A shows a typical standard curve for the hybridization of an human apoE probe to the template DNA from which it was synthesized; Figure 4A shows a similar standard curve for an apoA-I probe. Hybridization is linear in both cases over the entire range of input standard DNA. Figure 3B shows the hybridization of the apoE probe to RNA isolated from human Hep G2 cells. By comparison to the standard curve, 1 µg of RNA hybridized to 1.7 pg of apoE probe. Correction for the size of the probe (157 bases) compared to apoE mRNA (1150 bases) indicates that apoE mRNA is present at 12.4 pg/µg RNA. Similarly, Fig. 4B shows that 1 µg of RNA hybridized to 3.3 pg of apoA-I probe. Correction for the size of the probe (191 bases) compared to apoA-I mRNA (1100 bases) yields 19 pg/µg RNA.

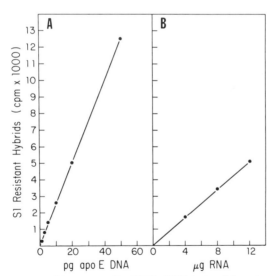

FIG. 3. Measurement of human apoE mRNA. RNA isolated from human Hep G2 cells was hybridized with a single-stranded apoE cDNA probe as described in the text. (A) The hybridization of probe to the M13 template DNA from which it was synthesized. Input DNA is expressed in terms of the apoE cDNA insert. (B) The hybridization of probe to Hep G2 RNA.

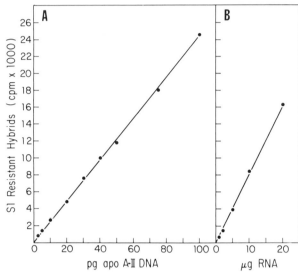

FIG. 4. Measurement of human apoA-I mRNA. RNA isolated from human Hep G2 cells was hybridized with single stranded apoA-I cDNA probe as described in the text. (A) The hybridization of probe to the M13 template DNA from which it was synthesized. Input DNA is expressed in terms of the apoA-I cDNA fragment in the probe. (B) The hybridization of probe to Hep G2 RNA.

When probe is made by replacement synthesis, a pure single-stranded hybridization standard may not be available. In this case the data yield relative values within an assay. To compare between assays, one RNA sample is selected as a standard and run with each assay. The calculation of absolute mRNA values must be made by the first method described above. A third method for assay standardization involves the synthesis of pure mRNA sequence using an appropriate cloning vector based on the Sp6 phage promoter.[22]

4. Assay sensitivity is determined by (1) probe specific activity, (2) background S1 nuclease resistance of the probe, (3) and the quantity of RNA that is assayed. As an example, with a specific activity of 500 cpm/pg and a typical S1 nuclease resistant background of 1%, the background from 100 pg of probe is 500 cpm. The minimum reliable hybridization value (2× background) of 1000 cpm corresponds to 2 pg of probe or 15 pg of mRNA in the case of apoE mRNA. This represents 24×10^6 molecules of apoE mRNA. If this value were obtained with 200 μg of RNA repre-

[22] M. Green, T. Maniatis, and D. Melton, *Cell* **32,** 681 (1983).

senting 20×10^6 cells, the minimum reliable value is approximately 1 mRNA molecule/cell. Similarly, if 20 or 2 μg of RNA was assayed, the minimum reliable values would be 10 mRNA molecules/cell or 100 mRNA molecules/cell, respectively. For most purposes a sensitivity of a few mRNA molecules per cell seems adequate. However, if it were necessary to increase the sensitivity, this may be done by decreasing the S1 nuclease resistance of the probe or scaling up the assay to measure more RNA. With probes made in M13 the S1 nuclease resistance is generally 0.5–1%. Probe made by replacement synthesis and two rounds of preparative hybridization frequently shows less than 0.1% resistance to S1 nuclease thus affording a 10-fold increase in sensitivity. Similarly, the specific activity of the probe can be increased at least 10-fold although in most cases it is not the absolute amount of radioactivity that limits the sensitivity of the assay.

Normalization of mRNA Values

The assay described above yields absolute mRNA values per unit of total cellular RNA. Expression of these values on a tissue weight or per cell basis requires knowledge of the RNA and DNA contents of the tissue or cells from which the RNA is isolated. Calculations for the normalization of mRNA values to tissue weight, DNA content, and molecules/cell are illustrated in Table IV. Tissue DNA and RNA contents are measured by colorimetric assays after selectively solubilizing RNA and DNA components according to the following protocol.

Step 1. A weighed tissue fragment (50–100 mg) or cell pellet is homogenized in 2 ml 5% cold trichloroacetic acid (TCA) using a motor-driven ground glass tissue grinder (Kontes, Duall grinder) until the homogenate is uniform and milky in appearance. The homogenate is quantitatively transferred to a 12 ml graduated centrifuge tube with 2 or more 1 ml washes of the homogenizer with 5% TCA. The sample is centrifuged at 2000 g for 10 min, and the supernatant is discarded.

Step 2. Suspend the pellet in 2–4 ml 5% TCA, and vortex vigorously until well mixed. Wash down the walls of the tube with additional 5% TCA, and centrifuge as above. Discard the supernatant, and repeat the suspension and washing of the insoluble pellet 4 more times. Drain the pellet well after the last centrifugation.

Step 3. Suspend the pellet in 0.5 ml 0.3 N KOH, cover with parafilm, and incubate at 37° for 12–15 hr to hydrolyze RNA. Cool the sample on ice, and add 0.5 ml 10% TCA. After an additional 15 min on ice, centrifuge the sample as above and remove the supernatant for the determination of ribose by the orcinol procedure.

TABLE IV
CALCULATION OF ApoE mRNA VALUES

Parameter	Value	Source
1. Typical values from the liver of the cynomolgus monkey, *Macaca fascicularis*		
ApoE mRNA	20×10^{-6} g/g RNA	Hybridization assay
Tissue RNA	6.2×10^{-3} g/g tissue	Colorimetric assay
Tissue DNA	2.0×10^{-3} g/g tissue	Colorimetric assay
ApoE mRNA molecular weight	3.7×10^5	mRNA sequence[a]
Diploid DNA content	5.45×10^{-12} g/cell	[b]
2. Normalization to tissue weight 20×10^{-6} g mRNA/g RNA $\times$ 6.2×10^{-3} g RNA/g tissue = 124×10^{-9} g mRNA/g tissue		
3. Normalization to DNA content 20×10^{-6} g mRNA/g RNA $\times$ 6.2 g RNA/2.0 g DNA = 62×10^{-6} g mRNA/g DNA		
4. Normalization to molecules/cell 62×10^{-6} g mRNA/g DNA $\times$ 5.45×10^{-12} g DNA/cell $\times$ 1.63×10^{18} molecules mRNA/g mRNA = 550 mRNA molecules/cell		

[a] V. I. Zannis, J. McPherson, G. Goldberger, S. K. Karathanasis, and J. L. Breslow, *J. Biol. Chem.* **259**, 5495 (1984).

[b] M. G. Manfredi Romanini, *Proc. Int. Congr. Primatol.*, 3rd **2**, 141 (1971).

Step 4. Suspend the pellet in 2 ml 5% TCA, wash down the tube walls with 2–4 ml 5% TCA, centrifuge as above, and discard the supernatant.

Step 5. Suspend the pellet in 1 ml 5% TCA and heat for 20 min at 90°. A marble or inverted glass ampule may be used as a condensor to minimize evaporation. Avoid boiling or bumping may occur. Cool the sample on ice, add additional 5% TCA to a volume of 1.5 ml, and centrifuge as above. Remove the supernatant for the determination of deoxyribose by the diphenylamine procedure.

Orcinol procedure: as described.[23]

Reagents

1. 60 ml concentrated sulfuric acid plus 40 ml water
2. 1.6 g orcinol (Sigma Chemical Company) in 100 ml water prepared just before use
3. 1 vol reagent 2 plus 7.5 vol reagent 1

[23] E. A. Kabat and M. M. Mayer, "Experimental Immunochemistry." Thomas, Springfield, Ill., 1961.

Procedure

1. Add 2.12 ml reagent 3 to 0.25 ml supernatant from step 3
2. Heat 15 min at 80°
3. Measure absorbance at 540 nm.

Standards. Prepare a standard curve with 5–150 μg RNA introduced at the beginning of step 3 in the above procedure. The RNA content of tissues is usually in the range of 1–6 mg/g.

Diphenylamine procedure: as described.[24]

Reagents

1. 1.5 g diphenylamine (Sigma) in 100 ml glacial acetic acid plus 1.5 ml concentrated sulfuric acid; store in the dark
2. 0.1 ml acetaldehyde in 5 ml water
3. 0.1 ml reagent 2 plus 20 ml reagent 1; reagents 2 and 3 are prepared just before use

Procedure

1. Add 1.4 ml reagent 3 to 0.7 ml supernatant from step 5
2. Incubate 16–20 hr at 30°
3. Measure absorbance at 600 nm

Standard. Prepare a standard curve with 5–100 μg DNA introduced at the beginning of step 5. The DNA content of tissues is usually in the range of 0.3–3 mg/g.

Acknowledgments

We thank Dr. Jan Breslow for kindly providing the cDNA clones for human apoA-I and apoE. Excellent technical assistance was provided by Salvatore Mungal and Penelope Strockbine. This research was supported by grants AM 18171 and HL 32868 from the National Institutes of Health. G.S.S. and D.A.G. are predoctoral trainees in Pharmacological Sciences (NIH GM 07518).

[24] K. Burton, *Biochem. J.* **62**, 315 (1956).

[40] Intra- and Extracellular Modifications of Apolipoproteins

By VASSILIS I. ZANNIS, SOTIRIOS K. KARATHANASIS, GAYLE M. FORBES, and JAN L. BRESLOW

Introduction

The plasma lipoproteins are spherical particles with cores of nonpolar neutral lipid consisting of cholesteryl ester and triglycerides and coats of relatively polar materials consisting of phospholipid, free cholesterol, and proteins.[1-5] The protein components of lipoproteins are called apolipoproteins and have been designated apoA-I, apoA-II, apoA-IV, apoB, apoC-I, apoC-II, apoC-III, apoD, and apoE.[6] The metabolism of lipoproteins is a complex pathway which contains several steps and is apparently linked tightly to the metabolism of individual apolipoproteins.

In this chapter we will review the methods used to study the intra- and extracellular modifications of apolipoproteins from the time of their synthesis to the time of their catabolism. Some of the apolipoprotein modifications may constitute important, but still poorly understood, signals which control various aspects of lipoprotein metabolism. Emphasis will be given to human apolipoprotein A-I and apolipoprotein E.

Materials

The materials, solutions and apparatus used for one- and two-dimensional gel electrophoresis, as well as the various antisera used for immunoprecipitation, are described by Zannis elsewhere in this volume [49].

[^{35}S]Methionine (300 Ci/mmol) was obtained from New England Nuclear. X-Ray film Cronex-4 was purchased from DuPont. *Clostridium*

---

[1] D. Atkinson, M. A. F. Davis, and R. B. Leslie, *Proc. R. Soc. London Ser. B* **186,** 165 (1974).
[2] P. Laggner, G. M. Kostner, U. Rakusch, and D. Worcester, *J. Biol. Chem.* **255,** 11832 (1981).
[3] L. C. Smith, J. H. Pownall, and A. M. Gotto, Jr., *Annu. Rev. Biochem.* **47,** 751 (1978).
[4] P. N. Herbert, G. Assmann, A. M. Gotto, Jr., and D. S. Frederickson, "The Metabolic Basis of Inherited Disease" (J. B. Stanbury, J. B. Wyngaarden, D. S. Frederickson, J. L. Goldstein, and M. D. Brown, eds.), 5th ed., pp. 589, 1982.
[5] A. M. Scanu, R. E. Byrne, and M. Mihovilovic, *Crit. Rev. Biochem.* **13,** 109 (1982).
[6] P. Alaupovic, "Protides of the Biological Fluids" (H. Peeters, ed.) p. 9. Pergamon, Oxford, 1971.

perfrigens neuraminidase, deoxycholate, Triton X-100, and hydrogen peroxide (30% w/w) were purchased from Sigma. Methionine-free MEM was purchased from Grand Island Biological, Co. [^{14}C]Proline (293 mCi/mmol), [^{3}H]lysine (68 Ci/mmol), [^{3}H]arginine (18.3 Ci/mmol), and [^{14}C]histidine (13.8 Ci/mmol) were purchased from New England Nuclear. IgG-Sorb was purchased from Enzyme Center, Boston, MA. The dog pancreatic membranes were a gift from J. A. Majzoub (Massachusetts General Hospital, Boston). Tunicamycin was obtained from Calbiochem. All other materials were the purest grade commercially available.

Methods

Detection of Signal Peptide Sequences in Apolipoproteins

Signal peptide sequences in apolipoproteins can be ascertained by comparison of the cell-free translation products of a specific apolipoprotein mRNA with the corresponding plasma apolipoprotein forms. The apolipoprotein forms which result from cell-free translation differ from the plasma forms by size and charge and can be easily identified by two-dimensional gel electrophoresis and autoradiography as explained below.

The precise signal peptide sequences of apolipoproteins can be derived from the DNA sequences either of full length cDNA clones or from genomic clones carrying the apolipoprotein genes. These procedures are outlined by Karathanasis *et al.* (this volume [41]). Alternatively, they can be determined by microsequence analysis of the primary translation product of apolipoprotein mRNA that was radiolabeled with ^{3}H- or ^{14}C-containing amino acids as described by Gordon *et al.*[7,8]

Synthesis of Apolipoproteins by Cell Cultures or Organ Cultures and by Cell-Free Translation of mRNA: Comparison of the Newly Synthesized with the Plasma Forms

mRNA Isolation and Cell-Free Translation. The isolation of mRNA from human liver and HepG2 cells has been described.[9,10] After optimization for mRNA concentration the translation was performed in a total

---

[7] J. I. Gordon, D. P. Smith, R. Andy, D. H. Alpers, G. Schonfeld, and A. W. Strauss, *J. Biol. Chem.* **257,** 971 (1982).
[8] J. I. Gordon, K. A. Budelier, H. F. Sims, C. Edelstein, A. M. Scanu, and A. W. Strauss, *J. Biol. Chem.* **258,** 14054 (1983).
[9] V. I. Zannis, J. McPherson, G. Goldberger, S. K. Karathanasis, and J. L. Breslow, *J. Biol. Chem.* **259,** 5495 (1984).
[10] V. I. Zannis, S. F. Cole, C. L. Jackson, D. M. Kurnit, and S. K. Karathanasis, *Biochemistry*, submitted (1984).

volume of 25 µl containing 60 µCi [^{35}S]methionine for 1 hr at 37°, using a rabbit reticulocyte lysate system.[11] In some experiments the translation was performed in the presence of dog pancreatic membranes.[12]

Growth and Labeling of HepG2 and Hep3B Cells. The human hepatoma cell lines HepG2 and Hep3B[13] were grown in 100-mm-diameter petri dishes with minimum essential medium (MEM) supplemented with 10% fetal bovine serum. Prior to labeling, 90% confluent monolayer cultures were washed twice in sterile Earle's balanced salt solution (EBSS) and incubated for 5 min to 8 hr with methionine-free medium containing 0.10 mCi of [^{35}S]methionine/ml medium.

Organ Cultures of Fetal Human Tissues. Fetal tissues were obtained from human abortuses under approved protocols as described previously.[14] After examination, small pieces of fetal liver and small intestine were obtained and placed in methionine-free MEM containing 10% dialyzed fetal calf serum and saturated with 95% O_2 and 5% CO_2. Within 10 min of delivery, the specimens were brought to the laboratory and dissected into pieces with a diameter of less than 0.5 mm in the case of the liver and 2.0 mm in the case of the small intestine. The dissected tissues were placed onto a stainless-steel grid and positioned in a plastic organ culture dish (Falcon, Division of Becton-Dickinson, Co., Cockeysville, MD). Tissue culture medium, which consisted of methionine-free MEM containing 10% dialyzed fetal calf serum and supplemented with 2 mM glutamine, 100 units/ml of penicillin, and 50 µg/ml of streptomycin, was added to the organ culture until it reached the lower surface of the grid. A drop of medium was then added on the top of the grid to cover the tissue along with 250 µCi/dish of [^{35}S]methionine (300 Ci/mmol). The organ cultures were then placed in Torbal anaerobic jars and aerated with 95% O_2, 5% CO_2. The jars were closed, incubated at 36.5°, reaerated with 95% O_2 and 5% CO_2 every 2 hr, and the incubation terminated after 7 hr. Similar methods have been used for organ cultures of adult human and monkey tissues.[15,16]

Immunoprecipitation of Radiolabeled Apolipoproteins Synthesized by Cells or Tissues in Culture and by Cell-Free Translation of mRNA. The culture medium or the cells or tissue homogenates were mixed with carrier plasma apolipoproteins and immunoprecipitated with the appropriate

[11] H. R. B. Pelham, and R. J. Jackson, *Eur. J. Biochem.* **67**, 247 (1976).
[12] J. A. Majzoub, M. Rosenblatt, B. Fennick, R. Maunus, H. M. Kronenberg, J. T. Potts, Jr., and J. F. Habener, *J. Biol. Chem.* **255**, 11478 (1980).
[13] B. B. Knowles, C. C. Howe, and D. P. Aden, *Science* **209**, 497 (1980).
[14] V. I. Zannis, D. Kurnit, and J. L. Breslow, *J. Biol. Chem.* **257**, 536 (1982).
[15] V. I. Zannis, J. L. Breslow, and A. J. Katz, *J. Biol. Chem.* **255**, 8612 (1980).
[16] R. J. Nicolosi, and V. I. Zannis, *J. Lipid Res.* **25**, 879–887 (1984).

antibodies. The apolipoproteins synthesized by cell-free translation of mRNA were precipitated in a similar fashion. The above techniques allow immunoprecipitation of either a single apolipoprotein when we employ a monospecific antiserum or several apolipoproteins when we employ a combination of monospecific antisera. The optimum amount of antibodies needed for immunoprecipitation was determined as follows: a fixed amount of serum was mixed with 5, 10, 20, 40, 80 and 160 μl of antiserum. Each mixture was adjusted to a final concentration of 10 mM sodium phosphate, pH 7 and all samples were adjusted to the same final volume with normal saline. After 3 days at 4°, the immunoprecipitate was collected by centrifugation for 3 min in a microcentrifuge and washed twice with cold normal saline and once with water. The preparation was then ready for one- or two-dimensional polyacrylamide gel electrophoresis (PAGE).

Preparation of Protein Samples for One- and Two-dimensional Gel Electrophoresis. Prior to electrophoresis the immunoprecipitate was dissolved either in the one-dimensional SDS–PAGE buffer (Tris-HCl 0.0625 M, pH 6.8, 10% v/v glycerol, 5% v/v β-mercaptoethanol, and 2% w/v SDS) or the two-dimensional PAGE buffer [9.5 M urea, 2% v/v nonidet P-40, 2.1% ampholines (1.38% pH 5 to 8, 0.16% pH 2.5 to 4.0, 0.46% pH 4 to 6) and 5% v/v β-mercaptoethanol)] and then used for one- and two-dimensional gel electrophoresis, respectively.

Immunoprecipitation of Radiolabeled Apolipoproteins by Staphylococcal Protein A Preparation. An alternate method of immunoprecipitation of the newly synthesized apolipoproteins utilizes a staphylococcal protein A (IgG-Sorb) preparation.[17] The lyophilized staph A preparation was reconstituted in 9.8 ml of double distilled water and was sonicated for 10 min in a Bransonic 12 water bath sonicator. The suspension was distributed in 0.5-ml aliquots into microcentrifuge tubes and was centrifuged for 30 sec. The pellet was washed three times by resuspension and centrifugation, as above, in a solution of 10 mM sodium phosphate pH 7.2, 85 mM NaCl, 5 mM KCl, 0.5% deoxycholate (w/v), 1.0% Triton X-100, 1% sodium dodecyl sulfate, and 1 mg/ml bovine serum albumin. The content of each tube was resuspended in 0.5 ml of the same buffer; this provides a 10% suspension of IgG-Sorb. The cell cultures or the organ cultures were lysed in a solution containing 10 mM sodium phosphate, pH 7.2, 85 mM NaCl, 5 mM KCl, 0.50% deoxycholate, and 1% Triton X-100. The cell lysate was diluted 1:1 with a solution of 10 mM sodium phosphate, pH 7.0, 85 mM NaCl, 5 mM KCl, 1% sodium dodecyl sulfate, centrifuged for

---

[17] V. I. Zannis, S. K. Karathanasis, H. Keutmann, G. Goldberger, and J. L. Breslow, *Proc. Natl. Acad. Sci. U.S.A.* **80**, 2574 (1983).

5 min at 4° in a microcentrifuge, and the supernatant was retained. The culture medium was similarly adjusted to a final concentration of 10 mM sodium phosphate, pH 7.2, 85 mM NaCl, 5 mM KCl, 0.25% deoxycholate, 0.5% Triton X-100, and 0.5% sodium dodecyl sulfate. An aliquot of 5 to 10 μl of apolipoprotein antiserum was added to 200 to 1000 μl of cell lysate or culture medium prepared as described above and the mixture was incubated at 4° overnight.

A 50- to 75-μl aliquot of IgG-Sorb suspension was added for an additional hour of incubation at 4° and the mixture was centrifuged for 30 sec in a microcentrifuge. The pellet was washed once by resuspension and 30 sec centrifugation in a buffer containing 10 mM sodium phosphate, pH 7.2, 85 mM NaCl, 5 mM KCl, 0.50% deoxycholate, 1.0% Triton X-100, 1.0% sodium dodecyl sulfate, and 1 mg/ml bovine serum albumin, then four times in the same buffer without albumin. The pellet was suspended in the one- or two-dimensional PAGE buffer centrifuged for 60 sec and the supernatant was used for one- or two-dimensional gel analysis.

One- or Two-Dimensional Polyacrylamide Gel Electrophoresis. These procedures are outlined in detail by Zannis elsewhere in this volume [49] and in references 18–20. The gel obtained from PAGE, following immunoprecipitation of the radiolabeled samples when stained for protein, will show the position of the carrier plasma apolipoproteins. The autoradiogram will show the position of the newly synthesized radiolabeled apolipoproteins. The MW and isoelectric point relationship of plasma apolipoproteins to the newly synthesized forms can be obtained by superimposing the autoradiogram on the corresponding one- or two-dimensional gel stained for protein.

Intracellular Modifications of Apolipoproteins: Pulse-Chase Experiments

Pulse-chase experiments of apoE synthesized by HepG2 cells were performed as follows: HepG2 cells were placed in 30-mm petri dishes. When the cultures reached approximately 90% confluency, they were washed with sterile EBSS and incubated for 5 to 20 min with 1 ml of methionine-free MEM containing 0.2 mCi of [^{35}S]methionine (pulse). The [^{35}S]methionine-containing medium was removed at the appropriate times. The cultures were washed once with MEM (chase) and incubated

[18] V. I. Zannis and J. L. Breslow, *Biochemistry* **20**, 1033 (1981).
[19] V. I. Zannis, J. L. Breslow, T. R. SanGiacomo, D. P. Aden, and B. B. Knowles, *Biochemistry* **20**, 7089 (1981).
[20] V. I. Zannis, V. Rooney, P. Fraser, and J. L. Breslow "CRC Handbook of Electrophoresis" (L. A. Lewis, ed.), Vol. 3, p. 319. CRC Press, Boca Raton, Fla., 1983.

in 2 ml of the same medium for various times. At the end of the chase period the medium was collected and the cells were lysed as described above. The cell lysate and the culture medium were immunoprecipitated with antihuman apoE antibodies and analyzed by one- and two-dimensional gel electrophoresis and autoradiography. Some of the cultures were treated with 2.5 µg/ml tunicamycin for 6 hr prior to labeling with [^{35}S]methionine and with either 2.5 or 7.5 µg/ml tunicamycin during the labeling period.

Extracellular Modification of Apolipoproteins

The extracellular modifications of apoE and apoA-I are assessed by comparison of the isoprotein composition of the newly secreted and the plasma apolipoprotein forms, as well as by comparison of the sequences of the newly secreted to the plasma forms are described below.

Other Procedures

Isolation and Sequence of Radiolabeled Apolipoproteins Synthesized by HepG2 Cells. To obtain apoA-I isoprotein 2 (apoA-I$_2$) and sialated apoE (apoE$_s$) with radiolabeled proline, lysine, arginine, or histidine residues, two monolayers were washed with EBSS and incubated for 4 hr with 18 ml of serum-free Eagle's medium free of proline, lysine, arginine, or histidine, respectively, and containing 0.125 mCi of [^{14}C]proline, 0.5 mCi of [^{3}H]lysine, 0.5 mCi of [^{3}H]arginine, or 0.125 mCi of [^{14}C]histidine, respectively. The medium was collected and centrifuged at 30,000 g for 25 min. The supernatant was then dialyzed, lyophilized, dissolved in two-dimensional PAGE buffer, and subjected to two-dimensional polyacrylamide gel electrophoresis. The resultant gel was fixed and dried, and, after autoradiography, the area corresponding to the radiolabeled apoA-I$_2$ and apoE$_s$ was excised from the gel. ApoA-I$_2$ and apoE$_s$ were extracted from these gel pieces by vigorous shaking overnight at room temperature in a solution containing 0.125 M Tris–HCl, pH 6.8, 10% (v/v) glycerol, 0.1% sodium dodecyl sulfate, and 1 mM EDTA. The recovery of the radiolabeled apolipoproteins was 50 to 90%.[9,17] Automated Edman degradations were carried out with the Beckman model 890C sequencer with the 0.1 M Quadrol program. ApoA-I$_2$ labeled with [^{14}C]proline (100,000 cpm), [^{3}H]lysine (300,000 cpm), [^{3}H]arginine (50,000 cpm), or [^{14}C]histidine (200,000 cpm) and apoE$_s$ [^{14}C]proline (14,000 cpm) were dissolved in anhydrous trifluoroacetic acid. Bovine serum albumin (2 mg) was added as carrier. The butyl chloride effluent containing the thiazolinone from each cycle was divided into aliquots for scintillation counting and for conversion to the phenylthiohydantoin to determine relative yield of the bovine serum albumin carrier.

Quantification of Radiolabeled Apolipoproteins. Radiolabeled apoA-I_2 and apo E isoproteins were excised from one- or two-dimensional polyacrylamide gels and solubilized in 2 ml of 30% w/w H_2O_2 at 54° by an overnight incubation in a scintillation vial. The solubilized acrylamide was then mixed with 10 ml scintillation fluid and counted.[17]

Treatment of ApoE with Neuraminidase. Lyophilized VLDL at a concentration of 0.5 mg/ml was dissolved in 0.5 ml of 0.1 M sodium acetate buffer, pH 5. The solution was treated with two units of *Clostridium perfringens* neuraminidase at 37° for 2 hr. After treatment, the samples were dialyzed against water, lyopholized, and used for electrophoretic analysis.[18]

Quantitation of the Plasma Forms of ApoA-I and ApoE Isoproteins. The relative concentrations of the plasma apoA-I and apoE isoproteins can be estimated from the intensity of the Coomassie brilliant blue dye eluted from specific isoprotein spots of the polyacrylamide gel.[21,22] For this analysis the protein spots were cut from the two-dimensional gels, extracted overnight with 25% pyridine, and the optical density at 605 nm was determined. The volume of the solvent used for extraction was estimated to give an A_{605} in the range of 0.05 to 0.25 units. Control experiments in which known amounts of purified human apoA-I and apoE were eluted from acrylamide gels under these same conditions showed linearity of A_{605} nm vs protein concentration.

Results and Discussion

The Signal Peptide Sequences of Apolipoproteins

Most secreted proteins synthesized in cell-free translation systems have been shown to contain a 16–26 amino acid-long (usually) NH_2-terminal signal peptide sequences.[23–25] Blobel and Dobberstein proposed that these leader sequences direct the cotranslational translocation of secreted proteins across the membrane of the rough endoplasmic reticulum. In

---

[21] C. Fenner, R. R. Traut, D. T. Mason, and J. Wikman-Coffelt, *Anal. Biochem.* **63**, 595 (1975).
[22] V. I. Zannis, A. M. Lees, R. S. Lees, and J. L. Breslow, *J. Biol. Chem.* **257**, 4978 (1982).
[23] G. Blobel, P. Walter, C. N. Chang, B. Goldman, A. H. Erickson, and V. R. Lingappa, *Symp. Soc. Exp. Biol.* **33**, 9 (1979).
[24] M. Inouye and S. Halegoua, *Crit. Rev. Biochem.* **7**, 339 (1980).
[25] G. Von Heijne, *Eur. J. Biochem.* **133**, 17 (1983).

THE SIGNAL PEPTIDE SEQUENCES OF HUMAN APOLIPOPROTEINS

	Sequence	References
ApoA-I	Met Lys Ala Ala Val Leu Thr Leu Ala Val Leu Phe Leu Thr Gly Ser Gln Ala	17, 29
ApoA-II	Met Lys Leu Leu Ala Ala Thr Val Leu Leu Leu Thr Ile Cys Ser Leu Glu Gly	8, 34
ApoA-IV	Met Phe[a] Leu Lys[a] Ala Val Val Leu Thr[a] Leu Ala Leu Val Ala Val Ala Gly[a,b] Ala[a] Arg[a] Ala	32
ApoC-I	Met Arg Leu Phe Leu Ser Leu Pro Val Leu Val Val Val Leu Ser Ile Val Leu Glu Gly Pro Ala Pro Ala Gln Gly	33
ApoC-II	Met Gly Thr Arg Leu Leu Pro Ala Leu Phe Leu Val Leu Leu Val Leu Gly Phe Glu Val Gln Gly	34
ApoC-III	Met Gln Pro Arg Val Leu Leu Val Val Ala Leu Leu Ala Leu Leu Ala Ser Ala Arg Ala	31, 34
ApoE	Met Lys Val Leu Trp Ala Ala Leu Leu Val Thr Phe Leu Ala Gly Cys Gln Ala	9

[a] The identity of these residues was derived from the cDNA sequence of a human apoA-I cDNA probe (Karathanasis, S. K., Yunis, I., and Zannis, V. I., *Biochemistry*, in press (1986).
[b] Reference 32 has Leu in this position.

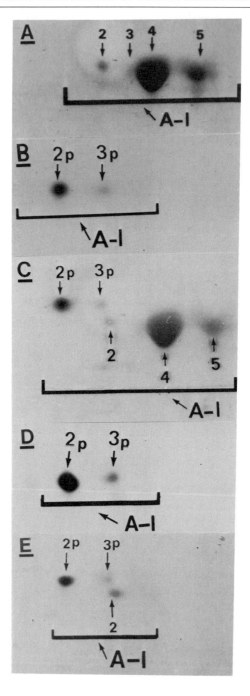

most,[26] but not all,[27] cases the signal peptide is cleaved cotranslationally by a membrane bound enzyme (signal peptidase) of the rough endoplasmic reticulum.[23,26] As expected, all apolipoproteins contain signal peptide sequences which range from 18 to 26 amino acids in length.[7,9,17,28–34] These sequences have been deduced from the DNA sequences of full length apolipoprotein cDNA clones or genomic clones and are shown in the table. The existence of signal peptide sequences can be ascertained by comparison of the apolipoprotein forms synthesized in a cell-free translation system in the presence and absence of dog pancreatic membranes to their plasma counterparts. The comparison of apoA-I and apoE isoproteins synthesized in a cell-free translation system to the corresponding plasma forms are shown in Figs. 1A–E and 2A–D, respectively.

[26] G. Blobel and B. Dobberstein, *J. Cell Biol.* **67**, 852 (1975).
[27] V. R. Lingappa, J. R. Lingappa, and G. Blobel, *Nature (London)* **281**, 117 (1979).
[28] J. I. Gordon, D. P. Smith, D. H. Alpers, and A. W. Strauss, *J. Biol. Chem.* **257**, 8418 (1982).
[29] J. I. Gordon, H. F. Sims, S. R. Lentz, C. Edelstein, A. M. Scanu, and A. W. Strauss, *J. Biol. Chem.* **258**, 4037 (1983).
[30] J. W. McLean, C. Fukazawa, and J. M Taylor, *J. Biol. Chem.* **258**, 8993 (1983).
[31] S. K. Karathanasis, V. I. Zannis, and J. L. Breslow, *Lipid Res.* **26**, 451 (1985).
[32] J. I. Gordon, C. L. Bisgaier, H. F. Sims, O. P. Sachder, R. M. Glickman, and A. W. Strauss, *J. Biol. Chem.* **259**, 468 (1984).
[33] T. J. Knott, M. E. Robertson, L. M. Priestley, M. Urdea, S. Wallis, and J. Scott, *Nucleic Acids Res.* **12**, 3904 (1984).
[34] C. R. Sharpe, A. Sidoli, C. S. Shelley, M. A. Lucero, C. C. Shoulders, and F. E. Baralle, *Nucleic Acids Res.* **12**, 3917 (1984).

FIG. 1. Two-dimensional polyacrylamide gel electrophoresis and autoradiography of proteins immunoprecipitated from the translation cocktail with specific antihuman apoA-I. A 50-μl aliquot of the translation cocktail of adult human liver mRNA was immunoprecipitated with antihuman apoA-I and IgG-Sorb. The apoA-I antibody complex was extracted with two-dimensional PAGE buffer and analyzed by two-dimensional polyacrylamide gel electrophoresis and autoradiography. (A) Gel stained for protein. The position of the plasma apoA-I isoproteins is indicated. (B) Autoradiogram of the gel in (A). The position of the immunoprecipitated ^{35}S-labeled apolipoproteins apo-A-I$_{2p(3p)}$ is indicated. ApoA-I$_{2p}$ and 3$_p$ represent the primary translation product of apoA-I mRNA. (C) Autoradiogram in B superimposed on the gel in (A). (D) Autoradiogram obtained after similar analysis of the translation products of HepG2 mRNA. (E) Autoradiogram obtained after similar analysis of the translation products of human liver mRNA processed cotranslationally with dog pancreatic membranes. Note the conversion of apoA-I$_{2p(3p)}$ to apoA-I$_2$. In this and subsequent figures, only the area of the gel or autoradiogram in the vicinity of apoA-I (or other protein of interest) is shown.

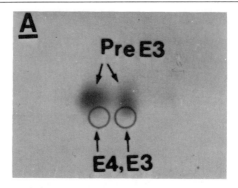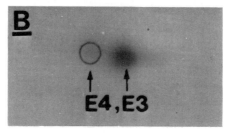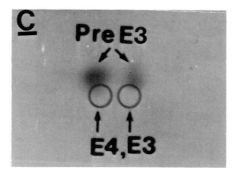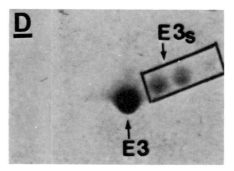

Detection of Differences between the Secreted and the Plasma Forms of Apolipoproteins

Plasma apoA-I is composed of several isoproteins with approximately the same apparent MW but different isoelectric points. These isoproteins were initially designated with numbers 2 to 6. The major plasma form is isoprotein 4 whereas isoprotein 2 represents 1 to 2% of the total plasma apoA-I (Fig. 3A).[14,16,22] In contrast, the apoA-I which is synthesized by HepG2 and Hep3B cells and by organ cultures of human intestine and liver consist primarily of isoprotein 2 (Fig. 3B–D). ApoA-I isoprotein 2 (ApoA-I_2) differs by +2 charges from isoprotein 4 (ApoA-I_4). Microsequence analysis of ApoA-I_2 synthesized by HepG2 cells and labeled as proline, lysine, histidine, and arginine residues[17] indicated that the secreted ApoA-I_2 extends 6 amino acids beyond the amino terminal of the plasma ApoA-I_4.[17,28] The microsequence analysis in combination with nucleotide sequence analysis of full length apoA-I cDNA clone showed that the sequence of the N-terminal hexapeptide of ApoA-I_2 is Arg-His-Phe-Trp-Gln-Gln. This finding establishes the molecular differences between secreted ApoA-I_2 (proapoA-I) and the major plasma form (ApoA-I_4). In addition, it explains the charge differences between these two ApoA-I isoproteins.

Similar studies have shown that the majority apoE, which is secreted by HepG2 and Hep3B cells and by hepatic organ cultures, consists mainly of isoproteins which are more acidic and have higher apparent MW than the corresponding major plasma apoE form[9,13,14] (Fig. 4A and C). These acidic isoproteins are also minor constituents of the plasma apoE form and are converted to this form by treatment with *C. perfrigens* neuraminidase (Fig. 4B and D). These findings indicate that the secreted form of

---

FIG. 2. Analysis of ^{35}S-labeled apoE isoproteins synthesized by cell-free translation of total mRNA obtained from HepG2 cells. (A–C) The analysis by two-dimensional PAGE and autoradiography of proteins immunoprecipitated from the translation mixture with specific antihuman apoE. An aliquot of 50 μl of the translation mixture of HepG2 mRNA was mixed with 100 μg of human very low-density lipoprotein and immunoprecipitated with antihuman apoE as explained in the text. The immunoprecipitate was dissolved in two-dimensional PAGE buffer and analyzed by two dimensional PAGE and autoradiography. (A) The autoradiogram obtained from this analysis is shown. The position of plasma apoE4 and apoE3 indicated by open circles was established by superimposing the autoradiogram on the corresponding two-dimensional slab gel that was stained for protein as explained in Fig. 1. (B) An autoradiogram obtained after similar analysis of the translation products of HepG2 mRNA processed cotranslationally with dog pancreatic membranes. Note the conversion of proapoE3 to apoE3. (C) An autoradiogram obtained after similar analysis of the translation products of total mRNA obtained from human liver. (D) For comparison purposes two-dimensional gel electrophoresis of 25 μg of VLDL apoE obtained from a normal human subject.

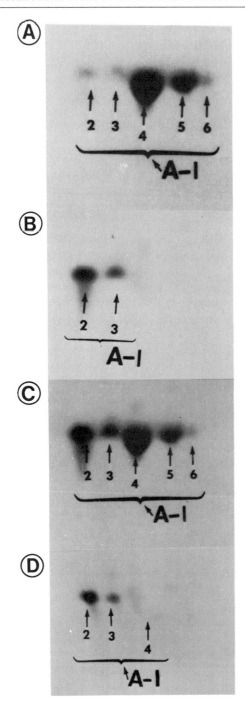

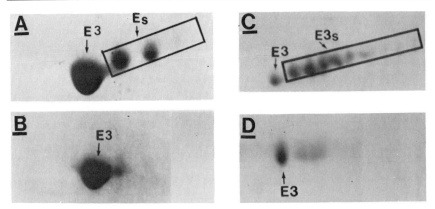

FIG. 4. Comparison of plasma and newly synthesized human apoE patterns observed by two-dimensional gel electrophoresis: (A) 35 μg of VLDL apoE; (B) 35 μg of VLDL apoE after treatment with *Clostridium perfrigens* neuraminidase; (C) apoE synthesized by the human hepatoma cell line Hep3B grown in the presence of [^{35}S]methionine; (D) apoE synthesized by human hepatoma cell line Hep3B after treatment with *Clostridium perfrigens* neuraminidase. E3 indicates asialo and E_s indicates asialo apoE isoproteins.

apoE is modified by carbohydrate chains containing sialic acid. The different forms of apoE isoproteins have been designated as described previoulsy.[35] The unmodified (asialo) form obtained from an E 3/3 phenotype has been designated apoE3 and the modified (sialo) forms have been designated apoE3$_{s1}$, apoE3$_{s2}$, apoE$_{s3}$, etc.[35] Microsequence analysis of the sialo apoE$_{s3}$ forms secreted by HepG2 cells showed that in contrast to apoA-I, the secreted sialo apoE isoproteins have the same number of amino acids as the corresponding plasma asialo apoE3 form.[9,36]

The studies described above also showed that HepG2 and hep3B cultures as well as the hepatic organ cultures synthesize a low molecular

[35] V. I. Zannis, J. L. Breslow, G. Utermann, R. W. Mahley, K. H. Weisgraber, R. J. Havel, J. L. Goldstein, M. S. Brown, G. Schonfeld, W. R. Hazzard, and C. B. Blum, *J. Lipid Res.* **23**, 911 (1982).
[36] S. C. Rall, K. H. Weisgraber, and R. W. Mahley, *J. Biol. Chem.* **257** 4171 (1981).

FIG. 3. Comparison of normal plasma to hepatic apoA-I isoproteins by two-dimensional gel electrophoresis and autoradiography. An aliquot of the hepatic organ culture medium was mixed with 30 μg of human HDL and immunoprecipitated with antihuman apoA-I. The immunoprecipitate was subjected to two-dimensional gel analysis. (A) The gel stained for protein. The positions of plasma apoA-I isoproteins are indicated. (B) The autoradiogram of the gel in (A). The positions of the immunoprecipitable ^{35}S-labeled apoA-I isoproteins are indicated. (C) The autoradiogram of (B) superimposed on the gel stained for protein. (D) An autoradiogram obtained after similar analysis of the intestinal organ culture medium. Note that ApoA-I$_2$ differs by +2 charges from apoA-I$_4$.

weight protein designated X.[13,14,20] Subsequent studies by Gordon *et al.* established that a low molecular weight apolipoprotein represents a pro-apoA-II form which contains a five amino acid N-terminal extension with the sequence Ala-Leu-Val-Arg-Arg.[8] Earlier studies suggested that 45% of the apoA-II prosegment is cleaved intracellularly and 55% extracellularly.[8] However, recent pulse-chase experiments of hepatoma cell cultures (HepG2) have shown that the cleavage of the apoA-II prosegment occurs mostly extracellularly.

Intracellular Modification of Apolipoproteins

The experiments presented in Figs. 2 and 4 suggest that following cleavage of the signal peptide, apoE is modified with carbohydrate chains containing sialic acid and is secreted as sialo apoE. The time course of intracellular apoE modification and secretion has been assessed by pulse-chase experiments of HepG2 cultures. In these studies HepG2 cells were

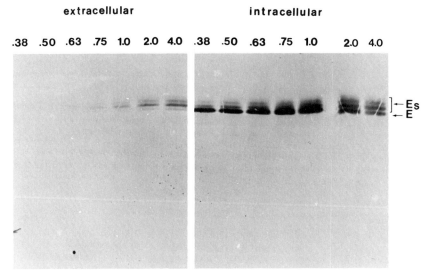

FIG. 5. Time course of apoE modification and secretion. Monolayer cultures of HepG2 cells grown in 30-mm petri dishes were pulsed with 0.2 Ci/ml [^{35}S]methionine and chased as explained in the text. The cell lysate and the culture medium were collected. The intracellular and extracellular apoE was immunoprecipitated and analyzed by one-dimensional SDS–PAGE and autoradiography. This figure shows the autoradiogram obtained from this analysis. Numbers on the top show the combined pulse and chase time in hours. The position of apoE and apoE$_s$ is indicated. Note that secreted apoE consists mainly of E$_s$ whereas the sialation of intracellular apoE is a function of pulse time. Apo E secretion is first detected at 0.5 hr.

pulsed with [^{35}S]methionine and chased. The intracellular and extracellular apoE was immunoprecipitated and analyzed by one- and two-dimensional polyacrylamide gel electrophoresis and autoradiography. The apoE and apoE$_s$ proteins were excised from the gels solubilized and counted. A typical pulse-chase experiment is shown in Fig. 5. Secretion of apoE was detected 30 min after pulse. The secreted apoE was detected 30 min after pulse. The secreted apoE is totally modified; in contrast, as shown in Figs. 5 and 6, the intracellular modification of apoE is a function of pulse time and can be detected after 16 min pulse. Tunicamycin did not affect the modification and secretion of apoE (Figs. 7 and 8). The tunicamycin treatment inhibited protein synthesis by 10% and [^{3}H]mannose incorporation into proteins by 93%. This finding suggests that the carbohydrate chains are attached to apoE by O-glycosidic linkages. Furthermore, the kinetics of appearance of the apoE$_{s2}$, apoE$_{s4}$, etc. suggest that the different sialated forms of apoE do not have a precursor product relationship.

Earlier studies had shown that ApoB is a glycoprotein containing 8 to 10% of high mannose and complex oligosaccharide chains.[37-39] Labeling of apoB with [^{3}H]mannose or [^{3}H]glucosamine showed that the carbohydrate is added to apoB cotranslationally in two stages corresponding to when the native polypeptide has acquired approximately 34 and 80% of its length.[39]

Extracellular Modifications of Apolipoproteins

As discussed earlier and shown in Figs. 3 and 4, the secreted forms of apoA-I and apoE differ from the corresponding plasma forms. Thus the secreted form of apoA-I consists of approximately 80% isoprotein 2. In contrast, the predominant plasma apoA-I form is isoprotein 4 (approximately 80%) whereas isoprotein 2 represents approximately 1 to 2% of the total plasma apoA-I.[22] Similarly, the secreted forms of apoE consist of approximately 90% sialo apoE and 10% asialo apoE (Fig. 8) whereas the plasma forms consist of approximately 24% sialo and 76% asialo apoE.[40] These observations suggest that following secretion the apoA-I and apoE are further modified in the plasma compartment.

The extracellular modification of the human apoA-I involves proteolytic removal of the hexapeptide Arg-His-Phe-Trp-Gln-Gln. The amino acid

---

[37] P. Lee and W. C. Breckenridge, *Can. J. Biochem.* **53**, 829 (1976).
[38] N. Swaminathan and F. Aladjem, *Biochemistry* **15**, 1516 (1976).
[39] P. Siuta-Mangano, S. C. Howard, W. J. Lennarz, and M. D. Lane, *J. Biol. Chem.* **257**, 4292 (1982).
[40] V. I. Zannis, C. Blum, R. Lees, and J. L. Breslow, *Circulation* **66**, 2 (1982).

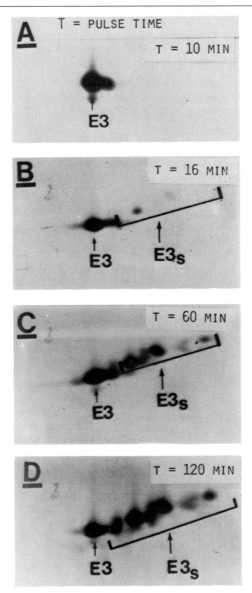

FIG. 6. Time course of intracellular modification of apoE. Cultures grown as explained in Fig. 5 were pulsed with 0.2 Ci/ml [^{35}S]methionine continuously for 10, 16, 60, and 120 min (A, B, C, and D, respectively). ApoE was immunoprecipitated from the cell lysate and analyzed by two-dimensional gel electrophoresis and autoradiography.

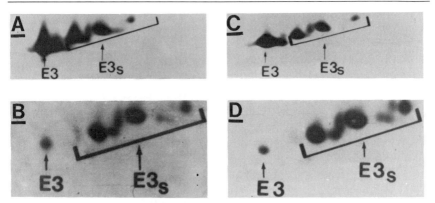

FIG. 7. Effect of tunicamycin on ApoE modification. HepG2 cultures grown as explained in Fig. 5 were pulsed continuously for 1 hr with 0.2 Ci/ml [^{35}S]methionine in the presence and absence of tunicamycin. ApoE was immunoprecipitated from the culture medium and the cell lysate and analyzed by two-dimensional PAGE. (A, B) Intracellular and secreted apoE obtained from untreated cultures. (C, D) Intracellular and secreted apoE obtained from cells treated with tunicamycin. The tunicamycin concentration was 2.5 µg/ml for 6 hr prior to labeling and 7.5 µg/ml during the pulse period.

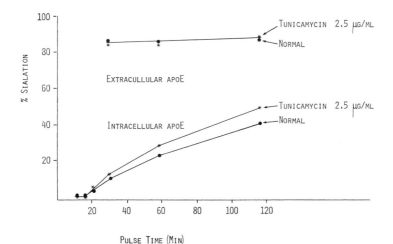

FIG. 8. HepG2 cultures grown as explained in Fig. 5 were pulse continuously with 0.2 Ci/ml [^{35}S]methionine in the presence and absence of 2.5 µg/ml tunicamycin. ApoE was immunoprecipitated from the culture medium and the cell lysate analyzed by one-dimensional SDS PAGE. The protein bands corresponding to apoE and apoE$_s$ (see Fig. 5) were cut from the gel solubilized and counted. The percentage sialation of intracellular and secreted apoE is plotted as a function of pulse time. The immunoprecipitable ^{35}S-labeled apoE present in the cell lysate and culture medium of the untreated and treated cultures was very similar. This figure indicates that under the conditions of the experiment tunicamycin did not affect the synthesis, the modification, or the secretion of apoE.

sequence around the scissile bond is Try-Gln-Gln-Asp-Glu. Preliminary studies have shown that proapoA-I secreted by HepG2 cells is converted partially to the plasma form by a proteolytic activity present in plasma and lymph.[41] This activity cleaved at the Gln-Asp bond of proapoA-I. Although this proteolytic activity has the expected features of the pro-apoA-I to plasma apoA-I converting protease, the rate of the conversion is slow and the proform is converted only partially to the plasma form under the experimental conditions used.[41] The exact nature of this activity will be clarified when the enzyme is purified and we achieve inhibition of its function *in vivo* with drugs or antibodies.

The physiological significance of proapoA-I to plasma apoA-I conversion remains a subject of speculation. Tangier disease is characterized by increases in the relative concentration of the plasma proapoA-I[22] as well as low levels of plasma apoA-I and HDL.[4] Although several explanations may account for the abnormalities of apoA-I and HDL observed in these patients it is possible that the proapoA-I to plasma apoA-I conversion is necessary for the stability of the apoA-I and HDL particle. Rall *et al.* have reported recently normal activation of lecithin:cholesterol acyltransferase by the proapoA-I form.[42]

The extracellular modification of apoE involves removal of the sialic acid and probably the entire oligosaccharide molecule. Nothing is known about the activities involved in this modification or its physiological significance. It is possible, though, that desialation of apoE may be required for the transfer of this protein from one lipoprotein particle to another and its recognition by cellular receptors.[43,44] Figure 9 shows a schematic representation of the intra- and extracellular modifications of apoA-I and apoE.

As discussed, the extracellular modifications of apoA-II also involves extracellular proteolytic cleavage of amino terminal pentapeptide Ala-Leu-Val-Arg-Arg.[8] The processing enzyme is a thiol protease which is synthesized and secreted by human hepatoma cell cultures (HepG2) cells.[45]

---

[41] C. Edelstein, J. I. Gordon, K. Toscas, H. F. Sims, A. W. Strauss, and A. M. Scanu, *J. Biol. Chem.* **258,** 11430 (1983).

[42] S. C. Rall, K. H. Weisgraber, R. W. Mahley, Y. Ogawa, C. H. Fielding, G. Utermann, J. Haas, A. Steinmetz, H. J. Menzel, and G. Assmann, *Fed. Proc.* **43,** 1815 (1984).

[43] J. L. Goldstein and M. S. Brown, "The Metabolic Basis of Inherited Disease" (J. B. Stanbury, J. B. Wyngaarden, D. S. Frederickson, J. L. Goldstein, and M. S. Brown, eds.), 5th ed., p. 672. McGraw-Hill, New York, 1982.

[44] D. Y. Hui, T. L. Innerarity, and R. W. Mahley, *J. Biol. Chem.* **256,** 5646 (1981).

[45] J. I. Gordon, H. F. Sims, C. Edelstein, A. M. Scanu, and A. W. Strauss, *J. Biol. Chem.* **258,** 15556 (1984).

A

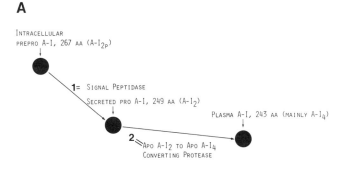

B

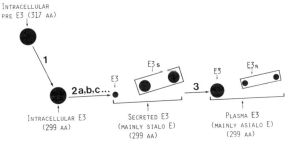

FIG. 9. Schematic presentation of the intra- and extracellular modification of human apoA-I (A) and apoE (B). The numbers indicate (in A) signal peptidase; (2) apoA-I_2 to apoA-I_4 converting protease present in plasma, lymph, and/or on cell surfaces; (in B) (1) signal peptidase; (2) a, b, c, etc., putative glycosyl transferases; (3) desialating enzyme present in plasma and/or on cell surfaces.

Other Modifications of Apolipoproteins

As shown in Fig. 1A, human apoA-I has several other isoproteins besides isoprotein 2 (proapoA-I) and isoprotein 4 (plasma apoA-I). Particularly prominent are isoproteins 5 and 6 which comprise approximately 15% of the plasma apoA-I. These isoproteins apparently result from successive modifications of proapoA-I. The molecular changes which lead to the formation of these proteins and their physiological signficance are not known. There are several other observations of apolipoprotein isoproteins, in human and animal species, which are apparently the products of extracellular modification of a precursor apolipoprotein form. For instance the isoprotein composition of human apoC-III shown in Fig. 10A suggests that apoC-III-0 (and occasionally apoC-III-I) contain more than

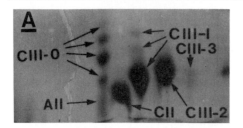
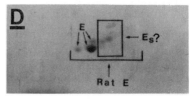
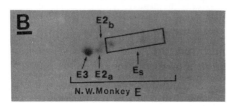
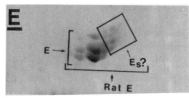
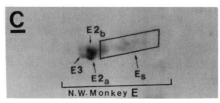

FIG. 10. Isoprotein forms of human apoC-III and of monkey and rat apoE. (A) Two-dimensional PAGE analysis of 250 μg of human VLDL. Only the area of the gel in the vicinity of apoC-III is shown. Note the different forms of apoC-III-0 and apoC-III-1. (B) Two-dimensional analysis of a fresh lipoprotein fraction of $d = 1.019-1.042$ obtained from squirrel monkey (New World monkey); only the area of the gel in the vicinity of apoE is shown. The apoE consists of asialo ($apoE_3$) and sialo ($apoE_s$) isoproteins, as well as two minor components designated $E2_a$ and $E2_b$. (C) Analysis of the same fraction after it was stored at 4° for 2 months. The storage results in the disappearance of the apoE designated by the circle and the increase in concentration of the two other isoprotein forms $apoE2_a$ and $apoE2_b$. This conversion is inhibited by storage of the sample in the presence of 0.1 mM DTNB. The isoelectric point relationship of apoE3 with the new forms was established with mixing experiments.[18] (D) Two-dimensional analysis of a lipoprotein fraction of $d = 1.068-1.096$ ml obtained from fresh rat plasma. Only the area of the gel in the vicinity of apoE is shown. Note that rat plasma apoE consists of several isoproteins with MW differences which are not affected by neuraminidase treatment. (E) Two-dimensional analysis of the rat lipoprotein fraction of $d = 1.068-1.096$ after storage for 30 days at 4°. This storage increases the number and the concentration of the higher MW apoE forms.

one isoprotein. It has been established previously that apoC-III-0, apoC-III-I, and apoC-III-II are modified with carbohydate chains which contain 0, 1, and 2 molecules sialic acid, respectively.[46] However, the molecular

[46] H. B. Brewer, R. Shulman, P. Herbert, and R. Ronan, *J. Biol. Chem.* **249**, 4975 (1974).

basis of the different apoC-III-0 and apoC-III-I isoproteins remains unknown. Figure 10B shows the isoprotein composition of squirrel monkey apoE obtained from fresh plasma. The pattern of apoE isoproteins resembles that of human plasma apoE and is composed of asialo and sialo apoE isoproteins. Storage of the plasma apoE at 4° results in the gradual conversion of the major asialo apoE isoproteins to two new forms which differ in size and/or charge from the original forms (Fig. 10C). This conversion is inhibited *in vitro* by 0.1 mM DTNB.[47]

Finally, rat plasma and newly secreted apoE contain several isoproteins which differ in size and charge (Fig. 10D–E). Storage of rat plasma lipoproteins at 4° results in changes in plasma apoE isoprotein patterns. These changes are similar to those seen in monkeys. The new patterns contain several more isoproteins with higher molecular weight as compared to the original pattern. The molecular nature and the physiological significance of these isoproteins require further clarification.

Summary

This chapter outlined the methods used to study intra- and extracellular modifications of apolipoproteins.

These and other related studies have shown that several of the apolipoproteins undergo a series of intra- and extracellular modifications as follows:

1. All apolipoproteins studied contain an 18–26 long signal peptide which is cleaved cotranslationally by the signal peptidase of the rough endoplasmic reticulum.

2. ApoE is further modified intracellularly with carbohydrate chains containing sialic acid and is secreted in the modified form designated apoE$_s$. The modified apoE is subsequently desialated in plasma.

3. ApoA-I is secreted in a proapoA-I form, which consists of 249 amino acids. The N-terminal hexapeptide of proapoA-I is cleaved extracellularly by a proapoA-I to plasma apoA-I converting protease. This cleavage generates the plasma apoA-I form which consists of 243 amino acids.

4. Other known apolipoprotein modifications include the modification of apoB, apoC-III, and apoD with carbohydrate chains that contain sialic acid and the proteolytic cleavage of the proapoA-II segment.

At the present time we are able to distinguish several isoprotein forms for a particular apolipoprotein. In addition, we began to understand the biochemical changes which lead to a few of these isoproteins.

[47] V. I. Zannis, R. J. Nicolosi, E. Jensen, K. C. Hayes, and J. L. Breslow, *J. Lipid Res.* **26**, 1421–1430 (1985).

Future research should be directed toward a better understanding not only of the structure but most importantly of the physiological significance of the different apolipoprotein forms.

Acknowledgments

This work was sponsored by grants from the National Institutes of Health (HL33952), the National Science Foundation (DCB840017), the March of Dimes Birth Defects Foundation (1-817), and the American Heart Association (83-963). Drs. Vassilis I. Zannis and Jan L. Breslow are Established Investigators of the American Heart Association.

We would like to thank Martin Ross, Johanna vanderSpek, and Elizabeth Walsh for technical assistance.

[41] Characterization of the Apolipoprotein A-I–C-III Gene Complex

By SOTIRIOS K. KARATHANASIS, VASSILIS I. ZANNIS, and JAN L. BRESLOW

Introduction

The human apolipoprotein A-I (apoA-I) and apolipoprotein C-III (apoC-III) genes have been recently isolated,[1,2] and it has been shown that they are physically linked.[3] This apoA-I–apoC-III gene complex is located on human chromosome 11[4,5] and has been structurally characterized by restriction endonuclease mapping analysis. Structural comparisons of this gene complex between normal individuals and patients with combined apoA-I–apoC-III deficiency and premature atherosclerosis[6] in-

---

[1] C. C. Shoulders, A. R. Kornblihtt, B. S. Munro, and F. E. Baralle, *Nucleic Acids Res.* **11**, 2827 (1983).

[2] S. K. Karathanasis, J. McPherson, V. I. Zannis, and J. L. Breslow, *Nature (London)* **304**, 371 (1983).

[3] S. K. Karathanasis, V. I. Zannis, and J. L. Breslow, *Proc. Natl. Acad. Sci. U.S.A.* **80**, 6147 (1983).

[4] P. Cheung, F. T. Kao, M. L. Law, C. Jones, T. T. Puck, and L. Chan, *Proc. Natl. Acad. Sci. U.S.A.* **81**, 508 (1984).

[5] G. A. P. Bruns, S. K. Karathanasis, and J. L. Breslow, *Arteriosclerosis* **4**, 97 (1984).

[6] R. A. Norum, J. B. Lakier, S. Goldstein, A. Angel, R. B. Goldberg, W. D. Block, D. K. Noffze, P. J. Dolphin, J. Edelglass, D. D. Bogorad, and P. Alaupovic, *N. Engl. J. Med.* **306**, 1513 (1982).

dicated the presence of a DNA insertion in the coding regions of the apoA-I gene of these patients.[7] This DNA insertion segregates as a typical autosomal Mendelian allele among the first degree relatives of these patients, and homozygosity for this insertion correlates with the development of the disease.[8] In this chapter, we will describe the methodology employed to carry out these studies.

Isolation and Characterization of apoA-I and ApoC-III cDNA Clones

ApoA-I and apoC-III are relatively abundant proteins in human plasma,[9-11] and are predominantly synthesized by intestine and liver.[12-16] ApoA-I represents about 1% of the total protein synthesized in cell-free translational systems programmed with poly(A) selected human liver RNA.[17] These observations indicate that apoA-I and apoC-III cDNA clones should occur with a relatively high frequency in cDNA libraries constructed from human liver mRNA. To identify these clones, we used the previously reported apoA-I and apoC-III amino acid sequences to deduce partial nucleotide sequences for the corresponding mRNAs as indicated in Fig. 1. These oligonucleotides were synthesized by a solid-phase phosphate triester method using the previously described reaction conditions and procedures.[18-20]

Briefly, for the apoA-I oligonucleotide, a sample of 25 mg of functionalized silica gel, charged with 1.3 μmol of 5-O-dimethoxytritylthymidine attached by its 3'-OH group through ester linkage to the solid phase, was unblocked at the 5'-OH group with a Lewis acid (saturated $ZnBr_2$/aqueous CH_3NO_2). The sample was condensed with a mixture of 10 mg each

[7] S. K. Karathanasis, V. I. Zannis, and J. L. Breslow, *Nature* (London) **305**, 823 (1984).
[8] S. K. Karathanasis, R. A. Norum, V. I. Zannis, and J. L. Breslow, *Nature* (London) **301**, 718 (1983).
[9] G. Schonfeld, and B. Pfleger, *J. Clin. Invest.* **54**, 236 (1974).
[10] J. B. Karlin, D. J. Juhn, J. I. Starr, A. M. Scanu, and A. H. Rubinstein, *J. Lipid Res.* **17**, 30 (1976).
[11] J. J. Albers, P. W. Wahl, V. G. Cabana, W. R. Hazzard, and J. G. Hoover, *Metabolism* **25**, 633 (1976).
[12] A. L. Wu and H. G. Windmueller, *J. Biol. Chem.* **254**, 7316 (1979).
[13] P. H. R. Green, A. R. Tall, and R. M. Glickman, *J. Clin. Invest.* **61**, 528 (1978).
[14] G. Schonfeld, E. Bell, and D. H. Alpers, *J. Clin. Invest.* **61**, 1539 (1978).
[15] H. G. Windmueller and A. L. Wu, *J. Biol. Chem.* **256**, 3012 (1981).
[16] D. Rachmilewitz, J. J. Albers, D. R. Saunders, and M. Fainaru, *Gastroenterology* **75**, 677 (1978).
[17] D. G. Ross, V. I. Zannis, D. M. Kurnit, and J. L. Breslow, *Biophys. J.* **37**, 396 (1982).
[18] M. D. Matteucci and M. H. Caruthers, *J. Am. Chem. Soc.* **103**, 3185 (1981).
[19] M. D. Matteucci and M. H. Caruthers, *Tetrahedron Lett.* **21**, 719 (1980).
[20] S. L. Beaucage and M. H. Caruthers, *Tetrahedron Lett.* **22**, 1859 (1981).

of protected nucleoside phosphoramidites of cytosine and thymidine [5'-dimethoxytrityl-N,N,N,N-acrysoyldeoxycytidine-3'-(methoxy)diethylaminophosphine and 5'-dimethoxytrityl-thymidine-3'-(methoxy)diethylaminophosphine], which were activated with tetrazole. The nucleoside phosphoramidites used throughout this procedure were obtained from ChemGenes Corporation (Waltham, MA). Any unreacted 5'-OH of the silica gel-bound nucleoside was blocked subsequently by reaction with a large excess of a very reactive phosphite. The phosphites were next oxidized with iodine to phosphates as described.[18] The cycle was repeated with the appropriate nucleoside phosphoramidite(s) until the last condensation was performed, subsequent to which the 5'-OH was not unblocked. At each point where an ambiguity in the DNA sequence existed, a mixture of derivatized nucleosides was employed as indicated in Fig. 1A. All reactions were carried out in a filtration device fashioned from Teflon and stainless steel.

After the synthesis was completed, the methyl groups of the phosphodiesters were removed by treatment with thiophenol and the ester bond

A

A.A. Residue	105	106	107	108	109	
(NH_2-Terminus)	GLN	LYS	LYS	TRP	GLN	
mRNA 5'	A_G^A	AA_G^A	AA_G^A	UGG	CA_G^A	3'
DNA 3'	T_C^T	TT_C^T	TT_C^T	ACC	GT_C^T	5'

B

A.A. Residue	63	64	65	66	67	
(NH_2-Terminus)	GLU	PHE	TRP	ASP	LEU	
mRNA 5'	GA_G^A	UU_C^U	UGG	GA_C^U	$_C^U U$	3'
DNA 3'	CT_C^T	AA_G^A	ACC	CT_G^A	$_G^A A$	5'

FIG. 1. ApoA-I and apoC-III oligonucleotide probe sequences. ApoA-I (A) and apoC-III (B) residue number and the corresponding amino acid, mRNA, and oligonucleotide sequences are shown. Nucleotides above the mRNA and oligonucleotide sequences indicate the code ambiguity for the corresponding amino acids. At these ambiguity points the appropriate derivatized nucleoside mixtures were used during organic synthesis to generate oligonucleotide mixtures (see text).

joining the oligonucleotide to the support was cleaved by treatment with concentrated ammonia as were the base-protecting groups. The reaction products then were fractionated by preparative HPLC by using a Waters C_8 column. The sample was loaded in 0.1 M triethylammonium bicarbonate at pH 7.0 and eluted by a linear gradient up to 25% acetonitrile over 40 min. This procedure separates failure sequences from the desired trityl oligomers, which emerge at the top of the gradient. Detritylation in 80% acetic acid for 20 min at room temperature then was followed by a second preparative HPLC under the same conditions. A dominant peak emerging approximately halfway through the gradient proved to be the desired tetradecamer mixture. This peak was subjected to polynucleotide kinase labeling at the 5' terminus with [γ-^{32}P]ATP[21] and run on a 20% polyacrylamide gel. The gel then was subjected to autoradiography, and over 95% of the oligomers ran as 14-base-long nucleotides. The overall yield from starting material was approximately 5–10%.

A similar procedure was employed for the synthesis of the apoC-III oligonucleotide probe (Fig. 1B) and the resultant oligonucleotides were mixtures of 16 oligomers for both apoA-I and apoC-III (Fig. 1A and B). Thus the ambiguity of the oligonucleotide sequences corresponding to apoA-I and apoC-III mRNAs due to the degeneracy of the code was accounted for by preparing each oligonucleotide sequence as a mixture of oligomers as indicated in Fig. 1A and B.

It has been shown earlier that oligonucleotides 13–15 bases in length are sufficient for the detection of a unique gene in a genomic yeast DNA library.[22] However, isolation of unique genes from mammalian DNA genomic libraries with such short DNA oligomers may be more difficult because of the greater complexity of the DNA in these libraries. On the other hand, screening of mammalian cDNA libraries constructed from a tissue that expresses the gene of interest circumvents the problem of the high DNA complexity present in genomic libraries. Therefore, we screened a previously constructed[23] adult human liver cDNA library with the oligonucleotide probe described above. To establish the hybridization condition for the screening, we used empirical formulas[24] to estimate that the temperature maximum of a perfectly matched 14-base-long DNA oli-

---

[21] P. W. J. Rigby, M. Dieckmann, C. Rhodes, and P. Berg, *J. Mol. Biol.* **113**, 237 (1977).

[22] J. W. Szostak, J. I. Stiles, B. K. Tye, P. Chiu, F. Sherman, and R. Wu, this series, Vol. 68, p. 419.

[23] D. E. Woods, A. F. Markham, A. T. Ricker, G. Goldberger, and H. R. Colten, *Proc. Natl. Acad. Sci. U.S.A.* **79**, 5661 (1982).

[24] R. W. Davis, D. Botstein, and J. R. Roth, "Advanced Bacterial Genetics." Cold Spring Harbor Laboratory, Cold Spring Harbor, New York, 1980.

gomer with a 50% content of G-C should be 51°. This indicated that hybridization at 1 M salt at 31–36° should satisfy the opposing requirements for sensitivity (low stringency) and specificity (high stringency). The human liver library cDNA clones were plated on 82-mm petri dishes at a density of 1000 bacterial colonies per dish. After growth and chloramphenicol amplification, the colonies were transferred to nitrocellulose filters[25] and hybridized to the oligonucleotides which had been radiolabeled by phosphorylating their 5'-OH end with [γ-^{32}P]ATP (New England Nuclear, 3000 Ci/mmol) and kinase (Biolabs).[21] The hybridization mixture contained 0.75 M NaCl, 0.15 M Tris–HCl (pH 8), 10 mM EDTA, 0.1% bovine serum albumin, 0.1% polyvinylpyrrolidone, 0.1% Ficoll, 0.1% sodium pyrophosphate, 0.1% SDS, yeast tRNA (carrier) at 100 μg/ml, and 0.65 μg of 5'-labeled oligonucleotide probe with a specific activity of 5 × 10^8 cpm/μg. The filters were hybridized at 23° for 16 hr. The filters then were washed in 0.9 M NaCl, 0.09 M sodium citrate, 0.05% sodium pyrophosphate, and 1% SDS sequentially at 23, 30, 40, and 50°. As predicted, screening of the cDNA library with the mixture of labeled 14-base-long DNA oligomers at room temperature (low stringency) showed a high background of nonspecific hybridization. Washing of the filters at 30–40° (high stringency) increased the specificity without significantly affecting the sensitivity of the hybridization signal. Finally, washing of the filters at 50° totally eliminated the hybridization signal.

This methodology allowed the identification of putative apoA-I and apoC-III cDNA clones. The clones were collected and the screening procedure repeated several times until single bacterial colonies hybridizing to apoA-I and apoC-III oligonucleotide probes were obtained. These colony-purified clones were grown in 1 liter 2XYT broth[26] supplemented with 20 μg/ml tetracycline[14] at 37° for 15–20 hr with vigorous shaking. One milliliter of these large bacterial growths was mixed with 0.15 ml autoclaved glycerol and stored at −20° for clone preservation. The rest of the cultures were used to prepare the recombinant plasmids carried by these clones. The plasmids were isolated by the alkaline lysis method as detailed elsewhere[27] and were mapped with various restriction endonucleases. Restriction maps of several putative apoA-I and apoC-III cDNA plasmids are indicated in Fig. 2. Restriction mapping of these plasmids facilitated complete nucleotide sequencing of their cDNA inserts. Both

---

[25] M. Grunstein and D. S. Hogness, *Proc. Natl. Acad. Sci. U.S.A.* **72**, 3961 (1975).

[26] J. H. Miller, "Experiments in Molecular Genetics." Cold Spring Harbor Laboratory, Cold Spring Harbor, New York, 1972.

[27] T. Maniatis, F. E. Fritsch, and J. Sambrook, "Molecular Cloning, A Laboratory Manual." Cold Spring Harbor Laboratory, Cold Spring Harbor, New York, 1982.

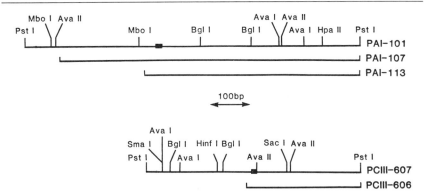

FIG. 2. Restriction maps of the cDNA inserts in putative apoA-I (pAI-101, pAI-107, pAI-113) and apoC-III (pCIII-606, pCIII-607) recombinant plasmids. Restriction sites are shown. Black boxes indicate regions homologous to the corresponding apoA-I and apoC-III oligonucleotide probes.

the chemical[28] and the enzymatic[29] methods for DNA sequencing were employed, and the obtained sequences have been reported elsewhere.[3,30,31]

In frame translation of these cDNA sequences showed that they correspond to apoA-I and apoC-III mRNAs and allowed the clarification of literature discrepancies in the amino acid sequences of these polypeptides.[2,31] In addition, these cDNA sequences showed that apoA-I and apoC-III are synthesized as a prepropeptide and a prepeptide, respectively.[2,31,32] We therefore conclude that design and synthesis of oligonucleotides deduced from apoA-I and apoC-III amino acid sequences allowed the isolation of the corresponding cDNA clones present in a human liver cDNA library.

Isolation and Characterization of the Human ApoA-I Gene

cDNA clones contain nucleotide sequences corresponding to mature mRNAs of a gene. Therefore, the structure of cDNA clone inserts does not represent the structure of the relevant gene as it occurs in the genome.

[28] A. M. Maxam and W. Gilbert, *Proc. Natl. Acad. Sci. U.S.A.* **74,** 560 (1977).
[29] F. Sanger, S. Nicklen, and A. R. Coulson, *Proc. Natl. Acad. Sci. U.S.A.* **74,** 5463 (1977).
[30] J. L. Breslow, D. Ross, J. McPherson, H. Williams, D. Kurnit, A. L. Nussbaum, S. K. Karathanasis, and V. I. Zannis, *Proc. Natl. Acad. Sci. U.S.A.* **79,** 6861 (1982).
[31] S. K. Karathanasis, V. I. Zannis, and J. L. Breslow, unpublished complete characterization of the human apoC-III gene, 1984.
[32] V. I. Zannis, S. K. Karathanasis, H. T. Keutmann, G. Goldberger, and J. L. Breslow, *Proc. Natl. Acad. Sci. U.S.A.* **80,** 2574 (1983).

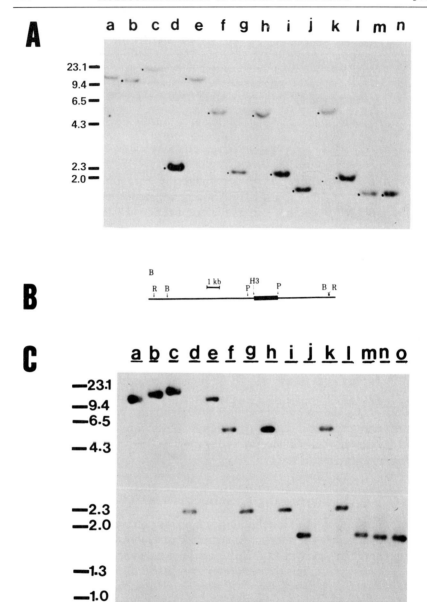

To obtain the gene structure, radiolabeled probes prepared from a cDNA clone DNA insert are used to carry out genomic blotting analysis of total chromosomal DNA. Specifically, to derive the restriction map of the human apoA-I gene, we prepared human DNA from peripheral lymphocytes as we have detailed elsewhere.[8] This DNA was digested with various restriction enzymes, and the resulting digests were electrophoresed on 1% agarose gels. The electrophoresed DNA was then transferred on to a nitrocellulose filter as described earlier[33] and hybridized with ^{32}P-labeled[30] pAI-113 cDNA clone insert (pAI-113 probe, Fig. 2). After washing the filter to remove the excess label, the filter was autoradiographed. The resulting autoradiogram (Fig. 3A) shows that this cDNA probe hybridizes to specific genomic restriction endonuclease fragments, the sizes of which allowed the construction of a restriction map of the human apoA-I gene (Fig. 3B). Since this restriction map was consistent when DNA isolated from a number of different individuals was used, we can conclude that this (Fig. 3B) restriction endonuclease organization represents the most frequent apoA-I gene allele. Clearly, deviations from this map would represent polymorphic alleles which may or may not be related to defects in apoA-I gene function.

Complete understanding of the structural organization of a gene can be accomplished by cloning it. Cloning allows a tremendous amplification of the relevant gene and its flanking DNA sequences. This amplified material can be used for detailed restriction endonuclease mapping and nucleotide sequencing analyses. In order to clone a gene, it is necessary to construct and screen a genomic DNA library. In contrast to cDNA libraries, which

[33] E. M. Southern, *J. Mol. Biol.* **98**, 503 (1975).

FIG. 3. Mapping of chromosomal and cloned human apoA-I gene. DNA prepared from human blood (A) or λ apoA-I #6 (C) was digested with restriction enzymes, electrophoresed on a 0.8% agarose gel, blotted, and hybridized with ^{32}P-labeled pAI-113 probe. The resulting autoradiograms are shown. Restriction enzymes used were as follows. Lanes: (a) *Eco*RI, (b) *Bam*HI, (c) *Hin*dIII, (d) *Pst*I, (e) *Eco*RI/*Bam*HI, (f) *Eco*RI/*Hin*dIII, (g) *Eco*RI/*Pst*I, (h) *Bam*HI/*Hin*dIII, (i) *Bam*HI/*Pst*I, (j) *Hin*dIII/*Pst*I, (k) *Eco*RI/*Bam*HI/*Hin*dIII, (l) *Eco*RI/*Bam*HI/*Pst*I, (m) *Eco*RI/*Hin*dIII/*Pst*I, (n) *Bam*HI/*Hin*dIII/*Pst*I, and (o) *Eco*RI/*Bam*HI/*Hin*dIII/*Pst*I. Size markers were *Hin*dIII-digested bacteriophage DNA (New England Biolabs) and are indicated on the left of the autoradiogram. (B) and (D) are restriction maps of the apoA-I gene in human blood chromosomal DNA and λ apoA-I #6 clone, respectively. These maps were constructed using the sizes of the hybridization bands obtained from the corresponding autoradiograms. The thick line in these restriction maps indicates the region hybridizing to the pAI-113 probe. *Eco*RI sites due to the charon 4A cloning vector are enclosed in boxes. Restriction sites are indicated as follows: (R) *Eco*RI, (B) *Bam*HI, (P) *Pst*I, and (H3) *Hin*dIII.

only contain cloned cDNA sequences corresponding to the mRNAs produced in the particular tissue used to construct the library, genomic libraries contain clones that carry every DNA sequence present in the entire genome. Therefore, genomic libraries constructed from any tissue DNA should contain clones carrying the relevant gene. A genomic library from human fetal liver DNA in the λ bacteriophage charon 4A vector has been previously reported.[34] Screening of this charon 4A genomic library with pAI-113 probe allowed the identification of putative apoA-I gene clones. One of these clones (λ apoA-I #6) was purified, grown in large amounts, and phage DNA was prepared. If this genomic clone DNA carries a human chromosomal fragment containing the apoA-I gene, it would then be expected that restriction endonuclease mapping analysis of λ apoA-I #6 would produce a map compatible with the apoA-I genomic map. Restriction mapping analysis showed that λ apoA-I #6 (Fig. 3C and D) and genomic (Fig. 3A and B) maps are identical. We therefore concluded that λ apoA-I #6 contains the human apoA-I gene and that this gene has the same structure in fetal liver (Fig. 3A and B) or adult lymphocyte (Fig. 3C and D) chromosomal DNAs. This cloned material allowed a detailed restriction mapping analysis and complete nucleotide sequencing of the human apoA-I gene which has been reported elsewhere.[2]

Linkage of Human Apo A-I and ApoC-III Genes

The methodology outlined for isolation and characterization of the apoA-I gene could have been applied to the isolation and characterization of the apoC-III gene. However, comparisons of apoA-I and apoC-III amino acid sequences had previously suggested that the corresponding genes may have been derived from a common evolutionary precursor by gene duplication.[35] In addition, we had previously observed that a defect in the apoA-I gene of patients with premature atherosclerosis was related to combined apoA-I–apoC-III deficiency in these patients.[7,8] These observations implied that apoA-I and apoC-III genes may be in close physical proximity in the human genome.

To examine this possibility, the cloned apoA-I gene (λ apoA-I #6) was digested with various restriction enzymes, blotted, and hybridized with ^{32}P-labeled insert of the pCIII-606 cDNA clone insert (pCIII-606 probe, Fig. 2). The resulting autoradiogram (Fig. 4A) indicated that the apoA-I gene containing genomic clone also contains sequences strongly hybridiz-

[34] R. M. Lawn, E. F. Fritsch, R. C. Parker, G. Blake, and T. Maniatis, *Cell* **15**, 1157 (1978).
[35] W. C. Barker and M. O. Dayhoff, *Comp. Biochem. Physiol.* **57B**, 309 (1977).

ing to apoC-III cDNA sequences. The sites of the hybridizing bands in Fig. 4A allowed the construction of a restriction map of the region in this cloned DNA that contains sequences homologous to apoC-III cDNA (Fig. 4B). Restriction mapping analysis of total human lymphocyte DNA with the apoC-III cDNA insert probe produced the autoradiogram shown in Fig. 4C, and the derived restriction map in Fig. 4D is in agreement with the cloned DNA map (Fig. 4B). We therefore conclude that apoA-I and apoC-III are physically linked and that both genes are present in the cloned fragment of the human genome contained in λ apoA-I #6. This observation facilitated detailed restriction mapping analysis and complete nucleotide sequencing of the human apoC-III gene.[31] Figure 5A shows a restriction map of the linked genes, their relative location, and structural features.

ApoA-I–ApoC-III Gene Complex in Patients with Combined
ApoA-I–ApoC-III Deficiency and Premature Atherosclerosis

Lipoprotein abnormalities can lead to atherosclerosis and the development of coronary heart disease.[36] ApoA-I activates LCAT[37-39] and participates as the major structural protein component in HDL.[9-11] Both HDL and LCAT are involved in the reverse cholesterol transport mechanism by which systemic cholesterol homeostasis is maintained.[40] It is therefore reasonable to speculate that apoA-I deficiency may result in cholesterol accumulation in peripheral tissues and coronary arteries resulting in premature atherosclerosis. This mechanism has been recently suggested to explain the development of premature atherosclerosis in patients with combined apoA-I–apoC-III deficiency.[6] Deficiencies in gene products can result from mutational inactivation of the corresponding genes. In this case depending upon the particular mutagenic event, it is possible that these mutations may lead to or be associated with restriction site polymorphisms in the relevant gene locus.

[36] P. N. Herbert, G. Assmann, A. M. Gotto, Jr., and D. S. Fredrickson, in "The Metabolic Basis of Inherited Disease" (J. B. Stanbury, J. B. Wyngaarden, D. S. Fredrickson, J. L. Goldstein, and M. S. Brown, eds.), 5th Ed., p. 589. McGraw-Hill, New York, 1982.
[37] C. J. Fielding, V. G. Shore, and P. D. Fielding, *Biochem. Biophys. Res. Commun.* **46**, 1943 (1972).
[38] A. Soutar, C. Garner, H. N. Baker, J. T. Sparrow, R. L. Jackson, A. M. Gotto, and L. C. Smith, *Biochemistry* **14**, 3057 (1975).
[39] S. Yokoyama, D. Fukushima, J. P. Kupferberg, F. J. Kezdy, and E. T. Kaiser, *J. Biol. Chem.* **255**, 7333 (1980).
[40] R. L. Jackson, A. M. Gotto, O. Stein, and Y. Stein, *J. Biol. Chem.* **250**, 7204 (1975).

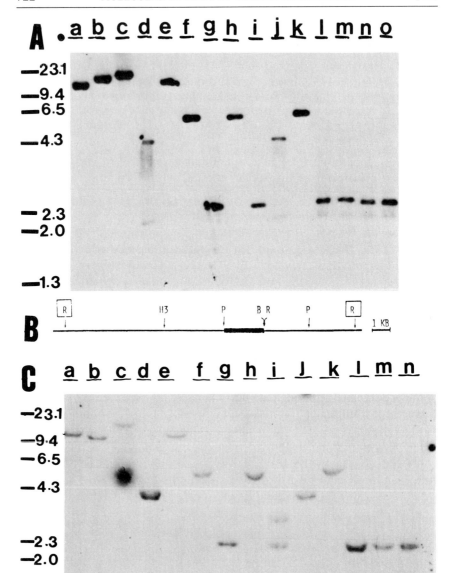

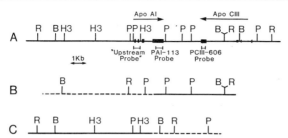

FIG. 5. Restriction maps of apoA-I–apoC-III gene complex in normal individuals and patients with apoA-I–apoC-III deficiency. (A) Restriction map of apoA-I–apoC-III gene complex in normal individuals has been derived from Figs. 3B, 3D, 4B, and 4D and additional characterization[2] of λ apoA-I #6. (B and C) Restriction maps derived by genomic blotting analysis of apoA-I–apoC-III-deficient patients blood DNA using pAI-113 and pCIII-606 probes or "upstream" probes, respectively. Arrows indicate direction of transcription and black boxes the coding regions of the genes. The corresponding positions of pAI-113, pCIII-606, and "upstream" probes in the gene complex are shown. Restriction sites are indicated as in Fig. 3.

To examine this possibility in the genome of these apoA-I–apoC-III deficient patients, we analyzed their chromosomal DNA by genomic blotting methods. DNA prepared from these patients' lymphocytes was digested with various restriction enzymes, electrophoresed, and, after blotting, hybridized with pAI-113 or pCIII-606 probes. The sizes of the hybridization bands in the resulting autoradiogram allowed the construction of a restriction map (Fig. 5B) of the apoA-I–apoC-III gene complex in these patients. Comparison between Fig. 5A and B shows that the restriction maps of the genomic DNA region between the pAI-113 and pCIII-606 probes is identical in normal individuals and apoA-I–apoC-III-deficient patients. However, the maps differ dramatically in the region to the 5' direction of pAI-113 probe (dotted line Fig. 5B). This difference is exemplified by the *Hin*dIII site normally occurring in the apoA-I gene (Fig. 5A) which is absent in the corresponding genomic region of these

---

FIG. 4. Mapping of chromosomal and cloned human apoC-III gene. DNA prepared from λ apoA-I #6 (A) or human blood (C) was digested with restriction enzymes, electrophoresed on a 0.8% agarose gel, blotted, and hybridized with ^{32}P-labeled pCIII-606 probe. The resulting autoradiograms are shown. The restriction enzymes used and size markers are as in Fig. 3. (B) and (D) are restriction maps of the apoC-III gene λ apoA-I #6 and human blood chromosomal DNA, respectively. These maps were constructed using the sizes of the hybridization bands from the corresponding autoradiograms. The thick line in these restriction maps indicates the region hybridizing to the pCIII-606 probe. *Eco*RI sites due to charon 4A cloning vector are enclosed in boxes. Restriction sites are indicated as in Fig. 3.

patients (Fig. 5B). Since single restriction site polymorphism could not explain these data, we speculated that this difference could have been generated by a DNA deletion or insertion in the apoA-I gene of the patients. Specifically, DNA deletion of the region in the vicinity of the 5' end of pAI-113 probe and extending to the 5' direction of the gene would have eliminated the *Hin*dIII site present in the normal apoA-I gene and would have produced new restriction sites in the 5' region of the gene. Alternatively, DNA insertion in the vicinity of the 5' end of pAI-113 probe would have brought new restriction sites in the area and moved this *Hin*dIII site further up to the 5' direction of the gene.

To distinguish between these two alternatives, we reasoned that if the region to the 5' direction of pAI-113 probe has been delected then DNA probes containing sequences to the 5' direction of the *Hin*dIII site present in the normal apoA-I gene should not produce hybridization bands when used for genomic blotting analysis of patients' chromosomal DNA. The 0.45 kb *Pst*I–*Hin*dIII fragment occurring to the 5' direction of the pAI-113 probe in the normal apoA-I gene (Fig. 5A) was subcloned in the bacteriophage M13 mp8 and single-stranded phage template was prepared.[41] About 0.1 pM of this template was primed with 2.5 pM of the M13 universal sequencing primer (Biolabs) by incubating 20 min at 65° and cooling at room temperature for an additional 10 min. This primed single-stranded template was then used to carry out enzymatic extension of the primer in the presence of 50 mM of γ-^{32}P-labeled (specific activity 700 Ci/mmol) deoxyribonucleotides (New England Nuclear) in 30 mM Tris (pH 7.5), 50 mM NaCl, 15 mM MgCl, 1 mM DTT with 0.5 units of *E. coli* DNA polymerase (Klenow fragment—Biolabs). The enzymatic reaction mixture was incubated 30 min on ice followed by 60 min at room temperature. The reaction was terminated by a phenol extraction and the incorporated label (probe) was purified by BioGel P60 (Bio-Rad) gel filtration. This cloned M13 derived probe contains sequences occurring from the *Hin*dIII site present in the normal apoA-I gene to the *Pst*I site occurring 0.45 kb to the 5' direction and is termed "upstream" probe (Fig. 5A).

Genomic blotting analysis of chromosomal DNA prepared from these apoA-I–apoC-III-deficient patients using this M13-derived upstream probe produced hybridization bands the sizes of which allowed the construction of a restriction map shown in Fig. 5C. Comparison of Fig. 5A and C shows that the restriction map to the 5' direction of this upstream probe is identical in normal individuals and patients with apoA-I–apoC-III deficiency, while these maps differ drastically at the 3' direction of this upstream probe (dotted line, Fig. 5C). We therefore conclude that the

[41] J. Messing, R. Crea, and P. Seeburg, *Nucleic Acids Res.* **5**, 1513 (1981).

apoA-I gene of these patients has been disrupted by a DNA insertion. This insertion has occurred in the region of the 3' direction of the *Hin*dIII site of the normal apoA-I gene and to the 5' direction of the 5' end of the pAI-113 probe, thus disrupting the fourth exon of the apoA-I gene.

Familial Segregation of the DNA Insertion in the ApoA-I Gene

If the DNA insertion found in the apoA-I gene of the apoA-I–apoC-III-deficient patients is responsible for apoA-I gene inactivation then it would be expected that homozygosity or heterozygosity for this insertion should be associated with apoA-I deficiency or half normal serum apoA-I levels, respectively. Since apoA-I serum levels of the first degree relatives of these patients is half normal,[6] we speculated that they may be heterozygous for this DNA insertion.

To examine this possibility, chromosomal DNA was prepared from peripheral lymphocytes of the relatives of these patients and after digestion with *Eco*RI was blotted and hybridized with the pAI-113 probe. The resulting autoradiogram in Fig. 6 shows that pAI-113 probe detects the expected (Fig. 5A and B) 13 and 6.5 kb genomic fragments in normal individuals and the apoA-I–apoC-III-deficient patients, respectively. However, both 13 and 6.5 kb hybridization bands are observed in DNA prepared from the first degree relatives of these patients. We therefore conclude that the DNA insertion in the apoA-I gene is hereditarily stable, transmitted in a typical Mendelian fashion, and is associated with apoA-I gene inactivation.

Conclusions

Structural comparisons of the apoA-I–apoC-III gene locus in normal individuals and apoA-I–apoC-III-deficient patients showed that a DNA insertion in the apoA-I gene of these patients is associated with apoA-I gene inactivation. In addition, preliminary observations have indicated that this DNA insertion is the result of a DNA rearrangement in the apoA-I–apoC-III gene locus of these patients and that this rearrangement has also interferred with apoC-III gene expression. It thus appears that a single mutagenic event may be responsible for both apoA-I and apoC-III gene inactivation. It is therefore clear that comparison of apolipoprotein gene structures in normal and diseased states can reveal mutations affecting gene expression which may be directly related to the development of the disease. In the future, it would be interesting to compare possible differences in factors controlling apolipoprotein gene expression in normal and diseased states. It is possible that mutational alterations in these

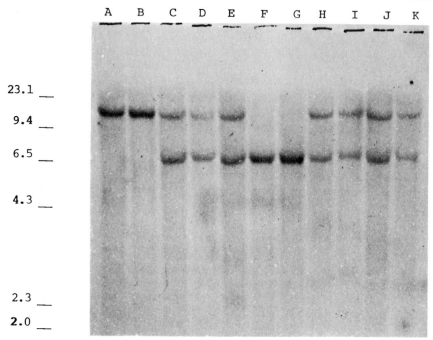

FIG. 6. Genomic blotting analysis of first-degree relatives of apoA-I–apoC-III-deficient patients. Chromosomal blood DNA was prepared, digested with EcoRI, electrophoresed, blotted, and hybridized with pAI-113 probe. The resulting autoradiogram is shown. The DNA used was derived from a normal individual (A), the maternal grandfather (B), father (C), mother (D), brother (E), apoA-I-deficient probands (F) and (G), son (H) and daughter (I) of proband (G), and the son (J) and daughter (K) of proband (F). Size markers are indicated as in Fig. 3.

factors may cause abnormal regulation of apolipoprotein gene expression which, in turn, could result in the development of lipoprotein abnormalities and coronary artery disease.

Acknowledgments

This work was supported by grants from the National Institutes of Health (HL32032 and HL33952), the National Science Foundation (DCB-8400173), the March of Dimes Birth Defects Foundation (1-817), and the Massachusetts Affiliate of the American Heart Association (13-517-5). Drs. S. Karathanasis, V. Zannis, and J. Breslow are Established Investigators of the American Heart Association. Dr. S. Karathanasis is a scholar of the Syntex Corporation.

[42] Genetic Polymorphism in the ApoA-I/C-III Complex

By CAROL C. SHOULDERS and F. E. BARALLE

Introduction

This chapter, rather than dealing with technical details, will concern itself with strategies that employ the analysis of DNA polymorphism in the study of the genetic disease. Particular emphasis will be placed on the identification of such polymorphic sequences that associate with lipoprotein metabolism disorders.

The hyperlipidemias, with hypertension, diabetes mellitus, and cigarette smoking, are among the major risk factors for the development of atheroma. Familial studies have shown that there is a strong inherited component to hyperlipidemia, but only in the minority of patients is the condition transmitted as a monogenic defect.

Conventionally, hyperlipidemia has been studied by looking for abnormalities in protein structure and function in the clearly defined disorders. In this manner, several genetic defects in the cell surface receptor that normally controls the degradation of LDL have been identified.[1,2] In addition, a mutant form of apolipoprotein A-I (apoA-I_{Milano}[3]) associated with hypertriglyceridemia and a polymorphic form of apolipoprotein E[4,5] associated with type III hyperlipoproteinemia have been demonstrated to exist in the population through variations in protein structure. More recently, in an attempt to identify biochemical defects underlying the less well-characterized forms of hyperlipidemia, large population samples of serum have been screened for variant forms of apoA-I, II, and IV. This approach has become possible by the development of methodology that allows the rapid isolation of these apolipoproteins, from serum samples, for two-dimensional gel analysis.[6] Thus, to-date, six variant forms of apoA-I and 1 variant form of apoA-IV have been identified.[7] This technol-

---

[1] M. S. Brown and J. L. Goldstein, *Science* **185**, 61 (1974).
[2] H. Tolleshaug, K. K. Hobgood, M. S. Brown, and J. L. Goldstein, *Cell* **32**, 941 (1983).
[3] K. H. Weisgraber, T. P. Bersot, R. W. Mahley, G. Franceschini, and C. R. Sirtori, *J. Clin. Invest.* **66**, 901 (1980).
[4] G. Utermann, M. Jaeschke, and J. Menzel, *FEBS Lett.* **56**, 352 (1975).
[5] K. H. Weisgraber, S. C. Rall, Jr., and R. W. Mahley, *J. Biol. Chem.* **256**, 9077 (1981).
[6] G. Utermann, G. Feussner, G. Franceschini, J. Haas, and A. Steinmetz, *J. Biol. Chem.* **257**, 501 (1982).
[7] H.-J. Menzel, G. A. Smann, S. C. Rall, K. H. Weisgraber, and R. Mahley, *J. Biol. Chem.* **259**, 3070 (1984).

ogy, however, is limited in that most variants require to differ in size or charge in order to be recognized. Perhaps therefore it is not surprising that in practice, these approaches have made little impact on identifying the genetic factors underlying the common types of hyperlipidemia. Moreover, one can envisage that these disorders do not necessarily occur as a result of an altered protein product, but rather as a result of subtle alterations in the level of expression of one or more of the many genes involved in lipid transport and metabolism.

To illustrate the difficulties involved in identifying any variation arising from the interaction of more than 1 gene product, let us consider an example of two unlinked loci, L and p, with their normal alleles, L1, L2, p1, and p2 and their mutant alleles L* and p*. The distribution of L and p alleles in an imaginary family is shown in Fig. 1. If one assumes that the hyperlipidemia is manifested as a monogenic defect and that the mutant L* gene is dominant, then it is straightforward to follow the mode of inheritance (Fig. 1A). However, consider the case where the hyperlipidemia is transmitted by the interaction of two genes, one recessive and one dominant; we would then observe the picture shown in Fig. 1B, where only one case of hyperlipidemia appears in three generations. If the expression of the pathological phenotype occurs only in the presence of environmental factors or by the interaction of more than two genes, it is clear that it will be extremely difficult to identify the genetic components in the pathogenesis of the disease by pedigree analysis. We believe this to be true for conditions such as cardiovascular disease and atherosclerosis, where the scores of genes involved in the regulation of lipid metabolism, arterial blood pressure and vessel wall quality, all interact to contribute to the development of these disorders.

Recombinant DNA technology[8] now provides a means by which the limitations associated with pedigree analysis can be minimized. This approach makes it possible to readily detect alterations in nucleotide sequence in the genome that are associated with pathological conditions. In some cases, these alterations occur in the region of the genome coding for a protein and therefore result in an altered protein product, while in others, they occur in noncoding regions. The latter may either cause differences in the regulation of its neighboring gene or, more likely be silent, although they may be physically linked to some other abnormality present in a neighboring region of the chromosome. Whatever the case, many of these sequence variations can be detected as changes in the length of characteristic fragments produced from the digestion of genomic

---

[8] This series, Vols. 99, 100, 101, several articles.

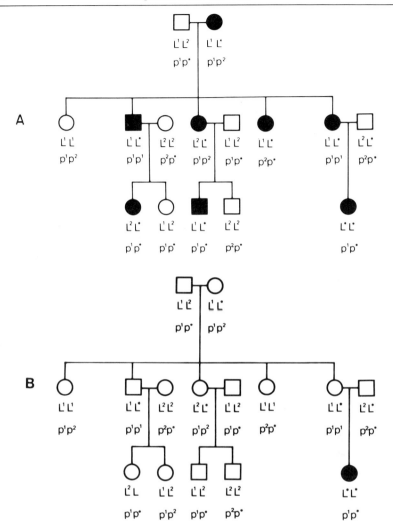

FIG. 1. Imaginary familial pedigree. L and p denotes two loci, L^1, L^2; p^1 and p^2 are normal alleles; L^* and p^* denote the mutant genes. The dark symbols denote the presence of the pathological condition. (A) Monogenic, L^* dominant; (B) digenic, L^* recessive, p^* dominant.

DNA with specific restriction enzymes (restriction fragment length polymorphism, RFLP) and, because they are inherited according to Mendelian principles, they serve as genetic markers in many human genetic disorders.

The Basic Procedure for Detecting RFLP

Genomic DNA is generally purified from peripheral blood leukocytes[9] and then specifically digested with a restriction endonuclease[10,11] appropriate for the gene(s) or DNA region under study. The cleaved DNA fragments are fractionated by size in an agarose gel, denatured, and then transferred onto a filter using the procedure of "genomic blotting."[12] The filter-bound DNA is incubated with a radioactive probe (generally prepared from cloned cDNA or genomic DNA) and then extensively washed so that the probe will remain bound only to those restriction fragment(s) containing complementary sequences; the latter are visualized by autoradiography.[13]

A RFLP will be detected when the region under study has either a rearrangement in its sequences (addition, deletion, translocation, inversion), or a substitution of a nucleotide such that either a new restriction enzyme site is created or a previously existing one is lost (see Fig. 2).

Preparation of Materials Necessary for the Detection of RFLPs

Isolation and Restriction of Genomic DNA

Genomic DNA can be isolated from various tissues or cultured cells. For population studies, the most convenient source is the leukocytes contained in peripheral blood. DNA (25–60 μg) can be obtained from each milliliter of whole blood, and therefore 10–20 ml of blood provides sufficient DNA for many analyses. The DNA which is easily purified from this source can be stored at $-20°$ reliably for many years.

Prior to searching for RFLPs in a population sample, one generally constructs a restriction enzyme map of the gene under investigation so that one is provided with some clues to the most appropriate restriction endonuclease to use for this purpose. A small amount of DNA from a normal individual (10–20 μg) is digested with several different restriction endonucleases in combination and singly, size fractionated, blotted, and then incubated with a suitable DNA probe to determine the length of the fragment(s) that contain the complementary sequences. In this type of analysis, it is also of interest to determine the orientation of the gene with

---

[9] L. M. Kunkel, K. D. Smith, S. J. Boyer, D. S. Borgaonkar, S. Wachtels, O. J. Miller, W. R. Breg, H. W. Jones, and J. M. Rary, *Proc. Natl. Acad. Sci. U.S.A.* **79**, 4381 (1977).
[10] H. O. Smith and D. Nathans, *J. Mol. Biol.* **81**, 419 (1973).
[11] R. J. Roberts, *Nucleic Acids Res.* **11**, r135 (1983).
[12] E. M. Southern, *J. Mol. Biol.* **38**, 503 (1975).
[13] R. A. Laskey and A. D. Mills, *FEBS Lett.* **82**, 314 (1977).

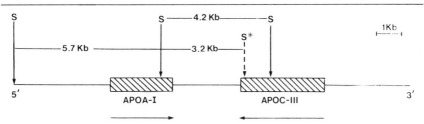

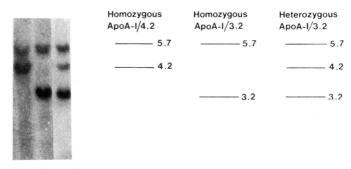

FIG. 2. RFLP analysis of DNA obtained from three unrelated individuals. The $SstI(s)$ restriction sites of the apoA-I/C-III complex have been analyzed. (For further detail, see section on Analysis of RFLPs associated with the apoA-I/C-III gene complex.) The upper part of the figure shows the relationship of these sites to the apoA-I/C-III genes together with the length of the fragments predicted from the presence or absence of the $SstI$ polymorphic site (s*). The lower part of the figure shows an autoradiograph of a Southern blot of $SstI$ restricted human DNA obtained from 3 unrelated hypertriglyceridemic Caucasian patients. The adjacent line diagram shows the genotype of the 3 individuals as defined by variations in restriction enzyme patterns.

respect to the restriction enzyme map. This can be readily achieved when the 5' and 3' ends of a gene probe hybridize to restriction fragments of different sizes that are derived from a single endonuclease digest of genomic DNA. For example, in the case of apoA-I, the 5' end of the gene probe is contained in a 5.7 kb $SstI$ restriction fragment, while the 3' end is contained in a 4.2 kb $SstI$ fragment (Fig. 2).

The restriction endonucleases used in "gene mapping" are often those that recognize tetra- or hexanucleotide sequences and thus on average cleave a random DNA molecule every 256 bp (4^4) and 4096 bp (4^6), respectively. Although with some particular enzyme digests the average size of fragments produced may deviate significantly from this theoretical value. In eukaryotic DNA, for example, the dinucleotide sequence CG is present at low frequency; thus, those restriction enzymes that recognize sequences containing the dimer CG produce larger than average fragments.

The choice of which restriction enzyme to use in the search for possible polymorphism associated with a particular gene in the population is dependent on several factors. For example, in the case of the apoA-I gene, an analysis of its restriction enzyme map (Fig. 2 and Ref. 14 for further details) revealed that for this purpose SstI would be an ideal enzyme to use for the following reasons. It isolates and cleaves a 9.9 kb region of the human genome that contains both the apoA-I structural gene and its 3' and 5' flanking regions into 3 fragments of 5.7 kb, 4.2 kb, and 35 bp, all of which contain part of the structural gene. The size of the two large fragments allows one to survey a large region of the genome containing the apoA-I gene with a single genomic blot, and yet to maintain high resolution for detection of variation in molecular weight. In addition, this enzyme was plentiful, relatively cheap (less than $1 US/individual), and known to reliably digest to completion DNA prepared from blood cells; all factors which favored its use for rapid screening of large numbers of individuals by an unskilled worker.

Construction and Isolation of cDNA and Genomic Clones

A variety of methods have now been described for the construction and isolation of cDNA clones and genomic clones.[15a,b] In principle, genomic DNA or cDNA (prepared from mRNA) is joined to a suitably prepared vector, placed in a bacterial host, and then amplified to produce a readily available source of DNA for use in both structural analysis and hybridization studies.

cDNA and genomic clones are now routinely isolated from cDNA and genomic libraries, respectively. Construction of total libraries makes it relatively easy to isolate any DNA or cDNA fragment for which a probe is available. Genomic libraries contain a complete collection of cloned fragments which comprise the entire genome of the organism used, while cDNA libraries contain the cDNA molecules produced from a heterogeneous mRNA preparation. The heterogeneity of cDNA clones in this

---

[14] C. C. Shoulders and F. E. Baralle, *Nucleic Acids Res.* **10**, 4873 (1982).

[15a] T. Maniatis, E. F. Fritsch, and J. Sambrook, "Molecular Cloning—A Laboratory Manual," p. 211. Cold Spring Harbor Laboratory, Cold Spring Harbor, New York, 1982.

[15b] U. Gubler and B. J. Hoffman, *Gene* **25**, 263 (1983).

library will, to a large extent, reflect the complexity of the original mRNA preparation used. Hence, this usually makes the isolation of clones corresponding to a relatively abundant mRNA species (e.g., albumin in a liver mRNA preparation) relatively easy, but the isolation of those corresponding to rare mRNA species more difficult. Although the isolation of clones from total libraries requires screening of a larger number of clones than is required when either purified DNA fragments or purified mRNA preparations are used, they give a larger flexibility since all sequences should be present and because they can readily be amplified and stored, they can be used whenever required.

Conventionally, most methods of producing cDNA libraries have relied on the ability of single-stranded cDNA molecules, produced from a mRNA preparation, to form a hair-pin loop at their 3' end, which is then exploited to prime the synthesis of their second strand. Subsequently, S1 nuclease is used to open and digest the single-stranded loop before the double-stranded cDNA molecules are inserted into plasmid vector via "homopolymer tailing" or molecular linkers. In practice, it is often found that the "homopolymer tailing" of, or addition of molecular linkers to, the cDNA and the vector DNA molecules prior to recombination is the most technically demanding and time consuming step of these procedures. We have, therefore, used an alternative approach to construct cDNA libraries which avoids this step.[14] The 5' overhanging ends of nuclease S1 treated double-stranded cDNA are repaired by "filling in" with the Klenow fragment of DNA polymerase I, and then directly inserted into as blunt-ended vector using T_4 ligase. This minor modification to the otherwise conventional procedure simplifies the construction of recombinant clones such that a complex cDNA library can be readily made within a few days from minimal amounts of mRNA.

Favorable conditions for directly screening cDNA libraries for a specific recombinant using synthetic oligonucleotides as hybridization probes are now well established.[16] This means in principle that, given a sufficiently complex library, a clone corresponding to any protein product can be isolated provided that 5 contiguous amino acids (preferably those with low degeneracy in their codons) of its sequence are known. Indeed, with this methodology, many of the apolipoprotein cDNA clones have now been isolated.[17,18]

[16] R. B. Wallace, M. J. Johnson, T. Hirose, T. Miyake, E. H. Kawashima, and K. Itakura, *Nucleic Acids Res.* **9,** 879 (1981).
[17] J. L. Breslow, D. Ross, J. McPherson, H. Williams, D. Kurnit, A. L. Nussbaum, S. K. Karathanasis, and V. I. Zannis, *Proc. Natl. Acad. Sci. U.S.A.* **79,** 6861 (1982).
[18] C. R. Sharpe, A. Sidoli, M. A. Lucero, C. S. Shelley, C. C. Shoulders, and F. E. Baralle, *Nucleic Acids Res.* **12,** 3917 (1984).

Preparation of Probes for Identification of Sequences within Digests of Total Eukaryotic DNA

The detection of particular sequences within a sample of DNA digested with one or more restriction enzymes is dependent on many factors. These include the quantity of DNA used, the binding and transfer efficiency of the individual fragments to a filter, the type of X-ray film used, and the specific radioactivity of the probe. In this section, this latter factor will be discussed; refs. 12 and 13 provide more detailed discussion of the other factors.

Currently, DNA is radioactively labeled *in vitro* by a variety of methods that reliably produce probes with a specific activity of 10^8–10^9 dpm/μg of DNA. For example, oligonucleotides are phosphorylated with polynucleotide kinase using γ-^{32}P-labeled nucleoside triphosphates (7000 Ci/mmol)[19] and double-stranded DNA by a nick-translation reaction,[20] using α-^{32}P-labeled nucleoside triphosphate (>3000 Ci/mmol). Another method involves making a radioactive copy of a DNA insert contained in a single-stranded cloning vector.[14] At present we favor the latter method. Under optimum conditions, it produces probes of higher specific activity than can be obtained using the more conventional method of nick translation. In the future, however, we can envisage that either fluorescent probes[21] or RNA probes prepared by transcription from the SP6 polymerase promoter[22] will eventually be the methods of choice.

cDNA clones and clones containing short fragments (approximately 0.2 to 1 kb) of genomic DNA are the usual hybridization probes used in the analysis of genomic DNA digests. However, direct detection of the genetic lesion can now be achieved using short oligonucleotide probes.[23] For example, this method has been used in the analysis of some of the defective β-globin alleles, where the structural mutations are known to be a single nucleotide substitution that may not create or destroy a cleavage site for any available restriction enzyme.[24] Individuals homozygous and heterozygous for such a trait can be identified using 2 synthetic oligonucleotides as hybridization probes in the genomic blot procedure. One of the oligonucleotides is exactly complementary to the normal DNA segment around the site of mutation while the other is complementary to the

---

[19] C. C. Richardson, *Proc. Natl. Acad. Sci. U.S.A.* **54**, 158 (1965).
[20] P. W. J. Rigby, M. Dieckmann, C. Rhodes, and P. Berg, *J. Mol. Biol.* **113**, 237 (1977).
[21] P. R. Langer-Safer, M. Levine, and D. C. Ward, *Proc. Natl. Acad. Sci. U.S.A.* **79**, 4381 (1982).
[22] M. R. Green, T. Maniatis, and D. A. Melton, *Cell* **32**, 681 (1983).
[23] B. J. Conner, A. A. Reyes, C. Morin, K. Itakura, R. L. Teplitz, and R. B. Wallace, *Proc. Natl. Acad. Sci. U.S.A.* **80**, 278 (1983).
[24] S. H. Orkin, A. F. Markham, and H. H. Kazazian, *J. Clin. Invest.* **71**, 775 (1983).

mutant sequences. After hybridization, stringent washing conditions are used such that a single mismatch will destabilize the hybrid and hence the "mutant probe" will give a signal only with the mutant gene but not with the normal gene and vice versa.

In principle, any given pair of alleles can be distinguished by the use of specific synthetic oligonucleotides, given that a nucleotide substitution is responsible for their creation and the DNA sequence around the site of mutation is known. Thus, this methodology could well provide a means of genotyping large population samples for the characterized polymorphic forms of apoA-I[25] and E.[26-28] Such analysis should prove to be rapid to carry out since one restriction enzyme digest of chromosomal DNA should be sufficient for the genotyping of several different combinations of alleles. This type of study would thus make it possible to examine whether any combination of particular alleles underlie certain forms of genetic hyperlipoproteinemia.

The Use of RFLP to Define Genetic Markers

Neutral polymorphisms that are scattered throughout the genome often give rise to RFLPs that can be used as genetic markers to follow the inheritance of defective alleles in linkage analysis studies. To be "informative," these markers need to be physically closely linked to the defective loci so that they are inherited together as a unit. Clearly, the further apart the RFLP site and the defective loci under study are on the chromosome, the more likely it is that recombination will occur in any one meiosis, and hence the less useful they will be in diagnosis of any given disorder.

Sickle cell anemia was the first disease to be diagnosed on the basis of a RFLP.[29] Linkage of this RFLP with the pathological condition was established by using a probe specific for the normal version of the known defective gene (see Fig. 3). Southern blot analysis of *Hpa*I restricted genomic DNA revealed that in the majority of normal West African individuals (<80%), this gene was contained in a 7.6 kb fragment, while in

---

[25] K. H. Weisgraber, S. C. Rall, Jr., T. P. Bersot, R. W. Mahley, G. Franceschini, and C. R. Sirtori, *J. Biol. Chem.* **258**, 2508 (1983).
[26] S. C. Rall, Jr., K. H. Weisgraber, T. L. Innerarity, and R. W. Mahley, *Proc. Natl. Acad. Sci. U.S.A.* **79**, 4696 (1982).
[27] S. C. Rall, Jr., K. H. Weisgraber, T. L. Innerarity, T. P. Bersot, R. W. Mahley, and C. B. Blum, *J. Clin. Invest.* **72**, 1288 (1983).
[28] J. W. Mclean, N. A. Elshourbagy, D. J. Chang, R. W. Mahley, and J. M. Taylor, *J. Biol. Chem.* **259**, 6498 (1984).
[29] Y. W. Kan and A. M. Dozy, *Proc. Natl. Acad. Sci. U.S.A.* **75**, 5631 (1978).

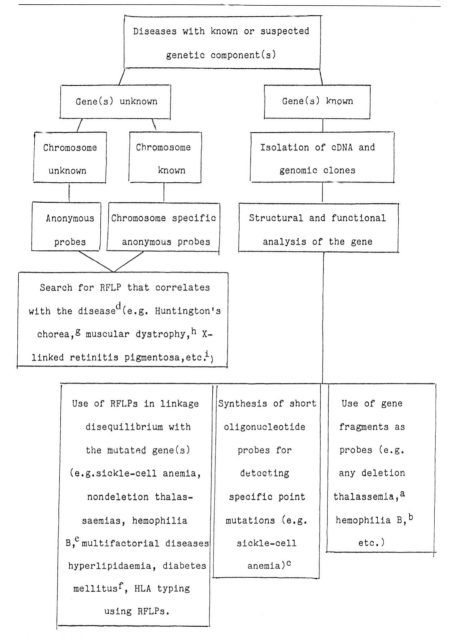

FIG. 3. Different approaches for looking for RFLPs. [a] R. A. Flavell, J. M. Kooter, E. de Boer, P. F. R. Little, G. Annison, and R. Williamson, *Nucleic Acids Res.* **6**, 2749 (1979). [b] F. Giannelli, K. H. Choo, D. J. G. Rees, Y. Boyd, C. R. Rizza, and G. G. Brownlee, *Nature*

individuals bearing the sickle cell trait, it was located in a fragment of 13.0 kb.

An alternative approach to establishing linkage between a RFLP and a pathological condition to that used for sickle cell anemia is shown in Fig. 3. Such methodology is based on calculations which suggest that if 500–1000 random DNA markers were available, it would be nearly certain that in a family study, at least one of them would show linkage to any given monogenic disorder.[30] The elegance of this approach means that genetic markers do not need to be defined by gene-specific probes, the only requirement being that the marker is closely linked to the functional mutation underlying the defect. This is clearly an advantage for disorders where all that is known is that they are genetic in character. To date, for example, anonymous probes have been found which define genetic markers in a number of families for Huntington's chorea,[31] muscular dystrophy,[32] and X-linked retinitis pigmentosa.[33] The very nature of this approach, however, means the inheritance of the genetic marker may have to be analyzed in many family members of the affected individual to determine which form of the genetic marker (i.e., which size of restriction fragment) is linked to the defective gene. Nevertheless, in spite of this limitation, the remarkable achievements described above undoubtedly demonstrate the vast potential of this approach. One can anticipate that,

---

[30] D. Botstein, R. L. White, M. Skolnick, and R. W. Davis, *Am. J. Hum. Genet.* **32**, 314 (1980).

[31] J. F. Gusella, N. S. Wexler, P. M. Conneally, S. L. Naylor, M. A. Anderson, R. E. Tanzi, P. C. Watkins, K. Ottina, M. R. Wallace, A. Y. Sakaguchi, A. B. Young, I. Shoulson, E. Bonilla, and J. B. Martin, *Nature (London)* **306**, 234 (1983).

[32] J. M. Murray, K. E. Davies, P. S. Harper, L. Meredith, C. R. Mueller, and R. Williamson, *Nature (London)* **300**, 69 (1982).

[33] S. S. Bhattacharya, A. F. Wright, J. F. Clayton, W. H. Price, C. I. Phillips, C. M. E. McKeown, M. Jay, A. C. Bird, P. L. Pearson, E. M. Southern, and H. J. Evans, *Nature (London)* **309**, 253 (1984).

(*London*) **303**, 181 (1983). [c] S. H. Orkin, A. F. Markham, and H. H. Kazazian, *J. Clin. Invest.* **71**, 775 (1983). [d] D. Botstein, R. L. White, M. Skolnick, and R. W. Davis, *Am. J. Hum. Genet.* **32**, 314 (1980). [e] F. Giannelli, D. S. Anson, K. H. Choo, D. J. G. Rees, P. R. Winship, N. Ferrari, C. R. Rizza, and G. G. Brownlee, *Lancet* **1**, 239 (1984). [f] G. I. Bell, J. H. Karam, and W. J. Rutter, *Proc. Natl. Acad. Sci. U.S.A.* **78**, 5759 (1981). [g] J. F. Gusella, N. S. Wexler, P. M. Conneally, S. L. Naylor, M. A. Anderson, R. E. Tanzi, P. C. Watkins, K. Ottina, M. R. Wallace, A. Y. Sakaguchi, A. B. Young, I. Shoulson, E. Bonilla, and J. B. Martin, *Nature (London)* **306**, 234 (1983). [h] J. M. Murray, K. E. Davies, P. S. Harper, L. Meredith, C. R. Mueller, and R. Williamson, *Nature (London)* **300**, 69 (1982). [i] J. M. Murray, K. E. Davies, P. S. Harper, L. Meredith, C. R. Mueller, and R. Williamson, *Nature (London)* **300**, 69 (1982).

in the future, it will be applied to extend linkage analysis in an attempt to (1) provide additional genetic markers so that diagnosis can be carried out in all affected families, and (2) isolate the specific region of the genome involved in causing the pathological phenotype.

Finally, of course, one must not neglect the fact that there are cases where the sequence variations give rise to RFLPs that are directly responsible for the disorder. In contrast to *neutral* polymorphisms, the majority of these *functional* polymorphisms will be intragenic, occurring in regions that affect either the transcriptional, translational regulation of the gene or the quality of the final product. In some cases, the particular restriction enzymes which will generate such RFLPs may only be identified when the defective gene has been fully characterized by sequence analysis.[34]

Limitations Associated with the RFLP Approach

Both technical and theoretical factors limit the potential of this approach to define genetic markers for human disease. The technical limitations are listed in the table, together with ways that these can, in some instances, be resolved. Two theoretical factors, namely the frequency of cross-over between the RFLP and the establishment of phase (i.e., which of the alleles at a given locus is linked to the mutant loci), have already been discussed in previous sections. A third limitation often encountered in the diagnosis of pathological conditions in both family studies and large population samples will be discussed in this section. For this purpose, we shall consider the genetically heterogeneous syndrome β-thalassemia.

In the majority of these disorders, no useful linkage disequilibrium exists between a RFLP and the β-thalassemic determinant. That is to say, although all β-thalassemic patients in a given population sample may possess a particular RFLP, it is present at too high a frequency in a control population of normal individuals to be of value in antenatal diagnosis. However, the extensive studies of Orkin *et al.*[35] have now suggested a means to minimize such a limitation. In brief, they subdivided 91 β-thalassemic patients into 9 major classes (haplotypes I–IX) on the basis of their genotype as defined by 7 RFLPs present in the β-globin gene locus. Sequence analysis of the β-globin gene, known to carry the functional mutation, from several individuals in each haplotype class revealed a coupling of specific thalassemic mutations to RFLPs haplotypes. For example, of 20 loci associated with types I–III and VII, 19 showed linkage

[34] R. F. Geever, L. B. Wilson, F. S. Nallaseth, P. F. Milner, M. Bittner, and J. T. Wilson, *Proc. Natl. Acad. Sci. U.S.A.* **78**, 5081 (1981).
[35] S. H. Orkin, H. H. Kazazian, Jr., S. E. Antonarakis, S. C. Goff, C. D. Boehm, J. P. Sexton, P. G. Waber, and P. J. V. Giardina, *Nature (London)* **296**, 627 (1982).

Minimization of the Practical Problems in the Use of RFLPs

Limitations	Ways to overcome the limitation
Restriction endonuclease analysis cannot directly detect a point mutation that does not affect a restriction endonuclease recognition site	In those cases where no linked RFLP(s) has been defined, design 2 oligonucleotides for use as hybridization probes; one should be a perfect match with the normal DNA sequence around the site of the mutation, and the other a perfect match with the mutant sequence[a]
Certain mutations can affect a restriction endonuclease site, but the fragments produced are too small to be detected or too similar in size to be easily separated from each other using the conventional Southern blot procedure	Fractionate DNA samples on either a higher percentage agarose gels of increased length or on acrylamide gels[b]; if using the former, gel electrophoresis and DNA transfer at 4° can yield better results[d]
The occurrence of similar genes in clusters may mask or confuse the results of the restriction fragment studies because one radioactive probe may identify fragments from all of the similar genes	If this cannot be corrected by high stringency hybridization or washing conditions, use an oligonucleotide or flanking region of the gene under investigation as hybridization probe
Methylation of cytidine bases can be a problem because these modifications of DNA within the recognition sequence can render the DNA refractory to digestion with certain endonucleases (usually those that have a CG in their recognition sequence) which do not recognize the specific cleavage if cytidine is methylated	Replicate the human DNA in *E. coli* strains that are methylase free or use isoschizomers that ignore the methylation[c]

[a] B. J. Conner, A. A. Reyes, C. Morin, K. Itakura, R. L. Teplitz, and R. B. Wallace, *Proc. Natl. Acad. Sci. U.S.A.* **80**, 278 (1983); S. H. Orkin, A. F. Markham, and H. H. Kazazian, Jr., *J. Clin. Invest.* **71**, 775 (1983).
[b] E. Frei, A. Levy, P. Gowland, and M. Noll, this series, Vol. 100, p. 309.
[c] C. Waalwijk and R. A. Flavell, *Nucleic Acids Res.* **5**, 3231 (1978).
[d] S. E. Y. Goodbourn, D. R. Higgs, J. B. Clegg, and D. J. Weatherall, *Mol. Biol. Med.* **2**, 223 (1985).

of mutation to haplotype. In this manner, these workers were able to identify the most common functional mutation underlying β-thalassemia in Mediterranean countries, thus providing a means whereby oligonucleotide probes specific for this defect could be used in their diagnosis.[24]

Should the coupling of haplotypes and specific mutations at a given locus within a defined population prove to be a general phenomenon, the strategy described above will undoubtedly be of value for characterizing genes associated with other inherited diseases.

Analysis of RFLPs Associated with the Apo-AI/C-III Gene Complex

It has now been demonstrated that apoA-I and C-III genes are closely physically linked in the genome by characterizing the insert of a genomic clone, apoA-I 6, previously shown to contain apoA-I sequences[36] (see Fig. 2). DNA from this clone was specifically digested with several different restriction endonucleases in combination and singly, size fractionated, blotted onto a nitrocellulose filter, and then incubated with an apoA-I cDNA probe to determine the length of the fragments that contained complementary sequences. The filter was then extensively washed to remove all traces of the hybridizing apoA-I probe and then rehybridized with an apoC-III probe. In this way, the fragments resulting from restriction enzyme digestion that contained both apoA-I and C-III sequences could be detected and a partial restriction map of the two genes constructed. From such data it was possible to conclude that apoA-I and C-III genes are separated by 2.7 ± 0.1 kb. Further blotting experiments utilizing 5' and 3' ends of the two cloned cDNA inserts as hybridizing probes showed the two genes are convergently transcribed and thus in the chromosome lie on opposite strands of the DNA molecule. This section describes two different types of RFLPs associated with this gene complex and their related phenotypic abnormalities. In particular, emphasis is placed on the methods now available to the recombinant DNA technologist for determining precisely the molecular defect underlying such abnormalities.

Association of a Mutant ApoA-I Allele with Atherosclerosis in Two Sisters

A mutant apoA-I allele that is inherited as a Mendelian trait linked to premature atherosclerosis has recently been described in an affected family.[37] Preliminary blotting experiments, using a number of restriction enzyme digests, have shown that this allele is contained in different size fragments from those of the normal allele. Such results clearly demonstrate that the mutant allele has not arisen by the acquisition of a point mutation but rather by the acquisition or loss of sequences in the apoA-I gene locus. By more extensive genomic blotting experiments using a series of different hybridization probes spanning the normal apoA-I gene

---

[36] S. K. Karathanasis, J. McPherson, V. I. Zannis, and J. L. Breslow, *Nature* (London) **304**, 371 (1983).

[37] S. K. Karathanasis, R. A. Norum, V. I. Zannis, and J. L. Breslow, *Nature* (London) **301**, 718 (1983).

locus, it has been deduced that an insertion of 6.5 kb in the 3' end of the apoA-I gene is responsible for the creation of this mutant allele.[38]

To characterize the mutant allele further, a library of genomic clones containing DNA fragments from one of the affected individuals has been created and the clone containing apoA-I/C-III sequences isolated.[39] Analysis of the latter, by restriction enzyme mapping and some DNA sequence determination, has made it possible to show that at least part of the 6.5 kb insert in this mutant apoA-I gene resides in normal individuals approximately 5 kb downstream of the apoA-I gene. Furthermore, this insert is deleted from its normal position in the genome of patients having the mutant allele. Thus, at present, it seems probable that this deletion may have removed the promoter region of the apoC-III from its normal location and may well explain the molecular basis of plasma apoC-III deficiency in patients homozygous for this mutant apoA-I/C-III complex.

In the affected family, the homozygotes have skin and tendon xanthomas, corneal clouding, and severe premature atherosclerosis associated with very low plasma HDL levels and deficiencies of apoA-I and C-III.[40] The obligate heterozygote members of the family, in contrast, are asymptomatic, but have lower than average levels of plasma HDL, apoA-I, and C-III. These levels fall into the range normally associated with increased risk of cardiovascular disease.[41,42] However, whether the above described mutant allele is a common cause of low HDL in the general population is at present unknown. The parents of the homozygous probands are heterozygous for this allele and, while they are both descended from ancestors of Scottish origin, they are believed to be unrelated. Nevertheless as yet, this mutant allele has not been found in any other family or population groups analyzed to date.[43] Whatever the case, as seen from the preceding discussion this mutant allele can be readily detected in carriers by the genomic blotting procedure and therefore such methodology should prove useful in the genetic counselling of this affected family and its future progeny.

---

[38] S. K. Karathanasis, V. I. Zannis, and J. L. Breslow, *Nature (London)* **305**, 823 (1983).

[39] S. K. Karathanasis, V. I. Zannis, and J. L. Breslow, *Fed. Proc., Fed. Am. Soc. Exp. Biol.* **43** (Part 2), 1815 (Abstr.) (1984).

[40] R. A. Norum, J. B. Lakier, S. Goldstein, A. Angel, R. B. Goldberg, W. D. Block, D. K. Woffze, P. J. Dolphin, J. Edelglass, D. D. Bogorad, and P. Alanpovic, *N. Engl. J. Med.* **306**, 1513 (1982).

[41] G. J. Miller and N. E. Miller, *Lancet* **1**, 16 (1975).

[42] W. Willet, C. H. Hennekens, A. J. Siegel, M. M. Adner, and W. P. Castelli, *N. Engl. J. Med.* **303**, 1159 (1980).

[43] A. Rees, C. C. Shoulders, J. Stocks, D. J. Galton, and F. E. Baralle, *Lancet* **1**, 444 (1983).

Association of a RFLP with Hypertriglyceridemia

A RFLP has now been found that distinguishes between two alleles (apoA-I/4.2, apoA-I/3.2) of the apoA-I/C-III gene complex in the population.[43] On restriction with endonuclease Sst1, allele apoA-I/4.2 yields two fragments of 5.7 and 4.2 kb, which hybridize to an apoA-I cDNA probe, whereas allele apoA-I/3.2 yields fragments of 5.7 and 3.2 kb (Fig. 2). The finding of individuals homozygous and heterozygous for both of the alleles in the population rules out the possibility that the two alleles represented experimental artefacts, two common causes of which arise from partial digestion of DNA and the presence of contaminating plasmid DNA in the genomic DNA sample. Furthermore, it confirms that the sequence variation giving rise to the RFLP is indeed inherited and not a consequence of spontaneous *de novo* mutation.

The frequency of the two apoA-I alleles has been estimated for both normolipidemic and type IV/V hypertriglyceridemic Caucasians patients.[43,44] Interestingly, none of the normolipidemics ($n = 52$) analyzed have the apoA-I/3.2 allele, whereas 26 out of 74 of the hypertriglyceridemics did. Thus, in this population sample, the apoA-I/3.2 allele associates with certain forms of hypertriglyceridemia.

In contrast, however, in other racial groups, a vast difference is seen in the distribution of this allele in normolipidemic individuals. For example, 65% of normolipidemic Chinese and 35% of Japanese possess this allele.[44] Clearly, therefore, in these racial groups, the apoA-I/3.2 allele cannot be used as a genetic marker for hypertriglyceridemia.

In the Caucasian populations, in order to use the RFLP directly as a genetic marker for hypertriglyceridemia, it is important to establish that no Caucasian normolipidemic individual possesses the apoA-I/3.2 allele. This requirement will clearly be met if the sequence variation giving rise to the RFLP is directly responsible for the aetiology of the disorder. However, if this allele should prove to be a marker for some abnormality present in a neighboring region of the chromosome which directly causes the hypertriglyceridemic symptoms, the frequency of meiotic crossing over between the RFLP and the linked mutant locus will need to be considered. Meiotic crossing over is estimated to occur in about 1 per 10^8 bp per meiosis.[45] Therefore, if the mutant gene and the genetic marker are about 20 kb apart, the possibility of crossing over occurring between them and hence the diagnostic error rate will be in the order of 1–5000. Clearly,

---

[44] A. Rees, J. Stocks, C. R. Sharpe, M. A. Vella, C. C. Shoulders, J. Katz, F. E. Baralle, and D. J. Galton, *J. Clin. Invest.* **76,** 1090 (1985).
[45] V. A. McKusick and F. H. Ruddle, *Science* **196,** 390 (1977).

the further apart the two loci, the higher the possibility of crossover and the higher the diagnostic error rate.

At present, this genetic marker tells us nothing about the cause of the abnormal lipoprotein levels seen in those patients who bear it. In order to improve upon this situation, we have clarified the role of the sequence variation giving rise to this marker. Genomic blotting experiments using a number of different restriction enzymes ruled out the possibility that the apoA-I/3.2 allele arises from the deletion of sequences in the complex and provided good evidence that a polymorphic nucleotide 2.7 ± 0.1 kb downstream of the apoA-I gene was responsible for its creation. With these data, it was then possible to locate precisely the position of this polymorphic nucleotide in the apoA-I/3.2 allele, by sequence analysis of a genomic clone constructed from Caucasian DNA homozygous for this allele. It has turned out to be the substitution of a cytosine residue for a guanosine at the 40th position of the 3' noncoding region of the apoC-III gene (Sharpe *et al.*, unpublished). This substitution destroys a restriction enzyme site for *Bst*XI and thus provides a means of determining whether the polymorphic sequence giving rise to the apoA-I/3.2 allele in the normolipidemic non-Caucasian groups[44] is of the same nature.

The location of the apoA I/3.2 polymorphic nucleotide suggests that in itself it is neutral but constitutes a mutation present within either the apoA-I or C-III gene. This functional mutation could affect the level of expression, or cause some structural modification, in either of these apolipoproteins. Again, this latter possibility can be checked by the sequence analysis of relevant sections of genomic classes obtained from hypertriglyceridemic individuals who bear this allele. However, in the event that no differences can be discerned in any known functionally important regions of the two apoA-I alleles, it would then be necessary to examine their regulation of expression under physiological conditions, to investigate whether any other sequences in the apoA-I/C-III complex are important for optimal functioning. This investigation would be most readily carried out by inserting the gene complex from normal and hypertriglyceridemic individuals, of both genotypes, via a suitable vector into eukaryotic cells and evaluating the relative levels of expression of both apoA-I and C-III.[46] One can envisage that such studies may show, in liver and intestine cells, there are different but specific sequences in the apoA-I/C-III complex regulating the transcription of these two genes. Moreover, a difference in these specific sequences between the two allelic complexes may account for the differences observed in phenotype.

---

[46] "Transcription and Translation: A Practical Approach" (B. D. Hames and S. J. Higgins, eds.), several articles. Practical Approach Series, IRL Press, Oxford, 1984.

As described previously, however, a RFLP can act as a genetic marker for a defective loci up to 10 to 20 × 10^6 bp away and, therefore, clearly the above described methods may not find any functional differences between the two alleles of the apoA-I/C-III complex. In this case, to isolate the particular area of the genome associated with the defective phenotype, such techniques that allow one to "walk" or "jog" the chromosome, will be of value.[47] In principle, this procedure involves isolating a series of overlapping clones from a genomic library upstream and downstream from a given point, mapping them with respect to one another, and then isolating from the 5' and 3' extremities of the resulting cluster further fragments that can be used to isolate further overlapping clones. This procedure is repeated as many times as required to reach the desired destination.

Finally, in a broader context, the association of a RFLP with any given genetic disease can open up many avenues for future clinical research. It should now be possible, for example, to examine whether there is any significant difference in either the clearance rate of dietary triglyceride from the circulation, or in the severity and incidence of cardiovascular disease between hypertriglyceridemic patients of different genotypes. Previously, the results from such investigations may well have been masked by the heterogeneous nature of such a disorder. Moreover, by using two or more genetic markers, the polygenic origin of disorders such as atherosclerosis can be evaluated. Already in this respect, preliminary results from Jowett et al.[48] indicate that the severity of hypertriglyceridemia may depend upon the additive effect of two genetic defects. In their studies, they divided 33 hypertriglyceridemic patients into 6 categories on the basis of their genotype. Two genetic markers were used for this purpose; the RFLP associated with the apoA I/C-III complex, and the RFLP associated with the insulin gene (designated class 1 and 3[49]). Significantly, perhaps, in diabetic patients homozygous for the insulin class 3 allele, they noted a correlation between the severity of hypertriglyceridemia and the genotype of the apoA-I/C-III complex.

Conclusion

This chapter has described recent work in the apoA-I/C-III gene complex to illustrate the potential of RFLPs to identify genetic factors under-

---

[47] R. M. Lawn, E. F. Fritsch, R. C. Parker, G. Blake, and T. Maniatis, *Cell* **15**, 1157 (1978).
[48] N. I. Jowett, A. Rees, L. G. Williams, J. Stocks, M. A. Vella, G. A. Hitman, J. Katz, and D. J. Galton, *Diabetologia* **27**, 380 (1984).
[49] G. I. Bell, J. H. Karam, and W. J. Rutter, *Proc. Natl. Acad. Sci. U.S.A.* **78**, 5759 (1981).

lying lipid metabolism disorders. In the future, one can envisage the recognition of such RFLPs linked to other relevant genes (e.g., apolipoproteins A-II, C-II, E, etc., LDL receptor, and lipoprotein lipase) will increase our understanding of the pathophysiology of the common forms of hyperlipidemia. In this manner, the phenotype-based classification of such disorders could be supplanted by one based on genotype. Such a step forward will undoubtedly be of benefit in evaluating the relationship between different forms of hyperlipidemia and cardiovascular disease.

Acknowledgments

We should like to thank Dr. M. Myant, Dr. A. R. Kornblihtt, A. F. Dean, and C. S. Shelley for critical reading of the manuscript. The work was supported by a grant from the British Heart Foundation.

[43] Molecular Cloning and Sequence Analysis of Human Apolipoprotein A-II cDNA

By LAWRENCE CHAN, MARSHA N. MOORE, and YUAN-KAI TSAO

Introduction

ApoVLDL-II, the major apolipoprotein in avian very low-density lipoproteins, was the first eukaryotic apolipoprotein studied by modern recombinant DNA techniques. The cDNA for apoVLDL-II mRNA was first cloned in plasmid several years ago.[1,2] ApoVLDL-II mRNA was partially purified by various size fractionation techniques including sucrose gradient centrifugation and Sepharose 4B chromatography.[1] The enriched mRNA was used as a template for dscDNA synthesis. The dscDNA was tailed with 15–20 dC by terminal transferase. It was inserted into dG-tailed pBR322 by standard techniques.[1] Selection of ApoVLDL-II cDNA from the various colonies was accomplished by the method of hybrid-arrested cell-free translation.[3] The various clones were sequenced by the Maxam–Gilbert technique. The amino acid sequence of ApoVLDL-II

[1] L. Chan, A. Dugaiczyk, and A. R. Means, *Biochemistry* **19**, 5631 (1980).
[2] B. Wieringa, W. Roskam, A. Arnberg, J. vander Zwaag-Gerritsen, G. AB, and M. Gruber, *Nucleic Acids Res.* **7**, 2147 (1979).
[3] B. M. Paterson, B. E. Roberts, and E. L. Kuff, *Proc. Natl. Acad. Sci. U.S.A.* **74**, 4370 (1976).

protein was also determined. Direct sequence comparison confirmed the authenticity of the cloned ApoVLDL-II cDNA.[4]

The cloning of ApoVLDL-II cDNA was relatively straightforward, since the corresponding mRNA could be induced to high levels (20–30% of total mRNA population) by estrogen treatment,[5] and even without further purification, about 20–30% of the clones were expected to be ApoVLDL-II cDNAs. The size fractionation further increased the proportion to 60–70%. Such an approach is not feasible in the case of mammalian apolipoproteins, since their mRNAs generally constitute less than 1% of the total mRNA population. Alternative approaches, however, are available. These include specific immunoprecipitation of polysomes[6] and oligonucleotide hybridization.[7] In this chapter, we shall review our experience with the use of oligonucleotides as hybridization probes in the selection of specific cDNAs. We have successfully used this technique in the selection of cDNA clones for human apoA-I, apoA-II, apoE, apoC-II, and rat sterol carrier protein. We will use the cloning of human apoA-II cDNA as an example, though we will also draw on our experience with other apolipoprotein cDNAs to illustrate some general points.

Use of Oligonucleotide as Hybridization Probe

The use of oligonucleotides as hybridization probes has been developed in a number of laboratories. The procedure that we follow is a modification of the method of Suggs et al.[8] In the selection of oligonucleotide sequences, some amino acid sequence information must be available. The least degenerate portion of the known amino acid sequence is selected, and mixtures of oligonucleotides predicted from such regions are synthesized and used as hybridization probes. Mixtures of as many as 128 different oligonucleotides have been applied successfully for the selection of apolipoprotein clones.[9] The oligonucleotides are end-labeled and used for colony hybridization.

[4] A. Dugaiczyk, A. S. Inglis, P. M. Strike, R. W. Burley, W. G. Beattie, and L. Chan, *Gene* **14,** 175 (1981).
[5] J. Codina-Salada, J. P. Moore, and L. Chan, *Endocrinology* **113,** 1158 (1983).
[6] J. M. Taylor and T. P. P. Tse, *J. Biol. Chem.* **251,** 7461 (1976).
[7] S. V. Suggs, R. B. Wallace, T. Hirose, E. H. Kawashima, and K. Itakura, *Proc. Natl. Acad. Sci. U.S.A.* **78,** 6613 (1981).
[8] S. V. Suggs, T. Hirose, T. Miyake, E. H. Kawashima, M. J. Johnson, K. Itakura, and R. B. Wallace, *in* "Developmental Biology Using Purified Genes" (D. D. Brown and C. F. Fox, eds.), p. 683. Academic Press, New York, 1981.
[9] S. C. Wallis, S. Rogne, L. Gill, A. Markham, M. Edge, O. Woods, R. Williamson, and S. Humphries, *EMBO J.* **2,** 2369 (1983).

Reagents

TE buffer, 10 mM Tris–HCl, pH 8.0, 1 mM EDTA
T_4 polynucleotide kinase
DE52 ion-exchange resin (Whatman)

Protocol. 5'-End labeling of oligonucleotide probe. In a siliconized eppendorf microfuge tube, set up the following.

1. Oligonucleotide, 50 pmol
   Tris–HCl, pH 7.6, 50 mM
   $MgCl_2$, 10 mM
   Dithiothreitol, 5 mM
   Spermidine, 0.1 mM
   EDTA, 0.1 mM
   [γ-^{32}P]ATP (>1000 Ci/mmol), 500–1000 pmol
   T_4 polynucleotide kinase, 20 units
2. Incubate at 37° for 30 min
3. Stop reaction by addition of 450 μl TE buffer
4. Apply mixture to a 1 cm DE52 column equilibrated with TE
5. Wash column with TE until radioactivity in eluate levels toward baseline
6. Elute oligonucleotide with ~10 ml 0.5 M NaCl in TE
7. Use labeled oligonucleotide directly for hybridization.

Oligonucleotides are generally labeled on the day of hybridization. If they are not used within a few days, we generally repurify the labeled material on a DE52 column. The specific activity of the probes has generally been ~10^{10} cpm/μg for 14- and 17-mers.

Screening of cDNA Clones

For small cDNA libraries (for example, those consisting of less than 1000), recombinant cDNA clones are picked and put on fresh antibiotic plates [e.g., for cDNA cloned into the *Pst*I site of pBR322, tetracycline (25 μg/ml) plates are used] in an ordered array in duplicate. They are transferred to Whatman 541 filters according to the following protocol which is a modification of the method of Gergen *et al.*[10]

Reagents

6 × SSC (1 × SSC = 150 mM NaCl, 15 mM Na-citrate, pH 7.0)
LB agar
   Bactoagar, 10 g
   Bactotryptone, 10 g

---

[10] J. P. Gergen, R. H. Stern, and P. C. Wensink, *Nucleic Acids Res.* **7**, 2115 (1979).

NaCl, 5 g
Yeast extract, 5 g
Water to 1 liter
Whatman 541 filters (cut to fit plates).

Protocol. Colony transfer.

1. Label Whatman 541 filters with pencil. Autoclave filters.
2. Grow colonies to a diameter of 3–4 mm before transfer.
3. Place filters over colonies and incubate an additional 2 hr at 37°. Avoid air bubbles between agar surface and filter.
4. Lift colonies, and transfer filter, colony side up, onto LB agar containing 250 µg/ml chloramphemcol and incubate another 24 hr at 37°.
5. Air dry filters thoroughly.
6. Treat filters in 500 ml solutions of the following in succession: (1) 2 × 0.5 N NaOH, 5 min each, (2) 2 × 0.5 M Tris–HCl, pH 7.4, 5 min each, (3) 2 × 2 × SSC, 5 min each, (4) 2 × 95% ethanol rinse briefly. Air dry thoroughly.

For large libraries, the above method can also be used, though it is labor intensive. High density screening as described by Hanahan and Meselson[11] is more efficient.

Colony Hybridization Using ^{32}P-Labeled Oligonucleotides

The method we have used for colony hybridization is modified from that of Suggs *et al.*[8]

Reagents

6 × NET (1 × NET = 150 mM NaCl, 15 mM Tris–HCl, pH 7.5, 1 mM EDTA)
NP40
6 × SSC

Protocol. Colony hybridization on Whatman 541 filters.

1. Prehybridize filters at 55° for 2 hr in 6 × NET, 0.5% NP40, and 100 µg/ml *E. coli* DNA and 30 µg/ml pBR322 DNA (previously sonicated; heated to boiling and rapidly cooled).
2. Hybridization is done in a minimal volume (2–3 ml per filter) of 6 × NET, 0.5% NP40, 250 µg/ml tRNA, and 1 ng/ml ^{32}P-labeled oligonucleotide (for each individual sequence); allow at least 10^6 cpm/filter (minimum 10^6 cpm/ml). Temperature dependent on oligonucleotide.[8] (Calculation is based on the T_d. T_d is the temperature at which one-half of the

[11] D. Hanahan and M. Meselson, this series, Vol. 100, p. 333.

duplexes are dissociated under the conditions of the experiment.[12] Temperature of hybridization is usually 2° below the T_d. $T_d = 2° \times$ number of AT base pairs + 4° × number of GC base pairs. When mixtures are used, the lowest calculated T_d will be used.) Incubate submerged for 2-20 hr.

3. Wash filters: 4 washes 6 × SSC (250 ml each) at 0° 15 min each; 1 wash 6 × SSC at temperature of hybridization (1 min at temperature of hybridization, not longer).

4. Blot dry. Expose to Kodak XR-1 X-ray film between 2 intensifier screens for 1-5 hr.

5. Whatman 541 filters can be reused after another NaOH, Tris, and 2 × SSC wash (see protocol on colony transfer).

We have found this procedure to be effective. In individual cases, we have found the following modifications necessary.

1. Temperature of hybridization. The formula used to calculate the optimal temperature of hybridization is based on the empirical observations of Suggs et al.[8] and Wallace et al.[12] In some individual cases, we have had to lower the temperature, in a stepwise fashion, by 5° each time, to obtain good signals of hybridization.

2. Washing. Usually the last wash at the temperature of hybridization can be skipped without significant effects on the background.

3. Interpretation of positive hybridization signals. Often, different parts of the filter show varying degrees of background on exposure to X-ray films. We found it most helpful to compare the intensity of the signal with its immediate neighbors. The positives in Fig. 1 are quite distinct.

4. Rescreening of positive colonies. We have found that positive colonies should be picked, and rescreened on another filter, together with negative controls (e.g., pBR322 as control). The original positive filters can be included in rescreening. Only the colonies which show definite positive signals on repeat hybridization are further studied.

Molecular Cloning and Sequence Analysis of Human ApoA-II cDNA

We have used the procedures detailed above for the cloning of a number of cDNAs including human apoA-II cDNA. The general approach should be feasible with cDNAs for other proteins, as long as partial sequence information is available. We will describe our experience with human apoA-II as an illustration of some of the results expected from such an approach.

---

[12] R. B. Wallace, J. Shaffer, R. F. Murphy, J. Bonner, T. Hirose, and K. Itakura, *Nucleic Acids Res.* **6**, 3543 (1979).

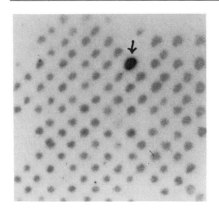 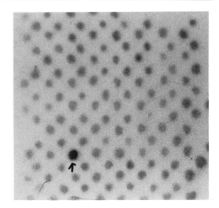

FIG. 1. Hybridization signals obtained with oligonucleotide probes. Colony hybridization was carried out by using a mixture of oligonucleotides as described in the text. The Whatman 541 filters were exposed to X-ray films for 16 hr. Both positive signals indicated by the arrows were subsequently proven to be authentic apoA-II cDNA clones by nucleotide sequencing.

To screen for apoA-II cDNA, we have used a mixture of oligonucleotides with the following sequences: 5′d[GTPyTGPuAAPuTAPyTGN(C/G)(A/T)]3′. These sequences are complementary to the possible mRNA sequences predicted from amino acids 12 to 17 (Ser-Gln-Tyr-Phe-Gln-Thr) of the sequence of apoA-II published by Brewer et al.[13] The 4 base wobble for Thr was dropped to reduce the total number of oligonucleotides. Our temperature of hybridization was 37° which was 3° lower than the calculated temperature. The signals we obtained using the protocols in this chapter are illustrated in Fig. 1.

Restriction mapping of 7 positive clones indicate that at least two of them have similar inserts. The longer one, pAII-1, was completely sequenced.[14] The sequencing strategy and complete nucleotide sequence of pAII-1 are shown in Fig. 2a and b. It should be noted that to ensure that the sequence was accurate, it was determined for the most part on both DNA strands. All the restriction sites on the map are accounted for by sequence analysis, and all of the restriction sites used to end-label DNA fragments, or for subcloning in M13, have been sequenced across by other restriction fragments. The clone contains a cDNA insert of 433 nucleo-

---

[13] H. B. Brewer, S. E. Lux, R. Ronan, and K. M. John, *Proc. Natl. Acad. Sci. U.S.A.* **69**, 1304 (1972).

[14] M. N. Moore, F. T. Kao, Y. K. Tsao, and L. Chan, *Biochem. Biophys. Res. Commun.* **123**, 1 (1984).

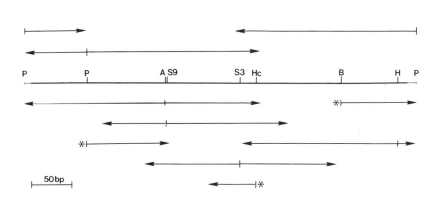

```
1   GTT ACC AAC ATG AAG CTG CTC GCA GCA ACT GTG CTA CTC CTC ACC ATC TGC AGC CTT GAA GGA GCT TTG GTT CGG AGA
                MET LYS LEU LEU ALA ALA THR VAL LEU LEU LEU THR ILE CYS SER LEU GLU GLY ALA LEU VAL ARG ARG
79  CAG GCA AAG GAG CCA TGT GTG GAG AGC CTG GTT TCT CAG TAC TTC CAG ACC GTG ACT GAC TAT GGC AAG GAC CTG ATG
    GLN ALA LYS GLU PRO CYS VAL GLU SER LEU VAL SER GLN TYR PHE GLN THR VAL THR ASP TYR GLY LYS ASP LEU MET
157 GAG AAG GTC AAG AGC CCA GAG CTT CAG GCC GAG GCC AAG TCT TAC TTT GAA AAG TCA AAG GAG CAG CTG ACA CCC CTG
    GLU LYS VAL LYS SER PRO GLU LEU GLN ALA GLU ALA LYS SER TYR PHE GLU LYS SER LYS GLU GLN LEU THR PRO LEU
235 ATC AAG AAG GCT GGA ACG GAA CTG GTT AAC TTC TTG AGC TAT TTC GTG GAA CTT GGA ACA CAG CCT GCC ACC CAG TGA
    ILE LYS LYS ALA GLY THR GLU LEU VAL ASN PHE LEU SER TYR PHE VAL GLU LEU GLY THR GLN PRO ALA THR GLN
313 AGT GTC CAG ACC ATT GTC TTC CAA CCC CAG CTG GCC TCT AGA ACA CCC ACT GGC CAG TCC TAG AGC TCC TGT CCC TAC
391 CCA CTC TTT GCT ACA ATA AAT GCT GAA TGA ATC CAA AAA AAA AA
```

FIG. 2. (a) Partial restriction map and sequence analysis of pAII-1. The arrows above the map indicate sequences obtained by the dideoxynucleotide chain termination method following subcloning of the PstI restriction fragments into mp8 or mp9. The arrows below the map indicate sequences obtained by the chemical degradation technique. Asterisk (*) indicates 5' end labeling. The other fragments were sequenced by 3' end labeling by reverse transcriptase. P, PstI; A, AvaII; S9, Sau96I; S3, Sau3AI; Hc, HincII; B, BalI; H, HinfI. (b) Nucleotide sequence of pAII-1. The deduced amino acid sequence is presented. The mature plasma protein starts with the first residue on the second line, i.e., Gln. The preprosegments are represented by the 23 residues on the first line, ending with Arg. The putative polyadenylation signal, AATAAA, is underlined.

tides which include the poly(A) tail, the complete 3'-nontranslated region of 115 nucleotides, the complete coding region of 300 nucleotides, and 8 nucleotides of the 5'-nontranslated region. The coding portion predicts a peptide of 100 amino acids. The mature plasma apo-AII peptide starts at residue 24, and consists of 77 amino acid residues. This part of the derived amino acid sequence completely matches the amino acid sequence published by Brewer et al.[13] with one exception. Residue 37 of the mature protein is predicted to the Glu rather than Gln as reported previously.[13] As this part of the DNA sequence was determined on both strands on four restriction fragments by the chemical degradation method, and also matched the sequence obtained by the M13 dideoxynucleotide chain termination method, we are confident that the codon for this amino acid is GAG. Whether the difference observed represents true sequence heterogeneity of human apoA-II remains to be determined. If it is proven to be the case, it may be of functional significance since it changes the charge properties of the apolipoprotein.

The first 23 residues of the derived amino acid sequence correspond to the prepropeptide of human apoA-II. This part of the sequence completely matches the partial sequence reported by Gordon et al.[15]

Examination of the codon usage of apoA-II mRNA reveals a marked preference for specific codons used for Lys (AAG, 10/10) and Gln (CAG, 7/7). Similar preference for these codons was also noted in apoA-I mRNA.[16] The significance of this observation is unclear.

Conclusion

We have presented our experience with the use of oligonucleotide probes to screen for apoA-II cDNA clones. If the procedures described are followed carefully, they can be used to screen for cDNAs of other proteins or apolipoproteins for which partial amino acid sequence information is available. The selection of cDNA clones is now a fairly straightforward undertaking. Once a clone is identified, various interesting studies can be performed (e.g., chromosomal mapping, restriction fragment length polymorphism, etc.), many of which are described in other chapters of this volume.

Acknowledgments

The work described in this chapter was supported by grants from the National Institutes of Health, HL-27341, and the March of Dimes Birth Defects Foundation. We thank Ms. Sundi Chen for expert technical assistance in the studies described in this chapter.

---

[15] J. I. Gordon, K. A. Budelier, H. F. Sims, C. Edelstein, A. M. Scanu, and A. W. Strauss, J. Biol. Chem. **258,** 14054 (1983).

[16] P. Cheung and L. Chan, Nucleic Acids Res. **11,** 3703 (1983).

[44] Rat Apolipoprotein A-IV: Application of Computational Methods for Studying the Structure, Function, and Evolution of a Protein[1]

By Mark S. Boguski, Nabil A. Elshourbagy, John M. Taylor, and Jeffrey I. Gordon

An increasingly common problem in molecular biology is that the primary structures of proteins, both real and hypothetical, are often available long before any definite function has been ascribed to them. There are many computational techniques that can be applied to this problem of discovering the functions of protein (and nucleic acid) sequences. These techniques fall into two overlapping categories: methods for examining potential evolutionary relationships among sequences and methods for predicting structural and functional characteristics of individual sequences. The purpose of the present chapter is not to provide a comprehensive survey of these techniques but rather to describe how we have used some of them to gain insight into the physiologic function of rat apoA-IV and its structural and evolutionary relationships to other apolipoproteins.

ApoA-IV is one of the major protein components of rat chylomicrons and HDL and is also associated with chylomicrons in humans and dogs.[2,3] Approximately 50% of plasma A-IV is not associated with lipoprotein particles and may undergo extensive redistribution among circulating lipoproteins and a free pool.[4,5] The cholesteryl ester content of HDL may regulate apoA-IV binding to HDL.[6,7] However the precise metabolic function of this protein remains obscure.

---

[1] This work was supported by Grants AM 30292 and AM 31615 from the National Institutes of Health. M.S.B. is supported by a Medical Scientist Training Program Grant GM 07200 and the Gerty T. Cori Predoctoral Fellowship from Sigma Chemical Company. This work was presented to the Faculty of the Graduate School of Arts and Sciences of Washington University in partial fulfillment of the requirements for the degree of Doctor of Philosophy for M.S.B. J.I.G. is an Established Investigator of the American Heart Association.
[2] J. B. Swaney, F. Braithwaite, and H. A. Eder, *Biochemistry* **16,** 271 (1977).
[3] K. H. Weisgraber, T. P. Bersot, and R. W. Mahley, *Biochem. Biophys. Res. Commun.* **85,** 287 (1978).
[4] U. Beisiegel and G. Utermann, *Eur. J. Biochem.* **93,** 601 (1979).
[5] N. H. Fidge, *Biochim. Biophys. Acta* **619,** 129 (1980).
[6] J. G. DeLamatre, C. A. Hoffmeier, and P. S. Roheim, *J. Lipid. Res.* **24,** 1578 (1983).
[7] C. H. Sloop, L. Dory, R. Hamilton, B. R. Krause, and P. S. Roheim, *J. Lipid Res.* **24,** 1429 (1983).

We have isolated, from a rat intestinal cDNA library, a full-length apoA-IV clone and determined its nucleotide sequence.[8] The derived amino acid sequence, for the primary translation product, contains 391 residues with a calculated molecular weight of 44,465. ApoA-IV contains at least 13 tandem repetitions of a highly conserved 22-amino-acid amphipathic segment which presumably arose by a series of intragenic unequal crossovers.[8] The structure and organization of these repeat units bear a striking similarity to repeated sequences in human apoA-I.

Methods for Searching Data Banks of Biological Sequences

Often the first or only clue to the biological function of a newly discovered sequence is similarity to another sequence of known function. However it should be noted that only a small fraction of all proteins and nucleic acids in nature have been sequenced to date and furthermore the collection of these sequences is biased: phylogenetic representation is very uneven and sequences that are abundant and easily purified are overrepresented. Thus, for newly discovered sequences, chances are small that database searching will reveal anything of biological significance in any individual case. Nevertheless database searching is the logical first step in sequence identification and analysis.

The two most complete and widely available databases in the United States are the Bolt, Beranek and Newman, Inc. GENETIC SEQUENCE DATA BANK[9] (GenBank) and the National Biomedical Research Foundation PROTEIN SEQUENCE DATABASE[10] (NBRF). GenBank (Release 35.0) contains over 5 megabases representing 7727 sequences and the NBRF database (Release 6.0, 28 August 1985) contains nearly three quarters of a million residues representing 3309 sequences. The information stored in these databases overlaps to some extent.

There are several available computer programs used to conduct database searches. For example, Wilbur and Lipman[11] developed SRCHN and SRCHGP for nucleic acid and protein searches, respectively. SRCHN and SRCHGP score matches on the basis of nucleotide or amino acid identity and allow the user to adjust various search parameters such as "window," "k-tuple," and "gap penalty." A k-tuple is defined as a

---

[8] M. S. Boguski, N. Elshourbagy, J. M. Taylor, and J. I. Gordon, *Proc. Natl. Acad. Sci. U.S.A.* **81,** 5021 (1984).

[9] GenBank, Research Systems Division, Bolt Beranek and Newman Inc., 10 Moulton St., Cambridge, MA 02238.

[10] National Biomedical Research Foundation, Georgetown University Medical Center, 3900 Reservoir Rd., Washington, D.C. 20007.

[11] W. J. Wilbur and D. J. Lipman, *Proc. Natl. Acad. Sci. U.S.A.* **80,** 726 (1983).

group of k consecutive sequence elements (bases or amino acids) where k is a small positive integer. All k-tuple matches increment the score by one; all gaps decrement the score by a user-specified gap penalty (that is independent of gap size). The positions of all k-tuple matches are identified and an optimal path through these matches is determined. To increase the speed at which the optimal path is located, the program considers only those matches that fall within a certain window space. For convenience, SRCHN and SRCHGP supply default values for the search parameters. For practical purposes, increasing k increases specificity and speed while decreasing sensitivity; low gap penalties maximize homology at the risk of introducing a biologically unreasonable number of sequence interruptions.

We used rat apoA-IV mRNA and derived amino acid sequences (Fig. 1) to search the GenBank and NBRF databases. The top matches for each database are presented in Table I. It is apparent that the best matches are to other apolipoprotein entries in the databases, specifically multiple apoA-I entries and apoE. There are also some other matches that seem meaningful in the biological context of lipid transport, namely mouse casein and rat serum albumin, but does apoA-IV really have any biologically significant relationship to bovine satellite DNA or an *E. coli* DNA replication initiation protein? Probably not, and this raises the issue of how to distinguish false positive results. Practically speaking, the best objective approach to assessing the significance of a sequence similarity is to apply a criterion of mutual consistency: do alternate search and comparison methods produce the same results?

Program SEARCH is available from the NBRF Protein Identification Resource[12] (PIR). It has the advantage of permitting alternate scoring systems besides simple amino acid identity. For example, one can use a "mutation data matrix" based on amino acid replacements tolerated by natural selection in protein families. This is a much more sensitive method for detecting distant relationships. When a search of the Protein Sequence Database (January 1984 Release) was conducted with this program, using various segments of rat apoA-IV as test sequences, apolipoproteins A-I and E (LPHUA1, LPGDA1, LPHUE) had the highest comparison scores ranging from 5.8 to 6.5 standard deviations above the mean. Serum albumins (ABHUS, ABBOS) also scored highly as did apolipoproteins A-II, C-I, C-II, and C-III (LPMQA2, LPHUA2, LPHUC1, LPHUC2, LPHUC3). That these latter apolipoproteins are *not* among the top 40

---

[12] This program and others are described in M. O. Dayhoff, "Atlas of Protein Sequence and Structure," Vol. 5, Suppl. 3. National Biomedical Research Foundation, Washington, D.C., 1978, and M. O. Dayhoff, W. C. Barker, and L. T. Hunt, this series, Vol. 91, p. 524.

```
     -81       -71       -61       -51       -41       -31       -21       -11        -1        10        20        30
ATCCTCACAGCGACACAGTGAACCATTTCTTCTGACGCTAGGAAACATCCAGTGTAGCTGAGGCTGTCCCAACCCAGTGAGGAGCCCAGGATGTTCCTGAAGGCTGTGGTGCTGACCGTG
                                                                                          MetPheLeuLysAlaValValLeuThrVal
                                                                                           ^signal peptide  .  .  .  .

      40        50        60        70        80        90       100       110       120       130       140       150
GCCCTGGTGGCCATCACCGGGACCCAGGCTGAGGTCACTTCCGACCAGGTGGCCAATGTGATGTGGGACTACTTCACCCAGCTAAGCAACAATGCCAAGGAGGCTGTGGAACAACTGCAG
AlaLeuValAlaIleThrGlyThrGlnAlaGluValThrSerAspGlnValAlaAsnValMetTrpAspTyrPheThrGlnLeuSerAsnAsnAlaLysGluAlaValGluGlnLeuGln
 .  .  .  .  .  .  .  .  .  .  .  .  ^mature plasma protein

     160       170       180       190       200       210       220       230       240       250       260       270
AAGACAGATGTCACTCAACAGCTCAATACCCTCTTCCAGGACAAACTTGGGAACATTAACACCTATGCCGATGACCTGCAGAACAAGCTGGTGCCCTTCGCCGTTCAACTGAGTGGACAT
LysThrAspValThrGlnGlnLeuAsnThrLeuPheGlnAspLysLeuGlyAsnIleAsnThrTyrAlaAspAspLeuGlnAsnLysLeuValProPheAlaValGlnLeuSerGlyHis
                                                                                             *
     280       290       300       310       320       330       340       350       360       370       380       390
CTAACCAAGGAAACAGAGAGGGTGAGGGAAGAGATCCAGAAGGAGCTCGAGGACCTACGGGCCAACATGATGCCCCATGCCAACAAAGTGAGCCAGATGTTCGGGGACAACGTGCAGAAG
LeuThrLysGluThrGluArgValArgGluGluIleGlnLysGluLeuGluAspLeuArgAlaAsnMetMetProHisAlaAsnLysValSerGlnMetPheGlyAspAsnValGlnLys
                                                                             *
     400       410       420       430       440       450       460       470       480       490       500       510
TTGCAGGAACACCTGAGGCCCTATGCCACGGACCTGCAAGCTCAGATCAACGCACAGACCCAGGATATGAAACGCCAGCTGACCCCCTACATCCAGCGCATGCAGACCACAATACAAGAC
LeuGlnGluHisLeuArgProTyrAlaThrAspLeuGlnAlaGlnIleAsnAlaGlnThrGlnAspMetLysArgGlnLeuThrProTyrIleGlnArgMetGlnThrThrIleGlnAsp
                  *                                                                     *
     520       530       540       550       560       570       580       590       600       610       620       630
AATGTGGAAAACCTGCAGTCCTCCATGGTGCCCTTTGCCAACGAGCTAAAGGAAAAGTTTAACCAGAATATGGAAGGGCTCAAGGGGCAACTAACCCCCCGTGCCAACGAGCTGAAGGCC
AsnValGluAsnLeuGlnSerSerMetValProPheAlaAsnGluLeuLysGluLysPheAsnGlnAsnMetGluGlyLeuLysGlyGlnLeuThrProArgAlaAsnGluLeuLysAla
                              *                                                                       *
     640       650       660       670       680       690       700       710       720       730       740       750
ACGATCGACCAGAACCTGGAAGACCTGCGCAGCAGACTGGCCCCTCTCGCGGAGGGTGTGCAGGAAAAACTCAACCATCAGATGGAAGGCTTGGCCTTCCAGATGAAGAAGAACGCTGAG
ThrIleAspGlnAsnLeuGluAspLeuArgSerArgLeuAlaProLeuAlaGluGlyValGlnGluLysLeuAsnHisGlnMetGluGlyLeuAlaPheGlnMetLysLysAsnAlaGlu
                                          *
     760       770       780       790       800       810       820       830       840       850       860       870
GAGCTCCATACCAAAGTCTCCACAAATATCGACCAGCTGCACAAGAATCTGGCCCCGCTGGTGGAAGATGTGCAAAGCAAGCTGAAAGGCAACACGGAAGGACTGCAGAAGTCTCTGGAA
GluLeuHisThrLysValSerThrAsnIleAspGlnLeuGlnLysAsnLeuAlaProLeuValGluAspValGlnSerLysLeuLysGlyAsnThrGluGlyLeuGlnLysSerLeuGlu
                                                    *
     880       890       900       910       920       930       940       950       960       970       980       990
GACCTGAACAAGCAGCTGGACCAGCAGGTGGAGGTGTTCCGGCGTGCTGTGGAGCCCCTGGGGGATAAGTTCAACATGGCTCTGGTGCAGCAGATGGAGAAGTTCAGGCAGCAGCTGGGC
AspLeuAsnLysGlnLeuAspGlnGlnValGluValPheArgArgAlaValGluProLeuGlyAspLysPheAsnMetAlaLeuValGlnGlnMetGluLysPheArgGlnGlnLeuGly
                                                    *
    1000      1010      1020      1030      1040      1050      1060      1070      1080      1090      1100      1110
TCCGATTCGGGGGACGTGGAAAGCCACTTGAGCTTCCTGGAGAAGAACCTGAGGGAAAAGGTCAGCTCCTTTATGAGCACCCTGCAAAAAAAGGGGAGCCCAGACCAGCCCCTAGCCCTC
SerAspSerGlyAspValGluSerHisLeuSerPheLeuGluLysAsnLeuArgGluLysValSerSerPheMetSerThrLeuGlnLysLysGlySerProAspGlnProLeuAlaLeu

    1120      1130      1140      1150      1160      1170      1180      1190      1200      1210      1220      1230
CCCCTCCCGGAGCAGGTTCAGGAACAGGTCCAGGAGCAGGTGCAGCCCAAACCTCTGGAGAGCTGAGCTGTCCCTGTGTCCTCGGCCCATCACAGCAGCAGACACCTGCCCTGCCCCACC
ProLeuProGluGlnValGlnGluGlnValGlnGluGlnValGlnProLysProLeuGluSer

    1240      1250      1260      1270      1280      1290      1300      1310      1320      1330
ACCTGTCTGTCACTCTGTTCCCAAGCACTTCTCGTACCAGCTGAGGACACATGTCCTGTGGGTGACGATACCTCCTCTCGTTACTCAATAAAGCATCTGAG
                                                                                     Poly(A)
                                                                                      signal
```

FIG. 1. Nucleotide sequence of rat apoA-IV cDNA and derived protein sequence. Proline residues that initiate repeated sequences are indicated by an asterisk (*).

TABLE I
RESULTS OF DATABASE SEARCHES WITH RAT ApoA-IV[a]

Matched sequence	Comparison score	Standard deviations above mean score
GenBank release 19.0 (1175 mammalian sequences)		
1. Human apoA-I gene (J00098)	26	12
2. Human apoA-I mRNA (J00100)	22	8.8
3. Mouse epsilon casein mRNA (J00379)	20	7.3
4. Human apoE gene (K00396)	20	7.3
5. Human insulin 3' repetitive sequence (J00268)	19	6.6
6. Human HLA-DR α-chain (J00204)	18	5.9
7. Bovine satellite DNA (J00039)	18	5.9
...		
12. Rat apoE mRNA (J00705)	17	5.1
NBRF January 1984 release (2511 total sequences)		
1. Human proapoA-I (LPHUA1)	19	4.2
2. Human apoE (LPHUE)	17	3.7
3. Dog apoA-I (LPDGA1)	16	3.5
4. *Bacillus subtilis* flagellin (FLBS68)	15	3.1
5. *E. coli* plasmid replication initiation protein (IDECRK)	15	3.1
6. Rat angiotensin precursor (ANRT)	15	2.9
7. Rat serum albumin precursor (ABRTS)	14	2.9

[a] For program SRCHN (nucleic acids), the k-tuple was 4, the window was 20, and the gap penalty was 4. For program SRCHGP (proteins), the k-tuple was 2, the window was 20, and the gap penalty was 10. Positions 8 to 11 in the rank order of matches for the GenBank search were occupied by multiple immunoglobulin entries. GenBank sequences are referenced by their accession numbers; NBRF sequences are referenced by their retrieval key codes.

matches using the Wilbur–Lipman program SRCHGP raises the issue of false negative results and emphasizes the importance of not relying on a single approach to the problem of assessing sequence similarities.

Recently, Lipman and Pearson[13] modified the sequence similarity algorithm used in SRCHN and SRCHGP to improve its sensitivity and efficiency. This algorithm was then implemented in the program FASTP which also uses the aforementioned mutation data (PAM250) matrix for scoring. FASTP can perform very rapid and sensitive protein similarity

[13] D. J. Lipman and W. R. Pearson, *Science* **227**, 1435 (1985).

searches. FASTP (and its DNA sequence database companion FASTN) represent significant advances over their earlier counterparts.

Methods for the Analysis of Sequence Similarities

As Doolittle has pointed out,[14] establishing a relationship between sequences is not merely a matter of finding, in a database, one sequence that resembles another. Once a similar sequence has been identified, rigorous comparisons must be carried out to distinguish authentic relationships from chance similarity or convergence. There are a number of methods for assessing the statistical and biological significance of a sequence similarity and we favor applying several methods to see if they produce mutually consistent results. It should be noted that, in general, DNA comparisons are less informative and subject to more background "noise."[14,15] Thus the analyses that follow employ the amino acid sequence of apoA-IV.

McLachlan[15-17] devised the *comparison matrix* method to quantitatively analyze the relationships among proteins. This method calculates *matching probability* scores for spans of amino acids based upon observed frequencies of amino acid replacements in homologous proteins. One sequence can be compared against itself or another sequence. A matrix of scores representing all possible alignments of the two sequences is constructed and all scores exceeding a user-specified threshold value are plotted. In this way, a graphical representation of the locations of related sequences within two proteins is obtained. The probability of chance occurrence can be estimated from the distribution of matching scores that result when randomly permuted sequences of the two proteins are used. The theoretical distribution for infinite sequences of the same amino acid composition can also be calculated.

The comparison matrix method was used to identify and study the repeated pattern of amphipathic docosapeptides in human apoA-I.[18] Because the database searches indicated a significant relationship between apolipoproteins A-I and A-IV, we used McLachlan's method to analyze this relationship. In proteins that are closely related but lack a repeating structure, the highest scores tend to fall on a single *main diagonal*. Re-

---

[14] R. F. Doolittle, *Science* **214**, 149 (1981).
[15] A. D. McLachlan, *J. Mol. Biol.* **169**, 15 (1983).
[16] A. D. McLachlan, *J. Mol. Biol.* **61**, 409 (1971).
[17] A. D. McLachlan, *J. Mol. Biol.* **72**, 417 (1972).
[18] A. D. McLachlan, *Nature (London)* **267**, 465 (1977).

peated sequences reveal themselves as shorter diagonals offset from the main diagonal. Figure 2 is the comparison matrix for human apoA-I vs rat apoA-IV. Immediately apparent are the strong repeats between residues 85-370 in apoA-IV and residues 70-255 in apoA-I. Another striking feature of the comparison matrix is the long main diagonal (md) shared by these two sequences. This result indicates a colinear relationship along nearly the entire length of the apoA-I sequence. The main diagonal fades away in several locations indicating variation in the strength of the homology between A-I and A-IV, but the pattern always reappears on the same diagonal indicating an absence of deletions or insertions. It is interesting to note in this context that the diagonal representing homology between the signal sequences (sp) does *not* resume on the main diagonal indicating a deletion in the amino terminus of apoA-IV. Based upon the observed number of amino acid identities, apoA-I and A-IV are most closely homologous in the 40 or so residues spanning positions 89-129 in A-I and 80-120 in A-IV (Fig. 2).

How significant is the relationship between human apoA-I and rat apoA-IV? Figure 3 shows the cumulative probability distributions of matching scores for real and shuffled sequences. The distribution of real scores begins to deviate from the distribution of shuffled scores at about 1 standard deviation above the mean. At about 4 standard deviations above the mean, there is a large "shoulder" (curve A) at scores ranging from about 143 to 157 indicating a large number of spans with similar scores. In infinite sequences of the same amino acid compositions, scores in this range have a frequency of chance occurrence of about 10^{-8} to 10^{-11} (Table II). Two spans reached a score of 169. For each of these spans, the frequency of observed scores exceeds that of expected scores by a factor of 261 million.

Program RELATE (NBRF/PIR) represents an alternative approach to assessing the significance of a sequence similarity. In much the same way as McLachlan's method, RELATE compares all possible segments (spans) of a user-specified length from one sequence with all segments of the same length from a second sequence. Segment scores are determined using the mutation data matrix, already mentioned in connection with program SEARCH. For each segment, RELATE generates (1) a score, (2) the positions of the first residues of the segments from each sequence, and (3) the difference between the first residue positions (the displacement). In sequences composed of multiple tandem repeats, displacements of the high-scoring segments tend to be multiples of the repeat length. RELATE also calculates a *segment comparison score* based on a comparison of the mean score for the real sequences and the mean scores from

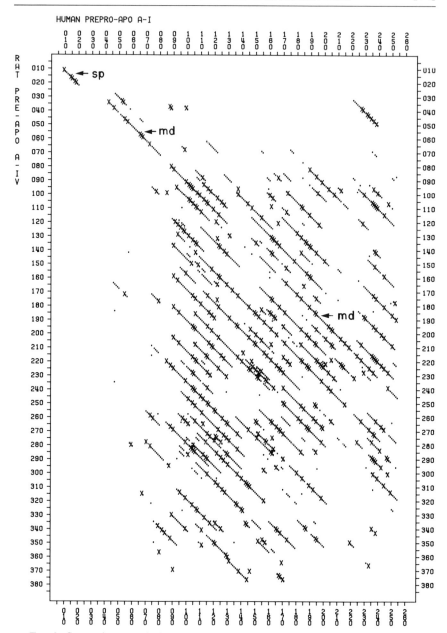

FIG. 2. Comparison matrix for human preproapoA-I vs rat preapoA-IV. A comparison (or matching) score is the unweighted sum of similarity scores across a span. A span of 29 residues was used. (The span must be an odd number and it is convenient to choose a length slightly greater than any suspected repeat unit.) "X" and "·" indicate the centers of spans

TABLE II
COMPARISON STATISTICS FOR RAT ApoA-IV VS HUMAN ApoA-I[a]

Score (S)	Number of spans of score S or higher	Observed frequency	Expected frequency	Ratio observed/expected
87	51,065	0.489	0.583	0.84
100	22,352	0.214	0.167	1.28
110	10,381	0.994×10^{-1}	0.328×10^{-1}	3.03
125	2,353	0.225×10^{-1}	0.795×10^{-3}	28.3
132	1,036	0.992×10^{-2}	0.958×10^{-5}	1030
143	167	0.160×10^{-2}	0.673×10^{-8}	237,000
157	23	0.220×10^{-3}	0.648×10^{-11}	33,900,000
169	2	0.191×10^{-4}	0.732×10^{-14}	261,000,000

[a] Comparison parameters were the same as in Figs. 2 and 3 and the same as those used previously to demonstrate the repeated sequences in human apoA-I (ref. 18). The number of elements in a comparison matrix is the product of the number of residues in each of the two sequences being compared. Thus the apoA-IV vs A-I matrix contained $391 \times 267 = 104,397$ spans. The mean score for shuffled sequences of the same length and amino acid composition was 87 with a standard deviation of 14.

multiple comparisons of randomized sequences. This segment comparison score, expressed in SD units,[19] can be used to estimate the probability of this degree of chance similarity between the sequences. The companion program DOTMATRIX generates graphic output very similar to McLachlan's comparison matrix.

Program RELATE can be used to compare a sequence with itself and in this way we have analyzed the repeated sequences within apoA-IV. The results are summarized in Table III. Every one of the segment displacements is a multiple of 11, leaving no doubt that the length of the repeat unit is 11 residues. The top scores are clustered in a narrow range (58 to 64.88) nearly 4 standard deviations above the mean score indicating

[19] A score in SD units is defined as the number of standard deviations by which the maximum score for the real two sequences exceeds the average maximum score for a large number (e.g., 100) of random permutations of the two sequences.

with exceed the threshold value. In addition, "X" indicates an amino acid identity between the sequences. All 2353 spans with comparison scores of 125 or greater (Table II) are displayed. A score of 125 exceeds the mean score by more than 3 standard deviations (Fig. 3, Table II) and has a frequency of chance occurrence of less than 10^{-3} in infinite random sequences of the same amino acid composition. The arrows labeled "sp" and "md" indicate signal peptide homology and the main diagonal, respectively.

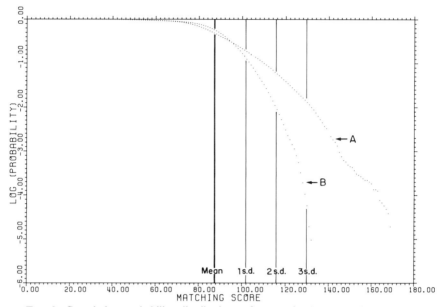

FIG. 3. Cumulative probability distributions of scores in the comparison matrix for human preproapoA-I vs rat preapoA-IV. Curve A is the distribution for the real sequences; curve B is the distribution for randomized sequences. The mean score was 87 with a standard deviation of 14.

TABLE III
INTRASEQUENCE COMPARISON OF RAT PreapoA-IV:
FREQUENCIES OF THE TOP 177 SCORES WITH THE
SAME DISPLACEMENT[a]

Displacement	Frequency	Average score
−44	73	64.88
−22	35	63.23
−110	23	60.61
−66	22	61.36
−88	13	60.00
−132	6	58.83
−99	4	58.25
−55	1	58.00

[a] Program RELATE (NBRF/PIR) was used to compare rat apoA-IV with itself. The mutation data matrix was used for scoring. The fragment length was 29. The resultant scores ranged from −61 to +79 with a mean score of −13 and a standard deviation of 19.

TABLE IV
SEGMENT COMPARISON SCORES FOR RAT ApoA-IV WITH VARIOUS PROTEINS[a]

Sequence 1	Sequence 2	Score (SD units)	Chance probability
Rat A-IV	Rat A-I	10.604	$< 10^{-23}$
Rat A-IV	Rat E	16.187	$<< 10^{-23}$
Rat A-IV	Human A-I	17.336	$<< 10^{-23}$
Rat A-IV	Human E3	11.730	$<< 10^{-23}$
Rat A-IV	Human A-II	4.839	$\sim 0.287 \times 10^{-6}$
Rat A-IV	Human C-I	3.796	$\sim 0.317 \times 10^{-4}$
Rat A-IV	Human C-II	2.484	$\sim 0.621 \times 10^{-2}$
Rat A-IV	Human C-III	0.936	~ 0.159
Rat A-IV	Rat serum albumin	1.491	$\sim 0.668 \times 10^{-1}$
Rat A-IV	Rat casein	−1.960	—

[a] This is a portion of the output from program RELATE (segment comparison score summary). Sequence 1 is compared with Sequence 2. The mutation data matrix was used for scoring and the fragment length was 11. The scores were calculated from 100 runs of randomized sequences. The probabilities of obtaining these scores by chance were estimated from the normal distribution as described in ref. 12. A negative score indicates that the mean score for the randomized sequences exceeded the mean score for the real sequences. The sources of the sequences were as follows: rat preapoA-IV (ref. 8), rat preproapoA-I (ref. 29), rat preapoE (ref. 20). All other sequences were from the NBRF database.

that the repeat units are highly conserved with respect to each other. We previously observed that the repeated sequence block of apoA-IV appears to consist of at least 13 tandem repetitions of a 22 amino acid segment that was itself composed of two related 11-mers.[8] Furthermore, each 11-mer is more similar to the 11-mer once removed that to its adjacent 11 amino acid segment. Thus it seems that a 22 amino acid segment was the more common evolutionary unit and may be the functional unit of apoA-IV as well.

We next used RELATE to assess the significance of sequence similarities between rat apoA-IV and other proteins identified in the database searches. Because the length of the basic repeat unit was determined to be 11 residues, the fragment (span) length parameter was set to 11. The results are summarized in Table IV. It is apparent the apolipoproteins, A-I, A-IV, and E form a group of very closely related sequences. The odds against their being related by chance are astronomical. In contrast, the scores for apolipoproteins A-II, C-I, C-II, and C-III are many orders

[20] J. W. McLean, C. Fukazawa, and J. M. Taylor, *J. Biol. Chem.* **258**, 8993 (1983).

of magnitude less significant.[21] The cutoff score for relatedness is about 3.0 SD units.[12] ApoA-II and C-I exceed this score but apoC-II and C-III do not. However, pairwise comparisons among these four sequences generally revealed significant relationships (data not shown). This apparent paradox is resolved by the realization that statistically significant similarity may not be demonstrable between every pair of sequences in a protein family.[12] Recent results on the structure of apolipoprotein genes have provided further insight into the relationships among members of this family. The possibility of limited gene conversion suggests that global sequence comparisons can be somewhat misleading and thus analyses using individual apolipoprotein exons may be more suitable to detect subtle relationships.[22]

The final step in the analysis of similarity between two sequences is to generate some kind of "optimal alignment." Program ALIGN (NBRF/PIR) will do this for either protein or nucleic acid sequences; PRTALN and NUCALN[11] operate on protein and nucleic acid sequences, respectively. ALIGN can use either the now familiar mutation data matrix or a "unitary matrix" based only upon identity between sequence elements. PRTALN and NUCALN are restricted to scoring identities alone. All of these programs allow the user to select a number of alignment parameters (scoring matrix, matrix bias, and break penalty for ALIGN; k-tuple, window, and gap penalty for PRTALN and NUCALN). For example, one can elect to maximize the number of matches or minimize the number of gaps. If one were aligning the sequence of an interrupted gene with that of its cognate mRNA, a low gap penalty would be necessary to allow for intervening sequences. Different "optimal alignments" are obtained with different sets of parameters. Also, alignments for the same sequence may differ depending upon whether nucleotides or amino acids are used and which scoring matrix is chosen. Thus it is up to the investigator to decide which alignment represents reality in terms of the structural organization and/or the evolutionary history of a pair of sequences.

Interpreting optimal alignments can be problematical for a number of reasons including injudicious choice of alignment parameters and attempting to align two sequences which differ greatly in length. Another difficulty arises when one is trying to align a pair of sequences that contain

---

[21] Because apolipoproteins A-I, A-IV, and E are much larger than apolipoproteins A-II, C-I, C-II, and C-III, we thought that perhaps this size discrepancy was affecting the results. However experiments with length-adjusted control sequences did not significantly alter the conclusions.

[22] M. S. Boguski, E. H. Birkenmeier, N. A. Elshourbagy, J. M. Taylor, and J. I. Gordon, *J. Biol. Chem.*, in press (1986).

multiple internal duplications which have arisen via unequal crossing over. This mechanism of intragenic duplication can scramble an orderly array of repeated sequences. None of the alignment algorithms in these programs will allow transpositions of elements, or groups of elements, in the linear sequence. Thus one must be very careful when using optimal alignments to reconstruct a series of genetic events. The comparison matrix method or program RELATE is a more appropriate tool for this type of analysis. Fitch *et al.* have developed specific techniques for inferring the history of a duplication.[23]

Methods for Predicting Secondary Structure

Apolipoproteins have proven refractory to crystallization. Thus techniques other than X-ray crystallography have been used to obtain information about apolipoprotein structure. In addition to direct experimental methods, such as circular dichroism and the synthesis of model peptides, methods for the empirical prediction of protein structure have been useful.

The most popular prediction method is that of Chou and Fasman[24] which we have used to analyze structural potentials of the repeated sequences in apoA-IV. Figure 4 is a diagrammatic Chou–Fasman prediction for the entire preapoA-IV sequence. The predicted amounts of α-helix and β-sheet were about 56 and 15%, respectively. These values are in good agreement with the values estimated by Swaney *et al.* from circular dichroic measurements (52% α, 11% β).[2] The diagram shows a periodic pattern of α-helices punctuated by other backbone conformations. The regular pattern of β-turn-forming residues is particularly notable. There are regions with significant β-sheet potential and other regions would appear to exist as random coils. It is conceivable that randomly coiled regions might be induced to assume α-helical conformations upon binding to lipid. The helical content of apolipoproteins is known to increase under such conditions (reviewed in ref. 25). Alternatively, randomly coiled regions may be required for tertiary conformational flexibility.

The Chou–Fasman method has been criticized because it accurately predicts the correct structures only about 50% of the time.[26] Kabsch and Sander have made some sobering observations relevant to the dangers of

---

[23] W. M. Fitch, T. Smith, and J. Breslow, this volume [45].
[24] P. Y. Chou and G. D. Fasman, *Annu. Rev. Biochem.* **47**, 251 (1978).
[25] J. D. Morrisett, R. L. Jackson, and A. M. Gotto, *Biochim. Biophys. Acta* **472**, 93 (1977).
[26] W. Kabsch and C. Sander, *FEBS Lett.* **155**, 179 (1983).

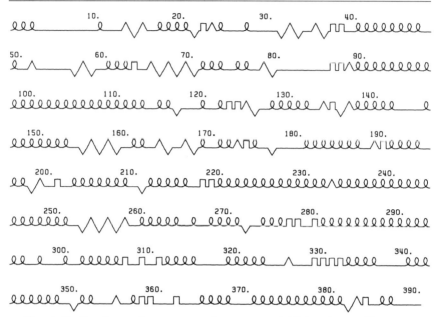

FIG. 4. Predicted secondary structure for rat preapoA-IV based upon Chou–Fasman rules. ℓ, α-helix; ⊓, β-turn; ∿, β-pleated sheet. Each symbol represents one amino acid residue. Flat lines indicate areas where no prediction could be made.

overzealous structure prediction.[27] They found many cases of identical pentapeptides that have completely different conformations in different proteins. The method of Garnier et al. seems to offer some practical advantages, especially for apolipoproteins, because it is possible to improve the accuracy of a prediction by introducing decision constants based upon circular dichroism data.[28] Clearly, until crystallization of an apolipoprotein has been achieved, empirical predictions will remain a mainstay of structural analysis. However, such predictions must be interpreted with great caution, especially in the absence of supporting experimental evidence.

Methods for Locating Functional Domains

Thus far we have established that (1) rat apoA-IV has intimate evolutionary relationships with apolipoproteins A-I and E, (2) apoA-IV has evolved by intragenic duplication of segments that are multiples of 11

[27] W. Kabsch and C. Sander, *Proc. Natl. Acad. Sci. U.S.A.* **81,** 1075 (1984).
[28] J. Garnier, D. J. Osguthorpe, and B. Robson, *J. Mol. Biol.* **120,** 97 (1978).

amino acids, and (3) apoA-IV has a predominantly α-helical structure. In addition, we have shown previously that apoA-IV has a high degree of structural similarity to both apoA-IV and apoE.[8,29] How can one objectively assess the functional implications of these findings?

Molecular model building has provided important insights about potential mechanisms of lipid–protein interaction,[30] but there are obvious limitations to the large scale application of this approach. Segrest and Feldmann[31] developed a computer algorithm that made it possible to search protein databases for amphipathic helices. "Edmundson wheel" diagrams, although originally developed for a different purpose,[32] have also been popular in analyzing sequences for lipid-binding potential. However this method is at best qualitative and the results are subject to biased interpretations.[33] Calculation of hydrophobic moments[34] has provided some interesting results when applied to the apolipoproteins.[35,36] However, the hydrophobic moment is a vector sum and its calculation thus requires a knowledge of protein structure. When this information is lacking, structural assumptions, based upon (possibly) unreliable empirical predictions, are necessary.

We have found the method of Kubota et al.[37] very useful in studying the periodic structure of apoA-IV.[8] With this technique, homology in protein sequences can be expressed by correlation coefficients. First the proteins or peptides of interest are converted to sequences of numerical values. These values can be any one of a number of quantitative properties of amino acids (e.g., relative mutability, bulkiness, polarity, hydrophobicity, α-helical potential). Alternatively any desired combination of individual properties can be used. In this case, the use of averaged values improves the ratio of signal to noise.[37] The shorter sequence is compared, in a sliding fashion, with the longer sequence. At increments of one residue, the computer calculates a correlation coefficient and plots this value. When the shorter sequence is aligned with a highly homologous region, a strong positive correlation occurs and one gets a sharp peak on the plot.

We searched the rat apoA-IV sequence with the sequence of a synthetic docosapeptide that had been designed as a prototypic LCAT activa-

[29] M. S. Boguski, N. Elshourbagy, J. M. Taylor, and J. I. Gordon, *Proc. Natl. Acad. Sci. U.S.A.* **82,** 992 (1985).
[30] J. P. Segrest, R. L. Jackson, J. D. Morrisett, and A. M. Gotto, *FEBS Lett.* **38,** 247 (1974).
[31] J. P. Segrest and R. J. Feldmann, *Biopolymers* **16,** 2053 (1977).
[32] M. Schiffer, and A. B. Edmundson, *Biophys. J.* **7,** 121 (1967).
[33] C. Flinta, G. Heline, and J. Johansson, *J. Mol. Biol.* **168,** 193 (1983).
[34] D. Eisenberg, *Annu. Rev. Biochem.* **53,** 595 (1984).
[35] H. J. Pownall, R. D. Knapp, A. M. Gotto, and J. B. Massey, *FEBS Lett.* **159,** 17 (1983).
[36] K. E. Krebs and M. C. Phillips, *FEBS Lett.* **175,** 263 (1984).
[37] Y. Kubota, S. Takahashi, K. Nishikawa, and T. Ooi, *J. Theor. Biol.* **91,** 347 (1981).

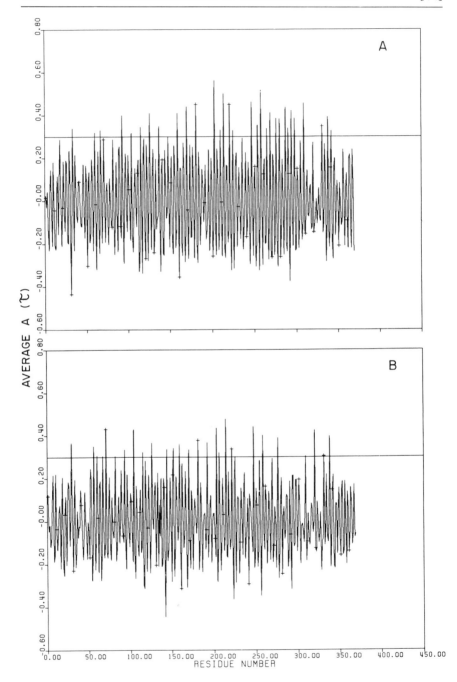

tor[38] to determine if apoA-IV had the structural requirements for LCAT activation. The results are presented in Fig. 5A. Multiple overlapping regions of significant homology are apparent. We also searched the rat apoA-IV sequence with a sequence that has been functionally defined as the apoB,E (LDL) receptor-binding domain in human apoE3.[39,40] It has been demonstrated previously that apoA-IV and apoE diverged from a common ancestral gene and furthermore that the apoB,E receptor-binding domain evolved via the functional differentiation of a repeated sequence.[22,29] The peptide representing residues 140 to 160 in human apoE3 was normalized to a length of 22 residues by the inclusion of one flanking residue (Ser-139). This docosapeptide was then used as a test sequence and the results are presented in Fig. 5B. Once again, multiple overlapping homologous regions are seen. As above, distances between the peaks generally correspond to multiples of 11 amino acids.

How does one interpret these data in light of what is known about LCAT activation and apoB,E receptor binding? Are the magnitudes of the correlations sufficient to predict that apoA-IV will activate LCAT and bind to the apoB,E (LDL) receptor? This depends entirely on the precise physical–chemical and structural determinants of these activities. We will return to this question after examining the structures of our test sequences.

Kyte and Doolittle developed a method for displaying the hydropathic character of a protein.[41] This method is based upon a unified hydropathy scale derived from chemical and statistical properties of the amino acids. The hydropathic profile of a protein is usually calculated from hydropathy values averaged over some user-specified span (window). For example, a

[38] S. Yokoyama, D. Fukushima, J. P. Kupferberg, F. J. Kezdy, and E. T. Kaiser, *J. Biol. Chem.* **255**, 7333 (1980).
[39] T. L. Innerarity, E. J. Friedlander, S. C. Rall, K. H. Weisgraber, and R. W. Mahley, *J. Biol. Chem.* **258**, 12341 (1983).
[40] K. H. Weisgraber, T. L. Innerarity, K. J. Harder, R. W. Mahley, R. W. Milne, Y. L. Marcel, and J. T. Sparrow, *J. Biol. Chem.* **258**, 12348 (1983).
[41] J. Kyte and R. F. Doolittle, *J. Mol. Biol.* **157**, 105 (1982).

FIG. 5. Correlograms for test sequences with preapoA-IV. The small horizontal lines perpendicular to and intersecting some peaks are spacing marks which occur every 5 residues. The horizontal line that intersects the vertical axis at 0.3 represents a level of statistical significance based on the results of Kubota *et al.* for several proteins of demonstrated periodicity.[37] (A) Synthetic LCAT activator peptide[38]: Pro-Lys-Leu-Glu-Glu-Leu-Lys-Glu-Lys-Leu-Lys-Glu-Leu-Leu-Glu-Lys-Leu-Lys-Glu-Lys-Leu-Ala; (B) apoB,E receptor-binding domain[39,40]: Ser-His-Leu-Arg-Lys-Leu-Arg-Lys-Arg-Leu-Leu-Arg-Asp-Ala-Asp-Asp-Leu-Gln-Lys-Arg-Leu-Ala.

window of 19 residues is most effective in locating potential membrane-spanning segments in proteins.[41] However averaging over a span obscures fine structural details. In the docosapeptide repeat units of apolipoproteins A-I and A-IV, the hydropathic character can change drastically over one or two residues. Thus the hydropathy profiles of the "LCAT activator" peptide and the apoB,E receptor-binding domain were plotted with a window of one residue (Fig. 6).

The hydropathy profiles of these two peptides are remarkably similar and conform to the definition of an amphipathic structure—a pattern which alternates regularly between hydrophobicity and hydrophilicity is clearly evident. The most obvious difference between these profiles is that the two central hydrophobic peaks are transposed. Also the magnitudes of the hydrophilic valleys are somewhat greater in the apoE se-

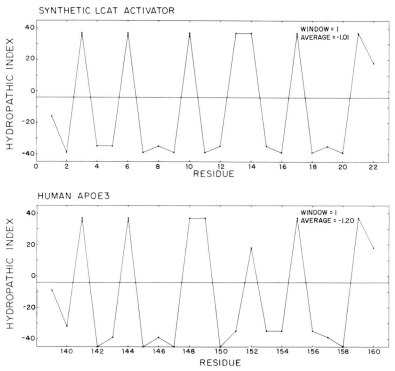

FIG. 6. Hydropathy profiles. The LCAT activator peptide (top panel) was designed to form an α-helix. The apoB,E receptor-binding domain (bottom panel) has a predominant α-helical potential by Chou–Fasman rules. In the apoE3 sequence, positively charged residues (particularly Arg-145 and Arg-158) are thought to be critical for receptor-binding activity.[43]

quence. Despite these differences, the apoB,E receptor-binding domain bears a striking similarity to the basic 22 amino repeat unit in apolipoproteins A-I and A-IV,[8] supporting our earlier conclusion that this domain is the derivative of an ancestral repeated sequence.[22,29]

Based on studies of chain length–function correlation in amphiphilic peptides, Fukushima et al. concluded that the physiological functions of apoA-I (including LCAT activation) are expressed by a multiplicity of weak, cooperative interactions rather than by a single active site.[42] In contrast, single amino acid substitutions in human apoE variants can result in almost total inactivity with respect to receptor binding (reviewed in ref. 43) indicating a very specific interaction at a single site. When the Kubota correlograms (Fig. 5) are interpreted in this light, it seems reasonable to conclude that apoA-IV has the structural requirements for LCAT activation but that it would not be expected to interact with the apoB,E receptor. In the latter case, the magnitudes of the correlation coefficients are simply too low. Still, this result demonstrates the important concept that the apoE receptor-binding domain is evolutionarily related to the repeated sequences in other apolipoproteins.

So far we have ignored any consideration of the influence of tertiary structure because it is simply indeterminable by present predictive methods. However, insofar as one can assume that segments with the greatest local hydrophilicity are found on the surfaces of proteins, program PRPLOT (NBRF/PIR) may be of some utility in assessing the likelihood that a region may form, for example, a receptor-binding site. This program uses an algorithm developed to predict antigenic determinants from amino acid sequence data based on the premise that most antigenic determinants are surface features of proteins.[44] To cite a single example, PRPLOT identifies the apoB,E receptor-binding site as a locally hydrophilic region of human apoE and thus this domain is a likely surface component of the protein. This may also be true when apoE is bound to lipid. The receptor-binding domain is an amphipathic structure (Fig. 6). Doolittle has observed that the surface coverings of many globular proteins are composed of amphipathic helices with the hydrophobic face oriented toward the interior of the protein.[14] The surface coverings of lipoprotein particles could be similarly organized.

---

[42] D. Fukushima, S. Yokoyama, D. J. Kroon, F. J. Kezdy, and T. K. Kaiser, *J. Biol. Chem.* **255**, 10651 (1980).
[43] R. W. Mahley and T. L. Innerarity, *Biochim. Biophys. Acta* **737**, 197 (1983).
[44] T. P. Hopp and K. R. Woods, *Proc. Natl. Acad. Sci. U.S.A.* **78**, 3824 (1981).

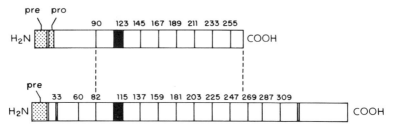

FIG. 7. Structural alignment of human preproapoA-I (top) with rat preapoA-IV (bottom). Residue numbers are for the primary translation products. This diagram represents the structural organization of amphipathic docosapeptides based upon landmark proline residues (Fig. 1). Optimal sequence alignments using high gap penalties and the unitary scoring matrix are consistent with this organization.

Summary

There are a great variety of computational methods available to study protein and nucleic acid sequences. The choice of a computer program appropriate to a particular problem and the critical interpretation of the results can lead to specific, and experimentally testable, predictions of a protein's structure and function and may yield insights into its evolution and the location of its gene.

We have shown that rat apoA-IV bears a striking structural similarity to human apoA-I (summarized in Fig. 7). Statistical analyses of homologies between apolipoproteins A-I, A-IV, and E demonstrate conclusively that all three sequences diverged from a common ancestral gene. That apoA-IV largely composed of 22 amino acid amphipathic segments with α-helical potential suggests that it possesses the structural requirements for LCAT activation.[45] Analysis of the apoB,E receptor-binding domain of human apoE3 has demonstrated that it evolved from an ancestral repeated sequence. Assuming that the genes for these proteins evolved as a result of a series of intra- and intergenic unequal crossovers, it is likely that their genetic loci were at one time linked. The repeated sequences of which these genes are composed have propagated themselves in an expansionary manner. Given this fact, the existence of other genes or pseudogenes based upon this repeated sequence motif is a distinct possibility.

---

[45] A. Steinmetz and G. Utermann (*J. Biol. Chem.* **260**, 2258, 1985) have shown that human apoA-IV is indeed a potent LCAT activator *in vitro*.

Acknowledgments

We thank Jens Birktoft, David Eisenberg, Walter Fitch, Mark Freeman, Dean Goddette, Larry Holden, David Lipman, A. D. McLachlan, and William Pearson for providing computer programs. We thank Walter Fitch, Henry Pownall, and Armin Steinmetz from making results available to us prior to publication. We thank Luis Glaser and A. W. Strauss for critical readings of the manuscript.

[45] Detecting Internally Repeated Sequences and Inferring the History of Duplication

By WALTER M. FITCH, TEMPLE SMITH, and JAN L. BRESLOW

Introduction

The problem of identifying sequence repeats resulting from ancient gene duplications and/or unequal crossings over is not fundamentally different from the problem of identifying sequence similarities between any two genetic sequences. The best understood algorithms are the dynamic programming methods first introduced into molecular biology[1] by Needelman and Wunch in 1970 and mathematically formalized by Sellers in 1974. These algorithms are readily implemented on nearly any computer system and run in time proportional to the square of the total sequence length. The original identification of amino acid tandem repeats in the apolipoproteins[2,3] resulted from use of dynamic programming related algorithms to search for maximum similarity between segments of these sequences and the entire sequences. A number of generalizations of dynamic programming methods have appeared in the literature, among which is the common subsequence algorithm[4] which is of particular interest in the identification of imperfect repeats. This algorithm generates a ranked set of alignments between maximally similar segments of the sequences (or within a single sequence) being compared.

The interest in identifying internal repeats, particularly tandem repeats, is 2-fold, the understanding of the structural implications and of the

---

[1] Kruskal (1983) recently reviewed the history of the uses of such sequence comparison algorithms across a wide spectrum of the life sciences: J. B. Kruskal, *SIAM Rev.* **25**, 201 (1983).
[2] Barker, C. Winona and M. O. Dayhoff, *in* "Atlas of Protein Sequences," Vol. 5, Suppl. 2, p. 254.
[3] W. M. Fitch, *Genetics* **86**, 623 (1977).
[4] T. F. Smith and M. S. Waterman, *J. Mol. Biol.* **147**, 195 (1981).

evolutionary implications. In the myosin heavy chain case, for example, the structure of two intertwined α-helical polypeptides imposes a periodic constraint on these molecules. A similar periodic (heptad) repeat is observed in the triple-stranded coiled-coil of the influenza hemagglutinin molecule.[5] This particular repeat was predicted theoretically.[6] Thus one would expect the long-term maintenance of periodic or tandem repeats given various structural constraints. Such sequence repeats are most easily generated through combinations of gene duplication and unequal crossings over.[7] In the apolipoprotein case used in this chapter, for example, the hydrophobicity and charge patterns are highly conserved even though there has been considerable amino acid replacement. It is of some interest to note that considerable sequence similarity (even at the nucleotide level) is detected between the human apolipoprotein A-I consensus repeat[8] and rat cardiac myosin. This similarity may have been imposed by structural constraints (analogy), rather than any shared genetic ancestry (homology).

Method for Detecting Similar Repeat Segments

Dynamic programming methods require the calculation of an $n \times m$ matrix, H, where n and m are the lengths of the sequences being compared. The calculation of any matrix element, H_{ij}, is a single function of the similarity of the element i in one sequence to element j in the other. To find the maximally similar segments shared between the two molecular sequences $A = a_1 a_2 \ldots a_n$ and $B = b_1 b_2 \ldots b_m$, we first need a measure of similarity $s(a, b)$ between the sequence elements a_i and b_j and a weight for deletions of length k, W_k. Next we calculate the matrix H where H_{ij} is the maximum similarity among all alignments of two segments ending with a_i and b_j, respectively. These values are obtained from the recursive relationship:

$$H_{ij} = \max\{H_{i-1,j-1} + s(a_i, b_j), \max_k(H_{i-k,j} - W_k), \max_h(H_{i,j-h} - W_h), 0\}$$

for $1 < i < n$ and $1 < j < m$, with initial values $H_{k0} = H_{0h} = 0$ for $0 < k < n$ and $0 < h < m$.

---

[5] A. W. Wilson, J. J. Skehel, and D. C. Wiley, *Nature (London)* **289**, 366 (1981).
[6] A. D. McLachlan and M. Stewart, *J. Mol. Biol.* **98**, 293 (1975).
[7] T. F. Smith, M. S. Waterman, and J. Sadler, *Nucleic Acids Res.* **11**, 2205 (1982).
[8] S. K. Karathanasis, V. I. Zannis, and J. L. Breslow, *Proc. Natl. Acad. Sci. U.S.A.* **80**, 6147 (1983).

The formula for H_{ij} follows by considering the possibilities for ending the segments at any a_i and b_j.
1. If a_i and b_j are associated, the similarity is $H_{i-1,j-1} + s(a_i,b_j)$.
2. If a_i is at the end of a deletion of length k, the similarity is $H_{i-k,j} - W_k$.
3. If b_j is at the end of a deletion of length l, the similarity is

$$H_{i,j-h} - W_h.$$

4. Finally, a zero is included to prevent calculating a negative similarity, indicating no similarity of interest up to a_i and b_j.

The pair of segments with maximum similarity is found by first locating the maximum element of H. The other matrix elements leading to this maximum value are then sequentially determined with a traceback procedure ending with an element of H equal to zero. The pair of segments with the next best similarity is found by applying the traceback procedure to the second largest element of H not associated with the first traceback. This procedure identifies the segments as well as produces the corresponding alignment. In order to utilize this algorithm for identifying internal repeats, a sequence is run against itself and the diagonal elements, H_{ii} are set to zero.

This procedure not only results in the identification of potential repeats but generates a measure of their relative relatedness via the locally maximal H matrix similarity values, associated with each pair of aligned segments or repeats. There are, however, two cautionary considerations required in interpreting such data. The first is that one needs an estimate of the statistical significance of any given implied relationship between segments. Traditionally, in the absence of an analytical solution, Monte Carlo simulations were carried out. Mean expected values and standard deviation about this mean were then calculated. But recent work on the statistics of runs of exact matches[9] and on maximum runs of matches containing fixed numbers of mismatches[10] allows one to estimate the expected length of the longest run, R, of similarity between random sequences. These analyses show that the expected length of maximally similar segments by chance is on the order of

$$R \simeq 2 \ln(n)/\ln(\lambda) + k - \ln(k!)/\ln(\lambda) - O(1)$$

Here n is the length of the sequence under investigation, k is the number of accepted mismatches, and λ is one over the probability of a match. For internal repeats, the term $2 \ln(n)$ must be replaced by $\ln[n(n-1)/2]$. The

---

[9] S. Karlin, Ghandour, Ghassau, Ost, Friedemann, Tavare, Simon, Korn, and J. Laurence, *Proc. Natl. Acad. Sci. U.S.A.* **80**, 5660 (1983).
[10] L. Gordon, M. F. Schilling, and M. S. Waterman, Submitted (1984).

variance on this mean is independent of n as n becomes large and has the form $\pi^2[6 \ln(\lambda)]$. This means for nucleotide sequences that the standard deviation is about 1.3.

Using apolipoprotein E as an example, one does not expect, by chance alone, to find any pair of segments possessing as many as 11 [$R = \ln(711 \times 710/2)/\ln(1/0.346) - 1 = 10.7$] consecutive exact nucleotide matches. The value of 0.346 is the average of the ΣP_i^2 for the three codon positions (see Table II) and is considerably larger than the 0.302 obtained if the coding positions are undifferentiated. This materially affects the results as R would in that case have been 9.4 and we would expect to have seen no segments sharing as many as 10 consecutive exact matches rather than 11, thereby overestimating the improbability of an observation.

One can also examine strings composed only of the first (or second or third) codon positions.[11] For positions 2 and 3 of apolipoprotein E, the values of R are 7.0 and 11.4, respectively ($n = 237$). This shows the great effect that the nucleotide composition may have on one's expectations on the length of the longest run of successive matches occurring by chance, for we do not expect to see matching sequences as long as 7 in the former but they must be 12 long before we do not expect to see them in the latter. The standard deviation on the latter is 1.6 and two standard deviations above the mean would imply that an observation of 15 [$>11.4 + 2(1.6)$] consecutive exact matches would have a probability of occurring by chance of ~0.01. For the former [$(9 - 7.0)/0.8$] would be 2.5 standard deviations above the mean with a probability of ~0.006 of occurring by chance.

The second cautionary consideration in interpreting the results from the above dynamic programming similarity method is that one cannot directly infer the history of the generation of the repeats from their pairwise similarity values. This arises in large part from the fact that their most likely method of generation, unequal crossings over, will both expand and contract any series of tandem repeats, thereby mixing various subregions of the repeating elements. Therefore, the standard molecular taxonomic approach so useful in reconstructing probable histories of a single gene cannot be directly employed. The recognition of this motivated the method of reconstruction discussed later in this chapter.

Before leaving the dynamic programming methods we should note that they are of rather general utility. They can be used in a search for rather distant repeats which are no longer of the same length or are not tandemly arranged. If, however, the repeats are tandem and of the same (or nearly so) length they are generally easily identified by the fact that all the

[11] T. F. Smith, W. M. Fitch, and M. S. Waterman, Submitted (1984).

segments found to be similar to any given one will differ in their relative position along the sequence by multiples of some fixed length.

A related although nonmathematical method of searching for similarity among or within genetic sequences is the graphic "dot matrix." Here a matrix is plotted in which the elements M_{ij} are either left blank if, beginning with the ith element k successive elements of a sequence do not match, within some criterion, to the k successive elements beginning with the jth element of the other sequence (or itself in the case of internal repeats), or filled with a dot if the elements do match. This method has the advantage of providing visualization of the relationships among closely similar regions. However, unless the expected run length statistics mentioned above are employed, success is far from certain. McLachlan[12] was able to identify the 28 residue repeat in the myosin heavy chains using this method but Cheung and Chan[13] failed to confirm, at both the DNA and amino acid levels, the 11 residue repeat previously reported in the human apoA-I molecule. This failure is important to note since the above dynamic programming method identified all 12 11-residue repeats at both the nucleic acid and protein levels as being more than two standard deviations above that expected by chance. This was even the case when rather low mismatch and deletion penalties were employed [$s(a,b)$ for $a \neq b$ of minus 0.9 compared with $s(a,b) = 1.0$ for $a = b$ and $W_k = -1.0 - 0.9k$].

Methods for Detecting Tandem Repeats

The following method is designed to identify tandem repeats in cases like the apolipoproteins where there is extensive tandem repetition and to provide a significance level as to that identification.

The method depends upon the assumption of tandem repetition. If, for example, a protein has 237 amino acids and a repeating unit of 11 amino acids, then any 2 amino acids 11 apart (and there are $237 - 11 + 1 = 227$ of them) represent a test of the hypothesis that the sequence was originally created from a repeating set of 11 amino acids. We need only to sum the amount of difference over all 227 pairs of amino acids, 11 apart, and compare that to the sum expected distance for 227 random drawings from a pool of the same composition as that of the protein being examined. Knowing the variance of the latter permits the computation of the one-tailed probability that the observed distance was that much less than the expected distance.

[12] A. D. McLachlan, *Nature (London)* **267**, 465 (1977).
[13] P. Cheung and L. Chan, *Nucleic Acids Res.* **11**, 3703 (1983).

One can use any kind of sequence data available although we shall restrict ourselves here to three kinds, nucleotides, condons, and amino acids. One can also use any suitable measure of distance (dissimilarity) although we shall restrict ourselves here to base differences (or minimum base differences, MBD, in the case of amino acids).

For the case of nucleotide sequences, let n = the length of the sequence and $n = \Sigma n_i (i = 1,4)$ where i represents the four nucleotides. The total ways of drawing (without replacement) two nucleotides is $n(n - 1)$ and the ways that they may be identical nucleotides is $\Sigma n_i(n_i - 1)$ summed over all i. The probability that they are identical is

$$p = [\Sigma n_i(n_i - 1)]/[n(n - 1)] \tag{1}$$

Let r = the repeat length. There will be $n - r + 1$ pairs of nucleotides in the test sample and hence the number of differences one would expect to see in a random sequence of length n is

$$e = (n - r + 1)(1 - p) \tag{2}$$

and the variance is

$$v = (n - r + 1)(1 - p)p \tag{3}$$

If one has observed a total of d differences in examining the $n - r + 1$ pairs of nucleotides r apart, one can obtain the standard measure

$$s = (d - e)/\sqrt{v} \tag{4}$$

The probability that a d is as low as that observed is obtained by finding the one-tailed probability of the value of s in the normal distribution provided m is negative. If s is positive, the probability is greater than 0.5.

As will be seen in the results section, the above method is subject to bias that tends to exaggerate significance levels when there is a periodic difference in nucleotide composition. This bias is likely to be readily apparent when the sequence is a protein coding sequence where there is a known three nucleotide periodic pattern.[7] This bias is corrected by the analysis of coding sequences presented now.

Let n' be the length of the sequence in codons and r' the repeat length in codons. Let p_j equal the probability that two randomly drawn nucleotides match when drawn from a pool having the same composition as that of the jth position of the codons ($j = 1,3$). Finally, let n_{ij} = the number of nucleotides of type i in the jth position of the codons. A similar logic applies and so

$$p_j = [\Sigma n_{ij}(n_{ij} - 1)]/n'(n' - 1)] \tag{5}$$
$$e_j = (n' - r' + 1)(1 - p_j) \tag{6}$$

and
$$v_j = (n' - r' + 1)(1 - p_j)p_j \tag{7}$$

Treating these as three independent distributions gives
$$e' = \Sigma e_j \tag{8}$$
$$v' = \Sigma v_j \tag{9}$$

and
$$s = (e'-d)/\sqrt{v'} \tag{10}$$

The third case, MBD and amino acid sequence, is only more complicated in that one uses a tetranomial rather than a binomial distribution. This arises because the base differences range from zero to three rather than from zero to one.

Since the lengths in amino acids equal the lengths in codons, we retain n' and r' from the previous case. Let n_a and n_b be the number of amino acids of kind a and b in the sequence. Let $m(a,b)$ be the MBD between amino acids of kind a and b. Finally, let p_k equal the probability that two randomly selected amino acids have k minimum base differences.[14]

$$p_0 = [\Sigma n_a(n_a - 1)]/[n(n' - 1)] \tag{11}$$
$$p_k = 2\Sigma n_a n_b/[n'(n' - 1)] \quad \text{given } m(a,b) = k \neq 0 \tag{12}$$
$$\mu = p_1 + 2p_2 + 3p_3 \tag{13}$$
$$e' = (n' - r' + 1)\mu \tag{14}$$
$$v' = p_1^2 + 4p_2^2 + 9p_3^2 - \mu^2 \tag{15}$$

with s being defined by (10).

Repeat Results. In Table I is shown the analysis of human apolipoprotein E coding sequence[15] as if it were a simple nucleotide sequence with potential repeat lengths from 7 to 30 being examined. The principal result is that for every 1 of the 10 r values that are evenly divisible by three, the standard measure is negative, but in only 2 of the other 20 is it negative and even then it is barely negative. This arises because the probability of a nucleotide match is greater in frame and less out of frame than for two nucleotides chosen randomly from the general population. These probabilities are shown in Table II.

In Table III is shown the analysis of the same sequence using the different nucleotide composition in each of three successive codon positions (left) and on the basis of minimum base differences in the encoded amino acids with potential repeat lengths of 7 to 30 codons or amino acids.

---

[14] W. M. Fitch, *J. Mol. Biol.* **16**, 9 (1966).
[15] V. I. Zannis, J. McPherson, G. Goldberger, S. K. Karathanasis, and J. L. Breslow, *J. Biol. Chem.* (1984).

TABLE I
TEST FOR REPEATED SEQUENCES IN HUMAN
APOLIPOPROTEIN E SEQUENCES[a]

Repeat length	Nucleotide			
	Substitutions		Standard measure	Probability
	Exp.	Fnd.		
7	491.7	504	1.01	8.4E-1
8	491.0	509	1.48	9.3E-1
9	490.3	432	−4.80	8.2E-7
10	489.6	499	.77	7.8E-1
11	488.9	496	.58	7.2E-1
12	488.2	450	−3.15	8.2E-4
13	487.5	489	.12	5.5E-1
14	486.8	529	3.48	1.0E-2
15	486.1	479	−.59	2.8E-1
16	485.4	500	1.21	8.9E-1
17	484.7	491	.52	7.0E-1
18	484.0	454	−2.48	6.5E-3
19	483.3	479	−.36	3.6E-1
20	482.6	498	1.28	9.0E-1
21	481.9	435	−3.89	5.0E-5
22	481.2	482	.06	5.3E-1
23	480.5	499	1.54	9.4E-1
24	479.8	431	−4.06	2.5E-5
25	479.1	487	.66	7.4E-1
26	478.4	499	1.71	9.6E-1
27	477.7	459	−1.56	5.9E-2
28	477.0	484	.58	7.2E-1
29	476.3	483	.56	7.1E-1
30	475.6	452	−1.97	2.4E-2

[a] Exp. and Fnd. are the number of nucleotide substitutions expected and found summed over all the pairs of nucleotides that are the repeat length apart. The probability notation 8.4E-1 is to be read 8.4×10^{-1}, the E being understood to imply exponent of 10.

The probabilities shown are as if only one repeat length had been sampled. Since there are 24 repeat lengths examined, it would be conservative to multiply the stated probabilities by 24. Upon doing so, no repeat has a lower number of substitutions than expected at the 0.05 level except for repeat lengths of 11 and 22. For these latter two lengths the p values run from 1.6×10^{-4} down to 7×10^{-7}. The p values are more than one order of magnitude lower for a length of 11 than for 22, indicating that the cross-

TABLE II
NUCLEOTIDE FREQUENCIES IN HUMAN APOLIPOPROTEIN E BY CODON POSITION[a]

Position	Frequency				Fraction (P_i)				ΣP_i^2	σ
	A	C	G	U	A	C	G	U		
1	34	85	103	15	0.143	0.359	0.435	0.063	0.343	1.20
2	77	49	55	56	0.325	0.207	0.232	0.236	0.258	0.81
3	12	92	126	7	0.051	0.388	0.532	0.029	0.437	1.55
Sum	123	226	284	78	0.173	0.318	0.399	0.110	0.302	1.07

[a] The probability of two nucleotides matching if drawn at random, with replacement, from the total pool is the sum of the squares of the bottom four fractions, viz. $0.173^2 + 0.318^2 + 0.399^2 + 0.11^2 = 0.302$. Thus the sequence as a whole is sufficiently CG rich to cause 30% matches for random sequences of that composition. To find the expectation for comparisons in frame, the same computation for each of the rows for positions 1 to 3 must be calculated. These are 0.343, 0.258, and 0.437 and their average, 0.346, is the probability that nucleotides of two random sequences will match if they have this particular positional composition and are aligned in frame. If the alignment is out of frame, the probabilities of a match are 0.237, 0.227, and 0.380 when position 1 is aligned with position 2, position 2 with 3, and 3 with 1, respectively, for an average of 0.281 over the whole alignment. The standard deviation (σ) was calculated as the square root of $\pi^2/[6(\ln(1/\Sigma P_i^2))^2]$, and is the σ for the length of the longest run of consecutive matches.[9]

over unit was more often 11 than 22 codons. The p values are lower for the amino acid test than the codon test, indicating that third position variability is contributing noise to the analysis that is being filtered out by using the amino acid sequence. Table IV shows a similar result for human apolipoprotein A-I using the sequence of Karathanasis et al.[8]

Method for Inferring the History of Duplication

In 1974, Smith[16] showed how a set of tandem sequences might arise and evolve starting with the gene duplication and followed by many unequal crossings over. The procedure to be presented now attempts to work backward through those temporal events. The original procedure for this was given by Fitch[3] and assumed that one could divide the sequence into segments each of one repeat length.

That earlier method had one assumption that is biologically unrealistic, namely that all crossings over occurred at the points where the repeats were arbitrarily divided to create the segments. In fact, the multiple repeats can arise without any limitation on where the crossover point

[16] G. P. Smith, *Cold Spring Harbor Symp. Quant. Biol.* **38**, 507 (1974).

TABLE III
TEST FOR REPEATED SEQUENCES IN HUMAN APOLIPOPROTEIN E SEQUENCES[a]

Repeat length	Codon				Amino acid			
	Substitutions		Standard measure	Probability	Substitutions		Standard measure	Probability
	Exp.	Fnd.			Exp.	Fnd.		
7	453.4	435	−1.50	6.7E-2	319.4	300	−1.95	2.6E-2
8	451.5	431	−1.66	4.8E-2	318.0	308	−1.01	1.6E-1
9	449.5	459	0.78	7.8E-1	316.6	317	0.04	5.2E-1
10	447.5	452	0.37	6.4E-1	315.2	332	1.70	9.6E-1
11	445.6	382	−5.20	9.7E-8[b]	313.8	260	−5.45	2.5E-8[b]
12	443.6	464	1.68	9.5E-1	312.4	334	2.19	9.9E-1
13	441.6	445	0.28	6.1E-1	311.0	325	1.42	9.2E-1
14	439.6	406	−2.77	2.8E-3	309.6	290	−2.00	2.3E-2
15	437.7	428	−0.80	2.1E-1	308.2	292	−1.66	4.8E-1
16	435.7	434	−0.14	4.4E-1	306.8	315	0.84	8.0E-1
17	433.7	442	0.69	7.5E-1	305.4	316	1.08	8.6E-1
18	431.8	428	0.31	3.8E-1	304.1	305	0.10	5.4E-1
19	429.8	433	0.27	6.1E-1	302.6	316	1.38	9.2E-1
20	427.8	439	0.94	8.3E-1	301.3	309	0.80	7.9E-2
21	425.8	455	2.44	9.9E-1	299.9	321	2.19	9.9E-1
22	423.9	372	−4.36	6.7E-6[b]	298.5	253	−4.73	1.1E-6[b]
23	421.9	423	0.09	5.4E-1	297.1	304	0.72	7.6E-1
24	419.9	421	0.09	5.4E-1	295.7	306	1.08	8.6E-1
25	418.0	402	−1.35	8.9E-2	294.3	287	−0.76	2.2E-1
26	416.0	391	−2.12	1.7E-2	292.9	275	−1.88	3.0E-2
27	414.0	426	1.02	8.5E-1	291.5	303	1.21	8.9E-1
28	412.0	429	1.44	9.3E-1	290.1	306	1.68	9.5E-1
29	410.1	409	−0.09	4.6E-1	288.7	287	−0.18	4.3E-1
30	408.1	398	−0.86	1.9E-1	287.3	287	−0.03	4.9E-1

[a] Exp. and Fnd. are the number of nucleotide substitutions expected and found summed over all the pairs of nucleotides that are the repeat length (in codons) apart. For the amino acids, the minimum base differences were used. The probability notation of 2.6E-2 is to be read 2.6×10^{-2}, the E being understood to imply exponent of 10.

[b] Those probabilities that are less than 0.05 after multiplication by 24.

occurs provided only that the misalignment is by a multiple of the repeat unit.

A simple method is available to permit one to place the crossover at any point along the sequence and arises from the repeat detection method above. As one moves down the length of the sequence examining pairs of nucleotides $(k, k + r)$ one simply keeps track of the number of differences, d, in the last r pairs. This number is the minimum number of

TABLE IV
TEST FOR REPEATED SEQUENCES IN HUMAN APOLIPOPROTEIN A-I SEQUENCES[a]

Repeat length	Codon				Amino acid			
	Substitutions		Standard measure	Probability	Substitutions		Standard measure	Probability
	Exp.	Fnd.			Exp.	Fnd.		
7	538.6	506	−2.52	5.8E-3	382.4	366	−1.53	6.3E-2
8	536.5	536	−0.04	4.8E-1	381.0	374	−0.65	2.6E-1
9	534.5	535	0.04	5.2E-1	379.5	386	0.61	7.3E-1
10	532.4	544	0.90	8.2E-1	378.0	393	1.40	9.2E-1
11	530.3	456	−5.80	8.8E-9[b]	376.6	308	−6.42	1.2E-10[b]
12	528.3	553	1.93	9.7E-1	375.1	396	1.96	9.7E-1
13	526.2	518	−0.64	2.6E-1	373.6	389	1.44	9.3E-1
14	524.1	526	0.15	5.6E-1	372.2	372	−0.02	4.9E-1
15	522.1	486	−2.84	2.3E-3	370.7	346	−2.33	9.9E-3
16	520.0	523	0.24	5.9E-1	369.2	379	0.92	8.2E-1
17	518.0	529	0.87	8.1E-1	367.8	382	1.35	9.1E-1
18	515.9	499	−1.34	9.1E-2	366.3	347	−1.83	3.3E-2
19	513.8	490	−1.89	2.9E-2	364.8	358	−0.65	2.6E-1
20	511.8	526	1.13	8.7E-1	363.4	387	2.25	9.9E-1
21	509.7	515	0.42	6.6E-1	361.9	373	1.06	8.6E-1
22	507.6	437	−5.63	3.3E-9[b]	360.5	300	−5.79	3.6E-9[b]
23	505.6	522	1.31	9.1E-1	359.0	375	1.54	9.4E-1
24	503.5	521	1.40	9.2E-1	357.5	384	2.54	9.9E-1
25	501.4	504	0.20	5.8E-1	356.1	353	−0.29	3.8E-1
26	499.4	497	−0.19	4.2E-1	354.6	349	−0.54	2.9E-1
27	497.3	485	−0.99	1.6E-1	353.1	343	−0.98	1.6E-1
28	495.3	489	−0.50	3.1E-1	351.7	352	0.03	5.1E-1
29	493.2	470	−1.88	3.0E-2	350.2	321	−2.84	2.3E-3
30	491.1	463	−2.28	1.1E-2	348.7	333	−1.53	6.3E-2

[a] See footnote *a* Table III for explanation of notation.
[b] See footnote *b*, Table III.

nucleotide substitutions required if one were to postulate that the segments ending at k and $k + r$ arose by an unequal crossover increasing the length by r. Similar counts must be kept for multiples of r.

With such a table, one looks for the smallest value of d per unit of r on the assumption that the two most similar segments represent the most recent result of the crossover process. In case of ties for d per unit r, the largest multiple of r is chosen. One then replaces these two segments with a single segment whose sequence has the property that it contains all sequences, and only those sequences, capable of giving rise after duplication to the two removed segments in a number of nucleotide substitutions equal to the observed number of differences. The method for doing this is found in Fitch.[17,18]

The replacement of two segments (probably r long but possibly a multiple thereof) by one segment reduces the length of the original sequence by the length of the segment. On that reduced sequence, the process may be repeated to find the presumably next most recent unequal crossing over. Indeed, the process may be repeated until the reduced sequence is less than $2r$ long or the d values are in the range of random expectation. The latter case should occur whenever the original sequence has been reduced to a single repeat unit plus other 3' and 5' (or N-terminal and C-terminal) segments that did not arise by duplication of the repeat unit. The sum of the d values is the minimum number of nucleotide substitutions required to account for descent of the reduplicated segments from a single repeat unit if the historic course (order and location) of the crossovers was as reconstructed.

There is a situation where one would expect the above procedure to err but which is correctable. Figure 1 shows four hypothetical segments and their history of duplication with numbers indicating nucleotide substitution in the appropriate time intervals. Since $d_{AB} = 40$ is less than any other, we correctly choose A and B as the segments arising from the most recent duplication. This will lead to a new reduced sequence composed of segments ECD. But E has fewer differences with either C or D than C has with D. Thus the preceding algorithm would not cause C to be joined properly with D. This problem arises because the distances are not all being counted from the same point in time. This is corrected by maintaining a vector for every nucleotide position that contains an additional distance value to be added when that nucleotide is now part of an ancestral sequence. The ancestral nucleotide's added value is half (or some other fraction were it desirable) the sum of the values for the two descendant nucleotides that arose from it plus either zero or one half. A half is added if the two descendant nucleotides contain no nucleotide in common, zero otherwise. (For example, if one descendant nucleotide possessed a pyrimidine and the other a guanine, there is no nucleotide in common.) This is equivalent to making the distance to any ancestor equal to half the changes on each of its two lines of descent.

Phylogenetic Results. An example of the product of the preceding method is presented in Fig. 2 where the C-terminal 146 residues of human apolipoprotein A-I[19] are given a phylogenetic interpretation which may be

[17] W. M. Fitch, *Syst. Zool.* **20,** 406 (1971).
[18] W. M. Fitch, *Am. Nat.* **111,** 223 (1977).
[19] H. N. Baker, A. M. Gotto, and R. L. Jackson, *J. Biol. Chem.* **250,** 2725 (1975).

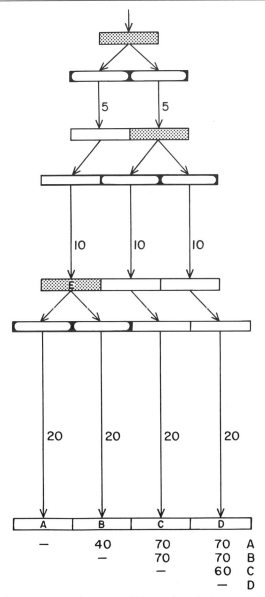

FIG. 1. Example of an ancestral sequence E being closer to two present day sequences, C and D (50 units of change), than either of those present day sequences are to each other (60 units of change). The duplicating segment is shown hatched, the two resulting duplicated segments are shown with solid rounded corners. The amount of change between duplications is shown on the Figure.

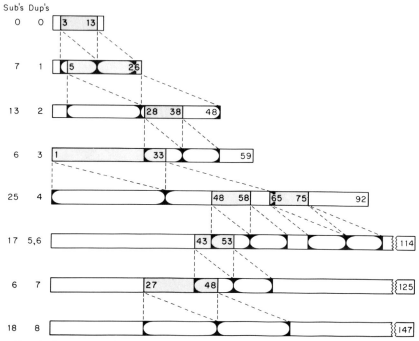

FIG. 2. A representative history of human apolipoprotein A-I duplication. Only the C-terminal 146 amino acids are shown. The sequence as published by Baker et al.[19] was used to permit comparison with the result of an earlier method.[3] That sequence was postulated to have a gap in it which accounts for the length of the sequence being shown as 147 in the bottom line of the figure. The gap was later filled by an amino acid that was missed in the original sequence analysis. There are eight duplications, one of length 33 (numbered 5), two of length 22 (numbered 1 and 4), the rest of length 11. There is one pair of duplications that have no residues in common (numbers 5 and 6) and hence are shown on the same line as their order of occurrence is indeterminate. The right-most of the 5, 6 pair of duplications could also have occurred at any time more recently than shown since none of its residues was involved in any subsequent duplications. The representation of duplication is the same as in Fig. 1 except that the first and last codon of the duplicating repeat is shown within the segments and the length of the segment is shown on the far right. There are also 92 more amino acids in the N-terminus of the mature apolipoprotein that are not represented. The duplications (dup's on figure) are numbered from the most ancient although that is the reverse of the order in which they are recognized. The number of required nucleotide substitutions (sub's in the figure) following a duplication before the next duplication occurs is recorded on the line where the next duplication will arise and sums to 92 substitutions.

contrasted to the older result found in Fig. 10 of Fitch[3] which used the same residues. The older, fixed-segment method required 104 nucleotide substitutions to explain the sequence. The above method, without weighting ancestral sequences, found a solution requiring 94 substitutions. When weighting was used, the result shown in Fig. 2 was obtained and requires 93 substitutions. This is a reduction of more than 10% in the number of nucleotide substitutions required and represents a significant improvement in the method of phylogenetic reconstruction of sequence repeats. But, as we shall discuss shortly, the correspondence between the history of those repeats and their reconstructed phylogeny is nevertheless problematical.

Discussion. The method for detecting tandem repeats is quite powerful. Barker and Dayhoff,[2] Fitch,[3] and McLachlan[12] all independently discovered the repeat of human apolipoprotein A-I with McLachlan and Fitch providing probabilities that it arose by chance of $\leqslant 10^{-4}$ and 2×10^{-8}, respectively. Nevertheless, Cheung and Chan[13] claimed, on the basis of being unable to observe the repeats in a dot matrix, "that there is no pattern of internally repeated segments in apoA-I, either at the amino acid or at the DNA level." But failure to prove something's existence is not proof of nonexistence or, put more statistically, failure to reject a null hypothesis does not prove the null hypothesis. What their work really illustrates is that dot matrices are not the most sensitive analytical tool for detecting repeats although, in the case of apolipoprotein A-I, a different choice of parameters for the dot matrix procedure would have revealed the repeats. The procedure used here generates a probability that the repeat structure is a random event of 10^{-10} for minimum base differences at the amino acid level and 10^{-8} for codons at the nucleic acid level. In addition to these statistical indications, there is the fact that the vast majority of amino acid replacements preserve the charge and hydrophobicity of the amphipathic helical repeat.[3,20]

The sensitivity of the method in any particular case is dependent upon (1) the number of repeats, (2) the fraction of the sequence that is composed of repeats, and (3) the evolutionary conservatism (resistance to change) in the repeated sequences following duplication, with the sensitivity increasing as each of those three factors increases. For rare or nontandem repeats, one is better served by the method of Smith and Waterman.[4]

The phylogenetic reconstruction method given here has several shortcomings. At each round of duplication, it picks the repeat with the fewest required substitutions. But there may be more than one of them. In the example shown in Fig. 2, there were seven different optimal choices for the first repeat and two for the third to give, ostensibly, 14 possible trees. In our particular case, one of the seven repeat choices in the first round still remained and was used at the fourth round, but this does not greatly affect the total trees one might examine. In practice, the total number of alternatives can be much higher and the choices made among the early

[20] J. P. Segrest, R. L. Jackson, J. D. Morrisett, and A. M. Gotto, *FEBS Lett.* **38,** 247 (1974).

alternatives can alter the choices available among the later alternatives so that the total number of trees, T, one might examine can be very large even if choices are restricted to those that are optimal in each round. Worst yet, assuming one desires the most parsimonious tree (which need not be historically correct), finding that tree may well depend upon making a suboptimal choice in an early round and thus not be among the many examined.

The second point is that many of the above T trees may be barely distinguishable from the best one(s). Even the difference between weighting and unweighting for prior substitutions may not make much difference, as in the apolipoprotein A-I case. Clearly, trees that differ by only one or two substitutions are not significantly different.

The third point is that the procedure provides no way of inferring contractions (as opposed to expansions) resulting from the crossover process, yet such contractions surely occur.

The appropriate inference from a result such as that portrayed in Fig. 1 is that it is a representative tree whose proposed ancestral history may well approximate the protein's actual history in total number and kind of different substitutions. In detail (where a specific duplication occurred or how many substitutions occurred between any two duplications) it is likely to be wide of the mark.

[46] Isolation of cDNA and Genomic Clones for Apolipoprotein C-II

By CYNTHIA L. JACKSON, GAIL A. P. BRUNS, and JAN L. BRESLOW

Introduction

Recombinant DNA technology allows examination of the regulation of gene expression by transcriptional as well as translational mechanisms. For apolipoproteins, one can now study the determinants of a functional protein by examining its gene structure and regulation as well as at the level of translation and posttranslational modification. This has been and will continue to be important in the definition of the molecular basis for some reported clinical lipoprotein disorders and,[1] in addition, will contrib-

[1] J. L. Goldstein and M. S. Brown, in "The Metabolic Basis of Inherited Disease" (J. B. Stanbury, J. B. Wyngaarden, D. S. Fredrickson, J. L. Goldstein, and M. S. Brown, eds.), 5th Ed., p. 622. McGraw-Hill, New York, 1983.

ute to our understanding of lipoprotein metabolism in normal individuals. In this chapter, we describe the application of recombinant DNA technology to apolipoprotein C-II. We have chosen techniques which take advantage of known information about apolipoprotein C-II in order to isolate and characterize cDNA and genomic clones. The isolation of cDNA and genomic clones encoding apoC-II is the initial work in the genetic study of this apolipoprotein, to better understand the normal regulation of apoC-II synthesis and defects that might occur in this process at the molecular level.

Methods

Isolation of cDNA Clones Encoding ApoC-II

Preparation of the Oligonucleotide Probe. Since the protein sequence for apoC-II is known,[2] a mixture of oligonucleotides was chemically synthesized that corresponded to all possible mRNAs that could code for a portion of the protein. We identified a region of 5 amino acids each of which could be specified by only one or two codons to minimize the number of oligomers to be synthesized. Examination of the protein sequence revealed that amino acid residues 6–10 were Gln-Asp-Glu-Met-Pro. Figure 1A shows the mRNA and corresponding cDNA sequence for these amino acids. The 14 nucleotide oligomer synthesized extends through the first two nucleotides of the codon for proline.

The oligonucleotide was synthesized using the solid phase phosphate triester method as described previously.[3,4] A protected nucleoside phosphoramidite (Chem Gene Corp) of thymidine [5'-dimethoxytritylthymidine-3'-(methoxy)diethylaminophosphine] was activated with tetrazole. This was condensed with functionalized silica gel to which DMTrG has been attached by its 3'-OH groups and the blocked 5' end subsequently unblocked prior to reaction with the activated thymidine phosphoramidite. Any remaining unreacted 5'-OH of the silica gel bound nucleoside was removed by reaction with a phosphite which was subsequently oxidized to phosphate. This cycle was repeated with the appropriate nucleoside phosphoramidite or mixture where an ambiguity in the sequence existed. When the synthesis was complete, the methyl groups of the phosphodiester were removed by treatment with thiophenol and the oli-

[2] R. L. Jackson, H. N. Baker, E. B. Gilliam, and A. M. Gotto, *Proc. Natl. Acad. Sci. U.S.A.* **74,** 1942 (1977).
[3] M. D. Matteucci and M. H. Caruthers, *J. Am. Chem. Soc.* **103,** 3185 (1981).
[4] M. D. Matteucci and M. H. Caruthers, *Tetrahedron Lett.* **21,** 719 (1980).

A

A.A. RESIDUE		6	7	8	9	10	
		GLN	ASP	GLU	MET	PRO	
mRNA	5'	CA^A_G	GA^U_C	GA^A_G	AUG	CC	3'
cDNA	3'	GT^T_C	CT^A_G	CT^T_C	TAC	GG	5'

B

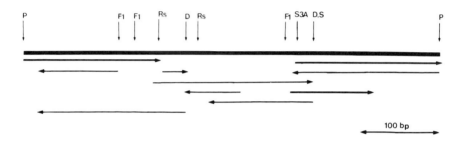

C

```
                                                   LeuValLeuLeuValLeuGlyPheGluValGlnGly
                                                   CTTGTCCTCCTGGTATTGGGATTTGAGGTCCAGGGG

ThrGlnGlnProGlnGlnAspGluMetProSerProThrPheLeuThrGlnValLysGluSerLeuSerSerTyr
ACCCAACAGCCCCAGCAAGATGAGATGCCTAGCCCGACCTTCCTCACCCAGGTGAAGGAATCTCTCTCCAGTTAC

TrpGluSerAlaLysThrAlaAlaGlnAsnLeuTyrGluLysThrTyrLeuProAlaValAspGluLysLeuArg
TGGGAGTCAGCAAAGACAGCCGCCCAGAACCTGTACGAGAAGACATACCTGCCCGCTGTAGATGAGAAACTCAGG

AspLeuTyrSerLysSerThrAlaAlaMetSerThrTyrThrGlyIlePheThrAspGlnValLeuSerValLeu
GACTTGTACAGCAAAAGCACAGCAGCCATGAGCACTTACACAGGCATTTTTACTGACCAAGTTCTTTCTGTGCTG

LysGlyGluGlu
AAGGGAGAGGAGTAACAGCCAGACCCCCCATCAGTGGACAAGGGGAGAGTCCCCTACTCCCCTGATCCCCCAGGT

TCAGACTGAGCTCCCCCTTCCCAGTAGCTCTTGCATCCTCCTCCCAACTCTAGCCTGAATTCTTTTCAATAAAAA
ATACAATTC
```

FIG. 1. (A) ApoC-II oligonucleotide probe used to screen the human cDNA library. (B) The sequencing strategy used for the DNA sequence analysis. Only the position of the relevant restriction sites are indicated: *Hin*fI (FI), *Rsa*I (Rs), *Dde*I (D), *Sau*3A (S3A), *Sac*I (S), *Pst*I (P). The horizontal arrows indicate the direction and extent of sequence determination. Calibration bar = 100 bp. (C) Complete nucleotide sequence of clone pCII-711.

gonucleotide was released from the support using ammonia. The trityloligomers were separated from incomplete reaction products using HLPC and elution with a linear gradient of acetonitrile from 0 to 25%. Detritylation was done in 80% acetic acid and the oligonucleotides repurified on HPLC using identical conditions. The resulting tetradecamer mixture was ready to be radiolabeled for use as a specific apoC-II probe. The oligonucleotide was labeled at the 5' end with [γ-^{32}P]ATP by incubating for 30 min at 37° with polynucleotide kinase (NEB) in a buffer containing 50 mM Tris (pH 7.4), 10 mM MgCl$_2$, 1 mM spermidine, and 1 mM EDTA.[5] The extent of labeling was determined by electrophoresis on 20% polyacrylamide gel electrophoresis.

Screening of the Adult Human Liver cDNA Library. An adult human liver cDNA library[6] in the vector PKT218 was used to isolate the apoC-II cDNA clones. Approximately 2000 colonies were plated per 10 cm nitrocellulose filter and grown overnight. The colonies were replica plated, grown for a few hours at 37°, and then amplified overnight with chloramphenicol. The duplicate filters were treated using a modification of the colony hybridization procedures of Grunstein and Hogness.[7] The colonies were lysed by placing the filters, colony side up, on Whatman sheets soaked in 1.5 mg/ml lysozyme, 25% sucrose, 50 mM Tris (pH 8.0), 1 mM EDTA, followed by 0.5 N NaOH, 0.2% Triton. The filters were neutralized by three successive transfers into 0.5 M Tris (pH 7.5), 1.5 M NaCl. After air drying, the filters were wet with 100 μg/ml proteinase K in 1× SET, finally rinsed in 1×SET. The filters were baked in a vacuum oven for 2 hr at 80° and then prewashed for several hours at 65° in 3× SSC, 1% SDS. Prehybridization was done in a solution which contained 5× SET, 0.2% BSA, 0.2% Ficol, 0.2% polyvinylpyrollidine (PVP), 0.1 M NaHPO$_4$ (pH 7.0), 0.1% SDS, 0.1% sodium pyrophosphate, and 50 μg/ml heat denatured salmon sperm DNA at 65° for 2–3 hr.

The filters were hybridized in a solution which contained in addition to the radioactive oligonucleotide (1 × 10^6 cpm/filter), 10% dextran sulfate, 5× SET, 0.1% BSA, 0.1% Ficol, 0.1% PVP, 50 mM NaHPO$_4$ (pH 7.0), 0.1% SDS, 0.1% sodium pyrophosphate, and 50 μg/ml salmon sperm DNA, overnight at 25°. Following hybridization, the filters were washed with several changes of 6× SSC, 1% SDS at room temperature followed by a wash at 37°. The filters were then exposed to Kodak XAR film overnight.

[5] A. M. Maxam and W. Gilbert, this series, Vol. 65, p. 499.
[6] D. E. Woods, A. F. Markham, A. I. Ricker, G. Goldberger, and H. R. Colten, *Proc. Natl. Acad. Sci. U.S.A.* **79**, 5661 (1982).
[7] M. Grunstein and D. Hogness, *Proc. Natl. Acad. Sci. U.S.A.* **72**, 3961 (1975).

Colonies which gave a duplicate positive signal were then picked and transferred to a gridded nitrocellulose filter to be retested by the above procedure. Colonies which hybridized strongly a second time were colony purified and grown in liquid culture. Small amounts of plasmid DNA were isolated using the alkaline lysis method[8] and the size of the cDNA insert examined after digestion with the enzyme *Pst*I. Most of the clones contained inserts roughly 500 base pairs in size, which is consistent with the known protein size allowing for the presence of 5' and 3' untranslated regions and the possibility of a signal peptide as well as the nucleotides added in the cloning procedure.

Identification of the ApoC-II Clone. The most unequivocal way to identify apoC-II cDNA clones was DNA sequencing, since the translated DNA sequence can be directly compared to the known protein sequence. In order to carry out DNA sequencing using the Maxam and Gilbert method[5] both a large amount of plasmid DNA and knowledge of some restriction sites within the cDNA are necessary.

Bacterial clones were grown in double-strength YT broth, supplemented with tetracycline (20 μg/ml) and plasmid DNA was isolated by the alkaline lysis method.[8] When necessary, plasmid DNA was further purified on CsCl gradients. Restriction endonucleases were purchased from New England Biolabs and used according to the conditions recommended by the manufacturer. Plasmid or isolated insert was digested with various restriction endonucleases, and the results were analyzed by agarose and polyacrylamide gel electrophoresis.

The apoC-II cDNA clone pCII-711 was identified and sequenced by digesting the plasmid with *Pst*I and labeling at the 3' end with ^{32}P-labeled cordycepin and terminal transferase (NEN) in a solution containing 0.1 M Na-cacodylate (pH 7.3), 1 mM 2-mercaptoethanol at 37° for 30 min. The labeled DNA was then subjected to secondary restriction digestion separated by polyacrylamide gel electrophoresis, electroeluted, and precipitated from ethanol. Once identified, the entire cDNA insert was sequenced. In order to determine the DNA sequence of both strands, fragments generated with other restriction enzymes were labeled at their 5' end with polynucleotide kinase after treatment with bacterial alkaline phosphatase. The sequencing strategy is shown in Fig. 1B.

Results

The complete DNA sequence of pCII-711 is shown in Fig. 1C. This clone is ~500 bp long and contains sequences corresponding to the

---

[8] H. C. Bierboim and J. Doly, *Nucleic Acids Res.* **1**, 1513 (1979).

mRNA coding for the entire plasma form of apoC-II. From the known NH_2-terminal amino acid threonine to the stop codon TAA, it appears that mature apoC-II is 79 amino acids in length. In addition, upstream from the NH_2-terminus, there are 36 bp which, when read in frame, code for the amino acid sequence Leu-Val-Leu-Leu-Val-Leu-Gly-Phe-Glu-Val-Gln-Gly. Furthermore, pCII also contains a 144 bp region that corresponds to the 3′ untranslated region of the apoC-II mRNA, which includes a standard polyadenylation signal A-A-T-A-A-A, the last A of which is 12 bp upstream from the polyadenylation site. pCII-711 also includes a portion of the poly(A) tail of apoC-II mRNA.

Once a cDNA clone has been isolated, one has a specific tool that can be used to examine the regulation of gene expression, the structure and location of the chromosomal gene, as well as genetic variation in that gene.

We have used the apoC-II cDNA clone to map the apoC-II gene and to isolate a cloned apoC-II gene from a human library. In addition, we have used this clone to determine on which human chromosome the apoC-II gene is located.[9]

Isolation of ApoC-II Genomic Clones

The apoC-II gene was isolated from human genomic library cloned in charon 4A as described previously.[10,11] Briefly, the phages were plated on 15-cm dishes at a density of 10,000 plaques per dish. The phages were transferred to nitrocellulose, denatured, neutralized, and baked according to the procedure of Benton and Davis.[12] The filters were prehybridized for 3 hr at 65° in the same solution as was used for the cDNA screening. The filters were hybridized in 1 ml per filter of prehybridization solution containing 10% dextran sulfate and 1×10^6 cpm/ml of radioactively labeled pCII-711 insert DNA.

Several strongly hybridizing clones were identified, and one of these designated λ apoC-II #1 was plaque purified and grown in large amounts in 500 ml of double-strength YT media[13] with the bacterial strain LE392 as

---

[9] C. L. Jackson, G. A. P. Bruns, and J. L. Breslow, *Proc. Natl. Acad. Sci. U.S.A.*, in press.

[10] T. Maniatis, K. C. Hardson, E. Lacy, J. Lauer, C. O'Connell, D. Quon, D. K. Sim, and A. Efstratiadis, *Cell* **15**, 687 (1978).

[11] S. K. Karathanasis, V. I. Zannis, and J. L. Breslow, *Proc. Natl. Acad. Sci. U.S.A.* **80**, 6147 (1983).

[12] W. D. Benton and R. W. Davis, *Science* **196**, 180 (1977).

[13] J. H. Miller, in "Experiments in Molecular Genetics," p. 433. Cold Spring Harbor Laboratory, Cold Spring Harbor, New York, 1972.

host. The recombinant phage was precipitated with polyethylene glycol and purified on CsCl step gradients. Phage DNA was prepared by phenol extraction and ethanol precipitation. To map the gene for apoC-II in λ apoC-II #1, the phage DNA was digested by various restriction enzymes and then separated by electrophoresis on 0.7% agarose gels. After Southern transfer[14] to a nitrocellulose filter, hybridization was carried out with the PstI insert from cDNA clone pCII-711 that had been radiolabeled by nick translation.[15] The filter was then washed for 1 hr at 65° in 0.3 M NaCl, 0.03 M Na-citrate (pH 7.0) containing 0.1% SDS. The identical experiment was performed with DNA isolated from peripheral blood leukocytes of normal individuals.

In order to characterize the gene further, it was digested with restriction enzymes and the resulting fragments were subcloned into plasmid vectors pBR322 or pUC9. Initially, a 5.8 kb BamHI fragment spanning the region which hybridized to pCII-711 was subcloned. The λ apoC-II #1 was digested with BamHI and the digestion products separated by electrophoresis on low melt agarose. The fragment of interest was cut out of the gel and purified by phenol extraction and precipitation with ethanol. The DNA was ligated with BamHI digested pBR322 in 50 mM Tris (pH 7.4), 10 mM MgCl$_2$, 10 mM dithiothreitol, 2 mM spermidine HCl, and 0.5 mM rATP using T$_4$ DNA ligase (NEB). Smaller fragments were isolated for restriction mapping and sequencing by further subcloning of the original BamHI subclone.

Results

The enzymatic digestion of λ apoC-II #1 DNA singly with EcoRI, BamHI, HindIII, and PstI produced single hybridization bands of 4.2, 5.8, 10.8, and 4.0 kb, respectively. These single, as well as multiple, enzymatic digests allowed the construction of a restriction endonuclease map of the human apoC-II gene in λ apoC-II #1 (Fig. 2). This map was in complete agreement with the map derived from the Southern transfer analysis of the apoC-II gene in the genome. These studies also showed that the size of the human DNA insert in λ apoC-II #1 is approximately 14 kb and the sequence of the apoC-II gene that hybridizes to pCII-711 is roughly in the middle of the human DNA insert. The DNA sequencing strategy and information derived from the further characterization of the apoC-II gene within this insert is shown in Fig. 3. The gene contains at least 2 introns, one is approximately 350 nucleotides in length and inter-

[14] E. Southern, *J. Mol. Biol.* **98**, 503 (1975).
[15] P. W. J. Rigby, M. Dieckmann, C. Rhodes, and P. Berg, *J. Mol. Biol.* **113**, 237 (1977).

A

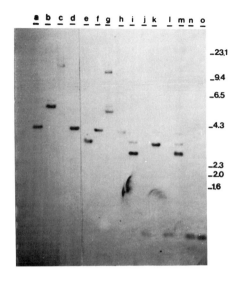

B

FIG. 2. Restriction analysis of clone λ C-II #1. (A) One microgram of DNA was digested with restriction enzyme, first singly, then in various combinations. The DNA was separated by electrophoresis on 0.7% agarose gels, transferred to nitrocellulose fibers, and hybridized with ^{32}P-labeled pCII-711 insert. The resulting autoradiogram is shown in A. Size markers were HindIII digested DNA and HinfI digested pBR322. The restriction enzymes used were (a) EcoRI, (b) BamHI, (c) HindIII, (d) PstI, (e) BamHI + EcoRI, (f) HindIII + EcoRI, (g) HindIII + BamHI, (h) PstI + HindIII, (i) PstI + BamHI, (j) PstI + EcoRI, (k) HindIII + BamHI + EcoRI, (l) PstI + HindIII + EcoRI, (m) PstI + HindIII + BamHI, (n) PstI + BamHI + EcoRI, (o) PstI + HindIII + BamHI + EcoRI. The faint bands present are partial digestion products. (B) This map was constructed using the results obtained in A. The thick line indicates the region hybridizing to the pCIII—711 probe. Restriction sites are indicated as follows: EcoRI (R), BamHI (B), PstI (P), and HindIII (H3).

rupts the sequence at the codon specifying arginine at position 50 in the mature apoC-II. The other intron of unknown size interrupts the sequence in the signal peptide region between amino acid −3 and −4 in the protein sequence. This preliminary information needs to be confirmed by sequencing from the signal peptide region into the 5' area of the intron.

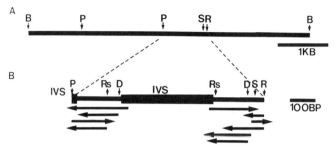

FIG. 3. Restriction map of clone λ C-II #1, subclones and sequence analysis strategy. (A) Restriction map of the λ C-II #1 BamHI subclone which contains the apoC-II (B) A blow-up of the region containing the subclone which contains most of the mature apoC-II sequence. The thick lines correspond to the apoC-II introns. Solid arrows indicate the direction and extent of nucleotide sequence determination.

Chromosomal Localization Utilizing Somatic Cell Hybrids

A panel of DNA obtained from human–rodent somatic cell hybrids was used. The primary hybrid clones were derived from fusion to hypoxanthine phosphoribosyltransferase (HPRT)-deficient Chinese hamster E36 cells or mouse RAG cells with leukocytes or fibroblasts from four unrelated individuals. Two of the leukocyte donors were female carriers of different, reciprocal X/19 translocation chromosomes. These included the X/19W translocation t(X;19)(q23–25::q13) and the X/19B translocation [(X;19)(q1::?p/q13)]. In the latter translocation, the breakpoint on chromosome 19 has not been more precisely determined because of the absence of easily distinguishable landmarks on this chromosome. The hybrid clones have been characterized extensively for human chromosome complements by analysis of human isozyme markers characteristic of each chromosome and by cytogenetic techniques.[16,17]

In addition, cloned DNA probes have been used to monitor 19 of the human autosomes and the X chromosome in the DNAs of the mapping panels used in the present study.[18–20] The $19q^+$ and Xq^- translocation chromosomes were monitored in the hybrid clones by isozyme and cyto-

[16] P. H. Ingram, G. A. P. Bruns, V. N. Regina, R. E. Eiseman, and P. S. Gerald, *Biochem. Genet.* **15,** 455 (1977).
[17] G. A. P. Bruns, B. J. Mintz, A. C. Teary, V. N. Regina, and P. S. Gerald, *Cytogenet. Cell Genet.* **22,** 182 (1978).
[18] D. M. Kurnit, B. W. Philipp, and G. A. P. Bruns, *Cytogenet. Cell Genet.* **34,** 282 (1983).
[19] A. S. Whitehead, G. A. P. Bruns, A. P. Markham, H. R. Colten, and D. E. Woods, *Science* **221,** 69 (1983).
[20] N. Kanda, F. Alt, R. R. Schreck, G. A. P. Bruns, D. Baltimore, and S. A. Latt, *Proc. Natl. Acad. Sci. U.S.A.* **80,** 4069 (1983).

genetic techniques[16,17] and with the cloned X-chromosome DNAs.[21] The chromosome 19 isozyme markers were phosphohexose isomerase, α-mannosidase, lysosomal DNase, and peptidase D; those for the X chromosome were hypoxanthine phosphoribosyltransferase, glucose-6-phosphate dehydrogenase, phosphoglycerate kinase, and α-galactosidase.[22] DNA from parental and hybrid cells was digested to completion with BamHI, separated by electrophoresis on 0.8% agarose gels, and transferred to nitrocellulose. Prehybridization and hybridization were carried out as described[21] with the nick-translated insert from pCII-711 as the probe.

Results

Under the hybridization conditions used, the 5.8 kb BamHI fragment was easily distinguished from the mouse parental cell RAG (Fig. 4) or did not hybridize in the case of the hamster E36 cells. The 5.8 kb component characteristic of human DNA exhibited concordant segregation with chromosome 19 in 7 clones from fusion of RAG cells with leukocytes or fibroblasts of two normal males, with the 19q$^+$B translocation chromosome in 7 clones, and with the 19q$^+$W translocation chromosome in all but one of 14 clones (see the table). The single discordant clone had a rearranged 19q$^+$ translocation chromosome by cytogenetic analysis, it did not express the chromosome by cytogenetic analysis, and it did not express the chromosome 19 isozyme markers phosphohexose isomerase or α-mannosidase. As the 19q$^+$ translocation chromosomes contained an overlapping region of the X chromosome (q24-qter), the segregation of pCII-711 was analyzed in the 7 hybrid clones derived from karyotypically normal males. In 6 of these clones, pCII-711 segregation was independent of the X chromosome. These observations suggest that the apoC-II gene may be assignable to chromosome 19 and regionalized to 19pter-q13. The hybridization pattern of pCII-711 was discordant with the segregation of the other autosomes and the sex chromosomes in the somatic cell hybrids studied. The discordancy indices varied from 0.25 to 0.86 (see the table).

Discussion

Apolipoprotein CII is important as a cofactor for lipoprotein lipase (LPL; triacylglycerol acylhydrolase, EC 3.1.1.3) which catalyzes the hydrolysis of triglycerides in chylomicrons and very low-density lipopro-

---

[21] G. A. P. Bruns, J. F. Gusella, C. Keys, A. C. Leary, D. Housman, and P. S. Gerald, *Adv. Exp. Med. Biol.* **154,** 60 (1982).
[22] T. B. Shows and P. J. McAlpine, *Cytogenet. Cell Genet.* **32,** 221 (1982).

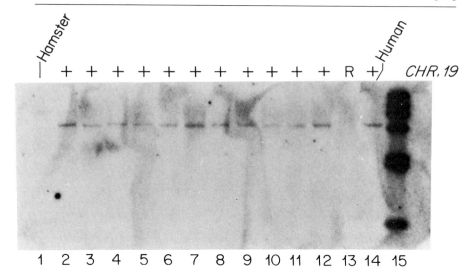

Fig. 4. Hybridization of pCII-711 with DNAs from human–hamster somatic cell hybrids. The DNAs are lane 1, hamster E36 cells; lanes 2–12, human–hamster hybrid clones that express human PHI, MANB, and have the $19q^+W$ translocation chromosomes (19pter-q13::Xq24-qter)[23,24] (these clones also express human HPRT and G6PD); lane 13, human–hamster hybrid from the same hybrid series that has an extensively rearranged $19q^+$ translocation chromosome and does not express PHI or MANB (pCII-711 does not hybridize with DNA from this clone); lane 14, human DNA; lane 15, T_4 polymerase labeled HindIII digest of λ DNA. The $19q^+$ translocation chromosome is the only chromosome shared by the hybrids in lanes 2–12. Segregation of pCII-711 with the X chromosome was excluded by analysis of other clones (see the table).

teins (VLDL).[23–25] The isolation of cDNA and genomic clones for apoC-II described here will provide the foundation for studies of normal apoC-II synthesis and mutations affecting this process.[26] From the DNA sequence of the cDNA clone, it is possible to deduce the amino acid sequence of the entire mature or plasma form of apoC-II. The amino acid sequence predicted differs somewhat from the previously determined amino acid sequence.[2] The DNA sequence specifies a polypeptide of 79 amino acids rather than the 78 previously reported. In addition, amino acids 2 and 17 are glutamine not glutamic acid, amino acid 27 is glutamic acid not glu-

[23] R. J. Havel, V. G. Shore, B. Shore, and D. M. Bier, *Circ. Res.* **33**, 595 (1970).
[24] J. C. LaRosa, R. I. Levy, P. Herbert, S. E. Lux, and D. S. Fredrickson, *Biochem. Biophys. Res. Commun.* **41**, 57 (1970).
[25] P. Nilsson-Ehle, A. S. Garfinkel, and M. C. Schotz, *Annu. Rev. Biochem.* **49**, 667 (1980).
[26] W. C. Breckenridge, J. A. Little, G. Steiner, A. Chow, and M. Poapst, *N. Engl. J. Med.* **298**, 1265 (1978).

HYBRIDIZATION PATTERN OF ApoC-II PROBE WITH DNA MAPPING PANELS

Hybrid-ization	Chromo-some	\multicolumn{24}{c}{Human Chromosomes}																						
		1	2	3	4	5	6	7	8	9	10	11	12	13	14	15	16	17	18	19[a]	20	21	22	X[b]	Y
+	+	8	5	9	11	9	15	10	6	11	12	10	13	11	14	12	8	6	10	21	13	15	12	3	1
−	−	5	4	3	2	5	6	3	4	7	4	4	3	3	2	4	4	4	3	6	4	5	4	1	5
+	−	13	16	12	10	12	6	11	15	10	9	11	8	10	7	9	13	13	10	0	8	6	9	18	20
−	+	2	3	2	4	2	1	4	3	0	3	3	3	2	5	3	2	1	2	1[c]	3	2	1	6	0
Discordant fraction		0.54	0.68	0.54	0.52	0.50	0.25	0.54	0.64	0.36	0.43	0.50	0.39	0.46	0.43	0.43	0.56	0.58	0.52	0.04	0.46	0.29	0.38	0.86	0.77

[a] For the 21 hybrid clones derived from fusions with WBC from the 2 different X/19 translocation carriers, this column represents the 19q+ translocation chromosome.
[b] The X column represents the intact X and the 2 different Xq− translocation chromosomes. As the 19q+ translocation chromosomes included a region of the X (q24-qter or q1-ter), the segregation of apoC-II was shown to be discordant with that of an intact X chromosome in 6 of the 7 clones derived from fusions with karyotypic normal males.
[c] This single discordant clone has a rearranged 19q+ translocation chromosome by cytogenetic analysis and does not express phosphohexose isomerase or α-mannosidase, 2 isozyme markers for chromosome 19.[29]

tamine, and amino acids 20 through 26 are Glu-Ser-Leu-Ser-Ser-Tyr-Trp not the reported Glu-Trp-Leu-Ser-Ser-Tyr. Recently, Hospattankar et al. reported the amino acid sequence of apoC-II from normal and hyperlipoproteinemic subjects and it is compatible with the DNA derived sequence from clone pCII-711.[27] ApoC-II appears to contain a signal peptide, since the 12 amino acids upstream of the NH_2-terminal threonine are characteristic of signal peptide.[28] However, the pCII-711 does not specify the entire signal peptide because it does not include an ATG to initiate translation. Recently, Sharpe et al. reported the DNA sequence of an apoC-II cDNA clone that specifies a 22 amino acid signal peptide.[29] The DNA sequence derived by these investigators is in complete agreement with the one derived for clone pCII-711.

The gene structure of apoC-II, determined to date, conforms to the structure of the other apolipoprotein genes.[11,30] There is an intron in the gene which interrupts the codon for amino acid 50. This is similar to IVS-3 in apoA-I, which interrupts at the codon specifying amino acid 43, and IVS-3 in apoE, which interrupts the codon specifying amino acid 61. The apoC-II gene also has an intron interrupting the region coding for the signal peptide at codons -3 to -4, which is similar to the IVS-2's for both the apoA-I and apoE genes.

In addition, mapping of the apoC-II gene to chromosome 19 places it on the same autosome as the apoE gene,[30] as well as the gene involved in the autosomal dominant lipoprotein disorder, familial hypercholesterolemia.[31,32] The exact relationship of the apoC-II locus to these other important loci specifying aspects of lipoprotein metabolism remains to be determined. Finally, the gene for myotonic dystrophy, a condition of undetermined etiology, has also been assigned to chromosome 19.[33] It may be possible to use polymorphisms in the apolipoprotein genes to facilitate the study of the transmission of the myotonic dystrophy locus.

---

[27] A. V. Hospattankar, T. Fairwell, R. Ronan, and H. B. Brewer, Jr., *J. Biol. Chem.* **259**, 318 (1984).

[28] G. Blobel, P. Walter, C. N. Chang, B. Goldman, A. H. Erickson, and V. R. Lingappa, *Symp. Soc. Exp. Biol. (G.B.)* **33**, 9 (1979).

[29] C. R. Sharpe, A. Sidoli, C. S. Shelley, M. A. Lucero, C. C. Shoulders, and F. E. Baralle, *Nucleic Acids Res.* **12**, 3917 (1984).

[30] J. McPherson, G. A. P. Bruns, S. K. Karathanasis, and J. L. Breslow, *J. Biol. Chem.* submitted (1984).

[31] J. Ott, H. G. Schrrott, J. F. Goldstein, W. R. Hazzard, F. H. Allen, C. T. Falk, and A. G. Motulsky, *Am. J. Hum. Genet.* **26**, 598 (1974).

[32] K. Berg and A. Heiberg, *Cytogenet. Cell Genet.* **22**, 621 (1978).

[33] K. E. Davies, J. Jackson, R. Williamson, P. S. Harper, S. Bau, M. Sarfarazi, L. Meredith, and G. Feg, *J. Med. Genet.* **20**, 259 (1983).

[47] Cloning of the cDNA for Rat and Human Apolipoprotein E mRNA

By JOHN M. TAYLOR, ROBERT W. MAHLEY, and CHIKAFUSA FUKAZAWA

Recombinant DNA technology has made it possible to determine the structure and to characterize the regulation of apolipoprotein mRNAs. The synthesis and cloning of cDNAs to these mRNAs constitute important steps toward a molecular understanding of apolipoprotein mRNAs. A variety of experimental strategies, which are undergoing continuous refinement, are feasible in the synthesis and cloning of these cDNAs. Those methods relevant to the cloning of rat and human apolipoprotein E (apoE) mRNAs[1,2] will be described here.

Principle of the Method

The basic strategy for cloning the rat apoE cDNA is to screen a rat liver cDNA library with cDNA hybridization probes prepared against liver mRNA fractions that are enriched or deficient in apoE mRNA.[1] The mRNA fractions are prepared by sucrose gradient centrifugation and analyzed by cell-free translation in a mRNA-dependent protein-synthesizing system. The cDNAs prepared from these fractions permit the selection of a cloned recombinant plasmid that is a candidate for containing the apoE sequence. The cloned apoE cDNA is identified initially by selection of the corresponding mRNA from total liver mRNA by hybridization to immobilized plasmid DNA. The specifically bound mRNA is eluted and translated in a mRNA-dependent protein-synthesizing system, and the translation products are analyzed by immunoprecipitation and gel electrophoresis. The identity of the hybridization-selected recombinant plasmid is verified by determining the nucleotide sequence of the cloned cDNA.[1]

To obtain a cloned cDNA corresponding to human apoE mRNA, the expected sequence homology between the rat and human species is exploited. A portion of the cloned rat cDNA is employed as a hybridization probe to screen a human liver cDNA library at a reduced stringency.[2] The identity of the cloned human apoE cDNA is established by nucleotide sequence determination.[2]

---

[1] J. W. McLean, C. Fukazawa, and J. M. Taylor, *J. Biol. Chem.* **258,** 8993 (1983).
[2] J. W. McLean, N. A. Elshourbagy, D. J. Chang, R. W. Mahley, and J. M. Taylor, *J. Biol. Chem.* **259,** 6498 (1984).

Cloning and Characterization of the Rat ApoE cDNA

Isolation and Characterization of Total Poly(A)-Containing RNA. The quality of the poly(A)-containing RNA that is to be used for cDNA synthesis, cloning, and mRNA translation is critical to the cloning of apoE cDNA. Undegraded mRNA is required for translation and is essential for the synthesis of full-length cDNAs that will be of benefit for both library construction and screening. Therefore, detailed protocols are provided for this fundamental technique.

Preparation of Total Liver RNA. Total cellular RNA is prepared essentially as described by Chirgwin *et al.*[3] In this method, tissues are dissolved by homogenizing them in guanidine thiocyanate, a strong chaotropic agent that inactivates ribonuclease. Then, RNA is selectively precipitated from the homogenate. The protocol described here for liver RNA extraction[4] has proven to be effective for the preparation of RNA from many different tissues of various animal species.

Liver is frozen in liquid nitrogen as quickly as possible after excision, and 7 g of frozen tissue is pulverized under liquid nitrogen. The tissue is transferred to a plastic centrifuge bottle containing 100 ml of 4 M guanidine thiocyanate, 25 mM sodium citrate at pH 7.0, 0.25% N-lauryl sarcosine, 0.1 M 2-mercaptoethanol, and 1.0% Sigma Antifoam A emulsion (Sigma Chemical Co., St. Louis, Missouri). The tissue is homogenized with a Polytron mixer (Brinkman, Westbury, NY) for 1 min. The homogenate is centrifuged at 6000 rpm in a Sorvall GSA rotor for 15 min at $-10°$, and the supernatant fluid is collected without disturbing the small remaining pellet. Add 2.5 ml of 1.0 M acetic acid, mix, and add 75 ml of ethanol at $-20°$. Allow the RNA to precipitate at $-20°$ for at least 6 hr or overnight. The RNA is collected by centrifugation, and the firm pellet is dissolved in 50 ml 7.5 M guanidine hydrochloride containing 25 mM sodium citrate and 5 mM dithiothreitol. Add 1.25 ml of 1.0 M acetic acid, mix, and add 25 ml of ethanol. The RNA is precipitated as above. For each successive step, sterile technique and sterile reagents are used, except for ethanol and the guanidine solution, which are treated as sterile. The RNA is collected, redissolved, and reprecipitated as above, with a twofold reduction in each solution volume added. The final RNA pellet is dispersed in 100 ml of absolute ethanol to extract residual guanidine hydrochloride. The ethanol is removed by centrifugation, and the RNA pellet is thoroughly drained. The RNA is dissolved in water, adjusted to 0.25 M potas-

[3] J. M. Chirgwin, A. E. Przybyla, R. J. MacDonald, and R. J. Rutter, *Biochemistry* **18,** 5294 (1979).

[4] G. A. Ricca, R. W. Hamilton, J. W. McLean, A. Conn, J. E. Kalinyak, and J. M. Taylor, *J. Biol. Chem.* **256,** 10362 (1981).

sium acetate at pH 5.5 and 70% ethanol, and precipitated overnight at −20°. The RNA can be stored as a precipitate under ethanol at −20° or dissolved in water and frozen at −70°.

Preparation of Poly(A)-Containing RNA. The preparation of undegraded poly(A)-containing RNA requires a strict adherence to sterile technique, including autoclaving of solutions (except for formamide solutions) and glassware to destroy ribonuclease. Nonautoclaved items can be treated with 0.3% diethyl pyrocarbonate in 5% ethanol for ~1 hr, followed by rinsing with sterile water. Since fingers are a major source of contaminating ribonuclease, gloves are helpful. Disposable sterile plastic tubes and pipets and separate RNA-only glassware will also minimize problems.

Total RNA is bound to an affinity ligand in high salt and eluted in a salt-free buffer by either elevated temperature or by the inclusion of formamide. The most commonly used affinity ligand supports employed are oligo(dT)-cellulose or poly(U)-Sepharose (Pharmacia, Piscataway, New Jersey), and either matrix works effectively. In both cases, after a single binding cycle, the eluted poly(A)-containing RNA usually contains an equivalent amount or more of ribosomal RNA. If further purification of the poly(A)-containing RNA is desired, a second adsorption cycle is necessary. An additional purification of the RNA by sedimentation through 5.7 M CsCl[3,5] can be included as well before the second affinity step. However, these extra purification steps are usually not required for most experimental uses.

This protocol applies to the use of the poly(U)-Sepharose affinity matrix.[4] Total RNA is dissolved in 20 mM HEPES at pH 7.2 containing 10 mM EDTA and 1% sodium dodecyl sulfate (SDS), heated at 65° for 10 min, then quickly cooled to room temperature. This heat treatment appears to dissociate mismatched interstrand associations between RNA molecules that might form during the extraction procedure. The RNA solution is diluted with an equal volume of water and adjusted to 300 mM NaCl. It is added to poly(U)-Sepharose (Pharmacia) that has been equilibrated in RNA binding buffer (300 mM NaCl, 0.5% SDS, 5 mM EDTA, and 10 mM HEPES at pH 7.2) at a ratio of 100 A_{260} units of RNA per 0.1 g of poly(U)-Sepharose. The slurry is mixed gently at room temperature for 1 hr, added to a sterile 25-cm column, and the solution is drained from the Sepharose matrix. The matrix is washed with 5 column volumes of binding buffer. Poly(A)-containing RNA is eluted with 4 bed volumes of 70% formamide (previously deionized) containing 5 mM HEPES and 2 mM EDTA at pH 7.2 by successive additions of 1 bed volume each. The eluate

---

[5] V. Glisin, R. Crkvenjakov, and C. Byus, *Biochemistry* **13**, 2633 (1974).

is adjusted to 0.25 M NaCl, and the RNA is precipitated by 2.5 vol of ethanol at $-20°$ overnight. Before reusing the column, it is treated with 90% formamide in 5 mM HEPES and 2 mM EDTA at pH 7.2 that has been warmed to 37°. The matrix is rinsed with 0.5× binding buffer, and it can be stored in this buffer.

Fractionation of RNA by Sucrose Gradient Sedimentation. The isolated RNA is fractionated according to size by sedimentation in an isokinetic sucrose gradient.[6] Gradients of 5 to 29.9% sucrose, using ribonuclease-free sucrose (available commercially), are prepared with gradient buffer (1.0% SDS, 5 mM EDTA, and 50 mM HEPES at pH 7.5) in centrifuge tubes for the Beckman SW41 rotor. The poly(A)-containing RNA is dissolved in gradient buffer, adjusted to an approximate concentration of 5 mg/ml, heated at 65° for 10 min, and quickly cooled to room temperature. Approximately 400 μg of the RNA sample should be loaded on each gradient and sedimented at 27,000 rpm for 17 hr at 20°. Collect 0.4-ml gradient fractions, using an automated collection device if available, and monitor the A_{260} of the fractions. Approximately 28 fractions are obtained, with most of the mRNA located in the central region of the gradient. Adjust the fractions to 0.3 M in sodium acetate (pH 5.5) and 70% ethanol, and precipitate the RNA overnight. Repeat the precipitation at least once to be sure that traces of SDS are removed. The RNA samples are dissolved in water and used directly for translation analysis or cDNA synthesis.

Translation of Poly(A)-Containing RNA. Total rat liver poly(A)-containing RNA, as well as the individual RNA fractions collected from sucrose gradient sedimentation, is translated in a cell-free mRNA-dependent protein-synthesizing system.[1] The translation system, derived from the rabbit reticulocyte lysate, is prepared and used according to the method of Pelham and Jackson.[7] Details of the method as well as technical considerations have recently been reviewed.[8] Lysate and translation kits that can be used with this system are commercially available from various suppliers. The translation products of exogenous mRNAs are usually labeled with [^{3}H]leucine or [^{35}S]methionine. However, [^{3}H]arginine also makes an effective label for apoE,[1] which is an arginine-rich protein.[9]

Apolipoprotein E mRNA translation products are identified by immunoprecipitation with rabbit antibodies that are prepared against puri-

[6] K. S. McCarty, Jr., R. T. Vollmer, and K. S. McCarty, *Anal. Biochem.* **61,** 165 (1974).
[7] H. R. Pelham and R. J. Jackson, *Eur. J. Biochem.* **67,** 247 (1976).
[8] W. J. Merrick, this series, Vol. 101, p. 606.
[9] R. W. Mahley, *Klin. Wochenschr.* **61,** 225 (1983).

fied rat plasma apoE.[1] Following protein synthesis, 30-μl aliquots of the final reaction mixture are added to 120 μg of apoE antibody in the presence of 1.0% Nonidet P-40, 1.0% sodium deoxycholate, 0.1% SDS, 100 mM NaCl, 50 mM LiCl, 5 mM EDTA, 1 mM phenylmethylsulfonyl fluoride, 0.02% sodium azide, and 50 mM Tris–HCl at pH 7.4. The mixtures are incubated at 37° for 30 min, then at 4° overnight. Formaldehyde-fixed *Staphylococcus aureus* (IgGSORB, The Enzyme Center, Boston) that is sufficient to bind 190 μg of IgG is added and mixed for 30 min at room temperature. Insoluble material is collected by centrifugation, and the pellets are washed four times with 300 μl of the above buffer through repeated suspension by vortexing, sonication, and centrifugation. The immunoprecipitated proteins are dissolved by resuspending the pellets at pH 6.8 in 63 mM Tris–HCl containing 2% SDS, 5% mercaptoethanol, and 2 mM dithiothreitol and heating them at 100° for 5 min.

Total translation products and immunoprecipitates are analyzed[1] by electrophoresis in 10 to 18% polyacrylamide gradient gels containing 0.1% SDS.[10] After the gels are electrophoresed, they are impregnated with a commercial fluor, dried under vacuum onto filter paper, and exposed to X-ray film at −70° for several days to develop a fluorogram of the gel. Typical results of a translation analysis of rat liver mRNA are shown in Fig. 1. Immunoprecipitation analysis identifies apoE as a major primary translation product of total liver mRNA, with a $M_r = 34,000$. As shown by the analysis of the RNA gradient fractions, apoE mRNA is enriched in two gradient fractions (Fractions 12 and 13), and it is deficient in neighboring gradient fractions. Thus, cDNA hybridization probes can be synthesized from the mRNA in these fractions to permit a differential screening of cDNA libraries to select candidate clones for apoE mRNA. In addition, the mRNA from the apoE mRNA-enriched fractions can be used as templates for the synthesis of cDNA libraries to enhance the selection of an apoE clone.

Construction of cDNA Libraries. The initial cDNA library for the cloning of apoE cDNA[1] was prepared from size-fractionated rat liver mRNA with reverse transcriptase. Briefly, first-strand cDNA synthesis is primed by oligo(dT), which hybridizes to the 3'-terminal poly(A) segment of the mRNA. The synthesis of the second cDNA strand relies upon the ability of single-stranded cDNA to fold back upon itself and form a hairpin structure at its 3'-terminal region.[11] This structure allows self-priming for second-strand cDNA synthesis. Before the resultant double-stranded cDNA (ds-cDNA) can be inserted into a plasmid for cloning, the single-

[10] U. K. Laemmli, *Nature (London)* **227**, 680 (1970).
[11] T. Maniatis, S. G. Kee, A. Efstratiadis, and F. D. Kafatos, *Cell* **8**, 163 (1976).

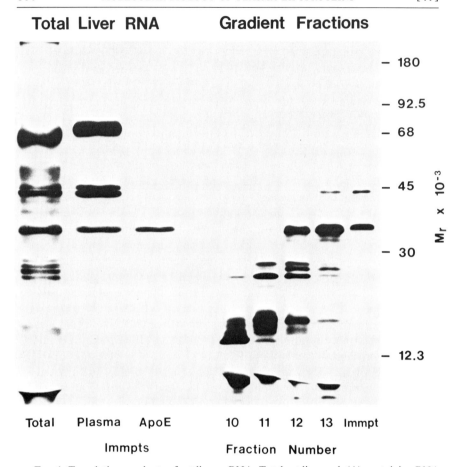

FIG. 1. Translation products of rat liver mRNA. Total rat liver poly(A)-containing RNA and fractions of this RNA that were obtained by sucrose gradient sedimentation were translated in a cell-free system in the presence of [^{3}H]leucine. The total translation products were reacted with antibodies to total plasma proteins and apoE, and gradient Fraction 13 translation products were reacted with antibody to apoE. Immunoprecipitates and total translation products were electrophoresed on 10 to 18% polyacrylamide gradient gels, which were examined by fluorography. The sizes of molecular weight markers are indicated.[2]

stranded loop of the hairpin is opened by digestion with S1 nuclease, resulting in a variable loss of sequences that corresponds to the 5'-terminal region of mRNA. Thus, the first cloned apoE cDNA (pALE124) that was selected and characterized was found to be missing the 5' region of its corresponding mRNA.[1]

To obtain cloned cDNAs that contained the missing 5' sequences, 1 μg of a 100-base pair restriction endonuclease fragment from the 5' end of the

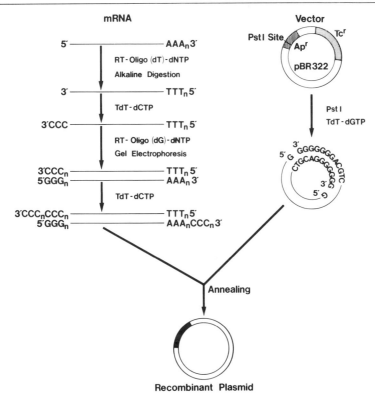

FIG. 2. Strategy for cloning full-length cDNAs. Messenger RNA is incubated with oligo(dT), which binds to the 3'-terminal poly(A) segment and serves as a primer for reverse transcriptase (RT) in the presence of deoxynucleotide (dNTP) substrates for the synthesis of a single-stranded cDNA. The mRNA is removed by digestion with alkali. The remaining cDNA serves as a primer for the 3'-terminal addition of an oligo(dC) segment by terminal deoxynucleotidyl transferase (TdT). The binding of oligo(dG) to the tailed cDNA primes second-strand cDNA synthesis by reverse transcriptase. After size fractionation by gel electrophoresis, the double-stranded cDNA serves as a primer for oligo(dC) addition. The tailed double-stranded cDNA is annealed to pBR322 vector DNA that has been digested by *Pst*I in the ampicillin resistance gene (Ap^r) and oligo(dG) tailed. The annealed DNA is transfected into a bacterial host, and recombinant plasmids are selected by ampicillin sensitivity and tetracycline resistance (Tc^r).

pALE124 cDNA insert is hybridized to 10 μg of total liver poly(A)-containing RNA for the priming of cDNA synthesis by reverse transcriptase. Double-stranded cDNA is synthesized by the method of Land *et al.*, which has been described in detail.[12] In this method, which is illustrated in Fig. 2, the 3' end of the single-stranded cDNA is extended with an oli-

---

[12] H. Land, M. Grez, H. Hauser, W. Lindenmaier, and G. Schutz, this series, Vol. 100, p. 285.

go(dC) tail by terminal deoxynucleotidyltransferase. Then, second-strand cDNA synthesis is primed by oligo(dG). In the more general case, when oligo(dT)-primed first-strand cDNAs are copied in this manner, full-length ds-cDNAs are usually obtained if the mRNA template has not been degraded.

For all of the apoE cDNA cloning, the plasmid pBR322 is employed as the cloning vector, and *E. coli* strain RR1 is the host.[13] Bacteria are transfected by the $CaCl_2$ method of Dagert and Ehrlich[14] without modification. Since insertion of ds-cDNA into the *Pst*I site of pBR322 results in the inactivation of the ampicillin-resistance gene, recombinant plasmids are selected on the basis of ampicillin sensitivity and tetracycline resistance.

Screening of cDNA Libraries. Transfected bacteria are grown on agar plates containing LB medium[15] with 50 µg/ml of ampicillin, and then they are transferred to gridded 85-mm nitrocellulose filters that have been placed on similar fresh plates. The colonies are transferred with applicator sticks that have been dipped in sterile 1% aqueous eosin Y to provide a color marker for the gridded filters. Plates are incubated at 37° overnight, and the filters are removed. The colonies are lysed, and the DNA is denatured and fixed to the filters essentially as described.[16]

Recombinant plasmids from a rat liver cDNA library are screened by colony filter hybridization[17] with ^{32}P-labeled single-stranded cDNAs[15] that are synthesized from poly(A)-containing RNA fractions that are enriched or deficient in apoE mRNA (Fig. 1, Fractions 13 and 11, respectively). Hybridization and filter-washing conditions are essentially as described.[16] Since apoE mRNA appears to be the most abundant RNA species in the Fraction 13 (Fig. 1) material, a single colony that hybridizes intensely with Fraction 13 cDNA but not with Fraction 11 cDNA has been selected for further characterization.

The cDNA insert from this selected plasmid has been subcloned into bacteriophage M13mp7 to facilitate further study of apoE mRNA.[1] The single-stranded recombinant bacteriophage progeny containing the mRNA coding strand provides a convenient template for cDNA synthesis, and the complementary strand bacteriophage DNA facilitates the hybridization selection of the corresponding mRNA. The methods for

[13] F. Bolivar and K. Backman, this series, Vol. 68, p. 245.
[14] M. Dagert and S. D. Ehrlich, *Gene* **6,** 23 (1979).
[15] T. Maniatis, E. F. Fritsch, and J. Sambrook, "Molecular Cloning: A Laboratory Manual." Cold Spring Harbor Laboratory, Cold Spring Harbor, New York, 1982.
[16] C. A. Reardon, Y.-K. Paik, D. J. Chang, G. E. Davies, R. W. Mahley, Y.-F. Lau, and J. M. Taylor, this volume [48].
[17] M. Grunstein and D. S. Hogness, *Proc. Natl. Acad. Sci. U.S.A.* **72,** 3961 (1975).

subcloning DNA into M13 bacteriophage vectors, as well as a description of improved M13 vectors, have been reviewed in detail.[18] To identify M13 recombinants that contain coding strand inserts, 1-μl aliquots of bacteriophage are spotted onto gridded nitrocellulose filters, baked for 2 hr at 80° under vacuum, and hybridized to ^{32}P-labeled cDNA synthesized from total liver mRNA. Complementary strand inserts can be identified by hybridization to the coding strand bacteriophage DNA.[18]

Characterization of Rat ApoE cDNA Clones. The rat apoE cDNA clone was initially identified by hybridization selection of the mRNA by immobilized plasmid DNA.[1] The identity of the cloned cDNA has been confirmed by nucleotide sequence analysis.

Hybridization Selection of mRNA. Five micrograms of M13 bacteriophage DNA containing a complementary strand insert is adjusted to 3 M NaCl and 0.3 M sodium citrate in a final volume of 20 μl, applied to a 1-cm^2 nitrocellulose filter, and baked at 80° for 2 hr under vacuum. The filter is cut into small pieces and incubated in 0.5 ml of hybridization buffer (70% formamide previously deionized) containing 0.3 M NaCl, 0.2% SDS, 5 mM EDTA, and 10 mM HEPES, pH 7.4, at 55° for 5 min. The solution is discarded, and the filter is incubated again with four successive changes of buffer. Total liver poly(A)-containing RNA (40 μg) in 150 μl of hybridization buffer is incubated for 6 hr at a decreasing temperature ranging from 55 to 44° and for 16 hr at 44°. The filters are washed five times for 15 min each at 65° in 75 mM NaCl, 7.5 mM sodium citrate, 0.5% SDS, and twice at 65° in 2.5 mM EDTA at pH 7.2. The hybridized RNA is eluted by 150 μl of 90% formamide containing 0.1% SDS, 1 mM EDTA, and 10 mM HEPES at pH 7.4, supplemented with 20 μg of *E. coli* tRNA, adjusted to 0.3 M sodium acetate at pH 5.5, and precipitated overnight at −20° with 2 vol of ethanol. This mRNA is translated in the cell-free protein-synthesizing system, and a portion of the reaction mixture is reacted with antibodies to rat apoE. The mRNA selected by the cloned cDNA encodes a protein of M_r = 34,000 that reacts with the antibody to apoE as demonstrated by polyacrylamide gel electrophoresis and fluorography analysis of the translation products and immunoprecipitate.

The identity of the cloned cDNA has been confirmed by nucleotide sequence analysis of the cDNA inserts of pALE124, which contains three-fourths of the corresponding mRNA sequence, and a second plasmid, pALE9, which contains the remainder of the apoE coding sequence.[1] Comparison of the derived rat amino acid sequence to that of the previously determined human amino acid sequence[19] has shown that the

---

[18] J. Messing, this series, Vol. 101, p. 20.
[19] S. C. Rall, Jr., K. H. Weisgraber, and R. W. Mahley, *J. Biol. Chem.* **257**, 4171 (1982).

two proteins are identical in 69% of their residue positions. The amino acid identities are clustered in two broad domains—an amino-terminal domain of 173 residues where 80% are identical, and a carboxyl-terminal domain of 84 residues where 70% are identical—separated by a short region of nonhomology. These two domains may be associated with specific functional roles in apoE.

Cloning and Characterization of the Human ApoE cDNA

The strategy for the cloning of the human apoE cDNA is based on the homology to the rat sequence. Since several regions of extended amino acid sequence of known identity are located throughout the two proteins, one would anticipate that the corresponding nucleotide positions would be highly conserved in these regions. Therefore, a human liver cDNA library is screened with a fragment from a rat apoE cloned cDNA. Hybridization conditions should be at reduced stringency to account for the possibility of a limited sequence mismatch in codon third-base positions.

A human liver cDNA library has been constructed essentially as described by Land et al.,[12] using liver RNA obtained from an adult trauma victim.[2] To enhance the library for recombinant plasmids containing long cDNA inserts, the ds-cDNA is electrophoresed in a 6.5% polyacrylamide gel, and material larger than 500 base pairs is extracted for subsequent cloning.

The human liver cDNA library is screened with a single-stranded recombinant M13 bacteriophage DNA containing a portion of the rat apoE cDNA that corresponds to the 3'-terminal 264 nucleotides of rat liver mRNA.[1,2] Filters containing the immobilized DNA of transfected bacterial colonies are prepared as described above for the rat cDNA library. The hybridization buffer is essentially the same as described,[16] but the formamide concentration is reduced from 50 to 25% to lower the strin-

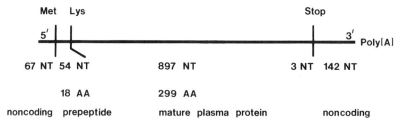

FIG. 3. Structure of human apoE mRNA. The structure of human apoE mRNA was derived from the nucleotide sequence analysis of the corresponding cDNA and a structural analysis of the gene.[2,20] Major structural regions of the mRNA are indicated.

gency of the reaction. In addition, the incubation temperature is reduced from 42 to 35° to lower further the hybridization stringency. After the filters are hybridized, they are washed three times (20 min each wash) at 20° in 360 mM NaCl, 0.1% SDS, 2 mM EDTA, and 20 mM sodium phosphate at pH 7.0, followed by four 15 min washes at 35° in the same buffer. Three recombinant plasmids that are candidates for containing an apoE cDNA should be present on examination of the filters by autoradiography.

The identity of the human apoE cloned cDNAs has been established by nucleotide sequence determination[2] and by comparison of the derived amino acid analysis to the previously determined sequence of plasma apoE.[19] The corresponding human apoE mRNA has a structure (Fig. 3) that is typical of most eukaryotic mRNAs. It contains noncoding regions at both the 5'- and 3'-terminal regions of the molecule. The exact length of the 5'-terminal noncoding segment has been determined by S1 nuclease analysis of a human apoE cloned genomic DNA fragment.[20] The availability of cloned cDNAs for apoE mRNAs and knowledge of their complete sequences will make possible a detailed correlation of specific structural domains of the proteins with their functional properties.

[20] Y.-K. Paik, D. J. Chang, C. A. Reardon, G. E. Davies, R. W. Mahley, and J. M. Taylor, *Proc. Natl. Acad. Sci. U.S.A.* **82**, 3445 (1985).

[48] Cloning and Expression of the Human Apolipoprotein E Gene

By CATHERINE A. REARDON, YOUNG-KI PAIK, DAVID J. CHANG, GLENN E. DAVIES, ROBERT W. MAHLEY, YUN-FAI LAU, and JOHN M. TAYLOR

Detailed analyses of gene structure and the determination of specific sequences that are required for gene expression and regulation have been made possible by the ability to isolate single genes from cloned libraries of genomic DNA in recombinant vectors. Derivatives of the bacteriophage λ are widely employed as vectors that commonly permit the replacement of nonessential bacteriophage genes by foreign DNA of about 15–20 kb in length.[1-4] Bacteriophage libraries made from the human genome, which has a sequence complexity of about 3×10^9 base pairs, require the screening of $0.8-1 \times 10^6$ recombinants to have a 99% probability of selecting a unique gene (i.e., the apolipoprotein (apo) E gene).[4,5] To increase the probability of isolating an intact gene along with functional domains or linked genes that might be present in flanking sequences, cosmids have been developed as cloning vehicles. Cosmids permit insertion of 30- to 45-kb-long segments of foreign DNA into a vector that contains the λ cos site for packaging into bacteriophage heads, a plasmid origin of replication, and an antibiotic-resistance selection marker gene.[6,7] The longer inserts reduce the number of recombinants that typically need to be screened for the selection of a unique gene.

An efficient system for characterizing the expression of cloned mammalian genes is DNA-mediated gene transfer into cultured heterologous cells.[8,9] Both stable and transient (acute) transformation of cells by the foreign gene are employed for studying gene expression. For transient expression, viral enhancers are often used to increase gene expression, and selectable gene markers for mammalian cells usually are employed to facilitate the isolation of stable transformants.[10-12] Cosmid vectors that contain these elements have been constructed and permit the direct use of an isolated clone in gene transfection without additional manipulation.

The human apoE gene has been isolated from available genomic libraries that have been constructed in both λ and cosmid vectors.[13] The protocols employed for this purpose are outlined here.

[1] D. Tiemeier, L. Enquist, and P. Leder, *Nature (London)* **263**, 526 (1976).
[2] F. R. Blattner, B. G. Williams, A. E. Blechl, K. Denniston-Thompson, H. E. Faber, L. A. Furlong, D. J. Grunwald, D. O. Kiefer, D. D. Moore, J. W. Schumm, E. L. Sheldon, and O. Smithies, *Science* **196**, 161 (1977).
[3] T. Maniatis, R. C. Hardison, E. Lacy, J. Lauer, C. O'Connell, D. Quon, G. K. Sim, and A. Efstratiadis, *Cell* **15**, 687 (1978).
[4] T. Maniatis, E. F. Fritsch, and J. Sambrook, "Molecular Cloning: A Laboratory Manual." Cold Spring Harbor Laboratory, Cold Spring Harbor, New York, 1982.
[5] L. Clarke and J. Carbon, *Proc. Natl. Acad. Sci. U.S.A.* **72**, 4361 (1975).
[6] J. Collins and B. Hohn, *Proc. Natl. Acad. Sci. U.S.A.* **75**, 4242 (1978).
[7] Y.-F. Lau and Y. W. Kan, *Proc. Natl. Acad. Sci. U.S.A.* **80**, 5225 (1983).
[8] A. Pellicer, D. Robins, B. Wold, R. Sweet, J. Jackson, I. Lowy, J. M. Roberts, G.-K. Sim, S. Silverstein, and R. Axel, *Science* **209**, 1414 (1980).
[9] R. C. Mulligan, B. H. Howard, and P. Berg, *Nature (London)* **277**, 108 (1979).
[10] P. Mellon, V. Parker, Y. Gluzman, and T. Maniatis, *Cell* **27**, 279 (1981).
[11] M. Wigler, S. Silverstein, L.-S. Lee, A. Pellicer, Y. Cheng, and R. Axel, *Cell* **11**, 223 (1977).
[12] R. C. Mulligan and P. Berg, *Science* **209**, 1422 (1980).
[13] Y.-K. Paik, D. J. Chang, C. A. Reardon, G. E. Davies, R. W. Mahley, and J. M. Taylor, *Proc. Natl. Acad. Sci. U.S.A.* **82**, 3445 (1985).

Screening of a Human Genome Library in Bacteriophage λ

Properties of the Library. The construction and characterization of a human genome library in the bacteriophage λ cloning vector, Charon 4A, has been described previously by Lawn et al.[14] and made available through the generosity of Dr. Tom Maniatis. In brief, human fetal liver DNA was digested partially with the restriction endonucleases HaeIII and AluI, and fragments of 15–20 kb in length were isolated by sucrose gradient centrifugation. The fragments were made resistant to EcoRI by treatment with EcoRI methylase, then ligated to EcoRI synthetic linkers. Cohesive fragment ends were generated by digestion with EcoRI for insertion into the EcoRI site of Charon 4A. The final genome library contained about 1×10^6 independently derived bacteriophages that had been amplified 10^6-fold.

Principle of the Screening Method. The human genome library was screened for recombinant clones containing the apoE gene according to the plaque hybridization method of Benton and Davis.[15] In this method, host bacteria are infected with the recombinant bacteriophages, and plaques are allowed to develop in a bacterial lawn. The bacteriophages and bacteriophage DNA present in the plaques are transferred to a nitro cellulose filter by adsorption. The plaque DNA is denatured, fixed to the filter, and hybridized to a probe that is ^{32}P labeled by nick translation[16] or by random priming.[17] In the case of apoE, the probe consisted of an AvaI–HinfI restriction endonuclease fragment that was 990 nucleotides in length and was derived from a full-length cloned human apoE cDNA.[18]

Preparing Plaque Replica Filters. To screen the library for apoE gene recombinants, 1×10^6 bacteriophages in aliquots of 50,000 per 140-mm-diameter culture plate are examined.[4] Each aliquot is incubated with ~5 $\times 10^8$ *E. coli* strain LE392 at 37° for 15 min in SM buffer (100 mM NaCl, 10 mM MgSO$_4$, 0.01% gelatin, and 50 mM Tris–HCl at pH 7.4). To each aliquot, add 6 ml of 0.7% molten top agar (equilibrated at 50°) and pour onto dry culture plates containing 1.5% bottom agar in NZYDT medium (GIBCO, Grand Island, New York). Allow the top agar to solidify, and incubate the plates at 37° until plaques reach 0.1–0.2 mm in diameter. Cool the plates at 4° for approximately an hour to harden the agar, then allow the plates to reach room temperature. Carefully place a 137-mm-

---

[14] R. M. Lawn, E. F. Fritsch, R. C. Parker, G. Blake, and T. Maniatis, *Cell* **15**, 1157 (1978).
[15] W. D. Benton and R. W. Davis, *Science* **196**, 180 (1977).
[16] P. W. Rigby, M. Dieckmann, C. Rhodes, and P. Berg, *J. Mol. Biol.* **113**, 237 (1977).
[17] A. P. Feinberg and B. Vogelstein, *Anal. Biochem.* **132**, 6 (1983).
[18] J. W. McLean, N. A. Elshourbagy, D. J. Chang, R. W. Mahley, and J. M. Taylor, *J. Biol. Chem.* **259**, 6498 (1984).

diameter nitrocellulose filter on the agar surface for 1 min. Make unambiguous identification marks by stabbing the filter with a needle containing India ink. Remove the filter and set it aside. Add a second filter for 1.5 min, and make duplicate identification marks. Store the plates covered at 4°.

To denature and fix the DNA to the filters, place the filters (plaque-side up) for 5 min on three sheets of Whatman 3MM paper saturated with 0.5 M NaOH containing 1.5 M NaCl. Neutralize the filters by two treatments of 5 min each on 3MM paper saturated with 1.0 M Tris–HCl (pH 7.5) containing 1.5 M NaCl. Rinse the filters for 5 min in 2× SSC (0.3 M NaCl and 0.03 M sodium citrate), and allow them to dry at room temperature. Sandwich each filter between sheets of 3MM paper, and bake for 2 hr at 80° in a vacuum oven.

Hybridization of Filter-Bound DNA. Prepare the filters for hybridization by incubating them at 42° in a prehybridization solution for 4–6 hr in a 140-mm-diameter glass culture dish sealed with Parafilm (American Can Co., Greenwich, Connecticut). Each dish can contain up to 20 filters, and the solution should cover the filters. Alternatively, the filters can be incubated in sealed plastic bags. During incubation, the filters are placed on a gently rotating platform. The prehybridization solution[4] contains 50% formamide (deionized previously), 5× SSPE (0.9 M NaCl, 50 mM NaH$_2$PO$_4$, 5 mM EDTA, pH 7.4), 1.0 mg/ml Ficoll, 1.0 mg/ml polyvinylpyrrolidone, 1.0 mg/ml bovine serum albumin, 0.1% sodium dodecyl sulfate, and 100 µg/ml denatured, sheared herring sperm DNA. For hybridization, an apoE cDNA probe that has been ^{32}P labeled to ~1 × 10^8 cpm/µg is denatured (heat at 100° for 5 min) and added to the prehybridization solution at 10^6 cpm/ml. The incubation of the filters is continued for ~24 hr at 42° with gentle shaking.

Following hybridization, the filters are washed at room temperature in four changes (15 min each) in 2× SSPE containing 0.1% sodium dodecyl sulfate, then at 55° in four changes (15 min each) of 0.1× SSPE containing 0.1% sodium dodecyl sulfate. The washing is done with gentle mixing and with a volume of ~100 ml per filter. The filters are dried at room temperature, placed on sheets of 3MM paper, covered with plastic food wrap, and exposed to X-ray film for 1 to 3 days at −70°.

Plaques that hybridize specifically to the apoE probe appear as superimposable spots on the autoradiograms of duplicate filters. The positions of the colonies are marked on the X-ray films and used to locate the corresponding positions on the original culture plates. A plug of agar medium containing the area of each positive plaque is taken, placed in 1 ml of SM buffer containing a drop of chloroform, and allowed to stand at room temperature for ~2 hr to allow the bacteriophages to diffuse into the solution. The bacteriophages from each plaque area are rescreened as described above at a density of about 1000 plaques per plate. Positive plaques may be screened again at a density of about 200 plaques per plate to obtain individual reconfirmed recombinants that are specific for the apoE probe. An isolated, single plaque is taken and used for the preparation of a plate stock and for subsequent preparation of DNA.[4]

Characterization of the Cloned ApoE Gene. In our studies, the screening of 1×10^6 recombinant bacteriophages of the human genome library has yielded one positive clone for the apoE gene, and the clone has been isolated as described above. A preliminary characterization of the human DNA insert is required to determine that no rearrangements in the structure of the apoE gene have occurred upon cloning. Bacteriophage DNA is prepared and examined by restriction endonuclease mapping and by hybridization of Southern blots of this DNA to ^{32}P-labeled apoE cDNA. The recombinant bacteriophage contains about 20 kb of inserted human DNA, which includes the intact apoE gene. A comparison is made between the restriction endonuclease digestion patterns of the apoE gene region of the insert DNA and the Southern blots of human genome DNA that have been digested by the same enzymes. In both cases, fragments of identical size hybridize to the apoE cDNA probe, suggesting that the overall structure of the cloned apoE gene and the native gene are the same. In addition, the restriction endonuclease mapping indicates that the mRNA-encoding part of the apoE gene consists of four exons that are interrupted by three introns. To facilitate further studies, the human DNA insert of the isolated bacteriophage is subcloned into the *Eco*RI and *Bam*HI sites of the plasmid pUC9.

The complete nucleotide sequence of the human apoE gene, as well as 856 nucleotides of the 5'-flanking region and 629 nucleotides of the 3'-flanking region, have been determined.[13] The structure of the gene and its relationship to the structures of the corresponding mRNA and protein product are illustrated in Fig. 1. The nucleotide sequence of the mRNA and the amino acid sequence of the protein have been determined previously.[18,19] The lengths of the exons are 44, 66, 193, and 860 nucleotides, and the intron lengths are 760, 1092, and 582 nucleotides in their 5' to 3' order. In comparison to the corresponding mRNA sequence, the first intron occurs in the 5'-noncoding region 23 nucleotides upstream from the translation initiation codon, the second intron occurs in the codon for Gly_{-15} of the signal peptide region, and the third intron occurs in the codon for Arg_{-61} of the mature plasma protein region. Additional structural features of the gene have been characterized.[13]

---

[19] S. C. Rall, Jr., K. Weisgraber, and R. W. Mahley, *J. Biol. Chem.* **257,** 4171 (1982).

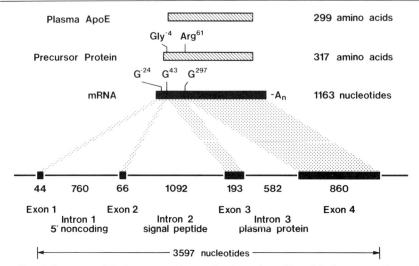

FIG. 1. Structure of the human apoE gene. A schematic outline of the human apoE gene is shown in relation to the mRNA which it encodes, the primary translation product of the mRNA, and the mature plasma protein product of the gene.

Screening of a Human Genome Library in a Cosmid Vector

Properties of the Library. The construction of the cosmid vector, pCV108 (Fig. 2), and the preparation of a human genomic cosmid library have been described previously.[7] Briefly, cosmid vector pJB8[20] is modified by the insertion of a fragment from an SV40 hybrid gene (SV40-neo)[21] containing a neomycin-resistance gene, an SV40 origin of replication, and an SV40 enhancer sequence. The neomycin-resistance marker allows for the selection of stable transformants of mammalian cells in the presence of the antibiotic G418. Genomic fragments of human DNA are prepared by the partial digestion of human leukocytic DNA with *Mbo*I and isolation of 30- to 45-kb fragments on sucrose gradients.[7] These fragments are inserted into the *Bam*HI site in pCV108, and the ligated DNA is packaged *in vitro* into λ bacteriophage heads and transduced into *E. coli* ED8767. The final library contains about 5×10^5 independent recombinants that have been amplified about 10-fold.

High-Density Bacterial Colony Screening. The cosmid library is screened according to the high-density method of Hanahan and Meselson.[22] Aliquots of the library are spread onto 137-mm nitrocellulose

[20] D. Ish-Horowicz and J. F. Burke, *Nucleic Acids Res.* **9**, 2989 (1981).
[21] P. J. Southern and P. Berg, *J. Mol. Appl. Genet.* **1**, 327 (1982).
[22] D. Hanahan and M. Meselson, this series, Vol. 100, p. 333.

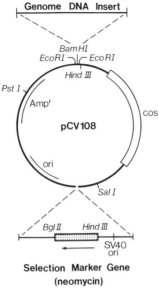

FIG. 2. Structure of the pCV108 cosmid vector. The human genome insert is contained within a unique BamHI restriction endonuclease site. The locations of the λ-derived *cos* gene, the SV40 origin of replication, the neomycin-resistance selection marker gene, the plasmid origin of replication, the ampicillin-resistance gene, and certain restriction endonuclease sites are indicated.

filters (Millipore, HATF, 0.45 μm; Millipore Corp., Bedford, Massachusetts) on agar plates of LB medium containing 50 μg/ml of ampicillin (LA plates) at a density of about 50,000 colonies per filter. The bacteria are grown at 37° until they reach a size of 0.1–0.2 mm. These master filters are lifted off the agar plates, and placed colony side up on three sheets of 3MM paper. A nitrocellulose filter that has been wetted by placing it on an agar plate is placed over the master filter and the two filters are pressed together firmly. The two filters are keyed to each other by unambiguous marking with a needle containing India ink. Two sets of replicate filters are prepared from each of the master filters, and the master filters are stored temporarily at 4°. The cosmids on the replica filters are grown at 37° for 6 hr and amplified by growing overnight on LA plates containing 200 μg/ml chloramphenicol.

To lyse the colonies and to denature and fix the DNA to the filters, place the filters (colony side up) for 5 min on three sheets of 3MM paper saturated with 0.5 M NaOH. Neutralize the filters by two 5-min treatments on 3MM paper saturated with 1.0 M Tris–HCl containing 1.5 M NaCl. Rinse the filters for 5 min in 0.5 M Tris–HCl containing 1.5 M

NaCl. Submerge the filters in a dish containing 2× SSC and 0.1% sodium dodecyl sulfate, and gently remove cellular debris from the filters by wiping gently with a tissue. Rinse the filters for 5 min in 2× SSC, allow them to dry at room temperature, place them between sheets of 3MM paper, and bake at 80° for 2 hr in a vacuum oven.

The filters are hybridized to the apoE cDNA probe and treated as described for screening the bacteriophage library. Colonies that screen as positive for the apoE sequence are sampled from the master filters, grown overnight in 5 ml of LB medium containing ampicillin, and rescreened at a density of about 1000 colonies per filter. Positive colonies may be screened again at a density of about 200 colonies per filter to obtain single recombinants that hybridize to the apoE probe.

For long-term storage, master filters are incubated on LA plates containing 15% glycerol at 37° for 3 hr. Then, the filters are replicated through to the identification mark keying step. The filters are left together and placed between two layers, three sheets per layer, of dry 3MM paper. This sandwich is sealed in a plastic bag along with one sheet of wet 3MM paper and stored at −70°.

Characterization of Human ApoE Gene-Containing Cosmid Clones

Preliminary Examination of the Cosmid DNA. Approximately 1×10^6 colonies from the gene library were screened with the human apoE cDNA probe, which yielded 16 positive recombinants for the apoE gene. Preparations of cosmid DNA were made for each recombinant, according to the alkaline-sodium dodecyl sulfate method for plasmid DNA isolation.[23] The DNA was analyzed by restriction endonuclease digestion, agarose gel electrophoresis, Southern blotting of the gels to nitrocellulose filters, and hybridization of the filters to apoE cDNA and cosmid vector probes. The cDNA hybridized to fragments of cosmid insert DNA that had a size consistent with the known structure of the apoE gene. These mapping studies showed that the apoE gene was located near either end of the human genomic DNA insert in all cases. The cosmids were grouped into two classes, and a representative of each class was selected for further characterization. Cosmid pCLE1, which represents one class, contains about 35,000 nucleotides of human genomic DNA flanking the 5' terminus of the apoE gene. Cosmid pCLE271, which represents the other class, contains about 35,000 nucleotides of human genomic DNA flanking the 3' terminus of the apoE gene.

[23] H. C. Birnboim, this series, Vol. 100, p. 243.

Stable Transfection of Cultured Mouse L Cells. To determine whether a functional apoE gene has been isolated, its expression upon transfection into heterologous cultured cells is examined. Mouse L cells[24] that can be adapted to grow in serum-free medium are employed to facilitate the subsequent characterization of the apoE gene products. The day prior to transfection, cells are seeded at 1×10^6 cells per 100-mm tissue culture dish and allowed to grow until they reach a 75% confluence density (~24 hr).

Human apoE gene-containing cosmid DNA is introduced into mouse L cells by a modification of the calcium phosphate coprecipitation method described by Graham and van der Eb.[25] Each tissue culture dish of cultured cells is transfected with 5 μg of cosmid DNA in a sterile 1.0-ml suspension of a calcium phosphate coprecipitate. To prepare the coprecipitate, 5 μl of a 1 mg/ml cosmid DNA solution is mixed with 500 μl of 274 mM NaCl, 1 mM KCl, 1 mM NaH$_2$PO$_4$, 0.2% dextrose, and 42 mM HEPES at pH 7.05 in a sterile 15-ml test tube. A sterile, cotton-plugged Pasteur pipet is placed into the solution, and air is gently bubbled through the solution. While this solution is bubbling, 495 μl of 0.252 M CaCl is added dropwise (slowly), after which the pipet is removed. The calcium phosphate–DNA coprecipitate is allowed to form for 30 min at room temperature, and then is mixed vigorously by vortexing for 30 sec. The coprecipitate is added to 10 ml of medium in the culture dish with simultaneous mixing to avoid a pH shock, and then the cells are incubated overnight.

To increase the efficiency of DNA uptake by the cells, they are then subjected to a glycerol shock. The culture medium is removed, and 3 ml of 15% glycerol in PBS (140 mM NaCl containing 8 mM NaH$_2$PO$_4$, 1.5 mM KH$_2$PO$_4$, and 3 mM KCl, at pH 7.5) is gently added to the cells. After 2 min, the glycerol solution is removed, and the cells are washed twice with PBS. Then, 10 ml of fresh culture medium is added and the cells are incubated at 37° for 48 hr, or until they reach confluence. Stable transfectants are selected by growing the cells in culture medium containing the antibiotic G418 as described below.

Transiently transfected cells can also be examined for the expression of the apoE gene. Cultured cells are treated as above, except that 5×10^6 cells are seeded into a 75-cm^2 tissue culture flask, and the cells are transfected with 25 μg of DNA. Optimum expression of the exogenous gene usually occurs about 48 hr after the glycerol shock.

[24] D. J. Merchant and K. B. Hellman, *Proc. Soc. Exp. Biol. Med.* **110,** 194 (1962).
[25] F. L. Graham and A. J. van der Eb, *Virology* **52,** 456 (1973).

Selection of Stable Transfected Cells. Confluent monolayers of cultured cells that have been transfected with the apoE gene are treated with trypsin, and 1/20 of the cells are seeded into 100-mm tissue culture dishes containing culture medium (Dulbecco's modified Eagle's medium plus 10% fetal calf serum) supplemented with 400 µg/ml G418 (GIBCO). The medium is changed every 4 days, with the loss of unstable transfected cells becoming noticeable after about 1 week. By about 2 weeks, individual colonies of G418-resistant cells can be observed as small, round, white spots. These colonies are isolated by removing the medium, gently scraping the colonies off the dish with a pipet tip, and rinsing the area with 5–10 µl of medium. Each colony is transferred to small culture dishes containing 1 ml of medium with G418 and cultured for about 4 weeks, or until stable cells are obtained. Individual cell lines are maintained in culture medium containing 200 µg/ml G418.

Analysis of the Expression of the Tranfected ApoE Gene

Stable cell lines that are transformed with the human apoE gene can be analyzed for the expression of this exogenous gene by examining its RNA transcripts and protein product.

Preparation of RNA. Total cellular RNA is isolated from the transformed cell lines by a modification of the procedure of Chirgwin *et al.*[26] Culture medium is removed from tissue culture flasks, and RNA extraction buffer (4.0 M guanidinium thiocyanate, 0.25% N-lauryl sarcosine, 25 mM sodium citrate at pH 7.0, 100 mM 2-mercaptoethanol, and 1.0% Sigma Antifoam A emulsion (Sigma Chem. Co., St. Louis, Missouri) is added at 1.0 ml per 75-cm^2 flask. After the cells are dissolved, the DNA is sheared by passing the solution five times through a 23-gauge needle. The solution is overlaid on a 2.0-ml cushion of 5.7 M CsCl containing 100 mM EDTA at pH 7.0 in a Beckman SW55 centrifuge tube (Beckman Instruments, Fullerton, California), and sedimented at 25,000 rpm at 15° for 18 hr. The supernatant fluid is aspirated thoroughly without contaminating the RNA pellet with cell homogenate. The pellet is rinsed with 70% ethanol, and the RNA is dissolved in 25 mM Tris–HCl containing 10 mM EDTA at pH 8.0.

Dot Blot Analysis. Analysis of human apoE mRNA concentrations in transfected cells is performed with a template manifold apparatus (Schleicher and Schuell, Inc., Keene, NH) to assure uniform dot size. Total cellular RNA is applied to the apparatus in four different amounts (3.0,

[26] J. M. Chirgwin, A. E. Przybyla, R. J. MacDonald, and R. J. Rutter, *Biochemistry* **18**, 5294 (1979).

2.0, 1.0, and 0.5 μg), and supplemented with yeast RNA to provide a final total amount of 3 μg of RNA/sample. The RNA samples are denatured by adjusting them to 1.0 M formaldehyde, 0.9 M NaCl, and 0.09 M sodium citrate, and heating them at 55° for 15 min. The sample is diluted into 20 vol of 3 M NaCl containing 0.3 M sodium citrate and applied to nitrocellulose filters under a gentle vacuum. The sample wells are washed with additional diluent, and the filter is baked at 80° for 2 hr under vacuum and then hybridized with a ^{32}P-labeled human apoE cDNA. Human liver RNA dilutions over a 100-fold concentration range are included on each blot as reference standards. Autoradiograms of the filters are analyzed by quantitative scanning densitometry.

In our studies, the level of human apoE mRNA expressed by the exogenous gene in different stable cell lines has been found to vary greatly, from undetectable to amounts comparable to that observed in the human liver. In no case is the endogenous host mouse apoE gene activated by the transfection of the human apoE gene, as measured by the criteria described here. A similar range of expression also has been observed in transiently transfected cells.

Northern Blot Analysis. The human apoE RNA transcripts synthesized by the transfected cells are examined also by Northern blot analysis.[27] Typically, 5 to 20 μg of total cellular RNA is adjusted to 10% glyoxal (deionized) and 25 mM NaH$_2$PO$_4$ at pH 6.5 and incubated at 50° for 1 hr. The denatured RNA samples are electrophoresed in a 1.1% agarose gel in 25 mM NaH$_2$PO$_4$ at pH 6.5 with a recirculating buffer system for 4 hr at 5 V/cm. The RNA is transferred overnight to a nitrocellulose filter by blotting, and the filter is baked for 2 hr at 80° under vacuum. To completely remove the glyoxal adduct, the filter is soaked overnight in 0.3 M sodium citrate containing 3.0 M NaCl, or placed in 20 mM Tris–HCl at pH 8.0 at 100° and allowed to cool to room temperature. The filters are hybridized with a human apoE cDNA probe, and an autoradiogram is obtained. Typically, the apoE mRNA that is synthesized by transfected cells is the same size as that in normal human liver.

Characterization of Secreted ApoE by Transfected Cells

Because apolipoproteins are not normal secretion products of mouse L cells, it is of particular interest to determine whether the transfected cells are capable of secreting human apoE into the culture medium. To eliminate the potential effects of exogenous lipoproteins, the cells are adapted to grow in a defined medium (HB101, Hana Biologicals, Inc.,

---

[27] P. S. Thomas, this series, Vol. 100, p. 255.

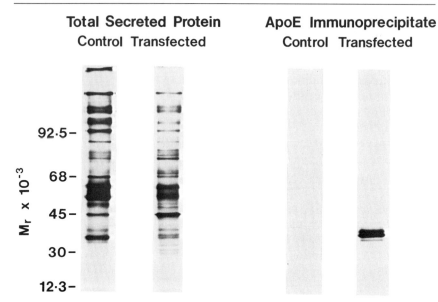

FIG. 3. Secretion of human apoE by cultured mouse L cells transfected by the isolated human apolipoprotein E gene. Mouse L cells were transfected with the DNA from cosmid pCLE1. The cells, in 75-cm^2 flasks, were labeled in serum-free medium containing 100 μCi/ml of [^{35}S]methionine for 6 hr at 37°, and the medium was concentrated 10-fold. Aliquots of the medium containing 2 × 10^5 cpm trichloroacetic acid-precipitable material were immunoprecipitated with anti-human apoE antibodies as described.[28] Total acid-precipitable material from the medium (2 × 10^4 cpm) and the immunoprecipitates were electrophoresed in 5–20% polyacrylamide gradient gels containing 0.1% sodium dodecyl sulfate, and a fluorogram prepared essentially as described.[28]

Berkeley, CA) in the absence of serum. The cells are passed through two cycles of medium changes before further analysis. Semiconfluent monolayers of cells are incubated for 18 hr in serum-free medium containing 100 μCi/ml of [^{35}S]methionine (specific activity >600 Ci/mmol) and 40 μM methionine. The culture medium is concentrated in the presence of 0.1 mM phenylmethylsulfonyl fluoride to inhibit proteases. Both total medium proteins and the immunoprecipitate from reaction of antihuman apoE are examined by electrophoresis in polyacrylamide gels. A typical analysis of a particular transfected mouse L cell line is shown in Fig. 3. Human apoE is found to be secreted by all transfected cells that contained the corresponding mRNA, and the secreted apoE is identical in electrophoretic properties to that of normal human plasma apoE.

[28] J. W. McLean, C. Fukazawa, and J. M. Taylor, *J. Biol. Chem.* **258**, 8993 (1983).

The availability of the cloned human apoE gene and knowledge of its structure will facilitate a direct examination of the molecular basis of its regulation. Expression of the isolated gene in transfected cells will permit the biological assay of specific reconstructions designed to correlate particular sequences with their functions.

Acknowledgments

The authors would like to thank James X. Warger and Norma Jean Gargasz for graphics, Linda Harris Odumade for manuscript preparation, and Barbara Allen and Sally Gullatt Seehafer for editorial assistance.

[49] Genetic Polymorphism in Human Apolipoprotein E

By VASSILIS I. ZANNIS

Introduction

Apolipoprotein E (apoE) was first identified in 1973[1] and has subsequently been found in lipoproteins of all mammalian species studied.[2,3] It is a single polypeptide composed of 299 amino acids of known sequence.[4] Although liver may be one of the major tissues capable of synthesizing apoE, recent studies have shown that this apolipoprotein may also be synthesized by numerous other tissues.[5-11] Newly synthesized apoE is

---

[1] V. G. Shore and B. Shore, *Biochemistry* **12**, 502 (1973).
[2] V. I. Zannis and J. L. Breslow, *J. Mol. Cell. Biochem.* **42**, 3 (1982).
[3] R. W. Mahley, *in* "Disturbances in Lipid and Lipoprotein Metabolism" (J. Dietschy, A. M. Gotto, Jr., and J. A. Ontko, eds.), p. 181. Clinical Physiology Series, American Physiology Society, Bethesda, Maryland, 1978.
[4] S. C. Rall, K. H. Weisgraber, and R. W. Mahley, *J. Biol. Chem.* **257**, 4171 (1981).
[5] T. E. Felker, M. Fainaru, R. L. Hamilton, and R. J. Havel, *J. Lipid Res.* **18**, 4654 (1977).
[6] R. L. Hamilton, *in* "Disturbances in Lipid and Lipoprotein Metabolism" (J. Dietschy, A. M. Gotto, Jr., and J. A. Ontko, eds.), p. 155. Clinical Physiology Series, American Physiology Society, Bethesda, Maryland, 1978.
[7] A. L. Wu and H. B. Windmueller, *J. Biol. Chem.* **254**, 7316 (1979).
[8] S. K. Basu, M. S. Brown, Y. K. Ho, R. J. Havel, and J. L. Goldstein, *Proc. Natl. Acad. Sci. U.S.A.* **78**, 7545 (1981).
[9] S. K. Basu, Y. K. Ho, M. S. Brown, D. W. Bilheimer, R. G. W. Anderson, and J. L. Goldstein, *J. Biol. Chem.* **257**, 9788 (1982).
[10] M. L. Blue, D. L. Williams, S. Zucker, S. A. Khan, and C. B. Blum, *Proc. Natl. Acad. Sci. U.S.A.* **80**, 283 (1983).
[11] V. I. Zannis, S. F. Cole, G. Forbes, D. Kurnit, and S. Karathanasis, *Biochemistry* **24**, 4450–4455 (1985).

incorporated into the high-density lipoprotein region and is subsequently transferred to triglyceride-rich lipoproteins.[12] Recent studies have shown that human plasma apoE consists of several isoproteins which differ in size and/or charge. This complexity of human apoE is the result of both genetic variation of apoE in the human population and posttranslational modification of apoE with carbohydrate chains containing sialic acid.[13,14] Present evidence indicates that apoE is synthesized as sialo apoE and is subsequently desialated in plasma.[2,14–16] ApoE desialation may play an important, but as yet undefined, role in lipoprotein metabolism.[17] Other studies have shown that apoE can mediate lipoprotein catabolism by both extrahepatic[18–20] and hepatic[21–23] tissues. Extrahepatic tissues catabolize lipoproteins containing apoE by the LDL (apoB/E) receptor.[18–20] Hepatic tissues catabolize lipoproteins containing apoE by the LDL (apoB/E) receptor as well as by the chylomicron remnant (apoE) receptor.[21–23] Structural mutations in apoE dramatically affect its recognition and catabolism by lipoprotein receptors.[24–29] A structural mutation in the apoE gene is believed to underlie type III hyperlipoproteinemia (type III HLP),[13,14,27–32] a condition which leads to premature arteriosclerosis.[33]

[12] C. B. Blum, *J. Lipid Res.* **23,** 1308 (1982).
[13] V. I. Zannis, P. W. Just, and J. L. Breslow, *Am. J. Hum. Genet.* **33,** 11 (1981).
[14] V. I. Zannis and J. L. Breslow, *Biochemistry* **20,** 1033 (1981).
[15] V. I. Zannis, J. L. Breslow, T. R. San Giacomo, D. P. Aden, and B. B. Knowles, *Biochemistry* **20,** 7089 (1981).
[16] V. I. Zannis, D. Kurnit, and J. L. Breslow, *J. Biol. Chem.* **257,** 536 (1982).
[17] V. I. Zannis, C. Blum, R. Lees, and J. L. Breslow, *Circulation* **66,** 170 (1982).
[18] T. P. Bersot, R. W. Mahley, M. S. Brown, and J. L. Goldstein, *J. Biol. Chem.* **251,** 2395 (1976).
[19] T. L. Innerarity and R. W. Mahley, *Biochemistry* **18,** 1440 (1978).
[20] R. E. Pitas, T. L. Innerarity, K. S. Arnold, and R. W. Mahley, *Proc. Natl. Acad. Sci. U.S.A.* **76,** 2311 (1979).
[21] M. Carrella and A. D. Cooper, *Proc. Natl. Acad. Sci. U.S.A.* **76,** 338 (1979).
[22] B. C. Sherrill, T. L. Innerarity, and R. W. Mahley, *J. Biol. Chem.* **255,** 1804 (1980).
[23] D. Y. Hui, T. L. Innerarity, and R. W. Mahley, *J. Biol. Chem.* **256,** 5646 (1981).
[24] R. J. Havel, Y. S. Chao, E. E. Windler, L. Kotite, and L. S. S. Guo, *Proc. Natl. Acad. Sci. U.S.A.* **77,** 4349 (1980).
[25] M. E. Gregg, L. A. Zech, E. J. Schaefer, and M. B. Brewer, Jr., *Science* **211,** 584 (1981).
[26] W. J. Schneider, P. T. Kovanen, M. S. Brown, J. L. Goldstein, G. Utermann, W. Weber, R. J. Havel, L. Kotite, J. P. Kane, T. L. Innerarity, and R. W. Mahley, *J. Clin. Invest.* **68,** 1075 (1981).
[27] K. H. Weisgraber, T. L. Innerarity, and R. W. Mahley, *J. Biol. Chem.* **257,** 2518 (1982).
[28] S. C. Rall, Jr., K. H. Weisgraber, T. L. Innerarity, and R. W. Mahley, *Proc. Natl. Acad. Sci. U.S.A.* **79,** 4696 (1982).
[29] K. H. Weisgraber, S. C. Rall, Jr., and R. W. Mahley, *J. Biol. Chem.* **256,** 9077 (1981).
[30] G. Utermann, M. Jaeschke, and J. Menzel, *FEBS Lett.* **56,** 352 (1975).
[31] G. Utermann, M. Hees, and A. Steinmetz, *Nature (London)* **269,** 604 (1977).

Finally, cholesterol feeding in animals results in the accumulation in plasma of lipoprotein particles designated HDL_c, $\beta VLDL$, and IDL which are enriched in cholesteryl ester and apoE[2,34-38] and have a high affinity for the LDL and chylomicron remnant receptors.[18-23] The above information suggests that this apolipoprotein plays a central role in lipoprotein and cholesterol metabolism. In this chapter, we will describe the methodologies used to study the polymorphism of human apoE. We will also review our current knowledge of the genetics of human apoE and the relationship between structural mutations in the apoE gene and type III HLP.

Study of ApoE Polymorphism by Two-Dimensional Polyacrylamide Gel Electrophoresis (2D-PAGE) and One-Dimensional Isoelectric Focusing

Principle

The principle of two-dimensional polyacrylamide gel electrophoresis is shown in Fig. 1A. This technique, developed by O'Farrel in 1975,[39] separates proteins on the basis of their isoelectric point on the x axis and by their molecular weight on the y axis. When a mixture of apolipoproteins is analyzed simultaneously in this system, we obtain a reproducible two-dimensional map of this group of proteins (Fig. 1A). In this map, each apolipoprotein or isoprotein has a defined position relative to the other apolipoproteins.

There are two methods to screen for apolipoprotein variations: (1) by running two gels, the first gel will contain the unknown apolipoproteins and the second will contain a mixture of approximately equal concentrations of the unknown sample and a sample containing all the normal

---

[32] J. L. Breslow, V. I. Zannis, T. R. San Giacomo, J. L. H. C. Third, T. Tracy, and C. J. Glueck, *J. Lipid Res.* **23,** 1224 (1982).

[33] D. S. Fredrickson, J. L. Goldstein, and M. D. Brown, in "The Metabolic Basis of Inherited Disease" (J. B. Stanbury, J. D. Wyngaarden, and D. S. Fredrickson, eds.), p. 604. McGraw-Hill, New York, 1978.

[34] R. W. Mahley, K. H. Weisgraber, T. Innerarity, H. B. Brewer, Jr., and G. Assman, *Biochemistry* **14,** 2817 (1975).

[35] R. W. Mahley, K. H. Weisgraber, and T. Innerarity, *Biochemistry* **15,** 2979 (1976).

[36] J. L. Rodriguez, G. C. Ghiselli, D. Torreggiani, and C. R. Sirtoni, *Atherosclerosis* **23,** 73 (1976).

[37] L. L. Rudel, R. Shah, and D. G. Greene, *J. Lipid Res.* **20,** 55 (1979).

[38] L. S. S. Guo, R. L. Hamilton, J. P. Kane, C. J. Fielding, and G. C. Chen, *J. Lipid Res.* **23,** 531 (1982).

[39] P. H. O'Farrel, *J. Biol. Chem.* **250,** 4007 (1975).

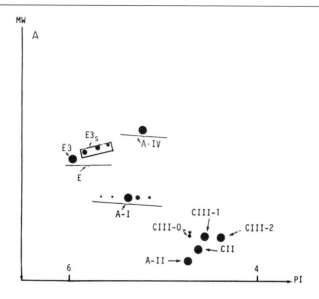
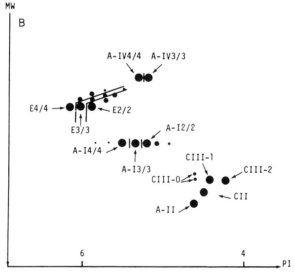

FIG. 1. (A) Schematic presentation of the 2D-PAGE patterns of normal human apolipoproteins. The apolipoproteins and their corresponding isoproteins are designated by dots. (B) Schematic presentation of a 2D-PAGE patterns of a mixture of normal and variant human apolipoproteins. The variant apolipoproteins focus in a different position than those of normal apolipoproteins and can be recognized by comparison of the patterns in A and B. In this and subsequent figures, cathode is on the left and anode is on the right.

apolipoproteins (Fig. 1B); and (2) by running a single gel that will contain internal isoelectric point markers.

Materials

Bovine albumin, ovalbumin, lysozyme, trypsin inhibitor, Tris, glycine, and agarose were purchased from Sigma Chemical Company. Ampholines, pH 2.5–4 and 5–8, were obtained from LKB. Nonidet P-40 was purchased from Particle Data Laboratories, Ltd. Sodium dodecyl sulfate (SDS), manufactured by British Drughouse Chemical, Ltd., was purchased from Gallard-Schleisinger. Acrylamide, bis(acrylamide), N,N,N',N'-tetramethylenediamine, ammonium persulfate, Coomassie brilliant blue, bromophenol blue, and Biolyte (ampholines, pH 4–6) were obtained from Bio-Rad. Urea, ultrapure grade, was supplied by Schwarz/Mann. Antihuman apoA-II (sheep) was donated by Dr. Peter Herbert, Miriam Hospital, Providence, RI. Antihuman apoE (goat) and antihuman apoA-I (goat) was donated by Atlantic antibodies, Scarborough, ME. All other materials were of the purest grade commercially available.

Solutions

The following solutions were used for one-dimensional isoelectric focusing and (2D-PAGE):
1. Equilibration buffer
   Tris–HCl 0.5 M, pH 6.8, 125 ml
   10% SDS pure, 200 ml
   Glycerol, 100 ml
   Deionized H_2O, 525 ml
   2-Mercaptoethanol, 50 ml
   Total volume, 1000 ml
2. Polyacrylamide solution—Isoelectric focusing 10× (recipe for 10 gels)
   Urea, ultra pure, 6.6 g
   30% Acrylamide (Bio-Rad), 1.58 ml
   NP 40, 4.8 ml
   Mix while warming solution under running warm H_2O
   Ampholines 5–8, 0.43 ml
   Biolyte 4–6, 0.144 ml
   Ampholines 2.5–4, 0.050 ml
   Degas for 20 min with gentle stirring
   10% ammonium persulfate, 0.021 ml
   TEMED, 0.019 ml

The suggested combination of ampholines provides a pH gradient in the region of 4 to 6.7

3. 30% Polyacrylamide solution for tube gels
   Acrylamide (Bio-Rad), 56.76 g
   Bis(acrylamide), 3.24 g
   Dissolve to 200 ml deionized H_2O and filter through No. 1 Whatman filter
4. Lysis buffer
   Urea, ultra pure, 28.5 g
   NP 40, 1.0 g
   β-mercaptoethanol, 2.5 ml
   Mix while warming solution under running warm H_2O
   Ampholines 5–8, 1.5 ml
   Biolyte 4–6, 0.6 ml
   Ampholines 2.5–4, 0.4 ml
   Bring up to final volume of 50 ml with deionized water
5. Overlay buffer
   1:1 dilution of lysis buffer with deionized water
6. 0.01 M NaOH
   This solution is made fresh just prior to use
   Eight pellets (0.8 g) NaOH are dissolved in 1 liter
7. 0.01 M H_3PO_4
   A stock of 1 M H_3PO_4 is prepared by diluting 80.60 ml of 85% w/v H_3PO_4 to a final volume of 1 liter deionized water. The stock is then diluted 1/100 with deionized water to a final concentration of 0.01 M
8. 30% polyacrylamide solution for slab gels
   Acrylamide, 292 g
   Bis(acrylamide), 8 g
   Dissolve in deionized water to a final volume of 1 liter. The solution is stored at 4° in an amber bottle
9. Running gel (recipe for 10 gels)
   30% polyacrylamide, 100 ml
   Tris–HCl buffer 0.5 M, pH 8.8, 62.5 ml
   Deionized water, 83.75 ml
   10% SDS, pure, 2.50 ml
   10% ammonium persulfate, 1.0 ml
   TEMED, 0.3 ml
10. Sealer gel (recipe for 5 gels)
    Running gel, 11.5 ml
    10% ammonium persulfate, 0.176 ml
    TEMED, 0.020 ml

11. Stacking gel (recipe for 10 gels)
    30% polyacrylamide, 15 ml
    Tris–HCl 0.5 M, pH 6.8, 25 ml
    Deionized water, 58 ml
    10% SDS, pure, 1 ml
    10% ammonium persulfate, 0.3 ml
    TEMED, 0.2 ml
12. Running buffer
    Glycine, 14.35 g
    Tris base, 3 g
    Bring to 990 ml with water
    Add SDS 10%, 10 ml
13. Fixing (destaining) solution
    Deionized water, 2900 ml
    Glacial acetic acid, 600 ml
    Methanol, 3500 ml
14. Staining solution
    10 g Coomassie brilliant blue dye in 4 liters of fixing solution
15. 1% w/v agarose
    Tris–HCl 0.5 M, pH 6.8, 12.5 ml
    10% SDS, pure, 20.0 ml
    Glycerol, 10.0 ml
    Deionized water, 52.5 ml
    Agarose, 1 g
    Heat solution to dissolve agarose, then add 5 ml mercaptoethanol. Store in $-20°$ in 10-ml aliquots
16. Sample buffer for SDS slab gel electrophoresis
    Tris–HCl, 0.5 M, pH 6.8, 12.5 ml
    10% SDS, pure, 20.0 ml
    Glycerol, 10.0 ml
    Deionized water, 52.5 ml
    β-mercaptoethanol, 5.0 ml
17. Molecular weight markers
    Chymotrypsinogen (5 mg/ml), 200 μl
    Sample buffer, for SDS slab gels, 200 μl
    Boil for 1 min
    BSA (5 mg/ml), 200 μl
    Ovalbumin (5 mg/ml), 200 μl
    HDL (5 mg/ml), 200 μl
    Trypsin inhibitor (5 mg/ml), 200 μl
    Egg white lysozyme (5 mg/ml), 200 μl
    Aldolase (20 mg/ml), 50 μl

Sample buffer, 1050 μl
Bromophenol blue (0.1% w/v), 100 μl
Boil the mixture for 2 min; store in $-20°$
18. Acrylamide solution—slab gel isoelectric focusing
    Acrylamide gel 30%, 6 ml
    Urea, ultra pure, 19.2 g
    Deionized water, 15 ml
    Ampholine 4%, pH 4–6, 2 ml
    Bring the volume to 40 ml with deionized water
    Degas for 20 min with gentle stirring, then add 10% ammonium persulfate, 0.16 ml
    TEMED, 0.040 ml
19. Fixing solution for slab gel isoelectric focusing
    Trichloroacetic acid, 57.5 g
    Sulfosalicylic acid, 17.3 g
    Deionized water to a final volume of 100 ml
20. Sample buffer for slab gel isoelectric focusing
    1 M Tris–HCl, pH 8.2, 1 ml
    Urea, ultra pure, 48 g
    0.1 M dithiothreitol, 10 ml
    Deionized water to a final volume of 100 ml

Sample Preparation

Very low-density lipoprotein is obtained by ultracentrifugation at 104,000 g at 4° for 20 hr using a 40.3 rotor. The VLDL is dialyzed against distilled water overnight, lyophilized and dissolved in either the lysis buffer for the two-dimensional analysis or the sample buffer for the one-dimensional isoelectric focusing analysis. Immunoprecipitation of proteins from plasma is performed as follows: 15 μl of plasma is mixed with 60 μl of antihuman apoA-I, 40 ml of antihuman apoA-II, and 115 μl 0.01 potassium phosphate buffer pH 7.1, 0.15 M NaCl. The mixture is left at 4° for 2–3 days. The immunoprecipitate is collected by centrifugation for 2 min in a microfuge, and is washed twice with ice cold normal saline and once with distilled water. The immunoprecipitate is then dissolved in the appropriate buffer and is used for one- or two-dimensional gel analysis. The samples which will be used for slab gel isoelectric focusing are dialyzed in 0.01 M ammonium bicarbonate overnight, lyophilized, and delipidated with 2:1 chloroform–methanol. The delipidated protein is collected by centrifugation at 3000 rpm for 30 min, washed with ether, dried under N_2, and dissolved in sample buffer.

Procedures

Isoelectric Focusing (First Dimension). Isoelectric focusing is performed in glass tubes 13 cm long and 3 mm in diameter. Before each run, the tubes are soaked overnight in 1/50 dilution of 7× detergent, rinsed with tap water, soaked in a solution of dichrol (an acid dichromate cleaning solution) for a minimum of 1 hr, rinsed with tap water, and dried well in a drying oven. To prepare the cylindrical gels, the tubes are sealed on one side with a strip of parafilm and placed vertically in a tube holder. A 20-ml syringe, fitted with a 20 × 1.5 g needle and 6 in. of 15-mm tubing is used to fill the tubes with acrylamide solution. The tip of the tubing is inserted at the bottom of the cylindrical tubes and is moved upward as the tube is filled with acrylamide. This motion prevents the trapping of air bubbles. The tubes are filled to within 1 cm from the top, and topped with overlay buffer. After the acrylamide has polymerized, the parafilm is removed from the bottom and the overlay buffer is removed. The tube gels are positioned in the isoelectric focusing apparatus (Bio-Rad). The lower chamber of the apparatus is filled with a 0.01 N phosphoric acid (approximately 1.8 liter). The lower chamber is filled to within 1–2 cm of the top to prevent overheating of the tube gels. Such overheating results in detachment of the acrylamide gel from the glass tubes. When all 18 tubes are placed in the apparatus the lower chamber is filled with the 0.01 N phosphoric acid. The upper portion of the apparatus is rotated within the buffer-filled lower chamber to dislodge air bubbles from underneath the gels. Lyophilized samples or immunoprecipitated protein pellets are resuspended in lysis buffer and a maximum of 50 μl is loaded into the tube with a pipetman. The sample is topped with 20 μl of overlay buffer and filled to the top with 10–15 μl of 0.01 N NaOH. If the sample aliquot is less than 50 μl, lysis buffer is added to bring the volume to 50 μl. When all the samples are loaded, the upper chamber is filled with 0.01 N NaOH. The apparati are connected with the power supply units and electrophoresed for 1 hr at 150 V, 12 hr at 500 V, and 2 hr at 1000 V for a total of 8150 V-hr. When the run is completed, the tube gels are removed, six at a time, from the apparatus and extruded from the tubes using a 10-ml syringe fitted with a small piece of rubber tubing. The syringe is filled with warm water and the acid end of the tube is inserted into the tubing. The gel is warmed slightly under running hot water and forced from the tube onto a piece of parafilm by applying gentle pressure with the syringe. The gel is then transferred to a labeled 15-ml screw-cap plastic culture tube which is then filled with equilibration buffer, capped, frozen in an ethanol–dry ice slurry, and stored at $-70°$ until they are run on the second dimension. Alternatively, the tube gels may be fixed in a solution of 11.5% TCA–3.4% sulfosalicylic acid, stained for 30 min, and destained with fixing solution and water as explained below. This treatment will show the one-dimensional isoelectric focusing patterns.

SDS–Polyacrylamide Gel Electrophoresis (Second Dimension). The plates used are 27 cm in length and 19 cm in width. Notches on the plates are 2 cm long and 1 cm wide. Plastic rulers 0.75 mm thick are cut 1 cm wide and 30 cm long and used as spacers. One side of each plate is washed with ethanol and the spacers are greased lightly with vacuum pump oil on both sides. Straight (rectangular) plates are placed, clean side up, on the bench top and spacers are positioned along either edge. The notched plates are laid over the straight plates with the clean side facing down, and the two plates are clamped together with 4 small binder clips 1/3″ and 2/3″ from the bottom of the plate. The bottom of the plate is sealed with saran wrap which is then clamped to the glass plate. The plates are placed in white polypropylene test tube racks, taking care to avoid tearing the saran wrap seal.

The gels are sealed with 4 ml sealer acrylamide solution. Following polymerization of the sealer layer, the remainder of the running gel is poured into the plates until it reaches 5 cm from the top. The acrylamide layer is overlayed with water using a spray bottle. When the gels have polymerized, a sharp line will appear on the acrylamide–water interface. The water is removed and the acrylamide surface is blotted with 3 mm Whatman paper. Next, the stacking gel is poured until it reaches the top of the notched plates and a 1.5-mm-thick spacer is placed across the exposed surface of the stacking gel and clamped in place with two binder clips. Care should be taken to avoid trapping air bubbles between the spacer and the acrylamide surface. Sharp interfaces are very important for good results. A small spacer (3 cm long, 2 mm wide) is inserted close to one of the edges of the gel. This will be used to load molecular weight protein standards. When the stacking gel has polymerized, the clips, saran wrap, and the top spacer are removed. The slab gel is positioned at a 30° angle with the notched plates facing up. The tubes containing the cylindrical gels are thawed quickly with warm water. The gels are placed on a piece of parafilm and oriented with the acid side on the right and the basic on the left. The acid side of the tube gel usually appears as an undistorted, perfectly clear region, whereas the basic side is most often slightly tapered, swollen, discolored, or cloudy.

The tube gel is allowed to slide off the parafilm sheet so that it rests in the notch of the gel plate on the surface of the stacking gel. The cylindrical gel is cemented in place with agarose, taking care to avoid trapping air bubbles between the cylindrical gel and the stacking gel.

The SDS gel carrying the cylindrical gel on top is placed on the gel apparatus and the upper and the lower reservoirs are filled with running buffer. Bubbles are removed from the bottom of the gel with a rubber policeman. The apparati are connected in parallel (a maximum of nine gels can be run on a single Buchler 3-1500 constant power supply). The gels are run at 50 V for 1 hr and 110 V overnight (until the dye front reaches the bottom of the gel).

Staining and Destaining. The slab gels are removed from the apparati, put into plastic boxes containing 200 ml of fixing solution, and placed on the shaker. The gels must remain in the fixing solution for a minimum of 1 hr; however, they can be left there for several days. The gels are then stained with 200 ml of staining solution for 1 hr with continuous shaking to prevent deposition of precipitated stain on the gel surface. The stain is then replaced with water. Good destaining is accomplished by exposing the stained gel alternately to fixing solution and water. After each successive period of destaining the gels should be placed in water for at least 30 min to rehydrate. Generally, 1 hr exposure to the fixing solution is sufficient for complete destaining. However, this time will vary depending on the thickness of the gel, staining time, rehydration time, etc. After destaining, the gels are either dried and stored or photographed and discarded.

One-Dimensional Slab Gel Isoelectric Focusing. The plates used for slab gel isoelectric focusing are 13 cm in length and 14 cm wide. The spacers used are 1.5 mm thick. Apparati similar to those described above are used. The samples are loaded and overlayed with 0.01–NaOH. The upper chamber is filled with 0.01 N NaOH and the lower chamber is filled with 0.01 N H_3PO_4. The slab gel is electrophoresed at 250 V for 16 hr. When the run is completed, the slab gel is fixed for 30 min to remove the ampholines, stained for 30 min, and destained for 30–60 min as described above.

Determination of ApoE Phenotypes

Determination of ApoE Phenotypes by 2D-PAGE
  Using Mixing Experiments

2D-PAGE has shown that most of the apolipoproteins consist of several isoprotein components which differ in molecular weight and/or charge. Our discussion will concentrate on human apoE. However, the same procedures can be applied to characterize polymorphisms which occur in other apolipoproteins.

In the early stages of our studies, we observed two distinct electropho-

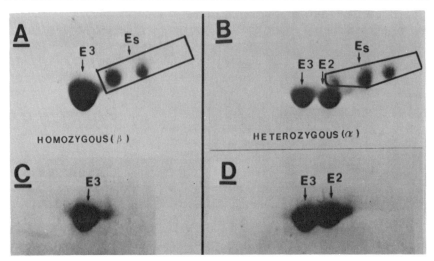

FIG. 2. Two-dimensional gel electrophoresis patterns of two easily distinguishable α and β apoE phenotypes are shown. (A) The homozygous (β) and (B) the heterozygous (α) apoE phenotypes, respectively. Note the multiplicity of apoE isoproteins which comprise the α and β apoE phenotypes. (C and D) Homozygous (β) and heterozygous (α) phenotypes following treatment with *C. perfringens* neuraminidase.[14] In this and subsequent figures, only the area of the gel in the vicinity of apoE is shown.

retic patterns of apoE which were designated apoE class β and α[13,14,40] (Fig. 2A and B). Following a recently adopted uniform apoE nomenclature,[41] the apoE classes (and subclasses) we had described are now referred to as apoE phenotypes. The α phenotype has two major asialo apoE isoproteins of apparent MW 38K (determined by SDS–polyacrylamide gel electrophoresis in 12% gels) and at least six sialo apoE (apoE$_s$) isoproteins. In contrast, the β phenotype has only one major asialo apoE isoprotein of apparent MW 38K and at least three sialo apoE isoproteins. The sialo apoE isoproteins, which are more acidic and have slightly higher molecular weight than the asialo apoE isoproteins, are converted to the corresponding asialo isoprotein forms upon treatment of apoE with *C. perfringens* neuraminidase[14–16] (Fig. 2C and D).

[40] V. I. Zannis and J. L. Breslow, *J. Biol. Chem.* **255**, 1759 (1980).
[41] V. I. Zannis, J. L. Breslow, G. Utermann, R. W. Mahley, K. H. Weisgraber, R. J. Havel, J. L. Goldstein, M. S. Brown, G. Schonfeld, W. R. Hazzard, and C. B. Blum, *J. Lipid Res.* **23**, 911 (1982). The following correspondence exists between the early nomenclature of apoE phenotypes and genotypes and the one adopted in ref. 41: apoE alleles, ε2 = εIV, ε3 = εIII, ε4 = εII; apoE phenotypes, E4/4 = βII, E3/3 = βIII, E2/2 = βIV, E4/3 = αII, E3/2 = αIII, E4/2 = αIV.

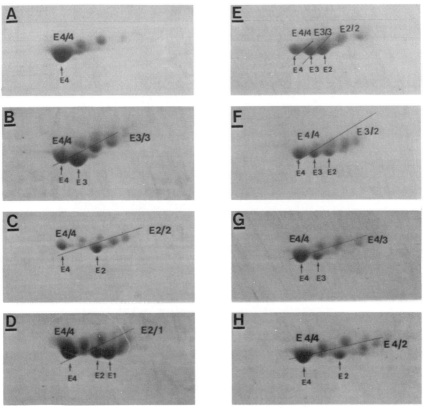

FIG. 3. Two-dimensional gel electrophoresis of mixtures of VLDL fractions obtained from individuals with different apoE phenotypes: (A) 35 μg of equal mixture obtained from two different subjects with E4/4 phenotypes; (B) 25 μg of E4/4 and 30 μg of E3/3; (C) 15 μg of E4/4 and 20 μg of E2/2; (D) 30 μg of E4/4 and 35 μg of E2/1; (E) 20 μg of E4/4, 20 μg of E3/3, and 20 μg of E2/2; (F) 20 μg of E4/4 and 20 μg of E3/2; (G) 25 μg of E4/4 and 25 μg of E4/3; (H) 25 μg E4/4 and 25 μg E4/2. The phenotypes that were mixed and the major asialo isoproteins of each phenotype are indicated.

Comparison of the apoE patterns in a large number of human subjects revealed additional genetic heterogeneity within the heterozygous and homozygous apoE phenotypes. A total of six common and one rare apoE phenotypes were established as follows: equal quantities of VLDL obtained from subjects with unknown and known apoE phenotypes (usually E4/4) were mixed, and the mixture was analyzed by 2D-PAGE. Typical mixing experiments are shown in Fig. 3A–H. These experiments allowed us to distinguish three common homozygous and three common heterozy-

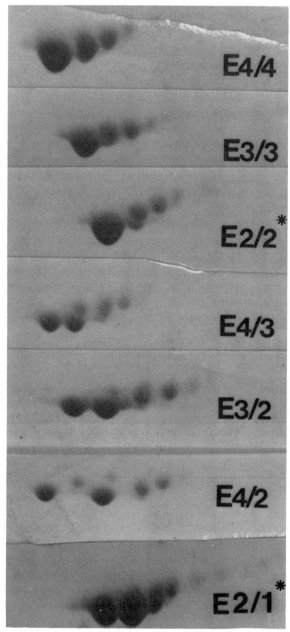

FIG. 4. Two-dimensional electrophoretic patterns of the six common E4/4, E3/3, E2/2, E4/3, E3/2, E4/2, and one rare E2/1 apoE phenotype observed in humans. * indicates phenotypes associated with type III HLP.

gous phenotypes of apoE (Fig. 4). The common homozygous phenotypes have been designated E4/4, E3/3, E2/2, and the common heterozygous phenotypes have been designated E4/3, E3/2, E4/2.[13,14,41] The rare heterozygous apoE phenotype has been designated E2/1. Other laboratories have similarly described E2/1 and E3/1 phenotypes.[42,43]

Determination of ApoE Phenotypes by 2D-PAGE Using Internal Isoelectric Point Markers

The apoE phenotypes can be obtained in a single two-dimensional run utilizing internal isoelectric point markers. In order to obtain the apoE phenotypes by this procedure, apoA-I and apoA-II immunoprecipitated from plasma, and lyophilized VLDL or apoE are immunoprecipitated from plasma, are dissolved in lysis buffer, mixed together, and analyzed by 2D-PAGE. This analysis provides a two-dimensional map that contains apoA-I, apoA-II, apoA-IV, apoC-II, apoC-III-0, apoC-III-1, apoC-III-2, and apoE, along with the heavy (H) and light (L) immunoglobulin chain (Fig. 5). Two proteins in the IgG heavy chain region designated X and Y along with the apoA-I isoprotein 2 (proapoA-I) provide isoelectric point markers which allow in most instances accurate determination of the apoE phenotypes (Fig. 6A–H). Occasionally, when the concentration of isoelectric point markers is low mixing experiments are necessary in order to determine unequivocally the apoE phenotype. The essential reagent required for this analysis is the antihuman apoE antibodies. HDL can be used as the source of apoA-I and proapoA-I that serves as an isoelectric point marker.

Determination of ApoE Phenotypes by One-Dimensional Isoelectric Focusing

One-dimensional isoelectric focusing using tube gels has been employed extensively in various laboratories for determination of apoE phenotypes.[30,31,44–50] The six common apoE phenotypes obtained by this pro-

---

[42] K. H. Weisgraber, S. C. Rall, Jr., T. L. Innerarity, and R. W. Mahley, *J. Clin. Invest.* 1024 (1984).
[43] R. E. Gregg, G. Ghiselli, and J. B. Brewer, Jr., *J. Clin. Endocrinol. Metab.* **57,** 969 (1983).
[44] A. Pagnan, R. J. Havel, J. P. Kane, and L. Kotite, *J. Lipid Res.* **18,** 613 (1977).
[45] S. W. Weigman, B. Suarez, J. M. Falko, J. L. Witztum, J. Kolar, M. Raben, and G. Schonfeld, *J. Lab. Clin. Med.* **549** (1979).
[46] G. R. Warnick, C. Mayfield, J. J. Albers, and W. R. Hazzard, *Clin. Chem.* **25,** 279 (1979).
[47] G. Utermann, A. Steinmetz, and W. Weber, *Hum. Genet.* **60,** 344 (1982).
[48] M. R. Wardell, P. A. Suckling, and E. D. Janus, *J. Lipid Res.* **23,** 1174 (1982).
[49] D. Bouthillier, C. F. Sing, and J. Davignon, *J. Lipid Res.* **24,** 1060 (1983).
[50] G. Utermann, U. Langenback, U. Beisiegel, and W. Weber, *Am. J. Hum. Genet.* **32,** 339 (1980).

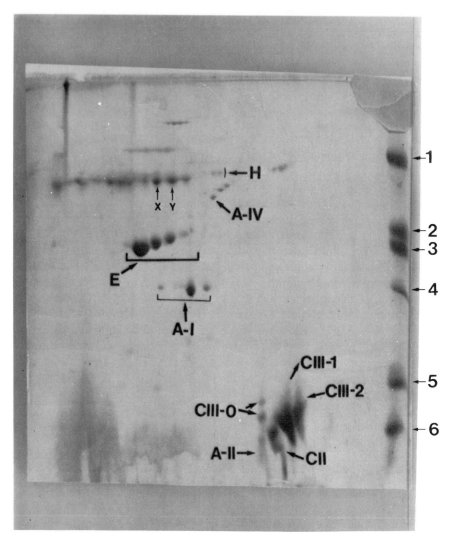

Fig. 5. Two-dimensional gel electrophoresis of a mixture of VLDL and apoA-I and apoA-II immunoprecipitated with specific antibodies is shown. The various apolipoproteins are indicated. H indicates heavy IgG chains. The numbers 1–6 at the right side indicate protein molecular weight markers as follows: 1, bovine serum albumin (68,000); 2, ovalbumin (43,000); 3, aldolase (40,000); 3, human apoA-I (28,000); 6, trypsin inhibitor (19,000); 5, egg white lysozyme (14,300).

Fig. 6. ApoE phenotypes determined by 2D-PAGE using internal isoelectric point markers. A–G contain the following phenotypes: A, E4/4; B, E3/3; C, E2/2; D, E3/2; E, E4/3; F, E4/2; G, E2/1. H contains a mixture of the three homozygous (E4/4, E3/3, and E2/2) phenotypes. The isoelectric point markers X and Y focus over and slightly to the right of apoE2 and apoE1, respectively. The proapoA-I (apoA-I$_2$) focuses under and slightly to the right of apoE2.

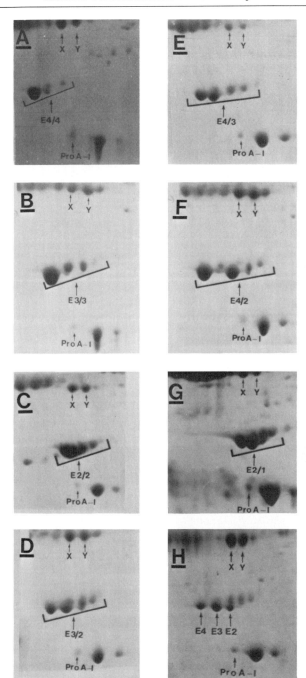

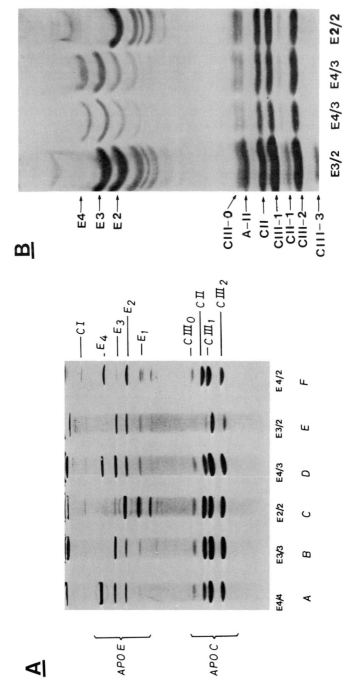

FIG. 7. (A) The six apolipoprotein E phenotypes as determined by one-dimensional isoelectric focusing on cylindrical polyacrylamide gels (see Ref. 49). Reproduced with the permission of the authors and the *Journal of Lipid Research*. (B) ApoE phenotypes determined by one-dimensional slab gel isoelectric focusing. Reproduced with the permission of Dr. Frank Sacks. The apoE isoproteins and phenotypes as well as the apoA-II, apoC-II, and apoC-III isoproteins are indicated.

cedure by Bouthillier et al.[49] are shown in Fig. 7A. The phenotypes may also be determined in most instances by one-dimensional slab gel isoelectric focusing (Fig. 7B). If an ambiguity arises, then the pattern may be clarified further by one of the two-dimensional techniques described above.

Comparative Advantages of the One- and Two-Dimensional Methods

The two-dimensional technique separates the proteins both on the basis of molecular weight and isoelectric point. The main advantage of this method is that it allows unequivocal determination of the phenotypes regardless of the apoE3:apoE2 ratio, the degree of sialation of apoE, or the presence of other protein contaminants such as human serum albumin and the immunoglobulin heavy and light chains. The samples used in this analysis do not require extensive purification such as double centrifugation and delipidation of VLDL. As explained previously, the samples employed in our experiments were obtained by a single 18 hr ultracentrifugation of plasma or by immunoprecipitation of apolipoproteins from plasma.

The immunoprecipitation method eliminates the need for time-consuming ultracentrifugations and allows determination of apoE phenotypes in a laboratory setting that lacks ultracentrifuges. The simplifications introduced in the preparation of the samples coupled with the scaling up of the two-dimensional technique enables a skilled technician to phenotype an average of 30 samples per week. The main disadvantage of the 2D-PAGE is the relative complexity of the technique which renders it impractical for a clinical laboratory setting.

The main advantage of the one-dimensional technique is its relative simplicity. Its main disadvantages are that it requires two ultracentrifugations and delipidation of the samples. Misinterpretation of the apoE patterns on one-dimensional isoelectric focusing gels may occur when the VLDL sample is contaminated by human serum albumin HSA (Fig. 8A–D), when the concentration of the sialo apoE (apoE$_s$) is increased (Fig. 8E), and when the apoE pattern has been degenerated due to aging of the sample (Fig. 9A–G).

Artifacts

During the course of our experiments, we have observed the following two predominant artifacts in the apoE phenotypes. (1) Prolonged storage of the VLDL solution at 4° causes smearing of the apoE pattern and gradual shift of the apoE isoproteins toward more acidic isoelectric points (Fig. 9A–E). The smearing can be prevented if VLDL is stored at −20° in

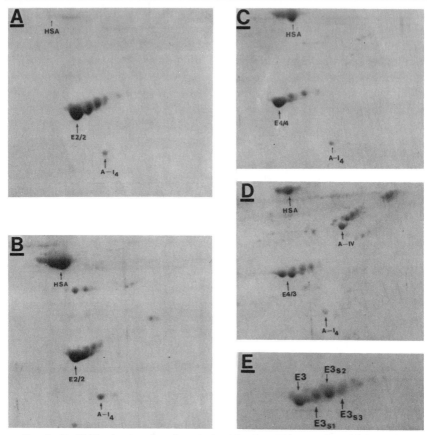

Fig. 8. (A–C) The isoelectric point relationship of apoE isoproteins with human serum albumin (HSA). (D) Protein contaminants found in VLDL obtained by a single ultracentrifugation. HSA focuses in the apoE4 and apoE3 region. Although HSA and the other protein contaminants do not affect the two-dimensional patterns, it may cause artifacts in the one-dimensional patterns. For instance, the pattern of B and C would mimic an E3/2 and E4/3 phenotype, respectively, on one-dimensional isoelectric focusing gels. (E) A pattern with increased sialation of apoE. This pattern occurs in patients with Tangier disease.[17] The pattern mimics an E3/1 phenotype on one-dimensional isoelectric focusing gels.

a lyophilized form. (2) Prolonged storage of plasma causes degradation of apoE. This degradation may lead to a total disappearance of the apoE isoproteins or it may generate fragments of various sizes (Fig. 9E). Limited degradation produces patterns like those shown in Fig. 9F and G. These patterns, in addition to the main apoE isoprotein, contain one apoE fragment of 35K which differs by three positive charges from the original apoE isoproteins.

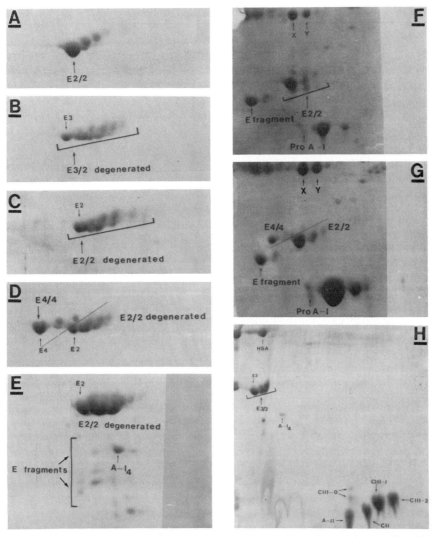

FIG. 9. (A–H) Alterations in the apoE patterns caused by aging of the samples. (A) Normal E2/2 pattern obtained from fresh VLDL. (B and C) Degenerated E3/2 and E2/2 patterns, respectively. The degeneration has been caused by storage of VLDL for 2 weeks at 4°. Note the smearing of the patterns in B and C and the generation of more acidic isoproteins. (D) A mixture of degenerated E2/2 with normal E4/4 phenotype. (E) Degenerated E2/2 phenotype and degradation products of apoE. (F) Partial degradation of apoE. The degradation was caused by storage of plasma at 4° for 10 days. (G) A mixture of normal E4/4 phenotype with the sample analyzed in F. This panel shows that the major degradation product of apoE has an apparent MW of 35K and differs by approximately +3 charges from the apoE isoprotein of origin. (H) A shift of the apolipoprotein patterns toward the cathode. This shift results from incomplete dialysis of the samples. Note the compression of the basic proteins (apoE, HSA) and the improved resolution of the acidic apolipoproteins (apoCs and apoA-II).

Finally increased salt concentration in the lyophilized sample, due to incomplete dialysis of the sample, causes a shift of the apolipoprotein patterns toward the cathode (Fig. 9H). This results in the compression of the apoE and HSA patterns and an improved resolution of the apoCs and apoA-II.

Recommended Methods

The fastest approach for the determination of the apoE phenotype in our laboratory is the immunoprecipitation of the apoE from plasma followed by 2D-PAGE analysis of the sample. The proapoA-I (added to the sample in the form of HDL or immunoprecipitated apoA-I) and proteins X and Y in the heavy chain region of IgG allow, in most instances, accurate determination of the apoE phenotypes (Figs. 5 and 6A–H). For laboratories where the 2D-PAGE is impractical, the patterns may be obtained by one-dimensional isoelectric focusing using well-purified VLDL samples. We recommend that whenever there is ambiguity in the patterns determined by one- or two-dimensional techniques, that these patterns be verified by appropriate mixing experiments (Fig. 3A–H). Finally, it is important to use fresh samples for VLDL fractionation or for apoE immunoprecipitation. When VLDL is not used immediately, it should be stored in lyophilized form at $-20°$. The VLDL samples used should be dialyzed overnight against water to achieve good separation of the apoE isoproteins.

The Genetic Basis of ApoE Polymorphism

The Genetic Model of ApoE Inheritance

The observation that the heterozygous phenotypes could be mimicked by mixing two different homozygous phenotypes (Fig. 3B and C) suggested that the apoE phenotypes were genetically determined and that the different phenotypes represented heterozygosities and homozygosities for different apoE alleles. To substantiate this hypothesis and determine the genetics of human apoE, we have studied the inheritance of apoE phenotypes in 34 families where both parents and a total of 84 children were phenotyped. These studies are compatible with the following genetic model (Fig. 10A). The common apoE phenotypes are specified at a single structural gene locus with three common alleles: $\varepsilon 4$, $\varepsilon 3$, $\varepsilon 2$. Individuals homozygous for alleles $\varepsilon 4$, $\varepsilon 3$, and $\varepsilon 2$ have the apoE phenotypes E4/4, E3/3, and E2/2, respectively. Individuals heterozygous for apoE alleles

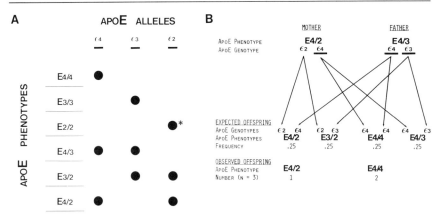

FIG. 10. (A) Schematic presentation of the single structural locus three-allele model of apoE inheritance. Closed circles represent the major asialo apoE isoproteins. (B) Schematic presentation of the inheritance of apoE phenotypes based on the single structural locus three-allele model of apoE inheritance.

have the apoE phenotypes E4/3, E3/2, and E4/2 which correspond to genotypes ε4/ε3, ε3/ε2, and ε4/ε2, respectively. A representative example of inheritance of apoE phenotypes is shown in Fig. 10B.

Frequency of ApoE Alleles and Phenotypes

We have analyzed the apoE phenotypes of 152 normal volunteers (individuals without hyperlipidemia or atherosclerosis). The phenotype frequencies were E4/4 = 0.03, E3/3 = 0.58, E2/2 = 0.013, E4/3 = 0.14, E3/2 = 0.22, E4/2 = 0.02. Therefore, in this population, the apoE alleles occurred with frequencies of $\varepsilon 4 = 0.12$, $\varepsilon 3 = 0.76$, $\varepsilon 2 = 0.13$.[32] If we assume a Hardy-Weinberg distribution of the apoE alleles, the apoE phenotype frequencies should be E4/4 = 0.013, E3/3 = 0.58, E2/2 = 0.018, E4/3 = 0.18, E3/2 = 0.20, E4/2 = 0.03. The genetic model we proposed has been verified in independent studies by several other groups.[47–50] The apoE allele frequencies have also been assessed by others. In Germany, apoE phenotypes were assessed in 1031 blood bank donors and the apoE allele frequencies were $\varepsilon 4 = 0.15$, $\varepsilon 3 = 0.77$, $\varepsilon 2 = 0.08$.[47] An analysis of 426 blood bank donors in New Zealand gave apoE allele frequencies of $\varepsilon 4 = 0.16$, $\varepsilon 3 = 0.72$, $\varepsilon 2 = 0.12$.[48] The minor differences in apoE allele frequencies among these studies may be due to selection bias or may reflect true genetic differences between the populations studied.

ApoE Phenotypes Have Resulted from Structural Mutations in the ApoE Gene

Amino acid and DNA sequence analysis of apoE protein and gene, respectively, have shown that the apoE phenotypes are the result of structural mutations in the apoE gene. The protein sequence studies have identified five polymorphic amino acid sites at positions 112, 127, 145, 146, and 158. In the most common apoE polypeptide (apoE3) specified by the ε3 allele, these sites have the following amino acids: 112 Cys, 127 Gly, 145 Arg, 146 Lys, and 158 Arg.[28] The polypeptide one charge unit more basic than apoE3, specified by the ε4 allele, differs from apoE3 by a 112 Cys to Arg substitution.[29,51] The apoE isoproteins one charge unit more acidic than apoE3 differ from apoE3 by either a 158 Arg to Cys, a 145 Arg to Cys or a 146 Lys to Gln substitution. These alleles have been designated ε2, ε2*, and ε2**, respectively.[29,51] Individuals may possess the E2/2 phenotype with any combination of the ε2, ε2*, and ε2** alleles. These data show that within a given apoE phenotype there can be genetic heterogeneity. This concept is further supported by other data that show the association of apoE phenotypes other than E2/2 with type III HLP.[32] Finally, the apoE1 isoprotein differs from apoE3 by a 127 Gly to Asp and a 158 Arg to Cys substitution.[42] In addition to structural mutations, other significant variations in the apoE gene are suggested by the observation of an individual with undetectable amounts of plasma apoE who presented with a type III HLP phenotype.[52]

In recent studies, DNA sequence analysis of apoE cDNA and phage genomic clones carrying the apoE gene supports the concept that the apoE alleles are due to mutations in the apoE structural gene. These studies showed that at the polymorphic amino acid sites, the codons specifying the most common apoE allele ε3 were residue 112 TGC, 127 GGC, 145 CGT, 146 AAG, and 158 CGC. A single base substitution in the first or second nucleotide of these codons could account for the amino acid substitutions observed by protein sequence analysis of the variant apoE alleles.[53] The specific nucleotide and amino acid changes which lead to the different apoE phenotypes and genotypes are shown in Fig. 11.

---

[51] S. C. Rall, K. H. Weisgraber, T. L. Innerarity, T. P. Bersot, and R. W. Mahley, *J. Clin. Invest.* **72,** 1288 (1983).

[52] G. Ghiselli, E. J. Schaefer, P. Gascon, and H. B. Brewer, Jr., *Science* **214,** 1239 (1981).

[53] J. L. Breslow, J. McPherson, A. L. Nussbaum, H. W. Williams, F. Lofquist-Kahl, S. K. Karathanasis, and V. I. Zannis, *J. Biol. Chem.* **257,** 14639 (1982).

ALLELE	CODON AT THE POLYMORPHIC SITE					BASE SUBSTITUTION RELATIVE TO ε3 ALLELE	AMINO ACID SUBSTITUTION RELATIVE TO E3/3 PHENOTYPE
	112	127	145	146	158		
ε3	TGC	GGC	CGT	AAG	CGC	NONE	NONE
ε4	CGC	SAME	SAME	SAME	SAME	T C	112 Cys→Arg
ε2	SAME	SAME	SAME	SAME	TGC	C T	158 Arg→Cys
ε2*	SAME	SAME	TGT	SAME	SAME	C T	145 Arg→Cys
ε2**	SAME	SAME	SAME	CAG	SAME	A C	146 Lys→Gln
ε1	SAME	GAC	SAME	SAME	TGC	{G A, C T}	[127 Gly→Asp, 158 Arg→Cys]

Diagram of ApoE Sequence Variations:

- ε3: Cys-112 — Gly-127 — Arg-145 — Lys-146 — Arg-158
- ε4: Arg-112 — Gly-127 — Arg-145 — Lys-146 — Arg-158
- ε2: Cys-112 — Gly-127 — Arg-145 — Lys-146 — Cys-158
- ε2*: Cys-112 — Gly-127 — Cys-145 — Lys-146 — Arg-158
- ε2**: Cys-112 — Gly-127 — Arg-145 — Gln-146 — Arg-158
- ε1: Cys-112 — Asp-127 — Arg-145 — Lys-146 — Cys-158

FIG. 11. Schematic presentation of the nucleotide and amino acid substitutions which correspond to the different apoE alleles.

Association of ApoE Phenotype E2/2 with Type III HLP

Familial type III HLP, also called familial dysbetalipoproteinemia, broad β, or floating β disease, is characterized by premature atherosclerosis, xanthomas, elevated cholesterol and triglyceride levels, cholesterol enriched βVLDL and IDL particles, and increased plasma apoE levels.[33,54–58] The most reliable criterion used in the past for diagnosis of this disease was an increase in the ratio of VLDL cholesterol to total triglyceride ($r \geq 0.30$) and a triglyceride concentration between 150 and 1000 mg/dl.[56,59,60] The frequency of the disease was estimated to be 0.1 to 0.01% in the population.[33,61]

In 1975, Utermann and colleagues described an apparent apoE4 and apoE3 isoprotein deficiency in patients with type III hyperlipoproteinemia.[30,31] Subsequent work established that the apoE phenotype observed in type III HLP patients (E2/2) is not deficient in any isoprotein but rather is shifted to a more acidic isoelectric point as a result of a structural mutation that changes the net charge of apoE.[14,15,40] In a recent study, we determined the clinical symptoms and lipoprotein patterns in 17 individuals with type III HLP and their relatives and spouses, and used the apoE phenotype E2/2 as a molecular marker to study the transmission as well as the phenotypic expression of the disease.[32] We found that the apoE phenotype E2/2 occurred in 15 type III HLP probands and the apoE phenotype E4/2 in 2 probands. In another study still in progress of the families of 17 additional probands, we found that the E2/2 phenotype occurred in 16 and the E3/2 phenotype in one proband. Thus, in these studies, the apoE phenotype E2/2 was found in 91% of probands with type III HLP where it would be expected to occur in only 1 to 2% of the normal population.[32] These studies and others show a very strong association of the E2/2 phenotype with type III HLP. However, there are individuals who apparently have this condition who do not have this apoE2/2 phenotype. It is reasonable to expect that some of these patients have other mutations in a functional domain of apoE which do not change the net charge of this apolipoprotein, but affect its biological function(s).

[54] D. S. Fredrickson, R. I. Levy, and R. S. Lees, *New Engl. J. Med.* **276**, 34 (1967).
[55] D. S. Fredrickson, R. I. Levy, and F. R. Lindgren, *J. Clin. Invest.* **47**, 2446 (1968).
[56] W. R. Hazzard, D. Forte, Jr., and E. L. Bierman, *Metabolism* **21**, 1009 (1972).
[57] C. B. Blum, L. Aron, and R. Sciacca, *J. Clin. Invest.* **66**, 1240 (1980).
[58] R. Havel, J. L. Kotite, J. L. Vigne, J. P. Kane, J. Tun, N. Phillips, and G. C. Chen, *J. Clin. Invest.* **6**, 1351 (1980).
[59] M. Mishkel, D. J. Nazir, and S. Crother, *Clin. Chim. Acta* **58**, 121 (1975).
[60] D. S. Fredrickson, J. Morganroth, and R. I. Levy, *Ann. Intern. Med.* **82**, 150 (1975).
[61] J. A. Morrison, K. Kelly, R. Horwitz, P. Khoury, P. M. Laskarzewski, M. J. Mellies, and C. J. Glueck, *Metabolism* **31**, 158 (1982).

Reduced Binding to Lipoprotein Receptors of ApoE Derived from Individuals with the E2/2 Phenotype May Underlie Type III HLP

ApoE is one of the apolipoproteins which, as a component of lipoprotein particles and liposomes, binds to cell surface receptors and mediates the catabolism of lipoproteins by hepatic and extrahepatic tissues.[18-23] Turnover studies showed that the catabolism of ^{125}I-labeled apoE derived from an individual with type III HLP and the E2/2 phenotype was slower than when apoE from normal individuals was used.[24,25] In addition to these *in vivo* studies, extensive experiments involving fibroblast cultures and membrane preparations have shown that apoE of different phenotypes in phospholipid complexes display variable degrees of competition for the LDL receptor.[26] ApoE from individuals with the E3/3 and E4/4 phenotypes display the same competition for the LDL receptor as described previously.[27] However, apoE derived from individuals with the E2/2 phenotype do not compete as well for the LDL receptor.[28] Mahley and colleagues have noted functional heterogeneity in this regard within the E2/2 phenotype.[28] These studies have shown that apoE which has arisen from a 158 Arg to Cys substitution competes very inefficiently, whereas apoE which has arisen from a 145 Arg to Cys substitution competes almost normally for the LDL receptor.[28] These binding experiments are consistent with earlier observations showing accumulation of remnant lipoproteins in the plasma of patients with type III HLP which are enriched in cholesteryl esters and apoE.[57,58,62-64] These apoE-rich lipoprotein remnants are apparently the result of slow clearance *in vivo* of the apoE-containing lipoproteins due to the described structural defect in apoE.

Factors Affecting the Phenotypic Expression of Type III HLP

Numerous studies suggest that the E2/2 phenotype alone is not sufficient to cause type III HLP and that other genetic factors may be required for the phenotypic expression of the disease.[32,33,65-72] All of these studies

---

[62] W. R. Hazzard and E. L. Bierman, *Metabolism* **25**, 777 (1976).
[63] A. Chait, W. R. Hazzard, J. J. Albers, R. P. Kushwaha, and J. D. Brunzell, *Metabolism* **27**, 1055 (1978).
[64] A. Chait, J. D. Brunzell, J. J. Albers, and W. R. Hazzard, *Lancet* **1**, 1176 (1977).
[65] D. S. Fredrickson and R. I. Levy, in "The Metabolic Basis of Inherited Disease" (J. B. Stanbury, J. B. Wyngaarden, and D. S. Fredrickson, eds.), 3rd Ed., p. 545. McGraw-Hill, New York, 1972.
[66] J. Morganroth, R. I. Levy, and D. S. Fredrickson, *Ann. Intern. Med.* **82**, 158 (1975).
[67] W. R. Hazzard, T. F. O'Donnell, and Y. L. Lee, *Ann. Intern. Med.* **82**, 141 (1975).
[68] J. J. C. Marien, H. A. M. Hulsmans, and C. M. van Gent, *Acta Med. Scand.* **196**, 149 (1974).

indicate that, in most cases, type III HLP is expressed in families with a tendency to hypertriglyceridemia based on environmental or other genetic factors or some combination of the two. The concept that other genetic and/or environmental factors may be required for the expression of type III HLP has received direct biochemical support from recent studies of Rall et al.[73] These investigators have shown that apoE from subjects with the apoE2/2 phenotype who are normolipidemic, and even hypolipidemic, behave the same in competition experiments and, presumably, has the same molecular defect as apoE from patients with the E2/2 phenotype (158 Cys) who express type III HLP. In conclusion, the genetic and biochemical data involving human apoE cannot completely account for the lipid and lipoprotein abnormalities observed in patients with type III HLP. Future studies should be directed toward other genetic and environmental factors which trigger the onset of this disease. Such studies will require a long-term follow-up of asymptomatic subjects with the E2/2 phenotype in order to assess what environmental factors may trigger the onset of the disease. In addition, it will require suitable genetic markers to assess the contribution of other monogenic hyperlipidemias to the expression of type III HLP.

Summary

1. This chapter provides the methodologies employed to study the polymorphism of human apoE.

2. These and other related studies have advanced our understanding of the structure and function of this protein as follows:
   a. The complex array of human apoE observed by two-dimensional gel electrophoresis results from genetic variation and posttranslational modification.
   b. The genetic polymorphism of apoE is explained by the existence of three common alleles ($\varepsilon 4$, $\varepsilon 3$, $\varepsilon 2$) at a single structural gene locus. Combinations of above alleles can generate three homozygous (E4/4, E3/3, E2/2) and three heterozygous (E4/3, E3/2, E4/2) apoE phenotypes.

---

[69] P. O. Kwiterovich, C. Neill, S. Margolis, M. Thamer, and P. Bachorik, *Clin. Res.* **23**, 262A (1975).

[70] G. Utermann, N. Pruin, and A. Steinmetz, *Clin. Genet.* **15**, 63 (1979).

[71] G. Utermann, K. H. Vogelberg, A. Steinmetz, W. Schoenborn, N. Pruia, M. Jaeschke, M. Hees, and W. Canzler, *Clin. Genet.* **15**, 37 (1979).

[72] W. R. Hazzard, G. R. Warnick, G. Utermann, and J. J. Albers, *Metabolism* **30**, 79 (1981).

[73] S. C. Rall, Jr., K. H. Weisgraber, T. L. Innerarity, and R. W. Mahley, *J. Clin. Invest.* **71**, 1023 (1983).

c. The apoE phenotype E2/2 is found in 91% of patients with type III hyperlipoproteinemia and can be used as a molecular marker for the diagnosis of this disease. However, other rare or common apoE phenotypes have been observed in patients with type III HLP.
d. ApoE originating from E2/2 phenotype (Arg 158 to Cys 158 substitution) has reduced affinity for the LDL receptor. This property of apoE2 can account partially for the accumulation of apoE-rich lipoprotein remnants in the plasma of patients with type III HLP. However, other genetic or environmental factors are necessary for the phenotypic expression of the disease.

Acknowledgments

This work was supported by grants from the National Institutes of Health (HL32339), the March of Dimes Birth Defects Foundation (1-817), and the American Heart Association (83-963). Dr. Vassilis I. Zannis is an Established Investigator of the American Heart Association.

We would like to thank Mr. Marty Ross and Johanna vanderSpek for technical assistance and Ms. Anne Gibbons for typing this manuscript.

[50] Genetic Mapping of Apolipoprotein Genes in the Human Genome by Somatic Cell Hybrids

By FA-TEN KAO and LAWRENCE CHAN

There are undoubtedly large numbers of genes controlling mammalian lipoprotein metabolism. Some of these include the genes that determine the expression and primary structure, the intracellular processing, posttranslational modification, assembly, packaging, and secretion of the individual apolipoproteins. Finally, the catabolism and receptor- or nonreceptor-mediated uptake of the various apolipoproteins are also under genetic control. These genes interact with environmental factors in determining plasma lipoprotein levels and composition, and are important in the hereditary component of atherosclerosis.

Mapping of genes involved in lipoprotein metabolism by using pedigree analysis of families segregating relevant traits in man has made significant but relatively slow progress. For example, by such techniques, the gene for apoE has been linked to the complement component *C3* locus on chromosome 19[1]; the gene that is associated with familial hypercholesterolemia (*FHC*) has also been localized on chromosome 19.[2] A gene order of *FHC-C3-APOE* has been suggested in such studies.[3] Furthermore, Utermann *et al.*[4] found that genetic apoE phenotypes and the apoA-I mutant (apoA-I$_{\text{Marburg}}$) segregated independently, indicating that the structural gene loci for apoE and apoA-I are not linked.

While linkage analysis is well suited to studies in the mouse (see Chapter [52], this volume) it is time-consuming and labor intensive in man. Recently, advances in cytogenetics and somatic cell and molecular genetics have greatly facilitated gene mapping. Specifically, these recent innovations include the use of somatic cell hybrids, cloned eukaryotic genes, and *in situ* nucleic acid hybridization for chromosomal localization of specific genes, including various apolipoproteins. In this chapter, we will discuss the use of somatic cell hybrids in apolipoprotein gene mapping. In Chapter [51] the technique of *in situ* nucleic acid hybridization on metaphase chromosomes will be described in detail.

Introductory Remarks

By family studies and classical linkage analysis, between 1911 and 1967, while nearly 100 genes were shown to be sex linked, only a few autosomal genes were assigned to specific chromosomes. A breakthrough in human gene mapping occurred in 1967 by using a totally different approach, namely somatic cell hybrid method. Since then, rapid progress has been made in mapping of autosomal as well as X-linked genes. More than 800 genes have now been assigned to various human chromosomes.[5] The development of recombinant DNA technology has also made important contributions to gene mapping, especially by combining cell hybrids and cloned gene probes for chromosomal and regional assignment of many structural genes that may not be expressed in cultured cells or cell hybrids.

Establishment of Human–Rodent Somatic Cell Hybrids

Somatic cells grown in culture can be fused together either spontaneously at low frequency or induced by Sendai virus or polyethylene glycol.

[1] B. Olaisen, P. Teisberg, and T. Gedde-Dahl, Jr., *Hum. Genet.* **62**, 233 (1982).
[2] K. Berg and A. Heiberg, *Cytogenet. Cell Genet.* **22**, 621 (1978).
[3] K. Berg, J. O. Julsrud, A. L. Borresen, G. Fey, and S. E. Humphries, *Cytogenet. Cell Genet.* **37**, 417 (1984).
[4] G. Utermann, H. Steinmetz, R. Paetzgold, J. Wilk, G. Feussner, E. Kaffarnik, C. Mueller-Eckhardt, D. Seidel, K. H. Vogelberg, and F. Aimmer, *Hum. Genet.* **61**, 329 (1982).
[5] *Hum. Gene Map. 8, Cytogenet. Cell Genet.* **40**, 1 (1985).

The resulting cell hybrids usually retain chromosomes from one parental genome and lose chromosomes from the other genome. Weiss and Green were the first to demonstrate the preferential loss of human chromosomes from human–mouse hybrids.[6] Thus, it has become possible to identify human chromosomes that are retained or lost in the hybrids and to correlate them with the presence or absence of a specific human gene product. For example, if a particular human enzyme is always segregating together with a specific human chromosome, we may conclude that the gene coding for this enzyme is located on that human chromosome. When two genes are located on the same chromosome regardless of their distance and their locations on the same or different arms, they are termed "syntenic." The synteny analysis provides the essence of gene mapping using somatic cell hybrid method.[7]

Reagents

Saline G: Dissolve the following in 1 liter (final volume) of triple-distilled water, sterilize, and store at 4°: NaCl 8.0 g, KCl 0.4 g, $MgSO_4 \cdot 7H_2O$ 0.154 g, $CaCl_2 \cdot 2H_2O$ 0.016 g, $Na_2HPO_4 \cdot 7H_2O$ 0.29 g, KH_2PO_4 0.15 g, glucose 1.10 g, Phenol Red 0.0012 g[8]

Nonselective medium: F12[8]

Selective medium: if using one of the CHO-Kl auxotrophic mutants as parent, the selective medium is F12D.[8]

Protocol. A typical fusion experiment using Sendai virus as the fusing agent is as follows.[9–11]

1. Mix rodent mutant cells with normal human lymphocytes, 10^6 cells each, in a tube containing serum-free Saline G at a final volume of 0.5–1.0 ml.

2. Add 500 HAU (haemagglutinating units) of UV-inactivated Sendai virus and mix the tube well.

3. Place the tube at 4° for 20 min to facilitate cell clumping.

4. Place the tube at 37° for another 20 min for cell fusion.

5. Dispense cells into large dishes containing nonselective medium and allow growth for 2–3 days before changing to selective medium.

6. Frequently change medium to remove human lymphocytes.

7. Hybrid colonies develop after 10–20 days in selective medium and are isolated for further characterization.

[6] M. C. Weiss and H. Green, *Proc. Natl. Acad. Sci. U.S.A.* **58**, 1104 (1967).
[7] F. T. Kao, *Int. Rev. Cytol.* **85**, 109 (1983).
[8] F. T. Kao and T. T. Puck, *Methods Cell Biol.* **8**, 23 (1974).
[9] F. T. Kao, R. T. Johnson, and T. T. Puck, *Science* **164**, 312 (1969).
[10] F. T. Kao and T. T. Puck, *Nature (London)* **228**, 329–332 (1970).
[11] C. Jones, F. T. Kao, and R. T. Taylor, *Cytogenet. Cell Genet.* **28**, 181 (1980).

Human cells including fibroblasts can be selectively eliminated in the fusion experiment by adding 1×10^{-7} M ouabain which kills human cells but not rodent cells or cell hybrids.[12]

Characterization of Human Chromosome Content in Cell Hybrids

Although many assays can be performed to identify particular genetic markers that have been assigned to particular human chromosomes, the following two analyses are the most commonly and conveniently used, especially when large numbers of cell hybrids will have to be analyzed for each of the 24 different human chromosomes.

Isozyme Analysis

The identity of human chromosomes retained in the hybrids can be determined by the presence of an isozyme marker previously assigned to a specific human chromosome. All human chromosomes except the Y have at least one isozyme marker that is convenient to assay, thus making the isozyme analysis a very useful procedure for revealing the presence of particular human chromosomes in the hybrids.

The general procedure for isozyme analysis involves preparation of crude cell extracts from hybrid cells and their parents, application of cell extracts to Cellogel or starch gel electrophoresis for separating isozymes of different species origin, histochemical staining of the enzyme activity *in situ,* and identification of the human isozyme in the hybrids. The positive reaction of a human isozyme marker in the hybrids indicates the presence of a specific human chromosome, at least in part if not whole, to which the human isozyme marker is assigned. For large human chromosomes, it is helpful to assay for isozyme markers that are located near the distal end of each of the two arms to ensure the intactness of the chromosome. The Handbook by Harris and Hopkinson[13] on the isozyme analysis can be consulted for specifics and details of each individual isozyme assay. A typical isozyme analysis using Cellogel electrophoresis is described below.

Reagents

Saline G

Lysis buffer: 5 mM sodium phosphate buffer, pH 6.4 containing 1 mM Na$_2$EDTA, 1 mM dithiothreitol, and 20 μM NADP

[12] M. L. Law and F. T. Kao, *Somat. Cell Genet.* **4,** 465 (1978).
[13] H. Harris and D. A. Hopkinson, "Handbook of Enzyme Electrophoresis in Human Genetics." North Holland Publ., Amsterdam, 1976.

Protocol

1. Grow cells to large quantities and harvest by trypsinization.
2. Count cell number.
3. Transfer cells to Eppendorf centrifuge tube and wash cells once with Saline G.
4. Centrifuge and remove supernatant. Keep cell pellet in freezer ($-79°$) if not used immediately.
5. To prepare cell extract, add deionized water to cell pellet to make final cell density of 10^8 cells per ml. In some cases, use "lysis buffer" instead of water for better protection of enzymes in the cell extract.
6. Freeze and thaw 3 times in dry ice/acetone bath. Mix cell pellet after each thawing. Centrifuge. Use supernatant as crude cell extract for isozyme assay.
7. Apply cell extract to Cellogel (Kalex Scientific Co., Manhasset, NY) using sample applicator (Gilman).
8. Run electrophoresis using buffers, voltage, and running time according to each isozyme.[13-15] Usually the gel is electrophoresed at 200 V for 2–4 hr.
9. For histochemical staining of the enzyme activity, mix equal volumes of staining solution and 1% agar gel. Place Cellogel on staining agar. Incubate at 37° for appropriate time to develop the stain. Photograph if desired.
10. Wash off agar from gel. Store gel in 5% acetic acid.
11. The histochemical staining solution is prepared fresh according to each isozyme.[13-15] For the isozymes that are revealed by staining with fluorescent reagent, gels are placed under UV illuminator for visualization and photography.

Cytogenetic Analysis

Chromosome banding techniques[16] have been used to identify human chromosomes in the hybrids. The addition of Giemsa-11 differential staining techniques[17,18] has increased the reliability of the identification of certain human chromosomes retained in cell hybrids. This technique is particularly valuable in revealing translocations between human and rodent

---

[14] P. Meera Khan, *Arch. Biochem. Biophys.* **145**, 470 (1971).
[15] H. Van Someren, H. B. Ban Henegouwen, W. Los, E. Wurzer-Figurelli, B. Doppert, M. Vervloet, and P. Meera Khan, *Humangenetik* **25**, 189 (1974).
[16] T. Caspersson, L. Zech, and C. Hohansson, *Exp. Cell Res.* **60**, 315 (1970).
[17] M. Bobrow, K. Madan, and P. L. Pearson, *Nature (London) New Biol.* **238**, 122 (1972).
[18] B. Alhadeff, M. Velivasakis, and M. Siniscalco, *Cytogenet. Cell Genet.* **19**, 236 (1977).

chromosomes,[19] and also in confirming human chromosomes with specific deletions. The power of cytogenetic analysis can be further increased by performing both chromosome banding and Giemsa-11 staining techniques on the same chromosome slide preparations.[20] In this sequential staining procedure, the chromosome slide is treated with trypsin banding and Giemsa-11 differential staining in sequential steps and the human chromosomes (intact, partial, or translocated) can be reliably identified by both staining techniques. Cytogenetic analysis of the human chromosome content in the hybrids not only can identify translocations and the intactness of human chromosomes, but also reveals the proportion of hybrid cells that contains a particular human chromosome. Such information is helpful in making reliable synteny analysis. The various cytogenetic techniques used in determining human chromosome content in cell hybrids are briefly described in the following.

Reagents

Stock trypsin solution: dissolve 1.25 g of Difco 1 : 250 trypsin in 200 ml of deionized water, adjust to pH 2.5 with 5 N HCl; store at $-10°$

Giemsa staining solution for trypsin banding: add 2 ml of Gurr's improved Giemsa to 50 ml of Gurr's buffer solution (1 Gurr's buffer tablet, pH 6.8, to 1 liter of deionized water)

Stock Giemsa-11 staining solution: use Giemsa stain powder (Fisher G-146) and prepared in glycerol according to the procedures of Alhadeff *et al.*[18] Store in a brown bottle, sealed tightly and wrapped in aluminum foil

Protocol: Trypsin Banding Technique

1. Air-dried chromosome slides are immersed in clean Coplin jars containing working trypsin solution for 5–20 sec. Working trypsin solution is prepared as 1% stock trypsin solution.
2. Rinse twice in Gurr's buffer solution.
3. Stain with Giemsa staining solution for 2–5 min.
4. Rinse in deionized water. Air dry.

Alkaline Giemsa-11 Differential Staining Technique

1. Air-dried chromosome slides are soaked for 2 hr in distilled water at 60°.
2. Slides are immersed in the staining solution for a specified time. The staining solution is prepared by adding 3 ml of stock Giemsa-11

---

[19] K. K. Friend, S. Chen, and F. H. Ruddle, *Somat. Cell Genet.* **2**, 183 (1976).
[20] H. G. Morse, D. Patterson, and C. Jones, *Mamm. Chromosomes Newslett.* **23**, 127 (1982).

staining solution to 47 ml of 0.05 M Na$_2$HPO$_4$ buffer, pH 11.3, previously warmed at 37°. Wipe off the precipitate on the surface of diluted stain with tissue paper.

3. At the end of staining, the Coplin jar with slides is placed under running distilled water to flush away the stain.

4. The slides are removed and dried by the air jet. The chromosome slides are prepared according to usual procedures but at low density (50 or fewer cells per low-power field). The age of the slide is not important, but the metaphases should be well spread. After staining, chromosomes that are understained appear uniformly blue and those overstained appear uniformly magenta. At a critical point of differential staining, human chromosomes stain a light blue and rodent chromosomes magenta. Some human chromosomes may display bright red regions at centromeres and secondary constriction regions.

Sequential Staining Technique

1. Photograph well-banded metaphase cells from trypsin-banding slides and record the location.
2. Destain the slide in 3 : 1 methanol : acetic acid.
3. Soak the slide in distilled water for 2 hr at 60°.
4. Restain with Giemsa-11 differential staining procedure as previously described.

Construction of Hybrid Clone Panel

Unless a particular gene to be mapped can only be assayed in hybrids derived from particular combinations of special parental cells, a general set of hybrids can be assembled whose human chromosome content has already been reliably determined. This hybrid clone panel can be used in a very general and convenient way to assign many genes to specific human chromosomes.

Theoretically, five hybrids with unique combinations of all human chromosomes are adequate for this purpose.[21] In practice, however, additional hybrids are analyzed to substantiate the assignment. Ideally, final confirmation can come from cell hybrids that contain only one human chromosome of interest.[22,23] Due to the relative instability of various human chromosomes retained in different hybrids, it is necessary to monitor the hybrids regularly to ensure the correct human chromosome content. It

---

[21] F. H. Ruddle and R. P. Creagan, *Annu. Rev. Genet.* **9,** 407 (1975).
[22] F. T. Kao, C. Jones, and T. T. Puck, *Proc. Natl. Acad. Sci. U.S.A.* **73,** 193 (1976).
[23] E. C. Lai, F. T. Kao, M. L. Law, and S. L. C. Woo, *Am. J. Hum. Genet.* **35,** 385 (1983).

should also be done with hybrids that are thawed from frozen stocks. If the proportion of a particular human chromosome in a hybrid falls to a low level, for example, 10–20%, this hybrid will not be used in synteny analysis for that particular chromosome. For the human chromosomes that can be stably retained under selective conditions, such as auxotrophic markers[24] or the HPRT marker,[25] periodic cytogenetic analysis will ensure the intactness of the retained human chromosomes. When the hybrids are grown to large quantities for preparing DNA or cell extracts, it is important to monitor the human chromosome content in the hybrid cells at the final stage of expansion. This analysis will reveal the representation of the human chromosomes in the DNAs or cell extracts prepared for mapping uses.

Use of Hybrid Clone Panel for Gene Mapping

The following describes mapping procedures using hybrid clone panels that have been established in a number of laboratories.

Gene Mapping by Assaying Gene Products

Gene mapping by assaying the gene products in the hybrids has been generally used before cloned genes become available. The prerequisite for such analysis is that the species difference of the gene products can be demonstrated by techniques including those that depend on certain physical or chemical properties of the enzymes or proteins, affinity for specific antibodies, etc. Since many tissue-specific functions tend to be turned off in cultured cells or in cell hybrids,[26] specific induction systems or special parental cell types will have to be devised for mapping studies. However, gene mapping by assaying gene products offers unique opportunities for mapping genes that have regulatory functions, and also for mapping genes for which cloned gene probes are not yet available.

Mapping of Structural Genes Using Cloned Gene Sequences and Southern Blot Analysis

Since the first human gene coding for β-globin was isolated by recombinant DNA techniques in 1977, large numbers of cloned human gene sequences, either genomic or cDNA, have been used for gene mapping purposes. The development of Southern blotting method has greatly in-

---

[24] T. T. Puck and F. T. Kao, *Annu. Rev. Genet.* **16**, 25 (1982).
[25] J. W. Littlefield, *Science* **145**, 709 (1964).
[26] N. R. Ringertz and R. E. Savage, "Cell Hybrids." Academic Press, New York, 1976.

creased the sensitivity and accuracy in hybridization experiments involving gene probes and the total genomic DNA. Using this method, the first apolipoprotein gene mapped to a specific human chromosome is apolipoprotein A-I (apoA-I), in the long arm of chromosome 11 region 11q13-qter.[27] By virtue of close proximity of the genes apoA-I and apoC-III, the latter can also be assigned to the same chromosomal region as apoA-I.

Restriction Enzyme Digestion, Electrophoretic Separation, and Southern Blot Analysis of DNA Fragments. In a typical mapping analysis using cloned genes and the hybrid clone panel, a suitable restriction enzyme is first selected which can produce restriction fragments of different lengths for human and rodent genomic DNA. This enzyme will be used to digest genomic DNA from the hybrid clone panel. The following steps are generally included in a Southern blot hybridization experiment for synteny analysis.

Reagents

Restriction enzyme buffers (10×): these are prepared according to manufacturer's recommendations
SSC (standard saline citrate): 150 mM NaCl, 15 mM Na-citrate, pH 7.0. Tris-acetate buffer: 40 mM Tris-acetate, pH 8.0, containing 2 mM Na$_2$EDTA
SDS (sodium dodecyl sulfate, specially pure from BDH Chemicals. Distributed in the U.S. by Gallard-Schlesinger Chemical Manufacturing Company)
Denhardt's reagent (50×): add to 500 ml of H$_2$O Ficoll 5 g, poly(vinylpyrrolidone) 5 g, BSA 5 g; filter and store at −20°

Protocol

1. Restriction enzyme digest of hybrid DNA: Hybrid cell DNA is digested under conditions recommended by the suppliers. An excess of enzyme (e.g., 3–5 units/μg DNA, incubation for 6 hr) is used to ensure complete digestion. λ phage DNA can be included as internal controls.
2. Apply 10 μg of digested DNA per hybrid to 0.75% agarose gel.
3. Run electrophoresis overnight in Tris-acetate buffer at 30 V.
4. Stain the gel with ethidium bromide (10 μg/ml) to visualize the digested DNA with UV light.
5. Transfer the DNA from gel to nitrocellulose filter by the method of Southern.[28]

---

[27] P. Cheung, F. T. Kao, M. L. Law, C. Jones, T. T. Puck, and L. Chan, *Proc. Natl. Acad. Sci. U.S.A.* **81,** 508 (1984).
[28] E. Southern, *J. Mol. Biol.* **98,** 503 (1975).

6. Nick-translate the cDNA or genomic DNA probe with [α-^{32}P]dCTP to high specific activity (greater than 10^8 cpm per μg DNA).[29]

7. Prehybridize the filter in a sealed plastic bag at 42° for at least 4 hr or overnight. The prehybridization solution includes 50% formamide, 5× Denhardt's reagent, 0.05 M Na-phosphate, pH 6.5, 500 μg/ml denatured salmon sperm DNA, 5× SSC.

8. Hybridize the filter by adding the labeled probe (heated to denature) to the plastic bag containing freshly prepared 50% formamide, 1× Denhardt's reagent, 0.02 M Na-phosphate, pH 6.5, 100 μg/ml denatured salmon sperm DNA, 5× SSC, 10% dextran sulfate.[30] Continue hybridization for 16–24 hr at 42°.

9. Carefully remove the filter from the bag. Rinse in 2× SSC + 0.1% SDS.

10. Wash the filter three times at room temperature at 5 min each in 2× SSC + 0.1% SDS, followed by another three washes at 55° at 15 min each in 0.1× SSC + 0.1% SDS. Finally, wash the filter in 0.1× SSC at 55° until virtually no radioactivity is detectable.

11. Dry the filter on 3MM paper for at least 1 hr before exposed to Dupont Cronex-4 X-ray film at −70° for 16 hr to 1 week, with a Dupont Lightning Plus intensifying screen. As an example, Figs. 1 and 2 show the results of Southern blot analysis of the human apoA-I gene.[27]

Interpretation of Results. In the Southern analysis, if a lane containing a specific hybrid DNA exhibits both human and rodent hybridization bands, it is interpreted as possessing a specific human chromosome which contains the hybridizable gene probe sequences. The hybrid exhibiting only the rodent band(s) should possess no such human chromosome. This analysis is analogous to that used in isozyme assays to establish synteny relationships. It should be pointed out that due to the presence of only one of the two homologous chromosomes that is usually retained in the hybrids, the hybridization intensity of the human bands detected in cell hybrids may be weaker than that observed in the human parental DNA which contains two copies of the gene. If heterologous probes are used, less stringent hybridization conditions and longer exposure time of the X-ray film may be helpful in developing hybridization bands with human DNA sequences.

Pros and Cons of DNA Blotting versus Gene Product Analysis. The unique features in gene mapping using cloned gene probes and Southern blot analysis include the following: (1) the structural gene is mapped in the human genome; (2) gene probes from any species can be used provided

---

[29] P. Rigby, M. Dieckman, C. Rhodes, and P. Berg, *J. Mol. Biol.* **113**, 237 (1977).
[30] G. M. Wahl, M. Stern, and G. R. Stark, *Proc. Natl. Acad. Sci. U.S.A.* **76**, 3683 (1979).

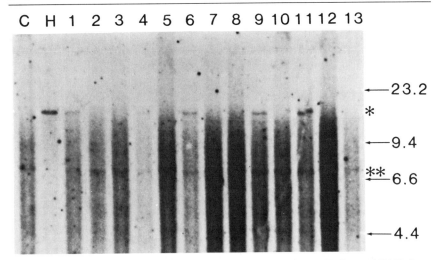

FIG. 1. Hybridization of cloned human apoA-I cDNA to HincII-digested DNA from human/Chinese hamster somatic cell hybrids and their parental cells. Lanes: C, Chinese hamster cell CHO-K1; H, human cell HT-1080; 1–13, HincIII-digested DNA from various somatic cell hybrids. Molecular weight standards consisted of HindIII-digested λ phage DNA. *Position of human-specific band hybridizing to the ^{32}P-labeled DNA probe; **position of CHO-specific band cross-hybridizing to the probe. Lanes 1, 4, 6, 9, and 11 represent positive hybrids showing the presence of the human band as defined in Lane H. From Cheung et al.[27]

that they cross-hybridize to the human gene sequences; (3) gene mapping can be carried out regardless of its expression in cell hybrids; (4) random human DNA fragments with unknown functions can also be mapped in the human genome, and if the random DNA fragments are polymorphic, they can be used as genetic markers for establishing linkage relationships with human traits or inherited diseases.[31] However, gene mapping using live cell hybrids for assaying gene products will continue to be useful, particularly in situations where gene probes are not yet available, and also for mapping genes or DNA sequences with regulatory functions.

Regional and Fine Structure Mapping

Cell hybrids containing well-defined deletions or translocations of a particular human chromosome can be used for regional mapping of genes assigned to that chromosome. However, such a practice will rely on the

[31] D. Botstein, R. White, M. Skolnick, and R. Davis, Am. J. Hum. Genet. **32**, 314 (1980).

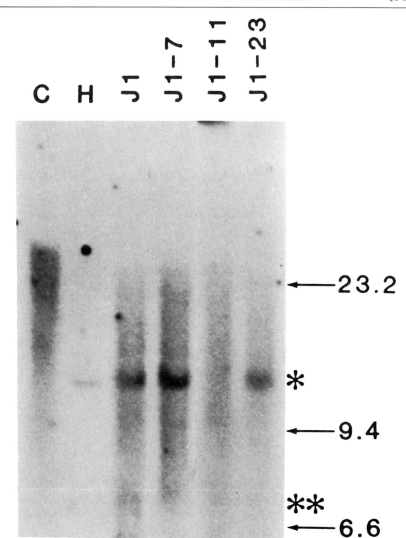

FIG. 2. Hybridization of cloned human apoA-I cDNA to HincII-digested DNA from cell hybrids containing intact human chromosome 11 (J1) or clones containing chromosome 11 with deletions (J1-7, J1-11, and J1-23). The specific chromosomal regions retained are J1-7, p11 → qter; J1-23, p13 → qter; and J1-11, pter → q13. Lanes: C, CHO-K1; H, human cell HT-1080. This autoradiogram represents a 24 hr exposure of the X-ray film to the filter. *Position of human specific band hybridizing to ^{32}P-labeled apoA-I cDNA; **position of CHO-specific band cross-hybridizing to the probe. The hamster band showed up on all the lanes except H on more prolonged exposure of the X-ray film (96 hr) to the filter. The lanes containing J1 (intact chromosome 11), J1-7, and J1-23 DNAs are positive, whereas the lane containing J1-11 is negative. This experiment permits the localization of human apoA-I/C-III gene complex to region q13 → qter of chromosome 11. From Cheung et al.[27]

availability of relevant structural changes of the chromosome. A more general method will be the use of *in situ* nucleic acid hybridization which can map a gene to a specific region of the chromosome limited only by the spread of the silver grain around the gene locus. This technique is described in Chapter [51] of this volume. For further refined mapping within a small region of the chromosome, other techniques need to be developed. The use of radiation-induced chromosome breakage for segregating closely linked genes as demonstrated by Goss and Harris[32] may provide estimates of relative distances and the order of genes that are so closely located that cytogenetics or *in situ* hybridization cannot resolve.[33] The use of species-specific, middle repetitive sequence probes, as demonstrated by Gusella *et al.*[34] and Law *et al.*,[35] can provide additional genetic markers for defining multiple sites on a particular chromosome. Such DNA sequence markers can be effectively used for fine structure mapping of specific regions of the chromosome in which various deletions produced by chromosomal breaks and localized in a small region may not be resolvable by current cytogenetic or *in situ* hybridization techniques.

Acknowledgments

Experiments performed in the authors' laboratories described in this chapter were supported in part by grants from the National Institutes of Health, HD-02080, HL-27341, and the March of Dimes Birth Defects Foundation.

[32] S. Goss and H. Harris, *Nature (London)* **255,** 680 (1975).
[33] S. Goss, *Cytogenet. Cell Genet.* **25,** 161 (1979).
[34] J. F. Gusella, C. Jones, F. T. Kao, D. Housman, and T. T. Puck, *Proc. Natl. Acad. Sci. U.S.A.* **79,** 7804 (1982).
[35] M. L. Law, J. N. Davidson, and F. T. Kao, *Proc. Natl. Acad. Sci. U.S.A.* **79,** 7390 (1982).

[51] Chromosomal Fine Mapping of Apolipoprotein Genes by *in Situ* Nucleic Acid Hybridization to Metaphase Chromosomes

By MARY E. HARPER and LAWRENCE CHAN

Over the past 7 years, classical genetic linkage analyses have localized only two genes related to lipoprotein metabolism, namely, the genes for familial hypercholesterolemia (FHC)[1] and for apolipoprotein E (apoE)[2] to

[1] K. Berg and A. Heiberg, *Cytogenet. Cell Genet.* **22,** 621 (1978).
[2] B. Olaisen, P. Teisberg, and T. Gedde-Dahl, Jr., *Hum. Genet.* **62,** 233 (1982).

human chromosome 19. Development of somatic cell hybrids (see Chapter [50]) and molecular cloning of specific apolipoprotein genes have accelerated the mapping of apolipoprotein genes considerably. In the last 2 years, as this review goes to press, at least five apolipoprotein genes have been localized to specific human chromosomes, namely, the apoA-I/C-III gene complex to chromosome 11,[3] apoE and apoC-II structural genes to chromosome 19,[2-7] and the apoA-II gene to chromosome 1.[5] In the case of apoA-I/C-III genes, regional mapping was possible by use of somatic cell hybrid lines containing specific deletions of human chromosome 11. Using such hybrids, Cheung et al. have localized the A-I/C-III gene complex to 11q13 → 11qter.[3] In addition, the low-density lipoprotein receptor (LDLR) has been mapped to chromosome 19 by somatic cell hybrid analysis.[6]

Somatic cell hybrids, including well-characterized lines containing chromosome deletions, thus constitute an important tool for mapping apolipoprotein genes. However, at present limited numbers of such somatic cell lines are available. Most are also unstable and unless carefully recharacterized periodically, misinterpretations might result from the use of such cell hybrids. Furthermore, the chromosomal regions defined by such deletions are generally fairly large and, unless multiple nonoverlapping deletions are available, do not allow for fine gene mapping. Therefore, while somatic cell hybrids represent a breakthrough in gene mapping, another technique, *in situ* nucleic acid hybridization to mitotic chromosome preparations, has been developed recently to localize single-copy genes (such as apolipoprotein genes). This method is rapid, generally allows higher resolution, and is a powerful tool for fine chromosomal gene mapping.

Introductory Remarks

Following the development of *in situ* hybridization methodology in 1969,[7,8] its application to mapping of a variety of repetitive sequences was

[3] P. Cheung, F. T. Kao, M. L. Law, C. Jones, T. T. Puck, and L. Chan, *Proc. Natl. Acad. Sci. U.S.A.* **81**, 508 (1984).
[4] C. L. Jackson, G. A. P. Bruns, and J. L. Breslow, *Proc. Natl. Acad. Sci. U.S.A.* **81**, 2945 (1984).
[5] M. N. Moore, F. T. Kao, Y. K. Tsao, and L. Chan, *Biochem. Biophys. Res. Commun.* **123**, 1 (1984).
[6] U. Francke, M. S. Brown, and J. L. Goldstein, *Proc. Natl. Acad. Sci. U.S.A.* **81**, 2826 (1984).
[7] J. G. Gall and M. L. Pardue, *Proc. Natl. Acad. Sci. U.S.A.* **63**, 378 (1969).
[8] H. A. John, M. L. Birnstiel, and K. W. Jones, *Nature (London)* **223**, 582 (1969).

quickly realized. However, single- or low-copy genes, such as structural genes for apolipoproteins, could not be detected. Factors responsible for the low sensitivity included low hybridization efficiencies and limitation to soft β-emitters such as ^{3}H and ^{125}I. In order to increase the sensitivity of detecting unique gene hybrids, it was found that the effective length of the probe molecules could be increased.[9] Dextran sulfate, an anionic polymer, greatly accelerates the renaturation of DNA in solution by volume exclusion.[10] Wahl et al. found that 10% dextran sulfate also accelerates the hybridization rate of DNA probes to immobilized nucleic acids and, furthermore, promotes formation of probe networks when the DNA is initially randomly cleaved and single stranded.[11] Thus, when DNA is radiolabeled by nick-translation[12] and subsequently denatured, partially complementary DNA fragments can reanneal to each other to form DNA networks.

In applying this approach to the detection of single-copy sequences on human mitotic chromosomes, cloned sequences were labeled with [^{3}H]dNTPs. Use of ^{3}H results in clean cytological backgrounds and probes with long half-lives, and it was found that adequate specific activities for relatively rapid detection could be obtained.[9] It was also determined that careful control of pH and temperature preserves chromosome morphology and allows G-banding of autoradiographed chromosomes by several series of staining–destaining–restaining with Wright stain.[9,13] Thus labeled, G-banded chromosomes can be visualized directly in the microscope and grain locations directly and accurately analyzed. The method is now a powerful tool for localizing single-copy and low-repeat DNA sequences in normal cells as well as in cells with abnormal karyotypes.

Mitotic Chromosome Preparation

For accurate analyses of well banded chromosomes, high quality preparations consisting of large numbers of well spread, reasonably contracted mitotic spreads are required. In studies utilizing normal chromosomes, high quality preparations can be consistently obtained from methotrexate-synchronized peripheral blood lymphocyte cultures.[14,15]

[9] M. E. Harper and G. F. Saunders, *Chromosoma* **83**, 431 (1981).
[10] J. G. Wetmur, *Biopolymers* **14**, 2517 (1975).
[11] G. M. Wahl, M. Stern, and G. R. Stark, *Proc. Natl. Acad. Sci. U.S.A.* **76**, 3683 (1979).
[12] P. W. J. Rigby, M. Dieckmann, C. Rhodes, and P. Berg, *J. Mol. Biol.* **119**, 237 (1977).
[13] M. E. Chandler and J. J. Yunis, *Cytogenet. Cell Genet.* **22**, 352 (1978).
[14] J. J. Yunis, *Science* **191**, 1268 (1976).
[15] J. J. Yunis and M. E. Chandler, *Prog. Clin. Pathol.* **7**, 267 (1977).

With this synchronization technique, the wave of mitotic cells may be collected approximately 5 hr after release from the block following short exposure to a low dose of colcemid. By use of appropriate harvest conditions, it is possible to observe a mitotic index of 8–12%, with the majority of mitotic cells in early metaphase and the remainder in prometaphase or in mid-metaphase. Early metaphase chromosomes are selected for (550 band stage; ISCN[16]) due to their ease of analysis, although prometaphase and prophase chromosomes (880 bands and higher), which are more difficult to obtain on a routine basis, result in higher resolution and more refined gene localization. Other methods have also been described for obtaining early mitotic chromosomes, such as those involving addition of ethidium bromide,[17] actinomycin D,[18] or BUdR[19] to cultures before harvest or BUdR synchronization.[20] However, comparison with chromosomes containing unsubstituted DNA is recommended since lower hybridization efficiencies have been observed with several of these procedures.

G-banding with Wright stain requires chromosome preparations that exhibit very low refractive qualities, i.e., "flat" and gray with no light halos when visualized by phase contrast microscopy. This type of preparation can be routinely obtained if relative humidity is considered during the spreading procedure, as described below.

Reagents

Heparin, preservative-free, 1000 units/ml
RPMI 1640 media
Fetal calf serum
Penicillin–streptomycin, 10,000 units/ml penicillin, 10,000 µg/ml streptomycin
Glutamine, 200 mM, frozen in aliquots
Hanks' balanced salt solution (BSS)
Methotrexate sodium parenteral (Lederle #4554)
  Prepare 10^{-5} M solution in Hanks' BSS and freeze in aliquots
Thymidine (Sigma #T-9250)
  Prepare 10^{-3} M solution in Hanks' BSS and freeze in aliquots

---

[16] ISCN (1981): An international system for human cytogenetic nomenclature—High resolution banding (1981). *Cytogenet. Cell Genet.* **31,** 1 (1981).
[17] T. Ikeuchi and M. Sasaki, *Proc. Jpn. Acad.* **55,** 15 (1979).
[18] J. Ryback, A. Tharapel, S. Robinett, M. Garcia, C. Mankinen, and M. Freeman, *Hum. Genet.* **60,** 328 (1982).
[19] J. V. Shah, R. S. Verma, J. Rodriguez, and H. Dosik, *J. Med. Genet.* **20,** 452 (1983).
[20] B. U. Zabel, S. L. Naylor, A. Y. Sakaguchi, G. I. Bell, and T. B. Shows, *Proc. Natl. Acad. Sci. U.S.A.* **80,** 6932 (1983).

Phytohemagglutinin, M form (Gibco #670-0576)
Colcemid, 10 µg/ml (Gibco #120-5210)
KCl
Methanol, 500 ml bottles
Acetic acid, 500 ml bottles

Protocol

1. Draw 3 ml peripheral blood in syringe containing 0.1 ml heparin. Mix by inversion for 5 min.
2. Prepare complete media consisting of RPMI 1640 media, 20% fetal calf serum, 1% penicillin–streptomycin, and 1% glutamine. For 4 pellets, prepare 40 ml complete media, add 32 µl heparin and 1.6 ml phytohemagglutinin, and distribute to 8 T-25 culture flasks.
3. Add 300 µl blood to each flask. Swirl to mix, cap tightly, and incubate in an upright position at 37° for 72 hr.
4. Swirl flasks to resuspend cells. Add 50 µl methotrexate (final 10^{-7} M) to each flask. Swirl again to mix, cap tightly, and return to 37° incubator for 17 hr.
5. Swirl flasks to mix. Pour contents of flasks into four sterile 15 ml, round-bottom culture tubes, two flasks per tube. Centrifuge at 200 g in clinical centrifuge (1100 rpm in tabletop model) for 8 min. Remove all but approximately 0.5 ml supernatant without disturbing pellets.
6. Swirl tubes lightly to resuspend cells and add 10 ml RPMI 1640 media at room temperature to each tube. Cap tubes, invert several times to resuspend completely.
7. Centrifuge again for 8 min. Repeat rinse.
8. After second rinse, resuspend contents of each tube in 10 ml complete media containing 1% thymidine (final 10^{-5} M). Invert to mix.
9. Pour contents of each tube into a new T-25 flask. Cap tightly and incubate in upright position at 37° for 5 hr to 5 hr 10 min. Since we recommend harvesting only 2 pellets at one time, we generally begin harvest of two pellets after 5 hr release time, and the other two flasks after 5 hr 10 min.
10. To begin harvest, swirl flasks to resuspend cells and add 60 µl colcemid to each flask. Swirl to mix and incubate at 37° for 10 min. This step as well as all subsequent steps should be timed exactly.
11. Pour contents of each flask into 15-ml disposable centrifuge tube. Spin at 200 g for 8 min.
12. Remove supernatant, leaving about 0.5 ml. Gently tap each tube several times to partially resuspend cells. Add 8 ml 75 mM KCl, freshly prepared and prewarmed to 37°. Mix gently and only until cell pellet is completely resuspended. Place in 37° water bath for 10 min.

13. Spin at 200 g for 5 min. Remove all supernatant except for about 1/3 ml.

14. Prepare 3:1 methanol:acetic acid fixative immediately before use. Mix and draw up full pasteur pipet of fixative. Pick up one tube, vigorously flick bottom of tube two times to *partially* resuspend cells. Immediately add fixative *drop-by-drop* while shaking tube *very vigorously*. Keep adding fixative in this manner until 1 ml has been added, then lower pipet and draw liquid and cells up and down to mix. Mix very well until any clumps have disappeared; add more fixative to reach 6 ml. Mix again, cap tightly and let set; repeat with second tube, then with second pair of tubes, etc.

15. After 20–30 min, spin tubes at 200 g for 5 min. Remove supernatant and add 2 ml freshly prepared fixative to each tube. Mix, cover, and spin. Repeat rinse four more times without rest between, using freshly prepared fixative each time.

16. After final spin, remove supernatant and resuspend in fresh fixative (0.3–1.0 ml, depending upon size of pellet). Drop onto slides (previously soaked in 70% ethanol and wiped dry) from distance of approximately 2.5 ft. Slide should be placed at about 30° angle. Drop 2 drops of cells, then stand upright to dry. If humidity is low, cold wet slides should be used. If humidity is high, room temperature, dry slides will be optimal. Intermediate humidities often require intermediate degrees of temperature and humidity: room temperature, wet or steamed over 65° water bath for 5–10 sec, etc. Very humid conditions may require transfer of slides to incubator at 37–55° immediately after dropping. Chromosomes as well as interphase nuclei should exhibit low refractive qualities and appear very flat.

17. Use phase microscopy to determine if cell density is optimal and adjust accordingly. Preparations are usually best if spread the same day as harvest. If storage in refrigerator is necessary, increase amount of fixative and cap tightly. Before making preparations, bring suspension to room temperature and rinse several times with fresh fixative.

18. Store slides at room temperature or desiccated at 4°.

^{3}H Labeling of DNA Probes by Nick-Translation

Radiolabeling by nick-translation[12] results in reasonably high activity probes when multiple dNTPs of high specific activity are incorporated. In addition, this reaction results in random nicking of the DNA, an important component of probe amplification by dextran sulfate. However, the extent of nicking should be minimized in order to maximize single-stranded DNA length. This is most easily accomplished by initial characterization of the activity of each lot of DNase. In this regard, a nick-translation reaction is started, aliquots are removed at time points, e.g., 40, 60, 80, 100, 120 min, and percentages of ^{3}H incorporation are determined by TCA precipitation (see Steps 6–9 below). The time point at which incorporation begins to plateau (25–35%) is then used in subsequent reactions. Sizing by denaturing gel electrophoresis (e.g., a formamide–formaldehyde gel[21]) should indicate single-stranded lengths greater than 500 bases. The procedure described below is adapted from Lai et al.[22]

Reagents

[^{3}H]dCTP (40–60 Ci/mmol, in Tricine, New England Nuclear #NET-601A)

[^{3}H]TTP (90–110 Ci/mmol, in Tricine, New England Nuclear #NET-520A)

dATP, dGTP; solution of 375 μM dATP (Sigma #D-6500), 375 μM dGTP (Sigma #D-4010) in H_2O and frozen in aliquots

DNase I (Worthington #LS00 06330), reconstituted in 10 mM $CaCl_2$, pH 5–6, and frozen in aliquots

DNA Polymerase I (5–10 units/μl, New England Nuclear #NEE-100)

2-Mercaptoethanol (Bio-Rad #161-0710), stored at $-20°$

Salmon sperm DNA, sheared by sonication or depurination to approximately 300 bp and purified

ProtoSol Tissue Solubilizer (New England Nuclear #NEF-935)

Econofluor Scintillation Solution (New England Nuclear #NEF-969)

NACS Prepac Convertible Column (BRL #1525NP)

Capillary pipets, 1 μl to contain (CMS #326-173)

Protocol

1. Prepare 10× buffer: 500 mM Tris–HCl, pH 7.5, 75 mM $MgCl_2$, 100 mM 2-mercaptoethanol, 500 μg/ml bovine serum albumin.

2. Dilute DNase from 1 mg/ml to 50 ng/ml with chilled H_2O.

3. Mix in Eppendorf tube on ice: 10 μl [^{3}H]TTP (25 μCi), 5 μl [^{3}H]dCTP (12.5 μCi), 2.5 μl 10× buffer, 2 μl dATP, dGTP, 2 μl DNA (0.5 μg), 1.5 μl H_2O, 1 μl DNase (use capillary pipet), and 1 μl Polymerase I.

4. Incubate reaction tube at 14° for predetermined period (60–120 min).

5. Stop reaction by adding 5 μl of 100 μM EDTA; bring volume to 50 μl.

6. Transfer 1 μl with capillary pipet to 49 μl H_2O.

---

[21] H. Lehrach, D. Diamond, J. M. Wozney, and H. Boedtker, *Biochemistry* **16**, 4763 (1977).
[22] E. C. Lai, S. L. C. Woo, A. Dugaiczyk, and B. W. O'Malley, *Cell* **16**, 201 (1979).

7. TCA precipitate 10 µl of diluted reaction mix in the presence of 300 µg BSA with 400 µl cold 10% TCA. Vortex, and set on ice for at least 30 min. Spin in microcentrifuge for 5 min, pour off supernatant, rinse with 400 µl 10% TCA, centrifuge again, pour off supernatant, and tap out remaining supernatant. Dissolve pellet completely in 300 µl Protosol, which may require heating for 15–30 min at 37–45°. Add to 5 or 10 ml Econofluor in scintillation vial.

8. Add 10 µl of diluted reaction mix directly to 5 or 10 ml Econofluor containing 300 µl Protosol.

9. Count in liquid scintillation counter. Total TCA precipitable cpm (or dpm) are estimated to be incorporated into 0.5 µg of DNA. Efficiency of incorporation is equal to total TCA precipitable cpm divided by total cpm in reaction. Activity should be at least 10^7 cpm/0.5 µg DNA.

10. Chromatograph through NACS Prepac column. Count 1 µl of first two elution fractions in Econofluor containing Protosol. To these fractions, add 5 µg sheared salmon sperm DNA and ethanol precipitate. Resuspend in total of 100 µl H_2O, count 1 µl, and calculate total cpm recovered. Percentage of 0.5 µg DNA recovered (total recovered cpm divided by TCA precipitable cpm) is used to calculate final concentration of labeled DNA.

NOTE: Chromatography through Sephadex G-75 can also be used to purify ^{3}H-labeled DNA from unincorporated dNTPs.

In Situ Hybridization

As previously discussed, sensitivity of the *in situ* hybridization technique is greatly increased by addition of 10% dextran sulfate to the hybridization reaction. We also modified several hybridization parameters to preserve chromosome morphology, including careful control of pH throughout all steps, addition of phosphate buffer to the hybridization reaction, rapid denaturation in 70% formamide–2× SSC at 70°, and use of formamide to lower the hybridization temperature to 37°.[9,13] These improved conditions in combination with direct staining of autoradiographed preparations with Wright stain[15] permit G-banding of labeled chromosomes and therefore more thorough and accurate analyses.

Other factors may also affect chromosome morphology and subsequent banding. Chromosome preparations should be aged at least 1 week, either at room temperature or desiccated at 4°. Several hours rest after denaturation before hybridization appears to improve morphology, as does minimal duration of hybridization, e.g., 8–12 hr.

In regard to hybridization efficiency, slides stored at room temperature for over 4 weeks appear to exhibit less chromosomal label. However, we and other investigators (K. Davies, personal communication) have found that slides stored at 4° under desiccation retain adequate hybridization efficiency for at least several months. Following denaturation, slides are usually hybridized within 6 hr and not more than 18 hr.

Reagents

RNase A, from bovine pancreas (Sigma #R-4875)
Salmon sperm DNA (Sigma #D-1626), sheared by sonication or depurination to approximately 300 bp and purified
Formamide, reagent grade or nucleic acid grade
Dextran sulfate (Pharmacia #17-0340-01)
Rubber cement, Carter

Protocol

1. Grind dextran sulfate to fine powder in mortar with pestle. Prepare 20% dextran sulfate in formamide by stirring or shaking for several hours. Filter through Versapor 1200 membrane filter (1.2 μm pore, 25 mm diameter, Gelman #66393) using syringe filter holder (Gelman #4320). Prepare weekly.

2. Prepare 10× SSCP: 1.2 M NaCl, 0.15 M Na-citrate, 0.2 M NaPO$_4$ (add 2/5 volume 0.5 M NaPO$_4$ buffer, pH 6.0). The pH of the solution should be 5.5–6.5. Prepare mock hybridization mix without probe and carrier DNAs by mixing 5 parts 20% dextran sulfate in formamide: 2 parts 10× SSCP: 3 parts H$_2$O. The pH should be 7.0–7.2. If necessary, adjust pH of 0.5 M NaPO$_4$ and prepare new 10× SSCP.

3. Prepare 10× SSC: 1.5 M NaCl, 0.15 M Na-citrate. Adjust pH to 6.5 with HCl; the pH of 2× SSC should be 7.0–7.1.

4. Dissolve RNase in 2× SSC at 1 mg/ml. Heat in boiling water bath for 10 min; freeze in aliquots.

5. Locate best areas on slide preparations using phase microscopy. Mark area on back of each slide with diamond pencil. An 18 × 18 mm area is usually adequate; if mitoses are sparse, mark 22 × 30 mm or 22 × 40 mm areas.

6. Dilute RNase from 1 mg/ml to 100 μl/ml with 2× SSC. Place 200 μl diluted RNase on each slide and cover with 24 × 50-mm coverslip. Place in moist chamber (e.g., tray or dish containing large tubing strips to support slides above water, covered) and incubate at 37° for 1 hr.

7. Remove coverslips by gently sliding each off slides. Rinse well with agitation in 4 changes 2× SSC in coplin jars at room temperature, 2 min each. Dehydrate in 70, 80, and 95% ethanol, 2 min each, and dry well with air jet.

8. Prepare denaturing solution of 70% formamide, 2× SSC. Adjust

the pH to 7.0 with HCl, pour into coplin jar, place in water bath at 50–55°, and heat bath to 73–75°. Immerse each slide individually in denaturing solution at 70.5°. Solution will immediately cool to 70°; time for 2 min. Transfer quickly to 70% ethanol, rinse for 1 min with agitation, and continue dehydration in 80, 90, and 100% ethanol, 1 min each. Dry with air jet.

9. Hybridization mix is prepared immediately before use: 5 parts 20% dextran sulfate in formamide; 2 parts 10× SSCP; 2 parts ^{3}H-labeled probe DNA at 5× final concentration; 1 part sheared salmon sperm DNA at 10× final concentration (final carrier DNA concentration is 500× final probe concentration). Final concentration of probe DNA is generally 10–200 ng/ml for plasmid probes and 1–5 µg/ml for phage probes. Use of probe concentrations that are too high will result in marked, nonspecifically located grains and/or clusters of grain throughout the karyotype, on interphase nuclei and between cells. An 18 × 18 mm coverslip requires 10 µl probe; 22 × 30 mm requires 25 µl, etc. Mix very well since solutions are viscous. Denature hybridization mixes at 70° for 5 min; quickly cool in ice bath.

10. Place hybridization mix on slides and cover with coverslips. Seal edges of coverslips with rubber cement using pasteur pipet cut to larger bore. Incubate in moist chamber at 37° for 8–16 hr.

11. Remove rubber cement with forceps and rinse slides well with constant agitation in three changes of 50% formamide, 2× SSC, pH 7.0 at 39°, 3 min each. (Coverslips should slide off in first rinse.) Rinse further in five changes 2× SSC at 39°, 2 min each, with agitation. Dehydrate in 70, 80, and 95% ethanol, 2 min each, and dry with air jet.

Autoradiography

In our laboratory, autoradiography is carried out in a completely darkened room. If desired, safelight filter Wratten No. 2 may be used with Eastman NTB2 emulsion. A drying box with fan, described in Schmid[23] and also commercially available from Oncor (#52030), is optional. However, if autoradiography is routinely performed on a high volume of slides, and/or if relative humidity is high, a drying box should be considered.

Reagents

Nuclear track emulsion, Kodak type NTB2 (Eastman Kodak DC Special Products #165-4433)
Pipette filler, 10 ml (Markson #D-13666)

---

[23] W. Schmid, in "Human Chromosome Methodology" (J. J. Yunis, ed.), p. 91. Academic Press, New York, 1965.

Protocol

1. In darkened room, open emulsion box and place emulsion bottle in water bath at 44–45°. Water should extend to neck of bottle. Start timers preset for 20 and 30 min.

2. After 20 min, remove bottle and open cap. There will be a solidified plug of emulsion at the top. Carefully insert a clean wooden stick or pipet through the plug several times until plug sinks to bottom. Return bottle to water bath for last 10 min. At that time, the emulsion should be completely liquified and warm.

3. Place 20-ml aliquots of emulsion into disposable 50-ml centrifuge tubes by use of a 10-ml pipet and 10-ml pipet filler. The first tube should contain 20 ml of warm distilled water; return to water bath for use immediately after aliquoting. Place other filled tubes into light-proof specimen containers and set aside.

4. Dip clean, blank slide several times in diluted emulsion to mix. Blot end of slide on gauze for about 10 sec and set in blank slide rack (e.g., Fisher #12-587-20). Dip second blank slide in emulsion, almost to bottom; pause about 1 sec, remove in smooth motion, blot, and set in blank slide rack. Dip third blank slide in similar manner, blot, and set in second rack called the drying rack (this slide is called the test slide; it will be exposed and developed with the first hybridized slides).

5. Dip each hybridized slide (in prearranged order), blot, and set in drying rack. Place slides in rack from back to front.

6. Place drying rack in drying box for 1 hr. If not available, place in light-proof box or cabinet for 1–2 hr.

7. Dip fourth blank slide, blot, and set in blank slide rack.

8. Cap diluted emulsion and place in light-proof container. Tape container, as well as containers of undiluted emulsion, with black electrical tape.

9. When slides are secure, turn on light. Examine blank slides, second and fourth will illustrate homogeneity of emulsion. Fourth will indicate final level and quality of remaining emulsion.

10. Prepare black boxes by placing two blank slides in each, about 1″ from ends. Roll Drierite desiccant in gauze packages and place two in each box at ends. Decide front of box for both top and bottom so that particular slides may be removed later. Label each box.

11. After drying period is over, place slides in middle sections of boxes, according to prearranged number per box. After all slides are boxed, tape each with several rounds of black electrical tape. Ensure that there are no wrinkles and seal tape well by pressing.

12. Place in air-tight boxes containing Drierite and expose at 4°.

13. Diluted emulsions can be used repeatedly but must be well mixed after liquifying.

Development is carried out in Kodak Dektol, which minimizes the production of unexposed background grains on NTB emulsions compared with other developers.[24] Following development and fixation, autoradiographs must be partially rinsed before exposure to light since grains are not stable to light in the presence of fixer. After complete rinsing, slides should be thoroughly dried in air for at least several hours before staining.

Reagents

Dektol developer (Kodak #146-4726)
Kodafix solution (Kodak #146–4080)

Protocol

1. Place three staining dishes (glass or plastic) in tray or pan. Fill dishes with Dektol developer diluted 1:1 with distilled H_2O, distilled H_2O, and Kodafix diluted 1:3 with distilled H_2O. Cool solutions to 15°.

2. In dark room, preset three timers to 2, 5, and 5 min. After darkening, untape boxes and transfer slides to slide tray.

3. Place tray in developer for 2 min, lifting up and down very gently several times to agitate in approximately 30 sec intervals.

4. Transfer to H_2O stop bath for 20–25 sec, agitating every few seconds.

5. Transfer to fixer for 5 min, again agitating gently every 30 sec.

6. Transfer to large beaker of water to partially rinse fixer. If running water is available, keep water flowing at very gentle rate. Time for 5 min before admitting light. During this time, retape boxes containing slides for longer exposure.

7. After admitting light, rinse for 5 additional min in gently running tap water. Dry in air slowly and completely.

G-Banding and Data Analysis

The Wright staining technique produces sharp, highly contrasted and consistent G-bands without the need for pretreatment.[15] This procedure is simple and easily reproduced provided protocol details are followed.

Reagents

Wright stain, MC/B Manufacturing Chemists (only brand recommended) available through Curtis-Matheson (#WX-25)
Methanol, 500 ml bottles

---

[24] J. E. Neeley and J. W. Combs, *J. Histochem. Cytochem.* **24,** 1057 (1976).

Protocol

1. To prepare stain, pour 500 ml methanol from new bottle into clean bottle reserved only for mixing stain. Start to stir. Slowly add 1.25 g Wright stain to final concentration of 0.25%. Cap bottle and stir for 60 min. Filter through double Whatman #1 filter paper into original dark methanol bottle. Tightly cap and store in dark area at room temperature for at least 10–14 days.

2. Prepare 0.06 M phosphate buffer, pH 6.8, by mixing 51 parts 0.06 M KH_2PO_4 with 49 parts 0.06 M Na_2HPO_4. Adjust pH to 6.8 if necessary.

3. To stain, pipet 3 ml buffer into tube. Add 1 ml stain with pipet, pipet up and down two times to mix, and pour immediately onto slide placed horizontally on staining rack. Time for 5 min when staining hybridized slides. To rinse, lift end of slide at the same time as applying water so that the stain runs off slide and is not trapped in emulsion. Dry immediately with air jet.

4. Check staining by microscopy; this is most easily done by use of a high-power, high dry objective, e.g., Zeiss Epiplan 80× STM. If chromosomes are too dark, rinse in running water for about 10–15 sec, then dry and check again. Repeat as necessary. Rinsing also helps reduce cytoplasmic and emulsion-trapped stain.

5. If chromosomes are too light or not banded, destain and restain. Destaining consists of the following: 95% ethanol for 1.5 min, 95% ethanol containing 1% HCl for 35 sec, and methanol for 1.5 min. Agitate slide continuously in vertical motion, particularly in EtOH-HCl, and dry slide between solutions with air jet. Change solutions daily or as needed.

6. Restain slide as described above. Appearance of darkly stained chromosomes, which often occurs after one destaining–restaining series, is a positive sign. Rinse with water as described above in order to elicit a more differentiated pattern. If chromosomes are still unbanded, one more series of destaining–restaining can be carried out; more than two series is usually detrimental, but not always.

7. Unhybridized chromosome preparations usually require a staining time of only 1.5–2.5 min. They do not rinse easily with water; therefore, determination of the exact time for differentiated G-bands is usually recommended.

Data are then compiled directly by microscopic observation of grain locations in each cell. Grains are drawn exactly as observed on the ideogram of human metaphase chromosomes.[16] Cells are preselected for preserved morphology and good spreading under low magnification and after observation at high power (80× or 100× objective) must be included in the analysis. Criteria for chromosomal grains should be set, e.g., all grains which lie on top of a chromosome or appear to be touching a chromo-

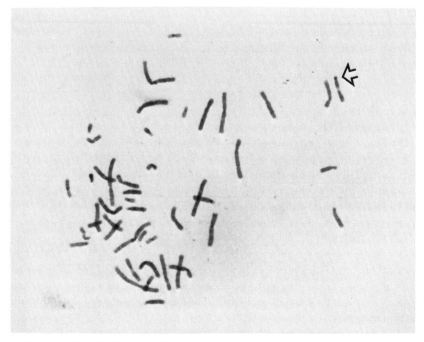

FIG. 1. *In situ* hybridization of an apoA-I-specific probe to human mitotic cell, showing localization of the apoA-I gene to the long arm of chromosome 11. Cloned apoA-I cDNA (pA1-3) was ^{3}H labeled by nick-translation to 3.4×10^7 cpm/μg and hybridized at 20 ng/ml; slides were exposed for 15 days. Analysis of 30 cells indicated significant clustering of grains around bands 11q22-q23 (M. E. Harper, P. Cheung, and L. Chan, in preparation, 1986).

some. Grain locations are compiled and the statistical significance of labeling of one (or several) chromosome bands or segments is evaluated by either the percentage of grains or percentage of labeled sites (each labeled site may comprise one to several grains). Label on the particular chromosomal segment should be significant throughout the entire karyotype and not only along the relevant chromosome.

An example of a human metaphase cell hybridized with a human apoA-I cDNA probe is shown in Fig. 1.

Acknowledgments

MH gratefully acknowledges Grady F. Saunders for collaboration and support in the development of this method. MH was supported by a National Research Service Award from the National Institutes of Health (GM 07992). Experiments performed in LC's laboratory described in this chapter were supported by grants from the National Heart, Lung and Blood Institute, HL-23741, and the March of Dimes Birth Defects Foundation.

[52] Genetic Control of Plasma Lipid Transport: Mouse Model

By ALDONS J. LUSIS and RENEE C. LEBOEUF

Introduction

This chapter introduces the principles of mouse genetics and discusses the advantages of the mouse as a model for identifying genetic factors in mammalian plasma lipid transport. It does not describe the biochemistry of mouse plasma lipoproteins and related enzymes; for this, the reader is referred to the chapter by Chapman[1] in the present volume and to several recent reports on mouse plasma lipoproteins and lipases.[2-5] The organization of the chapter is as follows. The first section discusses the concept that the mouse provides a useful model for biochemical–genetic studies of mammalian lipoproteins. The next section provides a description of how genetic variants in the mouse are identified and analyzed. The third section illustrates the kinds of questions to which biochemical–genetic approaches can be applied. And the last section provides a short list of sources and procedures.

Of Mice and Men

There are a large number of genes (probably hundreds) that participate in determining the levels and structures of plasma lipoproteins, and the gross genetic defects studied thus far in humans and a few animal models probably constitute a small subset of the various genes that control plasma lipids.[6] Unfortunately, only limited aspects of the genetic regulation of lipoproteins can be examined directly in humans. An important

---

[1] M. J. Chapman, this volume [3].
[2] R. C. LeBoeuf, D. L. Puppione, V. N. Schumaker, and A. J. Lusis, *J. Biol. Chem.* **258,** 5063 (1983).
[3] M.-C. Camus, M. J. Chapman, P. Forgez, and P. M. Laplaud, *J. Lipid Res.* **24,** 1210 (1983).
[4] O. Ben-Zeev, A. J. Lusis, R. C. LeBoeuf, J. Nikazy, and M. C. Schotz, *J. Biol. Chem.* **258,** 13632 (1983).
[5] K. L. Reue, D. H. Quon, K. A. O'Donnell, G. J. Dizikes, G. C. Fareed, and A. J. Lusis, *J. Biol. Chem.* **259,** 2100 (1984).
[6] D. S. Fredrickson and P. N. Herbert, *in* "Disturbances in Lipid and Lipoprotein Metabolism" (J. M. Dietschy, A. M. Gotto, Jr., and J. A. Outko, eds.), p. 199. Williams & Wilkins, Baltimore, 1978.

complication in humans is the interaction between environmental and genetic factors, and genetic analysis in humans is often restricted to family histories. Moreover, the high degree of genetic heterogeneity among human populations makes it difficult to identify the effects of individual genes unless those effects are relatively gross. Nevertheless, combined biochemical–genetic studies have been enormously informative in examining the functional and dynamic aspects of lipoprotein expression. This is best exemplified by the studies of Brown, Goldstein, and co-workers on the receptor mediated uptake of LDL.

Such biochemical–genetic approaches can profitably be extended to the mouse. The mouse is the classical mammal for genetic studies and as such it offers a variety of advantages over other animal models. First, there are hundreds of different inbred strains of mice available, each strain representing a unique gene pool in which natural polymorphisms have been fixed by inbreeding. Variations[7] are identified by individually surveying the various strains. The relatively brief surveys conducted thus far for plasma lipoproteins and related functions have revealed a wealth of useful variants (Table I). Second, the mouse map contains hundreds of genetic markers covering most regions of the 20 pairs of mouse chromosomes. This permits the researcher to map to precise chromosomal sites new variations that are identified.[8] Third, mouse geneticists have developed a variety of special tools which facilitate genetic analysis, including congenic strains and recombinant inbred strains (discussed below). Finally, over the years a number of mutant mice have been identified which may prove useful as models for human disease (Table II).

Biochemical–genetic studies with mice will clearly be of general significance, clarifying plasma lipid transport in humans and other mammals. Although the relative amounts of the density classes of plasma lipoproteins differ considerably between mice and humans, their lipoproteins and apolipoproteins are structurally and functionally similar.[2,3] Also, linkage of genes on chromosomes is roughly conserved (in blocks) among mammals; for example, the regions of mouse chromosome 9 and human chromosome 11 that contain the structural gene for apolipoprotein A-I contain other homologous markers.[9] Thus, mapping studies with mice will have implications for the locations of genes in humans. Finally, on the basis of

---

[7] The term "variation" is often used in place of "mutation" when referring to a naturally occurring polymorphism.

[8] At present, mapping with the use of somatic hybrid cells generally permits localization only to a chromosome or subregion of a chromosome.

[9] T. K. Antonucci, O. H. von Deimling, B. B. Rosenblum, L. C. Skow, and M. H. Meisler, *Genetics* **107**, 463 (1984).

TABLE I
PHENOTYPIC VARIATIONS AFFECTING PLASMA LIPID TRANSPORT
AND RELATED FUNCTIONS IN MICE

Function	Structural (S) or quantitative (Q) variation
ApoA-I[a]	S
ApoA-II[a]	S,Q
ApoB[b]	Q
ApoC-III[b,c]	Q
ApoE[c]	S,Q
HDL[a,d]	S,Q
LDL/VLDL[d]	S,Q
LPL (heart, adipose)[d]	S,Q
Plasma cholesterol (HDL,LDL/VLDL)[e]	Q
Amyloid proteins[f]	S,Q
Acetyl-LDL receptor[g]	Q
Lecithin : cholesterol acyltransferase[h]	Q
Diet-induced atherosclerosis[d]	—
Obesity[i]	—
Lipid deposition[i]	—

[a] A. J. Lusis, B. A. Taylor, R. W. Wangenstein, and R. C. LeBoeuf, *J. Biol. Chem.* **258**, 5071 (1983).
[b] R. C. LeBoeuf and A. J. Lusis, unpublished, 1983.
[c] R. C. LeBoeuf, D. L. Puppione, V. N. Schumaker, and A. J. Lusis, *J. Biol. Chem.* **258**, 5063 (1983).
[d] R. C. LeBoeuf, A. J. Lusis, K. Reue, and B. Paigen, unpublished, 1984.
[e] S. Steppan, R. C. LeBoeuf, A. J. Lusis, and M. C. Schotz, unpublished, 1984.
[f] B. A. Taylor and L. Rowe, *Mol. Gen. Genet.* **195**, 491 (1984).
[g] L. Metza and A. J. Lusis, unpublished, 1983.
[h] C.-H. Chen and R. C. LeBoeuf, unpublished, 1983.
[i] A. A. Kandutsch and D. C. Coleman, in "Biology of the Laboratory Mouse" (E. L. Green, ed.), p. 377. McGraw-Hill, New York, 1966.

recent results (discussed below) we are sanguine about the possibility that studies with mice will provide insights into genetic aspects of atherosclerosis and other disorders involving lipoproteins.

Identification and Analysis of Plasma Lipid Transport Genes

Genetic variations are usually identified by screening individual inbred strains of mice. The methods of choice for the identification of structural

TABLE II
Genes Affecting Lipid Transport or Related Functions in the Mouse

Gene symbol	Gene name	Chromosome
Alp-1	ApoA-I (structure)[a]	9
Alp-2	ApoA-II (structure, quantity)[a,b]	1
A[y]	Yellow (accompanied by obesity)[c]	2
Ad	Adult obesity and diabetes[d]	7
ald	Adrenocorticoid lipid depletion[c]	1
db	Diabetes[e]	4
db[ad]	Adipose (increase lipogenesis)[c]	19
fat	Fat[f]	—
fm	Foam-cell reticulosis[c]	—
Hdl-1	High-density lipoprotein structure[a]	1
hyt	Hypothyroid (accompanied by elevated serum cholesterol)[g]	12
jp[msd]	Myelin synthesis deficiency[h]	X
Lpl-1	Lipoprotein lipase[i]	—
mld	Myelin deficient[j]	—
Nil	Neonatal intestinal lipidosis[k]	7
ob	Obese[c]	6
obl	Obese like[l]	—
oed	Oedematous[m]	—
pu	Pudgy[c]	7
Saa	Serum amyloid protein[n]	7
t	T locus (affects lipase activities)[o]	17
tub	Tubby[p]	7

[a] A. J. Lusis, B. A. Taylor, R. W. Wangenstein, and R. C. LeBoeuf, *J. Biol. Chem.* **258**, 5071 (1983).

[b] R. C. LeBoeuf and A. J. Lusis, unpublished, 1983.

[c] A. A. Kandutsch and D. C. Coleman, in "Biology of the Laboratory Mouse" (E. L. Green, ed.), p. 377. McGraw-Hill, New York, 1966.

[d] M. E. Wallace and F. M. MacSwiney, *J. Hyg.* **82**, 309 (1979).

[e] D. Coleman, *Diabetalogia* **14**, 141 (1978).

[f] K. P. Hummel and D. Coleman, *Mouse News Lett.* **50**, 43 (1974); personal communication.

[g] W. G. Beamer, E. M. Eicher, L. J. MacTais, and J. L. Southard, *Science* **212**, 61 (1981).

[h] H. Meier and A. D. MacPike, *Exp. Brain Res.* **10**, 512 (1970).

[i] O. Ben-Zeev, A. J. Lusis, R. C. LeBoeuf, J. Nikazy, and M. C. Schotz, *J. Biol. Chem.* **258**, 13632 (1983).

[j] D. P. Doolittle and K. M. Schweikart, *J. Hered.* **68**, 331 (1977).

[k] M. E. Wallace and B. M. Herbertson, *J. Med. Genet.* **6**, 361 (1966).

[l] J. L. Guenet, *Mouse News Lett.* **67**, 30 (1982); personal communication.

[m] M. B. Schiffman, M. L. Santorineou, S. E. Lewis, H. A. Turchin, and S. Gluecksohn-Waelsch, *Genetics* **81**, 525 (1975).

[n] B. A. Taylor and L. Rowe, *Mol. Gen. Genet.* **195**, 491 (1984).

[o] J. R. Paterniti, Jr., W. V. Brown, and H. W. Ginsberg, *Science* **221**, 167 (1983).

[p] D. L. Coleman, E. M. Eicher, and J. L. Southard, *Mouse News Lett.* **59**, 25 (1978); personal communication.

variations at the level of protein are electrophoresis and heat stability.[10] If mRNA or genomic sequences for the gene of interest have been cloned, the most sensitive method for identification of variation at the structural locus is to screen for restriction fragment length polymorphisms. Quantitative variations are usually recognized by enzymatic activity or various immunoassays. Typically, most quantitative variations differ by a factor of only 2 to 3, although many larger variations have been identified. Thus, it is not normally feasible to identify variations comparable to the various rare dyslipoproteinemias in human populations; nevertheless, the less dramatic variations have been extremely informative in a variety of studies.[10,11]

Once a variation has been identified, the responsible genes can be characterized by examining progeny of genetic crosses. Whether a variation shows codominant or recessive/dominant inheritance is determined by examining F_1 progeny. The number and location of genes involved have classically been determined by examining segregation patterns in backcross and F_2 generation progeny. Since a large number of genetic markers covering most of the mouse genome is available, it is possible to map genes with considerable precision.

Traditional backcross analysis is time consuming, expensive, and laborious. These problems can be avoided through the use of recombinant inbred (RI) strains, a relatively recent tool in mouse genetics that is largely replacing backcross analysis.[12,13] RI strains are constructed by inbreeding F_2 progeny derived from two different preexisting progenitor strains (Fig. 1). Several RI strains are derived independently, each strain forming a stable segregant population consisting of a unique mixture of genes derived from the two parental strains. The set of RI strains permits linkage analysis, since alleles for linked genes tend to become fixed among the RI strains in the same combinations as in the progenitor strains, while unlinked genes become randomized with respect to one another. An important advantage of RI strains for linkage studies is that data for the segregation of genetic markers are cumulative, and each RI strain needs to be typed for a marker only once. Thus, to map a new genetic variation, one need only compare the distribution of alleles for the gene with the distributions of previously typed markers. Concordant dis-

---

[10] K. Paigen, *Annu. Rev. Genet.* **13**, 417 (1979).
[11] M. Negishi and D. W. Nebert, *J. Biol. Chem.* **254**, 11015 (1979).
[12] B. A. Taylor, in "Origins of Inbred Mice" (M. D. Morse, III, ed.), p. 423. Academic Press, New York, 1978.
[13] D. W. Bailey, in "The Mouse in Biomedical Research: History, Genetics and Wild Mice" (H. L. Foster, J. D. Small, and J. G. Fox, eds.), Vol. 1, p. 223. Academic Press, New York, 1981.

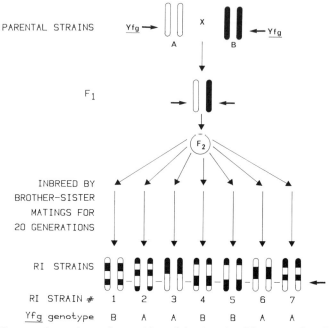

FIG. 1. Construction and use of recombinant inbred strains. The parental strains, A and B, have different allelic forms of "Your favorite gene," *Yfg*, shown with an arrow. Using these parental strains, a set of 7 recombinant inbred (RI) strains was constructed by first making F_1 hybrids, which carry both the A and B forms of *Yfg*, and then crossing these animals to yield the F_2 generation. At this point, F_2 mating pairs are chosen. The F_2 mother and her offspring are isolated and the offspring are brother–sister mated through 20 generations to establish new inbred strains. Each RI strain contains a unique mixture of genes derived from the 2 parental strains, A and B, which have become scrambled due to crossing-over between chromosomes derived from either the A or B parent and fixed due to extensive inbreeding. Each strain of the RI set is typed to determine which of the parental strain alleles has been fixed and this generates a strain distribution pattern characteristic of the *Yfg* locus.

tributions are indicative of linkage. RI strains provide the added advantage that multiple observations can be made on single recombinant genotypes, which is particularly important when examining small quantitative differences or complex phenotypes such as susceptibility to atherosclerosis. In order to utilize RI strains it is necessary that the gene of interest differ between progenitors of a set of RI strains; since dozens of RI sets have now been constructed[13] this is possible in many cases.

Congenic strains are another tool which have been invaluable in mouse genetics, particularly in the analysis of histocompatibility and immune regulation.[13] Their genomes contain a small chromosomal region derived from one strain that has been transferred, by repeated backcrossing, onto the genetic background of a second strain. Congenic strains provide a means of cleanly examining the effects of individual genes by separating them, through the genetic transfer, from other genes that may influence the phenotype. The studies of *Hdl-1* and *Alp-2* discussed below provide a good illustration of the use of congenic strains.

Applications of Mouse Genetics to Problems of Lipid Transport

Gene Mapping. As yet, only a small number of genes that directly control components of the plasma lipid transport system have been mapped in the mouse. Most of these involve HDL, the most abundant density class of lipoproteins in the mouse. Charge variations for apolipoproteins A-I and A-II were identified by isoelectric focusing and were used to map the structural genes for the proteins by examining strain distributions of alleles among recombinant inbred strains.[14] The structural gene for apoA-I, designated *Alp-1*, is linked to the marker *Lap-1* on mouse chromosome 9, and the structural gene for apoA-II, designated *Alp-2*, is tightly linked to the lymphocyte alloantigen locus *Lym-20* on chromosome 1 (Table III). Variations controlling the structure of intact HDL particles were identified by native polyacrylamide gel electrophoresis, and genetic studies showed that one such variation is determined by a single Mendelian gene (designated *Hdl-1*) that is tightly linked to the apoA-II structural locus on chromosome 1.[14] Also linked to the *Alp-2* locus are a gene controlling the levels of circulating apoA-II,[15] a gene controlling the levels of serum amyloid P-component,[16] and a gene determining susceptibility to diet-induced atherosclerosis.[17] Thus, there may be a cluster of genes involved in lipid transport and related functions on mouse chromosome 1; alternatively, some of these phenotypic variations may be determined by a common gene. A family of genes encoding serum amyloid proteins (which are acute phase reactants induced by inflammatory stimuli and incorporated into HDL) is located on chromosome 7.[18] Genes which independently control the levels of lipoprotein lipase activity in heart and adipose tissue have been identified but have not yet been mapped; however, they are unlinked to all of the above loci (Tables II and III).

---

[14] A. J. Lusis, B. A. Taylor, R. W. Wangenstein, and R. C. LeBoeuf, *J. Biol. Chem.* **258**, 5071 (1983).
[15] R. C. LeBoeuf and A. J. Lusis, unpublished, 1983.
[16] R. F. Mortensen and B. A. Taylor, *Fed. Proc., Fed. Am. Soc. Exp. Biol.* **43**, 2988 (1984).
[17] R. C. LeBoeuf, A. J. Lusis, K. Reue, and B. Paigen, unpublished, 1984.
[18] B. A. Taylor and L. Rowe, *Mol. Gen. Genet.* **195**, 491 (1984).

TABLE III
SEGREGATION OF LIPOPROTEIN AND LIPOPROTEIN LIPASE
VARIATIONS AMONG RI STRAINS DERIVED FROM C57BL/6(B) and
BALB/c (C)

Locus[a]	RI strain						
	D	E	G	H	I	J	K
Alp-1[b]	C	C	C	B	B	B	C
Lap-1[b]	C	C	C	B	B	B	C
Alp-2[b]	C	B	C	B	B	C	C
Lym-20[b]	C	B	C	B	B	C	C
Hdl-1[b]	C	B	C	B	B	C	C
LPL activity (heart)[c]	B	?	C	B	?	C	C
LPL activity (adipose)[c]	B	B	B	C	B	C	C

[a] *Alp-1, Alp-2,* and *Hdl-1* are genes controlling the structures of apoA-I, apoA-II, and high-density lipoproteins and are described in the text. *Lap-1* is a structural locus for leucine amino peptidase, a marker on mouse chromosome 9, and *Lym-20* is a locus for a lymphocyte alloantigen, a marker on mouse chromosome 1. Lipoprotein lipase (LPL) activity in adipose appears to be determined by a single locus (designated *Lpl-1*) while LPL activity in heart may be determined by multiple loci. Alleles resembling those carried by C57BL/6 or BALB/c indicated with a B or C, respectively.

[b] A. J. Lusis, B. A. Taylor, R. W. Wangenstein, and R. C. LeBoeuf, *J. Biol. Chem.* **258,** 5071 (1983).

[c] O. Ben-Zeev, A. J. Lusis, R. C. LeBoeuf, J. Nikazy, and M. C. Schotz, *J. Biol. Chem.* **258,** 13632 (1983).

Given the numerous variations that have already been identified (Table I), the list of mapped genes controlling lipid transport in mice should increase rapidly. It is not always possible to identify genetic variants among inbred strains at the level of protein; for example, an exhaustive search for apoE charge variants in over 60 inbred and wild stocks was unsuccessful.[15] In such cases, it is still quite likely that variation can be identified and the structural gene mapped using cloned probes to analyze restriction fragment length polymorphisms. If variations cannot be identified at the level of either protein or DNA, one could resort to somatic cell hybrid approaches.

The presence of a group of linked genes on mouse chromosome 1 raises the possibility that certain lipid transport genes may be clustered on chromosomes, as is the case for some gene families. There is some evidence for this from human studies (involving somatic cell hybrid or family history approaches), indicating that structural genes for apoA-I and apoC-III are tightly linked on chromosome 11[19,20] and that the genes for apoE, apoC-II, and the LDL receptor reside on chromosome 19.[21-23]

An important aspect of mapping genes for lipid transport is in suggesting relationships between different components or functions. For example, the fact that HDL charge and size in mice are controlled by a gene at the apoA-II structural locus suggested a possible role for apoA-II in HDL structure, which has now been confirmed (see below). Similarly, variation at the apoA-I/apoC-III locus in humans has been correlated with hypertriglyceridemia, suggesting that those genes determine the levels of certain lipoproteins.[24] Among the various morphological mutants in mice which are affected in functions such as obesity and abnormal lipid deposition, none appears to map near the plasma lipid transport genes identified thus far (Table II).

Relationships among Components of the Lipid Transport System. Genetic variants for LPL in mice have proved useful in examining the relationship between the enzyme present in different tissues. Studies of the activity and heat stability of LPL among different strains and among one set of recombinant inbred strains showed no correlation for enzyme from heart and adipose tissues (Table II). This indicates that the enzymes expressed in the two tissues are regulated independently and that they may well be derived from separate structural genes.[4] Similarly, a structural variation for apoA-I was useful in demonstrating that the multiple isoforms of the apoA in plasma are all derived from a common structural gene (Fig. 2).[25]

Lipoprotein Assembly and Metabolism. Electrophoretic examination of lipoproteins from various strains of mice has revealed extensive variation for all density classes (Fig. 3).[14,17] Biochemical and genetic analysis of these variations should permit the identification of genetic factors con-

---

[19] G. A. P. Bruns, S. K. Karathanasis, and J. L. Breslow, *Arteriosclerosis* **4**, 97 (1984).

[20] S. W. Law, G. Gray, H. B. Brewer, Jr., A. Y. Sakaguchi, and S. L. Naylor, *Biochem. Biophys. Res. Commun.* **118**, 934 (1984).

[21] A. M. Cumming and F. W. Robertson, *J. Med. Genet.* **19**, 417 (1982).

[22] C. L. Jackson, G. A. P. Bruns, and J. L. Breslow, *Fed. Proc., Fed. Am. Soc. Exp. Biol.* **43**, 1641 (1984).

[23] K. Berg, J. O. Julsrud, A. L. Borresen, G. Fey, and S. Humphries, *Proc. Hum. Gene Map. Meet.* **7**, 25 (1983).

[24] A. Rees, C. C. Shoulders, J. Stocks, D. J. Galton, and F. E. Barello, *Lancet,* **Feb. 26,** 44 (1983).

[25] K. A. O'Donnell and A. J. Lusis, *Biochem. Biophys. Res. Commun.* **114**, 275 (1983).

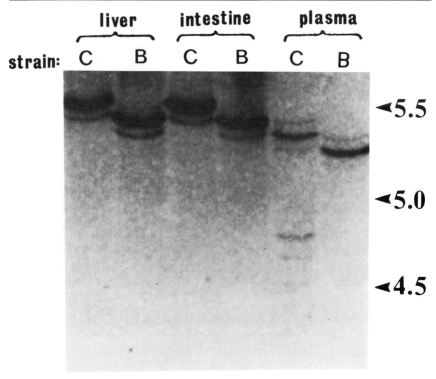

FIG. 2. Structural mutation of apoA-I simultaneously alters the charge isoforms derived from both liver and intestine. ApoA-I from strains C57BL/6 (B) or BALB/c (C), which contain apoA-I structural alleles determining differently charged proteins, was subjected to isoelectric focusing under denaturing conditions. The left four lanes are a fluorograph of apoA-I translated *in vitro* in the presence of [^{35}S]methionine from mRNA isolated from liver or intestine. Shown for comparison at the right are the patterns for plasma apoA-I stained for protein. The pH gradient is shown alongside. The difference in charge between the *in vitro* translation product and the plasma protein is due to a signal sequence and/or a prosegment that is removed from the plasma protein.

trolling lipoprotein size, density, and composition. One variation for HDL, affecting particle size and charge, has now been examined using these approaches. The variation is due to a single Mendelian gene, termed *Hdl-1*, which maps at the apoA-II structural locus on chromosome 1.[14] LeBoeuf has shown that the variation is associated with an altered content of apoA-II in HDL, suggesting that a structural or regulatory variation of A-II may be responsible for the observed structural changes of HDL.[26]

[26] R. C. LeBoeuf, unpublished, 1983.

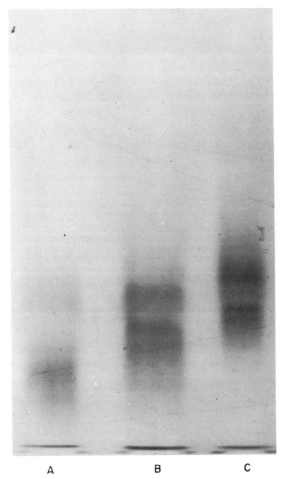

FIG. 3. Phenotypes of HDL from different inbred strains. Plasma lipoproteins isolated from different inbred strains by density centrifugation were subjected to nondenaturing gel electrophoresis in 5% polyacrylamide, and HDL were stained for protein with Coomassie brilliant blue. Electrophoresis was from top to bottom. Three major phenotypes, differing in mobility, could be distinguished. The most rapid migrating form (termed phenotype A) was unique to strain Peru. Some other strains had intermediate mobility (termed B) while others had slow mobility (termed C). The difference between the B and C phenotypes is determined by the *HDL-1* gene on chromosome 1.

The studies of *Alp-2* and *Hdl-1* have been simplified and clarified by the availability of a congenic strain (B6.C-*H-25^c*) which carries the minor histocompatibility allele *H-25^c*, transferred from BALB/cBy onto the genetic background of C57BL/6By by 14 successive backcrosses.[27] *H-25*

[27] D. W. Bailey, *Immunogenetics* **2**, 249 (1975).

was found to map very near the *Alp-2* locus, and examination of B6.C-*H-25*^{c} showed that the *Alp-2* and *Hdl-1* loci of BALB/cBy were transferred to the C57BL/6By background along with *H-25*.[14] The congenic not only confirmed the tight linkage of *Alp-2* and *Hdl-1*, but it also made it possible to examine the effects of the different *Alp-2* and *Hdl-1* alleles on a common genetic background.

Lipoprotein–Receptor Functions. The genetic disorder, familial hypercholesterolemia, has been crucial in clarifying the function of the (B,E)–receptor and, similarly, variations for lipoprotein receptors in mice will undoubtedly be informative in terms of understanding their functions. As yet, the only receptor for which strain surveys have been performed is the acetyl-LDL receptor on macrophages, and variations of 2- to 3-fold have been detected.[28]

Role of Lipoproteins in Atherosclerosis, Obesity, and Host Defense. One of the most exciting aspects of the mouse genetic studies relates to the possibility of identifying genetic factors in diseases and in host defense associated with lipid transport. Thompson[29] made the original observation that strains of mice differ in their susceptibility to diet-induced atherosclerosis, but, subsequent attempts to examine the inheritance of susceptibility using traditional backcross analysis showed multigenic inheritance that appeared to be difficult to analyze further. However, recent advances in characterizing the lesions and in developing improved diets, combined with the use of recombinant inbred strains rather than backcross analysis, suggest that it should be possible to identify individual genes determining susceptibility.[17,30] Strains of mice also differ in susceptibility to obesity (Table II) and in various host defense functions such as resistance to infection and immune regulation. Whether lipoproteins are involved in any of these differences is unknown, but with the elucidation of lipoprotein variations it may be possible to identify genetic correlations between the functions.

Sources and Procedures

Mice. Mice can be purchased from a variety of suppliers, but the largest and most widely used is the Jackson Laboratory, Bar Harbor, Maine. The source of mice is important for genetic studies since a given strain from different suppliers may not be identical. A list of the charac-

---

[28] L. Metza and A. J. Lusis, unpublished, 1983.
[29] T. S. Thompson, *J. Atherosclerosis Res.* **10**, 113 (1969).
[30] J. D. Morrisett, H. S. Kim, J. R. Patsch, S. R. Datta, and J. J. Trentin, *Arteriosclerosis* **2**, 312 (1982).

teristics, origins, and suppliers of common inbred strains is given by Statts.[31] A list of recombinant inbred strains is given by Bailey.[13] Not all strains are commercially available, but they can usually be obtained from the scientists who maintain them. The standard reference for the maintenance, handling, and use of mice is *Biology of the Laboratory Mouse*.[32] References 33 and 34 also provide valuable background information. A source for current information concerning new variations, strains, and linkage data is provided by the *Mouse News Letter* (subscription information from Miss Joan Staats, Jackson Laboratory, Bar Harbor, Maine, 04609).

Diet. Given the many dietary factors influencing lipid metabolism, it is important to monitor and standardize the diet. We normally maintain our mice on standard laboratory chow but have noticed that the types of fatty acids present vary from batch to batch (this may be due to the fact that the "rendered fat" is obtained from a variety of sources). For studies of diet-induced atherosclerosis, an "atherogenic" diet was developed by Thompson.[29] It contains 30% casein, 5% cholesterol, 2% sodium cholate, 5% alphacol, 4% vitamin mixture, 4% salt mixture, 6.5% sucrose, 6.5% dextrose, 6.5% dextrin, 0.5% choline chloride, and 30% cocoa butter. This diet can be commercially purchased from Teklad Laboratories (Madison, WI). In contrast to earlier reports,[30,35] we have found that this diet is atherogenic when mixed with normal mouse chow in a ratio of 75% mouse chow and 25% atherogenic diet.[17]

Screening Mice for Genetic Variation. The first step is to develop a sensitive, rapid, and reproducible assay for the function of interest. The most widely used assays are discussed above. When developing an assay for genetic variation, a common strain such as C57BL/6 or BALB/c should be chosen for initial studies. When screening for quantitative variations, the effects of age, sex, diet, and perhaps diurnal cycle should be examined.

Typically, between 10 and 50 strains reflecting a wide range of lineages are surveyed (for example, see ref. 14). For quantitative studies between 3 and 10 mice per strain are usually examined. Since recombinant inbred

---

[31] J. Statts, *Cancer Res.* **40,** 2083 (1980).
[32] A. A. Kandutsch and D. C. Coleman, *in* "Biology of the Laboratory Mouse" (E. L. Green, ed.), p. 377. McGraw-Hill, New York, 1966.
[33] "Origins of Inbred Mice" (M. C. Morse, III, ed.). Academic Press, New York, 1978.
[34] "The Mouse in Biomedical Research: History, Genetics and Wild Mice" (H. L. Foster, J. D. Small, and J. G. Fox, eds.), Vols. 1–4. Academic Press, New York, 1981.
[35] A. Roberts and J. S. Thompson, *in* "Atherosclerosis Drug Discovery" (C. E. Day, ed.), p. 313. Plenum, New York, 1976.

strains provide a tremendous aid in genetic analysis and mapping, progenitors of available recombinant inbred strain sets should be included. Also, the genetic pool from which variations can be identified has been increased greatly during the past decade by the construction of a number of new stocks from various wild populations of mice.

When screening strains of mice or examining progeny of genetic crosses it is important to minimize nongenetic influences, particularly in the case of quantitative variations. Thus, mice are usually matched as closely as possible in age (generally 2–3 months), sex, diet, maintenance conditions, and time of sacrifice or bleeding. Since feeding can have a large effect on lipoproteins, we usually fast mice overnight before taking samples.

Detailed record-keeping and care in the maintenance of mice are especially important in genetic studies.

Genetic Analysis. The inheritance patterns for a variation are examined using conventional backcross analysis or recombinant inbred strains (see above). Among the questions of interest are the following: Does the variation exhibit codominant or recessive/dominant inheritance? How many genes are responsible for the variation between strains? What is the chromosomal location of the gene determining the variation?

Dominance relationships can be determined by examining F_1 progeny. F_1 progeny are the direct descendants of two parental strains and are usually labeled as (female parent strain × male parent strain)F_1. Structural gene variations usually exhibit codominant inheritance. For example, in the case of a structural gene variation resolved by electrophoresis, F_1 animals would be expected to exhibit both parental electrophoretic forms. Regulatory or processing gene variations can exhibit recessive/dominant or codominant inheritance, depending on the nature of the variation. In the case of quantitative variations, if F_1 progeny exhibit a level that is intermediate between the parental levels, the inheritance is termed "additive."

To determine the number of genes controlling a phenotypic variation, backcrosses (between F_1 and one or the other progenitor strains) or F_2 crosses can be constructed. A genetic variation which is determined by a single Mendelian gene will exhibit a 1 : 1 (parental : F_1) ratio of phenotypes in backcross progeny and a 1 : 2 : 1 (parent 1 : F_1 : parent 2) ratio of phenotypes in F_2 progeny. Multiple genes, depending on the nature of their interactions, will give rise to altered ratios or nonparental phenotypes. When one says that a particular character is "determined by a single gene," it is important to realize that the statement refers only to the genes differing between the strains being examined. For example, there are obviously numerous genes that participate in determining the size and density of HDL, but the *difference* in HDL structure between strains C57BL/6 and BALB/c is determined by a single gene, *Hdl-1*.

Mapping genes to chromosomes using backcross analysis is time consuming, expensive, and laborious. A large pool of F_2 or backcross animals (usually 50 or more) must first be produced. Second, the animals must be assayed for the variation of interest, as well as for other biochemical markers whose chromosomal map positions are already known. One then determines map positions by looking for concordant segregation between previously mapped biochemical marker phenotypes and those of the new variant. The distance between genes is expressed in terms of the percentage recombination in a backcross (1% recombination is termed a centimorgan and is equivalent to roughly one million base pairs of DNA).

As discussed above, genetic analysis is enormously simplified by the use of RI strains. In addition to gene mapping (see above), RI strains are useful in determining the number of genes responsible for a variation. If the variation of interest is controlled by one major Mendelian gene, then only parental phenotypes are expected in the RI strains in approximately 1 : 1 (parent 1 : parent 2) ratios. This is because RI strains are homozygous for one or the other parental allele at each gene. If two or more unlinked genes control the variation, nonparental phenotypes generally result. Complete distributions for markers segregating in RI strains are available from scientists maintaining the stocks.

Isolation of Lipoproteins and Apolipoproteins. Lipoproteins are usually isolated by sequential ultracentrifugation based on their flotation properties in salt solutions of increasing density.[36] However, the density fractions commonly used to isolate human lipoproteins ($d < 1.006$ g/ml, $d = 1.006–1.063$ g/ml and $d = 1.063–1.21$ g/ml) do not completely resolve mouse lipoproteins.[2] Specifically, we find that the LDL fraction ($d = 1.006–1.063$ g/ml) contains α and pre-β migrating lipoproteins as well as LDL (which exhibit β mobility), and that apoA-I and E are present in addition to apoB. Based on density gradient ultracentrifugation,[2,3] a more appropriate density interval for LDL is 1.02–1.06 g/ml, although this fraction may still contain a small amount of apoA-I. We isolate lipoproteins from the pooled plasma of 10 to 60 animals using either SW 60 or SW 41 rotors. Alternatively, mouse lipoproteins can be isolated with density gradients.[37] To prevent bacterial growth and lipoprotein degradation, we usually include EDTA, sodium azide, gentamycin, and glutathione in solutions.[36]

---

[36] V. N. Schumaker and D. L. Puppione, this series, Vol. 128.
[37] B. H. Chung, J. P. Segrest, J. T. Cone, J. Pfau, J. C. Geer, and L. A. Duncan, *J. Lipid Res.* **22**, 1003 (1981).

Precipitation of mouse lipoproteins with heparin-manganese is useful for separating VLDL/LDL from HDL.[38,39] This method is particularly convenient for measurements of cholesterol and triglyceride present in VLDL/LDL and HDL fractions, but it does not completely separate lipoproteins from other mouse plasma proteins. We routinely begin with 50 μl to 1 ml of mouse plasma in an Eppendorf tube (1.5 ml) to which heparin (5% in 0.15 M NaCl) and $MnCl_2$ are added to final concentrations of 0.2% and 0.05 M, respectively. The solution is mixed gently and after 15 min incubation at room temperature, the precipitate is removed by centrifugation in a microfuge for 3 min. The supernate, containing HDL and plasma proteins, is carefully decanted, and the pellet, containing VLDL/LDL, is brought to 100 μl with 10% sodium bicarbonate. After incubation at room temperature for 15 min, the solution is again centrifuged and the supernate (containing VLDL/LDL) is decanted. Both fractions can be assayed directly for cholesterol or triglyceride content.

Mouse apolipoproteins can be isolated using conventional column chromatographic procedures or by preparative polyacrylamide gel electrophoresis. HPLC should also prove useful in mouse apolipoprotein isolations,[40] although this has not yet been examined. The most abundant apolipoprotein in mouse plasma is apoA-I (about 1 mg/ml plasma). It can be easily isolated from delipidated (2 times, with ethanol–ether) mouse HDL by gel filtration on a column (1.5 × 60 cm) of Sephacryl S-200 equilibrated with 0.1 M Tris–HCl, pH 8.6, containing 1 mM EDTA and 6 M urea.[2] Unlike rat HDL, mouse HDL contains a negligible amount of apoE and, thus, the apoA-I obtained with this single isolation step is essentially pure. ApoA-II (about 0.2 mg/ml plasma) is also resolved on this column, but it must be separated from small amounts of contaminating C apolipoproteins by DEAE-cellulose column chromatography.[2] ApoE and C-III can be isolated from delipidated mouse VLDL as described by Lim and Scanu.[41]

We have found it convenient to isolate apoB and certain less abundant apolipoproteins (A-IV from VLDL, C-peptides from HDL or VLDL) by preparative gel electrophoresis. We routinely run 3-mm-thick polyacrylamide slab gels (10 × 16 cm) in the presence of SDS. Intact lipoproteins are dissolved directly in sample buffer,[42] and up to 3 mg of total protein can be applied to each gel. The gels are stained briefly in Coomassie blue (45 min in 0.25% Coomassie brilliant blue R-250 in 50% methanol and 10% acetic acid) and destained first in 30% methanol and 5% acetic acid (30

[38] A. Montcalm and R. C. LeBoeuf, unpublished, 1983.
[39] M. Burstein, H. R. Scholnick, and R. Morfin, *J. Lipid Res.* **11**, 583 (1970).
[40] D. Pfaffinger, C. Edelstein, and A. M. Scanu, *J. Lipid Res.* **24**, 796 (1983).
[41] C. T. Lim and A. M. Scanu, *Artery* **2**, 483 (1976).
[42] U. K. Laemmli, *Nature (London)* **227**, 680 (1970).

min) and then overnight in water. Bands are sliced out, minced, and either homogenized with water for use as antigens in antibody preparations, or electroeluted from the gel (elution buffer is 0.05 M Tris-acetate, pH 7.8, containing 0.1% SDS) in a dialysis bag or an elution chamber (C.B.S. Scientific, Del Mar, CA).

Analysis of Plasma Lipids, Lipoproteins, and Apolipoproteins. Plasma cholesterol values for different mouse strains show about a 3-fold variation (from 59 mg/dl plasma to 169 mg/dl).[43] When maintained on diets containing elevated levels of cholesterol and saturated fats, these values can increase up to 5-fold. We have found that enzymatic procedures are unsuitable for plasma cholesterol determinations in plasma of mice on high lipid diets due to concomitant increases in plasma triglyceride levels, which interfere with optical absorption measurements. Thus, we generally use the Rudel and Morris[44] procedure to measure plasma and lipoprotein cholesterol levels. Triglyceride and phospholipid analyses are performed as described.[2]

"Western" blotting (or immunoblotting") is particularly useful for identifying quantitative and structural variations in mouse plasma proteins and has also proved useful for examining lipase directly in tissue homogenates.[26] We generally solubilize 1–4 μl of mouse plasma, or appropriate amounts of tissue homogenates, directly in either SDS sample buffer (for SDS gels) or isoelectric focusing sample buffer containing 9.5 M urea, 2% Triton X-100, and 2% ampholines (for isoelectric focusing gels).[2] After electrophoresis the proteins are electrophoretically transferred to nitrocellulose filters and treated with antibody as described by Burnette.[45] Although probably less quantitative than radioimmunoassay, an advantage of Western blotting is that it is possible to examine several proteins simultaneously (by mixing antibodies to different proteins) and to rapidly survey a large number of individual mice.

Addendum

There has been rapid progress in identifying genetic factors controlling transport of lipids. In particular, three unlinked loci containing the structural genes for apolipoproteins have now been identified in both mice and humans. In humans these loci are located on the p21-qter region of chromosome 1 (apoA-II),[46,47] the q13 region of chromosome 11 (apoA-1 and

[43] S. Steppan, R. C. LeBoeuf, A. J. Lusis, and M. C. Schotz, unpublished, 1984.
[44] L. L. Rudel and M. D. Morris, *J. Lipid Res.* **14,** 364 (1973).
[45] W. M. Burnette, *Anal. Biochem.* **112,** 195 (1981).
[46] K. J. Lackner, S. W. Law, H. B. Brewer, Jr., A. Y. Sakaguchi, and S. L. Naylor, *Biochem. Biophys. Res. Commun.* **122,** 877 (1984).
[47] T. J. Knott, R. L. Eddy, M. E. Robertson, L. M. Priestley, J. Scott, and T. B. Shows, *Biochem. Biophys. Res. Commun.* **125,** 299 (1984).

apoC-III,[48,49] and the centromere-qter region of chromosome 19 (apoC-I, apoC-II, apoE).[47,50–52] In mice, three homologous loci, showing conservation of flanking genetic markers, have been located on chromosomes 1 (apoA-II),[14] 7 (apoE),[53] and 9 (apoA-I).[14] Recently, we have isolated cDNA clones for human apoB, and somatic cell genetic studies indicate that the gene for apoB resides on chromosome 2, unlinked to the loci for the other apolipoproteins.[54] In addition, the gene for the LDL receptor has been mapped to the short arm of human chromosome 19[55] and the gene for 3-hydroxy-3-methylglutaryl coenzyme A reductase has been mapped to human chromosome 5.[56]

Acknowledgments

We are grateful to our colleagues, especially V. Schumaker, M. Schotz, and B. Paigen, for advice and encouragement. The authors' laboratories are supported by grants from the American Heart Association, Greater Los Angeles Affiliate (816-F1-1), and the National Institutes of Health (HL28481).

---

[48] S. W. Law, G. Gray, H. B. Brewer, Jr., A. Y. Sakaguchi, and L. S. Naylor, *Biochem. Biophys. Res. Commun.* **118**, 934 (1984).

[49] G. A. P. Bruns, S. K. Karathanasis, and J. L. Breslow, *Arteriosclerosis* **4**, 97 (1984).

[50] J. A. Donald, S. C. Wallis, A. Kessling, P. Tippett, E. B. Robson, S. Ball, K. E. Davies, P. Scambler, K. Berg, A. Neiberg, R. Williamson, and S. E. Humphries, *Hum. Genet.* **69**, 39 (1985).

[51] C. L. Jackson, G. A. P. Bruns, and J. L. Breslow, *Proc. Natl. Acad. Sci. U.S.A.* **81**, 2945 (1984).

[52] A. J. Lusis, C. Heinzmann, R. S. Sparkes, R. Geller, M. C. Sparkes, T. Mohandas, *8th Int. Workshop Hum. Gene Map., Cytogenet. Cell Genet.*, in press (1985).

[53] A. J. Lusis, B. Taylor, D. Quon, and R. C. LeBoeuf, unpublished.

[54] M. Mehrabian, T. Mohandas, R. Sparkes, V. Schumaker, C. Heinzmann, S. Zollman, and A. Lusis, *Som. Cell Mol. Genet.* in press (1986).

[55] U. Francke, M. S. Brown, and J. L. Goldstein, *Proc. Natl. Acad. Sci. U.S.A.* **81**, 2826 (1984).

[56] T. Mohandas, C. Heinzmann, R. S. Sparkes, P. Edwards, and A. Lusis, *Som. Cell Mol. Genet.* in press (1986).

[53] Molecular Cloning of Bovine LDL Receptor cDNAs

By DAVID W. RUSSELL and TOKUO YAMAMOTO

Introduction

The low-density lipoprotein (LDL) receptor is a cell surface protein that plays a central role in the metabolism of cholesterol in humans and animals.[1] The LDL receptor pathway is of interest to the cell biologist as a model system for the study of receptor-mediated endocytosis, a process by which macromolecules enter cells after binding to receptors in coated pits on the cell surface.[2] It is of interest to the membrane biochemist as an integral membrane protein that mediates endocytosis of the cholesterol-rich LDL particle via protein–protein interactions with the apolipoproteins B and E present in LDL.[1] Finally, the LDL receptor is of interest to the molecular biologist in that expression of the receptor gene is regulated by a feedback mechanism involving cholesterol, and because mutations in the gene affecting its structure and function give rise to one of the most prevalent human genetic diseases, familial hypercholesterolemia.[3]

The application of the powerful techniques of molecular biology to the study of this interesting protein should greatly enhance our knowledge of its structure and function. The primary sequence of the receptor protein could be obtained by DNA sequence analysis of cloned complementary DNAs (cDNAs). This information could be used to derive testable models of the orientation of the LDL receptor in the plasma membrane, to predict domains within the protein involved in ligand binding and receptor recycling, and in the generation of domain-specific probes in the form of antipeptide antibodies. Site-specific mutation of the cloned receptor cDNA followed by expression of the mutant gene would allow precise testing of the roles of various domains in endocytosis. In addition, the availability of cloned cDNA probes would allow direct study of the expression of the LDL receptor gene and perhaps outline mechanisms involved in feedback regulation. Finally, insight into the molecular genetics of the receptor gene could be obtained using cloned cDNAs to characterize mutations occurring at this locus in patients with familial hypercholesterolemia.

---

[1] J. L. Goldstein and M. S. Brown, *Annu. Rev. Biochem.* **46,** 897 (1977).
[2] J. L. Goldstein, R. G. W. Anderson, and M. S. Brown, *Nature (London)* **279,** 679 (1979).
[3] J. L. Goldstein and M. S. Brown, *Annu. Rev. Genet.* **13,** 259 (1979).

As a first step in the application of molecular cloning techniques to the study of the LDL receptor, we describe here the methods employed to obtain partial cDNA clones of the bovine LDL receptor mRNA.[4]

Principle of the Method

The LDL receptor is a trace protein of cultured cells and animal tissues, accounting for less than 0.01% of the total membrane protein.[5] As first approximations, membrane proteins constitute about 10% of total cell protein and levels of an mRNA encoding a given protein roughly parallel its abundance in the cell. Thus, the LDL receptor mRNA may represent as little as 0.001% of the total cell mRNA. To facilitate cloning of this low abundance mRNA (Fig. 1), the techniques of polyribosome immune purification are first used to enrich for the receptor mRNA species. Poly(A)$^+$ RNA isolated from purified polyribosomes is employed as a template in the construction of a cDNA library using the techniques and vectors of Okayama and Berg.[6] Synthetic oligodeoxyribonucleotide probes derived from the amino acid sequence of an internal cyanogen bromide fragment of the bovine receptor are then used to identify cDNA clones in the library derived from the LDL receptor mRNA. Final identification of a receptor cDNA clone is based on DNA sequence and Northern blotting analyses.

Reagents, Buffers, Tissues, and Cells

Bovine adrenal glands and liver are obtained from a local slaughterhouse within 5 min of sacrifice, trimmed of excess fat, placed immediately in liquid nitrogen, powdered in a Waring blender, and stored at $-70°$. Human epidermoid carcinoma A-431 cells are grown in Dulbecco's modified Eagle's essential medium containing 10% fetal calf serum. Induction and suppression of LDL receptors in these cells are as described.[7]

Heparin (catalogue No. H-3125), Triton X-100 (catalogue No. T-6878), and phenylmethylsulfonylfluoride (catalogue No. P-7626) are obtained from Sigma. Trichodermin is a kind of gift of Dr. W. O. Godtfredsen, Leo Pharmaceuticals, DK-2750, Ballerup, Denmark. RNase-free sucrose and Nonidet P-40 are from Bethesda Research Laboratory. Sodium deoxy-

---

[4] D. W. Russell, T. Yamamoto, W. J. Schneider, C. J. Slaughter, M. S. Brown, and J. L. Goldstein, *Proc. Natl. Acad. Sci. U.S.A.* **80**, 7501 (1983).
[5] W. J. Schneider, U. Beisiegel, J. L. Goldstein, and M. S. Brown, *J. Biol. Chem.* **257**, 2664 (1982).
[6] H. Okayama and P. Berg, *Mol. Cell. Biol.* **2**, 161 (1982).
[7] R. G. W. Anderson, M. S. Brown, and J. L. Goldstein, *J. Cell Biol.* **88**, 441 (1981).

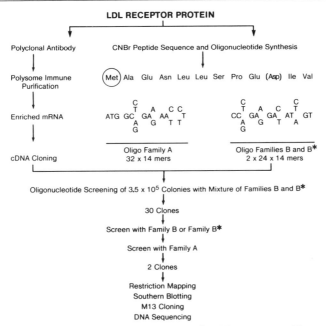

FIG. 1. Strategy for cloning a cDNA for the bovine LDL receptor. The methods employed in each step are described in the text.

cholate is from Fisher. Protein A sepharose is from Pharmacia. Oligo(dT)-cellulose (type 7) is obtained from P-L Biochemicals. Synthetic oligonucleotide mixtures are a kind gift of Dr. Michael Smith, University of British Columbia.

Buffers

Buffer A. 25 mM Tris–HCl, pH 7.5, 25 mM NaCl, 5 mM $MgCl_2$, 2% Triton X-100 (v/v). After autoclaving, heparin is added to 0.3 mg/ml from a sterile-filtered 40 mg/ml stock solution in H_2O, trichodermin is added to 1 µg/ml from a 1 mg/ml stock solution in ethanol, and phenylmethylsulfonylfluoride is added to 60 µg/ml from a 20 mg/ml stock solution in dimethyl sulfoxide.

Buffer B. 0.5 M sucrose, 25 mM Tris–HCl, pH 7.5, 25 mM NaCl, 5 mM $MgCl_2$, 0.5% Triton X-100 (v/v). After autoclaving, heparin is added to 0.2 mg/ml and trichodermin to 1 µg/ml as described above.

Buffer C. 1.5 M NaCl, 50 mM $MgCl_2$, and 1% (v/v) Nonidet P-40.

Buffer D. 25 mM Tris–HCl, pH 7.5, 150 mM NaCl, 5 mM $MgCl_2$, 0.1% Nonidet P-40 (v/v), 0.2 mg/ml heparin, and 1 µg/ml trichodermin.

Methods

Isolation of Polysomes from Bovine Adrenal Glands

Polysomes are isolated using a modification of the $MgCl_2$ precipitation method of Palmiter.[8] All buffers are autoclaved and all glassware is baked overnight at 175° before use.

Four 10-g aliquots of frozen powdered bovine adrenal glands are weighed into plastic dishes on dry ice. One aliquot is added to 42.5 ml of Buffer A in a baked beaker and homogenized immediately at speed 3.5 for 2 min and then speed 8 for 25 sec in a Brinkman Instruments polytron. The pink homogenate is quickly transferred to two 40-ml sterile screw cap centrifuge tubes and centrifuged for 1 min at 4° in a Sorval SA-600 rotor at 12,000 rpm. The rotor is stopped as quickly as possible, the yellow fatty surface layer is removed by aspiration and discarded, and the dark red supernatants are decanted into a 50-ml screw cap tube on ice containing 5 ml of 1 M $MgCl_2$. The final $MgCl_2$ concentration is 100 mM. This polysome–$MgCl_2$ suspension is left on ice while the 3 subsequent aliquots of adrenal powder are brought to this point. Once the supernatant from the last aliquot is brought to 100 mM $MgCl_2$ it is incubated 1 hr on ice. The processing of each aliquot to the $MgCl_2$ precipitation step requires about 20 min; thus the $MgCl_2$–polysome supernatant from the first tissue aliquot incubates on ice for a total of 2 hr while the last aliquot incubates for 1 hr. After this incubation, 20-ml aliquots of the polysome precipitates are carefully overlayed on 20 ml of Buffer B and centrifuged in 40-ml screw cap tubes in a Sorvall SA-600 rotor for 15 min at 15,000 rpm at 4°. After centrifugation, the tubes are placed on ice and the pink supernatant above the sucrose cushion is removed by aspiration and discarded. The sides of the tube are rinsed with 10 ml of sterile H_2O. The H_2O rinse is then removed followed by the sucrose cushion. Care is taken to avoid the large brown translucent polysome pellet on the side of the tube. After removal of the sucrose cushion, 1 ml of sterile 25 mM Tris–HCl, pH 7.5 is added to each tube and the polysome pellet is dislodged using a P1000 Pipetman with a sterile tip. The pellets are broken up by repeated pipetings and transferred to a Potter Elvehjem homogenization tube. The centrifuge tubes are rinsed with an additional 1 ml of 25 mM Tris–HCl, pH 7.5 and the rinses are transferred to the homogenization tube. The polysome pellets, in a volume of about 20 ml, are resuspended with 5 to 10 strokes at 4° with a tight fitting pestle. After resuspension, the solution is brought to 0.5% sodium deoxycholate with the addition of 2 ml of sterile-filtered 5%

---

[8] R. D. Palmiter, *Biochemistry* **13**, 3606 (1974).

detergent solution (w/v) in 25 mM Tris–HCl, pH 7.5, and then homogenized further with 2 to 3 strokes of the pestle. The exact volume of the polysome homogenate is measured and 0.1 volume of Buffer C is added, together with heparin to 0.1 mg/ml and trichodermin to 1 μg/ml, followed by gentle mixing. The sample is clarified by centrifugation in a baked 30-ml Corex tube for 10 min at 15,000 rpm at 4° in a Sorvall SA-600 rotor. The supernatant containing the polysomes is decanted to a 50-ml plastic tube, a 0.3-ml aliquot is removed for sucrose gradient analysis, and the remainder is quick-frozen in liquid nitrogen and stored at $-70°$.

In general, 40 g of bovine adrenal powder yields approximately 1000 A_{260} units of material at a concentration of 30 to 50 A_{260} units/ml. Analysis of 3 to 5 A_{260} units on 5 ml 10 to 40% linear sucrose gradients indicates that 50 to 70% of the A_{260} material sediments faster than 80 S monosomes. The remainder is present at the top and in the 80 S monosome region of the gradient.

Preparation of Polyclonal Antibody for Use in Polysome Immune Purification

An antibody to highly purified bovine LDL receptor is prepared by immunization of rabbits as described previously.[9] Following cardiac puncture, immune serum is passed over a protein A-sepharose column to obtain an IgG fraction.[10] Nonimmune rabbit IgG is isolated and purified in a similar manner. Purified IgG fractions are assayed for gross RNase contamination by incubating 5 A_{260} units of polysomes with 30 μg of IgG fraction for 1 hr on ice in Buffer D followed by centrifugation through 5 ml 10 to 40% linear sucrose gradients. None of the IgG fractions alters the polysome sedimentation profile, and as such is considered suitable for polysome immune purification.

Polysome Immune Purification of mRNA

One thousand A_{260} units of polysomes is clarified by centrifugation for 10 min at 12,000 rpm at 4° in a Sorvall SA-600 and then diluted to 15 A_{260} per ml by the addition of Buffer D. Polyclonal antibody is added at a ratio of 1 mg per 160 A_{260} units of polysomes[11] and the solution is stirred slowly on ice for 1 hr. The polysome–antibody slurry is then chromatographed

---

[9] U. Beisiegel, T. Kita, R. G. W. Anderson, W. J. Schneider, M. S. Brown, and J. L. Goldstein, *J. Biol. Chem.* **256,** 4071 (1981).

[10] U. Beisiegel, W. J. Schneider, J. L. Goldstein, R. G. W. Anderson, and M. S. Brown, *J. Biol. Chem.* **256,** 11923 (1981).

[11] J. P. Kraus and L. E. Rosenberg, *Proc. Natl. Acad. Sci. U.S.A.* **79,** 4015 (1982).

twice on a 5 cm³ column of protein A-Sepharose[12] (0.7 × 13 cm) equilibrated in Buffer D at a flow rate of 8 to 10 ml/hr at 4°. The column is washed overnight with ~120 ml of Buffer D. Bound polysomes are eluted at maximal flow rate with 20 ml of 25 mM Tris–HCl, pH 7.5 and 20 mM EDTA. The eluted fraction containing dissociated ribosomes and messenger RNA is heated 5 min at 65°, brought to 0.5 M NaCl by the addition of 0.6 g solid baked NaCl, and to 0.2% SDS by the addition of 0.4 ml of sterile-filtered 10% SDS (w/v).[13] The solution is cooled to room temperature in a beaker of H$_2$O and chromatographed at maximum flow rate through a 1 ml column (0.8 × 2.3 cm) of oligo(dT)-cellulose equilibrated in 10 mM Tris–HCl, pH 7.5, 0.5 M NaCl. The column is washed with 20 ml of this buffer and poly(A)$^+$ RNA is eluted with 5 ml of 10 mM Tris–HCl, pH 7.5. Fifty micrograms of RNase-free yeast carrier tRNA is added, and the RNA is precipitated twice with NaOAc and ethanol. The immunopurified RNA is resuspended in H$_2$O (20 μl) for use in *in vitro* translation assays and cDNA cloning reactions. The very small amounts of RNA obtained with this procedure preclude its exact quantitation; however, it is estimated that less than 2 μg of poly(A)$^+$ RNA is obtained from 1000 A_{260} units of polysomes.

In Vitro Translation

To assay qualitatively the degree of enrichment of LDL receptor RNA by the polysome immune purification procedures, small aliquots are incubated in H$_2$O with 2.5 mM CH$_3$HgOH for 10 min and translated in rabbit reticulocyte lysates prepared as described by Pelham and Jackson[14] and supplemented with 80 mM KOAc, 1 mM Mg(OA$_c$)$_2$, 19 amino acids (− methionine) at 16 $\mu$$M$ each, and [^{35}S]methionine at 0.2 mCi/ml. The final concentration of CH$_3$HgOH in the translation reaction is 0.3 mM. Translation products are analyzed by electrophoresis on SDS 7% polyacrylamide gels.

Previous biosynthetic studies in a variety of species indicate that the LDL receptor is initially synthesized as a precursor protein with an apparent molecular weight of about 120,000, as determined by SDS–gel electrophoresis.[15] In general, translation of 1 μl of the immune-selected poly(A)$^+$ RNA yielded a putative ~120 kDa LDL receptor protein band following fluorography and exposure to Kodak XAR-5 film for 16 to 24 hr (Fig. 2, lane 3). Because of its large size and the scarcity of translation products in

[12] S. Z. Shapiro and J. R. Young, *J. Biol. Chem.* **256**, 1495 (1981).
[13] A. J. Korman, P. J. Knudsen, J. F. Kaufman, and J. L. Strominger, *Proc. Natl. Acad. Sci. U.S.A.* **79**, 1844 (1982).
[14] H. R. B. Pelham and R. J. Jackson, *Eur. J. Biochem.* **67**, 247 (1976).
[15] H. Tolleshaug, J. L. Goldstein, W. J. Schneider, and M. S. Brown, *Cell* **30**, 715 (1982).

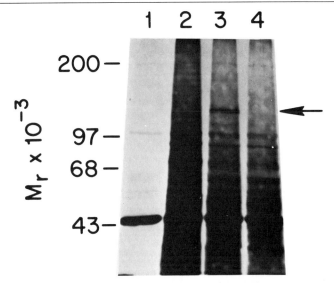

FIG. 2. *In vitro* translation of poly(A)⁺ RNA in rabbit reticulocyte lysates. Additions to the lysates were lane 1, 2, μl of H$_2$O; lane 2, 0.1 μg of total poly(A)⁺ RNA extracted from bovine adrenal glands; lane 3, 2 μl (10%) of poly(A)⁺ RNA derived from polysomes selected with an antireceptor IgG; lane 4, 2 μl (10%) of poly(A)⁺ RNA derived from polysomes selected with a nonimmune IgG. The fluorographed SDS/7% polyacrylamide gel is exposed to film for 23 hr at −70°. The arrow marks the putative LDL receptor precursor described in the text.

this region of the gel, this band is visible in the total translation products synthesized from the immune selected RNA. It accounts for about 1% of the total absorbance present on the autoradiogram as determined by densitometry. Despite numerous attempts with different protocols and antireceptor antibodies, it has not been possible to immune precipitate this putative LDL receptor protein following *in vitro* translation. To obtain indirect evidence that the 120 kDa protein is in fact the LDL receptor and not an artifact of the immune purification procedure, poly(A)⁺ RNA is purified as described above except that a preimmune antisera is employed. Translation of this RNA does not yield the 120 kDa translation product (Fig. 2, lane 4); however, all background bands are visualized. The presence of the 120 kDa band following translation of immune-selected RNA and its absence following translation of preimmune-selected RNA are consistent with enrichment of the LDL receptor mRNA.

cDNA Cloning

The vectors and methods described by Okayama and Berg[6] are employed in the construction of a cDNA library from the immune-selected

RNA. Two characteristics of the cloning system are relevent here. First, the method lends itself to use with very small amounts of starting mRNA. In the cloning of the LDL receptor cDNA, tiny amounts of starting template RNA are obtained by the polysome immune purification procedure. In the Okayama–Berg cloning method the mass of the dT-tailed plasmid primer used in the initial cloning reaction facilitates recovery of cDNA in the many subsequent ethanol precipitation steps. Second, the Okayama–Berg cloning system results in plasmids containing a higher proportion of longer cDNAs than is usually obtained by other methods. In the cloning of the LDL receptor cDNA we did not know the location of the CNBr fragment in the receptor protein whose sequence is used to derive the oligonucleotide probes (Fig. 1). Thus it is possible that a near full-length cDNA copy of the LDL receptor mRNA would be required for detection by the oligonucleotide probes.

To clone the LDL receptor cDNA, immune-selected RNA derived from 2000 A_{260} units of polysomes is precipitated with ethanol in the presence of 1.4 µg of dT-tailed vector primer. The nucleic acid pellet is resuspended directly in the reverse transcription buffer of Okayama and Berg. All subsequent reactions are performed exactly as described by them with the exception that reverse transcriptase and terminal deoxytransferase are obtained from Life Sciences. *E. coli* DNA ligase is from P-L, and *E. coli* DNA Polymerase I (endonuclease-free) is from Boehringer Mannheim. With the above amounts of input RNA (estimated at 4 µg), a cDNA library which contains about 5×10^5 clones is obtained.

Identification of LDL Receptor cDNA Clones

Portions of the cDNA library generated by the above methods are used to transform *Escherichia coli* strain RR1 to ampicillin resistance by the CaCl$_2$ shock procedure.[16] To obtain a large representative cDNA library, it is critical that the transformation frequency of the competent cells in control reactions be no less than 5×10^7 transformants per µg of supercoiled plasmid pBR322 DNA. For screening of the cDNA library, approximately 10,000 ampicillin-resistant transformants are plated per 15-cm-diameter nitrocellulose filter and grown overnight at 37° on Luria Broth agar plates containing 25 µg/ml ampicillin. Two replica filters are made from the master filter and prepared for hybridization after amplification with chloramphenicol. To reduce background hybridization from nonnucleic acid material present in the lysed colonies, baked filters are washed overnight in 50 m*M* Tris–HCl, pH 8.0, 1 m*M* EDTA, 1 *M* NaCl,

---

[16] T. Maniatis, E. F. Fritsch, and J. Sambrook, "Molecular Cloning: A Laboratory Manual," p. 1. Cold Spring Harbor Laboratory, Cold Spring Harbor, New York, 1982.

and 0.1% SDS (500 ml per 10 filters) at 37°. Prehybridization is done at 65° for >3 hr in 4 × SSC, 10 × Denhardt's solution, and 100 μg/ml sonicated and denatured *E. coli* DNA. Hybridization is performed overnight in fresh solution containing ^{32}P 5'-end-labeled oligonucleotide mixtures (6 × 10^6 cpm per pmol) at 1 pmol/ml. The hybridization temperature is calculated from the empirical formula $T_m = 2°$ (number of dA · dT bp) + 4° (number of dG · dC bp).[17] In the case of the mixed oligonucleotide probes used here, in which the exact base at a wobble position is unknown, an A · T base pair is used to calculate the T_m. Probe A is used at 36° and probe B at 38° (Fig. 1). Following hybridization, filters are washed in batches of 10 in 4 × SSC; 0.1% SDS at the hybridization temperature for periods of 5 to 30 min until the background radioactivity is substantially reduced as judged with a hand-held Geiger counter. Washed filters are air dried and subjected to autoradiography at −70° with Kodak XAR-5 film and Dupont Cronex Lightning Plus intensifying screens. Positive clones are readily visualized after overnight exposures under these conditions. The use of replica filters controls for occasional false signals due to nonspecific hybridization or contaminated film cassettes and intensifying screens.

Identification of candidate LDL receptor cDNA clones is accomplished with two different oligonucleotide probes derived from the sequence of a single cyanogen bromide fragment (Fig. 1). Initially, the cDNA library is screened with a probe that is nearer the carboxyl-terminus (probe B). Positive clones are then rescreened with the slightly upstream second probe (probe A). Approximately 3.5 × 10^5 colonies are screened with probe B, resulting in the isolation of 30 positive clones. Of these 30 clones, 2 hybridize with probe A and are considered to be strong candidates for an LDL receptor cDNA. The use of this type of cascade hybridization involving two essentially adjacent oligonucleotide probes allows an almost unambiguous identification of LDL receptor cDNA clones. Here, the failure to hybridize to the second slightly upstream probe A eliminates 28 clones which must share limited (spanning probe B) but not extensive (spanning both probes) sequence homology with the receptor.

The two putative LDL receptor clones are characterized by restriction mapping and the sizes of their cDNA inserts are calculated. Both plasmids have identical 2.8 kb inserts and are thus sibling clones (Fig. 3). The regions of homology with the oligonucleotide probes in the plasmid (pLDLR-1) are determined by Southern blotting analysis of DNA after digestion with restriction enzymes. For this purpose ~0.5 μg of restricted

---

[17] M. Smith, in "Methods for DNA and RNA Sequencing" (S. M. Weissmann, ed.), p. 1. Praeger, New York, 1983.

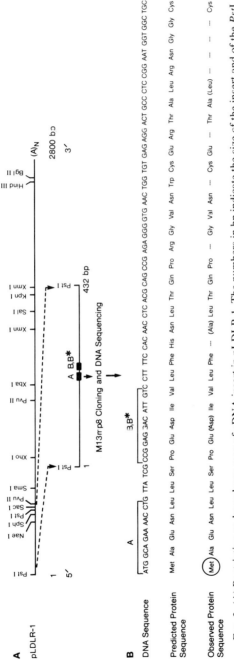

FIG. 3. (A) Restriction endonuclease map of cDNA insert in pLDLR-1. The numbers in bp indicate the size of the insert and of the *Pst*I DNA fragment containing regions of homology with oligonucleotide probes A and B. (B) Comparison of the nucleotide sequence of a region of the pLDLR-1 cDNA and the amino acid sequence of a CNBr peptide of the bovine LDL receptor. The 108-bp nucleotide DNA sequence corresponds to the central portion of the 432-bp *Pst*I fragment. Overlines indicate the regions corresponding to the A and B oligonucleotide probes. In the observed protein sequence, the amino acids in parentheses indicate tentative assignments based on data obtained in only one of the two sequenator runs. Dashes in the observed protein sequence correspond to amino acid residues whose identity was not determined in either of the sequenator runs. The circled methionine was inferred on the basis of the preference of CNBr to cleave after this residue.

plasmid DNA is size-fractionated by electrophoresis in agarose gels and transferred to nitrocellulose by capillary blotting. Baked filters are prehybridized for >6 hr at 37° in 6 × SSC, 10 × Denhardt's solution, 1 mM NaPP, and 100 µg/ml of yeast tRNA or sonicated and denatured *E. coli* DNA. Hybridization is carried out for >6 hr at temperatures calculated as described above in fresh solution containing ^{32}P 5'-end-labeled oligonucleotides (6 × 10^6 cpm/pmol) at 0.17 pmol/ml. Each filter is then washed three times for 5 min each at the hybridization temperature in 6 × SSC, 1 mM NaPP, 0.1% SDS. Following air drying, filters are wrapped with Saran Wrap and exposed to Kodak XAR-5 film for >30 min. These experiments localize the region of oligonucleotide homology within a 432 bp *Pst*I fragment corresponding to the 5' end of the cDNA insert in plasmid pLDLR-1 (Fig. 3).

This fragment is subjected to DNA sequence analysis[18] following cloning into the bacteriophage M13 mp8 vector.[19] Preliminary data from these M13 clones indicates that the region of oligonucleotide homology is located in the center of the 432 bp *Pst*I fragment in regions of the sequencing gel that are difficult to read unambiguously, possibly due to the high G + C content of the DNA. The structure of this difficult region is determined by primer extension of ^{32}P 5'-end-labeled oligonucleotides on the above M13 template followed by chemical sequencing.[20]

For this primer extension sequencing, 100 pmol of oligonucleotide probe B is 5' end-labeled in a 10 µl reaction volume containing 50 mM Tris–HCl, pH 7.5, 10 mM MgCl$_2$, 0.5 mM dithiothreitol, 2–10 units of T$_4$ polynucleotide kinase, and 250 µCi of [γ^{32}P]ATP (7000 Ci/mmol) at 37° for 45 min. The kinase enzyme is inactivated by heating the reaction at 65° for 10 min. The radiolabeled oligonucleotide mixture is combined with 0.1 pmol of M13 template DNA containing the complementary strand of the 432 base *Pst*I fragment, and is annealed at 55° for 5 min. The four unlabeled deoxynucleoside triphosphates are then added to a final concentration of 0.25 mM followed by 1 unit of the Klenow fragment of *E. coli* DNA polymerase I (obtained from Boehringer-Mannheim). Primer extension is allowed to occur for 5 to 10 min at 22° after which time the Klenow enzyme is inactivated by heat treatment at 65° for 10 min. Sodium chloride is added to a final concentration of 50 mM followed by 50 units of *Pst*I restriction endonuclease. In a 60 min incubation at 37°, *Pst*I cleaves the M13 DNA at the upstream cloning site rendered double stranded by the extension of the oligonucleotide primer. This 5'-end labeled primer-ex-

---

[18] F. Sanger, S. Nicklen, and A. R. Coulson, *Proc. Natl. Acad. Sci. U.S.A.* **74**, 5463 (1977).
[19] J. Messing, this series, Vol 101, p. 20.
[20] A. M. Maxam and W. Gilbert, this series, Vol. 65, p. 499.

tended fragment is released from the M13 template by the addition of 10 μl of 90% formamide, 0.02% bromophenol blue, 0.02% xylene cyanol, 25 mM EDTA, and incubation at 100° for 4 min. Electrophoresis on a 7 M urea–5% polyacrylamide gel in 50 mM Tris-borate, pH 8.3, 1 mM EDTA resolves a fragment of approximately 240 nucleotides corresponding to the extended oligonucleotide. This fragment is excised from the gel, eluted, and subjected to DNA sequence analysis as described by Maxam and Gilbert.[20] The results provide an unambiguous sequence upstream of the oligonucleotide B homology, including a region of DNA precisely complementary to one of the components of the oligonucleotide A probe.

Figure 3 shows the sequence of a 108 nucleotide portion of the cDNA in the region of complementarity to the A and B oligonucleotide probes. The sequence is derived from a combination of the above strategies. The predicted protein sequence agrees well with that determined from the CNBr peptide of the LDL receptor. The above primer extension–chemical sequencing method is useful in the rapid screening of candidate cDNA clones selected by a single oligonucleotide probe, provided that some amino acid sequence is known in the amino- or carboxy-terminal direction. Although the template employed in the above primer extension reaction is a single-stranded M13 subclone, the method works well on double-stranded DNAs if they are first denatured by NaOH treatment. In addition, the appropriate oligonucleotide and M13 template can be used to generate fragments for chemical sequencing from DNAs replete with secondary structure and thus more difficult to sequence by the dideoxy method.

The results in Fig. 3 indicating concordance between the nucleotide and amino acid sequence are a strong indication that pLDLR-1 contains a cDNA for the bovine LDL receptor. Further evidence for this identity is obtained in Northern blotting experiments.[21] Previous surveys of LDL receptor enzymatic activity indicate that the adrenal cortex has approximately 10-fold more receptors than the liver.[22] As such, the relative amounts of LDL receptor mRNA in these two tissues should reflect this difference. Total RNA is extracted from powdered adrenal glands and liver by the guanidinium isothiocyanate/CsCl method[16] and purified by two cycles of oligo(dT)-cellulose chromatography. For blotting analysis, 2 to 20 μg of poly(A)$^+$ RNA is incubated for 10 min at 65° in a volume of 50 μl containing 1 M glyoxal, 36% dimethyl sulfoxide (v/v), and 40 mM 3-N-morpholinopropanesulfonic acid (pH 7.0). Ten microliters of 75% glycerol (v/v), 40 mM 3-N-morpholinopropanesulfonic acid (pH 7.0), and 0.04%

---

[21] P. S. Thomas, this series, Vol. 100, p. 255.
[22] M. S. Brown, P. T. Kovanen, and J. L. Goldstein, *Recent Prog. Horm. Res.* **35**, 215 (1979).

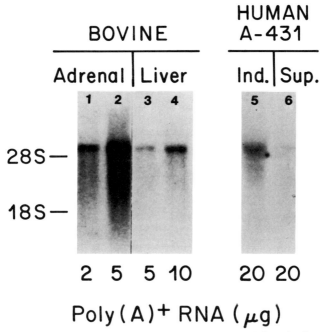

FIG. 4. Hybridization of ^{32}P-labeled pLDLR-1 to poly(A)$^+$ RNA from bovine tissues and human cells. Amounts and sources of poly(A)$^+$ RNA are as follows: lanes 1 and 2, 2 and 5 μg from bovine adrenal; lanes 3 and 4, 5 and 10 μg from bovine liver; lane 5, 20 μg from human A-431 cells induced for LDL receptor expression; and lane 6, 20 μg from human A-431 cells suppressed for LDL receptor expression. Gels are exposed to film for 24 hr (lanes 1–4) or 48 hr (lanes 5 and 6) at −70°. The positions to which bovine 18 S and 28 S ribosomal RNAs migrate are indicated.

bromophenol blue are then added and samples are electrophoresed at 30 V for 16 hr through 1.5% agarose gels in the above buffer which is recirculated. The size-fractionated RNA samples are then transferred to Zeta Probe membranes (Bio-Rad) in 20 × SSC for 24 hr, baked 2 hr *in vacuo* at 80°, prehybridized for 16 hr in 50% formamide (v/v), 5 × SSC, 5 × Denhardt's solution, 0.1% SDS, 100 μg/ml denatured salmon sperm DNA, and 1 μg/ml poly(A). Hybridization is carried out in fresh solution containing 1 × 10^6 cpm/ml of ^{32}P-labeled[23] pLDLR-1 (10^8 cpm/μg) for 24 hr. Hybridized filters are washed 2 times for 10 min in 2 × SSC, 0.1% SDS at 22° and then once at 53° for 1 hr in 0.1 × SSC, 0.1% SDS.

As indicated in Fig. 4, increasing amounts of adrenal gland RNA (lanes 1 and 2) yield a progressively stronger hybridization signal corre-

[23] P. W. J. Rigby, M. Dieckmann, C. Rhodes, and P. Berg, *J. Mol. Biol.* **113**, 237 (1977).

sponding to a unique mRNA of ~5.5 kb. Densitometric scanning shows that the signal obtained with a given amount of RNA from the adrenal is 9-fold more intense than that obtained with an identical amount of liver RNA (lanes 3 and 4). These results are in good agreement with the amount of LDL receptor activity measured by LDL binding in these two tissues and provide further evidence in support of the identity of pLDLR-1 as a cDNA clone of the LDL receptor mRNA. Furthermore, the finding that pLDLR-1 hybridizes to an mRNA of ~5.5 kb indicates that the 2.8 kb insert in this clone represents about one-half of the bovine LDL receptor mRNA.

Summary

Using methods described above, a partial cDNA clone for the bovine LDL receptor has been isolated. DNA sequence analysis and Northern blotting experiments are used to confirm the identity of pLDLR-1. Further DNA sequence analysis of pLDLR-1 reveals that the partial cDNA insert encodes 264 amino acids corresponding to the carboxy-terminal 25% of the bovine LDL receptor.[24] Antipeptide antibodies directed against regions of the predicted protein sequence specifically recognize the purified bovine receptor. These findings provide an independent confirmation of the identity of pLDLR-1.

Future Directions: The Human LDL Receptor

The cloning of a bovine receptor cDNA is the first step in a long-term strategy aimed at the molecular genetics of the human LDL receptor.[3] Central to this approach is the ability of the bovine cDNA to cross-hybridize with the human receptor mRNA. When poly(A)$^+$ RNA is isolated from human A-431 cells grown in the absence of sterols (receptor-induced) or presence of sterols (receptor-suppressed) and analyzed by Northern blotting with the bovine pLDLR-1 probe (Fig. 4), a strong hybridization signal from an mRNA of ~5.5 kb is detected in the induced RNA (lane 5). This signal is reduced in amount by more than 90% in the suppressed RNA (lane 6). The regulation of this mRNA by sterols in the medium is consistent with it encoding the human LDL receptor.[1,7] Based on these results, the bovine cDNA has been used as a probe to screen a human genomic DNA library[25] at reduced stringency and a fragment of

---

[24] D. W. Russell, W. J. Schneider, T. Yamamoto, K. L. Luskey, M. S. Brown, and J. L. Goldstein, *Cell* **37**, 577 (1984).

[25] R. M. Lawn, E. F. Fritsch, R. C. Parker, G. Blake, and T. Maniatis, *Cell* **15**, 1157 (1978).

the human gene encoding the COOH-terminus of the receptor has been isolated. Exon probes from this clone in turn have been used to isolate a full-length cDNA clone of the human LDL receptor. It is hoped that these clones will be useful probes for the study of receptor-mediated endocytosis[2] and the molecular genetics of familial hypercholesterolemia.[3]

Acknowledgments

This work is supported by grants from the National Institutes of Health (HL 31346 and HL 20948) and the Robert A. Welch Foundation (I-971). D. W. R. is the recipient of an NIH Research Career Development Award (HL 01287). We wish to thank Drs. Michael Brown, Joseph Goldstein, and Wolfgang Schneider for perspicacious advice, encouragement, and reagents.

Author Index

Numbers in parentheses are footnote reference numbers and indicate that an author's work is referred to although the name is not cited in the text.

A

Ab, G., 675, 745
Abermann, R., 461
Abrams, M., 544
Abreu, E., 38, 647, 652(7)
Aburatani, H., 536
Adams, D. J., 389, 395(9)
Adamson, A. W., 390, 399(11)
Adamson, G. L., 112(a), 113, 167, 456
Addelman, R., 503
Aden, D. P., 692, 694, 701(13), 704(13), 824, 834(15), 848(15)
Adner, M. M., 741
Adolphson, J. L., 74, 77(55), 141(55)
Adrian, M., 471
Agarwal, K. L., 44
Aggerbeck, L. P., 434, 452, 456, 458, 461(7), 462(7), 465, 466, 467(18), 468
Agostini, B., 116(p), 117
Ah, F. W., 59
Aimmer, F., 852
Akanuma, Y., 536, 552
Aladjem, F., 107, 433, 434(11)
Aladjam, F., 250, 258(28), 705
Alaupovic, P., 5, 75, 76, 81(o), 83, 92(w), 93, 94, 96(A), 112(b), 113, 126(57), 127(58), 263, 288, 297, 298, 299(3), 300(2, 3), 301(3), 302, 303(3), 304(4), 305(4), 310(1), 548, 690, 712, 721(6), 725(6), 741
Albers, J. J., 18, 21, 74, 77(55), 124, 125(177, 178), 141(55), 201, 208, 224, 251, 256, 300, 301(9), 303(9), 417, 431(3), 564, 566(e), 567, 568, 578, 713, 721(11), 837, 849(72), 850
Albert, W., 433, 434(10)
Alexander, C., 72, 85(30)
Alexander, C. A., 21, 22, 67
Alhadeff, B., 855, 856
Alexander, A. E., 392
Alexander, H., 126

Alexander, M., 116(B), 118
Alix, J. F., 301, 309
Allen, F. H., 800
Allen, J. K., 504
Allen, T. M., 658
Allerhand, A., 481, 490(33), 491, 493(33, 35), 495(35), 497(41), 498(34, 41), 499(34), 500(34, 35), 501(34, 35), 502, 503(34, 35), 504(34), 514(33, 35)
Alley, M. C., 529, 530(15), 534(15), 549
Alper, C., 153
Alpers, D. H., 53, 57(67), 61(67, 95), 62(118), 131(z), 134(v), 138, 139, 141, 240, 691, 699(7), 701(28), 713, 716(14)
Alt, F., 796
Amara, S. G., 56
Ameryckx, J. P., 136(Z^{11}), 140, 312
Anantharamaiah, G. M., 629, 633, 643
Anderson, D. W., 115(b), 117, 169, 223, 457
Anderson, L.-O., 327
Anderson, M. A., 736(g), 737
Anderson, R. G. W., 26, 35, 578, 609, 610(6), 823, 895, 896, 899, 908(7), 909(2)
Anderson, R. J., 59
Andrews, A. L., 109, 584
Andy, R., 61, 62(118), 141, 691, 699(7)
Anfinsen, C. B., 534
Angel, A., 712, 721(6), 725(6), 741
Angelin, B., 30, 36(104), 274, 281, 282(3), 286(3)
Annison, G., 736
Anson, D. S., 737
Antonarakis, S. E., 738
Antoniewicz, R., 124
Antonucci, T. K., 878
Anwar, M. K., 633
Apella, E., 235
Applegate, K., 422, 451, 455, 578
Arai, M., 70(8), 71
Arnberg, A., 745
Arnold, K. S., 33, 528, 824, 825(20), 849(20)

Aron, L., 18, 848, 849(57)
Ashbrook, J. D., 328, 330, 332, 333(29, 33), 334, 335(27, 33), 336(33)
Ashraf, M., 581
Ashworth, J. N., 72, 214
Askinazi, L., 651, 655(17)
Askonas, B. A., 535
Assman, G., 339(17), 340, 496, 501(44)
Assmann, G., 50, 63, 79(m), 82, 98(h), 100, 106, 129(h, j, o), 135(Z^1), 137, 140, 216, 276, 373, 374(18), 402, 455, 476, 478(10), 481(10), 558, 562(r, s), 563, 690, 708(4), 721
Astrup, H. N., 127
Atassi, M. Z., 542
Atkinson, D., 81(I), 83, 101, 105(106), 109, 113(v), 114, 116(x), 118, 121, 122(149, 158), 250(3), 251, 252(31), 448, 452(10), 584, 586(20), 587(20), 588(20), 592, 593(20), 594(20), 595(20), 596(20, 34), 597, 598(36), 601(34, 36), 603, 604(20, 36, 41), 605(20, 36, 41), 690
Attie, A. D., 33, 35(121), 70(10), 81, 107
Aune, K. C., 562(w), 563, 574
Austin, J. P., 109
Aveldano, M. I., 323
Avigan, J., 214, 215(15)
Avila, E. M., 58, 489, 490(33), 493(33), 495, 497(41), 498(41), 514(33)
Avitable, J., 501
Avogaro, P., 433
Axel, R., 812
Ayrault Jarrier, M., 301, 309
Azuma, J., 106, 110(k), 111, 121, 133(I), 139

B

Baccarana-Contri, M., 94, 98(f), 99, 115(l), 117, 120, 121(144)
Bachmann, L., 459, 461
Bachorik, P., 849(69), 850
Bachorik, P. S., 151, 153(1), 251
Bachorik, P. W., 35
Backman, K., 808
Bailey, D. W., 881, 882(13), 887, 889
Baird, M., 48
Baker, C., 51

Baker, D. P., 36
Baker, H. N., 134(Q), 139, 294, 542(55, 56), 543, 721, 784, 786, 789
Baker, H. W., 129(h), 137
Baker, N., 65, 66(142), 76, 131(D), 138, 141(65), 248, 262
Bakke, H., 125
Baldwin, W. D., 50
Ball, S., 894
Balmer, J., 19
Baltimore, D., 796
Ban Henegouwen, H. B., 855
Bank, A., 48
Banker, G. A., 270
Banks, J. M., 354, 357(4)
Baralle, F. E., 8, 47, 49, 56, 224, 235, 697(34), 699, 712, 732, 733(14), 734(14), 741, 742(43), 743(44), 799(29), 800
Baranov, O. K., 123, 124(168)
Barany, G., 632
Barello, F. E., 885
Barenholz, Y., 595
Bargoot, F. G., 482, 483, 485(21)
Barker, R., 260
Barker, W. C., 76, 720, 755, 764(12)
Barnes, B. A., 3
Barnhart, C. L., 218, 250, 254(26), 255(26), 376, 377(4), 584, 589(19), 590(19)
Barnhart, R. L., 581
Barr, S. I., 142, 549, 550(94)
Barratt, M. D., 109, 584
Barter, P., 85, 106(84)
Barter, P. J., 31, 103, 105
Barth, R. K., 56, 57(88), 70, 141(7)
Bartlett, G. R., 257
Barto, K. P., 37
Basu, S. K., 26, 37, 58, 59, 63, 135(Z^2), 140, 578, 608, 611(1), 823
Bates, S., 178
Battersby, J. E., 12
Bau, S., 800
Bausserman, L. L., 220, 312, 313(11), 314(11, 15), 315, 367(7), 368, 379
Bautovich, G. J., 419
Baxter, C. F., 232
Bayard, F., 288(10), 289
Beamer, W. G., 880
Beato, M., 52, 54(64)
Beattie, W. G., 746
Beaubatie, L., 78, 79(c), 80(u), 82, 84(76),

88(76), 92(i', k, l), 93, 95(n, o, p) 97, 98 (q, r, s), 101(76, 77), 127(76, 90), 143(76, 77, 90)
Beaucage, S. L., 713
Beaudry, B. A., 327
Beaudru, K., 611
Becker, E. D., 474, 478(1)
Beg, Z. H., 66
Beisiegel, U., 8, 131(y), 138, 223, 240(7), 241(7), 753, 837, 845(50), 896, 899
Belcher, J., 29
Bell, E., 713, 716(14)
Bell, G. I., 736(f), 737, 744, 866
Bell, R. M., 21
Bell-Quint, J., 68, 81(A), 83, 90, 92(o), 93, 96(s), 99(u), 101(96), 109(96), 110(i), 111, 112(m), 114, 116(t, z), 117, 118, 119(96)
Belyaev, D. K., 124
Benassayag, C., 336
Benditt, E. P., 136(Z^{11}, Z^{12}), 140, 311, 312(6), 313(6, 10), 316(4), 317(4), 319(14)
Bengtson-Olivecrona, G., 30, 39(107)
Bengtsson, G., 127
Bensadoun, A., 389
Benson, L., 36
Benton, W. D., 793, 813
Ben-Zeev, O., 877, 880, 884, 885(4)
Berde, C. B., 327, 337(19)
Berenson, G. S., 81(C, F, H, L), 83, 88, 106, 107
Berg, K., 50, 74, 77(52), 124, 125(179, 181), 141(52), 800, 852, 863, 885, 894
Berg, P., 63, 677, 715, 716(21), 734, 812, 813, 816, 860, 865, 868(12), 896, 907
Berg, T., 20, 37, 38(62), 40(62)
Bergseth, M., 87, 99(w), 100, 340, 341(9), 342(9)
Bergstrand, A., 67
Berlin, E., 121, 122(148)
Berliner, J. A., 36
Bersot, T. P., 129(P), 131(y, A), 133(N), 134(R, V), 138, 236, 278, 287, 727, 735, 753, 824, 825(18), 846, 849(18)
Bertram, P. D., 578
Betscholtz, C., 51
Beubler, E., 309
Bhatnager, P. K., 543, 545, 550(66)
Bhattacharya, S. S., 737

Bhown, A. S., 629
Bidallier, M., 87, 99(w), 100
Bieber, L. L., 388, 570
Bier, D. M., 17, 798
Bierboim, H. C., 792
Bierman, E. L., 40, 848, 849
Bigbee, W. L., 534
Biggs, H. G., 188
Bilheimer, D. W., 26, 35, 36, 397, 578, 823
Billington, T., 31, 546
Bina-Stein, M., 671
Bird, A. C., 737
Birdsall, N. J. M., 485
Birkenmeier, E. H., 764, 769(22), 771(22)
Birnboim, H. C., 818
Birnstiel, M. L., 864
Bisgaier, C. L., 24, 131(y), 138, 697(32), 699
Bishop, C. A., 288, 339(10), 340, 341(10), 352
Bishop, J. O., 671
Bissell, D. M., 37
Bittner, M., 48, 538, 738
Bittolo, G., 433
Bjurulf, C., 365
Blake, G., 720, 744, 813, 908
Blanche, P. J., 417, 457, 564, 565(22), 566(f), 567, 571(22), 572(22), 581, 651, 654(15), 655
Blanchette, L., 76, 548
Blanchetto, L., 263
Blaton, V., 81(N, P), 83, 92(x), 93, 96(D), 99 (C, F), 100, 104(g), 105, 120(126), 122(127), 129(k, n), 130(u, x), 137, 138, 142(124, 125), 365, 367, 368(4, 6), 370(4)
Blattner, F. R., 812
Blaufass, M. C., 53, 57(67), 61(67), 134(V), 139, 141
Blechl, A. E., 812
Bligh, E. G., 257
Blix, G., 72, 144
Blobel, G., 61, 63, 696, 699(23)
Block, W. D., 712, 721(6), 725(6), 741
Bloemendal, H., 344
Blomberg, F., 515
Blomhoff, J. P., 92(j), 93, 95(l)
Blomhoff, R., 37
Bloth, B., 165
Blue, M.-L., 26, 58, 823
Blum, C., 273, 280(1), 287, 705
Blum, C. B., 26, 58, 223, 647, 655(1), 735,

823, 824, 834, 837(41), 842(17), 848, 849(57)
Blum, J., 50
Blumenthal, L., 367(9), 368
Blumenthal, R., 657, 658, 659(1), 660(1, 24), 663(1, 12), 665(12, 14), 666(12), 667(12), 668(12, 14)
Bobrow, M., 855
Boedtker, H., 869
Boehm, C. D., 738
Boehni, P., 60
Bogorad, D. D., 712, 721(6), 725(6), 741
Boguski, M. S., 47, 240, 754, 763(8, 29), 764, 767(8), 769(22, 29), 771(8, 22, 29)
Bohlen, P., 256
Böhlen, P., 555, 570(2)
Bohmer, T., 86(*e*), 87, 90, 91(93), 92(*h*), 93, 95(*j*)
Bojanovski, D., 63, 75, 81(*O*), 83, 92(*w*), 93, 96(*A*), 127(58), 224, 303
Bojanovsky, M., 433
Bolivar, F., 808
Bollen, A., 534
Boman, H., 124, 125(179, 181)
Bond, M., 55
Bondjers, G., 218, 223, 300, 301(8), 303(8)
Bongiovanni, G., 647, 650(6), 652(6)
Bonilla, E., 736(g), 737
Bonner, J., 749
Borchardt, R. A., 64, 66, 131(*D*), 138
Borensztajn, J., 65, 131(*D*), 138, 262
Borensztajn, S., 247
Borgaonkar, D. S., 730
Børresen, A. L., 50, 124, 125(182, 183, 184), 852, 885
Bosinger, C. D., 634, 640(17)
Bostrom, K., 218
Botstein, D., 715, 737, 861
Bouthillier, D., 837, 840(49), 841, 845(49)
Bowman, B. H., 50
Boyd, Y., 736
Boyer, S. J., 730
Boyles, J., 37
Boyles, J. K., 26
Bradley, W. A., 34, 36, 70, 88(2), 141(1, 2), 142(213), 151, 158, 219, 247, 379, 528, 529(3), 583
Bradshaw, R. A., 543, 546(57)
Bragdon, J. H., 4, 72, 73(22), 155, 170, 181, 227

Brainard, J. R., 19, 476, 479(9), 480(9), 487(9), 499, 502, 513
Braithwaite, F., 128(*b*), 131(*z*), 135(Z^1), 137, 138, 140, 753, 765(2)
Brandt, J., 327
Brasure, E. B., 478, 479(12)
Breathnach, P., 46
Brecher, P. I., 649, 651(13)
Breckenridge, H. C., 98(*j*), 100
Breckenridge, W. C., 17, 90, 705, 798
Breg, R., 730
Breslow, J. L., 8, 47, 49, 50(26, 27, 28, 41, 44), 56(26), 62, 64, 69(41), 135(*z*), 139, 224, 227, 273, 280(1), 282(15), 295(16), 296, 339(16, 18), 340, 672, 674(6), 675(7), 688, 691, 692, 693, 694, 695(9, 17), 696(18), 697(9, 17, 31), 699(9, 17), 701(9, 14, 17, 22), 703(9), 704(14, 20), 705, 708(22), 711, 712, 713, 717(2, 3), 719(8, 30), 720(2, 7, 8), 721(31), 723(2), 733, 740, 741, 765, 774, 779, 781(8), 793, 800(11), 823, 824(2, 32), 825(2), 834(13, 14, 15, 16), 837(13, 14, 41), 842(17), 845(32), 846(32), 848(14, 15, 32, 40), 849(32), 864, 885, 894
Brewer, H. B., 74, 77(54), 81(*M*), 83, 123(54), 126, 128(*a'*), 129(*h'*, *j*, *n*), 130(*t*), 134(*U*), 136(Z^7, Z^9), 137, 138, 139, 140, 146, 710, 750, 751
Brewer, H. B., Jr., 825, 846, 885, 893, 894
Brewer, H. Bryan, Jr., 213, 223, 224, 226, 227, 228(28), 235, 236, 238(27, 42), 240(42), 241(28), 245(27), 288, 295(17), 296, 306, 307(20), 308(20), 309(20), 342, 375, 377(3), 378, 381, 382, 383(9), 386, 476, 478(10), 481(10), 496, 501(44), 542, 543, 544(59), 546, 562(*r*), 563, 570
Brewer, H. B., Jr., 31, 49, 50, 56, 62, 63, 66
Brewer, J. B., Jr., 837
Brewer, M. B., Jr., 824, 849(25)
Brewster, M. E., 19
Bridoux, A. M., 86(*i*), 87
Brinster, R. L., 52
Brinton, E. A., 40
Britten, R. J., 683
Brock, M., 54, 56
Bronzert, T. J., 128(*a'*), 137, 223, 224, 381, 386, 543, 544(59)
Brouillette, C. G., 201, 513, 559, 564(17),

571(17), 572(17), 574(17), 629, 640, 643, 646
Brouillette, G. G., 404(9), 405, 410(9)
Brown, F. B., 34
Brown, G., 540
Brown, J. H., 544
Brown, J. R., 336
Brown, M. D., 824(33), 825, 848(33), 849(33)
Brown, M. F., 501
Brown, M. S., 8, 19, 20(57), 26, 33(23), 34, 35, 36(26), 37(137), 46, 48, 50, 58, 59, 60, 63, 64(18), 65, 70(9), 71, 85(9), 86(j), 87, 90(85), 92(m), 93, 95(q), 102(85), 105(85), 110(h), 111, 112(j), 113, 135(Z^2), 140, 273, 280(1), 524, 525, 578, 581, 599, 602(38), 606, 607(38, 44), 608, 609, 610(4, 5, 6, 7, 8, 11), 611(1), 612(4, 5), 613(4), 703, 708, 727, 788, 823, 824, 825(18), 834, 837(41), 849(18, 26), 864, 894, 895, 896, 899, 900, 906, 908(1, 7), 909(2, 3)
Brown, R. K., 3
Brown, W. V., 5, 169, 217, 220, 288, 367, 368(3), 370(3), 371, 372(16), 379, 562(q, v), 563, 880
Brownlee, G. G., 737
Bruck, C., 534
Bruns, G. A. P., 47, 49, 50(26, 28, 44), 56(26), 295(16), 296, 712, 793, 796, 797(16, 17), 800, 864, 885, 894
Brunzell, J. D., 849
Bryant, L. R., 81(J), 83
Buchko, M. K., 8
Budelier, K. A., 62, 235, 691, 697(8), 704(8), 708(8), 752
Buell, G. N., 44
Bugaisky, G. E., 56
Bull, H. B., 388
Bullock, B. C., 142
Bumpus, F. M., 146
Burke, D. J., 169, 204, 417, 424(2), 425(2), 456
Burke, J. F., 45, 816
Burke, W. F., 618
Burley, R. W., 746
Burnette, W. M., 893
Burnette, W. N., 538
Burns, A. T. H., 671
Burns, C. H., 8
Burns, C. P., 323

Burnstein, M., 145, 306, 309(18), 892
Burrous, M., 151
Burton, K., 689
Butbul, E., 33
Butler, D. P., 79(l), 82
Butler, R., 549, 551(96)
Buttress, N., 80(w, y), 83, 85, 86(k), 87, 92(n), 93, 95(r), 99(t), 112(l), 114, 116(s), 117
Byers, M., 51
Byers, S. O., 647
Byrne, R. E., 151, 690
Byus, C., 681, 803

C

Cabana, V. G., 256, 578, 713, 721(11)
Cachera, C., 530
Calame, K., 55
Calandra, S., 79(h), 82, 94, 95(g), 97, 98(f), 99, 101, 102(107), 115(l), 117, 120, 121(144), 128(b, f), 137
Calhoun, W. K., 79(l), 82
Calvert, G. D., 103, 113(P), 114
Camejo, G., 5, 115(j), 117, 215, 216, 388, 392, 393, 395, 584, 585
Campbell, I. D., 493
Camus, M. C., 73, 76(41), 78(41), 79(e), 82, 92(c), 93, 94, 95(d), 96, 98(d), 99, 101(41), 102(41), 103(41), 110(d), 111, 112(c), 113, 115(c), 117, 120(41), 123(41), 128(c), 129(q), 132(E), 135(w, Z^2), 137, 138, 139, 140, 141(67), 877, 878(3), 891(3)
Cancro, M. P., 550
Canet, D., 498
Canzler, W., 849(71), 850
Capecchi, M. R., 51
Carbon, J., 812
Cardin, A. D., 151, 159, 218, 247, 250, 254(26), 255(26), 376, 377(4), 581, 584, 589, 590
Carlson, L. A., 110(b), 111, 455
Carrella, M., 824, 825(21), 849(21)
Carrington, A., 474
Caruthers, M. H., 713, 714(18), 789
Casadaban, M. J., 48
Casey, J., 679
Casey, M. L., 8, 33(23), 46

Caspersson, T., 855
Castelli, W. P., 223, 741
Castellino, F. J., 121, 260
Caster, H., 99(F), 100, 105, 120(126), 129(n), 130(x), 137, 558, 562(s), 563
Cazzolato, G., 433
Chacko, G. K., 40
Chaikoff, I. L., 90
Chait, A., 37, 849
Chajek, I., 300, 301(7), 303(7)
Chalvardjian, A., 72(33), 73
Chambers, R. E., 315
Chambon, P., 46
Chan, L., 10, 49, 50, 56, 57, 70, 88(2, 3), 141(1, 2), 142(3), 224, 235, 712, 745, 746, 750, 752, 777, 787, 859, 860(27), 861(27), 862(27), 864, 876
Chan, S. I., 487
Chan, T.-M., 511
Chandler, M. E., 865, 870(13), 874(15)
Chang, B. C., 561(d), 563
Chang, B. H., 404(9), 405, 410(9)
Chang, C. N., 696, 699(23)
Chang, D. J., 46, 47, 61(17), 735, 801, 806(2), 808, 810(2, 20), 811, 812, 813, 815(13, 18)
Chang, J. Y., 142, 549, 550(94)
Chang, M. K., 354
Chantler, S. M., 528
Chao, J., 151, 159, 247
Chao, Y.-S., 34, 64, 824, 849(24)
Chapman, D., 485, 486
Chapman, M. J., 71, 73, 74, 75(20), 71(20, 41), 78(41, 43), 79(e), 81(G, Q), 83, 84(50), 85(20), 86(d), 87, 88(43), 90(20), 92(c, g, u, y), 93, 94, 95(b, d, i), 96(x, B, E), 97, 98(d), 99(A), 101(41, 50), 102(20, 41), 103(41), 105, 106(20), 107, 110(d, g, l, m), 111, 112(c, h), 113(p, u, w), 114, 115(c), 116(A), 117, 118, 120(20, 41, 43, 50), 123(41, 43, 50), 126(20), 127, 128(c), 129(i, n, q), 130(x), 132(E, F), 133(I, J, K, L, O), 135(w, Z^2, Z^4), 137, 138, 139, 140, 141(20, 50, 67), 129, 164), 142(45, 82), 153, 171, 249, 877, 878(3), 891(3)
Chapman, M. J. S., 182, 543, 549(62), 551
Chappell, D. A., 36
Chataing, B., 38
Chatterjee, S., 108

Chen, C., 422, 564
Chen, C. H., 18, 224, 566(e), 567, 879
Chen, G. C., 128(d), 135(Z^3), 137, 140, 142, 263, 519, 520, 521(4), 522, 523(4), 524, 525, 526, 527(11), 609, 610(8), 825, 848, 849(58)
Chen, H. Y., 52
Chen, J. R., 26
Chen, J.-S., 543, 546(57)
Chen, S., 856
Chen, P. S., 556
Chen, R., 658
Chen, R. F., 322
Chen, T.-C., 513
Cheng, S.-L., 37
Cheng, Y., 812
Cheng, F., 81(A), 83, 92(o), 93, 96(s), 99(u), 110(i), 111, 112(m), 114, 116(t), 117
Cheung, F., 90, 101(96), 109(96), 119(96)
Cheung, M., 201
Cheung, M. C., 21, 300, 301(9), 303(9), 417, 431(3)
Cheung, P., 10, 49, 56, 224, 712, 752, 777, 787, 859, 860(27), 861, 862, 864, 876
Chick, H., 144
Child, J., 37
Child, J. S., 250, 258(30), 544
Childers, R. F., 489, 490(33), 491, 493(33, 35), 495(35), 500(35), 501(35), 502, 503(35), 514(33, 35)
Chin, D. J., 8, 59, 60, 63
Chin, J. C., 58
Chin, R. C., 232
Chiovetti, R., Jr., 24, 657
Chirgwin, J. M., 42, 681, 802, 803(3), 820
Chiu, P., 715
Chobanian, J. V., 649, 651(13)
Choo, K. H., 737
Chou, J., 48
Chou, P. Y., 584, 765
Chow, A., 17, 798
Chrambach, A., 418, 419
Christeff, N., 336
Christiansen, N. J., 21
Chun, P. W., 249, 252(22), 253(22), 260(22)
Chung, B. H., 24, 92(d), 93, 98(g), 134(s), 139, 182, 188(8), 190, 194(10), 195, 201, 513, 559, 564(17), 571(17), 572(17), 574(17), 611, 629, 640, 643, 646, 891

Chung, H., 128(e), 137, 224
Chung, J., 310, 339, 341(8)
Church, E. C., 220, 379
Cistola, D. P., 327, 333(25), 337(25)
Clark, A., 89, 91(92), 105(92), 119(92), 120(92)
Clark, S. B., 448, 452(10)
Clark, W. A., 79(k), 82, 92(d), 93, 95(e), 98(g), 112(e), 113, 115(g), 117
Clarke, C. F., 8, 59
Clarke, J., 658
Clarke, J. T. R., 108, 327, 337(18)
Clarke, L., 812
Clarke, S., 260
Clarkson, T. B., 124
Clayton, J. F., 737
Cleland, L. G., 658
Clevidence, B. A., 81(H), 83
Clinton, G. M., 66, 131(d), 138
Clouse, A. O., 503
Clute, O. L., 337, 338(40)
Codina-Salada, J., 746
Coffin, S., 485, 486
Cohen, C., 126
Cohen, L. W., 48
Cohn, E. J., 3, 72, 214
Colacicco, G., 392, 393, 399
Cole, S. F., 691, 823
Cole, T. G., 529, 535, 543(39), 545, 547(84), 548(7), 549(7, 39, 84), 550(84), 551(7)
Coleman, D., 880
Coleman, D. C., 879, 880, 889
Coleman, D. L., 880
Coleman, R., 21
Colescott, R. L., 634, 640(17)
Colgan, V., 53
Collen, D., 543
Collins, F. D., 79(i), 82
Collins, J., 812
Colten, H. R., 715, 791, 796
Combs, J. W., 874
Cone, J. T., 190, 194(10), 195, 201, 611, 891
Conn, A., 802, 803(4)
Conneally, P. M., 736(g), 737
Conner, B. J., 734, 739
Conrad, B., 673
Cook, H. W., 327, 337(18)
Cook, P. I., 634, 640(17)
Cook, W. H., 74, 110(k), 111, 116(z), 118
Cook, R. P., 257

Cooper, A. D., 20, 90, 824, 825(21), 849(21)
Cordes, E. H., 476, 479(9), 480(9), 487(9), 490, 491, 493(33, 35), 495(35, 38), 496, 497(38, 41, 43), 498(38, 41), 500(35), 501(35), 502(38), 503(35, 38, 42), 504(38, 42, 43), 505, 506, 507(43), 514(33, 35)
Corey-Gibson, J., 169
Coseo, M. C., 550
Costello, M. J., 459
Cotman, C. W., 270
Coulson, A. R., 44, 717, 902
Cowburn, D., 485, 486
Cox, A. C., 116(y), 118
Cox, D. J., 260
Cox, D. W., 17
Cox, R. A., 681
Crary, D. D., 126
Crea, R., 724
Creagan, R. P., 857
Crespi, H. L., 156
Crkvenjakov, R., 681, 803
Cross, C. E., 80(x), 83, 112(l), 114, 116(s), 117, 127, 452
Cross, T. T., 515
Crother, S., 848
Cryer, A., 28, 29(95), 80(t), 82
Cuatrecasa, P., 534
Cuatrecasas, P., 299
Culwell, A., 367(9), 368
Cumming, A. M., 50, 885
Cump, E., 568
Cumper, C. W. N., 392
Curry, M. D., 303
Curtiss, L. K., 354, 357(3), 358(3), 359, 360(3, 10), 361(10), 529, 530(14), 534, 535(14), 538(14), 543(14), 545(28), 549, 551(96)
Cushley, R. J., 513, 514(73)
Cysyk, R. L., 658

D

Dabach, Y., 39
Dagert, M., 808
Dahlen, G. H., 74, 247
Dairman, W., 256, 555, 570(2)
Dalal, K. B., 263
Dalbadie-McFarland, G., 48
Daley, M. J., 358

Dalferes, E. R., 81(C), 83, 87
Dalton, D. J., 49
Damen, J., 432
Dangerfield, W. G., 72(35), 73, 85(35)
Daniel, L. W., 327
Dargar, R., 529, 534(12), 535(12), 543(39), 544(12), 549(12, 39), 551, 552(101)
Darnell, J. E., 54
Das, H. K., 47, 50(28)
Dash, M. J., 419
Datta, S. K., 103, 115(f), 117, 120(110)
Datta, S. R., 888, 889(30)
David, J. A. K., 604
Davidson, E. O., 156
Davidson, J. N., 863
Davidson, N., 679
Davies, J. M., 529, 534(12), 535(12), 543(39), 544(12), 548(7), 549(7, 12, 39), 551(7)
Davies, G. E., 47, 613, 808, 810(20), 811, 812, 815(13)
Davies, K. E., 736(h, i), 737, 800, 894
Davies, R. S., 52
Davignon, J., 76, 87, 99(w), 100, 263, 548, 837, 840(49), 841(49), 845(49)
Davis, B. J., 536
Davis, C. G., 8, 33(23), 46
Davis, L. D., 78, 94, 109(75), 110(a), 111, 165
Davis, L. G., 478, 479(11), 481(11)
Davis, M., 55
Davis, M. A. F., 116(y), 118, 690
Davis, R., 861
Davis, R. A., 64, 66, 131(D), 138
Davis, R. W., 715, 737, 793, 813
Dawson, G., 108
Dawson, J. R., 250, 259(23)
Dawson, P. A., 26
Dawson, R., 584, 589(12), 590(12)
Day, C. E., 72, 85(30)
Dayhoff, M. O., 76, 720, 755, 764(12), 773, 787
Dearborn, D. G., 389
DeBault, L. E., 326
deBoer, E., 736
Deckelbaum, R., 33
Deckelbaum, R. J., 103, 122, 507, 561(f), 563, 574, 601, 608
Dedieu, J. C., 458, 461(7), 462(7)
Deeley, R., 53, 671

Deganello, S., 115(a), 117
Degovic, G., 113(P), 114, 119, 121(143), 122(143)
DeLalla, O., 144
DeLalla, O. F., 72, 73(21), 155
DeLamatre, J. G., 753
Delcourt, R., 72, 122(29)
DeLean, A., 547
Delpierre, C., 530
DeLucia, A. J., 81(J), 83
Denniston-Thompson, K., 812
Dennis, W. H., 658
Denis, M. J., 459
Denning, G. M., 326
De Parscau, L., 25
dePury, G. G., 79(i), 82
Derouaux, G., 3
Deufel, T., 250, 258(29)
Deutsch, D. G., 542
Deykin, D., 18
Diamond, D., 869
Dickson, J., 109
DiDomenico, B. J., 56
Dieckman, M., 860
Dieckmann, M., 677, 715, 716(21), 734, 794, 813, 865, 868(12), 907
Dietrich, W., 484
Dietschy, J. M., 33, 35, 36(122)
Dinh, D. M., 223
DiVerdi, J. A., 515
Dixon, H. B. F., 544
Dizikes, G. J., 56, 877
Dobberstein, B., 699
Dobbins, W. O., 68
Dobberstein, B., 61
Dobelier, K. A., 25
Dobson, C. M., 493
Doddrell, D., 490, 498(34), 499(34), 500, 501(34), 503(34), 504(34)
Doi, Y., 106
Dolphin, P. J., 67, 543, 712, 721(6), 725(6), 741
Doly, J., 792
Donahue, T. F., 52
Donald, J. A., 894
Donaldson, V. H., 151, 159, 247
Doolittle, D. P., 880
Doolittle, R. F., 758, 769, 770(41), 771(14)
Doppert, B., 855

Doria, G., 535
Dory, L., 27, 574, 753
Dosik, H., 866
Douglas, J. F., 116(B), 118
Doumas, B. T., 188
Douste-Blazy, Ph., 529, 544(8), 545(8)
Doyle, J. T., 223
Dozy, A. M., 735
Dragsten, P., 657, 659(1), 660(1), 663(1)
Drake, S. J., 122, 123(163)
Dray, S., 124, 125(177, 178)
Drengler, S. M., 109, 559, 561(h), 563, 570(13), 574(13)
Dresel, H. A., 37
Drevon, C. A., 20, 37, 38(62), 40(62)
Driscoll, C., 48
Driscoll, D. M., 57, 58(94), 59(99), 64
DuBien, L. H., 73, 74(40), 78(40), 112(f), 113, 115(h), 117, 120(40)
Dubochet, J., 471
Dugaiczyk, A., 745, 746, 869
Duncan, L. A., 190, 194(10), 891
Dunn, G. D., 105
Dyer, W. J., 257

E

Early, P., 55
Eatough, D. J., 327, 332(23)
Eckel, R. H., 28
Eckerson, M. L., 79(k), 82, 89, 91(92), 92(c), 93, 95(e), 98(g), 105(92), 112(e), 113, 115(g), 117, 119(92), 120(92)
Eddy, R., 51, 893, 894(47)
Edelglass, J., 712, 721(6), 725(6), 741
Edelhoch, H., 217
Edelstein, C., 5, 14, 25, 62, 99(B), 100, 104(i), 105, 116(D), 118, 121, 129(l, n), 130(w), 131(z), 137, 138, 146, 151, 153(3), 154, 156, 168(14), 171, 182, 215, 224, 235, 236, 339(15, 20), 340, 341(2, 3, 4, 7, 15), 342(2, 20), 343(3, 15), 345, 347, 375, 377(1), 379, 383, 434, 458, 570, 572(35), 691, 697(8, 29), 699, 704(8), 708(8), 752, 892
Edelstein, D., 79(j), 82, 115(n), 117, 128(e), 130(v), 137, 138
Eden, H. A., 227

Eder, H. A., 4, 40, 72, 73(22), 79(j), 82, 128(b), 131(z), 135(Z^1), 137, 138, 139, 140, 155, 170, 181, 753, 765(2)
Edge, M., 746
Edge, S. B., 236, 238(42), 240(42)
Edgington, T. S., 359, 360(10), 361(10), 529, 530(14), 534, 535(14), 538(14), 543(14), 545(28)
Edmundson, A. B., 767
Edner, O. J., 482, 483, 485(21)
Edwards, P. A., 8, 36, 37, 59, 250, 258(30), 544, 894
Efstratiadis, A., 793, 805, 812
Eggena, P., 433, 434(11)
Ehnholm, C., 165, 279, 287(14)
Ehrlich, S. D., 808
Eicher, E. M., 880
Eiseman, R. E., 796, 797(16)
Eisenberg, D., 17, 402, 403, 406, 767
Eisenberg, S., 25, 33, 78, 95(f), 97, 98(i), 100, 102(74), 103(74), 105(74), 112(g), 113, 115(i), 117, 120(74), 128(a), 137, 156, 397
Elam, R., 354, 357(4)
Elbein, R. C., 90, 91(95), 92(p), 93
Elbrecht, A., 53
Elder, E., 354
Elliott, H., 4
Elliott, H. A., 213
Elliott, R. W., 56, 57(88), 70, 141(7)
Ellman, G. L., 254
Ellsworth, J. L., 38
Elovson, J., 65, 66(142), 76, 131(D), 138, 141(65), 248, 250, 259(24), 261(53), 262
Elshourbagy, N. A., 26, 46, 61(17), 47, 240, 754, 763(8, 29), 764, 767(8), 769(22, 29), 771(8, 22, 29), 801, 806(2), 810(2), 813, 815(18)
Elson, C. E., 125
Emes, A. V., 435
Emond, D., 580
Endemann, G., 116(z), 118
Enders, G. H., 23, 29, 35(71)
Engelhorn, S. C., 534
Engelke, D. R., 51
Enquist, L., 812
Entenman, C., 90
Eoff, S. P., 81(z), 83
Erickson, A. H., 696, 699(23)

Erickson, S., 20
Erickson, S. K., 90
Ericsson, J. L., 67
Ericsson, L. H., 136(Z^{11}), 140, 311, 312(6), 313(6)
Eriksen, N., 136(Z^{11}, Z^{12}), 140, 311, 312(6), 313(6, 10), 316(4), 317(4), 319(14)
Ertel-Miller, J. C., 56, 57(88), 70, 141(7)
Escaig, Jacques, 471
Eskild, W., 37
Evans, H. J., 737
Evans, L., 459
Evans, R. M., 56
Evans, M. T. A., 389, 395(9), 399
Evans, S., 432
Ewens, S. L., 300, 301(9), 303(9)
Ewing, A. M., 155
Ey, P. L., 534

F

Faber, H. E., 812
Faergeman, O., 25, 28(80), 31(80), 85, 90, 92(e), 93, 106(84), 110(f), 111
Fager, G., 223, 300, 301(8), 303(8), 310
Fahey, J. L., 534, 535
Fainaru, M., 25, 223, 263, 713, 823
Fairclough, G. F., Jr., 159, 218, 249, 379
Fairwell, T., 50, 128(a'), 137, 224, 226, 235, 295(17), 296, 543, 544(59)
Falck, J., 610, 611(14)
Falk, C. T., 800
Falko, J. M., 536, 837
Faloona, G. R., 81(z), 83
Farber, E., 68
Fareed, G. C., 56, 877
Farr, N. J., 253, 256(40), 555, 570(1a), 602, 611
Farrar, T. C., 474
Fasman, G. D., 576, 584, 604, 765
Fathman, C. G., 529, 530(15), 534(15)
Faust, J. R., 59, 60, 599, 602(38), 607(38), 609, 610(4), 612(4), 613(4)
Feeney, J., 511
Feg, G., 800
Feinberg, A. P., 813
Felber, B. K., 55
Feldhoff, R. C., 337, 338(40)

Feldman, E. B., 448, 452(10)
Feldmann, R. J., 627, 767
Felgner, P., 662
Felker, T. E., 823
Fellows, R. E., 130(v), 138, 236
Felski, C., 79(n), 82
Felts, J. M., 4
Fenner, C., 696
Fennick, B., 692
Feramisco, J. R., 44
Ferenz, C. R., 628(5), 629
Ferguson, K. A., 270
Fernando-Warnakulasuriya, G. J. P., 79(k), 82, 89, 91(92), 92(d), 93, 95(e), 98(g), 105(92), 112(e), 113, 115(g), 117, 119(92), 120(92)
Ferrari, N., 737
Ferreri, L. F., 90, 91(95), 92(P), 93
Ferretti, J. A., 474, 478(1)
Feussner, G., 227, 241(31), 727, 852
Fey, G., 852, 885
Fiddes, J. C., 44
Fidge, N., 74, 92(r), 93, 99(x), 101(47), 121, 123(47, 153, 240), 241(48), 546
Fidge, N. G., 31
Fidge, N. H., 94, 96(u), 131(B), 138, 753
Fielding, C., 14
Fielding, C. H., 708
Fielding, C. J., 28, 38, 57, 90, 126, 128(d), 135(Z^3), 137, 140, 142, 223, 300, 301(7), 303(7), 578, 579(53), 721, 825
Fielding, P. D., 721
Fielding, P. E., 25, 38, 126, 223, 303
Finer, E. G., 109, 121, 122(154), 482, 483(22), 485(22), 487(22), 489(22), 497(22), 584
Finer-Moore, J., 60
Fink, G. R., 52
Finlayson, J. S., 306, 308(21)
Finlayson, R., 72(35), 73, 85(35)
Firestone, R., 610, 611(14)
Fischer-Dzoga, K., 178
Fish, P. A., 19
Fisher, H. G., 540
Fisher, M., 354, 355(4), 357, 358(7)
Fisher, W. R., 247, 249(5), 250(32), 251(5), 252(5, 22), 253(22, 38), 254(25), 256, 258(5, 38), 259, 260(22), 584, 590
Fitch, W. M., 765, 773, 776, 779, 784, 786, 787

Flanagan, S. A., 79(o), 82, 92(f), 93, 95(h), 98(k), 105, 127(116)
Flavell, R. A., 736, 739
Flavin, M., 658
Fless, G., 458
Fless, G. M., 71, 74(17), 77(17), 96(y, z), 97, 101, 102(109), 104(h), 105, 106, 107, 113(r, s, t), 114, 120(17), 121(109, 131), 141(17), 142(122), 178
Fletcher, J. E., 321, 322, 328(7), 329(7), 332, 333(7, 33), 334, 335(7, 27, 33), 336(33), 337(7), 338(7)
Flint, S. J., 671
Flinta, C., 767
Fogelman, A. M., 8, 36, 37, 59, 250, 258(30), 544
Folch, J., 555
Folk, G. E., 80(s), 82
Forbes, G., 823
Foreman, J. R., 171, 182, 542
Forester, G. P., 24, 647, 655(1)
Forester, L., 647, 650(6), 652(6)
Forgez, P., 73, 74, 76(41), 78(41), 79(e), 81(Q), 82, 83, 84(50), 92(c, y), 93, 94, 95(d), 96(E), 98(d), 99, 101(41, 50), 102(41), 103(41), 107, 110(d, m), 111, 112(c), 113(w), 114, 117, 120(41, 50), 123(41, 50), 128(C), 129(i, n, q), 130(x), 132(E), 133(J, L), 135(w, Z^2), 137, 138, 139, 140, 141(50, 67) 877, 878(3), 891(3)
Formisano, S., 382, 570
Fornieri, C., 79(h), 82, 95(g), 97, 101, 102(107)
Forte, D. Jr., 848
Forte, G. M., 73, 116(r, u), 117, 120(37), 155, 157(2), 447
Forte, T., 68, 73, 112(K), 114, 116(g), 117, 195, 223, 313, 448, 451, 452(10), 455, 458
Forte, T. M., 80(x), 81(A), 83, 90, 92(o), 93, 96(s), 99(u), 101(96), 109(96), 110(i), 111, 112(b', l, m), 113, 114, 115(b), 116(s, t, z), 117, 118, 119(96), 127, 169, 251, 442, 452(3), 455, 456, 457, 564, 565(22), 566(f), 567, 568, 571(22), 572(22), 578, 581, 583, 651, 654(15), 655
Foster, H. L., 889
Fountain, M. W., 657
Fox, J. G., 889
Fox, R. R., 126

Francecchini, G., 227, 241(31), 278, 622, 727, 735
Francke, U., 50, 864, 894
Fraenkel-Conrat, H., 544
Franklin, E. C., 311, 314(3)
Franklin, F. A., 35
Frase, S., 19
Fraser, P., 694, 704(20)
Fredrickson, D. S., 4, 5, 17, 22, 23(69) 139, 147, 155, 157, 167(1), 193, 217, 220(27), 251, 288, 294, 321, 322(2), 379, 455, 543, 708(4), 721, 798, 824(33), 825, 848(33), 849(33), 877
Freeman, M., 866
Freeman, N. K., 155(7, 8, 9, 10), 156
Freeze, H. H., 107
Frei, E., 739
Frey-Wyssling, A., 457
Fried, M., 81(z), 83, 620, 621, 623
Friedberg, S. J., 623
Friedlander, E. J., 528, 581, 769
Friedman, H. S., 153
Friedman, M., 647
Friend, K. K., 856
Fries, E., 585
Fritsch, E. F., 43, 676, 679(14), 680(14), 720, 732, 744, 902, 903(16), 906(16), 908
Fritsch, F. E., 716, 808, 812, 813(4), 814(4), 815(4)
Fritz, R., 106, 107
Frosi, T. G., 622
Fruchart, J.-C., 530
Fujita, H., 386
Fukami, A., 447
Fukazawa, C., 56, 58(89), 70, 135(Z^1), 140, 141(4), 763, 801, 804(1), 805(1), 806(1), 808(1), 809(1), 810(1), 822
Fukushima, D., 404(7, 8), 405, 408(8), 628(4), 629, 636, 644(4), 721, 769, 771
Furlong, L. A., 812
Furman, R. H., 5, 288

G

Gage, S. H., 19
Gaines, G. L., Jr., 390, 391(10), 399(10)
Gainsborough, H., 213
Galfre, G., 530, 534

Gall, J. G., 864
Gallagher, J. G., 74, 112(*o*), 114, 119(48), 123(48), 133(*I*), 139, 562(*w*), 563, 574
Galster, W., 85
Galton, D. J., 741, 742(43), 743(44), 744, 885
Galzigna, L., 588
Gambhir, P. N., 474, 478(1)
Garcia, L., 5
Garcia, M., 866
Gardiner, S., 214
Gardner, J. A., 213
Garfinkel, A. S., 17, 29(45), 798
Garner, C., 721
Garner, C. W., 129(*h*), 137
Garnier, J., 766
Gartside, P. S., 290
Gascon, P., 846
Gasser, S., 60
Gaubatz, J. W., 74, 248, 504
Gazith, J., 260
Gedde-Dahl, T., 49, 50(43)
Gedde-Dahl, T., Jr., 852, 863, 864(2)
Geer, J. C., 92(*d*), 93, 98(*g*), 182, 188(8), 190, 194(10), 891
Geever, R. F, 48, 738
Gefter, N., 51
Geller, R., 894
Gerald, P. S., 796, 797(16, 17)
Gerding, J. J. T., 344
Gergen, J. P., 747
Gerhards, R., 484
Gerrity, R. G., 79(*P*), 82
Gershoni, J. M., 535, 537(38)
Getz, G. S., 42, 57, 58(94), 59(99), 61(91), 64, 65, 66, 73, 74(40), 78(40), 112(*f*), 113, 120(40), 131(*D*), 138, 247
Getz, G. V., 115(*h*), 117
Gherardi, E., 79(*h*), 82, 95(*g*), 97, 101, 102(107)
Ghesquiere, J.-C., 530
Ghiselli, G. C., 104, 141, 142(213), 224, 226, 228(28), 241(28), 825, 837, 846
Ghisellini, M., 128(*b*, *f*), 137
Gianfranceschi, G., 622
Giannelli, F., 737
Gianturco, H., 151
Gianturco, S. H., 34, 36, 158, 219, 247, 379, 528, 529(3), 546
Giardina, P. J. V., 738
Gidez, L. I., 129(*P*), 133(*N*), 134(*R*, *V*), 138

Gifter, M. L., 358
Gil, G., 59, 60, 63
Gilbert, W., 44, 717, 791, 792(5), 905, 906
Gill, L., 50, 746
Gillespie, J. M., 3
Gilliam, E. B., 134(*Q*), 139, 151, 219, 379, 542(56), 543, 789
Gilman-Sachs, A., 124, 125(180)
Gingsberg, H. N., 169
Ginsberg, H. W., 880
Ginsburg, G. S., 511, 595, 596(34), 597, 598(36), 601(34, 36), 604(36), 605(36)
Giotas, C., 427
Girard-Globa, A., 624
Gishelli, G., 62
Gishelli, G. C., 105, 142(119)
Gjone, E., 422, 451, 455, 578
Glass, C., 568, 581
Glass, C. R., 39, 40
Glasscock, M. A., 195, 201
Glauert, A. M., 442
Glauman, H., 67
Glenner, G. G., Jr., 315
Glickman, J., 24
Glickman, R. M., 24, 131(*y*), 138, 223, 240, 241(47), 578, 697(32), 699, 713
Glines, L. A., 169
Glineur, C., 534
Glisin, V., 681, 803
Glomset, J. A., 4, 17, 38, 422, 451, 455, 578, 580
Glonek, T., 476, 477, 478(8), 479(8, 11, 12), 481(11)
Glueck, C. J., 496, 497(43), 503, 504(43), 507(43), 824(32), 825, 845(32), 846(32), 848(32), 849(32)
Gluecksohn-Waelsch, S., 880
Glushko, V., 490, 498(34), 499(34), 500(34), 501(34), 503(34), 504(34)
Gluzman, Y., 812
Goerke, J., 520, 649, 658
Goff, S. C., 738
Gofman, J. W., 4, 72, 73(21), 144, 155, 213
Goh, E. H., 122
Goldberg, R. B., 712, 721(6), 725(6), 741
Goldberger, G., 62, 688, 691, 693, 695(9, 17), 697(9, 17), 699(9, 17), 701(9, 17), 703(9), 715, 717, 779, 791
Goldberger, R. F., 671
Goldblatt, P. J., 68

Goldman, B., 696, 699(23)
Goldman, P., 53
Goldstein, D., 182
Goldstein, J. F., 800
Goldstein, J. L., 8, 19, 20(57), 26, 33(23), 34, 35, 36(26), 37(137), 46, 48, 50, 58, 59, 60, 63, 64(18), 65, 70(9), 71, 85(9), 86(*j*), 87, 90(85), 92(*m*), 93, 95(*q*), 102(85), 105(85), 110(*h*), 111, 112(*j*), 113, 135(Z^2), 140, 273, 280(1), 524, 525, 578, 581, 599, 602(38), 606, 607(38, 44), 608, 609, 610(4, 5, 6, 7, 8, 11), 611(1), 612(4, 5), 613(4), 703, 708, 727, 788, 823, 824(33), 825(18), 834, 837(41), 848(33), 849(18, 26, 33), 864, 894, 895, 896, 899, 900, 906, 908(1, 7), 909(2, 3)
Goldstein, S., 74, 81 (*G, Q*), 83, 84(50), 88(43), 92(*u, y*), 93, 95(*b*), 96(*x, B, E*), 97, 99(A), 101(50), 106, 110(*l, m*), 111, 113(*p, u, w*), 114, 116(A), 118, 120(43, 50), 123(43, 50), 129(*n*), 130(*x*), 133(*I, J, K, L*), 137, 139, 141(50, 129, 164), 171, 543, 549(62), 551, 712, 721(6), 725(6), 741
Gollan, J., 338
Gonen, B., 529, 534(12), 535(12), 544(12), 549(12)
Gong, E. C., 447
Gong, E. L., 251, 417, 457, 564, 565(22), 566(*f*), 567, 568, 571(22), 572(22), 578, 581, 651, 654(15, 16), 655
Gooding, K. M., 343
Goodman, D. S., 328, 329(28), 332, 333(28, 30), 335(28)
Gorden, J. L., 631
Gordon, J., 361, 423, 438, 440(20), 531, 535, 536(35)
Gordon, J. I., 25, 47, 53, 57(67), 61(67, 91, 95), 62(118), 131(*y, z*), 134(*v*), 138, 139, 141, 151, 153(3), 224, 235, 240, 671, 691, 697(8, 29, 32), 699(7), 701(28), 704(8), 708(8), 752, 754, 763(8, 29), 764, 767(8), 729(22, 29), 771(8, 22, 29)
Gordon, J. S., 26
Gordon, L., 775
Gordon, R. S., Jr., 321, 322(2)
Gordon, T., 223
Gordon, V., 74, 285, 520, 521(4), 522, 523(4)
Gorini, G., 535
Gorman, C. M., 60

Gorrissen, H., 513, 514(73)
Goss, S., 863
Gotto, A. M., 7, 13, 16, 18, 22, 23(69), 25, 32, 35(117), 41, 42(1), 74, 99(*c*), 100, 104(*g*), 105, 112(*o*), 114, 119(48), 121(121), 122(121, 155), 123(48), 129(*h, k*), 130(*u*), 133(*I*), 134(*Q*), 137, 138, 139, 141, 142(119, 213), 145, 147, 247, 248, 288, 294, 339(10), 340, 341(6, 10), 350(6), 351, 352, 367(8, 9, 10), 368(6), 370(12), 371, 372(15), 456, 583, 584, 604, 627, 628(6), 629, 639(6), 642(6), 645(6), 721, 765, 767, 784, 786(19), 787, 789
Gotto, A. M., Jr., 10, 11, 12, 30, 34, 36, 37, 39(107, 108), 151, 158, 219, 377, 379, 403, 404, 405(6), 406(2), 408, 409, 410(6), 411(15, 16), 476, 479(9), 480(9), 487(9), 499, 502, 513, 528, 529(3), 542(56), 543, 545, 546, 550(66), 558, 559, 561(*g, o*), 562(*p, w, x, y*), 563, 564, 566(*g*), 567, 574, 581, 609, 610(6), 625, 628(5), 629, 690, 708(4), 721
Gowland, P., 739
Graham, D. E., 393, 395(18), 401
Graham, F. L., 819
Graham, P., 68
Granda, J. L., 584
Grande, F., 80(*r*), 82
Granot, E., 103
Gray, G., 49, 224, 894
Green, A. A., 72, 79(*a'*), 82
Green, H., 853
Green, M., 686
Green, M. R., 734
Green, N., 126
Green, P. H. R., 24, 223, 240, 241(47), 578, 647, 648(4), 650(4), 651(4), 652(4), 653(4), 654(4), 713
Greene, D. G., 74, 77(56), 92(*v*), 93, 141(56), 825
Greenfield, N., 576, 604
Gregg, M. E., 824, 849(25)
Gregg, R. E., 31, 63, 224, 236, 238(42), 240(42), 546, 837
Gregoriadis, G., 658, 661
Gresham, G. A., 105, 142(124)
Grez, M., 807, 810(12)
Grimes, P. A., 659
Gronenborn, B., 675
Groot, P. H. E., 30, 171, 180, 182

Gross, E., 544, 609
Gross, R. H., 44
Groudine, M., 53
Grow, T. E., 620, 621, 623
Gruber, M., 675, 745
Grummer, R. H., 124
Grunstein, M., 44, 716, 791, 808
Grunwald, D. J., 812
Guarante, L., 44, 48(14)
Gubler, U., 732
Guenet, J. L., 880
Gulberg, J. E., 155(9), 156
Gulbrandsen, C. L., 223
Gulik, A., 458, 461(7), 462
Gulik-Krzywicki, T., 456, 458, 459, 461(7), 462(7), 466, 467
Gunther, R. A., 80(x), 83, 112(l), 114, 116(s), 117, 127, 452
Guo, L., 283
Guo, L. S., 57, 649
Guo, L. S. S., 20, 128(d), 135(Z^3, Z^4), 137, 139, 142, 520, 521(4), 522, 523(4), 658, 824, 825, 849(24)
Gurd, F. R.-N., 3, 214
Gurusiddaappa, S., 633
Gusella, J. F., 736(g), 737, 797, 863
Gustasfson, A., 5, 214, 217(13), 288, 300, 301(6), 303(6), 310, 609
Gutowsky, H. S., 514
Guyer, W., 481
Gwynne, J. T., 40

H

Ha, Y. C., 103
Haas, J., 708, 727
Haas, S. M., 388, 570
Habener, J.F., 692
Haberland, M. E., 36
Hagel, P., 344
Hagenbuche, O., 56
Hagins, W. A., 657
Hahm, K.-S., 535, 543(39), 545, 547(84), 549(39, 84), 550(84)
Hahn, P. F., 4
Halari, M., 383, 570, 572(35)
Halegoua, S., 696
Halfman, C. J., 336
Hall, C. E., 457

Halperin, G., 39
Hamer, D. H., 55
Hames, C. G., 223
Hamilton, J. A., 101, 105(106), 327, 333(25), 337(25), 448, 452(10), 486, 487, 488(31), 489(31), 490(33), 491, 492, 493(33, 35), 495(35, 38), 496, 497(38, 41, 43), 498(38, 41), 499(50), 500, 501(35), 502(38), 503(35, 38, 42), 504(38, 42, 43), 505, 506, 507(43), 509, 511, 512, 513, 514(33, 35, 40)
Hamilton, R., 658, 753
Hamilton, R. J., 649
Hamilton, R. L., 14, 19, 20, 21, 22(66), 23(58), 24(58), 25, 27, 28(80), 29, 31(80), 35(71), 57, 67, 78, 85, 86(j), 87, 90(85), 92(e, m), 93, 94, 95(q), 102(85), 105(85), 109(75), 111, 112(j), 113, 128(d), 135(Z^3), 137, 140, 142, 165, 169, 216, 219, 247, 263, 349, 448, 520, 521(4), 522, 523(4), 526, 527(11), 578, 579(53), 823, 825
Hamilton, R. W., 802, 803(4)
Hanahan, D., 748, 816
Hancock, W. S., 12, 288, 339(10), 340, 341(6, 10), 350, 351, 352
Hansma, H., 86(c), 87, 98(m), 100, 115(m), 117, 142
Hanson, R. H., 136(Z^{12}), 140, 311
Hanssum, H., 498, 499(49)
Hara, I., 145
Harder, K. J., 543, 769
Harding, D. R. K., 12, 288, 339(10), 340, 341(10), 352
Hardison, R. C., 812
Hardman, D. A., 31, 65, 76, 106(63), 131(c), 138, 141(63), 247, 262, 263, 264, 523, 529, 536(10), 586
Hardson, K. C., 793
Hardy, W. B., 214
Harmony, J. A. K., 19, 38, 58, 495, 497(41), 498(41)
Harper, M. E., 865, 870(9), 876
Harper, P. S., 736(h, i), 737, 800
Harper, R. W., 223
Harrington, W. F., 260
Harris, H., 550, 854, 855(13), 863
Hart, R. G., 134(p), 135(Z^6), 139, 140
Haschemeyer, R. H., 159, 218, 249, 379
Hasegawa, M., 216, 584, 586

Hasler, J., 124
Hasler-Rapacz, J., 124
Haslinger, A., 52, 54(64)
Hass, J., 227, 241(31)
Hastie, N. D., 56, 57(88), 70, 141(7)
Hatch, F. T., 145, 155(11), 156, 167(11)
Haupt, H., 301, 302, 306, 308(17)
Hauser, H., 109, 399, 481, 487, 493, 495(39), 497(39), 498(39), 513, 584, 807, 810(12)
Havekes, L., 171, 180(6), 182
Havel, R. J., 4, 14, 17, 19, 20(57), 21, 22(66), 23, 24(60), 25(60), 26, 27, 28(80), 29, 30, 31(80, 99), 32(115), 33(115), 34, 35(71, 115, 116), 39(105), 40, 58, 64, 67, 72, 73(22), 85, 86(e, j), 87, 90(85), 91(93), 92(e, h, m), 93, 95(j, g), 101, 102(85, 105), 105(85), 106(84), 109(105), 110(a, a', f, h), 111, 112(j), 113, 134(Q), 135(Z^2), 139, 140, 155, 167, 169(29), 170, 181, 216, 223, 227, 263, 273, 280(1), 288, 292(5), 310, 349, 520, 523, 578, 579(53), 584, 649, 658, 703, 798, 823, 824, 834, 837(41), 849(24, 26, 58)
Hawkes, R., 531
Hawkins, H. L., 578, 579(51)
Hay, R. V., 42, 57, 60, 61(91)
Hayat, M. A., 443
Hayes, K. C., 71, 81(E), 83, 92(t), 93, 96(w), 99(z), 105(18), 116(c), 118, 120(18), 711
Hazzard, W. R., 74, 77(55), 81(K), 83, 141(55), 256, 273, 280(1) 703, 713, 721(11), 800, 834, 837(41), 848, 849(72), 850
Hees, M., 824, 837(31), 849(71), 850
Hegsted, D. M., 79(o), 82, 92(f), 93, 95(h), 98(k), 105, 127(116)
Heiberg, A., 800, 852, 863
Heide, K., 301, 302, 306, 308(17), 534
Heideman, C., 74, 248
Heimberg, H., 105
Heimberg, M., 122
Heimberger, N., 301, 302
Heinen, R. J., 220, 379
Heinrickson, R. L., 128(e), 130(v), 137, 138, 224, 236, 542
Heinzmann, C., 894
Helenius, A., 215, 249, 584, 585(17), 586, 599
Helfman, D. M., 44
Helgerud, P., 20, 37, 38(62), 40(62)

Heline, G., 767
Heller, D., 555
Hellman, K. B., 819
Henderson, L. O., 220, 379
Henderson, P. J. F., 386
Henderson, T. O., 476, 477, 478(8), 479(8, 11, 12), 481(11)
Henkart, P., 657, 659(1), 660(1), 663(1)
Hennekens, C. H., 741
Henri, I., 50
Henry, R., 109, 116(y), 118, 121, 122(154), 482, 483(22), 485(22), 487(22), 489(22), 497(22), 584
Hensley, W. J., 419
Herbert, E., 65
Herbert, P., 17, 798
Herbert, P. N., 22, 23(69), 129(p), 133(M, N), 134(R, V), 138, 139, 147, 220, 236, 251, 288, 294, 296, 312, 313(11), 314(11, 15), 315, 367(7), 368, 379, 455, 543, 564, 566(k), 567, 569(24), 580(24), 690, 708(4), 710, 721, 877
Herbert, R., 134(u), 139
Herbertson, B. M., 880
Hermans, J., Jr., 411
Herring, V., 213
Herrmann, S., H., 548
Heuser, J. E., 459
Hewitt, J., 213
Hickson, D., 408, 409, 562(z), 563
Hickson, D. L., 18
Highet, R. J., 496, 501(44)
Hijmans, W., 314
Hill, P., 116(B), 118
Hillyard, L. A., 90
Himmelfarb, S., 260
Hinton, R. H., 182
Hirose, T., 733, 746, 748(8), 749(8)
Hirs, C. H. W., 381, 381(18)
Hirz, R., 485, 486
Hitman, G. A., 744
Hjelmeland, L. M., 659, 660(24)
Hjortland, M. C., 223
Ho, W. K. K., 432
Ho, Y. K., 8, 26, 37, 58, 59, 135(Z^2), 140, 578, 581, 609, 610, 823
Hoak, J. C., 324, 325(14)
Hobgood, K. H., 65
Hobgood, K. K., 727
Hoeg, J. M., 236, 238(42), 240(42)

Hoekstra, D., 657
Hofer, E., 54
Hoff, H. F., 74, 112(o), 114, 119(48), 123(48), 133(I), 139
Hoffman, B. J., 732
Hoffman, J. S., 311
Hoffman, M. S., 81(z), 82
Hoffmeier, C. A., 753
Hofschneider, P. W., 675
Hogan, V., 651, 655(17)
Hogness, D., 44, 791
Hogness, D. S. 716, 808
Hogue, M., 141, 529, 535, 536(40), 550(11)
Hohansson, C., 855
Hohn, B., 812
Hokom, M., 544
Hokum, M., 37, 250, 258(30)
Holasek, A., 87, 88, 133(I), 134(S), 139, 142, 288, 309, 433, 434(10)
Holdsworth, G. 58, 544
Hollander, G., 39
Holloway, C. J., 433
Holme, R., 92(j), 93, 95(l)
Holmes, D. H., 223
Holmquist, L., 433
Holtgreve, H., 52, 54(64)
Hood, L., 55
Hoover, J. G., 713, 721(11)
Hopkinson, D. A., 854, 855(13)
Hopp, T. P., 771
Hoppe-Seyler, F., 213
Horne, R. W., 442
Hornick, C., 29
Hornick, C. A., 23, 35(71)
Horrocks, L. A., 323
Hort, I., 72
Horwitz, D. L., 77, 115(k), 117, 123(72)
Horwitz, R., 848
Houser, A., 128(a'), 137, 224, 543, 544(59)
Hospattankar, A. V., 295, 296
Housman, D., 797, 863
Howald, M. A., 79(n), 82
Howard, A. N., 105, 142(124)
Howard, B. H., 60, 812
Howard, C. H., 5
Howard, S. C., 64, 68(131), 705
Howe, C. C., 692, 701(13), 704(13)
Howell, K. E., 68
Hradek, G. T., 29
Hsia, J. C., 327

Hsu, J., 367(8, 9), 368, 370(12)
Huang, C., 557
Huang, L. H., 76, 124(68), 125(68), 126, 141(68)
Huang, L. S., 126, 132(H), 139
Huang, Y. O., 65, 66(142), 76, 131(D), 138, 141(65), 248, 262
Hudson, B. S., 327, 337(19)
Huff, M. W., 31
Hughes, L. B., 19, 124
Hughes, S. H., 44
Hughes, T. A., 629
Hughes, W. L., 72, 584
Hughes, W. L., Jr., 214
Hui, D. Y., 551, 708, 824, 825(23), 849(23)
Hulley, S. B., 223
Hulsmans, H. A. M., 849
Hummel, K. P., 880
Humphries, S., 50, 746
Humphries, S. E., 50, 295(18), 296, 852, 885, 894
Hunt, A., 481
Hunt, L. T., 755, 764(12)
Huo, W. H., 124
Hurn, B. A. L., 528
Husbands, D. R., 12
Husby, G., 311, 312
Hutton, W. C., 480
Hwang, S.-L. C., 34, 528, 529(3)
Hynd, B. A., 290, 552
Hynes, R. O., 55

I

Ihm, J., 19, 38
Ikai, A., 249, 536, 584, 585, 586, 623
Ikeda, I., 91
Ikeuchi, T., 866
Illingworth, D. R., 81(D), 83, 92(S), 93, 96(v), 99(y), 116(B), 118
Ilsemann, K., 129(o), 137, 339(17), 340
Imai, H., 121, 122(155)
Imaizumi, I., 223
Imaizumi, K., 25
Inglis, A. S., 746
Ingram, P. H., 796, 797(16)
Innerarity, T. L., 30, 33, 34, 36(104), 37, 51, 74, 77(54), 81(M), 83, 104, 105, 112(i), 113, 123(54), 126, 129(h, j), 130(t),

136(Z^7, Z^8), 137, 138, 140, 263, 273, 279, 287(14), 355, 528, 543, 544, 551(74), 562(*t*), 563, 581, 606, 607(45), 609, 658, 663(12), 665(12, 14), 666(12), 667(12), 668(12, 14), 708, 735, 769, 770(43), 771, 824, 825(19, 20, 22, 23), 837, 846(28, 42), 849(19, 20, 22, 23, 26, 27, 28), 850
Inouye, M., 696
Ish-Horowicz, D., 45, 816
Itakura, H., 536
Itakura, K., 48, 733, 734, 739, 746, 748(8), 749(8)

J

Jackson, C. L., 47, 50(26), 56(26), 691, 793, 864, 885, 894
Jackson, J., 800, 812
Jackson, R., 367(10), 368(6), 370(12), 372
Jackson, R. J., 804, 900
Jackson, R. L., 16, 17, 29(46), 30(46), 34, 39(46), 58, 71, 74, 99(*c*), 100, 103(14), 104(*g*), 105, 112(*o*), 114, 119(48), 121(121), 122(121), 155), 123(48), 126(14), 129(*h*, *k*), 130(*u*), 133(*I*), 134(*Q*, *S*), 137, 138, 139, 142(121), 151, 159, 218, 247, 250, 254(26), 255(26), 294, 295, 296, 376, 377(4), 403, 542(55, 56), 543, 544, 546, 562(*y*), 563, 581, 584, 589(19), 590(19), 627, 692, 721, 765, 767, 784, 786(19), 787, 789
Jacobs, J. C., 250, 259(24)
Jacobson, K., 657
Jacobson, S. F., 51
Jaeger, J. S., 76, 124(68), 125(68), 126(68), 132(*H*), 138, 141(68)
Jaeschke, M., 49, 727, 824, 837(30), 849(71), 850
Jahn, C. E., 224
Jakoby, W. B., 231
Jan, L., 459
Jan, Y., 459
Janado, M., 74, 106, 110(*K*), 111, 116(*z*), 118
Janero, D. R., 21, 23(68), 68
Janus, E. D., 837, 845(48)
Jaramillo, J. J., 36
Jardetzky, O., 474, 478(4), 490(13), 492(4), 497(13), 514(4)
Jauhiainen, M., 288(11, 12), 289

Jay, M., 737
Jeager, J., 37
Jen, L. C., 619
Jenkins, C. H., 142
Jenkin, C. R., 534
Jenkins, P. J., 223
Jensen, E., 711
Jensen, L. C., 167, 170, 182, 455
Jentoft, N., 389
Jerabek, L., 126
Joerns, S., 456
Johansson, J., 767
Johansson, M. B., 88
Johansson, M. B. N., 72(34), 73, 85(34)
Johansson, S., 327
John, H. A., 864
John, K., 129(*o*), 137, 322, 328(7), 329(7), 333(7), 335(7), 337(7), 338(7)
John, K. M., 236, 327, 336(17), 542, 750, 751(13)
Johnson, D. A., 126
Johnson, F. L., 251
Johnson, G. L., 616
Johnson, J. D., 581
Johnson, M. J., 733, 746, 748(8), 749(8)
Johnson, R. T., 853
Jonas, A., 109, 116(*w*), 117, 121, 122(150), 127, 130(*s*), 135(*x*), 138, 139, 529, 558, 559(8), 561(*e*, *h*, *i*, *j*, *n*), 562(*u*), 563, 564, 565(19, 23, 25), 566(*d*, *h*, *i*, *j*, *k*), 567, 568, 569(16, 23, 24), 570(13, 19), 571(16, 25, 26), 573(16), 574(13), 575(8, 10), 577(25), 578(43, 48), 579(16, 25), 580(18, 23, 24, 25, 26, 32), 582, 649, 652, 653, 654(18)
Jones, A., 128(*g*), 137
Jones, A. L., 29, 101, 102(105), 109(105), 110(a), 111, 167, 169(29)
Jones, C., 49, 712, 853, 856, 857, 859, 860(27), 861(27), 862(27), 863, 864
Jones, D. M., 80(*w*), 83, 85, 86(*k*), 87, 513, 559, 564(17), 571(17), 572(17), 574(17)
Jones, H. W., 730
Jones, J. L., 629, 640, 646
Jones, K. W., 864
Jones, R., 99(*B*), 100, 104(*i*), 116(*D*), 118, 121
Jones, S., 18
Jost, J. P., 671
Jost-Vu, E., 29

Jowett, N. I., 744
Juarez-Salinas, H., 534
Juhn, D. J., 171, 178, 182, 247, 713, 721(10)
Julsrud, J. O., 852, 885
Jung, R. W., 121, 122(150)
Junien, C., 50
Jurgens, G., 113(P), 114, 119, 121(143), 122(143)
Just, P. W., 49, 824, 834(13), 837(13)

K

Kabat, E. A., 688
Kabsch, W., 765, 766
Kaduce, T. L., 80(s), 82
Kafatos, F. D., 805
Kaffarnik, E., 852
Kagan, A., 223
Kahlon, T. S., 112(a'), 113, 169, 456
Kahnt, F. W., 3
Kai, A. I., 216
Kainosho, M., 487
Kaiser, E., 628(17), 634, 640(17)
Kaiser, E. T., 404(7, 8), 405, 408(8), 627, 628(3, 4), 629, 636, 644(4), 721, 769, 771
Kakis, G., 90, 98(j), 100
Kalab, M., 72
Kalinyak, J. E., 802, 803(4)
Kamio, A., 121, 122(155)
Kan, Y. W., 735, 812, 816(7)
Kanda, N., 796
Kandutsch, A. A., 879, 880, 889
Kane, J. P., 24, 30, 31, 32, 35(116), 39(105), 57, 65, 66, 76, 85, 86(j), 87, 90(85), 92(m), 93, 95(q), 102(85), 105(85), 106(63), 110(a', c, h), 111, 112(j), 113, 126(62), 128(d), 131(c), 134(Q), 135(Z^3), 137, 138, 139, 140, 141(62, 63), 142, 153, 156, 169, 216, 219, 247, 249, 262, 263, 264, 274, 288, 292(5), 310, 349, 519, 520, 521(4), 522, 523(4), 524, 525, 526, 527(11), 529, 536(10), 584, 586, 609, 610(8), 824, 825, 837, 848, 849(26, 58)
Kanellis, P., 404(9), 405, 410(9), 646
Kannan, R., 65, 66(142), 76, 131(D), 138, 141(65), 248, 262
Kao, F. T., 49, 50, 235, 712, 750, 853, 854, 857, 858, 859, 860(27), 861(27), 862(27), 863, 864

Kao, Y. J., 609, 610(6)
Kapadra, G., 418
Kappers, A., 344
Karam, J. H., 736(f), 737, 744
Karathanasis, S. K., 8, 47, 49, 50(27, 28, 41, 44), 56, 62, 69(41), 224, 296, 672, 674(6), 675(7), 688, 691, 693, 695(9, 17), 697(9, 17, 31), 699(9, 17), 701(9, 17), 703(9), 712, 713, 717(2, 3), 719(8, 30), 720(2, 7, 8), 721(31), 723(2), 733, 740, 741, 774, 779, 781, 793, 800(11), 846, 885, 894
Karin, M., 52, 54(64)
Karlin, J. B., 30, 34, 39(108), 142, 171, 182, 247, 528, 529(3), 713, 721(10)
Karlsson, B. W., 72(34), 73, 85(34), 88
Kashimura, N., 106, 121, 133(I), 139
Kashyap, M., 201
Kashyap, M. D., 30, 39(105)
Kashyap, M. L., 290, 552
Katz, A. J., 227, 339(16), 340, 692
Katz, F. N., 63
Katz, J., 742, 743(44), 744
Kaufman, J. F., 900
Kaufman, R. J., 59
Kawahara, K., 259
Kawai, Y., 377
Kawashima, E. H., 733, 746, 748(8), 749(8)
Kay, D. H., 443
Kay, L., 50, 227
Kay, L. L., 62, 288
Kayden, H. J., 310, 339, 341(8), 452
Kayushina, R., 113(P), 114, 119, 121(143), 122(143)
Kazazian, H. H., 734, 736(c), 737, 739(24)
Kazazian, H. H., Jr., 738
Kazmar, R. E., 529, 530(15), 534(15)
Kean, C., 36
Kee, S. G., 805
Keil, B., 153
Keim, P., 129(v), 138, 146, 474, 501(6), 514(6)
Kellems, R. F., 59
Keller, G. A., 40
Kelley, J. L., 75, 81(O), 83, 92(w), 93, 96(A), 127(58), 171, 174, 178
Kelly, J. L., 169, 303
Kelly, K., 848
Kelus, A. S., 124, 125(176)
Keniry, M. A., 514
Kennedy, L., 354

Kent, S. B. H., 634
Kercret, H., 513, 559, 564(17), 571(17), 572(17), 574(17), 640, 657
Kesaniemi, A., 357, 358(7)
Kessling, A., 894
Keutmann, H., 62, 693, 695(17), 697(17), 699(17), 701(17), 717
Keys, C., 797
Kezdy, F., 14
Kezdy, F. J., 122, 627, 628(3, 4), 629, 636, 644(4), 721, 769, 771
Kezdy, F. T., 404(7, 8), 405, 408(8)
Khan, S. A., 26, 58, 633, 823
Khoury, P., 848
Kida, Y., 91
Kiefer, D. O., 812
Kilgore, L. L., 249, 250(32), 251, 252(22), 253(22, 38), 256, 258(38), 260(22), 584
Kim, H. S., 98(e), 99, 103, 115(f), 117, 120(110), 888, 889(30)
Kimber, B. J., 511
Kind, L., 327
King, W. C., 422, 451, 455, 578
Kingsbury, R., 60
Kinnunen, P. K. J., 39
Kintzinger, J.-P., 514
Kirby, C., 658
Kirkeby, K. K., 79(d), 82
Kiser, R. E., 326
Kita, T., 33, 70(9), 71, 85(9), 86(j), 90(85), 92(m), 93, 95(q), 102(85), 105(85), 110(h), 111, 112(j), 113, 899
Kitchens, T., 529
Klausner, R. D., 313, 657, 658, 659(1), 660(1, 24), 663(1, 12), 665(12, 14), 666(12), 667(12), 668(12, 14)
Klausner, R. K., 658
Klein, M. P., 498
Kloer, K. U., 438, 440(19), 441(19)
Kluge, K., 529, 530(15), 534(15)
Knapp, R. D., 403, 406(2), 499, 502, 513, 767
Knight, K. L., 124, 125(180)
Knipping, G. M. T., 87, 88, 113(P), 114, 119, 121(143), 122(143), 133(I), 134(S), 139, 142, 288
Knobler, C. M., 110(c), 111
Knott, T. J., 51, 235, 697(33), 699, 893, 894(47)
Knowles, B. B., 692, 694, 701(13), 704(13), 824, 834(15), 848(15)
Knowles, P. F., 570, 572(33)
Knudsen, P. J., 900
Kodama, T., 536
Koga, S., 5, 77, 115(k), 117, 123(72), 339, 341(2), 342(2), 568
Koh, S.-W. M., 326
Kohne, D. E., 683
Kokatnur, M. G., 87
Kokatur, M. G., 81(C), 83
Kokum, M. M., 36
Kolar, J., 536, 837
Komano, T., 110(k), 111, 121, 133(I), 139
König, T., 447
Kooter, J. M., 736
Korda, N., 124
Koren, E., 302, 305
Korman, A. J., 900
Kornblitt, A. R., 8, 47, 224, 712
Kostner, G. M., 87, 88, 132(I), 136(Z^9, Z^{10}), 139, 140, 217, 220, 222, 301, 303, 306, 307(19), 308(19), 309(19), 433, 434(10), 493, 493(39), 497(39), 498(39), 690
Kotite, J. L., 848, 849(58)
Kotite, L., 32, 64, 85, 86(j), 87, 90(85), 92(m), 93, 95(q), 102(85), 105(85), 110(h), 111, 112(j), 113, 134(Q), 139, 288, 292(5), 310, 824, 837, 849(24, 26)
Kottke, B. A., 142, 223, 549, 550(94)
Kovanen, P. T., 34, 36, 824, 849(26), 906
Kowalski, J., 182
Krajnovich, D. J., 109, 559, 561(e, n), 563, 573
Kramer, G. C., 80(x), 83, 112(l), 114, 116(s), 117, 127, 452
Kraus, J. P., 44, 899
Krause, B. R., 27, 578, 753
Krauss, R. M., 112(b), 113, 119, 169, 195, 204, 417, 424(2), 425(2), 427, 456
Kraute, P., 52, 54(64)
Krebs, K. E., 392, 402, 403, 767
Krieger, M., 524, 525, 599, 602, 607, 609, 610(4, 5, 6, 7, 8, 11), 611(14), 612(4, 5), 613(4)
Kreil, G., 61
Kreissman, S. G., 44
Krishnaiah, K. V., 65, 131(D), 138, 247, 262
Kritchevsky, D., 8
Kritchewsky, D., 79(g), 82
Kritchevsky, G., 555
Kronenberg, H. M., 692

Kroon, D. J., 404(7), 405, 771
Kroon, P., 487, 504, 609
Krul, E. S., 252, 529, 545, 546(84), 548(7), 549(7, 84), 550(84), 551(7)
Kruskal, J. B., 773
Kruski, A. W., 71, 78(19), 105(19), 108, 117(E), 118, 120(19), 169, 171, 178(8), 452, 476, 477, 478(8), 479(8, 11), 481(11)
Kubasek, F. O. T., 80(y), 83, 92(n), 93, 95(r), 99(t), 112(l), 114, 116(s), 117
Kubota, Y., 767, 769(37)
Kuchinskiene, Z., 100(b), 111
Kuehl, K. S., 536
Kuff, E. L., 745
Kuiken, L., 578, 579(51)
Kuksis, A., 90, 98(j), 100, 105
Kumar, N., 658
Kumit, D. M., 691
Kummerow, F. A., 105, 121(121), 122(121, 155), 142(121)
Kummerow, P. A., 153
Kunitake, S. T., 78, 94, 99(v), 100, 109(75), 110(a, c), 111, 116(x), 118, 121, 122(149), 156, 165, 169, 219, 448
Kunkel, H. G., 72, 142(26)
Kunkel, L. M., 730
Kuo, W. H., 163, 249, 340, 589
Kupfenberg, J. P., 404(8), 405, 408(8), 628(4), 629, 636, 644(4), 721, 769
Kupferer, P., 538
Kurnit, D., 8, 56, 64, 672, 675(7), 692, 701(14), 704(14), 713, 717, 719(30), 733, 796, 823, 824, 834(16)
Kuroda, M., 70(8), 71
Kurt, G., 153
Kurtz, D. T., 51
Kushwaha, R. P., 849
Kusserow, S. K., 558, 559, 561(g, o), 562(z), 563
Kuskwaka, R. S., 81(K), 83
Kutter, E., 677
Kuusi, T., 39, 279, 287(14)
Kwiterovich, P. O., 35, 108, 849(69), 850
Kyte, J., 769, 770(41)

L

Lackner, K. J., 235, 236, 238(42), 240(42), 893
Lacy, E., 793, 812
Laemmli, U. K., 360, 805, 892
Laggner, P., 113(P), 114, 119, 121(143), 122(143), 504, 562(x), 563, 583, 690
Lagocki, P. A., 402
Lagrange, D., 74, 81(Q), 83, 84(50), 92(y), 93, 95(b), 96(E), 101(50), 110(m), 111, 113(w), 114, 120(50), 123(50), 129(n), 130(x), 133(L), 137, 139, 141(50), 171, 182
Lai, E. C., 857, 869
Laitinen, L. A., 39
Laitinen, M. V., 288(12), 289
Lakier, J. B., 712, 721(6), 725(6), 741
Lalich, J. J., 125
Lally, I. J., 105
Lally, J. S., 31
Lamischka, I. R., 55
Lamplugh, S. M., 288, 339(10), 340, 341(10), 352
Lan, S. F., 8, 59
Land, H., 807, 810
Lane, D. M., 21, 23(68), 63, 64(130), 68(131)
Lane, M. D., 705
Langdon, R. G., 250, 256, 478, 479(14), 481
Langenback, U., 837, 845(50)
Langer-Safer, P. R., 734
Lapanje, S., 260
LaPiana, M. J., 251
Laplaud, P. M., 73, 76(41), 78(41), 79(c, e), 80(u), 82, 84(76), 88(76), 92(c, i, k, l), 93, 94, 95(b, d, n, p), 97, 98(d, g, r, s,), 99, 101(41, 76, 77), 102(41), 103(41), 171, 182, 110(d), 111, 112(c), 113, 115(c), 117, 120(41), 123(41), 127(76, 90), 128(c), 129(q), 132(E), 135(Z^2), 137, 140, 143(76, 77, 90), 877, 878(3), 891(3)
LaRosa, J. C., 17, 798
LaRue, A., 128(a'), 137, 224, 543, 544(59)
Laskarzewski, P. M., 848
Laskey, R. A., 730
Lasser, N. L., 79(j), 82
Laties, A. M., 659
Latner, A. L., 435
Latt, S. A., 796
Lattier, G., 66
Lattier, G. R., 131(D), 138
Lau, Y.-F., 808, 812, 816(7)
Lauer, J., 793, 812

Laurell, S., 324
Law, M. L., 49, 712, 854, 857, 859, 860(27), 861(27), 862(27), 863, 864
Law, S. W., 49, 56, 224, 235, 885, 893, 894
Lawn, R. M., 720, 744, 813, 908
Layuez, J., 365
Lazier, C. B., 53
Lea, T., 312, 313(12)
Leary, A. C., 797
Leat, W. M. F., 80(w, y), 83, 85, 86(k), 87, 92(n), 93, 95(r), 99(t), 116(s), 117
Lebherz, H. G., 26
LeBoeuf, R. C., 70, 73(5), 79(f), 82, 92(b), 93, 95(c), 96, 98(c), 99, 102(5), 115(e), 117, 120(5, 6), 128(c'), 129(q), 132(E), 135(Z^2), 137, 138, 143(6), 877, 878(2), 879, 880, 883, 884(15), 885(4, 14, 17), 886(14), 888(14, 17), 889(14, 17), 891(2), 892(2), 893(2, 26), 894(14)
Leclercq, B., 81(P), 83, 92(x), 93, 96(D), 105, 122(127), 142(124)
Led, J. J., 509
Leder, P., 812
Ledford, J. H., 127, 132(F), 133(o), 135(Z^4), 139, 140
Lee, A. G., 485
Lee, D. M., 77, 94, 112(b), 113, 158, 163, 247, 249, 340, 589
Lee, J. A., 4, 105
Lee, L.-S., 812
Lee, N. S., 136(Z^9), 140, 306, 307(20), 308(20), 309(20), 377, 380, 381(16)
Lee, P., 705
Lee, Y. L., 849
Lees, A. M., 696, 701(22), 708(22)
Lees, M., 555
Lees, R. S., 4, 145, 155(11), 156, 157(1), 167(1, 11), 193, 411, 696, 701(22), 705, 708(22), 824, 842(17), 848
Leford, J. Holland, 213
Legmann, P., 145, 306, 309(18)
Lehrach, H., 869
Lehrer, S. L., 577
Lehrman, M. A., 48
Lelkes, P. I., 658, 659(21a), 661, 663(21a)
LeMaire, M., 462
Lennarz, W. J., 63, 64(130), 68(131), 705
Lent, W. M. F., 112(b), 114
Lentz, B. R., 595
Lentz, S. R., 697(29), 699

Lepault, J., 471
LeQuire, V. S., 65
Lerner, R. A., 126
Leserman, L. D., 657, 659(1), 660(1), 663(1)
Leslie, R. B., 109, 116(y), 118, 121, 122(154), 482, 483(22), 485(22), 486, 487(22), 489(22), 492(22), 584, 690
Lever, W. F., 3
Levine, B., 481
Levine, M., 734
Levine, Y. K., 485
Levitt, M., 406
Levy, G. C., 498, 499
Levy, R. I., 4, 5, 17, 77, 128(a), 137, 155, 156, 157(1), 167(1), 193, 217, 220(27), 251, 288, 379, 397, 568, 798, 848, 849
Levy-Wilson, B., 51
Lewis, I. H., 79(a'), 82
Lewis, J. T., 657
Lewis, L. A., 72, 146
Lewis, L. L., 104, 105, 115(n), 117, 128(e), 137
Lewis, S. B., 136(Z^{11}), 140, 310
Lewis, S. E., 880
Li, D., 124
Li, W., 659
Liao, W. S., 26
Lieber, E., 555
Lievens, J., 365, 371, 372(16), 373, 374(18, 19)
Lievens, M. J., 402, 558, 562(s), 563
Light, J. A., 62, 224, 226, 227, 228(28), 241(28)
Lijnen, H. R., 543
Lim, C. T., 127, 129(l), 130(s, v), 135(x), 137, 138, 139, 310, 339, 341(4, 7, 8), 452, 892
Limm, M., 485, 486
Lin, A. H.-Y., 34, 528, 529(3)
Lin, C. T., 57, 70, 88(3), 142(3)
Lindall, A. W., 80(r)
Lindenmaier, W., 807, 810(12)
Lindgren, F., 213
Lindgren, F. R., 848
Lindgren, F. T., 4, 112(a', b'), 113, 115(b), 117, 119, 155(7, 8, 9, 10, 13), 156, 159(167), 169, 170, 182, 195, 223, 455, 456, 457
Lindquist, S., 56
Lingappa, J. R., 699

Lingappa, V. R., 63, 696, 699(23)
Lin-Lee, C., 57
Lin-Lee, Y. C., 70, 88(2, 3), 141(1, 2), 142(3)
Lin-Su, M. H., 70, 141(1)
Lipman, D. J., 754, 757, 764(11)
Lisam, L., 60
Liscum, L., 8, 59, 68
Little, J. A., 17, 798
Little, P. F. R., 736
Littlefield, J. W., 858
Liu, B. W., 552
Liu, C. H., 3
Lodish, H. F., 63
Lofland, H. B., 120
Lofquist-Kjahl, F., 8, 672, 674(6), 846
Loh, E., 99(v), 100, 110(c), 111, 116(x), 118
Longacre, S. S., 671
Lokesh, B. R., 324
London, R. E., 501, 503
Long, J. A., 86(e), 87, 90, 91(93), 92(h), 93, 95(j)
Lopez, A., 216
Lopez, F., 216
Lorimer, A. R., 32, 35(117)
Los, W., 855
Lossow, W. J., 170, 182, 455
Lowell, C. A., 320
Lowry, M. A., 534
Lowry, O. H., 253, 256, 555, 570, 602, 611
Lowy, I., 812
Lucchini, G., 52
Lucero, M. A., 47, 235, 697(34), 699, 733, 799(29), 800
Luley, C., 438, 440, 441(19)
Lundberg, B., 598, 600(37), 606
Lund-Katz, S., 496, 503(45), 513
Lusis, A. J., 56, 70, 73(5), 79(f), 82, 92(b), 93, 98(c), 99, 102(5), 115(e), 117, 120(5, 6), 128(c'), 129(q), 132(E), 135(Z^2), 137, 138, 143(6), 877, 878(2), 879, 880, 883, 884(15), 885(4, 14, 17), 886(14), 888(14, 17), 889(14, 17), 891(2), 892(2), 893(2), 894(14)
Lusk, L. T., 73, 74(40), 78(40), 112(f), 113, 115(h), 117, 120(40)
Luskey, K. L., 8, 46, 59, 60, 63, 908
Lussier-Caean, S., 87, 99(w), 100
Lux, S. E., 17, 129(o), 137, 236, 542, 568, 750, 751(13), 798
Luz, Z., 514
Luzzati, V., 458, 465, 467(18), 468

M

McAdam, K. P. W. J., 312, 313(11), 314(11, 15), 315
McAlpine, P. J., 797
MacAndrews, M. I., Jr., 616
McBride, J. R., 81(F), 83, 87, 88
McCarty, K. S., 804
McCarty, K. S., Jr., 804
McCaslin, D. R., 585
McConathy, W. J., 75, 127(58), 297, 298, 299(3), 300(2, 3), 301(3), 302, 303(3), 304(4), 305(4), 435, 438(18), 440(18), 441(18)
McConnell, H. M., 657
McCullough, P. J., 131(B), 138
MacDonald, R. J., 42, 59, 681, 802, 803(3), 820
McDowall, A. W., 471
McElvain, S. M., 616
McFarlane, A. S., 214, 217(10)
McGahan, R., 404(9), 405, 410(9), 646
McGill, J. R., 50
McHugh, H. T., 564, 565(25), 566(i, j), 567, 571(25, 26), 577(25), 579(25), 580 (25, 26)
McIntosh, G. H., 103
MacKenzie, S. L., 434
McKeown, C. M. E., 737
McKnight, S. L., 60
McKusick, V. A., 742
McLachlan, A. D., 474, 758, 774, 777
McLean, J. W., 46, 56, 58(89), 61(17), 70, 135(Z^1), 140, 141(4), 735, 763, 801, 802, 803(4), 804(1), 805(1), 806(1, 2), 808(1), 809(1), 810(1, 2), 813, 815(18), 822
McLean, L. R., 18
McMullen, J. J., 79(l), 82
McNair, D. S., 509, 511(63)
McPhaul, M. J., 609, 610(5), 611(14), 612(5)
McPherson, J., 8, 47, 50(28), 56, 672, 674(6), 675(7), 688, 691, 695(9), 697(9), 699(9), 701(9), 703(9), 712, 717(2), 719(30), 720(2), 723(2), 733, 740, 779, 800, 846
MacPike, A. D., 880
MacRitchie, F., 388
MacSwiney, F. M., 880
McTaggart, F., 73, 74, 78(43), 81(G), 83, 88(43), 92(u), 93, 96(x), 99(A), 110(l), 111, 112(h), 113(P), 114, 166(A), 118, 120(43), 123(43), 133(J), 139, 142(45)

MacTais, L. J., 880
McVicar, J. P., 169, 219
Ma, S. K., 110(c), 111
Mach, B., 671
Macheboeuf, M. A., 3, 144, 213, 214
Maciejko, J. J., 223
Madan, K., 855
Maeda, H., 163, 249, 340, 549, 589
Magin, R. L., 658
Mahley, R. W., 26, 30, 33, 34, 36(104), 37, 46, 47, 51, 61(17), 63, 64, 71, 73(11), 74(11, 39), 75(13, 36), 77(39, 42, 49, 54), 79(m), 80(q), 81(M), 83, 84(11, 49), 95(m), 97, 98(h, p), 100, 102(11), 104, 105, 109(42), 112(l), 113, 115(o), 116(y), 117, 118, 120(11, 13), 123(42, 54), 126(11, 13), 127, 129(h', j, r), 130(t), 131(y, A), 132(G), 135(Z, Z^1), 136(Z^7, Z^8), 137, 138, 139, 140, 141(13), 142(11, 13, 36), 143, 169, 216, 263, 273, 274, 275(4), 276(7), 277, 278, 279, 280(1), 281, 282(3), 284, 285, 286(3), 287(14), 339(13), 340, 341(13), 342(13), 355, 528, 542, 543, 544(52), 551(74), 562(t), 563, 581, 606, 607(45), 609, 703, 708, 727, 735, 753, 769, 770(43), 771, 801, 804, 806(2), 808, 809, 810(2, 20), 811(19), 812, 813, 815(13, 18), 823, 824, 825(18, 19, 20, 22, 23), 834, 837(41), 846(28, 29, 42), 849(18, 19, 20, 22, 23, 26, 27, 28), 850
Mahoney, E. M., 354, 357(4)
Maine, G. T., 649
Majzoub, J. A., 692
Makela, O., 534
Malamud, D., 535, 537(37), 538(37)
Malcolm, B. R., 401
Malhotra, S., 79(g), 82, 104
Malley, M. J., 247
Mallinson, A., 182
Malloy, M. J., 32, 35(115), 263, 523
Malmendier, C. L., 136(Z^{11}), 140, 312
Malone-McNeal, M., 66, 131(D), 138
Manfredi Romanini, M. G., 688
Mangel, J., 49
Maniatis, T., 43, 51, 55, 676, 679(14), 680(14), 686, 716, 720, 732, 734, 744, 793, 805, 808, 812, 813(4), 814(4), 815(4), 902, 903(16), 906(16), 908
Mankinen, C., 866
Manley, J. L., 51

Mannickarottu, V., 154, 171
Manning, J. A., 327
Mantulin, W. W., 377, 411
Manty, W., 213
Manzoni, C., 622
Mao, S. J. T., 142, 223, 529, 530(15), 534(15), 543, 545, 549, 550(66, 94), 562(y), 563
Marcel, Y. L., 76, 87, 99(w), 100, 141, 263, 339(9), 340, 341(1, 9), 342(1, 9), 435, 438(18), 440(18), 441(18), 529, 535, 536(40), 542(13), 543, 544(8, 13), 545(8, 32), 548, 550(11), 580, 769
March, S., 299
Marchalonis, J. J., 531
Margolins, H. S., 247
Margolis, S., 250, 256, 434, 849(69), 850
Margolius, H. S., 151, 159
Marhaug, G., 312, 314, 315(17), 316(17)
Marien, J. J. C., 849
Markham, A., 50, 746
Markham, A. F., 295(18), 296, 715, 734, 736(c), 737, 739(24), 791, 796
Markley, J. L., 511
Markwell, M. K., 388, 570
Marlett, J. A., 79(o), 82, 92(e), 93, 95(h), 98(f), 105, 127(116)
Marsh, D., 570, 572(33)
Marsh, J. B., 76, 106, 107, 131(D), 138, 248, 339(11), 340, 341(11)
Martin, J. B., 736(g), 737
Martin, R. B., 478, 479(14), 480, 481
Martin, W. G., 72, 74, 110(j, k), 111, 116(z, B), 118
Martinez, H. M., 519, 523(1), 524(1)
Martinez-Arias, A., 48
Marzetta, C. A., 251
Marzotto, A., 544
Mason, D. T., 696
Mason, W. R., 559, 575(10)
Massey, J. B., 10, 11, 18, 367(8), 368, 370, 371, 372, 403, 406(2), 410, 411(15, 16), 558, 559, 561(g, o), 562(p, z), 563, 625, 767
Mathur, S. M., 110(e), 11, 115(d), 117
Mathur, S. N., 112(d), 113
Matteucci, M. D., 713, 714(18), 789
Matwiyoff, N. A., 503
Matz, C. E., 564, 565(19, 23), 566(d, h), 567, 568, 569(23), 570(19), 580(18, 23, 32)
Maunus, R., 692

Maurel, D., 78, 80(*u*), 82, 88, 82, 92(*k, l*), 93, 95(*n, o, p*), 97, 98(*q, r, s*), 101(77), 127(90), 143(77, 90)
Mauser, T., 358
Maxam, A., 44, 717, 791, 792(5), 905, 906
Mayer, M. M., 688
Mayfield, C., 837
Mayo, B. C., 481
Mayo, K. E., 52
Mazzone, T., 37
Mead, M. G., 72(35), 73, 85(35)
Means, A. R., 745
Means, E. G., 354
Medgysi, G. A., 535
Meek, R. L., 311
Meera Khan, P., 855
Mchl, T. D., 354
Mehrabian, M., 894
Meier, H., 880
Meierhofer, J., 616
Meisler, M. H., 49, 878
Meister, N., 459
Melin, M., 72, 214
Mellies, M. J., 848
Mellon, P., 812
Melnik, B., 129(*o*), 137
Melrik, B., 339(17), 340
Melton, D., 686, 734
Memod, J. J., 56
Meng, M., 50, 128(*d*), 135(Z^4), 137, 140, 227, 283, 288
Menzel, H. J., 708, 727
Menzel, J., 727, 824, 837(30)
Meravech, M., 44
Merchant, D. J., 819
Meredith, L., 736(h, i), 737, 800
Meredith, S. C., 627, 628(3)
Merimee, T. J., 354
Merrick, W. J., 804
Merrifield, R. B., 632, 633, 634
Meryman, H. T., 457
Mescher, M. F., 548
Meselson, M., 748, 816
Messing, J., 675, 724, 809, 905
Metcalfe, J. C., 485
Metza, L., 879, 888
Michel, A. R., 634
Middelhoff, G., 367, 368(3), 370(3), 562(*q, v*), 563
Mihovilovic, M., 690

Miljanich, P., 116(*z*), 118
Miller, G. J., 223, 741
Miller, J. H., 716, 793
Miller, J. S., 44, 49(11)
Miller, K. W., 513
Miller, N. E., 741
Miller, N. F., 223
Miller, N. H., 256
Miller, O. J., 730
Mills, A. D., 730
Mills, G. L., 72, 74, 77(23), 78(23), 79(*b*), 80(*b*), 81(*b, Q*), 82, 83, 84(23, 50), 85(23), 86(*a', d*), 87, 88(23), 89(23), 91(23), 92(*a, g, y*), 93, 95(*a, i*), 96(*E*), 98(*a, l*), 99(*a*), 101(50), 102, 105, 106, 107, 110(*g*), 111, 112(*h, l*), 113(*w*), 114, 119, 120(50), 123(50), 127, 129(*i, n*), 130(*x*), 132(*F*), 133(*J, L, O*), 135(Z^4), 137, 139, 140, 141(50, 129), 142(45, 82)
Milne, R. W., 76, 141, 263, 435, 438(18), 440(18), 441(18), 529, 535, 536(40), 542, 543, 544(8, 13), 545(8, 32), 548, 550(11), 769
Milner, P. F., 48, 738
Milstein, C., 530, 534
Minari, O., 623
Mintz, B. J., 796, 797(17)
Mishkel, M., 848
Mitchell, C. D., 451, 455, 578
Mitchell, J. R., 389, 395(9)
Mittleman, D., 3
Miyake, T., 733, 746, 748(8), 749(8)
Mjøs, O. D., 25, 28(80), 31(80), 90, 92(*e*), 93, 110(*f*), 111
Moffat, L. F., 60
Mohandas, T., 894
Moller, J. V., 462
Montaguti, M., 94, 98(*f*), 99, 115(*l*), 117, 120, 121(144)
Montcalm, A., 892
Mookeriea, S., 105
Moor, H., 457, 459
Moore, C. M., 50
Moore, D. D., 812
Moore, J. P., 746
Moore, M. N., 50, 235, 750, 864
Morello, A. M., 79(*o*), 81(*F'*), 83, 87, 88, 92(*f*), 93, 95(*h*), 98(*k*), 105, 120(89), 127(116)
Morfin, R., 892

Morganroth, J., 848, 849
Morin, C., 48, 734, 739
Morioka, H., 91
Morisett, J., 368
Morris, C. F., 538
Morris, M. D., 4, 105, 893
Morrisett, J. D., 16, 36, 74, 98(e), 99, 103, 120(110), 115(f), 117, 121, 122(155), 248, 294, 372, 403, 476, 479(9), 480(9), 487(9), 499, 502, 504, 542(55), 543, 562(w, x), 563, 574, 584, 604, 627, 765, 767, 787, 888, 889(30)
Morrisey, J. H., 246
Morrison, J. A., 848
Morrison, P., 85
Morrow, J. F., 320
Morse, H. G., 856
Morse, M. C., III, 889
Morsett, H., 648, 649(8), 650(8)
Mortelmans, J., 99(c), 100, 104(g), 129(k), 130(u), 137, 138
Mortensen, R. F., 883
Morton, R. E., 303
Motulsky, A. G., 800
Mougin-Schutz, A., 624
Mount, D. W., 673
Mouton, R. F., 3
Mova, R., 627, 628(3)
Mueller, C. R., 736(h, i), 737
Mueller, M., 459
Mueller-Eckhardt, C., 852
Muhlethaler, K., 457
Mukerjee, P., 332, 333(31)
Mulford, D. J., 72, 214
Muller, K. W., 583
Muller-Hill, B., 675
Mulligan, R. C., 812
Mullinix, K. P., 671
Munk, P., 260
Muñoz, V., 5, 388, 395
Munro, A., 32, 35(117)
Munro, B. S., 8, 47, 224, 712
Munson, P. J., 547
Muramatusu, M., 395
Murase, T., 552
Murchio, J. C., 170, 182, 455
Murphy, R. F., 749
Murray, J. M., 736(h, i), 737
Muskinski, J. F., 306, 308(21)
Musliner, T. A., 427
Myatt, G., 72(35), 73, 85(35)
Myher, J. M., 90, 98(j), 100
Mykelbost, O., 50, 295, 296
Myklebost, S. R., 295(18), 296

N

Nada, T., 546
Naglkerke, J. F., 37
Naito, H. K., 79(p), 82, 85, 104, 105, 115(n), 117, 123(79), 128(e), 137
Nakai, H., 51
Nakamura, M., 358
Nakamura, N., 549
Nakanishi, M., 677
Nakaya, Y., 309
Nallaseth, F. S., 48, 738
Narayan, K. A., 79(l), 82
Nathans, D., 730
Natvig, J. B., 311, 312, 313(12)
Naylor, S. L., 49, 50, 736(g), 737, 866, 885, 893, 894
Nazir, D. J., 848
Nebert, D. W., 881
Neeley, J. E., 874
Negishi, M., 881
Neiberg, A., 894
Neill, C., 849(69), 850
Nelson, B. J., 91
Nelson, C., 647, 655(1)
Nelson, C. A., 142, 551, 552
Nelson, D. J., 478, 490(13), 497(13)
Nelson, G. J., 104
Nermut, M. V., 471
Nestel, P., 546
Nestel, P. J., 31, 223, 240, 241(48)
Nestruck, A. C., 87, 99(w), 100, 339(9), 340, 341(1, 9), 342(1, 9), 433
Neurath, H., 388
Newhouse, Y., 277
Newman, T. C., 26
Ng, S. Y., 51
Ng, T. C., 201, 513, 559, 564(17), 571(17), 572(17), 574(17), 640, 646
N'guyen, T.-D., 435, 438(18), 440(18), 441(18)
Niazi, G.. 48
Niblack, G. D., 223
Nichols, A. V., 38, 73, 78, 84(78), 86(b), 87,

102(78), 105(78), 112(k), 114, 115(b), 116(q, r, u, z), 117, 118, 120(37, 78), 155(7, 8, 9,12), 156(4), 157(2), 169, 251, 417, 442, 447, 451, 452(3), 457, 458, 564, 565(22), 566(f), 567, 568, 571(22), 572(22), 578, 581, 583, 649, 651(14), 654(15, 16), 655
Nicklen, S., 44, 717, 905
Nicolosi, R. J., 71, 79(o), 81(E, F'), 83, 87, 88, 92(f, t), 93, 95(h), 96(w), 98(k), 99(z), 105(18), 116(c), 118, 120(18, 89), 127(116), 692, 701(16), 711
Niday, E., 531
Niedman, P. D., 433
Nielsen, T. B., 618
Nikazy, J., 877, 880, 884, 885(4)
Nikkila, E. A., 39
Nilsson, J., 154, 171
Nilsson-Ehle, P., 17, 29(45), 798
Nishida, J., 153
Nishida, T., 336, 643
Nishikawa, K., 767, 769(37)
Noble, R. P., 593, 597(33)
Noel, J. G., 544
Noffze, D. K., 712, 721(6), 725(6)
Noll, M., 739
Nordhausen, R. W., 116(z), 118
Northrop, C. A., 80(w), 83, 85, 86(k), 87
Norum, K. R., 20, 37, 38(62), 40(62), 422, 451, 455, 578
Norum, R. A., 712, 713, 719(8), 720(8), 721(6), 725(6), 740, 741
Noyes, B. E., 44
Noyes, C., 130(v), 138, 236
Nugent, P., 546
Nunez, E., 336
Nunez, J., 626
Nussbaum, A. L., 8, 56, 672, 674(6), 675(7), 717, 719(30), 733, 846
Nutik, R., 90

O

O'Brien, K., 617, 619, 624(11)
Ockner, R., 338
Ockner, R. K., 327
O'Connell, C., 793, 812
O'Donnell, K. A., 56, 877, 885
O'Donnell, T. F., 849
O'Farrel, P. H., 342, 677, 825
Ogawa, Y., 708
Oh, G. S., 128(a'), 137
Oh, S. Y., 355
Ohno, T., 671
Ohno, Y., 145
Ohtsuki, M., 458
Okayama, H., 63, 896
Okazaki, M., 145
Olaisen, B., 49, 50(43), 852, 863, 864(2)
Oldfield, E., 514
Olivecrona, T., 25, 33, 127
Olofsson, S. O., 218, 223, 298, 300, 301(6, 8), 303(6, 8), 304(4), 305(4), 310
O'Malley, B. W., 869
Oncley, J. L., 5, 214
Onodera, K., 106
Ooi, T., 767, 769(37)
Opella, S. J., 478, 490(13), 497(13), 515
Oppenheimer, N., 496, 503(42), 503, 504(42), 505, 506
Oram, J. F., 40
Orkin, S. H., 55, 734, 736(c), 737, 738, 739(24)
Osborne, J. C., 136(Z^9), 140, 146
Osborne, J. C., Jr., 213, 217, 224, 306, 307(20), 308(20), 309(20), 313, 342, 375, 377(3), 378, 380, 381(16), 382, 383(9), 386, 562(u), 563, 570, 577, 578(48)
Osborne, J. L., 126
Osborne, M. J., 593
Osborne, T. F., 59, 60
Osguthorpe, D. J., 766
Oshcry, Y., 33, 78, 95(f), 97, 98(i), 100, 102(74), 103(74), 105(74), 112(g), 113, 115(i), 117, 120(74)
Ostapchuk, P., 26, 672, 679(8), 681(8)
Osterman, D., 627, 628(3)
Ostrem, J., 92(j), 93
Ostwald, R., 20, 86(b, c), 87, 98(m), 100, 115(m), 117, 128(d), 135(Z^4), 137, 140, 142, 283
Ott, G. S., 126, 134(u), 139, 141(195), 288, 339(19), 340, 341(5), 348, 349
Ott, J., 800
Ottina, K., 736(g), 737
Otway, S., 248
Oudin, J., 123
Owens, N. L., 109, 584

P

Packard, C. J., 32, 35(117), 86(g), 87, 456
Padley, R. J., 65, 66
Paetzgold, R., 852
Pagano, R. E., 657
Page, A. L., 124
Page, I. A., 72
Page, I. H., 79(a'), 82
Pagnan, A., 32, 837
Paigen, B., 879, 881, 883, 885(17), 888(17), 889(17)
Paik, Y.-K., 47, 808, 810(20), 811, 812, 815(13)
Pajetta, P., 544
Palade, G., 66
Palade, G. E., 68, 535, 537(38)
Palmieri, E., 620
Palmiter, R. D., 52, 898
Panyim, S., 671
Pao, Q., 409, 562(z), 563, 564, 566(g), 567
Papahadopoulos, D., 657, 661
Papanadjopoulos, D., 647, 648(2)
Papenberg, J., 116(p), 117
Paralta, J. M., 535, 536(36)
Pardue, M. L., 864
Pargaonkar, P. S., 81(H), 83, 106, 107
Parham, P., 534, 548(23), 549(23)
Parikh, I., 299
Park, C. E., 105
Parker, F., 4
Parker, R. C., 720, 744, 813, 908
Parker, V., 812
Parks, J. S., 99(D), 100, 327, 333(25), 337(25), 487, 488(31), 489(31)
Parks, T. S., 129(m), 130(w), 134(T), 135(Y), 136(Z^{13}), 137, 139
Parmar, Y. I., 513, 514(73)
Parmelee, D. C., 136(Z^{11}), 140, 311, 312(6), 313(6)
Paroutaud, P., 136(Z^{11}), 140
Parrish, D. O., 657
Partetelle, D., 534
Pasquali-Ronchetti, I., 79(h), 82, 94, 95(g), 97, 98(f), 99, 101, 102(107), 115(l), 117, 120, 121(144)
Paterniti, J. R., 880
Patsch, J. R., 13, 25, 30, 34, 39(107, 108), 103, 105, 115(f), 117, 120(110), 121(121), 122(121), 142(121), 169, 476, 479(9), 480(9), 487(9), 499, 502, 546, 888, 889(30)
Patsch, W., 13, 34, 169, 528, 544, 549(1)
Patterson, B. W., 109, 127, 130(s), 135(x), 138, 139, 249, 250, 252(22), 253(38), 254(25), 256, 258(38), 259, 260(22), 559, 561(e, h), 563, 570(13), 574(13), 590
Patterson, D., 856
Pattniak, N. M., 478, 479(12)
Patthy, L., 544
Patton, J. G., 549
Paulus, H. E., 65, 76, 106(63), 131(c), 138, 141(63), 247, 262, 529, 536(10)
Paulus, H. F., 586
Payvar, F., 44
Pearlstein, E., 434
Pearson, G. D. N., 358
Pearson, P. L., 737, 855
Pearson, T., 534
Pearson, W. R., 320, 757
Pease, R. J., 51
Peat, I. R., 498, 499
Peary, D. L., 36, 158, 247
Pedersen, K. O., 3, 144
Peeters, H., 81(N, P), 83, 92(x), 93, 96(D), 99(F), 100, 104(g), 105, 120(126), 122(126), 129(k, n), 130(u, x), 137, 138, 142(124, 125), 365, 367(7), 368(3, 4), 370(3, 4, 12), 371, 372(16), 373, 374(19), 562(q), 563
Pelham, H. R. B., 52, 692, 804, 900
Pellicer, A., 812
Penttila, I. M., 288(12), 289
Pererira, L. V., 543
Perham, R. N., 544
Pescador, R., 86(h), 87
Peterfy, F., 534
Peters, T., Jr., 337, 338(40)
Petersen, S. B., 509
Peterson, J., 609, 610(7)
Petrie, G. E., 582
Pflugshaupt, R., 153
Pfaffinger, D., 339(15), 340, 341(15), 343(15), 347, 892
Pfau, J., 891
Pfau, P., 190, 194(10)
Pfleger, B., 34, 528, 529, 534(12), 535(12), 544(12), 549(1, 12), 550(6), 568, 713, 721(9)
Phelps, D. E., 498

Philipp, B. W., 796
Phillips, C. I., 737
Phillips, M. C., 18, 38, 109, 389, 392, 393, 395(9, 18), 396, 399, 401, 402, 403, 496, 503(45), 513, 584, 767
Phillips, M. L., 78, 94, 101, 105(106), 109(75), 110(a), 111, 165, 448
Phillips, N., 848, 849(58)
Pilch, P. F., 616
Pinon, J. C., 86(i), 87
Pinter, G. G., 131(z), 138
Pisano, J., 610, 611(14)
Pisano, J. J., 235
Pitas, R. E., 26, 34, 37, 287, 544, 562(t), 563, 581, 606, 607(45), 609, 824, 825(20), 849(20)
Pitman, R. L., 107
Pitner, T. P., 498, 499(48)
Pittman, R. C., 33, 35(121), 39, 40, 70(10), 71, 568, 581(31)
Pitts, L. L., 96(C), 97, 99(E), 142
Plasky, W. Z., 327, 332(23)
Ploplis, V. A., 121
Plow, E. F., 361
Poapst, M., 17, 798
Polacek, D., 151, 153(4), 339, 341(3), 343(3), 345
Polanovski, J., 301, 309
Pollard, H., 456, 544, 584
Polz, E., 136(Z^9, Z^{10}), 140, 301, 306, 307(19), 308(19), 309(19)
Ponsin, G., 409, 628(6), 629, 639(6), 642, 645(6)
Portman, O. W., 116(B), 118
Pottenger, L., 481
Potts, J. T., Jr., 692
Poupko, R., 514
Powell, G. M., 380, 381(16)
Powell, L. M., 51
Pownall, H. J., 10, 11, 18, 41, 42(1), 105, 121(121), 122(121, 155), 142(121), 339, 341(6), 350(6), 351, 367(8, 9, 10), 368, 370(12), 371, 372(15), 377, 403, 404, 405(6), 406(2), 408, 409, 410(6), 411(15, 16), 504, 558, 559, 561(g, o), 562(p, z), 563, 564, 566(g), 567, 576, 604, 609, 610(6), 625, 628(5, 6), 629, 639(6), 642(6), 645(6), 690, 767
Prasad, S., 30, 39(107)
Prasad, S. C., 34, 528, 529(3)

Preissner, K., 625
Price, W. H., 737
Priestly, L. M., 51, 235, 697(33), 699, 893, 894(47)
Privat, J.-P., 562(u), 563, 575, 577, 578(43, 48)
Protter, A. A., 26, 672, 677, 679(8), 681(8)
Proudfoot, N. J., 55
Proudfoot, N., 51
Prowse, S. J., 534
Pruia, N., 849(71), 850
Pruin, N., 849(70), 850
Prydz, H., 50
Przyalba, A. E., 42, 802, 803, 820
Przybyla, A. E., 681
Ptashne, M., 44, 48(14)
Puck, T. T., 49, 712, 853, 857, 858, 859, 860(27), 861(27), 862(27), 863, 864
Puhakainen, E. V., 288(12), 289
Puppione, D. L., 70, 71, 73(5, 15), 78(15), 79(f), 80(v), 81(B), 82, 83, 86(b), 87, 91(16), 92(b), 93, 94, 95(c), 96, 98(c), 99(v), 101, 102(5), 105(106), 109(75), 110(a), 111, 112(k), 114, 115(e), 116(q, r, u, x), 117, 118, 120(5, 16, 37), 121, 122(149), 128(c'), 129(q), 132(E), 135(Z^2), 137, 138, 155, 157, 165, 166, 250, 251, 259(24), 448, 877, 878(2), 879, 891(2), 892(2), 893(2)

Q

Quinet, E., 223, 240, 241(47)
Quinn, D. M., 504
Quon, D., 793, 812
Quon, D. H., 56, 877, 894

R

Raben, M., 536, 837
Rachmilewitz, D., 713
Radhakrishnamarthy, B., 81(C, F, H, L), 83, 87, 88, 106, 107
Ragland, J. B., 195, 578, 579(51)
Rahbani-Nobar, M., 435
Rakusch, U., 690
Rall, L. B., 51

Rall, S., 64
Rall, S. C., 51, 63, 127, 132(*E*), 135(*W, Z, Z*⁵), 136(*Z*⁷), 138, 139, 140, 141(67), 342, 703, 708, 727, 769, 823, 846
Rall, S. C., Jr., 33, 275, 276(7), 277, 278, 282, 284, 287(14), 528, 542, 543, 544(52), 581, 727, 735, 809, 811(19), 815, 824, 837, 846(28, 29, 42), 849(28), 850
Ralston, E., 657, 658, 659(1), 660(1, 24), 663(1, 12), 665(12), 666(12), 667(12), 668(12)
Ramirez, F., 48
Ranck, J. L., 458
Randall, R. J., 253, 256(40), 555, 570(1a), 602, 611
Randolph, A., 128(*e*), 137, 224
Rapacz, J., 123, 124(166), 125(166)
Rapport, M. M., 392, 393
Rary, J. M., 730
Ravida, C. A., 30, 36(104)
Ray, B. R., 156
Raymond, T. L., 81(*J*), 83
Reader, W., 544, 584
Reardon, C. A., 47, 57, 61(91), 64, 808, 810(20), 811, 812, 815(13)
Reardon, I., 128(*e*), 137, 224
Redgrave, T. G., 91, 109(104), 170, 171(2)
Reed, R. G., 337, 338(40)
Rees, A., 49, 741, 742(43), 743(44), 744, 885
Rees, D. J. G., 737
Rees, P. M., 389, 395(9)
Reese, T. S., 459
Reeves, H. C., 618
Regina, V. N., 796, 797(16, 17)
Regnier, F. E., 343
Regts, J., 648, 649(8), 650(8)
Rehfeld, S. J., 327, 332(23)
Reichelt, R., 447
Reijnoud, D.-J., 513
Reinhardt, W. O., 90
Reinhart, M. P., 535, 537(37), 538(37)
Renaud, G., 29
Renkonen, O., 159
Retegui, L., 529, 544(8), 545(8)
Reue, K. L., 56, 877, 879, 883, 885(17), 888(17), 889(17)
Reyes, A. A., 734, 739
Reynolds, G. A., 59
Reynolds, J. A., 142, 216, 218, 249, 250(18), 252(18), 254(18), 258(18), 259(18), 260(18), 375, 377(2), 386(2), 584, 585(15), 586, 589(22), 590, 591(31), 592(31), 603, 604(31), 623
Rhazin, A., 53
Rhoades, C., 677
Rhoads, G., 223
Rhodes, C., 715, 716(21), 734, 794, 813, 860, 865, 868(12), 907
Ricca, G. A., 802, 803(4)
Ricciardi, R. P., 44, 49(1)
Richards, E. G., 520, 521(4), 522, 523(4), 584
Richards, J. H., 48
Richards, R. I., 52, 54(64)
Richardson, C. C., 734
Ricker, A. I., 791
Ricker, A. T., 715
Riemen, M. W., 633
Rifici, V. A., 40
Rigby, P., 860
Rigby, P. W., 813
Rigby, P. W. J., 677, 715, 716(21), 734, 794, 865, 868(12), 907
Riggs, A. D., 48, 53
Riley, J. W., 223, 240, 241(47)
Ringertz, N. R., 858
Ritter, M. C., 568
Rivier, J., 56
Rizza, C. R., 737
Robards, A. W., 469
Robbins, R. J., 28
Roberts, A., 889
Roberts, B. E., 44, 49(11), 745
Roberts, D. C. K., 71, 73(13), 85(12), 142(12), 170, 171(2)
Roberts, G. C. K., 474, 478(4), 492(4), 514(4)
Roberts, J. M., 812
Roberts, R. J., 730
Roberts, R. M., 44, 48(14)
Robertson, E., 51
Robertson, F. W., 50, 885
Robertson, M. E., 697(33), 699, 893, 894(47)
Robertson, R. N., 109, 121, 122(154), 482, 483(22), 485(22), 487(22), 489(22), 497(22), 584
Robinett, S., 866
Robins, D., 812
Robinson, D. S., 248
Robinson, K., 290
Robinson, L. A., 602

Robinson, M. T., 549, 551(96)
Robson, B., 766
Robson, E. B., 894
Rodbard, D., 418, 419, 546, 547
Rodriquez, J., 866
Rodriquez, J. L., 104, 105, 142(119), 825
Roeder, R. G., 51
Rogers, J., 55, 295(18), 296
Rogers, J. L., 256
Rogne, S., 50, 746
Rohde, M. F., 377, 411, 513, 559, 562(p), 563, 564, 566(g), 567, 625
Roheim, P. S., 27, 79(j), 82, 131(z), 138, 142, 574, 753
Rolih, C. A., 71, 74(17), 77(17), 120(17), 141(17)
Ronan, R., 50, 62, 128(u'), 129(o), 134(u), 137, 139, 224, 236, 288, 295(17), 296, 542, 543, 544(59), 546, 710, 750, 751(13)
Rooke, J. A., 75, 78(59)
Rooney, V., 694, 704(20)
Rose, L., 103
Roseborough, N. J., 253, 256(40), 555, 570(1a), 602, 611
Rosen, J., 126
Rosenberg, L. E., 44, 899
Rosenblatt, M., 692
Rosenblum, B. B., 878
Rosenfeld, M. J., 56
Rosenthal, C. J., 311, 314(3)
Roskam, W., 745
Ross, D., 8, 56, 672, 675(7), 713, 717, 719(30), 733
Rosseneu, M., 50, 81(P), 83, 92(x), 93, 96(D), 99(C), 100, 104(g), 105, 122(127), 129(k), 130(u), 137, 138, 313, 365, 367(7), 368(3, 4, 5, 6), 370(3, 4, 12), 371(5), 372(5, 16), 373, 374(18, 19), 402, 558, 562(q, r), 563
Roth, J. R., 715
Roth, K., 511
Roth, R., 367(10), 368
Roth, R. I., 105, 121(121), 122(121), 142(121)
Rothblat, G. H., 838
Rouser, G., 555
Rowe, L., 879, 880, 883
Roy, M. J., 611
Roy, R., 568
Royden, G., 481
Rubalcaba, E., 546

Rubenstein, A. H., 142, 171, 182, 247, 713, 721(10)
Rubenstein, B., 432
Rubin, C. E., 21
Rubinstein, A., 169
Ruddle, F. H., 742, 856, 857
Rudel, L. L., 4, 26, 74, 77(56), 81(*I*), 83, 92(v), 93, 96(C), 97, 99(D, E), 100, 113(v), 114, 120, 121, 122(158), 129(m), 130(w), 134(T), 135(Y), 136(Z^{13}), 137, 139, 141(56), 142, 251, 825, 893
Rudman, R., 5
Russell, B., 448, 452(10)
Russell, D. W., 8, 33(23), 36(26), 46, 48, 63, 896, 908
Ruterjans, H., 515
Rutter, R. J., 820
Rutter, W. J., 42, 681, 736(f), 737, 744, 802, 803(3)
Ryback, J., 866

S

Saarinen, P., 39
Sabesin, S., 19
Sabesin, S. M., 195, 578, 579(51)
Saboureau, M., 78, 79(C), 82, 84(76), 88(76), 92(i'), 93, 101(76), 127(76), 143(76)
Sachder, O. P., 131(y), 138, 697(32), 699
Sachkelford, R. M., 124
Sadler, J., 774, 778(7)
Sakaguchi, A. Y., 49, 736(g), 737, 866, 885, 893, 894
Salpeter, M. M., 461
Sambrook, J., 43, 676, 679(14), 680(14), 716, 732, 808, 812, 813(4), 814(4), 815(4), 902, 903(16), 906(16)
Samulski, E. T., 514
Sanchez-Muniz, F. J., 438
Sanchez-Muntz, F. J., 186
Sander, C., 765, 766
Sandor, G., 214
Sandra, A., 657
Sanger, F., 44, 717, 905
Sanger, L., 68
SanGiacomo, T. R., 694, 824(32), 834(15), 845(32), 846(32), 848(15, 32), 849(32)
Sansone, G., 48

Santorineou, M. L., 880
Santos, E. C., 327, 330, 332, 333(29, 33), 335(33), 336(33)
Sardet, C., 86(*b, c*), 87, 98(*m*), 100, 115(*m*), 117, 142
Sarfarazi, M., 800
Saritelli, A. L., 313, 314(15)
Sarthy, A., 52
Sarvas, H., 534
Sasaki, M., 866
Sasaki, N., 58
Sass, M., 499
Sata, T., 101, 102(105), 109(105), 110(*a, a'*), 111, 167, 169(29), 216, 310, 349
Saudek, C. D., 223
Saunders, D. R., 713
Saunders, G. F., 865, 870(9)
Savage, R. E., 858
Savina, M. A., 124
Savu, L., 336
Sawchenko, P. G., 56
Sawyerr, A. M., 80(*t*), 82
Scambler, P., 894
Scanu, A., 5, 14
Scanu, A. M., 12, 25, 62, 71, 74(17), 77(17), 96(*y, z*), 97, 99(*B*), 100, 101, 102(109), 104(*h, i*), 105, 106, 107, 108, 113(*s, t*), 114, 115(*a', k, n*), 116(*D*), 117, 118, 120(17), 121(109, 131), 122, 123(72), 127, 128(*e*), 129(*l, n*), 130(*s, v, x*), 135(*x*), 137, 138, 139, 141(17), 142(122), 146, 151, 153(3, 4), 154, 156, 168(14), 171, 178, 182, 215, 223, 224, 235, 236, 240, 241(49), 247, 275, 310, 339(15, 20), 340, 341(2, 3, 4, 7, 8, 15), 342(2, 20), 343(3, 15), 345, 347, 375, 377(1), 379, 383, 391, 397(13), 402, 434, 452, 456, 458, 475, 476, 477, 478(8), 479(8, 11, 12), 481(11), 485, 486, 542, 544, 568, 570, 572(35), 584, 623, 647, 651(5), 690, 691, 697(8, 29), 699, 704(8), 708(8), 713, 721(10), 752, 892
Schaefer, E. J., 31, 62, 128(a), 137, 156, 223, 224, 226, 228(28), 241(28), 309, 380, 381(16), 824, 846, 849(25)
Scharschmidt, B. F., 29
Schechter, J., 250, 258(30)
Scheck, L. M., 30
Scheek, L. M., 171, 180(6), 182
Scheider, W., 324, 334(12), 338(12)

Scherphof, G., 648, 649(8), 650(8)
Schibler, U., 56
Schiffman, M. B., 880
Schilling, M. F., 775
Schimke, R. T., 44, 59
Schleicher, E., 250, 258(29)
Schmid, K., 3
Schmid, W., 872
Schmidt, C. F., 629, 662
Schmidt, P., 339(14), 340, 341(14)
Schmit, V. M., 58, 59(99)
Schmitt-Fumian, W. W., 459
Schmitz, G., 50, 63, 129(*o*), 137, 339(17), 340
Schneider, W. J., 8, 33(23), 36(26), 46, 48, 65, 824, 849(26), 896, 900, 908
Schoenborn, W., 849(71), 850
Scholnick, H. R., 892
Schonfeld, G., 13, 34, 53, 57(67), 61(67), 62(118), 65, 79(*n*), 82, 131(*D*), 134(*V*), 138, 139, 141, 247, 252, 262, 273, 280(1), 528, 529, 534(12), 535(12), 536, 543(39), 544(12), 545, 546(57), 547(84), 548(7), 549(1, 7, 12, 39, 84), 550(6, 84), 551(7), 552(101), 568, 691, 699(7), 703, 713, 716(14), 721(9), 834, 837(41)
Schotz, M. C., 17, 29(45), 798, 877, 879, 880, 884, 885(4), 893
Schousboe, I., 308, 309(22)
Schreck, R. R., 796
Schreiber, E., 513
Schreiber, J. R., 58, 59(99)
Schreiner, H., 48
Schroeder, F., 122
Schroeder, J. P., 616
Schrrott, H. G., 800
Schuerch, A. R., 671
Schuh, J., 159, 218, 249, 379
Schullery, E., 662
Schulman, R. S., 251
Schultz, A., 80(*r*), 82
Schultze, H. E., 301, 302, 306
Schumaker, V. N., 70, 73(5), 78, 79(*f*), 82, 92(*b*), 93, 94, 95(*c*), 96, 98(*c*), 99(*v*), 100, 101, 102(5), 105(106), 109(75), 110(*a, c*), 111, 115(*e*), 116(*x*), 117, 118, 120(5), 128(*c'*), 129(*q*), 132(*E*), 135(Z^2), 137, 138, 165, 168, 250, 251, 256, 259(24), 448, 549, 551, 590, 877, 878(2), 879, 891(2), 892(2), 893(2), 894

Schuman, J., 38
Schumm, J. W., 812
Schutz, G., 807, 810(12)
Schwarzbauer, J. E., 55
Schweikart, K. M., 880
Schwertner, H. A., 153
Schwick, H. G., 534
Sciacca, R., 848, 849(57)
Sciarratta, G., 48
Scoffone, E., 544
Scott, J., 51, 237, 697(33), 699, 893, 894(47)
Scott, L. W., 30, 39(108)
Scott, P. J., 113(*P*), 114
Seager, J., 250, 258(30), 544
Seeburg, P., 724
Segrest, J., 372
Segrest, J. P., 16, 24, 92(*d*), 93, 98(*g*), 182, 188(8), 190, 194(10), 195, 201, 208, 403, 404(9), 405, 410(9), 513, 559, 564(17), 571(17), 572(17), 574(17), 584, 611, 627, 629, 640, 643, 646, 657, 767, 787, 891
Segura, R., 74, 112(*o*), 114, 119(48), 123(48), 133(*I*), 139
Seherphof, G., 432
Seidel, D., 433, 852
Semis, H. F., 25
Seppala, I., 534
Servillo, L., 386
Sexton, J. P., 738
Seymour, J., 277
Shackelford, J. E., 26
Shaffer, J., 749
Shah, J. V., 866
Shah, R., 74, 77(56), 92(*v*), 93, 141(56), 825
Shahrokh, Z., 649, 651(14)
Shainoff, J. R., 104, 105, 115(*n*), 117, 127(*e*), 137
Shakir, K. M. M., 434
Shander, M., 55
Shander, M. H. M., 51
Shapira, S. K., 48
Shapiro, D., 54, 56
Shapiro, S. Z., 900
Sharon, I., 33
Sharpe, C. R., 47, 235, 697(34), 699, 733, 742, 743(44), 799(29), 800
Sharrow, S. O., 657
Shastry, B. S., 51
Shaw, D., 474
Shaw, P. H., 56, 57(88), 70, 141(7)

Shechter, I., 37, 544
Sheldon, E. L., 812
Shelley, C. S., 47, 235, 697(34), 699, 733, 799(29), 800
Shelness, G. S., 55, 672, 679(8), 681(8)
Shen, B. W., 14, 122, 156, 168(14), 339(20), 340, 342(20), 375, 377(1), 391, 397(13)
Shen, M. M. S., 112(*a'*, *b*), 113, 119, 195, 456
Shepherd, J., 32, 35(117), 86(*g*), 87, 456
Sherman, F., 715
Sherrill, B. C., 141, 142(213), 824, 825(22), 849(22)
Shimada, Y., 70(8), 71
Shinomiya, M., 544
Shipley, G. G., 14, 122, 411, 507, 561(*f*), 563, 574, 601, 608
Shiratori, T., 18
Shireman, R., 250(32), 251, 584
Shore, B., 5, 17, 134(*P*), 135(Z^6), 136(Z^{11}), 139, 140, 214, 288, 310, 339(12), 340, 341(12), 342(12), 348, 379, 536, 584, 798, 823
Shore, R., 86(*f*), 87
Shore, V., 5, 86(*f*), 87, 214, 348, 379, 536, 584
Shore, V. G., 17, 126, 128(*a'*), 134(*P*, *U*), 135(Z^6), 136(Z^{11}), 137, 139, 140, 141(195), 223, 288, 310, 339(12, 19), 340, 341(5, 12), 342(12), 348, 349, 721, 798, 823
Shore, V. S., 655
Shoulders, C., 56
Shoulders, C. C., 8, 47, 49, 224, 235, 697(34), 699, 712, 732, 733(14), 734(14), 741, 742(43), 743(44), 799(29), 800, 885
Shoulson, I., 736(g), 737
Shows, T. B., 51, 797, 866, 893, 894(47)
Shrewsburg, M. A., 90
Shulman, R. S., 129(*P*), 133(*M*, *N*), 134(*R*, *U*, *V*), 138, 139, 220, 236, 288, 294, 296, 313, 314(15), 379, 543, 710
Shuman, J., 647, 652(7)
Sidoli, A., 47, 235, 697(34), 699, 733, 799(29), 800
Siedel, D., 155, 157
Siegel, A. J., 741
Sigler, G., 12
Silverfield, J. C., 529, 530(15), 534(5)
Silverstein, S., 812

Sim, D. K., 793
Sim, G. K., 812
Simon, K., 249
Simoni, R. D., 327, 337(19)
Simons, A. R., 535, 536(36)
Simons, K., 165, 215, 584, 585(17), 586, 599
Simpson, R. B., 330, 333(29)
Sims, H. F., 62, 131(y), 138, 151, 153(3), 224, 235, 691, 697(8, 29, 32), 699, 704(8), 708(8), 752
Sing, C. F., 837, 840(49), 841(49), 845(49)
Singer, S. J., 393, 401
Singh, S., 158
Single, S., 247
Siniscalco, M., 855, 856(18)
Sipe, J. D., 314, 315
Sirtori, C., 622
Sirtori, C. R., 104, 105, 142(119), 278, 727, 735
Sirtoni, C. R., 825
Siuta-Mangano, P., 21, 23(68), 705
Sivanandaiah, K. M., 633
Skehel, J. J., 774
Skinner, E. R., 75, 78(59)
Skipski, V. P., 90, 91(100), 98(b), 99, 102(100), 104, 106, 108(100), 519
Sklar, L. A., 327, 337(19)
Skogen, B., 312, 313(12)
Skolnick, M., 737, 861
Skow, L. C., 878
Skrabal, P., 481
Slater, H. R., 86(g), 87
Slater, R.-J., 72, 142(26)
Slaughter, C. A., 550
Slaughter, C. J., 8, 36(26), 896
Sletten, K., 312, 313(12)
Sleytr, V. B., 469
Sliwkowski, M. B., 62
Sloane-Stanley, G. H., 555
Sloop, C. H., 27, 578, 753
Small, D. M., 14, 81(I), 83, 101, 105(106), 113(v), 114, 116(x), 118, 121, 122(149, 158), 327, 333(25), 337(25), 411, 448, 452(10), 487, 488(31), 489(31), 507, 511, 513, 561(f), 563, 574, 595, 596(34), 597, 598(36), 601(34, 36), 604(36), 605(36), 608, 649, 651, 655(17), 656
Small, J. D., 889
Smann, G. A., 727
Smith, C. C., 81(L), 83

Smith, D. G., 543
Smith, D. P., 57, 61(95), 62(118), 131(Z), 138, 141, 240, 691, 699(7), 701(28)
Smith, E. L., 544
Smith, G. D., 74, 94, 96(u), 101(47), 123(47)
Smith, G. P., 781
Smith, H. M., 109
Smith, H. O., 730
Smith, K. D., 730
Smith, L., 38
Smith, L. C., 10, 11, 12, 18, 30, 34, 39(108), 41, 42(1), 129(h), 134(s), 137, 139, 546, 576, 609, 610(6), 690, 721
Smith, M., 902
Smith, R., 250, 259, 332, 333(32), 339(32), 584, 589(12), 590(12)
Smith, T., 765
Smith, T. F., 773, 774, 776, 778(7), 787
Smithies, O., 812
Sobroff, J. M., 321
Socorro, L., 215, 216, 584, 585
Sodhi, H. S., 434
Soetewey, F., 81(P), 83, 92(x), 93, 96(D), 105, 122(127), 129(k), 137, 365, 367(7), 368(3, 4), 370(3, 4), 371, 372(16), 373, 374(19)
Sogard, M., 458
Sokoloski, E. A., 476, 478(10), 481(10), 496, 501(44)
Soltys, B. J., 327
Soronti, B., 126
Soutar, A., 12, 721
Southard, J. L., 880
Southern, E., 48, 859
Southern, E. M., 719, 730, 737, 794
Southern, P. J., 816
Spady, D. K., 35
Sparkes, M. C., 894
Sparkes, R. S., 894
Sparks, C. E., 76, 131(D), 138, 248, 339(11), 340, 341(11), 392, 396
Sparks, R. S., 549, 551(96)
Sparrow, J. T., 10, 11, 12, 288, 294, 339(10), 340, 341(6, 10), 350(6), 351, 352, 367(9), 404, 405(6), 409, 410(6), 513, 542(55), 543, 545, 550(66), 562(y, z), 563, 628(5, 6), 629, 639(6), 642(6), 645(6), 721, 769
Spatola, A. F., 633
Spaziani, E., 23, 29, 35(71)
Spector, A. A., 80(s), 82, 110(e), 111,

112(*d*), 113, 115(*d*), 117, 320, 321, 322(1), 323, 324, 325(1, 14), 326, 327, 328(7), 329(7), 330, 332, 333(1, 7, 29, 33), 334, 335(7, 27, 33), 336(1, 17, 33), 337(7), 338(7)
Spence, M. W., 327, 337(18)
Sprecher, D. L., 224, 238(27), 245(27)
Srinivasan, S. R., 81(*C, F, H, L*), 82, 87, 88, 106, 107
Staehelin, T., 361, 423, 438, 440(20), 535, 536(35)
Stalenhoef, A. F. H., 32, 35(116), 50
Stallcup, K. L., 548
Stange, E., 116(*P*), 117
Stange, E. F., 33, 36(122)
Stanker, L. H., 534
Stark, G. R., 613, 860, 865
Staros, J. V., 616
Starr, J. I., 713, 721(10)
Starzl, T. E., 223
Statts, J., 889
Stead, D., 92(*q*), 93, 96(*t*), 101, 102(108), 105(106), 112(*n*), 114, 116(*v*), 117, 123
Stearman, R. S., 320
Stedje, K., 544
Steele, J. C., 216, 218, 247, 249, 250(18), 252(18), 254(18), 258(18), 259(18), 260(18), 377, 584, 585(15), 589, 618
Steere, R. L., 457
Steim, J. M., 482, 483, 485(21)
Stein, O., 38, 39, 68, 581, 721
Stein, R., 44
Stein, S., 256, 555, 570(2)
Stein, Y., 38, 39, 68, 581, 721
Steinberg, D., 33, 35(121), 39, 40, 70(10), 71, 107, 568, 581(31)
Steinbrecher, U. P., 355, 357(6), 358(7)
Steinberg, D., 354, 357(4)
Steiner, G., 17, 432, 798
Steiner, P. M., 503
Steinmetz, A., 18, 227, 241(31), 280, 708, 727, 772, 824, 837(31), 845(47), 849(70, 71), 852
Stephens, R. E., 276
Steppan, S., 879, 893
Stern, M., 860, 865
Stern, O., 576
Stern, R. H., 747
Stevens, G. R., 170, 182, 455
Stewart, M., 774

Steyrer, E., 134(*s*), 139, 142, 288
Stiles, J. I., 715
Stiller, E., 5, 339, 341(2), 342(2), 568
Stocks, J., 49, 741, 742(43), 743(44), 744, 885
Stoffel, W., 513, 625
Stoltz, J. M., 108
Stone, R. A., 659
Stone, W. H., 124
Stonik, J. A., 66
Stout, C., 81(*O*), 83, 92(*w*), 93
Strauss, A. W., 25, 53, 57(67), 61(67, 95), 62(118), 131(*y, z*), 134(*v*), 138, 139, 141, 151, 153(3), 224, 235, 240, 691, 697(8, 29, 32), 699(7), 701(28), 704(8), 708(8), 752
Strauss, J. F., III, 40
Strike, P. M., 746
Strisower, B., 213
Strisower, E. H., 73, 116(*r, u*), 117, 120(37), 155, 156(2)
Strominger, J. L., 900
Strong, K., 409, 628(6), 629, 639(6), 642(6), 645(6)
Strong, L. E., 72, 214
Stroud, R. M., 60
Struck, D. K., 63, 64(130)
Suarez, B., 536, 837
Suarez, Z. M., 5
Suckling, P. A., 837, 845(48)
Sudhof, T. C., 48
Sugano, M., 91, 105
Sugenor, D. M., 3
Suggs, S. V., 746, 748, 749
Suita, P. B., 63, 64(130)
Suita-Mangano, P., 64, 68(131)
Sullivan, C. P., 548
Sundaram, G. S., 434
Suominen, L., 598, 600(37), 606
Suzue, G., 339, 341(1), 342(1), 580
Svanberg, U., 218
Svensson, H., 72, 144
Swaminathan, N., 107, 250, 258(28), 705
Swaney, J. B., 128(*b*), 129(*p*), 131(*z*), 133(*N*), 134(*R, V*), 135(*Z*^{1}), 137, 138, 139, 140, 536, 559, 561(*d, k, l, m*), 563, 617, 619, 620, 622(15), 624(11), 626, 753, 765(2)
Swanson, L. W., 56
Sweeny, S. A., 564, 566(*k*), 567, 569(24), 580(24), 582

Sweet, R., 812
Swift, L. L., 65, 66
Sybers, H. D., 34, 546
Szoka, F., 647, 648(2), 657, 661
Szostak, J. W., 715

T

Taam, L., 224, 238(27), 245(27)
Tajima, S., 377, 388
Takahashi, S., 767, 769(37)
Takaku, F., 536
Takats, J., 110(j), 111
Takeichi, M., 657
Talkowski, C., 489, 490(33), 491, 493(33, 35), 495(35), 500(35), 501(35), 502, 503(35), 514(33, 35)
Tall, A. P., 24
Tall, A. R., 38, 81(I), 83, 113(v), 114, 116(x), 118, 121, 122(149, 158), 223, 411, 527, 561(f), 563, 574, 578, 602, 647, 648(4), 649, 650(4, 6), 651(4, 13), 652(4, 6, 7), 653(4), 654(4), 655(1, 17), 656, 713
Tam, J. P., 633
Tamnkum, J. W., 55
Tan, B. H. A., 435
Tan, M. H., 30
Tan, T., 66, 131(D), 138
Tanaka, N., 116(B), 118
Tanaka, R. D., 8, 59
Tanaka, Y., 57, 70, 88(3), 105, 142(3)
Tandeter, H. B., 658, 659(21a), 663(21a)
Tanford, C., 116(y), 118, 250, 259(23), 260, 261, 332, 333(32), 339(32), 584, 585, 589(12), 590(12)
Tanzawa, K., 70(8), 71
Tanzi, R. E., 736(g), 737
Tardieu, A., 458, 465, 467(18), 468
Tartar, A., 530
Tarugi, P., 79(h), 82, 95(g), 97, 101, 102(107), 128(b, f), 137
Tarver, A. P., 504
Tasch, M. A., 551, 552(101)
Taso, Y. K., 50
Tata, F., 50
Tate, R. L., 381
Tauber, J.-P., 288(10), 289
Taunton, O. D., 34, 74, 112(o), 114, 119(48), 123(48), 129(h), 133(I), 134(S), 137, 139, 294, 456, 542(55), 543, 546
Taveirne, M. J., 50
Taylor, B. A., 70, 120(6), 143(6), 879, 880, 881, 883, 884, 885(14), 886(14), 888(14), 889(14), 894(14)
Taylor, E. W., 456
Taylor, H. L., 72, 214
Taylor, J. M., 26, 42, 46, 47, 56, 58(89), 61(17), 70, 135(Z^1), 140, 141(4), 240, 735, 746, 754, 763(8, 29), 764, 767(8), 769(22, 29), 771(8, 22, 29), 801, 802, 803(4), 804(1), 805(1), 806(1, 2), 808(1), 809(1), 810(1, 2, 20), 811, 812, 813, 815(13, 18), 822
Taylor, R. T., 853
Taylour, C. E., 72, 77(23), 78(23), 79(b), 80(b), 81(b), 82, 84(23), 85(23), 86(a'), 87, 88(23), 89(23), 91(23), 92(a), 93, 95(a), 98(a, l), 99(a), 102, 105, 112(l), 114, 119
Teary, A. C., 796, 797(17)
Teisberg, P., 49, 50(43), 852, 863, 864(2)
Teplitz, R. L., 734, 739
Terpsira, A. H. M., 186
Terpstra, A. M. M., 438
Terry, E. W., 534
Terwilliger, T. C., 17, 402, 403
Thamer, M., 849(69), 850
Tharapel, A., 866
Theolis, R., 141
Theolis, R., Jr., 529, 535, 536(40), 542(13), 544(13)
Theorell, A. H. T., 144
Third, J. L. H. C., 824(32), 825, 845(32), 846(32), 848(32), 849(32)
Thomas, G. P., 44
Thomas, J. K., 121
Thomas, P. S., 821, 906
Thompson, J. S., 889
Thompson, T. E., 595, 662
Thompson, T. S., 888, 889
Tibbling, G., 324
Tiemeier, D., 812
Tikkanen, M. J., 529, 534(12), 535(12), 543(39), 544(12), 545, 547(84), 548(7), 549(1, 12, 39, 84), 550(84), 551(7), 552(101)
Tillack, T. W., 662
Timasheff, S. N., 381, 382(18)

Tippett, P., 894
Tipping, L. R. H., 481
Tiselius, A., 72, 144
Titani, K., 136(Z^{11}), 140, 311, 312(6), 313(6)
Tivol, W., 433, 434(11)
Tjaden, M., 4
Tjoeng, F. S., 633
Tolbert, N. E., 388, 570
Tollefson, J. H., 18, 300, 301(9), 303(9)
Tolleshaug, H., 65, 727, 900
Toomey, M. L., 99(v), 100, 116(x), 118
Torain, B. F., 315
Toribara, T. Y., 556
Torregiani, D., 104, 105, 142(119), 825
Torsvik, H., 124, 125(179, 181)
Toscas, K., 25, 62, 151, 153(3), 224, 708
Toth, J., 5, 339, 341(2), 342(2), 568
Tournier, J.-F., 288(10), 289
Towbin, H., 361, 423, 438, 440(20), 535, 536(35)
Tracy, R., 121, 122(155)
Tracy, T., 824(32), 825, 845(32), 846(32), 848(32), 849(32)
Träuble, H., 562(v), 563
Traut, R. R., 696
Treisman, R., 55
Trentin, J. J., 103, 115(f), 117, 120(110), 888, 889(30)
Triplett, R. B., 247, 249(5), 251(5), 252(5), 258(5)
Troxler, R. F., 135(Z^5), 136(Z^7), 140, 141
Trurnit, H. J., 392
Tsang, V. C. W., 535, 536(36)
Tsao, B. P., 534, 545(28)
Tsao, Y. K., 235, 750, 864
Tse, T. P. P., 746
Tsujita, Y., 70(8), 71
Tun, P., 848, 849(58)
Tunggal, B., 513
Turchin, H. A., 880
Turkewitz, A. P., 548
Turkova, J., 548
Turtle, J. R., 419
Tye, B. K., 715

U

Udenfriend, S., 256, 555, 570(2)
Uhlenbruck, G., 106

Uhler, M., 65
Urdea, M., 235, 697(33), 699
Urdea, M. S., 51
Uroma, E., 3
Usher, D. C., 76, 124(68), 125(68), 126(68), 132(H), 139, 141(68)
Utermann, G., 49, 131(y), 138, 223, 227, 240(7), 241(7, 31), 273, 280(1), 282, 433, 703, 708, 727, 753, 772, 824, 834, 837(30, 31, 41), 845(47, 50), 849(26, 70, 71, 72), 850, 852
Uzawa, H., 549

V

Vaino, P., 39
Vaith, P., 106
Vale, W. W., 56
Valente, A. J., 163, 249, 340, 589
Vallette, G., 336
Van Berkel, T. J. C., 37
Van Brogt, P. H., 50
Vandamme, D., 81(P), 83, 92(x), 93, 96(D), 105, 122(127), 142(125)
Vandercasteele, N., 99(E), 100, 105, 120(126), 129(n), 130(x), 137
Van der Eb, A. J., 819
Vanderhoek, J., 581
VanderZwaag-Gerritsen, J., 745
Van Dyke, R. W., 29
van Gent, C. M., 849
van Gent, T., 39
Van Holde, K. E., 572
Van Landschoot, N., 105, 142(125)
Van Lenten, B. J., 36, 142
van Noort, W. L., 171, 182, 180(6)
Van Renswoude, J., 657
van Rijswijk, M. H., 314
Van Rollins, M., 323
Van Someren. H., 855
van't Hooft, F. M., 31, 39, 171, 180(6), 182
van Tol, A., 39
Van Tornout, P., 373, 374(18), 558, 562(s), 563
Van Tornout, R., 402
Van Winkle, W. B., 559, 562(p), 563, 564, 566(g), 567, 625
Varghese, M., 354

Varjo, P., 39
Velivasakis, M., 855, 856(18)
Vella, M. A., 742, 743(44), 744
Verbin, R. S., 68
Vercaemst, R., 81(P), 83, 92(x), 93, 96(D), 99(C, F), 100, 104(g), 105, 120(126), 122(127), 129(k, n), 130 (u, x), 137, 138, 367, 368(4, 6), 370(4), 371, 372(16), 373, 374(19), 558, 562(s), 563
Verdery, R. D., 529, 542(13), 544(13)
Verma, R. S., 866
Vervloet, M., 855
Vezina, C., 580
Via, D. P., 37
Vigne, J.-L., 28, 223, 848, 849(58)
Virgil, D. G., 35
Virtanen, J. A., 39
Vitek, M. P., 44
Vitello, L., 99(B), 100, 104(i), 115(a'), 116(D), 117, 118, 121
Vogelberg, K. H., 849(71), 850, 852
Vogelstein, B., 813
Vold, R. L., 498
Vollmer, R. J., 804
Volmer, M., 576
von Deimling, O. H., 878
Von Heijne, G., 696
von Hunger, K., 232

W

Waalwijk, C., 739
Waber, P. G., 738
Wachtels, S., 730
Wadsö, I., 365
Wahl, G. M., 860, 685
Wahl, P., 562(u), 563, 573, 575, 577, 578(43, 48)
Wahl, P. W., 713, 721(11)
Waite, M., 327, 432
Wakefield, T., 79(l), 82
Waldner, H., 457
Walker, L. F., 65, 73, 74(40), 78(40), 112(f), 113, 117, 120(40), 131(D), 138, 247, 262
Walker, S. M., 124, 125(179)
Walker, T. E., 503
Wall, R., 55
Wallace, M. E., 880

Wallace, M. R., 736(g), 737
Wallace, R. B., 733, 734, 739, 746, 748(8), 749(8)
Wallinder, L., 127
Wallis, S., 697(33), 699
Wallis, S. C., 746, 894
Walsh, K. A., 136(Z^{11}), 140, 311, 312(6), 313(6)
Walsh, M. T., 250(31), 251, 252(31), 584, 586(20), 587(20), 588(20), 592, 593(20), 594(20), 595(20), 596(20), 597, 598(36), 601(36), 603, 604(20, 36, 41), 605(20, 36, 41)
Walter, P., 696, 699(23)
Walton, R. K. W., 122, 123(163)
Wangenstein, R. W., 70, 120(6), 143(6), 879, 880, 883, 884, 885(14), 886(14), 888(14), 889(14), 894(14)
Wang, S.-Y., 672, 679(8), 681(8)
Wangermann, G., 447
Wanner, L., 49
Ward, D. C., 734
Wardell, M. R., 837, 845(48)
Warner, H., 556
Warnick, G. R. 837, 849(72), 850
Warnick, R., 208
Warren, R., 52
Wassall, S. R., 513, 514(73)
Watanabe, H., 438, 440(19), 441(19)
Watanabe, Y., 70(8, 9, 10), 71, 85(9)
Waterman, M. S., 773, 774, 775, 776, 778(7), 787(4)
Watkins, P. C., 736(g^7), 737
Watson, B., 31
Watson, W. A., 188
Watt, R. M., 216, 249, 529, 530(16), 534(16), 545(16), 550(16), 584, 586, 589(22), 590, 591(31), 592(31), 603, 604(31)
Watt, T. S., 529, 530(16), 534(16), 545(16), 550(16)
Watts, A., 570, 572(33)
Waugh, D., 121, 122(149)
Waugh, J. S., 498
Webber, L. S., 81(C), 83, 87
Weber, G., 573
Weber, K., 593
Weber, W., 280, 282, 824, 837, 845(47, 50), 849(26)
Weech, P. K., 76, 435, 438(18), 440(18), 441(18), 529, 548, 550(11)

Wegrzyn, J., 124
Wehrly, K., 133(M), 134(U), 139, 294, 296, 543
Wei, S.-P. L., 323
Weidman, S. W., 34, 528, 536, 549(1)
Weigman, S. W., 837
Weil, D., 50
Weinberg, R. B., 223, 240, 241(49), 275
Weinstein, D., 39, 107, 389, 568, 581(31)
Weinstein, J. N., 24, 649, 657, 658, 659(1), 660(1, 24), 663(1, 12), 665(12, 14), 666(12), 667(12), 668(12, 14)
Weintraub, H., 53
Weisbrod, S., 53
Weisgraber, K. H., 33, 64, 71, 73(11), 74(11, 39), 75(13), 77(39, 42, 49, 54), 79(m), 80(q), 81(M), 82, 83, 84(11, 49), 95(m), 97, 98(h, P), 100, 102(11), 109(42), 115(o), 116(y), 117, 118, 120(11, 13), 123(42, 54), 126(11, 13), 127, 129(h', j, r), 130(t), 131(y, A), 132(G), 135(Z, Z^1, Z^5), 136(Z^7, Z^8), 137, 138, 139, 140, 141(13), 142(11, 13), 169, 216, 273, 274, 275(4), 276, 277, 278, 279, 280(1), 282, 284, 285, 287(14), 339(13), 340, 341(13), 342, 355, 528, 542, 543, 544(52), 581, 609, 703, 708, 727, 735, 753, 769, 809, 811(19), 815, 823, 824, 825, 834, 837(41), 846(28, 29, 42), 849(27, 28), 850
Weisinger, R., 338
Weiss, M. C., 853
Weiss, R. M., 402, 403
Weissman, I. L., 126
Weisweiler, P., 339(14), 340, 341(14), 342(13)
Welch, P. K., 263
Welch, V. A., 92(q), 93, 96(t), 101, 102(108), 105(106), 112(n), 114, 116(v), 117, 123
Wells, M. A., 79(k), 82, 89, 91(92), 92(d), 93, 95(e), 98(g), 105(92), 112(e), 113, 115(g), 117, 119(92), 120(92)
Wensink, P. C., 747
Werb, Z., 26, 58
Werth, V. P., 656
West, C. E., 170, 171(2)
Westphal, H. M., 52, 54(64)
Wetmur, J. G., 865
Wetterau, J. R., 558, 559(8), 561(i, j), 563, 569(16), 571(16), 573(16), 575(8), 579(16)
Wexler, N. S., 736(g), 737
Whicher, J. T., 315
Whidby, J. F., 498, 499(48)
White, R., 861
White, R. M., 17
White, R. L., 737
Whitehead, A. S., 796
Wicken, M. P., 44
Wieland, H., 433
Wieland, O. H., 250, 258(29)
Wieland, V. H., 155, 157
Wieringa, B., 675, 745
Wigler, M., 812
Wiklund, O., 223, 300, 301(8), 303(8)
Wikman-Coffelt, J., 696
Wilbur, W. J., 754, 764(11)
Wilchek, M., 231
Wilcox, H. G., 81(z), 83, 105
Wiley, D. C., 774
Wilhelmsen, L., 223
Wilhelmsson, C., 300, 301(8), 303(8)
Wilhelmy, L., 390
Wilk, J., 852
Wilkinson, T., 92(d), 93, 98(g), 182, 188(8)
Willet, W., 741
Williams, B. G., 812
Williams, D. L., 26, 55, 58, 72, 672, 679, 681(8), 823
Williams, E., 489, 490(33), 491, 493(33, 35), 495(35), 500(35), 501(35), 502, 503(35), 514(33, 35)
Williams, H., 8, 56, 717, 719(30), 733
Williams, H. W., 672, 674(6), 675(7), 846
Williams, K. J., 647, 651(5), 656
Williams, L. G., 744
Williams, M., 14
Williams, M. C., 520, 578, 579(53)
Williams, R. H., 4
Williams, R. J. P., 481, 492(37), 493
Williamson, B., 295(18), 296
Williamson, R., 50, 736(h, i), 737, 746, 800, 894
Wills, R. D., 155(9), 156, 182
Wilschut, J. C., 648, 649(8), 650(8)
Wilson, A. W., 774
Wilson, D. M., 503, 546
Wilson, J. T., 48, 738

Wilson, L. B., 48, 738
Windler, E., 29, 34, 64, 824, 849(24)
Windmueller, H. G., 25, 62, 65, 76, 77, 129(P), 133(N), 134(R, V), 138, 139, 223, 236, 713, 823
Wing, L., 155(9), 156
Winona, C., 773, 787(2)
Winship, P. R., 737
Wissler, R. W., 99(B), 100, 104(i), 105, 113(r), 114, 116(D), 118, 121, 142 (122)
Witt, K. R., 218, 250, 254(26), 255(26), 376, 377(4), 584, 589(19), 590(19)
Witztum, J. L., 34, 354, 355, 357(3, 4, 6), 358(3, 7), 360(3), 528, 536, 549(1), 837
Wlodauer, A., 24
Woffze, D. K., 741
Wold, B., 812
Wold, F., 614
Wolf, R. H., 81(L), 83, 129(n), 130(w), 137
Wolff, J. A., 656
Wong, A. W., 167
Wong, T. N., 633
Woo, S. L. C., 857, 869
Woods, D. E., 295(18), 296, 715, 791, 796
Woods, K. R., 771
Woods, O., 746
Woodward, C. J. H., 186
Woodward, C. S. H., 438
Worcester, D., 690
Wough, D., 116(x), 118
Wozney, J. M., 869
Wrann, M., 324
Wright, A. F., 737
Wu, A. L., 25, 65, 76, 223, 713, 823
Wu, C.-S. C., 519, 523(1), 524(1), 525
Wu, R., 715
Wurm, H., 136(Z^{10}), 140, 301, 306, 307(19), 308(19), 309(19)
Wurzer-Figurelli, E., 855

X

Xavier, A. V., 492(37), 493

Y

Yachida, Y., 623
Yamamoto, A., 377, 388
Yamamoto, M., 105
Yamamoto, T., 8, 33(23), 36(26), 46, 896, 908
Yamanaka, W., 86(b), 87
Yanagita, Y., 623
Yang, F., 50
Yang, J. T., 519, 523(1), 524(1)
Yates, M., 456, 458, 466
Yatvin, M., 658
Yeagle, P. L., 478, 479(14), 480, 481
Yermolaev, V. I., 124
Yeshurun, D. L., 34, 546
Yokoyama, S., 377, 388, 404(7, 8), 405, 408, 552, 628(4), 629, 636(4), 644, 721, 769, 771
Yoshikami, S., 657
Young, A. B., 736(g), 737
Young, C., 121, 122(148)
Young, E. G., 213
Young, J. R., 900
Young, R. A., 56
Yunis, J. J., 865, 870(13), 874(15)

Z

Zabel, B. U., 866
Zaharko, D. S., 658
Zahn, H., 616
Zampighi, G., 586
Zannis, V. I., 8, 47, 49, 50(27, 41), 56, 62, 64, 69(41), 135(z), 139, 224, 227, 273, 280(1), 282(15), 296, 339(16, 18), 340, 672, 674(6), 675(7), 688, 691, 692, 693, 694, 695(9, 17), 696(18), 697(9, 17, 31), 699(9, 17), 701(9, 14, 16, 17, 22), 703(9), 704(14, 20), 705, 708(22), 711, 712, 713, 717(2, 3), 719(8, 30), 720(2, 7, 8), 721(31), 723(2), 733, 740, 741, 774, 779, 781(8), 793, 800(11), 823, 824(2, 32), 825(2), 834(13, 14, 15, 16), 837(13, 14, 41), 842(17), 845(32), 846(32), 848(14, 15, 32, 40), 849(32)
Zech, L., 855

Zech, L. A., 31, 62, 223, 380, 381(16), 824, 849(25)
Zechner, R., 134(S), 139, 142, 288
Ziegler, A., 534
Zierenberg, O., 513
Ziessow, D., 499
Ziff, E., 51
Zilversmit, D. B., 19, 89, 124, 214, 217(14), 303
Zimmerman, C., 235
Zimmerman, D., 81(z), 83
Zinsmeister, A. R., 223
Zipper, P., 113(P), 114, 119, 121(143), 122(143)
Zollman, S., 894
Zucker, S., 26, 58, 823
Zukel, W. J., 223

Subject Index

A

ABC nomenclature, 297
Abetalipoproteinemia, 5, 22–23, 68, 147, 157
Absorption spectroscopy, 516–517
ACAT. *See* Acyl:coenzyme A-*O*-acyltransferase
Acyl:coenzyme A-*O*-acyltransferase, activity, 18, 20, 31
Adrenal gland, synthesis of apolipoproteins, 26
Adrenocorticotropic hormone, in LDL receptor regulation, 36
Affinity chromatography, 78, 169
  for lipoprotein isolation, 10
African green monkey
  apoA-I, physical characteristics, 129
  apoA-II, physical characteristics, 130
  apoC-II, physical characteristics, 134
  apoC-III, physical characteristics, 135
  circulating lipoprotein levels, 81
  HDL, physical properties, 120
  HDL and subclasses, chemical composition, 99
  lipid metabolism, 142
  threonine-poor apoproteins, physical characteristics, 136
Agarose electrophoresis, 167
Aggregation, definition, 376
Air–water interface, quantitation of adsorption and desorption at, 389
Albumin
  addition of fatty acid, celite method, 324–326
  binding isotherm for fatty acid, 331
  comparison with plasma lipoproteins, 338–339
  delipidation, 322
  fatty acid binding, 321
    data analysis, 333
    data correction for fatty acid association, 333
    effect of fatty acid structure, 335–336
    measurement, 327–339
    multiple binding model, 334–336
      distribution analysis, 334–335
    Scatchard model, 333–334
    sites, 333, 337–338
      fluorescence spectroscopy, 337
      localization, 337
      proteolytic cleavage, 337
      tryptophan fluorescence quenching by, 336
  glucosylation, 362
  incubation with n-heptane solution containing fatty acid, 327
  lipid transport, 336–339
  solution
    addition of fatty acid salts, 322–324
    fatty acid sonification in, 327
    incubation with hormone-stimulated fat pad, 327
    incubation with triglyceride emulsion containing lipase, 327
  structure, 336
  unbound fatty acid association, 332
Alcohol–ether, solvent system, 215
Alcoholic hepatitis, 578
ALIGN, 764
Ammonium molybdate, 443
Amphipathic helical hypothesis, 403–404, 583–584, 627–629, 770
Amphipathic helices, 16–17, 771, 787
  self-association, 16
Amphiphilic peptide, chain length-function correlation, 771
α-Amylase, gene, 56
Amyloidogenesis, 311
Angled-head rotor, 196, 209
Anticoagulant, added to plasma samples, 153, 158, 218
Apes, circulating lipoprotein levels, 81
Apolipoprotein II, cDNA probe
  labeling, 680

preparation, 677–679
Apolipoprotein A, 5, 223, 837
Apolipoprotein A-I, 5, 24, 39, 40, 106, 109, 125, 142, 151, 154, 223, 279, 340, 654, 690, 838, 883
  absorption spectroscopy, extinction coefficients, 516
  adsorption kinetics, 402
  adsorption to egg phosphatidylcholine monolayer, 398
  amino acid analysis, 235–236
  amino acid sequence, 12, 141, 224
  analytical isoelectrofocusing, 234
  analytical SDS–gel electrophoresis, 232
  anion-exchange HPLC, 348–349
  and apoA-II, interfacial exchange, 402–403
  and apoA-IV, sequence similarities analysis, 758–761
  association enthalpy with synthetic lecithin, 368
  and CAD, 4
  cDNA clones
    characterization, 713–717
    isolation, 713–717
  cDNA probe, human metaphase cell hybridized with, data analysis, 875–876
  characterization, 232–235
  chromosomal localization, 859, 893–894
  in chylomicron synthesis, 23–24, 68–69
  clones, 8
  complexes with lipids, sodium cholate-mediated formation, 566
  complex with DMPC, 637
    enthalpy titration curve for, 373–374
  conformational change on lipid binding, 368–370
  consensus domain
    peptide analog, 630–631
    similarity to rat cardiac myosin, 774
  cross-linking experiments, 624–625
  cyanogen bromide fragmentation, 542
  distribution, 341
    in HDL classes, 441
  Edman amino-terminal analysis, 235
  enthalpy of α-helix formation, 370
  enthalpy titration curve, 373–374
  enzymatic cleavage, 543
  extracellular modification, 695, 705–709

  fluorescence properties, 517
  function, 223–224
  gene, 52, 800
    characterization, 717–720
    chromosomal location, 49–51
    DNA insertion in, familial segregation of, 725–726
    isolation, 717–720
    linkage, to apoC-III gene, in humans, 720–723
    Southern blot analysis, 860–862
  genomic clones, 47
  glucosylation, 361, 362
  in HDL, 74
  HPLC, 341
  human
    biosynthesis, 224
    compared to rat apoA-IV, statistics for, 761
  immunoblots, 234–235
  interaction with lipid vesicles, 558–560, 561
  iodination, 380
  isoelectric point, 341
  isoforms
    characterization, 232–235
    isolation, 231–232
    nomenclature, 226–227
    separation, 353
  isolation, 10–11, 203–205, 339
  isoprotein 2, isolation, 695
  isoproteins, 709
    plasma forms, quantitation, 696
    secreted vs. plasma forms, 698–701
  LCAT activation, 18, 627, 645–646
  and LCAT activity, 38
  lipid-binding segments, 374
  lipoprotein distribution, 9
  mammalian, 127
  modification, 709, 711
  molecular weight, 9, 341
  monoclonal antibodies, purification, 534
  mouse
    chromosomal localization, 894
    plasma level, 892
  mRNA
    measurement, 685–686
    specific cDNA probes for, 674–675
  mutant, 852
  nitration, 380

oligonucleotide probe sequences, 713–714
Ouchterlony immunodiffusion, 235
physical characteristics, 128–129
plasma, 709
plasma concentration, 9
polymorphic
  affinity chromatography, 231
  gel permeation chromatography, 228–229
  HPLC, 229
  isolation, 228–231
  preparative SDS–PAGE, 229–230
  size exclusion chromatography, 228–229
polymorphic forms, 735
prepro-form, 224
  amino acid sequence, 225
  proteolytic processing, 226
  vs. rat preapoA-IV, comparison matrix, 762
pro-form, 141–142, 224–226
  function, 62–63
  isolation, 228
  structure, 62
repeat sequences, 754
reversed-phase HPLC, 350
role in lipid transport, 13
secondary structure, effect of dissociation of oligomers, 383
secreted vs. plasma forms, 701–703
secretion, 25
self- and mixed associations, 377–378
separation method, 341
sequence analysis, 755–757
sequence repeat, 777, 787
  phylogenetic interpretation, 784–787
signal peptide, 61, 697
solubilization, 342
structural mutation, 885–886
structural transitions with association/aggregation, 378
surface chemistry, 397
synthesis, 223
  regulation, 57
  at sites other than enterocytes and hepatocytes, 26–27
tissue source, 9
two-dimensional gel electrophoresis, 226, 234, 244

urea polyacrylamide gel electrophoresis, 232–234
variant forms, 727
Apolipoprotein A-I$_{Marburg}$, 852
Apolipoprotein A-I$_{Milano}$, 622, 727
Apolipoprotein A-I–apoC-III complex
  characterization, 712–726
  genes, chromosomal localization, 864
  genetic polymorphism, 727–745
  location, 712
  in patients with combined apoA-I–apoC-III deficiency, 721–725
  RFLPs associated with, analysis, 740–744
  structure, 712–713
Apolipoprotein A-I–apoC-III deficiency, 428, 712, 720, 725–726
  apoA-I–apoC-III gene complex in, 721–725
Apolipoprotein A-I-containing lipoprotein, 300
  isolation, 227–228
Apolipoprotein A-II, 5, 151, 154, 201, 340, 690, 704, 838, 840, 883
  absorption spectroscopy, extinction coefficients, 516
  adsorption to egg phosphatidylcholine monolayer, 398
  amino acid sequence, 12, 141
  anion-exchange HPLC, 348–349
  association enthalpy with synthetic lecithin, 368
  association with DMPC, enthalpy of, 370, 372
  association with LMPC, enthalpy of, 370, 372
  catabolism, 40
  cDNA
    cloning, 745–746
    molecular cloning, 749–750
    sequence analysis, 750–752
  characterization, 240–241
  chromosomal localization, 893
  in chylomicron synthesis, 24
  complexes with lipids, sodium cholate-mediated formation, 566
  conformational change on lipid binding, 368–370
  cyanogen bromide fragmentation, 542
  distribution, 341

enzymatic cleavage, 543
extracellular modification, 708
fluorescence properties, 517, 518
function, 223
gene, chromosomal location, 50
glucosylation, 361, 362
HPLC, 341
human
    amino acid composition, 236
    synthesis, 235
interaction with lipid vesicles, 558, 560, 562
isoelectric point, 341
isoforms, 238–239
    characterization, 240–241
    isolation, 240
isolation, 11, 203–205, 339
lipoprotein distribution, 9
mammalian, 127
molecular weight, 9, 341
mouse
    chromosomal localization, 894
    isolation, 892
mRNA, 745
nomenclature, 238
physical characteristics, 129–130
plasma concentration, 9
polymorphic
    affinity chromatography, 239
    gel permeation chromatography, 238
    homogeneity, 238
    HPLC, 238
    ion-exchange chromatography, 239
    isolation, 238
    preparative SDS–PAGE, 238–239
    size exclusion chromatography, 238
prepro-form
    amino acid sequence, 235, 237
    proteolytic processing, 236, 237
pro-form
    conversion to mature form, 236, 237
    function, 62–63
    structure, 61–62
recombinants formed with, cross-linking experiments, 624–625
reversed-phase HPLC, 350
role in lipid transport, 13
secondary structure, effect of dissociation of oligomers, 383
self- and mixed associations, 377–378

separation methods, 341
sialylated forms, 238
signal peptide sequence, 697
solubilization, 342
and structural change in HDL, 886
structural transitions with association/aggregation, 378
structure, 236
synthesis, 25, 223
tissue source, 9
two-dimensional gel electrophoresis, 244
variant forms, 727
Apolipoprotein A-II(D_2), physical characteristics, 130
Apolipoprotein A-II-containing lipoprotein, 300
isolation, 238
Apolipoprotein A-IV, 223, 690, 769
activation of LCAT, 18
amino acid sequence, 12, 141, 754
analytical SDS–gel electrophoresis, 244
characterization, 243–244
in chylomicron synthesis, 24
consensus domain, 632
    peptide analog, 630–631
function, 223–224
functional domains, methods for locating, 766–772
gene
    chromosomal location, 49–50, 51
    functional organization, 47
human
    amino acid composition, 236
    isoforms, 241
    structure, 240
isoforms
    characterization, 243–244
    isolation, 243
isolation, 892
lipoprotein distribution, 9
mammalian, 127
metabolic function, 753
molecular weight, 9, 754
periodic structure, studying, 767
physical characteristics, 131
plasma, 753
plasma concentration, 9
polymorphic
    affinity antibody chromatography, 242

gel permeation chromatography, 242–243
heparin affinity chromatography, 242–243
HPLC, 242
isolation, 242–243
preparative SDS-PAGE, 242
size exclusion chromatography, 242
storage, 242
pre-form, 240
intrasequence comparison, 762
rat, 753–774
relationships with apoA-I and apoE, 766
results of database search for sequence analysis, 755–757
segment comparison with various proteins, 763–764
structural similarity to human apoA-I, 772
repeat units, 754
role in lipid transport, 13
secondary structure, methods for predicting, 765–766
signal peptide, 61, 697
synthesis, 223
regulation, 57
tissue source, 9
two-dimensional gel electrophoresis, 244
urea PAGE, 244
variant forms, 727
Apolipoprotein A-IV$_{Marburg}$, 241
Apolipoprotein A-IV-containing lipoprotein, isolation, 242
Apolipoprotein A-V, physical characteristics, 131
Apolipoprotein B, 5, 31, 74, 106, 126, 262, 584, 608, 690, 705
aggregates, role of sulfhydryl blocking, 590
aggregation, 377
amino acid analysis, 258
analysis, 255–257
assembly, 23
binding to LDL receptors, 34–35
biochemical properties, 263
biosynthesis, 65
and CAD, 4
CD spectra, 585

characterization, 257–260
chemical modification of amino acid residues, 544
circular dichroic measurements, 383
composition, 250
cyanogen bromide fragmentation, 542–543
delipidation, 215, 219, 221, 222, 252–255, 586, 587
conformation after, 258
considerations in, 249–250
organic solvent, 253–255
distribution, 341
enzymatic cleavage, 543
fluorescamine assay, 256
fluorescence properties, 517
in formation of triglyceride-rich lipoproteins, 21–22
gene, chromosomal location, 51
glucosylation, 354, 361, 362
glycosylation, 63–64
guanidine HCl solubilization, 589
HPLC, 341
human, cDNA clones, 894
hydrophobicity, 258
immunological cross-reactivity, 141
immunologic studies, 258
in intestinal lipoprotein assembly, 68
isoelectric point, 341
isolation, 11, 168, 353
large. See apoB-100
of large VLDL, conformation, 34
in LDL and reassembled complexes, secondary structure, 604–605
in LDL metabolism, 33
and lipid autoxidation, 218
lipid peroxidation, inhibition, 249
Lowry assay, 256
mammalian, 127
modification, 711
molecular weight, 341, 342, 377, 590
determination, effect of solution properties, 260–262
monoclonal antibodies, purification, 534
mouse, isolation, 892
nomenclature, 76
phosphorylation, 66
properties, 218
proteolytic cleavage, 247
purity, 255

quantitation, 256
receptor binding and uptake studies, 551–552
residual lipid, measurement, 257
sedimentation coefficient in guanidine–HCl, 258–259
sedimentation coefficient in surfactant solutions, 260
self-association, 263
separation methods, 341
size, 258
solubilization, 342, 585–590
  denaturants, 589–590
  detergents, 585–589
  ethanol–diethyl ether method, 589
  organic solvents, 589
solubilized with NaDC, SDS–PAGE, 587, 588
solution properties, 216, 221
speciation, metabolic significance, 263
structure, 378
synthesis, at sites other than enterocytes and hepatocytes, 26
usage, and method of isolation, 250
variants, 65–66
Apolipoprotein B_h, 65–66
Apolipoprotein B_l, 65–66
Apolipoprotein B,E receptor, 888. *See also* LDL receptor
Apolipoprotein B,E receptor-binding domain, 770–772
Apolipoprotein B-26, 151, 247, 248
Apolipoprotein B-48, 141, 247–248, 259, 261
  amino acid analysis, 267–269
  amino acid composition, 269
  in chylomicron synthesis, 23–24
  Ferguson plot, 271, 272
  lipoprotein distribution, 9
  metabolism of particles containing, 35
  molecular weight, 9, 269–272
  in nascent VLDL, 24
  particle size, 76
  physical characteristics, 131, 132, 133
  plasma concentration, 9
  purification, 263–267
    gel permeation chromatography, 264
    SDS–PAGE, 264–265
    ultracentrifugation, 264
  residence time in humans, 31
  role in lipid transport, 13
  SDS–PAGE, 269–270
    effect of protein load, 270–272
  tissue source, 9, 262
Apolipoprotein B-74, 151
Apolipoprotein B-76, 247, 248
Apolipoprotein B-100, 38, 40, 123, 141, 151, 261–263
  amino acid composition, 269
  Ferguson plot, 271, 272
  in LDL, 33
  lipoprotein distribution, 9
  metabolism of particles containing, 35
  molecular weight, 9, 272, 624
  in nascent VLDL, 24
  particle size, 76
  physical characteristics, 131–133
  in plasma, 247
  plasma concentration, 9
  role in lipid transport, 13
  tissue source, 9
Apolipoprotein B-DMPC complex, 592–594
  density gradient centrifugation, 593–594
  electron microscopy, 593–594
Apolipoprotein B-egg PC complex, 590–592
  gel filtration chromatography, 591, 592
Apolipoprotein C, 5, 23, 40, 340
  acquisition by nascent chylomicrons, 27
  binding to LDL receptors, 34
  chromatofocusing, 288
  components, 5
  DEAE-Sephacel chromatography, 290–292
  delipidation, 215
  elution profile, 352
  first isolated, 288
  glucosylation, 361
  HPLC, 288
  isolation, 11, 339
  loss during chylomicron remnant formation, 28–29
  mouse, isolation, 892
  preparative isoelectric focusing, 288
  purification, 288–294
  residence time in humans, 31
  solubilization, 342
  synthesis, 24
  in VLDL metabolism, 32, 35

Apolipoprotein C-I, 51, 690
  absorption spectroscopy, extinction
    coefficients, 516
  activation of LCAT, 18
  anion-exchange HPLC, 348–349
  association enthalpy with synthetic
    lecithin, 368
  complexes with lipids, sodium cholate-
    mediated formation, 566
  conformational change on lipid binding,
    369
  cyanogen bromide fragmentation, 542
  distribution, 341
    in HDL classes, 441
  enzymatic cleavage, 543
  fluorescence properties, 517
  gene, chromosomal location, 50
  glucosylation, 362
  HPLC, 341
  human
    amino acid composition, 295
    amino acid sequence, 294
  interaction with lipid vesicles, 562
  isoelectric point, 341, 432
  isolation, 288
  lipoprotein distribution, 9
  mammalian, 127
  molecular weight, 9, 294–296, 341
  physical characteristics, 133–134
  plasma concentration, 9
  properties, 294–296
  purity, 291
  role in lipid transport, 13
  secondary structure, effect of dissocia-
    tion of oligomers, 383
  self- and mixed associations, 377–378
  separation methods, 341
  signal peptide sequence, 697
  structural transitions with association/
    aggregation, 378
  tissue source, 9
Apolipoprotein C-II, 31, 51, 142, 201, 690,
  797, 837, 840
  absorption spectroscopy, extinction
    coefficients, 516
  activation of LPL, 17
  aggregation, 377
  amino acid sequence, 12, 798–800
  cDNA clones
    characterization, 792–793
    identification, 792
    isolation, 789–793
    uses, 793
  chemical modification of amino acid
    residues, 544
  chromosomal localization using somatic
    cell hybrids, 796–799
  complexes with lipids, sodium cholate-
    mediated formation, 566
  cyanogen bromide fragmentation, 542
  deficiency, 17
  distribution, 9, 341
    in HDL classes, 441
  elution profile, 291–292
  enzymatic cleavage, 543
  fluorescence properties, 517
  gene
    chromosomal location, 50
    structure, 800
  genomic clones, 47
    isolation, 793–796
    restriction analysis, 794–796
  HPLC, 293–294, 341
  human
    amino acid composition, 295
    amino acid sequence, 295
  isoelectric point, 341
  isolation, 288
  locus, 800
  mammalian, 127
  molecular weight, 9, 296, 341
  physical characteristics, 134
  plasma concentration, 9
  preparative isoelectric focusing, 292–294
  probe, hybridization pattern with DNA
    mapping panels, 797, 799
  purity, 292
  role in lipid transport, 13
  secondary structure, effect of dissocia-
    tion of oligomers, 383
  separation methods, 341
  signal peptide sequence, 697
  structural transitions with association/
    aggregation, 378
  tissue source, 9
Apolipoprotein C-II-1, 292
Apolipoprotein C-II deficiency, 147
Apolipoprotein C-III, 5, 690, 840
  absorption spectroscopy, extinction
    coefficients, 516

amino acid sequence, 12, 141
association enthalpy with synthetic lecithin, 368
cDNA clones
 characterization, 713–717
 isolation, 713–717
chromosomal localization, 859, 894
complexes with lipids, sodium cholate-mediated formation, 566
conformational change on lipid binding, 368–370
distribution, 9, 341
 in HDL classes, 441
enzymatic cleavage, 543
fluorescence properties, 517
gene, 52
 chromosomal location, 49–50, 51
 linkage, to apoA-I gene, in humans, 720–723
 mapping, 720–723
genomic clones, 47
glycosylation, 63–64
HPLC, 341
human
 amino acid composition, 295
 amino acid sequence, 296
interaction with lipid vesicles, 562
isoelectric point, 341
isolation, 288, 342
isoprotein composition, 709–711
mammalian, 127
modification, 711
molecular weight, 9, 296, 341
mouse, isolation, 892
oligonucleotide probe sequences, 713–714
physical characteristics, 134–135
plasma concentration, 9
role in lipid transport, 13
self-association, 377–378
separation methods, 341
signal peptide sequence, 697
space-filling model, 631
structure, 378
synthesis, regulation, 57
tissue source, 9
Apolipoprotein C-III$_1$, anion-exchange HPLC, 348–349
Apolipoprotein C-III$_2$, anion-exchange HPLC, 348–349

Apolipoprotein C-III-0, 292, 837
Apolipoprotein C-III-1, 292, 837
Apolipoprotein C-III-2, 292, 837
Apolipoprotein D, 75, 141, 297–304, 690
 amino acid composition, 301
 carbohydrate content, 301–302
 characterization, 300–301
 distribution, 9, 303
  in HDL classes, 441
 functional aspects, 303–304
 isoforms, 300–301
 isolation, 203–205, 298–299
 and LCAT activity, 38
 modification, 711
 molecular weight, 9, 301
 plasma concentration, 9, 303
 tissue source, 9
Apolipoprotein D-containing lipoprotein, isolation, 299–300
Apolipoprotein E, 23, 24, 31, 38–40, 126, 142, 151, 656–657, 690, 769, 776, 863, 864
 absorption spectroscopy, extinction coefficients, 516
 acquisition by nascent chylomicrons, 27
 alleles
  frequency, 845
  structure, 46
 amino acid analysis, 283
 amino acid sequence, 12, 141
 analytical isoelectric focusing, 280–281
 binding to LDL receptors, 34
 biosynthesis, 142
 cDNA cloning, principle, 801
 characterization, 279–287
 chemical modification of amino acid residues, 544
 in chylomicron processing, 28
 clones, 8
 CNBr peptide, isolation, 286–287
 conformational change on lipid binding, 369
 cyanogen bromide fragmentation, 542–543
 cysteamine treatment, 280–282
 decomposition, 279
 delipidation, 222
 desialation, 824
 distribution, 341
  in HDL classes, 441

enzymatic cleavage, 543
extracellular modification, 695, 705–709
first identified, 823
functional domains, 284–285
functional tests, 287
gene, 800, 851–852
   chromosomal location, 50, 51
   cloned, characterization, 815–816
   dot blot analysis, 820–821
   expression, analysis, 820–821
   human, structure, 815–816
   Northern blot analysis, 821
   screening genomic library for, 813–814
genetic polymorphism, 823–851
genomic clones, 47
glucosylation, 361, 362
glycosylation, 63–64
in HDL, 74
α-helical content, 285
HPLC, 341
human
   cDNA, cloning, 810–811
   gene, 811
   mRNA, structure, 810–811
   sequence repeats, 779–783
   test for repeated sequences, 782–783
immunological cross-reactivity, 141
inheritance, genetic model, 844–845
interaction with lipid vesicles, 562
intracellular modification, 704–707
   effect of tunicamycin, 705–707
   pulse-chase experiments, 694
   time course, 705–706
isoelectric focusing, 276–277
isoelectric point, 341
isoforms
   purification, 276–278
   separation, 353
isolation, 11, 216, 273–279
isoproteins, 710
   plasma forms, quantitation, 696
   secreted vs. plasma forms, 698–701
LDL-receptor binding domain, 769
mammalian, 127
modification, 709, 711
   and secretion, time course, 704–705
molecular seive chromatography, 275–276

molecular weight, 341
mouse
   chromosomal localization, 894
   isolation, 892
mRNA
   calculation, 688
   measurement, 685
   probe, synthesis, 675–676
   specific cDNA probes for, 674
mRNA translation products, 804–805
in nascent VLDL, 24
neuraminidase treatment, 281–282, 696
nomenclature, 273
patterns on one-dimensional isoelectric focusing gels, misinterpretation, 841–843
α phenotype, 834
β phenotype, 834
phenotype E2/2
   affinity for LDL receptor, 851
   association with type III hyperlipoprotenemia, 848
   and binding to lipoprotein receptors, 849
   as molecular marker for HLP III, 851
phenotypes, 835–837, 852
   determination, 833–844
      artifacts, 841–844
      by 2D-PAGE using internal isoelectric point markers, 837–839
      by 2D-PAGE using mixing experiments, 833–837
      by one-dimensional isoelectric focusing, 837–841
      comparison of one- and two-dimensional methods, 841
   frequency, 845
   genetic heterogeneity, 835
   result of structural gene mutations, 846–847
phenotypic categories, 280
phenotyping, 273, 279–280
physical characteristics, 135–136
polymorphic forms, 735
polymorphism, 273, 727
   genetic basis, 844–850
   one-dimensional isoelectric focusing, 825–833

two-dimensional PAGE, 825–833
preparative SDS–PAGE, 276
purification
    heparin-Sepharose chromatography, 278
    Immobiline gels, 277–278
rat
    cDNA, cloning, 802–809
    cDNA clones, characterization, 809
rat and human, sequence homology, 801
receptor binding, 287
receptor binding and uptake studies, 551–552
residence time in humans, 31
role in lipid transport, 13
role in lipoprotein and cholesterol metabolism, 825
SDS–PAGE, 282–283
secreted by transfected cells, characterization, 821–823
secreted vs. plasma forms, 701–703
self-association, 377–378
separation methods, 341
sequence analysis, 283–287, 755–757
sialated, isolation, 695
signal peptide, 61, 697
solubilization, 342
source, 273–274
storage, 279
structural mutations, 824
structural polymorphisms, 49
structure, 378
synthesis, 823–824
    in extrahepatic tissues, 58–59
    regulation, 57
    at sites other than enterocytes and hepatocytes, 26–27
thiopropyl chromatography, 278–279
variants, 285, 286
in VLDL metabolism, 33, 35
Apolipoprotein E-containing lipoprotein
    delipidation, 274
    isolation, 274
Apolipoprotein E-II
    lipoprotein distribution, 9
    molecular weight, 9
    plasma concentration, 9
    tissue source, 9
Apolipoprotein E-III
    lipoprotein distribution, 9
    molecular weight, 9
    plasma concentration, 9
    tissue source, 9
Apolipoprotein E-IV
    lipoprotein distribution, 9
    molecular weight, 9
    plasma concentration, 9
    tissue source, 9
Apolipoprotein F, 304–306
    amino acid analysis, 305
    amino acid composition, 301
    characterization, 305
    distribution, 306
    functional aspects, 306
    isoelectric point, 305, 432
    isolation, 304–305
    molecular weight, 305
Apolipoprotein G
    amino acid composition, 301
    isolation, 309
    molecular weight, 309
Apolipoprotein H, 306–309
    amino acid composition, 301, 308
    carbohydrate content, 302
    characterization, 308
    distribution, 9, 308–309
    function, 309
    isoforms, 308
    isolation, 306–308
    mammalian, 127
    molecular weight, 9, 308, 377
    physical characteristics, 136
    plasma concentration, 9
    structure, 377, 378
    tissue source, 9
Apolipoprotein–lipid complex
    analytical ultracentrifugation, 572
    characterization, 569–578
    chemical composition, 570–571
    discoidal
        apolipoprotein structure in, 574–568
        circular dichroism spectra, 575
        fluorescence spectra, 576–578
        lipid phase behavior, 574–575
        reconstituted, comparison with native HDL, 578–582
        sodium cholate-mediated formation, 564–567
    electron microscopy, 571–572

fluorescent conjugates, preparation with dansyl chloride, 573–574
gel filtration, 569, 571
gradient gel electrophoresis, 572
isolation, 568–569
isopycnic gradient density centrifugation, 569–570
neutron diffraction study, 574
nuclear magnetic resonance, 574
preparation, 558–569
  by detergent-mediated synthesis, 560–565
  by spontaneous interaction of apolipoproteins and lipid vesicles, 558–563
quasielastic light scattering, 574
shape determination, 571–574
size determination, 571–574
small angle X-ray scattering, 574
static fluorescence polarization, 573–574
Apolipoproteinopathies, 12
Apolipoprotein–phospholipid association. *See* Phospholipidapolipoprotein association
Apolipoproteins, 338
  absorption spectroscopy, extinction coefficients, 516
  acetylation, 389
  acquisition by triglyceride-rich lipoproteins, 23
  adsorption kinetics, 401–402
  adsorption to insoluble lipid monolayer, 399, 400
  aggregates, 376, 377, 385
  at air–water interface, 387–403
  amino acid sequences, 127
  analysis, for mouse strains, 893
  assembly, 69
  assignment of epitopes, 535–536
    dot immunoblotting, 535
    electrotransfer after SDS–PAGE or urea–PAGE, 536–537
    immunological detection on nitrocellulose, 538–539
    isoelectric focusing, 536
    protein transfer after IEF–PAGE, 537–538
    protein transfer blots, 535–536
  associating, 377–378
  binding to lipoprotein, 16
  biosynthesis, 61–63
  cDNA clone analysis, 127
  characteristics, in normal fasting humans, 9
  chemical modification of amino acid residues, 543–544
  cholesterol-induced, 142–143
  chromosomal localization, 10
  circular dichroism, 383, 765–766
  codispersion and interaction with lipids
    cosonication method, 554
    detergent-mediated methods, 554
    methods for, 554
    spontaneous reaction of vesicles with apolipoproteins, 554
  conformations, and function, 528
  covalently modified, repurification, 380
  covalent modification, 380
  cyanogen bromide fragmentation, 542–543
  definition, 8, 375–376
  displacement from air–water and lipid monolayer–water interface, experiments, 402–403
  enzymatic cleavage, 543
  epitopes
    antibody competition assay, 539–541
    antibody cotitration assay, 540–542
    assignment to specific regions, 542–545
    conserved through evolution, 551–552
    enumeration, 538–539
    identification, 538–539
  extracellular modification, 695, 705–709
  family concept, 626
  fluorescence measurements, 383
  fluorescence properties, 517–518
    wavelength maxima, 517
  free energy of transfer from water to phospholipid surface, 412
  free energy values for phospholipid association, 410
  function, 403
    linking epitopes to, 551–552
  gene mapping
    by assaying gene products, 858
    regional and fine structure, 861–863

using cloned gene sequences and Southern blot analysis, 858–859
genes
  chromosomal location, 49–51, 864
  genetic mapping, 851–863
  polymorphisms, 48–49
glucosylated. See also Protein, glucosylated
  amino acid composition analysis, 357
  antibody preparation, 357–358
  immunochemical identification of, 356
  trinitrobenzenesulfonic acid assay, 357
glucosylation
  by column chromatography, 362
  effect on function, 355–356
  by immunoadsorption, 362–363
  tandem assay, 363–364
  by Western blot analyses, 360–362
homology between human and other mammalian, 75–76
HPLC, 339–353
immunoblotting, 419
as immunogen, 529–530
interfacial exchange, 402–403
for interfacial studies, 388–389
intracellular modification, 704–707
  pulse-chase experiments, 694–695
ionization, enthalpy of, 372–374
isoelectric points, 341, 342
light scattering measurements, 384
mammalian
  classification, 71–75
  nomenclature, 71–75
  physicochemical characteristics, 126–142
mean residue helical amphipathic moments, 406
molecular properties, evaluation, 380–381
molecular species in solution, 376–378
mouse, isolation, 891–893
mRNA
  calculation of values from hybridization data, 684–686
  hybridization assay, 680–687
    sensitivity, 686
  measurement, 672–689

normalization of values, 687–688
probe preparation
  by replacement synthesis, 677–679
  strand separation, 679–680
  single-stranded probes, hydroxylapatite fractionation, 676–677
  specific cDNA probes for, 674–675
nomenclature, 75–76
nonassociating, 377
one- or two-dimensional PAGE, 693, 694
peptide analog
  18A, 628, 630, 631, 635, 637, 639, 641, 643
    complex with DMPC, 637
  18Aa, 628
  18A-Pro-18A, 628, 632, 635, 637, 639, 641, 643, 645
  advantages, 646
  amphipathic helical structures, 630
  Ap, 628–630, 636, 645
  18As, 628
  binding studies, 642–643
  C_0-LAP, 639
  C_4-LAP, 628, 639
  C_6-LAP, 643
  C_8-LAP, 628, 639, 645
  C_{12}-LAP, 628, 639
  C_{16}-LAP, 628, 639
  C_{18}-LAP, 628, 645
  Co-LAP, 628
  competition experiments, 642–644
  design, 629–632
  desVal^{10}18A, 628, 630–632, 635, 637, 639, 641, 643, 645
  disadvantages, 647
  displacement studies, 643–644
  LAP, 642
  LAP-16, 628, 629, 630, 638
  LAP-20, 628, 629, 630, 638
  LAP-24, 628, 629, 630
  LCAT activation, 644–646
  lipid interactions, 636
    circular dichroism study, 639
    differential scanning calorimetry, 640–642
    electron microscopy, 636–637
    intrinsic tryptophan fluorescence, 638–639

SUBJECT INDEX

nondenaturing gradient gel electrophoresis, 637–638
proton NMR, 640–641
reverse 18A, 628, 630, 631, 635, 637, 639, 641, 643, 645
synthesis, 632–635
polar amino acid residues, role in lipid–protein interaction, 16
polymorphisms, characterization, 833
polypeptide backbone refolding, 386
preparation, in lipid-free form, 146
primary translation products, post-translational processing, 141–142
purified, surface pressure–molecular area isotherms, 394
purity, assessment, 554
radioactive labeling, 389
radioiodinated, autoradiography, 419
radiolabeled
    immunoprecipation by staphylococcal protein A preparation, 693–694
    quantification, 696
    synthesized by HepG2 cells
        isolation, 695–696
        sequence, 695–696
    synthesized by cell or tissue culture and by, cell-free translation of mRNA, immunoprecipitation, 692–693
receptor binding and uptake studies, 551–552
for reconstitution of HDL, 554–555
reductive methylation, 389
refolding of polypeptide backbone, 381, 383
region specific immunoassays, 553
regions that have high affinity for lipid/water interface, 14
role in lipid transport, 13
secondary structure, methods for predicting, 765–766
secreted vs. plasma forms, 701–704
secretion, 19–27
sedimentation equilibrium measurements, 382, 384
    data analysis, 386
sedimentation velocity measurements, 382
self- and mixed association, 383
self-association, 385

study of, 619
separation
    general considerations, 340–343
    procedures, 339–341
signal peptides, 61, 696–701
    detection, 691
solubilization and dissociation, 379–380
in solution, quantitation, 555
solution properties, 375–387
    effect of experimental design, 385–386
    methods of study, 381–384
    relevance to native lipoprotein structure and metabolism, 386–387
    and sample purity, 385
spreading at air–water interface, 392–394
storage, 381
study, laboratory techniques for, 144–147
surface activity, 402
surface chemistry, 412–413
    interpretation of data, 399–403
    reagents for, 389–390
surface conformation, 399–401
surface potential at air–water interface, 397–399
surface-seeking properties, 16–17
synthesis, 19–27
    by cell cultures, 691–692
    by cell-free translation of mRNA, 691–692
    by organ cultures, 691–692
    regulation, 56–58
    at sites other than enterocytes and hepatocytes, 26–27
synthetic, helical moment analysis, 412
synthetic peptides, 544–545, 627–647
tertiary structure, 647
types, 5
urea-soluble, Sephadex G-75 gel filtration, 290
Apolipoprotein affinity chromatography, for purification of monoclonal antibody, 534
Apolipoprotein SAA
    amino acid content, 313
    amino acid sequence homology, 311
    characteristics, 312–314
    functions, 319–320

as indicator of tissue damage, 320
isotypes, 311, 312–313
molecular mass determinations, 313
preparation, 312
protein content, 314
Apolipoprotein SAA$_1$, 311
amino acid sequence, 313
Apolipoprotein SAA$_2$, 311
Apolipoprotein VLD-LII, avian, gene, 53, 55
Apoprotein, 5–8
Artherurus macrourus. *See* Porcupine
Artiodactyls, circulating lipoprotein levels, 80, 87
Ascites, murine, LDL, 119
Asinus hemionus. *See* Onager
Association
definition, 376
mixed, definition, 376
Ateles sp. *See* Spider monkey
Atherosclerosis
mutant apoA-I allele associated with, 740–741
premature, 712. *See also* Familial hypercholesterolemia apoA-I–apoC-III gene complex in, 721–725
regression, injection of phospholipids to produce, 656
role of lipoproteins, mouse genetic studies of, 888
Autoradiography, of apolipoprotein gene preparations, 872–874

B

Baboon, 89
apoA-I, physical characteristics, 129
apoA-II
physical characteristics, 130
structure, 236
apoB-100, physical characteristics, 133
apolipoprotein, 75, 127
physical characteristics, 141
circulating lipoprotein levels, 81
HDL
carbohydrate composition, 106
phospholipid content, 104
physical properties, 116, 120

HDL and subclasses, chemical composition, 99
LDL
carbohydrate composition, 107
chemical composition, 96
physical properties, 113
lipid metabolism, 142
lipoprotein distribution, 78
lipoprotein lipids, fatty acid profile, 105
lipoproteins
chemical composition, 91
immunologic properties, 123
Lp(a), 74
plasma apolipoprotein analogous to human apoD, 303
plasma VLDL, chemical composition, 92
serum lipoprotein profiles, 176–180
Bacteriophage λ, 45
as cloning vehicle, 812
human genome library in
properties, 813
screening, 813
plaque replica filter preparation, 813–814
Badger. *See also* European badger
apolipoproteins, 127
HDL and subclasses, chemical composition, 98
LDL, chemical composition, 95, 102
lipoprotein distribution, 78
lipoproteins, chemical composition, 91, 101
plasma VLDL, chemical composition, 92
Benzamidine, 153
Binding enthalpy, 365
Bis(sulfosuccinimidyl)suberate, 615, 616, 624
reaction with, 617
Black rhinoceros, circulating lipoprotein levels, 86
Blood
collection
for lipoprotein isolation, 151–155
preservative cocktail for, 152–153
procedure, 153–155
enzymatic degradations in, after collection, 151–152
Blot hybridization, 671

SUBJECT INDEX

Bottlenose dolphin, circulating lipoprotein levels, 80
Bovine. *See* Cattle
Bovine serum albumin. *See also* Serum albumin
  Ferguson plot, 271
  structure, 336, 337
Brain, synthesis of apolipoproteins, 26
Brij 36T, 216

C

CAD, 13. *See* Coronary artery disease
Calcitonin, gene, 56
Calf. *See* Cattle
California sea lion, circulating lipoprotein levels, 80
Callithrix jacchus. See Common marmoset
Calorimeter, types, 365
Camel
  circulating lipoprotein levels, 87
  HDL and subclasses, chemical composition, 99
  LDL, chemical composition, 95, 102
  lipoprotein profile, 84, 85
  plasma VLDL, chemical composition, 92
Camelus bactrianus. See Camel
Canis dingo. See Dingo
Canis familiaris. See Dog
Carboxyfluorescein
  characteristics, 658–660
  fluorescence self-quenching in small unilammellar vesicles, 658–659
  purification, 659–661
  release from liposomes
    monitoring, 662–663
    phase transition release, 663–664
    phase transition release profiles, 665–666
Carboxyfluorescein leakage assay, of lipoprotein–liposome interaction, 657–668
Carnivores
  circulating lipoprotein levels, 80
  lipoprotein profiles, 84
β-Carotene, as spectroscopic probe of order in lipoprotein lipids, 524–527
Carotenoids, in chylomicrons, 20
Casein, mouse, 755

Cat, circulating lipoprotein levels, 80
Cattle, 89
  apoA-I, physical characteristics, 128
  apoA-II(D_2), physical characteristics, 130
  apoC-III, physical characteristics, 135
  chylomicrons, 90
  circulating lipoprotein levels, 81
  HDL
    native structure, studies, 121
    physical properties, 115, 119
  HDL and subclasses, chemical composition, 99
  IDL
    chemical composition, 94
    physical properties, 109
  LDL
    chemical composition, 96, 102
    physical properties, 112, 119
  lipoprotein allotypes, 124
  lipoprotein distribution, 78
  lipoprotein lipids, fatty acid profile, 105
  lipoproteins
    antigenic sites, 122
    chemical composition, 91, 101
    physical properties, 109, 122
  plasma lipoprotein, 71
  plasma VLDL, chemical composition, 92
  VLDL, physical properties, 109, 110
Cavia porcellus. See Guinea pig
Complementary DNA
  clones, screening, 747–748
  cloning, 901–903
  contraction, 732–733
  full-length, cloning strategy, 807–808
  isolation, 732–733
  single-stranded probes
    preparation, 672–680
      with bacteriophage M13, 672–674
    selection, 672–673
Complementary DNA libraries, 903
  adult human liver, screening to isolate apoC-II cDNA clones, 791–792
  construction, 43–45, 732–733, 805–808
  human liver
    construction, 810
    screening, 810–811
  screening, 808–809

$C_{12}E_8$, solubilization of apoB, 587–589
Cebus albifrons. See Cebus monkey
Cebus apella. See Cebus monkey
Cebus monkey
  circulating lipoprotein levels, 81
  HDL, physical properties, 115, 120
  HDL and subclasses, chemical composition, 99
  LDL, chemical composition, 96
  lipoprotein lipids, fatty acid profile, 105
  plasma VLDL, chemical composition, 92
Cell lines, UT-1, compactin resistance, 59
Cellulose acetate gel electrophoresis, 433
Ceratotherium simum. See White rhinoceros
Cercocebus albigena. See Mangaby
Cercopithecus aethiops. See African green monkey
Cetaceans, circulating lipoprotein levels, 80
Cetyltrimethylammonium bromide, delipidation with, 215
Chimpanzee, 89
  apoA-I, physical characteristics, 129
  apoA-II, physical characteristics, 130
  apoB-100, physical characteristics, 133
  apolipoproteins, physical characteristics, 141
  circulating lipoprotein levels, 81
  HDL
    carbohydrate composition, 106
    phospholipid content, 104
    physical properties, 120
  HDL and subclasses, chemical composition, 99
  LDL
    carbohydrate composition, 107
    chemical composition, 96
    physical properties, 113
  lipid metabolism, 142
  lipoprotein lipids, fatty acid profile, 105
  lipoprotein profile, 84, 88
  lipoproteins
    chemical composition, 91, 101
    immunologic properties, 123
    native structure, studies, 121
  Lp(a), 74
  plasma VLDL, chemical composition, 92

VLDL, physical properties, 110
Chloroform-methanol, solvent system, 215, 217
Cholesterol
  ^{13}C NMR, 496
  flux between cell and medium, factor associated with, 8
  in LDL receptor regulation, 36
  metabolism, 895
  profile, obtained with vertical autoprofile procedure, 190–192
  in regulation of HMG CoA reductase, 60
  uptake, by liver, 30
Cholesterol feeding, 73, 825
  and apolipoprotein synthesis, 57–58
Cholesterol transport, 13, 20, 101–102. *See also* Reverse cholesterol transport
  pathway, 27
Cholesteryl esters
  microemulsion particles containing, circular dichroism study, 520–522
  NMR, spin-lattice relaxation times, 506
  transfer, by specific factors in plasma, 18–19
Cholesteryl linoleate, ^{13}C NMR, 504–505
Chromatofocusing, 339, 433, 441
Chromatography. *See also specific chromatographic procedure* of apolipoproteins, 383–384
Chromosome
  G-banding with Wright stain, 865–866, 874–875
  mitotic, preparation, 865–868
Chromosome breakage, radiation-induced, for segregating closed linked genes, 863
Chylomicron remnant, 28
  mammalian
    lipid composition, 89–90
    protein composition, 89–90
  receptor-mediated endocytosis, 28–29
Chylomicrons, 262, 263, 417, 753, 797
  apoB-48 purification from, 263
  assembly, 68
  catabolism, 29–31
  core lipids, chemical composition, 11
  cross-linking, 626
  density, 10
  electrophoretic definition, 10

formation, in intestines and liver, 19–24
freeze-fracture and freeze-etching
    electron microscopic studies, 458
¹H NMR, 487–488
human
    concentration range, 157
    density, 72, 157
    electrophoretic class, 157
    flotation rates, 157
    size range, 157
isolation, 242
lipids, hydrolysis, 17
mammalian
    lipid composition, 89–90
    protein composition, 89–90
molecular weight, 10
nascent, surface components, 23
particle size, 10
processing, 27–31
secretion, 68–69
    as function of fat absorption, 20
structure, 12
surface components, chemical composition, 11
Circular dichroism, 519
    of apolipoproteins, 383, 401
    contribution of lipids in UV wavelength region, 519–524
    to determine structural changes, 409
    intrinsic, of constituent lipids of lipoproteins, 519–520
    of lipoproteins, 108, 121
    of organized lipids in model systems, 520–524
Citellus mexicanus. See Ground squirrel
Coated pits, 29, 608
Colony hybridization, 748–750
Column chromatography, 145, 146
Common marmoset
    apoA-I, physical characteristics, 129
    apoB-100, physical characteristics, 133
    apoB-48, physical characteristics, 133
    circulating lipoprotein levels, 81
    HDL, 73
        physical properties, 115, 120
    HDL and subclasses, chemical composition, 99
    LDL
        carbohydrate composition, 107
        chemical composition, 96

physical properties, 113
lipoprotein distribution, 78
lipoprotein profile, 88
lipoproteins, immunologic properties, 123
plasma VLDL, chemical composition, 92
VLDL, physical properties, 110
Common zebra, circulating lipoprotein levels, 80
Compactin, in LDL receptor regulation, 36
Coronary artery disease, 3–4
Cosmid
    as cloning vehicle, 812
    DNA, examination, 818
    human genome library in
        properties, 816
        screening, 816–818
    pCV108, structure, 816–817
    pJB8, 816
Cosmid clones, human apoE gene-containing, characterization, 818–820
Cow. *See* Cattle
Cross-linked oligomers, molecular size, analysis, 618
Cross-linking
    analysis of results, 618–619
    applications, 620–626
    extent of modification, quantitation, 619
    interparticle, 619–620
    reaction performance, 616–617
Cross-linking reagent, criteria for selection, 614–616
Cryofixation, 457
    effect on sample solution, determination, 467–469
Crystallization, of apolipoprotein, 766
Cynomolgus monkey
    circulating lipoprotein levels, 81
    HDl, physical properties, 120
    HDL and subclasses, chemical composition, 99
    LDL
        chemical composition, 96
        native structure, studies, 121
        physical properties, 113
    lipid metabolism, 142
    lipoproteins, physicochemical properties, 122
    liver, apoE mRNA values, 688

D

D-2, 310
Decyl sulfate, 221
Deer, circulating lipoprotein levels, 81
Delipidation, 146, 213–222, 379
　advantages, 222
　of albumin, 322
　of apoB, 249–250, 252–255, 586, 587
　of apoE, 274
　of apoHDL, 343–345, 350
　of apoVLDL, 345–346, 349–352
　aqueous samples, 214–216, 220
　　apparatus, 220
　　detergents, 215–216
　　extraction procedure, 220
　　organic solvents, 214–215
　　reagents, 200
　dialysis, 219
　disadvantages, 222
　effect of apolipoprotein solution properties, 385
　effects of freezing, 217
　of HDL, 348–349
　isolation of lipoprotein fractions, 219
　of LDL, 589
　of lipoproteins, 549–550
　lyophilization, 219
　lyophilized samples, 216–218, 220–221
　　extraction procedure, 221
　　reagents, 220–221
　organic solvent, of apoB, 253–255
　procedure, 220–221
　resolubilization after, 221–222
　sample preparation, 218–219
　　solvent additives, 218–219
　of serum, effect of freezing, 214
　solvent systems, 213
　surfactant, 252–253
　of triglyceride-rich lipoproteins, 289–290
Density gradient ultracentrifugation
　in swinging bucket rotor, 170–181
　　advantages, 171–172, 180
　　analysis, 176
　　application, 176–177
　　centrifugation, 172–176
　　density gradient, 172–173
　　disadvantages, 180
　　effect of diffusion and centrifugation time on density gradient, 174–176
　　equipment, 173
　　five step NaCl/KBr density gradient, 171
　　four-step density gradient, 170
　　lipoprotein profiles, 174
　　NaCl step gradient, 171
　　preparation of gradient, 171–172
　　preparations of step gradients with different volumes of serum, 172
　　reproducibility, 176
　　sample collection, 172
　　sensitivity of UV monitor, 174
　　stock sodium bromide solution, 172
　in vertical rotors, 182
Deoxycholate, 221
　micellar dispersions of lipids, to produce discoidal HDL-like complexes, 564–565
Deoxyribose, determination, diphenylamine procedure, 688–689
Detergent–apoB complex, 584
DFDNB. See Difluorodinitrobenzene
DFP. See Diisopropylfluorophosphate
Diabetes, 354
Diceros bicornis. See Black rhinoceros
Didelphus virginiana. See Opossum
Diethyl ether, solvent for lipids, 214
Diethyl ether–methanol, solvent system, 255
Difference absorption spectra, of lipoproteins, 121
Differential scanning calorimetry, of lipoproteins, 108, 121
Difluorodinitrobenzene, 615–616, 619, 623, 624
　reaction, 617
　　with HDL, 620–621
Diimidoester reagents, 614–615, 619
Diisopropylfluorophosphate, in blood collected for lipoprotein isolation, 151
Dilauroyl phosphatidylcholine, interaction with apolipoproteins, 554, 559, 561
Dimethyl adipamate, 615
Dimethyl malonimidate, 615
Dimethyl suberimidate, 614–615, 624
　reaction, 616–617
　　with apoA-I and apoA-II, 619
Dimyristoyl phosphatidylcholine
　binding enthalpy to apoA-I and apoA-II, 367, 368

interaction with apolipoproteins, 561, 562
liposome preparation with, 661
mixed micelle with NaDC, preparation, 592–593
spontaneous reaction with apolipoproteins, 554
Dingo, circulating lipoprotein levels, 80
Dipalmitoyl phosphatidylcholine
interaction with apolipoproteins, 559–562, 566
liposome preparation with, 661
spontaneous reaction with apolipoproteins, 554
Diphenylhexatriene, fluorescence probe, 575, 578
Distearoyl phosphatidylcholine
interaction with apolipoproteins, 559
liposome preparation with, 661
Dithiobispropionimidate dihydrochloride, 623
DLPC. See Dilauroyl phosphatidylcholine
DMPC. See Dimyristoyl phosphatidylcholine
DMPC–apoA-I complex, cross-linking experiments, 624–625
DMPC–apoB complex, differential scanning calorimetry, 603
DMPC–cholesterol, interaction with apolipoproteins, 562
DMPC vesicles, interaction with apolipoproteins, 558–560
DMS. See Dimethyl suberimidate
DNA
digests of, probes for identification of sequences in, 734–735
filter-bound, hybridization, 814–815
genomic
isolation, 730–732
restriction, 730–732
restriction enzyme digestion, 859–860
RFLP analysis, 730–731
tissue content, measurement, 687–688
DNA : DNA reassociation
effect of fragment size, 683–684
effect of probe concentration, 682–683
DNA-excess solution hybridization, 671–672
DNA probe. ^{3}H labeling by nick-translation, 868–870

DNA sequencing, 45–46
n-Dodecyloctaethylene glycol monoether, 216. See also $C_{12}E_8$
Dog, 89
apoA-I, physical characteristics, 128
apoA-II, physical characteristics, 129
apoA-IV, physical characteristics, 131
apoB-48, physical characteristics, 132
apoB-100, physical characteristics, 132
apoE, physical characteristics, 135
apolipoproteins, physical characteristics, 141
chylomicrons, 90
circulating lipoprotein levels, 80
HDL, 73
phospholipid content, 104
physical properties, 115, 120
HDL and subclasses, chemical composition, 98
HDL subclasses, 120
IDL, chemical composition, 94
LDL
chemical composition, 95, 102
phospholipid content, 104
physical properties, 112
lipid metabolism, 142
lipoprotein lipids, fatty acid profile, 105
plasma lipoprotein, 71
plasma VLDL, chemical composition, 92
VLDL, physical properties, 109
Dolphin. See also Bottlenose dolphin; Pacific white-sided dolphin HDL, physical properties, 120
Dot blot analysis, 671
DOTMATRIX, 761
Dot matrix, 777, 787
DPPC. See Dipalmitoyl phosphatidylcholine
Drosophila, heat shock proteins, 56
DSPC. See Distearoyl phosphatidylcholine
du Nouy ring, 391, 393
Dynamic programing methods, 773–777
Dysbetalipoproteinemia, 428–429, 523. See also Hyperlipoproteinemia, type III
associated with genetic disorders, 7
conditions associated with secondary hyperlipoproteinemia, 7
laboratory definition, 7
Dyslipoproteinemia, 244–245
type III, 49

E

Edmundson wheel diagrams, 767
EDTA. *See* Ethylenediaminetetraacetic acid
Egg phosphatidylcholine
  complex with apoA-I, comparison to HDL, 580, 582
  interaction with apolipoproteins, 560, 562, 565–566
  liposomes, circular dichroism studies, 520–522
Electron microscopy, 442, 457
  of lipoproteins, 121
  low temperature, 471
  of negatively stained lipoproteins
    agreement with other techniques, 455–456
    application of sample, 447–448
    examination of sample, 448–450
    Formvar-carbon-coated grids, 445–446
    grids, 445
    specimen support films, 445
    thin carbon support films, 446–447
Electron spin resonance spectroscopy
  of lipoproteins, 108
  study of fatty acid binding to albumin, 327
Electrophoresis, 145
  according to particle size, 417–418
  effect of gel concentration, 270–272
cis-Eleostearic acid, 327
Elephant seal, circulating lipoprotein levels, 80
Endocytosis. *See* Receptor-mediated endocytosis
Endosomes, 29
Equilibrium dialysis, to determine constant of equilibrium, for phospholipid–apolipoprotein association, 408
Equus burchelli. See Common zebra
Equus caballus. See Horse
Equus przewalski. See Przewalski horse
Equus zebra. See Mountain zebra
Erinaceus Europaeus L. *See* Hedgehog
Erythrocebus patas. See Patas monkey
Ethanol–diethyl ether, solvent system, 215
Ethylenediaminetetraacetic acid, 153, 218
  in blood collected for lipoprotein isolation, 151
European badger
  lipid transport system, 143
  lipoprotein profile, 88
Exons, 45–47
Expression vectors, 44, 48

F

Fab fragments, purification, 535
Familial dysbetalipoproteinemia. *See* Hyperlipoproteinemia, type III
Familial hypercholesterolemia, 64, 69, 429, 800, 852, 863, 888, 895, 909
  diagnostic test for, 610
  mutations associated with, 47–48
Familial LCAT deficiency, 5
FASTN, 758
FASTP, 757–758
Fatty acid
  addition to aqueous solutions, 326–327
  binding to albumin, 320–321
    equilibrium dialysis measurement, 328
    equilibrium partition measurement, 327–328
    measurement, 327–339
  carried by albumin, 320
  fluorescent analogs, 327
  partition between hydrocarbon and aqueous solution, 330
  partition ratio, 330–331
  tissue uptake, 338
Fatty acid–albumin solution, preparation, 321–327
Felis domesticus. See Cat
Ferguson plots, 270–272
Ferret, circulating lipoprotein levels, 80
Fetal human tissue, organ cultures, 692
Fibrinopeptide A, 153
Fish eye disease, 455
Fluorescence polarization analysis, of lipoproteins, 121
Foam cells, 143
Formvar-carbon-coated grids, preparation, 445–446
Fox. *See* Red fox
Framingham Study, 3–4
Free fatty acid, 320–321
Freeze-drying, 457, 471
Freeze-etching, 457
Freeze-fracturing, 457

G

Gel electrophoresis, 339
Gel filtration, 169
  to determine constant of equilibrium, for phospholipid–apolipoprotein association, 408
  for lipoprotein isolation, 10
  purification of IgM monoclonal antibodies, 534
Gel filtration chromatography, study of fatty acid binding to albumin, 327
GenBank, 754
Gene mapping, somatic cell hybrid method, 852
Genes
  eukaryotic
    cloned, expression systems, 51
    organization and expression, 42–45
    promoter component, 45
    regulatory sequences, 45
    structural elements, 45–49
  expression
    characterization, 812
    regulation, 53–60
  regulation
    posttranscriptional, 54–56
    posttranslational, 65
    sequences required for, 51–53
  transcription, sequences required for, 51–53
GENETIC SEQUENCE DATA BANK. *See* GenBank
Genomic clones
  construction, 732–733
  isolation, 732–733
Gentamicin, added to plasma samples, 158
Geon–Pevicon block electrophoresis, 169
Giemsa-11 differential staining, 855–857
Glucosylation
  in vitro
    extent of lysine modification in, 357
    in presence of reducing agent, 356
    procedure, 356
  of proteins, 354
Glycine-serine-rich polypeptide, 310
β-Glycoprotein-I. *See* Apolipoprotein H
$β_2$-Glycoprotein I, physical characteristics, 136
$β_2$-Glycoprotein-II, 306
Glycosphingolipids, 106–108

Glycosylation, 63–64
Goat, lipoproteins, antigenic sites, 122
Golgi apparatus, glycosylation in, 63
Gradient gel electrophoresis, 167, 169, 456, 457
  electrophoretic blotting after, 430–431
  after immunoaffinity chromatography, 431
Ground squirrel, circulating lipoprotein levels, 79
Guanidine-hydrochloride, nonideality effects, 260–262
Guanidium chloride, 221, 222
Guinea pig, 89
  apoA-I, physical characteristics, 128
  apoB-100, physical characteristics, 132
  apoC-I, physical characteristics, 133
  apoE, physical characteristics, 135
  apoE mRNA, 57
  apolipoproteins, 127
  chylomicrons, 90
  circulating apolipoprotein levels, 142
  circulating lipoprotein levels, 86
  comigrating peptide, physical characteristics, 135
  HDL, physical properties, 115, 120
  HDL and subclasses, chemical composition, 98
  LDL
    chemical composition, 95, 102
    physical properties, 112
  lipid metabolism, 142
  lipoprotein lipids, fatty acid profile, 105
  lipoprotein profile, 84, 85
  lipoproteins
    chemical composition, 91
    immunologic properties, 123
  plasma VLDL, chemical composition, 92
  VLDL, physical properties, 110

H

Harbor seal
  circulating lipoprotein levels, 80
  HDL, physical properties, 115
  HDL subclasses, 120
HDL, 5, 142, 144, 155, 417, 647, 753
  apoA-I from, 227
  apoF, 306
  apoH, 308–309

apolipoprotein, 340, 342
  conformational change on lipid binding, 369
  interaction with DPPC vesicles, 667
  isolation, 11
  rate of adsorption at air–water interface, 394–395
  release of carboxyfluorescein, 664, 665
  reversed-phase HPLC, 350–353
  surface pressure–molecular area isotherm for spread monolayer of, 393
assembly, 69
catabolism, 37–40
charge-dependent separation of, 432, 433
in chylomicron processing, 30–31
^{13}C NMR, 490–491, 493–497
composition, 578
cross-linking experiments, 620–623
delipidation, 215, 343–345, 348–350
delipidized, preparation, 298
density, 170
discoidal, 24–25, 27
discoidal analogs, 584. *See also* HDL-like particles
electrophoretic blotting, 430–431
first isolated, 3
formation, 24–26
freeze-fracture and freeze-etching electron microscopic studies, 458
heterogeneity, on isoelectric focusing, 440–441
^{1}H NMR, 482–483, 485–487
human
  analysis, 167
  chemical composition, 168
  concentration range, 157
  density, 72, 157
  electrophoretic class, 157
  flotation rates, 157
  fractionation after centrifugation, 166
  normal plasma distribution, 423–425
  sequential flotation ultracentrifugation, 165
  size range, 157
  structural heterogeneity, 102
hydrolysis *in vitro* by phospholipase A$_2$, ^{31}P NMR study, 479
incubated with apolipoprotein monoclonal antibodies, immunoelectrotransfer blots, 538–539
interactions with cells and lipoproteins, 581–582
isoelectric focusing, 433–435
  on agarose gel, 435, 438–440
isolation, 77, 238
  by single vertical spin inverted rate zonal density gradient ultracentrifugation, 186–187
lipid content, 519
lipoprotein profile, 176
lyophilization, 217
mammalian, 84
  lipid and protein composition, 90–106
  phospholipid content, 103–106
  physical properties, 115–117, 119–121
model, 15
mouse, 892
  phenotypes from different strains, 887
  variant for, 886–887
nascent particles, 14, 23
native, CD of β-carotene as spectroscopic probe of order in, 526–527
negative correlation with CAD, 3–4
negatively stained, electron microscopy, 448
negative staining, 442, 443
NMR, 500–503
nonhuman, isolation, 168
particle size, 418
  in states of abnormal lipoprotein metabolism, 427–428
phospholipid-enriched species
  analysis, 651–652
  physical studies of, 653–654
  separation, 651–652
phospholipid enrichment, 652–653
phospholipids, consumed by LCAT reaction, 38–39
physical properties, 109
^{31}P NMR, 476, 478–479
  effects of paramagnetic reagents, 481
polymorphism, 125
protein constituents, 223

protein stoichiometry, cross-linking
  study of, 620–621
radiolabeling, 653
reactions with LCAT, 580
reconstitution, 553–582, 584
  lipids, 555–557
  preparation of materials, 554–555
in reverse cholesterol transport, 721
separation from albumin, by inverted
  rate-zonal density gradient ultracentrifugation, 198–201
single vertical spin ultracentrifugation,
  196, 198–201
size, 456–457
spin density gradient ultracentrifugation, 183
structure, 13, 578
  abnormal, cross-linking methodol-
    ogy to study, 622–623
subclasses, 73–74, 102–103
  in mammals, 120–121
subpopulations, 418
  identification, 417
  properties, 424
subspecies, 25–26
  separation, using single vertical spin
    ultracentrifugation, 201–205
SVS procedure, adaption to 70Ti rotor,
  202–205
transfer of phospholipids into, molecular basis, 654–655
VAP analyses, 201–204
HDL$_1$
  density, 170
  mammalian, physical properties, 120–
    121
  separation, 78
HDL$_2$, 154, 650
  core lipids, chemical composition, 11
  in coronary heart disease, 13
  cross-linking experiments with, 621–622
  degradation, 17
  density, 10, 170
  electrophoretic definition, 10
  fractionation, by SVS inverted rate-
    zonal density gradient ultracen-
    trifugation, 201–203
  isolation, by single vertical spin in-
    verted rate zonal density gradient
    ultracentrifugation, 187–188
  lipoprotein profile, 174

molecular weight, 10
particle size, 10
structure, 13
surface components, chemical composition, 11
HDL$_3$, 154, 650, 652
  core lipids, chemical composition, 11
  cross-linking experiments with, 621–622
  density, 10, 170
  electrophoretic definition, 10
  fractionation, by SVS inverted rate-
    zonal density gradient ultracen-
    trifugation, 201–203
  freeze-fracture/etching technique for,
    463–465
  isolation, by single vertical spin in-
    verted rate zonal density gradient
    ultracentrifugation, 187–188
  lipoprotein profile, 174
  molecular weight, 10
  particle size, 10
  ^{31}P NMR, 476–477
  separation from HDL$_2$, 171, 180
  structure, 13
  surface components, chemical composi-
    tion, 11
HDL$_c$, 142, 418, 523
HDL-like complexes
  composition, 579
  morphology, 579
  reactions with LCAT, 580
  spheroidal synthesis, 565–569
HDL-like particles
  interactions with cells and lipoproteins,
    581–582
  reconstitution, 554
Hedgehog
  apolipoproteins, 127
  circulating lipoprotein levels, 79
  HDL and subclasses, chemical compo-
    sition, 98
  LDL, chemical composition, 95, 102
  lipid transport system, 143
  lipoprotein, chemical composition, 101
  lipoprotein distribution, 78
  lipoprotein profile, 84, 88
  plasma VLDL, chemical composition,
    92
Helical amphipathic moment, 405–406
Helical hydrophobic moment, 17
Heparin, 145–146, 153

Heparin affinity chromatography, 75
Heparin chromatography, 169
Heparin–Sepharose chromatography, 78
  of apoE, 278
Hepatic lipase, 17, 39, 223–224
  activity, 33
  in chylomicron processing, 30
Hepatocellular disease, 157
Hepatocyte, assembly and secretion of VLDL, 21–22
Hepatoma cell lines
  Hep3B
    growth, 692
    labeling, 692
  HepG2
    growth, 692
    labeling, 692
n-Heptane, 217
Herbivores, lipoprotein profiles, 84
High-density lipoprotein. See HDL
High-performance hydroxylapatite chromatography, purification of IgM monoclonal antibodies, 534
High-performance liquid chromatography, 145, 146
  of apolipoproteins, 339–353, 380–381, 383–384
    advantages, 353
    disadvantages, 353
    methodology, 343–344
  column maintenance, 346–348
  gel permeation, of apolipoproteins, 343
  ion-exchange, of apolipoproteins, 348–350
  for lipoprotein isolation, 11
  of mouse apolipoproteins, 892
  reversed-phase, of apolipoproteins, 350–353
HMG-CoA reductase. See 3-Hydroxy-3-methylglutaryl coenzyme A reductase
Horse, See also Przewalski horse
  circulating lipoprotein levels, 80
  lipoprotein lipids, fatty acid profile, 105
  lipoproteins, antigenic sites, 122
Host defense, role of lipoproteins, mouse genetic studies of, 888
HPLC. See High-performance liquid chromatography
Huntington's chorea, 737
Hybrid clone panel
  construction, 857–858
  for genetic mapping, 858–861
Hybridization, oligonucleotide probes, 746–747
3-Hydroxy-3-methylglutaryl coenzyme A reductase, 8
  activity, 66
  biosynthesis, 63
  gene, chromosomal localization, 894
  regulation, 59–60
N-Hydroxysuccinimidyl-4-azidobenzoate, 615, 616
Hyperalphalipoproteinemia, 157
Hypercholesterolemia. See also Hyperlipoproteinemia, type IIa
  associated genetic disorders, 6
  conditions associated with secondary hyperlipoproteinemia, 6
  laboratory definition, 6
Hyperglycemia, protein glucosylation in, 354–355
Hyperlipemia
  endogenous. See also Hyperlipoproteinemia, type IV
    associated genetic disorders, 7
    conditions associated with secondary hyperlipoproteinemia, 7
    laboratory definition, 7
  exogenous. See also Hyperlipoproteinemia, type I
    associated genetic disorders, 6
    conditions associated with secondary hyperlipoproteinemia, 6
    laboratory definition, 6
  mixed. See also Hyperlipoproteinemia, type V
    associated genetic disorders, 7
    conditions associated with secondary hyperlipoproteinemia, 7
    laboratory definition, 7
Hyperlipidemia
  combined, 5–7, 727. See also Hyperlipoproteinemia, type IIb
    associated with genetic disorders, 6
    conditions associated with secondary hyperlipoproteinemia, 6
    laboratory definition, 6
  as monogenic defect, 727–729
Hyperlipoproteinemia
  human, plasma, VAP analysis, 192–193

phenotypes, 4–7
type I, 6. *See also* Hyperlipemia, exogenous
type IIa, 6. *See also* Hypercholesterolemia
type IIb, 6. *See also* Hyperlipidemia, combined
type III, 7, 274, 282, 429, 727, 824, 846. *See also* Dysbetalipoproteinemia
  association with apoE2/2, 848, 849
  phenotypic expression, factors affecting, 849–850
type IV, 7. *See also* Hyperlipemia, endogenous
type V, 7, 242, 274. *See also* Hyperlipidemia, mixed
Hyperthyroidism, 157
Hypertriglyceridemia, 49, 180, 274, 428, 429, 727
  RFLP associated with, 742–744
  severity, and additive effect of two genetic defects, 744
  type IV, 288
  type V, 288
Hypobetalipoproteinemia, 5, 147
Hypolipidemia, 5

I

IDL, 33, 35, 126, 155, 417
  core lipids, chemical composition, 11
  degradation, 17
  density, 10, 170
  electrophoretic definition, 10
  gel gradient for, 418
  human
    concentration range, 157
    density range, 157
    electrophoretic class, 157
    flotation rates, 157
    fractionation after centrifugation, 166
    sequential flotation ultracentrifugation, 164–165
    size range, 157
  isolation, 274
  mammalian, lipid and protein composition, 90–106
  molecular weight, 10

native, CD of β-carotene as spectroscopic probe of order in, 526
particle size, 10
  abnormal, 429
physical properties, 109–119
single vertical spin density gradient ultracentrifugation, 196–198
spin density gradient ultracentrifugation, 183
structure, 13
subpopulations, 427
surface components, chemical composition, 11
Immunoaffinity chromatography, 145
Immunoblotting, of mouse plasma proteins, 893
Immunochemistry, of lipoprotein structure, 527–553
Immunological techniques, for lipoproteins, 146
Indian rhinoceros, circulating lipoprotein levels, 86
Insectivores, circulating lipoprotein levels, 79
Insulin, and lipoprotein lipase activity, 29
Intermediate-density lipoprotein. *See* IDL
Intestine
  apolipoproteins from, 9
  formation of triglyceride-rich lipoproteins, 19–24
  lipid transport, 68
Introna, 45–47
Ion-exchange chromatography, 339
  of lipoproteins, 432–433
Isoelectric focusing, 121, 167
  on Agarose gel, 435–438
    conditions, 436
    optimization, 437–438
    pH range of ampholytes, 437–438
    of plasma lipoproteins, 438–440
    preparation of gel films, 435–436
    staining of films, 436–437
  on liquid phase, 433–434
  on polyacrylamide gel, 434–435
Isopycnic density gradient ultracentrifugation, 78, 154, 181
Isotachophoresis, 433, 441
Isothermal microcalorimetry, experimental set-up, 365–367

J

Jaguar
  lipoprotein distribution, 77
  lipoprotein profile, 89

K

Kallikrein inhibitor, for blood collected for lipoprotein isolation, 151
Kidney, synthesis of apolipoproteins, 26
Killer whale
  circulating lipoprotein levels, 80
  HDL, physical properties, 115, 120
  LDL, physical properties, 112

L

Lagenorhynchus obliquidens. See Pacific white-sided dolphin
Lagomorphs, circulating lipoprotein levels, 86
LAP-20, interaction with lipid vesicles, 562
LCAT. *See also* Familial LCAT deficiency; Lecithin-cholesterol acyltransferase
L cells, mouse
  apoE gene expression upon transfection into, 819
  secretion of human apoE, 821–823
  transfected with apoE genes, selection, 820
LDL, 5, 126, 142, 155, 262, 417
  for apoB preparation, screening for contaminants, 248
  apoB purity, assessment, 251–252
  apoF, 306
  apoH, 309
  and CAD, 3–4
  carbohydrate composition, 106, 107
  carboxyamidomethylation, 590
  carotenoids in, 524
  charge-dependent separation of, 432
  chimpanzee and human, identity, 123
  cholesterol in, 608
  ^{13}C NMR, 490–491, 493, 494
  core
    fluorescent probes reconstituted into, 610
    reconstitution
      preparation of starch tubes, 611
      procedure, 612–613
      reagents, 611
      spectroscopic probes reconstituted into, 610
      toxins reconstituted into, 610
  core lipids, chemical composition, 11
  cross-linking data, 623–624
  delipidation, 214, 215, 260, 589
    in presence of ion-exchange resin, 215–216
  density, 10, 170
  differential scanning calorimetry, 601–603
  electrophoretic blotting, 430–431
  electrophoretic definition, 10
  epitope expression, role of carbohydrates, 551
  extraction of endogenous core lipids, 612
  formation, VLDL in, 31–33
  fractionation, with spin density gradient ultracentrifugation, 183
  freeze-etching electron microscopy, 469–471
  freeze-fracture and freeze-etching electron microscopic studies, 458
  glucosylation, metabolism and regulatory activity of, 354–355
  ^{1}H NMR, 482–483, 485–487
  human
    analysis, 167
    chemical composition, 167
    concentration range, 157
    density, 72, 157
    electrophoretic class, 157
    flotation rates, 157
    fractionation after centrifugation, 166
    polymorphism, 123
    sequential flotation ultracentrifugation, 164–165
    size range, 157
  hydrophobic core
    composition, 608
    reconstitution, 608–613
  incubated with apolipoprotein monoclonal antibodies, immunoelectro-transfer blots, 538–539
  isoelectric focusing, 434
    on agarose gel, 438–440

isolation, 77, 250–251, 612–613
   by single vertical spin inverted rate zonal density gradient ultracentrifugation, 186–187
lipid content, 519
lipid extracts, CD spectra, 519–520
lipoprotein profile, 174, 176
lyophilization, 217–218, 612
mammalian
   cross-reactivities, 123
   lipid and protein composition, 90–106
   phospholipid content, 103–106
   physical properties, 112–113, 119
molecular weight, 10
morphology, 456
mouse, 892
   density, 891
   genetic variants for, 885
native, CD of β-carotene as spectroscopic probe of order in, 526
negatively stained, electron microscopy, 448
negative staining, 442
NMR, 500, 501, 503
nonhuman, isolation, 168
normal, thermodynamic characterization, 601–602
particle size, 10
   abnormal, 428–429
physical properties, 109
^{31}P NMR, 476, 478–479
   effects of paramagnetic reagents, 481
porcine, quantitation, 88
preparation, 611
   for apoB isolation, 250
protein component, 247
proteolysis, inhibition, 249
rabbit, allotypy, 125–126
rapid cryofixation, 468–469
reassembled, 595–600
   biological characterization, 606–607
   circular dichroism, 603–606
   physical characteristics, 600–606
   thermodynamic characterization, 601–603
reassembly, 582–608
   by cholesteryl ester replacement, in phospholipid–cholesteryl ester–apoB complex, 599–600

methods, 590–600
reconstituted
   biological activity, 610
   compared to native protein, 610
   properties, 610
   validation of β-carotene as spectroscopic probe for, 524–525
reconstitution
   with exogenous lipid, 612
   principle, 609
   removal of apolar core, 609–611
   single vertical spin density gradient ultracentrifugation, 196–198
size, 455–456
slow cryofixation, 467–468
solubilization, 612–613
solubilized with NaDC, gel filtration chromatography, 587
spin density gradient ultracentrifugation, 183
storage, 252
structure, 13
   temperature-dependent changes, 506–509
subpopulations, 418, 427
   identification, 417
   normal human distribution, 424–427
subspecies, separation, using single vertical spin ultracentrifugation, 204–207
surface components, chemical composition, 11
terminal catabolism, 33–37
LDL$_2$, freeze-fracture/etching technique for, 463–466
LDL-II, 74
LDL receptor, 33–35, 608
   apoE binding, 769
   biosynthesis, 64–65
   bovine, mRNA, partial cDNA clones, 896
   cDNA, cloning, 901–903
   cDNA clones
      identification, 902–908
      primer extension sequencing, 905–906
      restriction endonuclease mapping, 903–904
   chromosomal localization, 864
   cloning

buffers, 897
cells, 896–897
methods, 898–908
principle of method, 896–897
reagents, 896
tissues, 896–897
on fibroblast surface, 8
gene
chromosomal location, 50, 51, 894
functional organization, 47
human, molecular biology, 908–909
in lipoprotein metabolism, 69
mRNA, relative amounts in liver and adrenal cortex, 906–908
regulation, 35–36
RNA, *in vitro* translation, 900–901
role in cholesterol metabolism, 895
structure, 46
Lecithin
binding to apoHDL, enthalpy changes on, 371
long-chain, aqueous solution, 370
short-chain, aqueous solution, 370
zwitterionic, 20–21
Lecithin-cholesterol acyltransferase, 24, 39, 103
activation, 223–224, 627
polyclonal antibodies used to block, 552
by proapoA-I form, 708
study, using apolipoprotein peptide analogs, 644–646
activity, 4, 12, 14, 17–18, 31, 37–38, 223
and apoD, 303
isolation, 203–205
in reverse cholesterol transport, 721
Lecithin-cholesterol acyltransferase activator peptide, 770
Lecithin-cholesterol acyltransferase deficiency, 428, 455, 578
HDL in, 451
Light scattering, 456
of apolipoproteins, 384
Linkage analysis, 852, 863
Lion
lipoprotein distribution, 77
lipoprotein profile, 89
Lipase, *See* Hepatic lipase, Lipoprotein lipase

Lipid-apolipoprotein association, role of helical regions, 16
Lipid–protein interaction, 767
mechanism, structural determinants, 15–17
Lipid-transfer proteins, 18–19
Lipid transport, 338
genetic control of, mouse model for, 877–894
role of apolipoproteins, 13
Lipid transport systems, 143
auxiliary proteins, 309–310
Lipolytic enzymes, 17–19
Lipoprotein. *See also* Plasma lipoprotein; Triglyceride-rich lipoprotein
allelic variation, 550–551
analysis, for mouse strains, 893
antibody binding, 545
antibody affinity chromatography, 547–549
competitive binding assay, 545–547
assembly, in mice, 885–888
biosynthesis, 41–70
bovine, 73
catabolism, 27–40
apoE in, 824
CD spectra, correction for contribution of lipids, 522–524
characterization, 4
charge-dependent separations, 432–433
chemical precipitation techniques, 145–146
cholesterol-induced, 142–143
classes, 12
composition, 11–12
core lipids, chemical composition, 11
cryofixation, 459
sample preparation, 458
cryofixed, freeze-fracture/etching technique, 459–461
delipidation, 549–550
density classification, 72–73
electrophoretic classification, 72–73
enzymatic treatment, 548–550
epitope expression, 545–548
reconstituted models, 550
freeze-fracture/etching technique
artifacts, 465–466
cleaning replica, 461–462

interpretation of electron microscopic images, 462–465
observation of replica, 461–462
replication of exposed surface, 461
heterogeneity, 5
  charge difference as basis for, 432
human
  classes, 72
  family hypothesis, 75
as immunogen, 529
intracellular assembly and transport, 66–69
isolation, 10–11, 181
  blood collection for, 151–155
  from Golgi, 68
lipid composition, 11
lipid core, 583
lipids, β-carotene as probe for, 524–527
magnetically active nuclei in, 474, 475, 513–515
mammalian
  anomalies, 73–74
  carbohydrate composition, 106–108
  chemical properties, 89–108
  classification, 71–75
  distribution, anomalies in, 77–78
  isolation, 168
  nomenclature, 71–75
  phospholipid content, 103–106
  physical properties, 108–121
metabolism, 690
  genes involved in, 851
  in mice, 885–888
  physicochemical concept of, 404
α-migrating, 72
β-migrating, 72
molecular organization, 583
molecular weight determination, 430
mouse
  density fractions, 891
  isolation, 891–893
native
  CD of β-carotene as spectroscopic probe of order in, 525–527
  physicochemical studies, 121–122
negatively stained. See also Electron microscopy
  dialysis of samples, 444–445
negative staining, 442
  aggregation of particles, 451–453

artifacts, 450–455
clumping of particles, 450–451
distortion of particle shape, 451, 452
particle flattening, 451, 454–455
preparation of sample, 443–444
spotty background, 451–454
stains, 443
sucrose artifact, 444
nomenclature, 5–8
nonhuman, sequential flotation ultracentrifugation, 168–169
particle size distributions, in states of abnormal lipoprotein metabolism, 427–429
preparation, 649–650
preservative solutions for, 152
processing, 27–40
profiles, identification of species groups based on, 78–79
recombinant, cross-linking data, 624–626
region specific immunoassays, 553
secretion, 19–27
separation, 77
Shen/Kezdy model, 11
structure, 12–17, 583
  cross-linking reagents to study, 613–626
  immunochemistry of, 527–553
  in vitro perturbation, and analysis with antibodies, 548–550
study, laboratory techniques for, 144–147
subpopulations
  autoradiography, 419
  immunoblotting, 419
  localization and identification, 418–419
  protein vs. lipid staining, 418–419
surface components, chemical composition, 11
synthesis, 19–27
transfer factors, 14
with triglycerides with elevated proportions of saturated fatty acids, 78
two-dimensional analysis with agarose isoelectric focusing and other techniques, 440
variant forms, 192, 194
α-Lipoprotein, 157

isolation, 168–169
β-Lipoprotein, 123, 157
  isolation, 168–169
Lipoprotein affinity chromatography, for purification of monoclonal antibody, 534
Lipoprotein D
  isolation, 302
  lipid composition, 302
Lipoprotein F, 305
  characterization, 305–306
  isolation, 305–306
  lipid composition, 302
Lipoprotein lipase, 13, 127, 309, 797
  activation, polyclonal antibodies used to block, 552
  activity, 4, 17, 29–30
  genes, 883–884
  in LDL formation, 31
Lipoprotein–liposome interactions, 647–657
  biological relevance, 655–656
  carboxyfluorescein leakage assay, 657–668
    calculations, 664–665
    principle, 658
  cholesterol/phospholipid ration, effects, 649
  effects of liposome composition, 648–649
  incubation conditions, 650–651
    temperature, 650
    vesicle/HDL ratio, 650–651
    vesicle/plasma ration, 650–651
  lipoprotein preparation, 649–650
  phospholipid composition, effects, 649
  solute escape, fluorescence self-quenching method for monitoring, 657
  unilammelar vs. multilamellar liposomes, effects, 648
Lipoprotein receptor, cross-linking experiments with, 626
Lipoprotein-receptor functions, in mice, 888
Lipoprotein X, 5, 647, 651
  ^{1}H NMR, 487
  ^{31}P NMR, 479–480
Liposome
  containing phospholipids and cholesterol, circular dichroism, 520–522

  drug encapsulation in, 656
  preparation, 648, 661–662
  small unilamellar vesicles, preparation, 661–662
Liposome–apolipoprotein interaction, mechanisms, 667–668
Liver. *See also* Hepatocyte
  apolipoproteins from, 9
  apolipoprotein synthesis, 223
  cholesterol uptake, 30
  formation of triglyceride-rich lipoproteins, 19–24
  VLDL assembly, 68
Liver disease, obstructive, 5
Low density lipoprotein. *See* LDL
LMPC. *See* Lysomyristoyl phosphatidylcholine
Lp(a), 74, 77, 141, 154–155, 248
  contamination of LDL, 165
  isolation, 208
  nonhuman, isolation, 168
  plasma, isolation by VAP procedure, 192, 194
  single vertical spin density gradient ultracentrifugation, 196–198
  spin density gradient ultracentrifugation, 183
Lysolecithin
  aqueous solution, 370
  binding to apoHDL, enthalpy changes on, 371
  distribution in mammalian lipoproteins, 104
Lysomyristoyl phosphatidylcholine, association with apolipoprotein, 370
Lysozyme, Ferguson plot, 271

M

Macaca arctoides. See Stumptail macaque
Macaca fascicularis. See Cynomolgus monkey
Macaca mulatta. See Rhesus monkey
Macaca nemestrina. See Pigtail macaque
Macrophage, synthesis of apoE, 26
Malayan tapir, circulating lipoprotein levels, 80
Malondialdehyde, modification of LDL, 37
Mammals
  HDL, 78–85

LDL, 85–88
Man
  apoA-I, physical characteristics, 128
  apoA-II, physical characteristics, 129
  apoA-IV, physical characteristics, 131
  apoB-48, physical characteristics, 131
  apoB-100, physical characteristics, 131
  apoC-I, physical characteristics, 133
  apoC-II, physical characteristics, 134
  apoC-III, physical characteristics, 134
  apoE, physical characteristics, 135
  apoH, physical characteristics, 136
  circulating apolipoprotein levels, 142
  β_2-glycoprotein I, physical characteristics, 136
  HDL
    phospholipid content, 104
    physical properties, 115, 120
  HDL and subclasses, chemical composition, 98
  IDL, chemical composition, 94
  LDL
    carbohydrate composition, 107
    chemical composition, 95
    phospholipid content, 104
    physical properties, 112, 119
  lipoproteins
    antigenic sites, 122
    immunologic properties, 123
  loci containing structural genes for apolipoproteins, 893–894
  plasma VLDL, chemical composition, 92
  serum lipoprotein profiles, 179–180
  threonine-poor apoproteins, physical characteristics, 136
  VLDL
    phospholipid content, 104
    physical properties, 110
Mangaby, circulating lipoprotein levels, 81
Marsupials, circulating lipoprotein levels, 79
Mean helical hydrophobic moment, 402
Meles meles L. *See* European badger
Meles putorius furo L. *See* Ferret
Merianes unguiculatus. See Mongolian gerbil
Messenger RNA
  cell-free translation, 691–692
  hybridization selection, 809–810
  isolation, 691–692
  polysome immune purification, 899–900
  processing, 54–55
  rat liver, translation products, 805–806
Metallothionein, gene, regulation, 52, 54
Micelle, 558
Mink
  lipoprotein allotypes, 124–125
  lipoproteins, immunogenetic polymorphism, 125
Mirounga angustirostris. See Elephant seal
Molecular sieve chromatography, 4, 339
Mongolian gerbil
  apolipoproteins, 127
  circulating lipoprotein levels, 79
  HDL and subclasses, chemical composition, 98
  LDL, chemical composition, 95, 102
  lipoprotein lipids, fatty acid profile, 105
  plasma VLDL, chemical composition, 92
Monkey. *See also* New World monkeys; Old World monkeys; *specific species*
  HDL, 73
Monoclonal antibody
  to apolipoproteins, 146
  preparation, 357–358
  to apolipoproteins and lipoproteins
    detection of activity, 530
    dot immunoblotting, 531–534
    immunizations for, 529–530
    immunogens, 529
    production, 529–530
    purification, 534
    sandwich plate assay, 530–531
Mountain zebra, circulating lipoprotein levels, 80
Mouse, 89
  apoA-I, physical characteristics, 128
  apoA-II, physical characteristics, 129
  apoB, 262
  apoB-100, physical characteristics, 132
  apoC-III, physical characteristics, 135
  apoE, physical characteristics, 135
  apoE mRNA, 57
  apolipoproteins, physical characteristics, 141
  atherogenic diet, 889
  circulating lipoprotein levels, 79
  diet for, 889

genetics
    applications to lipid transport, 883–888
    backcross analysis, 881, 890–891
    congenic strains used in, 882–883
    recombinant inbred strains used in, 881–882, 890–891
  HDL, 73, 102–103
    physical properties, 115, 120
    variants, 885–887
  HDL and subclasses, chemical composition, 98
  IDL, chemical composition, 94
  LDL
    chemical composition, 95, 102
    physical properties, 112
  lipid transport genes, mapping, 883–885
  lipoprotein assembly and metabolism, 885–888
  lipoprotein distribution, 78
  lipoproteins
    chemical composition, 101
    immunologic properties, 123
  loci containing structural genes for apolipoproteins, 893–894
  model for biochemical-genetic study of mammalian lipoproteins, 877–879
  plasma lipid transport, phenotypic variations affecting, 878–879
  plasma lipid transport genes, 878, 880
    analysis, 879–883
    identification, 879–881
  plasma VLDL, chemical composition, 92
  RI strains, 881–882, 890–891
    lipoprotein and lipoprotein lipase variations among, 883–884
  screening for genetic variation, 889–890
  sources, 888–889
  threonine-poor apoproteins, physical characteristics, 136
  VLDL, physical properties, 110
Mouse News Letter, 889
Muscular dystrophy, 737
Mus musculus. See Mouse
Mutagenesis, oligonucleotide site-directed, 48
Myosin, 261
  Ferguson plot, 271, 272
Myotonic dystrophy, 800
Myristic acid
  binding to albumin, Scatchard plot, 331
  partition ratio relative to concentration in *n*-heptane, 330–331

N

NaN_3, added to plasma samples, 159
National Biomedical Research Foundation, PROTEIN SEQUENCE DATABASE. *See* NBRF
NBRF, 754
  Protein Identification Resource (PIR), 755
Negative staining, 442
New World monkeys
  circulating lipoprotein levels, 81, 87
  LDL, chemical composition, 102
Nonhuman primates
  circulating lipoprotein levels, 81
  plasma lipoprotein, 71
Nonideality, definition, 261
Nonidet P40, delipidation with, 215
Northern analysis, 671
NUCALN, 764
Nuclear magnetic resonance
  of apolipoprotein peptide analog–lipid interactions, 640–641
  ^{13}C, 475, 489–513
    advantages, 489–490
    Bruker 10 mm probe, 509–510
    chemical shifts and assignments, 492–497
    quantitation, 497–498
  chemical shift, 473, 492–497
    and assignments, 492–497
  continuous wave, 472–473
  ^1H, 475, 482–489
  ^2H, 475, 513–514
  linewidth, 473, 502–507
  of lipoproteins, 108, 121, 474
  model systems, 512–513
  ^{14}N, 475, 513, 515
  ^{15}N, 475, 513, 515
  ^{17}O, 475, 513–514
  optimizing sample volume, 511–512
  ^{31}P, 475–481
  peak intensity, 473

pulsed Fourier transform, 472–474
^{33}S, 475, 513–514
sample temperature, 507–511
signal/noise ratio, 511–512
spin-lattice relaxation time, 473, 498–502
    extreme narrowing region, 500
    fast inversion-recovery determination, 498
    study of fatty acid binding to albumin, 327
Nuclear Overhauser enhancement, 478–479, 497–499, 507
Nucleic acid hybridization, 671
  in situ, 870–872
    for chromosomal mapping of apolipoprotein genes, 863–876
    sensitivity, 865, 870

O

Obesity, role of lipoproteins, mouse genetic studies of, 888
Odobenus rosmarus. See Walrus
Odocoileus virgianus. See Deer
Old World monkeys
  circulating lipoprotein levels, 81
  LDL, chemical composition, 102
Oligomer, definition, 376
Oligonucleotide, ^{32}P-labeled, colony hybridization using, 748–750
Oligonucleotide probe, 746–747
  for apoC-II cDNA clones, preparation, 789–791
Omnivores, lipoprotein profiles, 84
Onager, circulating lipoprotein levels, 80
Opossum
  circulating lipoprotein levels, 79
  lipoprotein profile, 84
Optical rotatory dispersion, of lipoproteins, 108, 121
Orangutan, Lp(a), 74
Orcinus orca. See Killer whale
Oryctolagus cuniculus domesticus. See Rabbit
Ovalbumin, Ferguson plot, 271
Ovis aries. See Sheep
Owl monkey, HDL, physical properties, 120

P

Pacific white-sided dolphin, circulating lipoprotein levels, 80
Palmitoylpalmitoleoyl phosphatidylcholine, interaction with apolipoproteins, 560, 561, 566
Panthera leo. See Lion
Panthera onca. See Jaguar
Pan troglodytes. See Chimpanzee
Paper electrophoresis, 167
Papio anubis. See Baboon
Papio cynocephalus. See Baboon
Papio papio. See Baboon
cis-Parinaric acid, 327
Patas monkey
  apoA-I, physical characteristics, 129
  apoA-II, physical characteristics, 130
  apoE, physical characteristics, 136
  apolipoproteins, physical characteristics, 141
  circulating lipoprotein levels, 81
  HDL, carbohydrate composition, 106
  LDL, 74
    carbohydrate composition, 107
  lipid metabolism, 142
Pedigree analysis, 728–729, 851
Peripheral blood lymphocyte, methotrexate synchronization, 865–868
Perissodactyls, circulating lipoprotein levels, 80, 86
Phenylmethylsulfonylfluoride, 219, 379
Phoca vitulina. See Harbor seal
Phosphatidic acid, distribution in mammalian lipoproteins, 104
Phosphatidylcholine
  distribution in mammalian lipoproteins, 104
  multilamellar vesicles, 556–557
  for reconstitution of HDL, 555–557
  small unilamellar vesicles, 557
  transfer, by specific factors in plasma, 18–19
Phosphatidylcholine vesicles, spontaneous reaction with apolipoproteins, 554
Phosphatidylethanolamine, distribution in mammalian lipoproteins, 104
Phosphatidylinositol, distribution in mammalian lipoproteins, 104

Phosphatidylserine, distribution in mammalian lipoproteins, 104
Phospholipid, microemulsion particles containing, circular dichroism study, 520–522
Phospholipid–apoB complex, 590–594
Phospholipid–apolipoprotein association
  constant of equilibrium, calculation, 407–410
  contribution of ionic forces toward, 372
  enthalpy, 367–369
    changes from phospholipid crystalline structure, 370–372
  enthalpy of, 410, 411
  free energy of, enthalpic contribution to, 410
  free energy values, 410
    calculation, 407
    comparison of results, 410–413
  loss in conformational entropy with, 411
  spectroscopic changes with, 408–409
  thermodynamic parameters, measurement, 407
  thermodynamics, 403–413
Phospholipid–cholesteryl ester–apoB complex, 595–600
  electron microscopy, 597–598
  gel chromatography, 597–598
  gel filtration chromatography, 599–600
  microemulsion–apoB complex, 595–599
Phosphotungstate, 443
Pig, 89
  apoA-I, physical characteristics, 129
  apoB-100, physical characteristics, 133
  apoC-II, physical characteristics, 134
  apoE, physical characteristics, 136
  circulating apolipoprotein levels, 142
  HDL, 108
    carbohydrate composition, 106
    physical properties, 115, 120
  HDL and subclasses, chemical composition, 99
  IDL, chemical composition, 94
  LDL, 108
    carbohydrate composition, 107
    chemical composition, 96, 102
    physical properties, 112–113, 119
  lipoprotein allotypes, 124–125
  lipoprotein distribution, 77
  lipoprotein lipids, fatty acid profile, 105
  lipoprotein profile, 84, 85–88
  lipoproteins
    chemical composition, 101
    immunogenetic polymorphism, 125
    immunologic properties, 123
    native structure, studies, 121
    physical properties, 109
    physicochemical properties, 122
  plasma glycosphingolipids, 108
  plasma lipoprotein, 71
  plasma VLDL, chemical composition, 92
  VLDL, 108
    physical properties, 121
Pigtail macaque
  circulating lipoprotein levels, 81
  HDL, physical properties, 120
  Lp(a), 74
Pinnipeds, circulating lipoprotein levels, 80
Plasma
  handling, 279
  preparation, for polyacrylamide gradient gel electrophoresis, 420
  solvent additives, 378–379
  storage, 279
    degradation of apoE in, 842–843
  two-dimensional gel electrophoresis, 244–246
    first dimension, 245–246
    immunoblot, 246
    sample preparation, 245
    second dimension, 246
Plasma apolipoprotein
  amphipathic helical model, 403–404
  lipid-associating regions, peptide analogs, 404–405
Plasma lipid, analysis, for mouse strains, 893
Plasma lipoprotein
  allotypy, 123–126
  clinical importance, 3–4
  families, physical properties, 10
  first isolated, 3
  fractions, isolation, 219
  freeze-fracture and freeze-etching electron microscopic studies, 458
  human, 297, 417
    sequential flotation ultracentrifugation, 164–166

immunogenetic polymorphism, 123-126
in vivo communication, and solution
    properties of apolipoproteins, 387
lipid autoxidation, 218
lyophilization prior to delipidation, 217
mammalian
    comparative analysis, 70-143
    immunological properties, 122-126
    interspecies relationships, 122-123
    quaternary organization, 217
        factors affecting, 214
    studies, history of, 144-145
    three step gradient single-angled head
        spin fractionation, gradient conditions for, 198-201
Plasma triglycerides, clearance, 4
Polar bear, circulating lipoprotein levels, 80
Polyacrylamide gradient gel electrophoresis
    calibration, 418, 419
    densitometry, 422
    electrophoretic transfer to nitrocellulose sheets, 422-423
        autoradiography, 423
        fixation procedure, 422
        lipid stain, 422
        procedure, 421-422
        protein stain, 422
    reagents, 419-420
    sample preparation, 420-421
    spacer solution used in, 420, 421
    staining procedure, 422
    supplies, 419-420
Polyclonal antibody
    to apolipoproteins, 146
        preparation, 357
    for use in polysome immune purification, preparation, 899
Polyclonal antisera, production, 528
Polyglycerophosphatides, distribution in mammalian lipoproteins, 104
Polymorphism
    functional, 738
    neutral, 735, 738
Polysome, from bovine adrenal glands, isolation, 898-899
POMC. *See* Preproopiomelanocortitropin
POPC. *See* Palmitoylpalmitoleoyl phosphatidylcholine

Porcupine, circulating lipoprotein levels, 79
Precipitation, for lipoprotein isolation, 10
Preproopiomelanocorticotropin, 65
Proline-rich polypeptide, 310
Protein. *See also* Secretory protein
    adsorption to lipid monolayers, 399
    comparison matrix, 758
    glucosylated
        competitive binding assay, 359-360
        direct (noncompetitive) binding assay, 359
        solid-phase radioimmunoassay, 359-360
    hydropathic profile, 769-771
    immunological detection, on nitrocellulose, 538-539
    intracellular targeting, 60-61
    in vivo glucosylated, antigen preparation, 358-359
    posttranslational phosphorylation, 66
    synthesis, 60
Protein conformational change, enthalpy of, 368-370
Protein-lipid association, structural determinants regulating, 404
Protein-lipid interaction, in plasma, 213
Protomer, definition, 376
PRPLOT, 771
PRTTALN, 764
Przewalski horse, circulating lipoprotein levels, 80

R

Rabbit, 89. *See also* Watanabe-heritable hyperlipidemic rabbit
    apoA-I, physical characteristics, 128
    apoB-48, physical characteristics, 132
    apoB-100, physical characteristics, 132
    apoC-I, physical characteristics, 134
    apoE, physical characteristics, 135
    apolipoproteins, physical characteristics, 141
    chylomicrons, 90
    circulating lipoprotein levels, 86
    Dutch Belt
        circulating lipoprotein levels, 86
        lipoprotein profile, 85

Fauve de Bourgogne, circulating lipoprotein levels, 86
HDL
    native structure, studies, 121
    physical properties, 115
HDL and subclasses, chemical composition, 99
LDL
    allotypy, 125–126
    chemical composition, 95, 102
    native structure, studies, 121
    physical properties, 112
lipid metabolism, 142
lipoprotein allotypes, 124
lipoprotein lipids, fatty acid profile, 105
lipoprotein profile, 85
lipoproteins
    antigenic sites, 122
    immunogenetic polymorphism, 125
    physicochemical properties, 122
New Zealand, lipoprotein profile, 85
New Zealand white, circulating lipoprotein levels, 86
plasma lipoprotein, 71
plasma VLDL, chemical composition, 92
Red Burgundy, lipoprotein profile, 85
VLDL
    native structure, studies, 121
    phospholipid content, 103, 104
    physical properties, 110
Rat, 89
apoA-I, physical characteristics, 128
apoA-II, physical characteristics, 129
apoA-IV, 241, 753–774
    physical characteristics, 131
    precursor, 240
apoA-V, physical characteristics, 131
apoB, 76, 248, 262
    sedimentation coefficient, 259
apoB-48, physical characteristics, 132
apoB-100, physical characteristics, 131
apoC-I, physical characteristics, 133
apoC-II, physical characteristics, 134
apoC-III, physical characteristics, 134
apoE, physical characteristics, 135
apolipoproteins, physical characteristics, 141
chylomicrons, 90
circulating apolipoprotein levels, 142

circulating lipoprotein levels, 79
β_2-glycoprotein I, physical characteristics, 136
HDL, 73, 102–103
    carbohydrate composition, 106
    distribution, strain variation, 77
    phospholipid content, 104
    physical properties, 115, 119, 120
HDL and subclasses, chemical composition, 98
IDL, chemical composition, 94
LDL
    carbohydrate composition, 107
    chemical composition, 95, 102
    isolation, 168
    phospholipid content, 103, 104
    physical properties, 112, 119
lipid metabolism, 142
lipoprotein lipids, fatty acid profile, 105
lipoprotein profile, 84
lipoproteins
    chemical composition, 91
    native structure, studies, 121
plasma lipoprotein, 71
plasma VLDL, chemical composition, 92
VLDL
    phospholipid content, 103, 104
    physical properties, 109, 110
Rate zonal centrifugation, 78, 169, 183
Rattus norvegicus. *See* Rat
Receptor-mediated endocytosis, 895, 909
    of chylomicrons, 28–29
Recombinant DNA techniques, 728–729, 788, 852
Red fox
    apolipoproteins, 127
    circulating lipoprotein levels, 80
    HDL and subclasses, chemical composition, 98
    LDL, chemical composition, 95, 102
    plasma VLDL, chemical composition, 92
RELATE, 759–765
Restriction endonuclease, used in gene mapping, 731–732
Restriction fragment length polymorphism, 729, 881
    associated with apoA-I–apoC-III gene complex, analysis, 740–744

detection
   approaches, 736
   materials needed, 730–735
   procedure, 730–731
   limitations to approach, 738–739
   use to define gene markers, 735–738
Retinol-binding protein, 13
Retinyl esters, in chylomicrons, 20
Reverse cholesterol transport, 37–40, 721
Reverse transcriptase, 43
Rhesus monkey, 89
   apoA-I, physical characteristics, 129
   apoA-II
      physical characteristics, 130
      structure, 236
   apoB-100, physical characteristics, 133
   apolipoproteins, physical characteristics, 141
   circulating apolipoprotein levels, 142
   circulating lipoprotein levels, 81
   HDL
      carbohydrate composition, 106
      phospholipid content, 104
      physical properties, 115, 119, 120
   HDL and subclasses, chemical composition, 99
   LDL
      carbohydrate composition, 107
      chemical composition, 96
      native structure, studies, 121
      phospholipid content, 104
      physical properties, 113
   lipid metabolism, 142
   lipoprotein lipids, fatty acid profile, 105
   lipoproteins
      antigenic sites, 122
      chemical composition, 101
      immunologic properties, 123
      physical properties, 109
   Lp(a), 74
   plasma VLDL, chemical composition, 92
Rhinoceros. *See* Black rhinoceros; Indian rhinoceros; White rhinoceros
Rhinoceros unicornis. See Indian rhinoceros
Rhinocerotidae, lipoprotein profile, 85
Ribonuclease, adsorption at air–water interface, 394, 395

Ribose, determination, orcinol procedure, 687–689
RNA
   fractionation, by sucrose gradient sedimentation, 804
   poly(A)-containing
      for cDNA synthesis, 802
      preparation, 803–804
      translation, 804–805
   preparation, for apolipoprotein mRNA hybridization assay, 680–682
   tissue content, measurement, 687–688
   total liver, preparation, 802–803
RNA-excess solution hybridization, 671
Rodents, circulating lipoprotein levels, 79, 86
Rough endoplasmic reticulum
   glycosylation in, 63
   lipoprotein and apoprotein formation and secretion, 20–21
$R_0 t$ analysis, 671

S

Saimiri sciureus. See Squirrel
Salt, mass balance equation for, 160
Scavenger receptor, 36–37
SDS–PAGE, 146
   effect of protein load, 270–272
Seal. *See* Elephant seal; Harbor seal
SEARCH, 755, 759
Secretory protein, secretory pathway, 66–67
Sedimentation equilibrium ultracentrifugation, 456
Self-association, definition, 376
Sequence
   identification and analysis, database searching, 754–758
   similarities
      identification, 773
      methods for analysis, 758–765
   similarity, assessment, 755
   single-copy, detection on human mitotic chromosomes, 865
Sequence probes, 863
Sequence repeats
   identification, 773
   internal, identification, 773–774
   mapping, 864–865

methods for inferring history of duplication, 781–788
  phylogenetic interpretation, 784–788
  procedure for inferring contractions, 788
  similar, methods for detecting, 774–777
  tandem
    detection methods, 777–781, 787
    identification, 773–774
Sequential flotation ultracentrifugation, 155–170, 181, 207, 208
  addition of inhibitors, 158–159
  adjusting plasma density to given volume for, 156
  advantages, 156, 169
  analysis, 167–168
  caps, 163–164
  density adjustment for further centrifugation, 167
  disadvantages, 156, 169–170
  effect of temperature, 163
  fractionation, 166–167
  further purification after, 169
  of human plasma lipoproteins, 164–166
  isolation of nonhuman lipoproteins, 168–169
  precautions, 156–158
  rotor, 156–158, 162–163
  run time, 163
  salt solution, 159–160
  sample calculations, 160–162
  sample preparation, 158–162
  solutions required, 159–160
  swinging bucket rotor, 169
  tubes, 163–164
  zonal rotor, 169
Serine protease, 218–219
Serine protease inhibitor, 219, 379
Serum albumin. *See also* Bovine serum albumin
  physiologic functions, 320
  rat, 755
    rate of adsorption at air–water interface, 394, 395
Serum amyloid A, 311
  assay, 314–319
    general considerations, 314–316
    suggested method, 316–319
  double antibody radioimmunoassay, 316–319

immunoassays, 314
  measurement, background, 315–316, 318
  quantification, 314, 319
  radioimmunoassay, 314–315
Serum amyloid protein, genes, 883
Serum cholesterol, and coronary artery disease, 3
Shandon Southern preparative SDS–PAGE, 264–267
  elution system, 265–266
  separation of B proteins of triglyceride-rich lipoproteins with, 267
Sheep
  apolipoproteins, 127
  circulating lipoprotein levels, 80
  HDL
    phospholipid content, 104
    physical properties, 115, 120
  HDL and subclasses, chemical composition, 99
  LDL
    chemical compositions, 95, 102
    phospholipid content, 104
    physical properties, 112
  lipoprotein allotypes, 124
  lipoproteins
    antigenic sites, 122
    chemical composition, 90
  plasma VLDL, chemical composition, 92
Sickle cell anemia, 48, 735–737
Signal peptidase, 61
Signal peptide, 61
Singer equation, 394
Single vertical spin density gradient ultracentrifugation, 182
  advantages, 207–208
  analysis of fractions, 188–190
  combination with continuous flow analysis of cholesterol, 190–192, 208
  density adjustment of sample, 185
  density gradient, overlayering method for, 185
  disadvantages, 208–209
  discontinuous KBr density gradient formation, 185
  equipment, 184
  gradient conditions, 189

gradient preparation, 208
gradients, 183
lipoprotein collection from, 186–190
lipoprotein degradation in, 207
lipoprotein species and subspecies
    analyzed, 207–208
materials, 184
method, 185–188
plasma volume limit per rotor, 209
preparation of plasma samples, 185
principle, 182–184
procedures, 183
resolution, 208–209
samples per spin, 208
spin times, 183, 207
strategies for preparation of lipoprotein
    species, 196–201
strategies for preparation of lipoprotein
    subspecies, 201–205
VAP procedure, 208
vertical autoprofile procedure, 190–192,
    206–207
    for plasma, 205
    quantification by computer-assisted
        curve demonstration, 192–196
Skeletal muscle, synthesis of apolipoproteins, 26
Sodium azide, 153, 218, 379
Sodium bromide, stock solution, 171–172
Sodium cholate, micellar dispersions of
    lipids, to produce discoidal HDL-like
    complexes, 564–566
Sodium deoxycholate
    delipidation with, 215, 252
    solubilization of apoB, 586–587
Sodium dodecyl sulfate
    delipidation with, 215, 252
    solubilization of apoB, 585
Somatic cell hybrid
    human–mouse
        alkaline Giemsa-11 differential
            staining technique, 855–857
        sequential staining, 857
        trypsin banding technique, 856
    human–rodent
        characterization of human chromosome content in, 854–857
        cytogenetic analysis, 855–857
        establishment, 852–854
        isozyme analysis, 854–855

for mapping apolipoprotein genes, 864
Southern blotting, 48
Spectroscopic studies, of lipoproteins,
    515–518
    extrinsic methods, 515
    intrinsic methods, 515–516
Sphingomyelin, 21
    CD spectrum, 519
    distribution in mammalian lipoproteins,
        103
    interaction with apolipoproteins, 561,
        562
Spider monkey
    circulating lipoprotein levels, 87
    HDL
        carbohydrate composition, 106
        physical properties, 120
    LDL, carbohydrate composition, 107
    lipoprotein profile, 88
Spin density gradient ultracentrifugation,
    181–209
    advantages, 181
    rotor, 181
Spleen, synthesis of apolipoproteins, 26
Squirrel. *See* Ground squirrel
Squirrel monkey
    HDL, physical properties, 115, 120
    HDL and subclasses, chemical composition, 99
    IDL, chemical composition, 94
    LDL, chemical composition, 96
    lipoprotein, chemical composition, 91,
        101
    plasma VLDL, chemical composition,
        92
SRCHGP, 754, 757
SRCHN, 754–755
Steer. *See* Cattle
Stumptail macaque, circulating lipoprotein
    levels, 81
Sucrose density ultracentrifugation, purification of IgM monoclonal antibodies,
    534
Surface concentration
    of adsorbed films, 395–399
    of protein monolayer at air–water
        interface, measurement, 395–398
Surface potential, 397–399
Surface pressure
    of adsorbed monolayers, 392–395

measurement, 390
Surface pressure–molecular area isotherms, of spread films, measurement, 390–391
Surface radioactivity, 395–398
Swingout rotor, 181–182, 208
Swingout rotor density gradient ultracentrifugation, 207
Swiss mouse, lipoprotein profile, 78

T

Tamarin monkey, HDL, physical properties, 120
Tangier disease, 5, 62–63, 147, 455, 708
Tapirus indicus. See Malayan tapir
TATA box, 45, 51
Tetramethyl urea, 216
β-Thalassemia, 48, 738
Thalassemias, 55
Thimerosal, added to plasma samples, 159
Thiopropyl chromatography, of apoE, 278–279
Threonine-poor apoproteins, physical characteristics, 136
Threonine-poor/SAA (serum amyloid) apolipoprotein, 310
mammalian, 127
Thymidine kinase, herpes viral, 52
Tocopheryl esters, in chylomicrons, 20
Topogenic sequences, 60
Transfer factors, 17–19
Transfer proteins, plasma phospholipid, 647–648
isolation, 652
role in exchange and transfer of phospholipids between vesicles and HDL, 652
Transferrin
Ferguson plot, 271
gene, chromosomal location, 50
Transferrin receptor, gene, chromosomal location, 50
Trasylol, 151
Triglyceride-rich lipoprotein
assembly and secretion, 68–69
delipidation, 289–290
formation, in intestines and liver, 19–24
isolation, 288–289
Triglyceride-rich particles, negatively stained, halo, 455, 456

Triglyceride. See Plasma triglycerides
Triglyceride transport, pathway, 27
Triton WR 1339, in apoB isolation, 248
Triton X-100, 216, 221
delipidation of apoB, 252, 253
delipidation of LDL with, 260
solubilization of apoB, 585–586
Tunicamycin, and intracellular modification of apolipoprotein, 705–707
Turnip yellow mosaic virus, freeze-fracture replica, 465–466
Tursiops truncatus. See Bottlenose dolphin
Two-dimensional gel electrophoresis, 146
of apolipoproteins, 342
of plasma, 244–246

U

Ultracentrifugation, 457
analytical, 167
to determine constant of equilibrium for phospholipid–apolipoprotein association, 408
flotation, 4, 77–78
for lipoprotein isolation, 10
modes, 144
Ultrafiltration, to determine constant of equilibrium, for phospholipid–apolipoprotein association, 408
Uranyl acetate, 443
Uranyl formate, 443
Uranyl oxalate, 443
Urea gel electrophoresis, 146
Uroporphyrinogen I synthase, 49

V

Vertical rotor, 187
principle, 182
Vervet. See African green monkey
Very high density lipoprotein. See VHDL
Very low density lipoprotein. See VLDL
VHDL, human, density, 72
Vitellogenin
production, 56
synthesis, hormonal regulation, 53–54
VLDL, 5, 126, 155, 262, 288, 417, 797–798
apoB-48 and apoB-100, separation, 548
apoB in, 247
apoF, 306
apoH, 309

apolipoproteins from, reversed-phase HPLC, 351–353
assembly, 21–23, 66, 68, 69
CD spectra, correction for contribution of lipids, 522–523
charge-dependent separation of, 432–433
^{13}C NMR, 490–491, 494
core lipids, chemical composition, 11
cross-linking, 626
delipidation, 215, 345–346, 349–352
density, 10, 170
electrophoretic definition, 10
enzymatic treatment, 548
formation, 19–24
and formation of LDL, 31–33
freeze-fracture and freeze-etching electron microscopic studies, 458
freeze-fracture/etching technique for, 463–466
gel gradient for, 418
having LDL-receptor binding activity, 548
HPLC profile, 347
human
  analysis, 167
  chemical composition, 167
  concentration range, 157
  density, 72, 157
  electrophoretic class, 157
  flotation rates, 157
  fractionation after centrifugation, 166
  sequential flotation ultracentrifugation, 164–165
  size range, 157
incubated with apolipoprotein monoclonal antibodies, immunoelectrotransfer blots, 538–539
isoelectric focusing, 434, 435
  on agarose gel, 438–440
isoelectric points, 121
isolation, 242, 274
  by single vertical spin inverted rate zonal density gradient ultracentrifugation, 186–187
lipid content, 519
lipids, hydrolysis, 17
lipoprotein profile, 174
lyophilization, 217
mammalian

lipid and protein composition, 90–106
  phospholipid content, 103–106
metabolism, 31
molecular weight, 10
morphology, 455
mouse, 892
nascent, surface components, 23
negatively stained
  electron microscopy, 448, 449–450
  flattening, 455
negative staining, 443
NMR, 500
particle size, 10, 20
  abnormal, 428–429
physical properties, 109–111
^{31}P NMR, 476
secretion from hepatocytes, 21–22
single vertical spin density gradient ultracentrifugation, 196–197
  wall adherence in, 208–209
size, 455
spin density gradient ultracentrifugation, 183
storage, 841
structure, 12–13
subpopulations, 427, 546
surface components, chemical composition, 11
β-VLDL, 142, 274
  interaction with monocyte-macrophage, 143
  native, CD of β-carotene as spectroscopic probe of order in, 526–527
β-VLDL receptor, 36, 37
VLDL remnants, terminal catabolism, 33–37
Vulpes vulpes. *See* Red fox

W

Walrus, circulating lipoprotein levels, 80
Watanabe-heritable hyperlipidemic rabbit, 70
  chylomicrons, 90
  LDL, physical properties, 112
  lipoprotein profile, 85
  lipoproteins, chemical composition, 91
  VLDL, physical properties, 110
Western blotting, of mouse plasma proteins, 893

WHHL rabbit. *See* Watanabe-heritable hyperlipidemic rabbit
White rhinoceros, circulating lipoprotein levels, 86
Wilhelmy plate technique, 390–391, 393

X

Xenopus, liver, vitellogenin production, 56
X-linked retinitis pigmentosa, 737
X protein, synthesized by hepatoma cell lines, 703–704

X-ray scattering, 109
of lipoproteins, 121

Z

Zalophus californianus. *See* California sea lion
Zebra. *See* Common zebra; Mountain zebra
Zonal rotor, 209
Zonal ultracentrifugation, 207

219640